Information Theory

This enthusiastic introduction to the fundamentals of information theory builds from classical Shannon theory through to modern applications in statistical learning, equipping students with a uniquely well-rounded and rigorous foundation for further study. The book introduces core topics such as data compression, channel coding, and rate-distortion theory using a unique finite blocklength approach. With over 210 end-of-part exercises and numerous examples, students are introduced to contemporary applications in statistics, machine learning, and modern communication theory. This textbook presents information-theoretic methods with applications in statistical learning and computer science, such as f-divergences, PAC-Bayes and variational principle, Kolmogorov's metric entropy, strong data-processing inequalities, and entropic upper bounds for statistical estimation. Accompanied by additional stand-alone chapters on more specialized topics in information theory, this is the ideal introductory textbook for senior undergraduate and graduate students in electrical engineering, statistics, and computer science.

Yury Polyanskiy is a Professor of Electrical Engineering and Computer Science at the Massachusetts Institute of Technology, with a focus on information theory, statistical machine learning, error-correcting codes, wireless communication, and fault tolerance. He is the recipient of the 2020 IEEE Information Theory Society James Massey Award for outstanding achievement in research and teaching in Information Theory.

Yihong Wu is a Professor of Statistics and Data Science at Yale University, focusing on the theoretical and algorithmic aspects of high-dimensional statistics, information theory, and optimization. He is the recipient of the 2018 Sloan Research Fellowship in Mathematics.

"Polyanskiy and Wu's book treats information theory and various subjects of statistics in a unique ensemble, a striking novelty in the literature. It develops in depth the connections between the two fields, which helps to presenting the theory in a more complete, elegant and transparent way. An exciting and inspiring read for graduate students and researchers."

Alexandre Tsybakov, *CREST-ENSAE, Paris*

"Since the publication of Claude E. Shannon's *A Mathematical Theory of Communication* in 1948, information theory has expanded beyond its original focus on reliable transmission and storage of information to applications in statistics, machine learning, computer science, and beyond. This textbook, written by two leading researchers at the intersection of these fields, offers a modern synthesis of both the classical subject matter and these recent developments. It is bound to become a classic reference."

Maxim Raginsky, *University of Illinois at Urbana-Champaign*

"The central role of information theory in data science and machine learning is highlighted in this book, and will be of interest to all researchers in these areas. The authors are two of the leading young information theorists currently active. Their deep understanding of the area is evident in the technical depth of the treatment, which also covers many communication theory-oriented aspects of information theory."

Venkat Anantharam, *University of California, Berkeley*

"Written in a mathematically rigorous yet accessible style, this book offers information-theoretic tools that are indispensable for high-dimensional statistics. It also presents the classic topic of coding theorems in the modern one-shot (finite block-length) approach. To put it briefly, this is the information theory textbook of the new era."

Shun Watanabe, *Tokyo University of Agriculture and Technology*

Information Theory

From Coding to Learning

Yury Polyanskiy
Massachusetts Institute of Technology

Yihong Wu
Yale University, Connecticut

Shaftesbury Road, Cambridge CB2 8EA, United Kingdom

One Liberty Plaza, 20th Floor, New York, NY 10006, USA

477 Williamstown Road, Port Melbourne, VIC 3207, Australia

314–321, 3rd Floor, Plot 3, Splendor Forum, Jasola District Centre,
New Delhi – 110025, India

103 Penang Road, #05–06/07, Visioncrest Commercial, Singapore 238467

Cambridge University Press is part of Cambridge University Press & Assessment, a department of the University of Cambridge.

We share the University's mission to contribute to society through the pursuit of education, learning and research at the highest international levels of excellence.

www.cambridge.org
Information on this title: www.cambridge.org/highereducation/isbn/9781108832908

DOI: 10.1017/9781108966351

When citing this work, please include a reference to the DOI 10.1017/9781108966351

First published 2025

A catalogue record for this publication is available from the British Library

Library of Congress Cataloging-in-Publication Data
Names: Polyanskiy, Yury, 1982– author. | Wu, Yihong, 1984– author.
Title: Information theory : from coding to learning / Yury Polyanskiy, Massachusetts Institute of Technology, Yihong Wu, Yale University, Connecticut.
Description: Cambridge, United Kingdom ; New York, NY : Cambridge University Press, 2025. | Includes bibliographical references and index.
Identifiers: LCCN 2024004713 | ISBN 9781108832908 (hardback) | ISBN 9781108966351 (ebook)
Subjects: LCSH: Information theory – Textbooks.
Classification: LCC Q360 .P65 2025 | DDC 003/.54–dc23/eng/20240617
LC record available at https://lccn.loc.gov/2024004713

ISBN 978-1-108-83290-8 Hardback

Additional resources for this publication at www.cambridge.org/polyanskiy-wu.

Dedicated to

The memory of Gennady	(Y.P.)
My family	(Y.W.)

Contents

Preface

This book is a modern introduction to information theory. In the last two decades, the subject has evolved from a discipline primarily dealing with problems of information storage and transmission ("coding") to one focusing increasingly on information extraction and denoising ("learning"). This transformation is reflected in the title and content of this book.

It took us more than a decade to complete this work. It started as a set of lecture notes accumulated by the authors through teaching regular courses at MIT, University of Illinois, and Yale, as well as topics courses at EPFL (Switzerland) and ENSAE (France). Consequently, the intended usage of this is as a textbook for a first course on information theory for graduate and senior undergraduate students, or for a second (topics) course delving deeper into specific areas.

There are two aspects that make this textbook unusual. First is that, while being written by information-theoretic "insiders," the material is very much outward looking. While we do cover in depth the bread-and-butter results (coding theorems) of information theory, we also dedicate much effort to ideas and methods which have found influential applications in statistical learning, statistical physics, ergodic theory, computer science, probability theory, and more. The second aspect is that we cover both the time-tested classical material (such as connections to combinatorics, ergodicity, and functional analysis) along with the latest developments of very recent years (large alphabet distribution estimation, community detection, mixing of Markov chains, graphical models, PAC-Bayes, and generalization bounds).

It is hard to mention everyone who helped us start and finish this work, but some stand out especially. We owe our debt to Sergio Verdú, whose course at Princeton is responsible for our life-long admiration of the subject. His passion and pedagogy are reflected, if imperfectly, on these pages. For an undistorted view see his forthcoming comprehensive monograph [437].

Next, we were fortunate to have many bright students contribute to typing of the original lecture notes (precursor of this book), as well as to correcting and extending the content. Among them, we especially thank Ganesh Ajjanagadde, Austin Collins, Yuzhou Gu, Richard Guo, Alexander Haberman, Matthew Ho, Qingqing Huang, Yunus Inan, Reka Inovan, Jason Klusowski, Anuran Makur, Pierre Quinton, Aolin Xu, Sheng Xu, Pengkun Yang, Andrew Yao, and Junhui Zhang.

We thank many colleagues who provided valuable feedback at various stages of the book draft over the years, in particular, Lucien Birgé, Marco Dalai, Meir Feder, Bob Gallager, Bobak Nazer, Or Ordentlich, Henry Pfister, Maxim Raginsky, Sasha

Rakhlin, Philippe Rigollet, Mark Sellke, and Nikita Zhivotovskiy. Rachel Cohen (MIT) has been very kind with her time and helped in a myriad of different ways.

We are grateful for the support from our editors Chloe Mcloughlin, Elizabeth Horne, Sarah Strange, and Julie Lancashire at Cambridge University Press (CUP) and for CUP to allow us to keep a free version online. Our special acknowledgement is to Julie for providing that initial push and motivation in 2019, without which we would have never even considered to go beyond the initial set of chaotic online lecture notes. (Though, if we knew it would take five years ...)

The cover art and artwork were contributed by the talented illustrator Nastya Mukhanova [312], whom we cannot praise enough.

Y. P. would like to thank Olga for her unwavering patience and encouragement. Her loving sacrifice made the luxury of countless hours of extra time available to Y. P. to work on this book. Y. P. would also like to extend a literary hug to Yana, Alina, and Evan and thank them for brightening up his life.

Y. W. would like to thank his parents and his wife, Nanxi.

Y. Polyanskiy
Y. Wu

Introduction

What Is Information?

The Oxford English Dictionary lists 18 definitions of the word *information*, while the Merriam-Webster Dictionary lists 17. This emphasizes the diversity of meaning and contexts in which the word *information* may appear. This book, however, is only concerned with a precise mathematical understanding of information, independent of the application domain.

How can we measure something that we cannot even define well? Among the earliest attempts of quantifying information we can list R. A. Fisher's works on the uncertainty of statistical estimates ("confidence intervals") and R. Hartley's definition of information as the logarithm of the number of possibilities. Around the same time, Fisher [169] and others identified connection between information and thermodynamic *entropy*. This line of thinking culminated in Claude Shannon's magnum opus [379], where he formalized the concept of (what we call today) the Shannon information and forever changed the human language by accepting John Tukey's word *bit* as the unit of its measurement. In addition to possessing a number of elegant properties, Shannon information turned out to also answer certain rigorous mathematical questions (such as the optimal rate of data compression and data transmission). This singled out Shannon's definition as the right way of quantifying information. Classical information theory, as taught in [106, 111, 177], focuses exclusively on this point of view.

In this book, however, we take a slightly more general point of view. To introduce it, let us quote an eminent physicist, L. Brillouin [76]:

> We must start with a precise definition of the word "information". We consider a problem involving a certain number of possible answers, if we have no special information on the actual situation. When we happen to be in possession of some information on the problem, the number of possible answers is reduced, and complete information may even leave us with only one possible answer. Information is a function of the ratio of the number of possible answers before and after, and we choose a logarithmic law in order to insure additivity of the information contained in independent situations.

Note that only the last sentence specializes the more general term *information* to Shannon's special version. In this book, we think of information without that last sentence. Namely, for us information is a measure of *difference between two beliefs about the system state*. For example, it could be the amount of *change* in our worldview following an observation or an event. Specifically, suppose that initially the probability distribution P describes our understanding of the world (e.g. P allows us to answer questions such as how likely it is to rain today). Following an observation our distribution changes to Q (e.g. upon observing clouds or a clear sky). The amount of information in the observation is the *dissimilarity* between P and Q. How to quantify dissimilarity depends on the particular context. As argued by Shannon, in many cases the right choice is the Kullback–Leibler (KL) divergence $D(Q\|P)$, see Definition 2.1. Indeed, if the prior belief is described by a probability mass function $P = (p_1, \dots, p_k)$ on the set of k possible outcomes, then the observation of the first outcome results in the new (posterior) belief vector $Q = (1, 0, \dots, 0)$ giving $D(Q\|P) = \log \frac{1}{p_1}$, and similarly for other outcomes. Since the outcome i happens with probability p_i we see that the average dissimilarity between the prior and posterior beliefs is

$$\sum_{i=1}^{k} p_i \log \frac{1}{p_i},$$

which is precisely the Shannon entropy, see Definition 1.1.

However, it is our conviction that measures of dissimilarity (or "information measures") other than the KL divergence are needed for applying information theory beyond the classical realms. For example, the concepts of total variation, Hellinger distance, and χ^2-divergence (both prominent members of the f-divergence family) have found deep and fruitful applications in the theory of statistical estimation and probability, as well as contemporary topics in theoretical computer science such as communication complexity, estimation with communication constraints, and property testing (we discuss these in detail in Part VI). Therefore, when we talk about information measures in Part I of this book we do not exclusively focus on those of Shannon type, although the latter are justly given a premium treatment.

What Is Information Theory?

Similarly to *information*, the subject of *information theory* does not have a precise definition. In the narrowest sense, it is a scientific discipline concerned with optimal methods of transmitting and storing data. The highlights of this part of the subject are so called "coding theorems" showing existence of algorithms for compressing and communicating information across noisy channels. Classical results, such as Shannon's noisy channel coding theorem (Theorem 19.9), not only show existence of algorithms, but also quantify their performance and show that such performance is best possible. This part is, thus, concerned with identifying fundamental limits of practically relevant (engineering) problems. Consequently, this branch is sometimes called "IEEE[1]-style information theory," and it influenced or revolutionized much of information technology we witness today: digital communication, wireless (cellular and WiFi) networks, cryptography (Diffie–Hellman), data compression (Lempel–Ziv family of algorithms), and a lot more.

[1] For Institute of Electrical and Electronics Engineers; pronounced "Eye-triple-E."

This book, however, is not limited to the IEEE-style information theory, because the true scope of the field is much broader. Indeed, the Hilbert's thirteenth problem (for smooth functions) was illuminated and resolved by Arnold and Kolmogorov via the idea of metric entropy that Kolmogorov introduced following Shannon's rate-distortion theory [441]. The isomorphism problem for Bernoulli shifts in ergodic theory has been solved by introducing the Kolmogorov–Sinai entropy [388, 323]. In physics, the Landauer principle and other works on Maxwell demon have been heavily influenced by the information theory [268, 42]. Natural language processing (NLP), the idea of modeling text as a high-order Markov model, has seen spectacular successes recently in the form of GPT [321] and related models. Many more topics ranging from biology, neuroscience, and thermodynamics to pattern recognition, artificial intelligence, and control theory all regularly appear in information-theoretic conferences and journals.

It seems that objectively circumscribing the territory claimed by information theory is futile. Instead, we would like to highlight what we believe to be the recent developments that fascinate us and which motivated us to write this book.

First, information processing systems of today are much more varied compared to those of last century. A modern controller (robot) is not just reacting to a few-dimensional vector of observations, modeled as a linear time-invariant system. Instead, it has million-dimensional inputs (e.g. a rasterized image), delayed and quantized, which also need to be communicated across noisy links. The target of statistical inference is no longer a low-dimensional parameter, but rather a high-dimensional (possibly discrete) object with structure (e.g. a sparse matrix, or a social network between people with underlying community structure). Furthermore, observations arrive to a statistician from spatially or temporally separated sources, which need to be transmitted cognizant of rate limitations. Recognizing these new challenges, multiple communities simultaneously started re-investigating classical results (Chapter 29) on the optimality of maximum likelihood and the (optimal) variance bounds given by the Fisher information. These developments in high-dimensional statistics, computer science, and statistical learning depend on the mastery of the f-divergences (Chapter 7), the mutual information method (Chapter 30), and the strong version of the data-processing inequality (Chapter 33).

Second, since the 1990s technological advances have brought about a slew of new noisy channel models. While classical theory addresses the so-called memoryless channels, the modern channels, such as in flash storage, or urban wireless (multipath, multi-antenna) communication, are far from memoryless. In order to analyze these, the classical "asymptotic iid" theory is insufficient. The resolution is the so-called "one-shot" approach to information theory, in which all main results are developed while treating the channel inputs and outputs as abstract [211]. Only at the last step are those inputs given the structure of long sequences and the asymptotic values are calculated. This new "one-shot" approach has additional relevance for quantum information theory, where it is in fact necessary.

Third, following impressive empirical achievements in the 2010s there was an explosion in the interest of understanding the methods and limits of machine learning from data. Information-theoretic principles were instrumental for several discoveries in this area. As examples, we recall the concept of metric entropy (Chapter 27) that is a cornerstone of Vapnik's approach to supervised learning (known as empirical risk minimization), non-linear regression, and the theory of density estimation (Chapter 32). In machine learning density estimation is known as probabilistic generative modeling, a prototypical problem in unsupervised learning. At present the best algorithms were derived by applying information-theoretic ideas: Gibbs variational principle for Kullback–Leibler divergence (in variational autoencoders (VAEs), see Example 4.2) and variational characterization of Jensen–Shannon divergence (in generative adversarial networks (GANs), see Example 7.5). Another fascinating connection is that the optimal prediction performance of online learning algorithms is given by the maximum of the mutual information. This is shown through a deep connection between prediction and universal compression (Chapter 13), which led to the discovery of the multiplicative weight update algorithm [446, 104].

On the theoretical side, a common information-theoretic method known as strong data-processing inequality (Chapter 33) led to resolutions of a series of problems in distributed estimation, community detection (in graphs), and principal component analysis (spiked Wigner model). The PAC-Bayes method, rooted in the Donsker–Varadhan characterization of the Kullback–Leibler divergence, led to numerous breakthroughs in the theory of bounding the generalization error of learning algorithms and in understanding concentration and uniform convergence properties of empirical processes in high dimensions (Section 4.8*).

Fourth, theoretical computer science has been exchanging ideas with information theory as well. Classical connections include entropy and combinatorics (Chapter 8); entropy and randomness extraction (Chapter 9); von Neumann's computation with noisy gates (Section 33.1); Ising, Potts, and coloring models on trees and general graphs (Section 33.5); and communication complexity and Hellinger distance (Exercise **I.41**). More recently, skillful applications of the chain rule led to an elegant strengthening of Szemerédi's regularity lemma in graph theory by Tao (Exercise **I.63**) and to a breakthrough in the union-closed sets conjecture by Gilmer (Exercise **I.61**). The so-called I-MMSE identity (Section 3.7*) was applied to get a very short proof of stochastic localization (Exercise **I.66**). In the area of randomized sampling and counting, the method of spectral independence (Exercise **VI.26**) resolved multiple long-standing conjectures.

Why Another Book on Information Theory?

Our motivation for writing this book was two-fold. First, in our experience there is a need for a graduate-level textbook on information theory, developed at an acceptable level of generality (i.e. not restricted to discrete, or categorical, random variables) while not sacrificing any mathematical rigor. Second, we wanted to introduce readers to all the exciting (classical and new) connections between information theory and other disciplines that we surveyed in the previous section. We believe that topics like the f-divergences, the one-shot point of view, the connections with statistical learning and probability are not covered adequately in existing textbooks and are future-proof: their significance will only grow with time. Currently being relegated to specialized monographs, acquisition of this toolkit by an aspiring student is delayed.

There are two great classical textbooks that are unlikely to become irrelevant any time soon: [106] by Cover and Thomas and [111] by Csiszár and Körner (and the revised edition of the latter [115]). The former has been a primary textbook for the majority of undergraduate courses on information theory in the world. It manages to rigorously introduce the concepts of entropy, information, and divergence, and prove all the main results of the field, while also sampling several less standard topics, such as universal compression, gambling, and portfolio theory.

The textbook [111] spearheaded the combinatorial approach on information theory, known as "the method of types." While more mathematically demanding than [106], [111] manages to introduce stronger results such as sharp estimates of error exponents and, especially, rate regions in multi-terminal communication systems. However, both books are almost exclusively focused on asymptotics, Shannon-type information measures, and discrete (finite-alphabet) cases.

Focused on specialized topics, several monographs are available. For a communication-oriented reader, the classical [177] is still indispensable. The one-shot point of view is taken in [211]. Connections to statistical learning theory and learning on graphs (belief propagation) is beautifully covered in [288]. Ergodic theory is the central subject in [198]. Quantum information theory – a burgeoning field – is treated in the recent [452]. The only textbook dedicated to the connection between information theory and statistics is by Kullback [265], though restricted to large-sample asymptotics in hypothesis testing. In nonparametric statistics, application of information-theoretic methods is briefly (but elegantly) covered in [425].

Nevertheless, it is not possible to quilt this textbook from chapters of these excellent predecessors. A number of important topics are treated exclusively here, such as those in Chapters 7 (f-divergences), 18 (one-shot coding theorems), 22 (finite blocklength), 27 (metric entropy), 30 (mutual information method), 32 (entropic bounds on estimation), and 33 (strong data-processing inequalities). Furthermore, building up to these chapters requires numerous small innovations across the rest of the textbook and are not available elsewhere. In addition, the exercises explore works of the last few years.

Going to omissions, this book almost entirely skips the topic of multi-terminal information theory (with exception of Sections 11.7*, 16.5*, and 25.3*). This difficult subject captivated much of the effort in the post-Shannon "IEEE-style" theory. We refer to the classical [115] and the recent excellent textbook [147] containing an encyclopedic coverage of this area.

Another unfortunate omission is the connection between information theory and functional inequalities [106, Chapter 17]. This topic has seen a flurry of recent activity, especially in logarithmic Sobolev inequalities, isoperimetry, concentration of measure, Brascamp–Lieb inequalities, (Marton–Talagrand) information-transportation inequalities, and others. We only briefly mention these topics in Sections 3.7*, 3.8*, and associated exercises (e.g. I.47 and I.65). For a fuller treatment, see the monograph [354] and references therein.

Finally, this book will not teach one how to construct practical error-correcting codes or design modern wireless communication systems. Following our Part IV, which covers the basics, an interested reader is advised to master the tools from coding theory via [361] and multiple-antenna channels via [424].

A Note to Statisticians

The interplay between information theory and statistics is a constant theme in the development of both fields. Since its inception, information theory has been indispensable for understanding the fundamental limits of statistical estimation. The prominent role of information-theoretic quantities, such as mutual information, f-divergence, metric entropy, and capacity, in establishing the minimax rates of estimation has long been recognized since the seminal work of Le Cam [273], Ibragimov and Khas'minski [222], Pinsker [329], Birgé [53], Haussler and Opper [216], and Yang and Barron [465], among many others. In Part VI of this book we give an exposition to some of the most influential information-theoretic ideas and their applications in statistics. This part is not meant to be a thorough treatment of decision theory or mathematical statistics; for that purpose, we refer to the classics [222, 277, 68, 425] and the more recent monographs [78, 190, 447] focusing on high dimensions. Instead, we apply the theory developed in previous Parts I–V of this book to several concrete and carefully chosen examples of determining the minimax risk in both classical (fixed-dimensional, large-sample asymptotic) and modern (high-dimensional, non-asymptotic) settings.

At a high level, the connection between information theory (in particular, data transmission) and statistical inference is that both problems are defined by a conditional distribution $P_{Y|X}$, which is referred to as the *channel* for the former and the *statistical model* or *experiment* for the latter. In both disciplines the ultimate goal is to estimate X with high fidelity based on its noisy observation Y using computationally efficient algorithms. However, in data transmission the set of allowed values of X is typically discrete and restricted to a carefully chosen subset of inputs (called codebook), the design of which is considered to be the main difficulty. In statistics, however, the space or the distribution of allowed values of X (the parameter) is constrained by the problem setup (for example, requiring sparsity or low rank on X), not by the statistician. Despite this key difference, both disciplines in the end are all about estimating X based on Y and information-theoretic ideas are applicable in both settings.

Specifically, in Chapter 29 we show how the data-processing inequality can be used to deduce classical lower bounds in statistical estimation (Hammersley–Chapman–Robbins, Cramér–Rao, van Trees). In Chapter 30 we introduce the *mutual information method*, based on the reasoning in joint source–channel coding. Namely, by comparing the amount of information contained in the data and the amount of information required for achieving a given estimation accuracy, both measured in bits, this method allows us to apply the theory of capacity and rate-distortion function developed in Parts IV and V to lower-bound the statistical risk. Besides being principled, this approach also unifies the three popular methods for proving minimax lower bounds due to Le Cam, Assouad, and Fano respectively (Chapter 31).

It is a common misconception that information theory only supplies techniques for proving negative results in statistics. In Chapter 32 we present three *upper bounds* on statistical estimation risk based on *metric entropy*: Yang and Barron's construction inspired by universal compression, Le Cam and Birgé's tournament based on pairwise hypothesis testing, and Yatracos' minimum-distance approach. These powerful methods are responsible for some of the strongest and most general results in statistics and applicable to both high-dimensional and nonparametric problems. Finally, in Chapter 33 we introduce the method based on strong data-processing inequalities and apply it to resolve an array of contemporary problems including community detection on graphs, distributed estimation with communication constraints, and generating random tree colorings. These problems are increasingly captivating the minds of computer scientists as well.

How to Use This Textbook

Each Part of this book starts with a preface that briefly summarizes its contents. As a first step, we recommend reading through those to get an overall idea about the content of the book.

An introductory class on information theory aiming at advanced undergraduate or graduate students can proceed with the following sequence:

- Part I: Chapters 1–3, Sections 4.1, 5.1–5.3, 6.1, and 3.6, focusing only on discrete probability space and ignoring Radon–Nikodym derivatives. Some mention of applications in combinatorics and cryptography (Chapters 8, 9, and select exercises) is recommended.
- Part II: Chapter 10, Sections 11.1–11.5.
- Part III: Chapter 14, Sections 15.1–15.3, and 16.1.
- Part IV: Chapters 17–18, Sections 19.1–19.3, 19.7, 20.1–20.2, and 23.1.
- Part V: Sections 24.1–24.3, 25.1, 26.1, and 26.3.
- Conclude with a few applications of information theory outside the classical domain (Chapters 30 and 33).

A graduate-level class on information theory with a traditional focus on communication and compression can proceed faster through Part I (omitting f-divergences and other non-essential chapters), but then cover Parts II–V in depth, including strong converse, finite-blocklength regime, and communication with feedback, but omitting Chapter 27. It is important to work through exercises at the end of Part IV for this kind of class.

For a graduate-level class on information theory with an emphasis on statistical learning, start with Part I (especially Chapter 7), followed by Part II (especially Chapter 13) and Part III, from Part IV limit coverage to Chapters 17–19, and from Part V to Chapter 27 (especially, Sections 27.1–27.4). This should leave more than half of the semester for carefully working through Part VI. For example, for a good pace we suggest leaving at least five or six lectures for Chapters 32 and 33. These last chapters contain some bleeding-edge research results and open problems, hopefully welcoming students to work on them. For that we also recommend going over the exercises at the end of Parts I and VI.

Difficult sections are marked with asterisks (*) and can be skipped on a first reading as they may rely on material from future chapters or external sources.

An extensive index should help connect different topics together. For example, looking up "community detection" shows all the many occurrences of this interesting example across the chapters.

Frequently Used Notation

General Conventions

- The following abbreviations and acronyms are used in this book, perhaps without further explanation: a.e., almost every (almost everywhere); a.s., almost surely; CDF, cumulative distribution function; iff, if and only if; iid, independent and identically distributed; LHS, left-hand side; PDF, probability density function; PMF, probability mass function; RHS, right-hand side; s.t., so that (such that); w.p., with probability; w.r.t., with respect to.
- We adopt the French style so that x is positive means $x \geq 0$ and x is strictly positive means $x > 0$. Similar conventions apply to other quantifiers such as monotonically increasing, convex, etc.
- The symbol $\triangleq$ reads *defined as* and $\equiv$ *abbreviated as*.
- The set of real numbers and integers are denoted by $\mathbb{R}$ and $\mathbb{Z}$. Let $\mathbb{N} = \{1, 2, \ldots\}$, $\mathbb{Z}_+ = \{0, 1, \ldots\}$, and $\mathbb{R}_+ = \{x\colon x \geq 0\}$.
- For $n \in \mathbb{N}$, let $[n] = \{1, \ldots, n\}$.
- Throughout the book, $x^n \triangleq (x_1, \ldots, x_n)$ denotes an n-dimensional vector, $x_i^j \triangleq (x_i, \ldots, x_j)$ for $1 \leq i < j \leq n$, and $x_S \triangleq \{x_i\colon i \in S\}$ for $S \subset [n]$.
- Unless explicitly specified, the logarithm log and exponential exp are with respect to a generic common base. The natural logarithm is denoted by $\ln = \log_e$ and $\exp_e\{\cdot\} = e^{(\cdot)}$.
- We agree to take $\exp\{+\infty\} = +\infty, \exp\{-\infty\} = 0, \log(+\infty) = +\infty$, and $\log(0) = -\infty$. The function $x \mapsto x \log x$ is extended to $x = 0$ by taking $0 \cdot \log 0 = 0$. The bivariate function $(x, y) \mapsto \log \frac{x}{y}$ extended to $x = 0$ and $y = 0$ is denoted by $\mathrm{Log}\frac{x}{y}$ and has a special convention (2.10).
- $a \wedge b = \min\{a, b\}$ and $a \vee b = \max\{a, b\}$.
- For $p \in [0, 1], \bar{p} \triangleq 1 - p$.
- $x^+ = \max\{x, 0\}$.
- Right and left limits: $f(x+) \triangleq \lim\limits_{y \searrow x} f(y)$ and $f(x-) \triangleq \lim\limits_{y \nearrow x} f(y)$.
- Limit inferior and limit superior: $\liminf_{n\to\infty} g_n \triangleq \lim_{n\to\infty} \inf_{m \geq n} g_m$ and $\limsup_{n\to\infty} g_n \triangleq \lim_{n\to\infty} \sup_{m \geq n} g_m$.
- $w_{\mathrm{H}}(x)$ denotes the Hamming weight (number of ones) of a binary vector x. $d_{\mathrm{H}}(x, y) = \sum_{i=1}^n 1\{x_i \neq y_i\}$ denotes the Hamming distance between vectors x and y of length n.

- Standard big-O notation is used throughout the book: for example, for any positive sequences $\{a_n\}$ and $\{b_n\}$, $a_n = O(b_n)$ or equivalently, $a_n \lesssim b_n$ if there is an absolute constant $c > 0$ such that $a_n \le cb_n$; $a_n = \Omega(b_n)$ or $a_n \gtrsim b_n$ if $b_n = O(a_n)$; $a_n = \Theta(b_n)$ or $a_n \asymp b_n$ if both $a_n = O(b_n)$ and $a_n = \Omega(b_n)$; $a_n = o(b_n)$ or $b_n = \omega(a_n)$ if $a_n \le \epsilon_n b_n$ for some $\epsilon_n \to 0$. In addition, if there is a parameter p in the discussion and the constant c in the definition of $a_n = O(b_n)$ depends on p, then we emphasize this fact by writing $a_n = O_p(b_n)$.

Information Theory and Statistics

- $h(\cdot)$ is the binary entropy function, $H(\cdot)$ denotes general Shannon entropy.
- $d(\cdot\|\cdot)$ is the binary divergence function, $D(\cdot\|\cdot)$ denotes general Kullback–Leibler divergence.
- Standard channels BSC_δ, BEC_δ, and $\mathsf{BIAWGN}_{\sigma^2}$.
- Common divergences are $\chi^2(\cdot\|\cdot)$ (chi-squared), $D_\alpha(\cdot\|\cdot)$ (Rényi divergence), and $D_f(\cdot\|\cdot)$ (general f-divergence).
- Common statistical distances $\mathrm{TV}(\cdot,\cdot)$ (total variation), $H^2(\cdot,\cdot)$ (Hellinger-squared), and $W_1(\cdot,\cdot)$ (Wasserstein distance).
- Depending on context P_θ may denote a parametric class of distributions indexed by the parameter θ, or it can mean the law of random variable θ.

Analysis and Algebra

- For $x \in \mathbb{R}^d$ we denote $\|x\|_p = \left(\sum_{i=1}^p |x_i|^p\right)^{1/p}$ and $\|x\| = \|x\|_2$ the standard Euclidean norm.
- Let $\mathrm{int}(E)$ and $\mathrm{cl}(E)$ denote the interior and closure of a set E, namely, the largest open set contained in and smallest closed set containing E, respectively.
- Let $\mathrm{co}(E)$ denote the convex hull of E (without topology), namely, the smallest convex set containing E, given by $\mathrm{co}(E) = \left\{\sum_{i=1}^n \alpha_i x_i : \alpha_i \ge 0, \sum_{i=1}^n \alpha_i = 1, x_i \in E, n \in \mathbb{N}\right\}$.
- For subsets A, B of a real vector space and $\lambda \in \mathbb{R}$, denote the dilation $\lambda A = \{\lambda a : a \in A\}$ and the Minkowski sum $A + B = \{a + b : a \subset A, B \in B\}$.
- For a metric space $(\mathcal{X}, d)$, a function $f : \mathcal{X} \to \mathbb{R}$ is called C-Lipschitz if $|f(x) - f(y)| \le Cd(x, y)$ for all $x, y \in \mathcal{X}$. We set $\|f\|_{\mathrm{Lip}(\mathcal{X})} = \inf\{C : f \text{ is } C\text{-Lipschitz}\}$.
- Linear algebraic notation. For a matrix A with real entries we define:
  - trace $\mathrm{tr}\, A = \sum_i A_{i,i}$;
  - $\sigma_1(A) \ge \sigma_2(A) \ge \cdots$ its list of singular values sorted in decreasing order, and recall that $\sigma_j(A)$ is the square-root of the jth largest eigenvalue of $A^\top A$;
  - the operator norm $\|A\|_{op} = \sup_{v \ne 0} \frac{\|Av\|}{\|v\|} = \sigma_1(A)$;
  - the Frobenius norm $\|A\|_{\mathrm{F}}^2 = \sum_{i,j} |A_{i,j}|^2 = \mathrm{tr}\, A^\top A = \mathrm{tr}\, AA^\top = \sum_i \sigma_i^2(A)$;
  - we write $A \succeq 0$ to denote that A is positive semidefinite, and $A \succeq B$ to denote that $A - B \succeq 0$. Similarly, positive definiteness is denoted by $\succ$.
- $\mathrm{Hess}\, f(x)$ denotes the Hessian of f: $(\mathrm{Hess}\, f(x))_{i,j} = \frac{\partial^2 f}{\partial x_i \partial x_j}(x)$. $\Delta f(x) = \mathrm{tr}\,\mathrm{Hess}\, f(x) = \sum_i \frac{\partial^2}{\partial x_i^2} f(x)$ denotes the Laplacian.

- Convolution of two functions f, g on $\mathbb{R}^d$ is defined as $(f * g)(x) = \int_{\mathbb{R}^d} f(y)g(x - y)dy$ for all x where the (Lebesgue) integral exists. For two probability measures p, q on $\mathbb{R}^d$ we also define $p * q$ to be the law of $A + B$ where $A \perp\!\!\!\perp B$ and $A \sim p$, $B \sim q$.

Measure Theory and Probability

- The Lebesgue measure on Euclidean spaces is denoted by Leb and also by vol (volume).
- Throughout the book, all measurable spaces $(\mathcal{X}, \mathcal{E})$ are standard Borel spaces. Unless explicitly needed, we suppress the underlying σ-algebra $\mathcal{E}$.
- The collection of all probability measures on $\mathcal{X}$ is denoted by $\mathcal{P}(\mathcal{X})$. For finite spaces we abbreviate $\mathcal{P}_k \equiv \mathcal{P}([k])$, a $(k-1)$-dimensional simplex.
- Let P be absolutely continuous with respect to Q, denoted by $P \ll Q$. The Radon–Nikodym derivative of P with respect to Q is denoted by $\frac{dP}{dQ}$. For a probability measure P, if $Q = \text{Leb}$, $\frac{dP}{dQ}$ is referred to as the probability density function (PDF); if Q is the counting measure on a countable $\mathcal{X}$, $\frac{dP}{dQ}$ is the probability mass function (PMF).
- Let $P \perp Q$ denote their mutual singularity, namely, $P(A) = 0$ and $Q(A) = 1$ for some A.
- A^c denotes the complement of a set A.
- The *support* of a probability measure P, denoted by $\text{supp}(P)$, is the smallest closed set C such that $P(C) = 1$. An *atom* x of P is such that $P(\{x\}) > 0$. A distribution P is *discrete* if $\text{supp}(P)$ is a countable set (consisting of its atoms).
- Let X be a random variable taking values on $\mathcal{X}$, which is referred to as the *alphabet* of X. Its realizations are labeled by lower case letters, for example, x. Thus, upper case, lower case, and script case are matched to random variables, realizations, and alphabets, respectively (as in $X = x \in \mathcal{X}$). Oftentimes $\mathcal{X}$ and $\mathcal{Y}$ are automatically assumed to be the alphabet of X and Y, etc. We also write $X \in \mathcal{X}$ to mean that random variable X is $\mathcal{X}$-valued.
- Let P_X denote the distribution (law) of the random variable X, $P_{X,Y}$ the joint distribution of X and Y, and $P_{Y|X}$ the conditional distribution of Y given X.
- A conditional distribution $P_{Y|X}$ is also called a Markov kernel acting between spaces $\mathcal{X}$ and $\mathcal{Y}$, written as $P_{Y|X}\colon \mathcal{X} \to \mathcal{Y}$. Given a conditional distribution $P_{Y|X}$ and a marginal we can form a joint distribution, written as $P_X \times P_{Y|X}$, or simply $P_X P_{Y|X}$. Its marginal P_Y is denoted by a composition operation $P_Y \triangleq P_{Y|X} \circ P_X$.
- The independence of random variables X and Y is denoted by $X \perp\!\!\!\perp Y$, in which case $P_{X,Y} = P_X \times P_Y$. Similarly, $X \perp\!\!\!\perp Y|Z$ denotes their conditional independence given Z, in which case $P_{X,Y|Z} = P_{X|Z} \times P_{Y|Z}$.
- For measures P and Q, their product measure is denoted by $P \times Q$ or $P \otimes Q$. The n-fold product of P is denoted by P^n or $P^{\otimes n}$. Similarly, given a Markov kernel $P_{Y|X}\colon \mathcal{X} \to \mathcal{Y}$ the kernel that acts independently on each of n coordinates is denoted as $P^n_{Y|X}$ or $P^{\otimes n}_{Y|X}\colon \mathcal{X}^n \to \mathcal{Y}^n$.

- Throughout the book, $X^n \equiv X_1^n \triangleq (X_1, \ldots, X_n)$ denotes an n-dimensional random vector. We write $X_1, \ldots, X_n \stackrel{\text{iid}}{\sim} P$ if they are independently and identically distributed (iid) as P, in which case $P_{X^n} = P^n$.
- The empirical distribution of a sequence $x_1, \ldots, x_n$ is denoted by $\hat{P}_{x^n}$; the empirical distribution of a random sample $X_1, \ldots, X_n$ is denoted by $\hat{P}_n \equiv \hat{P}_{X^n}$.
- $\xrightarrow{\text{a.s.}}$, $\xrightarrow{\mathbb{P}}$, and $\xrightarrow{\text{d}}$ denote convergence almost surely, in probability, and in distribution (law), respectively. We define $\stackrel{\text{d}}{=}$ to mean equality in distribution.
- Occasionally, for clarity we use a self-explanatory notation $\mathbb{E}_{Y \sim Q}[\cdot]$ to mean that the expectation is taken with Y generated from distribution Q. We also use cues like $\mathbb{E}_C[\cdot]$ to signify that the expectation is taken over C.
- Some commonly used distributions are as follows:
  - $\text{Ber}(p)$ is the Bernoulli distribution with mean p.
  - $\text{Bin}(n, p)$ is the binomial distribution with n trials and success probability p.
  - $\text{Poisson}(\lambda)$ is the Poisson distribution with mean λ.
  - $\mathcal{N}(\mu, \sigma^2)$ is the Gaussian (normal) distribution on $\mathbb{R}$ with mean μ and σ^2. $\mathcal{N}(\mu, \Sigma)$ is the Gaussian distribution on $\mathbb{R}^d$ with mean μ and covariance matrix Σ. Denote the standard normal density by $\varphi(x) = \frac{1}{\sqrt{2\pi}} e^{-x^2/2}$, the cumulative distribution function (CDF) and complementary CDF by $\Phi(t) = \int_{-\infty}^{t} \varphi(x) dx$ and $Q(t) = \Phi^c(t) = 1 - \Phi(t)$. The inverse of Q is denoted by Q^{-1}.
  - $Z \sim \mathcal{N}_c(\mu, \sigma^2)$ denotes the complex-valued circular symmetric normal distribution with expectation $\mathbb{E}[Z] = \mu \in \mathbb{C}$ and $\mathbb{E}[|Z - \mu|^2] = \sigma^2$.
  - For a compact subset $\mathcal{X}$ of $\mathbb{R}^d$ with non-empty interior, $\text{Unif}(\mathcal{X})$ denotes the uniform distribution on $\mathcal{X}$, with $\text{Unif}(a, b) \equiv \text{Unif}([a, b])$ for interval $[a, b]$. We also use $\text{Unif}(\mathcal{X})$ to denote the uniform (equiprobable) distribution on a finite set $\mathcal{X}$.
- For a $\mathbb{R}^d$-valued random variable X we denote $\text{Cov}(X) = \mathbb{E}[(X - \mathbb{E}[X])(X - \mathbb{E}[X])^\top]$ its covariance matrix. A conditional version is denoted as $\text{Cov}(X|Y) = \mathbb{E}[(X - \mathbb{E}[X|Y])(X - \mathbb{E}[X|Y])^\top]$.
- For a set $E \subset \Omega$ we denote by $1_E(\omega)$ the function equal to 1 iff $\omega \in E$. Similarly, $1\{\text{boolean condition}\}$ denotes a random variable that is equal to 1 iff the "boolean condition" is satisfied and otherwise equals 0. Thus, for example, $\mathbb{P}[X > 1] = \mathbb{E}[1\{X > 1\}]$.

Part I

Information Measures

Information measures form the backbone of information theory. The first part of this book is devoted to an in-depth study of some of them, most notably, entropy, divergence, and mutual information, as well as their conditional versions (Chapters 1–3). In addition to basic definitions illustrated through concrete examples, we will also study various aspects including chain rules, regularity, tensorization, variational representation, local expansion, convexity, and optimization properties, as well as the data-processing principle (Chapters 4–6). These information measures will be imbued with operational meaning when we proceed to classical topics in information theory such as data compression and transmission, in subsequent parts of the book. This part also includes topics connecting information theory to other subjects, such as I-MMSE relation (estimation theory), entropy-power inequality (probability), PAC-Bayes bounds, and Gibbs variational principle (machine learning).

In addition to the classical (Shannon) information measures, Chapter 7 provides a systematic treatment of f-divergences, a generalization of (Shannon) measures introduced by Csíszar that plays an important role in many statistical problems (see Parts III and VI). Finally, toward the end of this part we will discuss two operational topics: random number generators in Chapter 9 and the application of the entropy method to combinatorics and geometry in Chapter 8.

Several contemporary topics are developed in exercises such as stochastic block model (Exercise **I.49**), Gilmer's method in combinatorics (Exercise **I.61**), Tao's proof of Szemerédi's regularity lemma (Exercise **I.63**), Eldan's stochastic localization (Exercise **I.66**), Gross' log-Sobolev inequality (Exercise **I.65**), and others.

1 Entropy

This chapter introduces the first information measure – Shannon entropy. After studying its standard properties (chain rule, conditioning), we will briefly describe how one could arrive at its definition. We discuss axiomatic characterization, the historical development in statistical mechanics, as well as the underlying combinatorial foundation ("method of types"). We close the chapter with Han's and Shearer's inequalities, that both exploit submodularity of entropy. After this chapter, the reader is welcome to explore the applications in combinatorics (Chapter 8) and random number generation (Chapter 9), which are independent of the rest of this part.

1.1 Entropy and Conditional Entropy

Definition 1.1. (Entropy) Let X be a discrete random variable with probability mass function (PMF) $P_X(x), x \in \mathcal{X}$. The *entropy* (or *Shannon entropy*) of X is

$$\begin{aligned} H(X) &= \mathbb{E}\left[\log \frac{1}{P_X(X)}\right] \\ &= \sum_{x \in \mathcal{X}} P_X(x) \log \frac{1}{P_X(x)} . \end{aligned}$$

When computing the sum, we agree that (by continuity of $x \mapsto x \log \frac{1}{x}$)

$$0 \log \frac{1}{0} = 0. \tag{1.1}$$

Since entropy only depends on the distribution of a random variable, it is customary in information theory to also write $H(P_X)$ in place of $H(X)$, which we will do freely in this book. The basis of the logarithm in Definition 1.1 determines the units of entropy:

$$\begin{aligned} \log_2 &\leftrightarrow \text{bits}, \\ \log_e &\leftrightarrow \text{nats}, \\ \log_{256} &\leftrightarrow \text{bytes}, \\ \log &\leftrightarrow \text{arbitrary units, base always matches exp.} \end{aligned}$$

Different units will be convenient in different cases and so most of the general results in this book are stated with "baseless" log/exp.

Definition 1.2. (Joint entropy) The *joint entropy* of n discrete random variables $X^n \triangleq (X_1, X_2, \ldots, X_n)$ is

$$H(X^n) = H(X_1, \ldots, X_n) = \mathbb{E}\left[\log \frac{1}{P_{X_1,\ldots,X_n}(X_1, \ldots, X_n)}\right],$$

which can also be written explicitly as a summation over a joint probability mass function (PMF):

$$H(X^n) = \sum_{x_1} \cdots \sum_{x_n} P_{X_1,\ldots,X_n}(x_1, \ldots, x_n) \log \frac{1}{P_{X_1,\ldots,X_n}(x_1, \ldots, x_n)}.$$

Note that joint entropy is a special case of Definition 1.1 applied to the random vector $X^n = (X_1, X_2, \ldots, X_n)$ taking values in the product space.

Remark 1.1. The name "entropy" originates from thermodynamics – see Section 1.3, which also provides combinatorial justification for this definition. Another common justification is to derive $H(X)$ as a consequence of natural axioms for any measure of "information content" – see Section 1.2. There are also natural experiments suggesting that $H(X)$ is indeed the amount of "information content" in X. For example, one can measure the time it takes for ant scouts to describe the location of food to worker ants. It was found that when the nest is placed at the root of a full binary tree of depth d and food at one of the leaves, the time was proportional to the entropy of a random variable describing the food location [359]. (It was also estimated that ants communicate with about 0.7–1 bit/min and that communication time reduces if there are some regularities in path description: paths like "left,right,left,right,left,right" are described by scouts faster).

Entropy measures the intrinsic randomness or uncertainty of a random variable. In the simple setting where X takes values uniformly over a finite set $\mathcal{X}$, the entropy is simply given by log-cardinality: $H(X) = \log|\mathcal{X}|$. In general, the more spread out (respectively concentrated) a probability mass function is, the higher (respectively lower) is its entropy, as demonstrated by the following example.

Example 1.1. Bernoulli

Let $X \sim \text{Ber}(p)$, with $P_X(1) = p$ and $P_X(0) = \bar{p} \triangleq 1 - p$. Then

$$H(X) = h(p) \triangleq p \log \frac{1}{p} + \bar{p} \log \frac{1}{\bar{p}}.$$

Here $h(\cdot)$ is called the *binary entropy function*, which is continuous, concave on $[0, 1]$, symmetric around $\frac{1}{2}$, and satisfies $h'(p) = \log \frac{\bar{p}}{p}$, with infinite slope at 0 and 1. The highest entropy is achieved at $p = \frac{1}{2}$ (uniform), while the lowest entropy is achieved at $p = 0$ or 1 (deterministic). It is instructive to compare the plot of the binary entropy function with the variance $p(1 - p)$.

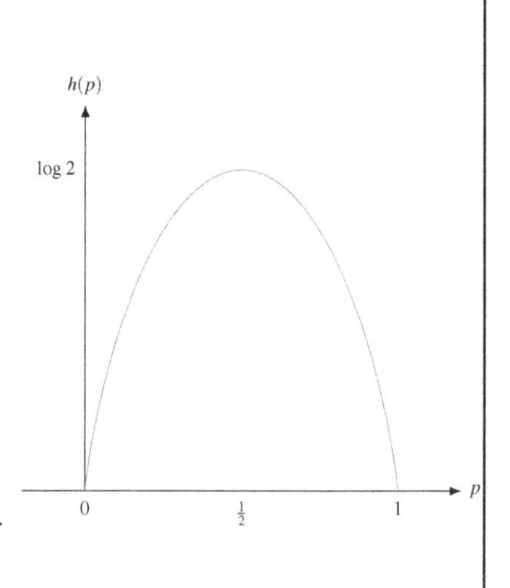

Example 1.2. Geometric

Let X be geometrically distributed, with $P_X(i) = p\bar{p}^i, i = 0, 1, \ldots$. Then $\mathbb{E}[X] = \frac{\bar{p}}{p}$ and

$$H(X) = \mathbb{E}\left[\log \frac{1}{p\bar{p}^X}\right] = \log \frac{1}{p} + \mathbb{E}[X] \log \frac{1}{\bar{p}} = \frac{h(p)}{p}.$$

Example 1.3. Infinite entropy

Is it possible that $H(X) = +\infty$? Yes, for example, $\mathbb{P}[X = k] \propto \frac{1}{k \ln^2 k}, k = 2, 3, \ldots$.

Many commonly used information measures have their conditional counterparts, defined by applying the original definition to a conditional probability measure followed by a further averaging. For entropy this is defined as follows.

Definition 1.3. (Conditional entropy) Let X be a discrete random variable and Y arbitrary. Denote by $P_{X|Y=y}(\cdot)$ or $P_{X|Y}(\cdot|y)$ the conditional distribution of X given $Y = y$. The conditional entropy of X given Y is

$$H(X|Y) = \mathbb{E}_{y \sim P_Y}[H(P_{X|Y=y})] = \mathbb{E}\left[\log \frac{1}{P_{X|Y}(X|Y)}\right].$$

Note that if Y is also discrete we can write out the expression in terms of joint PMF $P_{X,Y}$ and conditional PMF $P_{X|Y}$ as

$$H(X|Y) = \sum_x \sum_y P_{X,Y}(x, y) \log \frac{1}{P_{X|Y}(x|y)}.$$

Similar to entropy, conditional entropy measures the remaining randomness of a random variable when another is revealed. As such, $H(X|Y) = H(X)$ whenever Y is independent of X. But when Y depends on X, observing Y does lower the entropy of X. Before formalizing this in the next theorem, here is a concrete example.

Example 1.4. Conditional entropy and noisy channel

Let Y be a noisy observation of $X \sim \text{Ber}(1/2)$ as follows.

1 Let $Y = X \oplus Z$, where $\oplus$ denotes binary addition (XOR) and $Z \sim \text{Ber}(\delta)$ independently of X. In other words, Y agrees with X with probability δ and disagrees with probability $\bar{\delta}$. Then $P_{X|Y=0} = \text{Ber}(\delta)$ and $P_{X|Y=1} = \text{Ber}(\bar{\delta})$. Since $h(\delta) = h(\bar{\delta})$, $H(X|Y) = h(\delta)$. Note that when $\delta = \frac{1}{2}$, Y is independent of X and $H(X|Y) = H(X) = 1$ bit; when $\delta = 0$ or 1, X is completely determined by Y and hence $H(X|Y) = 0$.

2 Let $Y = X + Z$ be real-valued, where $Z \sim \mathcal{N}(0, \sigma^2)$. Then $H(X|Y) = \mathbb{E}[h(\mathbb{P}[X = 1|Y])]$, where $\mathbb{P}[X = 1|Y = y] = \frac{\varphi(\frac{y-1}{\sigma})}{\varphi(\frac{y}{\sigma}) + \varphi(\frac{y-1}{\sigma})}$ and $Y \sim \frac{1}{2}(\mathcal{N}(0, \sigma^2) + \mathcal{N}(1, \sigma^2))$. Figure 1.1 is a numerical plot of $H(X|Y)$ as a function of σ^2 which can be shown to be monotonically increasing from 0 to 1 bit. (Hint: Theorem 1.4(d).)

Figure 1.1 Conditional entropy of a Bernoulli X given its Gaussian noisy observation.

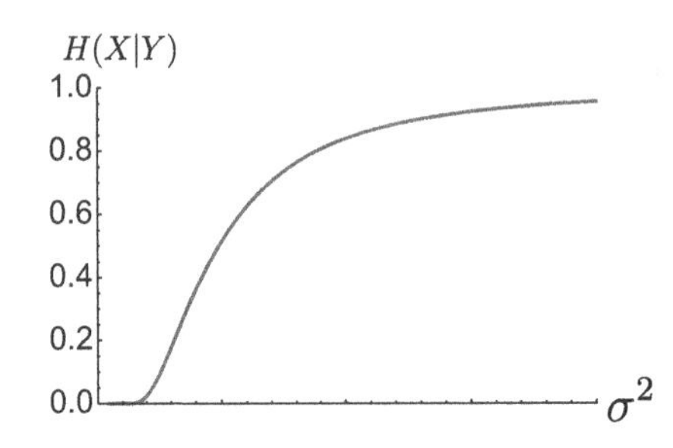

Before discussing various properties of entropy and conditional entropy, let us first review some relevant facts from convex analysis, which will be used extensively throughout the book.

Review: Convexity

- **Convex set**: A subset S of some vector space is *convex* if $x, y \in S \Rightarrow \alpha x + \bar{\alpha} y \in S$ for all $\alpha \in [0,1]$. (Recall: $\bar{\alpha} \triangleq 1 - \alpha$.)
  Examples: unit interval $[0,1]$; $S = \{\text{probability distributions on } \mathcal{X}\}$; $S = \{P_X : \mathbb{E}[X] = 0\}$.
- **Convex function**: $f : S \to \mathbb{R}$ is
  - *convex* if $f(\alpha x + \bar{\alpha} y) \le \alpha f(x) + \bar{\alpha} f(y)$ for all $x, y \in S, \alpha \in [0,1]$;
  - *strictly convex* if $f(\alpha x + \bar{\alpha} y) < \alpha f(x) + \bar{\alpha} f(y)$ for all $x \ne y \in S, \alpha \in (0,1)$;
  - *(strictly) concave* if $-f$ is (strictly) convex.

  Examples: $x \mapsto x \log x$ is strictly convex; the mean $P \mapsto \int x dP$ is convex but not strictly convex; variance is concave. (Question: is it strictly concave? Think of zero-mean distributions.)
- **Jensen's inequality**:
  For any S-valued random variable X,
  - f is convex $\Rightarrow f(\mathbb{E}X) \le \mathbb{E}f(X)$;
  - f is strictly convex $\Rightarrow f(\mathbb{E}X) < \mathbb{E}f(X)$, unless X is a constant ($X = \mathbb{E}X$ a.s.).

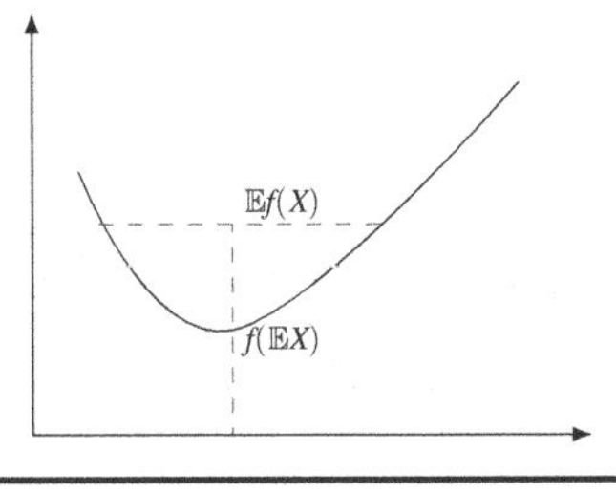

Theorem 1.4. (Properties of entropy)

(a) (Positivity) $H(X) \ge 0$ with equality iff X is a constant (no randomness).
(b) (Uniform distribution maximizes entropy) For finite $\mathcal{X}$, $H(X) \le \log|\mathcal{X}|$, with equality iff X is uniform on $\mathcal{X}$.
(c) (Invariance under relabeling) $H(X) = H(f(X))$ for any bijective f.
(d) (Conditioning reduces entropy) $H(X|Y) \le H(X)$, with equality iff X and Y are independent.
(e) (Simple chain rule)

$$H(X,Y) = H(X) + H(Y|X) \le H(X) + H(Y). \tag{1.2}$$

(f) *(Entropy under deterministic transformation)* $H(X) = H(X, f(X)) \geq H(f(X))$ *with equality iff* f *is one-to-one on the support of* P_X.
(g) *(Full chain rule)*

$$H(X_1, \ldots, X_n) = \sum_{i=1}^{n} H(X_i|X^{i-1}) \leq \sum_{i=1}^{n} H(X_i), \tag{1.3}$$

with equality iff $X_1, \ldots, X_n$ *are mutually independent.*

Proof.

(a) Since $\log \frac{1}{P_X(X)}$ is a positive random variable, its expectation $H(X)$ is also positive, with $H(X) = 0$ if and only if $\log \frac{1}{P_X(X)} = 0$ almost surely, namely, P_X is a point mass.
(b) Apply Jensen's inequality to the strictly concave function $x \mapsto \log x$:
$$H(X) = \mathbb{E}\left[\log \frac{1}{P_X(X)}\right] \leq \log \mathbb{E}\left[\frac{1}{P_X(X)}\right] = \log |\mathcal{X}|.$$
(c) $H(X)$ as a summation only depends on the values of P_X, not locations.
(d) Abbreviate $P(x) \equiv P_X(x)$ and $P(x|y) \equiv P_{X|Y}(x|y)$. Using $P(x) = \mathbb{E}_Y[P(x|Y)]$ and applying Jensen's inequality to the strictly concave function $x \mapsto x \log \frac{1}{x}$,
$$H(X|Y) = \sum_{x \in \mathcal{X}} \mathbb{E}_Y\left[P(x|Y) \log \frac{1}{P(x|Y)}\right] \leq \sum_{x \in \mathcal{X}} P(x) \log \frac{1}{P(x)} = H(X).$$
Additionally, this also follows from (and is equivalent to) Corollary 3.5 in Chapter 3 or Theorem 5.2 in Chapter 5.
(e) Telescoping $P_{X,Y}(X, Y) = P_{Y|X}(Y|X) P_X(X)$ and noting that both sides are positive $P_{X,Y}$-almost surely, we have
$$\begin{aligned}\mathbb{E}\left[\log \frac{1}{P_{X,Y}(X,Y)}\right] &= \mathbb{E}\left[\log \frac{1}{P_X(X) \cdot P_{Y|X}(Y|X)}\right] \\ &= \underbrace{\mathbb{E}\left[\log \frac{1}{P_X(X)}\right]}_{H(X)} + \underbrace{\mathbb{E}\left[\log \frac{1}{P_{Y|X}(Y|X)}\right]}_{H(Y|X)}.\end{aligned}$$
(f) The intuition is that $(X, f(X))$ contains the same amount of information as X. Indeed, $x \mapsto (x, f(x))$ is one-to-one. Thus by (c) and (e):
$$H(X) = H(X, f(X)) = H(f(X)) + H(X|f(X)) \geq H(f(X)).$$
The bound is attained iff $H(X|f(X)) = 0$ which in turn happens iff X is a *constant* given $f(X)$.
(g) Similar to (e), telescoping
$$P_{X_1 X_2 \cdots X_n} = P_{X_1} P_{X_2|X_1} \cdots P_{X_n|X^{n-1}}$$
and taking the logarithm prove the equality. The inequality follows from (d), with the case of equality occurring if and only if $P_{X_i|X^{i-1}} = P_{X_i}$ for $i = 1, \ldots, n$, namely, $P_{X^n} = \prod_{i=1}^{n} P_{X_i}$. □

To give a preview of the *operational meaning* of entropy, let us play the game of *20 Questions*. We are allowed to make queries about some unknown discrete random variable X by asking *yes–no* questions. The objective of the game is to guess the realized value of the random variable X. For example, $X \in \{a, b, c, d\}$ with $\mathbb{P}[X = a] = 1/2$, $\mathbb{P}[X = b] = 1/4$, and $\mathbb{P}[X = c] = \mathbb{P}[X = d] = 1/8$. In this case, we can ask "$X = a$?"; if not, proceed by asking "$X = b$?"; if not, ask "$X = c$?"; after which we will know for sure the realization of X. The resulting average number of questions is $1/2 + 1/4 \times 2 + 1/8 \times 3 + 1/8 \times 3 = 1.75$, which equals $H(X)$ in bits. An alternative strategy is to ask "$X = a, b$ or c, d" in the first round then proceed to determine the value in the second round, which always requires two questions and does worse on average.

It turns out (Section 10.3) that the minimal average number of yes–no questions to pin down the value of X is always between $H(X)$ bits and $H(X) + 1$ bits. In this special case the above scheme is optimal because (intuitively) it always splits the probability in half.

1.2 Axiomatic Characterization

One might wonder why entropy is defined as $H(P) = \sum p_i \log \frac{1}{p_i}$ and if there are other definitions. Indeed, the information-theoretic definition of entropy is related to entropy in statistical physics. Also, it arises as answers to specific operational problems, for example, the minimum average number of bits to describe a random variable as discussed above. Therefore it is fair to say that it is not pulled out of thin air.

Shannon in a 1948 paper has also showed that entropy can be defined *axiomatically*, as a function satisfying several natural conditions. Denote a probability distribution on m letters by $P = (p_1, \ldots, p_m)$ and consider a functional $H_m(p_1, \ldots, p_m)$. Assume H_m obeys the following axioms:

(a) Permutation invariance: $H_m(p_{\pi(1)}, \ldots, p_{\pi(m)}) = H_m(p_1, \ldots, p_m)$ for any permutation π on $[m]$.
(b) Expansibility: $H_m(p_1, \ldots, p_{m-1}, 0) = H_{m-1}(p_1, \ldots, p_{m-1})$.
(c) Normalization: $H_2(\frac{1}{2}, \frac{1}{2}) = \log 2$.
(d) Subadditivity: $H(X, Y) \leq H(X) + H(Y)$. Equivalently, $H_{mn}(r_{11}, \ldots, r_{mn}) \leq H_m(p_1, \ldots, p_m) + H_n(q_1, \ldots, q_n)$ whenever $\sum_{j=1}^n r_{ij} = p_i$ and $\sum_{i=1}^m r_{ij} = q_j$.
(e) Additivity: $H(X, Y) = H(X) + H(Y)$ if $X \perp\!\!\!\perp Y$. Equivalently, $H_{mn}(p_1 q_1, \ldots, p_m q_n) = H_m(p_1, \ldots, p_m) + H_n(q_1, \ldots, q_n)$.
(f) Continuity: $H_2(p, 1 - p) \to 0$ as $p \to 0$.

Then $H_m(p_1, \ldots, p_m) = \sum_{i=1}^m p_i \log \frac{1}{p_i}$ is the only possibility. The interested reader is referred to [115, Exercise **I.13**] and the references therein.

We note that there are other meaningful measure of randomness, including, notably, the *Rényi entropy of order* α introduced by Alfréd Rényi [357]:

$$H_\alpha(P) \triangleq \begin{cases} \frac{1}{1-\alpha} \log \sum_{i=1}^m p_i^\alpha, & \alpha \in (0,1) \cup (1,\infty), \\ \min_i \log \frac{1}{p_i}, & \alpha = \infty. \end{cases} \tag{1.4}$$

(The quantity H_∞ is also known as the *min-entropy*, or $H_{\min}$, in the cryptography literature). One can check that

1. $0 \le H_\alpha(P) \le \log m$, where the lower (respectively upper) bound is achieved when P is a point mass (respectively uniform);
2. $H_\alpha(P)$ is non-increasing in α and tends to the Shannon entropy $H(P)$ as $\alpha \to 1$;
3. Rényi entropy satisfies the above six axioms except for the subadditivity.

1.3 History of Entropy

In the early days of the industrial age, engineers wondered if it is possible to construct a perpetual motion machine. After many failed attempts, a law of conservation of energy was postulated: a machine cannot produce more work than the amount of energy it consumed from the ambient world. (This is also called the *first law* of thermodynamics.) The next round of attempts was then to construct a machine that would draw energy in the form of heat from a warm body and convert it to an equal (or approximately equal) amount of work. An example would be a steam engine. However, again it was observed that all such machines were highly inefficient. That is, the amount of work produced by absorbing heat Q was far less than Q. The remainder of the energy was dissipated to the ambient world in the form of heat. Again, after many rounds of attempting various designs Clausius and Kelvin proposed another law:

Second law of thermodynamics: There does not exist a machine that operates in a cycle (i.e. returns to its original state periodically), produces useful work and whose only other effect on the outside world is drawing heat from a warm body. (That is, every such machine should expend some amount of heat to some cold body too!)[1]

An equivalent formulation is as follows: "There does not exist a cyclic process that transfers heat from a cold body to a warm body." That is, every such process needs to be helped by expending some amount of external work; for example, air conditioners, sadly, will always need to use some electricity.

Notice that there is something annoying about the second law as compared to the first law. In the first law there is a quantity that is conserved, and this is somehow logically easy to accept. The second law seems a bit harder to believe in (and some engineers did not, and only their recurrent failures to circumvent it finally convinced them). So Clausius, building on an ingenious work of Carnot, figured out that there is an "explanation" to why any cyclic machine should expend heat. He proposed that there must be some hidden quantity associated with the machine, its entropy (initially described as "transformative content" or *Verwandlungsinhalt* in German),

[1] Note that the reverse effect (i.e. converting work into heat) is rather easy: friction is an example.

whose value must return to its original state. Furthermore, under any reversible (i.e. quasi-stationary, or "very slow") process operated on this machine the change of entropy is proportional to the ratio of absorbed heat and the temperature of the machine:

$$\Delta S = \frac{\Delta Q}{T}. \tag{1.5}$$

If heat Q is absorbed at temperature T_{hot} then, to return to the original state, one must return some amount of heat Q', where Q' can be significantly smaller than Q but never zero if Q' is returned at temperature $0 < T_{\text{cold}} < T_{\text{hot}}$. Further logical arguments can convince one that for an irreversible cyclic process the change of entropy at the end of the cycle can only be positive, and hence *entropy cannot reduce.*

There were a great many experimentally verified consequences that the second law produced. However, what is surprising is that the mysterious entropy did not have any formula for it (unlike, say, energy), and thus had to be computed indirectly on the basis of relation (1.5). This was changed with the revolutionary work of Boltzmann and Gibbs, who provided a microscopic explanation of the second law based on statistical physics principles and showed that, for example, for a system of n independent particles (as in an ideal gas) the entropy of a given macro-state can be computed as

$$S = kn \sum_{j=1}^{\ell} p_j \log \frac{1}{p_j}, \tag{1.6}$$

where k is the Boltzmann constant, and we assume that each particle can only be in one of ℓ molecular states (e.g. spin up/down, or if we quantize the phase volume into ℓ subcubes) and p_j is the fraction of particles in the jth molecular state.

More explicitly, their innovation was two-fold. First, they separated the concept of a micro-state (which in our example above corresponds to a tuple of n states, one for each particle) and the macro-state (a list $\{p_j\}$ of proportions of particles in each state). Second, they postulated that for experimental observations only the macro-state matters, but the multiplicity of the macro-state (number of micro-states that correspond to a given macro-state) is precisely the (exponential of the) entropy. The formula (1.6) then follows from the following explicit result connecting combinatorics and entropy.

Proposition 1.5. (Method of types) *Let $n_1, \dots, n_k$ be non-negative integers with $\sum_{i=1}^k n_i = n$, and denote the distribution $P = (p_1, \dots, p_k)$, $p_i = \frac{n_i}{n}$. Then the multinomial coefficient $\binom{n}{n_1, \dots, n_k} \triangleq \frac{n!}{n_1! \cdots n_k!}$ satisfies*

$$\frac{1}{(1+n)^{k-1}} \exp\{nH(P)\} \le \binom{n}{n_1, \dots, n_k} \le \exp\{nH(P)\}.$$

Proof. For the upper bound, let $X_1, \dots, X_n \overset{\text{iid}}{\sim} P$ and let $N_i = \sum_{i=1}^n 1\{X_j = i\}$ denote the number of occurrences of i. Then $(N_1, \dots, N_k)$ has a multinomial distribution,

$$\mathbb{P}[N_1 = n'_1, \ldots, N_k = n'_k] = \binom{n}{n'_1, \ldots, n'_k} \prod_{i=1}^{k} p_i^{n'_i},$$

for any non-negative integers n'_i such that $n'_1 + \cdots + n'_k = n$. Recalling that $p_i = n_i/n$, the upper bound follows from $\mathbb{P}[N_1 = n_1, \ldots, N_k = n_k] \le 1$. In addition, since $(N_1, \ldots, N_k)$ takes at most $(n+1)^{k-1}$ values, the lower bound follows if we can show that $(n_1, \ldots, n_k)$ is its mode. Indeed, for any n'_i with $n'_1 + \cdots + n'_k = n$, defining $\Delta_i = n'_i - n_i$ we have

$$\frac{\mathbb{P}[N_1 = n'_1, \ldots, N_k = n'_k]}{\mathbb{P}[N_1 = n_1, \ldots, N_k = n_k]} = \prod_{i=1}^{k} \frac{n_i!}{(n_i + \Delta_i)!} p_i^{\Delta_i} \le \prod_{i=1}^{k} n_i^{-\Delta_i} p_i^{\Delta_i} = 1,$$

where the inequality follows from $\frac{m!}{(m+\Delta)!} \le m^{-\Delta}$ and the last equality follows from $\sum_{i=1}^{n} \Delta_i = 0$. □

Proposition 1.5 shows that the multinomial coefficient can be approximated up to a polynomial (in n) term by $\exp(nH(P))$. More refined estimates can be obtained; see Exercise **I.2**. In particular, the binomial coefficient can be approximated using the binary entropy function as follows: Provided that $p = \frac{k}{n} \in (0, 1)$,

$$e^{-1/6} \le \frac{\binom{n}{k}}{\frac{1}{\sqrt{2\pi np(1-p)}} e^{nh(p)}} \le 1. \tag{1.7}$$

For more on combinatorics and entropy, see Exercises **I.1** and **I.3**, and Chapter 8. For more on the intricate relationship between statistical, mechanistic, and information-theoretic descriptions of the world, see Section 12.5* on Kolmogorov–Sinai entropy.

1.4* Submodularity

Recall that $[n]$ denotes a set $\{1, \ldots, n\}$, $\binom{S}{k}$ denotes subsets of S of size k, and 2^S denotes all subsets of S. A set-function $f: 2^S \to \mathbb{R}$ is called *submodular* if for any $T_1, T_2 \subset S$

$$f(T_1 \cup T_2) + f(T_1 \cap T_2) \le f(T_1) + f(T_2). \tag{1.8}$$

Submodularity is similar to concavity, in the sense that "adding elements gives diminishing returns." Indeed consider $T' \subset T$ and $b \notin T$. Then

$$f(T \cup b) - f(T) \le f(T' \cup b) - f(T').$$

Theorem 1.6. *Let X^n be a discrete random vector. Then $T \mapsto H(X_T)$ is submodular.*

Proof. Let $A = X_{T_1 \setminus T_2}, B = X_{T_1 \cap T_2}, C = X_{T_2 \setminus T_1}$. Then we need to show

$$H(A, B, C) + H(B) \le H(A, B) + H(B, C).$$

This follows from a simple chain

$$\begin{aligned} H(A,B,C) + H(B) &= H(A,C|B) + 2H(B) &(1.9)\\ &\le H(A|B) + H(C|B) + 2H(B) &(1.10)\\ &= H(A,B) + H(B,C). &(1.11) \end{aligned}$$

□

Note that entropy is not only submodular, but also monotone:

$$T_1 \subset T_2 \implies H(X_{T_1}) \le H(X_{T_2}).$$

So fixing n, let us denote by Γ_n the set of all non-negative, monotone, submodular set-functions on $[n]$. Note that via an obvious enumeration of all non-empty subsets of $[n]$, Γ_n is a closed convex cone in $\mathbb{R}_+^{2^n-1}$. Similarly, let us denote by Γ_n^* the set of all set-functions corresponding to distributions on $\mathcal{X}^n$. Let us also denote $\bar{\Gamma}_n^*$ the closure of Γ_n^*. It is not hard to show, see [474], that $\bar{\Gamma}_n^*$ is also a closed convex cone and that

$$\Gamma_n^* \subset \bar{\Gamma}_n^* \subset \Gamma_n.$$

The astonishing result of [475] is that

$$\begin{aligned} &\Gamma_2^* = \bar{\Gamma}_2^* = \Gamma_2, &(1.12)\\ &\Gamma_3^* \subsetneq \bar{\Gamma}_3^* = \Gamma_3, &(1.13)\\ &\Gamma_n^* \subsetneq \bar{\Gamma}_n^* \subsetneq \Gamma_n, \quad n \ge 4. &(1.14) \end{aligned}$$

This follows from the fundamental new information inequality not implied by the submodularity of entropy (and thus called *non-Shannon inequality*). Namely, [475] showed that for any 4-tuple of discrete random variables:

$$\begin{aligned} I(X_3;X_4) - I(X_3;X_4|X_1) - I(X_3;X_4|X_2) &\le \frac{1}{2}I(X_1;X_2) + \frac{1}{4}I(X_1;X_3,X_4)\\ &\quad + \frac{1}{4}I(X_2;X_3,X_4). \end{aligned}$$

Here we have used mutual information and conditional mutual information – notions that we will introduce later. However, the above inequality (with the help of Theorem 3.4) can be easily rewritten as a rather cumbersome expression in terms of entropies of sets of variables X_1, X_2, X_3, X_4. In conclusion, the work [475] demonstrated that the entropy set-function is more constrained than a generic submodular non-negative set-function even if one only considers linear constraints.

1.5* Han's Inequality and Shearer's Lemma

Theorem 1.7. (Han's inequality) *Let X^n be a discrete n-dimensional random vector and denote $\bar{H}_k(X^n) = \frac{1}{\binom{n}{k}} \sum_{T \in \binom{[n]}{k}} H(X_T)$ the average entropy of a k-subset of coordinates. Then $\frac{\bar{H}_k}{k}$ is decreasing in k:*

$$\frac{1}{n}\bar{H}_n \le \cdots \le \frac{1}{k}\bar{H}_k \le \cdots \le \bar{H}_1. \tag{1.15}$$

Furthermore, the sequence $\bar{H}_k$ is increasing and concave in the sense of decreasing slope:

$$\bar{H}_{k+1} - \bar{H}_k \le \bar{H}_k - \bar{H}_{k-1}. \tag{1.16}$$

Proof. Denote for convenience $\bar{H}_0 = 0$. Note that $\frac{\bar{H}_m}{m}$ is an average of differences:

$$\frac{1}{m}\bar{H}_m = \frac{1}{m}\sum_{k=1}^{m}(\bar{H}_k - \bar{H}_{k-1}).$$

Thus, it is clear that (1.16) implies (1.15) since increasing m by one adds a smaller element to the average. To prove (1.16) observe that from submodularity

$$H(X_1,\ldots,X_{k+1}) + H(X_1,\ldots,X_{k-1}) \le H(X_1,\ldots,X_k) + H(X_1,\ldots,X_{k-1},X_{k+1}).$$

Now average this inequality over all $n!$ permutations of indices $\{1,\ldots,n\}$ to get

$$\bar{H}_{k+1} + \bar{H}_{k-1} \le 2\bar{H}_k$$

as claimed by (1.16).

Alternative proof: Notice that by "conditioning decreases entropy" we have

$$H(X_{k+1}|X_1,\ldots,X_k) \le H(X_{k+1}|X_2,\ldots,X_k).$$

Averaging this inequality over all permutations of indices yields (1.16). □

Theorem 1.8. (Shearer's lemma) *Let X^n be a discrete n-dimensional random vector and let $S \subset [n]$ be a random variable independent of X^n and taking values in subsets of $[n]$. Then*

$$H(X_S|S) \ge H(X^n) \cdot \min_{i\in[n]} \mathbb{P}[i \in S]. \tag{1.17}$$

Remark 1.2. In the special case where S is uniform over all subsets of cardinality k, (1.17) reduces to Han's inequality $\frac{1}{n}H(X^n) \le \frac{1}{k}\bar{H}_k$. The case of $n = 3$ and $k = 2$ can be used to give an entropy proof of the following well-known geometry result that relates the size of a three-dimensional object to those of its two-dimensional projections: Place N points in $\mathbb{R}^3$ arbitrarily. Let N_1, N_2, and N_3 denote the number of distinct points projected onto the xy, xz, and yz planes, respectively. Then $N_1N_2N_3 \ge N^2$. For another application, see Section 8.2.

Proof. We will prove an equivalent (by taking a suitable limit) version: If $\mathcal{C} = (S_1,\ldots,S_M)$ is a list (possibly with repetitions) of subsets of $[n]$ then

$$\sum_j H(X_{S_j}) \ge H(X^n) \cdot \min_i \deg(i), \tag{1.18}$$

where $\deg(i) \triangleq \#\{j : i \in S_j\}$. Let us call $\mathcal{C}$ a *chain* if all subsets can be rearranged so that $S_1 \subseteq S_2 \subseteq \cdots \subseteq S_M$. For a chain, (1.18) is trivial, since the minimum on the right-hand side is either zero (if $S_M \neq [n]$) or equals the multiplicity of S_M in $\mathcal{C}$,[2] in which case we have

$$\sum_j H(X_{S_j}) \geq H(X_{S_M})\#\{j\colon S_j = S_M\} = H(X^n) \cdot \min_i \deg(i).$$

For the case of $\mathcal{C}$ not a chain, consider a pair of sets S_1, S_2 that are not related by inclusion and replace them in the collection with $S_1 \cap S_2, S_1 \cup S_2$. Submodularity (1.8) implies that the sum on the left-hand side of (1.18) does not increase under this replacement; values $\deg(i)$ are not changed. Notice that the total number of pairs that are not related by inclusion strictly decreases by this replacement: if T was related by inclusion to S_1 then it will also be related to at least one of $S_1 \cup S_2$ or $S_1 \cap S_2$; if T was related to both S_1, S_2 then it will be related to both of the new sets as well. Therefore, by applying this operation we must eventually arrive at a chain, for which (1.18) has already been shown. □

Remark 1.3. Han's inequality (1.16) holds for any submodular set-function. For Han's inequality (1.15) we also need $f(\emptyset) = 0$ (this can be achieved by adding a constant to all values of f). Shearer's lemma holds for any submodular set-function that is also non-negative.

Example 1.5. Non-entropy submodular function

Another submodular set-function is

$$S \mapsto I(X_S; X_{S^c}).$$

Han's inequality for this one reads

$$0 = \frac{1}{n} I_n \leq \cdots \leq \frac{1}{k} I_k \leq \cdots \leq I_1,$$

where $I_k = \frac{1}{\binom{n}{k}} \sum_{S\colon |S|=k} I(X_S; X_{S^c})$ measures the amount of k-subset coupling in the random vector X^n.

[2] Note that, consequently, for X^n without constant coordinates, and if $\mathcal{C}$ is a chain, (1.18) is only tight if $\mathcal{C}$ consists of only $\emptyset$ and $[n]$ (with multiplicities). Thus if degrees $\deg(i)$ are known and non-constant, then (1.18) can be improved, see [289].

2 Divergence

In this chapter we study divergence $D(P\|Q)$ (also known as information divergence, Kullback–Leibler (KL) divergence, or relative entropy), which is the first example of a dissimilarity (information) measure between a pair of distributions P and Q. As we will see later in Chapter 7, KL divergence is a special case of f-divergences. Defining KL divergence and its conditional version in full generality requires some measure-theoretic acrobatics (Radon–Nikodym derivatives and Markov kernels), that we spend some time on. (We stress again that all these abstractions can be ignored if one is willing to only work with finite or countably infinite alphabets.)

Besides definitions we prove the "main inequality" showing that KL divergence is non-negative. Coupled with the chain rule for divergence, this inequality implies the **data-processing inequality**, which is arguably the central pillar of information theory and this book. We conclude the chapter by studying local behavior of divergence when P and Q are close. In the special case when P and Q belong to a parametric family, we will see that divergence is locally quadratic with Hessian being the Fisher information, explaining the fundamental role of the latter in classical statistics.

2.1 Divergence and Radon–Nikodym Derivatives

Review: Measurability

For an exposition of measure-theoretic preliminaries, see [84, Chapters I and IV]. We emphasize two aspects. *First,* in this book we understand Lebesgue integration $\int_{\mathcal{X}} f d\mu$ as defined for measurable functions that are extended real-valued, that is, $f: \mathcal{X} \to \mathbb{R} \cup \{\pm\infty\}$. In particular, for negligible set E, that is, $\mu[E] = 0$, we have $\int_{\mathcal{X}} 1_E f d\mu = 0$ regardless of (possibly infinite) values of f on E, see [84, Chapter I, Proposition 4.13]. *Second,* we almost always assume that alphabets are standard Borel spaces. Some of the nice properties of standard Borel spaces:

- All complete separable metric spaces endowed with Borel σ-algebras are standard Borel. In particular, countable alphabets and $\mathbb{R}^n$ and $\mathbb{R}^\infty$ (space of sequences) are standard Borel.
- If $\mathcal{X}_i, i = 1, \ldots,$ are standard Borel, then so is $\prod_{i=1}^\infty \mathcal{X}_i$.

- Singletons $\{x\}$ are measurable sets.
- The diagonal $\{(x,x) : x \in \mathcal{X}\}$ is measurable in $\mathcal{X} \times \mathcal{X}$.

We now need to define the second central concept of this book: the *relative entropy*, or *Kullback–Leibler divergence*. Before giving the formal definition, we start with special cases. For that we fix some alphabet $\mathcal{A}$. The relative entropy between distributions P and Q on $\mathcal{X}$ is denoted by $D(P\|Q)$, defined as follows.

- Suppose $\mathcal{A}$ is a discrete (finite or countably infinite) alphabet. Then

$$D(P\|Q) \triangleq \begin{cases} \sum_{a\in\mathcal{A}:\, P(a),Q(a)>0} P(a)\log\frac{P(a)}{Q(a)}, & \mathrm{supp}(P) \subset \mathrm{supp}(Q), \\ +\infty, & \text{otherwise.} \end{cases} \tag{2.1}$$

- Suppose $\mathcal{A} = \mathbb{R}^k$, and P and Q have densities (PDFs) p and q with respect to the Lebesgue measure. Then

$$D(P\|Q) = \begin{cases} \int_{\{p>0,q>0\}} p(x)\log\frac{p(x)}{q(x)}dx, & \mathrm{Leb}\{p>0, q=0\} = 0, \\ +\infty, & \text{otherwise.} \end{cases} \tag{2.2}$$

These two special cases cover a vast majority of all cases that we encounter in this book. However, mathematically it is not very satisfying to restrict to these two special cases. For example, it is not clear how to compute $D(P\|Q)$ when P and Q are two measures on a manifold (such as a unit sphere) embedded in $\mathbb{R}^k$. Another problematic case is computing $D(P\|Q)$ between measures on the space of sequences (stochastic processes). To address these cases we need to recall the concepts of *Radon–Nikodym derivative* and *absolute continuity*.

Recall that for two measures P and Q, we say P is absolutely continuous w.r.t. Q (denoted by $P \ll Q$) if $Q(E) = 0$ implies $P(E) = 0$ for all measurable E. If $P \ll Q$, then the Radon–Nikodym theorem shows that there exists a function $f : \mathcal{X} \to \mathbb{R}_+$ such that for any measurable set E,

$$P(E) = \int_E f dQ \qquad \text{[change of measure]}. \tag{2.3}$$

Such f is called a *relative density* or a Radon–Nikodym derivative of P w.r.t. Q, denoted by $\frac{dP}{dQ}$. Note that $\frac{dP}{dQ}$ may not be unique. In the simple cases, $\frac{dP}{dQ}$ is just the familiar *likelihood ratio*:

- For discrete distributions, we can just take $\frac{dP}{dQ}(x)$ to be the ratio of PMFs.
- For continuous distributions, we can take $\frac{dP}{dQ}(x)$ to be the ratio of PDFs.

We can see that the two special cases of $D(P\|Q)$ were both computing $\mathbb{E}_P[\log\frac{dP}{dQ}]$. This turns out to be the most general definition that we are looking for. However, we will state it slightly differently, following the tradition.

Definition 2.1. (Kullback–Leibler (KL) divergence) Let P, Q be distributions on $\mathcal{A}$, with Q called the reference measure. The divergence (or relative entropy) between P and Q is

$$D(P\|Q) = \begin{cases} \mathbb{E}_Q\left[\frac{dP}{dQ}\log\frac{dP}{dQ}\right], & P \ll Q, \\ +\infty, & \text{otherwise}, \end{cases} \tag{2.4}$$

adopting again the convention from (1.1), namely, $0\log 0 = 0$.

Below we will show (Lemma 2.5) that the expectation in (2.4) is well defined (but possibly infinite) and coincides with $\mathbb{E}_P\left[\log\frac{dP}{dQ}\right]$ whenever $P \ll Q$.

To demonstrate the general definition in the case not covered by discrete/continuous specializations, consider the situation in which both P and Q are given as densities with respect to a common dominating measure μ, written as $dP = f_P d\mu$ and $dQ = f_Q d\mu$ for some non-negative f_P and f_Q. (In other words, $P \ll \mu$ and $f_P = \frac{dP}{d\mu}$.) For example, taking $\mu = P + Q$ always allows one to specify P and Q in this form. In this case, we have the following expression for divergence:

$$D(P\|Q) = \begin{cases} \int_{f_Q>0, f_P>0} d\mu\ f_P \log\frac{f_P}{f_Q}, & \mu(\{f_Q = 0, f_P > 0\}) = 0, \\ +\infty, & \text{otherwise}. \end{cases} \tag{2.5}$$

Indeed, first note that, under the assumption of $P \ll \mu$ and $Q \ll \mu$, we have $P \ll Q$ iff $\mu(\{f_Q = 0, f_P > 0\}) = 0$. Furthermore, if $P \ll Q$, then $\frac{dP}{dQ} = \frac{f_P}{f_Q}$ Q-a.e., in which case applying (2.3) and (1.1) reduces (2.5) to (2.4). Namely, $D(P\|Q) = \mathbb{E}_Q\left[\frac{dP}{dQ}\log\frac{dP}{dQ}\right] = \mathbb{E}_Q\left[\frac{f_P}{f_Q}\log\frac{f_P}{f_Q}\right] = \int d\mu f_P \log\frac{f_P}{f_Q} 1\{f_Q > 0\} = \int d\mu f_P \log\frac{f_P}{f_Q} 1\{f_Q > 0, f_P > 0\}$.

Note that $D(P\|Q)$ was defined to be $+\infty$ if $P \not\ll Q$. However, it can also be $+\infty$ even when $P \ll Q$. For example, $D(\text{Cauchy}\|\text{Gaussian}) = \infty$. However, it does not mean that there are somehow two different ways in which D can be infinite. Indeed, what can be shown is that in both cases there exists a sequence of (finer and finer) finite partitions Π of the space $\mathcal{A}$ such that evaluating KL divergence between the induced discrete distributions $P_{|\Pi}$ and $Q_{|\Pi}$ grows without a bound. This will be subject of Theorem 4.5 below.

Our next observation is that, generally, $D(P\|Q) \neq D(Q\|P)$ and, therefore, divergence is not a distance. We will see later that this is natural in many cases; for example it reflects the inherent *asymmetry* of hypothesis testing (see Part III and, in particular, Section 14.5). Consider the example of coin tossing where under P the coin is fair and under Q the coin always lands on the head. Upon observing HHHHHHH, one tends to believe it is Q but can never be absolutely sure; upon observing HHT, one knows for sure it is P. Indeed, $D(P\|Q) = \infty$, $D(Q\|P) = 1$ bit.

Having made these remarks we proceed to some examples. First, we show that D is unsurprisingly a generalization of entropy.

Theorem 2.2. (Entropy versus divergence) *If distribution P is supported on a finite set $\mathcal{A}$, then*

$$H(P) = \log|\mathcal{A}| - D(P\|U_{\mathcal{A}}),$$

where $U_{\mathcal{A}}$ is the uniform distribution on $\mathcal{A}$.

Proof. $D(P\|U_{\mathcal{A}}) = \mathbb{E}_P\left[\log\frac{P(X)}{1/|\mathcal{A}|}\right] = \log|\mathcal{A}| - H(P)$. □

Example 2.1. Binary divergence

Consider $P = \mathrm{Ber}(p)$ and $Q = \mathrm{Ber}(q)$ on $\mathcal{A} = \{0, 1\}$. Then

$$D(P\|Q) = d(p\|q) \triangleq p \log \frac{p}{q} + \bar{p} \log \frac{\bar{p}}{\bar{q}}. \tag{2.6}$$

Figure 2.1 shows how $d(p\|q)$ depends on p and q.

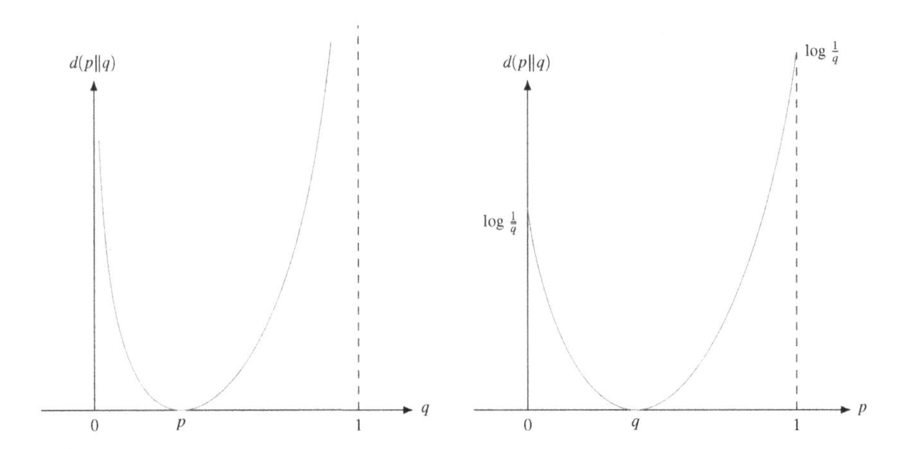

Figure 2.1 Binary divergence.

The following quadratic lower bound is easily checked:

$$d(p\|q) \geq 2(p - q)^2 \log e.$$

In fact, this is a special case of the famous Pinsker's inequality (Theorem 7.10).

Example 2.2. Real Gaussian

For two Gaussians on $\mathcal{A} = \mathbb{R}$,

$$D(\mathcal{N}(m_1, \sigma_1^2)\|\mathcal{N}(m_0, \sigma_0^2)) = \frac{\log e}{2} \frac{(m_1 - m_0)^2}{\sigma_0^2} + \frac{1}{2}\left[\log \frac{\sigma_0^2}{\sigma_1^2} + \left(\frac{\sigma_1^2}{\sigma_0^2} - 1\right) \log e\right]. \tag{2.7}$$

Here, the first and second terms compare the means and the variances, respectively.

Similarly, in the vector case of $\mathcal{A} = \mathbb{R}^k$ and assuming $\det \Sigma_0 \neq 0$, we have

$$\begin{aligned} D(\mathcal{N}(m_1, \Sigma_1)\|\mathcal{N}(m_0, \Sigma_0)) = {} & \frac{\log e}{2}(m_1 - m_0)^\top \Sigma_0^{-1}(m_1 - m_0) \\ & + \frac{1}{2}\left(\log\det \Sigma_0 - \log\det \Sigma_1 + \mathrm{tr}(\Sigma_0^{-1}\Sigma_1 - I) \log e\right). \end{aligned} \tag{2.8}$$

See Exercise **I.8** for the derivation.

Example 2.3. Complex Gaussian

The complex Gaussian distribution $\mathcal{N}_c(m,\sigma^2)$ with mean $m \in \mathbb{C}$ and variance σ^2 has density $\frac{1}{\pi\sigma^2}e^{-|z-m|^2/\sigma^2}$ for $z \in \mathbb{C}$. In other words, the real and imaginary parts are independent real Gaussians:

$$\mathcal{N}_c(m,\sigma^2) = \mathcal{N}\left(\begin{bmatrix}\mathrm{Re}(m) & \mathrm{Im}(m)\end{bmatrix}, \begin{bmatrix}\sigma^2/2 & 0 \\ 0 & \sigma^2/2\end{bmatrix}\right).$$

Then

$$D(\mathcal{N}_c(m_1,\sigma_1^2)\|\mathcal{N}_c(m_0,\sigma_0^2)) = \frac{\log e}{2}\frac{|m_1-m_0|^2}{\sigma_0^2} + \log\frac{\sigma_0^2}{\sigma_1^2} + \left(\frac{\sigma_1^2}{\sigma_0^2} - 1\right)\log e. \quad (2.9)$$

which follows from (2.8). More generally, for complex Gaussian vectors on $\mathbb{C}^k$, assuming $\det\Sigma_0 \neq 0$,

$$\begin{aligned} D(\mathcal{N}_c(m_1,\Sigma_1)\|\mathcal{N}_c(m_0,\Sigma_0)) &= (m_1-m_0)^H\Sigma_0^{-1}(m_1-m_0)\log e \\ &\quad + \log\det\Sigma_0 - \log\det\Sigma_1 + \mathrm{tr}(\Sigma_0^{-1}\Sigma_1 - I)\log e. \end{aligned}$$

2.2 Divergence: Main Inequality and Equivalent Expressions

Many inequalities in information can be attributed to the following fundamental result, namely, the non-negativity of divergence.

Theorem 2.3. (Information inequality)

$$D(P\|Q) \geq 0,$$

with equality iff $P = Q$.

Proof. In view of the definition (2.4), it suffices to consider $P \ll Q$. Let $\varphi(x) \triangleq x\log x$, which is strictly convex on $\mathbb{R}_+$. Applying Jensen's inequality:

$$D(P\|Q) = \mathbb{E}_Q\left[\varphi\left(\frac{dP}{dQ}\right)\right] \geq \varphi\left(\mathbb{E}_Q\left[\frac{dP}{dQ}\right]\right) = \varphi(1) = 0,$$

with equality iff $\frac{dP}{dQ} = 1$ Q-a.e., namely, $P = Q$. □

Here is a typical application of the previous result (variations of it will be applied numerous times in this book). This result is widely used in machine learning as it shows that minimizing average cross-entropy loss $\ell(Q,x) \triangleq \log\frac{1}{Q(x)}$ recovers the true distribution (Exercise **III.11**).

Corollary 2.4. *Let X be a discrete random variable with $H(X) < \infty$. Then*

$$\min_Q \mathbb{E}\left[\log\frac{1}{Q(X)}\right] = H(X),$$

and the unique minimizer is $Q = P_X$.

Proof. It is sufficient to prove that for any $Q \neq P_X$ we have

$$\mathbb{E}\left[\log \frac{1}{Q(X)}\right] > H(X).$$

If the LHS is infinite this is clear, so let us assume it is finite and hence $Q(x) > 0$ whenever $P_X(x) > 0$. Then subtracting $H(X)$ from both sides and using linearity of expectation we have

$$\mathbb{E}\left[\log \frac{1}{Q(X)}\right] - H(X) = \mathbb{E}\left[\log \frac{P_X(X)}{Q(X)}\right] = D(P_X \| Q) > 0,$$

where the inequality is via Theorem 2.3. □

Another implication of the proof of Theorem 2.3 is in bringing forward the reason for defining $D(P\|Q) = \mathbb{E}_Q\left[\frac{dP}{dQ}\log\frac{dP}{dQ}\right]$ as opposed to $D(P\|Q) = \mathbb{E}_P\left[\log\frac{dP}{dQ}\right]$. However, we still need to show that the two definitions are equivalent, which is what we do next. In addition, we will also unify the two cases ($P \ll Q$ versus $P \not\ll Q$) in Definition 2.1.

Lemma 2.5. *Let $P, Q, R \ll \mu$ and f_P, f_Q, f_R denote their densities relative to μ. Define a bivariate function $\mathrm{Log}\frac{a}{b}\colon \mathbb{R}_+ \times \mathbb{R}_+ \to \mathbb{R} \cup \{\pm\infty\}$ by*

$$\mathrm{Log}\frac{a}{b} = \begin{cases} -\infty, & a = 0, b > 0, \\ +\infty, & a > 0, b = 0, \\ 0, & a = 0, b = 0, \\ \log\frac{a}{b}, & a > 0, b > 0. \end{cases} \tag{2.10}$$

Then the following results hold:

- *First, the following expectation exists and equals*

$$\mathbb{E}_P\left[\mathrm{Log}\frac{f_R}{f_Q}\right] = D(P\|Q) - D(P\|R), \tag{2.11}$$

provided at least one of the divergences is finite.

- *Second, the expectation $\mathbb{E}_P\left[\mathrm{Log}\frac{f_P}{f_Q}\right]$ is well defined (but possibly infinite) and, furthermore,*

$$D(P\|Q) = \mathbb{E}_P\left[\mathrm{Log}\frac{f_P}{f_Q}\right]. \tag{2.12}$$

In particular, when $P \ll Q$ we have

$$D(P\|Q) = \mathbb{E}_P\left[\log\frac{dP}{dQ}\right]. \tag{2.13}$$

Remark 2.1. Note that ignoring the issue of dividing by or taking a log of 0, the proof of (2.12) is just the simple identity $\log\frac{dR}{dQ} = \log\frac{dRdP}{dQdP} = \log\frac{dP}{dQ} - \log\frac{dP}{dR}$. What permits us to handle zeros is the Log function, which satisfies several natural properties of the log: for every $a, b \in \mathbb{R}_+$

$$\mathrm{Log}\frac{a}{b} = -\mathrm{Log}\frac{b}{a}$$

and for every $c > 0$ we have

$$\mathrm{Log}\frac{a}{b} = \mathrm{Log}\frac{a}{c} + \mathrm{Log}\frac{c}{b} = \mathrm{Log}\frac{ac}{b} - \log(c)$$

except for the case $a = b = 0$.

Proof. First, suppose $D(P\|Q) = \infty$ and $D(P\|R) < \infty$. Then $P[f_R(Y) = 0] = 0$, and hence in computation of the expectation in (2.11) only the second part of convention (2.10) can possibly apply. Since also $f_P > 0$ P-almost surely, we have

$$\mathrm{Log}\frac{f_R}{f_Q} = \mathrm{Log}\frac{f_R}{f_P} + \mathrm{Log}\frac{f_P}{f_Q}, \tag{2.14}$$

with both logarithms evaluated according to (2.10). Taking expectation over P we see that the first term, equal to $-D(P\|R)$, is finite, whereas the second term is infinite. Thus, the expectation in (2.11) is well defined and equal to $+\infty$, as is the RHS of (2.11).

Now assume $D(P\|Q) < \infty$. This implies that $P[f_Q(Y) = 0] = 0$ and this time in (2.11) only the first part of convention (2.10) can apply. Thus, again we have identity (2.14). Since the P-expectation of the second term is finite, and of the first term non-negative, we again conclude that expectation in (2.11) is well defined, and equals the RHS of (2.11) (and both sides are possibly equal to $-\infty$).

For the second part, we first show that

$$\mathbb{E}_P\left[\min\left(\mathrm{Log}\frac{f_P}{f_Q}, 0\right)\right] \geq -\frac{\log e}{e}. \tag{2.15}$$

Let $g(x) = \min(x\log x, 0)$. It is clear $-\frac{\log e}{e} \leq g(x) \leq 0$ for all x. Since $f_P(Y) > 0$ for P-almost all Y, in convention (2.10) only the $\frac{1}{0}$ case is possible, which is excluded by the $\min(\cdot, 0)$ from the expectation in (2.15). Thus, the LHS in (2.15) equals

$$\begin{aligned}\int_{\{f_P > f_Q > 0\}} f_P(y)\log\frac{f_P(y)}{f_Q(y)}d\mu &= \int_{\{f_P > f_Q > 0\}} f_Q(y)\frac{f_P(y)}{f_Q(y)}\log\frac{f_P(y)}{f_Q(y)}d\mu \\ &= \int_{\{f_Q > 0\}} f_Q(y) g\left(\frac{f_P(y)}{f_Q(y)}\right)d\mu \\ &\geq -\frac{\log e}{e}.\end{aligned}$$

Since the negative part of $\mathbb{E}_P\left[\mathrm{Log}\frac{f_P}{f_Q}\right]$ is bounded, the expectation $\mathbb{E}_P\left[\mathrm{Log}\frac{f_P}{f_Q}\right]$ is well defined. If $P[f_Q = 0] > 0$ then it is clearly $+\infty$, as is $D(P\|Q)$ (since $P \not\ll Q$). Otherwise, let $E = \{f_P > 0, f_Q > 0\}$. Then $P[E] = 1$ and on E we have $f_P = f_Q \cdot \frac{f_P}{f_Q}$. Thus, with $\varphi(x) = x\log x$, we obtain

$$\mathbb{E}_P\left[\mathrm{Log}\frac{f_P}{f_Q}\right] = \int_E d\mu\, f_P \log\frac{f_P}{f_Q} = \int_E d\mu f_Q \varphi\left(\frac{f_P}{f_Q}\right) = \mathbb{E}_Q\left[1_E \varphi\left(\frac{f_P}{f_Q}\right)\right].$$

From here, we notice that $Q[f_Q > 0] = 1$ and on $\{f_P = 0, f_Q > 0\}$ we have $\varphi\left(\frac{f_P}{f_Q}\right) = 0$. Thus, the term 1_E can be dropped and we obtain the desired (2.12).

The final statement of the lemma follows from taking $\mu = Q$ and noticing that P-almost surely we have

$$\mathrm{Log}\frac{\frac{dP}{dQ}}{1} = \log\frac{dP}{dQ}. \qquad \square$$

2.3 Differential Entropy

The definition of $D(P\|Q)$ extends verbatim to measures P and Q (not necessarily probability measures), in which case $D(P\|Q)$ can be negative. A sufficient condition for $D(P\|Q) \geq 0$ is that P is a probability measure and Q is a sub-probability measure, that is, $\int dQ \leq 1 = \int dP$. The notion of *differential entropy* is simply the divergence with respect to the Lebesgue measure.

Definition 2.6. The differential entropy of a random vector X is

$$h(X) = h(P_X) \triangleq -D(P_X\|\mathrm{Leb}). \tag{2.16}$$

In particular, if X has probability density function (PDF) p, then $h(X) = \mathbb{E}\log\frac{1}{p(X)}$; otherwise $h(X) = -\infty$. The conditional differential entropy is $h(X|Y) \triangleq \mathbb{E}\log\frac{1}{p_{X|Y}(X|Y)}$ where $p_{X|Y}$ is a conditional PDF.

Example 2.4. Gaussian

For $X \sim N(\mu, \sigma^2)$,

$$h(X) = \frac{1}{2}\log(2\pi e\sigma^2). \tag{2.17}$$

More generally, for $X \sim N(\mu, \Sigma)$ in $\mathbb{R}^d$,

$$h(X) = \frac{1}{2}\log((2\pi e)^d \det \Sigma). \tag{2.18}$$

Warning: Even for continuous random variable X, $h(X)$ can be positive, negative, take values of $\pm\infty$ or even be undefined.[1] There are many crucial differences between the Shannon entropy and the differential entropy. For example, from Theorem 1.4 we know that deterministic processing cannot increase the Shannon entropy, that is, $H(f(X)) \leq H(X)$ for any discrete X, which is intuitively clear. However, this fails completely for differential entropy (e.g. consider scaling). Furthermore, for sums of independent integer-valued X and Y, $H(X+Y)$ is finite whenever $H(X)$ and $H(Y)$ are, because $H(X+Y) \leq H(X,Y) = H(X) + H(Y)$. This again fails for differential entropy. In fact, there exists real-valued X with finite $h(X)$ such that $h(X+Y) = \infty$ for any independent Y such that $h(Y) > -\infty$; there also exist X and Y with finite differential entropy such that $h(X+Y)$ does not exist (see [65, Section V]).

[1] For an example, consider a piecewise-constant PDF taking value $e^{(-1)^n n}$ on the nth interval of width $\Delta_n = \frac{c}{n^2}e^{-(-1)^n n}$.

Nevertheless, differential entropy shares many functional properties with the usual Shannon entropy. For a short application to Euclidean geometry see Section 8.4.

Theorem 2.7. (Properties of differential entropy) *Assume that all differential entropies appearing below exist and are finite (in particular all random variables have PDFs and conditional PDFs).*

(a) (Uniform distribution maximizes differential entropy) If $\mathbb{P}[X^n \in S] = 1$ then $h(X^n) \le \log \mathrm{Leb}(S)$, with equality iff X^n is uniform on S.
(b) (Scaling and shifting) $h(X^n + x) = h(X^n), h(\alpha X^n) = h(X^n) + n \log |\alpha|$ and, for an invertible matrix A, $h(AX^n) = h(X^n) + \log|\det A|$.
(c) (Conditioning reduces differential entropy) $h(X|Y) \le h(X)$. (Here Y is arbitrary.)
(d) (Chain rule) Let X^n have a joint probability density function. Then

$$h(X^n) = \sum_{k=1}^{n} h(X_k|X^{k-1}).$$

(e) (Submodularity) The set-function $T \mapsto h(X_T)$ is submodular.
(f) (Han's inequality) The function $k \mapsto \frac{1}{k\binom{n}{k}} \sum_{T \in \binom{[n]}{k}} h(X_T)$ is decreasing in k.

Proof. Parts (a), (c), and (d) follow from the similar argument in the proof (b), (d), and (g) of Theorem 1.4. Part (b) is by a change of variable in the density. Finally, (e) and (f) are analogous to Theorems 1.6 and 1.7, respectively. □

Interestingly, the first property is robust to small additive perturbations, see Exercise **I.6**. Regarding maximizing entropy under quadratic constraints, we have the following characterization of Gaussians.

Theorem 2.8. *Let $\mathrm{Cov}(X) = \mathbb{E}[XX^\top] - \mathbb{E}[X]\mathbb{E}[X]^\top$ denote the covariance matrix of a random vector X. For any $d \times d$ positive definite matrix Σ,*

$$\max_{P_X:\, \mathrm{Cov}(X) \preceq \Sigma} h(X) = h(N(0, \Sigma)) = \frac{1}{2}\log((2\pi e)^d \det \Sigma). \tag{2.19}$$

Furthermore, for any $a > 0$,

$$\max_{P_X:\, \mathbb{E}[\|X\|^2] \le a} h(X) = h\left(N\left(0, \frac{a}{d} I_d\right)\right) = \frac{d}{2}\log\frac{2\pi e a}{d}. \tag{2.20}$$

Proof. To show (2.19), without loss of generality, assume that $\mathbb{E}[X] = 0$. By comparing to a Gaussian, we have

$$\begin{aligned} 0 &\le D(P_X \| N(0, \Sigma)) \\ &= -h(X) + \frac{1}{2}\log((2\pi)^d \det(\Sigma)) + \frac{\log e}{2}\mathbb{E}[X^\top \Sigma^{-1} X] \\ &\le -h(X) + h(N(0, \Sigma)), \end{aligned}$$

where in the last step we apply $\mathbb{E}[X^\top \Sigma^{-1} X] = \mathrm{Tr}(\mathbb{E}[XX^\top]\Sigma^{-1}) \le \mathrm{Tr}(I)$ due to the constraint $\mathrm{Cov}(X) \preceq \Sigma$ and the formula (2.18). The inequality (2.20) follows analogously by choosing the reference measure to be $N(0, \frac{a}{d} I_d)$. □

Corollary 2.9. *The map $\Sigma \mapsto \log\det\Sigma$ is concave on the space of real positive definite $n \times n$ matrices.*

Proof. Let Σ_1, Σ_2 be positive definite $n \times n$ matrices. Let $Y \sim \mathrm{Ber}(1/2)$ and given $Y = 0$ we set $X \sim \mathcal{N}(0, \Sigma_1)$ and otherwise $X \sim \mathcal{N}(0, \Sigma_2)$. Let $\mathrm{Cov}(X) = \Sigma = \frac{1}{2}\Sigma_1 + \frac{1}{2}\Sigma_2$. Then we have $h(X|Y) \le h(X) \le \frac{1}{2}\log((2\pi e)^n \det\Sigma)$. For $h(X|Y)$ we apply (2.18) and after simplification obtain

$$\frac{1}{2}\log\det\Sigma_1 + \frac{1}{2}\log\det\Sigma_2 \le \log\det\left(\frac{1}{2}\Sigma_1 + \frac{1}{2}\Sigma_2\right),$$

which is exactly the claimed concavity. □

Finally, let us mention a connection between the differential entropy and the Shannon entropy. Let X be a continuous random vector in $\mathbb{R}^d$. Denote its discretized version by $X_m = \frac{1}{m}\lfloor mX \rfloor$ for $m \in \mathbb{N}$, where $\lfloor\cdot\rfloor$ is taken componentwise. Rényi showed that [358, Theorem 1], provided $H(\lfloor X \rfloor) < \infty$ and $h(X)$ is defined, we have

$$H(X_m) = d\log m + h(X) + o(1), \quad m \to \infty. \tag{2.21}$$

To interpret this result, consider, for simplicity, $d = 1, m = 2^k$ and assume that X takes values in the unit interval, in which case X_{2^k} is the k-bit uniform quantization of X. Then (2.21) suggests that for large k, the quantized bits behave as independent fair coin flips. The underlying reason is that for "nice" density functions, the restriction to small intervals is approximately uniform. For more on quantization see Section 24.1 (notably Section 24.1.5) in Chapter 24.

2.4 Markov Kernels

The main objects in this book are random variables and probability distributions. The main operation for creating new random variables, as well as for defining relations between random variables, is that of a *Markov kernel* (also known as a *transition probability kernel*).

Definition 2.10. A Markov kernel $K \colon \mathcal{X} \to \mathcal{Y}$ is a bivariate function $K(\cdot\,|\,\cdot)$, whose first argument is a measurable subset of $\mathcal{Y}$ and the second is an element of $\mathcal{X}$, such that:

1 For any $x \in \mathcal{X}$: $K(\cdot\,|x)$ is a probability measure on $\mathcal{Y}$.
2 For any measurable set A: $x \mapsto K(A|x)$ is a measurable function on $\mathcal{X}$.

The kernel K can be viewed as a random transformation acting from $\mathcal{X}$ to $\mathcal{Y}$, which draws Y from a distribution depending on the realization of X, including deterministic transformations as special cases. For this reason, we write $P_{Y|X} \colon \mathcal{X} \to$

$\mathcal{Y}$ and also $X \xrightarrow{P_{Y|X}} Y$. In information-theoretic context, we also refer to $P_{Y|X}$ as a *channel*, where X and Y are the channel input and output respectively. There are two ways of obtaining Markov kernels. The first way is defining them explicitly. Here are some examples of that:

1 Deterministic system: $Y = f(X)$. This corresponds to setting $P_{Y|X=x} = \delta_{f(x)}$.
2 Decoupled system: $Y \perp\!\!\!\perp X$. Here we set $P_{Y|X=x} = P_Y$.
3 Additive noise (convolution): $Y = X + Z$ with $Z \perp\!\!\!\perp X$. This time we choose $P_{Y|X=x}(\cdot) = P_Z(\cdot - x)$. The term "convolution" corresponds to the fact that the resulting marginal distribution $P_Y = P_X * P_Z$ is a convolution of measures.
4 Some of the most useful channels that we will use throughout the book are going to be defined starting in Examples 3.3, 3.4, and 3.6.

The second way is to *disintegrate* a joint distribution $P_{X,Y}$ by conditioning on X, which is denoted simply by $P_{Y|X}$. Specifically, we have the following result [84, Chapter IV, Theorem 2.18]:

Theorem 2.11. (Disintegration) *Suppose $P_{X,Y}$ is a distribution on $\mathcal{X} \times \mathcal{Y}$ with $\mathcal{Y}$ being standard Borel. Then there exists a Markov kernel $K\colon \mathcal{X} \to \mathcal{Y}$ so that for any measurable $E \subset \mathcal{X} \times \mathcal{Y}$ and any integrable f we have*

$$P_{X,Y}[E] = \int_{\mathcal{X}} P_X(dx)K(E^x|x), \quad E^x \triangleq \{y\colon (x,y) \in E\}, \tag{2.22}$$

$$\int_{\mathcal{X}\times\mathcal{Y}} f(x,y)P_{X,Y}(dx\,dy) = \int_{\mathcal{X}} P_X(dx) \int_{\mathcal{Y}} f(x,y)K(dy|x).$$

Note that above we have implicitly used the facts that the slices E^x of E are measurable subsets of $\mathcal{Y}$ for each x and that the function $x \mapsto K(E^x|x)$ is measurable (see [84, Chapter I, Propositions 6.8 and 6.9], respectively). We also notice that one joint distribution $P_{X,Y}$ can have many different *versions* of $P_{Y|X}$ differing on a measure-zero set of x's.

The operation of combining an input distribution on $\mathcal{X}$ and a kernel $K\colon \mathcal{X} \to \mathcal{Y}$ as we did in (2.22) is going to appear extensively in this book. We will usually denote it as *multiplication*: Given P_X and kernel $P_{Y|X}$ we can multiply them to obtain $P_{X,Y} \triangleq P_X P_{Y|X}$, which in the discrete case simply means that the joint PMF factorizes as the product of marginal and conditional PMFs:

$$P_{X,Y}(x,y) = P_{Y|X}(y|x)P_X(x),$$

and more generally is given by (2.22) with $K = P_{Y|X}$.

Another useful operation will be that of *composition* (marginalization), which we denote by $P_{Y|X} \circ P_X \triangleq P_Y$. In words, this means forming a distribution $P_{X,Y} = P_X P_{Y|X}$ and then computing the marginal P_Y, or, explicitly,

$$P_Y[E] = \int_{\mathcal{X}} P_X(dx)P_{Y|X}(E|x).$$

To denote this (linear) relation between the input P_X and the output P_Y we sometimes also write $P_X \xrightarrow{P_{Y|X}} P_Y$.

We must remark that technical assumptions such as restricting to standard Borel spaces are really necessary for constructing any sensible theory of disintegration/conditioning and multiplication. To emphasize this point we consider a (cautionary!) example involving a pathological measurable space $\mathcal{Y}$.

Example 2.5. $X \perp\!\!\!\perp Y$ but $P_{Y|X=x} \not\ll P_Y$ for all x

Consider $\mathcal{X}$ a unit interval with Borel σ-algebra and $\mathcal{Y}$ a unit interval with the σ-algebra $\sigma\mathcal{Y}$ consisting of all sets which are either countable or have a countable complement. Clearly $\sigma\mathcal{Y}$ is a sub-σ-algebra of a Borel one. We define the following kernel $K \colon \mathcal{X} \to \mathcal{Y}$:

$$K(A|x) \triangleq 1\{x \in A\}.$$

This is simply saying that Y is produced from X by setting $Y = X$. It should be clear that for every $A \in \sigma\mathcal{Y}$ the map $x \mapsto K(A|x)$ is measurable, and thus K is a valid Markov kernel. Letting $X \sim \text{Unif}(0,1)$ and using formula (2.22) we can define a joint distribution $P_{X,Y}$. But what is the conditional distribution $P_{Y|X}$? On the one hand, clearly we can set $P_{Y|X}(A|x) = K(A|x)$, since this was how $P_{X,Y}$ was constructed. On the other hand, we will show that $P_{X,Y} = P_X P_Y$, that is, $X \perp\!\!\!\perp Y$ and $X = Y$ at the same time! Indeed, consider any set $E = B \times C \subset \mathcal{X} \times \mathcal{Y}$. We always have $P_{X,Y}[B \times C] = P_X[B \cap C]$. Thus if C is countable then $P_{X,Y}[E] = 0$ and so is $P_X P_Y[E] = 0$. On the other hand, if C^c is countable then $P_X[C] = P_Y[C] = 1$ and $P_{X,Y}[E] = P_X P_Y[E]$ again. Thus, both $P_{Y|X} = K$ and $P_{Y|X} = P_Y$ are valid conditional distributions. But notice that since $P_Y[\{x\}] = 0$, we have $K(\cdot|x) \not\ll P_Y$ for every $x \in \mathcal{X}$. In particular, the value of $D(P_{Y|X=x} \| P_Y)$ can either be 0 or $+\infty$ for every x depending on the choice of the version of $P_{Y|X}$. It is, thus, advisable to stay within the realm of standard Borel spaces.

We will also need to use the following result extensively. We remind the reader that a σ-algebra is called separable if it is generated by a countable collection of sets. Any standard Borel space's σ-algebra is separable. The following is another useful result about Markov kernels, see [84, Chapter 5, Theorem 4.44]:

Theorem 2.12. (Doob's version of the Radon–Nikodym theorem) *Assume that $\mathcal{Y}$ is a measurable space with a separable σ-algebra. Let $P_{Y|X} \colon \mathcal{X} \to \mathcal{Y}$ and $R_{Y|X} \colon \mathcal{X} \to \mathcal{Y}$ be two Markov kernels. Suppose that for every x we have $P_{Y|X=x} \ll R_{Y|X=x}$. Then there exists a measurable function $(x,y) \mapsto f(y|x) \geq 0$ such that for every $x \in \mathcal{X}$ and every measurable subset E of $\mathcal{Y}$,*

$$P_{Y|X}(E|x) = \int_E f(y|x) R_{Y|X}(dy|x).$$

The meaning of this theorem is that the Radon–Nikodym derivative $\frac{dP_{Y|X=x}}{dR_{Y|X=x}}$ can be made *jointly* measurable with respect to (x,y).

2.5 Conditional Divergence, Chain Rule, Data-Processing Inequality

We aim to define the conditional divergence between two Markov kernels. Throughout this chapter we fix a pair of Markov kernels $P_{Y|X}: \mathcal{X} \to \mathcal{Y}$ and $Q_{Y|X}: \mathcal{X} \to \mathcal{Y}$, and also a probability measure P_X on $\mathcal{X}$. First, let us consider the case of discrete $\mathcal{X}$. We define the conditional divergence as

$$D(P_{Y|X}\|Q_{Y|X}|P_X) \triangleq \sum_{x\in\mathcal{X}} P_X(x)D(P_{Y|X=x}\|Q_{Y|X=x}).$$

In order to extend the above definition to more general $\mathcal{X}$, we need to first understand whether the map $x \mapsto D(P_{Y|X=x}\|Q_{Y|X=x})$ is even measurable.

Lemma 2.13. *Suppose that $\mathcal{Y}$ is standard Borel. The set $A_0 \triangleq \{x: P_{Y|X=x} \ll Q_{Y|X=x}\}$ and the function*

$$x \mapsto D(P_{Y|X=x}\|Q_{Y|X=x})$$

are both measurable.

Proof. Take $R_{Y|X} = \frac{1}{2}P_{Y|X} + \frac{1}{2}Q_{Y|X}$ and define $f_P(y|x) \triangleq \frac{dP_{Y|X=x}}{dR_{Y|X=x}}(y)$ and $f_Q(y|x) \triangleq \frac{dQ_{Y|X=x}}{dR_{Y|X=x}}(y)$. By Theorem 2.12 these can be chosen to be jointly measurable on $\mathcal{X} \times \mathcal{Y}$. Let us define $B_0 \triangleq \{(x,y): f_P(y|x) > 0, f_Q(y|x) = 0\}$ and its slice $B_0^x = \{y: (x,y) \in B_0\}$. Then note that $P_{Y|X=x} \ll Q_{Y|X=x}$ iff $R_{Y|X=x}[B_0^x] = 0$. In other words, $x \in A_0$ iff $R_{Y|X=x}[B_0^x] = 0$. The measurability of B_0 implies that of $x \mapsto R_{Y|X=x}[B_0^x]$ and thus that of A_0. Finally, from (2.12) we get that

$$D(P_{Y|X=x}\|Q_{Y|X=x}) = \mathbb{E}_{Y\sim P_{Y|X=x}}\left[\mathrm{Log}\frac{f_P(Y|x)}{f_Q(Y|x)}\right], \tag{2.23}$$

which is measurable, for example, [84, Chapter 1, Proposition 6.9]. □

With this preparation we can give the following definition.

Definition 2.14. (Conditional divergence) Assuming $\mathcal{Y}$ is standard Borel, define

$$D(P_{Y|X}\|Q_{Y|X}|P_X) \triangleq \mathbb{E}_{x\sim P_X}[D(P_{Y|X=x}\|Q_{Y|X=x})].$$

We observe that as usual in Lebesgue integration it is possible that a conditional divergence is finite even though $D(P_{Y|X=x}\|Q_{Y|X=x}) = \infty$ for some (P_X-negligible set of) x.

Theorem 2.15. (Chain rule) *For any pair of measures $P_{X,Y}$ and $Q_{X,Y}$ we have*

$$D(P_{X,Y}\|Q_{X,Y}) = D(P_{Y|X}\|Q_{Y|X}|P_X) + D(P_X\|Q_X), \tag{2.24}$$

regardless of the versions of conditional distributions $P_{Y|X}$ and $Q_{Y|X}$ one chooses.

Proof. First, let us consider the simplest case: $\mathcal{X}, \mathcal{Y}$ are discrete and $Q_{X,Y}(x,y) > 0$ for all x, y. Letting $(X,Y) \sim P_{X,Y}$ we get

$$
\begin{aligned}
D(P_{X,Y}\|Q_{X,Y}) &= \mathbb{E}\left[\log\frac{P_{X,Y}(X,Y)}{Q_{X,Y}(X,Y)}\right] = \mathbb{E}\left[\log\frac{P_X(X)P_{Y|X}(Y|X)}{Q_X(X)Q_{Y|X}(Y|X)}\right] \\
&= \mathbb{E}\left[\log\frac{P_{Y|X}(Y|X)}{Q_{Y|X}(Y|X)}\right] + \mathbb{E}\left[\log\frac{P_X(X)}{Q_X(X)}\right],
\end{aligned}
$$

completing the proof.

Next, let us address the general case. If $P_X \not\ll Q_X$ then $P_{X,Y} \not\ll Q_{X,Y}$ and both sides of (2.24) are infinity. Thus, we assume $P_X \ll Q_X$ and set $\lambda_P(x) \triangleq \frac{dP_X}{dQ_X}(x)$. Define $f_P(y|x)$, $f_Q(y|x)$, and $R_{Y|X}$ as in the proof of Lemma 2.13. Then we have $P_{X,Y}, Q_{X,Y} \ll R_{X,Y} \triangleq Q_X R_{Y|X}$, and for any measurable E

$$
P_{X,Y}[E] = \int_E \lambda_P(x) f_P(y|x) R_{X,Y}(dx\,dy), \quad Q_{X,Y}[E] = \int_E f_Q(y|x) R_{X,Y}(dx\,dy).
$$

Then from (2.12) we have

$$
D(P_{X,Y}\|Q_{X,Y}) = \mathbb{E}_{P_{X,Y}}\left[\mathrm{Log}\frac{f_P(Y|X)\lambda_P(X)}{f_Q(Y|X)}\right]. \tag{2.25}
$$

Note the following property of Log: For any $c > 0$

$$
\mathrm{Log}\frac{ac}{b} = \log(c) + \mathrm{Log}\frac{a}{b}
$$

unless $a = b = 0$. Now, since $P_{X,Y}[f_P(Y|X) > 0, \lambda_P(X) > 0] = 1$, we conclude that $P_{X,Y}$-almost surely

$$
\mathrm{Log}\frac{f_P(Y|X)\lambda_P(X)}{f_Q(Y|X)} = \log\lambda_P(X) + \mathrm{Log}\frac{f_P(Y|X)}{f_Q(Y|X)}.
$$

We aim to take the expectation of both sides over $P_{X,Y}$ and invoke linearity of expectation. To ensure that the issue of $\infty - \infty$ does not arise, we notice that the negative part of each term has finite expectation by (2.15). Overall, continuing (2.25) and invoking linearity we obtain

$$
D(P_{X,Y}\|Q_{X,Y}) = \mathbb{E}_{P_{X,Y}}[\log\lambda_P(X)] + \mathbb{E}_{P_{X,Y}}\left[\mathrm{Log}\frac{f_P(Y|X)}{f_Q(Y|X)}\right],
$$

where the first term equals $D(P_X\|Q_X)$ by (2.12) and the second $D(P_{Y|X}\|Q_{Y|X}|P_X)$ by (2.23) and the definition of conditional divergence. □

The chain rule has a number of useful corollaries, which we summarize below.

Theorem 2.16. (Properties of divergence) *Assume that $\mathcal{X}$ and $\mathcal{Y}$ are standard Borel. Then we have the following.*

(a) Conditional divergence can be expressed unconditionally:

$$
D(P_{Y|X}\|Q_{Y|X}|P_X) = D(P_X P_{Y|X}\|P_X Q_{Y|X}).
$$

(b) (Monotonicity) $D(P_{X,Y}\|Q_{X,Y}) \geq D(P_Y\|Q_Y)$.

(c) (Full chain rule)

$$
D(P_{X_1\cdots X_n}\|Q_{X_1\cdots X_n}) = \sum_{i=1}^n D(P_{X_i|X^{i-1}}\|Q_{X_i|X^{i-1}}|P_{X^{i-1}}). \tag{2.26}
$$

In the special case of $Q_{X^n} = \prod_{i=1}^n Q_{X_i}$,

$$D(P_{X_1 \cdots X_n} \| Q_{X_1} \cdots Q_{X_n}) = D(P_{X_1 \cdots X_n} \| P_{X_1} \cdots P_{X_n}) + \sum_{i=1}^n D(P_{X_i} \| Q_{X_i})$$
$$\geq \sum_{i=1}^n D(P_{X_i} \| Q_{X_i}), \tag{2.27}$$

where the inequality holds with equality if and only if $P_{X^n} = \prod_{j=1}^n P_{X_j}$.

(d) *(Tensorization)*

$$D\left(\prod_{j=1}^n P_{X_j} \middle\| \prod_{j=1}^n Q_{X_j}\right) = \sum_{j=1}^n D(P_{X_j} \| Q_{X_j}).$$

(e) *(Conditioning increases divergence) Given* $P_{Y|X}$, $Q_{Y|X}$, *and* P_X, *let* $P_Y = P_{Y|X} \circ P_X$ *and* $Q_Y = Q_{Y|X} \circ P_X$, *as represented by the diagram:*

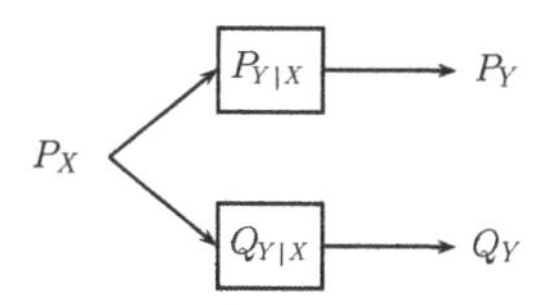

Then $D(P_Y \| Q_Y) \leq D(P_{Y|X} \| Q_{Y|X} | P_X)$, *with equality iff* $D(P_{X|Y} \| Q_{X|Y} | P_Y) = 0$.

We remark that as before without the standard Borel assumption even the first property can fail. For example, Example 2.5 shows an example where $P_X P_{Y|X} = P_X Q_{Y|X}$ but $P_{Y|X} \neq Q_{Y|X}$ and $D(P_{Y|X} \| Q_{Y|X} | P_X) = \infty$.

Proof.

(a) This follows from the chain rule (2.24) since $P_X = Q_X$.
(b) Apply (2.24), with X and Y interchanged and use the fact that conditional divergence is non-negative.
(c) By telescoping $P_{X^n} = \prod_{i=1}^n P_{X_i|X^{i-1}}$ and $Q_{X^n} = \prod_{i=1}^n Q_{X_i|X^{i-1}}$.
(d) Apply (c).
(e) The inequality follows from (a) and (b). To get conditions for equality, notice that by the chain rule for D:

$$D(P_{X,Y} \| Q_{X,Y}) = D(P_{Y|X} \| Q_{Y|X} | P_X) + \underbrace{D(P_X \| P_X)}_{=0}$$
$$= D(P_{X|Y} \| Q_{X|Y} | P_Y) + D(P_Y \| Q_Y). \qquad \square$$

Some remarks are in order:

- There is a nice interpretation of the full chain rule as a decomposition of the "distance" from P_{X^n} to Q_{X^n} as a sum of "distances" between intermediate distributions; see Exercise **I.43**.

- In general, $D(P_{X,Y}\|Q_{X,Y})$ and $D(P_X\|Q_X) + D(P_Y\|Q_Y)$ are incomparable. For example, if $X = Y$ under P and Q, then $D(P_{X,Y}\|Q_{X,Y}) = D(P_X\|Q_X) < 2D(P_X\|Q_X)$. Conversely, if $P_X = Q_X$ and $P_Y = Q_Y$ but $P_{X,Y} \neq Q_{X,Y}$ we have $D(P_{X,Y}\|Q_{X,Y}) > 0 = D(P_X\|Q_X) + D(P_Y\|Q_Y)$.

The following result, known as the *data-processing inequality (DPI)*, is an important principle in all of information theory. In many ways, it underpins the whole concept of information. The intuitive interpretation is that it is easier to distinguish two distributions using clean (respectively full) data as opposed to noisy (respectively partial) data. DPI is a recurring theme in this book, and later we will study DPI for other information measures such as mutual information and f-divergences (Chapters 3 and 7) as well as their strengthenings (Chapter 33).

Theorem 2.17. (DPI for KL divergence) *Let $P_Y = P_{Y|X} \circ P_X$ and $Q_Y = P_{Y|X} \circ Q_X$, as represented by the diagram:*

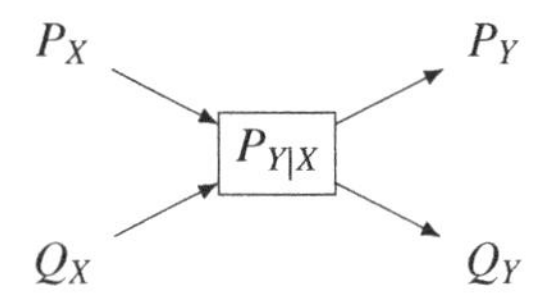

Then

$$D(P_Y\|Q_Y) \leq D(P_X\|Q_X), \tag{2.28}$$

with equality if and only if $D(P_{X|Y}\|Q_{X|Y}|P_Y) = 0$.

Proof. This follows from either the chain rule or monotonicity:

$$\begin{aligned} D(P_{X,Y}\|Q_{X,Y}) &= \underbrace{D(P_{Y|X}\|Q_{Y|X}|P_X)}_{=0} + D(P_X\|Q_X) \\ &= D(P_{X|Y}\|Q_{X|Y}|P_Y) + D(P_Y\|Q_Y). \end{aligned}$$ □

Corollary 2.18. (Divergence under deterministic transformation) *Let $Y = f(X)$. Then $D(P_Y\|Q_Y) \leq D(P_X\|Q_X)$, with equality if f is one-to-one.*

Note that $D(P_{f(X)}\|Q_{f(X)}) = D(P_X\|Q_X)$ does not imply that f is one-to-one; as an example, consider $P_X =$ Gaussian, $Q_X =$ Laplace, $Y = |X|$. In fact, the equality happens precisely when $f(X)$ is a *sufficient statistic* for testing P against Q; in other words, there is no loss of information in summarizing X into $f(X)$ as far as testing these two hypotheses is concerned. See Example 3.9 for details.

A particularly useful application of Corollary 2.18 is when we take f to be an indicator function:

Corollary 2.19. (Large-deviations estimate) *For any subset $E \subset \mathcal{X}$ we have*

$$d(P_X[E]\|Q_X[E]) \leq D(P_X\|Q_X),$$

where $d(\cdot\|\cdot)$ is the binary divergence function in (2.6).

Proof. Consider $Y = 1\{X \in E\}$. □

This method will be highly useful in large-deviations theory which studies rare events (Section 14.5 and Section 15.2), where we apply Corollary 2.19 to an event E which is highly likely under P but highly unlikely under Q.

2.6* Local Behavior of Divergence and Fisher Information

As we shall see in Section 4.5, KL divergence is in general not continuous. Nevertheless, it is reasonable to expect that in non-pathological cases the functional $D(P\|Q)$ vanishes when P approaches Q "smoothly." Due to the smoothness and strict convexity of $x\log x$ at $x = 1$, it is then also natural to expect that this functional decays "quadratically." In this section we examine this question first along the linear interpolation between P and Q, then, more generally, in smooth parameterized families of distributions. These properties will be extended to more general divergences later in Sections 7.10 and 7.11.

2.6.1* Local Behavior of Divergence for Mixtures

Let $0 \le \lambda \le 1$ and consider $D(\lambda P + \bar{\lambda}Q\|Q)$, which vanishes as $\lambda \to 0$. Next, we show that this decay is always sublinear.

Proposition 2.20. *When $D(P\|Q) < \infty$, the one-sided derivative at $\lambda = 0$ vanishes:*

$$\frac{d}{d\lambda}\bigg|_{\lambda=0} D(\lambda P + \bar{\lambda}Q\|Q) = 0.$$

If we exchange the arguments, the criterion is even simpler:

$$\frac{d}{d\lambda}\bigg|_{\lambda=0} D(Q\|\lambda P + \bar{\lambda}Q) = 0 \quad\iff\quad P \ll Q. \tag{2.29}$$

Proof.

$$\frac{1}{\lambda}D(\lambda P + \bar{\lambda}Q\|Q) = \mathbb{E}_Q\left[\frac{1}{\lambda}(\lambda f + \bar{\lambda})\log(\lambda f + \bar{\lambda})\right],$$

where $f = \frac{dP}{dQ}$. As $\lambda \to 0$ the function under expectation decreases to $(f-1)\log e$ monotonically. Indeed, the function

$$\lambda \mapsto g(\lambda) \triangleq (\lambda f + \bar{\lambda})\log(\lambda f + \bar{\lambda})$$

is convex and equals zero at $\lambda = 0$. Thus $\frac{g(\lambda)}{\lambda}$ is increasing in λ. Moreover, by the convexity of $x \mapsto x\log x$:

$$\frac{1}{\lambda}(\lambda f + \overline{\lambda})(\log(\lambda f + \overline{\lambda})) \le \frac{1}{\lambda}(\lambda f\log f + \overline{\lambda}1\log 1) = f\log f,$$

and by assumption $f\log f$ is Q-integrable. Thus the monotone convergence theorem applies.

To prove (2.29) first notice that if $P \not\ll Q$ then there is a set E with $p = P[E] > 0 = Q[E]$. Applying data processing for divergence to $X \mapsto 1_E(X)$, we get

$$D(Q\|\lambda P + \bar{\lambda}Q) \geq d(0\|\lambda p) = \log \frac{1}{1-\lambda p}$$

and the derivative is non-zero. If $P \ll Q$, then let $f = \frac{dP}{dQ}$ and notice simple inequalities

$$\log \bar{\lambda} \leq \log(\bar{\lambda} + \lambda f) \leq \lambda(f-1)\log e.$$

Dividing by λ and assuming $\lambda < \frac{1}{2}$ we get for some absolute constants c_1, c_2:

$$\left|\frac{1}{\lambda}\log(\bar{\lambda} + \lambda f)\right| \leq c_1 f + c_2.$$

Thus, by the dominated convergence theorem we get

$$\frac{1}{\lambda}D(Q\|\lambda P + \bar{\lambda}Q) = -\int dQ\left(\frac{1}{\lambda}\log(\bar{\lambda} + \lambda f)\right) \xrightarrow{\lambda\to 0} \int dQ(1-f) = 0. \qquad \square$$

Remark 2.2. More generally, under suitable technical conditions,

$$\frac{d}{d\lambda}\Big|_{\lambda=0} D(\lambda P + \bar{\lambda}Q\|R) = \mathbb{E}_P\left[\log\frac{dQ}{dR}\right] - D(Q\|R)$$

and

$$\frac{d}{d\lambda}\Big|_{\lambda=0} D(\bar{\lambda}P_1 + \lambda Q_1\|\bar{\lambda}P_0 + \lambda Q_0) = \mathbb{E}_{Q_1}\left[\log\frac{dP_1}{dP_0}\right] - D(P_1\|P_0) + \mathbb{E}_{P_1}\left[1 - \frac{dQ_0}{dP_0}\right]\log e.$$

See Exercise **I.22** for an example.

The main message of Proposition 2.20 is that the function

$$\lambda \mapsto D(\lambda P + \bar{\lambda}Q\|Q)$$

is $o(\lambda)$ as $\lambda \to 0$. In fact, in most cases it is quadratic in λ. To make a precise statement, we need to define the concept of χ^2-divergence – a version of f-divergence (see Chapter 7):

$$\chi^2(P\|Q) \triangleq \int dQ\left(\frac{dP}{dQ} - 1\right)^2.$$

This is a popular dissimilarity measure between P and Q, frequently used in statistics. It has many important properties, but we will only mention that χ^2 dominates KL divergence (see (7.34)):

$$D(P\|Q) \leq \log(1 + \chi^2(P\|Q)).$$

Our second result about the local behavior of KL divergence is the following (see Section 7.10 for generalizations):

Proposition 2.21. (KL is locally χ^2-like) *We have*

$$\liminf_{\lambda\to 0} \frac{1}{\lambda^2} D(\lambda P + \bar{\lambda} Q\|Q) = \frac{\log e}{2}\chi^2(P\|Q), \tag{2.30}$$

where both sides are finite or infinite simultaneously.

Proof. First, we assume that $\chi^2(P\|Q) < \infty$ and prove

$$D(\lambda P + \bar{\lambda} Q\|Q) = \frac{\lambda^2 \log e}{2}\chi^2(P\|Q) + o(\lambda^2), \qquad \lambda \to 0.$$

To that end notice that

$$D(P\|Q) = \mathbb{E}_Q\left[g\left(\frac{dP}{dQ}\right)\right],$$

where

$$g(x) \triangleq x\log x - (x-1)\log e.$$

Note that $x \mapsto \frac{g(x)}{(x-1)^2 \log e} = \int_0^1 \frac{sds}{x(1-s)+s}$ is decreasing in x on $(0,\infty)$. Therefore

$$0 \le g(x) \le (x-1)^2 \log e,$$

and hence

$$0 \le \frac{1}{\lambda^2} g\left(\bar{\lambda} + \lambda\frac{dP}{dQ}\right) \le \left(\frac{dP}{dQ} - 1\right)^2 \log e.$$

By the dominated convergence theorem (which is applicable since $\chi^2(P\|Q) < \infty$) we have

$$\lim_{\lambda\to 0} \frac{1}{\lambda^2}\mathbb{E}_Q\left[g\left(\bar{\lambda} + \lambda\frac{dP}{dQ}\right)\right] = \frac{g''(1)}{2}\mathbb{E}_Q\left[\left(\frac{dP}{dQ} - 1\right)^2\right] = \frac{\log e}{2}\chi^2(P\|Q).$$

Second, we show that unconditionally

$$\liminf_{\lambda\to 0} \frac{1}{\lambda^2} D(\lambda P + \bar{\lambda} Q\|Q) \ge \frac{\log e}{2}\chi^2(P\|Q). \tag{2.31}$$

Indeed, this follows from Fatou's lemma:

$$\liminf_{\lambda\to 0} \mathbb{E}_Q\left[\frac{1}{\lambda^2} g\left(\bar{\lambda} + \lambda\frac{dP}{dQ}\right)\right] \ge \mathbb{E}_Q\left[\liminf_{\lambda\to 0} g\left(\bar{\lambda} + \lambda\frac{dP}{dQ}\right)\right] = \frac{\log e}{2}\chi^2(P\|Q).$$

Therefore, from (2.31) we conclude that if $\chi^2(P\|Q) = \infty$ then so is the LHS of (2.30). □

2.6.2* Parametrized Family

Extending the setting of Section 2.6.1*, consider a parameterized set of distributions $\{P_\theta : \theta \in \Theta\}$, where the parameter space Θ is an open subset of $\mathbb{R}^d$. Furthermore, suppose that distributions P_θ are all given in the form of

$$P_\theta(dx) = p_\theta(x)\mu(dx),$$

where μ is some common dominating measure (e.g. Lebesgue or counting measure). If for each fixed x, the density $p_\theta(x)$ depends smoothly on θ, one can define the *Fisher information matrix* with respect to the parameter θ as

$$J_F(\theta) \triangleq \mathbb{E}_\theta\left[VV^\top\right], \quad V \triangleq \nabla_\theta \ln p_\theta(X), \tag{2.32}$$

where $\mathbb{E}_\theta$ is with respect to $X \sim P_\theta$. In particular, V is known as the *score*.

Under suitable regularity conditions, we have the identity

$$\mathbb{E}_\theta[V] = 0 \tag{2.33}$$

and several equivalent expressions for the Fisher information matrix:

$$\begin{aligned} J_F(\theta) &= \mathrm{Cov}_\theta(V) \\ &= 4\int \mu(dx)(\nabla_\theta\sqrt{p_\theta(x)})(\nabla_\theta\sqrt{p_\theta(x)})^\top \\ &= -\,\mathbb{E}_\theta[\mathrm{Hess}_\theta(\ln p_\theta(X))], \end{aligned}$$

where the last identity is obtained by differentiating (2.33) with respect to each θ_j.

The significance of the Fisher information matrix arises from the fact that it gauges the local behavior of divergence for smooth parametric families. Namely, we have (again under suitable technical conditions):[2]

$$D(P_{\theta_0}\|P_{\theta_0+\xi}) = \frac{\log e}{2}\xi^\top J_F(\theta_0)\xi + o(\|\xi\|^2), \tag{2.34}$$

which is obtained by integrating the Taylor expansion:

$$\ln p_{\theta_0+\xi}(x) = \ln p_{\theta_0}(x) + \xi^\top\nabla_\theta \ln p_{\theta_0}(x) + \frac{1}{2}\xi^\top \mathrm{Hess}_\theta(\ln p_{\theta_0}(x))\xi + o(\|\xi\|^2).$$

We will establish this fact rigorously later in Section 7.11. Property (2.34) is of paramount importance in statistics. We should remember it as: *Divergence is locally quadratic on the parameter space, with Hessian given by the Fisher information matrix.* Note that for the Gaussian location model $P_\theta = \mathcal{N}(\theta, \Sigma)$, (2.34) is in fact exact with $J_F(\theta) \equiv \Sigma^{-1}$ – see Example 2.2.

As another example, note that Proposition 2.21 is a special case of (2.34) by considering $P_\lambda = \bar{\lambda}Q + \lambda P$ parameterized by $\lambda \in [0,1]$. In this case, the Fisher information at $\lambda = 0$ is simply $\chi^2(P\|Q)$. Nevertheless, Proposition 2.21 is completely general while the asymptotic expansion (2.34) is not without regularity conditions (see Section 7.11).

Remark 2.3. Some useful properties of Fisher information are as follows:

[2] To illustrate the subtlety here, consider a scalar location family, that is, $p_\theta(x) = f_0(x-\theta)$ for some density f_0. In this case Fisher information $J_F(\theta_0) = \int \frac{(f_0{}')^2}{f_0}$ does not depend on θ_0 and is well defined even for compactly supported f_0, provided $f_0{}'$ vanishes at the endpoints sufficiently fast. But at the same time the left-hand side of (2.34) is infinite for any $\xi > 0$. Thus, a more general interpretation for Fisher information is as the coefficient in expansion $D(P_{\theta_0}\|\frac{1}{2}P_{\theta_0} + \frac{1}{2}P_{\theta_0+\xi}) = \frac{\xi^2}{8}J_F(\theta_0) + o(\xi^2)$. We will discuss this in more detail in Section 7.11.

- Reparameterization: It can be seen that if one introduces another parameterization $\tilde{\theta} \in \tilde{\Theta}$ by means of a smooth invertible map $\tilde{\Theta} \to \Theta$, then Fisher information matrix changes as

$$J_F(\tilde{\theta}) = A^\top J_F(\theta) A, \tag{2.35}$$

where $A = \frac{d\theta}{d\tilde{\theta}}$ is the Jacobian of the map. So we can see that J_F transforms similarly to the metric tensor in Riemannian geometry. This idea can be used to define a Riemannian metric on the parameter space Θ, called the *Fisher–Rao metric*, as is explored in a field known as information geometry [85, 17].

- Additivity: Suppose we are given a sample of n iid observations $X^n \overset{\text{iid}}{\sim} P_\theta$. As such, consider the parameterized family of product distributions $\{P_\theta^{\otimes n} : \theta \in \Theta\}$, whose Fisher information matrix is denoted by $J_F^{\otimes n}(\theta)$. In this case, the score is an iid sum. Applying (2.32) and (2.33) yields

$$J_F^{\otimes n}(\theta) = n J_F(\theta). \tag{2.36}$$

Example 2.6.

Let $P_\theta = (\theta_0, \dots, \theta_d)$ be a probability distribution on the finite alphabet $\{0, \dots, d\}$. We will take $\theta = (\theta_1, \dots, \theta_d)$ as the free parameter and set $\theta_0 = 1 - \sum_{i=1}^d \theta_i$. So all derivatives are with respect to $\theta_1, \dots, \theta_d$ only. Then we have

$$p_\theta(i) = \begin{cases} \theta_i, & i = 1, \dots, d, \\ 1 - \sum_{i=1}^d \theta_i, & i = 0, \end{cases}$$

and for the Fisher information matrix we get

$$J_F(\theta) = \operatorname{diag}\left(\frac{1}{\theta_1}, \dots, \frac{1}{\theta_d}\right) + \frac{1}{1 - \sum_{i=1}^d \theta_i} \mathbf{1}\mathbf{1}^\top, \tag{2.37}$$

where $\mathbf{1}$ is the $d \times 1$ vector of all ones. For future reference (see Sections 29.4 and 13.4*), we also compute the inverse and determinant of $J_F(\theta)$. By the matrix inversion lemma $(A + UCV)^{-1} = A^{-1} - A^{-1}U(C^{-1} + VA^{-1}U)^{-1}VA^{-1}$, we have

$$J_F^{-1}(\theta) = \operatorname{diag}(\theta) - \theta\theta^\top. \tag{2.38}$$

For the determinant, notice that $\det(A + xy^\top) = \det A \cdot \det(I + A^{-1}xy^\top) = \det A \cdot (1 + y^\top A^{-1} x)$, where we used the identity $\det(I + AB) = \det(I + BA)$. Thus, we have

$$\det J_F(\theta) = \prod_{i=0}^d \frac{1}{\theta_i}. \tag{2.39}$$

Example 2.7. Location family

In statistics and information theory it is common to talk about Fisher information of a (single) random variable or a distribution without reference to a parametric family. In such cases one is implicitly considering a location parameter. Specifically, for any density p_0 on $\mathbb{R}^d$ we define a *location family* of distributions on $\mathbb{R}^d$ by setting $P_\theta(dx) = p_0(x-\theta)dx, \theta \in \mathbb{R}^d$. Note that $J_F(\theta)$ here does not depend on θ. For this special case, we will adopt the standard notation: Let $X \sim p_0$ then

$$J(X) \equiv J(p_0) \triangleq \mathbb{E}_{X\sim p_0}[(\nabla \ln p_0(X))(\nabla \ln p_0(X))^\top] \tag{2.40}$$

$$= -\mathbb{E}_{X\sim p_0}[\text{Hess}(\ln p_0(X))], \tag{2.41}$$

where the second equality requires applicability of integration by parts. (See also (7.96) for a variational definition.)

3 Mutual Information

After the technical preparations in the previous chapters we define perhaps the most famous concept in the entire field of information theory, *mutual information*. It was originally defined by Shannon, although the name was coined later by Robert Fano.[1] It has two equivalent expressions (as a KL divergence and as a difference of entropies), both having their merits. In this chapter, we collect some basic properties of mutual information (non-negativity, chain rule, and the data-processing inequality). While defining conditional information, we also introduce the language of *directed graphical models*, and connect the equality case in the data-processing inequality with Fisher's concept of sufficient statistic.

So far in this book we have not yet attempted connecting information quantities to any operational concepts. The first time this will be done is in Section 3.6 where we relate mutual information to probability of error in the form of *Fano's inequality*, which states that whenever $I(X;Y)$ is small, one should not be able to predict X on the basis of Y with a small probability of error. As such, this inequality will be applied countless times in the rest of the book as a main workhorse for studying fundamental limits of problems in both information theory and statistics.

The connection between information and estimation is furthered in Section 3.7*, in which we relate mutual information and minimum mean-squared error in Gaussian noise (I-MMSE relation). From the latter we also derive the entropy-power inequality, which plays a central role in high-dimensional probability and concentration of measure.

3.1 Mutual Information

Mutual information was first defined by Shannon to measure the decrease in entropy of a random quantity following the observation of another (correlated) random quantity. Unlike the concept of entropy itself, which was well known by then in statistical mechanics, mutual information was new and revolutionary and had no analogs in science. Today, however, it is preferred to define mutual information in a different form (proposed in [379, Appendix 7]).

[1] Professor of Electrical Engineering at MIT, who developed the first course on information theory and as part of it formalized and rigorized much of Shannon's ideas. Most famously, he showed the "converse part" of the noisy channel coding theorem, see Section 17.4.

Definition 3.1. (Mutual information) The *mutual information* between a pair of random variables X and Y is

$$I(X;Y) = D(P_{X,Y} \| P_X P_Y).$$

The intuitive interpretation of mutual information is that $I(X;Y)$ measures the dependence between X and Y by comparing their joint distribution to the product of the marginals in the KL divergence, which, as we show next, is also equivalent to comparing the conditional distribution to the unconditional.

The way we defined $I(X;Y)$ it is a functional of the joint distribution $P_{X,Y}$. However, it is also rather fruitful to look at it as a functional of the pair $(P_X, P_{Y|X})$ – more on this in Section 5.1.

In general, the divergence $D(P_{X,Y} \| P_X P_Y)$ should be evaluated using the general definition (2.4). Note that $P_{X,Y} \ll P_X P_Y$ need not always hold. Let us consider the following examples, though.

Example 3.1.

If $X = Y \sim N(0,1)$ then $P_{X,Y} \not\ll P_X P_Y$ and $I(X;Y) = \infty$. This reflects our intuition that X contains an "infinite" amount of information requiring infinitely many bits to describe. On the other hand, if even one of X or Y is discrete, then we *always* have $P_{X,Y} \ll P_X P_Y$. Indeed, consider any $E \subset \mathcal{X} \times \mathcal{Y}$ measurable in the product σ-algebra with $P_{X,Y}(E) > 0$. Since $P_{X,Y}(E) = \sum_{x \in S} \mathbb{P}[(X,Y) \in S, X = x]$, there exists some $x_0 \in S$ such that $P_Y(E^{x_0}) \geq \mathbb{P}[X = x_0, Y \in E^{x_0}] > 0$, where $E^{x_0} \triangleq \{y \colon (x_0, y) \in E\}$ is a section of E (measurable for every x_0). But then $P_X P_Y(E) \geq P_X P_Y(\{x_0\} \times E^{x_0}) = P_X(\{x_0\}) P_Y(E^{x_0}) > 0$, implying that $P_{X,Y} \ll P_X P_Y$.

Theorem 3.2. (Properties of mutual information)

(a) (Mutual information as conditional divergence) Whenever $\mathcal{Y}$ is standard Borel,

$$I(X;Y) = D(P_{Y|X} \| P_Y | P_X). \tag{3.1}$$

(b) (Symmetry) $I(X;Y) = I(Y;X)$.
(c) (Positivity) $I(X;Y) \geq 0$, with equality $I(X;Y) = 0$ iff $X \perp\!\!\!\perp Y$.
(d) (Deterministic maps) For any function f we have

$$I(f(X);Y) \leq I(X;Y).$$

If f is one-to-one (with a measurable inverse), then $I(f(X);Y) = I(X;Y)$.
(e) (More data $\Rightarrow$ more information) $I(X_1, X_2; Z) \geq I(X_1; Z)$.

Proof.

(a) This follows from Theorem 2.16(a) with $Q_{Y|X} = P_Y$.
(b) Consider a Markov kernel K sending $(x,y) \mapsto (y,x)$. This kernel sends measure $P_{X,Y} \xrightarrow{K} P_{Y,X}$ and $P_X P_Y \xrightarrow{K} P_Y P_X$. Therefore, from the DPI Theorem 2.17 applied to this kernel we get

$$D(P_{X,Y} \| P_X P_Y) \geq D(P_{Y,X} \| P_Y P_X).$$

Applying this argument again shows that the inequality is in fact equality.

(c) This is just $D \geq 0$ from Theorem 2.3.

(d) Consider a Markov kernel K sending $(x, y) \mapsto (f(x), y)$. This kernel sends measure $P_{X,Y} \xrightarrow{K} P_{f(X),Y}$ and $P_X P_Y \xrightarrow{K} P_{f(X)} P_Y$. Therefore, from the DPI Theorem 2.17 applied to this kernel we get

$$D(P_{X,Y} \| P_X P_Y) \geq D(P_{f(X),Y} \| P_{f(X)} P_Y).$$

It is clear that the two sides correspond to the two mutual informations. For bijective f, simply apply the inequality to f and f^{-1}.

(e) Apply (d) with $f(X_1, X_2) = X_1$. □

Of the results above, the one we will use the most is (3.1). Note that it implies that $D(P_{X,Y} \| P_X P_Y) < \infty$ if and only if

$$x \mapsto D(P_{Y|X=x} \| P_Y)$$

is P_X-integrable. This property has a counterpart in terms of absolute continuity, as follows.

Lemma 3.3. *Let $\mathcal{Y}$ be standard Borel. Then*

$$P_{X,Y} \ll P_X P_Y \quad \iff \quad P_{Y|X=x} \ll P_Y \quad \textit{for } P_X\textit{-a.e. } x.$$

Proof. Suppose $P_{X,Y} \ll P_X P_Y$. We need to prove that *any* version of the conditional probability satisfies $P_{Y|X=x} \ll P_Y$ for almost every x. Note, however, that if we prove this for *some* version $\tilde{P}_{Y|X}$ then the statement for any version follows, since $P_{Y|X=x} = \tilde{P}_{Y|X=x}$ for P_X-a.e. x. (This measure-theoretic fact can be derived from the chain rule (2.24): since $P_X \tilde{P}_{Y|X} = P_{X,Y} = P_X P_{Y|X}$ we must have $0 = D(P_{X,Y} \| P_{X,Y}) = D(\tilde{P}_{Y|X} \| P_{Y|X} | P_X) = \mathbb{E}_{x \sim P_X}[D(\tilde{P}_{Y|X=x} \| P_{Y|X=x})]$, implying the stated fact.) So let $g(x, y) = \frac{dP_{X,Y}}{dP_X P_Y}(x, y)$ and $\rho(x) \triangleq \int_{\mathcal{Y}} g(x, y) P_Y(dy)$. Fix any set $E \subset \mathcal{X}$ and notice

$$P_X[E] = \int_{\mathcal{X} \times \mathcal{Y}} 1_E(x) g(x, y) P_X(dx)\, P_Y(dy) = \int_{\mathcal{X}} 1_E(x) \rho(x) P_X(dx).$$

On the other hand, we also have $P_X[E] = \int 1_E dP_X$, which implies $\rho(x) = 1$ for P_X-a.e. x. Now define

$$\tilde{P}_{Y|X}(dy|x) = \begin{cases} g(x, y) P_Y(dy), & \rho(x) = 1, \\ P_Y(dy), & \rho(x) \neq 1. \end{cases}$$

Directly plugging $\tilde{P}_{Y|X}$ into (2.22) shows that $\tilde{P}_{Y|X}$ does define a valid version of the conditional probability of Y given X. Since by construction $\tilde{P}_{Y|X=x} \ll P_Y$ for every x, the result follows.

Conversely, let $P_{Y|X}$ be a kernel such that $P_X[E] = 1$, where $E = \{x \colon P_{Y|X=x} \ll P_Y\}$ (recall that E is measurable by Lemma 2.13). Define $\tilde{P}_{Y|X=x} = P_{Y|X=x}$ if $x \in E$ and $\tilde{P}_{Y|X=x} = P_Y$ otherwise. By construction $P_X \tilde{P}_{Y|X} = P_X P_{Y|X} = P_{X,Y}$ and

$\tilde{P}_{Y|X=x} \ll P_Y$ for every x. Thus, by Theorem 2.12 there exists a jointly measurable $f(y|x)$ such that

$$\tilde{P}_{Y|X}(dy|x) = f(y|x)P_Y(dy),$$

and, thus, by (2.22)

$$P_{X,Y}[E] = \int_E f(y|x)P_Y(dy)P_X(dx),$$

implying that $P_{X,Y} \ll P_X P_Y$. □

3.2 Mutual Information as Difference of Entropies

As promised, we next introduce a different point of view on $I(X;Y)$, namely as a difference of entropies. This (conditional entropy) point of view of Shannon emphasizes that $I(X;Y)$ is also measuring the change in the spread or uncertainty of the distribution of X following the observation of Y.

Theorem 3.4. (Mutual information and entropy)

(a) $I(X;X) = \begin{cases} H(X), & X \textit{ discrete}, \\ +\infty, & \textit{otherwise}. \end{cases}$

(b) If X is discrete, then

$$I(X;Y) + H(X|Y) = H(X). \tag{3.2}$$

Consequently, either $H(X|Y) = H(X) = \infty$,[2] *or $H(X|Y) < \infty$ and*

$$I(X;Y) = H(X) - H(X|Y). \tag{3.3}$$

(c) If both X and Y are discrete, then

$$I(X;Y) + H(X,Y) = H(X) + H(Y),$$

so that whenever $H(X,Y) < \infty$ we have

$$I(X;Y) = H(X) + H(Y) - H(X,Y).$$

(d) Similarly, if X, Y are real-valued random vectors with a joint PDF, then

$$I(X;Y) = h(X) + h(Y) - h(X,Y)$$

provided that $h(X,Y) < \infty$. If X has a marginal PDF p_X and a conditional PDF $p_{X|Y}(x|y)$, then

$$I(X;Y) = h(X) - h(X|Y),$$

provided $h(X|Y) < \infty$.

[2] This is indeed possible if one takes $Y = 0$ (constant) and X from Example 1.3, demonstrating that (3.3) does not always hold.

(e) *If X or Y are discrete then $I(X;Y) \leq \min(H(X), H(Y))$, with equality iff $H(X|Y) = 0$ or $H(Y|X) = 0$, or, equivalently, iff one is a deterministic function of the other.*

Proof.

(a) By Theorem 3.2(a), $I(X;X) = D(P_{X|X} \| P_X | P_X) = \mathbb{E}_{x \sim X} D(\delta_x \| P_X)$. If P_X is discrete, then $D(\delta_x \| P_X) = \log \frac{1}{P_X(x)}$ and $I(X;X) = H(X)$. If P_X is not discrete, let $\mathcal{A} = \{x \colon P_X(x) > 0\}$ denote the set of atoms of P_X. Let $\Delta = \{(x,x) \colon x \notin \mathcal{A}\} \subset \mathcal{X} \times \mathcal{X}$. ($\Delta$ is measurable since it is the intersection of $\mathcal{A}^c \times \mathcal{A}^c$ with the diagonal $\{(x,x) \colon x \in \mathcal{X}\}$.) Then $P_{X,X}(\Delta) = P_X(\mathcal{A}^c) > 0$ but since

$$(P_X \times P_X)(E) \triangleq \int_{\mathcal{X}} P_X(dx_1) \int_{\mathcal{X}} P_X(dx_2) 1\{(x_1, x_2) \in E\}$$

we have by taking $E = \Delta$ that $(P_X \times P_X)(\Delta) = 0$. Thus $P_{X,X} \not\ll P_X \times P_X$ and thus by definition

$$I(X;X) = D(P_{X,X} \| P_X P_X) = +\infty.$$

(b) If $H(X|Y) = \infty$ then the statement holds (indeed, $H(X) = \infty$ because conditioning reduces entropy). Thus, we assume $H(X|Y) < \infty$. Since X is discrete there exists a countable set S such that $\mathbb{P}[X \in S] = 1$, and for any $x_0 \in S$ we have $\mathbb{P}[X = x_0] > 0$. Let λ be a counting measure on S and let $\mu = \lambda \times P_Y$, so that $P_X P_Y \ll \mu$. As shown in Example 3.1 we also have $P_{X,Y} \ll \mu$. Furthermore, $f_P(x,y) \triangleq \frac{dP_{X,Y}}{d\mu}(x,y) = p_{X|Y}(x|y)$, where the latter denotes the conditional PMF of X given Y (which is a proper PMF for almost every y, since $\mathbb{P}[X \in S | Y = y] = 1$ for a.e. y). We also have $f_Q(x,y) = \frac{dP_X P_Y}{d\mu}(x,y) = \frac{dP_X}{d\lambda}(x) = p_X(x)$, where the latter is an unconditional PMF of X. Note that by definition of Radon–Nikodym derivatives we have

$$\mathbb{E}[p_{X|Y}(x_0|Y)] = p_X(x_0). \tag{3.4}$$

Next, according to (2.12) we have

$$I(X;Y) = \mathbb{E}\left[\mathrm{Log}\frac{f_P(X,Y)}{f_Q(X,Y)}\right] = \mathbb{E}_{y \sim P_Y} \sum_{x \in S} \left[p_{X|Y}(x|y) \mathrm{Log}\frac{p_{X|Y}(x|y)}{p_X(x)}\right].$$

Note that $P_{X,Y}$-almost surely both $p_{X|Y}(X|Y) > 0$ and $P_X(x) > 0$, so we can replace Log with log in the above. On the other hand,

$$H(X|Y) = \mathbb{E}_{y \sim P_Y} \sum_{x \in S} \left[p_{X|Y}(x|y) \log \frac{1}{p_{X|Y}(x|y)}\right].$$

Adding these two expressions, we obtain

$$\begin{aligned} I(X;Y) + H(X|Y) &\overset{\text{(i)}}{=} \mathbb{E}_{y \sim P_Y} \sum_{x \in S} \left[p_{X|Y}(x|y) \log \frac{1}{p_X(x)}\right] \\ &\overset{\text{(ii)}}{=} \sum_{x \in S} \mathbb{E}_{y \sim P_Y}[p_{X|Y}(x|y)] \log \frac{1}{p_X(x)} \overset{\text{(iii)}}{=} \mathbb{E}\left[\log \frac{1}{P_X(X)}\right] \triangleq H(X), \end{aligned}$$

where in (i) we used linearity of the Lebesgue integral $\mathbb{E}_{P_Y} \sum_x$, in (ii) we interchange $\mathbb{E}$ and $\sum$ via Fubini, and (iii) holds due to (3.4).

(c) Simply add $H(Y)$ to both sides of (3.2) and use the chain rule for H from (1.2).
(d) These arguments are similar to the discrete case, except that counting measure is replaced with Lebesgue. We leave the details as an exercise.
(e) Follows from (b). □

From (3.2) we deduce the following result, which was previously shown in Theorem 1.4(d).

Corollary 3.5. (Conditioning reduces entropy) *For discrete X, $H(X|Y) \leq H(X)$, with equality iff $X \perp\!\!\!\perp Y$.*

Proof. If $H(X) = \infty$ then there is nothing to prove. Otherwise, apply (3.2). □

Thus, the intuition behind the last corollary (and an important innovation of Shannon) is to give meaning to the amount of entropy reduction (mutual information). It is important to note that conditioning reduces entropy *on average*, not per realization. Indeed, take $X = U \,\texttt{OR}\, Y$, where $U, Y \stackrel{\text{iid}}{\sim} \text{Ber}(1/2)$. Then $X \sim \text{Ber}(3/4)$ and $H(X) = h(\frac{1}{4}) < 1$ bit $= H(X|Y = 0)$, that is, conditioning on $Y = 0$ increases entropy. But *on average*, $H(X|Y) = \mathbb{P}[Y = 0]H(X|Y = 0) + \mathbb{P}[Y = 1]H(X|Y = 1) = \frac{1}{2}$ bit $< H(X)$, by the strong concavity of $h(\cdot)$.

Remark 3.1. (Information, entropy, and Venn diagrams) For discrete random variables, the following Venn diagram illustrates the relationship between entropy, conditional entropy, joint entropy, and mutual information from Theorem 3.4(b) and (c).

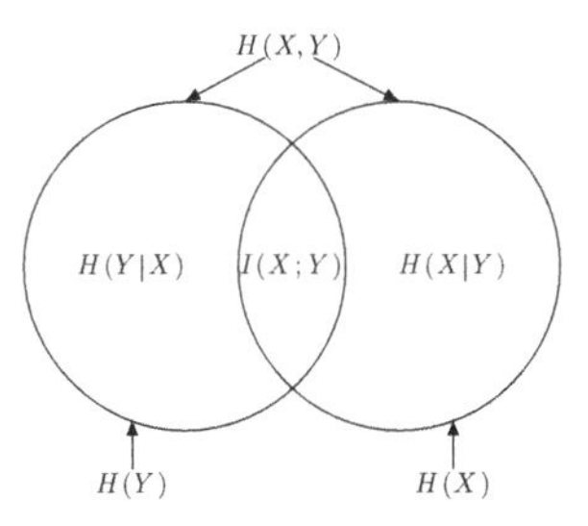

Applying analogously the inclusion–exclusion principle to three variables X_1, X_2, X_3, we see that the triple intersection corresponds to

$$H(X_1) + H(X_2) + H(X_3) - H(X_1, X_2) - H(X_2, X_3) - H(X_1, X_3) + H(X_1, X_2, X_3), \tag{3.5}$$

which is sometimes denoted by $I(X_1;X_2;X_3)$. It can be both positive and negative (why?).

In general, one can treat random variables as sets (so that X_i corresponds to set E_i and the pair (X_1, X_2) corresponds to $E_1 \cup E_2$). Then we can define a unique signed measure μ on the finite algebra generated by these sets so that every information quantity is found by replacing

$$I/H \to \mu \quad ; \to \cap \quad , \to \cup \quad | \to \backslash .$$

As an example, we have

$$H(X_1|X_2, X_3) = \mu(E_1 \setminus (E_2 \cup E_3)), \tag{3.6}$$

$$I(X_1, X_2; X_3|X_4) = \mu(((E_1 \cup E_2) \cap E_3) \setminus E_4). \tag{3.7}$$

By inclusion–exclusion, the quantity in (3.5) corresponds to $\mu(E_1 \cap E_2 \cap E_3)$, which explains why μ is not necessarily a positive measure. For an extensive discussion, see [110, Chapter 1.3].

3.3 Examples of Computing Mutual Information

Below we demonstrate how to compute I in both continuous and discrete settings.

Example 3.2. Bivariate Gaussian

Let X, Y be jointly Gaussian. Then

$$I(X;Y) = \frac{1}{2} \log \frac{1}{1 - \rho_{X,Y}^2}, \tag{3.8}$$

where $\rho_{X,Y} \triangleq \frac{\mathbb{E}[(X-\mathbb{E}X)(Y-\mathbb{E}Y)]}{\sigma_X \sigma_Y} \in [-1, 1]$ is the correlation coefficient; see Figure 3.1 for a plot.

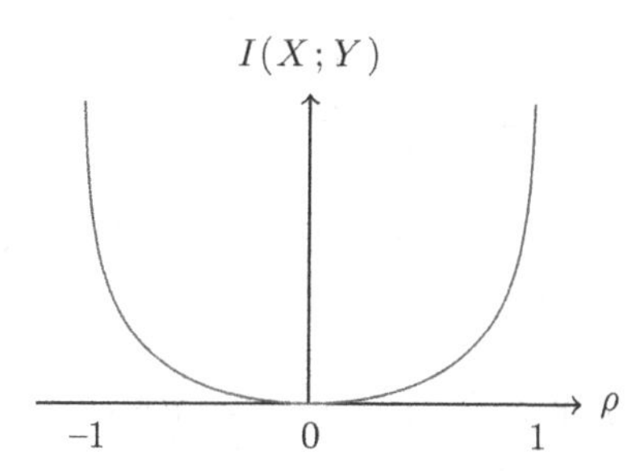

Figure 3.1 Mutual information between correlated Gaussians.

To show (3.8), by shifting and scaling if necessary, we can assume without loss of generality that $\mathbb{E}X = \mathbb{E}Y = 0$ and $\mathbb{E}X^2 = \mathbb{E}Y^2 = 1$. Then $\rho \equiv \rho_{X,Y} = \mathbb{E}XY$. By joint Gaussianity, $Y = \rho X + Z$ for some $Z \sim \mathcal{N}(0, 1 - \rho^2) \perp\!\!\!\perp X$. Then using the divergence formula for Gaussians (2.7), we get

$$\begin{aligned}
I(X;Y) &= D(P_{Y|X} \| P_Y | P_X) \\
&= \mathbb{E} D(\mathcal{N}(\rho X, 1 - \rho^2) \| \mathcal{N}(0, 1)) \\
&= \mathbb{E}\left[\frac{1}{2} \log \frac{1}{1 - \rho^2} + \frac{\log e}{2} \left((\rho X)^2 + 1 - \rho^2 - 1\right)\right] \\
&= \frac{1}{2} \log \frac{1}{1 - \rho^2}.
\end{aligned}$$

Alternatively, we can use the differential entropy representation in Theorem 3.4(d) and the entropy formula (2.17) for Gaussians:

$$\begin{aligned}I(X;Y) &= h(Y) - h(Y|X)\\ &= h(Y) - h(Z)\\ &= \frac{1}{2}\log(2\pi e) - \frac{1}{2}\log(2\pi e(1-\rho^2)) = \frac{1}{2}\log\frac{1}{1-\rho^2}.\end{aligned}$$

Here the second equality follows $h(Y|X) = h(Y - X|X) = h(Z|X) = h(Z)$ applying the shift invariance of h and the independence between X and Z.

Similar to the role of mutual information, the correlation coefficient also measures the dependence between random variables which are real-valued (or, more generally, valued in an inner product space) in a certain sense. In contrast, mutual information is invariant to bijections and much more general as it can be defined not just for numerical but for arbitrary random variables.

Example 3.3. AWGN channel

The additive white Gaussian noise (AWGN) channel is one of the main examples of Markov kernels that we will use in this book. This kernel acts from $\mathbb{R}$ to $\mathbb{R}$ by taking an input x and setting $K(\cdot|x) \sim \mathcal{N}(x, \sigma_N^2)$, or in equation form we write $Y = X + N$, with $X \perp\!\!\!\perp N \sim \mathcal{N}(0, \sigma_N^2)$. Pictorially, we can think of it as follows:

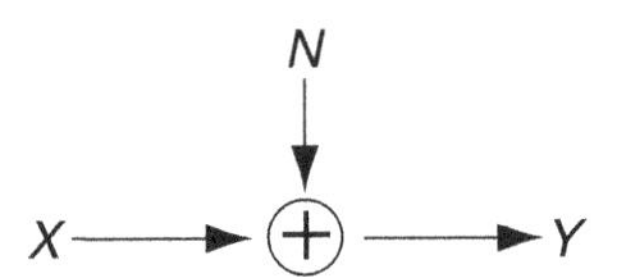

Now, suppose that $X \sim \mathcal{N}(0, \sigma_X^2)$, in which case $Y \sim \mathcal{N}(0, \sigma_X^2 + \sigma_N^2)$. Then by invoking (2.17) twice we obtain

$$I(X;Y) = h(Y) - h(Y|X) = h(X+N) - h(N) = \frac{1}{2}\log\left(1 + \frac{\sigma_X^2}{\sigma_N^2}\right),$$

where $\frac{\sigma_X^2}{\sigma_N^2}$ is frequently referred to as the *signal-to-noise ratio* (SNR). See Figure 3.2 for an illustration. Note that in engineering it is common to express SNR in decibels (dB), so that SNR in dB equals $10\log_{10}$ (SNR). Later, we will define the AWGN channel more formally in Definition 20.10.

Example 3.4. BI-AWGN channel

In communication and statistical applications one also often encounters a situation where the AWGN channel's input is restricted to $X \in \{\pm 1\}$. This Markov kernel is denoted $\mathsf{BIAWGN}_{\sigma_N^2} : \{\pm 1\} \to \mathbb{R}$ and acts by setting

$$Y = X + N, \qquad X \perp\!\!\!\perp N \sim \mathcal{N}(0, \sigma_N^2).$$

If we set $X \sim \text{Ber}(1/2)$ then in this case it is more convenient to calculate mutual information by a decomposition different from the AWGN case. Indeed, we have

$$I(X;Y) = H(X) - H(X|Y) = \log 2 - H(X|Y).$$

To compute $H(X|Y = y)$ we simply need to evaluate the posterior distribution given observation $Y = y$. In this case we have $\mathbb{P}[X = +1|Y = y] = \frac{e^{y/\sigma_N^2}}{e^{-y/\sigma_N^2} + e^{y/\sigma_N^2}}$. Thus, after some algebra we obtain the following expression

$$I(X;Y) = \log 2 - \int_{-\infty}^{\infty} \frac{1}{\sqrt{2\pi}} e^{-\frac{z^2}{2}} \log(1 + e^{-\frac{2}{\sigma^2} + \frac{2}{\sigma} z})\, dz.$$

(One can verify that $H(X|Y)$ here coincides with that in Example 1.4(2) with σ replaced by 2σ.) For this channel, the SNR is given by $\frac{\mathbb{E}[X^2]}{\mathbb{E}[N^2]} = \frac{1}{\sigma_N^2}$. We compare mutual informations of AWGN and BI-AWGN as a function of the SNR on Figure 3.2. Note that for low SNR restricting to binary input results in virtually no loss of information – a fact underpinning the role played by the BI-AWGN channel in many real-world communication systems.

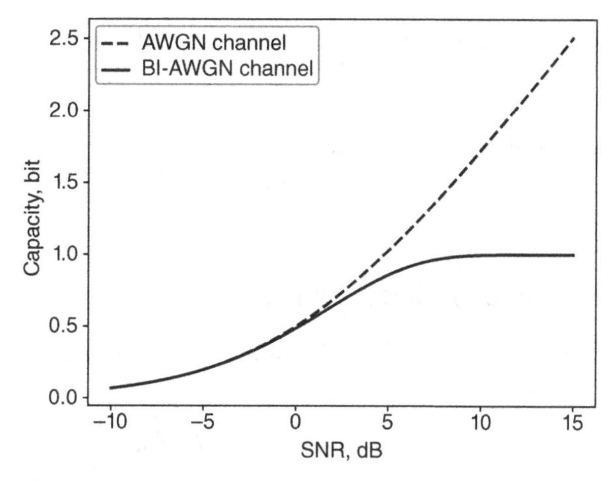

Figure 3.2 Comparing mutual information for the AWGN and BI-AWGN channels (see Examples 3.3 and 3.4). It will be shown later in this book that these mutual informations coincide with the capacities of respective channels.

Example 3.5. Gaussian vectors

Let $\mathbf{X} \in \mathbb{R}^m$ and $\mathbf{Y} \in \mathbb{R}^n$ be jointly Gaussian. Then

$$I(\mathbf{X};\mathbf{Y}) = \frac{1}{2} \log \frac{\det \Sigma_{\mathbf{X}} \det \Sigma_{\mathbf{Y}}}{\det \Sigma_{[\mathbf{X},\mathbf{Y}]}},$$

where $\Sigma_{\mathbf{X}} \triangleq \mathbb{E}\left[(\mathbf{X} - \mathbb{E}\mathbf{X})(\mathbf{X} - \mathbb{E}\mathbf{X})^\top\right]$ denotes the covariance matrix of $\mathbf{X} \in \mathbb{R}^m$, and $\Sigma_{[\mathbf{X},\mathbf{Y}]}$ denotes the covariance matrix of the random vector $[\mathbf{X}, \mathbf{Y}] \in \mathbb{R}^{m+n}$.

In the special case of additive noise, $\mathbf{Y} = \mathbf{X} + \mathbf{N}$ for $\mathbf{N} \perp\!\!\!\perp \mathbf{X}$, we have

$$I(\mathbf{X};\mathbf{X}+\mathbf{N}) = \frac{1}{2}\log\frac{\det(\Sigma_{\mathbf{X}} + \Sigma_{\mathbf{N}})}{\det \Sigma_{\mathbf{N}}}$$

since $\det \Sigma_{[\mathbf{X},\mathbf{X}+\mathbf{N}]} = \det \begin{pmatrix} \Sigma_{\mathbf{X}} & \Sigma_{\mathbf{X}} \\ \Sigma_{\mathbf{X}} & \Sigma_{\mathbf{X}}+\Sigma_{\mathbf{N}} \end{pmatrix} \overset{\text{why?}}{=} \det \Sigma_{\mathbf{X}} \det \Sigma_{\mathbf{N}}$.

Example 3.6. Binary symmetric channel

Recall the setting in Example 1.4(1). Let $X \sim \text{Ber}(1/2)$ and $N \sim \text{Ber}(\delta)$ be independent. Let $Y = X \oplus N$; or equivalently, Y is obtained by flipping X with probability δ.

As shown in Example 1.4(1), $H(X|Y) = H(N) = h(\delta)$ and hence

$$I(X;Y) = \log 2 - h(\delta).$$

The corresponding conditional distribution $P_{Y|X}$ (Markov kernel) is called the binary symmetric channel (BSC) with parameter δ and denoted by BSC_δ.

Example 3.7. Addition over finite groups

Generalizing Example 3.6, let X and Z take values on a finite abelian group G. If X is uniform on G and independent of Z, then

$$I(X;X+Z) = \log|G| - H(Z),$$

which simply follows from the fact that $X + Z$ is uniform on G regardless of the distribution of Z. The same holds for non-abelian groups, but then $+$ should be replaced with the group operation $\circ$ and the channel action is $x \mapsto x \circ Z$, $Z \sim P_Z$.

3.4 Conditional Mutual Information and Conditional Independence

Definition 3.6. (Conditional mutual information) If $\mathcal{X}$ and $\mathcal{Y}$ are standard Borel, then we define

$$I(X;Y|Z) \triangleq D(P_{X,Y|Z} \| P_{X|Z}P_{Y|Z}|P_Z) \tag{3.9}$$

$$= \mathbb{E}_{z\sim P_Z}[I(X;Y|Z=z)], \tag{3.10}$$

where the product $P_{X|Z}P_{Y|Z}$ is a conditional distribution such that $(P_{X|Z}P_{Y|Z})(A\times B|z) = P_{X|Z}(A|z)P_{Y|Z}(B|z)$, under which X and Y are independent conditioned on Z.

Denoting $I(X;Y)$ as a functional $I(P_{X,Y})$ of the joint distribution $P_{X,Y}$, we have $I(X;Y|Z) = \mathbb{E}_{z\sim P_Z}[I(P_{X,Y|Z=z})]$. As such, $I(X;Y|Z)$ is a linear functional in P_Z. Measurability of the map $z \mapsto I(P_{X,Y|Z=z})$ is not obvious, but follows from Lemma 2.13.

To further discuss the properties of conditional mutual information, let us first introduce the notation for conditional independence. A family of joint distributions can be represented by a directed acyclic graph (DAG) encoding the dependence structure of the underlying random variables. We do not intend to introduce formal definitions here and refer to the standard monograph for full details [272]. But in short, every problem consisting of finitely (or countably infinitely) many random variables can be depicted as a DAG. Nodes of the DAG correspond to random variables and incoming edges into the node U simply describe which variables need to be known in order to generate U. A simple example is a Markov chain (path graph) $X \to Y \to Z$, which represents distributions that factor as $\{P_{X,Y,Z} \colon P_{X,Y,Z} = P_X P_{Y|X} P_{Z|Y}\}$. We have the following equivalent descriptions:

$$\begin{aligned}
X \to Y \to Z \quad &\Leftrightarrow \quad P_{X,Z|Y} = P_{X|Y} \cdot P_{Z|Y} \\
&\Leftrightarrow \quad P_{Z|X,Y} = P_{Z|Y} \\
&\Leftrightarrow \quad P_{X,Y,Z} = P_X \cdot P_{Y|X} \cdot P_{Z|Y} \\
&\Leftrightarrow \quad X, Y, Z \text{ form a Markov chain} \\
&\Leftrightarrow \quad X \perp\!\!\!\perp Z|Y \\
&\Leftrightarrow \quad X \leftarrow Y \to Z, P_{X,Y,Z} = P_Y \cdot P_{X|Y} \cdot P_{Z|Y} \\
&\Leftrightarrow \quad Z \to Y \to X.
\end{aligned}$$

There is a general method for obtaining these equivalences for general graphs, known as *d-separation*, see [272]. We say that a variable V is a *collider* on some undirected path if it appears on the path as

$$\text{collider:} \qquad \cdots \to V \leftarrow \cdots. \tag{3.11}$$

Otherwise, V is called a non-collider (and hence appears as $\to V \to$, $\leftarrow V \leftarrow$, or $\leftarrow V \to$). A pair of collections of variables A and B are *d-connected* by a collection C if there exists an undirected path from some variable in A to some variable in B such that (a) there are no non-colliders in C and (b) every collider is either in C or has a descendant in C. The concept of d-connectedness is important because it characterizes conditional independence. Specifically, $A \perp\!\!\!\perp B|C$ in *every distribution* satisfying a given graphical model if and only if A and B are *not* d-connected by C. It is rather useful for many information-theoretic considerations to master this criterion. However, in our book we will not formally require this apparatus beyond the basic equivalences for a linear Markov chain listed above. We do recommend practicing these, however, for example, by doing Exercises **I.26–I.30**.

Theorem 3.7. (Further properties of mutual information) *Suppose that all random variables are valued in standard Borel spaces. Then:*

(a) $I(X;Z|Y) \geq 0$, *with equality iff* $X \to Y \to Z$.

(b) (Simple chain rule)[3]

$$I(X,Y;Z) = I(X;Z) + I(Y;Z|X)$$
$$= I(Y;Z) + I(X;Z|Y).$$

(c) (DPI for mutual information) If $X \to Y \to Z$, *then*

$$I(X;Z) \leq I(X;Y), \tag{3.12}$$

with equality iff $X \to Z \to Y$.

(d) If $X \to Y \to Z \to W$, *then* $I(X;W) \leq I(Y;Z)$.

(e) (Full chain rule)

$$I(X^n;Y) = \sum_{k=1}^{n} I(X_k;Y|X^{k-1}).$$

(f) (Permutation invariance) If f *and* g *are one-to-one (with measurable inverses), then* $I(f(X);g(Y)) = I(X;Y)$.

Proof.

(a) By definition and Theorem 3.2(c).

(b) First, notice that from (3.1) we have (with self-evident notation):

$$I(Y;Z|X=x) = D(P_{Y|Z,X=x} \| P_{Y|X=x} | P_{Z|X=x}).$$

Taking expectation over X here we get

$$I(Y;Z|X) = D(P_{Y|X,Z} \| P_{Y|X} | P_{X,Z}).$$

On the other hand, from the chain rule for D, (2.24), we have

$$D(P_{X,Y,Z} \| P_{X,Y} P_Z) = D(P_{X,Z} \| P_X P_Z) + D(P_{Y|X,Z} \| P_{Y|X} | P_{X,Z}),$$

where in the second term we noticed that conditioning on X, Z under the measure $P_{X,Y} P_Z$ results in $P_{Y|X}$ (independent of Z). Putting together the preceding two displays completes the proof.

(c) Apply Kolmogorov identity to $I(Y,Z;X)$:

$$I(Y,Z;X) = I(X;Y) + \underbrace{I(X;Z|Y)}_{=0}$$
$$= I(X;Z) + I(X;Y|Z).$$

(d) Several applications of the DPI: $I(X;W) \leq I(X;Z) \leq I(Y;Z)$.

(e) Recursive application of Kolmogorov identity.

(f) Apply DPI to f and then to f^{-1}. □

[3] Also known as "Kolmogorov identities."

Remark 3.2. In general, $I(X;Y|Z)$ and $I(X;Y)$ are incomparable. Indeed, consider the following examples:

- $I(X;Y|Z) > I(X;Y)$: We need to find an example of X, Y, Z which do not form a Markov chain. To that end notice that there is only one directed acyclic graph non-isomorphic to $X \to Y \to Z$, namely $X \to Y \leftarrow Z$. With this idea in mind, we construct $X, Z \stackrel{\text{iid}}{\sim} \text{Ber}(1/2)$ and $Y = X \oplus Z$. Then $I(X;Y) = 0$ since $X \perp\!\!\!\perp Y$; however, $I(X;Y|Z) = I(X;X \oplus Z|Z) = H(X) = 1$ bit.
- $I(X;Y|Z) < I(X;Y)$: Simply take X, Y, Z to be any random variables on finite alphabets and $Z = Y$. Then $I(X;Y|Z) = I(X;Y|Y) = H(Y|Y) - H(Y|X, Y) = 0$ by a conditional version of (3.3).

Remark 3.3. (Chain rule for $I \Rightarrow$ Chain rule for H) Set $Y = X^n$. Then $H(X^n) = I(X^n;X^n) = \sum_{k=1}^n I(X_k;X^n|X^{k-1}) = \sum_{k=1}^n H(X_k|X^{k-1})$, since $H(X_k|X^n, X^{k-1}) = 0$.

Remark 3.4. (DPI for divergence $\Longrightarrow$ DPI for mutual information) We proved DPI for mutual information in Theorem 3.7 using Kolmogorov's identity. In fact, DPI for mutual information is *implied by* that for divergence in Theorem 2.17:

$$I(X;Z) = D(P_{Z|X}\|P_Z|P_X) \leq D(P_{Y|X}\|P_Y|P_X) = I(X;Y),$$

where we note that for each x, we have $P_{Y|X=x} \xrightarrow{P_{Z|Y}} P_{Z|X=x}$ and $P_Y \xrightarrow{P_{Z|Y}} P_Z$. Therefore if we have a bivariate functional of distributions $\mathcal{D}(P\|Q)$ which satisfies DPI, then we can define a "mutual information-like" quantity via $I_{\mathcal{D}}(X;Y) \triangleq \mathcal{D}(P_{Y|X}\|P_Y|P_X) \triangleq \mathbb{E}_{x\sim P_X}\mathcal{D}(P_{Y|X=x}\|P_Y)$ which will satisfy DPI on Markov chains. A rich class of examples arises by taking $\mathcal{D} = D_f$ (an f-divergence – see Chapter 7).

Remark 3.5. (Strong data-processing inequalities) For many channels $P_{Y|X}$, it is possible to strengthen the data-processing inequality (2.28) as follows: For any P_X, Q_X we have

$$D(P_Y\|Q_Y) \leq \eta_{\text{KL}} D(P_X\|Q_X),$$

where $\eta_{\text{KL}} < 1$ and depends on the channel $P_{Y|X}$ only. Similarly, this gives an improvement in the data-processing inequality for mutual information in Theorem 3.7(c): For any $P_{U,X}$ we have

$$U \to X \to Y \quad \Longrightarrow \quad I(U;Y) \leq \eta_{\text{KL}} I(U;X).$$

For example, for $P_{Y|X} = \mathsf{BSC}_\delta$ we have $\eta_{\text{KL}} = (1 - 2\delta)^2$. Strong data-processing inequalities (SDPIs) quantify the intuitive observation that noise intrinsic to the channel $P_{Y|X}$ must reduce the information that Y carries about the data U, regardless of how we optimize the encoding $U \mapsto X$. We explore SDPI further in Chapter 33 as well as their ramifications in statistics.

In addition to the case of strict inequality in DPI, it is also worth taking a closer look at the case of equality. If $U \to X \to Y$ and $I(U;X) = I(U;Y)$, intuitively it means that, as far as U is concerned, there is no loss of information in summarizing

X into Y. In statistical parlance, we say that Y is a sufficient statistic of X for U. This is the topic for the next section.

3.5 Sufficient Statistic and Data Processing

Much later in the book we will be interested in estimating parameters θ of probability distributions of X. To that end, one often first tries to remove unnecessary information contained in X. Let us formalize the setting as follows:

- Let P_X^θ be a collection of distributions of X parameterized by $\theta \in \Theta$.
- Let $P_{T|X}$ be some Markov kernel. Let $P_T^\theta \triangleq P_{T|X} \circ P_X^\theta$ be the induced distribution on T for each θ.

Definition 3.8. (Sufficient statistic) We say that T is a *sufficient statistic* **of** X **for** θ if there exists a transition probability kernel $P_{X|T}$ so that $P_X^\theta P_{T|X} = P_T^\theta P_{X|T}$, that is, $P_{X|T}$ can be chosen to not depend on θ.

The intuitive interpretation of T being sufficient is that, with T at hand, one can ignore X; in other words, T contains all the relevant information to infer about θ. This is because X can be simulated on the sole basis of T without knowing θ. As such, X provides no extra information for identification of θ. Any one-to-one transformation of X is sufficient, however, this is not the interesting case. In the interesting cases the dimensionality of T will be much smaller (typically equal to that of θ) than that of X. See examples below.

Observe also that the parameter θ need not be a random variable, as Definition 3.8 does not involve any distribution (prior) on θ. This is a so-called *frequentist* point of view on the problem of parameter estimation.

Theorem 3.9. *Let θ, X, T be as in the setting above. Then the following are equivalent:*

(a) T is a sufficient statistic of X for θ.
(b) $\forall P_\theta$, $\theta \to T \to X$.
(c) $\forall P_\theta$, $I(\theta;X|T) = 0$.
(d) $\forall P_\theta$, $I(\theta;X) = I(\theta;T)$, that is, the data-processing inequality for mutual information holds with equality.

Proof. We omit the details, which amount to either restating the conditions in terms of conditional independence, or invoking equality cases in the properties of mutual information stated in Theorem 3.7. □

The following result of Fisher provides a criterion for verifying sufficiency:

Theorem 3.10. (Fisher's factorization theorem) *For all $\theta \in \Theta$, let P_X^θ have a density p_θ with respect to a common dominating measure μ. Let $T = T(X)$ be a deterministic function of X. Then T is a sufficient statistic of X for θ iff*

$$p_\theta(x) = g_\theta(T(x))h(x)$$

for some measurable functions g_θ and h and all $\theta \in \Theta$.

Proof. We only give the proof in the discrete case where p_θ represents the PMF. (The argument for the general case is similar replacing $\sum$ by $\int d\mu$). Let $t = T(x)$.

"$\Rightarrow$": Suppose T is a sufficient statistic of X for θ. Then $p_\theta(x) = P_\theta(X = x) = P_\theta(X = x, T = t) = P_\theta(X = x|T = t)P_\theta(T = t) = \underbrace{P(X = x|T = T(x))}_{h(x)}\underbrace{P_\theta(T = T(x))}_{g_\theta(T(x))}$.

"$\Leftarrow$": Suppose the factorization holds. Then

$$P_\theta(X = x|T = t) = \frac{p_\theta(x)}{\sum_x 1\{T(x) = t\}p_\theta(x)} = \frac{g_\theta(t)h(x)}{\sum_x 1\{T(x) = t\}g_\theta(t)h(x)}$$
$$= \frac{h(x)}{\sum_x 1\{T(x) = t\}h(x)},$$

free of θ. □

Example 3.8. Independent observations

In the following examples, a parameterized distribution generates an independent sample of size n, which can be summarized into a scalar-valued sufficient statistic. These can be verified by checking the factorization of the n-fold product distribution and applying Theorem 3.10.

- *Normal mean model.* Let $\theta \in \mathbb{R}$ and observations $X_1, \dots, X_n \overset{\text{iid}}{\sim} \mathcal{N}(\theta, 1)$. Then the sample mean $\bar{X} = \frac{1}{n}\sum_{j=1}^n X_j$ is a sufficient statistic of X^n for θ.
- *Coin flips.* Let $B_i \overset{\text{iid}}{\sim} \text{Ber}(\theta)$. Then $\sum_{i=1}^n B_i$ is a sufficient statistic of B^n for θ.
- *Uniform distribution.* Let $U_i \overset{\text{iid}}{\sim} \text{Unif}(0, \theta)$. Then $\max_{i\in[n]} U_i$ is a sufficient statistic of U^n for θ.

Example 3.9. Sufficient statistic for hypothesis testing

Let $\Theta = \{0, 1\}$. Given $\theta = 0$ or 1, $X \sim P_X$ or Q_X, respectively. Then Y – the output of $P_{Y|X}$ – is a sufficient statistic of X for θ iff $D(P_{X|Y}\|Q_{X|Y}|P_Y) = 0$, that is, $P_{X|Y} = Q_{X|Y}$ holds P_Y-a.s. Indeed, the latter means that for kernel $Q_{X|Y}$ we have

$$P_X P_{Y|X} = P_Y Q_{X|Y} \quad \text{and} \quad Q_X P_{Y|X} = Q_Y Q_{X|Y},$$

which is precisely the definition of sufficient statistic when $\theta \in \{0, 1\}$. This example explains the condition for equality in the data processing for divergence in Theorem 2.17. Then assuming $D(P_Y\|Q_Y) < \infty$ we have:

$$D(P_X\|Q_X) = D(P_Y\|Q_Y) \iff Y \text{ is a sufficient statistic for testing } P_X \text{ versus } Q_X.$$

Proof. Let $Q_{X,Y} = Q_X P_{Y|X}$, $P_{X,Y} = P_X P_{Y|X}$, then

$$
\begin{aligned}
D(P_{X,Y}\|Q_{X,Y}) &= \underbrace{D(P_{Y|X}\|Q_{Y|X}|P_X)}_{=0} + D(P_X\|Q_X) \\
&= D(P_{X|Y}\|Q_{X|Y}|P_Y) + D(P_Y\|Q_Y) \\
&\geq D(P_Y\|Q_Y)
\end{aligned}
$$

with equality iff $D(P_{X|Y}\|Q_{X|Y}|P_Y) = 0$, which is equivalent to Y being a sufficient statistic for testing P_X versus Q_X as desired. □

3.6 Probability of Error and Fano's Inequality

Let W be a random variable and $\hat{W}$ be our prediction of it. Depending on the information available for producing $\hat{W}$ we can consider three types of problems:

1 Random guessing: $W \perp\!\!\!\perp \hat{W}$.
2 Guessing with data: $W \to X \to \hat{W}$, where $X = f(W)$ is a deterministic function of W.
3 Guessing with noisy data: $W \to X \to Y \to \hat{W}$, where $X \to Y$ is given by some noisy channel.

Our goal is to draw converse statements of the following type: If the uncertainty of W is too high or if the information provided by the data is too scarce, then it is difficult to guess the value of W. In this section we formalize these intuitions using (conditional) entropy and mutual information.

Theorem 3.11. *Let $|\mathcal{X}| = M < \infty$. Then for any $\hat{X} \perp\!\!\!\perp X$,*

$$H(X) \leq F_M(\mathbb{P}[X = \hat{X}]) \tag{3.13}$$

where

$$F_M(x) \triangleq (1-x)\log(M-1) + h(x), \quad x \in [0,1] \tag{3.14}$$

and $h(x) = x\log\frac{1}{x} + (1-x)\log\frac{1}{1-x}$ is the binary entropy function (Example 1.1).

If $P_{\max} \triangleq \max_{x\in\mathcal{X}} P_X(x)$, then

$$H(X) \leq F_M(P_{\max}) = (1-P_{\max})\log(M-1) + h(P_{\max}), \tag{3.15}$$

with equality iff $P_X = (P_{\max}, \frac{1-P_{\max}}{M-1}, \ldots, \frac{1-P_{\max}}{M-1})$.

The function $F_M(\cdot)$ is shown in Figure 3.3. Notice that due to its non-monotonicity the statement (3.15) does not imply (3.13), even though $\mathbb{P}[X = \hat{X}] \leq P_{\max}$.

Proof. To show (3.13) consider an auxiliary (product) distribution $Q_{X,\hat{X}} = U_X P_{\hat{X}}$, where U_X is uniform on $\mathcal{X}$. Then $Q[X = \hat{X}] = 1/M$. Denoting $P[X = \hat{X}] \triangleq P_S$, applying the DPI for divergence to the data processor $(X, \hat{X}) \mapsto 1\{X = \hat{X}\}$ yields $d(P_S\|1/M) \leq D(P_{X\hat{X}}\|Q_{X\hat{X}}) = \log M - H(X)$.

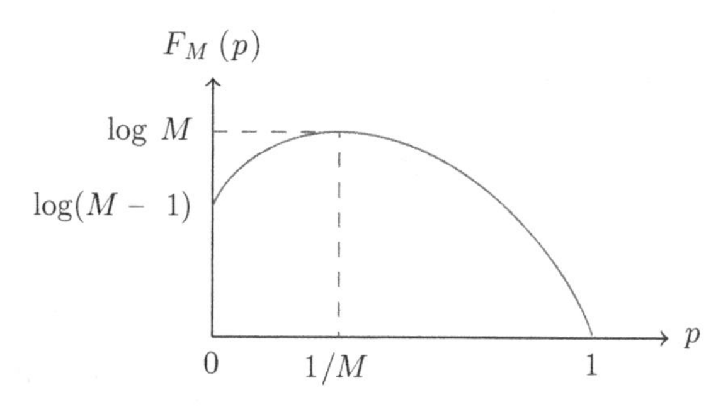

Figure 3.3 The function F_M in (3.14) is concave with maximum $\log M$ at maximizer $1/M$, but not monotone.

To show the second part, suppose one is trying to guess the value of X without any side information. Then the best bet is obviously the most likely outcome (mode) and the maximal probability of success is

$$\max_{\hat{X} \perp\!\!\!\perp X} \mathbb{P}[X = \hat{X}] = P_{\max}. \tag{3.16}$$

Thus, applying (3.13) with $\hat{X}$ being the mode yields (3.15). Finally, suppose that $P = (P_{\max}, P_2, \ldots, P_M)$ and introduce $Q = (P_{\max}, \frac{1-P_{\max}}{M-1}, \ldots, \frac{1-P_{\max}}{M-1})$. Then the difference of the right and left sides of (3.15) equals $D(P\|Q) \geq 0$, with equality iff $P = Q$. □

Remark 3.6. Let us discuss the unusual proof technique. Instead of studying directly the probability space $P_{X,\hat{X}}$ given to us, we introduced an auxiliary one: $Q_{X,\hat{X}}$. We then drew conclusions about the target metric (probability of error) for the auxiliary problem (the probability of error $= 1 - \frac{1}{M}$). Finally, we used DPI to transport the statement about Q to a statement about P: if $D(P\|Q)$ is small, then the probabilities of the events (e.g. $\{X \neq \hat{X}\}$) should be small as well. This is a general method, known as *meta-converse*, that we develop in more detail later in this book for channel coding (see Section 22.3). For the specific result (3.15), however, there are much more explicit ways to derive it – see Exercise **I.25**.

Similar to the Shannon entropy H, $P_{\max}$ is also a reasonable measure for randomness of P. In fact, recall from (1.4) that

$$H_\infty(P) = \log \frac{1}{P_{\max}} \tag{3.17}$$

is the Rényi entropy of order ∞, see (1.4). In this regard, Theorem 3.11 can be thought of as our first example of a *comparison of information measures*: it compares H and H_∞. We will study such comparisons systematically in Section 7.4.

Next we proceed to the setting of Fano's inequality where the estimate $\hat{X}$ is made on the basis of some observation Y correlated with X. We will see that the proof of the previous theorem trivially generalizes to this new case of possibly randomized estimators. Though not needed in the proof, it is worth mentioning that the best estimator minimizing the probability of error $\mathbb{P}[X \neq \hat{X}]$ is the maximum posterior (MAP) rule, that is, the posterior mode: $\hat{X}(y) = \mathrm{argmax}_x P_{X|Y}(x|y)$.

Theorem 3.12. (Fano's inequality) *Let* $|\mathcal{X}| = M < \infty$ *and* $X \to Y \to \hat{X}$. *Let* $P_e = \mathbb{P}[X \neq \hat{X}]$, *then*

$$H(X|Y) \leq F_M(1 - P_e) = P_e \log(M-1) + h(P_e). \tag{3.18}$$

Furthermore, if $P_{\max} \triangleq \max_{x \in \mathcal{X}} P_X(x) > 0$, *then regardless of* $|\mathcal{X}|$,

$$I(X;Y) \geq (1 - P_e) \log \frac{1}{P_{\max}} - h(P_e). \tag{3.19}$$

Proof. To show (3.18) we apply data processing (for divergence) to $P_{X,Y,\hat{X}} = P_X P_{Y|X} P_{\hat{X}|Y}$ versus $Q_{X,Y,\hat{X}} = U_X P_Y P_{\hat{X}|Y}$ and the data processor (kernel) $(X, Y, \hat{X}) \mapsto \{X \neq \hat{X}\}$ (note that $P_{\hat{X}|Y}$ is identical for both).

To show (3.19) we apply data processing (for divergence) to $P_{X,Y,\hat{X}} = P_X P_{Y|X} P_{\hat{X}|Y}$ versus $Q_{X,Y,\hat{X}} = P_X P_Y P_{\hat{X}|Y}$ and the data processor (kernel) $(X, Y, \hat{X}) \mapsto \{X \neq \hat{X}\}$ to obtain:

$$\begin{aligned} I(X;Y) &= D(P_{X,Y,\hat{X}} \| Q_{X,Y,\hat{X}}) \geq d(\mathbb{P}[X = \hat{X}] \| \mathbb{Q}[X = \hat{X}]) \\ &\geq -h(P_e) + (1 - P_e) \log \frac{1}{\mathbb{Q}[X = \hat{X}]} \geq -h(P_e) - (1 - P_e) \log P_{\max}, \end{aligned}$$

where the last step follows from $\mathbb{Q}[X = \hat{X}] \leq P_{\max}$ since $X \perp\!\!\!\perp \hat{X}$ under $\mathbb{Q}$. (Again, we refer to Exercise **I.25** for a direct proof.) □

The following corollary of the previous result emphasizes its role in providing converses (or impossibility results) for statistics and data transmission.

Corollary 3.13. (Lower bound on average probability of error) *Let* $W \to X \to Y \to \hat{W}$, *where* W *is uniform on* $[M] \triangleq \{1, \ldots, M\}$. *Then*

$$P_e \triangleq \mathbb{P}[W \neq \hat{W}] \geq 1 - \frac{I(X;Y) + h(P_e)}{\log M} \tag{3.20}$$

$$\geq 1 - \frac{I(X;Y) + \log 2}{\log M}. \tag{3.21}$$

Proof. Apply Theorem 3.12 and the data processing inequality for mutual information: $I(W;\hat{W}) \leq I(X;Y)$. □

3.7* Estimation Error in Gaussian Noise (I-MMSE)

In the previous section we considered estimating a discrete random variable X on the basis of observation Y and showed bounds on the probability of reconstruction error. Here we consider the case of $X \in \mathbb{R}^d$ and a quadratic loss, which is also known in signal processing as the mean-squared error (MSE). Specifically, whenever $\mathbb{E}[\|X\|^2] < \infty$ we define

$$\mathsf{mmse}(X|Y) \triangleq \mathbb{E}[\|X - \mathbb{E}[X|Y]\|^2],$$

where MMSE or mmse(...) stands for minimum MSE (which follows from the fact that the best estimator of X given Y is precisely $\mathbb{E}[X|Y]$). Just like Fano's inequality one can derive *inequalities* relating $I(X;Y)$ and $\mathsf{mmse}(X|Y)$. For example, from Tao's inequality (see Corollary 7.11) one can easily get for the case where $X \in [-1,1]$ that

$$0 \le \mathrm{Var}(X) - \mathsf{mmse}(X|Y) \le \frac{2}{\log e} I(X;Y),$$

which shows that the variance reduction of X due to Y is at most proportional to their mutual information. (Simply notice that $\mathbb{E}[|\,\mathbb{E}[X|Y] - \mathbb{E}[X]|^2] = \mathrm{Var}(X) - \mathsf{mmse}(X|Y)$.)

However, this section is not about such inequalities. Here we show a remarkable *equality* for the special case when Y is an observation of X corrupted by Gaussian noise. As applications of this identity we will derive *stochastic localization* in Exercise **I.66** and the entropy-power inequality in Theorem 3.16.

Theorem 3.14. (I-MMSE [205]) *Let $X \in \mathbb{R}^d$ be independent of $Z \sim \mathcal{N}(0, I_d)$. If $\mathbb{E}[\|X\|^2] < \infty$ then for all $\gamma > 0$ we have*

$$\frac{d}{d\gamma} I(X;\sqrt{\gamma}X + Z) = \frac{\log e}{2} \mathsf{mmse}(X|\sqrt{\gamma}X + Z), \tag{3.22}$$

$$\frac{d^2}{d\gamma^2} I(X;\sqrt{\gamma}X + Z) = -\frac{\log e}{2}\, \mathbb{E}[\|\Sigma_\gamma(\sqrt{\gamma}X + Z)\|_F^2], \tag{3.23}$$

where $\Sigma_\gamma(y) = \mathrm{Cov}(X|\sqrt{\gamma}X + Z = y)$ and $\|\cdot\|_F$ is the Frobenius norm of the matrix.

As a simple example, for Gaussian X, one may verify (3.22) by combining the mutual information in Example 3.3 with the MMSE in Example 28.1.

Before proving Theorem 3.14 we start with some notation and preliminary results. Let $I \subset \mathbb{R}$ be an open interval, μ a (positive) measure on $\mathbb{R}^d$, and $K, L\colon \mathbb{R}^d \times I \to \mathbb{R}$ such that the following conditions are met: (a) $\int K(x,\theta)\mu(dx)$ exists for all $\theta \in I$; (b) $\int_{\mathbb{R}^d} \mu(dx) \int_I d\theta |L(x,\theta)| < \infty$; (c) $t \mapsto \int_{\mathbb{R}^d} \mu(dx) L(x,t)$ is continuous; and (d) we have

$$\frac{\partial}{\partial\theta} K(x,\theta) = L(x,\theta)$$

for all x, θ. Then

$$\frac{\partial}{\partial\theta} \int_{\mathbb{R}^d} K(x,\theta)\, dx = \int_{\mathbb{R}^d} L(x,\theta)\, dx. \tag{3.24}$$

(To see this, take $\theta > \theta_0 \in I$ and write $K(x,\theta) = K(x,\theta_0) + \int_{\theta_0}^{\theta} dt L(x,t)$. Now we can integrate this over x and interchange the order of integrals to get $\int dx K(x,\theta) =$ constant $+ \int_{\theta_0}^{\theta} g(t)dt$, where $g(t) = \int dx L(x,t)$ is continuous.) Note that in the case of finite interval I both conditions (b) and (c) are implied by the condition: (e) for all $t \in I$ we have $|L(x,t)| \le \ell(x)$ and ℓ is μ-integrable.

Let $\phi_a(x) = \frac{1}{(2\pi a)^{d/2}} e^{-\|x\|^2/(2a)}$ be the density of $\mathcal{N}(0, aI_d)$. Suppose p is some probability distribution, and f is a function, then we denote by $p * f(x) = \mathbb{E}_{X' \sim p}[f(x - X')]$, which coincides with the usual convolution if p is a density. In particular, the Gaussian convolution $p * \phi_a$ is known as a *Gaussian mixture* with *mixing distribution* p. For any differential operator $D = \frac{\partial^k}{\partial x_{i_1} \cdots \partial x_{i_k}}$ we have

$$D(p * \phi_a) = p * (D\phi_a). \tag{3.25}$$

For $D = \frac{\partial}{\partial x_1}$ this follows from (3.24) by taking $\mu = p$, $K(x,\theta) = \phi_a(y-x)$ where $y = (\theta, y_2, \ldots, y_d)$. Conditions for $L(x,\theta)$ follow from the fact that $D\phi_a$ is uniformly bounded. The case of general D follows by induction.

As a next application we will show that

$$\frac{\partial}{\partial a} p * \phi_a(y) = p * \left(\frac{\partial}{\partial a}\phi_a\right)(y) = \frac{1}{2}\Delta(p * \phi_a)(y), \tag{3.26}$$

where $\Delta f = \mathrm{tr}(\mathrm{Hess}\ f)$ is the Laplacian. Notice that the second equality follows from (3.25) and the easily checked identity

$$\frac{\partial}{\partial a}\phi_a(x) = \frac{1}{2a^2}(\|x\|^2 - ad)\phi_a(x) = \frac{1}{2}\Delta\phi_a(x).$$

Thus, we only need to justify the first equality in (3.26). To that end, we use (3.24) with $\mu = p$, $K(x,a) = \phi_a(y-x)$ and $L(x,a) = \frac{\partial}{\partial a}K(x,a)$. Note that by the previous calculation we have $\sup_x |\frac{\partial}{\partial a}\phi_a(x)| < \infty$, and thus condition (e) of (3.24) applies and so (3.24) implies (3.26).

The next lemma shows a special property of Gaussian convolution (derivatives of log-convolution correspond to conditional moments).

Lemma 3.15. *Let $Y = X + \sqrt{a}Z$, where $X \perp\!\!\!\perp Z$, $X \sim p$, and $Z \sim \mathcal{N}(0, I_d)$. Then:*

1 If $\mathbb{E}[\|X\|] < \infty$ we have

$$\nabla \ln(p * \phi_a)(y) = \frac{1}{a}(\mathbb{E}[X|Y=y] - y) \tag{3.27}$$

and also

$$\|\nabla \ln p * \phi_a(x)\| \le \frac{3}{a}\|x\| + \frac{4}{a}\mathbb{E}[\|X\|]. \tag{3.28}$$

2 If $\mathbb{E}[\|X\|^2] < \infty$ we have

$$\mathrm{Hess}\ \ln(p * \phi_a)(y) = \frac{1}{a^2}\mathrm{Cov}[X|Y=y] - \frac{1}{a}I_d. \tag{3.29}$$

Proof. Notice that since $Y \sim p * \phi_a$ we have

$$\mathbb{E}[X|Y=y] = \frac{1}{p * \phi_a(y)}\int_{\mathbb{R}^d} xp(x)\phi_a(y-x)dx = \frac{1}{p * \phi_a(y)}\int_{\mathbb{R}^d}(y-x)p(y-x)\phi_a(x)dx$$
$$\stackrel{(a)}{=} y + \frac{a}{p * \phi_a(y)}\int p(y-x)(\nabla\phi_a(x)) \stackrel{(b)}{=} y + \frac{a}{p * \phi_a(y)}\nabla(p * \phi_a(x)),$$

where (a) follows from the fact that $\nabla\phi_a(x) = -\frac{x}{a}\phi_a(x)$ and (b) from (3.25). The proof of (3.27) is completed after noticing $\frac{1}{p*\phi_a(y)}\nabla(p * \phi_a(x)) = \nabla \ln(p * \phi_a)(y)$. Technical estimate (3.28) is shown in [345, Proposition 2].

The identity (3.29) is shown entirely similarly. □

With these preparations we are ready to prove the theorem.

Proof of Theorem 3.14. For simplicity, in this proof we compute all informations and entropies with natural base, so $\log = \ln$. With these preparations we can show (3.22). First, let $a = 1/\gamma$ and notice

$$I(X;\sqrt{\gamma}X + Z) = I(X;X + \sqrt{a}Z) = h(X + \sqrt{a}Z) - \frac{d}{2}\ln(2\pi ea),$$

where we computed differential entropy of the Gaussian via Theorem 2.8. Thus, the proof is completed if we show

$$\frac{d}{da}h(X + \sqrt{a}Z) = \frac{d}{2a} - \frac{1}{2a^2}\mathsf{mmse}(X|Y_a), \tag{3.30}$$

where we defined $Y_a = X + \sqrt{a}Z$. Let the law of X be p. Conceptually, the computation is just a few lines:

$$\begin{aligned}
-\frac{d}{da}h(X + \sqrt{a}Z) &= \frac{d}{da}\int (p * \phi_a)(x)\ln(p * \phi_a)(x)dx \\
&\overset{(a)}{=} \int \frac{\partial}{\partial a}[(p * \phi_a)(x)\ln(p * \phi_a)(x)]dx \\
&\overset{(b)}{=} \frac{1}{2}\int (1 + \ln p * \phi_a)\Delta(p * \phi_a)dx \\
&\overset{(c)}{=} \frac{1}{2}\int (p * \phi_a)\Delta(\ln p * \phi_a)dx \\
&\overset{(d)}{=} \frac{1}{2}\int (p * \phi_a)(y)\left(\frac{1}{a^2}\mathsf{mmse}(X|Y_a = y) - \frac{d}{a}\right)dy,
\end{aligned}$$

where (a) and (c) will require technical justifications, while (b) is just (3.26) and (d) is by taking the trace of (3.29). Note that (a) is just interchange of differentiation and integration, while (c) is simply the "self-adjointness" of the Laplacian.

We proceed to justifying (a). We will apply (3.24) with $\mu = \text{Leb}$, $I = (a_1, a_2)$ some finite interval, $K(x,a) = (p * \phi_a)(x)\ln(p * \phi_a)(x)$, and

$$L(x,a) = \frac{\partial}{\partial a}K(x,a) = \frac{1}{2}(1 + \ln(p * \phi_a)(x))(p * \Delta\phi_a)(x),$$

where we again used (3.26).

Integrating (3.28) we get

$$|\ln p * \phi_a(x) - \ln p * \phi_a(0)| \le \frac{3}{2a}\|x\|^2 + \frac{4}{a}\|x\|\,\mathbb{E}[\|X\|].$$

Since $p * \phi_a(0) \le \phi_a(0)$ we get that for all $a \in (a_1, a_2)$ we have for some $c > 0$:

$$|\ln p * \phi_a(x)| \le c(1 + \|x\| + \|x\|^2). \tag{3.31}$$

From this estimate we note that

$$K(x,a) \le c(1 + \|x\| + \|x\|^2)(p * \phi_a)(x).$$

The integral of the right-hand side over x is simply $c(1 + \mathbb{E}[\|Y_a\| + \|Y_a\|^2]) < \infty$, which confirms condition (a) of (3.24).

Next, we notice that for any differential operator D we have $D\phi_a(x) = f(x)\phi_a(x)$ where f is some polynomial in x. Since for $a < a_2$ we have $\sup_x \frac{f(x)\phi_a(x)}{\phi_{a_2}(x)} < \infty$ we have that for some constant c' and all $a < a_2$ and all x we have

$$|D(p * \phi_a)(x)| \le c' p * \phi_{a_2}(x), \tag{3.32}$$

where we used (3.25) as well. Thus, for $L(x,a)$ we can see that the first term is bounded by (3.31) and the second by the previous display, so that overall

$$L(x,a) \le \frac{cc'}{2} |2 + \|x\| + \|x\|^2| (p * \phi_{a_2})(x).$$

Since again the right-hand side is integrable, we see that condition (e) of (3.24) applies and thus the interchange of differentiation and integration in step (a) is valid.

Finally, we justify step (c). To that end we prove an auxiliary result first: If u, v are two univariate twice-differentiable functions with (i) $\int_{\mathbb{R}} |u''v|$ and $\int_{\mathbb{R}} |v''u|$ both finite and (ii) $\int |u'v'| < \infty$ then

$$\int_{\mathbb{R}} u''v = \int_{\mathbb{R}} v''u. \tag{3.33}$$

Indeed, from condition (ii) there must exist a sequence $c_n \to +\infty$, $b_n \to -\infty$ such that $|u'(c_n)v'(c_n)| + |u'(b_n)v'(b_n)| \to 0$. On the other hand, from condition (i) we have

$$\lim_{n\to\infty} \int_{b_n}^{c_n} u''v = \int_{\mathbb{R}} u''v,$$

and similarly for $\int v''u$. Now applying integration by parts we have

$$\int_{b_n}^{c_n} u''v = u'(c_n)v'(c_n) - u'(b_n)v'(b_n) + \int_{b_n}^{c_n} v''u,$$

and the first two terms vanish with n.

Next, consider multivariate twice-differentiable functions U, V with (i) $\int_{\mathbb{R}^d} |V \frac{\partial^2}{\partial x_i^2} U|$ and $\int_{\mathbb{R}^d} |U \frac{\partial^2}{\partial x_i^2} V|$ both finite and (ii) $\int_{\mathbb{R}^d} \|\nabla U\| \, \|\nabla V\| < \infty$, then

$$\int_{\mathbb{R}^d} V\Delta U = \int_{\mathbb{R}^d} U\Delta V. \tag{3.34}$$

We write $x = (x_1, x_2^d)$ by grouping the last $(d-1)$ coordinates together. Fix x_2^d and define $u(x_1) = U(x_1, x_2, \ldots, x_d)$ and $v(x_1) = V(x_1, x_2, \ldots, x_d)$. For Lebesgue-a.e. x_2^d we see that u, v satisfy conditions for (3.33). Thus, we obtain that for such x_2^d we have

$$\int_{\mathbb{R}} V(x) \frac{\partial^2}{\partial x_1^2} U(x)\, dx_1 = \int_{\mathbb{R}} U(x) \frac{\partial^2}{\partial x_1^2} V(x)\, dx_1.$$

Integrating this over x_2^d we get

$$\int_{\mathbb{R}^d} V(x) \frac{\partial^2}{\partial x_1^2} U(x)\, dx = \int_{\mathbb{R}^d} U(x) \frac{\partial^2}{\partial x_1^2} V(x)\, dx.$$

Now, to justify step (c) we have to verify that $U(x) = 1 + \ln(p * \phi_a)(x)$ and $V(x) = p * \phi_a(x)$ satisfy the conditions of the previous result. To that end, notice that from (3.29) we have $|\frac{\partial^2}{\partial y_i^2} U(y)| \le \frac{1}{a^2} \mathbb{E}[X_i^2 | Y_a = y] + \frac{1}{a}$ and thus

$$\int_{\mathbb{R}^d} |V \frac{\partial^2}{\partial x_i^2} U| = O_a(\mathbb{E}[X_i^2]) < \infty.$$

On the other hand, from (3.25) and (3.32) we have $|\frac{\partial^2}{\partial y_i^2}V(y)| \le c'p * \phi_{a_2}(y)$. From (3.31) then we obtain

$$\int_{\mathbb{R}^d} \left| U\frac{\partial^2}{\partial x_i^2}V \right| \le cc' \int (1+\|x\|+\|x\|^2)p * \phi_{a_2}(x) = cc'\,\mathbb{E}[1+\|Y_{a_2}\|+\|Y_{a_2}\|^2] < \infty.$$

Finally, for showing $\int_{\mathbb{R}^d} \|\nabla U\|\|\nabla V\| < \infty$ we apply (3.28) to estimate $\|\nabla U\| \lesssim_a 1+\|y\|$ and use (3.32) to estimate $\|\nabla V\| \lesssim_a p * \phi_{a_2}(x)$. Thus, we have

$$\int_{\mathbb{R}^d} \|\nabla U\|\|\nabla V\| \lesssim_a \mathbb{E}[1+\|Y_{a_2}\|] < \infty.$$

This completes the verification of the conditions and we conclude $\int U\Delta V = \int V\Delta U$ as required for step (c).

The identity (3.23) is obtained by differentiating function $\gamma \mapsto \mathsf{mmse}(X|\sqrt{\gamma}+Z)$ using very similar methods. We refer to [206] for full justification. □

Remark 3.7. (Tensorization of I-MMSE) We proved the I-MMSE identity for a d-dimensional vector directly. However, it turns out that the one-dimensional version implies the d-dimensional version. Specifically, suppose the one-dimensional version of (3.22) is already proven. Let us denote $X = (X_1,\ldots,X_d)$, $Y = (Y_1,\ldots,Y_d)$, $Z = (Z_1,\ldots,Z_d)$ as in (3.22). However, now let $\underline{\gamma} = (\gamma_1,\ldots,\gamma_d)$ be a vector and we set $Y_j = \sqrt{\gamma_j}X_j + Z_j$. We are interested in computing the derivative of $I(X;Y)$ along the diagonal $\underline{\gamma} = (\gamma,\ldots,\gamma)$. To that end, denote by $Y_{\sim j} = \{Y_i\colon i \ne j\}$ and notice that by the chain rule we have

$$\frac{\partial}{\partial\gamma_j}I(X;Y) = \frac{\partial}{\partial\gamma_j}I(X_j;Y_j|Y_{\sim j}). \tag{3.35}$$

Similarly, notice that $\mathbb{E}_{y_{\sim j}\sim P_{Y_{\sim j}}}[\mathsf{mmse}(X_j|Y_j, Y_{\sim j} = y_{\sim j})] = \mathsf{mmse}(X_j|Y)$. Thus, applying the one-dimensional version of (3.22) we get

$$\frac{\partial}{\partial\gamma_j}I(X;Y) = \frac{\log e}{2}\mathsf{mmse}(X_j|Y).$$

Now since $\mathsf{mmse}(X|Y) = \sum_j \mathsf{mmse}(X_j|Y)$ by summing (3.35) over j we obtain the d-dimensional version of (3.22). Note that we computed the derivative in a scalar parameter γ by introducing a vector one $\underline{\gamma}$ and then using the chain rule to simplify partial derivatives. This idea is the basis of the *area theorem* in information theory [361, Lemma 3] and *Guerra interpolation* in statistical physics [411].

3.8* Entropy-Power Inequality

As an application of the last section's result we demonstrate an important relation between the additive structure of $\mathbb{R}^d$ and entropy. To state the result, recall that from (2.18) an iid Gaussian vector Z^d with coordinates of power σ^2 have differential entropy $h(Z^d) = \frac{d}{2}\log(2\pi e\sigma^2)$. Correspondingly, for any $\mathbb{R}^d$-valued random variable X we define its *entropy power* to be

$$N(X) \triangleq \frac{1}{2\pi e} \exp\left\{\frac{2}{d} h(X)\right\}.$$

Note that by Theorem 2.8 the entropy power is maximized (under the second moment constraint) by an iid Gaussian vector. Thus, the result that we prove next can be interpreted as a statement that *convolution increases Gaussianity*.[4] The result was conjectured by Shannon and proved by Stam.

Theorem 3.16. (Entropy-power inequality (EPI) [400]) *Suppose $A_1 \perp\!\!\!\perp A_2$ are independent $\mathbb{R}^d$-valued random variables with finite second moments $\mathbb{E}[\|A_i\|^2] < \infty$, $i \in \{1, 2\}$. Then*

$$N(A_1 + A_2) \geq N(A_1) + N(A_2).$$

Remark 3.8. (Costa's EPI) Consider the case of $A_2 = \sqrt{a}Z, Z \sim \mathcal{N}(0, I_d)$. Then the EPI is equivalent to the statement that $\frac{d}{da} N(A_1 + \sqrt{a}Z) \geq 1$. For this special case, Costa [100] established a much stronger property that $a \mapsto N(A_1 + \sqrt{a}Z)$ is concave. A further improvement for this case, in terms of F_I-curve (see Definition 16.5) is proposed in [103].

Proof. We present an elegant proof of [438]. First, an observation of Lieb [281] shows that EPI is equivalent to proving: For all $U_1 \perp\!\!\!\perp U_2$ and $\alpha \in [0, 2\pi)$ we have

$$h(U_1 \cos\alpha + U_2 \sin\alpha) \geq h(U_1) \cos^2\alpha + h(U_2) \sin^2(\alpha). \tag{3.36}$$

(To see that (3.36) implies EPI simply take $\cos^2\alpha = \frac{N(A_1)}{N(A_1)+N(A_2)}$ and $U_1 = A_1/\cos\alpha$, $U_2 = A_2/\sin\alpha$.)

Next, we claim that proving (3.36) for general U_i is equivalent to proving it for their "smoothed" versions, that is, $\tilde{U}_i = U_i + \sqrt{\epsilon} Z_i$, where $Z_i \sim \mathcal{N}(0, I_d)$ is independent of U_1, U_2. Indeed, this technical continuity result follows, for example, from [345, Proposition 1], which shows that whenever $\mathbb{E}[\|U_i\|^2] < \infty$ then $\epsilon \mapsto h(U_i + \sqrt{\epsilon} Z_i)$ is continuous and in fact $h(U_i + \sqrt{\epsilon} Z) = h(U_i) + O(\sqrt{\epsilon})$ as $\epsilon \to 0$.

In other words, to prove Lieb's EPI it is sufficient to prove for all $\epsilon > 0$

$$h(X + \sqrt{\epsilon}Z) \geq h(U_1 + \sqrt{\epsilon}Z_1) \cos^2\alpha + h(U_2 + \sqrt{\epsilon}Z_2) \sin^2(\alpha),$$

where we also defined $X \triangleq U_1 \cos\alpha + U_2 \sin\alpha, Z \triangleq Z_1 \cos\alpha + Z_2 \sin\alpha$. Since the above inequality is scale-invariant, we can equivalently show for all $\gamma \geq 0$ the following:

$$h(\sqrt{\gamma}X + Z) \geq h(\sqrt{\gamma}U_1 + Z_1) \cos^2\alpha + h(\sqrt{\gamma}U_2 + Z_2) \sin^2(\alpha).$$

As a final simplification, we replace differential entropies by mutual informations. That is, the proof is completed if we show

$$I(X; \sqrt{\gamma}X + Z) \geq I(U_1; \sqrt{\gamma}U_1 + Z_1) \cos^2\alpha + I(U_2; \sqrt{\gamma}U_2 + Z_2) \sin^2(\alpha).$$

[4] Another deep manifestation of this phenomenon is in the context of the central limit theorem (CLT). Barron's *entropic CLT* states that for iid X_i's with zero mean and unit variance, the KL divergence $D(\frac{X_1+\cdots+X_n}{\sqrt{n}} \| \mathcal{N}(0, 1))$, whenever finite, converges to zero. This convergence is in fact monotonic as shown in [27, 102].

This last inequality clearly holds for $\gamma = 0$. Thus it is sufficient to prove the same inequality for derivatives (in γ) of both sides and then integrate from 0 to γ. Computing derivatives via (3.22) it amounts to show

$$\mathsf{mmse}(X|\sqrt{\gamma}X+Z) \geq \mathsf{mmse}(U_1|\sqrt{\gamma}U_1+Z_1)\cos^2\alpha+\mathsf{mmse}(U_2|\sqrt{\gamma}U_2+Z_2)\sin^2(\alpha).$$

But this latter inequality is very simple to argue, since clearly

$$\mathsf{mmse}(X|\sqrt{\gamma}X+Z) \geq \mathsf{mmse}(X|\sqrt{\gamma}U_1+Z_1,\sqrt{\gamma}U_2+Z_2).$$

On the other hand, for the right-hand side X is a sum of two conditionally independent terms and thus

$$\mathsf{mmse}(X|\sqrt{\gamma}U_1+Z_1,\sqrt{\gamma}U_2+Z_2) = \mathsf{mmse}(U_1|\sqrt{\gamma}U_1+Z_1)\cos^2\alpha + \mathsf{mmse}(U_2|\sqrt{\gamma}U_2+Z_2)\sin^2(\alpha). \qquad \square$$

In Corollary 2.9 we have already seen how properties of differential entropy can be translated to properties of positive definite matrices. Here is another application:

Corollary 3.17. (Minkowski inequality) *Let Σ_1, Σ_2 be real positive definite $n \times n$ matrices. Then*

$$\det(\Sigma_1+\Sigma_2)^{\frac{1}{n}} \geq \det(\Sigma_1)^{\frac{1}{n}} + \det(\Sigma_2)^{\frac{1}{n}}.$$

Proof. Take $A_i \sim \mathcal{N}(0,\Sigma_i)$, use (2.18) and the EPI. □

EPI is a corner stone of many information-theoretic arguments: for example, it was used to establish the capacity region of the Gaussian broadcast channel [45]. However, its significance extends throughout geometry and analysis, having deep implications for high-dimensional probability, convex geometry, and concentration of measure. As an example see Exercise **I.65** which derives the *log-Sobolev inequality* of Gross. Further discussions are outside of the scope of this book, but we recommend reading [106, Chapter 16].

4 Variational Characterizations and Continuity of Information Measures

In this chapter we collect some results on variational characterizations of information measures. It is a well-known method in analysis to study a functional by proving a variational characterization of the form $F(x) = \sup_{\lambda \in \Lambda} f_\lambda(x)$ or $F(x) = \inf_{\mu \in M} g_\mu(x)$. Such representations can be useful for multiple purposes:

- Convexity: the pointwise supremum of convex functions is convex.
- Regularity: the pointwise supremum of lower semicontinuous (lsc) functions is lsc.
- Bounds: the upper/lower bound on F follows by choosing any λ (μ) and evaluating f_λ (g_μ).

We will see in this chapter that divergence has two different sup-characterizations (over partitions and over functions). The mutual information is more special. In addition to inheriting the ones from KL divergence, it possesses two extra: an inf-representation over (centroid) measures Q_Y and a sup-representation over Markov kernels.

As applications of these variational characterizations, we discuss the *Gibbs variational principle*, which serves as the basis of many modern algorithms in machine learning, including the *expectation-maximization (EM) algorithm* and *variational autoencoders*; see Section 4.4. An important theoretical construct in machine learning is the idea of PAC-Bayes bounds (Section 4.8*).

From the information-theoretic point of view variational characterizations are important because they address the problem of continuity. We will discuss several types of continuity in this chapter. First is the continuity in discretization. This is related to the issue of *computation*. For complicated P and Q direct computation of $D(P\|Q)$ might be hard. Instead, one may want to discretize the infinite alphabet and compute numerically the finite sum. Does this approximation work, that is, as the quantization becomes finer, are the resulting finite sums guaranteed to converge to the true value of $D(P\|Q)$? The answer is positive and this continuity with respect to discretization is the content of Theorem 4.5.

Second is the continuity under change of the distribution. For example, this arises in the problem of *estimating information measures*. In many statistical setups, oftentimes we do not know P or Q, and we estimate the distribution by $\hat{P}_n$ using n iid observations sampled from P (in discrete cases we may set $\hat{P}_n$ to be simply the empirical distribution). Does $D(\hat{P}_n\|Q)$ provide a good estimator for $D(P\|Q)$? Does $D(\hat{P}_n\|Q) \to D(P\|Q)$ if $\hat{P}_n \to P$? The answer is delicate – see Section 4.5.

Third, there is yet another kind of continuity: continuity "in the σ-algebra." Despite the scary name, this one is useful even in the most "discrete" situations. For example, imagine that $\theta \sim \mathrm{Unif}(0,1)$ and $X_i \overset{\text{iid}}{\sim} \mathrm{Ber}(\theta)$. Suppose that you observe a sequence of X_i's until the random moment τ equal to the first occurrence of the pattern 0101. How much information about θ did you learn by time τ? We can encode these observations as

$$Z_j = \begin{cases} X_j, & j \le \tau, \\ ?, & j > \tau, \end{cases}$$

where ? designates the fact that we do not know the value of X_j on those times. Then the question we asked above is to compute $I(\theta;Z^\infty)$. We will show in this chapter that

$$I(\theta;Z^\infty) = \lim_{n\to\infty} I(\theta;Z^n) = \sum_{n=1}^{\infty} I(\theta;Z_n|Z^{n-1}), \tag{4.1}$$

thus reducing computation to evaluating an infinite sum of simpler terms (not involving infinite-dimensional vectors). Thus, even in this simple question about biased coin flips we have to understand how to safely work with infinite-dimensional vectors.

4.1 Geometric Interpretation of Mutual Information

Mutual information (MI) can be understood as a weighted "distance" from the conditional distributions to the marginal distribution. Indeed, for discrete X, we have

$$I(X;Y) = D(P_{Y|X}\|P_Y|P_X) = \sum_{x\in\mathcal{X}} D(P_{Y|X=x}\|P_Y)P_X(x).$$

Furthermore, it turns out that P_Y, similar to the center of gravity, minimizes this weighted distance and thus can be thought of as the best approximation for the "center" of the collection of distributions $\{P_{Y|X=x}: x \in \mathcal{X}\}$ with weights given by P_X. We formalize these results in this section and start with the proof of a "golden formula." Its importance is in bridging the two points of view on mutual information. Recall that on the one hand we had Fano's Definition 3.1, and on the other hand for discrete cases we had Shannon's definition (3.3) as difference of entropies. Then the next result (4.3) presents MI as the difference of *relative* entropies in the style of Shannon, while retaining applicability to continuous spaces in the style of Fano.

Theorem 4.1. (Golden formula) *For any Q_Y we have*

$$D(P_{Y|X}\|Q_Y|P_X) = I(X;Y) + D(P_Y\|Q_Y). \tag{4.2}$$

Thus, if $D(P_Y \| Q_Y) < \infty$, then

$$I(X;Y) = D(P_{Y|X} \| Q_Y | P_X) - D(P_Y \| Q_Y). \tag{4.3}$$

Proof. In the discrete case and ignoring the possibility of dividing by zero, the argument is really simple. We just need to write

$$I(X;Y) \stackrel{(3.1)}{=} \mathbb{E}_{P_{X,Y}}\left[\log \frac{P_{Y|X}}{P_Y}\right] = \mathbb{E}_{P_{X,Y}}\left[\log \frac{P_{Y|X} Q_Y}{P_Y Q_Y}\right]$$

and then expand $\log \frac{P_{Y|X} Q_Y}{P_Y Q_Y} = \log \frac{P_{Y|X}}{Q_Y} - \log \frac{P_Y}{Q_Y}$. The argument below is a rigorous implementation of this idea.

First, notice that by Theorem 2.16(e) we have $D(P_{Y|X} \| Q_Y | P_X) \geq D(P_Y \| Q_Y)$ and thus if $D(P_Y \| Q_Y) = \infty$ then both sides of (4.2) are infinite. Thus, we assume $D(P_Y \| Q_Y) < \infty$ and in particular $P_Y \ll Q_Y$. Rewriting the LHS of (4.2) via the chain rule (2.24) we see that the theorem amounts to proving

$$D(P_{X,Y} \| P_X Q_Y) = D(P_{X,Y} \| P_X P_Y) + D(P_Y \| Q_Y).$$

The case of $D(P_{X,Y} \| P_X Q_Y) = D(P_{X,Y} \| P_X P_Y) = \infty$ is clear. Thus, we can assume at least one of these divergences is finite, and, hence, also $P_{X,Y} \ll P_X Q_Y$.

Let $\lambda(y) = \frac{dP_Y}{dQ_Y}(y)$. Since $\lambda(Y) > 0$ P_Y-a.s., applying the definition of Log in (2.10), we can write

$$\mathbb{E}_{P_Y}[\log \lambda(Y)] = \mathbb{E}_{P_{X,Y}}\left[\mathrm{Log} \frac{\lambda(Y)}{1}\right]. \tag{4.4}$$

Notice that the same $\lambda(y)$ is also the density $\frac{dP_X P_Y}{dP_X Q_Y}(x, y)$ of the product measure $P_X P_Y$ with respect to $P_X Q_Y$. Therefore, the RHS of (4.4) by (2.11) applied with $\mu = P_X Q_Y$ coincides with

$$D(P_{X,Y} \| P_X Q_Y) - D(P_{X,Y} \| P_X P_Y),$$

while the LHS of (4.4) by (2.13) equals $D(P_Y \| Q_Y)$. Thus, we have shown the required

$$D(P_Y \| Q_Y) = D(P_{X,Y} \| P_X Q_Y) - D(P_{X,Y} \| P_X P_Y). \qquad \square$$

By dropping the second term in (4.2) we obtain the following result.

Corollary 4.2. (Mutual information as center of gravity) *For any Q_Y we have*

$$I(X;Y) \leq D(P_{Y|X} \| Q_Y | P_X)$$

and, consequently,

$$I(X;Y) = \min_{Q_Y} D(P_{Y|X} \| Q_Y | P_X). \tag{4.5}$$

If $I(X;Y) < \infty$, the unique minimizer is $Q_Y = P_Y$.

Remark 4.1. The variational representation (4.5) is useful for upper-bounding mutual information by choosing an appropriate Q_Y. Indeed, often each distribution in the collection $P_{Y|X=x}$ is simple, but their mixture, P_Y, is very hard to work with.

In these cases, choosing a suitable Q_Y in (4.5) provides a convenient upper bound. As an example, consider the AWGN channel $Y = X + Z$ in Example 3.3, where $\mathrm{Var}(X) = \sigma^2$, $Z \sim \mathcal{N}(0, 1)$. Then, choosing the best possible Gaussian Q and applying the above bound, we have

$$I(X;Y) \le \inf_{\mu \in \mathbb{R}, s \ge 0} \mathbb{E}[D(\mathcal{N}(X,1)\|\mathcal{N}(\mu,s))] = \frac{1}{2}\log(1+\sigma^2),$$

which is tight when X is Gaussian. For more examples and statistical applications, see Chapter 30.

Theorem 4.3. (Mutual information as distance to product distributions)

$$I(X;Y) = \min_{Q_X, Q_Y} D(P_{X,Y}\|Q_X Q_Y)$$

with the unique minimizer $(Q_X, Q_Y) = (P_X, P_Y)$.

Proof. We only need to use the previous corollary and the chain rule (2.24):

$$D(P_{X,Y}\|Q_X Q_Y) \stackrel{(2.24)}{=} D(P_{Y|X}\|Q_Y|P_X) + D(P_X\|Q_X) \ge I(X;Y). \qquad \square$$

Interestingly, the point of view in the previous result extends to conditional mutual information as follows: We have

$$I(X;Z|Y) = \min_{Q_{X,Y,Z}:\, X \to Y \to Z} D(P_{X,Y,Z}\|Q_{X,Y,Z}), \tag{4.6}$$

where the minimization is over all $Q_{X,Y,Z} = Q_X Q_{Y|X} Q_{Z|Y}$, see Section 3.4. Showing this characterization is very similar to the previous theorem. By repeating the same argument as in (4.2) we get

$$\begin{aligned}
&D(P_{X,Y,Z}\|Q_X Q_{Y|X} Q_{Z|Y}) \\
&= D(P_{X,Y,Z}\|P_X P_{Y|X} P_{Z|Y}) + D(P_X\|Q_X) + D(P_{Y|X}\|Q_{Y|X}|P_X) + D(P_{Z|Y}\|Q_{Z|Y}|P_Y) \\
&= D(P_{X,Y,Z}\|P_Y P_{X|Y} P_{Z|Y}) + D(P_X\|Q_X) + D(P_{Y|X}\|Q_{Y|X}|P_X) + D(P_{Z|Y}\|Q_{Z|Y}|P_Y) \\
&= \underbrace{D(P_{XZ|Y}\|P_{X|Y} P_{Z|Y}|P_Y)}_{I(X;Z|Y)} + D(P_X\|Q_X) + D(P_{Y|X}\|Q_{Y|X}|P_X) + D(P_{Z|Y}\|Q_{Z|Y}|P_Y) \\
&\ge I(X;Z|Y).
\end{aligned}$$

Characterization (4.6) can be understood as follows. The most general directed graphical model for the triplet (X, Y, Z) is a 3-clique (triangle).

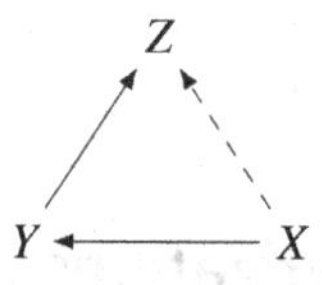

What is the information flow on the dashed edge $X \to Z$? To answer this, notice that removing this edge restricts the joint distribution to a Markov chain $X \to Y \to Z$. Thus, it is natural to ask what is the minimum (KL divergence) distance between a given $P_{X,Y,Z}$ and the set of all distributions $Q_{X,Y,Z}$ satisfying the Markov chain constraint. By the above calculation, the optimal choice is $Q_{X,Y,Z} = P_Y P_{X|Y} P_{Z|Y}$

and hence the distance is $I(X;Z|Y)$. For this reason, we may interpret $I(X;Z|Y)$ as the amount of information flowing through the $X \to Z$ edge.

In addition to inf-characterization, mutual information also has a sup-characterization.

Theorem 4.4. *For any Markov kernel $Q_{X|Y}$ such that $Q_{X|Y=y} \ll P_X$ for P_Y-a.e. y we have*

$$I(X;Y) \geq \mathbb{E}_{P_{X,Y}}\left[\log \frac{dQ_{X|Y}}{dP_X}\right].$$

If $I(X;Y) < \infty$ then

$$I(X;Y) = \sup_{Q_{X|Y}} \mathbb{E}_{P_{X,Y}}\left[\log \frac{dQ_{X|Y}}{dP_X}\right], \tag{4.7}$$

where the supremum is over Markov kernels $Q_{X|Y}$ as in the first sentence.

Remark 4.2. Similar to how (4.5) is used to upper-bound $I(X;Y)$ by choosing a good approximation to P_Y, this result is used to lower-bound $I(X;Y)$ by selecting a good (but computable) approximation $Q_{X|Y}$ to usually a very complicated posterior $P_{X|Y}$. See Section 5.6 for applications.

Proof. Since modifying $Q_{X|Y=y}$ on a negligible set of y's does not change the expectations, we will assume that $Q_{X|Y=y} \ll P_Y$ for every y. If $I(X;Y) = \infty$ then there is nothing to prove. So we assume $I(X;Y) < \infty$, which implies $P_{X,Y} \ll P_X P_Y$. Then by Lemma 3.3 we have that $P_{X|Y=y} \ll P_X$ for almost every y. Choose any such y and apply (2.11) with $\mu = P_X$ and noticing $\mathrm{Log}\frac{dQ_{X|Y=y}/dP_X}{1} = \log \frac{dQ_{X|Y=y}}{dP_X}$ we get

$$\mathbb{E}_{P_{X|Y=y}}\left[\log \frac{dQ_{X|Y=y}}{dP_X}\right] = D(P_{X|Y=y}\|P_X) - D(P_{X|Y=y}\|Q_{X|Y=y}),$$

which is applicable since the first term is finite for a.e. y by (3.1). Taking expectation of the previous identity over y we obtain

$$\mathbb{E}_{P_{X,Y}}\left[\log \frac{dQ_{X|Y}}{dP_X}\right] = I(X;Y) - D(P_{X|Y}\|Q_{X|Y}|P_Y) \leq I(X;Y), \tag{4.8}$$

implying the first part. The equality case in (4.7) follows by taking $Q_{X|Y} = P_{X|Y}$, which satisfies the conditions on Q when $I(X;Y) < \infty$. □

4.2 Variational Characterizations of Divergence: Gelfand–Yaglom–Perez

The point of the following theorem is that divergence on general alphabets can be defined via divergence on finite alphabets and discretization. Moreover, as the quantization becomes finer, we approach the true divergence.

Theorem 4.5. (Gelfand–Yaglom–Perez [182]) *Let P, Q be two probability measures on $\mathcal{X}$ with σ-algebra $\mathcal{F}$. Then*

$$D(P\|Q) = \sup_{\{E_1,\ldots,E_n\}} \sum_{i=1}^{n} P[E_i] \log \frac{P[E_i]}{Q[E_i]}, \tag{4.9}$$

where the supremum is over all finite $\mathcal{F}$-measurable partitions: $\bigcup_{j=1}^{n} E_j = \mathcal{X}, E_j \cap E_i = \emptyset$, and $0 \log \frac{1}{0} = 0$ and $\log \frac{1}{0} = \infty$ per our usual convention.

Remark 4.3. This theorem, in particular, allows us to prove all general identities and inequalities for the cases of discrete random variables and then pass to the general case. In the case of mutual information $I(X;Y) = D(P_{X,Y}\|P_X P_Y)$, the partitions of $\mathcal{X}$ and $\mathcal{Y}$ can be chosen separately, see (4.29).

Proof. "$\geq$": Fix a finite partition $E_1, \ldots, E_n$. Define a function (quantizer) $f\colon \mathcal{X} \to \{1, \ldots, n\}$ as follows: For any x, let $f(x)$ denote the index j of the set E_j to which x belongs. Let X be distributed according to either P or Q and set $Y = f(X)$. Applying the data-processing inequality for divergence yields

$$\begin{aligned} D(P\|Q) &= D(P_X\|Q_X) \\ &\geq D(P_Y\|Q_Y) \\ &= \sum_i P(E_i) \log \frac{P(E_i)}{Q(E_i)}. \end{aligned} \tag{4.10}$$

"$\leq$": To show $D(P\|Q)$ is indeed achievable, first note that if $P \not\ll Q$, then by definition, there exists B such that $Q(B) = 0 < P(B)$. Choosing the partition $E_1 = B$ and $E_2 = B^c$, we have $D(P\|Q) = \infty = \sum_{i=1}^{2} P[E_i] \log \frac{P[E_i]}{Q[E_i]}$. In the sequel we assume that $P \ll Q$ and let $X = \frac{dP}{dQ}$. Then $D(P\|Q) = \mathbb{E}_Q[X \log X] = \mathbb{E}_Q[\varphi(X)]$ by (2.4). Note that $\varphi(x) \geq 0$ if and only if $x \geq 1$. By the monotone convergence theorem, we have $\mathbb{E}_Q[\varphi(X)1\{X < c\}] \to D(P\|Q)$ as $c \to \infty$, regardless of the finiteness of $D(P\|Q)$.

Next, we construct a finite partition. Let $n = c/\epsilon$ be an integer and for $j = 0, \ldots, n-1$, let $E_j = \{j\epsilon \leq X \leq (j+1)\epsilon\}$ and $E_n = \{X \geq c\}$. Define $Y = \epsilon \lfloor X/\epsilon \rfloor$ as the quantized version. Since φ is uniformly continuous on $[0, c]$, for any $x, y \in [0, c]$ such that $|x - y| \leq \epsilon$, we have $|\varphi(x) - \varphi(y)| \leq \epsilon'$ for some $\epsilon' = \epsilon'(\epsilon, c)$ such that $\epsilon' \to 0$ as $\epsilon \to 0$. Then $\mathbb{E}_Q[\varphi(Y)1\{X < c\}] \geq \mathbb{E}_Q[\varphi(X)1\{X < c\}] - \epsilon'$. Moreover,

$$\begin{aligned} \mathbb{E}_Q[\varphi(Y)1\{X < c\}] &= \sum_{j=0}^{n-1} \varphi(j\epsilon) Q(E_j) \leq \epsilon' + \sum_{j=0}^{n-1} \varphi\left(\frac{P(E_j)}{Q(E_j)}\right) Q(E_j) \\ &\leq \epsilon' + Q(X \geq c) \log e + \sum_{j=0}^{n} P(E_j) \log \frac{P(E_j)}{Q(E_j)}, \end{aligned}$$

where the first inequality applies the uniform continuity of φ since $j\epsilon \leq \frac{P(E_j)}{Q(E_j)} < (j+1)\epsilon$, and the second applies $\varphi \geq -\log e$. As $Q(X \geq c) \to 0$ as $c \to \infty$, the proof is completed by first sending $\epsilon \to 0$ then $c \to \infty$. □

4.3 Variational Characterizations of Divergence: Donsker–Varadhan

The following is perhaps the most important variational characterization of divergence. We remind the reader of our convention $\exp\{-\infty\}=0$, $\log 0=-\infty$.

Theorem 4.6. (Donsker–Varadhan [134]) *Let P, Q be probability measures on $\mathcal{X}$ and denote a class of functions $\mathcal{C}_Q = \{f : \mathcal{X} \to \mathbb{R} \cup \{-\infty\} : 0 < \mathbb{E}_Q[\exp\{f(X)\}] < \infty\}$. Then*

$$D(P\|Q) = \sup_{f\in\mathcal{C}_Q} \mathbb{E}_P[f(X)] - \log \mathbb{E}_Q[\exp\{f(X)\}]. \tag{4.11}$$

In particular, if $D(P\|Q) < \infty$ then $\mathbb{E}_P[f(X)]$ is well defined and $< \infty$ for every $f \in \mathcal{C}_Q$. The identity (4.11) holds with $\mathcal{C}_Q$ replaced by the class of all $\mathbb{R}$-valued simple functions. If $\mathcal{X}$ is a normal topological space (e.g. a metric space) with the Borel σ-algebra, then also

$$D(P\|Q) = \sup_{f\in\mathcal{C}_b} \mathbb{E}_P[f(X)] - \log \mathbb{E}_Q[\exp\{f(X)\}], \tag{4.12}$$

where $\mathcal{C}_b$ is the class of all bounded continuous functions.

Proof. "$D \ge \sup_{f\in\mathcal{C}_Q}$": We can assume for this part that $D(P\|Q) < \infty$, since otherwise there is nothing to prove. Then fix $f \in \mathcal{C}_Q$ and define a probability measure Q^f (*tilted version* of Q) via

$$Q^f(dx) = \exp\{f(x) - \psi_f\}Q(dx), \quad \psi_f \triangleq \log \mathbb{E}_Q[\exp\{f(X)\}]. \tag{4.13}$$

Then, $Q^f \ll Q$. We will apply (2.11) next with reference measure $\mu = Q$. Note that according to (2.10) we always have $\mathrm{Log}\frac{\exp\{f(x)-\psi_f\}}{1} = f(x) - \psi_f$ even when $f(x) = -\infty$. Thus, we get from (2.11)

$$\mathbb{E}_P[f(X)] - \psi_f = \mathbb{E}_P\left[\mathrm{Log}\frac{dQ^f/dQ}{1}\right] = D(P\|Q) - D(P\|Q^f) \le D(P\|Q).$$

Note that (2.11) also implies that if $D(P\|Q) < \infty$ and $f \in \mathcal{C}_Q$ the expectation $\mathbb{E}_P[f]$ is well defined.

"$D \le \sup_f$" with supremum over all simple functions: The idea is to just take $f = \log\frac{dP}{dQ}$; however, to handle all cases we proceed more carefully. First, notice that if $P \not\ll Q$ then for some E with $Q[E] = 0 < P[E]$ and $c \to \infty$ taking $f = c1_E$ shows that both sides of (4.11) are infinite. Thus, we assume $P \ll Q$. For any partition of $\mathcal{X} = \bigcup_{j=1}^n E_j$ we set $f = \sum_{j=1}^n 1_{E_j} \log\frac{P[E_j]}{Q[E_j]}$. Then the right-hand sides of (4.11) and (4.9) evaluate to the same value and hence by Theorem 4.5 we obtain that the supremum over simple functions (and thus over $\mathcal{C}_Q$) is at least as large as $D(P\|Q)$.

Finally, to show (4.12), we show that for every simple function f there exists a continuous bounded f' such that $\mathbb{E}_P[f'] - \log \mathbb{E}_Q[\exp\{f'\}]$ is arbitrarily close to the same functional evaluated at f. To that end we first show that for any $a \in \mathbb{R}$ and measurable $A \subset \mathcal{X}$ there exists a sequence of continuous bounded f_n such that

$$\mathbb{E}_P[f_n] \to aP[A] \quad \text{and} \quad \mathbb{E}_Q[\exp\{f_n\}] \to \exp\{a\}Q[A] \tag{4.14}$$

hold *simultaneously*, that is, $f_n \to a1_A$ in the sense of approximating both expectations. We only consider the case of $a > 0$ below. Let compact F and open U be such that $F \subset A \subset U$ and $\max(P[U] - P[F], Q[U] - Q[F]) \le \epsilon$. Such F and U exist whenever P and Q are so-called regular measures. Without going into details, we just notice that finite measures on Polish spaces are automatically regular. Then by Urysohn's lemma there exists a continuous function $f_\epsilon \colon \mathcal{X} \to [0, a]$ equal to a on F and 0 on U^c. Then we have

$$aP[F] \le \mathbb{E}_P[f_\epsilon] \le aP[U],$$
$$\exp\{a\}Q[F] \le \mathbb{E}_Q[\exp\{f_\epsilon\}] \le \exp\{a\}Q[U].$$

Subtracting $aP[A]$ and $\exp\{a\}Q[A]$ for each of these inequalities, respectively, we see that taking $\epsilon \to 0$ indeed results in a sequence of functions satisfying (4.14).

Similarly, if we want to approximate a general simple function $g = \sum_{i=1}^n a_i 1_{A_i}$ (with A_i disjoint and $|a_i| \le a_{\max} < \infty$) we fix $\epsilon > 0$ and define functions $f_{i,\epsilon}$ approximating $a_i 1_{A_i}$ as above with sets $F_i \subset A_i \subset U_i$, so that $S \triangleq \bigcup_i (U_i \setminus F_i)$ satisfies $\max(P[S], Q[S]) \le n\epsilon$. We also have

$$\left| \sum_i f_{i,\epsilon} - g \right| \le a_{\max} \sum_i 1_{U_i \setminus F_i} \le n a_{\max} 1_S.$$

We then clearly have $|\mathbb{E}_P[\sum_i f_{i,\epsilon}] - \mathbb{E}_P[g]| \le a_{\max} n^2 \epsilon$. On the other hand, we also have

$$\begin{aligned}\sum_i \exp\{a_i\}Q[F_i] &\le \mathbb{E}_Q\left[\exp\left\{\sum_i f_{i,\epsilon}\right\}\right] \\ &\le \mathbb{E}_Q[\exp\{g\}1_{S^c}] + \exp\{n a_{\max}\}Q[S] \\ &\le \mathbb{E}_Q[\exp\{g\}] + \exp\{n a_{\max}\}n\epsilon.\end{aligned}$$

Hence taking $\epsilon \to 0$ the sum $\sum_i f_{i,\epsilon} \to \sum_i a_i 1_{A_i}$ in the sense of both $\mathbb{E}_P[\cdot]$ and $\mathbb{E}_Q[\exp\{\cdot\}]$. □

Remark 4.4.

(a) What is the Donsker–Varadhan representation useful for? By setting $f(x) = \epsilon \cdot g(x)$ with $\epsilon \ll 1$ and linearizing exp and log we can see that when $D(P\|Q)$ is small, expectations under P can be approximated by expectations over Q (change of measure): $\mathbb{E}_P[g(X)] \approx \mathbb{E}_Q[g(X)]$. This holds for all functions g with finite exponential moment under Q. Total variation distance provides a similar bound, but for a narrower class of bounded functions:

$$|\mathbb{E}_P[g(X)] - \mathbb{E}_Q[g(X)]| \le \|g\|_\infty \mathrm{TV}(P, Q).$$

(b) More formally, the inequality $\mathbb{E}_P[f(X)] \le \log \mathbb{E}_Q[\exp f(X)] + D(P\|Q)$ is useful in estimating $\mathbb{E}_P[f(X)]$ for complicated distribution P (e.g. over high-dimensional X with weakly dependent coordinates) by making a smart choice of Q (e.g. with iid components).

(c) In Chapter 5 we will show that $D(P\|Q)$ is convex in P (in fact, in the pair). A general method of obtaining variational formulas like (4.11) is via the Young–Fenchel duality, which we review below in (7.84). Indeed, (4.11) is exactly that inequality since the Fenchel–Legendre conjugate of $D(\cdot\|Q)$ is given by a convex map $f \mapsto \psi_f$. For more details, see Section 7.13.

(d) Donsker–Varadhan should also be seen as an "improved version" of the DPI. For example, one of the main applications of the DPI in this book is in obtaining estimates like

$$P[A] \log \frac{1}{Q[A]} \leq D(P\|Q) + \log 2, \tag{4.15}$$

which is the basis of the large-deviations theory (Corollary 2.19 and Chapter 15) and Fano's inequality (Theorem 3.12). The same estimate can be obtained by applying (4.11) with $f(x) = 1\{x \in A\} \log \frac{1}{Q[A]}$.

4.4 Gibbs Variational Principle

As we remarked before the Donsker–Varadhan characterization can be seen as a way of expressing the convex function $P \mapsto D(P\|Q)$ as the supremum of linear functions $P \mapsto \mathbb{E}_P[f] - \psi_f$, where $\psi_f = \log \mathbb{E}_Q[\exp\{f\}]$, in which case ψ_f is known as the convex conjugate. Now, by the general convex duality theory one would expect then that the function $f \mapsto \psi_f$ is convex and should have a similar characterization as the supremum of linear functions of f. In this section we derive it and show several of its (quite influential) classical and modern applications.

Proposition 4.7. (Gibbs variational principle) *Let $f: \mathcal{X} \to \mathbb{R} \cup \{-\infty\}$ be any measurable function and Q a probability measure on $\mathcal{X}$. Then*

$$\log \mathbb{E}_Q[\exp\{f(X)\}] = \sup_P \mathbb{E}_P[f(X)] - D(P\|Q),$$

where the supremum is taken over all P with $D(P\|Q) < \infty$. If the left-hand side is finite then the unique maximizer of the right-hand side is $P = Q^f$, a tilted version of Q defined in (4.13).

Proof. Let $\psi_f \triangleq \log \mathbb{E}_Q[\exp\{f(X)\}]$. First, if $\psi_f = -\infty$, then Q-a.s. $f = -\infty$ and hence P-a.s. also $f = -\infty$, so that both sides of the equality are $-\infty$. Next, assume $-\infty < \psi_f < \infty$. Then by Donsker–Varadhan (4.11) we get

$$\psi_f \geq \mathbb{E}_P[f(X)] - D(P\|Q).$$

On the other hand, setting $P = Q^f$ we obtain an equality. To show uniqueness, notice that $\mathrm{Log}\frac{dQ^f/dQ}{1} = f - \psi_f$ even when $f = -\infty$. Thus, from (2.11) we get whenever $D(P\|Q) < \infty$ that

$$\mathbb{E}_P[f(X) - \psi_f] = D(P\|Q) - D(P\|Q^f).$$

From here we conclude that $\mathbb{E}_P[f(X)] < \infty$ and hence we can rewrite the above as

$$\mathbb{E}_P[f(X)] - D(P\|Q) = \psi_f - D(P\|Q^f),$$

which shows that $\mathbb{E}_P[f(X)] - D(P\|Q) = \psi_f$ implies $P = Q^f$.

Next, suppose $\psi_f = \infty$. Let us define $f_n = f \wedge n$, $n \geq 1$. Since $\psi_{f_n} < \infty$ we have by the previous characterization that there is a sequence P_n such that $D(P_n\|Q) < \infty$ and as $n \to \infty$

$$\mathbb{E}_{P_n}[f(X) \wedge n] - D(P_n\|Q) = \psi_{f_n} \nearrow \psi_f = \infty.$$

Since $\mathbb{E}_{P_n}[f(X) \wedge n] \leq \mathbb{E}_{P_n}[f(X)]$, we have

$$\mathbb{E}_{P_n}[f(X)] - D(P_n\|Q) \geq \psi_{f_n} \to \infty,$$

concluding the proof. □

We now briefly explore how Proposition 4.7 has been applied over the last century. We start with the example from statistical physics and graphical models. Here the key idea is to replace sup over all distributions P with a subset that is easier to handle. This idea is the basis of much of variational inference [448].

Example 4.1. Mean-field approximation for Ising model

Suppose that we have a complicated model for a distribution of a vector $\tilde{X} \in \{0,1\}^{n+m}$ given by an *Ising model*

$$P_{\tilde{X}}(\tilde{x}) = \frac{1}{\tilde{Z}} \exp\{\tilde{x}^\top \tilde{A}\tilde{x} + \tilde{b}^\top \tilde{x}\},$$

where $\tilde{A} \in \mathbb{R}^{(n+m)\times(n+m)}$ is a symmetric *interaction matrix* with zero diagonal, $\tilde{b}$ is a vector of *external fields*, and $\tilde{Z}$ is a normalization constant. We note that often $\tilde{A}$ is very sparse with non-zero entries occurring only for those few variables x_i and x_j that are considered to be interacting (or adjacent in some graph). We decompose the vector $\tilde{X} = (X, Y)$ into two components: the last m coordinates are observables and the first n coordinates are hidden (latent), whose values we want to infer; in other words, our goal is to evaluate $P_{X|Y=y}$ upon observing y. It is clear that this conditional distribution is still an Ising model, so that

$$P_{X|Y}(x|y) = \frac{1}{Z} \exp\{x^\top Ax + b^\top x\}, \qquad x \in \{0,1\}^n,$$

where A is the $n \times n$ leading minor of $\tilde{A}$, and b and Z depend on y. Unfortunately, computing even a single value $P[X_1 = 1]$ is known to be generally computationally infeasible [395, 175], since evaluating Z requires summation over the 2^n values of x.

Let us denote $f(x) = x^\top Ax + b^\top x$ and by Q the uniform distribution on $\{0,1\}^n$. Applying Proposition 4.7 we obtain

$$\log Z - n\log 2 = \log \mathbb{E}_Q[\exp\{f\}] = \sup_{P_{X^n}} \mathbb{E}_P[f(X^n)] - D(P\|Q).$$

Then by Theorem 2.2 we get

$$\log Z = \sup_{P_{X^n}} \mathbb{E}_P[f(X^n)] + H(P_{X^n}).$$

As we said, exact computation of $\log Z$, though, is not tractable. An influential idea is instead to search for the maximizer in the class of product distributions $P_{X^n} = \prod_{i=1}^n \mathrm{Ber}(p_i)$. In this case, this supremization can be solved almost in closed form:

$$\sup_p p^\top A p + b^\top p + \sum_i h(p_i),$$

where $p = (p_1, \ldots, p_n)$. Since the objective function is strongly concave (Exercise **I.37**), we only need to solve the first-order optimality conditions (or *mean-field equations*), which is a set of n non-linear equations in n variables:

$$p_i = \sigma\left(b_i + 2\sum_{j=1}^n a_{i,j} p_j\right), \quad \sigma(x) \triangleq \frac{1}{1+\exp(-x)}.$$

These are solved by iterative message-passing algorithms [448]. Once the values of p_i are obtained, the *mean-field approximation* is to take

$$P_{X|Y=y} \approx \prod_{i=1}^n \mathrm{Ber}(p_i).$$

We stress that the mean-field idea is not only to approximate the value of Z, but also to consider the corresponding maximizer (over a restricted class of product distributions) as the approximate posterior distribution.

To get another flavor of examples, let us consider a more general setting, where we have some parametric collection of distributions $P_{X,Y}^{(\theta)}$ indexed by $\theta \in \mathbb{R}^d$. Often, the joint distribution is such that $P_X^{(\theta)}$ and $P_{Y|X}^{(\theta)}$ are both "simple," but $P_Y^{(\theta)}$ and $P_{X|Y}^{(\theta)}$ are "complex" or even intractable (e.g. in sparse linear regression and community detection, Section 30.3). As in the previous example, X is the latent (unobserved) and Y is the observed variable.

For a moment we will omit writing θ and consider the problem of evaluating $P_Y(y)$ – a quantity (known as evidence) showing how extreme the observed y is. Note that

$$P_Y(y) = \mathbb{E}_{X\sim P_X}[P_{Y|X}(y|x)].$$

Although by assumption P_X and $P_{Y|X}$ are both easy to compute, this marginalization may be intractable. As a workaround, we invoke Proposition 4.7 with $f(x) = \log P_{Y|X}(y|x)$ and $Q = P_X$ to get

$$\log P_Y(y) = \sup_R \mathbb{E}_R[f(X)] - D(R\|P_X) = \sup_R \mathbb{E}_{X\sim R}\left[\log \frac{P_{X,Y}(X,y)}{R(X)}\right], \tag{4.16}$$

where R is an arbitrary distribution. Note that the right-hand side only involves a simple quantity $P_{X,Y}$ and hence all the complexity of computation is moved to optimization over R. Expression (4.16) is known as *evidence lower bound* (ELBO) since for any fixed value of R we get a provable lower bound on $\log P_Y(y)$. Typically, one optimizes the choice of R over some convenient set of distributions to get the best (tightest) lower bound in that class.

One such application leads to the famous iterative (EM) algorithm, see (5.33) below. Another application is a modern density estimation algorithm, which we describe next.

Example 4.2. Variational autoencoders [246]

A canonical problem in unsupervised learning is density estimation: given a sample $Y_1, \ldots, Y_n \overset{\text{iid}}{\sim} P_Y$ estimate the true P_Y on $\mathbb{R}^d$. We describe a modern solution of [246]. First, they propose a latent parametric model (a *generative model*) for P_Y. Namely, Y is generated by first sampling a latent variable $X \sim \mathcal{N}(0, I_{d'})$ and then setting Y to be conditionally Gaussian:

$$Y = \mu(X; \theta) + D(X; \theta)Z, \quad X \perp\!\!\!\perp Z \sim \mathcal{N}(0, I_d),$$

where vector $\mu(\cdot; \theta)$ and diagonal matrix $D(\cdot; \theta)$ are deep neural networks with input $(\cdot)$ and weights θ. (See [246, Appendix C.2] for a detailed description.) The resulting distribution $P_Y^{(\theta)}$ is a (complicated) location-scale Gaussian mixture. To find the best density [246] aims to maximize the likelihood by solving

$$\max_\theta \sum_i \log P_Y^{(\theta)}(y_i).$$

Since the marginalization to obtain P_Y is intractable, we replace the objective (by an appeal to ELBO (4.16)) with

$$\max_\theta \sup_{R_{X|Y}} \sum_i \mathbb{E}_{X \sim R_{X|Y=y_i}} \left[\log \frac{p_{X,Y}^{(\theta)}(X, y_i)}{r_{X|Y}(X|y_i)} \right], \tag{4.17}$$

where we denoted the PDFs of $P_{X,Y}$ and $R_{X|Y}$ by lower case letters. Now in this form the algorithm is simply the EM algorithm (as we discuss below in Section 5.6). What brings the idea to the twenty-first century is restricting the optimization to the set of $R_{X|Y}$ which are again defined via

$$X = \tilde{\mu}(Y; \phi) + \tilde{D}(Y; \phi)\tilde{Z}, \quad Y \perp\!\!\!\perp \tilde{Z} \sim \mathcal{N}(0, I_{d'}),$$

where $\tilde{\mu}(Y; \phi)$ and diagonal covariance matrix $\tilde{D}(Y; \phi)$ are output by some neural network with parameter ϕ. The conditional distribution under this auxiliary model (*recognition model*), denoted by $R_{X|Y}^{(\phi)}$, is Gaussian. Since the ELBO (4.16) is achieved by the posterior $P_{X|Y}^{(\theta)}$, what this amounts to is to approximate the true posterior under the generative model by a Gaussian. Replacing also the expectation over $R_{X|Y=y_i}$ with its empirical version (by generating $\tilde{Z}_{ij} \overset{\text{iid}}{\sim} \mathcal{N}(0, I_{d'})$), we obtain the following:

$$\max_\theta \max_\phi \sum_i \sum_j \left[\log \frac{p_{X,Y}^{(\theta)}(x_{i,j}, y_i)}{r_{X|Y}^{(\phi)}(x_{i,j}|y_i)} \right], \quad x_{i,j} = \tilde{\mu}(y_i; \phi) + \tilde{D}(y_i; \phi)\tilde{Z}_{ij}. \tag{4.18}$$

Now plugging in the Gaussian form of the densities p_X, $p^{(\theta)}_{Y|X}$, and $r^{(\phi)}_{X|Y}$ one gets an expression whose gradients ∇_θ and ∇_ϕ can be easily computed by automatic differentiation software.[1] In fact, since $r^{(\phi)}_{X|Y}$ and $r^{(\theta)}_{Y|X}$ are both Gaussian, we can use less Monte Carlo approximation than (4.18), because the objective in (4.17) equals

$$\mathbb{E}_{X\sim R_{X|Y=y_i}}\left[\log \frac{p^{(\theta)}_{X,Y}(X,y_i)}{r^{(\phi)}_{X|Y}(X|y_i)}\right] = -D(R_{X|Y=y_i}\|P_X) + \mathbb{E}_{X\sim R_{X|Y=y_i}}\left[\log p^{(\theta)}_{Y|X}(y_i|X)\right],$$

where $P_X = \mathcal{N}(0, I'_d)$, $R_{X|Y=y_i} = \mathcal{N}(\tilde{\mu}(y_i;\ \phi);\ \tilde{D}^2(y_i;\ \phi))$, and $P^{(\theta)}_{Y|X=x} = \mathcal{N}(\mu(x;\ \theta);\ D^2(x;\ \theta))$ so that the first Gaussian KL divergence is in closed form (Example 2.2) and we only need to apply Monte Carlo approximation to the second term. For both versions, the optimization proceeds by (stochastic) gradient ascent over θ and ϕ until convergence to some (θ^*, ϕ^*). Then $P^{(\theta^*)}_Y$ can be used to generate new samples from the learned distribution, $R^{(\phi^*)}_{X|Y}$ to map ("encode") samples to the latent space and $P^{(\theta^*)}_{Y|X}$ to "decode" a latent representation into a target sample. We refer the reader to Chapters 3 and 4 in the survey [247] for other encoder and decoder architectures.

4.5 Continuity of Divergence

For a finite alphabet $\mathcal{X}$ it is easy to establish the continuity of entropy and divergence:

Proposition 4.8. *Let $\mathcal{X}$ be finite. Fix a distribution Q on $\mathcal{X}$ with $Q(x) > 0$ for all $x \in \mathcal{X}$. Then the map $P \mapsto D(P\|Q)$ is continuous. In particular, $P \mapsto H(P)$ is continuous.*

Warning: Divergence is never continuous in the pair, even for finite alphabets. For example, as $n \to \infty$, $d(\frac{1}{n}\|2^{-n}) \not\to 0$.

Proof. Notice that

$$D(P\|Q) = \sum_x P(x) \log \frac{P(x)}{Q(x)}$$

and each term is a continuous function of $P(x)$. □

Our next goal is to study continuity properties of divergence for general alphabets. We start with a negative observation.

Remark 4.5. In general, $D(P\|Q)$ is *not* continuous in either P or Q. For example, let $X_1, \ldots, X_n$ be iid and equally likely to be $\{\pm 1\}$. Then by the central limit theorem,

[1] An important part of the contribution of [246] is the "reparameterization trick." Namely, since [453] a standard way to compute $\nabla_\phi \mathbb{E}_{Q^{(\phi)}}[f]$ in machine learning is to write

$$\nabla_\phi \mathbb{E}_{Q^{(\phi)}}[f] = \mathbb{E}_{Q^{(\phi)}}[f(X)\nabla_\phi \ln q^{(\phi)}(X)]$$

and replace the latter expectation by its empirical approximation. However, in this case a much better idea is to write $\mathbb{E}_{Q^{(\phi)}}[f] = \mathbb{E}_{Z\sim\mathcal{N}}[f(g(Z;\phi))]$ for some explicit g and then move the gradient inside the expectation before computing the empirical version.

$S_n = \frac{1}{\sqrt{n}}\sum_{i=1}^n X_i \xrightarrow{\mathrm{d}} \mathcal{N}(0,1)$ as $n \to \infty$. But $D(P_{S_n}\|\mathcal{N}(0,1)) = \infty$ for all n because S_n is discrete. Note that this is also an example for strict inequality in (4.19).

Nevertheless, there is a very useful semicontinuity property.

Theorem 4.9. (Lower semicontinuity of divergence) *Let $\mathcal{X}$ be a metric space with Borel σ-algebra $\mathcal{H}$. If P_n and Q_n converge weakly to P and Q, respectively,*[2] *then*

$$D(P\|Q) \le \liminf_{n\to\infty} D(P_n\|Q_n). \tag{4.19}$$

On a general space if $P_n \to P$ and $Q_n \to Q$ pointwise[3] *(i.e. $P_n[E] \to P[E]$ and $Q_n[E] \to Q[E]$ for every measurable E) then* (4.19) *also holds.*

Proof. This simply follows from (4.12) since $\mathbb{E}_{P_n}[f] \to \mathbb{E}_P[f]$ and $\mathbb{E}_{Q_n}[\exp\{f\}] \to \mathbb{E}_Q[\exp\{f\}]$ for every $f \in C_b$. □

4.6* Continuity under Monotone Limits of σ-Algebras

Our final and somewhat delicate topic is to understand the (so far neglected) dependence of D and I on the implicit σ-algebra of the space. Indeed, the definition of divergence $D(P\|Q)$ implicitly (via the Radon–Nikodym derivative) depends on the σ-algebra $\mathcal{F}$ defining the measurable space $(\mathcal{X}, \mathcal{F})$. To emphasize the dependence on $\mathcal{F}$ we will write in this section only the underlying σ-algebra explicitly as follows:

$$D(P_{\mathcal{F}}\|Q_{\mathcal{F}}).$$

Our main results are the continuity under monotone limits of σ-algebras. Recall that a sequence of nested σ-algebras, $\mathcal{F}_1 \subset \mathcal{F}_2 \subset \cdots$, is written as $\mathcal{F}_n \nearrow \mathcal{F}$ when $\mathcal{F} \triangleq \sigma\left(\bigcup_n \mathcal{F}_n\right)$ is the smallest σ-algebra containing $\bigcup_n \mathcal{F}_n$ (the union of σ-algebras may fail to be a σ-algebra and hence needs completion). Similarly, a sequence of nested σ-algebras, $\mathcal{F}_1 \supset \mathcal{F}_2 \supset \cdots$, is written as $\mathcal{F}_n \searrow \mathcal{F}$ if $\mathcal{F} = \bigcap_n \mathcal{F}_n$ (the intersection of σ-algebras is always a σ-algebra). We will show in this section that we always have:

$$\mathcal{F}_n \nearrow \mathcal{F} \quad \Longrightarrow \quad D(P_{\mathcal{F}_n}\|Q_{\mathcal{F}_n}) \nearrow D(P_{\mathcal{F}}\|Q_{\mathcal{F}}), \tag{4.20}$$

$$\mathcal{F}_n \searrow \mathcal{F} \quad \Longrightarrow \quad D(P_{\mathcal{F}_n}\|Q_{\mathcal{F}_n}) \searrow D(P_{\mathcal{F}}\|Q_{\mathcal{F}}). \tag{4.21}$$

For establishing the first result, it will be convenient to extend the definition of the divergence $D(P_{\mathcal{F}}\|Q_{\mathcal{F}})$ to (a) any *algebra* of sets $\mathcal{F}$ and (b) two positive additive (not necessarily σ-additive) set-functions P, Q on $\mathcal{F}$. We do so following the Gelfand–Yaglom–Perez variational representation of divergence (Theorem 4.5).

[2] Recall that a sequence of random variables X_n converges in distribution to X if and only if their laws P_{X_n} converge weakly to P_X.

[3] Pointwise convergence is weaker than convergence in total variation and stronger than weak convergence.

Definition 4.10. (KL divergence over an algebra) Let P and Q be two positive, additive (not necessarily σ-additive) set-functions defined over an algebra $\mathcal{F}$ of subsets of $\mathcal{X}$ (not necessarily a σ-algebra). We define

$$D(P_{\mathcal{F}}\|Q_{\mathcal{F}}) \triangleq \sup_{\{E_1,\dots,E_n\}} \sum_{i=1}^{n} P[E_i]\log\frac{P[E_i]}{Q[E_i]},$$

where the supremum is over all finite $\mathcal{F}$-measurable partitions: $\bigcup_{j=1}^n E_j = \mathcal{X}, E_j \cap E_i = \emptyset$, and $0\log\frac{1}{0} = 0$ and $\log\frac{1}{0} = \infty$ per our usual convention.

Note that when $\mathcal{F}$ is not a σ-algebra or P, Q are not σ-additive, we do not have the Radon–Nikodym theorem and thus our original definition of KL divergence is not applicable.

Theorem 4.11. (Measure-theoretic properties of divergence) *Let P, Q be probability measures on the measurable space $(\mathcal{X}, \mathcal{H})$. Assume all algebras below are sub-algebras of $\mathcal{H}$. Then:*

- *(Monotonicity) If $\mathcal{F} \subseteq \mathcal{G}$ are nested algebras then*

$$D(P_{\mathcal{F}}\|Q_{\mathcal{F}}) \le D(P_{\mathcal{G}}\|Q_{\mathcal{G}}). \tag{4.22}$$

- *Let $\mathcal{F}_1 \subseteq \mathcal{F}_2 \subseteq \cdots$ be an increasing sequence of algebras and let $\mathcal{F} = \bigcup_n \mathcal{F}_n$ be their limit, then*

$$D(P_{\mathcal{F}_n}\|Q_{\mathcal{F}_n}) \nearrow D(P_{\mathcal{F}}\|Q_{\mathcal{F}}).$$

- *If $\mathcal{F}$ is $(P+Q)$-dense in $\mathcal{G}$ then*[4]

$$D(P_{\mathcal{F}}\|Q_{\mathcal{F}}) = D(P_{\mathcal{G}}\|Q_{\mathcal{G}}). \tag{4.23}$$

- *(Monotone convergence theorem) Let $\mathcal{F}_1 \subseteq \mathcal{F}_2 \subseteq \cdots$ be an increasing sequence of algebras and let $\mathcal{F} = \bigvee_n \mathcal{F}_n$ be the smallest σ-algebra containing all of $\mathcal{F}_n$. Then we have*

$$D(P_{\mathcal{F}_n}\|Q_{\mathcal{F}_n}) \nearrow D(P_{\mathcal{F}}\|Q_{\mathcal{F}})$$

and, in particular,

$$D(P_{X^\infty}\|Q_{X^\infty}) = \lim_{n\to\infty} D(P_{X^n}\|Q_{X^n}).$$

Proof. The first two items are straightforward applications of the definition. The third follows from the following fact: if $\mathcal{F}$ is dense in $\mathcal{G}$ then any $\mathcal{G}$-measurable partition $\{E_1,\dots,E_n\}$ can be approximated by an $\mathcal{F}$-measurable partition $\{E_1',\dots,E_n'\}$ with $(P+Q)[E_i\Delta E_i'] \le \epsilon$. Indeed, first we set E_1' to be an element of $\mathcal{F}$ with $(P+Q)(E_1\Delta E_1') \le \frac{\epsilon}{2n}$. Then, we set E_2' to be an $\frac{\epsilon}{2n}$-approximation of $E_2 \setminus E_1'$, etc. Finally, $E_n' = (\bigcup_{j\le 1} E_j')^c$. By taking $\epsilon \to 0$ we obtain $\sum_i P[E_i']\log\frac{P[E_i']}{Q[E_i']} \to \sum_i P[E_i]\log\frac{P[E_i]}{Q[E_i]}$.

[4] Recall that $\mathcal{F}$ is μ-dense in $\mathcal{G}$ if $\forall E \in \mathcal{G}, \epsilon > 0\, \exists E' \in \mathcal{F}$ s.t. $\mu[E\Delta E'] \le \epsilon$.

The last statement follows from the previous one and the fact that any algebra $\mathcal{F}$ is μ-dense in the σ-algebra $\sigma\{\mathcal{F}\}$ it generates for any bounded μ on $(\mathcal{X}, \mathcal{H})$ (see [142, Lemma III.7.1]). □

Finally, we address the continuity under the decreasing σ-algebra, that is, (4.21).

Proposition 4.12. *Let $\mathcal{F}_n \searrow \mathcal{F}$ be a sequence of decreasing σ-algebras with intersection $\mathcal{F} = \bigcap_n \mathcal{F}_n$; let P, Q be two probability measures on $\mathcal{F}_0$. If $D(P_{\mathcal{F}_0} \| Q_{\mathcal{F}_0}) < \infty$ then we have*

$$D(P_{\mathcal{F}_n} \| Q_{\mathcal{F}_n}) \searrow D(P_{\mathcal{F}} \| Q_{\mathcal{F}}). \tag{4.24}$$

The condition $D(P_{\mathcal{F}_0} \| Q_{\mathcal{F}_0}) < \infty$ cannot be dropped; see the example after (4.32).

Proof. Let $X_{-n} = \left.\frac{dP}{dQ}\right|_{\mathcal{F}_n}$. Since $X_{-n} = \mathbb{E}_Q\left[\frac{dP}{dQ}\middle|\mathcal{F}_n\right]$, we have that $(\ldots, X_{-1}, X_0)$ is a uniformly integrable martingale. By the martingale convergence theorem in reversed time, see [84, Theorem 5.4.17], we have almost surely

$$X_{-n} \to X_{-\infty} \triangleq \left.\frac{dP}{dQ}\right|_{\mathcal{F}}. \tag{4.25}$$

We need to prove that

$$\mathbb{E}_Q[X_{-n} \log X_{-n}] \to \mathbb{E}_Q[X_{-\infty} \log X_{-\infty}].$$

We will do so by decomposing $x \log x$ as follows:

$$x \log x = x \log^+ x + x \log^- x,$$

where $\log^+ x = \max(\log x, 0)$ and $\log^- x = \min(\log x, 0)$. Since $x \log^- x$ is bounded, we have from the bounded convergence theorem:

$$\mathbb{E}_Q[X_{-n} \log^- X_{-n}] \to \mathbb{E}_Q[X_{-\infty} \log^- X_{-\infty}].$$

To prove a similar convergence for $\log^+$ we need to notice two things. First, the function

$$x \mapsto x \log^+ x$$

is convex. Second, for any non-negative convex function ϕ s.t. $\mathbb{E}[\phi(X_0)] < \infty$ the collection $\{Z_n = \phi(\mathbb{E}[X_0|\mathcal{F}_n]) : n \geq 0\}$ is uniformly integrable. Indeed, we have from Jensen's inequality

$$\mathbb{P}[Z_n > c] \leq \frac{1}{c}\mathbb{E}[\phi(\mathbb{E}[X_0|\mathcal{F}_n])] \leq \frac{\mathbb{E}[\phi(X_0)]}{c}$$

and thus $\mathbb{P}[Z_n > c] \to 0$ as $c \to \infty$. Therefore, we have again by Jensen's inequality

$$\mathbb{E}[Z_n 1\{Z_n > c\}] \leq \mathbb{E}[\phi(X_0) 1\{Z_n > c\}] \to 0, \qquad c \to \infty.$$

Finally, since $X_{-n} \log^+ X_{-n}$ is uniformly integrable, we have from (4.25)

$$\mathbb{E}_Q[X_{-n} \log^+ X_{-n}] \to \mathbb{E}_Q[X_{-\infty} \log^+ X_{-\infty}]$$

and this concludes the proof. □

4.7 Variational Characterizations and Continuity of Mutual Information

Again, similarly to Proposition 4.8, it is easy to show that in the case of finite alphabets mutual information is always continuous on finite-dimensional simplices of distributions.[5]

Proposition 4.13.

(a) If $\mathcal{X}$ and $\mathcal{Y}$ are both finite alphabets, then $P_{X,Y} \mapsto I(X;Y)$ is continuous.

(b) If $\mathcal{X}$ is finite, then $P_X \mapsto I(X;Y)$ is continuous.

(c) Without any assumptions on $\mathcal{X}$ and $\mathcal{Y}$, let P_X range over the convex hull $\Pi = \mathrm{co}(P_1, \ldots, P_n) = \{\sum_{i=1}^n \alpha_i P_i : \sum_{i=1}^n \alpha_i = 1, \alpha_i \geq 0\}$. If $I(P_j, P_{Y|X}) < \infty$ (using the notation $I(P_X, P_{Y|X}) = I(X;Y)$) for all $j \in [n]$, then the map $P_X \mapsto I(X;Y)$ is continuous.

Proof. For the first statement, apply the representation

$$I(X;Y) = H(X) + H(Y) - H(X,Y)$$

and the continuity of entropy in Proposition 4.8.

For the second statement, take $Q_Y = \frac{1}{|\mathcal{X}|} \sum_{x \in \mathcal{X}} P_{Y|X=x}$. Note that

$$D(P_Y \| Q_Y) = \mathbb{E}_{Q_Y}\left[f\left(\sum_x P_X(x) h_x(Y) \right) \right],$$

where $f(t) = t \log t$ and $h_x(y) = \frac{dP_{Y|X=x}}{dQ_Y}(y)$ are bounded by $|\mathcal{X}|$ and non-negative. Thus, from the bounded convergence theorem we have that $P_X \mapsto D(P_Y \| Q_Y)$ is continuous. The proof is complete since by the golden formula

$$I(X;Y) = D(P_{Y|X} \| Q_Y | P_X) - D(P_Y \| Q_Y),$$

and the first term is linear in P_X.

For the third statement, form a Markov chain $Z \to X \to Y$ with $Z \in [n]$ and $P_{X|Z=j} = P_j$. Without loss of generality, assume that $P_1, \ldots, P_n$ are distinct extreme points of $\mathrm{co}(P_1, \ldots, P_n)$. Then there is a linear bijection between P_Z and $P_X \in \Pi$. Furthermore, $I(X;Y) = I(Z;Y) + I(X;Y|Z)$. The first term is continuous in P_Z by the previous claim, whereas the second one is simply linear in P_Z. Thus, the map $P_Z \mapsto I(X;Y)$ is continuous and so is $P_X \mapsto I(X;Y)$. □

Further properties of mutual information follow from $I(X;Y) = D(P_{X,Y} \| P_X P_Y)$ and corresponding properties of divergence, for example:

1 Donsker–Varadhan for mutual information: By the definition of mutual information

$$I(X;Y) = \sup_f \mathbb{E}[f(X,Y)] - \log \mathbb{E}[\exp\{f(X,\bar{Y})\}], \tag{4.26}$$

[5] Here we only assume that topology on the space of measures is compatible with the linear structure, so that all linear operations on measures are continuous.

where $\bar{Y}$ is a copy of Y, independent of X and the supremum can be taken over any of the classes of (bivariate) functions as in Theorem 4.6. Notice, however, that for mutual information we can also get a stronger characterization:[6]

$$I(X;Y) \geq \mathbb{E}[f(X,Y)] - \mathbb{E}[\log \mathbb{E}[\exp\{f(X,\bar{Y})\}|X]], \tag{4.27}$$

from which (4.26) follows by moving the outer expectation inside the log. Both of these can be used to show that $\mathbb{E}[f(X,Y)] \approx \mathbb{E}[f(X,\bar{Y})]$ as long as the dependence between X and Y (as measured by $I(X;Y)$) is weak, see Exercise **I.55**.

2 If $(X_n, Y_n) \xrightarrow{d} (X, Y)$ converge in distribution, then

$$I(X;Y) \leq \liminf_{n\to\infty} I(X_n;Y_n). \tag{4.28}$$

- Example of strict inequality: $X_n = Y_n = \frac{1}{n}Z$. In this case $(X_n, Y_n) \xrightarrow{d} (0,0)$ but $I(X_n;Y_n) = H(Z) > 0 = I(0;0)$.
- An even more impressive example: Let (X_p, Y_p) be uniformly distributed on the unit ℓ_p-ball on the plane: $\{(x,y) \in \mathbb{R}^2 : |x|^p + |y|^p \leq 1\}$. Then as $p \to 0$, $(X_p, Y_p) \xrightarrow{d} (0,0)$, but $I(X_p;Y_p) \to \infty$ (see Exercise **I.16**).

3 Mutual information as supremum over partitions:

$$I(X;Y) = \sup_{\{E_i\}\times\{F_j\}} \sum_{i,j} P_{X,Y}[E_i \times F_j] \log \frac{P_{X,Y}[E_i \times F_j]}{P_X[E_i]P_Y[F_j]}, \tag{4.29}$$

where the supremum is over finite partitions of spaces $\mathcal{X}$ and $\mathcal{Y}$.[7]

4 (Monotone convergence I):

$$I(X^\infty;Y) = \lim_{n\to\infty} I(X^n;Y), \tag{4.30}$$

$$I(X^\infty;Y^\infty) = \lim_{n\to\infty} I(X^n;Y^n). \tag{4.31}$$

This implies that the full amount of mutual information between two processes X^∞ and Y^∞ is contained in their finite-dimensional projections, leaving nothing in the tail σ-algebra. Note also that applying the (finite-n) chain rule to (4.30) recovers (4.1).

5 (Monotone convergence II): Recall that for any random process $(X_1, \ldots)$ we define its tail σ-algebra as $\mathcal{F}_{tail} = \bigcap_{n\geq 1} \sigma(X_n^\infty)$, where $X_n^\infty = (X_n, \ldots)$. Let X_{tail} be a random variable such that $\sigma(X_{tail}) = \bigcap_{n\geq 1} \sigma(X_n^\infty)$. Then

$$I(X_{tail};Y) = \lim_{n\to\infty} I(X_n^\infty;Y), \tag{4.32}$$

whenever the right-hand side is finite. This is a consequence of Proposition 4.12. Without the finiteness assumption the statement is incorrect. Indeed, consider $X_j \overset{iid}{\sim} \text{Ber}(1/2)$ and $Y = X_0^\infty$. Then each $I(X_n^\infty;Y) = \infty$, but $X_{tail} =$ constant a.e. by Kolmogorov's 0–1 law, and thus the left-hand side of (4.32) is zero.

[6] Just apply Donsker–Varadhan to $D(P_{Y|X=x_0} \| P_Y)$ and average over $x_0 \sim P_X$.

[7] To prove this from (4.9) one needs to notice that the algebra of measurable rectangles is dense in the product σ-algebra. See [129, Section 2.2].

4.8* PAC-Bayes

A deep implication of Donsker–Varadhan and Gibbs principle is a method, historically known as PAC-Bayes,[8] for bounding suprema of empirical processes. Here we present the key idea together with two applications: one in high-dimensional probability and the other in statistical learning theory.

But first, let us agree that in this section ρ and π will denote distributions on Θ and we will write $\mathbb{E}_\rho[\cdot]$ and $\mathbb{E}_\pi[\cdot]$ to mean integration over *only* the θ variable over the respective prior, that is,

$$\mathbb{E}_\rho[f_\theta(x)] \triangleq \mathbb{E}_{\theta\sim\rho}[f_\theta(x)] = \int_\Theta f_\theta(x)\rho(d\theta)$$

denotes a function of x. Similarly, $\mathbb{E}_{P_X}[f_\theta(X)]$ will denote expectation *only* over $X \sim P_X$. The following estimate is a workhorse of the PAC-Bayes method.

Proposition 4.14. (PAC-Bayes inequality) *Consider a collection of functions* $\{f_\theta : \mathcal{X} \to \mathbb{R}, \theta \in \Theta\}$ *such that* $(\theta, x) \mapsto f_\theta(x)$ *is measurable. Fix a random variable* $X \in \mathcal{X}$ *and prior* $\pi \in \mathcal{P}(\Theta)$. *Then with probability at least* $1 - \delta$ *we have for all* $\rho \in \mathcal{P}(\Theta)$:

$$\mathbb{E}_\rho[f_\theta(X) - \psi(\theta)] \le D(\rho\|\pi) + \log\frac{1}{\delta}, \qquad \psi(\theta) \triangleq \log \mathbb{E}_{P_X}[\exp\{f_\theta(X)\}]. \quad (4.33)$$

Furthermore, for any joint distribution $P_{\theta,X}$ *we have*

$$\mathbb{E}[f_\theta(X) - \psi(\theta)] \le I(\theta;X) \le D(P_{\theta|X}\|\pi|P_X). \quad (4.34)$$

Proof. We will prove the following result, known in this area as an *exponential inequality*:

$$\mathbb{E}_{P_X}\left[\sup_\rho \exp\{\mathbb{E}_\rho[f_\theta(X) - \psi(\theta)] - D(\rho\|\pi)\}\right] \le 1, \quad (4.35)$$

where the supremum inside is taken over all ρ such that $D(\rho\|\pi) < \infty$. Indeed, from it (4.33) follows via the Markov inequality. Notice that this supremum is taken over uncountably many values and hence it is not a priori clear whether the function of X under the outer expectation is even measurable. We will show the latter together with the exponential inequality.

To that end, we apply the Gibbs principle (Proposition 4.7) to the function $\theta \mapsto f_\theta(X) - \psi(\theta)$ and base measure π. Notice that this function may take value $-\infty$, but nevertheless we obtain

$$\sup_\rho \mathbb{E}_\rho[f_\theta(X) - \psi(\theta)] - D(\rho\|\pi) = \log \mathbb{E}_\pi[\exp\{f_\theta(X) - \psi(\theta)\}],$$

where the right-hand side is a measurable function of X. Exponentiating and taking expectation over X we obtain

[8] For "probably approximately correct" (PAC), as developed by Shawe-Taylor and Williamson [383], McAllester [299], Maurer [298], Catoni [83], and many others; see [16] for a survey.

$$\mathbb{E}_{P_X}\left[\sup_{\rho} \exp\{\mathbb{E}_\rho[f_\theta(X) - \psi(\theta)] - D(\rho\|\pi)\}\right] = \mathbb{E}_{P_X}\left[\mathbb{E}_\pi[\exp\{f_\theta(X) - \psi(\theta)\}]\right].$$

We claim that the right-hand side equals $\pi[\psi(\theta) < \infty] \le 1$, which completes the proof of (4.35). Indeed, let $E = \{\theta\colon \psi(\theta) < \infty\}$. Then for any $\theta \in E$ we have $\mathbb{E}_{P_X}[\exp\{f_\theta(X) - \psi(\theta)\}] = 1$, or in other words for all θ:

$$\mathbb{E}_{P_X}[\exp\{f_\theta(X) - \psi(\theta)\}]1\{\theta \in E\} = 1\{\theta \in E\}.$$

Now applying $\mathbb{E}_\pi$ to both sides here and invoking Fubini we obtain

$$\mathbb{E}_{P_X}[\mathbb{E}_\pi[\exp\{f_\theta(X) - \psi(\theta)\}]1\{\theta \in E\}] = \pi[E].$$

Finally, notice that $1\{\theta \in E\}$ can be omitted since for $\theta \in E^c$ we have $\exp\{f_\theta(X) - \infty\} = 0$ by agreement.

To show (4.34), for each x take $\rho = P_{\theta|X=x}$ in (4.35) to get

$$\mathbb{E}_{x\sim P_X}\left[\exp\{\mathbb{E}[f_\theta(X) - \psi(\theta)|X = x] - D(P_{\theta|X=x}\|\pi)\}\right] \le 1.$$

By Jensen's inequality we can move the outer expectation inside the exponent and obtain the right-most inequality in (4.34). To get the bound in terms of $I(\theta;X)$ take $\pi = P_\theta$ and recall (4.5). □

4.8.1 Uniform Convergence

As stated the PAC-Bayes inequality is too general to appreciate its importance. Its significance and depth are only revealed once the art of applying it is mastered. We first give such an example in the context of uniform convergence and high-dimensional probability.

Example 4.3. Norms of sub-Gaussian vectors [477]

Suppose $X \sim \mathcal{N}(0, \Sigma)$ takes values in $\mathbb{R}^d$. What is the magnitude of $\|X\| \equiv \|X\|_2$? First, by Jensen's inequality we get

$$\mathbb{E}[\|X\|] \le \sqrt{\mathbb{E}[\|X\|^2]} = \sqrt{\operatorname{tr}\Sigma}.$$

But what about a typical value of $\|X\|$, that is, can we show an upper bound on $\|X\|$ that holds with high probability? In order to see how PAC-Bayes could be useful here, notice that $\|X\| = \sup_{\|v\|=1} v^\top X$. Thus, we aim to use (4.33) to bound this supremum. For any v let $\rho_v = \mathcal{N}(v, \beta^2 I_d)$ and notice $v^\top X = \mathbb{E}_{\rho_v}[\theta^\top X]$. We also take $\pi = \mathcal{N}(0, \beta^2 I_d)$ and $f_\theta(x) = \lambda\theta^\top x$, $\theta \in \mathbb{R}^d$, where $\beta, \lambda > 0$ are parameters to be optimized later. Taking the base of log to be e, we compute explicitly $\psi(\theta) = \frac{1}{2}\lambda^2\theta^\top\Sigma\theta$, $\mathbb{E}_{\rho_v}[\psi(\theta)] = \frac{\lambda^2}{2}(v^\top\Sigma v + \beta^2 \operatorname{tr}\Sigma)$, and $D(\rho_v\|\pi) = \frac{\|v\|^2}{2\beta^2}$ via (2.8). Thus, using (4.33) restricted to ρ_v with $\|v\| = 1$ we obtain that with probability $\ge 1 - \delta$ we have for all v with $\|v\| = 1$

$$\lambda v^\top X \le \frac{\lambda^2}{2}(v^\top\Sigma v + \beta^2 \operatorname{tr}\Sigma) + \frac{1}{2\beta^2} + \ln\frac{1}{\delta}.$$

Now, we can optimize the right-hand side over β by choosing $\beta^2 = 1/\sqrt{\lambda^2 \operatorname{tr} \Sigma}$ to get

$$\lambda v^\top X \le \frac{\lambda^2}{2}(v^\top \Sigma v) + \lambda\sqrt{\operatorname{tr} \Sigma} + \ln\frac{1}{\delta}.$$

Finally, estimating $v^\top \Sigma v \le \|\Sigma\|_{op}$ and optimizing λ we obtain the resulting high-probability bound:

$$\|X\| \le \sqrt{\operatorname{tr} \Sigma} + \sqrt{2\|\Sigma\|_{op} \ln\frac{1}{\delta}}.$$

Although this result can be obtained using the standard technique of Chernoff bound (Section 15.1) – see [271, Lemma 1] for a stronger version or [69, Example 5.7] for general norms based on sophisticated Gaussian concentration inequalities – the advantages of the PAC-Bayes proof are that (a) it is not specific to Gaussians and holds for any X such that $\psi(\theta) \le \frac{\lambda^2}{2}\theta^\top \Sigma \theta$ (similar to sub-Gaussian random variables introduced below) and (b) its extensions can be used to analyze the concentration of sample covariance matrices [477].

To present further applications, we need to introduce a new concept.

Definition 4.15. (Sub-Gaussian random variables) A random variable X is called σ^2-sub-Gaussian if

$$\mathbb{E}[e^{\lambda(X-\mathbb{E}[X])}] \le e^{\frac{\sigma^2\lambda^2}{2}}, \qquad \forall \lambda \in \mathbb{R}.$$

Here are some useful observations:

- $\mathcal{N}(0, \sigma^2)$ is σ^2-sub-Gaussian. In fact it satisfies the condition with equality and explains the origin of the name.
- If $X \in [a, b]$, then X is $\frac{(b-a)^2}{4}$-sub-Gaussian. This is the well-known Hoeffding's lemma (see Exercise **III.22** for a proof).
- If X_i are iid and σ^2-sub-Gaussian then the empirical average $S_n = \frac{1}{n}\sum_{i=1}^n X_i$ is (σ^2/n)-sub-Gaussian.

There are many equivalent ways to define sub-Gaussianity [439, Proposition 2.5.2], including by requiring the tails of X to satisfy $\mathbb{P}[|X - \mathbb{E}[X]| > t] \le 2e^{-\frac{t^2}{2\sigma^2}}$. However, for us the most important property is the consequence of the two observations above: the empirical average of independent bounded random variables is $O(1/n)$-sub-Gaussian.

The concept of sub-Gaussianity is used in the PAC-Bayes method as follows. Suppose we have a collection of functions $\mathcal{F}$ from $\mathcal{X}$ to $\mathbb{R}$ and an iid sample $X^n \overset{\text{iid}}{\sim} P_X$. One of the main questions of *empirical process theory* and uniform convergence is to get a high-probability bound on

$$\sup_{f\in\mathcal{F}} \mathbb{E}[f(X)] - \hat{\mathbb{E}}_n[f(X)], \quad \hat{\mathbb{E}}_n[f(X)] \triangleq \frac{1}{n}\sum_{i=1}^n f(X_i).$$

Suppose that each $f \in \mathcal{F}$ takes values in $[0, 1]$. Then $(\mathbb{E} - \hat{\mathbb{E}}_n)f$ is $\frac{1}{4n}$-sub-Gaussian and applying PAC-Bayes inequality to functions $\lambda(\mathbb{E} - \hat{\mathbb{E}}_n)f(X)$ we get that with probability $\ge 1 - \delta$ for any ρ on $\mathcal{F}$ we have

$$\mathbb{E}_{f\sim\rho}[(\mathbb{E}-\hat{\mathbb{E}}_n)f(X)] \le \frac{\lambda}{8n} + \frac{1}{\lambda}\left(D(\rho\|\pi) + \ln\frac{1}{\delta}\right),$$

where π is a fixed prior on $\mathcal{F}$. This method can be used to get interesting bounds for countably infinite collections $\mathcal{F}$ (see Exercises **I.55** and **I.56**). However, the real power of this method shows when $\mathcal{F}$ is uncountable (as in the previous example for Gaussian norms).

We remark that bounding the supremum of a random process (e.g. empirical or Gaussian process) indexed by a continuous parameter is a vast subject [139, 430, 432]. The usual method is based on discretization and approximation (with a more advanced version known as chaining; see (27.22) and Exercise **V.28** for the counterpart of Gaussian processes). The PAC-Bayes inequality offers an alternative which often allows for sharper results and shorter proofs. There are also applications to high-dimensional probability (the small-ball probability and random matrix theory); see [318, 310]. In those works, PAC-Bayes is applied with π being the uniform distribution on a small ball.

Remark 4.6. (PAC-Bayes versus Rademacher complexity) Note that PAC-Bayes bounds the supremum of an empirical process. Indeed, for any value Y we can think of $\theta \mapsto f_\theta(Y)$ as a vector (random process) indexed by $\theta \in \Theta$. Each $\rho \in \mathcal{P}(\Theta)$ defines a linear function $F(\rho) \triangleq \mathbb{E}_{\theta\sim\rho}[f_\theta(Y)]$ of this vector. A modern method [38] of bounding the supremum of a collection of functions is to use *Rademacher complexity*. It turns out that PAC-Bayes can be rather naturally understood in that framework: if $|f_\theta(\cdot)| \le M$ then the set $\{F(\rho)\colon D(\rho\|\pi) \le C < \infty\}$ has Rademacher complexity $O(M\sqrt{\frac{C}{n}})$; see [240]. In fact they show that any set $\{F(\rho)\colon G(\rho) \le C\}$ satisfies this bound as long as G is strongly convex. Recall that $\rho \mapsto D(\rho\|\pi)$ is indeed strongly convex by Exercise **I.37**.

4.8.2 Generalization Bounds in Statistical Learning Theory

The original purpose of introducing PAC-Bayes was for the analysis of learning algorithms. We first introduce the standard (PAC) setting of learning theory.

Suppose one has access to a training sample $X^n = (X_1, \ldots, X_n) \overset{\text{iid}}{\sim} P$ and there is a space of parameters Θ. The goal is to find $\theta \in \Theta$ that minimizes the estimation error (loss). Specifically, the loss incurred by an estimator indexed by some parameter $\theta \in \Theta$ on the data point X_i is given by $\ell_\theta(X_i)$. Learning algorithms aim to find the minimizer of the *test loss* (also known as test error or generalization risk), that is,

$$L(\theta) \triangleq \mathbb{E}[\ell_\theta(X_{new})], \quad X_{new} \sim P.$$

Note that since P is unknown direct minimization of $L(\theta)$ is impossible. This setting encompasses many problems in machine learning and statistics. For example, we could have $X_i = (Y_i, Z_i)$, where Z_i is a feature (covariate) and Y_i is the label (response) with the loss being $1\{Y_i = f_\theta(Z_i)\}$ (in classification) or $(Y_i - f_\theta(Z_i))^2$ (in regression) and where f_θ is, for instance, a linear predictor $\theta^\top Z_i$ or a deep neural net with weights θ. Or (as in Example 4.2) we could be trying to estimate distribution P itself by optimizing cross-entropy loss $\log\frac{1}{p_\theta(X)}$ over a parametric class of densities.

As we mentioned, the value of $L(\theta)$ is not computable by the learner. What is computable is the *training error* (or empirical risk), namely

$$\hat{L}_n(\theta) \triangleq \frac{1}{n}\sum_{i=1}^{n} \ell_\theta(X_i).$$

So how does one select a good estimate $\hat{\theta}$ given training sample X^n? Here are two famous options:

- Empirical risk minimization (ERM): $\hat{\theta}_{\mathrm{ERM}} = \mathrm{argmin}_{\theta\in\Theta}\, \hat{L}_n(\theta)$.
- Gibbs sampler: Fix some $\lambda > 0$. Given some prior distribution π on Θ, draw $\hat{\theta}$ from the "posterior"

$$\rho(d\theta) \propto \pi(d\theta) e^{-\lambda \hat{L}_n(\theta)}. \tag{4.36}$$

  In the limit of $\lambda \to \infty$, this reduces to the ERM. Note that in this case $\hat{\theta}$ is a randomized function of X^n.

However, many other choices exist and are invented daily for specific problems.

The main issue that is to be addressed by theory is the following: The choice $\hat{\theta}$ is guided by the sample X^n and thus the value $\hat{L}_n(\hat{\theta})$ is probably not representative of the value $L(\hat{\theta})$ that the estimator will attain on a fresh sample X_{new} because of overfitting to the training sample. To gauge the amount of overfitting one seeks to prove an estimate of the form:

$$L(\hat{\theta}) \le \hat{L}_n(\hat{\theta}) + \text{small error terms}$$

with high probability over the sample X^n. Note that here $\hat{\theta}$ can be either deterministic (as in ERM) or a randomized (as in Gibbs) function of the training sample. In either case, it is convenient to think about the estimator as a θ drawn from a data-dependent distribution ρ, in other words, a channel from X^n to θ, so that we always understand $L(\hat{\theta})$ as $\mathbb{E}_\rho[L(\theta)]$ and $\hat{L}_n(\hat{\theta})$ as $\mathbb{E}_\rho[\hat{L}_n(\theta)]$. Note that in either case the values $L(\hat{\theta})$ and $\hat{L}_n(\hat{\theta})$ are random quantities depending on the sample X^n.

A specific kind of bound we will show is going to be a high-probability bound that holds *uniformly over all data-dependent* ρ, specifically

$$\mathbb{P}\left[\forall \rho\colon \mathbb{E}_{\theta\sim\rho}[L(\theta)] \le \mathbb{E}_{\theta\sim\rho}[\hat{L}_n(\theta)] + \text{excess risk}(\rho)\right] \ge 1-\delta,$$

for some excess risk depending on (n, ρ, δ). We emphasize that here the probability is with respect to $X^n \overset{\text{iid}}{\sim} P$ and the quantifier "$\forall\rho$" is inside the probability. Having a uniform bound like this suggests selecting that ρ which minimizes the right-hand side of the inequality, thus making the second term serve as a *regularization term* preventing overfitting.

The main theorem of this section is the following version of the *generalization error bound* of McAllester [299]. Many other similar bounds exist, for example see Exercise **I.54**.

Theorem 4.16. *Fix a reference distribution π on Θ and suppose that for all θ, x the loss $\ell_\theta(x) \in [0,1]$. Then for any $\delta \le e^{-1}$*

$$\mathbb{P}\left[\forall \rho\colon \mathbb{E}_{\theta\sim\rho}[L(\theta)] \le \mathbb{E}_{\theta\sim\rho}[\hat{L}_n(\theta)] + \frac{5}{4}\sqrt{\frac{D(\rho\|\pi) + \ln\frac{1}{\delta}}{2n}} + \frac{1}{\sqrt{10n}}\right] \ge 1-\delta. \tag{4.37}$$

The same result holds if (instead of being bounded) $\ell_\theta(X)$ is $\frac{1}{4}$-sub-Gaussian for each θ.

Before proving the theorem, let us consider a finite class Θ and argue that

$$\mathbb{P}\left[\forall \rho\colon \mathbb{E}_{\theta\sim\rho}[L(\theta)] \le \mathbb{E}_{\theta\sim\rho}[\hat{L}_n(\theta)] + \sqrt{\frac{1}{2n}\ln\frac{M}{\delta}}\right] \ge 1-\delta. \tag{4.38}$$

Indeed, by linearity, it suffices to only consider point mass distributions $\rho = \delta_\theta$. For each θ the random variable $\hat{L}_n(\theta) - L(\theta)$ is zero-mean and $\frac{1}{4n}$-sub-Gaussian (Hoeffding's lemma). Thus, applying Markov's inequality to $e^{\lambda(L(\theta)-\hat{L}_n(\theta))}$ we have for any $\lambda > 0$ and $t > 0$:

$$\mathbb{P}[L(\theta) - \hat{L}_n(\theta) \ge t] \le e^{\frac{\lambda^2}{8n}-\lambda t}.$$

Thus, setting t so that the right-hand side equals $\frac{\delta}{M}$ from the union bound we obtain that with probability $\ge 1-\delta$ simultaneously for all θ we have

$$L(\theta) - \hat{L}_n(\theta) \le \frac{\lambda}{8n} + \frac{1}{\lambda}\ln\frac{M}{\delta}.$$

Optimizing $\lambda = \sqrt{8n\ln\frac{M}{\delta}}$ yields (4.38). On the other hand, if we apply (4.37) with $\pi = \mathrm{Unif}(\Theta)$ and observe that $D(\rho\|\pi) \le \log M$, we recover (4.38) with only a slightly worse estimate of excess risk.

We can see that just like in the previous subsection, the core problem in showing Theorem 4.16 is that the union bound only applies to finite Θ and we need to work around that problem by leveraging the PAC-Bayes inequality.

Proof. First, we fix λ and apply the PAC-Bayes inequality to functions $f_\theta(X^n) = \lambda(L(\theta) - \hat{L}_n(\theta))$. By Hoeffding's lemma we know $\hat{L}_n(\theta)$ is $\frac{1}{8n}$-sub-Gaussian and thus $\psi(\theta) \le \frac{\lambda^2}{8n}$. Thus, we have with probability $\ge 1-\delta$ simultaneously for all ρ:

$$\mathbb{E}_\rho[L(\theta) - \hat{L}_n(\theta)] \le \frac{\lambda}{8n} + \frac{D(\rho\|\pi) + \ln\frac{1}{\delta}}{\lambda}. \tag{4.39}$$

Let us denote for convenience $b(\rho) = D(\rho\|\pi) + \ln\frac{1}{\delta}$. Since $\delta < e^{-1}$, we see that $b(\rho) \ge 1$. We would like to optimize λ in (4.39) by setting $\lambda = \lambda^* \triangleq \sqrt{8nb(\rho)}$. However, of course, λ cannot depend on ρ. Thus, instead we select a countable grid $\lambda_i = \sqrt{2n}2^i, i \ge 1$, and apply the PAC-Bayes inequality separately for each λ_i with probability chosen to be $\delta_i = \delta 2^{-i}$. Then from the union bound we have for all ρ and all $i \ge 1$ simultaneously:

$$\mathbb{E}_\rho[L(\theta) - \hat{L}_n(\theta)] \le \frac{\lambda_i}{8n} + \frac{b(\rho) + i\ln 2}{\lambda}.$$

Let $i^* = i^*(\rho)$ be chosen so that $\lambda^*(\rho) \le \lambda_{i^*} < 2\lambda^*(\rho)$. From the latter inequality we have $\sqrt{2n}2^{i^*} < 2\sqrt{8b(\rho)n}$ and thus

$$i^* \ln 2 < \ln 4 + \frac{1}{2}\ln(b(\rho)) \le \ln 4 - 1/2 + \frac{1}{2}b(\rho).$$

Therefore, choosing $i = i^*$ in the bound, upper-bounding $\lambda_{i^*} \le 2\lambda^*$ and $\frac{1}{\lambda_{i^*}} \le \frac{1}{\lambda^*}$ we get that for all ρ

$$\mathbb{E}_\rho[L(\theta) - \hat{L}_n(\theta)] \le \sqrt{\frac{b(\rho)}{8n}}\left(2 + \frac{3}{2}\right) + \frac{\ln 4 - 1/2}{\sqrt{8nb(\rho)}}.$$

Finally, bounding the last term by $\frac{1}{\sqrt{10n}}$ (since $b \ge 1$), we obtain the theorem. □

We remark that although we proved an optimized (over λ) bound, the intermediate result (4.39) is also quite important. Indeed, it suggests choosing ρ (the randomized estimator) based on minimizing the regularized empirical risk $\mathbb{E}_\rho \hat{L}_n(\theta) + \frac{1}{\lambda}D(\rho\|\pi)$. The minimizing ρ is just the Gibbs sampler (4.36) due to Proposition 4.7, which justifies its popularity.

PAC-Bayes bounds are often criticized on the following grounds. Suppose we take a neural network and train it (perhaps by using a version of gradient descent) until it finds some choice of weight matrices $\hat{\theta}$ that results in an acceptably low value of $\hat{L}_n(\hat{\theta})$. We would like now to apply PAC-Bayes Theorem 4.16 to convince ourselves that also the test loss $L(\hat{\theta})$ would be small. But notice that the weights of the neural network are non-random, that is, $\rho = \delta_{\hat{\theta}}$ and hence for any continuous prior π we will have $D(\rho\|\pi) = \infty$, resulting in a vacuous bound. For a while this was considered to be an unavoidable limitation, until an elegant work of [478]. There the authors argue that in the end weights of neural networks are stored as finite-bit approximations (floating point) and we can use $\pi(\theta) = 2^{-\text{length}(\theta)}$ as a prior. Here $\text{length}(\theta)$ represents the total number of bits in a compressed representation of θ. As we will learn in Part II this indeed defines a valid probability distribution (for any choice of the lossless compressor). In this way, the idea of [478] bridges the area of data compression and generalization bounds: if the trained neural network has highly compressible θ (e.g. has many zero weights) then it has smaller excess risk and thus is less prone to overfitting.

Before closing this section, let us also apply the "in expectation" version of the PAC-Bayes (4.34). Namely, again suppose that losses $\ell_\theta(x)$ are in $[0, 1]$ and suppose the learning algorithm (given X^n) selects ρ and then samples $\hat{\theta} \sim \rho$. This creates a joint distribution $P_{\hat{\theta},X^n}$. From (4.34), as in the preceding proof, for every $\lambda > 0$ we get

$$\mathbb{E}[L(\hat{\theta}) - \hat{L}_n(\hat{\theta})] \le \frac{\lambda}{8n} + \lambda I(\hat{\theta};X^n).$$

Optimizing over λ we obtain the bound

$$\mathbb{E}[L(\hat{\theta}) - \hat{L}_n(\hat{\theta})] \le \sqrt{\frac{1}{2n}I(\hat{\theta};X^n)}.$$

This version of McAllester's result [370, 462] provides a useful intuition: the algorithm's propensity to overfit can be gauged by the amount of information leaking from X^n into $\hat{\theta}$. For applications, though, a version with a flexible reference prior π, as in Theorem 4.16, appears more convenient.

5 Extremization of Mutual Information: Capacity Saddle Point

There are four fundamental optimization problems arising in information theory:

- Information projection (or I-projection): Given Q minimize $D(P\|Q)$ over convex class of P's. (See Chapter 15.)
- Maximum likelihood: Given P minimize $D(P\|Q)$ over some class of Q. (See Section 29.3.)
- Rate distortion: Given P_X minimize $I(X;Y)$ over a convex class of $P_{Y|X}$. (See Chapter 26.)
- Capacity: Given $P_{Y|X}$ maximize $I(X;Y)$ over a convex class of P_X. (This chapter.)

In this chapter we show that all these problems have convex/concave objective functions, discuss iterative algorithms for solving them, and study the capacity problem in more detail. Specifically, we will find that the supremum over input distributions P_X can also be written as the infimum over the output distributions P_Y and the resulting minimax problem has a saddle point. This will lead to understanding of capacity as the information radius of a set of conditional distributions $\{P_{Y|X=x}: x \in \mathcal{X}\}$ measured in KL divergence.

5.1 Convexity of Information Measures

Theorem 5.1. *The map $(P, Q) \mapsto D(P\|Q)$ is convex.*

Proof. Let $P_X = Q_X = \text{Ber}(\lambda)$ and define two conditional kernels:

$$P_{Y|X=0} = P_0, \quad P_{Y|X=1} = P_1,$$
$$Q_{Y|X=0} = Q_0, \quad Q_{Y|X=1} = Q_1.$$

An explicit calculation shows that

$$D(P_{X,Y}\|Q_{X,Y}) = \bar{\lambda}D(P_0\|Q_0) + \lambda D(P_1\|Q_1).$$

Therefore, from the DPI (monotonicity) we get:

$$\begin{aligned}\bar{\lambda}D(P_0\|Q_0) + \lambda D(P_1\|Q_1) &= D(P_{X,Y}\|Q_{X,Y}) \geq D(P_Y\|Q_Y)\\ &= D(\bar{\lambda}P_0 + \lambda P_1\|\bar{\lambda}Q_0 + \lambda Q_1).\end{aligned}$$

□

Remark 5.1. The proof shows that for an arbitrary measure of similarity $\mathcal{D}(P\|Q)$, the convexity of $(P,Q) \mapsto \mathcal{D}(P\|Q)$ is *equivalent* to the "conditioning increases divergence" property of $\mathcal{D}$. Convexity can also be understood as "mixing decreases divergence."

Remark 5.2. (Strict and strong convexity) There are a number of alternative arguments possible. For example, $(p,q) \mapsto p \log \frac{p}{q}$ is convex on $\mathbb{R}_+^2$, which is a manifestation of a general phenomenon: for a convex $f(\cdot)$ the *perspective* function $(p,q) \mapsto qf\left(\frac{p}{q}\right)$ is convex too. Yet another way is to invoke the Donsker–Varadhan variational representation of Theorem 4.6 and notice that the supremum of convex functions is convex. Our proof, however, allows us to immediately notice that the map $(P,Q) \mapsto D(P\|Q)$ is not *strictly convex*. Indeed, the gap in the DPI that we used in the proof is equal to $D(P_{X|Y}\|Q_{X|Y}|P_Y)$, which can be zero. For example, this happens if P_0, Q_0 have common support, which is disjoint from the common support of P_1, Q_1. At the same time the map $P \mapsto D(P\|Q)$, whose convexity was so crucial in the previous chapter, turns out to not only be strictly convex but in fact *strongly convex* with respect to total variation, see Exercise **I.37**. This strong convexity is crucial for the analysis of the *mirror descent* algorithm, which is a first-order method for optimization over probability measures (see [40, Examples 9.10 and 5.27]).

Theorem 5.2. *The map $P_X \mapsto H(X)$ is concave. Furthermore, if $P_{Y|X}$ is any channel, then $P_X \mapsto H(X|Y)$ is concave. If $\mathcal{X}$ is finite, then $P_X \mapsto H(X|Y)$ is continuous.*

Proof. For the special case of the first claim, when P_X is on a finite alphabet, the proof is complete by $H(X) = \log|\mathcal{X}| - D(P_X\|U_X)$. More generally, we prove the second claim as follows. Let $f(P_X) = H(X|Y)$. Introduce a random variable $U \sim \text{Ber}(\lambda)$ and define the transformation

$$P_{X|U} = \begin{cases} P_0, & U = 0, \\ P_1, & U = 1. \end{cases}$$

Consider the probability space $U \to X \to Y$. Then we have $f(\lambda P_1 + (1-\lambda)P_0) = H(X|Y)$ and $\lambda f(P_1) + (1-\lambda) f(P_0) = H(X|Y,U)$. Since $H(X|Y,U) \le H(X|Y)$, the proof is complete. Continuity follows from Proposition 4.13. □

Recall that $I(X;Y)$ is a function of $P_{X,Y}$, or equivalently $(P_X, P_{Y|X})$. Denote $I(P_X, P_{Y|X}) = I(X;Y)$.

Theorem 5.3. (Mutual information)

- *For fixed $P_{Y|X}$, $P_X \mapsto I(P_X, P_{Y|X})$ is concave.*
- *For fixed P_X, $P_{Y|X} \mapsto I(P_X, P_{Y|X})$ is convex.*

Proof. There are several ways to prove the first statement, all having their merits.

- *First proof*: Introduce $\theta \in \text{Ber}(\lambda)$. Define $P_{X|\theta=0} = P_X^0$ and $P_{X|\theta=1} = P_X^1$. Then $\theta \to X \to Y$. Then $P_X = \bar{\lambda} P_X^0 + \lambda P_X^1$. Also $I(X;Y) = I(X,\theta;Y) = I(\theta;Y) + I(X;Y|\theta) \ge I(X;Y|\theta)$, which is our desired $I(\bar{\lambda} P_X^0 + \lambda P_X^1, P_{Y|X}) \ge \bar{\lambda} I(P_X^0, P_{Y|X}) + \lambda I(P_X^1, P_{Y|X})$.

- *Second proof*: $I(X;Y) = \min_Q D(P_{Y|X}\|Q|P_X)$, which is a pointwise minimum of affine functions in P_X and hence concave.
- *Third proof*: Pick a Q and use the golden formula: $I(X;Y) = D(P_{Y|X}\|Q|P_X) - D(P_Y\|Q)$, where $P_X \mapsto D(P_Y\|Q)$ is convex, as the composition of the maps $P_X \mapsto P_Y$ (affine) and $P_Y \mapsto D(P_Y\|Q)$ (convex).

To prove the second (convexity) statement, simply notice that

$$I(X;Y) = D(P_{Y|X}\|P_Y|P_X).$$

The argument P_Y is a linear function of $P_{Y|X}$ and thus the statement follows from the convexity of D in the pair. □

Review: Minimax and Saddle Point

Suppose we have a bivariate function f. Then we always have the *minimax inequality*:

$$\inf_y \sup_x f(x,y) \geq \sup_x \inf_y f(x,y).$$

When does it hold with equality?

1. It turns out that minimax equality is implied by the existence of a saddle point (x^*, y^*), that is,

$$f(x,y^*) \leq f(x^*,y^*) \leq f(x^*,y) \qquad \forall x,y.$$

   Furthermore, minimax equality also implies the existence of a saddle point if inf and sup are achieved for all x, y (see [49, Section 2.6]).
2. There are a number of known criteria establishing

$$\inf_y \sup_x f(x,y) = \sup_x \inf_y f(x,y).$$

   They usually require some continuity of f, compactness of domains, and concavity in x and convexity in y. One of the most general versions is due to M. Sion [390].
3. The mother result of all this minimax theory is a theorem of von Neumann on bilinear functions: Let A and B have finite alphabets, and $g(a,b)$ be arbitrary, then

$$\min_{P_A} \max_{P_B} \mathbb{E}[g(A,B)] = \max_{P_B} \min_{P_A} \mathbb{E}[g(A,B)].$$

   Here $(x,y) \leftrightarrow (P_A, P_B)$ and $f(x,y) \leftrightarrow \sum_{a,b} P_A(a)P_B(b)g(a,b)$.
4. A more general version is: if $\mathcal{X}$ and $\mathcal{Y}$ are compact convex domains in $\mathbb{R}^n$, $f(x,y)$ continuous in (x,y), concave in x, and convex in y, then

$$\max_{x\in\mathcal{X}} \min_{y\in\mathcal{Y}} f(x,y) = \min_{y\in\mathcal{Y}} \max_{x\in\mathcal{X}} f(x,y).$$

5.2 Saddle Point of Mutual Information

The following result is a cornerstone of analyzing the maximum of mutual information over convex sets of input distributions. We will see applications immediately in this section, as well as much later in the book (in channel coding, sequential prediction, universal data compression, and density estimation).

Theorem 5.4. (Saddle point) *Let $\mathcal{P}$ be a convex set of distributions on $\mathcal{X}$. Suppose there exists $P_X^* \in \mathcal{P}$, called a* capacity-achieving input distribution, *such that*

$$\sup_{P_X \in \mathcal{P}} I(P_X, P_{Y|X}) = I(P_X^*, P_{Y|X}) \triangleq C.$$

Let $P_Y^* = P_{Y|X} \circ P_X^*$, called a capacity-achieving output distribution. *Then for all $P_X \in \mathcal{P}$ and for all Q_Y, we have*

$$D(P_{Y|X} \| P_Y^* | P_X) \leq D(P_{Y|X} \| P_Y^* | P_X^*) \leq D(P_{Y|X} \| Q_Y | P_X^*). \tag{5.1}$$

Proof. The right-hand inequality in (5.1) follows from $C = I(P_X^*, P_{Y|X}) = \min_{Q_Y} D(P_{Y|X} \| Q_Y | P_X^*)$, where the latter is (4.5).

The left-hand inequality in (5.1) is trivial when $C = \infty$. So assume that $C < \infty$, and hence $I(P_X, P_{Y|X}) < \infty$ for all $P_X \in \mathcal{P}$. Let $P_{X_\lambda} = \lambda P_X + \bar{\lambda} P_X^* \in \mathcal{P}$ and $P_{Y_\lambda} = P_{Y|X} \circ P_{X_\lambda}$. Clearly, $P_{Y_\lambda} = \lambda P_Y + \bar{\lambda} P_Y^*$, where $P_Y = P_{Y|X} \circ P_X$.

We have the following chain then:

$$\begin{aligned} C \geq I(X_\lambda; Y_\lambda) &= D(P_{Y|X} \| P_{Y_\lambda} | P_{X_\lambda}) \\ &= \lambda D(P_{Y|X} \| P_{Y_\lambda} | P_X) + \bar{\lambda} D(P_{Y|X} \| P_{Y_\lambda} | P_X^*) \\ &\geq \lambda D(P_{Y|X} \| P_{Y_\lambda} | P_X) + \bar{\lambda} C \\ &= \lambda D(P_{X,Y} \| P_X P_{Y_\lambda}) + \bar{\lambda} C, \end{aligned}$$

where inequality is by the right part of (5.1) (already shown). Thus, subtracting $\bar{\lambda} C$ and dividing by λ we get

$$D(P_{X,Y} \| P_X P_{Y_\lambda}) \leq C$$

and the proof is completed by taking $\liminf_{\lambda \to 0}$ and applying the lower semicontinuity of divergence (Theorem 4.9). □

Corollary 5.5. *In addition to the assumptions of Theorem 5.4, suppose $C < \infty$. Then the capacity-achieving output distribution P_Y^* is unique. It satisfies the property that for any P_Y induced by some $P_X \in \mathcal{P}$ (i.e. $P_Y = P_{Y|X} \circ P_X$) we have*

$$D(P_Y \| P_Y^*) \leq C < \infty \tag{5.2}$$

and in particular $P_Y \ll P_Y^*$.

Proof. The statement is: $I(P_X, P_{Y|X}) = C \Rightarrow P_Y = P_Y^*$. Indeed:

$$\begin{aligned} C = D(P_{Y|X} \| P_Y | P_X) &= D(P_{Y|X} \| P_Y^* | P_X) - D(P_Y \| P_Y^*) \\ &\leq D(P_{Y|X} \| P_Y^* | P_X^*) - D(P_Y \| P_Y^*) \\ &= C - D(P_Y \| P_Y^*) \Rightarrow P_Y = P_Y^*. \end{aligned}$$

Statement (5.2) follows from the left-hand inequality in (5.1) and the "conditioning increases divergence" property in Theorem 2.16. □

Remark 5.3.

- The finiteness of C is necessary for Corollary 5.5 to hold. For a counterexample, consider the identity channel $Y = X$, where X takes values on integers. Then any distribution with infinite entropy is a capacity-achieving input (and output) distribution.
- Unlike the output distribution, a capacity-achieving input distribution need not be unique. For example, consider $Y_1 = X_1 \oplus Z_1$ and $Y_2 = X_2$ where $Z_1 \sim \mathrm{Ber}(1/2)$ is independent of X_1. Then $\max_{P_{X_1 X_2}} I(X_1, X_2; Y_1, Y_2) = \log 2$, achieved by $P_{X_1 X_2} = \mathrm{Ber}(p) \times \mathrm{Ber}(1/2)$ for any p. Note that the capacity-achieving *output* distribution is unique: $P^*_{Y_1 Y_2} = \mathrm{Ber}(1/2) \times \mathrm{Ber}(1/2)$.

Applying Theorem 5.4 to conditional divergence gives the following result.

Corollary 5.6. (Minimax) *Under the assumptions of Theorem 5.4, we have*

$$\begin{aligned}\max_{P_X \in \mathcal{P}} I(X;Y) &= \max_{P_X \in \mathcal{P}} \min_{Q_Y} D(P_{Y|X} \| Q_Y | P_X) \\ &= \min_{Q_Y} \sup_{P_X \in \mathcal{P}} D(P_{Y|X} \| Q_Y | P_X).\end{aligned}$$

Proof. This follows from the standard property of saddle points: Maximizing/minimizing the leftmost/rightmost sides of (5.1) gives

$$\begin{aligned}\min_{Q_Y} \sup_{P_X \in \mathcal{P}} D(P_{Y|X} \| Q_Y | P_X) &\le \max_{P_X \in \mathcal{P}} D(P_{Y|X} \| P^*_Y | P_X) = D(P_{Y|X} \| P^*_Y | P^*_X) \\ &\le \min_{Q_Y} D(P_{Y|X} \| Q_Y | P^*_X) \le \max_{P_X \in \mathcal{P}} \min_{Q_Y} D(P_{Y|X} \| Q_Y | P_X),\end{aligned}$$

but by definition min max $\ge$ max min. Note that we were careful to only use max and min for the cases where we know the optimum is achievable. □

Review: Radius and Diameter

Let (X, d) be a metric space. Let A be a bounded subset.

1 *Radius* (also known as Chebyshev radius) of A: the radius of the smallest ball that covers A, that is,

$$\mathrm{rad}(A) = \inf_{y \in X} \sup_{x \in A} d(x, y). \tag{5.3}$$

2 *Diameter* of A:

$$\mathrm{diam}(A) = \sup_{x, y \in A} d(x, y). \tag{5.4}$$

Note that the radius and the diameter both measure the massiveness of a set.

3 From the definition and triangle inequality we have

$$\frac{1}{2}\text{diam}(A) \le \text{rad}(A) \le \text{diam}(A). \tag{5.5}$$

The lower and upper bounds are achieved when A is, for example, a Euclidean ball and the Hamming space, respectively.

4 In many special cases, the upper bound in (5.5) can be improved:
- A result of Bohnenblust [67] shows that in $\mathbb{R}^n$ equipped with any norm we always have $\text{rad}(A) \le \frac{n}{n+1}\text{diam}(A)$.
- For $\mathbb{R}^n$ with Euclidean distance Jung [239] proved $\text{rad}(A) \le \sqrt{\frac{n}{2(n+1)}}\text{diam}(A)$, attained by simplex. The best constant is sometimes called the Jung constant of the space.
- For $\mathbb{R}^n$ with ℓ_∞-norm the situation is even simpler: $\text{rad}(A) = \frac{1}{2}\text{diam}(A)$; such spaces are called centerable.

5.3 Capacity as Information Radius

The next simple corollary shows that the capacity of a channel (Markov kernel) is just the radius of a (finite) collection of distributions $\{P_{Y|X=x} : x \in \mathcal{X}\}$ when distances are measured by divergence (although, we recall that divergence is not a metric).

Corollary 5.7. *For any finite $\mathcal{X}$ and any kernel $P_{Y|X}$, the maximal mutual information over all distributions P_X on $\mathcal{X}$ satisfies*

$$\begin{aligned}\max_{P_X} I(X;Y) &= \max_{x\in\mathcal{X}} D(P_{Y|X=x}\|P_Y^*)\\ &= D(P_{Y|X=x}\|P_Y^*) \qquad \forall x\colon P_X^*(x) > 0.\end{aligned}$$

The last corollary gives a geometric interpretation to capacity: It equals the radius of the smallest divergence "ball" that encompasses all distributions $\{P_{Y|X=x} : x \in \mathcal{X}\}$. Moreover, the optimal center P_Y^* is a convex combination of some $P_{Y|X=x}$ and is *equidistant* to those.

The following is the information-theoretic version of "radius $\le$ diameter" (in KL divergence) for an arbitrary input space (see Theorem 32.4 for a related representation):

Corollary 5.8. *Let $\{P_{Y|X=x} : x \in \mathcal{X}\}$ be a set of distributions. Then*

$$C = \sup_{P_X} I(X;Y) \le \underbrace{\inf_{Q}\sup_{x\in\mathcal{X}} D(P_{Y|X=x}\|Q)}_{\text{radius}} \le \underbrace{\sup_{x,x'\in\mathcal{X}} D(P_{Y|X=x}\|P_{Y|X=x'})}_{\text{diameter}}.$$

Proof. By the golden formula, Corollary 4.2, we have

$$\begin{aligned}I(X;Y) &= \inf_Q D(P_{Y|X}\|Q|P_X) \le \inf_Q \sup_{x\in\mathcal{X}} D(P_{Y|X=x}\|Q)\\ &\le \inf_{x'\in\mathcal{X}}\sup_{x\in\mathcal{X}} D(P_{Y|X=x}\|P_{Y|X=x'}).\end{aligned}$$

□

5.4* Existence of Capacity-Achieving Output Distribution (General Channel)

In the previous section we have shown that the solution to

$$C = \sup_{P_X \in \mathcal{P}} I(X;Y)$$

can be (a) interpreted as a saddle point and (b) written in the minimax form; and also (c) that the capacity-achieving output distribution P_Y^* is unique. This was all done under the extra assumption that the supremum over P_X is attainable. It turns out that properties (b) and (c) can be shown without that extra assumption.

Theorem 5.9. (Kemperman [244]) *For any $P_{Y|X}$ and a convex set of distributions $\mathcal{P}$ such that*

$$C = \sup_{P_X \in \mathcal{P}} I(P_X, P_{Y|X}) < \infty, \tag{5.6}$$

there exists a unique P_Y^* with the property that

$$C = \sup_{P_X \in \mathcal{P}} D(P_{Y|X} \| P_Y^* | P_X). \tag{5.7}$$

Furthermore,

$$\begin{aligned} C &= \sup_{P_X \in \mathcal{P}} \min_{Q_Y} D(P_{Y|X} \| Q_Y | P_X) && (5.8)\\ &= \min_{Q_Y} \sup_{P_X \in \mathcal{P}} D(P_{Y|X} \| Q_Y | P_X) && (5.9)\\ &= \min_{Q_Y} \sup_{x \in \mathcal{X}} D(P_{Y|X=x} \| Q_Y), \qquad (\textit{if } \mathcal{P} = \{\textit{all } P_X\}). && (5.10) \end{aligned}$$

Note that condition (5.6) is automatically satisfied if there exists a Q_Y such that

$$\sup_{P_X \in \mathcal{P}} D(P_{Y|X} \| Q_Y | P_X) < \infty. \tag{5.11}$$

Example 5.1. Non-existence of capacity-achieving input distribution

Let $Z \sim \mathcal{N}(0, 1)$ and consider the problem

$$C = \sup_{P_X:\ \substack{\mathbb{E}[X]=0, \mathbb{E}[X^2]=P \\ \mathbb{E}[X^4]=s}} I(X; X+Z). \tag{5.12}$$

Without the constraint $\mathbb{E}[X^4] = s$, the capacity is uniquely achieved at the input distribution $P_X = \mathcal{N}(0, P)$; see Theorem 5.11. When $s \neq 3P^2$, such P_X is no longer feasible. However, for $s > 3P^2$ the maximum

$$C = \frac{1}{2} \log(1 + P)$$

is still attainable. Indeed, we can add a small "bump" to the Gaussian distribution as follows:

$$P_X = (1-p)\mathcal{N}(0,P) + p\delta_x,$$

where $p \to 0$ and $x \to \infty$ such that $px^2 \to 0$ but $px^4 \to s - 3P^2 > 0$. This shows that for the problem (5.12) with $s > 3P^2$, the capacity-achieving input distribution does not exist, but the capacity-achieving output distribution $P_Y^* = \mathcal{N}(0, 1+P)$ exists and is unique as Theorem 5.9 shows.

Proof of Theorem 5.9. Let P'_{X_n} be a sequence of input distributions achieving C, that is, $I(P'_{X_n}, P_{Y|X}) \to C$. Let $\mathcal{P}_n$ be the convex hull of $\{P'_{X_1}, \ldots, P'_{X_n}\}$. Since $\mathcal{P}_n$ is a finite-dimensional simplex, the (concave) function $P_X \mapsto I(P_X, P_{Y|X})$ is continuous (Proposition 4.13) and attains its maximum at some point $P_{X_n} \in \mathcal{P}_n$, that is,

$$I_n \triangleq I(P_{X_n}, P_{Y|X}) = \max_{P_X \in \mathcal{P}_n} I(P_X, P_{Y|X}).$$

Denote by P_{Y_n} the output distribution induced by P_{X_n}. We have then:

$$D(P_{Y_n} \| P_{Y_{n+k}}) = D(P_{Y|X} \| P_{Y_{n+k}} | P_{X_n}) - D(P_{Y|X} \| P_{Y_n} | P_{X_n}) \tag{5.13}$$

$$\le I(P_{X_{n+k}}, P_{Y|X}) - I(P_{X_n}, P_{Y|X}) \tag{5.14}$$

$$\le C - I_n, \tag{5.15}$$

where in (5.14) we applied Theorem 5.4 to $(\mathcal{P}_{n+k}, P_{Y_{n+k}})$. The crucial idea is to apply comparison of KL divergence (which is not a distance) with a true distance known as *total variation* defined in (7.3) below. Such comparisons are going to be the topic of Chapter 7. Here we take for granted the validity of Pinsker's inequality (see Theorem 7.10). According to that inequality and since $I_n \nearrow C$, we conclude that the sequence P_{Y_n} is Cauchy in total variation:

$$\sup_{k \ge 1} \mathrm{TV}(P_{Y_n}, P_{Y_{n+k}}) \to 0, \qquad n \to \infty.$$

Since the space of all probability distributions on a fixed alphabet is complete in total variation, the sequence must have a limit point $P_{Y_n} \to P_Y^*$. Convergence in TV implies weak convergence, and thus by taking a limit as $k \to \infty$ in (5.15) and applying the lower semicontinuity of divergence (Theorem 4.9) we get

$$D(P_{Y_n} \| P_Y^*) \le \lim_{k \to \infty} D(P_{Y_n} \| P_{Y_{n+k}}) \le C - I_n,$$

and therefore $P_{Y_n} \to P_Y^*$ in the (stronger) sense of $D(P_{Y_n} \| P_Y^*) \to 0$. By Theorem 4.1,

$$D(P_{Y|X} \| P_Y^* | P_{X_n}) = I_n + D(P_{Y_n} \| P_Y^*) \to C. \tag{5.16}$$

Take any $P_X \in \bigcup_{k \ge 1} \mathcal{P}_k$. Then $P_X \in \mathcal{P}_n$ for all sufficiently large n and thus by Theorem 5.4

$$D(P_{Y|X} \| P_{Y_n} | P_X) \le I_n \le C, \tag{5.17}$$

which, by the lower semicontinuity of divergence and Fatou's lemma, implies

$$D(P_{Y|X} \| P_Y^* | P_X) \le C. \tag{5.18}$$

To prove that (5.18) holds for arbitrary $P_X \in \mathcal{P}$, we may repeat the argument above with $\mathcal{P}_n$ replaced by $\tilde{\mathcal{P}}_n = \mathrm{conv}(\{P_X\} \cup \mathcal{P}_n)$, denoting the resulting sequences by $\tilde{P}_{X_n}$, $\tilde{P}_{Y_n}$ and the limit point by $\tilde{P}_Y^*$, and obtain

$$D(P_{Y_n} \| \tilde{P}_{Y_n}) = D(P_{Y|X} \| \tilde{P}_{Y_n} | P_{X_n}) - D(P_{Y|X} \| P_{Y_n} | P_{X_n}) \tag{5.19}$$

$$\leq C - I_n, \tag{5.20}$$

where (5.20) follows from (5.18) since $P_{X_n} \in \tilde{\mathcal{P}}_n$. Hence taking the limit as $n \to \infty$ we have $\tilde{P}_Y^* = P_Y^*$ and therefore (5.18) holds.

To see the uniqueness of P_Y^*, assuming there exists Q_Y^* that fulfills $C = \sup_{P_X \in \mathcal{P}} D(P_{Y|X} \| Q_Y^* | P_X)$, we show $Q_Y^* = P_Y^*$. Indeed,

$$C \geq D(P_{Y|X} \| Q_Y^* | P_{X_n}) = D(P_{Y|X} \| P_{Y_n} | P_{X_n}) + D(P_{Y_n} \| Q_Y^*) = I_n + D(P_{Y_n} \| Q_Y^*).$$

Since $I_n \to C$, we have $D(P_{Y_n} \| Q_Y^*) \to 0$. Since we have already shown that $D(P_{Y_n} \| P_Y^*) \to 0$, we conclude $P_Y^* = Q_Y^*$ (this can be seen, for example, from Pinsker's inequality and the triangle inequality $\mathrm{TV}(P_Y^*, Q_Y^*) \leq \mathrm{TV}(P_{Y_n}, Q_Y^*) + \mathrm{TV}(P_{Y_n}, P_Y^*) \to 0$).

Finally, to see (5.9), note that by definition capacity as a max-min is at most the min-max, that is,

$$C = \sup_{P_X \in \mathcal{P}} \min_{Q_Y} D(P_{Y|X} \| Q_Y | P_X) \leq \min_{Q_Y} \sup_{P_X \in \mathcal{P}} D(P_{Y|X} \| Q_Y | P_X)$$

$$\leq \sup_{P_X \in \mathcal{P}} D(P_{Y|X} \| P_Y^* | P_X) = C$$

in view of (5.16) and (5.17). □

Corollary 5.10. *Let $\mathcal{X}$ be countable and $\mathcal{P}$ a convex set of distributions on $\mathcal{X}$. If $\sup_{P_X \in \mathcal{P}} H(X) < \infty$ then*

$$\sup_{P_X \in \mathcal{P}} H(X) = \min_{Q_X} \sup_{P_X \in \mathcal{P}} \sum_x P_X(x) \log \frac{1}{Q_X(x)} < \infty$$

and the optimizer Q_X^* exists and is unique. If $Q_X^* \in \mathcal{P}$, then it is also the unique maximizer of $H(X)$.

Proof. Just apply Kemperman's theorem (Theorem 5.9) to the identity channel $Y = X$. □

Example 5.2. Max entropy

Assume that $f : \mathbb{Z} \to \mathbb{R}$ is such that $Z(\lambda) \triangleq \sum_{n \in \mathbb{Z}} \exp\{-\lambda f(n)\} < \infty$ for all $\lambda > 0$. Then

$$\max_{X:\ \mathbb{E}[f(X)] \leq a} H(X) \leq \inf_{\lambda > 0} \{\lambda a + \log Z(\lambda)\}.$$

This follows from taking

$$Q_X(n) = Z(\lambda)^{-1} \exp\{-\lambda f(n)\} \tag{5.21}$$

in Corollary 5.10. Distributions of this form are known as *Gibbs distributions* for the energy function f. This bound is often tight and achieved by $P_X(n) = Z(\lambda^*)^{-1}\exp\{-\lambda^* f(n)\}$ with λ^* being the minimizer, see Exercise **III.27**. (Note that Proposition 4.7 discusses the Lagrangian version of the same problem.)

5.5 Gaussian Saddle Point

For the additive-noise channel there is another curious saddle-point relation that we rigorously present in the next result. The proofs are based on applying the inf- and sup-characterizations of mutual information in (4.5) and (4.7), respectively. Note that we have already seen that the Gaussian distribution is extremal under covariance constraints (Theorem 2.8).

Theorem 5.11. *Let $X_g \sim \mathcal{N}(0,\sigma_X^2)$, $N_g \sim \mathcal{N}(0,\sigma_N^2)$, and $X_g \perp\!\!\!\perp N_g$. Then we have the following.*

1 *"Gaussian capacity":*

$$C = I(X_g;X_g + N_g) = \frac{1}{2}\log\left(1 + \frac{\sigma_X^2}{\sigma_N^2}\right).$$

2 *"Gaussian input is the best for Gaussian noise": For all $X \perp\!\!\!\perp N_g$ and $\mathrm{Var}\, X \le \sigma_X^2$,*

$$I(X;X + N_g) \le I(X_g;X_g + N_g), \tag{5.22}$$

with equality iff $X \stackrel{\mathrm{d}}{=} X_g$.

3 *"Gaussian noise is the worst for Gaussian input": For all N s.t. $\mathbb{E}[X_g N] = 0$ and $\mathbb{E}N^2 \le \sigma_N^2$,*

$$I(X_g;X_g + N) \ge I(X_g;X_g + N_g),$$

with equality iff $N \stackrel{\mathrm{d}}{=} N_g$ and independent of X_g.

This result encodes extremality properties of the normal distribution: for the AWGN channel, Gaussian input is the most favorable (attains the maximum mutual information, or capacity), while for a general additive-noise channel the least favorable noise is Gaussian. For a vector version of the former statement see Exercise **I.9**.

Proof. Without loss of generality, assume all random variables have zero mean. Let $Y_g = X_g + N_g$. Define

$$\begin{aligned} f(x) &\triangleq D(P_{Y_g|X_g=x}\|P_{Y_g}) = D(\mathcal{N}(x,\sigma_N^2)\|\mathcal{N}(0,\sigma_X^2+\sigma_N^2)) \\ &= \underbrace{\frac{1}{2}\log\left(1+\frac{\sigma_X^2}{\sigma_N^2}\right)}_{=C} + \frac{\log e}{2}\frac{x^2-\sigma_X^2}{\sigma_X^2+\sigma_N^2}. \end{aligned}$$

1 Compute $I(X_g;X_g+N_g) = \mathbb{E}[f(X_g)] = C$.

2 Recall the inf-representation (Corollary 4.2): $I(X;Y) = \min_Q D(P_{Y|X}\|Q|P_X)$. Then

$$I(X;X+N_g) \le D(P_{Y_g|X_g}\|P_{Y_g}|P_X) = \mathbb{E}[f(X)] \le C < \infty.$$

Furthermore, if $I(X;X+N_g) = C$, then the uniqueness of the capacity-achieving output distribution, see Corollary 5.5, implies $P_Y = P_{Y_g}$. But $P_Y = P_X * \mathcal{N}(0,\sigma_N^2)$, where $*$ denotes convolution. Then it must be that $X \sim \mathcal{N}(0,\sigma_X^2)$ simply by considering its characteristic function $\Psi_X(t) = \mathbb{E}[e^{itX}]$:

$$\Psi_X(t)\cdot e^{-\frac{1}{2}\sigma_N^2 t^2} = e^{-\frac{1}{2}(\sigma_X^2+\sigma_N^2)t^2} \Rightarrow \Psi_X(t) = e^{-\frac{1}{2}\sigma_X^2 t^2} \Longrightarrow X \sim \mathcal{N}(0,\sigma_X^2).$$

3 Let $Y = X_g + N$ and let $P_{Y|X_g}$ be the respective kernel. Note that here we only assume that N is *uncorrelated* with X_g, that is, $\mathbb{E}\left[NX_g\right] = 0$, not necessarily independent. Then

$$I(X_g;X_g+N) \ge \mathbb{E}\log\frac{dP_{X_g|Y_g}(X_g|Y)}{dP_{X_g}(X_g)} \tag{5.23}$$

$$= \mathbb{E}\log\frac{dP_{Y_g|X_g}(Y|X_g)}{dP_{Y_g}(Y)} \tag{5.24}$$

$$= C + \frac{\log e}{2}\mathbb{E}\left[\frac{Y^2}{\sigma_X^2+\sigma_N^2} - \frac{N^2}{\sigma_N^2}\right] \tag{5.25}$$

$$= C + \frac{\log e}{2}\frac{\sigma_X^2}{\sigma_X^2+\sigma_N^2}\cdot\left(1 - \frac{\mathbb{E}N^2}{\sigma_N^2}\right) \tag{5.26}$$

$$\ge C, \tag{5.27}$$

where

- (5.23) follows from (4.7),
- (5.24) $\frac{dP_{X_g|Y_g}}{dP_{X_g}} = \frac{dP_{Y_g|X_g}}{dP_{Y_g}}$,
- (5.26) $\mathbb{E}[X_g N] = 0$ and $\mathbb{E}[Y^2] = \mathbb{E}[N^2] + \mathbb{E}[X_g^2]$,
- (5.27) $\mathbb{E}N^2 \le \sigma_N^2$.

Finally, the conditions for equality in (5.23) (see (4.8)) require

$$D(P_{X_g|Y}\|P_{X_g|Y_g}|P_Y) = 0.$$

Thus, $P_{X_g|Y} = P_{X_g|Y_g}$, that is, X_g is conditionally Gaussian: $P_{X_g|Y=y} = \mathcal{N}(by, c^2)$ for some constants b and c. In other words, under $P_{X_g Y}$, we have

$$X_g = bY + cZ, \quad Z \sim \text{Gaussian} \perp\!\!\!\perp Y.$$

But then Y must be Gaussian itself by Cramér's theorem [107] or simply by considering characteristic functions:

$$\Psi_Y(t)\cdot e^{ct^2} = e^{c't^2} \Rightarrow \Psi_Y(t) = e^{c''t^2} \Longrightarrow Y \text{ is Gaussian.}$$

Therefore, (X_g, Y) must be jointly Gaussian and hence $N = Y - X_g$ is Gaussian. Thus we conclude that it is only possible to attain $I(X_g;X_g+N) = C$ if N is Gaussian of variance σ_N^2 and independent of X_g. □

5.6 Iterative Algorithms: Blahut–Arimoto, Expectation-Maximization, Sinkhorn

Although the optimization problems that we discussed above are convex (and thus would be considered algorithmically "easy" in finite dimensions), there are still clever ideas used to speed up their numerical solutions. The main underlying principle is the following *alternating minimization algorithm*:

- Optimization problem: $\min_t f(t)$.
- Assumption I: $f(t) = \min_s F(t,s)$ (i.e. f can be written as a minimum of some other function F).
- Assumption II: There exist two solvers $t^*(s) = \operatorname{argmin}_t F(t,s)$ and $s^*(t) = \operatorname{argmin}_s F(t,s)$.
- Iterative algorithm:
  - –Step 0: Fix some s_0, t_0.
  - –Step $2k-1$: $s_k = s^*(t_{k-1})$.
  - –Step $2k$: $t_k = t^*(s_k)$.

Note that there is a steady improvement at each step (the value $F(s_k, t_k)$ is decreasing), so it can be often proven that the algorithm converges to a local minimum, or even a global minimum under appropriate conditions (e.g. the convexity of f). Below we discuss several applications of this idea, and refer to [113] for proofs of convergence. In general, this class of iterative methods for maximizing and minimizing mutual information are called *Blahut–Arimoto algorithms* for their original discoverers [24, 62]. Unlike gradient ascent/descent that proceeds by small ("local") changes of the decision variable, algorithms in this section move by large ("global") jumps and hence converge much faster.

The basis of all these algorithms is the Gibbs variational principle (Proposition 4.7): for any function $c\colon \mathcal{Y} \to \mathbb{R}$ and any Q_Y on $\mathcal{Y}$, under the integrability condition $Z = \int Q_Y(dy)\exp\{-c(y)\} < \infty$, the minimum

$$\min_{P_Y} D(P_Y \| Q_Y) + \mathbb{E}_{Y \sim P_Y}[c(Y)] \tag{5.28}$$

is attained at $P_Y^*(dy) = \frac{1}{Z} Q_Y(dy)\exp\{-c(y)\}$. For simplicity below we mostly consider the case of discrete alphabets $\mathcal{X}, \mathcal{Y}$.

Maximizing Mutual Information (Capacity) We have a fixed $P_{Y|X}$ and the optimization problem

$$C = \max_{P_X} I(X;Y) = \max_{P_X} \max_{Q_{X|Y}} \mathbb{E}_{P_{X,Y}}\left[\log \frac{Q_{X|Y}}{P_X}\right],$$

where in the second equality we invoked (4.7). This results in the iterations

$$Q_{X|Y}(x|y) \leftarrow \frac{1}{Z(y)} P_X(x) P_{Y|X}(y|x),$$

$$P_X(x) \leftarrow Q'(x) \triangleq \frac{1}{Z}\exp\left\{\sum_y P_{Y|X}(y|x)\log Q_{X|Y}(x|y)\right\},$$

where $Z(y)$ and Z are normalization constants. To derive this, notice that for a fixed P_X the optimal $Q_{X|Y} = P_{X|Y}$. For a fixed $Q_{X|Y}$, we can see that

$$\mathbb{E}_{P_{X,Y}}\left[\log \frac{Q_{X|Y}}{P_X}\right] = \log Z - D(P_X \| Q'),$$

and thus the optimal $P_X = Q'$.

Denoting P_n to be the value of P_X at the nth iteration, we observe that

$$I(P_n, P_{Y|X}) \leq C \leq \sup_x D(P_{Y|X=x} \| P_{Y|X} \circ P_n). \tag{5.29}$$

This is useful since at every iteration not only do we get an estimate of the optimizer P_n, but also the gap to optimality $C - I(P_n, P_{Y|X}) \leq C -$ RHS. It can be shown, furthermore, that both the RHS and LHS in (5.29) monotonically converge to C as $n \to \infty$; see [113] for details.

Minimizing Mutual Information (Rate Distortion) We have a fixed P_X, a cost function $c(x, y)$, and the optimization problem

$$R = \min_{P_{Y|X}} I(X;Y) + \mathbb{E}[d(X,Y)] = \min_{P_{Y|X}, Q_Y} D(P_{Y|X} \| Q_Y | P_X) + \mathbb{E}[d(X,Y)], \tag{5.30}$$

where in the second equality we invoked (4.5). This minimization problem is the basis of lossy compression and will be discussed extensively in Part V. Using (5.28) we derive the iterations:

$$P_{Y|X}(y|x) \leftarrow \frac{1}{Z(x)} Q_Y(y) \exp\{-d(x,y)\},$$
$$Q_Y \leftarrow P_{Y|X} \circ P_X.$$

A sandwich bound similar to (5.29) holds here, see (5.32), so that one gets two computable sequences converging to R from above and below, as well as $P_{Y|X}$ converging to the argmin in (5.30).

EM Algorithm (Convex Case) Proposed in [121], the expectation-maximization (EM) algorithm is a heuristic for solving the maximum likelihood problem. It is known to converge to the global maximizer for convex problems. We first consider this special case (with a general one to follow next). Given a distribution P_X our goal is to minimize the divergence with respect to the mixture $Q_X = Q_{X|Y} \circ Q_Y$:

$$L = \min_{Q_Y} D(P_X \| Q_{X|Y} \circ Q_Y), \tag{5.31}$$

where $Q_{X|Y}$ is a given channel. This is a problem arising in the maximum likelihood estimation for mixture models where Q_Y is the unknown mixing distribution and $P_X = \frac{1}{n}\sum_{i=1}^n \delta_{x_i}$ is the empirical distribution of the sample $(x_1, \ldots, x_n)$. To derive an iterative algorithm for (5.31), we write

$$\min_{Q_Y} D(P_X \| Q_X) = \min_{Q_Y} \min_{P_{Y|X}} D(P_{X,Y} \| Q_{X,Y}).$$

(Note that taking $d(x,y) = -\log \frac{dQ_{X|Y}(x|y)}{dP_X(x)}$ shows that this problem is equivalent to (5.30).) By the chain rule, thus, we find the iterations

$$P_{Y|X} \leftarrow \frac{1}{Z(x)} Q_Y(y) Q_{X|Y}(x|y),$$
$$Q_Y \leftarrow P_{Y|X} \circ P_X.$$

Denote by $Q_Y^{(n)}$ the value of Q_Y at the nth iteration and $Q_X^{(n)} = Q_{X|Y} \circ Q_Y^{(n)}$. Notice that for any n and all Q_Y we have from Jensen's inequality,

$$D(P_X \| Q_X^{(n)}) - D(P_X \| Q_X) = \mathbb{E}_{X\sim P_X}\left[\log \mathbb{E}_{Y\sim Q_Y} \frac{dQ_{X|Y}}{dQ_X^{(n)}}\right] \le \mathrm{gap}(Q_X^{(n)}),$$

where $\mathrm{gap}(Q_X) \triangleq \log \mathrm{esssup}_y \mathbb{E}_{X\sim P_X}\left[\frac{dQ_{X|Y=y}}{dQ_X}\right]$. In all, we get the following sandwich bound:

$$D(P_X \| Q_X^{(n)}) - \mathrm{gap}(Q_X^{(n)}) \le L \le D(P_X \| Q_X^{(n)}), \tag{5.32}$$

and it can be shown that as $n \to \infty$ both sides converge to L, see, for example, [112, Theorem 5.3].

EM Algorithm (General Case) The EM algorithm is also applicable more broadly than (5.31), in which the quantity $Q_{X|Y}$ is fixed. In general, we consider the model where both $Q_Y^{(\theta)}$ and $Q_{X|Y}^{(\theta)}$ depend on the unknown parameter θ and the goal (see Section 29.3) is to maximize the total log-likelihood $\sum_{i=1}^n \log Q_X^{(\theta)}(x_i)$ over θ. A canonical example (which was one of the original motivations for the EM algorithm) is the k-component Gaussian mixture $Q_X^{(\theta)} = \sum_{j=1}^k w_j \mathcal{N}(\mu_j, 1)$; in other words, $Q_Y = (w_1, \ldots, w_k)$, $Q_{X|Y=j} = \mathcal{N}(\mu_j, 1)$, and $\theta = (w_1, \ldots, w_k, \mu_1, \ldots, \mu_k)$. If the centers μ_j's are known and only the weights w_j's are to be estimated, then we get the simple convex case in (5.31). Otherwise, we need to jointly optimize the log-likelihood over the centers and the weights, which is a non-convex problem.

Here, one way to approach the problem is to apply the ELBO (4.16) as follows:

$$\log Q_X^{(\theta)}(x_i) = \sup_P \mathbb{E}_{Y\sim P}\left[\log \frac{Q_{X,Y}^{(\theta)}(x_i, Y)}{P(Y)}\right].$$

Thus the maximum likelihood can be written as a double maximization problem,

$$\sup_\theta \sum_i \log Q_X^{(\theta)}(x_i) = \sup_\theta \sup_{P_{Y|X}} F(\theta, P_{Y|X}),$$

where

$$F(\theta, P_{Y|X}) = \sum_i \mathbb{E}_{Y\sim P_{Y|X=x_i}}\left[\log \frac{Q_{X,Y}^{(\theta)}(x_i, Y)}{P_{Y|X}(Y|x_i)}\right].$$

Thus, the iterative algorithm is to start with some θ and update according to

$$P_{Y|X} \leftarrow Q_{Y|X}^{(\theta)}, \quad \text{E-step},$$
$$\theta \leftarrow \operatorname*{argmax}_\theta F(\theta, P_{Y|X}), \quad \text{M-step}. \tag{5.33}$$

In general, if the log-likelihood function is non-convex in θ, EM iterations may not converge to the global optimum even with infinite sample size (see [234] for an example for three-component Gaussian mixtures). Furthermore, for certain problems in the E-step $Q^{(\theta)}_{Y|X}$ may be intractable to compute. In those cases one performs an approximate version of EM where the step $\max_{P_{Y|X}} F(\theta, P_{Y|X})$ is solved over a restricted class of distributions; see Examples 4.1 and 4.2.

Sinkhorn's Algorithm This algorithm [389] is very similar to, but not exactly the same as, the ones above. We fix $Q_{X,Y}$, two marginals V_X, V_Y, and solve the problem

$$S = \min\{D(P_{X,Y} \| Q_{X,Y}) \colon P_X = V_X, P_Y = V_Y)\}.$$

From the results of Chapter 15 (see Theorem 15.16 and Example 15.2) it is clear that the optimal distribution $P_{X,Y}$ is given by

$$P^*_{X,Y} = A(x) Q_{X,Y}(x,y) B(y),$$

for some $A, B \geq 0$. In order to find functions A, B we notice that under a fixed B the value of A that makes $P_X = V_X$ is given by

$$A(x) \leftarrow \frac{V_X(x) Q_{X,Y}(x,y) B(y)}{\sum_y Q_{X,Y}(x,y) B(y)}.$$

Similarly, to fix the Y marginal we set

$$B(y) \leftarrow \frac{A(x) Q_{X,Y}(x,y) V_Y(y)}{\sum_x A(x) Q_{X,Y}(x,y)}.$$

Sinkhorn's algorithm alternates the A and B updates until convergence.

The original version in [389] corresponds to $V_X = V_Y = \mathrm{Unif}([n])$, and the goal there was to show that any matrix $\{C_{x,y}\}$ with non-negative entries can be transformed into a doubly stochastic matrix $\{A(x) C_{x,y} B(y)\}$ by only rescaling rows and columns. The renewed interest in this classical algorithm arose from an observation that taking a jointly Gaussian $Q_{X,Y}(x,y) = c \exp\{-\|x-y\|^2/\epsilon\}$ produces a coupling $P_{X,Y}$ which resembles and approximates (as $\epsilon \to 0$) the optimal transport coupling required for computing the Wasserstein distance $W_2(V_X, V_Y)$; see [117] for more.

6 Tensorization and Information Rates

In this chapter we start with explaining the important property of mutual information known as tensorization (or single-letterization), which allows one to maximize and minimize mutual information between two high-dimensional vectors. Next, we extend the information measures discussed in previous chapters for random variables to random processes by introducing the concepts of *entropy rate* (for a stochastic process) and *mutual information rate* (for a pair of stochastic processes). For the former, it is shown that two stochastic processes that can be coupled well (i.e. have small *Ornstein's distance*) have close entropy rates – a fact to be used later in the discussion of ergodicity (see Section 12.5*). For the latter we give a simple expression for the information rate between a pair of stationary Gaussian processes in terms of their joint spectral density. This expression will be crucial much later, when we study Gaussian channels with colored noise (Section 20.6*).

6.1 Tensorization (Single-Letterization) of Mutual Information

For many applications we will be dealing with *memoryless* channels or memoryless sources. The following result is critical for extremizing mutual information in those cases.

Theorem 6.1. (Joint versus marginal mutual information)

(a) If the channel is memoryless, that is, $P_{Y^n|X^n} = \prod P_{Y_i|X_i}$, *then*

$$I(X^n;Y^n) \le \sum_{i=1}^{n} I(X_i;Y_i), \tag{6.1}$$

with equality iff $P_{Y^n} = \prod P_{Y_i}$. *Consequently, the (unconstrained) capacity is additive for memoryless channels:*

$$\max_{P_{X^n}} I(X^n;Y^n) = \sum_{i=1}^{n} \max_{P_{X_i}} I(X_i;Y_i).$$

(b) If the source is memoryless, that is, $X_1 \perp\!\!\!\perp \cdots \perp\!\!\!\perp X_n$, *then*

$$I(X^n;Y) \ge \sum_{i=1}^{n} I(X_i;Y), \tag{6.2}$$

with equality iff $P_{X^n|Y} = \prod P_{X_i|Y}$ P_Y*-almost surely.*[1] *Consequently,*

$$\min_{P_{Y^n|X^n}} I(X^n;Y^n) = \sum_{i=1}^{n} \min_{P_{Y_i|X_i}} I(X_i;Y_i).$$

Proof.

(a) Use $I(X^n;Y^n) - \sum I(X_i;Y_i) = D(P_{Y^n|X^n} \| \prod P_{Y_i|X_i} | P_{X^n}) - D(P_{Y^n} \| \prod P_{Y_i})$.
(b) Reverse the role of X and Y: $I(X^n;Y) - \sum I(X_i;Y) = D(P_{X^n|Y} \| \prod P_{X_i|Y} | P_Y) - D(P_{X^n} \| \prod P_{X_i})$. □

In short, we see that:

- For a product channel, the input maximizing the mutual information is a product distribution.
- For a product source, the channel minimizing the mutual information is a product channel.

This type of result is often known as *single-letterization* in information theory. It tremendously simplifies the optimization problem over a high-dimensional (multi-letter) problem to a scalar (single-letter) problem. For example, in the simplest case where X^n, Y^n are binary vectors, optimizing $I(X^n;Y^n)$ over P_{X^n} and $P_{Y^n|X^n}$ entails optimizing over 2^n-dimensional vectors and $2^n \times 2^n$ matrices, whereas optimizing each $I(X_i;Y_i)$ individually is easy. In analysis, the effect when some quantities extend additively to tensor powers is called *tensorization*. One of the most famous such examples is a log-Sobolev inequality; see Exercise **I.65** or [200]. Since forming a product of channels or distributions is a form of tensor power, the first part of the theorem shows that the capacity tensorizes.

Example 6.1.

Let us complement Theorem 6.1 with the following examples.

- *Expression (6.1) fails for non-product channels.* Let $X_1 \perp\!\!\!\perp X_2 \sim \text{Ber}(1/2)$. Let $Y_1 = X_1 + X_2$ (binary addition) and $Y_2 = X_1$. Then $I(X_1;Y_1) = I(X_2;Y_2) = 0$ but $I(X^2;Y^2) = 2$ bits.
- *Strict inequality in (6.1).* Consider $Y_k = X_k = U \sim \text{Ber}(1/2)$ for all k. Then $I(X_k;Y_k) = 1$ bit and $I(X^n;Y^n) = 1$ bit $< \sum I(X_k;Y_k) = n$ bits.
- *Strict inequality in (6.2).* Let $X_1 \perp\!\!\!\perp \cdots \perp\!\!\!\perp X_n$. Consider $Y_1 = X_2, Y_2 = X_3, \ldots, Y_n = X_1$. Then $I(X_k;Y_k) = 0$ for all k, and $I(X^n;Y^n) = \sum H(X_i) > 0 = \sum I(X_k;Y_k)$.

[1] That is, if $P_{X^n,Y} = P_Y \prod_{i=1}^n P_{X_i|Y}$ as joint distributions.

6.2* Gaussian Capacity via Orthogonal Symmetry

In this section we revisit the "Gaussian saddle-point" result from Theorem 5.11. There it was derived by an explicit argument. Here we demonstrate how tensorization can be used to show the extremality of Gaussian input/noise without any explicit calculations.

We start with the maximization of mutual information (capacity) question. In the notation of Theorem 5.11 we know that (for $Z \sim \mathcal{N}(0,1)$)

$$\max_{P_X:\ \mathbb{E}[X^2]\le\sigma_X^2} I(X;X+Z) = \frac{1}{2}\log\left(1+\sigma_X^2\right).$$

Note that from tensorization we also immediately get (for $Z^n \sim \mathcal{N}(0, I_n)$)

$$\max_{P_{X^n}:\ \mathbb{E}[\|X^n\|^2]\le n\sigma_X^2} I(X^n;X^n+Z^n) = \frac{n}{2}\log\left(1+\sigma_X^2\right).$$

Thus, the traditional way of solving n-dimensional problems is to solve a one-dimensional version by explicit (typically calculus of variations) computation and then apply tensorization. However, it turns out that sometimes directly solving the n-dimensional problem is magically easier and that is what we want to show next.

So, suppose that we are trying to directly solve

$$\max_{\mathbb{E}[\sum X_k^2]\le n\sigma_X^2} I(X^n;X^n+Z^n)$$

over the joint distribution P_{X^n}. By the tensorization property in Theorem 6.1(a) we get

$$\max_{\mathbb{E}[\sum X_k^2]\le n\sigma_X^2} I(X^n;X^n+Z^n) = \max_{\mathbb{E}[\sum X_k^2]\le n\sigma_X^2} \sum_{k=1}^n I(X_k;X_k+Z_k).$$

Given distributions $P_{X_1}\cdots P_{X_n}$ satisfying the constraint, form the "average of marginals" distribution $\bar{P}_X = \frac{1}{n}\sum_{k=1}^n P_{X_k}$, which also satisfies the single-letter constraint $\mathbb{E}[X^2] = \frac{1}{n}\sum_{k=1}^n \mathbb{E}[X_k^2] \le \sigma_X^2$. Then from the concavity in P_X of $I(P_X, P_{Y|X})$

$$I(\bar{P}_X;P_{Y|X}) \ge \frac{1}{n}\sum_{k=1}^n I(P_{X_k}, P_{Y|X}).$$

So $\bar{P}_X$ gives the same or better mutual information, which shows that the extremization above ought to grow linearly with n, that is,

$$\max_{\mathbb{E}[\sum X_k^2]\le n\sigma_X^2} I(X^n;X^n+Z^n) = n \max_{P_X:\ \mathbb{E}[X^2]\le\sigma_X^2} I(X;X+Z).$$

Next, let us return to $Y^n = X^n + Z^n$. Since an isotropic Gaussian is rotationally symmetric, for any orthogonal transformation $U \in O(n)$, $U\cdot(Z^n) \sim \mathcal{N}(0, I_n)$, so that $P_{UY^n|UX^n} = P_{Y^n|X^n}$, and

$$I(P_{X^n}, P_{Y^n|X^n}) = I(P_{UX^n}, P_{UY^n|UX^n}) = I(P_{UX^n}, P_{Y^n|X^n}).$$

Similarly to the "average of marginals" argument above, averaging over all orthogonal rotations U of X^n can only make the mutual information larger. Therefore, the optimal input distribution P_{X^n} can be chosen to be invariant under orthogonal transformations. Consequently, by Theorem 5.9, the (unique!) capacity-achieving output distribution $P^*_{Y^n}$ must be rotationally invariant. Furthermore, from the conditions for equality in (6.1) we conclude that $P^*_{Y^n}$ must have independent components. Since the only product distribution satisfying the power constraints and having rotational symmetry is an isotropic Gaussian, we conclude that $P_{Y^n} = (P^*_Y)^{\otimes n}$ and $P^*_Y = \mathcal{N}(0, 1 + \sigma_X^2)$. In turn, the only distribution P_X such that $P_{X+Z} = P^*_Y$ is $P_X = \mathcal{N}(0, \sigma_X^2)$ (this can be argued by considering characteristic functions).

The last part of Theorem 5.11 can also be handled similarly. That is, we can show that the minimizer in

$$\min_{P_N : \mathbb{E}[N^2]=1} I(X_G; X_G + N)$$

is necessarily Gaussian by going to a multi-dimensional problem and averaging over all orthogonal rotations.

The idea of "going up in dimension" (i.e. solving an $n = 1$ problem by going to an $n > 1$ problem first) as presented here is from [334] and only rederives something that we have already shown directly in Theorem 5.11. But the idea can also be employed for solving various non-convex differential entropy maximization problems; see [184].

6.3 Entropy Rate

Definition 6.2. The *entropy rate* of a random process $\mathbb{X} = (X_1, X_2, \ldots)$ is

$$H(\mathbb{X}) \triangleq \lim_{n\to\infty} \frac{1}{n} H(X^n) \tag{6.3}$$

provided the limit exists.

A sufficient condition for the entropy rate to exist is *stationarity*, which essentially means invariance with respect to time shift. Formally, $\mathbb{X}$ is stationary if $(X_{t_1}, \ldots, X_{t_n}) \stackrel{d}{=} (X_{t_1+k}, \ldots, X_{t_n+k})$ for any $t_1, \ldots, t_n, k \in \mathbb{N}$. This definition naturally extends to two-sided processes indexed by $\mathbb{Z}$.

Theorem 6.3. *For any stationary process $\mathbb{X} = (X_1, X_2, \ldots)$ the following hold:*

(a) $H(X_n|X^{n-1}) \le H(X_{n-1}|X^{n-2})$.
(b) $\frac{1}{n} H(X^n) \ge H(X_n|X^{n-1})$.
(c) $\frac{1}{n} H(X^n) \le \frac{1}{n-1} H(X^{n-1})$.
(d) *$H(\mathbb{X})$ exists and $H(\mathbb{X}) = \lim_{n\to\infty} \frac{1}{n} H(X^n) = \lim_{n\to\infty} H(X_n|X^{n-1})$. Both sequences converge to $H(\mathbb{X})$ from above.*
(e) *If $\mathbb{X}$ can be extended to a two-sided stationary process $\mathbb{X} = (\ldots, X_{-1}, X_0, X_1, X_2, \ldots)$, then $H(\mathbb{X}) = H(X_1|X^0_{-\infty})$ provided that $H(X_1) < \infty$.*

Proof.

(a) Further conditioning plus stationarity: $H(X_n|X^{n-1}) \le H(X_n|X_2^{n-1}) = H(X_{n-1}|X^{n-2})$.
(b) Using chain rule: $\frac{1}{n}H(X^n) = \frac{1}{n}\sum H(X_i|X^{i-1}) \ge H(X_n|X^{n-1})$.
(c) $H(X^n) = H(X^{n-1}) + H(X_n|X^{n-1}) \le H(X^{n-1}) + \frac{1}{n}H(X^n)$.
(d) $n \mapsto \frac{1}{n}H(X^n)$ is a decreasing sequence and lower-bounded by zero, hence has a limit $H(\mathbb{X})$. Moreover by chain rule, $\frac{1}{n}H(X^n) = \frac{1}{n}\sum_{i=1}^n H(X_i|X^{i-1})$. From here we claim that $H(X_n|X^{n-1})$ converges to the same limit $H(\mathbb{X})$. Indeed, from the monotonicity shown in part (a), $\lim_n H(X_n|X^{n-1}) = H'$ exists. Next, recall the following fact from calculus: if $a_n \to a$, then the Cesàro mean $\frac{1}{n}\sum_{i=1}^n a_i \to a$ as well. Thus, $H' = H(\mathbb{X})$.
(e) Assuming $H(X_1) < \infty$ we have from (4.30):

$$\lim_{n\to\infty} H(X_1) - H(X_1|X_{-n}^0) = \lim_{n\to\infty} I(X_1;X_{-n}^0) = I(X_1;X_{-\infty}^0)$$
$$= H(X_1) - H(X_1|X_{-\infty}^0). \qquad \square$$

Example 6.2. Stationary processes

Let us discuss some of the most standard examples of stationary processes.

(a) Memoryless source: If $\mathbb{X}$ is iid, then $H(\mathbb{X}) = H(X_1)$.
(b) An iid process is the simplest example of a stationary stochastic process. The next in complexity is a *mixed source*: Given two stationary (e.g. iid) processes $\mathbb{Y}$ and $\mathbb{Z}$, define another $\mathbb{X}$ as follows. Flip a coin with bias p. If head, set $\mathbb{X} = \mathbb{Y}$; if tail, set $\mathbb{X} = \mathbb{Z}$. Applying Theorem 3.4(b) yields $0 \le H(X^n) - (pH(Y^n) + \bar{p}H(Z^n)) \le \log 2$ for all n. Then $H(\mathbb{X}) = pH(\mathbb{Y}) + \bar{p}H(\mathbb{Z})$.
(c) Stationary Markov process: Let $\mathbb{X}$ be a Markov chain $X_1 \to X_2 \to X_3 \to \cdots$ with transition kernel $\mathbb{P}[X_2 = b|X_1 = a] = K(b|a)$ and initialized with an invariant distribution $X_1 \sim \mu$ (i.e. $\mu(b) = \sum_a K(b|a)\mu(a)$). Then $H(X_n|X^{n-1}) = H(X_n|X_{n-1})$ for all n and hence

$$H(\mathbb{X}) = H(X_2|X_1) = \sum_{a,b} \mu(a)K(b|a)\log\frac{1}{K(b|a)}.$$

See Exercise **I.31** for an example. This kind of process is what is called a *first-order* Markov process, since X_n depends only on X_{n-1}. There is an extension of that idea, where a kth-order Markov process is defined by a kernel $P_{X_n|X_{n-k}^{n-1}}$. Shannon classically suggested that such a process is a good model for natural language (with sufficiently large k), and a recent breakthrough in large language models [321] largely verified his vision.

Note that both of our characterizations of the entropy rate converge to the limit from above and thus evaluating $H(X_n|X^{n-1})$ or $\frac{1}{n}H(X^n)$ for arbitrary large n does not give any guarantees on the true value of $H(\mathbb{X})$ beyond an upper bound (in particular, we cannot even rule out $H(\mathbb{X}) = 0$). However, for a certain class of stationary processes, widely used in speech and language modeling, we can have a sandwich bound.

Definition 6.4. (Hidden Markov model (HMM)) Given a stationary Markov chain $\dots, S_{-1}, S_0, S_1, \dots$ on state space $\mathcal{S}$ and a Markov kernel $P_{X|S}\colon \mathcal{S} \to \mathcal{X}$, we define HMM as a stationary process $\dots, X_{-1}, X_0, X_1, \dots$ as follows. First a trajectory $S_{-\infty}^{\infty}$ is generated. Then, conditionally on it, we generate each $X_i \sim P_{X|S=S_i}$ independently. In other words, $\mathbb{X}$ is just $\mathbb{S}$ but observed over a stationary memoryless channel $P_{X|S}$ (called the *emission channel*).

One of the fundamental results in this area is due to Blackwell [60] who showed that a $\mathcal{P}(\mathcal{S})$-valued belief process $\mathbf{R}_n = (R_{s,n}, s \in \mathcal{S})$ given by $R_{s,n} \triangleq \mathbb{P}[S_n = s | X_{-\infty}^{n-1}]$ is in fact a stationary first-order Markov process. The common law μ of $\mathbf{R}_n$ (independent of n) is called the *Blackwell measure*. Although finding μ is very difficult even for the simplest processes (see example below), we do have the following representation of entropy rate in terms of μ:

$$H(\mathbb{X}) = \int_{\mathcal{P}(\mathcal{S})} \mu(d\mathbf{r})\, \mathbb{E}_{s\sim\mathbf{r}}[H(P_{X|S=s})].$$

That is, the entropy rate is an integral of a simple function $\mathbf{r} \mapsto \sum_s r_s H(P_{X|S=s})$ over μ.

Example 6.3. Gilbert–Elliott HMM [187, 151]

This is an HMM with binary states and binary emissions. Let $\mathcal{S} = \{0, 1\}$ and $\mathbb{P}[S_1 \neq S_0 | S_0] = \tau$, that is, the transition matrix of the $\mathbb{S}$ process is $\begin{pmatrix} 1-\tau & \tau \\ \tau & 1-\tau \end{pmatrix}$. Set $X_i = \mathsf{BSC}_\delta(S_i)$. In this case the Blackwell measure μ is supported on $[\tau, 1-\tau]$ and is the law of the random variable $\mathbb{P}[S_1 = 1 | X_{-\infty}^0]$ and the entropy rate can be expressed in terms of the binary entropy function h:

$$H(\mathbb{X}) = \int_0^1 \mu(dx) h(\delta\bar{x} + \bar{\delta}x),$$

where we remind the reader that $\bar{x} = 1 - x$, etc. In fact, we can express integration over μ in terms of the limit $\int f d\mu = \lim_{n\to\infty} K^n f(1/2)$, where K is the transition kernel of the belief process, which acts on functions $g\colon [0,1] \to \mathbb{R}$ as

$$Kg(x) = p(x) g\left(\frac{x\bar{\tau}\bar{\delta} + \bar{x}\tau\delta}{p(x)}\right) + \bar{p}(x) g\left(\frac{x\bar{\tau}\delta + \bar{x}\tau\bar{\delta}}{\bar{p}(x)}\right), \quad p(x) = 1 - \bar{p}(x) = \delta\bar{x} + \bar{\delta}x.$$

We can see that the belief process follows a simple fractional-linear update, but nevertheless the stationary measure μ is extremely complicated and can be either absolutely continuous or singular (fractal-like) [33, 32]. As such, understanding $H(\mathbb{X})$ as a function of (τ, δ) is a major open problem in this area. We remark, however, that if instead of the BSC we used $X = \mathsf{BEC}_\delta(S)$ then the resulting entropy rate is much easier to compute; see Exercise **I.32**.

Despite these complications, the entropy rate of HMM has a nice property: it can be tightly sandwiched between a monotonically increasing and a monotonically decreasing sequence. As we remarked above, such sandwich bound is not possible for general stationary processes.

Proposition 6.5. *Consider an HMM process $\mathbb{X}$ with state process $\mathbb{S}$. Then*

$$H(X_n|X_1^{n-1}, S_0) \le H(\mathbb{X}) \le H(X_n|X_1^{n-1}), \tag{6.4}$$

and both sides converge monotonically to $H(\mathbb{X})$ as $n \to \infty$.

Proof. The part about the upper bound we have already established. To show the lower bound, notice that

$$H(\mathbb{X}) = H(X_n|X_{-\infty}^{n-1}) \ge H(X_n|X_{-\infty}^{n-1}, S_0) = H(X_n|X_1^{n-1}, S_0),$$

where in the last step we used the Markov property $X_{-\infty}^0 \to S_0 \to X_1^n$. Next, we show that $H(X_n|X_1^{n-1}, S_0)$ is increasing in n. Indeed

$$H(X_{n+1}|X_1^n, S_0) = H(X_n|X_0^{n-1}, S_{-1}) \ge H(X_n|X_0^{n-1}, S_{-1}, S_0) = H(X_n|X_1^{n-1}, S_0),$$

where the first equality is by stationarity, the inequality is by adding conditioning (Theorem 1.4), and the last equality is due to the Markov property $(S_{-1}, X_0) \to S_0 \to X_1^n$.

Finally, we show that

$$H(\mathbb{X}) = \lim_{n\to\infty} H(X_n|X_1^{n-1}, S_0).$$

Indeed, note that by (4.30) we have

$$I(S_0;X_1^\infty) = \lim_{n\to\infty} I(S_0;X_1^n) \le H(S_0) < \infty,$$

and thus $I(S_0;X_1^n) - I(S_0;X_1^{n-1}) \to 0$. But we also have by the chain rule

$$I(S_0;X_1^n) - I(S_0;X_1^{n-1}) = I(S_0;X_n|X_1^{n-1}) = H(X_n|X_1^{n-1}) - H(X_n|X_1^{n-1}, S_0) \to 0.$$

Thus, the difference between the two sides of (6.4) vanishes with n. □

6.4 Entropy and Symbol (Bit) Error Rate

In this section we show that the entropy rates of two processes $\mathbb{X}$ and $\mathbb{Y}$ are close whenever they can be "coupled." Coupling of two processes means defining them on a common probability space so that the average distance between their realizations is small. In the following, we will require that the so-called *symbol error rate* (expected fraction of errors) is small, namely

$$\frac{1}{n}\sum_{j=1}^{n} \mathbb{P}[X_j \ne Y_j] \le \epsilon. \tag{6.5}$$

(The minimal such ϵ over all possible couplings is called *Ornstein's distance* between stochastic processes.) For a binary alphabet this quantity is known as the *bit error rate*, which is one of the performance metrics we consider for reliable data transmission in Part IV (see Section 17.1 and Section 19.6). Notice that if we define the Hamming distance as

$$d_H(x^n, y^n) \triangleq \sum_{j=1}^{n} 1\{x_j \neq y_j\} \tag{6.6}$$

then (6.5) corresponds to requiring $\mathbb{E}[d_H(X^n, Y^n)] \le n\epsilon$.

Before showing our main result, we show that Fano's inequality, Theorem 3.12, can be tensorized:

Proposition 6.6. *Let $X_1, \ldots, X_n$ take values on a finite alphabet of size M. Then*

$$H(X^n|Y^n) \le nF_M(1-\delta) = n(\delta \log(|\mathcal{X}|M - 1) + h(\delta)), \tag{6.7}$$

where the function F_M is defined in (3.14), and

$$\delta = \frac{1}{n}\mathbb{E}[d_H(X^n, Y^n)] = \frac{1}{n}\sum_{j=1}^{n} \mathbb{P}[X_j \neq Y_j].$$

Proof. For each $j \in [n]$, applying (3.18) to the Markov chain $X_j \to Y^n \to Y_j$ yields

$$H(X_j|Y^n) \le F_M(\mathbb{P}\left[X_j = Y_j\right]). \tag{6.8}$$

Then, upper-bounding the joint entropy by the sum of marginals, see (1.3), and combining with (6.8), we get

$$H(X^n|Y^n) \le \sum_{j=1}^{n} H(X_j|Y^n) \tag{6.9}$$

$$\le \sum_{j=1}^{n} F_M(\mathbb{P}[X_j = Y_j]) \tag{6.10}$$

$$\le nF_M\left(\frac{1}{n}\sum_{j=1}^{n} \mathbb{P}[X_j = Y_j]\right), \tag{6.11}$$

where in the last step we used the concavity of F_M and Jensen's inequality. □

Corollary 6.7. *Consider two processes $\mathbb{X}$ and $\mathbb{Y}$ with entropy rates $H(\mathbb{X})$ and $H(\mathbb{Y})$. If*

$$\mathbb{P}[X_j \neq Y_j] \le \epsilon$$

for every j and if $\mathbb{X}$ takes values on a finite alphabet of size M, then

$$H(\mathbb{X}) - H(\mathbb{Y}) \le F_M(1-\epsilon).$$

If both processes have alphabets of size M, then

$$|H(\mathbb{X}) - H(\mathbb{Y})| \le \epsilon \log M + h(\epsilon) \to 0 \qquad \text{as } \epsilon \to 0.$$

Proof. There is almost nothing to prove:

$$H(X^n) \le H(X^n, Y^n) = H(Y^n) + H(X^n|Y^n)$$

and apply (6.7). For the last statement just recall the expression for F_M. □

6.5 Mutual Information Rate

Extending the definition of entropy rate, the mutual information rate of two random processes $\mathbb{X} = (X_1, X_2, \ldots)$ and $\mathbb{Y} = (Y_1, Y_2, \ldots)$ is defined as follows.

Definition 6.8. (Mutual information rate)

$$I(\mathbb{X};\mathbb{Y}) = \lim_{n\to\infty} \frac{1}{n} I(X^n;Y^n)$$

provided the limit exists.

We provide an example in the context of Gaussian processes which will be useful in studying Gaussian channels with correlated noise (Section 20.6*).

Example 6.4. Gaussian processes

Consider $\mathbb{X}, \mathbb{N}$ two stationary Gaussian processes, independent of each other. Assume that their autocovariance functions are absolutely summable and thus there exist continuous power spectral density functions f_X and f_N. Without loss of generality, assume all means are zero. Let $c_X(k) = \mathbb{E}[X_1 X_{k+1}]$. Then f_X is the Fourier transform of the autocovariance function c_X, that is, $f_X(\omega) = \sum_{k=-\infty}^{\infty} c_X(k)e^{i\omega k}$, $|\omega| \le \pi$. Finally, assume $f_N \ge \delta > 0$. Then recall from Example 3.5:

$$\begin{aligned} I(X^n;X^n + N^n) &= \frac{1}{2}\log\frac{\det(\Sigma_{X^n} + \Sigma_{N^n})}{\det \Sigma_{N^n}} \\ &= \frac{1}{2}\sum_{i=1}^{n} \log \sigma_i - \frac{1}{2}\sum_{i=1}^{n} \log \lambda_i, \end{aligned}$$

where σ_i, λ_i are the eigenvalues of the covariance matrices $\Sigma_{Y^n} = \Sigma_{X^n} + \Sigma_{N^n}$ and Σ_{N^n}, which are all Toeplitz matrices, for example, $(\Sigma_{X^n})_{ij} = \mathbb{E}\left[X_i X_j\right] = c_X(i-j)$. By Szegö's theorem [199, Section 5.2]:

$$\frac{1}{n}\sum_{i=1}^{n} \log \sigma_i \to \frac{1}{2\pi}\int_{-\pi}^{\pi} \log f_Y(\omega)d\omega. \tag{6.12}$$

Note that $c_Y(k) = \mathbb{E}[(X_1 + N_1)(X_{k+1} + N_{k+1})] = c_X(k) + c_N(k)$ and hence $f_Y = f_X + f_N$. Thus, we have

$$\frac{1}{n}I(X^n;X^n + N^n) \to I(\mathbb{X};\mathbb{X} + \mathbb{N}) = \frac{1}{4\pi}\int_{-\pi}^{\pi} \log\frac{f_X(\omega) + f_N(\omega)}{f_N(\omega)}d\omega.$$

Maximizing this over f_X subject to a moment constraint leads to the famous *water-filling solution* $f_X^*(\omega) = |T - f_N(\omega)|^+$ – see Theorem 20.18.

7 f-Divergences

In Chapter 2 we introduced the KL divergence that measures dissimilarity between two distributions. This turns out to be a special case of a whole family of such measures, known as f-divergences, introduced by Csiszár [109]. Like KL divergence, f-divergences satisfy a number of useful properties:

- Operational significance: KL divergence forms a basis of information theory by yielding fundamental answers to questions in channel coding and data compression. Similarly, f-divergences such as χ^2, H^2, and TV have their foundational roles in parameter and hypothesis testing.
- Invariance to bijective transformations.
- Data-processing inequality.
- Variational representations (à la Donsker–Varadhan).
- Local behavior given by χ^2 (in nonparametric cases) or Fisher information (in parametric cases).

The purpose of this chapter is to establish these properties and prepare the ground for applications in subsequent chapters. The important highlight is a *joint range* theorem of Harremoës and Vajda [214], which gives the sharpest possible comparison inequality between arbitrary f-divergences and puts an end to a long sequence of results starting from Pinsker's inequality – Theorem 7.10. This material is especially relevant for those interested in "non-classical" applications of information theory, such as the ones we will explore in Part VI. Others can skim through this chapter and refer back to it upon need.

7.1 Definition and Basic Properties of f-Divergences

Definition 7.1. (f-Divergence) Let $f\colon (0,\infty) \to \mathbb{R}$ be a convex function with $f(1) = 0$. Let P and Q be two probability distributions on a measurable space $(\mathcal{X}, \mathcal{F})$. If $P \ll Q$ then the f-divergence is defined as

$$D_f(P\|Q) \triangleq \mathbb{E}_Q\left[f\left(\frac{dP}{dQ}\right)\right], \tag{7.1}$$

where $\frac{dP}{dQ}$ is a Radon–Nikodym derivative and $f(0) \triangleq f(0+)$. More generally, let $f'(\infty) \triangleq \lim_{x\downarrow 0} xf(1/x)$. Suppose that $Q(dx) = q(x)\mu(dx)$ and $P(dx) = p(x)\mu(dx)$ for some common dominating measure μ, then we have

$$D_f(P\|Q) = \int_{q>0} q(x)f\left(\frac{p(x)}{q(x)}\right)d\mu + f'(\infty)P[q=0] \tag{7.2}$$

with the agreement that if $P[q=0]=0$ the last term is taken to be zero regardless of the value of $f'(\infty)$ (which could be infinite).

Remark 7.1. For the discrete case, with $Q(x)$ and $P(x)$ being the respective PMFs, we can also write

$$D_f(P\|Q) = \sum_x Q(x)f\left(\frac{P(x)}{Q(x)}\right)$$

with the understanding that

- $f(0) = f(0+)$,
- $0f(\frac{0}{0}) = 0$, and
- $0f(\frac{a}{0}) = \lim_{x\downarrow 0} xf(\frac{a}{x}) = af'(\infty)$ for $a>0$.

Remark 7.2. A nice property of $D_f(P\|Q)$ is that the definition is invariant to the choice of the dominating measure μ in (7.2). This is not the case for other dissimilarity measures, such as the squared L_2-distance between the densities $\|p-q\|^2_{L^2(d\mu)}$ which is a popular loss function for density estimation in statistics literature (cf. Section 32.4).

The following are common f-divergences:

- **Kullback–Leibler (KL) divergence**: We recover the usual $D(P\|Q)$ in Chapter 2 by taking $f(x) = x\log x$.
- **Total variation**: $f(x) = \frac{1}{2}|x-1|$,

$$\mathrm{TV}(P,Q) \triangleq \frac{1}{2}\mathbb{E}_Q\left[\left|\frac{dP}{dQ}-1\right|\right] = \frac{1}{2}\int |dP-dQ| = 1-\int d(P\wedge Q). \tag{7.3}$$

  Moreover, $\mathrm{TV}(\cdot,\cdot)$ is a metric on the space of probability distributions.[1]
- **χ^2-divergence**: $f(x) = (x-1)^2$,

$$\chi^2(P\|Q) \triangleq \mathbb{E}_Q\left[\left(\frac{dP}{dQ}-1\right)^2\right] = \int\frac{(dP-dQ)^2}{dQ} = \int\frac{dP^2}{dQ}-1. \tag{7.4}$$

  Note that we can also choose $f(x) = x^2-1$. Indeed, f's differing by a linear term lead to the same f-divergence, see Proposition 7.2.
- **Squared Hellinger distance**: $f(x) = \left(1-\sqrt{x}\right)^2$,

$$H^2(P,Q) \triangleq \mathbb{E}_Q\left[\left(1-\sqrt{\frac{dP}{dQ}}\right)^2\right] = \int\left(\sqrt{dP}-\sqrt{dQ}\right)^2 = 2-2\int\sqrt{dPdQ}. \tag{7.5}$$

[1] In (7.3), $\int d(P\wedge Q)$ is the usual shorthand for $\int(\frac{dP}{d\mu}\wedge\frac{dQ}{d\mu})d\mu$ where μ is any dominating measure. The expressions in (7.4) and (7.5) are understood in the similar sense.

Here the quantity $B(P, Q) \triangleq \int \sqrt{dPdQ}$ is known as the *Bhattacharyya coefficient* (or Hellinger affinity) [52]. Note that $H(P, Q) = \sqrt{H^2(P, Q)}$ defines a metric on the space of probability distributions: indeed, the triangle inequality follows from that of $L_2(\mu)$ for a common dominating measure. Note, however, that

$$P \mapsto H(P, Q) \text{ is not convex.} \tag{7.6}$$

(This is because the metric H is not induced by a Banach norm on the space of measures.) For an explicit example, consider $p \mapsto H(\text{Ber}(p), \text{Ber}(0.1))$.

- **Le Cam divergence (distance)** [274, p. 47]: $f(x) = \frac{1-x}{2x+2}$,

$$\text{LC}(P, Q) = \frac{1}{2} \int \frac{(dP - dQ)^2}{dP + dQ}. \tag{7.7}$$

Moreover, $\sqrt{\text{LC}(P\|Q)}$ is a metric on the space of probability distributions [152], known as Le Cam distance.

- **Jensen–Shannon divergence**: $f(x) = x \log \frac{2x}{x+1} + \log \frac{2}{x+1}$,

$$\text{JS}(P, Q) = D\left(P \middle\| \frac{P+Q}{2}\right) + D\left(Q \middle\| \frac{P+Q}{2}\right). \tag{7.8}$$

Moreover, $\sqrt{\text{JS}(P\|Q)}$ is a metric on the space of probability distributions [152].

Remark 7.3. If $D_f(P\|Q)$ is an f-divergence, then it is easy to verify that $D_f(\lambda P + \bar{\lambda} Q\|Q)$ and $D_f(P\|\lambda P + \bar{\lambda} Q)$ are f-divergences for all $\lambda \in [0, 1]$. In particular, $D_f(Q\|P) = D_{\tilde{f}}(P\|Q)$ with $\tilde{f}(x) \triangleq x f(\frac{1}{x})$.

We start by summarizing some formal observations about f-divergences.

Proposition 7.2. (Basic properties) *The following hold:*

1. $D_{f_1+f_2}(P\|Q) = D_{f_1}(P\|Q) + D_{f_2}(P\|Q)$.
2. $D_f(P\|P) = 0$.
3. $D_f(P\|Q) = 0$ *for all* $P \neq Q$ *iff* $f(x) = c(x-1)$ *for some* c. *For any other* f *we have* $D_f(P\|Q) = f(0) + f'(\infty) > 0$ *for* $P \perp Q$.
4. *If* $P_{X,Y} = P_X P_{Y|X}$ *and* $Q_{X,Y} = P_X Q_{Y|X}$ *then the function* $x \mapsto D_f(P_{Y|X=x}\|Q_{Y|X=x})$ *is measurable and*

$$D_f(P_{X,Y}\|Q_{X,Y}) = \int_{\mathcal{X}} dP_X(x) D_f(P_{Y|X=x}\|Q_{Y|X=x}) \triangleq D_f(P_{Y|X}\|Q_{Y|X}|P_X), \tag{7.9}$$

the latter being referred to as the conditional f-divergence (similar to Definition 2.14 for conditional KL divergence).

5. *If* $P_{X,Y} = P_X P_{Y|X}$ *and* $Q_{X,Y} = Q_X P_{Y|X}$ *then*

$$D_f(P_{X,Y}\|Q_{X,Y}) = D_f(P_X\|Q_X). \tag{7.10}$$

In particular,

$$D_f(P_X P_Y\|Q_X P_Y) = D_f(P_X\|Q_X). \tag{7.11}$$

6 *Let* $f_1(x) = f(x) + c(x-1)$, *then*

$$D_{f_1}(P\|Q) = D_f(P\|Q) \qquad \forall P, Q.$$

In particular, we can always assume that $f \geq 0$ *and (if* f *is differentiable at 1) that* $f'(1) = 0$.

Proof. The first and second are clear. For the third property, verify explicitly that $D_f(P\|Q) = 0$ for $f = c(x-1)$. Next consider general f and observe that for $P \perp Q$, by definition we have

$$D_f(P\|Q) = f(0) + f'(\infty), \tag{7.12}$$

which is well defined (i.e. $\infty - \infty$ is not possible) since by convexity $f(0) > -\infty$ and $f'(\infty) > -\infty$. So all we need to verify is that $f(0) + f'(\infty) = 0$ if and only if $f = c(x-1)$ for some $c \in \mathbb{R}$. Indeed, since $f(1) = 0$, the convexity of f implies that $x \mapsto g(x) \triangleq \frac{f(x)}{x-1}$ is non-decreasing. By assumption, we have $g(0+) = g(\infty)$ and hence $g(x)$ is a constant on $x > 0$, as desired.

For property 4, let $R_{Y|X} = \frac{1}{2}P_{Y|X} + \frac{1}{2}Q_{Y|X}$. By Theorem 2.12 there exist jointly measurable $p(y|x)$ and $q(y|x)$ such that $dP_{Y|X=x} = p(y|x)dR_{Y|X=x}$ and $Q_{Y|X} = q(y|x)dR_{Y|X=x}$. We can then take μ in (7.2) to be $\mu = P_X R_{Y|X}$, which gives $dP_{X,Y} = p(y|x)d\mu$ and $dQ_{X,Y} = q(y|x)d\mu$ and thus

$$\begin{aligned}
&D_f(P_{X,Y}\|Q_{X,Y}) \\
&= \int_{\mathcal{X}\times\mathcal{Y}} d\mu 1\{y\colon q(y|x) > 0\}\, q(y|x) f\left(\frac{p(y|x)}{q(y|x)}\right) + f'(\infty)\int_{\mathcal{X}\times\mathcal{Y}} d\mu 1\{y\colon q(y|x) = 0\}\, p(y|x) \\
&\overset{(7.2)}{=} \int_{\mathcal{X}} dP_X \underbrace{\left\{\int_{\{y\colon q(y|x)>0\}} dR_{Y|X=x}\, q(y|x) f\left(\frac{p(y|x)}{q(y|x)}\right) + f'(\infty)\int_{\{y\colon q(y|x)=0\}} dR_{Y|X=x}\, p(y|x)\right\}}_{D_f(P_{Y|X=x}\|Q_{Y|X=x})},
\end{aligned}$$

which is the desired (7.9).

Property 5 follows from the observation: if we take $\mu = P_{X,Y} + Q_{X,Y}$ and $\mu_1 = P_X + Q_X$ then $\frac{dP_{X,Y}}{d\mu} = \frac{dP_X}{d\mu_1}$ and similarly for Q.

Property 6 follows from the first and the third. Note also that reducing to $f \geq 0$ is done by taking $c = f'(1)$ (or any subdifferential at $x = 1$ if f is not differentiable). □

7.2 Data-Processing Inequality; Approximation by Finite Partitions

Theorem 7.3. (Monotonicity)

$$D_f(P_{X,Y}\|Q_{X,Y}) \geq D_f(P_X\|Q_X). \tag{7.13}$$

Proof. Note that in the case $P_{X,Y} \ll Q_{X,Y}$ (and thus $P_X \ll Q_X$), the proof is a simple application of Jensen's inequality to Definition 7.1:

$$
\begin{aligned}
D_f(P_{X,Y}\|Q_{X,Y}) &= \mathbb{E}_{X\sim Q_X}\,\mathbb{E}_{Y\sim Q_{Y|X}}\left[f\left(\frac{dP_{Y|X}P_X}{dQ_{Y|X}Q_X}\right)\right] \\
&\geq \mathbb{E}_{X\sim Q_X}\left[f\left(\mathbb{E}_{Y\sim Q_{Y|X}}\left[\frac{dP_{Y|X}P_X}{dQ_{Y|X}Q_X}\right]\right)\right] \\
&= \mathbb{E}_{X\sim Q_X}\left[f\left(\frac{dP_X}{dQ_X}\right)\right].
\end{aligned}
$$

To prove the general case we need to be more careful. Let $R_X = \frac{1}{2}(P_X+Q_X)$ and $R_{Y|X} = \frac{1}{2}P_{Y|X} + \frac{1}{2}Q_{Y|X}$. It should be clear that $P_{X,Y}, Q_{X,Y} \ll R_{X,Y} \triangleq R_X R_{Y|X}$ and that for every x: $P_{Y|X=x}, Q_{Y|X=x} \ll R_{Y|X=x}$. By Theorem 2.12 there exist measurable functions p_1, p_2, q_1, q_2 so that

$$
dP_{X,Y} = p_1(x)p_2(y|x)dR_{X,Y}, \quad dQ_{X,Y} = q_1(x)q_2(y|x)dR_{X,Y},
$$

and $dP_{Y|X=x} = p_2(y|x)dR_{Y|X=x}$, $dQ_{Y|X=x} = q_2(y|x)dR_{Y|X=x}$. We also denote $p(x,y) = p_1(x)p_2(y|x)$, $q(x,y) = q_1(x)q_2(y|x)$.

Fix $t > 0$ and consider a supporting line to f at t with slope μ, so that

$$
f(u) \geq f(t) + \mu(u-t), \qquad \forall u \geq 0.
$$

Thus, $f'(\infty) \geq \mu$ and taking $u = \lambda t$ for any $\lambda \in [0,1]$ we have shown:

$$
f(\lambda t) + \bar{\lambda} t f'(\infty) \geq f(t), \qquad \forall t \geq 0, \lambda \in [0,1]. \tag{7.14}
$$

Note that we added the $t = 0$ case as well, since for $t = 0$ the statement is obvious (recall, though, that $f(0) \triangleq f(0+)$ can be equal to $+\infty$).

Next, fix some x with $q_1(x) > 0$ and consider the chain

$$
\begin{aligned}
&\int_{\{y:\, q_2(y|x)>0\}} dR_{Y|X=x}\, q_2(y|x) f\left(\frac{p_1(x)p_2(y|x)}{q_1(x)q_2(y|x)}\right) + \frac{p_1(x)}{q_1(x)} P_{Y|X=x}[q_2(Y|x)=0] f'(\infty) \\
&\overset{(a)}{\geq} f\left(\frac{p_1(x)}{q_1(x)} P_{Y|X=x}[q_2(Y|x)>0]\right) + \frac{p_1(x)}{q_1(x)} P_{Y|X=x}[q_2(Y|x)=0] f'(\infty) \\
&\overset{(b)}{\geq} f\left(\frac{p_1(x)}{q_1(x)}\right),
\end{aligned}
$$

where (a) is by Jensen's inequality and the convexity of f, and (b) by taking $t = \frac{p_1(x)}{q_1(x)}$ and $\lambda = P_{Y|X=x}[q_2(Y|x) > 0]$ in (7.14). Now multiplying the obtained inequality by $q_1(x)$ and integrating over $\{x\colon q_1(x) > 0\}$ we get

$$
\begin{aligned}
&\int_{\{q>0\}} dR_{X,Y}\, q(x,y) f\left(\frac{p(x,y)}{q(x,y)}\right) + f'(\infty) P_{X,Y}[q_1(X) > 0, q_2(Y|X) = 0] \\
&\quad \geq \int_{\{q_1>0\}} dR_X\, q_1(x) f\left(\frac{p_1(x)}{q_1(x)}\right).
\end{aligned}
$$

Adding $f'(\infty)P_X[q_1(X) = 0]$ to both sides we obtain (7.13) since both sides evaluate to the definition (7.2). □

The following is the main result of this section.

Theorem 7.4. (Data processing) *Consider a channel that produces Y given X based on the conditional law $P_{Y|X}$ (shown below).*

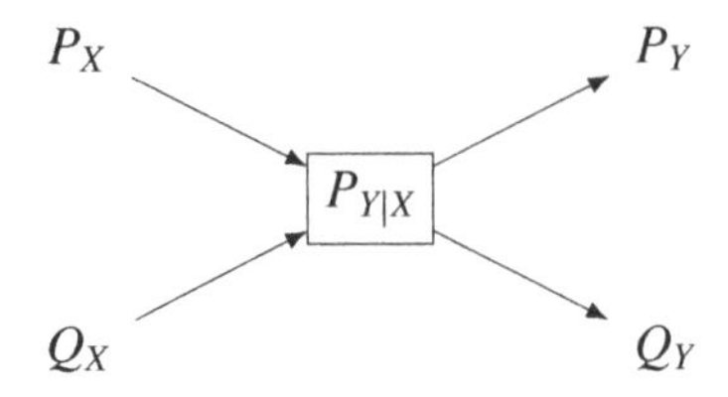

Let P_Y (respectively Q_Y) denote the distribution of Y when X is distributed as P_X (respectively Q_X). For any f-divergence $D_f(\cdot\|\cdot)$,

$$D_f(P_Y\|Q_Y) \le D_f(P_X\|Q_X). \tag{7.15}$$

Proof. This follows from the monotonicity property (7.13) and (7.10). □

Next we discuss some of the more useful properties of f-divergence that parallel those of KL divergence in Theorem 2.16.

Theorem 7.5. (Properties of f-divergences)

(a) Non-negativity: $D_f(P\|Q) \ge 0$. If f is strictly convex[2] *at 1, then $D_f(P\|Q) = 0$ if and only if $P = Q$.*
(b) Joint convexity: $(P, Q) \mapsto D_f(P\|Q)$ is a jointly convex function. Consequently, $P \mapsto D_f(P\|Q)$ and $Q \mapsto D_f(P\|Q)$ are also convex.
(c) Conditioning increases f-divergence: Let $P_Y = P_{Y|X} \circ P_X$ and $Q_Y = Q_{Y|X} \circ P_X$, or, pictorially,

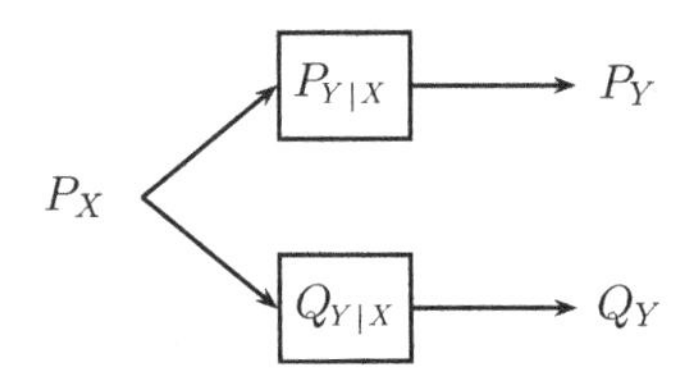

then

$$D_f(P_Y\|Q_Y) \le D_f\left(P_{Y|X}\|Q_{Y|X}|P_X\right).$$

Proof.

(a) Non-negativity follows from monotonicity by taking X to be unary. To show strict positivity, suppose for the sake of contradiction that $D_f(P\|Q) = 0$ for some $P \ne Q$. Then there exists some measurable A such that $p = P(A) \ne q = Q(A) > 0$. Applying the data-processing inequality (with $Y = 1\{X \in A\}$), we obtain $D_f(\text{Ber}(p)\|\text{Ber}(q)) = 0$. Consider two cases.

[2] By strict convexity at 1, we mean for all $s, t \in [0, \infty)$ and $\alpha \in (0, 1)$ such that $\alpha s + \bar{\alpha} t = 1$, we have $\alpha f(s) + \bar{\alpha} f(t) > f(1)$.

(i) $0 < q < 1$: Then $D_f(\mathrm{Ber}(p)\|\mathrm{Ber}(q)) = qf\left(\frac{p}{q}\right) + \bar{q}f\left(\frac{\bar{p}}{\bar{q}}\right) = f(1)$.

(ii) $q = 1$: Then $p < 1$ and $D_f(\mathrm{Ber}(p)\|\mathrm{Ber}(q)) = f(p) + \bar{p}f'(\infty) = 0$, that is, $f'(\infty) = \frac{f(p)}{p-1}$. Since $x \mapsto \frac{f(x)}{x-1}$ is non-decreasing, we conclude that f is affine on $[p, \infty)$.

Both cases contradict the assumed strict convexity of f at 1.

(b) Convexity follows from the DPI as in the proof of Theorem 5.1.

(c) Recall that the conditional divergence was defined in (7.9) and hence the inequality follows from monotonicity. Another way to see the inequality is as the result of applying Jensen's inequality to the jointly convex function $D_f(P\|Q)$. □

Remark 7.4. (Strict convexity) Just like for the KL divergence, f-divergences are never strictly convex in the sense that $(P, Q) \mapsto D_f(P\|Q)$ can be linear on an interval connecting (P_0, Q_0) to (P_1, Q_1). As in Remark 5.2 this is the case when (P_0, Q_0) have support disjoint from (P_1, Q_1). However, for f-divergences this can happen even with pairs with a common support. For example, $\mathrm{TV}(\mathrm{Ber}(p), \mathrm{Ber}(q)) = |p - q|$ is piecewise linear. In turn, strict convexity of f is related to certain desirable properties of f-information $I_f(X;Y)$; see Exercise **I.40**.

Remark 7.5. (g-Divergences) We note that, more generally, we may call functional $\mathcal{D}(P\|Q)$ a "g-divergence," or a generalized dissimilarity measure, if it satisfies the following properties: positivity, monotonicity (as in (7.13)), data-processing inequality (DPI, see (7.15)), and $\mathcal{D}(P\|P) = 0$ for any P. Note that the last three properties imply positivity by taking X to be unary in the DPI. In many ways g-divergence properties allow one to interpret it as a measure of information in the generic sense adopted in this book. We have seen that f-divergences satisfy two additional properties: conditioning increases divergence (CID) and convexity in the pair, the two being essentially equivalent (see proof of Theorem 5.1). CID and convexity do not necessarily hold for any f-divergence. Indeed, any monotone function of an f-divergence is a g-divergence, and of course those do not need to be monotone (see (7.6) for an example). Interestingly, there exist g-divergences which are not monotone transformations of any f-divergence, see [339, Section V]; the example there is in fact $\mathcal{D}(P\|Q) = \alpha - \beta_\alpha(P, Q)$ with β_α defined in (14.3) later in the book. On the other hand, for finite alphabets, [326] shows that any $\mathcal{D}(P\|Q) = \sum_i \phi(P_i, Q_i)$ is a g-divergence iff it is an f-divergence.

The following convenient property, a counterpart of Theorem 4.5, allows us to reduce any general problem about f-divergences to the problem on finite alphabets. The proof is in Section 7.14*.

Theorem 7.6. *Let P, Q be two probability measures on $\mathcal{X}$ with σ-algebra $\mathcal{F}$. Given finite $\mathcal{F}$-measurable partitions $\mathcal{E} = \{E_1, \ldots, E_n\}$ define the distribution $P_\mathcal{E}$ on $[n]$ by $P_\mathcal{E}(i) = P[E_i]$ and $Q_\mathcal{E}(i) = Q[E_i]$. Then*

$$D_f(P\|Q) = \sup_{\mathcal{E}} D_f(P_\mathcal{E}\|Q_\mathcal{E}), \tag{7.16}$$

where the supremum is over all finite $\mathcal{F}$-measurable partitions $\mathcal{E}$.

7.3 Total Variation and Hellinger Distance in Hypothisis Testing

As we will discover throughout the book, different f-divergences have different operational significance. For example, χ^2-divergence is useful in the study of Markov chains (see Example 33.8 and Exercise **VI.19**); in estimation the Bayes quadratic risk for a binary prior is determined by Le Cam divergence (7.7). Here we discuss the relation of TV and Hellinger H^2 to the problem of binary hypothesis testing. We will delve deep into this problem in Part III (and return to its composite version in Part VI). In this section, we only introduce some basics for the purpose of motivation and illustration.

The *binary hypothesis testing* problem is formulated as follows: One is given an observation (random variable) X, and it is known that either $X \sim P$ (a case referred to as null hypothesis H_0) or $X \sim Q$ (alternative hypothesis H_1). The goal is to decide, on the basis of X alone, which of the two hypotheses holds. In other words, we want to find a (possibly randomized) decision function $\phi\colon \mathcal{X} \to \{0, 1\}$ such that the sum of two types of probabilities of error,

$$P[\phi(X) = 1] + Q[\phi(X) = 0], \tag{7.17}$$

is minimized.

In this section we first show that optimization over ϕ naturally leads to the concept of TV. Subsequently, we will see that asymptotic considerations (when P and Q are replaced with $P^{\otimes n}$ and $Q^{\otimes n}$) leads to H^2. We start with the former case.

Theorem 7.7.

(a) sup-*representation of* TV*:*

$$\mathrm{TV}(P, Q) = \sup_E P(E) - Q(E) = \frac{1}{2} \sup_{f \in \mathcal{F}} \mathbb{E}_P[f(X)] - \mathbb{E}_Q[f(X)], \tag{7.18}$$

where the first supremum is over all measurable sets E, and the second is over $\mathcal{F} = \{f\colon \mathcal{X} \to \mathbb{R}, \|f\|_\infty \le 1\}$. In particular, the minimal total error probability in (7.17) is given by

$$\min_\phi \{P[\phi(X) = 1] + Q[\phi(X) = 0]\} = 1 - \mathrm{TV}(P, Q), \tag{7.19}$$

where the minimum is over all decision rules $\phi\colon \mathcal{X} \to \{0, 1\}$.[3]

(b) inf-*representation of* TV *[404]:*[4] *Provided that the diagonal $\{(x, x)\colon x \in \mathcal{X}\}$ is measurable,*

$$\mathrm{TV}(P, Q) = \min_{P_{X,Y}} \{P_{X,Y}[X \ne Y]\colon P_X = P, P_Y = Q\}, \tag{7.20}$$

where minimization is over joint distributions $P_{X,Y}$ with the property $P_X = P$ and $P_Y = Q$, which are called couplings of P and Q.

[3] The extension of (7.19) from simple to composite hypothesis testing is in (32.28).

[4] See Exercise **I.36** for another inf-representation.

Proof. Let p, q, μ be as in Definition 7.1. Then for any $f \in \mathcal{F}$ we have

$$\int f(x)(p(x) - q(x))d\mu \leq \int |p(x) - q(x)|d\mu = 2\mathrm{TV}(P, Q),$$

which establishes that the second supremum in (7.18) lower-bounds TV, and hence (by taking $f(x) = 2 \cdot 1_E(x) - 1$) so does the first. For the other direction, let $E = \{x : p(x) > q(x)\}$ and notice

$$0 = \int (p(x) - q(x))d\mu = \int_E + \int_{E^c} (p(x) - q(x))d\mu,$$

implying that $\int_{E^c}(q(x) - p(x))d\mu = \int_E(p(x) - q(x))d\mu$. But the sum of these two integrals precisely equals $2 \cdot \mathrm{TV}$, which implies that this choice of E attains equality in (7.18).

For the inf-representation, we notice that given a coupling $P_{X,Y}$, for any $\|f\|_\infty \leq 1$, we have

$$\mathbb{E}_P[f(X)] - \mathbb{E}_Q[f(X)] = \mathbb{E}[f(X) - f(Y)] \leq 2P_{X,Y}[X \neq Y],$$

which, in view of (7.18), shows that the inf-representation is always an upper bound. To show that this bound is tight one constructs X, Y as follows: with probability $\pi \triangleq \int \min(p(x), q(x))d\mu$ we take $X = Y = c$ with c sampled from a distribution with density $r(x) = \frac{1}{\pi}\min(p(x), q(x))$, whereas with probability $1 - \pi$ we take X, Y sampled independently from distributions $p_1(x) = \frac{1}{1-\pi}(p(x) - \min(p(x), q(x)))$ and $q_1(x) = \frac{1}{1-\pi}(q(x) - \min(p(x), q(x)))$ respectively. The result follows upon verifying that this $P_{X,Y}$ indeed defines a coupling of P and Q and applying the last identity of (7.3). □

Remark 7.6. (Variational representation) The sup-representation (7.18) of the total variation will be extended to general f-divergences in Section 7.13. However, only the TV has the representation of the form $\sup_{f \in \mathcal{F}} |\mathbb{E}_P[f] - \mathbb{E}_Q[f]|$ over a class of test functions. Distances of this form (for different classes of $\mathcal{F}$) are sometimes known as integral probability metrics (IPMs). And so TV is an example of an IPM for the class $\mathcal{F}$ of all bounded functions.

In turn, the inf-representation (7.20) has no analogs for other f-divergences, with the notable exception of Marton's d_2; see Remark 7.15. Distances defined via inf-representations over couplings are often called *Wasserstein distances*, and hence we may think of TV as the Wasserstein distance with respect to Hamming distance $d(x, x') = 1\{x \neq x'\}$ on $\mathcal{X}$. The benefit of variational representations is that choosing a particular coupling in (7.20) gives an upper bound on $\mathrm{TV}(P, Q)$, and choosing a particular f in (7.18) yields a lower bound.

Of particular relevance is the special case of testing with multiple observations, where the data $X = (X_1, \ldots, X_n)$ are iid drawn from either P or Q. In other words, the goal is to test

$$H_0 : X \sim P^{\otimes n} \quad \text{versus} \quad H_1 : X \sim Q^{\otimes n}.$$

By Theorem 7.7, the optimal total probability of error is given by $1-\mathrm{TV}(P^{\otimes n}, Q^{\otimes n})$. By the data-processing inequality, $\mathrm{TV}(P^{\otimes n}, Q^{\otimes n})$ is a non-decreasing sequence in n (and bounded by 1 by definition) and hence converges. One would expect that as $n \to \infty$, $\mathrm{TV}(P^{\otimes n}, Q^{\otimes n})$ converges to 1 and consequently the probability of error in the hypothesis test vanishes. It turns out that for fixed distributions $P \neq Q$, large-deviations theory (see Chapter 16) shows that $\mathrm{TV}(P^{\otimes n}, Q^{\otimes n})$ indeed converges to 1 as $n \to \infty$ and, in fact, exponentially fast:

$$\mathrm{TV}(P^{\otimes n}, Q^{\otimes n}) = 1 - \exp(-nC(P, Q) + o(n)), \tag{7.21}$$

where the exponent $C(P, Q) > 0$ is known as the *Chernoff information* of P and Q given in (16.2). However, as frequently encountered in high-dimensional statistical problems, if the distributions $P = P_n$ and $Q = Q_n$ depend on n, then the large-deviations asymptotics in (7.21) can no longer be directly applied. Since computing the total variation between two n-fold product distributions is typically difficult, understanding how a more tractable f-divergence is related to the total variation may give insight on its behavior. It turns out Hellinger distance is precisely suited for this task.

Shortly, we will show (Section 7.5.1) the following relation between TV and the Hellinger divergence:

$$\frac{1}{2}H^2(P, Q) \leq \mathrm{TV}(P, Q) \leq H(P, Q)\sqrt{1 - \frac{H^2(P, Q)}{4}} \leq 1. \tag{7.22}$$

Direct consequences of the bound (7.22) are the following.

- $H^2(P, Q) = 2$ if and only if $\mathrm{TV}(P, Q) = 1$. In this case, the probability of error is zero since essentially P and Q have disjoint supports.
- $H^2(P, Q) = 0$ if and only if $\mathrm{TV}(P, Q) = 0$. In this case, the smallest total probability of error is 1, meaning the best test is random guessing.
- Hellinger consistency is equivalent to TV consistency: we have

$$H^2(P_n, Q_n) \to 0 \iff \mathrm{TV}(P_n, Q_n) \to 0, \tag{7.23}$$

$$H^2(P_n, Q_n) \to 2 \iff \mathrm{TV}(P_n, Q_n) \to 1; \tag{7.24}$$

however, the speed of convergence need not be the same.

Theorem 7.8. *For any sequence of distributions P_n and Q_n, as $n \to \infty$,*

$$\mathrm{TV}(P_n^{\otimes n}, Q_n^{\otimes n}) \to 0 \iff H^2(P_n, Q_n) = o\left(\frac{1}{n}\right),$$

$$\mathrm{TV}(P_n^{\otimes n}, Q_n^{\otimes n}) \to 1 \iff H^2(P_n, Q_n) = \omega\left(\frac{1}{n}\right).$$

Proof. For convenience, let $X_1, X_2, \ldots, X_n \overset{\text{iid}}{\sim} Q_n$. Then

$$
\begin{aligned}
H^2(P_n^{\otimes n}, Q_n^{\otimes n}) &= 2 - 2\mathbb{E}\left[\sqrt{\prod_{i=1}^{n} \frac{P_n}{Q_n}(X_i)}\right] \\
&= 2 - 2\prod_{i=1}^{n} \mathbb{E}\left[\sqrt{\frac{P_n}{Q_n}(X_i)}\right] = 2 - 2\left(\mathbb{E}\left[\sqrt{\frac{P_n}{Q_n}}\right]\right)^n \\
&= 2 - 2\left(1 - \frac{1}{2}H^2(P_n, Q_n)\right)^n . \qquad (7.25)
\end{aligned}
$$

We now use (7.25) to conclude the proof. Recall from (7.23) that $\mathrm{TV}(P_n^{\otimes n}, Q_n^{\otimes n}) \to 0$ if and only if $H^2(P_n^{\otimes n}, Q_n^{\otimes n}) \to 0$, which happens precisely when $H^2(P_n, Q_n) = o(\frac{1}{n})$. Similarly, by (7.24), $\mathrm{TV}(P_n^{\otimes n}, Q_n^{\otimes n}) \to 1$ if and only if $H^2(P_n^{\otimes n}, Q_n^{\otimes n}) \to 2$, which is further equivalent to $H^2(P_n, Q_n) = \omega(\frac{1}{n})$. □

Remark 7.7. Property (7.25) is known as *tensorization*. More generally, we have

$$
H^2\left(\prod_{i=1}^{n} P_i, \prod_{i=1}^{n} Q_i\right) = 2 - 2\prod_{i=1}^{n}\left(1 - \frac{1}{2}H^2(P_i, Q_i)\right). \qquad (7.26)
$$

While some other f-divergences also satisfy tensorization, see Section 7.12, the H^2 has the advantage of a sandwich bound (7.22) making it the most convenient tool for checking asymptotic testability of hypotheses.

Remark 7.8. (Kakutani's dichotomy) Let $P = \prod_{i\geq 1} P_i$ and $Q = \prod_{i\geq 1} Q_i$, where $P_i \ll Q_i$. Kakutani's theorem shows the following dichotomy between these two distributions on the infinite sequence space:

- If $\sum_{i\geq 1} H^2(P_i, Q_i) = \infty$, then P and Q are mutually singular (i.e. $P \perp Q$).
- If $\sum_{i\geq 1} H^2(P_i, Q_i) < \infty$, then P and Q are equivalent (i.e. $P \ll Q$ and $Q \ll P$).

In the Gaussian case, say, $P_i = N(\mu_i, 1)$ and $Q_i = N(0, 1)$, the equivalence condition simplifies to $\sum \mu_i^2 < \infty$.

To understand Kakutani's criterion, note that by the tensorization property (7.26), we have

$$
H^2(P, Q) = 2 - 2\prod_{i\geq 1}\left(1 - \frac{H^2(P_i, Q_i)}{2}\right).
$$

Thus, if $\prod_{i\geq 1}(1 - \frac{H^2(P_i,Q_i)}{2}) = 0$, or equivalently $\sum_{i\geq 1} H^2(P_i, Q_i) = \infty$, then $H^2(P, Q) = 2$, which, by (7.22), is equivalent to $\mathrm{TV}(P, Q) = 0$ and hence $P \perp Q$. If $\sum_{i\geq 1} H^2(P_i, Q_i) < \infty$, then $H^2(P, Q) < 2$. To conclude the equivalence between P and Q, note that the likelihood ratio $\frac{dP}{dQ} = \prod_{i\geq 1} \frac{dP_i}{dQ_i}$ satisfies that either $Q(\frac{dP}{dQ} = 0) = 0$ or 1 by Kolmogorov's 0–1 law. See [143, Theorem 5.3.5] for details.

We end this section by discussing the related concept of *contiguity*. Note that if two distributions P_n and Q_n have vanishing total variation, then $P_n(A) = Q_n(A) + o(1)$ uniformly for all events A. Sometimes and especially for statistical applications we are only interested in comparing those events with probability close to 0 or 1. This leads us to the following definition.

Definition 7.9. (Contiguity and asymptotic separatedness) Let $\{P_n\}$ and $\{Q_n\}$ be sequences of probability measures. We say P_n is *contiguous* with respect to Q_n (denoted by $P_n \lhd Q_n$) if for any sequence $\{A_n\}$ of measurable sets, $Q_n(A_n) \to 0$ implies that $P_n(A_n) \to 0$. We say P_n and Q_n are *mutually contiguous* (denoted by $P_n \lhd\rhd Q_n$) if $P_n \lhd Q_n$ and $Q_n \lhd P_n$. We say that P_n is *asymptotically separated* from Q_n (denoted $P_n \triangle Q_n$) if $\limsup_{n\to\infty} \mathrm{TV}(P_n, Q_n) = 1$.

Note that when $P_n = P$ and $Q_n = Q$ these definitions correspond to $P \ll Q$ and $P \perp Q$, respectively, and thus should be viewed as their asymptotic versions. Clearly, $P_n \lhd\rhd Q_n$ is much weaker than $\mathrm{TV}(P_n, Q_n) \to 0$; for example, $P_n(A_n) = 1/2$ only guarantees $Q_n(A_n)$ is not tending to 0 or 1. In addition, if $P_n \lhd Q_n$, then any test that succeeds with high Q_n probability must fail with high P_n probability; in other words, P_n and Q_n cannot be distinguished perfectly so $\mathrm{TV}(P_n, Q_n) = 1 - \Omega(1)$. As such, contiguity and separatedness are mutually exclusive. Furthermore, often many interesting sequences of measures satisfy a dichotomy similar to Kakutani's: either $P_n \lhd\rhd Q_n$ or $P_n \triangle Q_n$; see [283].

Our interest in these notions arises from the fact that f-divergences are instrumental for establishing contiguity and separatedness. For example, from (7.24) we conclude that

$$P_n \triangle Q_n \iff \limsup_{n\to\infty} H^2(P_n, Q_n) = 2.$$

On the other hand, [386, Theorem III.10.1] shows

$$P_n \lhd Q_n \iff \lim_{\alpha\to 0+} \limsup_{n\to\infty} D_\alpha(P_n \| Q_n) = 0,$$

where D_α is the Rényi divergence (Definition 7.24). This criterion can be weakened to the following (commonly used) one: $P_n \lhd Q_n$ if $\chi^2(P_n \| Q_n) = O(1)$. Indeed, applying a change of measure and Cauchy–Schwarz, $P_n(A_n) = \mathbb{E}_{P_n}[1\{A_n\}] = \mathbb{E}_{Q_n}[\frac{dP_n}{dQ_n} 1\{A_n\}] \le \sqrt{1 + \chi^2(P_n \| Q_n)}\sqrt{Q_n(A_n)}$, which vanishes whenever $Q_n(A_n)$ vanishes. (See Exercise **I.49** for a concrete example in the context of community detection and random graphs.) In particular, a sufficient condition for mutual contiguity is the boundedness of likelihood ratio: $c \le \frac{P_n}{Q_n} \le C$ for some constants c, C.

7.4 Inequalities between f-Divergences and Joint Range

In this section we study the relationship, in particular, inequalities, between f-divergences. To gain some intuition, we start with the ad hoc approach by proving *Pinsker's inequality*, which bounds total variation from above in terms of the KL divergence.

Theorem 7.10. (Pinsker's inequality)

$$D(P \| Q) \ge (2 \log e) \mathrm{TV}^2(P, Q). \tag{7.27}$$

Proof. It suffices to consider the natural logarithm for the KL divergence. First we show that, by the data-processing inequality, it suffices to prove the result for Bernoulli distributions. For any event E, let $Y = 1\{X \in E\}$ which is Bernoulli with parameter $P(E)$ or $Q(E)$. By the DPI, $D(P\|Q) \geq d(P(E)\|Q(E))$. If Pinsker's inequality holds for all Bernoulli distributions, we have

$$\sqrt{\frac{1}{2}D(P\|Q)} \geq \mathrm{TV}(\mathrm{Ber}(P(E)), \mathrm{Ber}(Q(E)) = |P(E) - Q(E)|.$$

Taking the supremum over E gives $\sqrt{\frac{1}{2}D(P\|Q)} \geq \sup_E |P(E) - Q(E)| = \mathrm{TV}(P, Q)$, in view of Theorem 7.7.

The binary case follows easily from a second-order Taylor expansion (with integral remainder form) of $p \mapsto d(p\|q)$:

$$d(p\|q) = \int_q^p \frac{p-t}{t(1-t)}dt \geq 4\int_q^p (p-t)dt = 2(p-q)^2$$

and $\mathrm{TV}(\mathrm{Ber}(p), \mathrm{Ber}(q)) = |p - q|$. □

Pinsker's inequality has already been used multiple times in this book. Here is yet another implication that is further explored in Exercises **I.62** and **I.63** (Szemerédi regularity).

Corollary 7.11. (Tao's inequality [415]) *Let $Y \to X \to X'$ be a Markov chain with $Y \in [-1, 1]$. Then*

$$\mathbb{E}[|\,\mathbb{E}[Y|X] - \mathbb{E}[Y|X']|^2] \leq \frac{2}{\log e}I(Y;X|X'). \tag{7.28}$$

The same estimate holds for Y ranging over a unit ball in any normed vector space with $|\cdot|$ on the LHS being the norm.

Proof. If Y_1 and Y_2 are two random variables taking values in $[-1, 1]$ then by (7.20) there exists a coupling such that $\mathbb{P}[Y_1 \neq Y_2] \leq \mathrm{TV}(P_{Y_1}, P_{Y_2})$. Thus, $|\,\mathbb{E}[Y_1] - \mathbb{E}[Y_2]| \leq 2\mathrm{TV}(P_{Y_1}, P_{Y_2})$. Now, applying this to $P_{Y_1} = P_{Y|X=a}$ and $P_{Y_2} = P_{Y|X'=b}$ we obtain

$$|\,\mathbb{E}[Y|X=a] - \mathbb{E}[Y|X'=b]|^2 \leq 4\mathrm{TV}^2(P_{Y|X=a}, P_{Y|X'=b}) \leq \frac{2}{\log e}D(P_{Y|X=a}\|P_{Y|X'=b}),$$

where we applied Pinsker's inequality in the last step. The proof is completed by averaging over $(a, b) \sim P_{X,X'}$ and noticing that $D(P_{Y|X}\|P_{Y|X'}|P_{X,X'}) = I(Y;X|X')$ due to $P_{Y|X,X'} = P_{Y|X}$ by assumption. □

Pinsker's inequality and Tao's inequality are both sharp in the sense that the constants cannot be improved. For example, for (7.27) we can take $P_n = \mathrm{Ber}(1/2+1/n)$ and $Q_n = \mathrm{Ber}(1/2)$ and compute that $\frac{D(P_n\|Q_n)}{\mathrm{TV}^2(P_n,Q_n)} \to 2\log e$ as $n \to \infty$. (This is best seen by inspecting the local quadratic behavior in Proposition 2.21.) Nevertheless, this does not mean that the inequality (7.27) is not improvable, as the RHS can be replaced by some other function of $\mathrm{TV}(P, Q)$ with additional higher-order terms. Indeed, several such improvements of Pinsker's inequality are known. But what is the best inequality? In addition, another natural question is the reverse inequality: can we upper-bound $D(P\|Q)$ in terms of $\mathrm{TV}(P,Q)$?

Settling these questions rests on characterizing the *joint range* (the set of possible values) of a given pair of f-divergences. This systematic approach to comparing f-divergences (as opposed to the ad hoc proof of Theorem 7.10 we presented above) is the subject of the rest of this section.

Definition 7.12. (Joint range) Consider two f-divergences $D_f(P\|Q)$ and $D_g(P\|Q)$. Their joint range is a subset of $[0,\infty]^2$ defined by

$$\mathcal{R} \triangleq \{(D_f(P\|Q), D_g(P\|Q)) \colon P, Q \text{ are probability measures on some measurable space}\}.$$

In addition, the joint range over all k-ary distributions is defined as

$$\mathcal{R}_k \triangleq \{(D_f(P\|Q), D_g(P\|Q)) \colon P, Q \text{ are probability measures on } [k]\}.$$

As an example, Figure 7.1 gives the joint range $\mathcal{R}$ between the KL divergence and the total variation. By definition, the lower boundary of the region $\mathcal{R}$ gives the optimal refinement of Pinsker's inequality:

$$D(P\|Q) \geq F(\mathrm{TV}(P,Q)), \quad F(\epsilon) \triangleq \inf_{(P,Q)\colon \mathrm{TV}(P,Q)=\epsilon} D(P\|Q) = \inf\{s \colon (\epsilon, s) \in \mathcal{R}\}.$$

Also from Figure 7.1 we see that it is impossible to bound $D(P\|Q)$ from above in terms of $\mathrm{TV}(P,Q)$ due to the lack of upper boundary.

The joint range $\mathcal{R}$ may appear difficult to characterize since we need to consider P, Q over all measurable spaces; on the other hand, the region $\mathcal{R}_k$ for small k is easy to obtain (at least numerically). Revisiting the proof of Pinsker's inequality in Theorem 7.10, we see that the key step is the reduction to Bernoulli distributions. It is natural to ask: To obtain full joint range is it possible to reduce to the binary case? It turns out that it is always sufficient to consider quaternary distributions, or the convex hull of that of binary distributions.

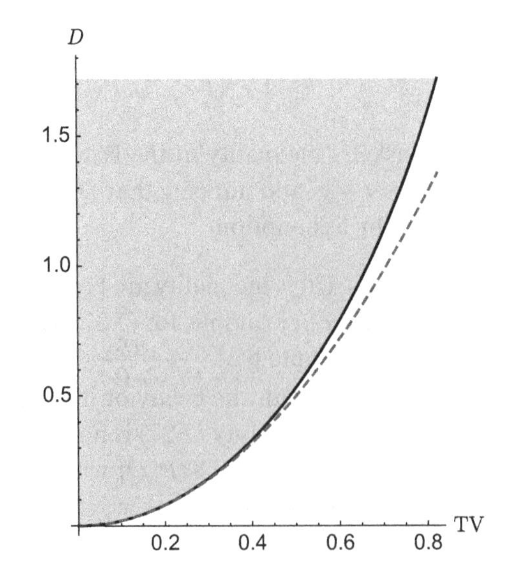

Figure 7.1 Joint range of TV and KL divergence. The dashed line is the quadratic lower bound given by Pinsker's inequality (7.27).

Theorem 7.13. (Harremoës and Vajda [214])

$$\mathcal{R} = \text{co}(\mathcal{R}_2) = \mathcal{R}_4,$$

where co *denotes the convex hull with a natural extension of convex operations to* $[0, \infty]^2$.

We will rely on the following famous result from convex analysis (see, e.g., [145, Chapter 2, Theorem 18]).

Lemma 7.14. (Fenchel–Eggleston–Carathéodory theorem) *Let* $S \subseteq \mathbb{R}^d$ *and* $x \in \text{co}(S)$. *Then there exists a set of* $d+1$ *points* $S' = \{x_1, x_2, \ldots, x_{d+1}\} \in S$ *such that* $x \in \text{co}(S')$. *If* S *has at most* d *connected components, then* d *points are enough.*

Proof. Our proof will consist of three claims:

- *Claim 1:* $\text{co}(\mathcal{R}_2) \subset \mathcal{R}_4$.
- *Claim 2:* $\mathcal{R}_k \subset \text{co}(\mathcal{R}_2)$.
- *Claim 3:* $\mathcal{R} = \mathcal{R}_4$.

Note that Claims 1 and 2 prove the most interesting part: $\bigcup_{k=1}^{\infty} \mathcal{R}_k = \text{co}(\mathcal{R}_2)$. Claim 3 is more technical and its proof can be found in [214]. However, the approximation result in Theorem 7.6 shows that $\mathcal{R}$ is the closure of $\bigcup_{k=1}^{\infty} \mathcal{R}_k$. Thus for the purpose of obtaining inequalities between D_f and D_g, Claims 1 and 2 are sufficient.

We start with Claim 1. Given any two pairs of distributions (P_0, Q_0) and (P_1, Q_1) on some space $\mathcal{X}$ and given any $\alpha \in [0, 1]$, define two joint distributions of the random variables (X, B) where $P_B = Q_B = \text{Ber}(\alpha)$, $P_{X|B=i} = P_i$, and $Q_{X|B=i} = Q_i$ for $i = 0, 1$. Then by (7.9) we get

$$D_f(P_{X,B} \| Q_{X,B}) = \bar{\alpha} D_f(P_0 \| Q_0) + \alpha D_f(P_1 \| Q_1),$$

and similarly for D_g. Thus, $\mathcal{R}$ is convex. Next, notice that

$$\mathcal{R}_2 = \tilde{\mathcal{R}}_2 \cup \{(pf'(\infty), pg'(\infty)) \colon p \in (0, 1]\} \cup \{(qf(0), qg(0)) \colon q \in (0, 1]\},$$

where $\tilde{\mathcal{R}}_2$ is the image of $(0, 1)^2$ of the continuous map

$$(p, q) \mapsto (D_f(\text{Ber}(p) \| \text{Ber}(q)), D_g(\text{Ber}(p) \| \text{Ber}(q))).$$

Since $(0, 0) \in \tilde{\mathcal{R}}_2$, we see that regardless of which off $f(0), f'(\infty), g(0), g'(\infty)$ are infinite, the set $\mathcal{R}_2 \cap \mathbb{R}^2$ is connected. Thus, by Lemma 7.14 any point in $\text{co}(\mathcal{R}_2 \cap \mathbb{R}^2)$ is a combination of two points in $\mathcal{R}_2 \cap \mathbb{R}^2$, which, by the argument above, is a subset of $\mathcal{R}_4$. Finally, it is not hard to see that $\text{co}(\mathcal{R}_2) \backslash \mathbb{R}^2 \subset \mathcal{R}_4$, which concludes the proof of $\text{co}(\mathcal{R}_2) \subset \mathcal{R}_4$.

Next, we prove Claim 2. Fix P, Q on $[k]$ and denote their PMFs by (p_j) and (q_j), respectively. Note that without changing either $D_f(P \| Q)$ or $D_g(P \| Q)$ (but perhaps, by increasing k by 1), we can make $q_j > 0$ for $j > 1$ and $q_1 = 0$, which we thus assume. Denote $\phi_j = \frac{p_j}{q_j}$ for $j > 1$ and consider the set

$$\mathcal{S} = \left\{ \tilde{Q} = (\tilde{q}_j)_{j \in [k]} \colon \tilde{q}_j \geq 0, \sum \tilde{q}_j = 1, \tilde{q}_1 = 0, \sum_{j=2}^{k} \tilde{q}_j \phi_j \leq 1 \right\}.$$

We also define a subset $\mathcal{S}_e \subset \mathcal{S}$ consisting of points $\tilde{Q}$ of two types:

1 $\tilde{q}_j = 1$ and $\phi_j \le 1$ for some $j \ge 2$.
2 $\tilde{q}_{j_1} + \tilde{q}_{j_2} = 1$ and $\tilde{q}_{j_1}\phi_{j_1} + \tilde{q}_{j_2}\phi_{j_2} = 1$ for some $j_1, j_2 \ge 2$.

It can be seen that $\mathcal{S}_e$ contains all the extreme points of $\mathcal{S}$. Indeed, any $\tilde{Q} \in \mathcal{S}$ with $\sum_{j\ge 2} \tilde{q}_j\phi_j < 1$ with more than one non-zero atom cannot be extremal (since there is only one active linear constraint $\sum_j \tilde{q}_j = 1$). Similarly, $\tilde{Q}$ with $\sum_{j\ge 2} \tilde{q}_j\phi_j = 1$ can only be extremal if it has one or two non-zero atoms.

We next claim that any point in $\mathcal{S}$ can be written as a convex combination of finitely many points in $\mathcal{S}_e$. This can be seen as follows. First, we can view $\mathcal{S}$ and $\mathcal{S}_e$ as subsets of $\mathbb{R}^{k-1}$. Since $\mathcal{S}$ is clearly closed and convex, by the Krein–Milman theorem (see [12, Theorem 7.68]), $\mathcal{S}$ coincides with the closure of the convex hull of its extreme points. Since $\mathcal{S}_e$ is compact (hence closed), so is $\text{co}(\mathcal{S}_e)$ [12, Corollary 5.33]. Thus we have $\mathcal{S} = \text{co}(\mathcal{S}_e)$ and, in particular, there are probability weights $\{\alpha_i : i \in [m]\}$ and extreme points $\tilde{Q}_i \in \mathcal{S}_e$ so that

$$Q = \sum_{i=1}^{m} \alpha_i \tilde{Q}_i. \tag{7.29}$$

Next, to each $\tilde{Q}$ we associate $\tilde{P} = \tilde{P}(\tilde{Q}) = (\tilde{p}_j)_{j\in[k]}$ as follows:

$$\tilde{p}_j = \begin{cases} \phi_j \tilde{q}_j, & j \in \{2, \dots, k\}, \\ 1 - \sum_{j=2}^{k} \phi_j \tilde{q}_j, & j = 1. \end{cases}$$

We then have that

$$\tilde{Q} \mapsto D_f(\tilde{P} \| \tilde{Q}) = \sum_{j\ge 2} \tilde{q}_j f(\phi_j) + f'(\infty)\tilde{p}_1$$

affinely maps $\mathcal{S}$ to $[0, \infty]$ (note that $f(0)$ or $f'(\infty)$ can equal ∞). In particular, if we denote $\tilde{P}_i = \tilde{P}(\tilde{Q}_i)$ corresponding to $\tilde{Q}_i$ in decomposition (7.29), we get

$$D_f(P\|Q) = \sum_{i=1}^{m} \alpha_i D_f(\tilde{P}_i \| \tilde{Q}_i),$$

and similarly for $D_g(P\|Q)$. We are left to show that $(\tilde{P}_i, \tilde{Q}_i)$ are supported on at most two points, which verifies that any element of $\mathcal{R}_k$ is a convex combination of k elements of $\mathcal{R}_2$. Indeed, for $\tilde{Q} \in \mathcal{S}_e$ the set $\{j \in [k] : \tilde{q}_j > 0 \text{ or } \tilde{p}_j > 0\}$ has cardinality at most two (for the second type of extremal points we notice $\tilde{p}_{j_1} + \tilde{p}_{j_2} = 1$ implying $\tilde{p}_1 = 0$). This concludes the proof of Claim 2. □

7.5 Examples of Computing Joint Range

In this section we show how to apply the method of Harremoës and Vajda for proving the best possible comparison inequalities between various f-divergences.

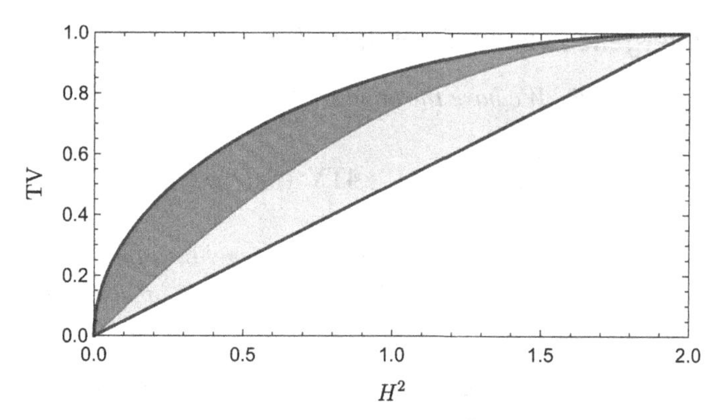

Figure 7.2 The joint range $\mathcal{R}$ of TV and H^2 is characterized by (7.22), which is the convex hull of the dark gray region $\mathcal{R}_2$.

7.5.1 Hellinger Distance versus Total Variation

The joint range $\mathcal{R}_2$ of H^2 and TV over binary distributions is simply:

$$\mathcal{R}_2 = \left\{(2(1 - \sqrt{pq} - \sqrt{\bar{p}\bar{q}}), |p - q|) : 0 \le p \le 1, 0 \le q \le 1\right\}.$$

This is shown as the non-convex dark gray region in Figure 7.2. By Theorem 7.13, the full joint range $\mathcal{R}$ is the convex hull of $\mathcal{R}_2$, which turns out to be exactly described by the sandwich bound (7.22) shown earlier in Section 7.3. This means that (7.22) is not improvable. Indeed, with t ranging from 0 to 1,

- the upper boundary is achieved by $P = \text{Ber}(\frac{1+t}{2})$, $Q = \text{Ber}(\frac{1-t}{2})$,
- the lower boundary is achieved by $P = (1 - t, t, 0)$, $Q = (1 - t, 0, t)$.

7.5.2 KL Divergence versus Total Variation

The joint range between KL and TV was previously shown in Figure 7.1. Although there is no known closed-form expression, the following parametric formula of the lower boundary (see Figure 7.1) is known [163, Theorem 1]:

$$\begin{cases} \text{TV}_t = \frac{1}{2}t\left(1 - \left(\coth(t) - \frac{1}{t}\right)^2\right), \\ D_t = -t^2\text{csch}^2(t) + t\coth(t) + \log(t\,\text{csch}(t)), \end{cases} \quad t \ge 0, \tag{7.30}$$

where we take the natural logarithm. Here is a corollary (weaker bound) due to [428]:

$$D(P\|Q) \ge \log\frac{1 + \text{TV}(P,Q)}{1 - \text{TV}(P,Q)} - \frac{2\text{TV}(P,Q)}{1 + \text{TV}(P,Q)}\log e. \tag{7.31}$$

Both bounds are stronger than Pinsker's inequality (7.27). Note the following consequences:

- $D \to 0 \Rightarrow \text{TV} \to 0$, which can be deduced from Pinsker's inequality.
- $\text{TV} \to 1 \Rightarrow D \to \infty$ and hence $D = O(1)$ implies that TV is bounded away from one. This can be obtained from (7.30) or (7.31), but not Pinsker's inequality.

7.5.3 χ^2-Divergence versus Total Variation

Proposition 7.15. *We have the bound*

$$\chi^2(P\|Q) \geq f(\mathrm{TV}(P,Q)) \geq 4\mathrm{TV}^2(P,Q), \quad f(t) = \begin{cases} 4t^2, & t \leq \frac{1}{2}, \\ \frac{t}{1-t}, & t \geq \frac{1}{2}, \end{cases} \tag{7.32}$$

where the function f is a convex increasing bijection of $[0,1)$ onto $[0,\infty)$. Furthermore, for every $s \geq f(t)$ there exists a pair of distributions (P,Q) such that $\chi^2(P\|Q) = s$ and $\mathrm{TV}(P,Q) = t$.

Proof. We claim that the binary joint range is convex. Indeed,

$$\mathrm{TV}(\mathrm{Ber}(p),\mathrm{Ber}(q)) = |p-q| \triangleq t, \quad \chi^2(\mathrm{Ber}(p)\|\mathrm{Ber}(q)) = \frac{(p-q)^2}{q(1-q)} = \frac{t^2}{q(1-q)}.$$

Given $|p-q| = t$, let us determine the possible range of $q(1-q)$. The smallest value of $q(1-q)$ is always 0 by choosing $p = t, q = 0$. The largest value is $1/4$ if $t \leq 1/2$ (by choosing $p = 1/2 - t$, $q = 1/2$). If $t > 1/2$ then we can at most get $t(1-t)$ (by setting $p = 0$ and $q = t$). Thus we get $\chi^2(\mathrm{Ber}(p)\|\mathrm{Ber}(q)) \geq f(|p-q|)$ as claimed. The convexity of f follows since its derivative is monotonically increasing. Clearly, $f(t) \geq 4t^2$ because $t(1-t) \leq \frac{1}{4}$. □

7.6 A Selection of Inequalities between Various Divergences

This section presents a collection of useful inequalities. For a more complete treatment, consider [374] and [425, Section 2.4]. Most of these inequalities are joint ranges, which means they are tight.

- KL versus TV: see (7.30). For discrete distributions there is partial comparison in the other direction ("reverse Pinsker," see [374, Section VI]):

$$D(P\|Q) \leq \log\left(1 + \frac{2}{Q_{\min}}\mathrm{TV}(P,Q)^2\right) \leq \frac{2\log e}{Q_{\min}}\mathrm{TV}(P,Q)^2, \quad Q_{\min} = \min_x Q(x).$$

- KL versus Hellinger:

$$D(P\|Q) \geq 2\log\frac{2}{2 - H^2(P,Q)} \geq \log e \cdot H^2(P,Q). \tag{7.33}$$

The first inequality gives the joint range and is attained at $P = \mathrm{Ber}(0)$, $Q = \mathrm{Ber}(q)$. For a fixed H^2, in general $D(P\|Q)$ has no finite upper bound, as seen from $P = \mathrm{Ber}(p)$, $Q = \mathrm{Ber}(0)$.

There is a partial result in the opposite direction (log-Sobolev inequality for Bonami–Beckner semigroup, see [122, Theorem A.1] and Exercise **I.64**):

$$D(P\|Q) \leq \frac{\log(\frac{1}{Q_{\min}} - 1)}{1 - 2Q_{\min}}\left(1 - (1 - H^2(P,Q))^2\right), \quad Q_{\min} = \min_x Q(x).$$

Another partial result is in Exercise **I.59**.

- KL versus χ^2:

$$0 \le D(P\|Q) \le \log(1 + \chi^2(P\|Q)) \le \log e \cdot \chi^2(P\|Q). \tag{7.34}$$

The left-hand inequality states that no lower bound on KL in terms of χ^2 is possible.

- TV versus Hellinger: see (7.22). A useful simplified bound from [186] is the following:

$$\mathrm{TV}(P,Q) \le \sqrt{-2\ln\left(1 - \frac{H^2(P,Q)}{2}\right)}.$$

- Le Cam versus Hellinger [274, p. 48]:

$$\frac{1}{2}H^2(P,Q) \le \mathrm{LC}(P,Q) \le H^2(P,Q). \tag{7.35}$$

- Le Cam versus Jensen–Shannon [423]:

$$\mathrm{LC}(P,Q)\log e \le \mathrm{JS}(P,Q) \le \mathrm{LC}(P,Q) \cdot 2\log 2. \tag{7.36}$$

- χ^2 versus TV: The full joint range is given by (7.32). Two simple consequences are

$$\mathrm{TV}(P,Q) \le \frac{1}{2}\sqrt{\chi^2(P\|Q)}, \tag{7.37}$$

$$\mathrm{TV}(P,Q) \le \max\left\{\frac{1}{2}, \frac{\chi^2(P\|Q)}{1+\chi^2(P\|Q)}\right\}, \tag{7.38}$$

where the second is useful for bounding TV away from one.

- JS versus TV: The full joint region is given by

$$2d\left(\frac{1-\mathrm{TV}(P,Q)}{2}\Big\|\frac{1}{2}\right) \le \mathrm{JS}(P,Q) \le \mathrm{TV}(P,Q)\cdot 2\log 2. \tag{7.39}$$

The lower bound is a consequence of Fano's inequality. For the upper bound notice that for $p, q \in [0,1]$ and $|p-q| = \tau$ the maximum of $d(p\|\frac{p+q}{2})$ is attained at $p = 0, q = \tau$ (from the convexity of $d(\cdot\|\cdot)$) and, thus, the binary joint range is given by $\tau \mapsto d(\tau\|\tau/2) + d(1-\tau\|1-\tau/2)$. Since the latter is convex, its concave envelope is a straight line connecting endpoints at $\tau = 0$ and $\tau = 1$.

7.7 Divergences between Gaussians

To get a better feel for the behavior of f-divergences, here we collect expressions (as well as asymptotic expansions) of divergences between Gaussian distributions.

1 Total variation:

$$\begin{aligned}\mathrm{TV}(\mathcal{N}(0,\sigma^2), \mathcal{N}(\mu,\sigma^2)) &= 2\Phi\left(\frac{|\mu|}{2\sigma}\right) - 1 = \int_{-\frac{|\mu|}{2\sigma}}^{\frac{|\mu|}{2\sigma}} \varphi(x)dx \\ &= \frac{|\mu|}{\sqrt{2\pi}\sigma} + O(\mu^2), \quad \mu \to 0.\end{aligned} \tag{7.40}$$

2 Hellinger distance:

$$H^2(\mathcal{N}(0,\sigma^2)\|\mathcal{N}(\mu,\sigma^2)) = 2 - 2e^{-\frac{\mu^2}{8\sigma^2}} = \frac{\mu^2}{4\sigma^2} + O(\mu^3), \quad \mu \to 0. \tag{7.41}$$

More generally,

$$H^2(\mathcal{N}(\mu_1,\Sigma_1)\|\mathcal{N}(\mu_2,\Sigma_2)) = 2 - 2\frac{|\Sigma_1|^{\frac{1}{4}}|\Sigma_2|^{\frac{1}{4}}}{|\bar{\Sigma}|^{\frac{1}{2}}} \exp\left\{-\frac{1}{8}(\mu_1-\mu_2)^\top \bar{\Sigma}^{-1}(\mu_1-\mu_2)\right\},$$

where $\bar{\Sigma} = \frac{\Sigma_1+\Sigma_2}{2}$.

3 KL divergence:

$$D(\mathcal{N}(\mu_1,\sigma_1^2)\|\mathcal{N}(\mu_2,\sigma_2^2)) = \frac{1}{2}\log\frac{\sigma_2^2}{\sigma_1^2} + \frac{1}{2}\left(\frac{(\mu_1-\mu_2)^2}{\sigma_2^2} + \frac{\sigma_1^2}{\sigma_2^2} - 1\right)\log e. \tag{7.42}$$

For a more general result see (2.8).

4 χ^2-divergence:

$$\chi^2(\mathcal{N}(\mu,\sigma^2)\|\mathcal{N}(0,\sigma^2)) = e^{\frac{\mu^2}{\sigma^2}} - 1 = \frac{\mu^2}{\sigma^2} + O(\mu^3), \quad \mu \to 0, \tag{7.43}$$

$$\chi^2(\mathcal{N}(\mu,\sigma^2)\|\mathcal{N}(0,1)) = \begin{cases} \frac{e^{\mu^2/(2-\sigma^2)}}{\sigma\sqrt{2-\sigma^2}} - 1, & \sigma^2 < 2, \\ \infty, & \sigma^2 \geq 2. \end{cases} \tag{7.44}$$

5 χ^2-divergence for Gaussian mixtures [225] (see also Exercise **I.48** for the Ingster–Suslina method applicable to general mixture distributions):

$$\chi^2(P*\mathcal{N}(0,\Sigma)\|\mathcal{N}(0,\Sigma)) = \mathbb{E}[e^{\langle\Sigma^{-1}X,X'\rangle}] - 1, \quad X \perp\!\!\!\perp X' \sim P. \tag{7.45}$$

7.8 Mutual Information Based on *f*-Divergence

Given an f-divergence D_f, we can define *f-information*, an extension of mutual information, as follows:

$$I_f(X;Y) \triangleq D_f(P_{X,Y}\|P_X P_Y). \tag{7.46}$$

Theorem 7.16. (Data processing) *For $U \to X \to Y$, we have $I_f(U;Y) \leq I_f(U;X)$.*

Proof. Note that $I_f(U;X) = D_f(P_{U,X}\|P_U P_X) \geq D_f(P_{U,Y}\|P_U P_Y) = I_f(U;Y)$, where we applied the data-processing theorem (Theorem 7.4) to the (possibly stochastic) map $(U,X) \mapsto (U,Y)$. See also Remark 3.4. □

A useful property of mutual information is that $X \perp\!\!\!\perp Y$ iff $I(X;Y) = 0$. A generalization of it is the property that for $X \to Y \to Z$ we have $I(X;Y) = I(X;Z)$ iff $X \to Z \to Y$. Both of these may or may not hold for I_f depending on the strict convexity of f, see Exercise **I.40**.

Another often used property of the standard mutual information is *subadditivity*: If $P_{A,B|X} = P_{A|X}P_{B|X}$ (i.e. A and B are conditionally independent given X), then

$$I(X;A,B) \leq I(X;A) + I(X;B). \tag{7.47}$$

However, other notions of f-information have complicated relationship with subadditivity:

1 The f-information corresponding to the χ^2-divergence,

$$I_{\chi^2}(X;Y) \triangleq \chi^2(P_{X,Y} \| P_X P_Y), \tag{7.48}$$

is not generally subadditive. There are two special cases when I_{χ^2} is subadditive: if either $I_{\chi^2}(X;A)$ or $I_{\chi^2}(X;B)$ is small [202, Lemma 26] or if $X \sim \text{Ber}(1/2)$ channels $P_{A|X}$ and $P_{B|X}$ are binary memoryless symmetric (BMS) (Section 19.4*); see [1].

2 The f-information corresponding to total variation $I_{\text{TV}}(X;Y) \triangleq \text{TV}(P_{X,Y}, P_X P_Y)$ is not subadditive. Furthermore, it has a counterintuitive behavior of "getting stuck." For example, take $X \sim \text{Ber}(1/2)$ and $A = \mathsf{BSC}_\delta(X)$, $B = \mathsf{BSC}_\delta(X)$ – two independent observations of X across the BSC. A simple computation (Exercise **I.35**) shows

$$I_{\text{TV}}(X;A,B) = I_{\text{TV}}(X;A) = I_{\text{TV}}(X;B).$$

In other words, an additional observation does not improve TV information at all. This is the main reason for the famous herding effect in economics [30].

3 The *symmetric KL-information*

$$I_{\text{SKL}}(X;Y) \triangleq D(P_{X,Y} \| P_X P_Y) + D(P_X P_Y \| P_{X,Y}), \tag{7.49}$$

the f-information corresponding to the *symmetric KL divergence* (also known as the *Jeffreys divergence*)

$$D_{\text{SKL}}(P,Q) \triangleq D(P\|Q) + D(Q\|P), \tag{7.50}$$

satisfies, quite amazingly [266], the *additivity property*:

$$I_{\text{SKL}}(X;A,B) = I_{\text{SKL}}(X;A) + I_{\text{SKL}}(X;B). \tag{7.51}$$

Let us prove this in the discrete case. First notice the following equivalent expression for I_{SKL}:

$$I_{\text{SKL}}(X;Y) = \sum_{x,x'} P_X(x) P_X(x') D(P_{Y|X=x} \| P_{Y|X=x'}). \tag{7.52}$$

From (7.52) we get (7.51) by the additivity $D(P_{A,B|X=x} \| P_{A,B|X=x'}) = D(P_{A|X=x} \| P_{A|X=x'}) + D(P_{B|X=x} \| P_{B|X=x'})$. To prove (7.52) first consider the obvious identity:

$$\sum_{x,x'} P_X(x) P_X(x') [D(P_Y \| P_{Y|X=x'}) - D(P_Y \| P_{Y|X=x})] = 0,$$

which is rewritten as

$$\sum_{x,x'} P_X(x) P_X(x') \sum_y P_Y(y) \log \frac{P_{Y|X}(y|x)}{P_{Y|X}(y|x')} = 0. \tag{7.53}$$

Next, by definition,

$$I_{\text{SKL}}(X;Y) = \sum_{x,y} [P_{X,Y}(x,y) - P_X(x)P_Y(y)] \log \frac{P_{X,Y}(x,y)}{P_X(x)P_Y(y)}.$$

Since the marginals of $P_{X,Y}$ and $P_X P_Y$ coincide, we can replace $\log \frac{P_{X,Y}(x,y)}{P_X(x)P_Y(y)}$ by any $\log \frac{P_{Y|X}(y|x)}{f(y)}$ for any f. We choose $f(y) = P_{Y|X}(y|x')$ to get

$$I_{\text{SKL}}(X;Y) = \sum_{x,y} [P_{X,Y}(x,y) - P_X(x)P_Y(y)] \log \frac{P_{Y|X}(y|x)}{P_{Y|X}(y|x')}.$$

Now averaging this over $P_X(x')$ and applying (7.53) to get rid of the second term inside the bracket above, we obtain (7.52). For another interesting property of I_{SKL} related to generalization, see Exercise **I.54**.

7.9 Empirical Distribution and χ^2-Information

Consider an arbitrary channel $P_{Y|X}$ and some input distribution P_X. Suppose that we have $X_i \stackrel{\text{iid}}{\sim} P_X$ for $i = 1, \ldots, n$. Let

$$\hat{P}_n = \frac{1}{n} \sum_{i=1}^{n} \delta_{X_i}$$

denote the empirical distribution corresponding to this sample. Let $P_Y = P_{Y|X} \circ P_X$ be the output distribution corresponding to P_X and $P_{Y|X} \circ \hat{P}_n$ be the output distribution corresponding to $\hat{P}_n$ (a random distribution). Note that for additive-noise channel $P_{Y|X=x}(\cdot) = \phi(\cdot - x)$, where ϕ is a fixed density, we can think of $P_{Y|X} \circ \hat{P}_n$ as a *kernel density estimator* (KDE), whose density is $\hat{p}_n(x) = (\phi * \hat{P}_n)(x) = \frac{1}{n}\sum_{i=1}^{n} \phi(X_i - x)$. Furthermore, using the fact that $\mathbb{E}[P_{Y|X} \circ \hat{P}_n] = P_Y$, we have

$$\mathbb{E}[D(P_{Y|X} \circ \hat{P}_n \| P_X)] = D(P_Y \| P_X) + \mathbb{E}[D(P_{Y|X} \circ \hat{P}_n \| P_Y)],$$

where the first term represents the bias of the KDE due to convolution and increases with the bandwidth of ϕ, while the second term represents the variability of the KDE and decreases with the bandwidth of ϕ. Surprisingly, the second term is sharply (within a factor of 2) given by the I_{χ^2}- information. More exactly, we prove the following result.

Proposition 7.17.

$$\mathbb{E}[D(P_{Y|X} \circ \hat{P}_n \| P_Y)] \le \log\left(1 + \frac{1}{n} I_{\chi^2}(X;Y)\right), \tag{7.54}$$

where $I_{\chi^2}(X;Y)$ is defined in (7.48). Furthermore,

$$\liminf_{n\to\infty} n\,\mathbb{E}[D(P_{Y|X} \circ \hat{P}_n \| P_Y)] \ge \frac{\log e}{2} I_{\chi^2}(X;Y). \tag{7.55}$$

In particular, $\mathbb{E}[D(P_{Y|X} \circ \hat{P}_n \| P_Y)] = O(1/n)$ if $I_{\chi^2}(X;Y) < \infty$ and $\omega(1/n)$ otherwise.

In Section 25.4* we will discuss an extension of this simple bound, in particular showing that in many cases about $n = \exp\{I(X;Y) + K\}$ observations are sufficient to ensure $D(P_{Y|X} \circ \hat{P}_n \| P_Y) = e^{-O(K)}$.

Proof. First, a simple calculation shows that

$$\mathbb{E}[\chi^2(P_{Y|X} \circ \hat{P}_n \| P_Y)] = \frac{1}{n} I_{\chi^2}(X;Y).$$

Then from (7.34) and Jensen's inequality we get (7.54).

To get the lower bound in (7.55), let $\bar{X}$ be drawn uniformly at random from the sample $\{X_1, \ldots, X_n\}$ and let $\bar{Y}$ be the output of the $P_{Y|X}$ channel with input $\bar{X}$. With this definition we have:

$$\mathbb{E}[D(P_{Y|X} \circ \hat{P}_n \| P_Y)] = I(X^n;\bar{Y}). \tag{7.56}$$

Next, apply (6.2) to get

$$I(X^n;\bar{Y}) \geq \sum_{i=1}^{n} I(X_i;\bar{Y}) = nI(X_1;\bar{Y}).$$

Finally, notice that

$$I(X_1;\bar{Y}) = D\left(\frac{n-1}{n} P_X P_Y + \frac{1}{n} P_{X,Y} \middle\| P_X P_Y\right)$$

and apply the local expansion of KL divergence (Proposition 2.21) to get (7.55). □

In the discrete case, by taking $P_{Y|X}$ to be the identity channel ($Y = X$) we obtain the following guarantee on the closeness between the empirical and the population distribution. This fact can be used to test whether the sample was truly generated by the distribution P_X.

Corollary 7.18. *Suppose P_X is discrete with support $\mathcal{X}$. If $\mathcal{X}$ is infinite, then*

$$\lim_{n\to\infty} n\,\mathbb{E}[D(\hat{P}_n \| P_X)] = \infty. \tag{7.57}$$

Otherwise, we have

$$\mathbb{E}[D(\hat{P}_n \| P_X)] \leq \log\left(1 + \frac{|\mathcal{X}| - 1}{n}\right) \leq \frac{\log e}{n}(|\mathcal{X}| - 1). \tag{7.58}$$

Proof. Simply notice that $I_{\chi^2}(X;X) = |\mathcal{X}| - 1$. □

Remark 7.9. For fixed P_X, the tight asymptotic result is

$$\lim_{n\to\infty} n\,\mathbb{E}[D(\hat{P}_n \| P_X)] = \frac{\log e}{2}(|\mathrm{supp}(P_X)| - 1). \tag{7.59}$$

See Lemma 13.2 below. See also Exercise **VI.10** for the results on estimating P_X under different loss functions by means other than using the empirical distribution.

Corollary 7.18 is also useful for the statistical application of *entropy estimation*. Given n iid observations, a natural estimator of the entropy of P_X is the empirical entropy $\hat{H}_{\text{emp}} = H(\hat{P}_n)$ (plug-in estimator). It is clear that the empirical entropy is an *underestimate*, in the sense that the bias

$$\mathbb{E}[\hat{H}_{\text{emp}}] - H(P_X) = -\mathbb{E}[D(\hat{P}_n \| P_X)]$$

is always negative. For fixed P_X, $\hat{H}_{\text{emp}}$ is known to be consistent even on countably infinite alphabets [22], although the convergence rate can be arbitrarily slow, which aligns with the conclusion of (7.57). However, for a large alphabet of size $\Theta(n)$, the upper bound (7.58) does not vanish (this is tight for, e.g., uniform distribution). In this case, one needs to de-bias the empirical entropy (e.g. on the basis of (7.59)) or employ different techniques in order to achieve consistent estimation. See Section 29.4 for more details.

7.10 Most f-Divergences Are Locally χ^2-Like

In this section we prove analogs of Proposition 2.20 and Proposition 2.21 for the general f-divergences.

Theorem 7.19. *Suppose that $D_f(P\|Q) < \infty$ and the derivative of $f(x)$ at $x = 1$ exists. Then,*

$$\lim_{\lambda \to 0} \frac{1}{\lambda} D_f(\lambda P + \bar{\lambda} Q \| Q) = (1 - P[\text{supp}(Q)]) f'(\infty),$$

where as usual we take $0 \cdot \infty = 0$ on the right-hand side.

Remark 7.10. Note that we do not need a separate theorem for $D_f(Q\|\lambda P + \bar{\lambda} Q)$ since the exchange of arguments leads to another f-divergence with $f(x)$ replaced by $x f(1/x)$.

Proof. Without loss of generality we may assume $f(1) = f'(1) = 0$ and $f \geq 0$. Then, decomposing $P = \mu P_1 + \bar{\mu} P_0$ with $\mu = P[\text{supp}(Q)]$, $P_0 \perp Q$ and $P_1 \ll Q$ we have

$$\frac{1}{\lambda} D_f(\lambda P + \bar{\lambda} Q \| Q) = \bar{\mu} f'(\infty) + \int dQ \frac{1}{\lambda} f\left(1 + \lambda\left(\mu \frac{dP_1}{dQ} - 1\right)\right).$$

Note that $g(\lambda) = f(1 + \lambda t)$ is positive and convex for every $t \in \mathbb{R}$ and hence $\frac{1}{\lambda} g(\lambda)$ is monotonically decreasing to $g'(0) = 0$ as $\lambda \searrow 0$. Since for $\lambda = 1$ the integrand is assumed to be Q-integrable, the dominated convergence theorem applies and we get the result. □

Theorem 7.20. *Let f be twice continuously differentiable on $(0, \infty)$ with*

$$\limsup_{x \to +\infty} f''(x) < \infty.$$

If $\chi^2(P\|Q) < \infty$, then $D_f(\bar{\lambda} Q + \lambda P\|Q) < \infty$ for all $0 \leq \lambda < 1$ and

$$\lim_{\lambda \to 0} \frac{1}{\lambda^2} D_f(\bar{\lambda} Q + \lambda P \| Q) = \frac{f''(1)}{2} \chi^2(P\|Q). \tag{7.60}$$

If $\chi^2(P\|Q) = \infty$ *and* $f''(1) > 0$ *then* (7.60) *also holds, that is,* $D_f(\bar{\lambda}Q + \lambda P\|Q) = \omega(\lambda^2)$.

Remark 7.11. The conditions of the theorem are satisfied by D, D_{SKL}, H^2, JS, LC, and all Rényi divergences of orders $\lambda < 2$ (with $f(x) = \frac{1}{\lambda-1}(x^\lambda - 1)$; see Definition 7.24). A similar result holds also for the case when $f''(x) \to \infty$ with $x \to +\infty$ (e.g. Rényi divergences with $\lambda > 2$), but then we need to make extra assumptions in order to guarantee the applicability of the dominated convergence theorem (often just the finiteness of $D_f(P\|Q)$ is sufficient).

Proof. Assuming that $\chi^2(P\|Q) < \infty$ we must have $P \ll Q$ and hence we can use (7.1) as the definition of D_f. Note that under (7.1) without loss of generality we may assume $f'(1) = f(1) = 0$ (indeed, for that we can just add a multiple of $(x-1)$ to $f(x)$, which does not change the value of $D_f(P\|Q)$). From the Taylor expansion we have then

$$f(1+u) = u^2 \int_0^1 (1-t) f''(1+tu)dt.$$

Applying this with $u = \lambda\frac{P-Q}{Q}$ we get

$$D_f(\bar{\lambda}Q + \lambda P\|Q) = \int dQ \int_0^1 dt(1-t)\lambda^2 \left(\frac{P-Q}{Q}\right)^2 f''\left(1 + t\lambda\frac{P-Q}{Q}\right). \quad (7.61)$$

Note that for any $\epsilon > 0$ we have $\sup_{x \geq \epsilon} |f''(x)| \triangleq C_\epsilon < \infty$. Note that $\frac{P-Q}{Q} \geq -1$ and, thus, for every λ the integrand is non-negative and bounded by

$$\left(\frac{P-Q}{Q}\right)^2 C_{1-\lambda}, \quad (7.62)$$

which is integrable over $dQ \times \mathrm{Leb}$ (by the finiteness of $\chi^2(P\|Q)$ and Fubini, which applies due to non-negativity). Thus, $D_f(\bar{\lambda}Q + \lambda P\|Q) < \infty$. Dividing (7.61) by λ^2 we see that the integrand is dominated by (7.62) and hence we can apply the dominated convergence theorem to conclude

$$\begin{aligned}\lim_{\lambda\to 0} \frac{1}{\lambda^2} D_f(\bar{\lambda}Q + \lambda P\|Q) &\overset{(a)}{=} \int_0^1 dt(1-t) \int dQ \left(\frac{P-Q}{Q}\right)^2 \lim_{\lambda \to 0} f''\left(1 + t\lambda\frac{P-Q}{Q}\right) \\ &= \int_0^1 dt(1-t) \int dQ \left(\frac{P-Q}{Q}\right)^2 f''(1) = \frac{f''(1)}{2}\chi^2(P\|Q),\end{aligned}$$

which proves (7.60).

We proceed to proving that $D_f(\lambda P + \bar{\lambda}Q\|Q) = \omega(\lambda^2)$ when $\chi^2(P\|Q) = \infty$. If $P \ll Q$ then this follows by replacing the equality in (a) with $\geq$ due to Fatou's lemma. If $P \not\ll Q$, we consider the decomposition $P = \mu P_1 + \bar{\mu}P_0$ with $P_1 \ll Q$ and $P_0 \perp Q$. From definition (7.2) we have $\left(\text{for } \lambda_1 = \frac{\lambda\mu}{1-\lambda\bar{\mu}}\right)$

$$D_f(\lambda P + \bar{\lambda}Q\|Q) = (1 - \lambda\bar{\mu})D_f(\lambda_1 P_1 + \bar{\lambda}_1 Q\|Q) + \lambda\bar{\mu}D_f(P_0\|Q) \geq \lambda\bar{\mu}D_f(P_0\|Q).$$

Recall from Proposition 7.2 that $D_f(P_0\|Q) > 0$ unless $f(x) = c(x-1)$ for some constant c and the proof is complete. □

7.11 *f*-Divergences in Parametric Families: Fisher Information

In Section 2.6.2* we have already previewed the fact that in parametric families of distributions, the Hessian of the KL divergence turns out to coincide with the Fisher information. Here we collect such facts and their proofs. These materials form the basis of sharp bounds on parameter estimation that we will study later in Chapter 29.

To start with an example, let us return to the Gaussian location model (GLM) $P_t \triangleq \mathcal{N}(t,1), t \in \mathbb{R}$. From the identities presented in Section 7.7 we obtain the following asymptotics:

$$\mathrm{TV}(P_t, P_0) = \frac{|t|}{\sqrt{2\pi}} + o(|t|), \quad H^2(P_t, P_0) = \frac{t^2}{4} + o(t^2),$$

$$\chi^2(P_t\|P_0) = t^2 + o(t^2), \qquad D(P_t\|P_0) = \frac{t^2}{2\log e} + o(t^2),$$

$$\mathrm{LC}(P_t, P_0) = \frac{1}{4}t^2 + o(t^2).$$

We can see that, with the exception of TV, other f-divergences behave quadratically under small displacement $t \to 0$. This turns out to be a general fact, and furthermore the coefficient in front of t^2 is given by the Fisher information (at $t = 0$). To proceed carefully, we need some technical assumptions on the family P_t.

Definition 7.21. (Regular single-parameter families) Fix $\tau > 0$, space $\mathcal{X}$, and a family P_t of distributions on $\mathcal{X}$, $t \in [0, \tau)$. We define the following types of conditions that we call regularity at $t = 0$:

(a) $P_t(dx) = p_t(x)\mu(dx)$, for some measurable $(t, x) \mapsto p_t(x) \in \mathbb{R}_+$ and a fixed measure μ on $\mathcal{X}$.

(b_0) There exists a measurable function $(s, x) \mapsto \dot{p}_s(x)$, $s \in [0, \tau), x \in \mathcal{X}$, such that for μ-almost every x_0 we have $\int_0^\tau |\dot{p}_s(x_0)|ds < \infty$ and

$$p_t(x_0) = p_0(x_0) + \int_0^t \dot{p}_s(x_0)ds. \tag{7.63}$$

Furthermore, for μ-almost every x_0 we have $\lim_{t\searrow 0} \dot{p}_t(x_0) = \dot{p}_0(x_0)$.

(b_1) We have $\dot{p}_t(x) = 0$ whenever $p_0(x) = 0$ and, furthermore,

$$\int_{\mathcal{X}} \mu(dx) \sup_{0\le t<\tau} \frac{(\dot{p}_t(x))^2}{p_0(x)} < \infty. \tag{7.64}$$

(c_0) There exists a measurable function $(s, x) \mapsto \dot{h}_s(x)$, $s \in [0, \tau), x \in \mathcal{X}$, such that for μ-almost every x_0 we have $\int_0^\tau |\dot{h}_s(x_0)|ds < \infty$ and

$$h_t(x_0) \triangleq \sqrt{p_t(x_0)} = \sqrt{p_0(x_0)} + \int_0^t \dot{h}_s(x_0)ds. \tag{7.65}$$

Furthermore, for μ-almost every x_0 we have $\lim_{t\searrow 0} \dot{h}_t(x_0) = \dot{h}_0(x_0)$.

(c_1) The family of functions $\{(\dot{h}_t(x))^2 : t \in [0, \tau)\}$ is uniformly μ-integrable.

Remark 7.12. Recall that the uniform integrability condition (c_1) is implied by the following stronger (but easier to verify) condition:

$$\int_{\mathcal{X}} \mu(dx) \sup_{0 \le t < \tau} (\dot{h}_t(x))^2 < \infty. \tag{7.66}$$

Impressively, if one also assumes the continuous differentiability of h_t then the uniform integrability condition becomes equivalent to the continuity of the Fisher information:

$$t \mapsto J_F(t) \triangleq 4 \int \mu(dx)(\dot{h}_t(x))^2. \tag{7.67}$$

We refer to [68, Appendix V] for this finesse.

Theorem 7.22. *Let the family of distributions $\{P_t \colon t \in [0, \tau)\}$ satisfy the conditions (a), (b_0), and (b_1) in Definition 7.21. Then we have*

$$\chi^2(P_t \| P_0) = J_F(0)t^2 + o(t^2), \tag{7.68}$$

$$D(P_t \| P_0) = \frac{\log e}{2} J_F(0)t^2 + o(t^2), \tag{7.69}$$

where $J_F(0) \triangleq \int_{\mathcal{X}} \mu(dx) \frac{(\dot{p}_0(x))^2}{p_0(x)} < \infty$ is the Fisher information at $t = 0$.

Proof. From assumption (b_1) we see that for any x with $p_0(x) = 0$ we must have $\dot{p}_t(x) = 0$ and thus $p_t(x) = 0$ for all $t \in [0, \tau)$. Hence, we may restrict all integrals below to the subset $\{x \colon p_0(x) > 0\}$, on which the ratio $\frac{(p_t(x) - p_0(x_0))^2}{p_0(x)}$ is well defined. Consequently, we have by (7.63)

$$\begin{aligned}
\frac{1}{t^2} \chi^2(P_t \| P_0) &= \frac{1}{t^2} \int \mu(dx) \frac{(p_t(x) - p_0(x))^2}{p_0(x)} \\
&= \frac{1}{t^2} \int \mu(dx) \frac{1}{p_0(x)} \left(t \int_0^1 du \dot{p}_{tu}(x) \right)^2 \\
&\overset{\text{(i)}}{=} \int \mu(dx) \int_0^1 du_1 \int_0^1 du_2 \frac{\dot{p}_{tu_1}(x) \dot{p}_{tu_2}(x)}{p_0(x)}.
\end{aligned}$$

Note that by the continuity assumption in (b_1) we have $\dot{p}_{tu_1}(x)\dot{p}_{tu_2}(x) \to \dot{p}_0^2(x)$ for every (u_1, u_2, x) as $t \to 0$. Furthermore, we also have $\left| \frac{\dot{p}_{tu_1}(x)\dot{p}_{tu_2}(x)}{p_0(x)} \right| \le \sup_{0 \le t < \tau} \frac{(\dot{p}_t(x_0))^2}{p_0(x_0)}$, which is integrable by (7.64). Consequently, application of the dominated convergence theorem to the integral in (i) concludes the proof of (7.68).

We next take a detour to show that for any f-divergence with twice continuously differentiable f (and in fact, without assuming (7.64)) we have:

$$\liminf_{t \to 0} \frac{1}{t^2} D_f(P_t \| P_0) \ge \frac{f''(1)}{2} J_F(0). \tag{7.70}$$

Indeed, similar to (7.61) we get

$$D_f(P_t \| P_0) = \int_0^1 dz (1-z) \, \mathbb{E}_{X \sim P_0} \left[f'' \left(1 + z \frac{p_t(X) - p_0(X)}{p_0(X)} \right) \left(\frac{p_t(X) - p_0(X)}{p_0(X)} \right)^2 \right]. \tag{7.71}$$

Dividing by t^2 notice that from (b_0) we have $\frac{p_t(X)-p_0(X)}{tp_0(X)} \xrightarrow{\text{a.s.}} \frac{\dot{p}_0(X)}{p_0(X)}$ and thus

$$f''\left(1+z\frac{p_t(X)-p_0(X)}{p_0(X)}\right)\left(\frac{p_t(X)-p_0(X)}{tp_0(X)}\right)^2 \to f''(1)\left(\frac{\dot{p}_0(X)}{p_0(X)}\right)^2.$$

Thus, applying Fatou's lemma yields (7.70).

Finally, plugging $f(x) = x\log x$ in (7.71) we obtain for the KL divergence

$$\frac{1}{t^2}D(P_t\|P_0) = (\log e)\int_0^1 dz\, \mathbb{E}_{X\sim P_0}\left[\frac{1-z}{1+z\frac{p_t(X)-p_0(X)}{p_0(X)}}\left(\frac{p_t(X)-p_0(X)}{tp_0(X)}\right)^2\right]. \tag{7.72}$$

The first fraction inside the bracket is between 0 and 1 and the second is bounded by $\sup_{0<t<\tau}\left(\frac{\dot{p}_t(X)}{p_0(X)}\right)^2$, which is P_0-integrable by (b_1). Thus, the dominated convergence theorem applies to the double integral in (7.72) and we obtain

$$\lim_{t\to 0}\frac{1}{t^2}D(P_t\|P_0) = (\log e)\int_0^1 dz\, \mathbb{E}_{X\sim P_0}\left[(1-z)\left(\frac{\dot{p}_0(X)}{p_0(X)}\right)^2\right],$$

completing the proof of (7.69). □

Remark 7.13. Theorem 7.22 extends to the case of multi-dimensional parameters as follows. Define the Fisher information matrix at $\theta \in \mathbb{R}^d$:

$$J_F(\theta) \triangleq \int \mu(dx)\nabla_\theta\sqrt{p_\theta(x)}\nabla_\theta\sqrt{p_\theta(x)}^\top. \tag{7.73}$$

Then (7.68) becomes $\chi^2(P_t\|P_0) = t^\top J_F(0)t + o(\|t\|^2)$ as $t \to 0$ and similarly for (7.69), which has previously appeared in (2.34).

Theorem 7.22 applies to many cases (e.g. to smooth subfamilies of exponential families, for which one can take $\mu = P_0$ and $p_0(x) \equiv 1$), but it is not without assumptions. To demonstrate the issue, consider the following example.

Example 7.1. Location families with compact support

We say that family P_t is a (scalar) location family if $\mathcal{X} = \mathbb{R}$, $\mu = \text{Leb}$, and $p_t(x) = p_0(x-t)$. Consider the following example, for $\alpha > -1$:

$$p_0(x) = C_\alpha \times \begin{cases} x^\alpha, & x \in [0,1], \\ (2-x)^\alpha, & x \in [1,2], \\ 0, & \text{otherwise,} \end{cases}$$

with C_α chosen from normalization. Clearly, here condition (7.64) is not satisfied and both $\chi^2(P_t\|P_0)$ and $D(P_t\|P_0)$ are infinite for $t > 0$, since $P_t \not\ll P_0$. But $J_F(0) < \infty$ whenever $\alpha > 1$ and thus one expects that a certain remedy should be possible. Indeed, one can compute those f-divergences that are finite for $P_t \not\ll P_0$ and find that for $\alpha > 1$ they are quadratic in t. As an illustration, we have

$$H^2(P_t, P_0) = \begin{cases} \Theta(t^{1+\alpha}), & 0 \le \alpha < 1, \\ \Theta(t^2 \log \frac{1}{t}), & \alpha = 1, \\ \Theta(t^2), & \alpha > 1 \end{cases} \tag{7.74}$$

as $t \to 0$. This can be computed directly, or from the more general result of [222, Theorem VI.1.1].[5] For a relation between Hellinger and Fisher information see also Eq. (VI.5) in Exercise **VI.5**.

The previous example suggests that quadratic behavior as $t \to 0$ can hold even when $P_t \not\ll P_0$, which is the case handled by the next (more technical) result, whose proof we placed in Section 7.14*. One can verify that condition (c_1) is indeed satisfied for all $\alpha > 1$ in Example 7.1, thus establishing the quadratic behavior. Also note that the stronger (7.66) only applies to $\alpha \ge 2$.

Theorem 7.23. *Given a family of distributions $\{P_t : t \in [0, \tau)\}$ satisfying the conditions (a), (c_0), and (c_1) of Definition 7.21, we have*

$$\chi^2(P_t \| \bar{\epsilon} P_0 + \epsilon P_t) = t^2 \bar{\epsilon}^2 \left(J_F(0) + \frac{1 - 4\epsilon}{\epsilon} J^{\#}(0) \right) + o(t^2), \qquad \forall \epsilon \in (0, 1), \tag{7.75}$$

$$H^2(P_t, P_0) = \frac{t^2}{4} J_F(0) + o(t^2), \tag{7.76}$$

where $J_F(0) = 4 \int \dot{h}_0^2 d\mu < \infty$ is the Fisher information and $J^{\#}(0) = \int \dot{h}_0^2 1\{h_0 = 0\} d\mu$ may be called the Fisher defect at $t = 0$.

Example 7.2. On Fisher defect

Note that in most cases of interest we will have the situation that $t \mapsto h_t(x)$ is actually differentiable for all t in some *two-sided* neighborhood $(-\tau, \tau)$ of 0. In such cases, $h_0(x) = 0$ implies that $t = 0$ is a local minimum and thus $\dot{h}_0(x) = 0$, implying that the defect $J^{\#}(0) = 0$. However, for other families this will not be so, sometimes even when $p_t(x)$ is smooth on $t \in (-\tau, \tau)$ (but not h_t). Here is such an example.

Consider $P_t = \text{Ber}(t^2)$. A straightforward calculation shows:

$$\chi^2(P_t \| \bar{\epsilon} P_0 + \epsilon P_t) = t^2 \frac{\bar{\epsilon}^2}{\epsilon} + O(t^4), \qquad H^2(P_t, P_0) = 2(1 - \sqrt{1 - t^2}) = t^2 + O(t^4).$$

Taking $\mu(\{0\}) = \mu(\{1\}) = 1$ to be the counting measure, we get the following:

$$h_t(x) = \begin{cases} \sqrt{1 - t^2}, & x = 0, \\ |t|, & x = 1, \end{cases} \qquad \dot{h}_t(x) = \begin{cases} \frac{-t}{\sqrt{1-t^2}}, & x = 0, \\ \text{sign}(t), & x = 1, t \ne 0, \\ 1, & x = 1, t = 0 \ \text{(just as an agreement)}. \end{cases}$$

[5] The statistical significance of this calculation is that if we were to estimate the location parameter t from n iid observations, then the precision δ_n^* of the optimal estimator up to constant factors is given by solving $H^2(P_{\delta_n^*}, P_0) \asymp \frac{1}{n}$, see [222, Chapter VI]. For $\alpha < 1$ we have $\delta_n^* \asymp n^{-\frac{1}{1+\alpha}}$ which is notably better than the empirical mean estimator (attaining precision of only $n^{-\frac{1}{2}}$). For $\alpha = 1/2$ this fact was noted by D. Bernoulli in 1777 as a consequence of his (newly proposed) maximum likelihood estimation.

Note that if we view P_t as a family on $t \in [0, \tau)$ for small τ, then conditions (a), (c_0), and (c_1) are clearly all satisfied ($\dot{h}_t$ is bounded on $t \in (-\tau, \tau)$). We have $J_F(0) = 4$ and $J^\#(0) = 1$ and thus (7.75) recovers the correct expansion for χ^2 and (7.76) for H^2.

Notice that the non-smoothness of h_t only becomes visible if we extend the domain to $t \in (-\tau, \tau)$. In fact, this issue is not seen in terms of densities p_t. Indeed, let us compute the density p_t and its derivative $\dot{p}_t$ explicitly too:

$$p_t(x) = \begin{cases} 1 - t^2, & x = 0, \\ t^2, & x = 1, \end{cases} \qquad \dot{p}_t(x) = \begin{cases} -2t, & x = 0, \\ 2t, & x = 1. \end{cases}$$

Clearly, p_t is continuously differentiable on $t \in (-\tau, \tau)$. Furthermore, the following expectation (typically equal to $J_F(t)$ in (7.67)),

$$\mathbb{E}_{X \sim P_t}\left[\left(\frac{\dot{p}_t(X)}{p_t(X)}\right)^2\right] = \begin{cases} 0, & t = 0, \\ 4 + \frac{4t^2}{1-t^2}, & t \neq 0, \end{cases}$$

is discontinuous at $t = 0$. To make things worse, at $t = 0$ this expectation does not match our definition of the Fisher information $J_F(0)$ in Theorem 7.23, and thus does not yield the correct small-t behavior for either χ^2 or H^2. In general, to avoid difficulties one should restrict to those families with $t \mapsto h_t(x)$ continuously differentiable in $t \in (-\tau, \tau)$.

7.12 Rényi Divergences and Tensorization

The following family of divergence measures introduced by Rényi is key in many applications involving product measures. Although these measures are not f-divergences, they are obtained as the monotone transformation of an appropriate f-divergence and thus satisfy DPI and other properties of f-divergences. Later, Rényi divergence will feature prominently in characterizing the optimal error exponents in hypothesis testing (see Section 16.1 and especially Remark 16.1), in approximating a channel output statistics (see Section 25.4*), and in non-asymptotic bounds for composite hypothesis testing (see Section 32.2.1).

Definition 7.24. For any $\lambda \in \mathbb{R} \setminus \{0, 1\}$, the *Rényi divergence* of order λ between probability distributions P and Q is defined as

$$D_\lambda(P\|Q) \triangleq \frac{1}{\lambda - 1} \log \mathbb{E}_Q\left[\left(\frac{dP}{dQ}\right)^\lambda\right],$$

where $\mathbb{E}_Q[(\frac{dP}{dQ})^\lambda]$ is formally understood as a $\mathrm{sign}(\lambda - 1) D_f(P\|Q) + 1$ with $f(x) = \mathrm{sign}(\lambda - 1)x^\lambda - 1)$ – see Definition 7.1. Extending Definition 2.14 of conditional KL divergence and assuming the same setup, the conditional Rényi divergence is defined as

$$\begin{aligned} D_\lambda(P_{X|Y}\|Q_{X|Y}|P_Y) &\triangleq D_\lambda(P_Y \times P_{X|Y}\|P_Y \times Q_{X|Y}) \\ &= \frac{1}{\lambda - 1} \log \mathbb{E}_{Y \sim P_Y} \int_{\mathcal{X}} (dP_{X|Y}(x))^\lambda (dQ_{X|Y}(x))^{1-\lambda}. \end{aligned}$$

Numerous properties of Rényi divergences are known, see [433]. Here we only note a few:

- Special cases of $\lambda = \frac{1}{2}, 1, 2$: Under mild regularity conditions $\lim_{\lambda\to 1} D_\lambda(P\|Q) = D(P\|Q)$. On the other hand, $D_2 = \log(1+\chi^2)$ and $D_{\frac{1}{2}} = -2\log\left(1 - \frac{H^2}{2}\right)$ are monotone transformations of the χ^2-divergence (7.4) and the Hellinger distance (7.5), respectively.
- For all $\lambda \in \mathbb{R}$ the map $\lambda \mapsto D_\lambda(P\|Q)$ is non-decreasing and the map $\lambda \mapsto (1-\lambda)D_\lambda(P\|Q)$ is concave.
- For $\lambda \in [0,1]$ the map $(P, Q) \mapsto D_\lambda(P\|Q)$ is convex.
- For $\lambda \geq 0$ the map $Q \mapsto D_\lambda(P\|Q)$ is convex.
- For Q uniform on a finite alphabet of size m, $D_\lambda(P\|Q) = \log m - H_\lambda(P)$, where H_λ is the Rényi entropy of order λ defined in (1.4). This recovers Theorem 2.2 as the special case of $\lambda = 1$.
- There is a version of the chain rule:

$$D_\lambda(P_{A,B}\|Q_{A,B}) = D_\lambda(P_B\|Q_B) + D_\lambda\left(P_{A|B}\|Q_{A|B}|P_B^{(\lambda)}\right), \tag{7.77}$$

  where $P_B^{(\lambda)}$ is the λ-tilting of P_B toward Q_B given by

$$P_B^{(\lambda)}(b) \triangleq P_B^\lambda(b)Q_B^{1-\lambda}(b)\exp\{-(\lambda-1)D_\lambda(P_B\|Q_B)\}. \tag{7.78}$$

- The key property is additivity under products, or *tensorization*:

$$D_\lambda\left(\prod_i P_{X_i}\middle\|\prod_i Q_{X_i}\right) = \sum_i D_\lambda(P_{X_i}\|Q_{X_i}), \tag{7.79}$$

  which is a simple consequence of (7.77). The D_λ's are the only divergences satisfying DPI and tensorization [311]. The most well-known special cases of (7.79) are for Hellinger distance (see (7.26)) and for χ^2:

$$1+\chi^2\left(\prod_{i=1}^n P_i\middle\|\prod_{i=1}^n Q_i\right) = \prod_{i=1}^n\left(1+\chi^2(P_i\|Q_i)\right).$$

  We can also obtain additive bounds for non-product distributions, see Exercises **I.42** and **I.43**.

The following consequence of the chain rule will be crucial in statistical applications later (see Section 32.2, in particular, Theorem 32.8).

Proposition 7.25. *Consider product channels $P_{Y^n|X^n} = \prod P_{Y_i|X_i}$ and $Q_{Y^n|X^n} = \prod Q_{Y_i|X_i}$. We have (with all optimizations over all possible distributions)*

$$\inf_{P_{X^n},Q_{X^n}} D_\lambda(P_{Y^n}\|Q_{Y^n}) = \sum_{i=1}^n \inf_{P_{X_i},Q_{X_i}} D_\lambda(P_{Y_i}\|Q_{Y_i}), \tag{7.80}$$

$$\sup_{P_{X^n},Q_{X^n}} D_\lambda(P_{Y^n}\|Q_{Y^n}) = \sum_{i=1}^n \sup_{P_{X_i},Q_{X_i}} D_\lambda(P_{Y_i}\|Q_{Y_i}) = \sum_{i=1}^n \sup_{x,x'} D_\lambda(P_{Y_i|X_i=x}\|Q_{Y_i|X_i=x'}). \tag{7.81}$$

In particular, for any collections of distributions $\{P_\theta : \theta \in \Theta\}$ *and* $\{Q_\theta : \theta \in \Theta\}$:

$$\inf_{P \in \text{co}\{P_\theta^{\otimes n}\}, Q \in \text{co}\{Q_\theta^{\otimes n}\}} D_\lambda(P\|Q) \geq n \inf_{P \in \text{co}\{P_\theta\}, Q \in \text{co}\{Q_\theta\}} D_\lambda(P\|Q), \tag{7.82}$$

$$\sup_{P \in \text{co}\{P_\theta^{\otimes n}\}, Q \in \text{co}\{Q_\theta^{\otimes n}\}} D_\lambda(P\|Q) \leq n \sup_{P \in \text{co}\{P_\theta\}, Q \in \text{co}\{Q_\theta\}} D_\lambda(P\|Q). \tag{7.83}$$

Remark 7.14. The mnemonic for (7.82) and (7.83) is that "mixtures of products are less distinguishable than products of mixtures." The former arise in statistical settings where iid observations are drawn from a single distribution whose parameter is drawn from a prior.

Proof. The second equality in (7.81) follows from the fact that D_λ is an increasing function of an f-divergence, and thus maximization should be attained at an extreme point of the space of probabilities, which are just the single-point masses. The main equalities (7.80) and (7.81) follow from (a) restricting optimizations to product distributions and invoking (7.79) and (b) the chain rule (7.77) for D_λ. For example for $n = 2$, we fix P_{X^2} and Q_{X^2}, which (via channels) induce joint distributions P_{X^2,Y^2} and Q_{X^2,Y^2}. Then we have

$$D_\lambda(P_{Y_1|Y_2=y}\|Q_{Y_1|Y_2=y'}) \geq \inf_{\tilde{P}_{X_1}, \tilde{Q}_{X_1}} D_\lambda(\tilde{P}_{Y_1}\|\tilde{Q}_{Y_1}),$$

since $P_{Y_1|Y_2=y}$ is a distribution induced by passing $\tilde{P}_{X_1} = P_{X_1|Y_2=y}$ through $P_{Y_1|X_1}$, and similarly for $Q_{Y_1|Y_2=y'}$. In all, we get

$$D_\lambda(P_{Y^2}\|Q_{Y^2}) = D_\lambda(P_{Y_2}\|Q_{Y_2}) + D_\lambda(P_{Y_1|Y_2}\|Q_{Y_1|Y_2}|P_{Y_2}^{(\lambda)}) \geq \sum_{i=1}^{2} \inf_{P_{X_i}, Q_{X_i}} D_\lambda(P_{Y_i}\|Q_{Y_i}),$$

as claimed. The case of sup is handled similarly.

From (7.80) and (7.81), we get (7.82) and (7.83) by taking $\mathcal{X} = \Theta$ and specializing the inf and sup to diagonal distributions P_{X^n} and Q_{X^n}, that is, those with the property that $P[X_1 = \cdots = X_n] = 1$ and $Q[X_1 = \cdots = X_n] = 1$. □

7.13 Variational Representation of *f*-Divergences

In Theorem 4.6 we presented a very useful variational representation of KL divergence due to Donsker and Varadhan. In this section we show how to derive such representations for other f-divergences in a principled way. The proofs are slightly technical and given in Section 7.14* at the end of this chapter.

Let $f\colon (0, +\infty) \to \mathbb{R}$ be a convex function. The convex conjugate $f^*\colon \mathbb{R} \to \mathbb{R} \cup \{+\infty\}$ of f is defined by:

$$f^*(y) = \sup_{x \in \mathbb{R}_+} xy - f(x), \quad y \in \mathbb{R}. \tag{7.84}$$

Denote the domain of f^* by $\text{dom}(f^*) \triangleq \{y\colon f^*(y) < \infty\}$. Two important properties of the convex conjugates are as follows.

1 f^* is also convex (which holds regardless of f being convex or not).
2 Biconjugation: $(f^*)^* = f$, which means

$$f(x) = \sup_y xy - f^*(y)$$

and implies the following (for all $x > 0$ and y)

$$f(x) + f^*(y) \geq xy.$$

Similarly, we can define a convex conjugate for any convex functional $\Psi(P)$ defined on the space of measures, by setting

$$\Psi^*(g) = \sup_P \int gdP - \Psi(P). \tag{7.85}$$

Under appropriate conditions (e.g. finite $\mathcal{X}$), biconjugation then yields the sought-after variational representation

$$\Psi(P) = \sup_g \int gdP - \Psi^*(g). \tag{7.86}$$

Next we will compute these conjugates for $\Psi(P) = D_f(P\|Q)$. It turns out to be convenient to first extend the definition of $D_f(P\|Q)$ to all finite signed measures P and then compute the conjugate. To this end, let $f_{\text{ext}}\colon \mathbb{R} \to \mathbb{R} \cup \{+\infty\}$ be an extension of f, such that $f_{\text{ext}}(x) = f(x)$ for $x \geq 0$ and f_{ext} is convex on $\mathbb{R}$. In general, we can always choose $f_{\text{ext}}(x) = \infty$ for all $x < 0$. In special cases, for example, $f(x) = |x-1|/2$ or $f(x) = (x-1)^2$, we can directly take $f_{\text{ext}}(x) = f(x)$ for all x. Now we can define $D_f(P\|Q)$ for all signed measures P in the same way as in Definition 7.1 using f_{ext} in place of f.

For each choice of f_{ext} we have a variational representation of f-divergence:

Theorem 7.26. *Let P and Q be probability measures on $\mathcal{X}$. Fix an extension f_{ext} of f and let f^*_{ext} be the conjugate of f_{ext}, that is, $f^*_{ext}(y) = \sup_{x\in\mathbb{R}} xy - f_{ext}(x)$. Denote* $\text{dom}(f^*_{ext}) \triangleq \{y\colon f^*_{ext}(y) < \infty\}$. *Then*

$$D_f(P\|Q) = \sup_{g\colon \mathcal{X}\to\text{dom}(f^*_{ext})} \mathbb{E}_P[g(X)] - \mathbb{E}_Q[f^*_{ext}(g(X))], \tag{7.87}$$

where the supremum can be taken either (a) over all simple g or (b) over all g satisfying $\mathbb{E}_Q[f^*_{ext}(g(X))] < \infty$.

We remark that when $P \ll Q$ both results (a) and (b) also hold for the supremum over $g\colon \mathcal{X} \to \mathbb{R}$, that is, without restricting to $g(x) \in \text{dom}(f^*_{\text{ext}})$.

As a consequence of the variational characterization, we get the following properties for f-divergences:

1 *Convexity*: First of all, note that $D_f(P\|Q)$ is expressed as a supremum of affine functions (since the expectation is a linear operation). As a result, we get that $(P, Q) \mapsto D_f(P\|Q)$ is convex, which was proved previously in Theorem 7.5 using a different method.

2 *Weak lower semicontinuity*: Recall the example in Remark 4.5, where $\{X_i\}$ are iid Rademachers (± 1), and

$$\frac{\sum_{i=1}^n X_i}{\sqrt{n}} \xrightarrow{d} \mathcal{N}(0,1)$$

by the central limit theorem; however, by Proposition 7.2, for all n,

$$D_f\left(\frac{P_{X_1+X_2+\cdots+X_n}}{\sqrt{n}} \middle\| \mathcal{N}(0,1)\right) = f(0) + f'(\infty) > 0,$$

since the former distribution is discrete and the latter is continuous. Therefore similar to the KL divergence, the best we can hope for f-divergence is semicontinuity. Indeed, if $\mathcal{X}$ is a nice space (e.g. Euclidean space), in (7.87) we can restrict the function g to continuous bounded functions, in which case $D_f(P\|Q)$ is expressed as a supremum of weakly continuous functionals (note that $f^* \circ g$ is also continuous and bounded since f^* is continuous) and is hence weakly lower semicontinuous, that is, for any sequence of distributions P_n and Q_n such that $P_n \xrightarrow{w} P$ and $Q_n \xrightarrow{w} Q$, we have

$$\liminf_{n\to\infty} D_f(P_n\|Q_n) \geq D_f(P\|Q).$$

3 *Relation to DPI:* As discussed in (4.15) variational representations can be thought of as extensions of the DPI. As an exercise, one should try to derive the estimate

$$|P[A] - Q[A]| \leq \sqrt{Q[A] \cdot \chi^2(P\|Q)}$$

via both the DPI and (7.91).

Example 7.3. Total variation and Hellinger

For total variation, we have $f(x) = \frac{1}{2}|x-1|$. Consider the extension $f_{\text{ext}}(x) = \frac{1}{2}|x-1|$ for $x \in \mathbb{R}$. Then

$$f^*_{\text{ext}}(y) = \sup_x \left\{xy - \frac{1}{2}|x-1|\right\} = \begin{cases} +\infty, & \text{if } |y| > \frac{1}{2}, \\ y, & \text{if } |y| \leq \frac{1}{2}. \end{cases}$$

Thus (7.87) gives

$$\text{TV}(P,Q) = \sup_{g:\, |g|\leq\frac{1}{2}} \mathbb{E}_P[g(X)] - \mathbb{E}_Q[g(X)], \tag{7.88}$$

which previously appeared in (7.18). A calculation for squared Hellinger yields $f^*_{\text{ext}}(y) = \frac{y}{1-y}$ with $y \in (-\infty, 1)$ and, thus, after changing from g to $h = 1-g$ in (7.87), we obtain

$$H^2(P,Q) = 2 - \inf_{h>0} \mathbb{E}_P[h] + \mathbb{E}_Q\left[\frac{1}{h}\right].$$

As an application, consider $f\colon \mathcal{X} \to [0,1]$ and $\tau \in (0,1)$, so that $h = 1 - \tau f$ satisfies $\frac{1}{h} \leq 1 + \frac{\tau}{1-\tau} f$. Then the previous characterization implies

$$\mathbb{E}_P[f] \leq \frac{1}{1-\tau}\mathbb{E}_Q[f] + \frac{1}{\tau}H^2(P,Q), \qquad \forall f\colon \mathcal{X} \to [0,1],\ \forall \tau \in (0,1).$$

Example 7.4. χ^2-divergence

For χ^2-divergence we have $f(x) = (x-1)^2$. Take $f_{\text{ext}}(x) = (x-1)^2$, whose conjugate is $f^*_{\text{ext}}(y) = y + \frac{y^2}{4}$. Applying (7.87) yields

$$\chi^2(P\|Q) = \sup_{g:\mathcal{X}\to\mathbb{R}} \mathbb{E}_P[g(X)] - \mathbb{E}_Q\left[g(X) + \frac{g^2(X)}{4}\right] \tag{7.89}$$

$$= \sup_{g:\mathcal{X}\to\mathbb{R}} 2\mathbb{E}_P[g(X)] - \mathbb{E}_Q[g^2(X)] - 1, \tag{7.90}$$

where the last step follows from a change of variable ($g \leftarrow \frac{1}{2}g - 1$).

To get another equivalent, but much more memorable, representation, we notice that (7.90) is not scale-invariant. To make it so, setting $g = \lambda h$ and optimizing over the $\lambda \in \mathbb{R}$ first we get

$$\chi^2(P\|Q) = \sup_{h:\,\mathcal{X}\to\mathbb{R}} \frac{(\mathbb{E}_P[h(X)] - \mathbb{E}_Q[h(X)])^2}{\mathrm{Var}_Q(h(X))}. \tag{7.91}$$

The statistical interpretation of (7.91) is as follows: If a test statistic $h(X)$ is such that the separation between its expectation under P and Q far exceeds its standard deviation, then this suggests the two hypotheses can be distinguished reliably. The representation (7.91) will turn out to be useful in statistical applications in Chapter 29 for deriving the Hammersley–Chapman–Robbins (HCR) lower bound as well as its Bayesian version, see Section 29.1.2, and ultimately the Cramér–Rao and van Trees lower bounds.

Example 7.5. Jensen–Shannon divergence and GANs

For the Jensen–Shannon divergence (7.8) we have $f(x) = x\log\frac{2x}{1+x} + \log\frac{2}{1+x}$. Computing the conjugate we obtain $f^*(s) = -\log(2 - \exp(s))$ with domain $s \in (-\infty, \log 2)$. We obtain from (7.87) the characterization

$$\mathrm{JS}(P,Q) = \sup_{g:\,\mathcal{X}\to(-\infty,\,\log 2)} \mathbb{E}_P[g] + \mathbb{E}_Q[\log(2 - \exp\{g(X)\})],$$

or after reparameterizing $h = \exp\{g\}/2$ we get

$$\mathrm{JS}(P,Q) = \sup_{h:\,\mathcal{X}\to(0,1)} 2\log 2 + \mathbb{E}_P[\log(h)] + \mathbb{E}_Q[\log(1-h)].$$

This characterization is the basis of an influential modern method of density estimation, known as *generative adversarial networks* (GANs) [193]. Here is its essence. Suppose that we are trying to approximate a very complicated distribution P on $\mathbb{R}^d$ by representing it as (the law of) a generator map $G\colon \mathbb{R}^m \to \mathbb{R}^d$ applied to a standard normal $Z \sim \mathcal{N}(0, I_m)$. The idea of [193] is to search for a good G by minimizing $\mathrm{JS}(P, P_{G(Z)})$. Due to the variational characterization we can equivalently formulate this problem as

$$\inf_G \sup_h \mathbb{E}_{X\sim P}[\log h(X)] + \mathbb{E}_{Z\sim \mathcal{N}}[\log(1-h(G(Z)))]$$

(and in this context the test function h is called a discriminator or, less often, a critic). Since the distribution P is only available to us through a sample of iid observations $x_1, \ldots, x_n \sim P$, we approximate this minimax problem by

$$\inf_G \sup_h \frac{1}{n}\sum_{i=1}^n \log h(x_i) + \mathbb{E}_{Z\sim \mathcal{N}}[\log(1-h(G(Z)))].$$

In order to be able to solve this problem another idea of [193] is to approximate the intractable optimizations over the infinite-dimensional function spaces of G and h by an optimization over neural networks. This is implemented via alternating gradient ascent/descent steps over the (finite-dimensional) parameter spaces defining the neural networks of G and h. Following the breakthrough of [193] variations on their idea resulted in finding $G(Z)$'s that yielded incredibly realistic images, music, videos, 3D scenery, and more.

Example 7.6. KL divergence

In this case we have $f(x) = x\log x$. Consider the extension $f_{\text{ext}}(x) = \infty$ for $x < 0$, whose convex conjugate is $f^*(y) = \frac{\log e}{e}\exp(y)$. Hence applying (7.87) with g replaced by $g + \log e$ yields

$$D(P\|Q) = \sup_{g\colon \mathcal{X}\to\mathbb{R}} \mathbb{E}_P[g(X)] - (\mathbb{E}_Q[\exp\{g(X)\}] - 1)\log e. \tag{7.92}$$

Note that in the last example, the variational representation (7.92) we obtained for the KL divergence is not the same as the Donsker–Varadhan identity in Theorem 4.6, that is,

$$D(P\|Q) = \sup_{g\colon \mathcal{X}\to\mathbb{R}} \mathbb{E}_P[g(X)] - \log \mathbb{E}_Q[\exp\{g(X)\}]. \tag{7.93}$$

In fact, (7.92) is weaker than (7.93) in the sense that for each choice of g, the obtained lower bound on $D(P\|Q)$ on the RHS is smaller. Furthermore, regardless of the choice of f_{ext}, the Donsker–Varadhan representation can never be obtained from Theorem 7.26 because, unlike (7.93), the second term in (7.87) is always linear in Q. It turns out that if we define $D_f(P\|Q) = \infty$ for all non-probability measures P, and compute its convex conjugate, we obtain in the next theorem a different type of variational representation, which, specialized to KL divergence in Example 7.6, recovers exactly the Donsker–Varadhan identity.

Theorem 7.27. *Consider the extension f_{ext} of f such that $f_{ext}(x) = \infty$ for $x < 0$. Let $S = \{x\colon q(x) > 0\}$ where q is as in (7.2). Then*

$$D_f(P\|Q) = f'(\infty)P[S^c] + \sup_g \mathbb{E}_P[g1_S] - \Psi^*_{Q,P}(g), \tag{7.94}$$

where

$$\Psi^*_{Q,P}(g) \triangleq \inf_{a\in\mathbb{R}} \mathbb{E}_Q[f^*_{ext}(g(X)-a)] + aP[S].$$

In the special case $f'(\infty) = \infty$, we have

$$D_f(P\|Q) = \sup_g \mathbb{E}_P[g] - \Psi_Q^*(g), \qquad \Psi_Q^*(g) \triangleq \inf_{a\in\mathbb{R}} \mathbb{E}_Q[f_{ext}^*(g(X) - a)] + a. \quad (7.95)$$

Remark 7.15. (Marton's divergence) Recall that in Theorem 7.7 we have shown both the sup- and inf-characterizations for the total variation. Do other f-divergences also possess inf-characterizations? The only other known example (to us) is due to Marton. Let

$$D_m(P\|Q) = \int dQ \left(1 - \frac{dP}{dQ}\right)_+^2,$$

which is clearly an f-divergence with $f(x) = (1-x)_+^2$. We have the following [69, Lemma 8.3]:

$$D_m(P\|Q) = \inf\{\mathbb{E}[P[X \neq Y|Y]^2] \colon X \sim P, Y \sim Q\},$$

where the infimum is over all couplings of P and Q. See Exercise **I.44**.

Marton's D_m divergence plays a crucial role in the theory of concentration of measure [69, Chapter 8]. Note also that while Theorem 7.20 does not apply to D_m, due to the absence of twice continuous differentiability, it does apply to the symmetrized Marton divergence $D_{sm}(P\|Q) \triangleq D_m(P\|Q) + D_m(Q\|P)$.

We end this section by describing some properties of Fisher information akin to those of f-divergences. In view of its role in the local expansion, we expect the Fisher information to inherit these properties such as monotonicity, data-processing inequality, and the variational representation. Indeed the first two can be established directly; see Exercise **I.46**. In [220] Huber introduced the following variational extension of the Fisher information (2.40) (in the location family) of a distribution on $\mathbb{R}$: for any $P \in \mathcal{P}(\mathbb{R})$, define

$$J(P) = \sup_h \frac{\mathbb{E}_P[h'(X)]^2}{\mathbb{E}_P[h(X)^2]}, \quad (7.96)$$

where the supremum is over all test functions $h \in C_c^1$ that are continuously differentiable and compactly supported such that $\mathbb{E}_P[h(X)^2] > 0$. Huber showed that $J(P) < \infty$ if and only if P has an absolutely continuous density p such that $\int (p')^2/p < \infty$, in which case (7.96) agrees with the usual definition (2.40).[6] This sup-representation can be anticipated by combining the variational representation (7.91) of χ^2-divergence and its local expansion (7.68) that involves the Fisher information. Indeed, setting aside regularity conditions, by Taylor expansion we have

$$\chi^2(P_t\|P) = \sup \frac{(\mathbb{E}[h(X+t) - h(X)])^2}{\mathbb{E}[h^2(X)]} = \sup \frac{\mathbb{E}[h'(X)]^2}{\mathbb{E}[h^2(X)]} \cdot t^2 + o(t^2),$$

[6] As an example in the reverse direction, $J(\text{Unif}(0,1)) = \infty$ which follows from choosing test functions such as $h(x) = \cos^2 \frac{x\pi}{\epsilon} 1\{|x| \leq \epsilon/2\}$ and $\epsilon \to 0$.

which is also $\chi^2(P_t\|P) = J(P)t^2 + o(t^2)$. A direct proof can be given by applying integration by parts and Cauchy–Schwarz: $(\int ph')^2 = (\int p'h)^2 \le \int h^2 p \int (p')^2/p$, which also shows that the optimal test function is given by the *score* function $h = p'/p$; for details, see [220, Theorem 4.2].

7.14* Technical Proofs: Convexity, Local Expansions, and Variational Representations

In this section we collect the proofs of some technical theorems from this chapter.

Proof of Theorem 7.23. By definition we have

$$L(t) \triangleq \frac{1}{\bar\epsilon^2 t^2}\chi^2(P_t\|\bar\epsilon P_0 + \epsilon P_t) = \frac{1}{t^2}\int_{\mathcal{X}} \mu(dx)\frac{(p_t(x)-p_0(x))^2}{\bar\epsilon p_0(x)+\epsilon p_t(x)} = \frac{1}{t^2}\int \mu(dx) g(t,x)^2, \tag{7.97}$$

where

$$g(t,x) \triangleq \frac{p_t(x)-p_0(x)}{\sqrt{\bar\epsilon p_0(x)+\epsilon p_t(x)}} = \phi(h_t(x);x), \qquad \phi(h;x) \triangleq \frac{h^2 - p_0(x)}{\sqrt{\bar\epsilon p_0(x)+\epsilon h^2}}.$$

By (c_0) the function $t \mapsto h_t(x) \triangleq \sqrt{p_t(x)}$ is absolutely continuous (for μ-a.e. x). Below we will show that $\|\phi(\cdot;x)\|_{\mathrm{Lip}} = \sup_{h\ge 0}|\phi'(h;x)| \le \frac{2-\epsilon}{(1-\epsilon)\sqrt{\epsilon}}$. This implies that $t \mapsto g(t,x)$ is also absolutely continuous and hence differentiable almost everywhere. Consequently, we have

$$g(t,x) = t\int_0^1 du\,\dot g(tu,x), \quad \dot g(t,x) \triangleq \phi'(h_t(x);x)\dot h_t(x).$$

Since $\phi'(\cdot;x)$ is continuous with

$$\phi'(h_0(x);x) = \begin{cases} 2, & x\colon h_0(x) > 0,\\ \frac{1}{\sqrt{\epsilon}}, & x\colon h_0(x) = 0 \end{cases} \tag{7.98}$$

(we verify these facts below too), we conclude that

$$\lim_{s\to 0}\dot g(s,x) = \dot g(0,x) = \dot h_0(x)\left(2\cdot 1\{h_0(x)>0\} + \frac{1}{\sqrt{\epsilon}}1\{h_0(x)=0\}\right), \tag{7.99}$$

where we also used continuity $\dot h_t(x) \to \dot h_0(x)$ by assumption (c_0).

Substituting the integral expression for $g(t,x)$ into (7.97) we obtain

$$L(t) = \int \mu(dx)\int_0^1 du_1 \int_0^1 du_2 \dot g(tu_1,x)\dot g(tu_2,x). \tag{7.100}$$

Since $|\dot g(s,x)| \le C|\dot h_s(x)|$ for some $C = C(\epsilon)$, we have from Cauchy–Schwarz

$$\int \mu(dx)|\dot g(s_1,x)\dot g(s_2,x)| \le C^2 \sup_t \int_{\mathcal{X}} \mu(dx)\dot h_t(x)^2 < \infty, \tag{7.101}$$

where the last inequality follows from the uniform integrability assumption (c_1). This implies that Fubini's theorem applies in (7.100) and we obtain

$$L(t) = \int_0^1 du_1 \int_0^1 du_2 G(tu_1, tu_2), \qquad G(s_1, s_2) \triangleq \int \mu(dx)\dot{g}(s_1, x)\dot{g}(s_2, x).$$

Notice that if a family of functions $\{f_\alpha(x) : \alpha \in I\}$ is uniformly square-integrable, then the family $\{f_\alpha(x) f_\beta(x) : \alpha \in I, \beta \in I\}$ is uniformly integrable simply because $|f_\alpha f_\beta| \le \frac{1}{2}(f_\alpha^2 + f_\beta^2)$. Consequently, from the assumption (c_1) we see that the integral defining $G(s_1, s_2)$ allows passing the limit over s_1, s_2 inside the integral. From (7.99) we get as $t \to 0$

$$\begin{aligned} G(tu_1, tu_2) \to G(0,0) &= \int \mu(dx)\dot{h}_0(x)^2 \left(4 \cdot 1\{h_0 > 0\} + \frac{1}{\epsilon} 1\{h_0 = 0\}\right) \\ &= J_F(0) + \frac{1 - 4\epsilon}{\epsilon} J^\#(0). \end{aligned}$$

From (7.101) we see that $G(s_1, s_2)$ is bounded and thus the bounded convergence theorem applies and

$$\lim_{t \to 0} \int_0^1 du_1 \int_0^1 du_2 G(tu_1, tu_2) = G(0,0),$$

which thus concludes the proof of $L(t) \to J_F(0)$ and of (7.75) assuming facts about ϕ. Let us verify those.

For simplicity, in the next paragraph we omit the argument x in $h_0(x)$ and $\phi(\cdot; x)$. A straightforward differentiation yields

$$\phi'(h) = 2h \frac{(1 - \frac{\epsilon}{2})h_0^2 + \frac{\epsilon}{2}h^2}{(\bar{\epsilon}h_0^2 + \epsilon h^2)^{3/2}}.$$

Since

$$\frac{h}{\sqrt{\bar{\epsilon}h_0^2 + \epsilon h^2}} \le \frac{1}{\sqrt{\epsilon}} \quad \text{and} \quad \frac{(1 - \frac{\epsilon}{2})h_0^2 + \frac{\epsilon}{2}h^2}{\bar{\epsilon}h_0^2 + \epsilon h^2} \le \frac{1 - \epsilon/2}{1 - \epsilon}$$

we obtain the finiteness of ϕ'. For the continuity of ϕ' notice that if $h_0 > 0$ then clearly the function is continuous, whereas for $h_0 = 0$ we have $\phi'(h) = \frac{1}{\sqrt{\epsilon}}$ for all h.

We next proceed to the Hellinger distance. Just like in the argument above, we define

$$M(t) \triangleq \frac{1}{t^2} H^2(P_t, P_0) = \int \mu(dx) \int_0^1 du_1 \int_0^1 du_2 \dot{h}_{tu_1}(x)\dot{h}_{tu_2}(x).$$

Exactly as above from Cauchy–Schwarz and $\sup_t \int \mu(dx)\dot{h}_t(x)^2 < \infty$ we conclude that Fubini applies and hence

$$M(t) = \int_0^1 du_1 \int_0^1 du_2 H(tu_1, tu_2), \qquad H(s_1, s_2) \triangleq \int \mu(dx)\dot{h}_{s_1}(x)\dot{h}_{s_2}(x).$$

Again, the family $\{\dot{h}_{s_1}\dot{h}_{s_2} : s_1 \in [0, \tau), s_2 \in [0, \tau)\}$ is uniformly integrable and thus from (c_0) we conclude that $H(tu_1, tu_2) \to \frac{1}{4}J_F(0)$. Furthermore, similar to (7.101) we see that $H(s_1, s_2)$ is bounded and thus

$$\lim_{t\to 0} M(t) = \int_0^1 du_1 \int_0^1 du_2 \lim_{t\to 0} H(tu_1, tu_2) = \frac{1}{4} J_F(0),$$

concluding the proof of (7.76). □

Proceeding to variational representations, we prove the counterpart of the Gelfand–Yaglom–Perez theorem (Theorem 4.5); see [185].

Proof of Theorem 7.6. The lower bound $D_f(P\|Q) \geq D_f(P_{\mathcal{E}}\|Q_{\mathcal{E}})$ follows from the DPI. To prove an upper bound, first we reduce to the case of $f \geq 0$ by property 6 in Proposition 7.2. Then define sets $S = \mathrm{supp}(Q)$, $F_\infty = \{\frac{dP}{dQ} = 0\}$, and for a fixed $\epsilon > 0$ let

$$F_m = \left\{\epsilon m \leq f\left(\frac{dP}{dQ}\right) < \epsilon(m+1)\right\}, \quad m = 0, 1, \ldots.$$

We have

$$\begin{aligned}\epsilon \sum_m m Q[F_m] \leq \int_S dQ f\left(\frac{dP}{dQ}\right) &\leq \epsilon \sum_m (m+1) Q[F_m] + f(0) Q[F_\infty] \\ &\leq \epsilon \sum_m m Q[F_m] + f(0) Q[F_\infty] + \epsilon. \end{aligned} \tag{7.102}$$

Notice that on the interval $I_m^+ = \{x > 1 \colon \epsilon m \leq f(x) < \epsilon(m+1)\}$ the function f is increasing and on $I_m^- = \{x \leq 1 \colon \epsilon m \leq f(x) < \epsilon(m+1)\}$ it is decreasing. Thus partition further every F_m into $F_m^+ = \{\frac{dP}{dQ} \in I_m^+\}$ and $F_m^- = \{\frac{dP}{dQ} \in I_m^-\}$. Then, we see that

$$f\left(\frac{P[F_m^\pm]}{Q[F_m^\pm]}\right) \geq \epsilon m.$$

Next, define the partition consisting of sets $\mathcal{E} = \{F_0^+, F_0^-, \ldots, F_n^+, F_n^-, F_\infty, S^c, \bigcup_{m>n} F_m\}$. For this partition we have, by the previous display:

$$D(P_{\mathcal{E}}\|Q_{\mathcal{E}}) \geq \epsilon \sum_{m\leq n} m Q[F_m] + f(0) Q[F_\infty] + f'(\infty) P[S^c]. \tag{7.103}$$

We next show that with sufficiently large n and sufficiently small ϵ the RHS of (7.103) approaches $D_f(P\|Q)$. If $f(0)Q[F_\infty] = \infty$ (and hence $D_f(P\|Q) = \infty$) then clearly (7.103) is also infinite. Thus, assume that $f(0)Q[F_\infty] < \infty$.

If $\int_S dQ f\left(\frac{dP}{dQ}\right) = \infty$, then the sum over m on the RHS of (7.102) is also infinite, and hence for any $N > 0$ there exists some n such that $\sum_{m\leq n} m Q[F_m] \geq N$, thus showing that the RHS of (7.103) can be made arbitrarily large. Thus, assume $\int_S dQ f\left(\frac{dP}{dQ}\right) < \infty$. Considering the LHS of (7.102) we conclude that for some large n we have $\sum_{m>n} m Q[F_m] \leq \frac{1}{2}$. Then, we must have again from (7.102)

$$\epsilon \sum_{m\leq n} m Q[F_m] + f(0) Q[F_\infty] \geq \int_S dQ f\left(\frac{dP}{dQ}\right) - \frac{3}{2}\epsilon.$$

Thus, we have shown that for arbitrary $\epsilon > 0$ the RHS of (7.103) can be made greater than $D_f(P\|Q) - \frac{3}{2}\epsilon$. □

Proof of Theorem 7.26. First, we show that for any $g\colon \mathcal{X} \to \mathrm{dom}(f^*_{\mathrm{ext}})$ we must have

$$\mathbb{E}_P[g(X)] \le D_f(P\|Q) + \mathbb{E}_Q[f^*_{\mathrm{ext}}(g(X))]. \tag{7.104}$$

Let $p(\cdot)$ and $q(\cdot)$ be the densities of P and Q. Then, from the definition of f^*_{ext} we have for every x s.t. $q(x) > 0$:

$$f^*_{\mathrm{ext}}(g(x)) + f_{\mathrm{ext}}\left(\frac{p(x)}{q(x)}\right) \ge g(x)\frac{p(x)}{q(x)}.$$

Integrating this over $dQ = q\,d\mu$ restricted to the set $\{q > 0\}$ we get

$$\mathbb{E}_Q[f^*_{\mathrm{ext}}(g(X))] + \int_{q>0} q(x) f_{\mathrm{ext}}\left(\frac{p(x)}{q(x)}\right) d\mu \ge \mathbb{E}_P[g(X)1\{q(X) > 0\}]. \tag{7.105}$$

Now, notice that

$$\sup\{y\colon y \in \mathrm{dom}(f^*_{\mathrm{ext}})\} = \lim_{x\to\infty} \frac{f_{\mathrm{ext}}(x)}{x} = f'(\infty). \tag{7.106}$$

Therefore, $f'(\infty)P[q(X) = 0] \ge \mathbb{E}_P[g(X)1\{q(X) = 0\}]$. Summing the latter inequality with (7.105) we obtain (7.104).

Next we prove that the supremum in (7.87) over simple functions g does yield $D_f(P\|Q)$, so that inequality (7.104) is tight. Armed with Theorem 7.6, it suffices to show (7.87) for finite $\mathcal{X}$. Indeed, for general $\mathcal{X}$, given a finite partition $\mathcal{E} = \{E_1, \dots, E_n\}$ of $\mathcal{X}$, we say a function $g\colon \mathcal{X} \to \mathbb{R}$ is $\mathcal{E}$-compatible if g is constant on each $E_i \in \mathcal{E}$. Taking the supremum over all finite partitions $\mathcal{E}$ we get

$$\begin{aligned} D_f(P\|Q) &= \sup_{\mathcal{E}} D_f(P_{\mathcal{E}}\|Q_{\mathcal{E}}) \\ &= \sup_{\mathcal{E}} \sup_{\substack{g\colon \mathcal{X}\to\mathrm{dom}(f^*_{\mathrm{ext}}) \\ g\ \mathcal{E}\text{-compatible}}} \mathbb{E}_P[g(X)] - \mathbb{E}_Q[f^*_{\mathrm{ext}}(g(X))] \\ &= \sup_{\substack{g\colon \mathcal{X}\to\mathrm{dom}(f^*_{\mathrm{ext}}) \\ g\ \text{simple}}} \mathbb{E}_P[g(X)] - \mathbb{E}_Q[f^*_{\mathrm{ext}}(g(X))], \end{aligned}$$

where the last step follows because the two suprema combined is equivalent to the supremum over all simple (finitely valued) functions g.

Next, consider finite $\mathcal{X}$. Let $S = \{x \in \mathcal{X}\colon Q(x) > 0\}$ denote the support of Q. We show the following statement:

$$D_f(P\|Q) = \sup_{g\colon S\to\mathrm{dom}(f^*_{\mathrm{ext}})} \mathbb{E}_P[g(X)] - \mathbb{E}_Q[f^*_{\mathrm{ext}}(g(X))] + f'(\infty)P(S^c), \tag{7.107}$$

which is equivalent to (7.87) by (7.106). By definition,

$$D_f(P\|Q) = \underbrace{\sum_{x\in S} Q(x) f_{\mathrm{ext}}\left(\frac{P(x)}{Q(x)}\right)}_{\triangleq \Psi(P)} + f'(\infty)P(S^c).$$

Consider the functional $\Psi(P)$ defined above where P takes values over all signed measures on S, which can be identified with $\mathbb{R}^S$. The convex conjugate of $\Psi(P)$ is as follows: for any $g\colon S \to \mathbb{R}$,

$$\begin{aligned}
\Psi^*(g) &= \sup_P \sum_x P(x)g(x) - Q(x) \left\{ \sup_{h \in \mathrm{dom}(f^*_{\mathrm{ext}})} \frac{P(x)}{Q(x)} h - f^*_{\mathrm{ext}}(h) \right\} \\
&= \sup_P \inf_{h\colon S \to \mathrm{dom}(f^*_{\mathrm{ext}})} \sum_x P(x)(g(x) - h(x)) + Q(x) f^*_{\mathrm{ext}}(h(x)) \\
&\overset{(a)}{=} \inf_{h\colon S \to \mathrm{dom}(f^*_{\mathrm{ext}})} \sup_P \sum_x P(x)(g(x) - h(x)) + \mathbb{E}_Q[f^*_{\mathrm{ext}}(h)] \\
&= \begin{cases} \mathbb{E}_Q[f^*_{\mathrm{ext}}(g(X))], & g\colon S \to \mathrm{dom}(f^*_{\mathrm{ext}}), \\ +\infty, & \text{otherwise}, \end{cases}
\end{aligned}$$

where (a) follows from the minimax theorem (which applies due to the finiteness of $\mathcal{X}$). Applying the convex duality in (7.86) yields the proof of the desired (7.107). □

Proof of Theorem 7.27. First we argue that the supremum on the right-hand side of (7.94) can be taken over all simple functions g. Then thanks to Theorem 7.6, it will suffice to consider finite alphabet $\mathcal{X}$. To that end, fix any g. For any δ, there exists a such that $\mathbb{E}_Q[f^*_{\mathrm{ext}}(g - a)] - aP[S] \le \Psi^*_{Q,P}(g) + \delta$. Since $\mathbb{E}_Q[f^*_{\mathrm{ext}}(g - a_n)]$ can be approximated arbitrarily well by simple functions we conclude that there exists a simple function $\tilde{g}$ such that simultaneously $\mathbb{E}_P[\tilde{g} 1_S] \ge \mathbb{E}_P[g 1_S] - \delta$ and

$$\Psi^*_{Q,P}(\tilde{g}) \le \mathbb{E}_Q[f^*_{\mathrm{ext}}(\tilde{g} - a)] - aP[S] + \delta \le \Psi^*_{Q,P}(g) + 2\delta.$$

This implies that restricting to simple functions in the supremization in (7.94) does not change the right-hand side.

Next consider finite $\mathcal{X}$. We proceed to compute the conjugate of Ψ, where $\Psi(P) \triangleq D_f(P\|Q)$ if P is a probability measure on $\mathcal{X}$ and $+\infty$ otherwise. Then for any $g\colon \mathcal{X} \to \mathbb{R}$, maximizing over all probability measures P we have

$$\begin{aligned}
\Psi^*(g) &= \sup_P \sum_{x \in \mathcal{X}} P(x)g(x) - D_f(P\|Q) \\
&= \sup_P \sum_{x \in \mathcal{X}} P(x)g(x) - \sum_{x \in S^c} P(x)g(x) - \sum_{x \in S} Q(x) f\left(\frac{P(x)}{Q(x)}\right) \\
&= \sup_P \inf_{h\colon S \to \mathbb{R}} \sum_{x \in S} P(x)[g(x) - h(x)] + \sum_{x \in S^c} P(x)[g(x) - f'(\infty)] + \sum_{x \in S} Q(x) f^*_{\mathrm{ext}}(h(x)) \\
&\overset{(a)}{=} \inf_{h\colon S \to \mathbb{R}} \left\{ \sup_P \left(\sum_{x \in S} P(x)[g(x) - h(x)] + \sum_{x \in S^c} P(x)[g(x) - f'(\infty)] \right) \right. \\
&\qquad \left. + \mathbb{E}_Q[f^*_{\mathrm{ext}}(h(X))] \right\} \\
&\overset{(b)}{=} \inf_{h\colon S \to \mathbb{R}} \left\{ \max\left(\max_{x \in S} g(x) - h(x), \max_{x \in S^c} g(x) - f'(\infty) \right) + \mathbb{E}_Q[f^*_{\mathrm{ext}}(h(X))] \right\} \\
&\overset{(c)}{=} \inf_{a \in \mathbb{R}} \left\{ \max\left(a, \max_{x \in S^c} g(x) - f'(\infty) \right) + \mathbb{E}_Q[f^*_{\mathrm{ext}}(g(X) - a)] \right\},
\end{aligned}$$

where (a) follows from the minimax theorem; (b) is due to P being a probability measure; and (c) follows since we can restrict to $h(x) = g(x) - a$ for $x \in S$, thanks to the fact that f^*_{ext} is non-decreasing (since $\text{dom}(f_{\text{ext}}) = \mathbb{R}_+$).

From convex duality we have shown that $D_f(P\|Q) = \sup_g \mathbb{E}_P[g] - \Psi^*(g)$. Notice that without loss of generality we may take $g(x) = f'(\infty) + b$ for $x \in S^c$. Interchanging the optimization over b with that over a we find that

$$\sup_b bP[S^c] - \max(a, b) = -aP[S],$$

which then recovers (7.94). To get (7.95) simply notice that if $P[S^c] > 0$, then both sides of (7.95) are infinite (since $\Psi^*_Q(g)$ does not depend on the values of g outside of S). Otherwise, (7.95) coincides with (7.94). □

8 Entropy Method in Combinatorics and Geometry

A commonly used method in combinatorics for bounding the number of certain objects from above involves a clever application of Shannon entropy. This method typically proceeds as follows: In order to count the cardinality of a given set $\mathcal{C}$, we draw an element uniformly at random from $\mathcal{C}$, whose entropy is given by $\log|\mathcal{C}|$. To bound $|\mathcal{C}|$ from above, we describe this random object by a random vector $X = (X_1, \ldots, X_n)$ then proceed to compute or upper-bound the joint entropy $H(X_1, \ldots, X_n)$ via one of the following methods:

- Marginal bound: $H(X_1, \ldots, X_n) \leq \sum_{i=1}^n H(X_i)$.
- Pairwise bound (Shearer's lemma) and generalization, see Theorem 1.8: $H(X_1, \ldots, X_n) \leq \frac{1}{n-1} \sum_{i<j} H(X_i, X_j)$.
- Chain rule (exact calculation): $H(X_1, \ldots, X_n) = \sum_{i=1}^n H(X_i|X_1, \ldots, X_{i-1})$.

We give three applications using each of the above three methods, respectively, in order of increasing difficulty: enumerating binary vectors of given average weight, counting triangles and other subgraphs, and Brégman's theorem.

Finally, to demonstrate how the entropy method can also be used for questions in Euclidean spaces, we prove the Loomis–Whitney and Bollobás–Thomason theorems based on analogous properties of *differential* entropy (Section 2.3).

8.1 Binary Vectors of Average Weight

Lemma 8.1. (Massey [294]) *Let $\mathcal{C} \subset \{0,1\}^n$ and let p be the average fraction of 1's in $\mathcal{C}$, that is,*

$$p = \frac{1}{|\mathcal{C}|} \sum_{x \in \mathcal{C}} \frac{w_{\mathrm{H}}(x)}{n},$$

where $w_{\mathrm{H}}(x)$ is the Hamming weight (number of 1's) of $x \in \{0,1\}^n$. Then $|\mathcal{C}| \leq \exp\{nh(p)\}$.

We emphasize that this result holds even if $p > 1/2$.

Proof. Let $X = (X_1, \ldots, X_n)$ be drawn uniformly at random from $\mathcal{C}$. Then

$$\log |\mathcal{C}| = H(X) = H(X_1, \ldots, X_n) \leq \sum_{i}^{n} H(X_i) = \sum_{i=1}^{n} h(p_i),$$

where $p_i = \mathbb{P}[X_i = 1]$ is the fraction of vertices whose ith bit is 1. Note that

$$p = \frac{1}{n}\sum_{i=1}^{n} p_i,$$

since we can either first average over vectors in $\mathcal{C}$ or first average across different bits. By Jensen's inequality and the fact that $x \mapsto h(x)$ is concave,

$$\sum_{i=1}^{n} h(p_i) \leq nh\left(\frac{1}{n}\sum_{i=1}^{n} p_i\right) = nh(p).$$

Hence we have shown that $\log |\mathcal{C}| \leq nh(p)$. □

As a consequence we obtain the following bound on the volume of the *Hamming ball*, which will be instrumental much later when we talk about metric entropy (Chapter 27).

Theorem 8.2.

$$\sum_{j=0}^{k} \binom{n}{j} \leq \exp\{nh(k/n)\}, \qquad k \leq n/2.$$

Proof. We take $\mathcal{C} = \{x \in \{0,1\}^n : w_{\mathrm{H}}(x) \leq k\}$ and invoke the previous lemma, which says that

$$\sum_{j=0}^{k} \binom{n}{j} = |\mathcal{C}| \leq \exp\{nh(p)\} \leq \exp\{nh(k/n)\},$$

where the last inequality follows from the fact that $x \mapsto h(x)$ is increasing for $x \leq 1/2$. □

For extensions to non-binary alphabets see Exercises **I.1** and **I.2**. Note that Theorem 8.2 also follows from the large-deviations theory in Part III:

$$\begin{aligned}\frac{\mathrm{LHS}}{2^n} &= \mathbb{P}(\mathrm{Bin}(n, 1/2) \leq k) \leq \exp\left\{-nd\left(\frac{k}{n}\middle\|\frac{1}{2}\right)\right\} \\ &= \exp\{-n(\log 2 - h(k/n))\} = \frac{\mathrm{RHS}}{2^n},\end{aligned}$$

where the inequality is the Chernoff bound on the binomial tail (see (15.19) in Example 15.1).

8.2 Shearer's Lemma and Counting Subgraphs

Recall that a special case of Shearer's lemma in Theorem 1.8 (Han's inequality) says:

$$H(X_1, X_2, X_3) \le \frac{1}{2}[H(X_1, X_2) + H(X_2, X_3) + H(X_1, X_3)].$$

A classical application of this result (see Remark 1.2) is to bound the cardinality of a set in $\mathbb{R}^3$ given the cardinalities of its projections.

For graphs H and G, define $N(H, G)$ to be the number of copies of H in G.[1] For example,

$$N(K_3, K_4) = 4, \quad N(P_3, K_4 - e) = 8.$$

If we know G has m edges, what is the maximal number of H that are contained in G? To study this quantity, we define

$$N(H, m) = \max_{G:\, |E(G)| \le m} N(H, G).$$

As an example, we show that the maximal number of triangles satisfies

$$N(K_3, m) \asymp m^{3/2}. \tag{8.1}$$

To show that $N(H, m) \gtrsim m^{3/2}$, consider $G = K_n$ which has $m = |E(G)| = \binom{n}{2} \asymp n^2$ and $N(K_3, K_n) = \binom{n}{3} \asymp n^3 \asymp m^{3/2}$.

To show the upper bound, fix a graph $G = (V, E)$ with m edges. Draw a labeled triangle uniformly at random and denote its vertices by (X_1, X_2, X_3). Then by Shearer's lemma,

$$\begin{aligned}\log(3!N(K_3, G)) = H(X_1, X_2, X_3) &\le \frac{1}{2}[H(X_1, X_2) + H(X_2, X_3) + H(X_1, X_3)] \\ &\le \frac{3}{2}\log(2m).\end{aligned}$$

Hence

$$N(K_3, G) \le \frac{\sqrt{2}}{3} m^{3/2}. \tag{8.2}$$

Remark 8.1. Interestingly, a linear algebra argument yields exactly the same upper bound as (8.2): Let A be the adjacency matrix of G with eigenvalues $\{\lambda_i\}$. Then

$$\begin{aligned}2|E(G)| &= \operatorname{tr}(A^2) = \sum \lambda_i^2, \\ 6N(K_3, G) &= \operatorname{tr}(A^3) = \sum \lambda_i^3.\end{aligned}$$

By Minkowski's inequality, $(6N(K_3, G))^{1/3} \le (2|E(G)|)^{1/2}$ which yields $N(K_3, G) \le \frac{\sqrt{2}}{3} m^{3/2}$.

[1] To be precise, here $N(H, G)$ is the number of subgraphs of G (subsets of edges) isomorphic to H. If we denote by $\operatorname{inj}(H, G)$ the number of injective maps $V(H) \to V(G)$ mapping edges of H to edges of G, then $N(H, G) = \frac{1}{|\operatorname{Aut}(H)|}\operatorname{inj}(H, G)$.

Using Shearer's lemma (Theorem 1.8), Friedgut and Kahn [173] obtained the counterpart of (8.1) for arbitrary H; this result was first proved by Alon [13]. We start by introducing the *fractional covering number* of a graph. For a graph $H = (V, E)$, define the fractional covering number as the value of the following linear program (LP):[2]

$$\rho^*(H) = \min_w \left\{ \sum_{e \in E} w(e) : \sum_{e \in E,\, v \in e} w(e) \geq 1 \; \forall v \in V, w(e) \in [0,1] \; \forall e \in E \right\}. \quad (8.3)$$

Theorem 8.3. *For any graph H there exist constants c_0 and c_1 such that*

$$c_0(H) m^{\rho^*(H)} \leq N(H, m) \leq c_1(H) m^{\rho^*(H)}. \quad (8.4)$$

For example, for triangles we have $\rho^*(K_3) = 3/2$ and Theorem 8.3 is consistent with (8.1).

Proof. *Upper bound*: Let $V(H) = [n]$ and let $w^*(e)$ be the solution for $\rho^*(H)$. For any G with m edges, draw a subgraph of G uniformly at random from all those that are isomorphic to H. Given such a random subgraph set $X_i \in V(G)$ to be the vertex corresponding to the ith vertex of H, $i \in [n]$. Now define a random 2-subset S of $[n]$ by sampling an edge e from $E(H)$ with probability $\frac{w^*(e)}{\rho^*(H)}$. By the definition of $\rho^*(H)$ we have for any $i \in [n]$ that $\mathbb{P}[i \in S] \geq \frac{1}{\rho^*(H)}$. We are now ready to apply Theorem 1.8:

$$\log N(H, G) = H(X) \leq H(X_S | S) \rho^*(H) \leq \log(2m) \rho^*(H),$$

where the last inequality is as before: if $S = \{v, w\}$ then $X_S = (X_v, X_w)$ takes one of $2m$ values. Overall, we get[3] $N(H, G) \leq (2m)^{\rho^*(H)}$.

Lower bound: This amounts to constructing a graph G with m edges for which $N(H, G) \geq c(H) |e(G)|^{\rho^*(H)}$. Consider the dual LP of (8.3),

$$\alpha^*(H) = \max_\psi \left\{ \sum_{v \in V(H)} \psi(v) : \psi(v) + \psi(w) \leq 1 \; \forall (vw) \in E, \psi(v) \in [0,1] \; \forall v \in V \right\}, \quad (8.5)$$

that is, the *fractional packing number*. By the duality theorem of LP, we have $\alpha^*(H) = \rho^*(H)$. The graph G is constructed as follows: For each vertex v of H, replicate it $m(v)$ times. For each edge $e = (vw)$ of H, replace it by a complete bipartite graph $K_{m(v), m(w)}$. Then the total number of edges of G is

$$|E(G)| = \sum_{(vw) \in E(H)} m(v) m(w).$$

[2] If the "$\in [0,1]$" constraints in (8.3) and (8.5) are replaced by "$\in \{0,1\}$," we obtain the covering number $\rho(H)$ and the independence number $\alpha(H)$ of H, respectively.

[3] Note that for $H = K_3$ this gives a bound weaker than (8.2). To recover (8.2) we need to take $X = (X_1, \ldots, X_n)$ to be uniform on all injective homomorphisms $H \to G$.

Furthermore, $N(G,H) \geq \prod_{v\in V(H)} m(v)$. To minimize the exponent $\frac{\log N(G,H)}{\log |E(G)|}$, fix a large number M and let $m(v) = \lceil M^{\psi(v)} \rceil$, where ψ is the maximizer in (8.5). Then

$$|E(G)| \leq \sum_{(vw)\in E(H)} 4M^{\psi(v)+\psi(w)} \leq 4M|E(H)|,$$

$$N(G,H) \geq \prod_{v\in V(H)} M^{\psi(v)} = M^{\alpha^*(H)},$$

and we are done. □

8.3 Brégman's Theorem

In this section, we present an elegant entropy proof due to Radhakrishnan [352] of Brégman's theorem [74], which bounds the number of *perfect matchings* (1-regular spanning subgraphs) in a bipartite graph.

We start with some definitions. The *permanent* of an $n \times n$ matrix A is defined as

$$\text{perm}(A) \triangleq \sum_{\pi\in S_n} \prod_{i=1}^{n} a_{i\pi(i)},$$

where S_n denotes the group of all permutations of $[n]$. For a bipartite graph G with n vertices on the left and right respectively, the number of perfect matchings in G is given by $\text{perm}(A)$, where A is the adjacency matrix of G. For example, we have

perm ( ) = 1, perm ( ) = 2.

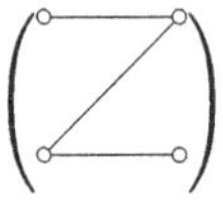

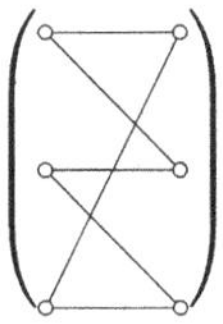

Theorem 8.4. (Brégman's theorem) *For any $n \times n$ bipartite graph with adjacency matrix A,*

$$\text{perm}(A) \leq \prod_{i=1}^{n} (d_i!)^{1/d_i},$$

where d_i is the degree of left vertex i (i.e. the sum of the ith row of A).

As an example, consider $G = K_{n,n}$. Then $\text{perm}(G) = n!$, which coincides with the RHS $[(n!)^{1/n}]^n = n!$. More generally, if G consists of n/d copies of $K_{d,d}$, then Brégman's bound is tight and $\text{perm}(A) = (d!)^{n/d}$.

Proof. If $\text{perm}(A) = 0$ then there is nothing to prove, so instead we assume $\text{perm}(A) > 0$ and some perfect matching exists. As a first attempt at proving Theorem 8.4 using the entropy method, we select a perfect matching uniformly at random which matches the ith left vertex to the X_ith right one. Let $X = (X_1, \ldots, X_n)$. Then

$$\log \operatorname{perm}(A) = H(X) = H(X_1, \ldots, X_n) \le \sum_{i=1}^{n} H(X_i) \le \sum_{i=1}^{n} \log(d_i).$$

Hence $\operatorname{perm}(A) \le \prod_i d_i$. This is worse than Brégman's bound by an exponential factor, since by Stirling's formula (I.2)

$$\prod_{i=1}^{n} (d_i!)^{1/d_i} \sim \left(\prod_{i=1}^{n} d_i \right) e^{-n}.$$

Here is our second attempt. The hope is to use the chain rule to expand the joint entropy and bound the conditional entropies more carefully. Let us write

$$H(X_1, \ldots, X_n) = \sum_{i=1}^{n} H(X_i | X_1, \ldots, X_{i-1}) \le \sum_{i=1}^{n} \mathbb{E}[\log N_i],$$

where N_i, as a random variable, denotes the number of possible values X_i can take conditioned on $X_1, \ldots, X_{i-1}$, that is, how many possible matchings for left vertex i given the outcome of where $1, \ldots, i-1$ are matched to. However, it is hard to proceed from this point as we only know the degree information, not the graph itself. In fact, since we do not know the relative positions of the vertices, there is no reason why we should order from 1 to n. The key idea is to *label the vertices randomly*, apply the chain rule in this random order, and then average.

To this end, pick π uniformly at random from S_n and independent of X. Then

$$\begin{aligned}
\log \operatorname{perm}(A) &= H(X) = H(X|\pi) \\
&= H(X_{\pi(1)}, \ldots, X_{\pi(n)} | \pi) \\
&= \sum_{k=1}^{n} H(X_{\pi(k)} | X_{\pi(1)}, \ldots, X_{\pi(k-1)}, \pi) \\
&= \sum_{k=1}^{n} H(X_k | \{X_j \colon \pi^{-1}(j) < \pi^{-1}(k)\}, \pi) \\
&\le \sum_{k=1}^{n} \mathbb{E} \log N_k,
\end{aligned}$$

where N_k denotes the number of possible matchings for vertex k given the outcomes of $\{X_j \colon \pi^{-1}(j) < \pi^{-1}(k)\}$ and the expectation is with respect to (X, π). The key observation is the next lemma.

Lemma 8.5. *N_k is uniformly distributed on $[d_k]$.*

Example 8.1.

As a concrete example for Lemma 8.5, consider the graph G on the right. For vertex $k = 1$, $d_k = 2$. Depending on the random ordering, if $\pi = 1{*}{*}$, then $N_k = 2$ w.p. 1/3; if $\pi = {*}{*}1$, then $N_k = 1$ w.p. 1/3; if $\pi = 213$, then $N_k = 2$ w.p. 1/3; if $\pi = 312$, then $N_k = 1$ w.p. 1/3. Combining everything, indeed N_k is equally likely to be 1 or 2.

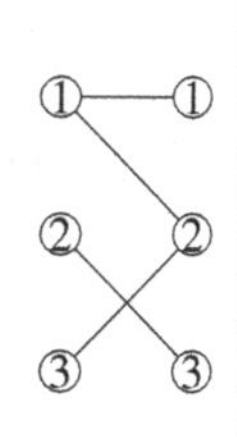

Applying Lemma 8.5,

$$\mathbb{E}_{(X,\pi)} \log N_k = \frac{1}{d_k} \sum_{i=1}^{d_k} \log i = \log(d_k!)^{1/d_k}$$

and hence

$$\log \operatorname{perm}(A) \le \sum_{k=1}^{n} \log(d_k!)^{1/d_k} = \log \prod_{i=1}^{n} (d_i!)^{1/d_i}. \qquad \square$$

Proof of Lemma 8.5. In fact, we will show that even conditioned on X^n the distribution of N_k is uniform. Indeed, if $d = d_k$ is the degree of the kth (right) node then let $J_1, \ldots, J_d$ be those right nodes that match with neighbors of k under the fixed perfect matching (one of the J_i's, say J_1, equals k). The random permutation π rearranges the J_i's in the order in which the corresponding right nodes are revealed. Clearly the induced order of the J_i's is uniform on the $d!$ possible choices. Note that if J_1 occurs in position $\ell \in \{1, \ldots, d\}$ then $N_k = d - \ell + 1$. Clearly ℓ and thus N_k are uniform on $[d] = [d_k]$. □

8.4 Euclidean Geometry: Bollobás–Thomason and Loomis–Whitney

The following famous result shows that n-dimensional rectangles simultaneously minimize the volumes of all coordinate projections.[4]

Theorem 8.6. (Bollobás–Thomason box theorem) *Let $K \subset \mathbb{R}^n$ be a compact set. For $S \subset [n]$, denote by $K_S \subset \mathbb{R}^S$ the projection of K onto those coordinates indexed by S. Then there exists a rectangle A s.t.* $\mathrm{Leb}(A) = \mathrm{Leb}(K)$ *and for all $S \subset [n]$:*

$$\mathrm{Leb}(A_S) \le \mathrm{Leb}(K_S).$$

Thus, rectangles are extremal objects from the point of view of maximizing the volumes of coordinate projections.

Proof. Let X^n be uniformly distributed on K. Then $h(X^n) = \log \mathrm{Leb}(K)$. Let A be a rectangle of size $a_1 \times \cdots \times a_n$ where

$$\log a_i = h(X_i | X^{i-1}).$$

Then, we have by Theorem 2.7(a):

$$h(X_S) \le \log \mathrm{Leb}(K_S).$$

On the other hand, by the chain rule and the fact that conditioning reduces differential entropy (recall Theorem 2.7(a) and (c)),

[4] Note that since K is compact, its projection and slices are all compact and hence measurable.

$$\begin{aligned} h(X_S) &= \sum_{i=1}^{n} 1\{i \in S\} h(X_i | X_{[i-1]\cap S}) \\ &\geq \sum_{i \in S} h(X_i | X^{i-1}) \\ &= \log \prod_{i \in S} a_i \\ &= \log \mathrm{Leb}(A_S). \end{aligned}$$

□

The following result is a continuous counterpart of Shearer's lemma (see Theorem 1.8 and Remark 1.2).

Corollary 8.7. (Loomis–Whitney) *Let K be a compact subset of $\mathbb{R}^n$ and let K_{j^c} denote the projection of K onto coordinates in $[n] \setminus j$. Then*

$$\mathrm{Leb}(K) \leq \prod_{j=1}^{n} \mathrm{Leb}(K_{j^c})^{\frac{1}{n-1}}. \tag{8.6}$$

Proof. Let A be a rectangle having the same volume as K. Note that

$$\mathrm{Leb}(K) = \mathrm{Leb}(A) = \prod_{j=1}^{n} \mathrm{Leb}(A_{j^c})^{\frac{1}{n-1}}.$$

By the previous theorem, $\mathrm{Leb}(A_{j^c}) \leq \mathrm{Leb}(K_{j^c})$. □

The meaning of the Loomis–Whitney inequality is best understood by introducing the average width of K in the jth direction: $w_j \triangleq \frac{\mathrm{Leb}(K)}{\mathrm{Leb}(K_{j^c})}$. Then (8.6) is equivalent to

$$\mathrm{Leb}(K) \geq \prod_{j=1}^{n} w_j,$$

that is, the volume of K is greater than or equal that of the rectangle of average widths.

9 Random Number Generators

In this chapter we consider the problem of creating high-quality random number generators. Given a stream of independent $\mathrm{Ber}(p)$ bits, with *unknown* p, we want to turn them into pure random bits, that is, independent $\mathrm{Ber}(1/2)$ bits. Our goal is to find a way of extracting as many fair coin flips as possible from possibly biased coin flips, without knowing the actual bias p.

In 1951 von Neumann [443] proposed the following scheme: Divide the stream into pairs of bits, then for each pair output 0 if input pair is 10, output 1 if it is 01, and otherwise do nothing and move to the next pair. Since both 01 and 10 occur with probability $p\bar{p}$ (where, we remind the reader that $\bar{p} = 1 - p$), regardless of the value of p, we obtain fair coin flips at the output. To measure the efficiency of von Neumann's scheme, note that, on average, we have $2n$ bits in and $2p\bar{p}n$ bits out. So the efficiency (rate) is $p\bar{p}$. The question is: Can we do better?

There are several choices to be made in the problem formulation. *Universal versus non-universal*: the source distribution can be unknown or partially known, respectively. *Exact versus approximately fair coin flips*: whether the generated coin flips are exactly fair or approximately, as measured by one of the f-divergences studied in Chapter 7 (e.g. total variation or KL divergence). In this chapter, we only focus on the universal generation of exactly fair coins. On the other extreme, in Part II we will see that optimal data compressors' output consists of almost purely random bits; however, those compressors are non-universal (need to know source statistics, e.g. bias p) and approximate.

For convenience, in this chapter we consider entropies measured in bits, that is, $\log = \log_2$ in this chapter.

9.1 Setup

Let $\{0,1\}^* = \bigcup_{k\geq 0}\{0,1\}^k = \{\emptyset, 0, 1, 00, 01, \ldots\}$ denote the set of all finite-length binary strings, where $\emptyset$ denotes the empty string. For any $x \in \{0,1\}^*$, $l(x)$ denotes the length of x.

Let us first introduce the definition of random number generator formally. If the input vector is X, denote the output (variable-length) vector by $Y \in \{0,1\}^*$. Then the desired property of Y is the following: Conditioned on the length of Y being k, Y is uniformly distributed on $\{0,1\}^k$.

Definition 9.1. (Randomness extractor) We say $\Psi\colon \{0,1\}^* \to \{0,1\}^*$ is an *extractor* if:

1 $\Psi(x)$ is a prefix of $\Psi(y)$ if x is a prefix of y;
2 for any n and any $p \in (0,1)$, if $X^n \overset{\text{iid}}{\sim} \text{Ber}(p)$, then $\Psi(X^n) \sim \text{Ber}(1/2)^k$ conditioned on $l(\Psi(X^n)) = k$ for each $k \geq 1$.

The efficiency of an extractor Ψ is measured by its *rate*:

$$r_\Psi(p) = \limsup_{n\to\infty} \frac{\mathbb{E}[l(\Psi(X^n))]}{n}, \quad X^n \overset{\text{iid}}{\sim} \text{Ber}(p).$$

In other words, Ψ consumes a stream of n coins with bias p and outputs on average $nr_\Psi(p)$ fair coins.

Note that the von Neumann scheme above defines a valid extractor Ψ_{vN} (with $\Psi_{\text{vN}}(x^{2n+1}) = \Psi_{\text{vN}}(x^{2n})$), whose rate is $r_{\text{vN}}(p) = p\bar{p}$. Clearly this is wasteful, because even if the input bits are already fair, we only get 25% in return.

9.2 Converse

We show that no extractor has a rate higher than the binary entropy function $h(p)$, even if the extractor is allowed to be non-universal (depending on p). The intuition is that the "information content" contained in each Ber(p) variable is $h(p)$ bits; as such, it is impossible to extract more than that. This is easily made precise by the data-processing inequality for entropy (since extractors are deterministic functions).

Theorem 9.2. *For any extractor Ψ and any $p \in (0,1)$,*

$$r_\Psi(p) \geq h(p) = p\log_2\frac{1}{p} + \bar{p}\log_2\frac{1}{\bar{p}}.$$

Proof. Let $L = \Psi(X^n)$. Then

$$nh(p) = H(X^n) \geq H(\Psi(X^n)) = H(\Psi(X^n)|L) + H(L) \geq H(\Psi(X^n)|L) = \mathbb{E}[L] \quad \text{bits},$$

where the last step follows from the assumption on Ψ that $\Psi(X^n)$ is uniform over $\{0,1\}^k$ conditioned on $L = k$. □

The rate of von Neumann's extractor and the entropy bound are plotted in Figure 9.1. Next we present two extractors, due to Elias [149] and Peres [328]

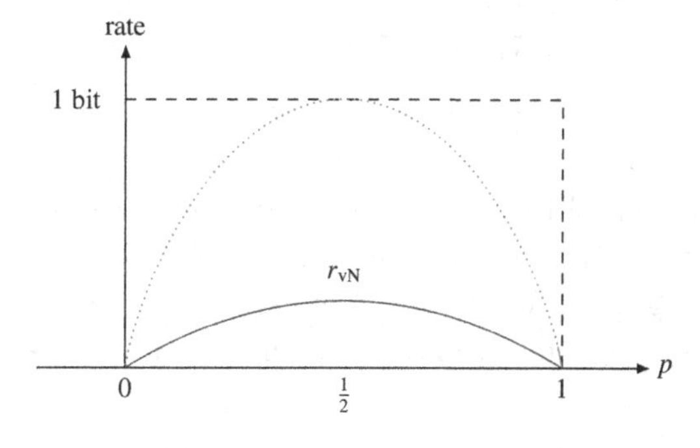

Figure 9.1 Rate function of von Neumann's extractor and the binary entropy function.

respectively, that attain the binary entropy function. (More precisely, both construct a sequence of extractors whose rate approaches the entropy bound).

9.3 Elias' Construction from Data Compression

The intuition behind Elias' scheme is the following:

1 For iid X^n, the probability of each string only depends on its *type*, that is, the number of 1's, see method of types in Exercise **I.1**. Therefore conditioned on the number of 1's being qn, X^n is uniformly distributed over the type class T_q. This observation holds universally for any value of the actual bias p.

2 Given a uniformly distributed random variable on some finite set, we can easily turn it into a *variable-length* string of fair coin flips. For example:
   - If U is uniform over $\{1, 2, 3\}$, we can map $1 \mapsto \emptyset, 2 \mapsto 0$, and $3 \mapsto 1$.
   - If U is uniform over $\{1, 2, \ldots, 11\}$, we can map $1 \mapsto \emptyset, 2 \mapsto 0, 3 \mapsto 1$, and the remaining eight numbers $4, \ldots, 11$ to three-bit strings.

   We will study the properties of these kinds of variable-length encoders later in Chapter 10.

Lemma 9.3. *Given U uniformly distributed on $[M]$, there exists $f\colon [M] \to \{0, 1\}^*$ such that conditioned on $l(f(U)) = k$, $f(U)$ is uniform over $\{0, 1\}^k$. Moreover,*

$$\log_2 M - 4 \le \mathbb{E}[l(f(U))] \le \log_2 M \quad \text{bits.}$$

Proof. We defined f by partitioning $[M]$ into subsets whose cardinalities are powers of 2, and assign elements in each subset to binary strings of that length. Formally, denote the binary expansion of M by $M = \sum_{i=0}^{n} m_i 2^i$, where the most significant bit $m_n = 1$ and $n = \lfloor \log_2 M \rfloor + 1$. Taking non-zero m_i's we can write $M = 2^{i_0} + \cdots + 2^{i_t}$ as a sum of distinct powers of 2 and thus define a partition $[M] = \bigcup_{j=0}^{t} M_j$, where $|M_j| = 2^{i_j}$. We map the elements of M_j to $\{0, 1\}^{i_j}$. Finally, notice that a uniform distribution conditioned on any subset is still uniform.

To prove the bound on the expected length, the upper bound follows from the same entropy argument $\log_2 M = H(U) \ge H(f(U)) \ge H(f(U)|l(f(U))) = \mathbb{E}[l(f(U))]$, and the lower bound follows from

$$\begin{aligned}\mathbb{E}[l(f(U))] &= \frac{1}{M}\sum_{i=0}^{n} m_i 2^i \cdot i = n - \frac{1}{M}\sum_{i=0}^{n} m_i 2^i (n-i) \ge n - \frac{2^n}{M}\sum_{i=0}^{n} 2^{i-n}(n-i) \\ &\ge n - \frac{2^{n+1}}{M} \ge n - 4,\end{aligned}$$

where the last step follows from $n \le \log_2 M + 1$. □

Elias' Extractor Fix $n \ge 1$. Let $w_{\mathrm{H}}(x^n)$ denote as before the Hamming weight (number of 1's) of a binary string x^n. Let $T_k = \{x^n \in \{0, 1\}^n\colon w_{\mathrm{H}}(x^n) = k\}$ define the Hamming sphere of radius k. For each $0 \le k \le n$, we apply the function f from Lemma 9.3 to T_k. This defines a mapping $\Psi_{\mathrm{E}}\colon \{0, 1\}^n \to \{0, 1\}^*$ and then we extend it to $\Psi_E : \{0,1\}^* \to \{0,1\}^*$ by applying the mapping per n-bit block and discard the last incomplete block. Then it is clear that the rate is given by $\frac{1}{n}\mathbb{E}[l(\Psi_E(X^n))]$. By Lemma 9.3, we have

$$\mathbb{E}\log\binom{n}{w_H(X^n)} - 4 \le \mathbb{E}[l(\Psi_E(X^n))] \le \mathbb{E}\log\binom{n}{w_H(X^n)}.$$

Using Stirling's approximation (see Exercise **I.1**) we can show

$$\log\binom{n}{w_H(X^n)} = nh(w_H(X^n)/n) + O(\log n).$$

Since $\frac{1}{n}w_H(X^n) = \frac{1}{n}\sum_{i=1}^n 1\{X_i = 1\}$, from the law of large numbers we conclude $\frac{w_H(X^n)}{n} \to p$ and since h is a continuous bounded function, we also have

$$\frac{1}{n}\mathbb{E}[l(\Psi_E(X^n))] = h(p) + O(\log n/n).$$

Therefore the extraction rate of Ψ_E approaches the optimum $h(p)$ as $n \to \infty$.

9.4 Peres' Iterated von Neumann's Scheme

The main idea is to recycle the bits thrown away in von Neumann's scheme and iterate. What von Neumann's extractor discarded are: (a) bits from equal pairs; and (b) the locations of the distinct pairs. To achieve the entropy bound, we need to extract the randomness out of these two parts as well.

First, some notation: Given x^{2n}, let $k = l(\Psi_{vN}(x^{2n}))$ denote the number of consecutive distinct bit pairs.

- Let $1 \le m_1 < \cdots < m_k \le n$ denote the locations such that $x_{2m_j} \ne x_{2m_j-1}$.
- Let $1 \le i_1 < \cdots < i_{n-k} \le n$ denote the locations such that $x_{2i_j} = x_{2i_j-1}$.
- $y_j = x_{2m_j}$, $v_j = x_{2i_j}$, $u_j = x_{2j} \oplus x_{2j+1}$.

Here y^k are the bits that von Neumann's scheme outputs and both v^{n-k} and u^n are discarded. Note that u^n is important because it encodes the location of the y^k and contains a lot of information. Therefore von Neumann's scheme can be improved if we can extract the randomness out of both v^{n-k} and u^n.

Peres' Extractor For each $t \in \mathbb{N}$, recursively define an extractor Ψ_t as follows:

- Set Ψ_1 to be von Neumann's extractor Ψ_{vN}, that is, $\Psi_1(x^{2n+1}) = \Psi_1(x^{2n}) = y^k$.
- Define Ψ_t by $\Psi_t(x^{2n}) = \Psi_t(x^{2n+1}) = (\Psi_1(x^{2n}), \Psi_{t-1}(u^n), \Psi_{t-1}(v^{n-k}))$.

As an example, consider input $x = 100111010011$ of length $2n = 12$. Then the output is determined recursively as follows:

$$\overbrace{(\underline{011})}^{y}\,\overbrace{(110100)}^{u}\,\overbrace{(101)}^{v}$$
$$(\underline{1})(010)(10)(\underline{0})$$
$$(\underline{1})(\underline{0})$$

Next we (a) verify Ψ_t is a valid extractor and (b) evaluate its efficiency (rate). Note that the bits that enter into the iteration are no longer iid. To compute the rate of Ψ_t, it is convenient to introduce the notion of exchangeability. We say X^n are *exchangeable* if the joint distribution is invariant under permutation, that is, $P_{X_1,\ldots,X_n} = P_{X_{\pi(1)},\ldots,X_{\pi(n)}}$ for any permutation π on $[n]$. In particular, if the X_i's are binary, then X^n are exchangeable if and only if the joint distribution only depends on the Hamming weight, that is, $P_{X^n}(x^n) = f(w_H(x^n))$ for some function f. Examples: X^n is iid Ber(p); X^n is uniform over the Hamming sphere T_k.

As an example, if X^{2n} are iid Ber(p), then conditioned on $L = k$, V^{n-k} are iid $\mathrm{Ber}(p^2/(p^2+\bar{p}^2))$, since $L \sim \mathrm{Binom}(n, 2p\bar{p})$ and

$$\begin{aligned}
\mathbb{P}[Y^k = y, U^n = u, V^{n-k} = v | L = k] &= \frac{p^{k+2m}\bar{p}^{n-k-2m}}{\binom{n}{k}(p^2+\bar{p}^2)^{n-k}(2p\bar{p})^k} \\
&= 2^{-k} \cdot \binom{n}{k}^{-1} \cdot \left(\frac{p^2}{p^2+\bar{p}^2}\right)^m \left(\frac{\bar{p}^2}{p^2+\bar{p}^2}\right)^{n-k-m} \\
&= \mathbb{P}[Y^k = y | L = k]\mathbb{P}[U^n = u | L = k] \\
&\quad \mathbb{P}[V^{n-k} = v | L = k],
\end{aligned}$$

where $m = w_H(v)$. In general, when X^{2n} are only exchangeable, we have the following:

Lemma 9.4. (Ψ_t preserves exchangeability) *Let X^{2n} be exchangeable and $L = \Psi_1(X^{2n})$. Then conditioned on $L = k$, Y^k, U^n, and V^{n-k} are independent, each having an exchangeable distribution. Furthermore, $Y^k \overset{iid}{\sim} \mathrm{Ber}(1/2)$ and U^n is uniform over T_k.*

Proof. It suffices to show that $\forall y, y' \in \{0,1\}^k, u, u' \in T_k$, and $v, v' \in \{0,1\}^{n-k}$ such that $w_H(v) = w_H(v')$, we have

$$\mathbb{P}[Y^k = y, U^n = u, V^{n-k} = v | L = k] = \mathbb{P}[Y^k = y', U^n = u', V^{n-k} = v' | L = k],$$

which implies that $\mathbb{P}[Y^k = y, U^n = u, V^{n-k} = v | L = k] = f(w_H(v))$ for some function f. Note that the string X^{2n} and the triple (Y^k, U^n, V^{n-k}) are in a one-to-one correspondence. Indeed, to reconstruct X^{2n}, simply read the k distinct pairs from Y and fill them according to the locations of the 1's in U and fill the remaining equal pairs from V. [For example, if $(y, u, v) = (01, 1100, 01)$, then $x = (10010011)$, and if $(y', u', v') = (11, 1010, 10)$, then $x' = (01110100)$.] Finally, note that u, y, v and u', y', v' correspond to two input strings x and x' of identical Hamming weight ($w_H(x) = k + 2w_H(v)$) and hence of identical probability due to the exchangeability of X^{2n}. □

Lemma 9.5. (Ψ_t is an extractor) *Let X^{2n} be exchangeable. Then $\Psi_t(X^{2n}) \overset{iid}{\sim} \mathrm{Ber}(1/2)$ conditioned on $l(\Psi_t(X^{2n})) = m$.*

Proof. Note that $\Psi_t(X^{2n}) \in \{0,1\}^*$. It is equivalent to show that for all $s^m \in \{0,1\}^m$,

$$\mathbb{P}[\Psi_t(X^{2n}) = s^m] = 2^{-m}\mathbb{P}[l(\Psi_t(X^{2n})) = m].$$

Proceed by induction on t. The base case of $t = 1$ follows from Lemma 9.4 (the distribution of the Y part). Assume Ψ_{t-1} is an extractor. Recall that $\Psi_t(X^{2n}) = (\Psi_1(X^{2n}), \Psi_{t-1}(U^n), \Psi_{t-1}(V^{n-k}))$ and write the length as $L = L_1 + L_2 + L_3$, where $L_2 \perp\!\!\!\perp L_3 | L_1$ by Lemma 9.4. Then

$$
\begin{aligned}
&\mathbb{P}[\Psi_t(X^{2n}) = s^m] \\
&\quad = \sum_{k=0}^{m} \mathbb{P}[\Psi_t(X^{2n}) = s^m | L_1 = k]\mathbb{P}[L_1 = k] \\
&\quad \overset{\text{Lemma 9.4}}{=} \sum_{k=0}^{m}\sum_{r=0}^{m-k} \mathbb{P}[L_1 = k]\mathbb{P}[Y^k = s^k | L_1 = k]\mathbb{P}[\Psi_{t-1}(U^n) = s_{k+1}^{k+r} | L_1 = k] \\
&\quad \times \mathbb{P}[\Psi_{t-1}(V^{n-k}) = s_{k+r+1}^{m} | L_1 = k] \\
&\quad \overset{\text{induction}}{=} \sum_{k=0}^{m}\sum_{r=0}^{m-k} \mathbb{P}[L_1{=}k]2^{-k}2^{-r}\mathbb{P}[L_2{=}r|L_1{=}k]2^{-(m-k-r)}\mathbb{P}[L_3{=}m-k-r|L_1{=}k] \\
&\quad = 2^{-m}\mathbb{P}[L = m]. \qquad \square
\end{aligned}
$$

Next we compute the rate of Ψ_t. Let $X^{2n} \overset{\text{iid}}{\sim} \text{Ber}(p)$. Then by the strong law of large numbers (SLLN), $\frac{1}{2n} l(\Psi_1(X^{2n})) \triangleq \frac{L_n}{2n}$ converges a.s. to $p\bar{p}$. Assume, again by induction, that $\frac{1}{2n} l(\Psi_{t-1}(X^{2n})) \xrightarrow{\text{a.s.}} r_{t-1}(p)$, with $r_1(p) = pq$. Then

$$
\frac{1}{2n} l(\Psi_t(X^{2n})) = \frac{L_n}{2n} + \frac{1}{2n} l(\Psi_{t-1}(U^n)) + \frac{1}{2n} l(\Psi_{t-1}(V^{n-L_n})).
$$

Note that $U^n \overset{\text{iid}}{\sim} \text{Ber}(2p\bar{p})$, $V^{n-L_n}|L_n \overset{\text{iid}}{\sim} \text{Ber}(p^2/(p^2+\bar{p}^2))$, and $L_n \xrightarrow{\text{a.s.}} \infty$. Then the inductive hypothesis implies that $\frac{1}{n} l(\Psi_{t-1}(U^n)) \xrightarrow{\text{a.s.}} r_{t-1}(2p\bar{p})$ and $\frac{1}{2(n-L_n)} l(\Psi_{t-1}(V^{n-L_n})) \xrightarrow{\text{a.s.}} r_{t-1}(p^2/(p^2+\bar{p}^2))$. We obtain the recursion:

$$
r_t(p) = p\bar{p} + \frac{1}{2} r_{t-1}(2p\bar{p}) + \frac{p^2+\bar{p}^2}{2} r_{t-1}\left(\frac{p^2}{p^2+\bar{p}^2}\right) \triangleq (Tr_{t-1})(p), \tag{9.1}
$$

where the operator T maps a continuous function on $[0,1]$ to another. Furthermore, T is monotone in the sense that if $f \le g$ pointwise then $Tf \le Tg$. Then it can be shown that r_t converges monotonically from below to the fixed point of T, which turns out to be exactly the binary entropy function h. Instead of directly verifying $Th = h$, here is a simple proof: Consider $X_1, X_2 \overset{\text{iid}}{\sim} \text{Ber}(p)$. Then $2h(p) = H(X_1, X_2) = H(X_1 \oplus X_2, X_1) = H(X_1 \oplus X_2) + H(X_1 | X_1 \oplus X_2) = h(2p\bar{p}) + 2p\bar{p}h(\frac{1}{2}) + (p^2+\bar{p}^2)h\left(\frac{p^2}{p^2+\bar{p}^2}\right)$.

The convergence of r_t to h is shown in Figure 9.2.

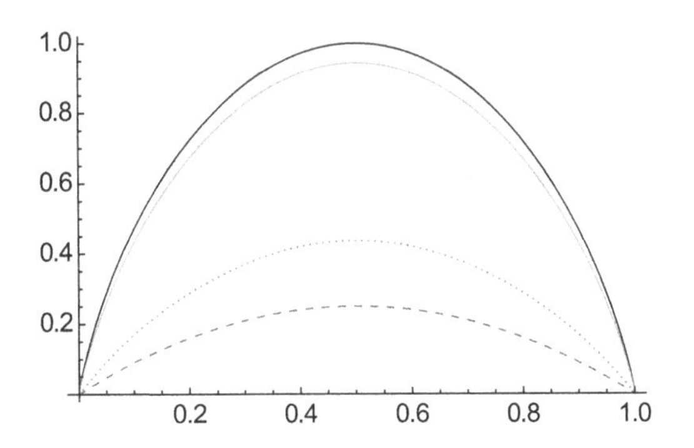

Figure 9.2 The rate function r_t of Ψ_t (by iterating von Neumann's extractor t times) versus the binary entropy function (black full line), for $t = 1$ (dashed line), $t = 4$ (dotted line), and $t = 10$ (gray full line).

9.5 Bernoulli Factory

Given a stream of Ber(p) bits with unknown p, for what kind of function $f\colon [0,1] \to [0,1]$ can we simulate iid bits from Ber($f(p)$)? Our discussion above deals with $f(p) \equiv \frac{1}{2}$. The most famous example is whether we can simulate Ber($2p$) from Ber(p), that is, $f(p) = 2p \wedge 1$. Keane and O'Brien [243] showed that all f that can be simulated are either constants or "polynomially bounded away from 0 or 1": for all $0 < p < 1$, $\min\{f(p), 1 - f(p)\} \geq \min\{p, 1-p\}^n$ for some $n \in \mathbb{N}$. In particular, doubling the bias is impossible.

The above result deals with what $f(p)$ can be simulated in principle. What type of computational devices are needed for such as task? Note that since $r_1(p)$ is quadratic in p, all rate functions r_t that arise from the iteration (9.1) are rational functions (ratios of polynomials), converging to the binary entropy function as Figure 9.2 shows. It turns out that for any rational function f that satisfies $0 < f < 1$ on $(0,1)$, we can generate independent Ber($f(p)$) from Ber(p) using either of the following schemes with finite memory [309]:

1 *Finite-state machine* (FSM): as shown in Table 9.1 this consists of an initial state (dark grey), intermediate states (white), and final states (light gray, output 0 or 1 then reset to initial state).
2 *Block simulation*: let A_0, A_1 be disjoint subsets of $\{0,1\}^k$. For each k-bit segment, output 0 if falling in A_0 or 1 if falling in A_1. If neither, discard and move to the next segment. The block size is at most the degree of the denominator polynomial of f.

Table 9.1 gives some examples of f that can be realized with these two architectures. (Exercise: How to generate $f(p) = 1/3$?)

It turns out that the only type of f that can be simulated using either FSM or block simulation is a rational function. For $f(p) = \sqrt{p}$, which satisfies Keane and O'Brien's characterization, it cannot be simulated by FSM or block simulation, but it can be simulated by the so-called pushdown automaton, which is an FSM operating with a stack (infinite memory) [309].

Table 9.1 Bernoulli factories realized by FSM or block simulation.

Goal	Block simulation	FSM
$f(p) = 1/2$	$A_0 = 10; A_1 = 01$	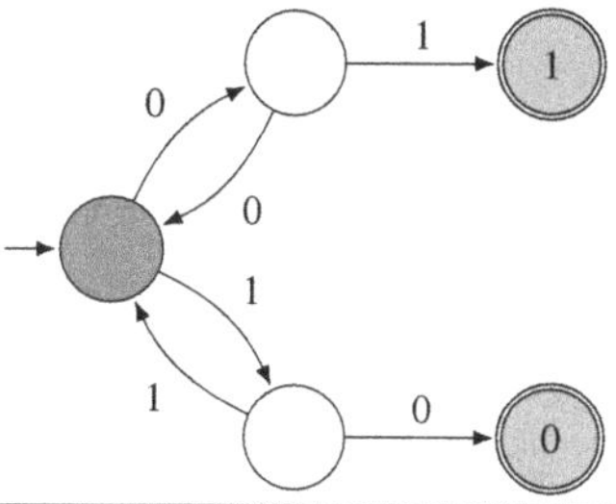
$f(p) = 2pq$	$A_1 = 00, 11; A_0 = 01, 10$	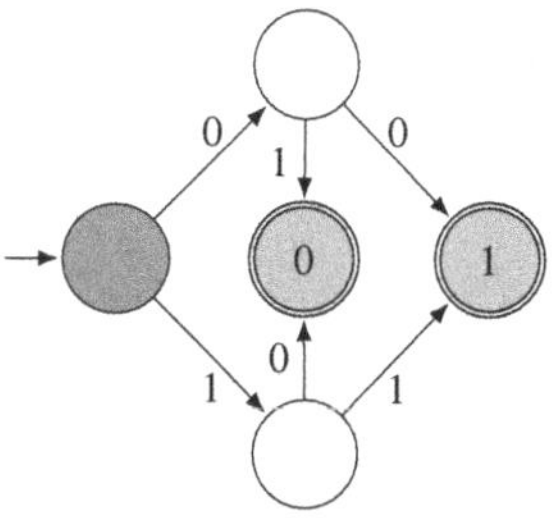
$f(p) = \frac{p^3}{p^3 + \bar{p}^3}$	$A_0 = 000; A_1 = 111$	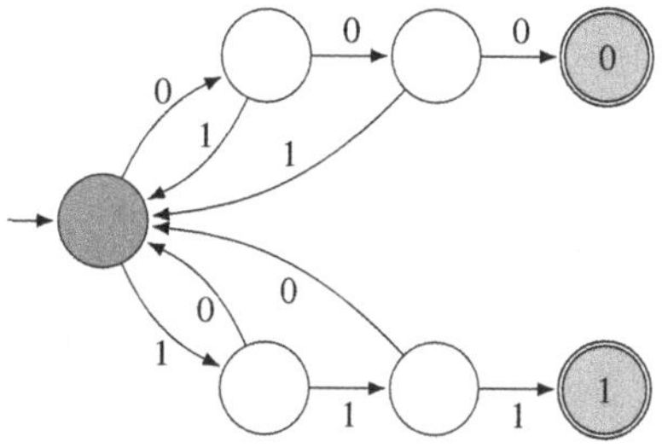

It is unknown how to find the optimal Bernoulli factory with the best rate. Clearly, a converse is the entropy bound $\frac{h(p)}{h(f(p))}$, which can be trivial (bigger than one).

Exercises for Part I

I.1 (Combinatorial meaning of entropy)

(a) Fix $n \geq 1$ and $0 \leq k \leq n$. Let $p = \frac{k}{n}$ and define $T_p \subset \{0,1\}^n$ to be the set of all binary sequences with fraction of 1's equal to p. Show that if $k \in [1, n-1]$ then

$$\frac{\exp\{nh(p)\}}{\sqrt{8k(n-k)/n}} \leq |T_p| = \binom{n}{k} \leq \frac{\exp\{nh(p)\}}{\sqrt{2\pi k(n-k)/n}}, \tag{I.1}$$

where $h(\cdot)$ is the binary entropy. Conclude that for all $0 \leq k \leq n$ we have

$$\log |T_p| = nh(p) + O(\log n).$$

(Hint: Stirling's approximation:

$$e^{\frac{1}{12n+1}} \leq \frac{n!}{\sqrt{2\pi n}(n/e)^n} \leq e^{\frac{1}{12n}}, \quad n \geq 1.) \tag{I.2}$$

(b) Let $Q^n = \mathrm{Ber}(q)^n$ be an iid Bernoulli distribution on $\{0,1\}^n$. Show that

$$\log Q^n[T_p] = -nd(p\|q) + O(\log n).$$

(c*) More generally, let $\mathcal{X}$ be a finite alphabet, $\hat{P}, Q$ distributions on $\mathcal{X}$, and $T_{\hat{P}}$ the set of all strings in $\mathcal{X}^n$ with composition $\hat{P}$. If $T_{\hat{P}}$ is non-empty (i.e. if $n\hat{P}(\cdot)$ is integral) then

$$\log |T_{\hat{P}}| = nH(\hat{P}) + O(\log n),$$
$$\log Q^n[T_{\hat{P}}] = -nD(\hat{P}\|Q) + O(\log n),$$

and furthermore, both $O(\log n)$ terms can be bounded as $|O(\log n)| \leq |\mathcal{X}| \log(n+1)$. (Hint: Show that the number of non-empty $T_{\hat{P}}$ is $\leq (n+1)^{|\mathcal{X}|}$.)

I.2 (Refined method of types) The following refines Proposition 1.5. Let $n_1, \ldots$ be non-negative integers with $\sum_i n_i = n$ and let k_+ be the number of non-zero n_i's. Then

$$\log \binom{n}{n_1, n_2, \ldots} = nH(\hat{P}) - \frac{k_+ - 1}{2} \log(2\pi n) - \frac{1}{2} \sum_{i:\, n_i > 0} \log \hat{P}_i - C_{k_+},$$

where $\hat{P}_i = \frac{n_i}{n}$ and $0 \le C_{k_+} \le \frac{\log e}{12}$. (Hint: Use (I.2).)

I.3 (Conditional entropy and Markov types)

(a) Fix $n \ge 1$, a sequence $x^n \in \mathcal{X}^n$, and define

$$N_{x^n}(a, b) = |\{(x_i, x_{i+1}) \colon x_i = a, x_{i+1} = b, i = 1, \ldots, n\}|,$$

where we define $x_{n+1} = x_1$ (cyclic continuation). Show that $\frac{1}{n} N_{x^n}(\cdot, \cdot)$ defines a probability distribution $P_{A,B}$ on $\mathcal{X} \times \mathcal{X}$ with equal marginals $P_A = P_B$. Conclude that $H(A|B) = H(B|A)$. Is $P_{A|B} = P_{B|A}$?

(b) Let $T^{(2)}_{x^n}$ (Markov type class of x^n) be defined as

$$T^{(2)}_{x^n} = \{\tilde{x}^n \in \mathcal{X}^n \colon N_{\tilde{x}^n} = N_{x^n}\}.$$

Show that elements of $T^{(2)}_{x^n}$ can be identified with cycles in the complete directed graph G on $\mathcal{X}$, such that for each $(a, b) \in \mathcal{X} \times \mathcal{X}$ the cycle passes $N_{x^n}(a, b)$ times through edge (a, b).

(c) Show that each such cycle can be uniquely specified by identifying the first node and by choosing at each vertex of the graph the order in which the outgoing edges are taken. From this and Stirling's approximation conclude that

$$\log |T^{(2)}_{x^n}| = nH(x_{T+1}|x_T) + O(\log n), \qquad T \sim \mathrm{Unif}([n]).$$

Check that $H(x_{T+1}|x_T) = H(A|B) = H(B|A)$.

(d) Show that for any time-homogeneous Markov chain X^n with $P_{X_1,X_2}(a_1, a_2) > 0 \ \forall a_1, a_2 \in \mathcal{X}$ we have

$$\log P_{X^n}(X^n \in T^{(2)}_{x^n}) = -nD(P_{B|A} \| P_{X_2|X_1} | P_A) + O(\log n).$$

I.4 (Maximum entropy)
Prove that for any X taking values on $\mathbb{N} = \{1, 2, \ldots\}$ such that $\mathbb{E}[X] < \infty$,

$$H(X) \le \mathbb{E}[X] h\left(\frac{1}{\mathbb{E}[X]}\right),$$

maximized uniquely by the geometric distribution. (Hint: Find an appropriate Q such that RHS – LHS = $D(P_X \| Q)$.)

I.5 (Finiteness of entropy) In Exercise **I.4** we have shown that the entropy of any $\mathbb{N}$-valued random variable with finite expectation is finite. Next let us improve this result.

(a) Show that $\mathbb{E}[\log X] < \infty \Rightarrow H(X) < \infty$.
Moreover, show that the condition of X being integer-valued is not superfluous by giving a counterexample.

(b) Show that if $k \mapsto P_X(k)$ is a decreasing sequence, then $H(X) < \infty \Rightarrow \mathbb{E}[\log X] < \infty$.

Moreover, show that the monotonicity assumption is not superfluous by giving a counterexample.

I.6 (Robust version of the maximal entropy) The maximal differential entropy among all densities supported on $[-b, b]$ is attained by the uniform distribution. Prove that as $\epsilon \to 0+$ we have

$$\sup\{h(M+Z) \colon M \in [-b,b], \mathbb{E}[Z] = 0, \operatorname{Var}[Z] \le \epsilon\} = \log(2b) + o(1),$$

where supremization is over all (not necessarily independent) random variables M, Z such that $M+Z$ possesses a density. (Hint: Ref. [162, Appendix C] proves the $o(1) = O(\epsilon^{1/3}\log\frac{1}{\epsilon})$ bound.)

I.7 (Maximum entropy under Hamming weight constraint)
For any $\alpha \le 1/2$ and $d \in \mathbb{N}$,

$$\max\{H(Y) \colon Y \in \{0,1\}^d, \mathbb{E}[w_{\mathrm{H}}(Y)] \le \alpha d\} = dh(\alpha),$$

achieved by the product distribution $Y \sim \mathrm{Ber}(\alpha)^{\otimes d}$. (Hint: Find an appropriate Q such that RHS − LHS = $D(P_Y \| Q)$.)

I.8 (Gaussian divergence)

(a) Under what conditions on $\mathbf{m}_0, \Sigma_0, \mathbf{m}_1, \Sigma_1$ is $D(\mathcal{N}(\mathbf{m}_1, \Sigma_1) \,\|\, \mathcal{N}(\mathbf{m}_0, \Sigma_0)) < \infty$?

(b) Compute $D(\mathcal{N}(\mathbf{m}, \Sigma) \,\|\, \mathcal{N}(0, I_n))$, where I_n is the $n \times n$ identity matrix.

(c) Compute $D(\mathcal{N}(\mathbf{m}_1, \Sigma_1) \,\|\, \mathcal{N}(\mathbf{m}_0, \Sigma_0))$ for non-singular Σ_0. (Hint: Think how the Gaussian distribution changes under shifts $\mathbf{x} \mapsto \mathbf{x} + \mathbf{a}$ and non-singular linear transformations $\mathbf{x} \mapsto \mathbf{A}\mathbf{x}$. Apply data processing to reduce to the previous case.)

I.9 (Water-filling solution) Let $M \in \mathbb{R}^{k \times n}$ be a fixed matrix, $X \perp\!\!\!\perp Z \sim \mathcal{N}(0, I_n)$.

(a) Let $M = U \Lambda V^\top$ be an SVD decomposition, so that U, V are orthogonal matrices and $\Lambda = \mathrm{diag}(\lambda_1, \ldots, \lambda_n)$ (with $\mathrm{rank}(M)$ non-zero λ_j's). Show that

$$\max_{P_X \colon \mathbb{E}[\|X\|^2] \le s^2} I(X; MX + Z) = \max_{P_X \colon \mathbb{E}[\|X\|^2] \le s^2} I(X; \Lambda X + Z).$$

(b) Conclude that

$$\max_{P_X \colon \mathbb{E}[\|X\|^2] \le s^2} I(X; MX + Z) = \frac{1}{2}\sum_{i=1}^{n} \log^+(\lambda_i^2 t),$$

where $\log^+ x = \max(0, \log x)$ and t is determined from solving $\sum_{i=1}^n |t - \lambda_i^{-2}|_+ = s^2$.

This distribution of the energy of X along singular vectors of M is known as *water-filling solution*; see Section 20.4.

I.10 (MIMO capacity) Let $M \in \mathbb{R}^{k \times n}$ be a *random, orthogonally invariant* matrix (i.e. $M \stackrel{(d)}{=} MU$ for any orthogonal matrix U). Let $X \perp\!\!\!\perp (Z, M)$ and $Z \sim \mathcal{N}(0, I_n)$. Show that

$$\max_{P_X:\ \mathbb{E}[\|X\|^2]\le s^2} I(X;MX+Z|M) = \frac{1}{2}\mathbb{E}\left[\log\det\left(I+\frac{s^2}{n}M^\top M\right)\right]$$
$$= \frac{1}{2}\sum_{i=1}^{n}\mathbb{E}\left[\log\left(1+\frac{s^2}{n}\sigma_i^2(M)\right)\right],$$

where $\sigma_i(M)$ are the singular values of M. (Hint: Average over rotations as in Section 6.2*.)

Note: In communication theory $M_{i,j} \overset{\text{iid}}{\sim} \mathcal{N}(0,1)$ (Ginibre ensemble) models a multi-input, multi-output (MIMO) channel with n transmit and k receive antennas. The matrix $MM^\top$ is a Wishart matrix and its spectrum, when n and k grow proportionally, approaches the Marchenko–Pastur distribution. The important practical consequence is that the capacity of a MIMO channel grows for high SNR as $\frac{1}{2}\min(n,k)\log \mathrm{SNR}$. This famous observation [419] is the reason modern WiFi and cellular systems employ multiple antennas.

I.11 (Conditional capacity) Consider a Markov kernel $P_{B,C|A}\colon \mathcal{A}\to\mathcal{B}\times\mathcal{C}$, which we will also understand as a collection of distributions $P^{(a)}_{B,C} \triangleq P_{B,C|A=a}$. Prove that

$$\inf_{Q_{C|B}}\sup_{a\in\mathcal{A}} D(P^{(a)}_{C|B}\|Q_{C|B}|P^{(a)}_B) = \sup_{P_A} I(A;C|B),$$

whenever the supremum on the right-hand side is finite and achievable by some distribution P^*_A. In this case, optimal $Q_{C|B} = P^*_{C|B}$ is found by disintegrating $P^*_{B,C} = \int P_{A^*}(da)P^{(a)}_{B,C}$. (Hint: Follow the steps of (5.1).)

I.12 Conditioned on $X=x$, let Y be Poisson with mean x, that is,

$$P_{Y|X}(k|x) = e^{-x}\frac{x^k}{k!},\quad k=0,1,2,\ldots.$$

Let X be an exponential random variable with unit mean. Find $I(X;Y)$.

I.13 (Information lost in erasures) Let X,Y be a pair of random variables with $I(X;Y)<\infty$. Let Z be obtained from Y by passing the latter through an erasure channel, that is, $X\to Y\to Z$ where

$$P_{Z|Y}(z|y) = \begin{cases} 1-\delta, & z=y,\\ \delta, & z=?,\end{cases}$$

where ? is a symbol not in the alphabet of Y. Find $I(X;Z)$.

I.14 (Information bottleneck) Let $X\to Y\to Z$ where Y is a discrete random variable taking values on a finite set $\mathcal{Y}$. Prove that

$$I(X;Z)\le\log|\mathcal{Y}|.$$

I.15 The *Hewitt–Savage 0–1 law* states that certain symmetric events have no randomness. Let $\{X_i\}_{i\ge1}$ be a sequence of iid random variables. Let E be an event determined by this sequence. We say E is exchangeable if it is invariant under permutation of finitely many indices in the sequence $\{X_i\}$, for example, the occurrence of E is unchanged if we permute the values of (X_1,X_4,X_7), etc.

Let us prove the Hewitt–Savage 0–1 law information-theoretically in the following steps:

(a) (Warm-up) Verify that $E = \{\sum_{i\geq 1} X_i \text{ converges}\}$ and $E = \left\{\lim_{n\to\infty} \frac{1}{n}\sum_{i=1}^n X_i = \mathbb{E}[X_1]\right\}$ are exchangeable events.

(b) Let E be an exchangeable event and $W = 1_E$ be its indicator random variable. Show that for any k, $I(W;X_1,\ldots,X_k) = 0$. (Hint: Use tensorization (6.2) to show that for arbitrary n, $nI(W;X_1,\ldots,X_k) \leq 1$ bit.)

(c) Since E is determined by the sequence $\{X_i\}_{i\geq 1}$, we have by continuity of mutual information:

$$H(W) = I(W;X_1,\ldots) = \lim_{k\to\infty} I(W;X_1,\ldots,X_k) = 0.$$

Conclude that E has no randomness, that is, $P(E) = 0$ or $P(E) = 1$.

(d) (Application to random walk) Often after the application of Hewitt–Savage, further efforts are needed to determine whether the probability is 0 or 1. As an example, suppose that X_i's are iid ± 1 and $S_n = \sum_{i=1}^n X_i$ denotes the symmetric random walk. Verify that the event $E = \{S_n = 0 \text{ finitely often}\}$ is exchangeable. Now show that $P(E) = 0$.

(Hint: Consider $E^+ = \{S_n > 0 \text{ eventually}\}$ and E^- similarly. Apply Hewitt–Savage to them and invoke symmetry.)

I.16 Let (X,Y) be uniformly distributed in the unit ℓ_p-ball $B_p \triangleq \{(x,y)\colon |x|^p + |y|^p \leq 1\}$, where $p \in (0,\infty)$. Also define the ℓ_∞-ball $B_\infty \triangleq \{(x,y)\colon |x| \leq 1, |y| \leq 1\}$.

(a) Compute $I(X;Y)$ for $p = 1/2$, $p = 1$, and $p = \infty$.

(b*) Determine the limit of $I(X;Y)$ as $p \to 0$.

I.17 Suppose $Z_1,\ldots,Z_n$ are independent Poisson random variables with mean λ. Show that $\sum_{i=1}^n Z_i$ is a sufficient statistic of $(Z_1,\ldots,Z_n)$ for λ.

I.18 Suppose $Z_1,\ldots,Z_n$ are independent and uniformly distributed on the interval $[0,\lambda]$. Show that $\max_{1\leq i\leq n} Z_i$ is a sufficient statistic of $(Z_1,\ldots,Z_n)$ for λ.

I.19 (Divergence of order statistic) Given $x^n = (x_1,\ldots,x_n) \in \mathbb{R}^n$, let $x_{(1)} \leq \cdots \leq x_{(n)}$ denote the ordered entries. Let P, Q be distributions on $\mathbb{R}$ and $P_{X^n} = P^n, Q_{X^n} = Q^n$.

(a) Prove that

$$D(P_{X_{(1)},\ldots,X_{(n)}} \| Q_{X_{(1)},\ldots,X_{(n)}}) = nD(P\|Q). \tag{I.3}$$

(b) Show that

$$D(\text{Bin}(n,p)\|\text{Bin}(n,q)) = nd(p\|q).$$

I.20 (Continuity of entropy on a finite alphabet) We have shown that on a finite alphabet entropy is a continuous function of the distribution. Quantify this continuity by explicitly showing

$$|H(P) - H(Q)| \leq h(\text{TV}(P,Q)) + \text{TV}(P,Q)\log(|\mathcal{X}| - 1)$$

for any P and Q supported on $\mathcal{X}$.

(Hint: Use Fano's inequality and the inf-representation (over coupling) of total variation in Theorem 7.7(a).)

I.21 (a) For any X such that $\mathbb{E}[|X|] < \infty$, show that

$$D(P_X \| \mathcal{N}(0,1)) \geq \frac{(\mathbb{E}[X])^2}{2} \quad \text{nats}.$$

(b) For $a > 0$, find the value and minimizer of

$$\min_{P_X :\, \mathbb{E}X \geq a} D(P_X \| \mathcal{N}(0,1)).$$

Is the minimizer unique?

I.22 Suppose $D(P_1 \| P_0) < \infty$ then show

$$\frac{d}{d\lambda}\Big|_{\lambda=0} D(\lambda P_1 + \bar{\lambda} Q \| \lambda P_0 + \bar{\lambda} Q) = 0.$$

This extends Proposition 2.20.

I.23 (Metric entropy and capacity) Let $\{P_{Y|X=x} : x \in \mathcal{X}\}$ be a set of distributions and let $C = \sup_{P_X} I(X;Y)$ be its capacity. For every $\epsilon \geq 0$, define[1]

$$N(\epsilon) = \min\{k : \exists Q_1, \ldots, Q_k : \forall x \in \mathcal{X}, \min_j D(P_{Y|X=x} \| Q_j) \leq \epsilon^2\}. \quad \text{(I.4)}$$

(a) Prove that

$$C = \inf_{\epsilon \geq 0} \left(\epsilon^2 + \log N(\epsilon)\right). \quad \text{(I.5)}$$

(Hint: When is $N(\epsilon) = 1$? See Theorem 32.4.)

(b) Similarly, show

$$I(X;Y) = \inf_{\epsilon \geq 0} (\epsilon + \log N(\epsilon; P_X)),$$

where the average-case covering number is

$$N(\epsilon; P_X) = \min\{k : \exists Q_1, \ldots, Q_k : \mathbb{E}_{x \sim P_X}[\min_j D(P_{Y|X=x} \| Q_j)] \leq \epsilon\}. \quad \text{(I.6)}$$

Comment: These estimates are useful because $N(\epsilon)$ for small ϵ roughly speaking depends on local (differential) properties of the map $x \mapsto P_{Y|X=x}$, unlike C which is global.

I.24 Consider the channel $P_{Y^m|X} : [0,1] \mapsto \{0,1\}^m$, where given $x \in [0,1]$, Y^m are iid Ber(x).

(a) Using the upper bound from Exercise **I.23** prove

$$C(m) \triangleq \max_{P_X} I(X;Y^m) \leq \frac{1}{2}\log m + O(1), \qquad m \to \infty.$$

(Hint: Find a covering of the input space.)

[1] $N(\epsilon)$ is the minimum number of radius-ϵ (in divergence) balls that cover the set $\{P_{Y|X=x} : x \in \mathcal{X}\}$. Thus, $\log N(\epsilon)$ is a metric entropy – see Chapter 27.

(b) Show a lower bound to establish

$$C(m) \geq \frac{1}{2}\log m + o(\log m), \qquad m \to \infty.$$

(Hint: Show that for any $\epsilon > 0$ there exists $K(\epsilon)$ such that for all $m \geq 1$ and all $p \in [\epsilon, 1-\epsilon]$ we have $|H(\text{Bin}(m,p)) - \frac{1}{2}\log m| \leq K(\epsilon)$.)

I.25 This exercise shows other ways of proving Fano's inequality in its various forms.

(a) Prove (3.15) as follows. Given any $P = (P_{\max}, P_2, \ldots, P_M)$, apply a random permutation π to the last $M-1$ atoms to obtain the distribution P_π. By comparing $H(P)$ and $H(Q)$, where Q is the average of P_π over all permutations π, complete the proof.

(b) Prove (3.15) by directly solving the convex optimization $\max\{H(P): 0 \leq p_i \leq P_{\max}, i = 1, \ldots, M, \sum_i p_i = 1\}$.

(c) Prove (3.19) as follows. Let $P_e = \mathbb{P}[X \neq \hat{X}]$. First show that

$$I(X;Y) \geq I(X;\hat{X}) \geq \min_{P_{Z|X}}\{I(P_X, P_{Z|X}) : \mathbb{P}[X = Z] \geq 1 - P_e\}.$$

Notice that the minimum is non-zero unless $P_e = P_{\max}$. Second, solve the stated convex optimization problem. (Hint: Look for invariants that the matrix $P_{Z|X}$ must satisfy under permutations $(X, Z) \mapsto (\pi(X), \pi(Z))$ then apply the convexity of $I(P_X, \cdot)$.)

I.26 Show that $P_{Y_1\cdots Y_n|X_1\cdots X_n} = \prod_{i=1}^n P_{Y_i|X_i}$ if and only if the Markov chain $Y_i \to X_i \to (X_{\setminus i}, Y_{\setminus i})$ holds for all $i = 1, \ldots, n$, where $X_{\setminus i} = \{X_j : j \neq i\}$.

I.27 (Distributions and graphical models)

(a) Draw all possible directed acyclic graphs (DAGs, or directed graphical models) compatible with the following distribution on $X, Y, Z \in \{0, 1\}$:

$$P_{X,Z}(x,z) = \begin{cases} 1/6, & x = 0, z \in \{0,1\}, \\ 1/3, & x = 1, z \in \{0,1\}, \end{cases} \qquad (I.7)$$

$$Y = X + Z \qquad (\text{mod } 2). \qquad (I.8)$$

You may include only the minimal DAGs (recall that a DAG is minimal for a given distribution if removal of any edge leads to a graphical model incompatible with the distribution).[2]

(b) Draw the DAG describing the set of distributions $P_{X^nY^n}$ satisfying $P_{Y^n|X^n} = \prod_{i=1}^n P_{Y_i|X_i}$.

(c) Recall that two DAGs G_1 and G_2 are called equivalent if they have the same vertex sets and each distribution factorizes w.r.t. G_1 if and only if it does so w.r.t. G_2. For example, it is well known that

$$X \to Y \to Z \iff X \leftarrow Y \leftarrow Z \iff X \leftarrow Y \to Z.$$

[2] *Note:* $\{X \to Y\}$, $\{X \leftarrow Y\}$, and $\{X \quad Y\}$ are the three possible directed graphical models for two random variables. For example, the third graph describes the set of distributions for which X and Y are independent: $P_{XY} = P_X P_Y$. In fact, $P_X P_Y$ factorizes according to any of the three DAGs, but $\{X \quad Y\}$ is the unique minimal DAG.

Consider the following two DAGs with countably many vertices:

$$X_1 \to X_2 \to \cdots \to X_n \to \cdots$$
$$X_1 \leftarrow X_2 \leftarrow \cdots \leftarrow X_n \leftarrow \cdots$$

Are they equivalent?

I.28 Give a necessary and sufficient condition for $A \to B \to C$ for jointly Gaussian (A, B, C) in terms of correlation coefficients. For discrete (A, B, C) denote $x_{abc} = P_{ABC}(a, b, c)$ and write the Markov chain condition as a list of degree-two polynomial equations in $\{x_{abc}, a \in \mathcal{A}, b \in \mathcal{B}, c \in \mathcal{C}\}$.

I.29 Let A, B, C be discrete with $P_{C|B}(c|b) > 0\ \forall b, c$.

(a) Show that

$$\begin{matrix} A \to B \to C \\ A \to C \to B \end{matrix} \implies A \perp\!\!\!\perp (B, C).$$

Discuss implications for sufficient statistic.

(b*) For binary (A, B, C) characterize all counterexamples.

Comment: Thus, a popular positivity condition $P_{A,B,C} > 0$ allows one to infer conditional independence relations, which are not true in general. In other words, a set of distributions satisfying certain (conditional) independence relations does not coincide with the closure of its intersection with $\{P_{A,B,C} > 0\}$, see [367] for more.

I.30 Consider the implication

$$I(A;C) = I(B;C) = 0 \implies I(A, B;C) = 0. \tag{I.9}$$

(a) Show (I.9) for jointly Gaussian (A, B, C).

(b) Find a counterexample for general (A, B, C).

(c) Prove or disprove: (I.9) also holds for arbitrary finite-cardinality discrete (A, B, C) under the positivity condition $P_{A,B,C}(a, b, c) > 0\ \forall a, b, c$.

I.31 Find the entropy rate of a stationary ergodic Markov chain with transition probability matrix

$$\mathbf{P} = \begin{bmatrix} \frac{1}{2} & \frac{1}{4} & \frac{1}{4} \\ 0 & \frac{1}{2} & \frac{1}{2} \\ 1 & 0 & 0 \end{bmatrix}.$$

I.32 (Solvable HMM) Similar to the Gilbert–Elliott process (Example 6.3) let $S_j \in \{\pm 1\}$ be a stationary two-state Markov chain with $\mathbb{P}[S_j = -S_{j-1}|S_{j-1}] = 1 - \mathbb{P}[S_j = S_{j-1}|S_{j-1}] = \tau$. Let $E_j \overset{\text{i.i.d.}}{\sim} \text{Ber}(\delta)$, with $E_j \in \{0, 1\}$ and let $X_j = \text{BEC}_\delta(S_j)$ be the observation of S_j through the binary erasure channel (BEC) with erasure probability δ, that is, $X_j = S_j E_j$. Find the entropy rate of X_j (you can give your answer in the form of a convergent series). Evaluate at $\tau = 0.11$, $\delta = 1/2$ and compare with $H(X_1)$.

I.33 Consider a binary symmetric random walk X_n on $\mathbb{Z}$ that starts at zero. In other words, $X_n = \sum_{j=1}^{n} B_j$, where $(B_1, B_2, \ldots)$ are independent and equally likely to be ± 1.

(a) When $n \gg 1$ does knowing X_{2n} provide any information about X_n? More exactly, prove that

$$\liminf_{n\to\infty} I(X_n;X_{2n}) > 0.$$

(Hint: Lower semicontinuity and central limit theorem.)

(b) Compute the exact value of $\lim_{n\to\infty} I(X_n;X_{2n})$.

I.34 (Entropy rate and contiguity) Theorem 2.2 states that if a distribution on a finite alphabet $\mathcal{X}$ is almost uniform then its entropy must be close to $\log|\mathcal{X}|$. This exercise extends this observation to random processes.

(a) Let Q_n be the uniform distribution on $\mathcal{X}^n$. Show that, if $\{P_n\} \lhd \{Q_n\}$ (see Definition 7.9), then $H(P_n) = H(Q_n) + o(n) = n\log|\mathcal{X}| + o(n)$, or equivalently $D(P_n\|Q_n) = o(n)$.

(b) Show that for non-uniform Q_n, $P_n \lhd\rhd Q_n$ does not imply $H(P_n) = H(Q_n) + o(n)$. (Hint: For a counterexample, consider the mixed source in Example 6.2(b).)

I.35 Let $I_{\mathrm{TV}}(X;Y) = \mathrm{TV}(P_{X,Y}, P_X P_Y)$. Let $X \sim \mathrm{Ber}(1/2)$ and conditioned on X generate A and B independently by setting them equal to X or $1 - X$ with probabilities $1 - \delta$ and δ, respectively (i.e. $A \leftarrow X \rightarrow B$). Show

$$I_{\mathrm{TV}}(X;A,B) = I_{\mathrm{TV}}(X;A) = \left|\frac{1}{2} - \delta\right|.$$

This means the second observation of X is "uninformative" (in the I_{TV} sense).

Similarly, show that when $X \sim \mathrm{Ber}(\delta)$ for $\delta < 1/2$ there exists joint distribution $P_{X,Y}$ so that $\mathrm{TV}(P_{Y|X=0}, P_{Y|X=1}) > 0$ (thus $I_{\mathrm{TV}}(X;Y)$ and $I(X;Y)$ are strictly positive), but at the same time $\min_{\hat{X}(Y)} \mathbb{P}[X \neq \hat{X}] = \delta$. In other words, observation Y is informative about X, but does not improve the probability of error.

Note: This effect is the basis of an interesting economic effect of herding [30].

I.36 Prove the following variational representation of the total variation:

$$\mathrm{TV}(P_0, P_1) = \frac{1}{2}\inf_{Q}\sqrt{\int dQ\left(\frac{d(P_0 - P_1)}{dQ}\right)^2}. \tag{I.10}$$

I.37 Show that the map $P \mapsto D(P\|Q)$ is strongly convex, that is, for all $\lambda \in [0,1]$ and all P_0, P_1, Q we have

$$\lambda D(P_1\|Q) + \bar{\lambda} D(P_0\|Q) - D(\lambda P_1 + \bar{\lambda} P_0\|Q) \geq 2\lambda\bar{\lambda}\,\mathrm{TV}(P_0, P_1)^2 \log e.$$

(Hint: Write LHS as $I(X;Y)$ for $X \sim \mathrm{Ber}(\lambda)$ and apply Pinsker's inequality.)

I.38 (Rényi divergences and Blackwell order) Let $p_\epsilon = \frac{e^\epsilon}{1+e^\epsilon}$. Show that for all $\epsilon > 0$ and all $\alpha > 0$ we have

$$D_\alpha(\mathrm{Ber}(p_\epsilon)\|\mathrm{Ber}(1-p_\epsilon)) < D_\alpha(\mathcal{N}(\epsilon,1)\|\mathcal{N}(0,1)).$$

Yet, for small enough ϵ we have

$$\mathrm{TV}(\mathrm{Ber}(p_\epsilon),\mathrm{Ber}(1-p_\epsilon)) > \mathrm{TV}(\mathcal{N}(\epsilon,1),\mathcal{N}(0,1)).$$

Note: This shows that domination under all Rényi divergences does not imply a similar comparison for other f-divergences [132]. On the other hand, we have the equivalence [311]:

$$\begin{aligned} &\forall\alpha > 0\colon D_\alpha(P_1\|P_0) \le D_\alpha(Q_1\|Q_0) \\ \iff \quad &\exists n_0 \forall n \ge n_0 \forall f\colon D_f(P_1^{\otimes n}\|P_0^{\otimes n}) \le D_f(Q_1^{\otimes n}\|Q_0^{\otimes n}). \end{aligned}$$

(The latter is also equivalent to the existence of a kernel K_n such that $K_n \circ P_i^{\otimes n} = Q_i^{\otimes n}$ – a so-called Blackwell order on pairs of measures, also known as channel degradation.)

I.39 (Rényi divergence as KL [384]) Show for all $\alpha \in \mathbb{R}$:

$$(1-\alpha)D_\alpha(P\|Q) = \inf_R\left(\alpha D(R\|P) + (1-\alpha)D(R\|Q)\right). \tag{I.11}$$

Whenever the LHS is finite, derive the explicit form of a unique minimizer R.

I.40 For an f-divergence, consider the following statements:

(i) If $I_f(X;Y) = 0$, then $X \perp\!\!\!\perp Y$.

(ii) If $X - Y - Z$ and $I_f(X;Y) = I_f(X;Z) < \infty$, then $X - Z - Y$.

Recall that $f\colon (0,\infty) \to \mathbb{R}$ is a convex function with $f(1) = 0$.

(a) Choose an f-divergence which is not a multiple of the KL divergence (i.e. f cannot be of form $c_1 x\log x + c_2(x-1)$ for any $c_1, c_2 \in \mathbb{R}$). Prove both statements for I_f.

(b) Choose an f-divergence which is non-linear (i.e. f cannot be of form $c(x-1)$ for any $c \in \mathbb{R}$) and provide examples that violate (i) and (ii).

(c) Choose an f-divergence. Prove that (i) holds, and provide an example that violates (ii).

I.41 (Hellinger and interactive protocols [31]) In the area of interactive communication Alice has access to X and outputs bits $A_i, i \ge 1$, whereas Bob has access to Y and outputs bits $B_i, i \ge 1$. The communication proceeds in rounds, so that at the ith round Alice and Bob see the previous messages of each other. This means that the conditional distribution of the *protocol* is given by

$$P_{A^n,B^n|X,Y} = \prod_{i=1}^n P_{A_i|A^{i-1},B^{i-1},X} P_{B_i|A^{i-1},B^{i-1},Y}.$$

Denote for convenience $\Pi_{x,y} \triangleq P_{A^n,B^n|X=x,Y=y}$. Show the following:

(a) (Cut-and-paste lemma) $H^2(\Pi_{x,y},\Pi_{x',y'}) = H^2(\Pi_{x,y'},\Pi_{x',y})$. Are there any other f-divergences with this property?

(b) $H^2(\Pi_{x,y}, \Pi_{x',y}) + H^2(\Pi_{x,y'}, \Pi_{x',y'}) \leq 2H^2(\Pi_{x,y}, \Pi_{x',y'})$.

I.42 (Chain rules I)

(a) Show using (I.11) and the chain rule for KL that

$$(1-\alpha)D_\alpha(P_{X^n}\|Q_{X^n}) \geq \sum_{i=1}^n \inf_a (1-\alpha)D_\alpha(P_{X_i|X^{i-1}=a}\|Q_{X_i|X^{i-1}=a}).$$

(b) Derive two special cases:

$$1 - \frac{1}{2}H^2(P_{X^n}, Q_{X^n}) \leq \prod_{i=1}^n \sup_a \left(1 - \frac{1}{2}H^2(P_{X_i|X^{i-1}=a}, Q_{X_i|X^{i-1}=a})\right),$$

$$1 + \chi^2(P_{X^n}\|Q_{X^n}) \leq \prod_{i=1}^n \sup_a (1 + \chi^2(P_{X_i|X^{i-1}=a}\|Q_{X_i|X^{i-1}=a})).$$

I.43 (Chain rules II)

(a) Show that the chain rule for divergence can be restated as

$$D(P_{X^n}\|Q_{X^n}) = \sum_{i=1}^n D(P_i\|P_{i-1}),$$

where $P_i = P_{X^i}Q_{X^n_{i+1}|X^i}$, with $P_n = P_{X^n}$ and $P_0 = Q_{X^n}$. The identity above shows how KL distance from P_{X^n} to Q_{X^n} can be traversed by summing distances between intermediate P_i's.

(b) Using the same path and the triangle inequality show that

$$\mathrm{TV}(P_{X^n}, Q_{X^n}) \leq \sum_{i=1}^n \mathbb{E}_{P_{X^{i-1}}} \mathrm{TV}(P_{X_i|X^{i-1}}, Q_{X_i|X^{i-1}}).$$

(c) Similarly, show for the Hellinger distance H:

$$H(P_{X^n}, Q_{X^n}) \leq \sum_{i=1}^n \sqrt{\mathbb{E}_{P_{X^{i-1}}} H^2(P_{X_i|X^{i-1}}, Q_{X_i|X^{i-1}})}.$$

See also [230, Theorem 7] for a deeper result, where for a universal $C > 0$ it is shown that

$$H^2(P_{X^n}, Q_{X^n}) \leq C\sum_{i=1}^n \mathbb{E}_{P_{X^{i-1}}} H^2(P_{X_i|X^{i-1}}, Q_{X_i|X^{i-1}}).$$

I.44 (a) Define *Marton's divergence*

$$D_m(P\|Q) = \int dQ\left(1 - \frac{dP}{dQ}\right)_+^2.$$

Prove that

$$D_m(P\|Q) = \inf_{P_{XY}}\{\mathbb{E}[P[X \neq Y|Y]^2] \colon P_X = P, P_Y = Q\},$$

where the infimum is over all couplings. $\left(\text{Hint: For one direction use the same coupling achieving TV. For the other direction notice that } \mathbb{P}[X \neq Y|Y] \geq 1 - \frac{P(Y)}{Q(Y)}.\right)$

(b) Define *symmetrized Marton's divergence*

$$D_{sm}(P\|Q) = D_m(P\|Q) + D_m(Q\|P).$$

Prove that

$$D_{sm}(P\|Q) = \inf_{P_{XY}}\{\mathbb{E}[\mathbb{P}[X \neq Y|Y]^2] + \mathbb{E}[\mathbb{P}[X \neq Y|X]^2]: P_X = P, P_Y = Q\}.$$

I.45 (Center of gravity under f-divergences) Recall from Corollary 4.2 the fact that $\min_{Q_Y} D(P_{Y|X}\|Q_Y|P_X) = I(X;Y)$ is achieved at $Q_Y = P_Y$. Prove the following versions for other f-divergences:

(a) Suppose that for P_X-a.e. x, $P_{Y|X=x} \ll \mu$ with density $p(y|x)$.[3] Then

$$\inf_{Q_Y} \chi^2(P_{Y|X}\|Q_Y|P_X) = \left(\int \mu(dy)\sqrt{\mathbb{E}[p_{Y|X}(y|X)^2]}\right)^2 - 1. \tag{I.12}$$

If the right-hand side is finite, the minimum is achieved at $Q_Y(dy) \propto \sqrt{\mathbb{E}[p(y|X)^2]}\mu(dy)$.

(b) Show that

$$\inf_{Q_Y} \chi^2(Q_Y\|P_{Y|X}|P_X) = \left(\int \mu(dy)\frac{1}{g(y)}\right)^{-1} - 1, \tag{I.13}$$

where $g(y) \triangleq \mathbb{E}[p_{Y|X}(y|X)^{-1}]$ and we follow the agreement $1/0 = \infty$ for all reciprocals. If the right-hand side is finite, then the minimum is achieved by $Q_Y(dy) \propto \frac{1}{g(y)}1\{g(y) < \infty\}\mu(dy)$.

(c) Show that

$$\inf_{Q_Y} D(Q_Y\|P_{Y|X}|P_X) = -\log \int \mu(dy)\exp(\mathbb{E}[\log p(y|X)]). \tag{I.14}$$

If the right-hand side is finite, the minimum is achieved at $Q_Y(dy) \propto \exp(\mathbb{E}[\log p(y|X)])\mu(dy)$.

Note: This exercise shows that the center of gravity with respect to other f-divergences need not be P_Y but its reweighted version. For statistical applications, see Exercises **VI.6**, **VI.9**, and **VI.10**, where (I.12) and (I.13) are used to determine the form of the Bayes estimator.

I.46 (DPI for Fisher information) Let $p_\theta(x,y)$ be a smoothly parameterized family of densities on $\mathcal{X} \otimes \mathcal{Y}$ (with respect to some reference measure $\mu_X \otimes \mu_Y$) where $\theta \in \mathbb{R}^d$. Let $J_F^{X,Y}(\theta)$ denote the Fisher information matrix of the joint distribution and similarly $J_F^X(\theta), J_F^Y(\theta)$ those of the marginals.

[3] Note that the results do not depend on the choice of μ, so we can take for example $\mu = P_Y$, in view of Lemma 3.3.

(a) (Monotonicity) Assume the interchangeability of derivative and integral, namely, $\nabla_\theta p_\theta(y) = \int \mu_X(dx)\nabla_\theta p_\theta(x,y)$ for every θ, y. Show that $J_F^Y(\theta) \preceq J_F^{X,Y}(\theta)$.

(b) (Data-processing inequality) Suppose, in addition, that $\theta \to X \to Y$. (In other words, $p_\theta(y|x)$ does not depend on θ.) Then show that $J_F^Y(\theta) \preceq J_F^X(\theta)$, with equality if Y is a sufficient statistic of X for θ.

I.47 (Fisher information inequality) Consider real-valued $A \perp\!\!\!\perp B$ with differentiable densities and finite (location) Fisher information $J(A)$, $J(B)$. Then Stam's inequality [400] shows that

$$\frac{1}{J(A+B)} \geq \frac{1}{J(A)} + \frac{1}{J(B)}. \tag{I.15}$$

(a) Show that Stam's inequality is equivalent to $(a+b)^2 J(A+B) \leq a^2 J(A) + b^2 J(B)$ for all $a, b > 0$.

(b) Let $X_1 = a\theta + A$, $X_2 = b\theta + B$. This defines a family of distributions of (X_1, X_2) parameterized by $\theta \in \mathbb{R}$. Show that its Fisher information is given by $J_F(\theta) = a^2 J(A) + b^2 J(B)$.

(c) Let $Y = X_1 + X_2$ and assume that conditions for the applicability of the DPI for Fisher information (Exercise **I.46**) hold. Conclude the proof of (I.15).

Note: A simple sufficient condition that implies (I.15) is that the densities of A and B are everywhere strictly positive on $\mathbb{R}$. For a direct proof in this case, see [58].

I.48 The *Ingster–Suslina formula* [225] computes the χ^2-divergence between a mixture and a simple distribution, exploiting the second moment nature of χ^2. Let P_θ be a family of probability distributions on $\mathcal{X}$ parameterized by $\theta \in \Theta$. Each distribution (prior) π on Θ induces a mixture $P_\pi \triangleq \int P_\theta \pi(d\theta)$. Assume that the P_θ's have a common dominating distribution Q.

(a) Show that

$$\chi^2(P_\pi \| Q) = \mathbb{E}[G(\theta, \tilde{\theta})] - 1,$$

where $\theta, \tilde{\theta}$ are two "replicas" independently drawn from π and $G(\theta, \tilde{\theta}) \triangleq \int dQ \frac{dP_\theta}{dQ} \frac{dP_{\tilde{\theta}}}{dQ}$.

(b) Show that for Gaussian mixtures (with $*$ denoting convolution)

$$\chi^2(\pi * \mathcal{N}(0, I) \| \mathcal{N}(0, I)) = \mathbb{E}[e^{\langle \theta, \tilde{\theta} \rangle}] - 1, \quad \theta, \tilde{\theta} \overset{\text{iid}}{\sim} \pi.$$

Deduce (7.45) from this result.

I.49 (Community detection) The Erdös–Rényi model, denoted by $\mathrm{ER}(n, p)$, is the distribution of a random graph with n nodes where each pair is connected independently with probability p. The stochastic block model (SBM), denoted by $\mathrm{SBM}(n, p, q)$, extends the homogeneous Erdös–Rényi model to incorporate community structure: Each node i is labeled by $\sigma_i \in \{\pm 1\}$ denoting its community membership; conditioned on $\sigma = (\sigma_1, \ldots, \sigma_n)$, each pair i and j are connected independently with probability p if $\sigma_i = \sigma_j$ and q otherwise.

(For example, the assortative case of $p > q$ models the scenario where individuals in the same community are more likely to be friends.) The problem of community detection asks whether $\mathrm{SBM}(n, p, q)$ is distinguishable from its Erdös–Rényi counterpart $\mathrm{ER}(n, \frac{p+q}{2})$ when n is large. Consider the sparse setting where $p = \frac{a}{n}$ and $q = \frac{b}{n}$ for fixed constants $a, b > 0$.

(a) Assume that the labels σ_i's are independent and uniform. Applying the Ingster–Suslina formula in Exercise **I.48**, show that as $n \to \infty$,

$$\chi^2\Big(\mathrm{SBM}(n, p, q) \Big\| \mathrm{ER}\Big(n, \frac{p+q}{2}\Big)\Big) = \begin{cases} \frac{1}{\sqrt{1-\tau}} + o(1), & \tau < 1, \\ \infty, & \tau \geq 1, \end{cases} \tag{I.16}$$

where $\tau \triangleq \frac{(a-b)^2}{2(a+b)}$.

(b) Assume that n is even and σ is uniformly distributed over the set of bisections $\{z \in \{\pm 1\}^n : \sum_{i=1}^n z_i = 0\}$, so that the two communities are equally sized. Show that (I.16) continues to hold.

Note: As a consequence of (I.16), we have the contiguity $\mathrm{SBM}(n, p, q) \triangleleft \mathrm{ER}(n, \frac{p+q}{2})$ whenever $\tau < 1$. In fact, they are mutually contiguous if and only if $\tau < 1$. This much more difficult result can be shown using the method of *small subgraph conditioning* developed by [365, 227]; see [308, Section 5].

I.50 (Sampling without replacement I [401]) Consider two ways of generating a random vector $X^n = (X_1, \ldots, X_n)$: under P, X^n are sampled from the set $[n] = \{1, \ldots, n\}$ without replacement; under Q, X^n are sampled from $[n]$ with replacement. Let's compare the joint distributions of the first k draws $X_1, \ldots, X_k$ for some $1 \leq k \leq n$.

(a) Show that

$$\mathrm{TV}(P_{X^k}, Q_{X^k}) = 1 - \frac{k!}{n^k}\binom{n}{k},$$

$$D(P_{X^k} \| Q_{X^k}) = -\log \frac{k!}{n^k}\binom{n}{k}.$$

Conclude that D and TV are $o(1)$ iff $k = o(\sqrt{n})$.

(b) Explain the specialness of $\sqrt{n}$ by finding an explicit test that distinguishes P and Q with high probability when $k \gg \sqrt{n}$. (Hint: Birthday paradox.)

I.51 (Sampling without replacement II [401]) Let $X_1, \ldots, X_k$ be a random sample of balls without replacement from an urn containing a_i balls of color $i \in [q]$, $\sum_{i=1}^q a_i = n$. Let $Q_X(i) = \frac{a_i}{n}$. In this exercise we compare the joint distributions of X^k with and without replacement by showing that

$$D(P_{X^k} \| Q_X^k) \leq c \frac{k^2(q-1)}{(n-1)(n-k+1)}, \qquad c = \frac{\log e}{2}.$$

(a) Let R_{m, b_0, b_1} be the distribution of the number of 1's in the first $m \leq b_0 + b_1$ coordinates of a randomly permuted binary string with b_0 0's and b_1 1's.

Show that

$$D(P_{X_{m+1}|X^m} \| Q_X | P_{X^m}) = \sum_{i=1}^{q} \mathbb{E}\left[\frac{a_i - V_i}{n-m} \log \frac{a_i - V_i}{p_i(n-m)}\right],$$

where $V_i \sim R_{m,n-a_i,a_i}$ and $p_i = a_i/n$.

(b) Show that the ith term above also equals $p_i \mathbb{E}[\log \frac{a_i - \tilde{V}_i}{p_i(n-m)}]$, $\tilde{V}_i \sim R_{m,n-a_i,a_i-1}$.

(c) Use Jensen's inequality to show that the ith term is upper-bounded by

$$p_i \log\left(1 + \frac{m}{(n-1)(n-m)} \frac{1-p_i}{p_i}\right).$$

(d) Use the bound $\log(1+x) \le x \log e$ to complete the proof.

I.52 (Effective de Finetti) We will show that for any distribution P_{X^n} invariant to permutation and $k < n$ there exists a mixture of iid distributions Q_{X^k} which approximates P_{X^k}:

$$\mathrm{TV}(P_{X^k}, Q_{X^k}) \le c\sqrt{\frac{k^2 H(X_1)}{n-k+1}}, \quad Q_{X_k} = \sum_{i=1}^{m} \lambda_i Q_i^{\otimes k}, \tag{I.17}$$

where $\sum_{i=1}^m \lambda_i = 1$, $\lambda_i \ge 0$, Q_i's are some distributions on $\mathcal{X}$ and $c > 0$ is a universal constant. Follow the steps:

(a) Show the identity (here P_{X^k} is arbitrary)

$$D\left(P_{X^k} \middle\| \prod_{j=1}^{k} P_{X_j}\right) = \sum_{j=1}^{k-1} I(X^j; X_{j+1}).$$

(b) Show that there must exist some $t \in \{k, k+1, \ldots, n\}$ such that

$$I(X^{k-1}; X_k | X_{t+1}^n) \le \frac{H(X^{k-1})}{n-k+1}.$$

(Hint: Expand $I(X^{k-1}; X_k^n)$ via the chain rule.)

(c) Show from steps (a) and (b) that

$$D\left(P_{X^k|T} \middle\| \prod P_{X_j|T} \middle| P_T\right) \le \frac{kH(X^{k-1})}{n-k+1},$$

where $T = X_{t+1}^n$.

(d) By Pinsker's inequality

$$\mathbb{E}_T\left[\mathrm{TV}\left(P_{X^k|T}, \prod P_{X_j|T}\right)\right] \le c\sqrt{\frac{kH(X^{k-1})|\mathcal{X}|}{n-k+1}}, \quad c = \frac{1}{\sqrt{2\log e}}.$$

Conclude (I.17) by the convexity of total variation.

Note: Another estimate [401, 123] is easy to deduce from Exercises **I.51** and **I.50**: there exists a mixture of iid Q_{X^k} such that

$$\mathrm{TV}(Q_{X^k}, P_{X^k}) \le \frac{k}{n} \min(2|\mathcal{X}|, k-1).$$

The bound (I.17) improves the above only when $H(X_1) \lesssim 1$.

I.53 (Wringing lemma [140, 420]) Prove that for any $\delta > 0$ and any (U^n, V^n) there exists an index set $I \subset [n]$ of size $|I| \le \frac{I(U^n;V^n)}{\delta}$ such that

$$I(U_t;V_t|U_I, V_I) \le \delta \qquad \forall t \in [n].$$

When $I(U^n;V^n) \ll n$, this shows that conditioning on a (relatively) few entries, one can make individual coordinates almost independent. (Hint: Show $I(A, B;C, D) \ge I(A;C) + I(B;D|A, C)$ first. Then start with $I = \emptyset$ and if there is any index t s.t. $I(U_t;V_t|U_I, V_I) > \delta$ then add it to I and repeat.)

I.54 (Generalization gap = I_{SKL}, [18]) A learning algorithm selects a parameter W based on observing (not necessarily independent) $S_1, \ldots, S_n$, where all S_i have a common marginal law P_S, with the goal of minimizing the loss on a fresh sample = $\mathbb{E}[\ell(W, S)]$, where $S^n \perp\!\!\!\perp S \sim P_S$ and ℓ is an arbitrary loss function.[4] Consider a Gibbs sampler (see Section 4.8.2) which chooses

$$W \sim P_{W|S^n}(w|s^n) = \frac{1}{Z(s^n)}\pi(w) \exp\left\{-\frac{\alpha}{n}\sum_{i=1}^{n} \ell(w, s_i)\right\},$$

where $\pi(\cdot)$ is a fixed prior on weights and $Z(\cdot)$ the normalization constant. Show that the generalization gap of this algorithm is given by

$$\mathbb{E}[\ell(W, S)] - \mathbb{E}\left[\frac{1}{n}\sum_{i=1}^{n} \ell(W, S_i)\right] = \frac{1}{\alpha} I_{\mathrm{SKL}}(W;S^n),$$

where I_{SKL} is the symmetric KL-information defined in (7.49).

I.55 Let (X, Y) be some dependent random variables. Suppose that for every x the random variable $h(x, Y)$ is ϵ^2-sub-Gaussian. Show that

$$\mathbb{E}_{P_{X,Y}}[h(X, Y)] - \mathbb{E}_{P_X \times P_Y}[h(X, Y)] \le \sqrt{2\epsilon^2 I(X;Y)}. \tag{I.18}$$

This allows one to control expectations of functions of dependent random variables by replacing them with independent pairs at the expense of the (square-root of the) mutual information slack.

I.56 ([370]) Let $A = \{A_j : j \in J\}$ be a countable collection of random variables and T a J-valued random index. Show that if each A_j is ϵ^2-sub-Gaussian, then

$$|\,\mathbb{E}[A_T]| \le \sqrt{2\epsilon^2 I(A;T)}.$$

I.57 (Divergence for mixtures [216, 250]) Let $\bar{Q} = \sum_i \pi_i Q_i$ be a mixture distribution.

(a) Prove that

$$D(P\|\bar{Q}) \le -\log\left(\sum_i \pi_i \exp(-D(P\|Q_i))\right),$$

[4] For example, if $S = (X, Y)$ we may have $\ell(w, (x, y)) = 1\{f_w(x) \ne y\}$ where f_w denotes a neural network with weights w.

improving over the simple convexity estimate $D(P\|\bar{Q}) \le \sum_i \pi_i D(P\|Q_i)$. (Hint: Prove that the function $Q \mapsto \exp\{-aD(P\|Q)\}$ is concave for every $a \le 1$.)

(b) Furthermore, for any distribution $\{\tilde{\pi}_j\}$ and any $\lambda \in [0,1]$ show that we have

$$\sum_j \tilde{\pi}_j D(Q_j\|\bar{Q}) + D(\pi\|\tilde{\pi}) \ge -\sum_i \pi_i \log \sum_j \tilde{\pi}_j e^{-(1-\lambda)D_\lambda(P_i\|P_j)}$$
$$\ge -\log \sum_{i,j} \pi_i \tilde{\pi}_j e^{-(1-\lambda)D_\lambda(P_i\|P_j)}.$$

(Hint: Prove $D(P_{A|B=b}\|Q_A) \ge -\mathbb{E}_{A|B=b}[\log \mathbb{E}_{A'\sim Q_A} \frac{g(A',b)}{g(A,b)}]$ via Donsker–Varadhan. Plug in $g(a,b) = P_{B|A}(b|a)^{1-\lambda}$, average over B, and use Jensen to bring outer $\mathbb{E}_{B|A}$ inside the log.)

I.58 (Mutual information and pairwise distances [216]) Suppose we have knowledge of pairwise distances $d_\lambda(x,x') \triangleq D_\lambda(P_{Y|X=x}\|P_{Y|X=x'})$, where D_λ is the Rényi divergence of order λ. What can be said about $I(X;Y)$? Let $X, X' \overset{\text{iid}}{\sim} P_X$. Using Exercise **I.57**, prove that

$$I(X;Y) \le -\mathbb{E}[\log \mathbb{E}[\exp(-d_1(X,X'))|X]]$$

and for every $\lambda \in [0,1]$

$$I(X;Y) \ge -\mathbb{E}[\log \mathbb{E}[\exp(-(1-\lambda)d_\lambda(X,X'))|X]].$$

See Theorem 32.5 for an application.

I.59 ($D \lesssim H^2 \log \frac{1}{H^2}$ trick) Show that for any $P, U, R, \lambda > 1$, and $0 < \epsilon < 2^{-5\frac{\lambda}{\lambda-1}}$ we have

$$D(P\|\epsilon U + \bar{\epsilon}R) \le 8(H^2(P,R) + 2\epsilon)\left(\frac{\lambda}{\lambda-1}\log\frac{1}{\epsilon} + D_\lambda(P\|U)\right).$$

Thus, a Hellinger ϵ-net for a set of P's can be converted into a KL $(\epsilon^2 \log \frac{1}{\epsilon})$-net; see Theorem 32.6 in Section 32.2.4.

(a) Start by proving the tail estimate for the divergence: For any $\lambda > 1$ and $b > e^{(\lambda-1)^{-1}}$

$$\mathbb{E}_P\left[\log\frac{dP}{dQ}\cdot 1\left\{\frac{dP}{dQ} > b\right\}\right] \le \frac{\log b}{b^{\lambda-1}}\exp\{(\lambda-1)D_\lambda(P\|Q)\}.$$

(b) Show that for any $b > 1$ we have

$$D(P\|Q) \le H^2(P,Q)\frac{b\log b}{(\sqrt{b}-1)^2} + \mathbb{E}_P\left[\log\frac{dP}{dQ}\cdot 1\left\{\frac{dP}{dQ} > b\right\}\right].$$

(Hint: Write $D(P\|Q) = \mathbb{E}_P[h(\frac{dQ}{dP})]$ for $h(x) = -\log x + x - 1$ and notice that $\frac{h(x)}{(\sqrt{x}-1)^2}$ is monotonically decreasing on $\mathbb{R}_+$.)

(c) Set $Q = \epsilon U + \bar{\epsilon}R$ and show that for every $\delta < e^{-\frac{1}{\lambda-1}} \wedge \frac{1}{4}$

$$D(P\|Q) \le \left(4H^2(P,R) + 8\epsilon + c_\lambda \epsilon^{1-\lambda}\delta^{\lambda-1}\right)\log\frac{1}{\delta},$$

where $c_\lambda = \exp\{(\lambda - 1)D_\lambda(P\|U)\}$. (Hint: notice that $H^2(P,Q) \le H^2(P,R) + 2\epsilon$ and $D_\lambda(P\|Q) \le D_\lambda(P\|U) + \log\frac{1}{\epsilon}$, and set $b = 1/\delta$.)

(d) Complete the proof by setting $\delta^{\lambda-1} = \frac{4H^2(P,R)+2\epsilon}{c_\lambda \epsilon^{\lambda-1}}$.

I.60 Let $G = (V, E)$ be a finite *directed* graph. Let

$$\Delta = \left|\left\{(x,y,z) \in V^3 : (x,y),(y,z),(z,x) \in E\right\}\right|,$$
$$\wedge = \left|\left\{(x,y,z) \in V^3 : (x,y),(x,z) \in E\right\}\right|.$$

Prove that $\Delta \le \wedge$.

(Hint: Prove $H(X,Y,Z) \le H(X) + 2H(Y|X)$ for random variables (X,Y,Z) distributed uniformly over the set of directed 3-cycles, that is, subsets $X \to Y \to Z \to X$.)

I.61 (Union-closed sets conjecture (UCSC)) Let X and Y be independent vectors in $\{0,1\}^n$. Show [88]

$$H(X \texttt{ OR } Y) \ge \frac{\bar{p}}{2\phi}(H(X) + H(Y)), \qquad \bar{p} \triangleq \min_i \min(\mathbb{P}[X_i = 0], \mathbb{P}[Y_i = 0]),$$

where $\texttt{OR}$ denotes coordinatewise logical-OR and $\phi = \frac{\sqrt{5}-1}{2}$. (Hint: Set $Z = X \texttt{ OR } Y$, use the chain rule $H(Z) \ge \sum_i H(Z_i|X^{i-1}, Y^{i-1})$, and the inequality for binary entropy $h(ab) \ge \frac{h(a)b+h(b)a}{2\phi}$.)

Comment: $\mathcal{F} \subset \{0,1\}^n$ is called a union-closed set if $x, y \in \mathcal{F} \implies (x \texttt{ OR } y) \in \mathcal{F}$. The UCSC states that $p = \max_i \mathbb{P}[X_i = 1] \ge 1/2$, where X is uniform over $\mathcal{F}$. Gilmer's method [189] applies the inequality above to Y taken to be an independent copy of X (so that $H(X \texttt{ OR } Y) \le H(X) = H(Y) = \log|\mathcal{F}|$) to prove that $p \ge 1 - \phi \approx 0.382$.

I.62 (Compression for regression) Let $Y \in [-1,1]$ and $X \in \mathcal{X}$ with $\mathcal{X}$ being finite (for simplicity). Auxiliary variables U, U' in this exercise are assumed to be deterministic functions of X. For simplicity assume $\mathcal{X}$ is finite (but giant). Let $\mathrm{cmp}(U)$ be a complexity measure satisfying $\mathrm{cmp}(U,U') \le \mathrm{cmp}(U) + \mathrm{cmp}(U')$, $\mathrm{cmp}(\text{constant}) = 0$, and $\mathrm{cmp}(\mathrm{Ber}(p)) \le \log 2$ for any p (think of $H(U)$ or $\log|\mathcal{U}|$). Choose U to be a maximizer of $I(U;Y) - \delta\,\mathrm{cmp}(U)$.

(a) Show that $\mathrm{cmp}(U) \le \frac{I(X;Y)}{\delta}$.

(b) For any U' show $I(Y;U'|U) \le \delta\,\mathrm{cmp}(U')$. (Hint: Check $U'' = (U,U')$.)

(c) For any event $S = \{X \in A\}$ show

$$|\mathbb{E}[(Y - \mathbb{E}[Y|U])1_S]| \le \sqrt{2\delta \ln 2}. \tag{I.19}$$

(Hint: By Cauchy–Schwarz you only need to show $\mathbb{E}[|\mathbb{E}[Y|U,1_S] - \mathbb{E}[Y|U]|^2] \lesssim \delta$, which follows by taking $U' = 1_S$ in (b) and applying Tao's inequality (7.28).)

(d) By choosing a proper S and applying the above to S and S^c conclude that

$$\mathbb{E}[|\mathbb{E}[Y|X] - \mathbb{E}[Y|U]|] \le 2\sqrt{2\delta \ln 2}.$$

(So any high-dimensional complex feature vector X can be compressed down to U whose cardinality is of order $I(Y;X)$ (and independent of $|\mathcal{X}|$) but which, nevertheless, is essentially as good as X for regression; see [51] for other results on *information distillation*.)

I.63 (IT version of Szemerédi regularity [415]) Fix $\epsilon, m > 0$ and consider random variables $Y, X = (X_1, X_2)$ with $Y \in [-1, 1]$, $\mathcal{X} = \mathcal{X}_1 \times \mathcal{X}_2$ finite (but enormous), and $I(X;Y) \le m$. In this exercise, all auxiliary random variables U have structure $U = (f_1(X_1), f_2(X_2))$ for some deterministic functions f_1, f_2. Thus U partitions $\mathcal{X}$ into product blocks and we call block $U = u_0$ *ϵ-regular* if

$$|\mathbb{E}[(Y - \mathbb{E}[Y|U])1_S|U = u_0]| \le \epsilon \qquad \forall S = \{X_1 \in A_1, X_2 \in A_2\}.$$

We will show there is $J = J(\epsilon, m)$ such that there exists a U taking values on $\mathcal{U}$ with $|\mathcal{U}| \le J$ and such that

$$\mathbb{P}[\text{block } U \text{ is not } \epsilon\text{-regular}] \le \epsilon. \tag{I.20}$$

(a) Suppose that we found random variables $Y \to X \to U' \to U$ such that (i) $I(Y;U'|U) \le \epsilon^4$ and (ii) for all S as above $I(Y;1_S|U') \le \frac{\epsilon^4}{|\mathcal{U}|^2}$. Then (I.20) holds with ϵ replaced by $O(\epsilon)$. (Hint: Define $g(u_0) = \mathbb{E}[|\,\mathbb{E}[Y|U'] - \mathbb{E}[Y|U]|\,|U = u_0]$ and show via (7.28) that $\mathbb{E}[g(U)] \lesssim \epsilon^2$. As in (I.19) argue that $\mathbb{E}[(Y - \mathbb{E}[Y|U'])1_S] \lesssim \frac{\epsilon^2}{|\mathcal{U}|}$. From the triangle inequality any u_0-block is $O(\epsilon)$-regular whenever $g(u_0) < \epsilon$ and $\mathbb{P}[U = u_0] > \frac{\epsilon}{|\mathcal{U}|}$. Finally, apply Markov inequality twice to show that the last condition is violated with $O(\epsilon)$ probability.)

(b) Show that such U', U indeed exist. (Hint: Construct a sequence $Y \to X \to \cdots \to U_j \to U_{j-1} \to \cdots \to U_0 = 0$ sequentially by taking U_{j+1} to be the maximizer of $I(Y;U) - \delta_{j+1} \log|\mathcal{U}|$ among all $Y \to X \to U \to U_j$ (compare with Exercise **I.62**) and $\delta_{j+1} = \frac{\epsilon^4}{|\mathcal{U}_j|^2}$. We take $(U', U) = (U_{n+1}, U_n)$ for the first pair that has $I(Y;U_{n+1}|U_n) \le \epsilon^4$. Show $n \le \frac{m}{\epsilon^4}$ and $|\mathcal{U}_n|$ is bounded by the nth iterate of the map $h \to h\exp\{mh^2/\epsilon^4\}$ started from $h = 1$.)

Remark: The point is that J does not depend on $P_{X,Y}$ or $|\mathcal{X}|$. For the classical version of Szemerédi's regularity lemma one takes X_1, X_2 to be uniformly sampled vertices of a bipartite graph and $Y = 1\{X_1 \sim X_2\}$ to be the incidence relation. An ϵ-regular block corresponds to an ϵ-regular bipartite subgraph, and the lemma decomposes an arbitrary graph into finitely many pairwise (almost) regular subgraphs.

I.64 (Entropy and binary convolution) Binary convolution is defined for $(a, b) \in [0, 1]^2$ by $a * b = a(1 - b) + (1 - a)b$ and describes the law of $\mathrm{Ber}(a) \oplus \mathrm{Ber}(b)$ where $\oplus$ denotes modulo-2 addition.

(a) (Mrs. Gerber's lemma, MGL[5]) Let $(U, X) \perp\!\!\!\perp Z \sim \mathrm{Ber}(\delta)$ with $X \in \{0, 1\}$. Show that

$$h(h^{-1}(H(X|U)) * \delta) \le H(X \oplus Z|U) \le H(X|U) + H(Z) - \frac{H(X|U)H(Z)}{\log 2}.$$

[5] Apparently, named after a landlady renting a house to Wyner and Ziv [458] at the time.

(Hint: Equivalently [458], we need to show that the parametric curve $(h(p), h(p * \delta)), p \in [0, 1/2]$ is convex.)

(b) Show that for any p, q the parametric curve $((1-2r)^2, d(p * r \| q * r)), r \in [0, 1/2]$ is convex. (Hint: See [368, Appendix A].)

MGL has various applications (Example 16.1 and Exercise **VI.21**), it tensorizes (see Exercise **III.32**), and its infinitesimal version (derivative in $\delta = 0+$) is exactly the log-Sobolev inequality for the hypercube [122, Section 4.1].

I.65 (Log-Sobolev inequality, LSI) Let X be an $\mathbb{R}^d$-valued random variable, $\mathbb{E}[\|X\|^2] < \infty$, and $X \perp\!\!\!\perp Z \sim \gamma$, where $\gamma = \mathcal{N}(0, I_d)$ is the standard Gaussian measure. Recall the notation for Fisher information matrix $J(\cdot)$ from (2.40).

(a) Show *de Bruijn's identity*:

$$\frac{d}{da} h(X + \sqrt{a}Z) = \frac{\log e}{2} \operatorname{tr} J(X + \sqrt{a}Z).$$

(Hint: Inspect the proof of Theorem 3.14.)

(b) Show that the entropy-power inequality implies

$$\frac{d}{da} \exp\{2h(X + \sqrt{a}Z)/d\} \geq 2\pi e.$$

(c) Conclude that Gaussians *minimize* the differential entropy among all X with bounded Fisher information $J(X)$, namely [400]

$$h(X) \geq \frac{n}{2} \log \frac{2\pi e n}{\operatorname{tr} J(X)}.$$

(d) Show the LSI of Gross [200]: For any f with $\int f^2 d\gamma = 1$, we have

$$\int f^2 \ln(f^2) d\gamma \leq 2 \cdot \int \|\nabla f\|^2 d\gamma.$$

(Hint: $P_X(dx) = f^2(x)\gamma(dx)$, prove $2\int (x^\top \nabla f) f \gamma(dx) = \mathbb{E}[\|X\|^2] - d$ and use $\ln(1+y) \leq y$.)

I.66 (Stochastic localization [148, 146]) Consider a discrete $X \sim \mu$ taking values in $\mathbb{R}^n$ and let $\rho = \mathbb{E}[\|X - \mathbb{E}[X]\|^2] = \sum_{i=1}^n \mathrm{Var}[X_i]$. We will show that for any $\epsilon > 0$ there exists a decomposition of $\mu = \mathbb{E}_\theta[\mu_\theta]$ as a mixture of measures μ_θ, which have similar entropy ($\mathbb{E}_\theta[H(\mu_\theta)] = H(\mu) - O(\rho/\epsilon)$) but have almost no pairwise correlations ($\mathbb{E}_\theta[\mathrm{Cov}(\mu_\theta)] \preceq \epsilon I_n$ and $\mathbb{E}_\theta[\|\mathrm{Cov}(\mu_\theta)\|_F^2] = O(\epsilon\rho)$). This has useful applications in statistical physics of Ising models.

(a) Let $Y_t = \sqrt{t}X + \sqrt{\epsilon}Z$, where $X \perp\!\!\!\perp Z \sim \mathcal{N}(0, I_d)$ and $t, \epsilon > 0$. Show that $\mathrm{Cov}(X|Y_t) \preceq \frac{\epsilon}{t} I_n$. (Hint: Consider the suboptimal estimator $\hat{X}(Y_t) = Y_t/\sqrt{t}$.)

(b) Show that $0 \leq H(X) - H(X|Y_t) \leq \frac{n}{2}\log(1 + \frac{t}{\epsilon n}\rho) \leq \frac{t \log e}{2\epsilon}\rho$. (Hint: Use (5.22).)

(c) Show that $\rho \geq \mathsf{mmse}(X|Y_1) - \mathsf{mmse}(X|Y_2) = \frac{1}{\epsilon}\int_1^2 \mathbb{E}[\|\Sigma_t(Y_t)\|_F^2] dt$, where $\Sigma_t(y) = \mathrm{Cov}[X|Y_t = y]$. (Hint: Use (3.23).)

Thus we conclude that for some $t \in [1, 2]$ decomposing $\mu = \mathbb{E}_{Y_t}[P_{X|Y_t}]$ satisfies the stated claims.

Part II

Lossless Data Compression

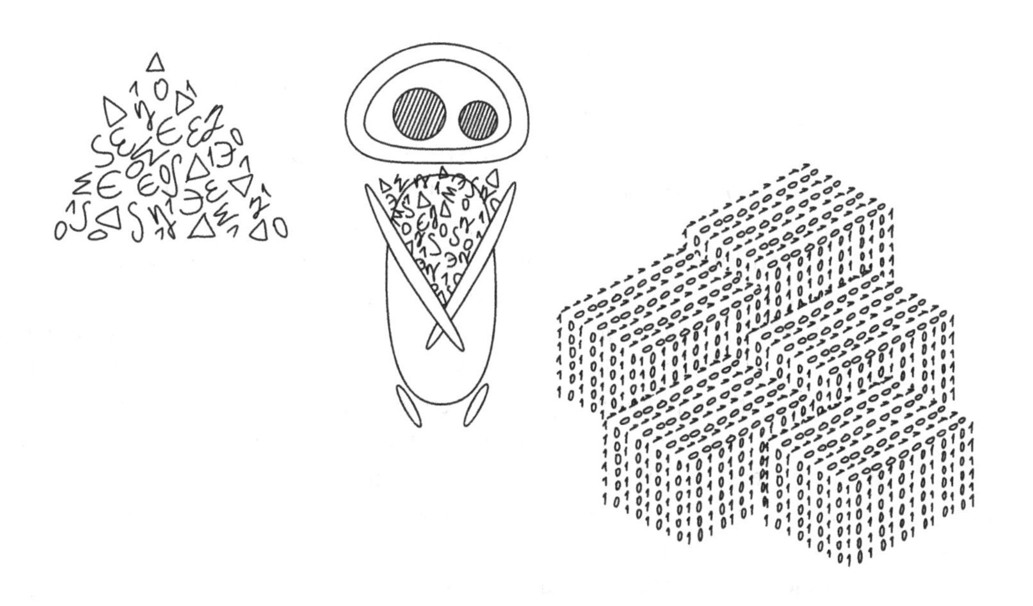

The principal goal of data compression is to represent a given sequence $a_1, a_2, \ldots, a_n$ produced by a source as a sequence of bits of minimal possible length under task-specific algorithmic constraints. Of course, reducing the number of bits is generally impossible, unless the source satisfies certain statistical restrictions, that is, if only a small subset of all sequences actually occur in practice. (Or, more precisely, only a small subset captures the majority of the overall probability distribution.) Is this the case for real-world data?

As a simple demonstration, one may take two English novels and compute empirical frequencies of occurrence of each letter. It will turn out to be the same for both novels (approximately). Thus, we can see that there is some underlying structure in English texts restricting possible output sequences. The structure goes beyond empirical frequencies of course, as further experimentation (involving digrams, word frequencies, etc.) may reveal. Thus, the main reason for the possibility of data compression is the *experimental (empirical) law: Real-world sources produce very restricted sets of sequences* of letters.

How do we model these restrictions? Further experimentation (with language, music, images) reveals that, frequently, the structure may be well described if we assume that sequences are generated probabilistically [379, Section III]. One of the lasting contributions of Shannon is another empirical law: *Real-world sequences may be described probabilistically with increasing precision starting from iid, first-order Markov, second-order Markov, etc.* Note that sometimes one needs to find an appropriate "basis" in which this "law" holds: for language you have a choice of representing the input as a sequence of either letters or words; for images you have a choice between using raw pixels or wavelets/local Fourier transforms. Indeed, a rasterized sequence of pixels does not exhibit any stable local structure whereas changing basis to wavelets and local Fourier transform reveals that structure. (Of course, one should not take these "laws" too far. In regards to language modeling, the (finite-state) Markov assumption is too simplistic to truly generate all proper sentences, see Chomsky [94]. However, astounding success of modern high-order Markov models, such as GPT-4 [321], shows that such models are very difficult to distinguish from true language.) Finding correct representations is practically very important, but in this book we assume this step has already been done and we are facing a simple stochastic process (iid, Markov, or ergodic). How do we represent it with the least number of bits?

In the beginning, we will simplify the problem even further and restrict attention to representing one random variable X in terms of (the minimal number of) bits. Later, X will be taken to be a large n-letter chunk of the target process, that is, $X = S^n = (S_1, \ldots, S_n)$. The types of compression we will consider in this book are:

- Variable-length lossless compression. Here we require $\mathbb{P}[X \neq \hat{X}] = 0$, where $\hat{X}$ is the decoded version. To make the question interesting, we compress X into a variable-length binary string. It will turn out that the optimal compression length is $H(X) - O(\log(1 + H(X)))$. If we further restrict attention to so-called prefix-free or uniquely decodable codes, then the optimal compression length is $H(X) + O(1)$. Applying these results to n-letter variables $X = S^n$ we see that the optimal compression length normalized by n converges to the entropy rate (Section 6.3) of the process $\{S_j\}$.

- Fixed-length almost lossless compression. Here, we allow some very small (or vanishing with $n \to \infty$ when $X = S^n$) probability of error, that is, $\mathbb{P}[X \neq \hat{X}] \leq \epsilon$. It turns out that under mild assumptions on the process $\{S_j\}$, here again we can compress to entropy rate but no more. This mode of compression permits various beautiful results in the presence of side information (Slepian–Wolf compression, Section 11.5).
- Lossy compression. Here we require only $\mathbb{E}[d(X, \hat{X})] \leq \epsilon$ where $d(\cdot, \cdot)$ is some loss function. This type of compression problem is the topic of Part V.

We also note that with $X = S^n$, it would be more correct to call the first two compression types above as "fixed-to-variable" and "fixed-to-fixed," because they take a fixed number of input letters and produce a variable or fixed number of output bits. There exist other types of compression algorithms, which we do not discuss, for example, a beautiful variable-to-fixed compressor of Tunstall [426].

10 Variable-Length Compression

In this chapter we consider a basic question: How does one describe a discrete random variable $X \in \mathcal{X}$ in terms of a variable-length bit string so that the description is the shortest possible? The basic idea, already used in the telegraph's Morse code, is completely obvious: shorter descriptions (bit strings) should correspond to more probable symbols. Later, however, we will see that this basic idea becomes a lot more subtle once we take X to mean a group of symbols. The discovery of Shannon was that compressing *groups* of symbols together (even if they are iid!) can lead to impressive savings in compressed length. That is, coding English text by first grouping 10 consecutive characters together is much better than doing so on a character-by-character basis. One should appreciate the boldness of Shannon's proposition since sorting all possible 26^{10} realizations of 10-letter English chunks in order of their decreasing frequency appears quite difficult. It is only later, with the invention of Huffman coding, arithmetic coding, and Lempel–Ziv compressors (decades after) that these methods became practical and ubiquitous.

In this chapter we discover that the minimal compression length of X is essentially equal to the entropy $H(X)$ for the single-shot, uniquely decodable, and prefix-free codes. These results are the first examples of *coding theorems* in our book, that is, results connecting an operational problem and an information measure. (For this reason, compression is also called *source coding* in information theory.) In addition, we also discuss the so-called Zipf law and how its widespread occurrence can be described information-theoretically.

10.1 Variable-Length Lossless Compression

The setting of lossless data compression is depicted as follows:

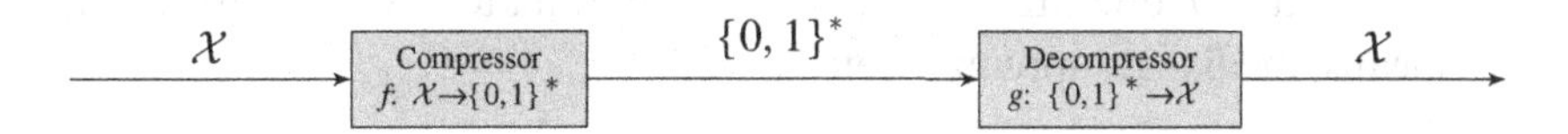

More formally, a function $f : \mathcal{X} \to \{0, 1\}^*$ is a variable-length single-shot lossless compressor of a random variable X if it satisfies the following properties:

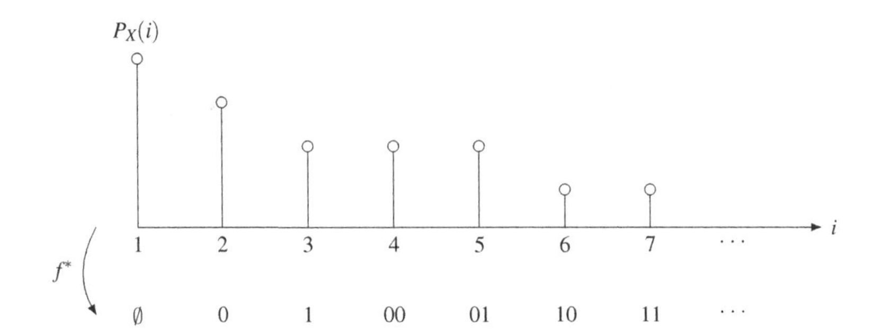

Figure 10.1 Illustration of the optimal variable-length lossless compressor f^*.

1 It maps each symbol $x \in \mathcal{X}$ into a variable-length string $f(x)$ in $\{0,1\}^* \triangleq \bigcup_{k\geq 0}\{0,1\}^k = \{\emptyset, 0, 1, 00, 01, \dots\}$. Each $f(x)$ is referred to as a *codeword* and the collection of codewords as the *codebook*.
2 It is *lossless* for X: there exists a decompressor $g\colon \{0,1\}^* \to \mathcal{X}$ such that $\mathbb{P}[X = g(f(X))] = 1$. In other words, f is injective on the support of P_X.

Notice that since $\{0,1\}^*$ is countable, lossless compression is only possible for discrete X. Also, since the structure of $\mathcal{X}$ is not important, we can relabel $\mathcal{X}$ such that $\mathcal{X} = \mathbb{N} = \{1, 2, \dots\}$ and sort the PMF decreasingly: $P_X(1) \geq P_X(2) \geq \cdots$. In *a single-shot compression* setting, see [252], we do not impose any additional constraints on the map f. Later in Section 10.3 we will introduce conditions such as prefix-freeness and unique decodability.

To quantify how good a compressor f is, we introduce the length function $l\colon \{0,1\}^* \to \mathbb{Z}_+$, for example, $l(\emptyset) = 0, l(01001) = 5$. We could consider different objectives for selecting the best compressor f, for example, minimizing any of $\mathbb{E}[l(f(X))]$, $\operatorname{esssup} l(f(X))$, or $\operatorname{median}[l(f(X))]$ appears reasonable. It turns out that there is a compressor f^* that minimizes all objectives simultaneously. As mentioned in the preface to this chapter, the main idea is to assign longer codewords to less likely symbols, and reserve the shorter codewords for more probable symbols. To make precise the optimality of f^*, let us recall the concept of stochastic dominance.

Definition 10.1. (Stochastic dominance) For real-valued random variables X and Y, we say Y stochastically dominates (or is stochastically larger than) X, denoted by $X \overset{\text{st.}}{\leq} Y$, if $\mathbb{P}[Y \leq t] \leq \mathbb{P}[X \leq t]$ for all $t \in \mathbb{R}$.

By definition, $X \overset{\text{st.}}{\leq} Y$ if and only if the CDF of X is larger than that of Y pointwise; in other words, the distribution of X assigns more probability to lower values than that of Y does. In particular, if X is dominated by Y stochastically, so are their means, medians, supremum, etc.

Theorem 10.2. (Optimal f^*) *Consider the compressor f^* defined (for a down-sorted PMF P_X) by $f^*(1) = \emptyset, f^*(2) = 0, f^*(3) = 1, f^*(4) = 00$, etc., assigning strings with increasing lengths to symbols $i \in \mathcal{X}$. (See Figure 10.1 for an illustration.) Then the following hold:*

1 *Length of codeword:*

$$l(f^*(i)) = \lfloor \log_2 i \rfloor.$$

2 *$l(f^*(X))$ is stochastically the smallest: For any lossless compressor $f\colon \mathcal{X} \to \{0,1\}^*$,*

$$l(f^*(X)) \overset{\text{st.}}{\leq} l(f(X)),$$

that is, for any k, $\mathbb{P}[l(f(X)) \leq k] \leq \mathbb{P}[l(f^*(X)) \leq k]$. As a result, $\mathbb{E}[l(f^*(X))] \leq \mathbb{E}[l(f(X))]$.

Proof. Note that

$$|A_k| \triangleq |\{x\colon l(f(x)) \leq k\}| \leq \sum_{i=0}^{k} 2^i = 2^{k+1} - 1 = |\{x\colon l(f^*(x)) \leq k\}| \triangleq |A_k^*|.$$

Here the inequality is because f is lossless so that $|A_k|$ can at most be the total number of binary strings of length up to k. Then

$$\mathbb{P}[l(f(X)) \leq k] = \sum_{x \in A_k} P_X(x) \leq \sum_{x \in A_k^*} P_X(x) = \mathbb{P}[l(f^*(X)) \leq k], \tag{10.1}$$

since $|A_k| \leq |A_k^*|$ and A_k^* contains all $2^{k+1} - 1$ most likely symbols. □

Having identified the optimal compressor the next question is to understand its average compression length $\mathbb{E}[\ell(f^*(X))]$. It turns out that one can in fact compute it exactly as an infinite series, see Exercise **II.1**. However, much more importantly, it turns out to be essentially equal to $H(X)$. Specifically, we have the following result.

Theorem 10.3. (Optimal average code length versus entropy [14])

$$H(X) \text{ bits} - \log_2[e(H(X)+1)] \leq \mathbb{E}[l(f^*(X))] \leq H(X) \text{ bits}.$$

Remark 10.1. (Source coding theorem) Theorem 10.3 is the first example of a *coding theorem* in this book, which relates the fundamental limit $\mathbb{E}[l(f^*(X))]$ (an operational quantity) to the entropy $H(X)$ (an information measure).

Proof. Define $L(X) = l(f^*(X)))$. For the upper bound, observe that since the PMFs are ordered decreasingly by assumption, $P_X(m) \leq 1/m$, so $L(m) \leq \log_2 m \leq \log_2(1/P_X(m))$. Taking expectation yields $\mathbb{E}[L(X)] \leq H(X)$.

For the lower bound,

$$\begin{aligned}
H(X) = H(X, L) = H(X|L) + H(L) &\overset{(a)}{\leq} \mathbb{E}[L] + H(L) \\
&\overset{(b)}{\leq} \mathbb{E}[L] + h\left(\frac{1}{1+\mathbb{E}[L]}\right)(1+\mathbb{E}[L]) \\
&= \mathbb{E}[L] + \log_2(1+\mathbb{E}[L]) + \mathbb{E}[L]\log_2\left(1+\frac{1}{\mathbb{E}[L]}\right) \qquad (10.2) \\
&\overset{(c)}{\leq} \mathbb{E}[L] + \log_2(1+\mathbb{E}[L]) + \log_2 e \\
&\overset{(d)}{\leq} \mathbb{E}[L] + \log_2(e(1+H(X))),
\end{aligned}$$

where in (a) we have used the fact that $H(X|L = k) \le k$ bits, because f^* is lossless, so that given $f^*(X) \in \{0,1\}^k$, X can take at most 2^k values; (b) follows by Exercise **I.4**; (c) is via $x \log(1 + 1/x) \le \log e, \forall x > 0$; and (d) is by the previously shown upper bound $H(X) \le \mathbb{E}[L]$. □

To give an illustration, we need to introduce an important method of going from a *single-letter* source to a *multi-letter* one, already alluded to in the preface. Suppose that $S_j \stackrel{\text{iid}}{\sim} P_S$ (this is called *a memoryless source*). We can group n letters of S_j together and consider $X = S^n$ as one super-letter. Applying our results to random variable X we obtain:

$$nH(S) \ge \mathbb{E}[l(f^*(S^n))] \ge nH(S) - \log_2 n + O(1).$$

In fact for memoryless sources, the exact asymptotic behavior is found in [409, Theorem 4]:

$$\mathbb{E}[l(f^*(S^n))] = \begin{cases} nH(S) + O(1), & P_S = \text{Unif}, \\ nH(S) - \frac{1}{2}\log_2 n + O(1), & P_S \ne \text{Unif}. \end{cases}$$

For the case of sources for which $\log_2 \frac{1}{P_S}$ has non-lattice distribution, it is further shown in [409, Theorem 3]:

$$\mathbb{E}[l(f^*(S^n))] = nH(S) - \frac{1}{2}\log_2(8\pi e V(S) n) + o(1), \tag{10.3}$$

where $V(S)$ is the *varentropy* of the source S:

$$V(S) \triangleq \text{Var}\left[\log_2 \frac{1}{P_S(S)}\right]. \tag{10.4}$$

Theorem 10.3 relates the *mean* of $l(f^*(X))$ to that of $\log_2 \frac{1}{P_X(X)}$ (entropy). It turns out that the distributions of these random variables are also closely related.

Theorem 10.4. (Code length distribution of f^*) $\forall \tau > 0, k \ge 0$,

$$\mathbb{P}\left[\log_2 \frac{1}{P_X(X)} \le k\right] \le \mathbb{P}\left[l(f^*(X)) \le k\right] \le \mathbb{P}\left[\log_2 \frac{1}{P_X(X)} \le k + \tau\right] + 2^{-\tau+1}.$$

Proof. Lower bound (achievability): Use $P_X(m) \le 1/m$. Then similarly as in Theorem 10.3, $L(m) = \lfloor \log_2 m \rfloor \le \log_2 m \le \log_2 \frac{1}{P_X(m)}$. Hence $L(X) \le \log_2 \frac{1}{P_X(X)}$ a.s.

Upper bound (converse): Consider the following chain

$$\begin{aligned} \mathbb{P}[L \le k] &= \mathbb{P}\left[L \le k, \log_2 \frac{1}{P_X(X)} \le k + \tau\right] + \mathbb{P}\left[L \le k, \log_2 \frac{1}{P_X(X)} > k + \tau\right] \\ &\le \mathbb{P}\left[\log_2 \frac{1}{P_X(X)} \le k+\tau\right] + \sum_{x \in \mathcal{X}} P_X(x) 1\left\{l(f^*(x)) \le k\right\} 1\left\{P_X(x) \le 2^{-k-\tau}\right\} \\ &\le \mathbb{P}\left[\log_2 \frac{1}{P_X(X)} \le k+\tau\right] + (2^{k+1} - 1) \cdot 2^{-k-\tau}. \end{aligned}$$

□

Remark 10.2. (Achievability versus converse) Traditionally, in information theory positive results ("compression length is smaller than ...") are called *achievability* and negative results ("compression length cannot be smaller than ...") are called *converse*.

So far our discussion applies to an arbitrary random variable X. Next we consider the source as a random process $(S_1, S_2, \dots)$ and introduce *blocklength n*. We apply our results to $X = S^n$, that is, by treating the first n symbols as a supersymbol. The following corollary states that the limiting behaviors, of $l(f^*(S^n))$ and $\log \frac{1}{P_{S^n}(S^n)}$ always coincide.

Corollary 10.5. *Let* $(S_1, S_2, \dots)$ *be a random process and* U, V *be real-valued random variables. Then*

$$\frac{1}{n}\log_2 \frac{1}{P_{S^n}(S^n)} \xrightarrow{\text{d}} U \quad \Leftrightarrow \quad \frac{1}{n} l(f^*(S^n)) \xrightarrow{\text{d}} U \tag{10.5}$$

and

$$\frac{1}{\sqrt{n}}\left(\log_2 \frac{1}{P_{S^n}(S^n)} - H(S^n)\right) \xrightarrow{\text{d}} V \quad \Leftrightarrow \quad \frac{1}{\sqrt{n}}(l(f^*(S^n)) - H(S^n)) \xrightarrow{\text{d}} V. \tag{10.6}$$

Proof. First recall that convergence in distribution is equivalent to convergence of the CDFs at all continuity points of the limiting CDF, that is, $U_n \xrightarrow{\text{d}} U \Leftrightarrow \mathbb{P}[U_n \le u] \to \mathbb{P}[U \le u]$ for all u at which the CDF of U is continuous (i.e. u is not an atom of U).

To get (10.5), apply Theorem 10.4 with $k = un$ and $\tau = \sqrt{n}$:

$$\mathbb{P}\left[\frac{1}{n}\log_2 \frac{1}{P_X(X)} \le u\right] \le \mathbb{P}\left[\frac{1}{n} l(f^*(X)) \le u\right] \le \mathbb{P}\left[\frac{1}{n}\log_2 \frac{1}{P_X(X)} \le u + \frac{1}{\sqrt{n}}\right] + 2^{-\sqrt{n}+1}.$$

To get (10.6), apply Theorem 10.4 with $k = H(S^n) + \sqrt{n}u$ and $\tau = n^{1/4}$:

$$\begin{aligned}\mathbb{P}\left[\frac{1}{\sqrt{n}}\left(\log \frac{1}{P_{S^n}(S^n)} - H(S^n)\right) \le u\right] &\le \mathbb{P}\left[\frac{l(f^*(S^n)) - H(S^n)}{\sqrt{n}} \le u\right] \\ &\le \mathbb{P}\left[\frac{1}{\sqrt{n}}\left(\log \frac{1}{P_{S^n}(S^n)} - H(S^n)\right) \le u + n^{-1/4}\right] \\ &\quad + 2^{-n^{1/4}+1}. \end{aligned} \tag{10.7}$$

□

Now let us particularize the preceding theorem to memoryless sources of iid S_j's. The important observation is that the log-likelihood becomes an iid sum:

$$\log \frac{1}{P_{S^n}(S^n)} = \sum_{i=1}^{n} \underbrace{\log \frac{1}{P_S(S_i)}}_{\text{iid}}.$$

This implies several results at once:

1 By the weak law of large numbers (WLLN), we know that $\frac{1}{n}\log\frac{1}{P_{S^n}(S^n)}\xrightarrow{\mathbb{P}}$ $\mathbb{E}\log\frac{1}{P_S(S)}=H(S)$. Therefore in (10.5) the limiting distribution U is degenerate, that is, $U=H(S)$, and we have the following result of fundamental importance:[1]

$$\frac{1}{n}l(f^*(S^n))\xrightarrow{\mathbb{P}}H(S).$$

That is, the optimal compression rate of an iid process converges to its entropy rate. This is a version of Shannon's source coding theorem, which we will also discuss in the subsequent chapter.

2 By the central limit theorem (CLT), if varentropy $V(S)<\infty$, then we know that V in (10.6) is Gaussian, that is,

$$\frac{1}{\sqrt{nV(S)}}\left(\log\frac{1}{P_{S^n}(S^n)}-nH(S)\right)\xrightarrow{\mathrm{d}}\mathcal{N}(0,1).$$

Consequently, we have the following Gaussian approximation for the probability law of the optimal code length

$$\frac{1}{\sqrt{nV(S)}}(l(f^*(S^n))-nH(S))\xrightarrow{\mathrm{d}}\mathcal{N}(0,1),$$

or, in shorthand,

$$l(f^*(S^n))\approx nH(S)+\sqrt{nV(S)}\mathcal{N}(0,1)\ \text{ in distribution.}$$

Gaussian approximation tells us the speed of convergence $\frac{1}{n}l(f^*(S^n))\to H(S)$ and also gives us a good approximation of the distribution of length at finite n.

Example 10.1. Ternary source

Next we apply our bounds to approximate the distribution of $l(f^*(S^n))$ in a concrete example. Consider a memoryless ternary source outputting iid n symbols from the distribution $P_S=[0.445, 0.445, 0.11]$. We first compare different results on the minimal expected length $\mathbb{E}[l(f^*(S^n))]$ in the following table:

Blocklength	Lower bound (10.3)	$\mathbb{E}[l(f^*(S^n))]$	$H(S^n)$ (upper bound)	Asymptotics (10.3)
$n=20$	21.5	24.3	27.8	$23.3+o(1)$
$n=100$	130.4	134.4	139.0	$133.3+o(1)$
$n=500$	684.1	689.2	695.0	$688.1+o(1)$

In all cases above $\mathbb{E}[l(f^*(S))]$ is close to a midpoint between the bounds.

[1] Recall that convergence to a constant in distribution is equivalent to that in probability.

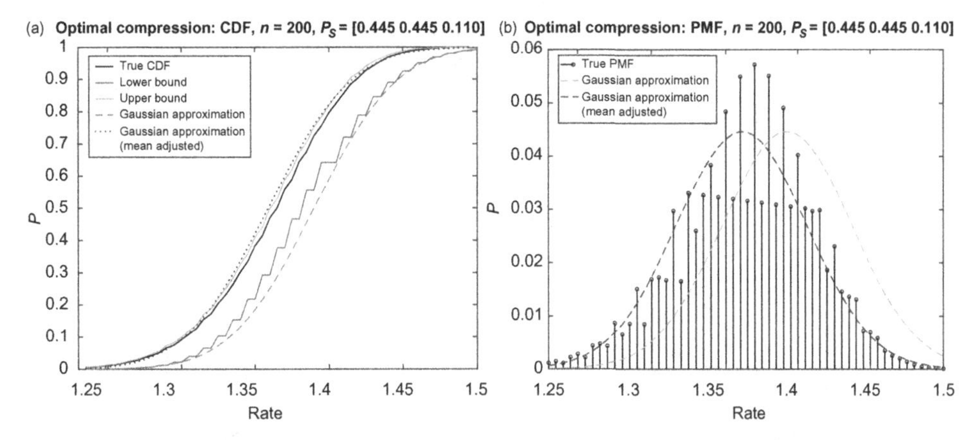

Figure 10.2 (a) Comparison of the true CDF of $l(f^*(S^n))$, bounds of Theorem 10.4 (optimized over τ), and the Gaussian approximations in (10.8) and (10.9). (b) PMF of the optimal compression length $l(f^*(S^n))$ and the two Gaussian approximations.

Next we consider the distribution of $l(f^*(S^n))$. Its Gaussian approximation is defined as

$$nH(S) + \sqrt{nV(S)}Z, \qquad Z \sim \mathcal{N}(0,1). \tag{10.8}$$

However, in view of (10.3) we also define the *mean-adjusted* Gaussian approximation as

$$nH(S) - \frac{1}{2}\log_2(8\pi e V(S)n) + \sqrt{nV(S)}Z, \qquad Z \sim \mathcal{N}(0,1). \tag{10.9}$$

Figure 10.2 compares the true distribution of $l(f^*(S^n))$ with the bounds and two Gaussian approximations.

10.2 Mandelbrot's Argument for Universality of Zipf's (Power) Law

Given a corpus of text it is natural to plot its *rank–frequency* table by sorting the word frequencies according to their rank i.e., we assume that words are indexed by integers so that their frequencies satisfy $p_1 \geq p_2 \geq \cdots$. The resulting tables, as noticed by Zipf [479], satisfy $p_r \asymp r^{-\alpha}$ for some value of α. Remarkably, this holds across various corpi of text in multiple different languages (and with $\alpha \approx 1$) – see Figure 10.3 for an illustration. Even more surprisingly, a lot of other similar tables possess the power-law distribution: "city populations, the sizes of earthquakes, moon craters, solar flares, computer files, wars, personal names in most cultures, number of papers scientists write, number of citations a paper receives, number of hits on web pages, sales of books and music recordings, number of species in biological taxa, people's incomes" (quoting from [316], which gives references for each study). This spectacular universality of the power law continues to provoke scientists from many disciplines to suggest explanations for its occurrence; see [306] for a survey of such. One of the earliest (in the context of natural language, following Zipf) is due to Mandelbrot [292] and is in fact intimately related to the topic of this chapter.

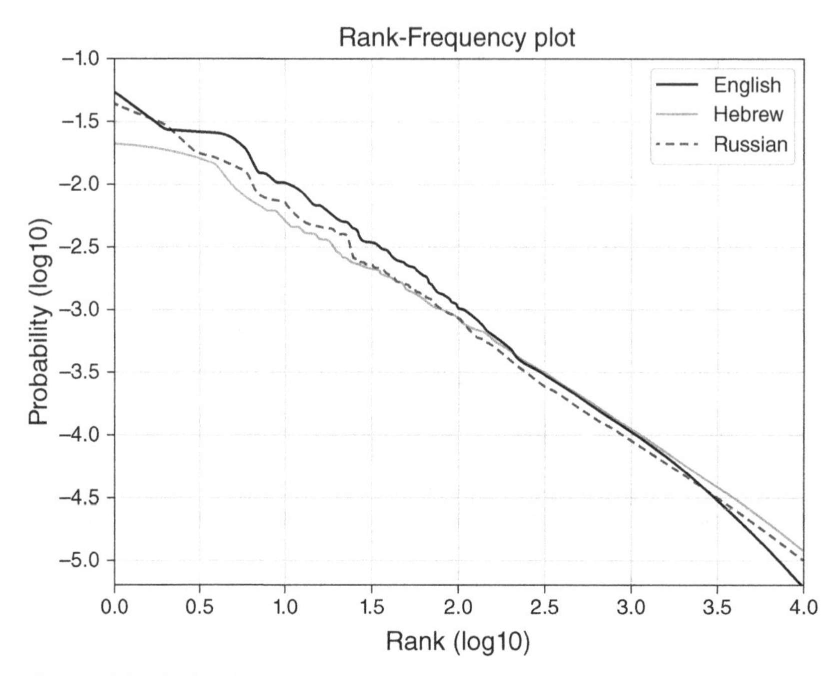

Figure 10.3 The log–log frequency–rank plots of the most used words in various languages exhibit a power-law tail with exponent close to 1, as popularized by Zipf [479]. Data from [399].

Let us go back to the question of minimal expected length of the representation of source X. We have shown bounds on this quantity in terms of the entropy of X in Theorem 10.3. Let us introduce the following function

$$\mathcal{H}(\Lambda) = \sup_{f, P_X} \{H(X) \colon \mathbb{E}[l(f(X))] \le \Lambda\},$$

where optimization is over lossless encoders and probability distributions $P_X = \{p_j \colon j = 1, \ldots\}$. Theorem 10.3 (or, more precisely, the intermediate result (10.2)) shows that

$$\Lambda \log 2 \le \mathcal{H}(\Lambda) \le \Lambda \log 2 + (1 + \Lambda) \log(1 + \Lambda) - \Lambda \log \Lambda.$$

It turns out that the upper bound is in fact tight. Furthermore, among all distributions the optimal tradeoff between entropy and minimal compression length is attained at power-law distributions.

To show that, notice that in computing $\mathcal{H}(\Lambda)$, we can restrict attention to sorted PMFs $p_1 \ge p_2 \ge \cdots$ (call this class $\mathcal{P}^{\downarrow}$), for which the optimal encoder is such that $l(f(j)) = \lfloor \log_2 j \rfloor$ (Theorem 10.2). Thus, we have shown

$$\mathcal{H}(\Lambda) = \sup_{P \in \mathcal{P}^{\downarrow}} \left\{ H(P) \colon \sum_j p_j \lfloor \log_2 j \rfloor \le \Lambda \right\}.$$

Next, let us fix the base of the logarithm of H to be 2, for convenience. (We will convert to arbitrary base at the end.) Applying Example 5.2 we obtain:

$$\mathcal{H}(\Lambda) \le \inf_{\lambda>0} \lambda\Lambda + \log_2 Z(\lambda), \tag{10.10}$$

where $Z(\lambda) = \sum_{n=1}^{\infty} 2^{-\lambda\lfloor\log_2 n\rfloor} = \sum_{m=0}^{\infty} 2^{(1-\lambda)m} = \frac{1}{1-2^{1-\lambda}}$ if $\lambda > 1$ and $Z(\lambda) = \infty$ otherwise. Clearly, the infimum over $\lambda > 0$ is a minimum attained at a value $\lambda^* > 1$ satisfying

$$\Lambda = -\left.\frac{d}{d\lambda}\right|_{\lambda=\lambda^*} \log_2 Z(\lambda).$$

Define the distribution

$$P_\lambda(n) \triangleq \frac{1}{Z(\lambda)} 2^{-\lambda\lfloor\log_2 n\rfloor}, \quad n \ge 1,$$

and notice that

$$\mathbb{E}_{P_\lambda}[\lfloor\log_2 X\rfloor] = -\frac{d}{d\lambda}\log_2 Z(\lambda) = \frac{2^{1-\lambda}}{1-2^{1-\lambda}},$$
$$H(P_\lambda) = \log_2 Z(\lambda) + \lambda\,\mathbb{E}_{P_\lambda}[\lfloor\log_2 X\rfloor].$$

Comparing with (10.10) we find that the upper bound in (10.10) is tight and attained by P_{λ^*}. From the first equation above, we also find $\lambda^* = \log_2 \frac{2+2\Lambda}{\Lambda}$. Altogether this yields

$$\mathcal{H}(\Lambda) = \Lambda \log 2 + (\Lambda+1)\log(\Lambda+1) - \Lambda\log\Lambda,$$

and the extremal distribution $P_{\lambda^*}(n) \asymp n^{-\lambda^*}$ is a power-law distribution with the exponent $\lambda^* \to 1$ as $\Lambda \to \infty$.

The Argument of Mandelbrot [292] The above derivation shows a special (extremality) property of the power law, but falls short of explaining its empirical ubiquity. Here is a way to connect the optimization problem $\mathcal{H}(\Lambda)$ to the evolution of the natural language. Suppose that there is a countable set $\mathcal{S}$ of elementary concepts that are used by the brain as building blocks of perception and communication with the outside world. As an approximation we can think that concepts are in one-to-one correspondence with language words. Now every concept x is represented internally by the brain as a certain pattern, in the simplest case as a sequence of zeros and ones of length $l(f(x))$ ([292] considers more general representations). Now we have seen that the number of sequences of concepts with a composition P grows exponentially (in length) with the exponent given by $H(P)$, see Proposition 1.5. Thus in the long run the probability distribution P over the concepts results in the rate of information transfer equal to $\frac{H(P)}{\mathbb{E}_P[l(f(X))]}$. Mandelbrot concludes that in order to transfer maximal information per unit, *language and brain representation co-evolve in such a way as to maximize this ratio*. Note that

$$\sup_{P,f} \frac{H(P)}{\mathbb{E}_P[l(f(X))]} = \sup_{\Lambda} \frac{\mathcal{H}(\Lambda)}{\Lambda}.$$

It is not hard to show that $\mathcal{H}(\Lambda)$ is concave and thus the supremum is achieved at $\Lambda = 0+$ and equals infinity. This appears to have not been observed by Mandelbrot. To fix this issue, we can postulate that for some unknown physiological reason there is a requirement of also having a certain minimal entropy $H(P) \geq h_0$. In this case

$$\sup_{P,f\colon H(P)\geq h_0} \frac{H(P)}{\mathbb{E}_P[l(f(X))]} = \frac{h_0}{\mathcal{H}^{-1}(h_0)},$$

and the supremum is achieved at a power-law distribution P. Thus, the implication is that *the frequency of word usage in human languages evolves until a power law is attained,* at which point it maximizes information transfer within the brain. That's the gist of the argument of [292]. It is clear that this does not explain the appearance of the power law in other domains, for which other explanations such as preferential attachment models are more plausible, see [306]. Finally, we mention that the P_λ distributions take discrete values $2^{-\lambda m - \log_2 Z(\lambda)}, m = 0, 1, 2, \ldots$ with multiplicities 2^m. Thus P_λ appears as a rather unsightly staircase on frequency–rank plots such as Figure 10.3. This artifact can be alleviated by considering non-binary brain representations with *unequal lengths* of signals.

10.3 Uniquely Decodable Codes, Prefix Codes and Huffman Codes

In the previous sections we have studied f^*, which achieves the stochastically (in particular, in expectation) shortest code length among all variable-length lossless compressors. Note that f^* is obtained by ordering the PMF and assigning shorter codewords to more likely symbols. In this section we focus on a specific class of compressors with good algorithmic properties which lead to low-complexity decoding and short delay when decoding from a stream of compressed bits. This part is more combinatorial in nature.

We start with a few definitions. Let $\mathcal{A}^+ = \bigcup_{n\geq 1} \mathcal{A}^n$ denote all non-empty finite-length strings consisting of symbols from the alphabet $\mathcal{A}$. Throughout this chapter $\mathcal{A}$ is a countable set.

Definition 10.6. (Extension of a code) The (symbol-by-symbol) extension of $f\colon \mathcal{A} \to \{0,1\}^*$ is $f\colon \mathcal{A}^+ \to \{0,1\}^*$ where $f(a_1, \ldots, a_n) = (f(a_1), \ldots, f(a_n))$ is defined by concatenating the bits.

Definition 10.7. (Uniquely decodable codes) $f\colon \mathcal{A} \to \{0,1\}^*$ is *uniquely decodable* if its extension $f\colon \mathcal{A}^+ \to \{0,1\}^*$ is injective.

Definition 10.8. (Prefix codes) $f\colon \mathcal{A} \to \{0,1\}^*$ is a *prefix code*[2] if no codeword is a prefix of another (e.g. 010 is a prefix of 0101).

[2] Also known as prefix-free/comma-free/self-punctuating/instantaneous code.

Example 10.2.

$\mathcal{A} = \{a, b, c\}$.

- $f(a) = 0, f(b) = 1, f(c) = 10$. Not uniquely decodable, since $f(ba) = f(c) = 10$.
- $f(a) = 0, f(b) = 10, f(c) = 11$. Uniquely decodable and a prefix code.
- $f(a) = 0, f(b) = 01, f(c) = 011, f(d) = 0111$ Uniquely decodable but not a prefix code, since as long as 0 appears, we know that the previous codeword has terminated.[3]

Remark 10.3.

1 Prefix codes are uniquely decodable and hence lossless, as illustrated in the following picture:

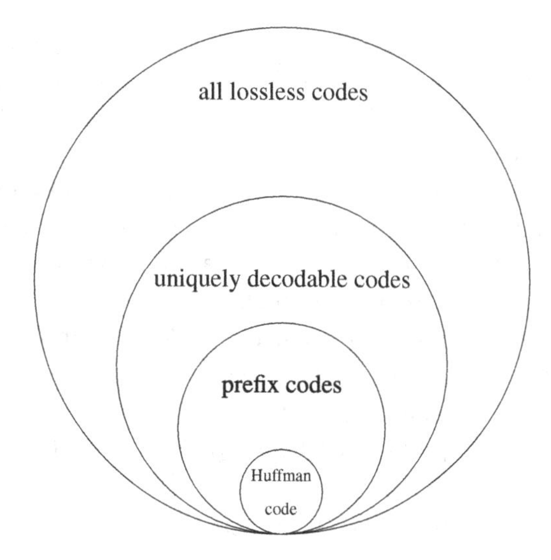

2 Similar to prefix-free codes, one can define suffix-free codes. Those are also uniquely decodable (one should start decoding in the reverse direction).
3 By definition, any uniquely decodable code does not have the empty string as a codeword. Hence $f: \mathcal{A} \to \{0, 1\}^+$ in both Definitions 10.7 and 10.8.
4 Unique decodability means that one can decode from a stream of bits without ambiguity, but one might need to look ahead in order to decide the termination of a codeword. (Think of the last example.) In contrast, prefix codes allow the decoder to decode instantaneously without looking ahead.
5 Prefix codes are in one-to-one correspondence with binary trees (with codewords at leaves). It is also equivalent to strategies to ask "yes/no" questions previously mentioned at the end of Section 1.1.

[3] In this example, if 0 is placed at the very end of each codeword, the code is uniquely decodable, known as the *unary code*.

Theorem 10.9. (Kraft–McMillan)

1 *Let $f\colon \mathcal{A} \to \{0,1\}^*$ be uniquely decodable. Set $l_a = l(f(a))$. Then f satisfies the* Kraft inequality

$$\sum_{a\in\mathcal{A}} 2^{-l_a} \le 1. \tag{10.11}$$

2 *Conversely, for any set of code length $\{l_a\colon a \in \mathcal{A}\}$ satisfying (10.11), there exists a prefix code f, such that $l_a = l(f(a))$. Moreover, such an f can be computed efficiently.*

Remark 10.4. The consequence of Theorem 10.9 is that as far as compression efficiency is concerned, we can ignore those uniquely decodable codes that are not prefix codes.

Proof. We prove the Kraft inequality for prefix codes and uniquely decodable codes separately. The proof for the former is probabilistic, following ideas in [15, Exercise **I.8**, p. 12]. Let f be a prefix code. Let us construct a probability space such that the LHS of (10.11) is the probability of some event, which cannot exceed one. To this end, consider the following scenario: Generate independent Ber(1/2) bits. Stop if a codeword has been written, otherwise continue. This process terminates with probability $\sum_{a\in\mathcal{A}} 2^{-l_a}$. The summation makes sense because the events that a given codeword is written are mutually exclusive, thanks to the prefix condition.

Now let f be a uniquely decodable code. The proof uses a *generating function* as a device for counting. (The analogy in coding theory is the weight enumerator function.) First assume $\mathcal{A}$ is finite. Then $L = \max_{a\in\mathcal{A}} l_a$ is finite. Let $G_f(z) = \sum_{a\in\mathcal{A}} z^{l_a} = \sum_{l=0}^{L} A_l(f) z^l$, where $A_l(f)$ denotes the number of codewords of length l in f. For $k \ge 1$, define $f^k\colon \mathcal{A}^k \to \{0,1\}^+$ as the symbol-by-symbol extension of f. Then $G_{f^k}(z) = \sum_{a^k\in\mathcal{A}^k} z^{l(f^k(a^k))} = \sum_{a_1}\cdots\sum_{a_k} z^{l_{a_1}+\cdots+l_{a_k}} = [G_f(z)]^k = \sum_{l=0}^{kL} A_l(f^k) z^l$. By the unique decodability of f, f^k is lossless. Hence $A_l(f^k) \le 2^l$. Therefore we have $G_f(1/2)^k = G_{f^k}(1/2) \le kL$ for all k. Then $\sum_{a\in\mathcal{A}} 2^{-l_a} = G_f(1/2) \le \lim_{k\to\infty}(kL)^{1/k} = 1$. If $\mathcal{A}$ is countably infinite, for any finite subset $\mathcal{A}' \subset \mathcal{A}$, repeating the same argument gives $\sum_{a\in\mathcal{A}'} 2^{-l_a} \le 1$. The proof is complete by the arbitrariness of $\mathcal{A}'$.

Conversely, given a set of code lengths $\{l_a\colon a \in \mathcal{A}\}$ s.t. $\sum_{a\in\mathcal{A}} 2^{-l_a} \le 1$, construct a prefix code f as follows: First relabel $\mathcal{A}$ to $\mathbb{N}$ and assume that $1 \le l_1 \le l_2 \le \cdots$. For each i, define

$$a_i \triangleq \sum_{k=1}^{i-1} 2^{-l_k}$$

with $a_1 = 0$. Then $a_i < 1$ by the Kraft inequality. Thus we define the codeword $f(i) \in \{0,1\}^+$ as the first l_i bits in the binary expansion of a_i. Finally, we prove that f is a prefix code by contradiction: Suppose that, for some $j > i$, $f(i)$ is the prefix of $f(j)$, since $l_j \ge l_i$. Then $a_j - a_i \le 2^{-l_i}$, since they agree on the most significant l_i bits. But $a_j - a_i = 2^{-l_i} + 2^{-l_{i+1}} + \cdots > 2^{-l_i}$, which is a contradiction. □

Remark 10.5. A conjecture of Ahlswede et al [7] states that for any set of lengths for which $\sum 2^{-l_a} \le \frac{3}{4}$ there exists a fix-free code (i.e. one which is simultaneously prefix-free and suffix-free). So far, existence has only been shown when the Kraft sum is $\le \frac{5}{8}$, see [467].

In view of Theorem 10.9, the optimal average code length among all prefix (or uniquely decodable) codes is given by the following optimization problem

$$\begin{aligned} L^*(X) \triangleq \min & \sum_{a\in\mathcal{A}} P_X(a) l_a \\ \text{s.t.} & \sum_{a\in\mathcal{A}} 2^{-l_a} \le 1, \\ & l_a \in \mathbb{N}. \end{aligned} \tag{10.12}$$

This is an *integer programming* (IP) problem, which, in general, is computationally hard to solve. It is remarkable that this particular IP can be solved in *near-linear* time, thanks to the Huffman algorithm. Before describing the construction of Huffman codes, let us give bounds to $L^*(X)$ in terms of entropy:

Theorem 10.10.

$$H(X) \le L^*(X) \le H(X) + 1 \text{ bits.} \tag{10.13}$$

Proof. Right inequality: Consider the following length assignment $l_a = \lceil \log_2 \frac{1}{P_X(a)} \rceil$,[4] which satisfies Kraft since $\sum_{a\in\mathcal{A}} 2^{-l_a} \le \sum_{a\in\mathcal{A}} P_X(a) = 1$. By Theorem 10.9, there exists a prefix code f such that $l(f(a)) = \left\lceil \log_2 \frac{1}{P_X(a)} \right\rceil$ and $\mathbb{E}l(f(X)) \le H(X) + 1$.

Left inequality: We give two proofs for this converse. One of the commonly used ideas to deal with combinatorial optimization is *relaxation*. Our first idea is to drop the integer constraints in (10.12) and relax it into the following optimization problem, which obviously provides a lower bound

$$\begin{aligned} L^*(X) \triangleq \min & \sum_{a\in\mathcal{A}} P_X(a) l_a \\ \text{s.t.} & \sum_{a\in\mathcal{A}} 2^{-l_a} \le 1. \end{aligned} \tag{10.14}$$

This is a nice *convex optimization* problem, with an affine objective function and a convex feasible set. Solving (10.14) by Lagrange multipliers (this is left as an exercise for the reader) yields that the minimum is equal to $H(X)$ (achieved at $l_a = \log_2 \frac{1}{P_X(a)}$).

Another proof is the following: For any f whose code lengths $\{l_a\}$ satisfy the Kraft inequality, define a probability measure $Q(a) = \frac{2^{-l_a}}{\sum_{a\in\mathcal{A}} 2^{-l_a}}$. Then

$$\mathbb{E}l(f(X)) - H(X) = D(P\|Q) - \log \sum_{a\in\mathcal{A}} 2^{-l_a} \ge 0. \qquad \square$$

[4] Such a code is called a Shannon code.

Next we describe the Huffman code, which achieves the optimum in (10.12). In view of the fact that prefix codes and binary trees are one-to-one, the main idea of the Huffman code is to build the binary tree from the bottom up: Given a PMF $\{P_X(a)\colon a \in \mathcal{A}\}$:

1 Choose the two least-probable symbols in the alphabet.
2 Delete the two symbols and add a new symbol (with combined probabilities). Add the new symbol as the parent node of the previous two symbols in the binary tree.

The algorithm terminates in $|\mathcal{A}|-1$ steps. Given the binary tree, the code assignment can be obtained by assigning 0/1 to the branches. Therefore the time complexity is $O(|\mathcal{A}|)$ (sorted PMF) or $O(|\mathcal{A}| \log |\mathcal{A}|)$ (unsorted PMF).

Example 10.3.

$\mathcal{A} = \{a, b, c, d, e\}, P_X = \{0.25, 0.25, 0.2, 0.15, 0.15\}$.

Huffman tree:

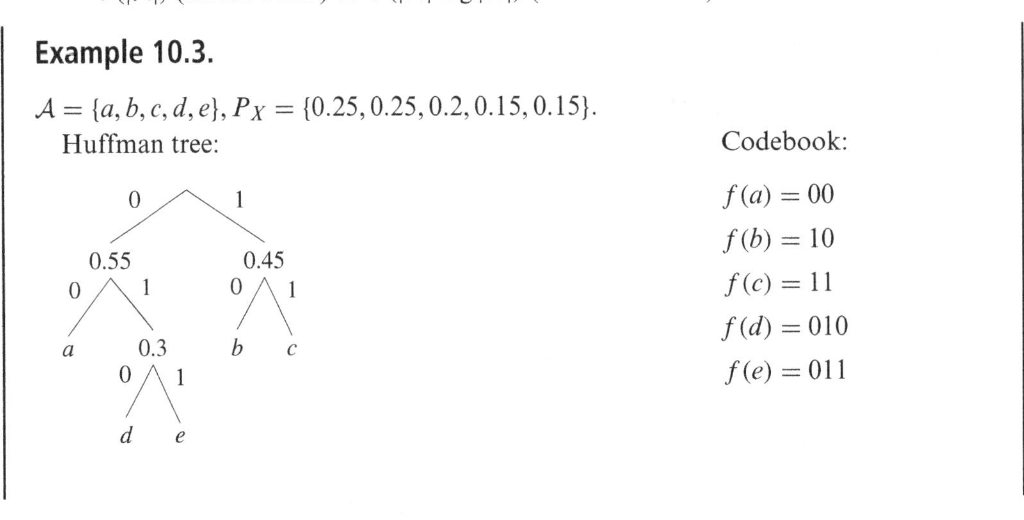

Codebook:

$f(a) = 00$
$f(b) = 10$
$f(c) = 11$
$f(d) = 010$
$f(e) = 011$

Theorem 10.11. (Optimality of Huffman codes) *The Huffman code achieves the minimal average code length (10.12) among all prefix (or uniquely decodable) codes.*

Proof. See [106, Section 5.8]. □

Remark 10.6. (Drawbacks of Huffman codes)

1 Constructing the Huffman code requires knowing the source distribution. This brings us the question: Is it possible to design a universal compressor which achieves entropy for a class of source distributions? And what is the price to pay? These questions are the topic of universal compression and will be addressed in Chapter 13.
2 To understand the main limitation of Huffman coding, we recall that (as Shannon pointed out), while Morse code already exploits the non-equiprobability of English letters, working with pairs (or more generally, n-grams) of letters achieves even more compression, since letters in a pair are not independent. In other words, to compress a block of symbols $(S_1, \ldots, S_n)$ by applying the Huffman code on a symbol-by-symbol basis one can achieve an average length of $\sum_{i=1}^{n} H(S_i) + n$ bits. But applying Huffman codes on a whole block

$(S_1, \ldots, S_n)$, that is the code designed for $P_{S_1,\ldots,S_n}$, allows one to exploit the memory in the source and achieve compression length $H(S_1, \ldots, S_n) + O(1)$. Due to (1.3) the joint entropy is smaller than $\sum_i H(S_i)$ (and usually *much* smaller). However, the drawback of this idea is that constructing the Huffman code has complexity $|\mathcal{A}|^n$ – exponential in the blocklength.

To resolve these problems we will later study other methods:

1 Arithmetic coding has a sequential encoding algorithm with complexity linear in the blocklength, while still attaining $H(S_1^n)$ length – Section 13.1.
2 The Lempel–Ziv algorithm also has low complexity and is even universal, provably optimal for all ergodic sources – Section 13.8.

As a summary of this chapter, we learned the following relationship between entropy and compression length of various codes:

$$H(X) - \log_2[e(H(X)+1)] \le \mathbb{E}[l(f^*(X))] \le H(X) \le \mathbb{E}[l(f_{\text{Huffman}}(X))] \le H(X)+1.$$

11 Fixed-Length Compression and Slepian–Wolf Theorem

In the previous chapter we introduced the concept of variable-length compression and studied its fundamental limits with and without the prefix-free condition. In some situations, however, one may desire that the output of the compressor always be of a fixed length, say, k bits. Unless k is unreasonably large, this will require relaxing the losslessness condition. This is the focus of this chapter: compression in the presence of (typically vanishingly small) probability of error. It turns out allowing even very small error enables several beautiful effects:

- The possibility to compress data via matrix multiplication over finite fields (linear compression or *hashing*).
- The possibility to reduce compression length from $H(X)$ to $H(X|Y)$ if side information Y is available at the decompressor (Slepian–Wolf).
- The possibility to reduce compression length below $H(X)$ if access to a compressed representation of side information Y is available at the decompressor (Ahlswede–Körner–Wyner).

All of these effects are ultimately based on the fundamental property of many high-dimensional probability distributions, the asymptotic equipartition (AEP), which we study in the context of iid distributions. Later we will extend this property to all ergodic processes in Chapter 12.

11.1 Source Coding Theorems

The coding paradigm in this section is illustrated as follows:

$\mathcal{X}$ → Compressor $f: \mathcal{X} \to \{0,1\}^k$ → $\{0,1\}^k$ → Decompressor $g: \{0,1\}^k \to \mathcal{X} \cup \{e\}$ → $\mathcal{X} \cup \{e\}$

Note that if we insist like in Chapter 10 that $g(f(X)) = X$ with probability one, then $k \geq \log_2 |\mathrm{supp}(P_X)|$ and no meaningful compression can be achieved. It turns out that by tolerating a small error probability, we can gain a lot in terms of code length! So, instead of requiring $g(f(x)) = x$ for all $x \in \mathcal{X}$, consider only lossless decompression for a subset $\mathcal{S} \subset \mathcal{X}$:

$$g(f(x)) = \begin{cases} x, & x \in \mathcal{S}, \\ \mathsf{e}, & x \notin \mathcal{S}, \end{cases}$$

and the probability of error is:

$$\mathbb{P}[g(f(X)) \neq X] = \mathbb{P}[g(f(X)) = \mathsf{e}] = \mathbb{P}[X \notin \mathcal{S}].$$

We summarize this formally next.

Definition 11.1. A compressor–decompressor pair (f, g) is called a fixed-length almost lossless (k, ϵ) source code for $X \in \mathcal{X}$, or (k, ϵ)-code for short, if:

$$f\colon\ \mathcal{X} \to \{0,1\}^k,$$
$$g\colon\ \{0,1\}^k \to \mathcal{X} \cup \{\mathsf{e}\},$$

such that $g(f(x)) \in \{x, \mathsf{e}\}$ for all $x \in \mathcal{X}$ and $\mathbb{P}[g(f(X)) = \mathsf{e}] \leq \epsilon$. The fundamental limit of fixed-length compression is simply the minimum probability of error and is defined as

$$\epsilon^*(X, k) \triangleq \inf\{\epsilon\colon \exists(k, \epsilon)\text{-code for } X\}.$$

The following result connects the respective fundamental limits of fixed-length almost lossless compression and variable-length lossless compression (Section 10.1):

Theorem 11.2. (Fundamental limit of fixed-length compression) *Recall the optimal variable-length compressor f^* defined in Theorem 10.2 and assume as before that $\mathcal{X} = \mathbb{N}$ and $P_X(1) \geq P_X(2) \geq \cdots$. Then*

$$\epsilon^*(X, k) = \mathbb{P}\left[l(f^*(X)) \geq k\right] = \sum_{x \geq 2^k} P_X(x).$$

Proof. Note that because of the assumption $\mathcal{X} = \mathbb{N}$ compressor must reserve one k-bit string for the error message even if $P_X(1) = 1$. The proof is essentially tautological. Note $1 + 2 + \cdots + 2^{k-1} = 2^k - 1$. Let $\mathcal{S}$ be the set of top $2^k - 1$ most likely (as measured by $P_X(x)$) elements $x \in \mathcal{X}$. Then

$$\epsilon^*(X, k) = \mathbb{P}[X \notin \mathcal{S}] = \mathbb{P}\left[l(f^*(X)) \geq k\right].$$

The last equality follows from (10.1). □

Comparing Theorems 10.2 and 11.2, we see that the optimal codes in these two settings work as follows:

- Variable-length: f^* encodes the $2^k - 1$ symbols with the highest probabilities to $\{\phi, 0, 1, 00, \ldots, 1^{k-1}\}$.
- Fixed-length: The optimal compressor f maps the elements of $\mathcal{S}$ into $(00\ldots00), \ldots, (11\ldots10)$ and the rest in $\mathcal{X}\backslash\mathcal{S}$ to $(11\ldots11)$. The decompressor g decodes perfectly except for outputting e upon receipt of $(11\ldots11)$.

Remark 11.1. (Detectable versus undetectable errors) In Definition 11.1 we required that the errors be always *detectable*, that is, $g(f(x)) = x$ or e. Alternatively, we can drop this requirement and allow *undetectable* errors, in which case we can of course do better since we have more freedom in designing codes. It turns out that we do not gain much by this relaxation. Indeed, if we define

$$\tilde{\epsilon}^*(X,k) = \inf\{\mathbb{P}[g(f(X)) \neq X] : f\colon \mathcal{X} \to \{0,1\}^k, g\colon \{0,1\}^k \to \mathcal{X} \cup \{\mathsf{e}\}\},$$

then $\tilde{\epsilon}^*(X,k) = \sum_{x>2^k} P_X(x)$. This follows immediately from $\mathbb{P}[g(f(X)) = X] = \sum_{x\in\mathcal{S}} P_X(x)$ where $\mathcal{S} \triangleq \{x\colon g(f(x)) = x\}$ satisfies $|\mathcal{S}| \leq 2^k$, because f takes no more than 2^k values. Compared to Theorem 11.2, we see that $\tilde{\epsilon}^*(X,k)$ and $\epsilon^*(X,k)$ only differ by $P_X(2^k) \leq 2^{-k}$. In particular, $\epsilon^*(X,k+1) \leq \tilde{\epsilon}^*(X,k) \leq \epsilon^*(X,k)$ and we can at most save a single bit in compressed strings.

These simple observations lead us to the first fundamental result of Shannon.

Corollary 11.3. (Shannon's source coding theorem) *Let S^n be iid discrete random variables. Then for any $R > 0$ and $\gamma \in \mathbb{R}$ asymptotically in blocklength n we have*

$$\lim_{n\to\infty} \epsilon^*(S^n, nR) = \begin{cases} 0, & R > H(S), \\ 1, & R < H(S). \end{cases}$$

If varentropy $V(S) < \infty$ then also

$$\lim_{n\to\infty} \epsilon^*(S^n, nH(S) + \sqrt{nV(S)}\gamma) = Q(\gamma),$$

where $Q(x) = \int_x^\infty \frac{1}{\sqrt{2\pi}} e^{-t^2/2} dt$ is the complementary CDF of $\mathcal{N}(0,1)$s.

Proof. Combine Theorem 11.2 with Corollary 10.5. □

This result demonstrates that if we are to compress an iid string S^n down to $k = k(n)$ bits then the minimal possible k enabling vanishing error satisfies $\frac{k}{n} = H(S)$, that is, we can compress to entropy rate of the iid process $\mathbb{S}$ and no more. Furthermore, if we allow a non-vanishing error ϵ then compression is possible down to

$$k = nH(S) + \sqrt{nV(S)}Q^{-1}(\epsilon)$$

bits. In the language of modern information theory, Corollary 11.3 derives both the asymptotic fundamental limit (minimal k/n) and the *normal approximation* under non-vanishing error.

The next desired step after understanding asymptotics is to derive finite-blocklength guarantees, that is, bounds on $\epsilon^*(X,k)$ in terms of the information quantities. As we mentioned above, the upper and lower bounds are typically called achievability and converse bounds. In the case of lossless compression such bounds are rather trivial corollaries of Theorem 11.2, but we present them for completeness next. For other problems in this part and other parts of the book, obtaining good finite-blocklength bounds is much more challenging.

Theorem 11.4. (Finite-blocklength bounds) *For all $\tau > 0$ and all $k \in \mathbb{Z}_+$ we have*

$$\mathbb{P}\left[\log_2 \frac{1}{P_X(X)} > k + \tau\right] - 2^{-\tau} \leq \tilde{\epsilon}^*(X,k) \leq \epsilon^*(X,k) \leq \mathbb{P}\left[\log_2 \frac{1}{P_X(X)} \geq k\right].$$

Proof. The argument for the lower (converse) bound is identical to the converse of Theorem 10.4. Indeed, considering the optimal (undetectable error) code let $\mathcal{S} = \{x\colon g(f(x)) = x\}$ and note

$$1-\tilde{\epsilon}^*(X,k) = \mathbb{P}[X \in \mathcal{S}] \le \mathbb{P}\left[\log_2 \frac{1}{P_X(X)} \le k+\tau\right] + \mathbb{P}\left[X \in \mathcal{S}, \log_2 \frac{1}{P_X(X)} > k+\tau\right].$$

For the second term we have

$$\mathbb{P}\left[X \in \mathcal{S}, \log_2 \frac{1}{P_X(X)} > k+\tau\right] = \sum_{x\in\mathcal{S}} P_X(x) 1\{P_X(x) < 2^{-k-\tau}\} \le |\mathcal{S}| 2^{-k-\tau} \le 2^{-\tau},$$

where we used the fact that $|\mathcal{S}| \le 2^k$. Combining the two inequalities yields the lower bound.

For the upper bound, without loss of generality we assume $P_X(1) \ge P_X(2) \ge \cdots$. Then by Theorem 11.2 we have

$$\epsilon^*(X,k) = \sum_{m \ge 2^k} P_X(m) \le \sum 1\left\{\frac{1}{P_X(m)} \ge 2^k\right\} P_X(m) = \mathbb{P}\left[\log_2 \frac{1}{P_X(X)} \ge k\right],$$

where $\le$ follows from the fact that the mth largest mass $P_X(m) \le \frac{1}{m}$. □

We now will do something strange. We will prove an upper bound that is weaker than that of Theorem 11.4 and furthermore the proof is much longer. However, this will be our first introduction to the technique of *random coding* (also known as the probabilistic method outside of information theory).[1] We will quickly find out that outside of the simplest setting of lossless compression, where the optimal encoder f^* was easy to describe, good encoders are very hard to find and thus random coding becomes indispensable. In particular the Slepian–Wolf theorem (Section 11.5 below), all of data transmission (Part IV), and lossy data compression (Part V) will be based on the method.

Theorem 11.5. (Random coding achievability) *For any $k \in \mathbb{Z}_+$ and any $\tau > 0$ we have*

$$\tilde{\epsilon}^*(X,k) \le \mathbb{P}\left[\log_2 \frac{1}{P_X(X)} > k - \tau\right] + 2^{-\tau}, \quad \forall \tau > 0, \tag{11.1}$$

that is, there exists a compressor–decompressor pair with the (possibly undetectable) error bounded by the right-hand side.

Proof. We first start by constructing a suboptimal decompressor g for a given f. Indeed, for a given compressor f, the optimal decompressor which minimizes the error probability is simply the maximum a posteriori (MAP) decoder, that is

$$g^*(w) = \operatorname*{argmax}_x P_{X|f(X)}(x|w) = \operatorname*{argmax}_{x\colon f(x)=w} P_X(x).$$

[1] These methods were discovered simultaneously by Shannon [379] and Erdös [153], respectively.

However, this decoder's performance is a little hard to analyze, so instead, we consider the following (suboptimal) decompressor g:

$$g(w) = \begin{cases} x, & \exists!\, x \in \mathcal{X} \text{ s.t. } f(x) = w \text{ and } \log_2 \frac{1}{P_X(x)} \le k - \tau, \\ & \text{(exists unique high-probability } x \text{ that is mapped to } w) \\ \mathsf{e}, & \text{otherwise.} \end{cases}$$

Note that $\log_2 \frac{1}{P_X(x)} \le k - \tau \iff P_X(x) \ge 2^{-(k-\tau)}$. We call those x "high-probability." (In the language of [106] and [115] these would be called "typical" realizations.)

Denote $f(x) = c_x$ and call the long vector $\mathcal{C} = [c_x : x \in \mathcal{X}]$ a codebook. It is instructive to think of $\mathcal{C}$ as a hashing table: it takes an object $x \in \mathcal{X}$ and assigns to it a k-bit hash value.

To analyze the error probability let us define

$$J(x, \mathcal{C}) \triangleq \left\{x' \in \mathcal{X} : c_{x'} = c_x, x' \ne x, \log_2 \frac{1}{P_X(x')} \le k - \tau\right\}$$

to be the set of high-probability inputs whose hashes collide with that of x. Then we have the following estimate for probability of error:

$$\begin{aligned} \mathbb{P}[g(f(X)) = \mathsf{e}] &= \mathbb{P}\left[\left\{\log_2 \frac{1}{P_X(X)} > k - \tau\right\} \cup \{J(X, \mathcal{C}) \ne \emptyset\}\right] \\ &\le \mathbb{P}\left[\log_2 \frac{1}{P_X(X)} > k - \tau\right] + \mathbb{P}[J(X, \mathcal{C}) \ne \emptyset]. \end{aligned}$$

The first term does not depend on the codebook $\mathcal{C}$, while the second term does. The idea now is to randomize over $\mathcal{C}$ and show that when we average over all possible choices of codebook, the second term is smaller than $2^{-\tau}$. Therefore there exists at least one codebook that achieves the desired bound. Specifically, let us consider $\mathcal{C}$ generated by setting each $c_x \overset{\text{iid}}{\sim} \mathrm{Unif}\left[\{0,1\}^k\right]$ and independently of X. Equivalently, since $\mathcal{C}$ can be represented by an $|\mathcal{X}| \times k$ binary matrix, whose rows correspond to codewords, we choose each entry to be an independent fair coin flip. Averaging the error probability (over $\mathcal{C}$ and over X), we have

$$\begin{aligned} \mathbb{E}_{\mathcal{C},X}[1\{J(X, \mathcal{C})\}] &= \mathbb{E}_{\mathcal{C},X}\left[1\left\{\exists x' \ne X : \log_2 \frac{1}{P_X(x')} \le k - \tau, c_{x'} = c_X\right\}\right] \\ &\le \mathbb{E}_{\mathcal{C},X}\left[\sum_{x' \ne X} 1\left\{\log_2 \frac{1}{P_X(x')} \le k - \tau\right\} 1\{c_{x'} = c_X\}\right] \quad \text{(union bound)} \\ &= 2^{-k}\mathbb{E}_X\left[\sum_{x' \ne X} 1\left\{P_X(x') \ge 2^{-k+\tau}\right\}\right] \\ &= 2^{-k} \sum_{x' \in \mathcal{X}} 1\left\{P_X(x') \ge 2^{-k+\tau}\right\} \\ &\le 2^{-k} 2^{k-\tau} = 2^{-\tau}, \end{aligned}$$

where the crucial penultimate step uses the fact that there can be at most $2^{k-\tau}$ values of x' with $P_X(x') > 2^{-k+\tau}$. □

Remark 11.2. (Why random coding works) The compressor $f(x) = c_x$ can be thought of as hashing $x \in \mathcal{X}$ to a random k-bit string $c_x \in \{0,1\}^k$, as illustrated below:

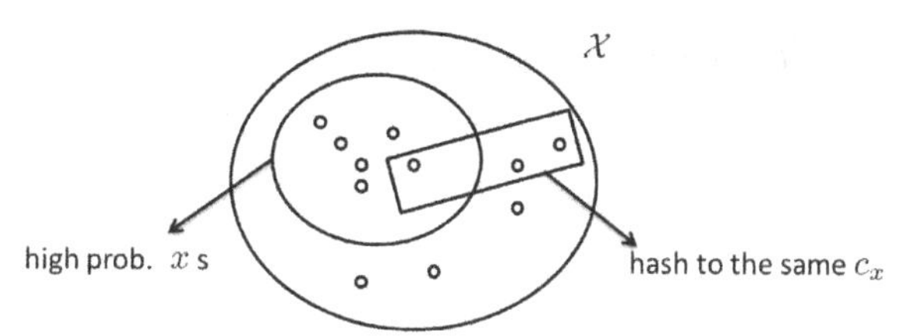

Here, we think of x as high probability iff $\log_2 \frac{1}{P_X(x)} \leq k - \tau$, or equivalently, $P_X(x) \geq 2^{-k+\tau}$. Therefore the number of those high-probability x's is at most $2^{k-\tau}$, which is far smaller (when we take $\tau \gg 1$) than 2^k, the total number of k-bit codewords. Hence the chance of hash-collision among high-probability x's is small.

Let us again emphasize that the essence of the random coding argument is the following. To prove the existence of an object with a certain a property, we construct a probability distribution (randomize) and show that on average the property is satisfied. Hence there exists at least one realization with the desired property. The downside of this argument is that it is not constructive, that is, it does not give us an algorithm to find the object. One may wonder whether there is a practical way to generate a large random hashing table and use it for compression. The problem is that generating such a table requires a lot of randomness and a lot of storage space (both are important resources). We will address this issue in Section 11.3, but for now let us make the following remark.

Remark 11.3. (Pairwise independence of codewords) In the proof we choose the random codebook to be uniform over all possible codebooks: $c_x \stackrel{\text{iid}}{\sim} \text{Unif}$. But a careful inspection (this is left as an exercise for the reader) shows that we only used pairwise independence, that is $c_x \perp\!\!\!\perp c_{x'}$ for any $x \neq x'$. This suggests that perhaps in generating the table we can use a lot fewer than $k|\mathcal{X}|$ random bits. Indeed, given two independent random bits B_1, B_2 we can generate three bits that are pairwise independent: $B_1, B_2, B_1 \oplus B_2$. This observation will lead us to the idea of linear compression studied in Section 11.3, where the codewords are independently generated not iid, but as elements of a random linear subspace.

11.2 Asymptotic Equipartition Property (AEP)

Finally, we address the following. Our random coding proof restricted attention only to those x's with sufficiently high probability $P_X(x) > 2^{-k+\tau}$. But it turns out that for iid sources we could have restricted attention only to what are called "typical" x's.

Proposition 11.6. (Asymptotic equipartition (AEP)) *Consider* iid S^n *and for any* $\delta >$ 0, *define the so-called entropy* δ*-typical set*

$$T_n^\delta \triangleq \left\{ s^n : \left| \frac{1}{n} \log \frac{1}{P_{S^n}(s^n)} - H(S) \right| \leq \delta \right\}.$$

Then the following properties hold:

1 $\mathbb{P}\left[S^n \in T_n^\delta\right] \to 1$ *as* $n \to \infty$.
2 $|T_n^\delta| \leq \exp\{(H(S) + \delta)n\}$.

For example if $S^n \stackrel{\text{iid}}{\sim} \text{Ber}(p)$, then $P_{S^n}(s^n) = p^{w_{\text{H}}(s^n)} \bar{p}^{n - w_{\text{H}}(s^n)}$, where $w_{\text{H}}(s^n)$ is the Hamming weight of the string (number of 1's). Thus the typical set corresponds to those sequences whose Hamming weight $\frac{1}{n} w_{\text{H}}(s^n)$ is close to the expected value of $p + O_p(\delta)$.

Proof. By the WLLN, we have

$$\frac{1}{n} \log \frac{1}{P_{S^n}(S^n)} \xrightarrow{\mathbb{P}} H(S). \tag{11.2}$$

Thus, $\mathbb{P}[S^n \in T_n^\delta] \to 1$. On the other hand, since for every $s^n \in T_n^\delta$ we have $P_{S^n}(s^n) > \exp\{-(H(S) + \delta)n\}$ there can be at most $\exp\{(H(S) + \delta)n\}$ elements in T_n^δ. □

To understand the meaning of the AEP, notice that it shows that the gigantic space $\mathcal{S}^n$ has almost all of probability P_{S^n} concentrated on an exponentially smaller subset T_n^δ. Furthermore, on this subset the measure P_{S^n} is approximately uniform: $P_{S^n}(s^n) = \exp\{-nH(S) \pm n\delta\}$.

To see how AEP is related to compression, let us give a third proof of Shannon's result, which, we remind, states that

$$\lim_{n \to \infty} \epsilon^*(S^n, nR) = \begin{cases} 0, & R > H(S), \\ 1, & R < H(S). \end{cases}$$

Indeed, let us consider an encoder f that enumerates (by strings in $\{0, 1\}^{nR}$) elements of T_n^δ. Then if $R > H(S) + \delta$ the decoding error happens with probability $\mathbb{P}[S^n \notin T_n^\delta] \to 0$. Hence any rate $R > H(S)$ results in a vanishing error. On the other hand, if $R < H(S)$ then it is clear that 2^{nR} bits cannot describe any significant portion of $|T_n^\delta|$ and since on the latter the measure P_{S^n} is almost uniform, the probability of error necessarily converges to 1 (in fact exponentially fast). There is a certain conceptual beauty in this way of the proving source coding theorem. For example, it explains why the optimal compressor's output should look almost like iid Ber(1/2):[2] after all it enumerates elements drawn almost uniformly from a set T_n^δ.

11.3 Linear Compression (Hashing)

So far we have seen three proofs of Shannon's theorem (Corollary 11.3), but unfortunately each of the proofs used methods that are not feasible to implement in

[2] This is the intuitive basis for why compressors can be used as random number generators; see Section 9.3.

practice. The first method required sorting all $|\mathcal{S}|^n$ realizations of the input data block, the second required constructing a $|\mathcal{S}|^n \times k$ hashing table, and the third enumerating the entropy-typical set $|T_n^\delta| = \exp\{nH(S) + o(n)\}$. In this section we show a fourth method, which is conceptually important and also results in a very simple compressor. (The decompressor that we describe is still going to be very impractical, but it can be made practical by leveraging efficient decoders of linear error-correcting codes.)

In this section we assume that the source takes the form $X = S^n$, where each coordinate is an element of a finite field (Galois field), that is, $S_i \in \mathbb{F}_q$, where q is the cardinality of $\mathbb{F}_q$. (This is only possible if $q = p^k$ for some prime number p and $k \in \mathbb{N}$.)

Definition 11.7. (Galois field) F is a finite set with operations $(+, \cdot)$ where the following hold.

- The addition operation $+$ is associative and commutative.
- The multiplication operation $\cdot$ is associative and commutative.
- There exist elements $0, 1 \in F$ s.t. $0 + a = 1 \cdot a = a$.
- $\forall a, \exists -a$, s.t. $a + (-a) = 0$.
- $\forall a \neq 0, \exists a^{-1}$, s.t. $a^{-1}a = 1$.
- Distributive: $a \cdot (b + c) = (a \cdot b) + (a \cdot c)$.

Simple examples of finite fields:

- $\mathbb{F}_p = \mathbb{Z}/p\mathbb{Z}$, where p is prime ("modulo-p arithmetic").
- $\mathbb{F}_4 = \{0, 1, x, x + 1\}$ with addition and multiplication as polynomials in $\mathbb{F}_2[x]$ modulo $x^2 + x + 1$.

A linear compressor is a linear function $H\colon \mathbb{F}_q^n \to \mathbb{F}_q^k$ (represented by a matrix $H \in \mathbb{F}_q^{k\times n}$) that maps each $x \in \mathbb{F}_q^n$ to its codeword $w = Hx$, namely

$$\begin{bmatrix} w_1 \\ \vdots \\ w_k \end{bmatrix} = \begin{bmatrix} h_{11} & \dots & h_{1n} \\ \vdots & & \vdots \\ h_{k1} & \dots & h_{kn} \end{bmatrix} \begin{bmatrix} x_1 \\ \vdots \\ x_n \end{bmatrix}.$$

Compression is achieved if $k < n$, that is, H is a fat matrix, which, again, is only possible in the almost lossless sense.

Theorem 11.8. (Achievability via linear codes) *Let $X \in \mathbb{F}_q^n$ be a random vector. For all $\tau > 0$, there exists a linear compressor $H \in \mathbb{F}_q^{n\times k}$ and decompressor $g\colon \mathbb{F}_q^k \to \mathbb{F}_q^n \cup \{\mathsf{e}\}$, s.t. its undetectable error probability is bounded by*

$$\mathbb{P}[g(HX) \neq X] \leq \mathbb{P}\left[\log_q \frac{1}{P_X(X)} > k - \tau\right] + q^{-\tau}.$$

Remark 11.4. Consider the Hamming space $q = 2$. In comparison with Shannon's random coding achievability, which uses $k2^n$ bits to construct a completely random codebook, here for linear codes we need kn bits to randomly generate the matrix H, and the codebook is a k-dimensional linear subspace of the Hamming space.

Proof. Fix τ. As pointed out in the proof of Shannon's random coding theorem (Theorem 11.5), given the compressor H, the optimal decompressor is the MAP decoder, that is, $g(w) = \operatorname{argmax}_{x\colon Hx=w} P_X(x)$, which outputs the most likely symbol that is compatible with the codeword received. Instead, as before we consider the following (suboptimal) decoder:

$$g(w) = \begin{cases} x, & \exists!\, x \in \mathbb{F}_q^n\colon w = Hx,\ x \text{ h.p.}, \\ \mathsf{e}, & \text{otherwise}, \end{cases}$$

where, as in the proof of Theorem 11.5, we denoted x to be "h.p." (high probability) whenever $\log_q \frac{1}{P_X(x)} \le k - \tau$.

Note that this decoder is the same as in the proof of Theorem 11.5. The proof is also mostly the same, except now hash collisions occur under the linear map H. Specifically, we have by applying the union bound twice:

$$\begin{aligned} \mathbb{P}[g(HX) \neq X|H] &\le \mathbb{P}[x \text{ not h.p.}] + \mathbb{P}\left[\exists x' \text{ h.p.}\colon x' \neq X, Hx' = HX\right] \\ &\le \mathbb{P}\left[\log_q \frac{1}{P_X(x)} > k - \tau\right] + \sum_x P_X(x) \sum_{x' \text{ h.p.}, x' \neq x} 1\{Hx' = Hx\}. \end{aligned}$$

Now we use random coding to average the second term over all possible choices of H. Specifically, choose H as a matrix independent of X where each entry is iid and uniform on $\mathbb{F}_q$. For distinct x_0 and x_1, the collision probability is

$$\begin{aligned} \mathbb{P}[Hx_1 = Hx_0] &= \mathbb{P}[Hx_2 = 0] && (x_2 \triangleq x_1 - x_0 \neq 0), \\ &= \mathbb{P}[H_1 \cdot x_2 = 0]^k && \text{(iid rows)}, \end{aligned}$$

where H_1 is the first row of the matrix H, and each row of H is independent. This is the probability that H_i is in the orthogonal complement of x_2. On $\mathbb{F}_q^n$, the orthogonal complement of a given non-zero vector has cardinality q^{n-1}. So the probability for the first row to lie in this subspace is $q^{n-1}/q^n = 1/q$, hence the collision probability is $1/q^k$. Averaging over H gives

$$\mathbb{E}_H\left[\sum_{x' \text{ h.p.}, x' \neq x} 1\{Hx' = Hx\}\right] = |\{x'\colon x' \text{ h.p.}, x' \neq x\}| q^{-k} \le q^{k-\tau} q^{-k} = q^{-\tau}.$$

This completes the proof. □

We remark that the bounds in Theorems 11.5 and 11.8 produce compressors with undetectable errors. However, the non-linear construction in the former is easy to modify to make all errors detectable (e.g. by increasing k by 1 and making sure the first bit is 1 for all $x = s^n$ with low probability). For the linear compressors, however, the errors cannot be made detectable.

Note that we restricted our theorem to inputs over $\mathbb{F}_q$. Can we loosen the requirements and produce compressors over an arbitrary commutative ring? In general, the answer is negative due to the existence of zero divisors in the commutative ring.

The latter ruin the key proof ingredient of low collision probability in the random hashing. Indeed, consider the following computation over $\mathbb{Z}/6\mathbb{Z}$:

$$\mathbb{P}\left[H\begin{bmatrix}1\\0\\\vdots\\0\end{bmatrix}=0\right]=6^{-k}\qquad\text{but}\qquad\mathbb{P}\left[H\begin{bmatrix}2\\0\\\vdots\\0\end{bmatrix}=0\right]=3^{-k},$$

since $0\cdot 2=3\cdot 2=0$ in $\mathbb{Z}/6\mathbb{Z}$.

11.4 Compression with Side Information at Both Compressor and Decompressor

We now move to discussing several variations of the compression problem when the data consists of a correlated pair $(X,Y)\sim P_{X,Y}$. The first variation is schematically depicted as follows:

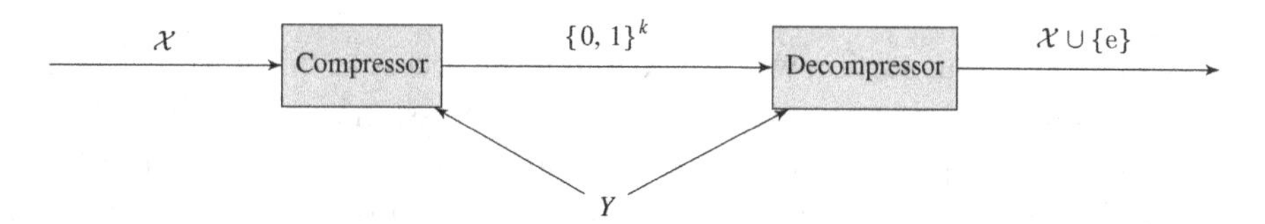

Formally, we make the following definition.

Definition 11.9. (Compression with side information) Given $P_{X,Y}$ we define the following:

- Compressor $f\colon \mathcal{X}\times\mathcal{Y}\to\{0,1\}^k$.
- Decompressor $g\colon \{0,1\}^k\times\mathcal{Y}\to\mathcal{X}\cup\{\mathsf{e}\}$.
- Probability of error $\mathbb{P}[g(f(X,Y),Y)\neq X]<\epsilon$. A code satisfying this property is called a (k,ϵ)-s.i. code.
- Fundamental limit: $\epsilon^*(X|Y,k)=\inf\{\epsilon\colon \exists(k,\epsilon)\text{-s.i. code}\}$.

Note that here unlike the source X, the side information Y need not be discrete. Conditioned on $Y=y$, the problem reduces to compression without side information studied in Section 11.1, but with the source X distributed according to $P_{X|Y=y}$. Since Y is known to both the compressor and decompressor, they can use the best code tailored for this distribution. Recall $\epsilon^*(X,k)$ defined in Definition 11.1, the optimal probability of error for compressing X using k bits, which can also be denoted by $\epsilon^*(P_X,k)$. Then we have the relationship

$$\epsilon^*(X|Y,k)=\mathbb{E}_{y\sim P_Y}[\epsilon^*(P_{X|Y=y},k)],$$

which allows us to apply various bounds developed before. In particular, we clearly have the following result.

Theorem 11.10.

$$\mathbb{P}\left[\log \frac{1}{P_{X|Y}(X|Y)} > k + \tau\right] - 2^{-\tau} \leq \epsilon^*(X|Y,k) \leq \mathbb{P}\left[\log_2 \frac{1}{P_{X|Y}(X|Y)} > k - \tau\right] + 2^{-\tau}, \quad \forall \tau > 0.$$

Corollary 11.11. *Let* $(X,Y) = (S^n, T^n)$ *where the pairs* $(S_i, T_i) \stackrel{\text{iid}}{\sim} P_{S,T}$. *Then*

$$\lim_{n\to\infty} \epsilon^*(S^n|T^n, nR) = \begin{cases} 0, & R > H(S|T), \\ 1, & R < H(S|T). \end{cases}$$

Proof. Indeed, note that from the WLLN we have

$$\frac{1}{n}\log \frac{1}{P_{S^n|T^n}(S^n|T^n)} = \frac{1}{n}\sum_{i=1}^{n} \log \frac{1}{P_{S|T}(S_i|T_i)} \xrightarrow{\mathbb{P}} H(S|T)$$

as $n \to \infty$. Thus, the result follows from setting $(X,Y) = (S^n, T^n)$ in the previous theorem. □

11.5 Slepian–Wolf: Side Information at Decompressor Only

In the previous section we learned that given access to side information at both compressor and decompressor the optimal compression rate is given by the conditional entropy $H(S|T)$. We now consider what happens if the side information $Y = T^n$ is not available at the compressor. This is demonstrated schematically as follows:

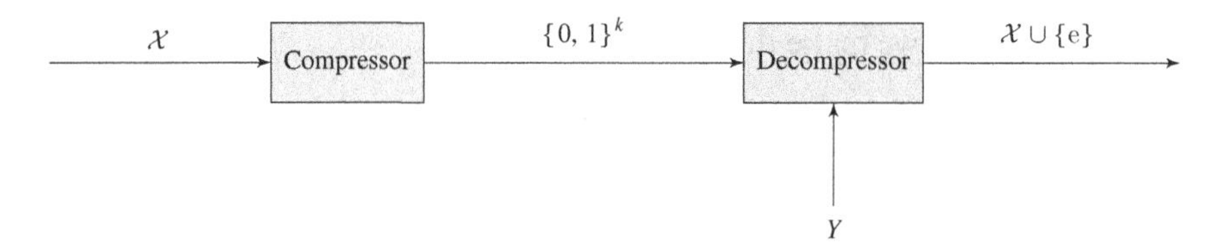

Formally, we make the following definition.

Definition 11.12. (Slepian–Wolf code) Given $P_{X,Y}$, we define a Slepian–Wolf coding problem as:

- Compressor $f\colon \mathcal{X} \to \{0,1\}^k$.
- Decompressor $g\colon \{0,1\}^k \times \mathcal{Y} \to \mathcal{X} \cup \{\mathsf{e}\}$.
- Probability of error $\mathbb{P}[g(f(X),Y) \neq X] \leq \epsilon$. A code satisfying this property is called a (k,ϵ)-Slepian–Wolf code.
- Fundamental limit: $\epsilon^*_{\mathrm{SW}}(X|Y,k) = \inf\{\epsilon : \exists (k,\epsilon)\text{-Slepian–Wolf code}\}$.

Here is the very surprising result of Slepian and Wolf,[3] which shows that the unavailability of side information at the compressor does not hinder the compression rate at all.

[3] This result is often informally referred to as "the most surprising result post-Shannon."

Theorem 11.13. (Slepian–Wolf [393])

$$\epsilon^*(X|Y,k) \le \epsilon^*_{\mathrm{SW}}(X|Y,k) \le \mathbb{P}\left[\log \frac{1}{P_{X|Y}(X|Y)} \ge k - \tau\right] + 2^{-\tau}.$$

From this theorem we will get by the WLLN the asymptotic result:

Corollary 11.14. (Slepian–Wolf [393])

$$\lim_{n\to\infty} \epsilon^*_{\mathrm{SW}}(S^n|T^n, nR) = \begin{cases} 0, & R > H(S|T), \\ 1, & R < H(S|T). \end{cases}$$

And we remark that the side information (T-process) is not even required to be discrete, see Exercise **II.9**.

Proof of Theorem 11.3. The LHS is obvious, since side information at the compressor and decompressor is better than only at the decompressor.

For the RHS, first generate a random codebook with iid uniform codewords: $\mathcal{C} = \{c_x \in \{0,1\}^k : x \in \mathcal{X}\}$ independently of (X, Y), then define the compressor and decoder as

$$f(x) = c_x,$$

$$g(w, y) = \begin{cases} x, & \exists!\, x\colon c_x = w, x \text{ h.p.}|y, \\ 0, & \text{otherwise}, \end{cases}$$

where we used the shorthand $x \text{ h.p.}|y \Leftrightarrow \log_2 \frac{1}{P_{X|Y}(x|y)} < k - \tau$. The error probability of this scheme, as a function of the codebook $\mathcal{C}$, is

$$\begin{aligned}
\mathbb{P}[X \ne g(f(X))|\mathcal{C}] &= \mathbb{P}\left[\log \frac{1}{P_{X|Y}(X|Y)} \ge k - \tau \text{ or } J(X, \mathcal{C}|Y) \ne \emptyset \Big| \mathcal{C}\right] \\
&\le \mathbb{P}\left[\log \frac{1}{P_{X|Y}(X|Y)} \ge k - \tau\right] + \mathbb{P}\left[J(X, \mathcal{C}|Y) \ne \emptyset \Big| \mathcal{C}\right] \\
&= \mathbb{P}\left[\log \frac{1}{P_{X|Y}(X|Y)} \ge k - \tau\right] + \sum_{x,y} P_{X,Y}(x,y) 1\{J(x, \mathcal{C}|y) \ne \emptyset\},
\end{aligned}$$

where $J(x, \mathcal{C}|y) \triangleq \{x' \ne x\colon x' \text{ h.p.}|y, c_x = c_{x'}\}$.

Now averaging over $\mathcal{C}$ we get

$$\begin{aligned}
\mathbb{P}[J(x, \mathcal{C}|y) \ne \emptyset] &\overset{(a)}{\le} \mathbb{E}_{\mathcal{C}}\left[\sum_{x' \ne x} 1\left\{x' \text{ h.p.}|y\right\} 1\{c_{x'} = c_x\}\right] \\
&\overset{(b)}{\le} 2^{k-\tau} \mathbb{P}[c_{x'} = c_x] \\
&\overset{(c)}{=} 2^{-\tau},
\end{aligned}$$

where (a) is a union bound, (b) follows from the fact that $|\{x'\colon x' \text{ h.p.}|y\}| \le 2^{k-\tau}$, and (c) is from $\mathbb{P}[c_{x'} = c_x] = 2^{-k}$ for any $x \ne x'$. □

Remark 11.5. (Undetectable error) Definition 11.12 allows the appearance of undetected errors. Now, we have seen that in all previous random coding results (except for the linear compression) we could always easily modify the compression algorithm to make all undetected errors detectable. However, Slepian–Wolf magic crucially depends on undetectable errors. Indeed, suppose we require that $g(f(x), y) = x$ or e for all x, y with $P_{X,Y}(x, y) > 0$. Suppose there is some $c \in \{0,1\}^k$ such that $f(x_1) = f(x_2) = c$. Then $g(c, y) = \mathsf{e}$ for all y, and the side information is not needed for such a c. On the other hand, if $c \in \{0,1\}^k$ is such that it has a unique x s.t. $f(x) = c$, then we can set $g(c, y) = x$ and ignore y again. Overall, we see that in either case side information is not useful at the decompressor and we can only compress down to $H(X)$ not $H(X|Y)$. Similarly, one can show that the Slepian–Wolf theorem does not hold in the setting of variable-length lossless compression: the minimal average compression length of any lossless algorithm is at least $H(X)$ (instead of $H(X|Y)$).

11.6 Slepian–Wolf: Compressing Multiple Sources

A simple extension of the previous result also covers another variation of the data-compression task, in which two correlated sources X and Y are compressed individually (possibly at two remote locations), but are decompressed jointly at the destination. This time, however, the goal is to reproduce both X and Y. This is depicted as follows:

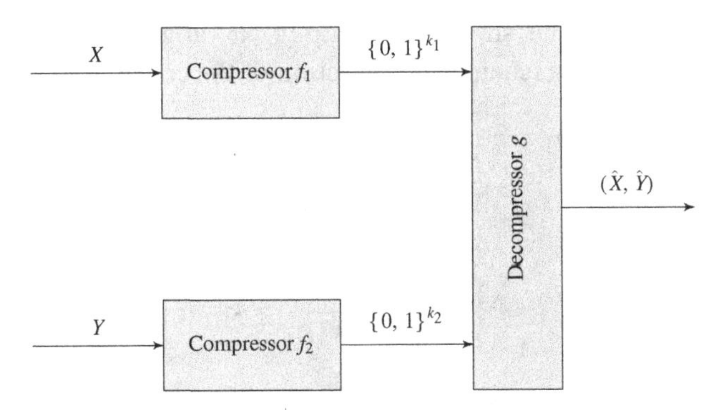

More formally, we have the following definition.

Definition 11.15. (Multi-terminal compression) Given $P_{X,Y}$ we define:

- Compressors $f_1 : \mathcal{X} \to \{0,1\}^{k_1}$, $f_2 : \mathcal{Y} \to \{0,1\}^{k_2}$.
- Decompressor g: $\{0,1\}_1^k \times \{0,1\}_2^k \to \mathcal{X} \times \mathcal{Y} \cup \{e\}$.
- Probability of error $\mathbb{P}[g(f_1(X), f_2(Y)) \neq (X, Y)] \leq \epsilon$. A code satisfying this property is called a (k_1, k_2, ϵ)-code (or multi-terminal Slepian–Wolf code).
- Fundamental limit: $\epsilon^*_{\mathrm{SW}}(X, Y, k_1, k_2) = \inf\{\epsilon : \exists (k_1, k_2, \epsilon)\text{-code}\}$.

The asymptotic fundamental limit here is given as follows.

Theorem 11.16. *Let* $(X, Y) = (S^n, T^n)$ *with* $(S_i, T_i) \overset{\text{iid}}{\sim} P_{S,T}$. *Then*

$$\lim_{n\to\infty} \epsilon^*_{\text{SW}}(S^n, T^n, nR_1, nR_2) = \begin{cases} 0, & (R_1, R_2) \in \text{int}(\mathcal{R}_{\text{SW}}), \\ 1, & (R_1, R_2) \notin \mathcal{R}_{\text{SW}}, \end{cases}$$

where $\mathcal{R}_{\text{SW}}$ *denotes the Slepian–Wolf rate region*

$$\mathcal{R}_{\text{SW}} = \left\{ (a, b) \colon \begin{array}{l} a \ge H(S|T), \\ b \ge H(T|S), \\ a + b \ge H(S, T). \end{array} \right.$$

The rate region $\mathcal{R}_{\text{SW}}$ typically looks like this:

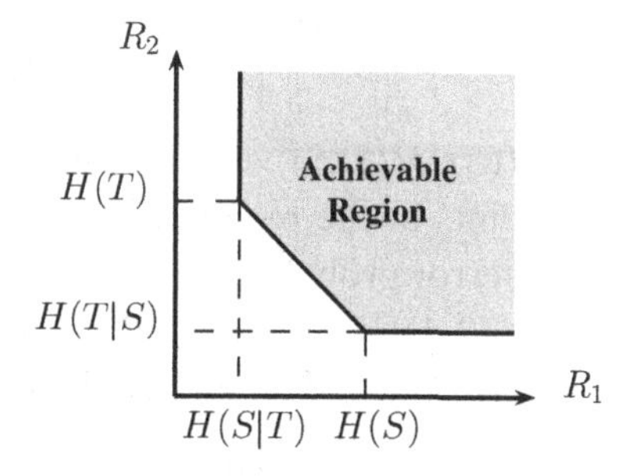

Since $H(T) - H(T|S) = H(S) - H(S|T) = I(S;T)$, the slope of the skewed line is -1.

Proof. *Converse*: Take $(R_1, R_2) \notin \mathcal{R}_{\text{SW}}$. Then one of three cases must occur:

1 $R_1 < H(S|T)$. In this case, even if the f_1 encoder and decoder had access to full T^n, we still cannot achieve vanishing error (Corollary 11.11).
2 $R_2 < H(T|S)$ (same).
3 $R_1 + R_2 < H(S, T)$. If this were possible, then we would be compressing the joint (S^n, T^n) at rate lower than $H(S, T)$, violating Corollary 11.3.

Achievability: First note that we can achieve the two corner points. The point $(H(S), H(T|S))$ can be approached by almost losslessly compressing S at entropy and compressing T with side information S at the decoder. To make this rigorous, let $k_1 = n(H(S) + \delta)$ and $k_2 = n(H(T|S) + \delta)$. By Corollary 11.3, there exist $f_1 \colon \mathcal{S}^n \to \{0, 1\}^{k_1}$ and $g_1 \colon \{0, 1\}^{k_1} \to \mathcal{S}^n$ s.t. $\mathbb{P}[g_1(f_1(S^n)) \neq S^n] \le \epsilon_n \to 0$. By Theorem 11.13, there exist $f_2 \colon \mathcal{T}^n \to \{0, 1\}^{k_2}$ and $g_2 \colon \{0, 1\}^{k_1} \times \mathcal{S}^n \to \mathcal{T}^n$ s.t. $\mathbb{P}[g_2(f_2(T^n), S^n) \neq T^n] \le \epsilon_n \to 0$. Since in our present setting the S^n is not available, we can feed the SW decompressor with an estimate of S^n given by $g_1(f_1(S^n))$ and define the joint decompressor by $g(w_1, w_2) = (g_1(w_1), g_2(w_2, g_1(w_1)))$ (see diagram below):

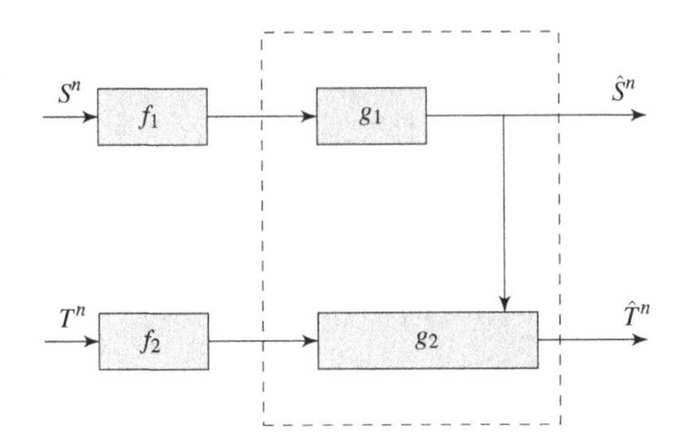

Apply union bound to get

$$\begin{aligned}
&\mathbb{P}\left[g(f_1(S^n), f_2(T^n)) \neq (S^n, T^n)\right] \\
&\quad = \mathbb{P}\left[g_1(f_1(S^n)) \neq S^n\right] + \mathbb{P}\left[g_2(f_2(T^n), g(f_1(S^n))) \neq T^n, g_1(f_1(S^n)) = S^n\right] \\
&\quad \leq \mathbb{P}\left[g_1(f_1(S^n)) \neq S^n\right] + \mathbb{P}\left[g_2(f_2(T^n), S^n) \neq T^n\right] \\
&\quad \leq 2\epsilon_n \to 0.
\end{aligned}$$

Similarly, the point $(H(S), H(T|S))$ can be approached.

To achieve other points in the region, use the idea of *time sharing*: If you can achieve with vanishing error probability any two points (R_1, R_2) and (R'_1, R'_2), then you can achieve for $\lambda \in [0, 1]$, $(\lambda R_1 + \bar{\lambda} R'_1, \lambda R_2 + \bar{\lambda} R'_2)$ by dividing the block of length n into two blocks of length λn and $\bar{\lambda} n$ and applying the two codes, respectively,

$$(S_1^{\lambda n}, T_1^{\lambda n}) \to \begin{bmatrix} \lambda n R_1 \\ \lambda n R_2 \end{bmatrix} \quad \text{using } (R_1, R_2) \text{ code,}$$

$$(S_{\lambda n+1}^{n}, T_{\lambda n+1}^{n}) \to \begin{bmatrix} \bar{\lambda} n R'_1 \\ \bar{\lambda} n R'_2 \end{bmatrix} \quad \text{using } (R'_1, R'_2) \text{ code.}$$

Therefore, all convex combinations of points in the achievable regions are also achievable, so the achievable region must be convex. □

11.7* Source Coding with a Helper (Ahlswede–Körner–Wyner)

In the Slepian–Wolf setting the goal was to compress/decompress X with decompressor having access to side information Y. A natural variation of the problem is to consider the case where access to Y itself is available over a rate-limited link, that is, we have the following setting:

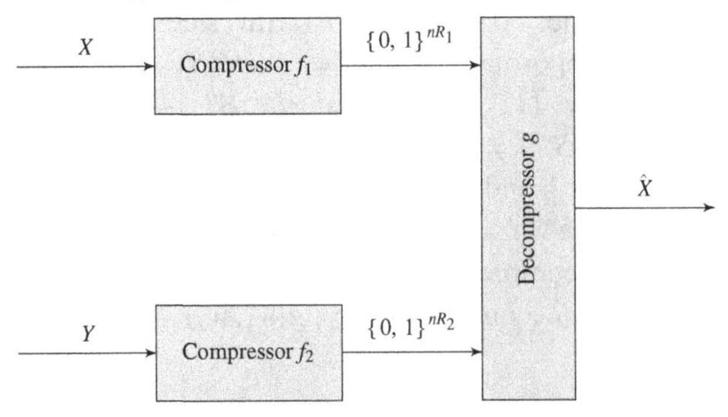

Note also that unlike the previous section, decompressor is only required to produce an estimate of X (not of Y), hence the name of this problem: compression with a (rate-limited) helper. The difficulty this time is that what needs to be communicated over this link from Y to the decompressor is not the information about Y but only that information in Y that is maximally useful for decompressing X. Despite similarity with the previous sections, this task is completely new and, consequently, characterization of rate pairs R_1, R_2 is much more subtle in this case. It was completed independently in two works [9, 460].

Theorem 11.17. (Ahlswede–Körner–Wyner) *Consider iid source $(X^n, Y^n) \sim P_{X,Y}$ with X discrete. The compressor produces message $W_1 = f_1(X^n)$ and the helper produces a message $W_2 = f_2(Y^n)$, with W_i consisting of at most nR_i bits, $i \in \{1,2\}$. The decompressor produces an estimate $\hat{X}^n = g(W_1, W_2)$. If rate pair (R_1, R_2) is achievable with vanishing probability of error $\mathbb{P}[\hat{X}^n \neq X^n] \to 0$, then there exists an auxiliary random variable U taking values on an alphabet of cardinality $|\mathcal{Y}| + 1$ such that $P_{X,Y,U} = P_{X,Y}P_{U|Y}$ (i.e. $X \to Y \to U$) and*

$$R_1 \geq H(X|U), \quad R_2 \geq I(Y;U). \tag{11.3}$$

Furthermore, for every such random variable U the rate pair $(H(X|U), I(Y;U))$ is achievable with vanishing error.

In other words, this time the set of achievable pairs (R_1, R_2) belongs to a region of $\mathbb{R}_+^2$ described as $\bigcup\{[H(X|U), +\infty) \times [I(Y;U), +\infty)\}$ with the union taken over all possible $P_{U|Y}\colon \mathcal{Y} \to \mathcal{U}$, where $|\mathcal{U}| = |\mathcal{Y}| + 1$. The boundary of the optimal (R_1, R_2) region is traced by an F_I-curve, a concept we will define later (Definition 16.5).

Proof. First, note that iterating over all possible random variables U (without cardinality constraint) the set of pairs (R_1, R_2) satisfying (11.3) is convex. Next, consider a compressor $W_1 = f_1(X^n)$ and $W_2 = f_2(Y^n)$. Then from Fano's inequality (3.19) assuming $\mathbb{P}[X^n \neq \hat{X}^n] = o(1)$ we have

$$H(X^n|W_1, W_2) = o(n).$$

Thus, from the chain rule and the fact that conditioning decreases entropy, we get

$$\begin{aligned} nR_1 &\geq I(X^n; W_1|W_2) \geq H(X^n|W_2) - o(n) && (11.4) \\ &= \sum_{k=1}^{n} H(X_k|W_2, X^{k-1}) - o(n) && (11.5) \\ &\geq \sum_{k=1}^{n} H(X_k|\underbrace{W_2, X^{k-1}, Y^{k-1}}_{\triangleq U_k}) - o(n). && (11.6) \end{aligned}$$

On the other hand, from (6.2) we have

$$nR_2 \ge I(W_2;Y^n) = \sum_{k=1}^{n} I\left(W_2;Y_k|Y^{k-1}\right) \tag{11.7}$$
$$= \sum_{k=1}^{n} I\left(W_2, X^{k-1};Y_k|Y^{k-1}\right) \tag{11.8}$$
$$= \sum_{k=1}^{n} I\left(W_2, X^{k-1}, Y^{k-1};Y_k\right), \tag{11.9}$$

where (11.8) follows from $I\left(W_2, X^{k-1};Y_k|Y^{k-1}\right) = I\left(W_2;Y_k|Y^{k-1}\right) + I(X^{k-1}; Y_k|W_2, Y^{k-1})$ and the fact that $(W_2, Y_k) \perp\!\!\!\perp X^{k-1}|Y^{k-1}$; and (11.9) from $Y^{k-1} \perp\!\!\!\perp Y_k$. Comparing (11.6) and (11.9) we notice that denoting $U_k = \left(W_2, X^{k-1}, Y^{k-1}\right)$ we have both $X_k \to Y_k \to U_k$ and

$$(R_1, R_2) \ge \frac{1}{n}\sum_{k=1}^{n}(H(X_k|U_k), I(U_k;Y_k))$$

and thus (from convexity) the rate pair must belong to the region spanned by all pairs $(H(X|U), I(U;Y))$.

To show that without loss of generality the auxiliary random variable U can be chosen to take at most $|\mathcal{Y}| + 1$ values, one can invoke Carathéodory's theorem (see Lemma 7.14). We omit the details.

Next, we show that for each U the mentioned rate pair is achievable. To that end, we first notice that if there were side information at the decompressor in the form of the iid sequence U^n correlated to X^n, then the Slepian–Wolf theorem implies that only rate $R_1 = H(X|U)$ would be sufficient to reconstruct X^n. Thus, the question boils down to creating a correlated sequence U^n at the decompressor by using the minimal rate R_2. One way to do it is to communicate U^n exactly by spending $nH(U)$ bits. However, it turns out that with $nI(U;X)$ bits we can communicate a "fake" $\hat{U}^n$ which nevertheless has conditional distribution $P_{X^n|\hat{U}^n} \approx P_{X^n|U^n}$ (such $\hat{U}^n$ is known as "jointly typical" with X^n). The possibility of producing such $\hat{U}^n$ is a result of independent prominence known as the covering lemma, which we will study much later – see Corollary 25.6. Here we show how to apply the covering lemma in this case.

By Corollary 25.6 and by Proposition 25.7 we know that for every $\delta > 0$ there exists a sufficiently large m and $\hat{U}^m = f_2(Y^m) \in \{0, 1\}^{mR_2}$ such that

$$X^m \to Y^m \to \hat{U}^m$$

and $I(X^m;\hat{U}^m) \ge m(I(X;U) - \delta)$. This means that $H(X^m|\hat{U}^m) \le mH(X|U) + m\delta$. We can now apply the Slepian–Wolf theorem to the block symbols $(X^m, \hat{U}^m)$. Namely, we define a new compression problem with $\tilde{X} = X^m$ and $\tilde{U} = \hat{U}^m$. These still take values on finite alphabets and thus there must exist (for sufficiently large ℓ) a compressor $W_1 = f_1(\tilde{X}^\ell) \in \{0, 1\}^{\ell\tilde{R}_1}$ and a decompressor $g(W_1, \tilde{U}^\ell)$ with a low probability of error and $\tilde{R}_1 \le H(\tilde{X}\,|\,\tilde{U}) + m\delta \le mH(X|U) + 2m\delta$. Now since the actual blocklength is $n = \ell m$ we get that the effective rate of this scheme is $R_1 = \frac{\tilde{R}_1}{m} \le H(X|U) + 2\delta$. Since $\delta > 0$ is arbitrary, the proof is completed. □

12 Entropy of Ergodic Processes

So far we have studied compression of iid sequence $\{S_i\}$, for which we demonstrated that the average compression length (for variable-length compressors) converges to the entropy $H(S)$ and that the probability of error (for fixed-length compressors) converges to zero or one depending on whether compression rate $R \lessgtr H(S)$. In this chapter, we shall examine similar results for a large class of processes with memory, known as *ergodic processes*. We start this chapter with a quick review of the main concepts of ergodic theory, then state our main results (Shannon–McMillan theorem, compression limit, and AEP). Subsequent sections are dedicated to proofs of the Shannon–McMillan and ergodic theorems. Finally, in the last section we introduce *Kolmogorov–Sinai entropy*, which associates to a fully deterministic transformation the measure of how "chaotic" it is. This concept plays a very important role in formalizing an apparent paradox: large mechanical systems (such as collections of gas particles) are on the one hand fully deterministic (described by Newton's laws of motion) and on the other hand have a lot of probabilistic properties (Maxwell distribution of velocities, fluctuations, etc.). Kolmogorov–Sinai entropy shows how these two notions can coexist. In addition the concept of Kolmogorov-Sinai entropy was used to resolve a long-standing open problem in dynamical systems regarding isomorphism of *Bernoulli shifts* [388, 323].

12.1 Bits of Ergodic Theory

Let us start with a dynamical system point of view on stochastic processes. Throughout this chapter we assume that all random variables are defined as functions on a common space of elementary outcomes $(\Omega, \mathcal{F})$.

Definition 12.1. (Measure-preserving transformation) $\tau\colon \Omega \to \Omega$ is a measure-preserving transformation, also known as a probability-preserving transformation (p.p.t.), if

$$\forall E \in \mathcal{F}, \quad P(E) = P(\tau^{-1}E).$$

The set E is called τ-invariant if $E = \tau^{-1}E$. The set of all τ-invariant sets forms a σ-algebra (this is left as an exercise for the reader) denoted $\mathcal{F}_{inv}$.

Definition 12.2. (Stationary process) A process $\{S_n, n = 0, \ldots\}$ is stationary if there exists a measure-preserving transformation $\tau\colon \Omega \to \Omega$ such that:

$$S_j = S_{j-1} \circ \tau = S_0 \circ \tau^j.$$

Therefore a stationary process can be described by the tuple $(\Omega, \mathcal{F}, \mathbb{P}, \tau, S_0)$ and $S_k = S_0 \circ \tau^k$.

Remark 12.1.

1 Alternatively, a random process $(S_0, S_1, S_2, \ldots)$ is stationary if its joint distribution is invariant with respect to shifts in time, that is, $P_{S_n^m} = P_{S_{n+t}^{m+t}}, \forall n, m, t$. Indeed, given such a process we can set $\Omega = \mathcal{S}^\infty$ and define a measure-preserving transformation as follows:

$$(s_0, s_1, \ldots) \xrightarrow{\tau} (s_1, s_2, \ldots). \tag{12.1}$$

So τ is a shift to the left.

2 An event $E \in \mathcal{F}$ is shift-invariant if

$$(s_1, s_2, \ldots) \in E \iff (s_0, s_1, s_2, \ldots) \in E, \quad \forall s_0$$

or, equivalently, $E = \tau^{-1}E$. Thus τ-invariant events are also called shift-invariant, when τ is interpreted as (12.1).

3 Some examples of shift-invariant events are $\{\exists n\colon x_i = 0, \forall i \geq n\}$, $\{\limsup x_i < 1\}$, etc. A non-shift-invariant event is $A = \{x_0 = x_1 = \cdots = 0\}$, since $\tau(1, 0, 0, \ldots) \in A$ but $(1, 0, \ldots) \notin A$.

4 Also recall that the tail σ-algebra is defined as

$$\mathcal{F}_{tail} \triangleq \bigcap_{n \geq 1} \sigma\{S_n, S_{n+1}, \ldots\}.$$

It is easy to check that all shift-invariant events belong to $\mathcal{F}_{tail}$. The inclusion is strict, as for example the event $\{\exists n\colon x_i = 0, \forall \text{ odd } i \geq n\}$ is in $\mathcal{F}_{tail}$ but not shift-invariant.

Proposition 12.3. (Poincaré recurrence) *Let τ be measure-preserving for $(\Omega, \mathcal{F}, \mathbb{P})$. Then for any measurable A with $\mathbb{P}[A] > 0$ we have*

$$\mathbb{P}\left[\bigcup_{k \geq 1} \tau^{-k} A \,\middle|\, A\right] = \mathbb{P}[\tau^k(\omega) \in A \text{ occurs infinitely often} | A] = 1.$$

Proof. Let $B = \bigcup_{k \geq 1} \tau^{-k} A$. It is sufficient to show that $\mathbb{P}[A \cap B] = \mathbb{P}[A]$ or equivalently

$$\mathbb{P}[A \cup B] = \mathbb{P}[B]. \tag{12.2}$$

To that end notice that $\tau^{-1} A \cup \tau^{-1} B = B$ and thus

$$\mathbb{P}[\tau^{-1}(A \cup B)] = \mathbb{P}[B],$$

but the left-hand side equals $\mathbb{P}[A \cup B]$ by the measure preservation of τ, proving (12.2). □

Consider τ mapping the initial state of the conservative (Hamiltonian) mechanical system to its state after passage of a given amount of time. It is known that τ preserves the Lebesgue measure in phase space (Liouville's theorem). Thus the Poincaré recurrence leads to rather counterintuitive conclusions. For example, opening the barrier separating two gases in a cylinder allows them to mix. Poincaré recurrence says that eventually they will return back to the original separated state (with each gas occupying its original half of the cylinder). Of course, the "paradox" is resolved by observing that it will take unphysically long for this to happen.

Definition 12.4. (Ergodicity) A transformation τ is ergodic if $\forall E \in \mathcal{F}_{inv}$ we have $\mathbb{P}[E] = 0$ or 1. A process $\{S_i\}$ is ergodic if all shift-invariant events are deterministic, that is, for any shift-invariant event E, $\mathbb{P}\left[S_1^\infty \in E\right] = 0$ or 1.

Here are some examples:

- $\{S_k = k^2\}$: ergodic but not stationary.
- $\{S_k = S_0\}$: stationary but not ergodic (unless S_0 is a constant). Note that the singleton set $E = \{(s, s, \ldots)\}$ is shift-invariant and $\mathbb{P}\left[S_1^\infty \in E\right] = \mathbb{P}[S_0 = s] \in (0, 1)$, not deterministic.
- $\{S_k\}$ iid is stationary and ergodic (by Kolmogorov's 0–1 law, tail events have no randomness).
- (Sliding-window construction of ergodic processes) If $\{S_i\}$ is ergodic, then $\{X_i = f(S_i, S_{i+1}, \ldots)\}$ is also ergodic. Such a process $\{X_i\}$ is called a *B-process* if S_i is iid.
- Here is an important example demonstrating how one can look at a simple iid process S_i in two ways via sliding window. Take $S_i \overset{\text{iid}}{\sim} \text{Ber}(1/2)$ and set $X_k = \sum_{n=0}^{\infty} 2^{-n-1} S_{k+n} = 2X_{k-1} \mod 1$. The marginal distribution of X_i is uniform on $[0, 1]$. Furthermore, *X_k's behavior is completely deterministic:* given X_0, all future X_k's can be computed exactly. This example shows that certain deterministic maps exhibit ergodic/chaotic behavior under iterative application: although the trajectory of X_k is completely deterministic, its time averages converge to expectations and in general "look random" since full determinism is only guaranteed if infinite-precision measurement of X_0 is available. Any discretization of the X_k's results in random behavior of the positive entropy rate – see more on this in Section 12.5*.
- There are also stronger conditions than ergodicity. Namely, we say that τ is mixing (or strongly mixing) if

$$\mathbb{P}[A \cap \tau^{-n} B] \to \mathbb{P}[A]\mathbb{P}[B].$$

We say that τ is weakly mixing if

$$\sum_{k=1}^{n} \frac{1}{n} \left| \mathbb{P}[A \cap \tau^{-k} B] - \mathbb{P}[A]\mathbb{P}[B] \right| \to 0.$$

Strong mixing implies weak mixing, which implies ergodicity (Exercise **II.12**).

- $\{S_i\}$: finite irreducible Markov chain with recurrent states is ergodic (in fact strong mixing), regardless of initial distribution.

  As a toy example, consider the kernel $P(0|1) = P(1|0) = 1$ with initial distribution $P(S_0 = 0) = 0.5$. This process only has two sample paths: $\mathbb{P}\left[S_1^\infty = (010101\ldots)\right] = \mathbb{P}\left[S_1^\infty = (101010\ldots)\right] = \frac{1}{2}$. It is easy to verify this process is ergodic (in the sense of Definition 12.4). Note however that in the Markov-chain literature a chain is called ergodic if it is irreducible, aperiodic, and recurrent. This example does not satisfy this definition (this clash of terminology is a frequent source of confusion).
- $\{S_i\}$: stationary zero-mean Gaussian process with autocovariance function $c(n) = \mathbb{E}[S_0 S_n^*]$.

$$\lim_{n\to\infty} \frac{1}{n+1} \sum_{t=0}^{n} c(t) = 0 \Leftrightarrow \{S_i\} \text{ ergodic} \Leftrightarrow \{S_i\} \text{ weakly mixing},$$
$$\lim_{n\to\infty} c(n) = 0 \Leftrightarrow \{S_i\} \text{ mixing}\,.$$

Intuitively speaking, an ergodic process may have infinite memory in general, but the memory should asymptotically vanish nevertheless. Indeed, we see that for a stationary Gaussian process ergodicity means the correlation dies (in the Cesàro-mean sense).

The *spectral measure* is defined as the (discrete-time) Fourier transform of the autocovariance sequence $\{c(n)\}$, in the sense that there exists a unique positive measure μ on $[-\pi, \pi]$ such that $c(n) = \frac{1}{2\pi}\int \exp(inx)\mu(dx)$. The spectral criteria can be formulated as follows:

$$\{S_i\} \text{ is ergodic} \Leftrightarrow \text{spectral measure has no atoms (CDF is continuous)},$$
$$\{S_i\} \text{ is a B-process} \Leftrightarrow \text{spectral measure has a density (power spectral density, see Example 6.4)}.$$

Detailed exposition on stationary Gaussian processes can be found in [135, Theorem 9.3.2, p. 474, Theorem 9.7.1, pp. 493–494].

12.2 Shannon–McMillan, Entropy Rate, and AEP

Equipped with the definitions of ergodicity we can state the three main results of this chapter. First is an analog of the law of large numbers for normalized log-likelihoods.

Theorem 12.5. (Shannon–McMillan–Breiman) *Let* $\mathbb{S} = \{S_1, S_2, \ldots\}$ *be a stationary and ergodic discrete process with entropy rate* $\mathcal{H} \triangleq H(\mathbb{S})$. *Then*

$$\frac{1}{n}\log\frac{1}{P_{S^n}(S^n)} \xrightarrow{\mathbb{P}} \mathcal{H}, \quad \textit{also a.s. and in } L_1. \tag{12.3}$$

Corollary 12.6. *Let $\{S_1, S_2, \dots\}$ be a discrete stationary and ergodic process with entropy rate $\mathcal{H}$ (in bits). Denote by f_n^* the optimal variable-length compressor for S^n and by $\epsilon^*(S^n, nR)$ the optimal probability of error of its fixed-length compressor with R bits per symbol (Definition 11.1). Then we have*

$$\frac{1}{n} l(f_n^*(S^n)) \xrightarrow{\mathbb{P}} \mathcal{H} \quad \text{and} \quad \lim_{n\to\infty} \epsilon^*(S^n, nR) = \begin{cases} 0, & R > \mathcal{H}, \\ 1, & R < \mathcal{H}. \end{cases} \tag{12.4}$$

Proof. By Corollary 10.5, the asymptotic distributions of $\frac{1}{n} l(f_n^*(S^n))$ and $\frac{1}{n}\log\frac{1}{P_{S^n}(S^n)}$ coincide. By the Shannon–McMillan–Breiman theorem (we only need convergence in probability) the latter converges to a constant $\mathcal{H}$. □

In Chapter 11 we learned the asymptotic equipartition property (AEP) for iid sources. Thanks to Shannon–McMillan–Breiman the same proof we did for the iid processes works for a general ergodic process.

Corollary 12.7. (AEP for stationary ergodic sources) *Let $\{S_1, S_2, \dots\}$ be a stationary and ergodic discrete process. For any $\delta > 0$, define the set*

$$T_n^\delta = \left\{ s^n : \left| \frac{1}{n} \log \frac{1}{P_{S^n}(s^n)} - \mathcal{H} \right| \le \delta \right\}.$$

Then

1. $\mathbb{P}\left[S^n \in T_n^\delta\right] \to 1$ *as* $n \to \infty$;
2. $\exp\{n(\mathcal{H} - \delta)\}(1 + o(1)) \le |T_n^\delta| \le \exp\{(\mathcal{H} + \delta)n\}(1 + o(1))$.

Some historical notes are in order. Convergence in probability for stationary ergodic Markov chains was already shown in [379]. The extension to convergence in L_1 for all stationary ergodic processes is due to McMillan in [302], and to almost sure convergence to Breiman [75].[1] A modern proof is in [11]. Note also that for a Markov chain, the existence of typical sequences and the AEP can be anticipated by thinking of a Markov process as a sequence of independent decisions regarding which transitions to take at each state. It is then clear that the Markov process's trajectory is simply a transformation of trajectories of an iid process, hence must concentrate similarly.

12.3 Proof of the Shannon–McMillan–Breiman Theorem

We shall show the L_1-convergence, which implies convergence in probability automatically. We will not prove a.s. convergence. To this end, let us first introduce Birkhoff–Khintchine's convergence theorem for ergodic processes, the proof of which is presented in the next subsection. The interpretation of this result is that time averages converge to the ensemble average.

[1] Curiously, both McMillan and Breiman left the field after these contributions. McMillan went on to head the US satellite reconnaissance program, and Breiman became a pioneer and advocate of the machine learning approach to statistical inference.

Theorem 12.8. (Birkhoff–Khintchine's ergodic theorem) *Let $\{S_i\}$ be a stationary and ergodic process. For any integrable function f, that is, $\mathbb{E}\,|f(S_1,\ldots)| < \infty$,*

$$\lim_{n\to\infty}\frac{1}{n}\sum_{k=1}^{n} f(S_k,\ldots) = \mathbb{E}\, f(S_1,\ldots) \quad \text{a.s. and in } L_1.$$

In the special case where f depends on finitely many coordinates, say, $f = f(S_1,\ldots,S_m)$,

$$\lim_{n\to\infty}\frac{1}{n}\sum_{k=1}^{n} f(S_k,\ldots,S_{k+m-1}) = \mathbb{E}\, f(S_1,\ldots,S_m) \quad \text{a.s. and in } L_1.$$

Example 12.1.

Consider $f = f(S_1)$. Then for an iid process Theorem 12.8 is simply the strong law of large numbers. On the extreme, if $\{S_i\}$ has constant trajectories, that is, $S_i = S_1$ for all $i \geq 1$, then such a process is non-ergodic and the conclusion of Theorem 12.8 fails (unless S_1 is a.s. constant).

We introduce an extension of the idea of the Markov chain.

Definition 12.9. (Finite-order Markov chain) $\{S_i\colon i \in \mathbb{N}\}$ is an mth-order Markov chain if $P_{S_{t+1}|S_1^t} = P_{S_{t+1}|S_{t-m+1}^t}$ for all $t \geq m$. It is called time-homogeneous if $P_{S_{t+1}|S_{t-m+1}^t} = P_{S_{m+1}|S_1^m}$.

Remark 12.2. Showing (12.3) for an mth-order time-homogeneous Markov chain $\{S_i\}$ is a direct application of Birkhoff–Khintchine. Indeed, we have

$$\begin{aligned}
\frac{1}{n}\log\frac{1}{P_{S^n}(S^n)} &= \frac{1}{n}\sum_{t=1}^{n}\log\frac{1}{P_{S_t|S^{t-1}}(S_t|S^{t-1})}\\
&= \frac{1}{n}\log\frac{1}{P_{S^m}(S^m)} + \frac{1}{n}\sum_{t=m+1}^{n}\log\frac{1}{P_{S_t|S_{t-m}^{t-1}}(S_t|S_{t-m}^{t-1})}\\
&= \underbrace{\frac{1}{n}\log\frac{1}{P_{S_1}(S_1^m)}}_{\to 0} + \underbrace{\frac{1}{n}\sum_{t=m+1}^{n}\log\frac{1}{P_{S_{m+1}|S_1^m}(S_t|S_{t-m}^{t-1})}}_{\to H(S_{m+1}|S_1^m)\text{ by Birkhoff–Khintchine}},
\end{aligned} \tag{12.5}$$

where we applied Theorem 12.8 with $f(s_1,s_2,\ldots) = \log\frac{1}{P_{S_{m+1}|S_1^m}(s_{m+1}|s_1^m)}$.

Now let us prove (12.3) for a general stationary ergodic process $\{S_i\}$ which might have infinite memory. The idea is to first approximate the distribution of that ergodic process by an mth-order Markov chain (finite memory) and make use of (12.5), then let $m \to \infty$ to make the approximation accurate. This is a highly influential contribution of Shannon to the theory of stochastic processes, known as *Markov approximation*.

Proof of Theorem 12.5 *in* L_1. To show that (12.3) converges in L_1, we want to show that

$$\mathbb{E}\left|\frac{1}{n}\log\frac{1}{P_{S^n}(S^n)}-\mathcal{H}\right|\to 0,\quad n\to\infty.$$

To this end, fix an $m\in\mathbb{N}$. Define the following auxiliary distribution for the process:

$$\begin{aligned}Q^{(m)}(S_1^\infty) &\triangleq P_{S_1^m}(S_1^m)\prod_{t=m+1}^{\infty}P_{S_t|S_{t-m}^{t-1}}(S_t|S_{t-m}^{t-1})\\ &= P_{S_1^m}(S_1^m)\prod_{t=m+1}^{\infty}P_{S_{m+1}|S_1^m}(S_t|S_{t-m}^{t-1}),\end{aligned}$$

where the second line applies stationarity. Note that under $Q^{(m)}$, $\{S_i\}$ is an mth-order time-homogeneous Markov chain.

By the triangle inequality,

$$\begin{aligned}\mathbb{E}\left|\frac{1}{n}\log\frac{1}{P_{S^n}(S^n)}-\mathcal{H}\right| \le\ & \underbrace{\mathbb{E}\left|\frac{1}{n}\log\frac{1}{P_{S^n}(S^n)}-\frac{1}{n}\log\frac{1}{Q_{S^n}^{(m)}(S^n)}\right|}_{\triangleq A}\\ &+\underbrace{\mathbb{E}\left|\frac{1}{n}\log\frac{1}{Q_{S^n}^{(m)}(S^n)}-H_m\right|}_{\triangleq B}+\underbrace{|H_m-\mathcal{H}|}_{\triangleq C},\end{aligned}$$

where $H_m\triangleq H(S_{m+1}|S_1^m)$.

We discuss each term separately next:

- $C=|H_m-\mathcal{H}|\to 0$ as $m\to\infty$ by Theorem 5.4 (recall that for stationary processes: $H(S_{m+1}|S_1^m)\to H$ from above).
- As shown in Remark 12.2, for any fixed m, $B\to 0$ in L_1 as $n\to\infty$, as a consequence of Birkhoff–Khintchine. Hence for any fixed m, $\mathbb{E}B\to 0$ as $n\to\infty$.
- For term A, applying the next Lemma 12.10,

$$\mathbb{E}[A]=\frac{1}{n}\mathbb{E}_P\left|\log\frac{dP_{S^n}}{dQ_{S^n}^{(m)}}\right|\le\frac{1}{n}D(P_{S^n}\|Q_{S^n}^{(m)})+\frac{2\log e}{en},$$

where

$$\begin{aligned}\frac{1}{n}D(P_{S^n}\|Q_{S^n}^{(m)}) &= \frac{1}{n}\mathbb{E}\left[\log\frac{P_{S^n}(S^n)}{P_{S^m}(S^m)\prod_{t=m+1}^{n}P_{S_{m+1}|S_m^1}(S_t|S_{t-m}^{t-1})}\right]\\ &=\frac{1}{n}(-H(S^n)+H(S^m)+(n-m)H_m)\\ &\to H_m-\mathcal{H}\quad \text{as } n\to\infty,\end{aligned}$$

with the second equality following from stationarity again.

Combining all three terms and sending $n\to\infty$, we obtain for any m,

$$\limsup_{n\to\infty}\mathbb{E}\left|\frac{1}{n}\log\frac{1}{P_{S^n}(S^n)}-\mathcal{H}\right|\le 2(H_m-\mathcal{H}).$$

Sending $m\to\infty$ completes the proof of the L_1-convergence. □

Lemma 12.10.

$$\mathbb{E}_P\left[\left|\log \frac{dP}{dQ}\right|\right] \leq D(P\|Q) + \frac{2\log e}{e}.$$

Proof. $|x\log x| - x\log x \leq \frac{2\log e}{e}, \forall x > 0$, since the LHS is zero if $x \geq 1$, and otherwise upper-bounded by $2\sup_{0\leq x\leq 1} x\log\frac{1}{x} = \frac{2\log e}{e}$. □

12.4* Proof of the Birkhoff–Khintchine Theorem

Proof of Theorem 12.8. For any function $\tilde{f} \in L_1$ and any ϵ, there exists a decomposition $\tilde{f} = f + h$ such that f is bounded and $h \in L_1$ with $\|h\|_1 = \mathbb{E}\,|h(S_1^\infty)| \leq \epsilon$.

Let us first focus on the bounded function f. Note that in the bounded domain $\mathcal{L}_1 \subset L_2$, thus $f \in L_2$. Furthermore, L_2 is a Hilbert space with inner product $(f,g) = \mathbb{E}[f(S_1^\infty)\overline{g(S_1^\infty)}]$.

For the measure-preserving transformation τ that generates the stationary process $\{S_i\}$, define the operator $T(f) = f \circ \tau$. Since τ is measure-preserving, we know that $\|Tf\|_2^2 = \|f\|_2^2$, thus T is a unitary and bounded operator.

Define the operator

$$A_n(f) = \frac{1}{n}\sum_{k=1}^{n} f \circ \tau^k.$$

Intuitively:

$$A_n = \frac{1}{n}\sum_{k=1}^{n} T^k = \frac{1}{n}(I - T^n)(I - T)^{-1}.$$

Then, if $f \perp \ker(I - T)$ we should have $A_n f \to 0$, since only components in the kernel can blow up. This intuition is formalized in the proof below.

Let us further decompose f into two parts $f = f_1 + f_2$, where $f_1 \in \ker(I - T)$ and $f_2 \in \ker(I - T)^\perp$. We make the following observations:

- If $g \in \ker(I - T)$, g must be a constant function. This is due to the ergodicity. Consider the indicator function 1_A: if $1_A = 1_A \circ \tau = 1_{\tau^{-1}A}$, then $\mathbb{P}[A] = 0$ or 1. For a general case, suppose $g = Tg$ and g is not constant, then at least some set $\{g \in (a,b)\}$ will be shift-invariant and have non-trivial measure, violating ergodicity.
- $\ker(I - T) = \ker(I - T^*)$. This is due to the fact that T is unitary:

  $$g = Tg \Rightarrow \|g\|^2 = (Tg, g) = (g, T^*g) \Rightarrow (T^*g, g) = \|g\|\|T^*g\| \Rightarrow T^*g = g,$$

  where in the last step we used the fact that Cauchy–Schwarz $(f,g) \leq \|f\| \cdot \|g\|$ only holds with equality for $g = cf$ for some constant c and in this case only $c = 1$ is possible.

- $\ker(I-T)^\perp = \ker(I-T^*)^\perp = [\mathrm{Im}(I-T)]$, where $[\mathrm{Im}(I-T)]$ denotes an L_2 closure of the image of the operator $(I-T)$.
- $g \in \ker(I-T)^\perp \iff \mathbb{E}[g] = 0$. Indeed, only zero-mean functions are orthogonal to constants.

With these observations, we know that $f_1 = m$ is a constant. Also, $f_2 \in [\mathrm{Im}(I-T)]$ so we further approximate it by $f_2 = f_0 + h_1$, where $f_0 \in \mathrm{Im}(I-T)$, namely $f_0 = g - g\circ\tau$ for some function $g \in L_2$, and $\|h_1\|_1 \le \|h_1\|_2 < \epsilon$. Therefore we have

$$A_n f_1 = f_1 = \mathbb{E}[f],$$
$$A_n f_0 = \frac{1}{n}(g - g\circ\tau^n) \to 0 \text{ a.s. and } L_1,$$

since $\mathbb{E}[\sum_{n\ge1}(\frac{g\circ\tau^n}{n})^2] = \mathbb{E}[g^2]\sum\frac{1}{n^2} < \infty$ and hence $\frac{1}{n}g\circ\tau^n \xrightarrow{\text{a.s.}} 0$ by Borel–Cantelli.
The proof is completed by showing

$$\mathbb{P}\left[\limsup_n A_n(h+h_1) \ge \delta\right] \le \frac{2\epsilon}{\delta}. \tag{12.6}$$

Indeed, then by taking $\epsilon \to 0$ we will have shown

$$\mathbb{P}\left[\limsup_{n\to\infty} A_n(f) \ge \mathbb{E}[f] + \delta\right] = 0$$

as required, and the opposite direction is shown analogously. □

The proof of (12.6) makes use of the maximal ergodic lemma stated as follows:

Theorem 12.11. (Maximal ergodic lemma) *Let $(\mathbb{P},\tau)$ be a probability measure and a measure-preserving transformation. Then for any $f \in L_1(\mathbb{P})$ we have*

$$\mathbb{P}\left[\sup_{n\ge1} A_n f > a\right] \le \frac{\mathbb{E}[f1\{\sup_{n\ge1} A_n f > a\}]}{a} \le \frac{\|f\|_1}{a},$$

where $A_n f = \frac{1}{n}\sum_{k=0}^{n-1} f\circ\tau^k$.

This is a so-called "weak L_1" estimate for a sublinear operator $\sup_n A_n(\cdot)$. In fact, this theorem is exactly equivalent to the following result:

Lemma 12.12. (Estimate for the maximum of averages) *Let $\{Z_n : n = 1,\ldots\}$ be a stationary process with $\mathbb{E}[|Z_1|] < \infty$. Then we have*

$$\mathbb{P}\left[\sup_{n\ge1}\frac{Z_1+\cdots+Z_n}{n} > a\right] \le \frac{\mathbb{E}[|Z_1|]}{a} \qquad \forall a > 0.$$

Proof. The argument for this lemma was originally quite involved, until a dramatically simple proof (below) was found by A. Garsia [180, Theorem 2.2.2]. Define

$$S_n = \sum_{k=1}^{n} Z_k,$$
$$L_n = \max\{0, Z_1, \ldots, Z_1 + \cdots + Z_n\},$$
$$M_n = \max\{0, Z_2, Z_2 + Z_3, \ldots, Z_2 + \cdots + Z_n\},$$
$$Z^* = \sup_{n \geq 1} \frac{S_n}{n}.$$

It is sufficient to show that

$$\mathbb{E}[Z_1 1\{Z^* > 0\}] \geq 0. \tag{12.7}$$

Indeed, applying (12.7) to $\tilde{Z}_1 = Z_1 - a$ and noticing that $\tilde{Z}^* = Z^* - a$ we obtain

$$\mathbb{E}[Z_1 1\{Z^* > a\}] \geq a\mathbb{P}[Z^* > a],$$

from which the lemma follows by upper-bounding the left-hand side with $\mathbb{E}[|Z_1|]$.

In order to show (12.7) we notice that

$$Z_1 + M_n = \max\{S_1, \ldots, S_n\}$$

and furthermore

$$Z_1 + M_n = L_n \qquad \text{on } \{L_n > 0\}.$$

Thus, we have

$$Z_1 1\{L_n > 0\} = L_n - M_n 1\{L_n > 0\},$$

where we do not need the indicator in the first term on the RHS, since $L_n = 0$ on $\{L_n > 0\}^c$. Taking expectation we get

$$\begin{aligned} \mathbb{E}[Z_1 1\{L_n > 0\}] &= \mathbb{E}[L_n] - \mathbb{E}[M_n 1\{L_n > 0\}] \\ &\geq \mathbb{E}[L_n] - \mathbb{E}[M_n] \\ &= \mathbb{E}[L_n] - \mathbb{E}[L_{n-1}] = \mathbb{E}[L_n - L_{n-1}] \geq 0, \end{aligned} \tag{12.8}$$

where we used $M_n \geq 0$, the fact that M_n has the same distribution as L_{n-1}, and $L_n \geq L_{n-1}$, respectively. Taking the limit as $n \to \infty$ in (12.8) and noticing that $\{L_n > 0\} \nearrow \{Z^* > 0\}$, we obtain (12.7). □

12.5* Sinai's Generator Theorem

As we mentioned in the introduction to this chapter, there is a classical conundrum in natural science. Our microscopic description of motions of atoms is fully deterministic (i.e. given positions and velocities of atoms at time t there is an operator τ that gives their positions at time $t + 1$). On the other hand, in many ways large systems behave probabilistically (as described by statistical mechanics, Gibbs distributions, etc.). An important conceptual bridge was built with the introduction of *Kolmogorov–Sinai entropy*, which in a nutshell attempts to resolve the conundrum by noticing that our way of describing a system at time t would typically involve only finitely many bits, and thus while τ is deterministic when acting on a full description of a state, from the point of view of any finite-bit description τ appears to act *stochastically*.

More formally, we associate to every probability-preserving transformation (p.p.t.) τ a number, called the Kolmogorov–Sinai entropy. This number is invariant to isomorphisms of p.p.t.'s (appropriately defined). Sinai's generator theorem then allows one to compute the Kolmogorov–Sinai entropy.

Definition 12.13. Fix a probability-preserving transformation τ acting on probability space $(\Omega, \mathcal{F}, \mathbb{P})$. The Kolmogorov–Sinai entropy of τ is defined as

$$\mathcal{H}(\tau) \triangleq \sup_{X_0} \lim_{n\to\infty} \frac{1}{n} H(X_0, X_0 \circ \tau, \dots, X_0 \circ \tau^{n-1}),$$

where the supremum is taken over all finitely valued random variables $X_0 \colon \Omega \to \mathcal{X}$ measurable with respect to $\mathcal{F}$.

Note that every random variable X_0 generates a stationary process adapted to τ, that is,

$$X_k \triangleq X_0 \circ \tau^k.$$

In this way, the Kolmogorov–Sinai entropy of τ equals the maximal entropy rate among all stationary processes adapted to τ. This quantity may be extremely hard to evaluate, however. Help comes in the form of the famous criterion of Y. Sinai. We need to elaborate on some more concepts first:

- The σ-algebra $\mathcal{G} \subset \mathcal{F}$ is $\mathbb{P}$-*dense* in $\mathcal{F}$, or sometimes we also say $\mathcal{G} = \mathcal{F} \mod \mathbb{P}$ or even $\mathcal{G} = \mathcal{F} \mod 0$, if for every $E \in \mathcal{F}$ there exists $E' \in \mathcal{G}$ s.t.

$$\mathbb{P}[E \Delta E'] = 0.$$

- Partition $\mathcal{A} = \{A_i \colon i = 1, 2, \dots\}$ measurable with respect to $\mathcal{F}$ is called *generating* if

$$\bigvee_{n=0}^{\infty} \sigma\{\tau^{-n}\mathcal{A}\} = \mathcal{F} \mod \mathbb{P}.$$

- Random variable $Y \colon \Omega \to \mathcal{Y}$ with a *countable* alphabet $\mathcal{Y}$ is called a *generator* of $(\Omega, \mathcal{F}, \mathbb{P}, \tau)$ if

$$\sigma\{Y, Y \circ \tau, \dots, Y \circ \tau^n, \dots\} = \mathcal{F} \mod \mathbb{P}.$$

Theorem 12.14. (Sinai's generator theorem) *Let Y be the generator of a p.p.t. $(\Omega, \mathcal{F}, \mathbb{P}, \tau)$. Let $H(\mathbb{Y})$ be the entropy rate of the process $\mathbb{Y} = \{Y_k = Y \circ \tau^k \colon k = 0, \dots\}$. If $H(\mathbb{Y})$ is finite, then $\mathcal{H}(\tau) = H(\mathbb{Y})$.*

Proof. Notice that since $H(\mathbb{Y})$ is finite, we must have $H(Y_0^n) < \infty$ and thus $H(Y) < \infty$. First, we argue that $\mathcal{H}(\tau) \geq H(\mathbb{Y})$. If Y has finite alphabet, then it is simply from the definition. Otherwise let Y be $\mathbb{Z}_+$-valued. Define a truncated version $\tilde{Y}_m = \min(Y, m)$, then since $\tilde{Y}_m \to Y$ as $m \to \infty$ we have from lower semicontinuity of mutual information, see (4.28), that

$$\lim_{m\to\infty} I(Y;\tilde{Y}_m) \geq H(Y).$$

and consequently for arbitrarily small ϵ and sufficiently large m

$$H(Y|\tilde{Y}) \leq \epsilon.$$

Then, consider the chain

$$\begin{aligned}
H(Y_0^n) &= H(\tilde{Y}_0^n, Y_0^n) = H(\tilde{Y}_0^n) + H(Y_0^n|\tilde{Y}_0^n) \\
&= H(\tilde{Y}_0^n) + \sum_{i=0}^{n} H(Y_i|\tilde{Y}_0^n, Y_0^{i-1}) \\
&\leq H(\tilde{Y}_0^n) + \sum_{i=0}^{n} H(Y_i|\tilde{Y}_i) \\
&= H(\tilde{Y}_0^n) + nH(Y|\tilde{Y}) \leq H(\tilde{Y}_0^n) + n\epsilon.
\end{aligned}$$

Thus, the entropy rate of $\tilde{\mathbb{Y}}$ (which is on a finite alphabet) can be made arbitrarily close to that of $\mathbb{Y}$, concluding that $\mathcal{H}(\tau) \geq \mathcal{H}(\mathbb{Y})$.

The bulk of the proof is to show that for any stationary process $\mathbb{X}$ adapted to τ the entropy rate is upper-bounded by $H(\mathbb{Y})$. To that end, consider $X\colon \Omega \to \mathcal{X}$ with finite $\mathcal{X}$ and define as usual the process $\mathbb{X} = \{X \circ \tau^k : k = 0, 1, \ldots\}$. By the generating property of $\mathbb{Y}$ we have that X (perhaps after modification on a set of measure zero) is a function of Y_0^∞. So are all X_k's. Thus

$$H(X_0) = I(X_0;Y_0^\infty) = \lim_{n\to\infty} I(X_0;Y_0^n),$$

where we used the continuity-in-σ-algebra property of mutual information, see (4.30). Rewriting the latter limit differently, we have

$$\lim_{n\to\infty} H(X_0|Y_0^n) = 0.$$

Fix $\epsilon > 0$ and choose m so that $H(X_0|Y_0^m) \leq \epsilon$. Then consider the following chain:

$$\begin{aligned}
H(X_0^n) &\leq H(X_0^n, Y_0^n) = H(Y_0^n) + H(X_0^n|Y_0^n) \\
&\leq H(Y_0^n) + \sum_{i=0}^{n} H(X_i|Y_i^n) \\
&= H(Y_0^n) + \sum_{i=0}^{n} H(X_0|Y_0^{n-i}) \\
&\leq H(Y_0^n) + m\log|\mathcal{X}| + (n-m)\epsilon,
\end{aligned}$$

where we used stationarity of (X_k, Y_k) and the fact that $H(X_0|Y_0^{n-i}) < \epsilon$ for $i \leq n - m$. After dividing by n and passing to the limit our argument implies

$$H(\mathbb{X}) \leq H(\mathbb{Y}) + \epsilon.$$

Taking here $\epsilon \to 0$ completes the proof.

Alternative proof: Suppose X_0 takes values on a finite alphabet $\mathcal{X}$ and $X_0 = f(Y_0^\infty)$. Then (this is a measure-theoretic fact) for every $\epsilon > 0$ there exists $m = m(\epsilon)$ and a function $f_\epsilon : \mathcal{Y}^{m+1} \to \mathcal{X}$ s.t.

$$\mathbb{P}[f(Y_0^\infty) \neq f_\epsilon(Y_0^m)] \leq \epsilon.$$

(This is just another way to say that $\bigcup_n \sigma(Y_0^n)$ is $\mathbb{P}$-dense in $\sigma(Y_0^\infty)$.) Define a stationary process $\tilde{\mathbb{X}}$ as

$$\tilde{X}_j \triangleq f_\epsilon(Y_j^{m+j}).$$

Notice that since $\tilde{X}_0^n$ is a function of Y_0^{n+m} we have

$$H(\tilde{X}_0^n) \leq H(Y_0^{n+m}).$$

Dividing by m and passing to the limit we conclude that the entropy rates satisfy

$$H(\tilde{\mathbb{X}}) \leq H(\mathbb{Y}).$$

Finally, to relate $\tilde{\mathbb{X}}$ to $\mathbb{X}$ notice that by construction for every j

$$\mathbb{P}[\tilde{X}_j \neq X_j] \leq \epsilon.$$

Since both processes take values on a fixed finite alphabet, from Corollary 6.7 we infer that

$$|H(\mathbb{X}) - H(\tilde{\mathbb{X}})| \leq \epsilon \log|\mathcal{X}| + h(\epsilon).$$

Altogether, we have shown that $H(\mathbb{X}) \leq H(\mathbb{Y}) + \epsilon \log|\mathcal{X}| + h(\epsilon)$. Taking $\epsilon \to 0$ concludes the proof. □

Some examples of Theorem 12.14 are as follows:

- Let $\Omega = [0, 1]$, $\mathcal{F}$ the Borel σ-algebra, $\mathbb{P} = \text{Leb}$, and

$$\tau(\omega) = 2\omega \mod 1 = \begin{cases} 2\omega, & \omega < 1/2, \\ 2\omega - 1, & \omega \geq 1/2. \end{cases}$$

  It is easy to show that $Y(\omega) = 1\{\omega < 1/2\}$ is a generator and that $\mathbb{Y}$ is an iid Bernoulli(1/2) process. Thus, we get that Kolmogorov–Sinai entropy is $\mathcal{H}(\tau) = \log 2$.

  Let us understand the significance of this example and Sinai's result. If we have a full "microscopic" description of the initial state of the system ω, then the future states of the system are completely deterministic: $\tau(\omega), \tau(\tau(\omega)), \ldots$. However, in practice we cannot possibly have a complete description of the initial state, and should be satisfied with some discrete (i.e. finite or countably infinite) measurement outcomes $Y(\omega)$, $Y(\tau(\omega))$, etc. What we infer from the previous result is that no matter how fine our discrete measurements are, they will still generate a process that will have finite entropy rate (equal to $\log 2$ bits per measurement). *This reconciles the apparent paradox between Newtonian (dynamical) and Gibbsian (statistical) points of view on large mechanical systems.* In more mundane terms, we may notice that Sinai's theorem tells us that much more complicated stochastic processes (e.g. the one generated by a ternary-valued measurement $Y'(\omega = 1\{\omega > 1/3\} + 1\{\omega > 2/3\})$ would still have an entropy rate the same as the simple iid Bernoulli(1/2) process.

- Let Ω be the unit circle $\mathbb{S}^1$, $\mathcal{F}$ the Borel σ-algebra, $\mathbb{P}$ the normalized length, and

$$\tau(\omega) = \omega + \gamma,$$

that is, τ is a rotation by the angle γ. (When $\frac{\gamma}{2\pi}$ is irrational, this is known to be an ergodic p.p.t.) Here $Y = 1\{|\omega| < 2\pi\epsilon\}$ is a generator for arbitrarily small ϵ and hence

$$\mathcal{H}(\tau) \le H(\mathbb{X}) \le H(Y_0) = h(\epsilon) \to 0 \qquad \text{as } \epsilon \to 0.$$

This is an example of a zero-entropy p.p.t.

Remark 12.3. Two p.p.t.'s $(\Omega_1, \tau_1, \mathbb{P}_1)$ and $(\Omega_0, \tau_0, \mathbb{P}_0)$ are called isomorphic if there exists $f_i \colon \Omega_i \to \Omega_{1-i}$ defined $\mathbb{P}_i$-almost everywhere and such that (1) $\tau_{1-i} \circ f_i = f_{1-i} \circ \tau_i$; (2) $f_i \circ f_{1-i}$ is the (identity on Ω_i (a.e.);) and (3) $\mathbb{P}_i[f_{1-i}^{-1} E] = \mathbb{P}_{1-i}[E]$. It is easy to see that Kolmogorov–Sinai entropies of isomorphic p.p.t.,s are equal. This observation was made by Kolmogorov in 1958. It was revolutionary, since it allowed one to show that p.p.t.,s corresponding to shifts of iid Ber(1/2) and iid Ber(1/3) processes are not isomorphic. Before, the only invariants known were those obtained from studying the spectrum of a unitary operator

$$U_\tau \colon L_2(\Omega, \mathbb{P}) \to L_2(\Omega, \mathbb{P}), \tag{12.9}$$

$$\phi(x) \mapsto \phi(\tau(x)). \tag{12.10}$$

However, the spectrum of τ corresponding to any non-constant iid process consists of the entire unit circle, and thus is unable to distinguish Ber(1/2) from Ber(1/3).[2]

[2] To see the statement about the spectrum, let X_i be iid with zero mean and unit variance. Then consider $\phi(x_1^\infty)$ defined as $\frac{1}{\sqrt{m}} \sum_{k=1}^m e^{i\omega k} x_k$. This ϕ has unit energy and as $m \to \infty$ we have $\| U_\tau \phi - e^{i\omega} \phi \|_{L_2} \to 0$. Hence every $e^{i\omega}$ belongs to the spectrum of U_τ.

13 Universal Compression

Unfortunately, the theory developed so far is not very helpful for anyone tasked with actually compressing a file of English text. Indeed, since the probability law governing text generation is not given to us, one cannot apply the compression results that we have discussed so far. In this chapter we will discuss how to produce compression schemes that do not require a priori knowledge of the distribution. For example, an n-letter input compressor maps $\mathcal{X}^n \to \{0,1\}^*$. There is no one fixed probability distribution P_{X^n} on $\mathcal{X}^n$, but rather a whole class of distributions. Thus, the problem of compression becomes intertwined with the problem of *distribution (density) estimation* and we will see that the optimal algorithms for both problems often coincide.

The plan for this chapter is as follows:

1. We will start by discussing the earliest example of a universal compression algorithm (of Fitingof). It does not talk about probability distributions at all. However, it turns out to be asymptotically optimal simultaneously for all iid distributions and with small modifications for all finite-order Markov chains.
2. The next class of universal compressors is based on assuming that the true distribution P_{X^n} belongs to a given class. These methods proceed by choosing a good model distribution Q_{X^n} serving as the minimax approximation to each distribution in the class. The compression algorithm for a single distribution Q_{X^n} is then designed as in previous chapters.
3. Finally, an entirely different idea are algorithms of Lempel–Ziv type. These automatically adapt to the distribution of the source, without any prior assumptions required.

Throughout this chapter, all logarithms are binary. Instead of describing each compression algorithm, we will merely specify some distribution Q_{X^n} and apply one of the following constructions:

- Sort all x^n in order of decreasing $Q_{X^n}(x^n)$ and assign values from $\{0,1\}^*$ as in Theorem 10.2; this compressor has lengths satisfying

$$\ell(f(x^n)) \le \log \frac{1}{Q_{X^n}(x^n)}.$$

- Set lengths to be

$$\ell(f(x^n)) \triangleq \left\lceil \log \frac{1}{Q_{X^n}(x^n)} \right\rceil$$

and apply Kraft's inequality from Theorem 10.9 to construct a prefix code.
- Use arithmetic coding (see next section).

The important conclusion is that in all these cases we have

$$\ell(f(x^n)) = \log \frac{1}{Q_{X^n}(x^n)} \pm \text{universal constant},$$

and in this way we may and will always replace lengths with $\log \frac{1}{Q_{X^n}(x^n)}$. In this architecture, the only task of a universal compression algorithm is to specify the Q_{X^n}, which is known as *universal probability assignment* in this context.

If one factorizes $Q_{X^n} = \prod_{t=1}^n Q_{X_t|X_1^{t-1}}$ then we arrive at a crucial conclusion: *universal compression* is equivalent to *sequential (online) prediction* under the log-loss, which in itself is simply a version of the density estimation task in learning theory. This exciting connection between compression and learning theory is explored in Section 13.6 and is a highlight of this chapter.

In turn, machine learning drives advances in universal compression. As of 2022 the best-performing text compression algorithms (see the leaderboard at [290]) use a deep neural network (specifically, a transformer model) that starts from a fixed initialization. As the input text is processed, parameters of the network are continuously updated via stochastic gradient descent causing progressively better prediction (and hence compression) performance.

This chapter, thus, can be understood as a set of results on both information theory (universal compression) and machine learning (online prediction/density estimation).

13.1 Arithmetic Coding

Constructing an encoder table from Q_{X^n} may require a lot of resources if n is large. Arithmetic coding provides a convenient workaround by allowing the encoder to output bits sequentially. *Notice that to do so, it requires that not only Q_{X^n} but also its marginalizations $Q_{X^1}, Q_{X^2}, \ldots$ be easily computable.* (This is not the case, for example, for Shtarkov distributions (13.12)–(13.14), which also have an additional problem of not being compatible for different n.)

Let us agree upon some ordering on the alphabet of $\mathcal{X}$ (e.g. $\mathsf{a} < \mathsf{b} < \cdots < \mathsf{z}$) and extend this order lexicographically to $\mathcal{X}^n$ (that is, for $x = (x_1, \ldots, x_n)$ and $y = (y_1, \ldots, y_n)$, we say $x < y$ if $x_i < y_i$ for the first i such that $x_i \neq y_i$, e.g. $\mathsf{baba} < \mathsf{babb}$). Then let

$$F_n(x^n) = \sum_{y^n < x^n} Q_{X^n}(y^n).$$

Associate to each x^n an interval $I_{x^n} = [F_n(x^n), F_n(x^n) + Q_{X^n}(x^n))$. These intervals are disjoint subintervals of $[0, 1)$. As such, each x^n can be represented uniquely by any point in the interval I_{x^n}. A specific choice is as follows. Encode

$$x^n \mapsto \text{largest dyadic interval } D_{x^n} \text{ contained in } I_{x^n} \tag{13.1}$$

and we agree to select the leftmost dyadic interval when there are two possibilities. Recall that dyadic intervals are intervals of the type $[m2^{-k}, (m+1)2^{-k})$ where m is an integer. We encode such an interval by the k-bit (zero-padded) binary expansion of the fractional number $m2^{-k} = 0.b_1 b_2 \ldots b_k = \sum_{i=1}^{k} b_i 2^{-i}$. For example, $[3/4, 7/8) \mapsto 110$, $[3/4, 13/16) \mapsto 1100$. We set the codeword $f(x^n)$ to be that string. The resulting code is a prefix code satisfying

$$\log_2 \frac{1}{Q_{X^n}(x^n)} \le \ell(f(x^n)) \le \left\lceil \log_2 \frac{1}{Q_{X^n}(x^n)} \right\rceil + 1. \tag{13.2}$$

(This is an exercise; see Exercise **II.13**.)

Observe that

$$F_n(x^n) = F_{n-1}(x^{n-1}) + Q_{X^{n-1}}(x^{n-1}) \sum_{y < x_n} Q_{X_n|X^{n-1}}(y|x^{n-1})$$

and thus $F_n(x^n)$ can be computed sequentially *if $Q_{X^{n-1}}$ and $Q_{X_n|X^{n-1}}$ are easy to compute.* This method is the method of choice in many modern compression algorithms because it allows one to dynamically incorporate the learned information about the data stream, in the form of updating $Q_{X_n|X^{n-1}}$ (e.g. if the algorithm detects that an executable file contains a long chunk of English text, it may temporarily switch to $Q_{X_n|X^{n-1}}$ modeling the English language).

We note that efficient implementation of an arithmetic encoder and decoder is a continuing research area. Indeed, performance depends on number-theoretic properties of the denominators of distributions $Q_{X_t|X^{t-1}}$, because as the encoder/decoder progress along the string, they need to periodically renormalize the current interval I_{x^t} to be $[0, 1)$ but this requires carefully realigning the dyadic boundaries. A recent idea of J. Duda, known as *asymmetric numeral system* (ANS) [138], led to such impressive computational gains that in less than a decade it was adopted by most compression libraries handling diverse data streams (e.g. the Linux kernel images, Dropbox, and Facebook traffic, etc.).

13.2 Combinatorial Construction of Fitingof

Fitingof [170] suggested that a sequence $x^n \in \mathcal{X}^n$ should be prescribed information $\Phi_0(x^n)$ equal to the logarithm of the number of all possible permutations obtainable from x^n (i.e. log-size of the type class containing x^n). As we have shown in Proposition 1.5:

$$\Phi_0(x^n) = nH(x_T) + O(\log n), \quad T \sim \text{Unif}([n]), \tag{13.3}$$

$$= nH(\hat{P}_{x^n}) + O(\log n), \tag{13.4}$$

where $\hat{P}_{x^n}$ is the empirical distribution of the sequence x^n:

$$\hat{P}_{x^n}(a) \triangleq \frac{1}{n}\sum_{i=1}^{n} 1\{x_i = a\}. \tag{13.5}$$

Then Fitingof argues that it should be possible to produce a prefix code with

$$\ell(f(x^n)) = \Phi_0(x^n) + O(\log n). \tag{13.6}$$

This can be done in many ways. In the spirit of what comes next, let us define

$$Q_{X^n}(x^n) \triangleq \exp\{-\Phi_0(x^n)\}c_n, \tag{13.7}$$

where the normalization constant c_n is determined by the number of types, namely, $c_n = 1/\binom{n+|\mathcal{X}|-1}{|\mathcal{X}|-1}$. Counting the number of different possible empirical distributions (types), we get

$$c_n = O(n^{-(|\mathcal{X}|-1)}),$$

and thus, by Kraft inequality, there must exist a prefix code with lengths satisfying (13.6).[1] Now taking expectation over $X^n \overset{\text{iid}}{\sim} P_X$ we get

$$\mathbb{E}[\ell(f(X^n))] = nH(P_X) + O(\log n),$$

for every iid source on $\mathcal{X}$.

Universal Compressor for all Finite-Order Markov Chains Fitingof's idea can be extended as follows. Define now the first-order information content $\Phi_1(x^n)$ to be the log of the number of all sequences, obtainable by permuting x^n with the extra restriction that the new sequence should have the same statistics on digrams. Asymptotically, Φ_1 is just the conditional entropy

$$\Phi_1(x^n) = nH(x_T|x_{T-1}) + O(\log n), \quad T \sim \text{Unif}([n]),$$

where $T-1$ is understood in the sense of modulo n. Again, it can be shown that there exists a code such that lengths

$$\ell(f(x^n)) = \Phi_1(x^n) + O(\log n).$$

This implies that for every first-order stationary Markov chain $X_1 \to X_2 \to \cdots \to X_n$ we have

$$\mathbb{E}[\ell(f(X^n))] = nH(X_2|X_1) + O(\log n).$$

This can be further continued to define $\Phi_2(x^n)$ leading to a universal code that is asymptotically optimal for all second-order Markov chains, and so on and so forth.

[1] Explicitly, we can do a two-part encoding: first describe the type class of x^n (takes $(|\mathcal{X}|-1)\log n$ bits) and then describe the element of the class (takes $\Phi_0(x^n)$ bits).

13.3 Optimal Compressors for a Class of Sources: Redundancy

So we have seen that we can construct compressor $f\colon \mathcal{X}^n \to \{0,1\}^*$ that achieves

$$\mathbb{E}[\ell(f(X^n))] \le H(X^n) + o(n),$$

simultaneously for all iid sources (or even all rth-order Markov chains). What should we do next? Krichevsky [260] suggested that the next barrier should be to minimize the regret, or *redundancy*,

$$\mathbb{E}[\ell(f(X^n))] - H(X^n)$$

simultaneously for all sources in a given class. We proceed to rigorous definitions.

Given a collection $\{P_{X^n|\theta} : \theta \in \Theta\}$ of sources, and a compressor $f\colon \mathcal{X}^n \to \{0,1\}^*$, we define its redundancy as

$$\sup_{\theta_0} \mathbb{E}[\ell(f(X^n))|\theta = \theta_0] - H(X^n|\theta = \theta_0).$$

Replacing code lengths with $\log \frac{1}{Q_{X^n}}$, we define the redundancy of the distribution Q_{X^n} as

$$\sup_{\theta_0} D(P_{X^n|\theta=\theta_0} \| Q_{X^n}).$$

Thus, the question of designing the best universal compressor (in the sense of optimizing the worst-case deviation of the average length from the entropy) becomes the question of finding the solution of:

$$Q^*_{X^n} = \operatorname*{argmin}_{Q_{X^n}} \sup_{\theta_0} D(P_{X^n|\theta=\theta_0} \| Q_{X^n}).$$

We therefore arrive at the following definition

Definition 13.1. (Redundancy in universal compression) Given a class of sources $\{P_{X^n|\theta=\theta_0} : \theta_0 \in \Theta, n = 1, \ldots\}$ we define its minimax redundancy as

$$\mathcal{R}^*_n \equiv \mathcal{R}^*_n(\Theta) \triangleq \min_{Q_{X^n}} \sup_{\theta_0 \in \Theta} D(P_{X^n|\theta=\theta_0} \| Q_{X^n}). \tag{13.8}$$

Assuming the finiteness of $\mathcal{R}^*_n$, Theorem 5.9 gives the maximin and capacity representation

$$\mathcal{R}^*_n = \sup_{\pi} \min_{Q_{X^n}} D(P_{X^n|\theta} \| Q_{X^n}|\pi) \tag{13.9}$$

$$= \sup_{\pi} I(\theta; X^n), \tag{13.10}$$

where optimization is over priors $\pi \in \mathcal{P}(\Theta)$ on θ. Thus redundancy is simply the capacity of the channel $\theta \to X^n$. This result, obvious in hindsight, was rather surprising in the early days of universal compression. It is known as *the capacity-redundancy theorem.*

Finding the exact Q_{X^n}-minimizer in (13.8) is a daunting task even for the simple class of all iid Bernoulli sources (i.e. $\Theta = [0,1]$, $P_{X^n|\theta} = \text{Ber}(\theta)^{\otimes n}$). In fact, for smooth parametric families the capacity-achieving input distribution is rather cumbersome: it is a discrete distribution with k_n atoms, k_n slowly growing as $n \to \infty$. A provocative conjecture was put forward by physicists [297, 2] that there is a certain universality relation:

$$\mathcal{R}_n^* = \frac{3}{4} \log k_n + o(\log k_n)$$

satisfied for all parametric families simultaneously. For the Bernoulli example this implies $k_n \asymp n^{2/3}$, but even this is open. However, as we will see below it turns out that these unwieldy capacity-achieving input distributions converge as $n \to \infty$ to a beautiful limiting law, known as the Jeffreys prior.

Remark 13.1. (Shtarkov, Fitingof, and individual sequence approach) There is a connection between the combinatorial method of Fitingof and the method of optimality for a class. Indeed, following Shtarkov we may want to choose distribution Q_{X^n} so as to minimize the worst-case redundancy *for each realization* x^n (not average!):

$$\mathcal{R}_n^{**}(\Theta) \triangleq \min_{Q_{X^n}} \max_{x^n} \sup_{\theta_0 \in \Theta} \log \frac{P_{X^n|\theta}(x^n|\theta_0)}{Q_{X^n}(x^n)}. \tag{13.11}$$

This minimization is attained at the *Shtarkov distribution* (also known as the *normalized maximal likelihood* (NML) code):

$$Q_{X^n}^{(S)}(x^n) = \frac{1}{Z} \sup_{\theta_0 \in \Theta} P_{X^n|\theta}(x^n|\theta_0), \tag{13.12}$$

where the normalization constant

$$Z = \sum_{x^n \in \mathcal{X}^n} \sup_{\theta_0 \in \Theta} P_{X^n|\theta}(x^n|\theta_0), \tag{13.13}$$

is called the *Shtarkov sum*. If the class $\{P_{X^n|\theta} : \theta \in \Theta\}$ is chosen to be all product distributions on $\mathcal{X}$ then

$$\text{(iid)} \quad Q_{X^n}^{(S)}(x^n) = \frac{\exp\{-nH(\hat{P}_{x^n})\}}{\sum_{x^n} \exp\{-nH(\hat{P}_{x^n})\}}, \tag{13.14}$$

where $H(\hat{P}_{x^n})$ is the empirical entropy of x^n. As such, compressing with respect to $Q_{X^n}^{(S)}$ recovers Fitingof's construction Φ_0 up to $O(\log n)$ differences between $nH(\hat{P}_{x^n})$ and $\Phi_0(x^n)$. If we take $P_{X^n|\theta}$ to be all first-order Markov chains, then we get construction Φ_1, etc. Note also that the problem (13.11) can be written as a minimization of the regret for each *individual sequence* (under the log-loss, with respect to a parameter class $P_{X^n|\theta}$):

$$\min_{Q_{X^n}} \max_{x^n} \left\{ \log \frac{1}{Q_{X^n}(x^n)} - \inf_{\theta_0 \in \Theta} \log \frac{1}{P_{X^n|\theta}(x^n|\theta_0)} \right\}. \tag{13.15}$$

In summary, using Shtarkov's distribution (minimizer of (13.15)) makes sure that any individual realization of x^n (whether it was or was not generated by $P_{X^n|\theta=\theta_0}$ for some θ_0) is compressed almost as well as the best compressor tailored for the class of $P_{X^n|\theta}$. Hence, if our model class $P_{X^n|\theta}$ approximates the generative process of

x^n well, we achieve nearly optimal compression. In Section 13.7 below we will also learn that $Q_{X_n|X^{n-1}}$ can be interpreted as an online estimator of the distribution of x_j's.

Remark 13.2. (Two redundancies) In the literature of universal compression, the quantity $\mathcal{R}_n^{**}$ is known as the *worst-case* or *pointwise* minimax redundancy, in comparison with the *average-case* minimax redundancy $\mathcal{R}_n^*$ in (13.8), which replaces $\max_{x^n}$ in (13.11) by $\mathbb{E}_{x^n \sim P_{X^n|\theta_0}}$. It is known that for many model classes, such as iid and finite-order Markov sources, $\mathcal{R}_n^*$ and $\mathcal{R}_n^{**}$ agree in the leading term as $n \to \infty$.[2] As $\mathcal{R}_n^* \leq \mathcal{R}_n^{**}$, typically the way one bounds the redundancies is to upper-bound $\mathcal{R}_n^{**}$ by bounding the pointwise redundancy (via combinatorial means) for a specific probability assignment and lower-bound $\mathcal{R}_n^*$ by applying (13.10) and bounding the mutual information for a specific prior; see Exercises **II.15** and **II.16** for an example and [112, Chapters 6 and 7] for more.

Remark 13.3. (Redundancy for single-shot codes) We note that any prefix code $f\colon \mathcal{X}^n \to \{0,1\}^*$ defines a distribution $Q_{X^n}(x^n) = 2^{-\ell(f(x^n))}$. (We assume the code's binary tree is full such that the Kraft sum equals one.) Therefore, our definition of redundancy in (13.8) assesses the excess of expected length $\mathbb{E}[\ell(f(X^n))]$ over $H(X^n)$ for the prefix codes. For single-shot codes (Section 10.1) without prefix constraints the optimal answers are slightly different, however. For example, the optimal universal code for all iid sources satisfies $\mathbb{E}[\ell(f(X^n))] \approx H(X^n) + \frac{|\mathcal{X}|-3}{2} \log n$ in contrast with $\frac{|\mathcal{X}|-1}{2} \log n$ for prefix codes, see [41, 257].

13.4* Asymptotic Maximin Solution: Jeffreys Prior

In this section we will only consider the simple setting of a class of sources consisting of all iid distributions on a given finite alphabet $|\mathcal{X}| = d+1$, which defines a d-parameter family of distributions. We will show that the prior, asymptotically achieving the capacity (13.10), is given by the Dirichlet distribution with parameters set to $1/2$. Recall that the Dirichlet distribution $\text{Dirichlet}(\alpha_0, \dots, \alpha_d)$ with parameters $\alpha_j > 0$ is a distribution for a probability vector $(\theta_0, \dots, \theta_d)$ such that $(\theta_1, \dots, \theta_d)$ has a joint density

$$c(\alpha_0, \dots, \alpha_d) \prod_{j=0}^{d} \theta_j^{\alpha_j - 1}, \tag{13.16}$$

and $\theta_0 = 1 - \sum_{j=1}^d \theta_j$, where $c(\alpha_0, \dots, \alpha_d) = \frac{\Gamma(\alpha_0 + \cdots + \alpha_d)}{\prod_{j=0}^d \Gamma(\alpha_j)}$ is the normalizing constant.

First, we give the formal setting as follows:

- Fix a finite alphabet $\mathcal{X}$ of size $|\mathcal{X}| = d+1$, which we will enumerate as $\mathcal{X} = \{0, \dots, d\}$.

[2] This, however, is not true in general. See Exercise **II.21** for an example where $\mathcal{R}_n^* < \infty$ but $\mathcal{R}_n^{**} = \infty$.

- As in Example 2.6, let $\Theta = \{(\theta_1, \ldots, \theta_d) \colon \sum_{j=1}^d \theta_j \le 1, \theta_j \ge 0\}$ parameterize the collection of all probability distributions on $\mathcal{X}$. Note that Θ is a d-dimensional simplex. We will also define

$$\theta_0 \triangleq 1 - \sum_{j=1}^d \theta_j.$$

- The source class is

$$P_{X^n|\theta}(x^n|\theta) \triangleq \prod_{j=1}^n \theta_{x_j} = \exp\left\{-n \sum_{a \in \mathcal{X}} \theta_a \log \frac{1}{\hat{P}_{x^n}(a)}\right\},$$

where as before $\hat{P}_{x^n}$ is the empirical distribution of x^n, see (13.5).

In order to find the (near-) optimal Q_{X^n}, we need to guess an (almost) optimal prior $\pi \in \mathcal{P}(\Theta)$ in (13.10) and take Q_{X^n} to be the mixture of $P_{X^n|\theta}$'s. We will search for π in the class of smooth densities on Θ and set

$$Q_{X^n}(x^n) \triangleq \int_\Theta P_{X^n|\theta}(x^n|\theta')\pi(\theta')d\theta'. \tag{13.17}$$

Before proceeding further, we recall the *Laplace method* of approximating exponential integrals. Suppose that $f(\theta)$ has a unique minimum at the interior point $\hat{\theta}$ of Θ and that Hessian Hess f is uniformly lower-bounded by a multiple of identity (in particular, $f(\theta)$ is strongly convex). Then taking the Taylor expansions of π and f we get

$$\int_\Theta \pi(\theta)e^{-nf(\theta)}d\theta = \int (\pi(\hat{\theta}) + O(\|t\|))e^{-n(f(\hat{\theta}) - \frac{1}{2}t^\top \text{Hess } f(\hat{\theta})t + o(\|t\|^2))}dt \tag{13.18}$$

$$= \pi(\hat{\theta})e^{-nf(\hat{\theta})} \int_{\mathbb{R}^d} e^{-x^\top \text{Hess } f(\hat{\theta})x} \frac{dx}{\sqrt{n^d}}(1 + O(n^{-1/2})) \tag{13.19}$$

$$= \pi(\hat{\theta})e^{-nf(\hat{\theta})} \left(\frac{2\pi}{n}\right)^{d/2} \frac{1}{\sqrt{\det \text{Hess } f(\hat{\theta})}}(1 + O(n^{-1/2})), \tag{13.20}$$

where in the last step we computed the Gaussian integral.

Next, we notice that

$$P_{X^n|\theta}(x^n|\theta') = \exp\{-n(D(\hat{P}_{x^n}\|P_{X|\theta=\theta'}) + H(\hat{P}_{x^n}))\},$$

and therefore, denoting

$$\hat{\theta}(x^n) \triangleq \hat{P}_{x^n},$$

we get from applying (13.20) to (13.17)

$$\log Q_{X^n}(x^n) = -nH(\hat{\theta}) + \frac{d}{2}\log\frac{2\pi}{n\log e} + \log\frac{P_\theta(\hat{\theta})}{\sqrt{\det J_F(\hat{\theta})}} + O(n^{-\frac{1}{2}}),$$

where we used the fact that $\text{Hess}_{\theta'} D(\hat{P}\|P_{X|\theta=\theta'})|_{\theta'=\hat{\theta}} = \frac{1}{\log e}J_F(\hat{\theta})$ with J_F being the Fisher information matrix introduced previously in (2.34). From here, using the fact that under $X^n \sim P_{X^n|\theta=\theta'}$ the random variable $\hat{\theta} = \theta' + O(n^{-1/2})$ we get by approximating $J_F(\hat{\theta})$ and $P_\theta(\hat{\theta})$

$$D(P_{X^n|\theta=\theta'}\|Q_{X^n}) = n(\mathbb{E}[H(\hat{\theta})] - H(X|\theta=\theta')) + \frac{d}{2}\log n \\ - \log\frac{P_\theta(\theta')}{\sqrt{\det J_F(\theta')}} + C + O(n^{-\frac{1}{2}}), \tag{13.21}$$

where C is some constant (independent of the prior P_θ or θ'). The first term is handled by the next result, refining Corollary 7.18.

Lemma 13.2. *Let $X^n \overset{\text{iid}}{\sim} P$ on a finite alphabet $\mathcal{X}$ such that $P(x) > 0$ for all $x \in \mathcal{X}$. Let $\hat{P} = \hat{P}_{X^n}$ be the empirical distribution of X^n, then*

$$\mathbb{E}[D(\hat{P}\|P)] = \frac{|\mathcal{X}|-1}{2n}\log e + o\left(\frac{1}{n}\right).$$

In fact, $nD(\hat{P}\|P) \to \frac{\log e}{2}\chi^2(|\mathcal{X}|-1)$ in distribution.

Proof. By the central limit theorem, $\sqrt{n}(\hat{P} - P)$ converges in distribution to $\mathcal{N}(0, \Sigma)$, where $\Sigma = \mathrm{diag}(P) - PP^\top$, where P is an $|\mathcal{X}|$-by-1 column vector. Thus, computing the second-order Taylor expansion of $D(\cdot\|P)$, see (2.34) and (2.37), we get the result. (To interchange the limit and the expectation, more formally we need to condition on the event $\hat{P}_n(x) \in (\epsilon, 1-\epsilon)$ for all $x \in \mathcal{X}$ to make the integrand function bounded. We leave these technical details as an exercise for the reader.) □

Continuing (13.21) we get in the end

$$D(P_{X^n|\theta=\theta'}\|Q_{X^n}) = \frac{d}{2}\log n - \log\frac{\pi(\theta')}{\sqrt{\det J_F(\theta')}} + \text{constant} + O(n^{-\frac{1}{2}}) \tag{13.22}$$

under the assumption of smoothness of prior π and that θ' is not on the boundary of Θ. Consequently, we can see that in order for the prior π to be the saddle-point solution, we should have

$$\pi(\theta') \propto \sqrt{\det J_F(\theta')},$$

provided that the right-hand side is integrable. The prior proportional to the square-root of the determinant of the Fisher information matrix is known as the *Jeffreys prior*. In our case, using the explicit expression for Fisher information (2.39), the Jeffreys prior π^* is found to be Dirichlet$(1/2, 1/2, \ldots, 1/2)$, with density

$$\pi^*(\theta) = c_d\frac{1}{\sqrt{\prod_{j=0}^d \theta_j}}, \tag{13.23}$$

where $c_d = \frac{\Gamma((d+1)/2)}{\Gamma(1/2)^{d+1}}$ is the normalization constant. The corresponding redundancy is then

$$\mathcal{R}_n^* = \frac{d}{2}\log\frac{n}{2\pi e} - \log\frac{\Gamma((d+1)/2)}{\Gamma(1/2)^{d+1}} + o(1). \tag{13.24}$$

Making the above derivation rigorous is far from trivial and was completed in [461]. (In Exercises **II.15** and **II.16** we analyze the $d = 1$ case and show $\mathcal{R}_n^* = \frac{1}{2}\log n + O(1)$.)

Overall, we see that the Jeffreys prior asymptotically maximizes (within $o(1)$) the $\sup_\pi I(\theta;X^n)$ and for this reason is called *asymptotically maximin solution.* Surprisingly [406], the corresponding mixture Q_{X^n}, that we denote $Q_{X^n}^{(\mathrm{KT})}$ (and study in detail in the next section), however, turns out to not give the asymptotically optimal redundancy. That is, we have for some $c_1 > c_2 > 0$ inequalities

$$\mathcal{R}_n^* + c_1 + o(1) \le \sup_{\theta_0} D\left(P_{X^n|\theta=\theta_0} \| Q_{X^n}^{(\mathrm{KT})}\right) \le \mathcal{R}_n^* + c_2 + o(1).$$

That is, $Q_{X^n}^{(\mathrm{KT})}$ is not asymptotically minimax (but it does achieve optimal redundancy up to $O(1)$ term). However, it turns out that patching the Jeffreys prior near the boundary of the simplex (or using a mixture of Dirichlet distributions) does result in asymptotically minimax universal probability assignments [461].

Extension to General Smooth Parametric Families The fact that the Jeffreys prior $\theta \sim \pi$ maximizes the value of mutual information $I(\theta;X^n)$ for general parametric families was conjectured in [46] in the context of selecting priors in Bayesian inference. This result was proved rigorously in [95, 96]. We briefly summarize the results of the latter.

Let $\{P_\theta : \theta \in \Theta_0\}$ be a smooth parametric family admitting a continuous and bounded Fisher information matrix $J_F(\theta)$ everywhere on the interior of $\Theta_0 \subset \mathbb{R}^d$. Then for every compact Θ contained in the interior of Θ_0 we have

$$\mathcal{R}_n^*(\Theta) = \frac{d}{2}\log\frac{n}{2\pi e} + \log\int_\Theta \sqrt{\det J_F(\theta)}d\theta + o(1). \tag{13.25}$$

Although the Jeffreys prior on Θ achieves (up to $o(1)$) the optimal value of $\sup_\pi I(\theta;X^n)$, to produce an approximate capacity-achieving output distribution Q_{X^n}, one needs to take a mixture with respect to a Jeffreys prior on a slightly larger set $\Theta_\epsilon = \{\theta : d(\theta,\Theta) \le \epsilon\}$ and take $\epsilon \to 0$ slowly with $n \to \infty$. This sequence of Q_{X^n}'s does achieve the optimal redundancy up to $o(1)$.

Remark 13.4. (Laplace's law of succession) In statistics Jeffreys prior is justified as being invariant to smooth reparameterization, as evidenced by (2.35). This is an important aspect of the objective Bayesianism. For example, in answering "will the sun rise tomorrow,"[3] Laplace proposed to estimate the probability by modeling sunrise as an iid Bernoulli process with a uniform prior on $\theta \in [0,1]$. However, this is clearly not very logical, as one may equally well postulate uniformity of $\alpha = \theta^{10}$ or $\beta = \sqrt{\theta}$. Jeffreys prior $\theta \sim \frac{1}{\sqrt{\theta(1-\theta)}}$ is invariant to reparameterization in the sense that if one computed $\sqrt{\det J_F(\alpha)}$ under α parameterization the result would be exactly the pushforward of the $\frac{1}{\sqrt{\theta(1-\theta)}}$ along the map $\theta \mapsto \theta^{10}$. Similarly, maximizers of $I(\theta;X^n)$ and $I(\alpha;X^n)$ are related by the pushforward.

[3] Interested readers should check *Laplace's rule of succession* and the sunrise problem; see [229, Chapter. 18] for a historical and philosophical account.

13.5 Sequential Probability Assignment: Krichevsky–Trofimov

From (13.23) it is not hard to derive the (asymptotically) optimal universal probability assignment Q_{X^n}. For simplicity we consider the Bernoulli case, that is, $d = 1$ and $\theta \in [0, 1]$ is the one-dimensional parameter. Then the Jeffreys prior and the resulting mixture distribution are given by[4]

$$P_\theta^* = \frac{1}{\pi\sqrt{\theta(1-\theta)}}, \tag{13.26}$$

$$Q_{X^n}^{(\mathrm{KT})}(x^n) = \frac{(2t_0-1)!!\cdot(2t_1-1)!!}{2^n n!}, \qquad t_a = \#\{j \le n\colon x_j = a\}. \tag{13.27}$$

This assignment can now be used to create a universal compressor via one of the methods outlined in the beginning of this chapter.

Note that $Q_{X^{n-1}}^{(\mathrm{KT})}$ coincides with the marginalization of $Q_{X^n}^{(\mathrm{KT})}$ to the first $n-1$ coordinates. This property is not specific to the KT distribution and holds for any Q_{X^n} that is given in the form $\int P_\theta(d\theta)P_{X^n|\theta}$ with P_θ not depending on n. What is remarkable, however, is that the conditional distribution $Q_{X_n|X^{n-1}}^{(\mathrm{KT})}$ has a rather elegant form:

$$Q_{X_n|X^{n-1}}^{(\mathrm{KT})}(1|x^{n-1}) = \frac{t_1+\frac{1}{2}}{n}, \qquad t_1 = \#\{j \le n-1\colon x_j = 1\}, \tag{13.28}$$

$$Q_{X_n|X^{n-1}}^{(\mathrm{KT})}(0|x^{n-1}) = \frac{t_0+\frac{1}{2}}{n}, \qquad t_0 = \#\{j \le n-1\colon x_j = 0\}. \tag{13.29}$$

This is the famous "add-1/2" rule of Krichevsky and Trofimov [261]. As mentioned in Section 13.1, this sequential assignment is very convenient for implementing an arithmetic coder.

Let $f_{\mathrm{KT}}\colon\{0,1\}^n \to \{0,1\}^*$ be the encoder assigning length $l(f_{\mathrm{KT}}(x^n)) = \left\lceil \log_2 \frac{1}{Q_{X^n}^{(\mathrm{KT})}(x^n)} \right\rceil$. Now from (13.24) we know that

$$\sup_{0\le\theta\le1}\{\mathbb{E}\left[l(f_{\mathrm{KT}}(S_\theta^n))\right] - nh(\theta)\} = \frac{1}{2}\log n + O(1).$$

Since (13.24) was not shown rigorously in Exercise **II.15** we prove the upper bound of this claim independently.

Remark 13.5. (Laplace "add-1" rule) A slightly less optimal choice of Q_{X^n} results from the Laplace prior (recall that Laplace advises to take P_θ to be uniform on $[0, 1]$). Then, in the case of binary (Bernoulli) alphabet we get

$$Q_{X^n}^{(\mathrm{Lap})} = \frac{1}{\binom{n}{w}(n+1)}, \qquad w = \#\{j\colon x_j = 1\}. \tag{13.30}$$

[4] We remind the reader that $(2a-1)!! = 1\cdot 3\cdots(2a-1)$. The expression for Q_{X^n} is obtained from the identity $\int_0^1 \frac{\theta^a(1-\theta)^b}{\sqrt{\theta(1-\theta)}}d\theta = \pi\frac{(2a-1)!!(2b-1)!!}{2^{a+b}(a+b)!}$ for integer $a, b \ge 0$, which in turn is derived by change of variable $z = \frac{\theta}{1-\theta}$ and using the standard keyhole contour on the complex plane.

The corresponding sequential probability assignment is given by

$$Q^{(\mathrm{Lap})}_{X_n|X^{n-1}}(1|x^{n-1}) = \frac{t_1+1}{n+1}, \quad t_1 = \#\{j \le n-1 \colon x_j = 1\}.$$

We notice two things. First, the distribution (13.30) is *exactly* the same as Fitingof's (13.7). Second, this distribution "almost" attains the optimal first-order term in (13.24). Indeed, when X^n is iid $\mathrm{Ber}(\theta)$ we have for the redundancy:

$$\mathbb{E}\left[\log\frac{1}{Q^{(\mathrm{Lap})}_{X^n}(X^n)}\right] - H(X^n) = \log(n+1) + \mathbb{E}\left[\log\binom{n}{W}\right] - nh(\theta), \quad W \sim \mathrm{Bin}(n,\theta). \tag{13.31}$$

From Stirling's expansion we know that as $n \to \infty$ this redundancy evaluates to $\frac{1}{2}\log n + O(1)$, uniformly in θ over compact subsets of $(0,1)$. However, for $\theta = 0$ or $\theta = 1$ the Laplace redundancy (13.31) clearly equals $\log(n+1)$. Thus, supremum over $\theta \in [0,1]$ is achieved close to endpoints and results in suboptimal redundancy $\log n + O(1)$. The Jeffreys prior (13.26) and the resulting KT compressor fixes the problem at the endpoints.

For a general (non-binary) alphabet the KT distribution is equally simple:

$$Q^{(\mathrm{KT})}_{X_n|X^{n-1}}(a|x^{n-1}) = \frac{t_a + \frac{1}{2}}{n + \frac{|\mathcal{X}|-2}{2}}, \qquad t_a = \#\{j \le n-1 \colon x_j = a\}. \tag{13.32}$$

In summary, to build a universal compressor for a class of all iid sources on a given finite alphabet $\mathcal{X}$ we can do the following:

1. Learner: Set $Q_{X_1} = \mathrm{Unif}[\mathcal{X}]$.
2. Arithmetic encoder: Subdivide $[0,1]$ interval using Q_{X_1}. Receive the first letter x_1 and select partition $Q_{X_1}(x_1)$.
3. Learner: Given x_1 compute $Q_{X_2|X_1=x_1}$ according to (13.32).
4. Arithmetic encoder: Subdivide currently selected partition according to $Q_{X_2|X_1=x_1}$. Receive the next letter x_2 and select partition $Q_{X_2|X_1=x_1}(x_2)$.
5. Etc.

For compression we only use the state of the arithmetic encoder to output corresponding $\{0,1\}$ bits. In the next section, however, we will see that the learner's output $Q_{X_n|X^{n-1}}$ is a very relevant quantity: it is the (regret-optimal) online density estimator under KL loss.

13.6 Online Prediction and Density Estimation

It turns out that the universal compression problem (or, more specifically, a universal probability assignment Q_{X^n} representing a whole class of distributions) automatically solves two other important problems: one in machine learning (online prediction) and the other in statistics (density estimation). For this reason, in information theory the problem of universal compression is also sometimes called "universal prediction." In the next two sections we will briefly explain this fundamental connection. For a full story we recommend an excellent textbook on the topic [86].

Let us fix $\mathcal{X}$ and a collection $P^{(\theta)}_{X^n}, \theta \in \Theta$ of measures on $\mathcal{X}^n$, which we will call a model class Θ. Although a case of continuous $\mathcal{X}$ is even more interesting (as we will explore in Section 32.1), for now we restrict our attention to discrete $\mathcal{X}$ and, thus, we understand $P^{(\theta)}_{X^n}(\cdot)$ as a PMF on $\mathcal{X}^n$. The most immediate choice is to have $P^{(\theta)}_{X^n}(x_1, \ldots, x_n) = \prod_{i=1}^n P^{(\theta)}_{X_1}(x_i)$, which would correspond to a class of iid sources, but we do not make this restriction.

Online Prediction Given a sequence $X_1, \ldots, X_n \sim P^{(\theta^*)}_{X^n}$ the learner sequentially observes $X_1, \ldots, X_{t-1}$ and outputs its prediction $Q_t(\cdot)$ about the next sample X_t. Once the next sample is revealed, the learner experiences a loss measured via log-loss:

$$\ell(Q_t, X_t) \triangleq \log \frac{1}{Q_t(X_t)}.$$

Given a sequence of predictors $\{Q_t\}$ we can assign average *cumulative* loss to it as follows:

$$\ell^{(a)}_n(\{Q_t\}, \theta^*) \triangleq \mathbb{E}_{P^{(\theta^*)}_{X^n}}\left[\sum_{t=1}^n \log \frac{1}{Q_t(X_t|X^{t-1})}\right].$$

The intimate connection between this problem of online prediction and universal compression is revealed by the following almost trivial observation. Notice that distribution $Q_t(\cdot)$ output by the learner at time t is in fact a function of observations $x_1, \ldots, x_{t-1}$. Therefore, we should be writing it more explicitly as $Q_t(\cdot;x^{t-1})$ to emphasize dependence on the (training) data. But then we can also think of $Q_t(x_t;x^{t-1})$ as a Markov kernel $Q_{X_t|X^{t-1}}(x_t|x^{t-1})$ and compute a joint probability distribution

$$Q_{X^n}(x^n) \triangleq \prod_{t=1}^n Q_t(x_t;x^{t-1}). \tag{13.33}$$

Conversely, any Q_{X^n} we can factorize sequentially in the form (13.33) and obtain an online predictor. It turns out that the choice of the optimal Q_{X^n} is precisely the same problem of *universal probability assignment* that is solved by the universal compression in (13.8). But before that we need to explain how to define optimality in the online prediction game.

Since the $\ell^{(a)}_n(\{Q_t\}, \theta^*)$ depends on the value of θ^* that governs the stochastics of the input sequence, our first instinct could be to try to minimize

$$\min_{\{Q_t\}} \sup_{\theta^* \in \Theta} \ell^{(a)}_n(\{Q_t\}, \theta^*).$$

However, this turns out to be a bad way to pose the problem. For example, if one of $P^{(\theta)}_{X^n} = \mathrm{Unif}[\mathcal{X}^n]$ then no predictor can achieve $\ell^{(a)}_n < n \log |\mathcal{X}|$. Furthermore, a trivial predictor that always outputs $Q_t = \mathrm{Unif}[\mathcal{X}]$ achieves this upper bound exactly. Thus in the minimax setting predicting $\mathrm{Unif}[\mathcal{X}]$ turns out to be optimal.

To understand how to work around this issue, let us first recall from Corollary 2.4 that if we have *oracle knowledge* about the true θ^* generating X_j's, then our choice would be to set Q_t to be the factorization of $P^{(\theta^*)}_{X^n}$. This achieves the loss

$$\ell_n^{(a)}(\{P^*_\theta\}, \theta^*) = H(P^{(\theta^*)}_{X^n}).$$

Thus, even if given the oracle information we cannot avoid the $H(P^{(\theta^*)}_{X^n})$ loss (this would also be called Bayes loss in machine learning). Note that for iid model class we have $H(P^{(\theta^*)}_{X^n}) = nH(P^{(\theta^*)}_{X_1})$ and the oracle loss is of order n. Consequently, the quality of the learning algorithm should be measured by the amount of excess of loss above the oracle loss. This quantity is known as *average regret* and defined as

$$\mathrm{AvgReg}_n(\{Q_t\}, \theta^*) \triangleq \mathbb{E}_{P^{(\theta^*)}_{X^n}}\left[\sum_{t=1}^n \log \frac{1}{Q_t(X_t|X^{t-1})}\right] - H(P^{(\theta^*)}_{X^n}).$$

Hence to design an optimal algorithm we want to minimize the worst regret, or in other words to solve the minimax problem

$$\mathrm{AvgReg}^*_n(\Theta) \triangleq \inf_{\{Q_t\}} \sup_{\theta^* \in \Theta} \mathrm{AvgReg}_n(\{Q_t\}, \theta^*). \tag{13.34}$$

This turns out to be completely equivalent to the universal compression problem, as we state next.

Theorem 13.3. *Recall the definition of compression redundancy $\mathcal{R}^*_n$ in (13.8). Then we have*

$$\mathrm{AvgReg}^*_n(\Theta) = \mathcal{R}^*_n(\Theta) \triangleq \min_{Q_{X^n}} \sup_{\theta^* \in \Theta} D(P^{(\theta^*)}_{X^n} \| Q_{X^n}),$$

where the minimum on the RHS is achieved and at a unique distribution $Q^*_{X^n}$. The optimal predictor is given by setting $Q_t(\cdot) = Q^*_{X_t|X^{t-1}=x^{t-1}}(\cdot)$. Furthermore, let $\theta \in \Theta$ have a prior distribution $\pi \in \mathcal{P}(\Theta)$. Then

$$\mathrm{AvgReg}^*_n(\Theta) = \sup_\pi I(\theta; X^n).$$

If there exists a maximizer π^* of the right-hand side maximization then the optimal estimator is found by factorizing $Q^*_{X^n} = \int \pi^*(d\theta) \prod_{i=1}^n P_\theta(x_i)$.

Proof. There is almost nothing to prove. We only need to rewrite the definition of average regret in terms of Q_{X^n} as follows:

$$\mathrm{AvgReg}_n(\{Q_t\}, \theta^*) = \mathbb{E}_{P^{(\theta^*)}_{X^n}}\left[\log \frac{P^{(\theta^*)}_{X^n}}{Q_{X^n}}\right] = D(P^{(\theta^*)}_{X^n} \| Q_{X^n}).$$

The rest of the claims follow from Theorem 5.9 (recall that $I(\theta; X^n) \le n \log |\mathcal{X}| < \infty$) and Theorem 5.4. □

As an application of this result, we see that the Krichevsky–Trofimov estimator achieves for any iid string $X^n \overset{\text{iid}}{\sim} P$ a log-loss

$$\sum_{t=1}^{n} \mathbb{E}\left[\log \frac{1}{Q^{(\mathrm{KT})}_{X_t|X^{t-1}}(X_t|X^{t-1})}\right] \le nH(P) + \frac{|\mathcal{X}|-1}{2}\log n + c_{\mathcal{X}},$$

where $c_{\mathcal{X}} < \infty$ is a constant independent of P or n. This excess above $nH(P)$ is optimal among all possible online estimators except possibly for a constant $c_{\mathcal{X}}$.

The problem we discussed may appear at first to be somewhat contrived, especially to someone who has been used to supervised learning/prediction tasks. Indeed, our prediction problem does not have any features to predict from! (We note that the featureless prediction is at the heart of developing large language models that are trained to predict sequences of words by minimizing log-loss [321]. Note that in this case the problem of loss minimization, as opposed to regret minimization, is meaningful due to non-iid nature of the sequence.) We can introduce a supervised learning version, where the prediction task is to estimate an unknown (label or quantity) Y_t given a correlated feature vector X_t. There is an analog of Theorem 13.3 for that case as well – see Exercises **II.20** and **II.22**.

Batch Regret In machine learning what we have defined above is known as cumulative (or online) regret, because we insisted that the estimator produce some prediction at every time step t. However, a more common setting is that of batch prediction, where the first $n-1$ observations are available as the training data and we only assess the loss on the new unseen sample (test data). This is called *batch loss* and the corresponding minimax regret is

$$\mathrm{BatchReg}^*_n(\Theta) \triangleq \inf_{Q_n(\cdot;x^{n-1})} \sup_{\theta^*\in\Theta} \underbrace{\mathbb{E}_{P^{(\theta^*)}_{X^n}}\left[\log\frac{1}{Q_n(X_n;X^{n-1})}\right] - H(P^{(\theta^*)}_{X_n|X^{n-1}})}_{D\left(P^{(\theta^*)}_{X_n|X^{n-1}} \,\middle\|\, Q_n(\cdot;X^{n-1})\right)}. \quad (13.35)$$

In other words, this is the optimal KL loss of predicting the next symbol by estimating its conditional distribution given the past, a central task in language models such as GPT [321]. Similar to Theorem 13.3 we can give a max-information formula for batch regret (Exercise **II.19**). However, it turns out that there is also a connection to universal compression. Indeed, we have the following inequalities:

$$\mathrm{AvgReg}^*_n(\Theta) - \mathrm{AvgReg}^*_{n-1}(\Theta) \le \mathrm{BatchReg}^*_n(\Theta) \le \frac{1}{n}\mathrm{AvgReg}^*_n(\Theta), \quad (13.36)$$

where the upper bound is only guaranteed to hold for iid models.[5] The inequality (13.36) is known as *online-to-batch conversion* or *estimation-compression inequality* [159, 241]; see Lemma 32.3 and Proposition 32.7 for a justification. The

[5] For stationary mth-order Markov models, the upper bound in (13.36) holds with $n-m$ in the denominator [213, Lemma 6].

estimator that achieves the above upper bound is very simple: it takes a probability assignment $Q^*_{X^n}$ and sets its predictor as

$$Q_n(x_n;x^{n-1}) \triangleq \frac{1}{n}\sum_{t=1}^{n} Q_{X_t|X^{t-1}}(x_n|x^{t-1}). \qquad (13.37)$$

However, unlike the cumulative regret, minimizers of the batch regret are distinct from those in universal compression. For example, for the model class of all iid distributions over k symbols, we know that (asymptotically) the "add-1/2" estimator of Krichevsky–Trofimov is optimal. However, for the batch loss it is not so (see the note at the end of Exercise **VI.10**). We also note that optimal batch regret in this case is $O(\frac{k-1}{n})$, but the online-to-batch rule only yields $O(\frac{(k-1)\log n}{n})$. On the other hand, for first-order Markov chains with $k \geq 3$ states, the online-to-batch upper bound turns out to be order optimal, as we have $\mathrm{BatchReg}^*_n \asymp \frac{1}{n}\mathrm{AvgReg}^*_n \asymp \frac{k^2}{n}\log\frac{n}{k^2}$ provided that $n \gg k^2$ [213]; however, proving this result, especially the lower bound, requires arguments native to Markov chains.

Density Estimation Consider now the following problem. Given a collection of (single-letter) distributions $P_X^{(\theta)}$ on $\mathcal{X}$ and $X_1,\ldots,X_{n-1} \overset{\text{iid}}{\sim} P_X^{(\theta)}$ we want to produce a density estimate $\hat{P}$ which minimizes the worst-case error as measured by KL divergence, that is, we seek to minimize

$$\sup_{\theta^*\in\Theta} \mathbb{E}_{X^{n-1}\overset{\text{iid}}{\sim} P_X^{(\theta^*)}}[D(\hat{P}\|P_X^{(\theta^*)})].$$

To connect to the previous discussion, we only need to notice that $\hat{P}(\cdot)$ can be interpreted as Q_n in the batch regret problem and we have an exact equality

$$\inf_{\hat{P}}\sup_{\theta^*\in\Theta} \mathbb{E}_{X^{n-1}\overset{\text{iid}}{\sim} P_X^{(\theta^*)}}[D(\hat{P}\|P_X^{(\theta^*)})] = \mathrm{BatchReg}^*_n(\Theta) \leq \frac{1}{n}\sup_{\pi} I(\theta;X^n).$$

Thus, we bound the minimax (KL divergence) density estimation rate by the capacity of a certain channel. The estimator achieving this bound is improper (i.e. $\hat{P} \neq P_X^{(\theta)}$ for any θ) and is given by (13.37). This is the basis of the Yang–Barron approach to density estimation, see Section 32.1 for more.

13.7 Individual Sequence and Worst-Case Regret

In the previous section we explained how the learner can predict stochastically generated strings X^n. Though very standard, this approach suffers from a common criticism: What if our model class Θ for the stochasticity of X^n is incorrect and the real data is not generated by any process in the class Θ? In this section, we show a somewhat surprising workaround: it turns out there can be a theory of predicting *any possible sequences* x^n even non-random and adversarially chosen! This exciting area is known in information theory as the individual sequence approach and we describe it next.

Consider the following problem: a sequence x^n is observed sequentially and our goal is to predict (by making a soft, or probabilistic, prediction) the next symbol given the past observations. The experiment proceeds as follows:

1 A string $x^n \in \mathcal{X}^n$ is selected by Nature.
2 Having observed $x_1, \ldots, x_{t-1}$ we are tasked to output a probability distribution $Q_t(\cdot|x^{t-1})$ on $\mathcal{X}$.
3 After that Nature reveals the next sample x_t and our loss for the tth prediction is evaluated via the log-loss:

$$\log \frac{1}{Q_t(x_t|x^{t-1})}.$$

The main objective is to find a sequence of predictors $\{Q_t\}$ that minimizes the *cumulative* loss:

$$\ell(\{Q_t\}, x^n) \triangleq \sum_{t=1}^{n} \log \frac{1}{Q_t(x_t|x^{t-1})}.$$

Consider first the naive goal of minimizing the worst-case loss:

$$\min_{\{Q_t\}_{t=1}^n} \max_{x^n} \ell(\{Q_t\}, x^n).$$

This is clearly hopeless. Indeed, at any step t the distribution Q_t must have at least one atom with weight at most $\frac{1}{|\mathcal{X}|}$, and hence for any predictor

$$\max_{x^n} \ell(\{Q_t\}, x^n) \geq n \log |\mathcal{X}|,$$

which is clearly achieved iff $Q_t(\cdot) \equiv \frac{1}{|\mathcal{X}|}$, that is, if the predictor simply makes uniform random guesses. This triviality is not surprising: In the absence of whatsoever prior information on x^n it is impossible to predict anything.

The exciting idea, originated by Feder, Merhav, and Gutman, see [161, 304], is to replace loss with *regret*, that is, the gap to the best possible *static oracle*. More precisely, suppose a non-causal oracle can examine the entire string x^n and output a constant $Q_t \equiv Q$. From the non-negativity of divergence this non-causal oracle achieves

$$\ell_{\text{oracle}}(x^n) = \min_{Q} \sum_{t=1}^{n} \log \frac{1}{Q(x_t)} = nH(\hat{P}_{x^n}).$$

Can a causal (but time-varying) predictor come close to this performance? In other words, we define *regret* of a sequential predictor as the excess risk over the static oracle,

$$\text{reg}(\{Q_t\}, x^n) \triangleq \ell(\{Q_t\}, x^n) - nH(\hat{P}_{x^n}),$$

and ask to minimize the worst-case regret:

$$\text{Reg}_n^* \triangleq \min_{\{Q_t\}} \max_{x^n} \text{reg}(\{Q_t\}, x^n). \tag{13.38}$$

Excitingly, non-trivial predictors emerge as solutions to the above problem, which furthermore do not rely on any assumptions on x^n.

We next consider the case of $\mathcal{X} = \{0,1\}$ for simplicity. To solve (13.38), first notice that designing a sequence $\{Q_t(\cdot|x^{t-1})\}$ is equivalent to defining one joint distribution Q_{X^n} and then factorizing the latter as $Q_{X^n}(x^n) = \prod_t Q_t(x_t|x^{t-1})$. Then the problem (13.38) becomes simply

$$\mathrm{Reg}_n^* = \min_{Q_{X^n}} \max_{x^n} \log \frac{1}{Q_{X^n}(x^n)} - nH(\hat{P}_{x^n}).$$

First, we notice that generally we have that optimal Q_{X^n} is the Shtarkov distribution (13.12), which implies that the regret coincides with the log of the Shtarkov sum (13.13). In the iid case we are considering, from (13.14) we get

$$\mathrm{Reg}_n^* = \log \sum_{x^n} \max_Q \prod_{i=1}^n Q(x_i) = \log \sum_{x^n} \exp\{-nH(\hat{P}_{x^n})\}.$$

This expression is, however, frequently not very convenient to analyze, so instead we consider upper and lower bounds. We may lower-bound the max over x^n with the average over the $X^n \sim \mathrm{Ber}(\theta)^n$ and obtain (also applying Lemma 13.2):

$$\mathrm{Reg}_n^* \geq \mathcal{R}_n^* + \frac{|\mathcal{X}|-1}{2} \log e + o(1),$$

where $\mathcal{R}_n^*$ is the universal compression redundancy defined in (13.8), whose asymptotics we derived in (13.24).

On the other hand, taking $Q_{X^n}^{(\mathrm{KT})}$ from Krichevsky–Trofimov (13.27) we find after some algebra and Stirling's expansion:

$$\max_{x^n} \log \frac{1}{Q_{X^n}^{(\mathrm{KT})}(x^n)} - nH(\hat{P}_{x^n}) = \frac{1}{2} \log n + O(1).$$

In all, we conclude that,

$$\mathrm{Reg}_n^* = \mathcal{R}_n^* + O(1) = \frac{|\mathcal{X}|-1}{2} \log n + O(1),$$

and remarkably, the per-letter regret $\frac{1}{n}\mathrm{Reg}_n^*$ converges to zero. That is, *there exists a causal predictor that can predict (under the log-loss) almost as well as any constant one, even if the latter is adapted to a particular sequence x^n non-causally.*

Explicit (asymptotically optimal) sequential prediction rules are given by the Krichevsky–Trofimov "add-1/2" rules (13.29). We note that the resulting rules are also independent of n ("horizon-free"). This is a very desirable property not shared by the optimal sequential predictors derived from factorizing the Shtarkov's distribution (13.12).

General Parametric Families The general definition of (cumulative) individual sequence (or worst-case) regret for a model class $\{P_{X^n|\theta=\theta_0} : \theta_0 \in \Theta\}$ is given by

$$\mathrm{Reg}_n^*(\Theta) = \min_{Q_{X^n}} \sup_{x^n} \log \frac{1}{Q_{X^n}(x^n)} - \inf_{\theta_0 \in \Theta} \log \frac{1}{P_{X^n|\theta=\theta_0}(x^n)}.$$

This regret can be *interpreted* as worst-case loss of a given estimator compared to the best possible one from a class $P_{X^n|\theta}$, when the latter is selected optimally for each sequence. In this sense, regret gives a uniform (in x^n) bound on the performance of an algorithm against a class.

It turns out that similarly to (13.25) the individual sequence redundancy for general d-parametric families (under smoothness conditions) can be shown to satisfy [363]:

$$\mathrm{Reg}_n^*(\Theta) = \mathcal{R}_n^*(\Theta) + \frac{d}{2}\log e + o(1) = \frac{d}{2}\log\frac{n}{2\pi} + \log\int_{\Theta}\sqrt{\det J_F(\theta)}d\theta + o(1).$$

In machine learning terms, we say that $\mathcal{R}_n^*(\Theta)$ in (13.8) is a cumulative sequential prediction regret under the well-specified setting (i.e. data X^n is generated by a distribution inside the model class Θ), while here $\mathrm{Reg}_n^*(\Theta)$ corresponds to a fully mis-specified setting (i.e. data is completely arbitrary). There are also interesting settings in between these extremes, for example, when data is iid but not from a model class Θ, see [162].

13.8 Lempel–Ziv Compressor

So given a class of sources $\{P_{X^n|\theta} : \theta \in \Theta\}$ we have shown how to produce an asymptotically optimal compressor by using Jeffreys prior. In the case of a class of iid processes, the resulting sequential probability of Krichevsky–Trofimov, see (13.32), had a very simple algorithmic description. When extended to more general classes (such as rth-order Markov chains), however, the sequential probability rules become rather complex. The Lempel–Ziv approach was to forgo the path "design Q_{X^n}, convert to $Q_{X_t|X^{t-1}}$, extract compressor" and attempt to directly construct a reasonable sequential compressor or, equivalently, derive an algorithmically simple sequential estimator $Q_{X_t|X^{t-1}}$. The corresponding joint distribution Q_{X^n} is hard to imagine, and the achieved redundancy is not easy to derive, but the algorithm becomes very transparent.

In principle, the problem is rather straightforward: As we observe a stationary process, we may estimate with better and better precision the conditional probability $\hat{P}_{X_n|X_{n-r}^{n-1}}$ and then use it as the basis for arithmetic coding. As long as $\hat{P}$ converges to the actual conditional probability, we will attain the entropy rate of $H(X_n|X_{n-r}^{n-1})$. Note that Krichevsky–Trofimov assignment (13.29) is clearly learning the distribution too: as n grows, the estimator $Q_{X_n|X^{n-1}}$ converges to the true P_X (provided that the sequence is iid). So in some sense the converse is also true: *any good universal compression scheme is inherently learning the true distribution.*

The main drawback of the learn-then-compress approach is the following. Once we extend the class of sources to include those with memory, we invariably are led to the problem of learning the joint distribution $P_{X_0^{r-1}}$ of r-blocks. However, the sample size required to obtain a good estimate of $P_{X_0^{r-1}}$ is exponential in r. Thus learning may proceed rather slowly. The Lempel–Ziv family of algorithms works around this in an ingeniously elegant way:

- First, estimating probabilities of rare substrings takes longest, but it is also the least useful, as these substrings almost never appear at the input.
- Second, *and the most crucial*, point is that an unbiased estimate of $P_{X^r}(x^r)$ is given by the reciprocal of the time since the last observation of x^r in the data stream.
- Third, there is a prefix code[6] mapping any integer n to a binary string of length roughly $\log_2 n$:

$$f_{int}\colon \mathbb{Z}_+ \to \{0,1\}^+, \qquad \ell(f_{int}(n)) = \log_2 n + O(\log\log n). \tag{13.39}$$

  Thus, by encoding the pointer to the last observation of x^r via such a code we get a string of length roughly $\log P_{X^r}(x^r)$ automatically.

There are a number of variations of these basic ideas, so we will only attempt to give a rough explanation of why it works, without analyzing any particular algorithm.

We proceed to formal details. First, we need to establish Kac's lemma.

Lemma 13.4. (Kac) *Consider a finite-alphabet stationary ergodic process* $\dots, X_{-1}, X_0, X_1, \dots$. *Let* $L = \inf\{t > 0\colon X_{-t} = X_0\}$ *be the last appearance of symbol* X_0 *in the sequence* $X_{-\infty}^{-1}$. *Then for any* u *such that* $\mathbb{P}[X_0 = u] > 0$ *we have*

$$\mathbb{E}[L|X_0 = u] = \frac{1}{\mathbb{P}[X_0 = u]}.$$

In particular, mean recurrence time $\mathbb{E}[L] = |\mathrm{supp}(P_X)|$.

Proof. Note that from stationarity the probability

$$\mathbb{P}[\exists t \geq k\colon X_t = u]$$

does not depend on $k \in \mathbb{Z}$. Thus by continuity of probability we can take $k = -\infty$ to get

$$\mathbb{P}[\exists t \geq 0\colon X_t = u] = \mathbb{P}[\exists t \in \mathbb{Z}\colon X_t = u].$$

However, the last event is shift-invariant and thus must have probability zero or one by ergodic assumption. But since $\mathbb{P}[X_0 = u] > 0$ it cannot be zero. So we conclude

$$\mathbb{P}[\exists t \geq 0\colon X_t = u] = 1. \tag{13.40}$$

Next, we have

[6] For this just notice that $\sum_{k\geq 1} 2^{-\log_2 k - 2\log_2 \log(k+1)} < \infty$ and use Kraft's inequality. See also Exercise **II.18**.

$$\mathbb{E}[L|X_0 = u] = \sum_{t\geq 1} \mathbb{P}[L \geq t | X_0 = u] \tag{13.41}$$

$$= \frac{1}{\mathbb{P}[X_0 = u]} \sum_{t\geq 1} \mathbb{P}[L \geq t, X_0 = u] \tag{13.42}$$

$$= \frac{1}{\mathbb{P}[X_0 = u]} \sum_{t\geq 1} \mathbb{P}[X_{-t+1} \neq u, \ldots, X_{-1} \neq u, X_0 = u] \tag{13.43}$$

$$= \frac{1}{\mathbb{P}[X_0 = u]} \sum_{t\geq 1} \mathbb{P}[X_0 \neq u, \ldots, X_{t-2} \neq u, X_{t-1} = u] \tag{13.44}$$

$$= \frac{1}{\mathbb{P}[X_0 = u]} \mathbb{P}[\exists t \geq 0 \colon X_t = u] \tag{13.45}$$

$$= \frac{1}{\mathbb{P}[X_0 = u]}, \tag{13.46}$$

where (13.41) is the standard expression for the expectation of a $\mathbb{Z}_+$-valued random variable, (13.44) is from stationarity, (13.45) is because the events corresponding to different t are disjoint, and (13.46) is from (13.40). □

The following result serves to explain the basic principle behind the operation of Lempel–Ziv methods.

Theorem 13.5. *Consider a finite-alphabet stationary ergodic process* $\ldots, X_{-1}, X_0, X_1, \ldots$ *with entropy rate* $\mathcal{H}$. *Suppose that* $X_{-\infty}^{-1}$ *is known to the decoder. Then there exists a sequence of prefix codes* $f_n(x_0^{n-1}, x_{-\infty}^{-1})$ *with expected length*

$$\frac{1}{n}\mathbb{E}[\ell(f_n(X_0^{n-1}, X_{\infty}^{-1}))] \to \mathcal{H}.$$

Proof. Let L_n be the last occurrence of the block x_0^{n-1} in the string $X_{-\infty}^{-1}$ (recall that the latter is known to the decoder), namely

$$L_n = \inf\{t > 0 \colon X_{-t}^{-t+n-1} = x_0^{n-1}\}.$$

Then, by Kac's lemma applied to the process $Y_t^{(n)} = X_t^{t+n-1}$ we have

$$\mathbb{E}[L_n | X_0^{n-1} = x_0^{n-1}] = \frac{1}{\mathbb{P}[X_0^{n-1} = x_0^{n-1}]}.$$

We now encode L_n using the code (13.39). Note that there is a crucial subtlety: even if $L_n < n$ and thus $[-t, -t+n-1]$ and $[0, n-1]$ overlap, the substring x_0^{n-1} can be decoded from the knowledge of L_n.

We have, by applying Jensen's inequality twice and noticing that $\frac{1}{n}H(X_0^{n-1}) \searrow \mathcal{H}$ and $\frac{1}{n}\log H(X_0^{n-1}) \to 0$, that

$$\frac{1}{n}\mathbb{E}[\ell(f_{int}(L_n))] \leq \frac{1}{n}\mathbb{E}\left[\log \frac{1}{P_{X_0^{n-1}}(X_0^{n-1})}\right] + o(1) \to \mathcal{H}.$$

From Kraft's inequality we know that for any prefix code we must have

$$\frac{1}{n}\mathbb{E}[\ell(f_{int}(L_n))] \geq \frac{1}{n}H(X_0^{n-1}|X_{-\infty}^{-1}) \to \mathcal{H}. \qquad \square$$

The result shown above demonstrates that the LZ algorithm has asymptotically optimal compression rate for every stationary ergodic process. Recall, however, that the previously discussed compressors also enjoyed non-stochastic (individual sequence) guarantees. For example, we have seen in Section 13.7 that the Krichevsky–Trofimov compressor achieves on *every input sequence* a compression ratio that is at most $O(\frac{\log n}{n})$ worse than the arithmetic encoder built with the best possible (for this sequence!) static probability assignment. It turns out that the LZ algorithm is also special from this point of view. In [332] (see also [160, Theorem 4]) it was shown that the LZ compression rate on *every input sequence* is better than that achieved by any finite state machine (FSM) up to correction terms $O\left(\frac{\log\log n}{\log n}\right)$. Consequently, investing via LZ achieves capital growth that is competitive against any possible FSM investor [160].

Altogether we can see that LZ compression enjoys certain optimality guarantees in both the stochastic and individual sequence senses.

Exercises for Part II

II.1 (Exact value of minimal compression length) Suppose $X \in \mathbb{N}$ and $P_X(1) \geq P_X(2) \geq \cdots$. Show that the optimal compressor f^*'s length satisfies

$$\mathbb{E}[l(f^*(X))] = \sum_{k=1}^{\infty} \mathbb{P}[X \geq 2^k].$$

II.2 (Mixed source) Consider a finite collection $\Pi = \{P_1, \ldots, P_m\}$ of distributions on $\mathcal{S}$. Mixed source $\{S_j\}$ is generated by sampling P_i from Π with probability π_i and then generating $S_j \overset{\text{iid}}{\sim} P_i$. Show that $\frac{1}{n} \log \frac{1}{P_{S^n}(S^n)}$ converges in distribution. Compute $\lim_{n\to\infty} \epsilon^*(S^n, nR)$.

II.3 Recall that an entropy rate of a process $\{X_j : j = 1, \ldots\}$ is defined as follows (provided the limit exists):

$$\mathcal{H} = \lim_{n\to\infty} \frac{1}{n} H(X^n).$$

Consider a four-state Markov chain with transition probability matrix

$$\begin{bmatrix} 0.89 & 0.11 & 0 & 0 \\ 0.11 & 0.89 & 0 & 0 \\ 0 & 0 & 0.11 & 0.89 \\ 0 & 0 & 0.89 & 0.11 \end{bmatrix}.$$

The distribution of the initial state is $[p, 0, 0, 1-p]$.

(a) Does the entropy rate of such a Markov chain exist? If it does, find it.

(b) Describe the asymptotic behavior of the optimum variable-length rate $\frac{1}{n}\ell(f^*(X_1, \ldots, X_n))$. Consider convergence in probability and in distribution.

(c) Repeat with transition matrix:

$$\begin{bmatrix} 0.89 & 0.11 & 0 & 0 \\ 0.11 & 0.89 & 0 & 0 \\ 0 & 0 & 0.5 & 0.5 \\ 0 & 0 & 0.5 & 0.5 \end{bmatrix}.$$

II.4 Consider a three-state Markov chain $S_1, S_2, \ldots$ with the following transition probability matrix:

$$\mathbf{P} = \begin{bmatrix} \frac{1}{2} & \frac{1}{4} & \frac{1}{4} \\ 0 & \frac{1}{2} & \frac{1}{2} \\ 1 & 0 & 0 \end{bmatrix}.$$

Compute the limit of $\frac{1}{n}\mathbb{E}[l(f^*(S^n))]$ when $n \to \infty$. Does your answer depend on the distribution of the initial state S_1?

II.5 (a) Let X take values on a finite alphabet $\mathcal{X}$. Prove that

$$\epsilon^*(X,k) \geq \frac{H(X) - k - 1}{\log(|\mathcal{X}| - 1)}.$$

(b) Deduce the following converse result: For a stationary process $\{S_k : k \geq 1\}$ on a finite alphabet $\mathcal{S}$,

$$\liminf_{n\to\infty} \epsilon^*(S^n, nR) \geq \frac{\mathcal{H} - R}{\log |\mathcal{S}|},$$

where $\mathcal{H} = \lim_{n\to\infty} \frac{H(S^n)}{n}$ is the entropy rate of the process.

II.6 *Run-length encoding* is a popular variable-length lossless compressor used in fax machines, image compression, etc. Consider compression of S^n, an iid Ber(δ) source with very small $\delta = \frac{1}{128}$ using run-length encoding: A chunk of consecutive $r \leq 255$ zeros (respectively ones) is encoded into a zero (respectively one) followed by an eight-bit binary encoding of r (if there are > 255 consecutive zeros then two or more nine-bit blocks will be output). Compute the average achieved compression rate

$$\lim_{n\to\infty} \frac{1}{n}\mathbb{E}[\ell(f(S^n)].$$

How does it compare with the optimal lossless compressor?

(Hint: Compute the expected number of nine-bit blocks output per chunk of consecutive zeros/ones; normalize by the expected length of the chunk.)

II.7 Draw n random points independently and uniformly from the vertices of the following square.

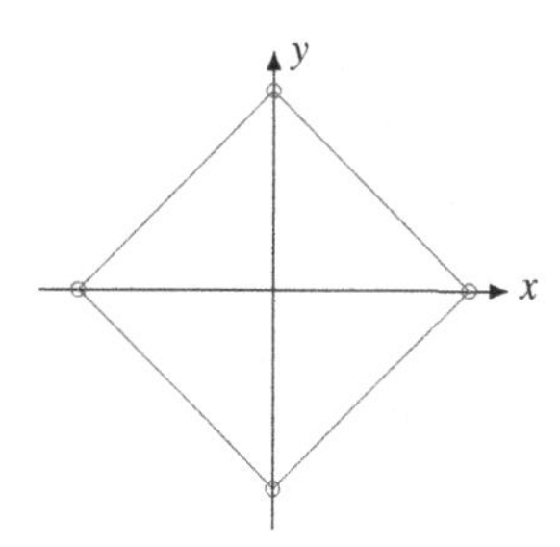

Denote the coordinates by $(X_1, Y_1), \ldots, (X_n, Y_n)$. Suppose Alice only observes X^n and Bob only observes Y^n. They want to encode their observation using R_X and R_Y bits per symbol respectively and send the codewords to Charlie who will be able to reconstruct the sequence of pairs.

(a) Find the optimal rate region for (R_X, R_Y).

(b) What if the square is rotated by 45°?

II.8 Consider a particle walking randomly on the graph of Exercise **II.7** (each edge is taken with equal probability; the particle does not stay in the same node). Alice observes the X coordinate and Bob observes the Y coordinate. How many bits per step (in the long run) does Bob need to send to Alice so that Alice will be able to reconstruct the particle's trajectory with vanishing probability of error? (Hint: You need to extend a certain theorem from Chapter 11 to the case of an ergodic Markov chain.)

II.9 Recall from Theorem 11.13 the upper bound on the probability of error for the Slepian–Wolf compression to k bits:

$$\epsilon^*_{\mathrm{SW}}(k) \le \min_{\tau>0} \mathbb{P}\left[\log_{|\mathcal{A}|} \frac{1}{P_{X^n|Y}(X^n|Y)} > k - \tau\right] + |\mathcal{A}|^{-\tau}. \quad \text{(II.1)}$$

Consider the following case, where $X^n = (X_1, \ldots, X_n)$ is uniform on $\{0, 1\}^n$ and

$$Y = (X_1, \ldots, X_n) + (N_1, \ldots, N_n),$$

where N_i are iid Gaussian with zero mean and variance 0.1. Let $n = 10$. Propose a method to numerically compute or approximate the bound (II.1) as a function of $k = 1, \ldots, 10$. Plot the results.

II.10 (Mismatched compression) Let P, Q be distributions on some discrete alphabet $\mathcal{A}$.

(a) Let $f_P^*: \mathcal{A} \to \{0, 1\}$ denote the optimal variable-length lossless compressor for $X \sim P$. Show that under Q,

$$\mathbb{E}_Q[l(f_P^*(X))] \le H(Q) + D(Q\|P).$$

(b) The Shannon code for $X \sim P$ is a prefix code f_P with the code length $l(f_P(a)) = \lceil \log_2 \frac{1}{P(a)} \rceil, a \in \mathcal{A}$. Show that if X is distributed according to Q instead, then

$$H(Q) + D(Q\|P) \le \mathbb{E}_Q[l(f_P(X))] \le H(Q) + D(Q\|P) + 1 \text{ bit}.$$

Comment: This can be interpreted as a robustness result for compression with model misspecification: When a compressor designed for P is applied to a source whose distribution is in fact Q, the suboptimality incurred by this mismatch can be related to divergence $D(Q\|P)$.

II.11 Consider a ternary fixed-length (almost lossless) compression $\mathcal{X} \to \{0,1,2\}^k$ with an additional requirement that the string in $w^k \in \{0,1,2\}^k$ should satisfy

$$\sum_{j=1}^{k} w_j \le \frac{k}{2}. \tag{II.2}$$

For example, $(0,0,0,0)$, $(0,0,0,2)$, and $(1,1,0,0)$ satisfy the constraint but $(0,0,1,2)$ does not.

Let $\epsilon^*(S^n,k)$ denote the minimum probability of error among all possible compressors of $S^n = \{S_j, j = 1,\dots,n\}$ with iid entries of finite entropy $H(S) < \infty$. Compute

$$\lim_{n\to\infty} \epsilon^*(S^n, nR)$$

as a function of $R \ge 0$.

(Hint: Relate to $\mathbb{P}[\ell(f^*(S^n)) \ge \gamma n]$ and use Stirling's formula (I.2) to find γ.)

II.12 Consider a probability measure $\mathbb{P}$ and a measure-preserving transformation $\tau\colon \Omega \to \Omega$. Prove that τ is ergodic if and only if for any measurable A, B we have

$$\frac{1}{n}\sum_{k=0}^{n-1} \mathbb{P}[A \cap \tau^{-k} B] \to \mathbb{P}[A]\mathbb{P}[B].$$

Comment: Thus ergodicity is a weaker condition than *mixing*: $\mathbb{P}[A\cap\tau^{-n}B] \to \mathbb{P}[A]\mathbb{P}[B]$.

II.13 (Arithmetic coding) We analyze the encoder defined by (13.1) for an iid source. Let P be a distribution on some ordered finite alphabet, say, $\mathtt{a} < \mathtt{b} < \cdots < \mathtt{z}$. For each n, define $p(x^n) = \prod_{i=1}^n P(x_i)$ and $q(x^n) = \sum_{y^n < x^n} p(y^n)$ according to the lexicographic ordering, so that $F_n(x^n) = q(x^n)$ and $|I_{x^n}| = p(x^n)$.

(a) Show that if $x^{n-1} = (x_1,\dots,x_{n-1})$, then

$$q(x^n) = q(x^{n-1}) + p(x^{n-1}) \sum_{\alpha < x_n} P(\alpha).$$

Conclude that $q(x^n)$ can be computed in $O(n)$ steps sequentially.

(b) Show that intervals I_{x^n} are disjoint subintervals of $[0,1)$.

(c) *Encoding*. Show that the code length $l(f(x^n))$ defined in (13.1) satisfies the constraint (13.2), namely, $\log_2 \frac{1}{p(x^n)} \le \ell(f(x^n)) \le \left\lceil \log_2 \frac{1}{p(x^n)} \right\rceil + 1$. Furthermore, verify that the map $x^n \mapsto f(x^n)$ defines a prefix code. *Warning:* This is not about checking Kraft's inequality.

(d) *Decoding*. Upon receipt of the codeword, we can reconstruct the interval D_{x^n}. Divide the unit interval according to the distribution P, that is, partition $[0, 1)$ into disjoint subintervals $I_{\mathsf{a}}, \ldots, I_{\mathsf{z}}$. Output the index that contains D_{x^n}. Show that this gives the first symbol x_1. Continue in this fashion by dividing I_{x_1} into $I_{x_1,\mathsf{a}}, \ldots, I_{x_1,\mathsf{z}}$, etc. Argue that x^n can be decoded losslessly. How many steps are needed?

(e) Suppose $P_X(\mathsf{e}) = 0.5, P_X(\mathsf{o}) = 0.3, P_X(\mathsf{t}) = 0.2$. Encode etoo (write the binary codewords) and describe how to decode.

(f) Show that the average length of this code satisfies

$$nH(P) \leq \mathbb{E}[l(f(X^n))] \leq nH(P) + 2 \quad \text{bits}.$$

(g) Assume that $X = (X_1, \ldots, X_n)$ is not iid but $P_{X_1}, P_{X_2|X_1}, \ldots, P_{X_n|X^{n-1}}$ are known. How would you modify the scheme so that we have

$$H(X^n) \leq \mathbb{E}[l(f(X^n))] \leq H(X^n) + 2 \quad \text{bits}.$$

II.14 (Enumerative codes) Consider the following simple universal compressor for binary sequences: Given $x^n \in \{0, 1\}^n$, denote by $n_1 = \sum_{i=1}^n x_i$ and $n_0 = n - n_1$ the number of ones and zeros in x^n. First encode $n_1 \in \{0, 1, \ldots, n\}$ using $\lceil \log_2(n + 1) \rceil$ bits, then encode the index of x^n in the set of all strings with n_1 number of ones using $\left\lceil \log_2 \binom{n}{n_1} \right\rceil$ bits. Concatenating two binary strings, we obtain the codeword of x^n. This defines a lossless compressor $f\colon \{0, 1\}^n \to \{0, 1\}^*$. Show that redundancy of this code is $\log n + O(1)$ by following steps:

(a) Verify that f is a prefix code.

(b) Let $\mathbb{E}_\theta$ be taken over $X^n \overset{\text{iid}}{\sim} \text{Ber}(\theta)$. Show that for any $\theta \in [0, 1]$,

$$\mathbb{E}_\theta[l(f(X^n))] \leq nh(\theta) + \log n + O(1),$$

where $h(\cdot)$ is the binary entropy function. On the other hand, show that

$$\sup_{0 \leq \theta \leq 1} \{\mathbb{E}_\theta[l(f(X^n))] - nh(\theta)\} \geq \log n + O(1).$$

[Optional: Explain why enumerative coding fails to achieve the optimal redundancy.]

(Hint: Stirling's approximation (I.2) might be useful.)

II.15 (Krichevsky–Trofimov codes). Consider the KT probability assignment for the binary alphabet (13.27) and its sequential form (13.29). Let f_{KT} be the encoder with length assignment $l(f(x^n)) = \left\lceil \log_2 \frac{1}{Q^{(\text{KT})}_{X^n}(x^n)} \right\rceil$ for all x^n.

(a) Prove that for any n and any $x^n \in \{0, 1\}^n$,

$$Q^{(\text{KT})}_{X^n}(x^n) \geq \frac{1}{2} \frac{1}{\sqrt{t_0 + t_1}} \left(\frac{t_0}{t_0 + t_1}\right)^{t_0} \left(\frac{t_1}{t_0 + t_1}\right)^{t_1},$$

where $t_i = t_i(x^n), i \in \{0, 1\}$ counts the number of i's occuring in x^{n-1}.

(Hint: Induction on n.)

(b) Conclude that the KT code length satisfies:

$$l(f_{\text{KT}}(x^n)) \leq nh\left(\frac{n_1}{n}\right) + \frac{1}{2}\log n + 2, \quad \forall x^n \in \{0, 1\}^n.$$

(c) Conclude that for KT codes the redundancy is bounded by, in the notation of Exercise **II.14b**,

$$\sup_{0\le\theta\le1} \{\mathbb{E}_\theta[l(f_{\mathrm{KT}}(X^n))] - nh(\theta)\} \le \frac{1}{2}\log n + O(1).$$

This establishes an upper-bound part of the $O(1)$ version of (13.24) for the binary alphabet. (In fact, part (b) shows the stronger result that the pointwise redundancy (see (13.11)) of the KT code satisfies $\max_{x^n\in\{0,1\}^n} \max_\theta \log \frac{P_{X^n|\theta}(x^n|\theta_0)}{Q^{(\mathrm{KT})}_{X^n}(x^n)} \le \frac{1}{2}\log n + O(1)$.)

II.16 (Redundancy lower bound: binary alphabet) In Exercise **II.15** we showed that the minimax average-case redundancy, see (13.8), for the iid Bernoulli model satisfies $\mathcal{R}^*_n([0,1]) \le \frac{1}{2}\log n + O(1)$. We show that this bound is tight by computing $I(\theta;X^n)$ for $\theta \sim \mathrm{Unif}([0,1])$. Thus, from the capacity-redundancy theorem (13.10), this provides a lower bound on the redundancy.

(a) Let $\hat\theta = \frac{1}{n}\sum_{i=1}^n X_i$ denote the empirical frequency of ones. Prove that

$$I(\theta;X^n) = I(\theta;\hat\theta).$$

(Hint: Section 3.5.)

(b) Show that the Bayes risk satisfies $\mathbb{E}[(\theta-\hat\theta)^2] = \frac{1}{6n}$.

(c) Compute the differential entropy $h(\theta)$.

(d) Justify each step in

$$h(\theta|\hat\theta) = h(\theta-\hat\theta|\hat\theta) \le h(\theta-\hat\theta) \le \frac{1}{2}\log\frac{2\pi e}{6n}.$$

(Hint: (2.20).)

(e) Assemble the above steps to conclude that $\mathcal{R}^*_n \ge \frac{1}{2}\log n + c_1$ and compute the value of c_1.

(f) Now Beta($\frac{1}{2}$) and redo the previous part. Do you get a better constant c_1?

Comment: We followed the strategy, introduced in [118], of lower-bounding $I(\theta;X^n)$ by guessing a good estimator $\hat\theta = \hat\theta(X_1,\dots,X_n)$ and bounding $I(\theta;\hat\theta)$ on the basis of the estimator error. The rationale is that if θ can be estimated well, then X^n needs to provide a large amount of information. We will further explore this idea in Chapter 30.

II.17 Consider the following collection of stationary ergodic Markov processes depending on parameter $\theta \in [0,1]$. The $X_1 \sim \mathrm{Ber}(1/2)$ and after that

$$X_t = \begin{cases} X_{t-1}, & \text{w.p. } 1-\theta, \\ 1-X_{t-1}, & \text{w.p. } \theta. \end{cases}$$

Denote the resulting Markov kernel as $P_{X^n|\theta}$.

(a) Compute $J_F(\theta)$.

(b) Prove that minimax redundancy $\mathcal{R}^*_n = (\frac{1}{2}+o(1))\log n$.

II.18 (Elias coding) In this problem we construct *universal* codes for integers. Namely, they compress any integer-valued (infinite-alphabet!) random variable almost to its entropy.

(a) Consider the following universal compressor for natural numbers: For $x \in \mathbb{N} = \{1, 2, \dots\}$, let $k(x)$ denote the length of its binary representation. Define its codeword $c(x)$ to be $k(x)$ zeros followed by the binary representation of x. Compute $c(10)$. Show that c is a prefix code and describe how to decode a stream of codewords.

(b) Next we construct another code using the one above: Define the codeword $c'(x)$ to be $c(k(x))$ followed by the binary representation of x. Compute $c'(10)$. Show that c' is a prefix code and describe how to decode a stream of codewords.

(c) Let X be a random variable on $\mathbb{N}$ whose probability mass function is decreasing. Show that $\mathbb{E}[\log(X)] \le H(X)$.

(d) Show that the average code length of c satisfies $\mathbb{E}[\ell(c(X))] \le 2H(X) + 2$ bit.

(e) Show that the average code length of c' satisfies $\mathbb{E}[\ell(c'(X))] \le H(X) + 2\log(H(X) + 1) + 3$ bit.

Comment: The two coding schemes are known as Elias γ-codes and δ-codes.

II.19 (Batch loss) Recall the definition of batch regret in online prediction in Section 13.6. Show that whenever the maximizer π^* exists we have

$$\mathrm{BatchReg}_n^*(\Theta) = \max_{\pi \in \mathcal{P}(\Theta)} I(\theta; X_n | X^{n-1}),$$

where optimization is over the distribution of $\theta \sim \pi$. (Hint: Apply Exercise **I.11**.)

II.20 (Supervised learning) Consider a possibly non-iid stochastic process $X^n = (X_1, \dots, X_n)$ and a parametric collection of conditional distributions $P_{Y|X}^{(\theta)}, \theta \in \Theta$, which we also understand as a kernel $P_{Y|X,\theta}$. Nature fixes θ^* and generates $Y_i \sim P_{Y|X=X_i}^{(\theta^*)}$ independently. These samples are sequentially fed to the learner who having observed (X^t, Y^{t-1}) outputs $\hat{Q}_t(\cdot) \in \mathcal{P}(\mathcal{Y})$ and experiences log-loss $-\log \hat{Q}_t(Y_t)$. The goal of supervised learning is to minimize the worst-case regret, that is, find the minimizer in

$$\mathrm{AvgReg}_n^*(\Theta) \triangleq \inf_{\{Q_t\}} \sup_{\theta^* \in \Theta} \mathbb{E}\left[\sum_{t=1}^{n} \log \frac{1}{\hat{Q}_t(Y_t)} - \log \frac{1}{P_{Y|X}^{(\theta^*)}(Y_t|X_t)}\right].$$

Here we show an analog of Theorem 13.3, namely that

$$\mathrm{AvgReg}_n^*(\Theta) = C_n \triangleq \max_{\pi} I(\theta; Y^n | X^n),$$

with optimization over $\pi \in \mathcal{P}(\Theta)$ of priors on θ. We assume the maximum is attained at some π^*.

(a) Let $D_n \triangleq \inf_{Q_{Y^n|X^n}} \sup_{\theta \in \Theta} D(P_{Y|X}^{(\theta)\otimes n} \| Q_{Y^n|X^n} | P_{X^n})$, where the infimum is over all conditional kernels $Q_{Y^n|X^n} : \mathcal{X}^n \to \mathcal{Y}^n$. Show $\mathrm{AvgReg}_n^*(\Theta) \le D_n$.

(b) Show that $C_n = D_n$ and that optimal $Q^*_{Y^n|X^n}(y^n|x^n) = \int \pi^*(d\theta) \prod_{t=1}^n P^{(\theta)}_{Y|X}(y_t|x_t)$ (Hint: Exercise **I.11**.)
(c) Show that we can always factorize $Q^*_{Y^n|X^n} = \prod_{t=1}^n Q_{Y_t|X^t,Y^{t-1}}$.
(d) Conclude that $Q^*_{Y^n|X^n}$ defines an optimal learner, who also operates without any knowledge of P_{X^n}.

Note: This characterization is mostly useful for upper-bounding regret (Exercise **II.22**). Indeed, the optimal learner requires knowledge of π^* which in turn often depends on P_{X^n}, which is not available to the learner. This shows why supervised learning is quite a bit more delicate than universal compression. Nevertheless, taking a "natural" prior π and factorizing the mixture $\int \pi(d\theta) P^{(\theta)\otimes n}_{Y|X}$ often gives very interesting and often almost optimal algorithms (e.g. exponential-weights update algorithm [446]).

II.21 (Average-case and worst-case redundancies are incomparable) This exercise provides an example where the worst-case minimax redundancy (13.11) is infinite but the average-case one (13.8) is finite. Take $n = 1$ and consider the class of distributions $\mathcal{P}_1 = \{P \in \mathcal{P}(\mathbb{Z}_+) : \mathbb{E}_P[X] \le 1\}$. Define

$$\mathcal{R}^* = \min_Q \sup_{P \in \mathcal{P}_1} D(P\|Q), \quad \mathcal{R}^{**} = \min_Q \max_{x \in \mathbb{Z}_+} \sup_{P \in \mathcal{P}_1} \log \frac{P(x)}{Q(x)}.$$

(a) Applying the capacity-redundancy theorem, show that $\mathcal{R}^* \le 2\log 2$. (Hint: Use Exercise **I.4** to bound the mutual information.)
(b) Prove that $\mathcal{R}^{**} = \infty$ if and only if the Shtarkov sum (13.13) is infinite, namely, $\sum_{x\in\mathbb{Z}_+} \sup_{P\in\mathcal{P}_1} P(x) = \infty$.
(c) Verify that

$$\sup_{P\in\mathcal{P}_1} P(x) = \begin{cases} 1, & x = 0, \\ 1/x, & x \ge 1, \end{cases}$$

and conclude $\mathcal{R}^{**} = \infty$. (Hint: Markov's inequality.)

II.22 (Linear regression) Let $X_i \overset{\text{iid}}{\sim} P_X$ on $\mathbb{R}^p$ with P_X being rotationally invariant (i.e. $X \overset{(d)}{=} UX$ for any orthogonal matrix U). Fix $\theta \in \mathbb{R}^p$ with $\|\theta\| \le s$ and given X_i generate $Y_i \sim \mathcal{N}(\theta^\top X_i, \sigma^2)$. Having observed Y^{t-1}, X^t (but not θ) the learner outputs a prediction $\hat{Y}_t$ of Y_t.

(a) Show that

$$\text{AvgReg}_n \triangleq \sup_{\|\theta\|\le s} \sum_{t=1}^n \mathbb{E}[(\hat{Y}_t - Y_t)^2] - n\sigma^2 \le \sigma^2 \sum_{i=1}^p \mathbb{E}\left[\ln\left(1 + \frac{s^2 n}{p}\lambda_i(\hat{\Sigma}_X)\right)\right],$$

where $\hat{\Sigma}_X = \frac{1}{n}\sum_{i=1}^n X_i X_i^\top$ is the sample covariance matrix. (Hint: Interpret the LHS as regret under log-loss and solve $\max_{\pi_\theta} I(\theta; Y^n|X^n)$ s.t. $\mathbb{E}[\|\theta\|^2] \le s^2$ via Exercise **I.10**.)

(b) Show that

$$\text{AvgReg}_n \le \sigma^2 \ln\det\left(I_p + \frac{s^2 n}{p}\Sigma_X\right) \le p\sigma^2 \ln\left(1 + \frac{s^2 n}{p^2}\mathbb{E}[\|X\|^2]\right).$$

(Hint: Jensen's inequality.)

Remark: Note that if $X_i \overset{\text{iid}}{\sim} \mathcal{N}(0, BI_p)$ the RHS is $p\sigma^2 \ln n + O(1)$. At the same time the prediction error of an ordinary least-squares estimate $\hat{Y}_t = \hat{\theta}^\top X_t$ for $n \geq p + 2$ is known to be exactly[7] $\mathbb{E}[(\hat{Y}_t - Y_t)^2] = \sigma^2(1 + \frac{p}{n-p-1})$ and hence achieves the optimal $p\sigma^2 \ln n + O(1)$ cumulative regret.

[7] This can be shown by applying Exercise **VI.3b** and evaluating the expected trace using [445, Theorem 3.1].

Part III

Hypothesis Testing and Large Deviations

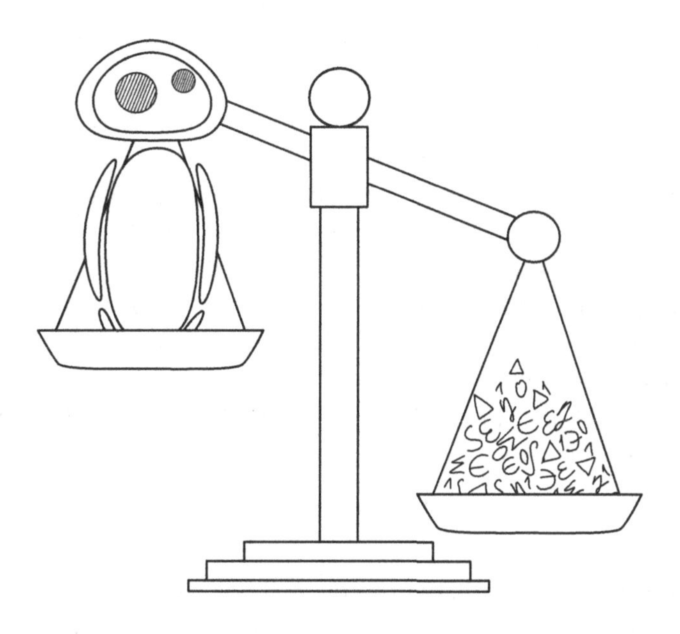

In this part we study the topic of binary hypothesis testing (BHT) which we first encountered in Section 7.3. This is an important area of statistics, with a definitive treatment given in [278]. Historically, there have been two schools of thought on how to approach this question. One is the so-called *significance testing* of Karl Pearson and Ronald Fisher. This is perhaps the most widely used approach in modern biomedical and social sciences. The concepts of null hypothesis, p-value, χ^2-test, and goodness-of-fit belong to this world. We will *not* be discussing these.

The other school was pioneered by Jerzy Neyman and Egon Pearson, and is our topic in this part. The concepts of type-I and type-II errors, likelihood-ratio tests, and Chernoff exponent are from this domain. This is, arguably, a more popular way of looking at the problem among the engineering disciplines (perhaps explained by its foundational role in radar and electronic signal detection).

The conceptual difference between the two is that in the first approach the full probabilistic model is specified *only* under the null hypothesis. (It still could be very specific, like $X_i \stackrel{\text{iid}}{\sim} \mathcal{N}(0,1)$, contain unknown parameters, like $X_i \stackrel{\text{iid}}{\sim} \mathcal{N}(\theta,1)$ with $\theta \in \mathbb{R}$ arbitrary, or be nonparametric, like $(X_i, Y_i) \stackrel{\text{iid}}{\sim} P_{X,Y} = P_X P_Y$ denoting that observables X and Y are statistically independent.) The main goal of the statistician in this setting is inventing a testing procedure that is able to find statistically significant deviations from the postulated null behavior. If such deviation is found then the null is rejected and (in scientific fields) a discovery is announced. The role of the alternative hypothesis (if one is specified at all) is to roughly suggest what features of the null are most likely to be violated and motivates the choice of test procedures. For example, if under the null $X_i \stackrel{\text{iid}}{\sim} \mathcal{N}(0,1)$, then both of the following are reasonable tests:

$$\frac{1}{n}\sum_{i=1}^{n} X_i \stackrel{?}{\approx} 0, \qquad \frac{1}{n}\sum_{i=1}^{n} X_i^2 \stackrel{?}{\approx} 1.$$

However, under the alternative, the first one would be preferred if "data has nonzero mean." and the second if "data has zero mean but variance not equal to one," Whichever of the alternatives is selected does not imply in any way the validity of the alternative. In addition, theoretical properties of the test are mostly studied under the null rather than the alternative. For this approach the null hypothesis (out of the two) plays a very special role.

The second approach treats hypotheses in complete symmetry. Exact specifications of probability distributions are required for both hypotheses and the precision of a proposed test is to be analyzed under both. This is the setting that is most useful for our treatment of the forthcoming topics of channel coding (Part IV) and statistical estimation (Part VI).

The outline of this part is the following. First, we define the performance metric $\mathcal{R}(P,Q)$ giving a full description of the BHT problem. A key result in this theory, the Neyman–Pearson lemma, determines the form of the optimal test and, at the same time, characterizes $\mathcal{R}(P,Q)$. We then specialize to the setting of iid observations and consider two types of asymptotics (as the sample size n goes to infinity):

Stein's regime (where type-I error is held constant) and Chernoff's regime (where errors of both types are required to decay exponentially). The fundamental limit in the former regime is simply a scalar (given by $D(P\|Q)$), while in the latter it is a region. To describe this region (in Chapter 16) we will first need to dive deep into another foundational topic: the theory of large deviations and information projection (Chapter 15).

14 Neyman–Pearson Lemma

In this chapter we formally define the problem of binary hypothesis testing between two simple hypotheses. We introduce the fundamental limit for this problem in the form of a region $\mathcal{R}(P, Q) \subset [0, 1]^2$, whose boundary is known as the *received operating characteristic (ROC) curve*. We will show how to compute this region/curve exactly (Neyman–Pearson lemma) and show the optimality of the likelihood-ratio tests in the process. However, for high-dimensional situations exact computation of the region is still too complex and we will also derive upper and lower bounds (as usual, we call them achievability and converse, respectively). Finally, we will conclude by introducing two different asymptotic settings: the Stein regime and the Chernoff regime. The answer in the former will be given completely (for iid distributions), while the answer for the latter will require further developments in subsequent chapters.

14.1 Neyman–Pearson Formulation

Consider the situation where we have two distributions P and Q on a space $\mathcal{X}$ one of which has generated our observation X. These two possibilities are summarized as a pair of hypotheses:

$$
\begin{aligned}
H_0 &: X \sim P, \\
H_1 &: X \sim Q,
\end{aligned}
$$

which states that, under hypothesis H_0 (the *null hypothesis*) X is distributed according to P, and under H_1 (the *alternative hypothesis*) X is distributed according to Q. A *test* (or decision rule) between the two distributions chooses either H_0 or H_1 based on the data X. We will consider deterministic tests and, more generally, randomized tests.

- Deterministic tests: $f : \mathcal{X} \to \{0, 1\}$, or equivalently, $f(x) = 1\{x \in E\}$, where E is known as a *decision region*.
- Randomized tests: $P_{Z|X} : \mathcal{X} \to \{0, 1\}$, so that $P_{Z|X}(1|x) \in [0, 1]$ is the probability of rejecting the null upon observing $X = x$.

Table 14.1 Expressions for common performance metrics of hypothesis tests.

Term	Expression
Type-I error, significance, size, false alarm rate, false positive	$1-\alpha$
Specificity, selectivity, true negative	α
Power, recall, sensitivity, true positive	$1-\beta$
Type-II error, missed detection, false negative	β
Accuracy	$\pi_1(1-\beta)+(1-\pi_1)\alpha$
F_1-score	$\frac{2\pi_1(1-\beta)}{1+\pi_1(1-\beta)-(1-\pi_1)\alpha}$
Bayesian error	$\pi_1\beta+(1-\pi_1)(1-\alpha)$
Positive predictive value (PPV), precision	$\frac{\pi_1(1-\beta)}{1-\pi_1\beta-(1-\pi_1)\alpha}$

Entries involving $\pi_1 = \mathbb{P}[H_1]$ correspond to the Bayesian setting where a prior probability on occurrence of H_1 is postulated.

Let $Z = 0$ denote that the test chooses P (accepting the null) and $Z = 1$ that the test chooses Q (rejecting the null).

This setting is called "testing simple hypothesis against simple hypothesis." Here "simple" refers to the fact that under each hypothesis there is only one distribution that could have generated the data. In comparison, a composite hypothesis postulates that $X \sim P$ for *some* P which is a given class of distributions; see Sections 16.4 and 32.2.1.

In order to quantify the performance of a test, we focus on two metrics. Let $\pi_{i|j}$ denote the probability of the test choosing i when the correct hypothesis is j, with $i,j \in \{0,1\}$. For every test $P_{Z|X}$ we associate a pair of numbers:

$$\alpha = \pi_{0|0} = P[Z=0] \quad \text{(probability of success given } H_0 \text{ is true)},$$
$$\beta = \pi_{0|1} = Q[Z=0] \quad \text{(probability of error given } H_1 \text{ is true)},$$

where $P[Z=0] = \int P_{Z|X}(0|x)P(dx)$ and $Q[Z=0] = \int P_{Z|X}(0|x)Q(dx)$. Depending on the field of study there are many different names (and transformations) that have been defined, see Table 14.1.

Because we have two performance metrics it is not easy to understand which one should be designated as the "best test." Consequently, there are several approaches:

- Bayesian: Assuming the prior distribution $\mathbb{P}[H_0] = \pi_0$ and $\mathbb{P}[H_1] = \pi_1$, we minimize the average probability of error:

$$P_b^* = \min_{P_{Z|X}:\, \mathcal{X}\to\{0,1\}} \pi_0\pi_{1|0} + \pi_1\pi_{0|1}. \tag{14.1}$$

- Minimax: Assuming there is an unknown prior distribution, we choose the test that performs the best for the worst-case prior

$$P_m^* = \min_{P_{Z|X}:\, \mathcal{X}\to\{0,1\}} \max\{\pi_{1|0}, \pi_{0|1}\}.$$

- Neyman–Pearson: Minimize the type-II error β subject to the condition that the success probability under the null is at least α.

In this part the Neyman–Pearson formulation is our choice. We formalize the fundamental performance limit as follows.

Definition 14.1. Given (P, Q), the Neyman–Pearson region consists of achievable points for all randomized tests

$$\mathcal{R}(P, Q) = \left\{(P[Z = 0], Q[Z = 0]) : P_{Z|X} : \mathcal{X} \to \{0, 1\}\right\} \subset [0, 1]^2. \tag{14.2}$$

In particular, its lower boundary is defined as (see Figure 14.1 for an illustration)

$$\beta_\alpha(P, Q) \triangleq \inf_{P[Z=0]\geq\alpha} Q[Z = 0]. \tag{14.3}$$

The Neyman–Pearson region encodes much useful information about the relationship between P and Q. In particular, the mutual singularity (see Figure 14.1) can be detected. Furthermore, every f-divergence can be computed from $\mathcal{R}(P, Q)$. For example, $\mathrm{TV}(P, Q)$ coincides with half the length of the longest vertical segment contained in $\mathcal{R}(P, Q)$ (Exercise **III.7**). In machine learning one of the most popular metrics used to characterize the quality of $\mathcal{R}(P, Q)$ is the area under the curve (AUC)

$$\mathrm{AUC}(P, Q) \triangleq 1 - \int_0^1 \beta_\alpha(P, Q) d\alpha.$$

We next prove several basic properties of $\mathcal{R}(P, Q)$.

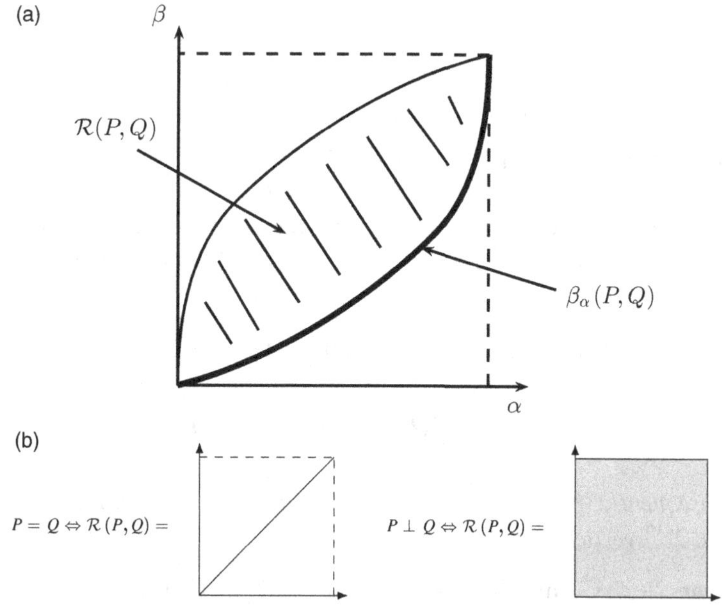

Figure 14.1 Illustration of the Neyman–Pearson regions: typical (a) and two extremal cases (b). Recall that P is mutually singular w.r.t. Q, denoted by $P \perp Q$, if $P[E] = 0$ and $Q[E] = 1$ for some E.

Theorem 14.2. (Properties of $\mathcal{R}(P, Q)$)

(a) $\mathcal{R}(P, Q)$ *is a closed, convex subset of* $[0, 1]^2$.
(b) $\mathcal{R}(P, Q)$ *contains the diagonal.*
(c) Symmetry: $(\alpha, \beta) \in \mathcal{R}(P, Q) \Leftrightarrow (1 - \alpha, 1 - \beta) \in \mathcal{R}(P, Q)$.

Proof.

(a) For convexity, suppose that $(\alpha_0, \beta_0), (\alpha_1, \beta_1) \in \mathcal{R}(P, Q)$, corresponding to tests $P_{Z_0|X}, P_{Z_1|X}$, respectively. Randomizing between these two tests, we obtain the test $\lambda P_{Z_0|X} + \bar{\lambda} P_{Z_1|X}$ for $\lambda \in [0, 1]$, which achieves the point $(\lambda\alpha_0 + \bar{\lambda}\alpha_1, \lambda\beta_0 + \bar{\lambda}\beta_1) \in \mathcal{R}(P, Q)$.

The closedness of $\mathcal{R}(P, Q)$ will follow from the explicit determination of all boundary points via the Neyman–Pearson lemma – see Remark 14.1. In more complicated situations (e.g. in testing against a composite hypothesis) simple explicit solutions similar to the Neyman–Pearson lemma are not available but closedness of the region can frequently be argued still. The basic reason is that the collection of bounded functions $\{g \colon \mathcal{X} \to [0, 1]\}$ (with $g(x) = P_{Z|X}(0|x)$) forms a weakly compact set and hence its image under the linear functional $g \mapsto (\int g dP, \int g dQ)$ is closed.

(b) Testing by random guessing, that is, $Z \sim \text{Ber}(1 - \alpha) \perp\!\!\!\perp X$, achieves the point (α, α).

(c) If $(\alpha, \beta) \in \mathcal{R}(P, Q)$ is achieved by $P_{Z|X}$, $P_{1-Z|X}$ achieves $(1 - \alpha, 1 - \beta)$. □

The region $\mathcal{R}(P, Q)$ consists of the operating points of all randomized tests, which include as special cases those of deterministic tests, namely

$$\mathcal{R}_{\text{det}}(P, Q) = \{(P(E), Q(E)) \colon E \text{ measurable}\}. \tag{14.4}$$

As the next result shows, the former is in fact the closed convex hull of the latter. Recall that $\text{cl}(E)$ (respectively $\text{co}(E)$) denote the closure (respectively convex hull) of a set E, namely, the smallest closed (respectively convex) set containing E. A useful example: For a subset E of a Euclidean space, and measurable functions $f, g \colon \mathbb{R} \to E$, we have $(\mathbb{E}[f(X)], \mathbb{E}[g(X)]) \in \text{cl}(\text{co}(E))$ for any real-valued random variable X.

Theorem 14.3. (Randomized test versus deterministic tests)

$$\mathcal{R}(P, Q) = \text{cl}(\text{co}(\mathcal{R}_{\text{det}}(P, Q))).$$

Consequently, if P and Q are on a finite alphabet $\mathcal{X}$, then $\mathcal{R}(P, Q)$ is a polygon of at most $2^{|\mathcal{X}|}$ vertices.

Proof. "$\supset$": Comparing (14.2) and (14.4), by definition, $\mathcal{R}(P, Q) \supset \mathcal{R}_{\text{det}}(P, Q)$, the former of which is closed and convex, by Theorem 14.2.

"$\subset$": Given any randomized test $P_{Z|X}$, define a measurable function $g \colon \mathcal{X} \to [0, 1]$ by $g(x) = P_{Z|X}(0|x)$. Then

$$P[Z=0] = \sum_x g(x)P(x) = \mathbb{E}_P[g(X)] = \int_0^1 P[g(X) \geq t]dt,$$

$$Q[Z=0] = \sum_x g(x)Q(x) = \mathbb{E}_Q[g(X)] = \int_0^1 Q[g(X) \geq t]dt,$$

where we applied the "area rule" that $E[U] = \int_{\mathbb{R}_+} \mathbb{P}[U \geq t]\,dt$ for any non-negative random variable U. Therefore the point $(P[Z=0], Q[Z=0]) \in \mathcal{R}$ is a mixture of points $(P[g(X) \geq t], Q[g(X) \geq t]) \in \mathcal{R}_{\text{det}}$, averaged according to t uniformly distributed on the unit interval. Hence $\mathcal{R} \subset \text{cl}(\text{co}(\mathcal{R}_{\text{det}}))$.

The last claim follows because there are at most $2^{|\mathcal{X}|}$ subsets in (14.4). □

Example 14.1. Testing Ber(p) versus Ber(q)

Assume that $p < \frac{1}{2} < q$. Using Theorem 14.3, note that there are $2^2 = 4$ events $E = \emptyset, \{0\}, \{1\}, \{0,1\}$. Then $\mathcal{R}(\text{Ber}(p), \text{Ber}(q))$ is given by the following diagram.

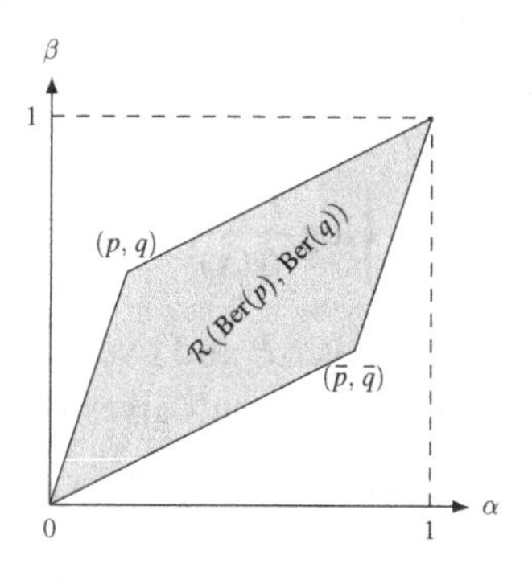

14.2 Likelihood-Ratio Tests

To define optimal hypothesis tests, we need to define the concept of the log-likelihood ratio (LLR). In the simple case when $P \ll Q$ we define the LLR $T(x) = \log \frac{dP}{dQ}(x)$ as a function $T\colon \mathcal{X} \to \mathbb{R} \cup \{-\infty\}$ by thinking of $\log 0 = -\infty$. In order to handle also the case of $P \not\ll Q$, we can leverage our concept of the Log function, see (2.10). Everywhere in this chapter expression $\log \frac{dP}{dQ}$ is to be understood in the sense of the following definition (even when log is not explicitly capitalized).

Definition 14.4. (Extended log-likelihood ratio) Assume that $dP = p(x)d\mu$ and $dQ = q(x)d\mu$ for some dominating measure μ (e.g. $\mu = P + Q$.) Recalling the definition of Log from (2.10) we define the extended LLR as

$$T(x) \triangleq \text{Log}\frac{p(x)}{q(x)} = \begin{cases} \log \frac{p(x)}{q(x)}, & p(x) > 0, q(x) > 0, \\ +\infty, & p(x) > 0, q(x) = 0, \\ -\infty, & p(x) = 0, q(x) > 0, \\ 0, & p(x) = 0, q(x) = 0. \end{cases}$$

Definition 14.5. (Likelihood-ratio test (LRT)) Given a binary hypothesis testing $H_0 : X \sim P$ versus $H_1 : X \sim Q$ the likelihood-ratio test (LRT) with threshold $\tau \in \mathbb{R} \cup \{\pm\infty\}$ is $1\{x : T(x) \le \tau\}$, in other words it decides

$$\mathrm{LRT}_\tau(x) = \begin{cases} \text{declare } H_0, & T(x) > \tau, \\ \text{declare } H_1, & T(x) \le \tau. \end{cases}$$

When $P \ll Q$ it is clear that $T(x) = \log \frac{dP}{dQ}(x)$ for P- and Q-almost every x. For this reason, everywhere in this part we abuse notation and write simply $\log \frac{dP}{dQ}$ to denote the *extended* (!) LLR as defined above. Notice that the LRT is a deterministic test, and that it does make intuitive sense: upon observing x, if $\frac{Q(x)}{P(x)}$ is large then Q is more likely and one should reject the null hypothesis P.

Note that for a discrete alphabet $\mathcal{X}$ and assuming $Q \ll P$ we can see

$$Q[T = t] = \exp(-t)P[T = t] \qquad \forall t \in \mathbb{R} \cup \{+\infty\}.$$

Indeed, this is shown by the following chain:

$$\begin{aligned} Q_T(t) &= \sum_{\mathcal{X}} Q(x) 1 \left\{ \log \frac{P(x)}{Q(x)} = t \right\} = \sum_{\mathcal{X}} Q(x) 1 \left\{ e^t Q(x) = P(x) \right\} \\ &= e^{-t} \sum_{\mathcal{X}} P(x) 1 \left\{ \log \frac{P(x)}{Q(x)} = t \right\} = e^{-t} P_T(t). \end{aligned}$$

We see that taking expectations over P and over Q are equivalent upon multiplying the expectant by $\exp(\pm T)$. The next result gives precise details in the general case.

Theorem 14.6. (Change of measure $P \leftrightarrow Q$) *The following hold:*

1 *For any $h : \mathcal{X} \to \mathbb{R}$ we have*

$$\mathbb{E}_Q[h(X) 1\{T > -\infty\}] = \mathbb{E}_P[h(X) \exp(-T)], \tag{14.5}$$

$$\mathbb{E}_P[h(X) 1\{T < +\infty\}] = \mathbb{E}_Q[h(X) \exp(T)]. \tag{14.6}$$

2 *For any $f \ge 0$ and any $-\infty < \tau < \infty$ we have*

$$\begin{aligned} \mathbb{E}_Q[f(X) 1\{T \ge \tau\}] &\le \mathbb{E}_P[f(X) 1\{T \ge \tau\}] \cdot \exp(-\tau), \\ \mathbb{E}_Q[f(X) 1\{T \le \tau\}] &\ge \mathbb{E}_P[f(X) 1\{T \le \tau\}] \cdot \exp(-\tau). \end{aligned} \tag{14.7}$$

Proof. We first observe that

$$Q[T = +\infty] = P[T = -\infty] = 0. \tag{14.8}$$

Then consider the chain

$$\begin{aligned} \mathbb{E}_Q[h(X) 1\{T > -\infty\}] &\overset{(a)}{=} \int_{\{-\infty < T(x) < \infty\}} d\mu\, q(x) h(x) \\ &\overset{(b)}{=} \int_{\{-\infty < T(x) < \infty\}} d\mu\, p(x) \exp(-T(x)) h(x) \\ &\overset{(c)}{=} \int_{\{-\infty < T(x) \le \infty\}} d\mu\, p(x) \exp(-T(x)) h(x) = \mathbb{E}_P[\exp(-T) h(T)], \end{aligned}$$

where in (a) we used (14.8) to justify restriction to finite values of T; in (b) we used $\exp(-T(x)) = \frac{q(x)}{p(x)}$ for $p, q > 0$; and (c) follows from the fact that $\exp(-T(x)) = 0$ whenever $T = \infty$. Exchanging the roles of P and Q proves (14.6).

The last part follows upon taking $h(x) = f(x)1\{T(x) \geq \tau\}$ and $h(x) = f(x)1\{T(x) \leq \tau\}$ in (14.5) and (14.6), respectively. □

The importance of the LLR is that it is a sufficient statistic for testing the two hypotheses (recall Section 3.5 and in particular Example 3.9), as the following result shows.

Corollary 14.7. *$T = T(X)$ is a sufficient statistic for testing P versus Q.*

Proof. For part 2 of Theorem 14.6, sufficiency of T would be implied by $P_{X|T} = Q_{X|T}$. For the case of $\mathcal{X}$ being discrete we have:

$$P_{X|T}(x|t) = \frac{P_X(x)P_{T|X}(t|x)}{P_T(t)} = \frac{P(x)1\left\{\frac{P(x)}{Q(x)} = \exp(t)\right\}}{P_T(t)}$$
$$= \frac{\exp(t)Q(x)1\left\{\frac{P(x)}{Q(x)} = \exp(t)\right\}}{P_T(t)} = \frac{Q_{XT}(xt)}{\exp(-T)P_T(t)} = \frac{Q_{XT}}{Q_T} = Q_{X|T}(x|t). \; \square$$

We leave the general case as an exercise for the reader.

From Theorem 14.3 we know that to obtain the achievable region $\mathcal{R}(P, Q)$, one can iterate over all decision regions and compute the region $\mathcal{R}_{\text{det}}(P, Q)$ first, then take its closed convex hull. But this is a formidable task if the alphabet is large or infinite. On the other hand, we know that the LLR is a sufficient statistic. Next we give bounds to the region $\mathcal{R}(P, Q)$ in terms of the statistics of the LLR. As usual, there are two types of statements:

- Converse (outer bounds): any point in $\mathcal{R}(P, Q)$ must satisfy certain constraints.
- Achievability (inner bounds): points satisfying certain constraints belong to $\mathcal{R}(P, Q)$.

14.3 Converse Bounds on $\mathcal{R}(P, Q)$

Theorem 14.8. (Weak converse) *For all $(\alpha, \beta) \in \mathcal{R}(P, Q)$, we have*

$$d(\alpha\|\beta) \leq D(P\|Q),$$
$$d(\beta\|\alpha) \leq D(Q\|P),$$

where $d(\cdot\|\cdot)$ is the binary divergence function in (2.6).

Proof. Use the data-processing inequality for KL divergence with $P_{Z|X}$; see Corollary 2.19. □

We will strengthen this bound with the aid of the following result.

Lemma 14.9. *For any test Z and any $\gamma > 0$ we have*

$$P[Z = 0] - \gamma Q[Z = 0] \leq P[T > \log\gamma],$$

where $T = \log \frac{dP}{dQ}$ *is understood in the extended sense of Definition 14.4.*

Note that we do not need to assume $P \ll Q$ precisely because $\pm\infty$ are admissible values for the (extended) LLR.

Proof. Defining $\tau = \log \gamma$ and $g(x) = P_{Z|X}(0|x)$ we get from (14.7):

$$P[Z = 0, T \leq \tau] - \gamma Q[Z = 0, T \leq \tau] \leq 0.$$

Decomposing $P[Z = 0] = P[Z = 0, T \leq \tau] + P[Z = 0, T > \tau]$ and similarly for Q we obtain then

$$\begin{aligned} P[Z = 0] - \gamma Q[Z = 0] &\leq P[T > \log \gamma, Z = 0] - \gamma Q[T > \log \gamma, Z = 0] \\ &\leq P[T > \log \gamma]. \end{aligned}$$

□

Theorem 14.10. (Strong converse) *For all* $(\alpha, \beta) \in \mathcal{R}(P, Q)$ *and all* $\gamma > 0$, *we have*

$$\alpha - \gamma\beta \leq P\left[\log \frac{dP}{dQ} > \log \gamma\right], \tag{14.9}$$

$$\beta - \frac{1}{\gamma}\alpha \leq Q\left[\log \frac{dP}{dQ} < \log \gamma\right]. \tag{14.10}$$

Proof. Apply Lemma 14.9 to (P, Q, γ) and $(Q, P, 1/\gamma)$. □

Theorem 14.10 provides an outer bound for the region $\mathcal{R}(P, Q)$ in terms of half-planes. To see this, fix $\gamma > 0$ and consider the line $\alpha - \gamma\beta = c$ by gradually increasing c from zero. There exists a maximal c, say c^*, at which point the line touches the lower boundary of the region. Then (14.9) says that c^* cannot exceed $P[\log \frac{dP}{dQ} > \log \gamma]$. Hence $\mathcal{R}(P, Q)$ must lie to the left of the line. Similarly, (14.10) provides bounds for the upper boundary. Altogether Theorem 14.10 states that $\mathcal{R}(P, Q)$ is contained in the intersection of an infinite collection of half-planes indexed by γ.

To apply the strong converse, Theorem 14.10, we need to know the CDF of the LLR, whereas to apply the weak converse, Theorem 14.8, we need only to know the expectation of the LLR, that is, the divergence. This is the usual pattern between the weak and strong converses in information theory.

14.4 Achievability Bounds on $\mathcal{R}(P, Q)$

Given the convexity of the set $\mathcal{R}(P, Q)$, it is natural to try to find all of its supporting lines (hyperplanes in dimension two), as it is well known that a closed convex set equals the intersection of all half-spaces that correspond to the supporting hyperplanes. We are thus led to the following problem: for $t > 0$,

$$\max\{\alpha - t\beta \colon (\alpha, \beta) \in \mathcal{R}(P, Q)\},$$

which is equivalent to minimizing the average probability of error in (14.1), with $t = \frac{\pi_1}{\pi_0}$. This can be solved without much effort. For simplicity, consider the discrete case. Then

$$\begin{aligned}\alpha^* - t\beta^* &= \max_{(\alpha,\beta)\in\mathcal{R}} (\alpha - t\beta) = \max_{P_{Z|X}} \sum_{x\in\mathcal{X}} (P(x) - tQ(x))P_{Z|X}(0|x) \\ &= \sum_{x\in\mathcal{X}} |P(x) - tQ(x)|^+,\end{aligned}$$

where the last equality follows from the fact that we are free to choose $P_{Z|X}(0|x)$, and the best choice is obvious:

$$P_{Z|X}(0|x) = 1\left\{\log \frac{P(x)}{Q(x)} \geq \log t\right\}.$$

Thus, we have shown that all supporting hyperplanes are parameterized by the LRT. This completely recovers the region $\mathcal{R}(P, Q)$ except for the points corresponding to the faces (flat pieces) of the region. The precise result is stated as follows:

Theorem 14.11. (Neyman–Pearson lemma) *For each α, β_α in (14.3) is attained by the following test:*

$$P_{Z|X}(0|x) = \begin{cases} 1, & \log \frac{dP}{dQ} > \tau, \\ \lambda, & \log \frac{dP}{dQ} = \tau, \\ 0, & \log \frac{dP}{dQ} < \tau, \end{cases} \tag{14.11}$$

where $\tau \in \mathbb{R}$ and $\lambda \in [0, 1]$ are the unique solutions to $\alpha = P\left[\log \frac{dP}{dQ} > \tau\right] + \lambda P\left[\log \frac{dP}{dQ} = \tau\right]$.

Proof of Theorem 14.11. Let $t = \exp(\tau)$. Given any test $P_{Z|X}$, let $g(x) = P_{Z|X}(0|x) \in [0, 1]$. We want to show that

$$\alpha = P[Z = 0] = \mathbb{E}_P[g(X)] = P\left[\frac{dP}{dQ} > t\right] + \lambda P\left[\frac{dP}{dQ} = t\right] \tag{14.12}$$

$$\Rightarrow \beta = Q[Z = 0] = \mathbb{E}_Q[g(X)] \overset{\text{goal}}{\geq} Q\left[\frac{dP}{dQ} > t\right] + \lambda Q\left[\frac{dP}{dQ} = t\right]. \tag{14.13}$$

Applying (14.7) with $f = g$ and $f = 1 - g$, we have

$$\begin{aligned}\beta &= \mathbb{E}_Q\left[g(X)1\left\{\frac{dP}{dQ} \leq t\right\}\right] + \mathbb{E}_Q\left[g(X)1\left\{\frac{dP}{dQ} > t\right\}\right] \\ &\geq \frac{1}{t}\mathbb{E}_P\left[g(X)1\left\{\frac{dP}{dQ} \leq t\right\}\right] + \mathbb{E}_Q\left[g(X)1\left\{\frac{dP}{dQ} > t\right\}\right] \\ &\overset{(14.12)}{=} \frac{1}{t}\left(\mathbb{E}_P\left[(1 - g(X))1\left\{\frac{dP}{dQ} > t\right\}\right] + \lambda P\left[\frac{dP}{dQ} = t\right]\right) + \mathbb{E}_Q\left[g(X)1\left\{\frac{dP}{dQ} > t\right\}\right] \\ &\geq \mathbb{E}_Q\left[(1 - g(X))1\left\{\frac{dP}{dQ} > t\right\}\right] + \lambda Q\left[\frac{dP}{dQ} = t\right] + \mathbb{E}_Q\left[g(X)1\left\{\frac{dP}{dQ} > t\right\}\right] \\ &= Q\left[\frac{dP}{dQ} > t\right] + \lambda Q\left[\frac{dP}{dQ} = t\right].\end{aligned}$$

□

Remark 14.1. As a consequence of the Neyman–Pearson lemma, all the points on the boundary of the region $\mathcal{R}(P, Q)$ are attainable. Therefore

$$\mathcal{R}(P, Q) = \{(\alpha, \beta) \colon \beta_\alpha \le \beta \le 1 - \beta_{1-\alpha}\}.$$

Since $\alpha \mapsto \beta_\alpha$ is convex on $[0, 1]$, hence continuous, the region $\mathcal{R}(P, Q)$ is a closed convex set, as previously stated in Theorem 14.2. Consequently, the infimum in the definition of β_α is in fact a minimum.

Furthermore, the lower half of the region $\mathcal{R}(P, Q)$ is the convex hull of the union of the following two sets:

$$\begin{cases} \alpha = P\left[\log \frac{dP}{dQ} > \tau\right], \\ \beta = Q\left[\log \frac{dP}{dQ} > \tau\right], \end{cases} \qquad \tau \in \mathbb{R} \cup \{\pm\infty\}$$

and

$$\begin{cases} \alpha = P\left[\log \frac{dP}{dQ} \ge \tau\right], \\ \beta = Q\left[\log \frac{dP}{dQ} \ge \tau\right], \end{cases} \qquad \tau \in \mathbb{R} \cup \{\pm\infty\}.$$

Therefore it does not lose optimality to restrict our attention to tests of the form $1\{\log \frac{dP}{dQ} \ge \tau\}$ or $1\{\log \frac{dP}{dQ} > \tau\}$. The convex combination (randomization) of the above two styles of tests lead to the achievability of the Neyman–Pearson lemma (Theorem 14.11).

Remark 14.2. The Neyman–Pearson test (14.11) is related to the LRT[1] as follows:

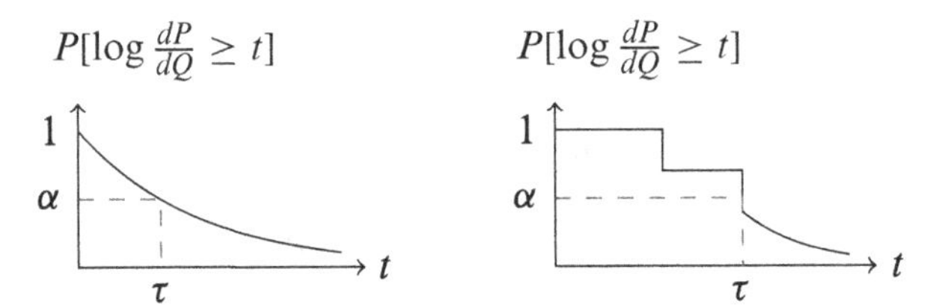

- Left figure: If $\alpha = P[\log \frac{dP}{dQ} > \tau]$ for some τ, then $\lambda = 0$, and (14.11) becomes the LRT $Z = 1\left\{\log \frac{dP}{dQ} \le \tau\right\}$.
- Right figure: If $\alpha \ne P[\log \frac{dP}{dQ} > \tau]$ for any τ, then we have $\lambda \in (0, 1)$, and (14.11) is equivalent to randomize over tests: $Z = 1\left\{\log \frac{dP}{dQ} \le \tau\right\}$ with probability $1 - \lambda$ or $1\left\{\log \frac{dP}{dQ} < \tau\right\}$ with probability λ.

Corollary 14.12. *For all $\tau \in \mathbb{R}$, there exists $(\alpha, \beta) \in \mathcal{R}(P, Q)$ such that*

$$\alpha = P\left[\log \frac{dP}{dQ} > \tau\right],$$
$$\beta \le \exp(-\tau) P\left[\log \frac{dP}{dQ} > \tau\right] \le \exp(-\tau).$$

[1] Note that it so happens that in Definition 14.4 the LRT is defined with an $\le$ instead of $<$.

Proof. For the case of discrete $\mathcal{X}$ it is easy to give an explicit proof:

$$Q\left[\log \frac{dP}{dQ} > \tau\right] = \sum Q(x) 1\left\{\frac{P(x)}{Q(x)} > \exp(\tau)\right\}$$
$$\leq \sum P(x) \exp(-\tau) 1\left\{\frac{P(x)}{Q(x)} > \exp(\tau)\right\}$$
$$= \exp(-\tau) P\left[\log \frac{dP}{dQ} > \tau\right].$$

The general case is just an application of (14.7). □

14.5 Asymptotics: Stein's Regime

Having understood how to compute and bound $\mathcal{R}(P, Q)$ we next proceed to the analysis of asymptotics. We will focus on iid observations in the large-sample asymptotics, that is, we will be talking about $\mathcal{R}(P^{\otimes n}, Q^{\otimes n})$ here. In other words, we consider

$$H_0: X_1, \ldots, X_n \overset{\text{iid}}{\sim} P,$$
$$H_1: X_1, \ldots, X_n \overset{\text{iid}}{\sim} Q, \tag{14.14}$$

where P and Q do not depend on n; this is a particular case of our general setting with P and Q replaced by their n-fold product distributions. We are interested in the asymptotics of the error probabilities $\pi_{0|1}$ and $\pi_{1|0}$ as $n \to \infty$ in the following two regimes:

- Stein's regime: When $\pi_{1|0}$ is constrained to be at most ϵ, what is the best exponential rate of convergence for $\pi_{0|1}$?
- Chernoff's regime: When both $\pi_{1|0}$ and $\pi_{0|1}$ are required to vanish exponentially, what is the optimal tradeoff between their exponents?

Recall that we are in the iid setting (14.14) and are interested in tests satisfying $1 - \alpha = \pi_{1|0} \leq \epsilon$ and $\beta = \pi_{0|1} \leq \exp(-nE)$ for some exponent $E > 0$. The motivation of this asymmetric objective is that often a "missed detection" ($\pi_{0|1}$) is far more disastrous than a "false alarm" ($\pi_{1|0}$). For example, a false alarm could simply result in extra computations (attempting to decode a packet when in fact only noise has been received), while missed detection results in a complete loss of the packet. The formal definition of the best exponent is as follows.

Definition 14.13. The *ϵ-optimal exponent in Stein's regime* is

$$V_\epsilon \triangleq \sup\{E : \exists n_0 \text{ s.t. } \forall n \geq n_0, \exists P_{Z|X^n} \text{ s.t. } \alpha > 1 - \epsilon, \beta < \exp(-nE)\},$$

and *Stein's exponent* is defined as $V \triangleq \lim_{\epsilon \to 0} V_\epsilon$.

It is an exercise for the reader to check the following equivalent definition:

$$V_\epsilon = \liminf_{n\to\infty} \frac{1}{n} \log \frac{1}{\beta_{1-\epsilon}(P_{X^n}, Q_{X^n})},$$

where β_α is defined in (14.3).

Here is the main result of this section.

Theorem 14.14. (Stein's lemma) *Consider the iid setting (14.14) where $P_{X^n} = P^n$ and $Q_{X^n} = Q^n$. Then*

$$V_\epsilon = D(P\|Q), \quad \forall \epsilon \in (0,1).$$

Consequently, $V = D(P\|Q)$.

The way to use this result in practice is the following. Suppose it is required that $\alpha \geq 0.999$, and $\beta \leq 10^{-40}$; what is the required sample size? Stein's lemma provides a rule of thumb: $n \geq -\frac{\log 10^{-40}}{D(P\|Q)}$.

Proof. We first assume that $P \ll Q$ so that $\frac{dP}{dQ}$ is well defined. Define the LLR

$$F_n = \log \frac{dP_{X^n}}{dQ_{X^n}} = \sum_{i=1}^n \log \frac{dP}{dQ}(X_i), \tag{14.15}$$

which is an iid sum under both hypotheses. As such, by the WLLN, under P, as $n \to \infty$,

$$\frac{1}{n} F_n = \frac{1}{n} \sum_{i=1}^n \log \frac{dP}{dQ}(X_i) \xrightarrow{\mathbb{P}} \mathbb{E}_P \left[\log \frac{dP}{dQ} \right] = D(P\|Q). \tag{14.16}$$

Alternatively, under Q, we have

$$\frac{1}{n} F_n \xrightarrow{\mathbb{P}} \mathbb{E}_Q \left[\log \frac{dP}{dQ} \right] = -D(Q\|P). \tag{14.17}$$

Note that both convergence results hold even if the divergence is infinite.

(Achievability) We show that $V_\epsilon \geq D(P\|Q) \equiv D$ for any $\epsilon > 0$. First assume that $D < \infty$. Pick $\tau = n(D - \delta)$ for some small $\delta > 0$. Then Corollary 14.12 yields

$$\begin{aligned} \alpha &= P(F_n > n(D-\delta)) \to 1, \quad \text{by (14.16)}, \\ \beta &\leq e^{-n(D-\delta)}. \end{aligned}$$

Then pick n large enough (depends on ϵ, δ) such that $\alpha \geq 1 - \epsilon$; we have that the exponent $E = D - \delta$ is achievable, $V_\epsilon \geq E$. Sending $\delta \to 0$ yields $V_\epsilon \geq D$. Finally, if $D = \infty$, the above argument holds for arbitrary $\tau > 0$, proving that $V_\epsilon = \infty$.

(Converse) We show that $V_\epsilon \leq D$ for any $\epsilon < 1$, to which end it suffices to consider $D < \infty$. As a warm-up, we first show a weak converse by applying Theorem 14.8 based on the data-processing inequality. For any $(\alpha, \beta) \in \mathcal{R}(P_{X^n}, Q_{X^n})$, we have

$$-h(\alpha) + \alpha \log \frac{1}{\beta} \leq d(\alpha\|\beta) \leq D(P_{X^n}\|Q_{X^n}). \tag{14.18}$$

For any achievable exponent $E < V_\epsilon$, by definition, there exists a sequence of tests such that $\alpha_n \geq 1-\epsilon$ and $\beta_n \leq \exp(-nE)$. Plugging this into (14.18) and using $h \leq \log 2$, we have $E \leq \frac{D(P\|Q)}{1-\epsilon} + \frac{\log 2}{n(1-\epsilon)}$. Sending $n \to \infty$ yields

$$V_\epsilon \leq \frac{D(P\|Q)}{1-\epsilon},$$

which is weaker than what we set out to prove. Nevertheless, this weak converse is tight for $\epsilon \to 0$, so that for Stein's exponent we have succeeded in proving the desired result of $V = \lim_{\epsilon\to 0} V_\epsilon \geq D(P\|Q)$. So the question remains: if we allow the type-I error to be $\epsilon = 0.999$, is it possible for the type-II error to decay faster? This is shown to be impossible by the strong converse next.

To this end, note that, in proving the weak converse, we only made use of the *expectation* of F_n in (14.18); we need to make use of the *entire distribution* (CDF) in order to obtain better results. Applying the strong converse in Theorem 14.10 to testing P_{X^n} versus Q_{X^n} and $\alpha = 1-\epsilon$ and $\beta = \exp(-nE)$, we have

$$1-\epsilon-\gamma\exp(-nE) \leq \alpha_n - \gamma\beta_n \leq P_{X^n}[F_n > \log\gamma].$$

Pick $\gamma = \exp(n(D+\delta))$ for $\delta > 0$. By the WLLN (14.16) the probability on the right-hand side goes to 0, which implies that for any fixed $\epsilon < 1$, we have $E \leq D+\delta$ and hence $V_\epsilon \leq D+\delta$. Sending $\delta \to 0$ completes the proof.

Finally, let us address the case of $P \not\ll Q$, in which case $D(P\|Q) = \infty$. By definition, there exists a subset A such that $Q(A) = 0$ but $P(A) > 0$. Consider the test that selects P if $X_i \in A$ for some $i \in [n]$. It is clear that this test achieves $\beta = 0$ and $1-\alpha = (1-P(A))^n$, which can be made less than any ϵ for large n. This shows $V_\epsilon = \infty$, as desired. □

Remark 14.3. (Non-iid data) Just like in Chapter 12 on data compression, Theorem 14.14 can be extended to stationary ergodic processes. Specifically, one can show that the Stein's exponent corresponds to the *relative* entropy rate, that is,

$$V_\epsilon = \lim_{n\to\infty}\frac{1}{n}D(P_{X^n}\|Q_{X^n}),$$

where $\{X_i\}$ is stationary and ergodic under both P and Q. Indeed, the counterpart of (14.16) based on the WLLN, which is the key for choosing the appropriate threshold τ, for ergodic processes is the Birkhoff–Khintchine convergence theorem (see Theorem 12.8).

The theoretical importance of Stein's exponent is in implications of the following type:

$$\forall E \subset \mathcal{X}^n,\ \ P_{X^n}[E] \geq 1-\epsilon \ \Rightarrow Q_{X^n}[E] \geq \exp(-nV_\epsilon + o(n)).$$

Thus knowledge of Stein's exponent V_ϵ allows one to prove exponential bounds on probabilities of arbitrary sets; this technique is known as "change of measure," which will be applied in large-deviations analysis in Chapter 15.

14.6 Chernoff Regime: Preview

We are still considering the iid setting (14.14), namely, testing

$$H_0 \colon X^n \sim P^n \qquad \text{versus} \qquad H_1 : X^n \sim Q^n,$$

but the objective in the Chernoff regime is to achieve exponentially small error probability of both types simultaneously. We say a pair of exponents (E_0, E_1) is *achievable* if there exists a sequence of tests such that

$$\begin{aligned} 1 - \alpha = \pi_{1|0} &\leq \exp(-nE_0), \\ \beta = \pi_{0|1} &\leq \exp(-nE_1). \end{aligned}$$

Intuitively, one exponent can be made large at the expense of making the other small. So the interesting question is to find their optimal tradeoff by characterizing the achievable region of (E_0, E_1). This problem was solved by [218, 61] and is the topic of Chapter 16. (See Figure 16.2 for an illustration of the optimal (E_0, E_1)-tradeoff.)

Let us explain what we already know about the region of achievable pairs of exponents (E_0, E_1).

First, Stein's regime corresponds to corner points of this achievable region. Indeed, Theorem 14.14 tells us that when fixing $\alpha_n = 1 - \epsilon$, namely $E_0 = 0$, picking $\tau = D(P\|Q) - \delta$ ($\delta \to 0$) gives the exponential convergence rate of β_n as $E_1 = D(P\|Q)$. Similarly, exchanging the roles of P and Q, we can achieve the point $(E_0, E_1) = (D(Q\|P), 0)$.

Second, we have shown in Section 7.3 that the minimum total error probability over all tests satisfies

$$\min_{(\alpha,\beta)\in\mathcal{R}(P^n,Q^n)} 1 - \alpha + \beta = 1 - \mathrm{TV}(P^n, Q^n) = \exp(-nE + o(n)).$$

That is, as $n \to \infty$, P^n and Q^n becomes increasingly distinguishable and their total variation converges to 1 exponentially, with exponent E given by $\max\min(E_0, E_1)$ over all achievable pairs. From the bounds (7.22) and tensorization of the Hellinger distance (7.25), we obtain

$$1 - \sqrt{1 - \exp(-2nE_H)} \leq 1 - \mathrm{TV}(P^n, Q^n) \leq \exp(-nE_H), \tag{14.19}$$

where we denoted

$$E_H \triangleq \log\left(1 - \frac{1}{2}H^2(P, Q)\right).$$

Thus, we can see that

$$E_H \leq E \leq 2E_H.$$

This characterization is valid even if P and Q depend on the sample size n which will prove useful later when we study *composite hypothesis testing* in Section 32.2.1. However, for fixed P and Q this is not precise enough. In order to determine the full set of achievable pairs, we need to make a detour into the topic of large deviations next. To see how this connection arises, notice that the (optimal) likelihood-ratio tests give us explicit expressions for both error probabilities:

$$1 - \alpha_n = P\left[\frac{1}{n}F_n \le \tau\right], \quad \beta_n = Q\left[\frac{1}{n}F_n > \tau\right],$$

where F_n is the LLR in (14.15). When τ falls in the range of $(-D(Q\|P), D(P\|Q))$, both probabilities are vanishing thanks to the WLLN, see (14.16) and (14.17), and we are interested in their exponential convergence rate. This falls under the purview of large-deviations theory.

15 Information Projection and Large Deviations

In this chapter we develop the tools needed for the analysis of the error exponents in hypothesis testing (Chernoff regime). We will start by introducing the concepts of large-deviations theory (log moment generating function (MGF) ψ_X, its convex conjugate ψ_X^*, known as the rate function, and revisit the idea of tilting). Then, we show that the probability of deviation of an empirical mean is governed by the solution of an information projection (also known as I-projection) problem:

$$\min_{Q:\, \mathbb{E}_Q[X]\geq\gamma} D(Q\|P) = \psi^*(\gamma).$$

Equipped with the information projection we will prove a tight version of the Chernoff bound. Specifically, for iid copies $X_1, \ldots, X_n$ of X, we show

$$\mathbb{P}\left[\frac{1}{n}\sum_{k=1}^{n} X_k \geq \gamma\right] = \exp\left(-n\psi^*(\gamma) + o(n)\right).$$

In the remaining sections we extend the simple information projection problem to a general minimization over convex sets of measures and connect it to empirical process theory (Sanov's theorem) and also show how to solve the problem under finitely many linear constraints (exponential families).

In the next chapter, we apply these results to characterize the achievable (E_0, E_1) region (as defined in Section 14.6) to get

$$(E_0(\theta) = \psi_P^*(\theta), \quad E_1(\theta) = \psi_P^*(\theta) - \theta),$$

with ψ_P^* being the rate function of $\log \frac{dP}{dQ}$ (under P). This gives us a complete (parametric) description of the sought-after tradeoff between the two exponents in the Chernoff regime.

15.1 Basics of Large-Deviations Theory

Let $X_1, \ldots, X_n$ be an iid sequence drawn from P and $\hat{P}_n = \frac{1}{n}\sum_{i=1}^{n} \delta_{X_i}$ their empirical distribution. The large-deviations theory focuses on establishing sharp exponential estimates of the kind

$$\mathbb{P}[\hat{P}_n \in \mathcal{E}] = \exp\{-nE + o(n)\}.$$

The full account of such theory requires delicate consideration of the topological properties of $\mathcal{E}$, and is the subject of classical treatments, for example, [120]. We focus here on a simple special case which, however, suffices for the purpose of establishing the Chernoff exponents in hypothesis testing, and also showcases all the relevant information-theoretic ideas. Our initial goal is to show the following result:

Theorem 15.1. *Consider a random variable X whose* log MGF $\psi_X(\lambda) = \log \mathbb{E}[\exp(\lambda X)]$ *is finite for all $\lambda \in \mathbb{R}$. Let $B = \operatorname{esssup} X$ and suppose $\mathbb{E}[X] < \gamma < B$. Then*

$$P\left[\sum_{i=1}^{n} X_i \geq n\gamma\right] = \exp\{-nE(\gamma) + o(n)\},$$

where $E(\gamma) = \sup_{\lambda \geq 0} \lambda\gamma - \psi_X(\lambda) = \psi_X^*(\gamma)$, known as the rate function.

The concepts of log MGF and the rate function will be elaborated in subsequent sections. We provide the proof below that should be revisited after reading the rest of the chapter.

Proof. Let us recall the usual Chernoff bound: For iid X^n, and any $\lambda \geq 0$, applying Markov's inequality yields

$$\begin{aligned}
\mathbb{P}\left[\sum_{i=1}^{n} X_i \geq n\gamma\right] &= \mathbb{P}\left[\exp\left(\lambda \sum_{i=1}^{n} X_i\right) \geq \exp(n\lambda\gamma)\right] \\
&\leq \exp(-n\lambda\gamma)\mathbb{E}\left[\exp\left(\lambda \sum_{i=1}^{n} X_i\right)\right] \\
&= \exp(-n\lambda\gamma + n\underbrace{\log \mathbb{E}[\exp(\lambda X)]}_{\psi_X(\lambda)}).
\end{aligned}$$

Optimizing over $\lambda \geq 0$ gives the following *non-asymptotic* upper bound (concentration inequality) which holds for any n:

$$\mathbb{P}\left[\sum_{i=1}^{n} X_i \geq n\gamma\right] \leq \exp\left\{-n \sup_{\lambda \geq 0}(\lambda\gamma - \psi_X(\lambda))\right\}. \tag{15.1}$$

This proves the upper bound part of Theorem 15.1.

To show the lower bound we need more tools that are going to be developed below. First, we will express $E(\gamma)$ as a certain KL minimization problem (see Theorem 15.9), known as information projection. Second, we will solve this problem (see (15.26)) to obtain the desired value of $E(\gamma)$. In the process of this proof we will also gain a deeper understanding of why the naive Chernoff bound turns out to be sharp. It will be seen that inequality (15.1) performs a change of measure to a new distribution Q, which is chosen to be the closest to P (in KL divergence) among all distributions Q with $\mathbb{E}_Q[X] \geq \gamma$. (This new distribution will turn out to be the tilted version of P, denoted by P_λ.) □

15.1.1 Log MGF and Rate Function

Definition 15.2. The log moment generating function (log MGF, also known as the *cumulant generating function*) of a real-valued random variable X is

$$\psi_X(\lambda) = \log \mathbb{E}[\exp(\lambda X)], \quad \lambda \in \mathbb{R}.$$

Per convention in information theory, we will denote $\psi_P(\lambda) = \psi_X(\lambda)$ if $X \sim P$.

As an example, for a standard Gaussian $Z \sim \mathcal{N}(0,1)$, we have $\psi_Z(\lambda) = \frac{\lambda^2}{2}$. Taking $X = Z^3$ yields a random variable such that $\psi_X(\lambda)$ is infinite for all non-zero λ.

In the remainder of the chapter, we shall make the following simplifying assumption, known as Cramér's condition.

Assumption 15.1. The random variable X is such that $\psi_X(\lambda) < \infty$ for all $\lambda \in \mathbb{R}$.

Most of the results we discuss in this chapter hold under a much weaker assumption of ψ_X having domain with non-empty interior. But proofs in this generality significantly obscure the elegance of the main ideas and we decided to avoid them. We note that Assumption 15.1 implies that all moments of X are finite.

Theorem 15.3. (Properties of ψ_X) *Under Assumption 15.1 we have the following:*

(a) ψ_X *is convex.*
(b) ψ_X *is continuous.*
(c) ψ_X *is infinitely differentiable and*

$$\psi'_X(\lambda) = \frac{\mathbb{E}[X\exp\{\lambda X\}]}{\mathbb{E}[\exp\{\lambda X\}]} = \exp\{-\psi_X(\lambda)\}\mathbb{E}[X\exp\{\lambda X\}].$$

In particular, $\psi_X(0) = 0$, $\psi'_X(0) = \mathbb{E}[X]$.

(d) If $a \le X \le b$ *a.s., then* $a \le \psi'_X \le b$.
(e) Conversely, if

$$A = \inf_{\lambda\in\mathbb{R}} \psi'_X(\lambda), \quad B = \sup_{\lambda\in\mathbb{R}} \psi'_X(\lambda),$$

then $A \le X \le B$ *a.s.*

(f) If X *is not a constant, then* ψ_X *is strictly convex, and consequently,* ψ'_X *is strictly increasing.*
(g) Chernoff bound:

$$P(X \ge \gamma) \le \exp(-\lambda\gamma + \psi_X(\lambda)), \quad \lambda \ge 0. \tag{15.2}$$

Remark 15.1. The slope of log MGF encodes the range of X. Indeed, Theorem 15.3(d) and (e) together show that the smallest closed interval containing the support of P_X equals (the closure of) the range of ψ'_X. In other words, A and B coincide with the essential infimum and supremum (min and max of a random variable in the probabilistic sense) of X respectively,

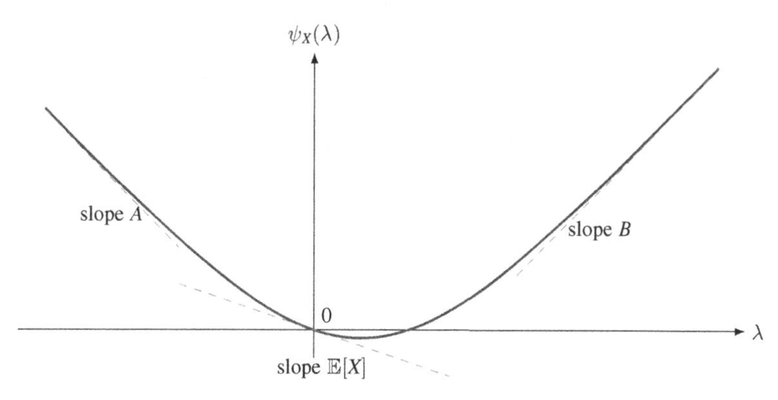

Figure 15.1 Example of a log MGF $\psi_X(\gamma)$ with P_X supported on $[A, B]$. The limiting maximal and minimal slopes are A and B respectively. The slope at $\gamma = 0$ is $\psi'_X(0) = \mathbb{E}[X]$. Here we plot for $X = \pm 1$ with $\mathbb{P}[X = 1] = 1/3$.

$$A = \operatorname{essinf} X \triangleq \sup\{a\colon X \geq a \text{ a.s.}\},$$
$$B = \operatorname{esssup} X \triangleq \inf\{b\colon X \leq b \text{ a.s.}\}.$$

See Figure 15.1 for an illustration.

Proof. For the proof we assume that the base of log and exp is e. Note that (g) is already proved in (15.1). The proof of (e)–(f) relies on Theorem 15.8 and can be revisited later.

(a) Fix $\theta \in (0, 1)$. Recall Hölder's inequality:

$$\mathbb{E}[|UV|] \leq \|U\|_p \|V\|_q, \quad \text{for } p, q \geq 1, \frac{1}{p} + \frac{1}{q} = 1,$$

where the L_p-norm of a random variable U is defined by $\|U\|_p = (\mathbb{E}|U|^p)^{1/p}$. Applying to $\mathbb{E}[e^{(\theta\lambda_1 + \bar{\theta}\lambda_2)X}]$ with $p = 1/\theta, q = 1/\bar{\theta}$, we get

$$\begin{aligned}\mathbb{E}[\exp((\lambda_1/p + \lambda_2/q)X)] &\leq \|\exp(\lambda_1 X/p)\|_p \|\exp(\lambda_2 X/q)\|_q \\ &= \mathbb{E}[\exp(\lambda_1 X)]^\theta \mathbb{E}[\exp(\lambda_2 X)]^{\bar{\theta}},\end{aligned}$$

that is, $e^{\psi_X(\theta\lambda_1 + \bar{\theta}\lambda_2)} \leq e^{\psi_X(\lambda_1)\theta} e^{\psi_X(\lambda_2)\bar{\theta}}$. Another proof is by expressing ψ''_X as a certain variance; see Theorem 15.8(c).

(b) By our assumptions on X, the domain of ψ_X is $\mathbb{R}$. By the fact that a convex function must be continuous on the interior of its domain, we conclude that ψ_X is continuous on $\mathbb{R}$.

(c) The subtlety here is that we need to be careful when exchanging the order of differentiation and expectation.

Assume without loss of generality that $\lambda \geq 0$. First, we show that $\mathbb{E}[|Xe^{\lambda X}|]$ exists. Since $e^{|X|} \leq e^X + e^{-X}$, we get

$$|Xe^{\lambda X}| \leq e^{|(\lambda+1)X|} \leq e^{(\lambda+1)X} + e^{-(\lambda+1)X}.$$

By assumption on X, both of the summands are absolutely integrable in X. Therefore by the dominated convergence theorem, $\mathbb{E}[|Xe^{\lambda X}|]$ exists and is continuous in λ.

Second, by the existence and continuity of $\mathbb{E}[|Xe^{\lambda X}|]$, $u \mapsto \mathbb{E}[|Xe^{uX}|]$ is integrable on $[0, \lambda]$, and we can switch the order of integration and differentiation as follows:

$$e^{\psi_X(\lambda)} = \mathbb{E}[e^{\lambda X}] = \mathbb{E}\left[1 + \int_0^\lambda Xe^{uX} du\right] \overset{\text{Fubini}}{=} 1 + \int_0^\lambda \mathbb{E}\left[Xe^{uX}\right] du$$
$$\Rightarrow \psi_X'(\lambda) e^{\psi_X(\lambda)} = \mathbb{E}[Xe^{\lambda X}],$$

thus $\psi_X'(\lambda) = e^{-\psi_X(\lambda)}\mathbb{E}[Xe^{\lambda X}]$ exists and is continuous in λ on $\mathbb{R}$.

Furthermore, using a similar application of the dominated convergence theorem we can extend to $\lambda \in \mathbb{C}$ and show that $\lambda \mapsto \mathbb{E}[e^{\lambda X}]$ is a holomorphic function. Thus it is infinitely differentiable.

(d) $A \le X \le B \Rightarrow \psi_X'(\lambda) = \frac{\mathbb{E}[Xe^{\lambda X}]}{\mathbb{E}[e^{\lambda X}]} \in [A, B]$.

(e) Suppose (for contradiction) that $P_X[X > B] > 0$. Then $P_X[X > B + 2\epsilon] > 0$ for some small $\epsilon > 0$. But then $P_\lambda[X \le B + \epsilon] \to 0$ for $\lambda \to \infty$ (see Theorem 15.8(d) below). On the other hand, we know from Theorem 15.8(b) that $\mathbb{E}_{P_\lambda}[X] = \psi_X'(\lambda) \le B$. This is not yet a contradiction, since P_λ might still have some very small mass at a very negative value. To show that this cannot happen, we first assume that $B - \epsilon > 0$ (otherwise just replace X with $X - 2B$). Next note that

$$\begin{aligned} B \ge \mathbb{E}_{P_\lambda}[X] &= \mathbb{E}_{P_\lambda}[X1\{X < B - \epsilon\}] + \mathbb{E}_{P_\lambda}[X1\{B - \epsilon \le X \le B + \epsilon\}] \\ &\quad + \mathbb{E}_{P_\lambda}[X1\{X > B + \epsilon\}] \\ &\ge \mathbb{E}_{P_\lambda}[X1\{X < B - \epsilon\}] + \mathbb{E}_{P_\lambda}[X1\{X > B + \epsilon\}] \\ &\ge -\mathbb{E}_{P_\lambda}[|X|1\{X < B - \epsilon\}] + (B + \epsilon)\underbrace{P_\lambda[X > B + \epsilon]}_{\to 1}. \end{aligned} \quad (15.3)$$

Therefore we will obtain a contradiction if we can show that $\mathbb{E}_{P_\lambda}[|X|1\{X < B - \epsilon\}] \to 0$ as $\lambda \to \infty$. To that end, notice that the convexity of ψ_X implies that $\psi_X' \nearrow B$. Thus, for all $\lambda \ge \lambda_0$ we have $\psi_X'(\lambda) \ge B - \epsilon/2$. Thus, we have for all $\lambda \ge \lambda_0$

$$\psi_X(\lambda) \ge \psi_X(\lambda_0) + (\lambda - \lambda_0)(B - \epsilon/2) = c + \lambda(B - \epsilon/2), \quad (15.4)$$

for some constant c. Then,

$$\begin{aligned} \mathbb{E}_{P_\lambda}[|X|1\{X < B - \epsilon\}] &= \mathbb{E}[|X|e^{\lambda X - \psi_X(\lambda)}1\{X < B - \epsilon\}] \\ &\le \mathbb{E}[|X|e^{\lambda X - c - \lambda(B - \epsilon/2)}1\{X < B - \epsilon\}] \\ &\le \mathbb{E}[|X|e^{\lambda(B-\epsilon) - c - \lambda(B - \epsilon/2)}] \\ &= \mathbb{E}[|X|]e^{-\lambda\epsilon/2 - c} \to 0, \quad \lambda \to \infty, \end{aligned}$$

where the first inequality is from (15.4) and the second from $X < B - \epsilon$. Thus, the first term in (15.3) goes to zero, implying the desired contradiction.

(f) Suppose ψ_X is not strictly convex. Since ψ_X is convex from part (a), ψ_X must be "flat" (affine) near some point. That is, there exists a small neighborhood of some λ_0 such that $\psi_X(\lambda_0+u) = \psi_X(\lambda_0)+ur$ for some $r \in \mathbb{R}$. Then $\psi_{P_\lambda}(u) = ur$ for all u in a small neighborhood of zero, or equivalently $\mathbb{E}_{P_\lambda}[e^{u(X-r)}] = 1$ for u small. The following Lemma 15.4 implies $P_\lambda[X = r] = 1$, but then $P[X = r] = 1$, contradicting the assumption $X \neq$ constant. □

Lemma 15.4. *If* $\mathbb{E}[e^{uS}] = 1$ *for all* $u \in (-\epsilon, \epsilon)$ *then* $S = 0$.

Proof. Expand in Taylor series around $u = 0$ to obtain $\mathbb{E}[S] = 0$, $\mathbb{E}[S^2] = 0$. Alternatively, we can extend the argument we gave for differentiating $\psi_X(\lambda)$ to show that the function $z \mapsto \mathbb{E}[e^{zS}]$ is holomorphic on the entire complex plane.[1] Thus by uniqueness, $\mathbb{E}[e^{uS}] = 1$ for all u. □

Definition 15.5. (Rate function) The rate function $\psi_X^*: \mathbb{R} \to \mathbb{R} \cup \{+\infty\}$ is given by the *Fenchel–Legendre conjugate* (convex conjugate) of the log MGF:

$$\psi_X^*(\gamma) = \sup_{\lambda \in \mathbb{R}} \lambda\gamma - \psi_X(\lambda). \tag{15.5}$$

Note that the maximization (15.5) is a convex optimization problem since ψ_X is strictly convex, so we can find the maximum by taking the derivative and finding the stationary point. In fact, ψ_X^* is precisely the convex conjugate of ψ_X; see (7.84).

The next result describes useful properties of the rate function. See Figure 15.2 for an illustration.

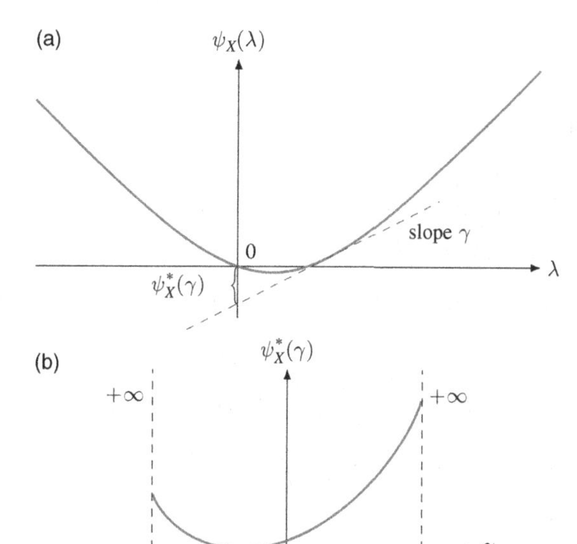

Figure 15.2 (a) Log MGF ψ_X and (b) its conjugate (rate function) ψ_X^* for X taking values in $[A, B]$, continuing the example in Figure 15.1.

[1] More precisely, if we only know that $\mathbb{E}[e^{\lambda S}]$ is finite for $|\lambda| \le 1$ then the function $z \mapsto \mathbb{E}[e^{zS}]$ is holomorphic in the vertical strip $\{z : |\mathrm{Re}\, z| < 1\}$.

Theorem 15.6. (Properties of ψ_X^*) *Assume that X is non-constant and satisfies Assumption 15.1.*

(a) Let $A = \operatorname{essinf} X$ and $B = \operatorname{esssup} X$. Then

$$\psi_X^*(\gamma) = \begin{cases} \lambda\gamma - \psi_X(\lambda) \text{ for } \lambda \text{ s.t. } \gamma = \psi_X'(\lambda), & A < \gamma < B, \\ \log \frac{1}{P(X=\gamma)}, & \gamma = A \text{ or } B, \\ +\infty, & \gamma < A \text{ or } \gamma > B. \end{cases}$$

(b) ψ_X^* is strictly convex and strictly positive except $\psi_X^*(\mathbb{E}[X]) = 0$.
(c) ψ_X^* is decreasing when $\gamma \in (A, \mathbb{E}[X]]$, and increasing when $\gamma \in [\mathbb{E}[X], B)$.

Proof. By Theorem 15.3(d), since $A \leq X \leq B$ a.s., we have $A \leq \psi_X' \leq B$. When $\gamma \in (A, B)$, the strictly concave function $\lambda \mapsto \lambda\gamma - \psi_X(\lambda)$ has a single stationary point which achieves the unique maximum. When $\gamma > B$ (respectively $< A$), $\lambda \mapsto \lambda\gamma - \psi_X(\lambda)$ increases (respectively decreases) without bounds. When $\gamma = B$, since $X \leq B$ a.s., we have

$$\begin{aligned} \psi_X^*(B) &= \sup_{\lambda\in\mathbb{R}} \lambda B - \log(\mathbb{E}[\exp(\lambda X)]) = -\log \inf_{\lambda\in\mathbb{R}} \mathbb{E}[\exp(\lambda(X - B))] \\ &= -\log \lim_{\lambda\to\infty} \mathbb{E}[\exp(\lambda(X - B))] = -\log P(X = B), \end{aligned}$$

by the monotone convergence theorem.

By Theorem 15.3(f), since ψ_X is strictly convex, the derivatives of ψ_X and ψ_X^* are inverses of each other. Hence ψ_X^* is strictly convex. Since $\psi_X(0) = 0$, we have $\psi_X^*(\gamma) \geq 0$. Moreover, $\psi_X^*(\mathbb{E}[X]) = 0$ follows from $\mathbb{E}[X] = \psi_X'(0)$. □

15.1.2 Tilted Distribution

As early as in Chapter 4, we have already introduced the concept of *tilting* in the proof of the Donsker–Varadhan variational characterization of divergence (Theorem 4.6). Let us formally define it now.

Definition 15.7. (Tilting) Given $X \sim P$ and $\lambda \in \mathbb{R}$, the tilted measure P_λ is defined by

$$P_\lambda(dx) = \frac{\exp\{\lambda x\}}{\mathbb{E}[\exp\{\lambda X\}]} P(dx) = \exp\{\lambda x - \psi_X(\lambda)\} P(dx). \tag{15.6}$$

In particular, if P has a PDF p, then the PDF of P_λ is given by $p_\lambda(x) = e^{\lambda x - \psi_X(\lambda)} p(x)$.

The set of distributions $\{P_\lambda : \lambda \in \mathbb{R}\}$ parameterized by λ is called a *standard (one-parameter) exponential family*, an important object in statistics [77]. Here are some examples:

- *Gaussian*: $P = \mathcal{N}(0, 1)$ with density $p(x) = \frac{1}{\sqrt{2\pi}} \exp(-x^2/2)$. Then P_λ has density $\frac{\exp(\lambda x)}{\exp(\lambda^2/2)} \frac{1}{\sqrt{2\pi}} \exp(-x^2/2) = \frac{1}{\sqrt{2\pi}} \exp(-(x - \lambda)^2/2)$. Hence $P_\lambda = \mathcal{N}(\lambda, 1)$.

- *Bernoulli*: $P = \text{Ber}(1/2)$. Then $P_\lambda = \text{Ber}\left(\frac{e^\lambda}{e^\lambda+1}\right)$ which puts more (respectively less) mass on 1 if $\lambda > 0$ (respectively < 0). Moreover, $P_\lambda \xrightarrow{d} \delta_1$ if $\lambda \to \infty$ or δ_0 if $\lambda \to -\infty$.
- *Uniform*: Let P be the uniform distribution on $[0, 1]$. Then P_λ is also supported on $[0, 1]$ with PDF $p_\lambda(x) = \frac{\lambda \exp(\lambda x)}{e^\lambda - 1}$. Therefore as λ increases, P_λ becomes increasingly concentrated near 1, and $P_\lambda \to \delta_1$ as $\lambda \to \infty$. Similarly, $P_\lambda \to \delta_0$ as $\lambda \to -\infty$.

In the above examples we see that P_λ shifts the mean of P to the right (respectively left) when $\lambda > 0$ (respectively < 0). Indeed, this is a general property of tilting.

Theorem 15.8. (Properties of P_λ) *Under Assumption 15.1 we have the following:*

(a) Log MGF:

$$\psi_{P_\lambda}(u) = \psi_X(\lambda + u) - \psi_X(\lambda).$$

(b) Tilting trades mean for divergence:

$$\mathbb{E}_{P_\lambda}[X] = \psi'_X(\lambda) \gtrless \mathbb{E}_P[X] \text{ if } \lambda \gtrless 0; \tag{15.7}$$

$$D(P_\lambda \| P) = \psi^*_X(\psi'_X(\lambda)) = \psi^*_X(\mathbb{E}_{P_\lambda}[X]). \tag{15.8}$$

(c) Tilted variance: $\text{Var}_{P_\lambda}(X) = \psi''_X(\lambda) \log e$.

(d) Finally

$$P(X > b) > 0 \Rightarrow \forall \epsilon > 0, P_\lambda(X \le b - \epsilon) \to 0 \text{ as } \lambda \to \infty;$$

$$P(X < a) > 0 \Rightarrow \forall \epsilon > 0, P_\lambda(X \ge a + \epsilon) \to 0 \text{ as } \lambda \to -\infty.$$

Therefore if $X_\lambda \sim P_\lambda$*, then* $X_\lambda \xrightarrow{d} \text{essinf}\, X = A$ *as* $\lambda \to -\infty$ *and* $X_\lambda \xrightarrow{d} \text{esssup}\, X = B$ *as* $\lambda \to \infty$.

Proof. Again for the proof we assume base e for exp and log.

(a) By definition.
(b) $\mathbb{E}_{P_\lambda}[X] = \frac{\mathbb{E}[X \exp(\lambda X)]}{\mathbb{E}[\exp(\lambda X)]} = \psi'_X(\lambda)$, which is strictly increasing in λ, with $\psi'_X(0) = \mathbb{E}_P[X]$.
$D(P_\lambda \| P) = \mathbb{E}_{P_\lambda} \log \frac{dP_\lambda}{dP} = \mathbb{E}_{P_\lambda} \log \frac{\exp(\lambda X)}{\mathbb{E}[\exp(\lambda X)]} = \lambda \mathbb{E}_{P_\lambda}[X] - \psi_X(\lambda) = \lambda \psi'_X(\lambda) - \psi_X(\lambda) = \psi^*_X(\psi'_X(\lambda))$, where the last equality follows from Theorem 15.6(a).
(c) $\psi''_X(\lambda) = \frac{\mathbb{E}[X^2 \exp(\lambda X)] - \mathbb{E}[X \exp(\lambda X)]^2}{\mathbb{E}[\exp(\lambda X)]^2} = \text{Var}_{P_\lambda}(X)$.
(d) We have

$$\begin{aligned} P_\lambda(X \le b - \epsilon) &= \mathbb{E}_P[e^{\lambda X - \psi_X(\lambda)} 1\{X \le b - \epsilon\}] \\ &\le \mathbb{E}_P[e^{\lambda(b-\epsilon) - \psi_X(\lambda)} 1\{X \le b - \epsilon\}] \\ &\le e^{-\lambda\epsilon} e^{\lambda b - \psi_X(\lambda)} \\ &\le \frac{e^{-\lambda\epsilon}}{P[X > b]} \to 0 \text{ as } \lambda \to \infty, \end{aligned}$$

where the last inequality is due to the usual Chernoff bound (Theorem 15.3(g)): $P[X > b] \le \exp(-\lambda b + \psi_X(\lambda))$. □

15.2 Large-Deviations Exponents and KL Divergence

Large-deviations problems deal with rare events by making statements about the tail probabilities of a sequence of distributions. Here, we are interested in the following special case: the speed of decay for $P\left[\frac{1}{n}\sum_{k=1}^n X_k \geq \gamma\right]$ for iid X_k when γ exceeds the mean.

In (15.1) we have used the Chernoff bound to obtain an upper bound on the exponent via the log MGF. Here we use a different method to give a formula for the exponent as a convex optimization problem involving the KL divergence, known as information projection (Section 15.3). Later in Section 15.4 we shall revisit the Chernoff bound after we have computed the value of the information projection.

Theorem 15.9. *Let $X_1, X_2, \ldots \stackrel{\text{iid}}{\sim} P$. Then for any $\gamma \in \mathbb{R}$,*

$$\lim_{n\to\infty} \frac{1}{n} \log \frac{1}{P\left[\frac{1}{n}\sum_{k=1}^n X_k > \gamma\right]} = \inf_{Q:\, \mathbb{E}_Q[X]>\gamma} D(Q\|P), \tag{15.9}$$

$$\lim_{n\to\infty} \frac{1}{n} \log \frac{1}{P\left[\frac{1}{n}\sum_{k=1}^n X_k \geq \gamma\right]} = \inf_{Q:\, \mathbb{E}_Q[X]\geq\gamma} D(Q\|P). \tag{15.10}$$

Furthermore, for every n we have the firm upper bound

$$P\left[\frac{1}{n}\sum_{k=1}^n X_k \geq \gamma\right] \leq \exp\left\{-n \cdot \inf_{Q:\, \mathbb{E}_Q[X]\geq\gamma} D(Q\|P)\right\}, \tag{15.11}$$

and similarly for $>$ in place of $\geq$.

Remark 15.2. (Subadditivity) One can argue from first principles that the limits (15.9) and (15.10) exist without computing their values. Indeed, note that the sequence $p_n \triangleq \log \frac{1}{P\left[\frac{1}{n}\sum_{k=1}^n X_k \geq \gamma\right]}$ satisfies $p_{n+m} \geq p_n p_m$ and hence $\log \frac{1}{p_n}$ is subadditive. As such, $\lim_{n\to\infty} \frac{1}{n}\log\frac{1}{p_n} = \inf_n \frac{1}{n}\log\frac{1}{p_n}$ by Fekete's lemma.

Proof. First note that if the events have zero probability, then both sides coincide with infinity. Indeed, if $P\left[\frac{1}{n}\sum_{k=1}^n X_k > \gamma\right] = 0$, then $P[X > \gamma] = 0$. Then $\mathbb{E}_Q[X] > \gamma \Rightarrow Q[X > \gamma] > 0 \Rightarrow Q \not\ll P \Rightarrow D(Q\|P) = \infty$ and hence (15.9) holds trivially. The case for (15.10) is similar.

In the sequel we assume both probabilities are non-zero. We start by proving (15.9). Set $P[E_n] = P\left[\frac{1}{n}\sum_{k=1}^n X_k > \gamma\right]$.

Lower bound on $P[E_n]$: Fix a Q such that $\mathbb{E}_Q[X] > \gamma$. Let X^n be iid. Then by the WLLN,

$$Q[E_n] = Q\left[\sum_{k=1}^n X_k > n\gamma\right] = 1 - o(1).$$

Now the data-processing inequality (Corollary 2.19) gives

$$d(Q[E_n]\|P[E_n]) \leq D(Q_{X^n}\|P_{X^n}) = nD(Q\|P).$$

And a lower bound for the binary divergence is

$$d(Q[E_n]\|P[E_n]) \geq -h(Q[E_n]) + Q[E_n]\log\frac{1}{P[E_n]}.$$

Combining the two bounds on $d(Q[E_n]\|P[E_n])$ gives

$$P[E_n] \geq \exp\left(\frac{-nD(Q\|P) - \log 2}{Q[E_n]}\right). \tag{15.12}$$

Optimizing over Q to give the best bound:

$$\limsup_{n\to\infty}\frac{1}{n}\log\frac{1}{P[E_n]} \leq \inf_{Q:\,\mathbb{E}_Q[X]>\gamma} D(Q\|P).$$

Upper bound on $P[E_n]$: The key observation is that given any X and any event E, $P_X(E) > 0$ can be expressed via the divergence between the conditional and unconditional distributions as: $\log\frac{1}{P_X(E)} = D(P_{X|X\in E}\|P_X)$. Define $\tilde{P}_{X^n} = P_{X^n|\sum X_i > n\gamma}$, under which $\sum X_i > n\gamma$ holds a.s. Then

$$\log\frac{1}{P[E_n]} = D(\tilde{P}_{X^n}\|P_{X^n}) \geq \inf_{Q_{X^n}:\,\mathbb{E}_Q[\sum X_i]>n\gamma} D(Q_{X^n}\|P_{X^n}). \tag{15.13}$$

We now show that the last problem "single-letterizes," that is, reduces to $n = 1$. Note that this is a special case of a more general phenomenon – see Exercise **III.12**. Consider the following two steps:

$$\begin{aligned} D(Q_{X^n}\|P_{X^n}) &\geq \sum_{j=1}^{n} D(Q_{X_j}\|P) \\ &\geq nD(\bar{Q}\|P), \qquad \bar{Q} \triangleq \frac{1}{n}\sum_{j=1}^{n} Q_{X_j}, \end{aligned} \tag{15.14}$$

where the first step follows from (2.27) in Theorem 2.16, after noticing that $P_{X^n} = P^n$, and the second step is by convexity of divergence (Theorem 5.1). From this argument we conclude that

$$\inf_{Q_{X^n}:\,\mathbb{E}_Q[\sum X_i]>n\gamma} D(Q_{X^n}\|P_{X^n}) = n\cdot\inf_{Q:\,\mathbb{E}_Q[X]>\gamma} D(Q\|P), \tag{15.15}$$

$$\inf_{Q_{X^n}:\,\mathbb{E}_Q[\sum X_i]\geq n\gamma} D(Q_{X^n}\|P_{X^n}) = n\cdot\inf_{Q:\,\mathbb{E}_Q[X]\geq\gamma} D(Q\|P). \tag{15.16}$$

In particular, (15.13) and (15.15) imply the required lower bound in (15.9).

Next we prove (15.10). First, notice that the lower bound argument (15.13) applies equally well, so that for each n we have

$$\frac{1}{n}\log\frac{1}{P\left[\frac{1}{n}\sum_{k=1}^{n} X_k \geq \gamma\right]} \geq \inf_{Q:\,\mathbb{E}_Q[X]\geq\gamma} D(Q\|P).$$

To get a matching upper bound we consider two cases:

- Case I: $P[X > \gamma] = 0$. If $P[X \geq \gamma] = 0$, then both sides of (15.10) are $+\infty$. If $P[X = \gamma] > 0$, then $P[\sum X_k \geq n\gamma] = P[X_1 = \cdots = X_n = \gamma] = P[X = \gamma]^n$. For the right-hand side, since $D(Q\|P) < \infty \implies Q \ll P \implies Q[X \leq \gamma] = 1$, the only possibility for $\mathbb{E}_Q[X] \geq \gamma$ is that $Q[X = \gamma] = 1$, that is, $Q = \delta_\gamma$. Then $\inf_{\mathbb{E}_Q[X]\geq\gamma} D(Q\|P) = \log \frac{1}{P(X=\gamma)}$.

- Case II: $P[X > \gamma] > 0$. Since $\mathbb{P}[\sum X_k \geq \gamma] \geq \mathbb{P}[\sum X_k > \gamma]$ from (15.9) we know that

$$\limsup_{n\to\infty} \frac{1}{n} \log \frac{1}{P\left[\frac{1}{n}\sum_{k=1}^n X_k \geq \gamma\right]} \leq \inf_{Q:\, \mathbb{E}_Q[X]>\gamma} D(Q\|P).$$

We next show that in this case

$$\inf_{Q:\, \mathbb{E}_Q[X]>\gamma} D(Q\|P) = \inf_{Q:\, \mathbb{E}_Q[X]\geq\gamma} D(Q\|P). \tag{15.17}$$

Indeed, let $\tilde{P} = P_{X|X>\gamma}$ which is well defined since $P[X > \gamma] > 0$. For any Q such that $\mathbb{E}_Q[X] \geq \gamma$, set $\tilde{Q} = \bar{\epsilon}Q + \epsilon\tilde{P}$ satisfying $\mathbb{E}_{\tilde{Q}}[X] > \gamma$. Then by convexity, $D(\tilde{Q}\|P) \leq \bar{\epsilon}D(Q\|P) + \epsilon D(\tilde{P}\|P) = \bar{\epsilon}D(Q\|P) + \epsilon \log \frac{1}{P[X>\gamma]}$. Sending $\epsilon \to 0$, we conclude the proof of (15.17). □

Remark 15.3. Note that the upper bound (15.11) also holds for independent *non-identically* distributed X_i. Indeed, we only need to replace the step (15.14) with $D(Q_{X^n}\|P_{X^n}) \geq \sum_{i=1}^n D(Q_{X_i}\|P_{X_i}) \geq nD(\bar{Q}\|\bar{P})$ where $\bar{P} = \frac{1}{n}\sum_{i=1}^n P_{X_i}$. This yields a bound (15.11) with P replaced by $\bar{P}$ in the right-hand side.

Example 15.1. Poisson-binomial tails

Consider X which is a sum of n independent Bernoulli random variables so that $\mathbb{E}[X] = np$. The distribution of X is known as *Poisson-binomial* [331, 414], including $\mathrm{Bin}(n,p)$ as a special case. Applying Theorem 15.9 (or Remark 15.3), we get the following tail bounds on X:

$$\mathbb{P}[X \geq k] \leq \exp\{-nd(k/n\|p)\}, \quad \frac{k}{n} > p, \tag{15.18}$$

$$\mathbb{P}[X \leq k] \leq \exp\{-nd(k/n\|p)\}, \quad \frac{k}{n} < p, \tag{15.19}$$

where for (15.18) we used the fact that $\min_{Q:\, \mathbb{E}_Q[X]\geq k/n} D(Q\|\mathrm{Ber}(p)) = \min_{q\geq k/n} d(q\|p) = d(\frac{k}{n}\|p)$ and similarly for (15.19). These bounds, in turn, can be used to derive various famous estimates:

- Multiplicative deviation from the mean (Bennett's inequality): We have

$$\mathbb{P}[X \geq u\,\mathbb{E}[X]] \leq \exp\{-\mathbb{E}[X]f(u)\} \quad \forall u > 1,$$
$$\mathbb{P}[X \leq u\,\mathbb{E}[X]] \leq \exp\{-\mathbb{E}[X]f(u)\} \quad \forall 0 \leq u < 1,$$

where $f(u) \triangleq u\log u - (u-1)\log e \geq 0$. These follow from (15.18) and (15.19) via the following useful estimate:

$$d(up\|p) \geq pf(u) \quad \forall p \in [0,1], u \in [0, 1/p]. \tag{15.20}$$

Indeed, consider the elementary inequality

$$x \log \frac{x}{y} \geq (x-y)\log e$$

for all $x, y \in [0, 1]$ (since the difference between the LHS and RHS is minimized over y at $y = x$). Using $x = 1 - up$ and $y = 1 - p$ establishes (15.20).

- Bernstein's inequality:

$$\mathbb{P}[X > np + t] \le e^{-t^2/(2(t+np))} \qquad \forall t > 0.$$

This follows from the previous bound for $u > 1$ by bounding

$$\frac{f(u)}{\log e} = \int_1^u \frac{u - x}{x} dx \ge \frac{1}{u} \int_1^u (u - x) dx = \frac{(u-1)^2}{2u}.$$

- Okamoto's inequality: For all $0 < p < 1$ and $t > 0$,

$$\mathbb{P}[\sqrt{X} - \sqrt{np} \ge t] \le e^{-t^2}, \tag{15.21}$$

$$\mathbb{P}[\sqrt{X} - \sqrt{np} \le -t] \le e^{-t^2}. \tag{15.22}$$

These simply follow from the inequality between KL divergence and Hellinger distance in (7.33). Indeed, we get $d(x\|p) \ge H^2(\mathrm{Ber}(x), \mathrm{Ber}(p)) \ge (\sqrt{x} - \sqrt{p})^2$. After Plugging $x = \frac{(\sqrt{np}+t)^2}{n}$ into (15.18) and (15.19) we obtain the result. We note that [317, Theorem 3] shows a stronger bound of e^{-2t^2} in (15.21).

Remarkably, the bounds in (15.21) and (15.22) do not depend on n or p. This is due to the variance-stabilizing effect of the square-root transformation for binomials: $\mathrm{Var}(\sqrt{X})$ is at most a constant for all n, p. In addition, $\sqrt{X} - \sqrt{np} = \frac{X - np}{\sqrt{X} + \sqrt{np}}$ is of a self-normalizing form: the denominator is on par with the standard deviation of the numerator. For more on self-normalizing sums, see [69, Problem 12.2].

15.3 Information Projection

The results of Theorem 15.9 motivate us to study the following general *information projection problem*: Let $\mathcal{E}$ be a convex set of distributions on some abstract space Ω, then for the distribution P on Ω, we want to compute

$$\inf_{Q \in \mathcal{E}} D(Q\|P).$$

Denote the minimizing distribution Q by Q^*. The next result shows that intuitively the "line" between P and optimal Q^* is "orthogonal" to $\mathcal{E}$ (see Figure 15.3).

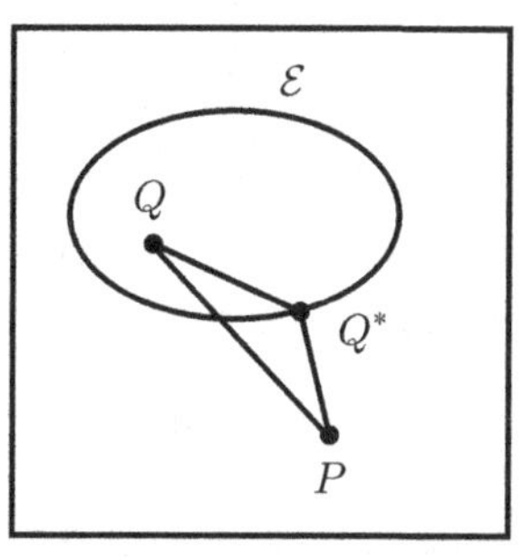

Figure 15.3 Illustration of information projection and the Pythagorean theorem.

Theorem 15.10. *Let $\mathcal{E}$ be a convex set of distributions. If there exists $Q^* \in \mathcal{E}$ such that $D(Q^*\|P) = \min_{Q\in\mathcal{E}} D(Q\|P) < \infty$, then $\forall Q \in \mathcal{E}$*

$$D(Q\|P) \geq D(Q\|Q^*) + D(Q^*\|P).$$

Proof. If $D(Q\|P) = \infty$, then there is nothing to prove. So we assume that $D(Q\|P) < \infty$, which also implies that $D(Q^*\|P) < \infty$. For $\lambda \in [0,1]$, form the convex combination $Q^{(\lambda)} = \bar{\lambda}Q^* + \lambda Q \in \mathcal{E}$. Since Q^* is the minimizer of $D(Q\|P)$, then

$$0 \leq \left.\frac{d}{d\lambda}\right|_{\lambda=0} D(Q^{(\lambda)}\|P) = D(Q\|P) - D(Q\|Q^*) - D(Q^*\|P).$$

The rigorous analysis requires an argument for interchanging derivatives and integrals (via the dominated convergence theorem) and is similar to the proof of Proposition 2.20. The details are in [114, Theorem 2.2]. □

Remark 15.4. If we view the picture above in the Euclidean setting, the "triangle" formed by P, Q^*, and Q (for Q^*, Q in a convex set, P outside the set) is always obtuse, and is a right triangle only when the convex set has a "flat face." In this sense, the divergence is similar to the squared Euclidean distance, and the above theorem is sometimes known as a "Pythagorean" theorem.

The relevant set $\mathcal{E}$ of Q's upon which we will focus next is the "half-space" of distributions $\mathcal{E} = \{Q : \mathbb{E}_Q[X] \geq \gamma\}$, where $X\colon \Omega \to \mathbb{R}$ is some fixed function (random variable). This is justified by the relationship to the large-deviations exponent in Theorem 15.9. First, we solve this I-projection problem explicitly.

Theorem 15.11. *Given a distribution P on Ω and $X\colon \Omega \to \mathbb{R}$ suppose Assumption 15.1 holds. We denote*

$$A = \inf \psi'_X = \operatorname{essinf} X \triangleq \sup\{a\colon X \geq a\ P\text{-a.s.}\}, \tag{15.23}$$
$$B = \sup \psi'_X = \operatorname{esssup} X \triangleq \inf\{b\colon X \leq b\ P\text{-a.s.}\}. \tag{15.24}$$

1 The information projection problem over $\mathcal{E} = \{Q : \mathbb{E}_Q[X] \geq \gamma\}$ has solution

$$\min_{Q\,:\,\mathbb{E}_Q[X]\geq\gamma} D(Q\|P) = \begin{cases} 0, & \gamma < \mathbb{E}_P[X], \\ \psi_P^*(\gamma), & \mathbb{E}_P[X] \leq \gamma < B, \\ \log\frac{1}{P(X=B)}, & \gamma = B, \\ +\infty, & \gamma > B, \end{cases} \tag{15.25}$$

$$= \psi_P^*(\gamma)1\{\gamma \geq \mathbb{E}_P[X]\}. \tag{15.26}$$

2 Whenever the minimum is finite, the minimizing distribution is unique and equal to the tilting of P along X, namely[2]

$$dP_\lambda = \exp\{\lambda X - \psi(\lambda)\} \cdot dP. \tag{15.27}$$

[2] Note that unlike the setting of Theorems *15.1* and *15.9* here P and P_λ are measures on an abstract space Ω, not necessarily on the real line.

3 *For all* $\gamma \in [\mathbb{E}_P[X], B)$ *we have*

$$\min_{\mathbb{E}_Q[X]\geq\gamma} D(Q\|P) = \inf_{\mathbb{E}_Q[X]>\gamma} D(Q\|P) = \min_{\mathbb{E}_Q[X]=\gamma} D(Q\|P).$$

Remark 15.5. Both Theorem 15.9 and Theorem 15.11 are stated for the right tail where the sample mean exceeds the population mean. For the left tail, simply apply these results to $-X_i$ to obtain for $\gamma < \mathbb{E}[X]$,

$$\lim_{n\to\infty}\frac{1}{n}\log\frac{1}{P\left[\frac{1}{n}\sum_{k=1}^n X_k < \gamma\right]} = \inf_{Q:\,\mathbb{E}_Q[X]<\gamma} D(Q\|P) = \psi_X^*(\gamma).$$

In other words, the large-deviations exponent is still given by the rate function (15.5) except that the optimal tilting parameter λ is negative.

Proof. We first prove (15.25).

- First case: Take $Q = P$.
- Fourth case: If $\mathbb{E}_Q[X] > B$, then $Q[X \geq B+\epsilon] > 0$ for some $\epsilon > 0$, but $P[X \geq B+\epsilon] = 0$, since $P[X \leq B] = 1$, by Theorem 15.3(e). Hence $Q \not\ll P \implies D(Q\|P) = \infty$.
- Third case: If $P(X = B) = 0$, then $X < B$ a.s. under P, and $Q \not\ll P$ for any Q s.t. $\mathbb{E}_Q[X] \geq B$. Then the minimum is ∞. Now assume $P[X = B] > 0$. Since $D(Q\|P) < \infty \implies Q \ll P \implies Q[X \leq B] = 1$, the only possibility for $\mathbb{E}_Q[X] \geq B$ is that $Q[X = B] = 1$, that is, $Q = \delta_B$. Then $D(Q\|P) = \log\frac{1}{P(X=B)}$.
- Second case: Fix $\mathbb{E}_P[X] \leq \gamma < B$, and find the unique λ such that $\psi_X'(\lambda) = \gamma = \mathbb{E}_{P_\lambda}[X]$ where $dP_\lambda = \exp(\lambda X - \psi_X(\lambda))dP$. This corresponds to tilting P far enough to the right to increase its mean from $\mathbb{E}_P X$ to γ, in particular $\lambda \geq 0$. Moreover, $\psi_X^*(\gamma) = \lambda\gamma - \psi_X(\lambda)$. Take any Q such that $\mathbb{E}_Q[X] \geq \gamma$, then

$$\begin{aligned}
D(Q\|P) &= \mathbb{E}_Q\left[\log\frac{dQdP_\lambda}{dPdP_\lambda}\right] \qquad (15.28)\\
&= D(Q\|P_\lambda) + \mathbb{E}_Q\left[\log\frac{dP_\lambda}{dP}\right]\\
&= D(Q\|P_\lambda) + \mathbb{E}_Q[\lambda X - \psi_X(\lambda)]\\
&\geq D(Q\|P_\lambda) + \lambda\gamma - \psi_X(\lambda)\\
&= D(Q\|P_\lambda) + \psi_X^*(\gamma)\\
&\geq \psi_X^*(\gamma), \qquad (15.29)
\end{aligned}$$

where the last inequality holds with equality if and only if $Q = P_\lambda$. In addition, this shows the minimizer is unique, proving the second claim. Note that even in the corner case of $\gamma = B$ (assuming $P(X = B) > 0$) the minimizer is a point mass $Q = \delta_B$, which is also a tilted measure (P_∞), since $P_\lambda \to \delta_B$ as $\lambda \to \infty$, see Theorem 15.8(d).

An alternative version of the solution, given by expression (15.26), follows from Theorem 15.6.

For the third claim, notice that there is nothing to prove for $\gamma < \mathbb{E}_P[X]$, while for $\gamma \geq \mathbb{E}_P[X]$ we have just shown

$$\psi_X^*(\gamma) = \min_{Q:\ \mathbb{E}_Q[X]\geq\gamma} D(Q\|P),$$

while from the next corollary we have

$$\inf_{Q:\ \mathbb{E}_Q[X]>\gamma} D(Q\|P) = \inf_{\gamma'>\gamma} \psi_X^*(\gamma').$$

The final step is to notice that ψ_X^* is increasing and continuous by Theorem 15.6, and hence the right-hand side infimum equals $\psi_X^*(\gamma)$. The case of $\min_{Q:\ \mathbb{E}_Q[X]=\gamma}$ D(Q||P) is handled similarly. □

Corollary 15.12. *For any Q with $\mathbb{E}_Q[X] \in (A, B)$, there exists a* unique $\lambda \in \mathbb{R}$ *such that the tilted distribution P_λ satisfies*

$$\mathbb{E}_{P_\lambda}[X] = \mathbb{E}_Q[X],$$
$$D(P_\lambda\|P) \leq D(Q\|P),$$

and furthermore the gap in the last inequality equals $D(Q\|P_\lambda) = D(Q\|P) - D(P_\lambda\|P)$.

Proof. Proceed as in the proof of Theorem 15.11, and find the unique λ s.t. $\mathbb{E}_{P_\lambda}[X] = \psi_X'(\lambda) = \mathbb{E}_Q[X]$. Then $D(P_\lambda\|P) = \psi_X^*(\mathbb{E}_Q[X]) = \lambda\mathbb{E}_Q[X] - \psi_X(\lambda)$. Repeat the steps (15.28) and (15.29) to obtain $D(Q\|P) = D(Q\|P_\lambda) + D(P_\lambda\|P)$. □

For any Q the previous result allows us to find a tilted measure P_λ that has the same mean as Q yet smaller (or equal) divergence distance to P. We will see that the same can be done under multiple linear constraints (Section 15.6*).

15.4 *I*-Projection and KL Geodesics

Figure 15.4 describes many properties of information projections, where we fix some baseline measure P. Then we have the following.

- Each set $\{Q\colon \mathbb{E}_Q[X] = \gamma\}$ is a slice of $\mathcal{P}(\mathbb{R})$, the space of probability distributions on $\mathbb{R}$. As γ varies from $-\infty$ to $+\infty$ the union of these slices fills the entire space of distributions with finite first moment.

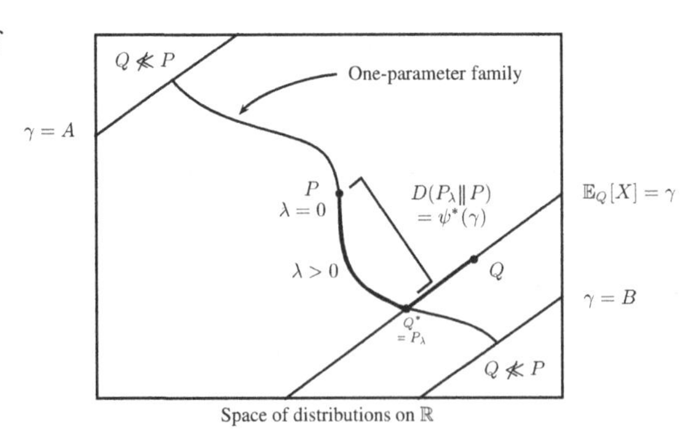

Figure 15.4 Illustration of information projections, tilting, and rate function.

- When $\gamma < A$ or $\gamma > B$, any distribution Q inside the slice satisfies $Q \not\ll P$.
- As γ varies between A and B the slices fill out the space of $\{Q\colon Q \ll P\}$. By Corollary 15.12, inside each slice there is one special distribution P_λ of the tilted form (15.6) that minimizes the divergence $D(Q\|P)$ to P.
- The set of P_λ's trace out a curve in the space of distributions. The "geodesic" distance from P to P_λ is measured by $\psi^*(\gamma) = D(P_\lambda\|P)$. This set of distributions is the one-parameter exponential family in Definition 15.7.

The key observation here is that the curve of this one-parameter family $\{P_\lambda : \lambda \in \mathbb{R}\}$ intersects each γ-slice $\mathcal{E} = \{Q\colon \mathbb{E}_Q[X] = \gamma\}$ "orthogonally" at the minimizing $Q^* \in \mathcal{E}$, and the distance from P to Q^* is given by $\psi^*(\lambda)$. To see this, note that applying Theorem 15.10 to the convex set $\mathcal{E}$ gives us $D(Q\|P) \geq D(Q\|Q^*) + D(Q^*\|P)$. Now thanks to Corollary 15.12, we in fact have an *equality* $D(Q\|P) = D(Q\|Q^*) + D(Q^*\|P)$ and $Q^* = P_\lambda$ for some tilted measure.

Let us give an intuitive (non-rigorous) justification for calling the curve $\{P_\mu, \mu \in [0,\lambda]\}$ the geodesic connecting $P = P_0$ to P_λ. Suppose there existed another curve $\{Q_\mu\}$ connecting P to P_λ and minimizing KL distance. Then the expectation $\mathbb{E}_{Q_\mu}[X]$ should continuously change from $\mathbb{E}_P[X]$ to $\mathbb{E}_{P_\lambda}[X]$. Now take any intermediate value γ' of the expectation $\mathbb{E}_{Q_\mu}[X]$. We know that on the slice $\{Q\colon \mathbb{E}_Q[X] = \gamma'\}$ the closet element to P is $P_{\mu'}$ for some $\mu' \in [0,\lambda]$. Thus, we could shorten the distance by traveling from P to $P_{\mu'}$ instead of to $Q_{\mu'}$.

Our treatment above is specific to distributions on $\mathbb{R}$. How do we find a geodesic between two arbitrary distributions $\tilde{P}$ and $\tilde{Q}$ on an abstract measurable space $\mathcal{X}$? To find the answer we notice that the "intrinsic" definition of the geodesic between P and P_λ above can be given as follows:

$$\frac{dP_\mu}{dP} = \frac{1}{Z(\mu)} \left(\frac{dP_\lambda}{dP}\right)^{\mu/\lambda},$$

where $Z(\mu) = \exp\{\psi(\mu)\}$ is a normalization constant. Correspondingly, we define the geodesic between $\tilde{P}$ and $\tilde{Q}$ as a parametric family $\{\tilde{P}_\mu, \mu \in [0,1]\}$ given by

$$\frac{d\tilde{P}_\mu}{d\tilde{P}} \triangleq \frac{1}{\tilde{Z}(\mu)} \left(\frac{d\tilde{Q}}{d\tilde{P}}\right)^{\mu}, \tag{15.30}$$

where the normalizing constant $\tilde{Z}(\mu) = \exp\{(\mu-1)D_\mu(\tilde{Q}\|\tilde{P})\}$ is given in terms of Rényi divergence. See also Exercise **III.25**.

Formal justification of (15.30) as a true geodesic in the sense of differential geometry was given by Cencov in [85, Theorem 12.3] for the case of finite underlying space $\mathcal{X}$. His argument was the following. To enable discussion of geodesics one needs to equip the space $\mathcal{P}([k])$ with a connection (or parallel transport map). It is natural to require the connection to be equivariant (or commute) with respect to some maps $\mathcal{P}([k]) \to \mathcal{P}([k'])$. Cencov lists (a) permutations of elements ($k = k'$); (b) embedding of a distribution $P \in \mathcal{P}([k])$ into a larger space by splitting atoms of $[k]$ (with specified ratio) into multiple atoms of $[k']$, so that $k < k'$; and (c) conditioning on an event ($k > k'$). It turns out there is a one-parameter family of connections satisfying (a) and (b), including the Riemannian (Levi-Civita) connection corresponding to a Fisher–Rao metric (2.35). However, there is only a unique connection satisfying all of (a)–(c). It is different from Fisher–Rao and its geodesics are exactly given by (15.30). Geodesically complete submanifolds in this metric are simply the exponential families (Section 15.6*). For more on this exciting area, see Cencov [85] and Amari and Nagaoka [17].

15.5 Sanov's Theorem

A corollary of the WLLN is that the empirical distribution $\hat{P}$ of n iid observations drawn from a distribution P converges weakly to P itself. The following theorem due to Sanov [371] quantifies the large-deviations behavior of this convergence.

Theorem 15.13. (Sanov's theorem) *Given $X_1, \ldots, X_n \overset{\text{iid}}{\sim} P$ on $\mathcal{X}$, denote the empirical distribution by $\hat{P} = \frac{1}{n}\sum_{j=1}^{n} \delta_{X_j}$. Let $\mathcal{E}$ be a set of distributions. Then under regularity conditions on $\mathcal{E}$ and P,*

$$\mathbb{P}[\hat{P} \in \mathcal{E}] = \exp\left\{-n \min_{Q \in \mathcal{E}} D(Q\|P) + o(n)\right\}.$$

Examples of regularity conditions in the above theorem include: (a) $\mathcal{X}$ is finite, P is fully supported on $\mathcal{X}$, and $\mathcal{E}$ is closed with non-empty interior – see Exercise **III.23** for a full proof in this case; and (b) $\mathcal{X}$ is a Polish space and $\inf_{Q \in \text{int}(\mathcal{E})} D(Q\|P) = \inf_{Q \in \text{cl}(\mathcal{E})} D(Q\|P)$ – this is the content of [120, Theorem 6.2.10]. Additionally, [120] contains full details about various other versions and extensions of Sanov's theorem to infinite-dimensional settings.

15.6* Information Projection with Multiple Constraints

We have considered so far the example of a single inequality $\mathbb{E}[X] \geq \gamma$. However, the entire theory can be extended to accommodate multiple constraints. Let P be a fixed distribution on some space $\mathcal{X}$ and let $\phi_1, \ldots, \phi_d : \mathcal{X} \to \mathbb{R}$ be arbitrary functions, which we will stack together into a vector-valued function $\phi : \mathcal{X} \to \mathbb{R}^d$. In this section we discuss the solution of the following I-projection problem, known as I-projection on a hyperplane:

$$F(\gamma) \triangleq \inf\{D(Q\|P) : \mathbb{E}_Q[\phi(X)] = \gamma\}, \qquad \gamma \in \mathbb{R}^d. \tag{15.31}$$

This problem arises in statistical physics, Gibbs variational principle, the exponential family, and many other fields. Note that taking P uniform corresponds to max-entropy problems.

In the case of $d = 1$ we have seen that whenever the value of the minimization is finite, the solution Q^* can be sought inside a single-parameter family of *tilted* measures P, see (15.27). For the more general case of $d > 1$ we define tilted measures as

$$P_\lambda(dx) \triangleq \exp\{\lambda^\top \phi(x) - \psi(\lambda)\}P(dx), \quad \lambda \in \mathbb{R}^d,$$

where the multi-dimensional log MGF of P (with respect to ϕ) is defined as

$$\psi(\lambda) \triangleq \mathbb{E}_{X\sim P}[\exp\{\lambda^\top \phi(X)\}]. \tag{15.32}$$

In order to discuss the solution of (15.31) we first make a simple observation analogous to Corollary 15.12:

Proposition 15.14. *If there exists λ such that $\psi(\lambda) < \infty$ and $\mathbb{E}_{X\sim P_\lambda}[\phi(X)] = \gamma$, then the unique minimizer of (15.31) is P_λ and for any Q with $\mathbb{E}_Q[\phi(X)] = \gamma$ we have*

$$D(Q\|P) = D(Q\|P_\lambda) + D(P_\lambda\|P). \tag{15.33}$$

Proof. Since $\log \frac{dP_\lambda}{dP} = \lambda^\top \phi(x) - \psi(\lambda)$ is finite everywhere we have that $D(P_\lambda\|P) = \lambda^\top \gamma - \psi(\lambda) < \infty$ and hence the solution of (15.31) is finite. The fact that P_λ is the unique minimizer follows from the identity (15.33) that we are to prove next. Take Q as in the statement and suppose that either $D(Q\|P)$ or $D(Q\|P_\lambda)$ is finite (otherwise there is nothing to prove). Since $P_\lambda \ll P$ this implies that $Q \ll P$ and so let us denote by $f_Q = \frac{dQ}{dP}$. From (2.11) we see that

$$\begin{aligned} D(Q\|P) - D(Q\|P_\lambda) &= \mathbb{E}_Q\left[\operatorname{Log}\frac{\exp\{\lambda^\top\phi(X) - \psi(\lambda)\}}{1}\right] \\ &= \mathbb{E}_Q\left[\log\exp\{\lambda^\top\phi(X) - \psi(\lambda)\}\right] \\ &= \lambda^\top\gamma - \psi(\lambda) = D(P_\lambda\|P), \end{aligned}$$

establishing the claim. □

Unfortunately, Proposition 15.14 is far from being able to completely resolve the problem (15.31) since it does not explain for what values $\gamma \in \mathbb{R}^d$ of the constraints it is possible to find a required tilting P_λ. For $d = 1$ we had a very simple characterization of the set of values that the means of P_λ's can achieve. Specifically, Theorem 15.8 showed (under Assumption 15.1)

$$\{\mathbb{E}_{X\sim P_\lambda}[\phi(X)] : \lambda \in \mathbb{R}\} = (A, B),$$

where A, B are the boundaries of the support of ϕ. To obtain a similar characterization for the case of $d > 1$, we let $\tilde{P}$ be the probability distribution on $\mathbb{R}^d$ of $\phi(X)$ when $X \sim P$, that is, $\tilde{P}$ is the pushforward of P along ϕ. The analog of (A, B) is then realized by the following concept:

Definition 15.15. (Convex support) Let $\tilde{P}$ be a probability measure on $\mathbb{R}^d$. We recall that support $\operatorname{supp}(\tilde{P})$ is defined as the intersection of all closed sets of full measure. The *convex support* of $\tilde{P}$ is defined as the intersection of all closed convex sets with full measure:

$$\operatorname{csupp}(\tilde{P}) \triangleq \bigcap\{S : \tilde{P}[S] = 1, S \text{ is closed and convex}\}.$$

It is clear that csupp is itself a closed convex set. Furthermore, it can be obtained by taking the convex hull of $\operatorname{supp}(\tilde{P})$ followed by the topological closure $\operatorname{cl}(\cdot)$, that is,

$$\mathrm{csupp}(\tilde{P}) = \mathrm{cl}(\mathrm{co}(\mathrm{supp}(\tilde{P}))). \tag{15.34}$$

(Indeed, $\mathrm{csupp}(\tilde{P}) \subset \mathrm{cl}(\mathrm{co}(\mathrm{supp}\,\tilde{P}))$ since the set on the right is convex and closed. On the other hand, for any closed half-space $H \supset \mathrm{csupp}(\tilde{P})$ of full measure, that is, $\tilde{P}[H] = 1$, we must have $\mathrm{supp}(\tilde{P}) \subset H$. Taking the convex hull and then the closure of both sides yields $\mathrm{cl}(\mathrm{co}(\mathrm{supp}(\tilde{P}))) \subset H$. Taking the intersection over all such H shows that $\mathrm{cl}(\mathrm{co}(\mathrm{supp}\,\tilde{P})) \subset \mathrm{csupp}(\tilde{P})$ as well.)

We are now ready to state the characterization of when I-projection is solved by a tilted measure.

Theorem 15.16. (I-projection on a hyperplane) *Suppose P and ϕ satisfy the following two assumptions: (a) the $d+1$ functions $(1, \phi_1, \ldots, \phi_d)$ are linearly independent P-a.s. and (b) the* log MGF *$\psi(\lambda)$ is finite for all $\lambda \in \mathbb{R}^d$. Then we have the following.*

1. *If there exist any Q such that $\mathbb{E}_Q[\phi] = \gamma$ and $D(Q\|P) < \infty$, we must have $\gamma \in \mathrm{csupp}(\tilde{P})$.*
2. *There is a solution λ to $\mathbb{E}_{P_\lambda}[\phi] = \gamma$ if and only if $\gamma \in \mathrm{int}(\mathrm{csupp}(\tilde{P}))$.*

Corollary 15.17. *Whenever $\gamma \in \mathrm{int}(\mathrm{csupp}(\tilde{P}))$ the I-projection problem (15.31) is solved by P_λ for some λ and the identity (15.33) holds. Furthermore, $F(\gamma) = \psi^*(\gamma)$ and $\lambda = \nabla F(\gamma)$.*

Remark 15.6. Assumption (b) of Theorem 15.16 can be relaxed to requiring only the domain of the log MGF to be an open set (see [85, Theorem 23.1] or [77, Theorem 3.6]). Applying Theorem 15.16, whenever $\gamma \in \mathrm{int}\,\mathrm{csupp}(\tilde{P})$ the I-projection can be sought in the tilted family P_λ and only in such a case. If $\gamma \notin \mathrm{csupp}(\tilde{P})$ then the I-projection is trivially impossible and every Q with the given expectation yields $D(Q\|P) = \infty$. When $\gamma \in \partial\,\mathrm{csupp}(\tilde{P})$ it could be that an I-projection (i.e. the minimizer of (15.31)) exists, is unique, and yields a finite divergence, but the minimizer is not given by a λ-tilting of P. It could also be that every feasible Q yields $D(Q\|P) = \infty$.

As a concrete example, consider $X \sim P = \mathcal{N}(0,1)$ and $\phi(x) = (x, x^2)$. Then $\mathrm{csupp}(\tilde{P}) = \{(\gamma_1, \gamma_2) \in \mathbb{R}^2 : \gamma_2 \geq \gamma_1^2\}$, consisting of all valid values of first and second moments (satisfying Cauchy–Schwarz) of distributions on $\mathbb{R}$. Then the solution to this I-projection problem is as follows.

- $\gamma_2 > \gamma_1^2$: the optimal Q is $\mathcal{N}(\gamma_1, \gamma_2 - \gamma_1^2)$, which is a tilted version of P along ϕ.
- $\gamma_2 = \gamma_1^2$: the only feasible Q is δ_{γ_1}, which results in $D(Q\|P) = \infty$.
- $\gamma_2 < \gamma_1^2$: there is no feasible Q.

Before giving the proof of the theorem we remind the reader about some of the standard and easy facts about *exponential families* of which P_λ is an example. In this context ϕ is called a vector of statistics and λ is the natural parameter. Note that all $P_\lambda \sim P$ are mutually absolutely continuous and hence we have from the linear independence assumption:

$$\mathrm{Cov}_{X \sim P_\lambda}[\phi(X)] \succ 0, \tag{15.35}$$

that is, the covariance matrix is (strictly) positive definite. Similarly to Theorem 15.3 we can show that $\lambda \mapsto \psi(\lambda)$ is a convex, infinitely differentiable function [77]. We want to study the map from natural parameter λ to the mean parameter μ:

$$\lambda \mapsto \mu(\lambda) \triangleq \mathbb{E}_{X \sim P_\lambda}[\phi(X)].$$

Specifically, we will show that the image $\mu(\mathbb{R}^d) = \operatorname{int}\operatorname{csupp}(\tilde{P})$. To that end note that, Similarly to Theorem 15.8(b) and (c), the first two derivatives give moments of ϕ as follows:

$$\mathbb{E}_{X \sim P_\lambda}[\phi(X)] = \nabla\psi(\lambda), \quad \operatorname{Cov}_{X \sim P_\lambda}[\phi(X)] = \operatorname{Hess}\,\psi(\lambda)\log e. \tag{15.36}$$

Together with (15.35) we see that then ψ is strictly convex and hence for any λ_1, λ_2 we have the strict monotonicity of $\nabla\psi$, that is,

$$(\lambda_1 - \lambda_2)^\top(\nabla\psi(\lambda_1) - \nabla\psi(\lambda_2)) > 0. \tag{15.37}$$

Additionally, from (15.35) we obtain that the Jacobian of the map $\lambda \mapsto \mu(\lambda)$ equals $\det \operatorname{Hess}\,\psi(\lambda) > 0$. Thus by the inverse function theorem the image $\mu(\mathbb{R}^d)$ is an open set in $\mathbb{R}^d$ and there is an infinitely differentiable inverse $\mu \mapsto \lambda = \mu^{-1}(\mu)$ defined on this set. Hence, the family P_λ can be equivalently reparameterized by μ's. What is non-trivial is that the image $\mu(\mathbb{R}^d)$ is convex and in fact coincides with $\operatorname{int}\operatorname{csupp}(\tilde{P})$.

Proof of Theorem 15.16. Throughout the proof we denote $C = \operatorname{csupp}(\tilde{P})$, $C_o = \operatorname{int}(\operatorname{csupp}(\tilde{P}))$.

Suppose there is a $Q \ll P$ with $t = \mathbb{E}_Q[\phi(X)] \notin C$. Then there is a (separating hyperplane) $b \in \mathbb{R}^d$ and $c \in \mathbb{R}$ such that $b^\top t < c \le b^\top p$ for any $p \in C$. Since $P[\phi(X) \in C] = 1$ we conclude that $Q[b^\top\phi(X) \ge c] = 1$. But then this contradicts the fact that $\mathbb{E}_Q[b^\top\phi(X)] < c$. This shows the first claim.

Next, we show that for any λ we have $\mu(\lambda) = \mathbb{E}_{P_\lambda}[\phi] \in C_o$. Indeed, by the previous paragraph we know $\mu(\lambda) \in C$. On the other hand, as we discussed the map $\lambda \to \mu(\lambda)$ is smooth and one-to-one, with smooth inverse. Hence the image of a small ball around λ is open and hence $\mu(\lambda) \in \operatorname{int}(C) = C_o$.

Finally, we prove the main implication that for any $\gamma \in C_o$ there must exist a λ such that $\mu(\lambda) = \gamma$. To that end, consider the unconstrained minimization problem

$$\inf_\lambda \psi(\lambda) - \lambda^\top\gamma. \tag{15.38}$$

If we can show that the minimum is achieved at some λ^*, then from the first-order optimality conditions we conclude the desired $\nabla\psi(\lambda^*) = \gamma$. Since the objective function is continuous, it is sufficient to show that the minimization without loss of generality can be restricted to a compact ball $\{\|\lambda\| \le R\}$ for some large R.

To that end, we first notice that if $\gamma \in C_o$ then for some $\epsilon > 0$ we must have

$$c_\epsilon \triangleq \inf_{v:\, \|v\|=1} P[v^\top(\phi(X) - \gamma) > \epsilon] > 0. \tag{15.39}$$

Indeed, suppose this is not the case. Then for any $\epsilon > 0$ there is a sequence v_k s.t.

$$P[v_k^\top(\phi(X) - \gamma) > \epsilon] \to 0.$$

Now by compactness of the sphere, $v_k \to \tilde{v}_\epsilon$ without loss of generality and thus we have, for every ϵ, some $\tilde{v}_\epsilon$ exists such that

$$P[\tilde{v}_\epsilon^\top(\phi(X) - \gamma) > \epsilon] = 0.$$

Again, by compactness there must exist convergent subsequence $\tilde{v}_\epsilon \to v_*$ and $\epsilon \to 0$ such that

$$P[v_*^\top(\phi(X) - \gamma) > 0] = 0.$$

Thus, $\mathrm{supp}(\tilde{P}) \subset \{x\colon v_*^\top \phi(x) \le v_*^\top \gamma\}$ and hence γ cannot be an interior point of $C = \mathrm{csupp}(\tilde{P})$.

Given (15.39) we obtain the following estimate, where we denote $v = \frac{\lambda}{\|\lambda\|}$:

$$\begin{aligned}\exp\{\psi(\lambda) - \lambda^\top\gamma\} &= \mathbb{E}_P[\exp\{\lambda^\top(\phi(X) - \gamma)\}] \\ &\ge \mathbb{E}_P[\exp\{\lambda^\top(\phi(X) - \gamma)\}1\{v^\top(\phi(X) - \gamma) > \epsilon\}] \\ &\ge c_\epsilon \exp\{\epsilon\|\lambda\|\}.\end{aligned}$$

Thus, returning to the minimization problem (15.38) we see that the objective function satisfies a lower bound

$$\psi(\lambda) - \lambda^\top\gamma \ge \log c_\epsilon + \epsilon\|\lambda\|.$$

Then it is clear that restricting the minimization to a sufficiently large ball $\{\|\lambda\| \le R\}$ is without loss of generality. As we explained this completes the proof. □

Example 15.2. Sinkhorn's problem

As an application of Theorem 15.16, consider a joint distribution $P_{X,Y}$ on finite $\mathcal{X} \times \mathcal{Y}$. Our goal is to solve a coupling problem:

$$\min\{D(Q_{X,Y}\|P_{X,Y})\colon Q_X = V_X, Q_Y = V_Y\},$$

where the marginals V_X and V_Y are given. As we discussed in Section 5.6, Sinkhorn identified an elegant iterative algorithm that converges to the minimizer. Here, we can apply our general I-projection theory to show that the minimizer has the form

$$Q^*_{X,Y}(x,y) = A(x)P_{X,Y}(x,y)B(y). \tag{15.40}$$

Specifically, let us assume $P_{X,Y}(x,y) > 0$ and consider $|\mathcal{X}| + |\mathcal{Y}|$ functions $\phi_a(x,y) = 1\{x = a\}$ and $\phi_b(x,y) = 1\{y = b\}$, $a \in \mathcal{X}, b \in \mathcal{Y}$. They are linearly independent. The set $\mathrm{csupp}(\tilde{P}) = \mathcal{P}(\mathcal{X}) \times \mathcal{P}(\mathcal{Y})$ in this case corresponds to all marginal distributions. Thus, whenever V_X, V_Y have no zeros they belong to $\mathrm{int}(\mathrm{csupp}(\tilde{P}))$ and the solution to the I-projection problem is a tilted version of $P_{X,Y}$ which is precisely of the form (15.40). In this case, it turns out that I-projection exists also on the boundary and even when $P_{X,Y}$ is allowed to have zeros but these cases are outside the scope of Theorem 15.16 and need to be treated differently, see [114].

16 Hypothesis Testing: Error Exponents

In this chapter our goal is to determine the achievable region of the exponent pairs (E_0, E_1) for the type-I and type-II error probabilities in Chernoff's regime when both exponents are strictly positive. Our strategy is to apply the achievability and (strong) converse bounds from Chapter 14 in conjunction with the large-deviations theory developed in Chapter 15. After characterizing the full tradeoff we will discuss an adaptive setting of hypothesis testing where, instead of committing ahead of time to testing on the basis of n samples, one can decide adaptively whether to request more samples or stop. We will find out that adaptivity greatly increases the region of achievable error exponents and will learn about the sequential probability ratio test (SPRT) of Wald. In the closing sections we will discuss relationships to more complicated settings in hypothesis testing: one with composite hypotheses and one with communication constraints.

16.1 (E_0, E_1)-Tradeoff

Recall the setting of the Chernoff regime introduced in Section 14.6, where the goal is to design tests satisfying

$$\pi_{1|0} = 1 - \alpha \leq \exp(-nE_0), \quad \pi_{0|1} = \beta \leq \exp(-nE_1).$$

To find the best tradeoff of E_0 versus E_1 we can define the following function:

$$\begin{aligned} E_1^*(E_0) &\triangleq \sup\{E_1 : \exists n_0, \forall n \geq n_0, \exists P_{Z|X^n} \text{ s.t. } \alpha > 1 - \exp(-nE_0), \beta < \exp(-nE_1)\} \\ &= \liminf_{n\to\infty} \frac{1}{n} \log \frac{1}{\beta_{1-\exp(-nE_0)}(P^n, Q^n)}, \end{aligned}$$

where β_α was defined in (14.3). This should be compared with Stein's exponent in Definition 14.13.

Define

$$T_k = \log \frac{dQ}{dP}(X_k), \quad k = 1, \dots, n,$$

which are iid copies of $T = \log \frac{dQ}{dP}(X)$. Then $\log \frac{dQ^n}{dP^n}(X^n) = \sum_{k=1}^n T_k$, which is an iid sum under both P and Q.

The log MGF of T under P (again assumed to be finite and also T is not a constant since $P \neq Q$) and the corresponding rate function are (see Definitions 15.2 and 15.5):

$$\psi_P(\lambda) = \log \mathbb{E}_P[\exp(\lambda T)], \quad \psi_P^*(\theta) = \sup_{\lambda \in \mathbb{R}} \theta\lambda - \psi_P(\lambda).$$

For discrete distributions, we have $\psi_P(\lambda) = \log \sum_x P(x)^{1-\lambda} Q(x)^\lambda$; in general, $\psi_P(\lambda) = \log \int d\mu \left(\frac{dP}{d\mu}\right)^{1-\lambda} \left(\frac{dQ}{d\mu}\right)^\lambda$ for some dominating measure μ.

Note that since $\psi_P(0) = \psi_P(1) = 0$, from the convexity of ψ_P (Theorem 15.3) we conclude that $\psi_P(\lambda)$ is finite on $0 \leq \lambda \leq 1$. Furthermore, assuming $P \ll Q$ and $Q \ll P$ we also have that $\lambda \mapsto \psi_P(\lambda)$ is continuous everywhere on $[0, 1]$. (The continuity on $(0, 1)$ follows from convexity, but for the boundary points we need more detailed arguments.) Although all results in this section apply under the (milder) conditions of $P \ll Q$ and $Q \ll P$, we will only present proofs under the (stronger) condition that log MGF exists for all λ, following the convention of the previous chapter. The following result determines the optimal (E_0, E_1)-tradeoff in a parametric form. For a concrete example, see Exercise **III.19** for testing two Gaussians.

Theorem 16.1. *Assume $P \ll Q$ and $Q \ll P$. Then*

$$E_0(\theta) = \psi_P^*(\theta), \quad E_1(\theta) = \psi_P^*(\theta) - \theta, \tag{16.1}$$

parameterized by $-D(P\|Q) \leq \theta \leq D(Q\|P)$, characterizes the upper boundary of the region of all achievable (E_0, E_1) pairs. (See Figure 16.1 for an illustration.)

Remark 16.1. (Rényi divergence) In Definition 7.24 we defined Rényi divergence D_λ. It turns out that D_λ's are intimately related to error exponents. Indeed, we have $\psi_P(\lambda) = (\lambda - 1)D_\lambda(Q\|P) = -\lambda D_{1-\lambda}(P\|Q)$. This provides another proof of why $\psi_P(\lambda)$ is negative for λ between 0 and 1, and also recovers the slope at endpoints: $\psi_P'(0) = -D(P\|Q)$ and $\psi_P'(1) = D(Q\|P)$. See also Exercise **I.39**.

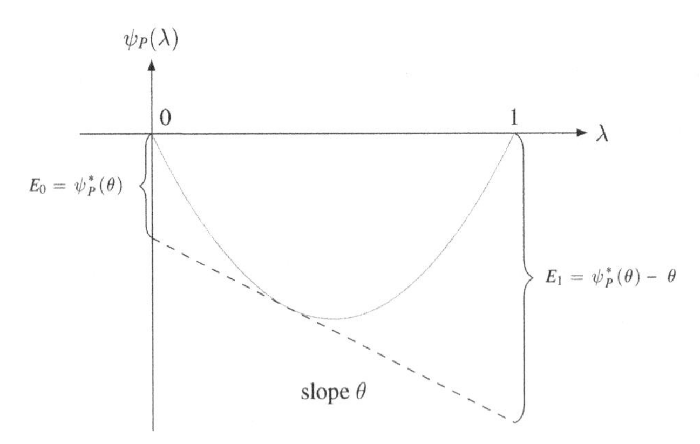

Figure 16.1 The geometric interpretation of Theorem 16.1 relies on the properties of $\psi_P(\lambda)$ and $\psi_P^*(\theta)$. Note that $\psi_P(0) = \psi_P(1) = 0$. Moreover, by Theorem 15.6, $\theta \mapsto E_0(\theta)$ is increasing, and $\theta \mapsto E_1(\theta)$ is decreasing.

Corollary 16.2. (Bayesian criterion) *Fix a prior* (π_0, π_1) *such that* $\pi_0 + \pi_1 = 1$ *and* $0 < \pi_0 < 1$. *Denote the optimal Bayesian (average) error probability by*

$$P_e^*(n) \triangleq \inf_{P_{Z|X^n}} \pi_0\pi_{1|0} + \pi_1\pi_{0|1}$$

with exponent

$$E \triangleq \lim_{n\to\infty} \frac{1}{n} \log \frac{1}{P_e^*(n)}.$$

Then

$$E = \max_\theta \min(E_0(\theta), E_1(\theta)) = \psi_P^*(0)$$

regardless of the prior, and

$$\psi_P^*(0) = -\inf_{\lambda\in[0,1]} \psi_P(\lambda) = -\inf_{\lambda\in[0,1]} \log \int (dP)^{1-\lambda}(dQ)^\lambda \triangleq C(P, Q) \tag{16.2}$$

is called the Chernoff exponent *or* Chernoff information.

Notice that from (14.19) we always have

$$\log\left(1 - \frac{1}{2}H^2(P, Q)\right) \le C(P, Q) \le 2\log\left(1 - \frac{1}{2}H^2(P, Q)\right)$$

and thus for small $H^2(P, Q)$ we have $C(P, Q) \asymp H^2(P, Q)$.

Remark 16.2. (Bhattacharyya distance) There is an important special case in which the Chernoff exponent simplifies. Instead of iid observations, consider independent, but not identically distributed, observations. Namely, suppose that two hypotheses correspond to two different strings x^n and $\tilde{x}^n$ over a finite alphabet $\mathcal{X}$. The hypothesis tester observes $Y^n = (Y_1, \ldots, Y_n)$ obtained by applying one of the two strings to the input of the memoryless channel $P_{Y|X}$; in other words, either $Y^n \sim \prod_{t=1}^n P_{Y|X=x_t}$ or $\prod_{t=1}^n P_{Y|X=\tilde{x}_t}$. (The alphabet $\mathcal{Y}$ does not need to be finite, but we assume this below.) Extending Corollary 16.2 it can be shown that in this case the optimal (average) probability of error $P_e^*(x^n, \tilde{x}^n)$ has (Chernoff) exponent[1]

$$E = -\inf_{\lambda\in[0,1]} \frac{1}{n} \sum_{t=1}^n \log \sum_{y\in\mathcal{Y}} P_{Y|X}(y|x_t)^\lambda P_{Y|X}(y|\tilde{x}_t)^{1-\lambda}.$$

If $|\mathcal{X}| = 2$ *and if the compositions (types) of* x^n *and* $\tilde{x}^n$ *are equal (!)*, the expression is invariant under $\lambda \leftrightarrow 1 - \lambda$ and thus from the convexity in λ we conclude that $\lambda = \frac{1}{2}$ is optimal,[2] yielding $E = \frac{1}{n} d_B(x^n, \tilde{x}^n)$, where

$$d_B(x^n, \tilde{x}^n) = -\sum_{t=1}^n \log \sum_{y\in\mathcal{Y}} \sqrt{P_{Y|X}(y|x_t)P_{Y|X}(y|\tilde{x}_t)} \tag{16.3}$$

[1] In short, this is because the optimal tilting parameter λ does not need to be chosen differently for different values of $(x_t, \tilde{x}_t)$.

[2] For another example where $\lambda = \frac{1}{2}$ achieves the optimal Chernoff information, see Exercise **III.30**.

is known as the *Bhattacharyya distance* between codewords x^n and $\tilde{x}^n$. (Compare with the Bhattacharyya coefficient defined after (7.5).) Without the two assumptions stated, $d_B(\cdot,\cdot)$ does not necessarily give the optimal error exponent. We do, however, always have the bounds, see (14.19):

$$\frac{1}{4}\exp\left(-2d_B(x^n,\tilde{x}^n)\right) \le P_e^*(x^n,\tilde{x}^n) \le \exp\left(-d_B(x^n,\tilde{x}^n)\right),$$

where the upper bound becomes tighter when the joint composition of $(x^n,\tilde{x}^n)$ and that of $(\tilde{x}^n, x^n)$ are closer.

Proof of Theorem 16.1. The idea is to apply the large-deviations theory to the iid sum $\sum_{k=1}^n T_k$. Specifically, let's rewrite the achievability and converse bounds from Chapter 14 in terms of T:

- Achievability (Neyman–Pearson): Applying Theorem 14.11 with $\tau = -n\theta$, the LRT achieves the following:

$$\pi_{1|0} = P\left[\sum_{k=1}^n T_k \ge n\theta\right], \quad \pi_{0|1} = Q\left[\sum_{k=1}^n T_k < n\theta\right]. \tag{16.4}$$

- Converse (strong): Applying Theorem 14.10 with $\gamma = \exp(-n\theta)$, any achievable $\pi_{1|0}$ and $\pi_{0|1}$ satisfy

$$\pi_{1|0} + \exp(-n\theta)\,\pi_{0|1} \ge P\left[\sum_{k=1}^n T_k \ge n\theta\right]. \tag{16.5}$$

For achievability, applying the non-asymptotic large–deviations upper bound in Theorem 15.9 (and Theorem 15.11) to (16.4), we obtain that for any n,

$$\pi_{1|0} = P\left[\sum_{k=1}^n T_k \ge n\theta\right] \le \exp\left(-n\psi_P^*(\theta)\right), \quad \text{for } \theta \ge \mathbb{E}_P T = -D(P\|Q),$$
$$\pi_{0|1} = Q\left[\sum_{k=1}^n T_k < n\theta\right] \le \exp\left(-n\psi_Q^*(\theta)\right), \quad \text{for } \theta \le \mathbb{E}_Q T = D(Q\|P).$$

Notice that by the definition of $T = \log\frac{dQ}{dP}$ we have

$$\psi_Q(\lambda) = \log\mathbb{E}_Q[e^{\lambda T}] = \log\mathbb{E}_P[e^{(\lambda+1)T}] = \psi_P(\lambda+1)$$
$$\Rightarrow \psi_Q^*(\theta) = \sup_{\lambda\in\mathbb{R}} \theta\lambda - \psi_P(\lambda+1) = \psi_P^*(\theta) - \theta.$$

Thus the pair of exponents $(E_0(\theta), E_1(\theta))$ in (16.1) is achievable.

For the converse, we aim to show that any achievable (E_0, E_1) pair must lie below the curve achieved by the above Neyman–Pearson test, namely $(E_0(\theta), E_1(\theta))$ parameterized by θ. Suppose $\pi_{1|0} = \exp(-nE_0)$ and $\pi_{0|1} = \exp(-nE_1)$ is achievable. Combining the strong converse bound (16.5) with the large–deviations lower bound, we have: for any fixed $\theta \in [-D(P\|Q), D(Q\|P)]$,

$$\begin{aligned}
&\exp(-nE_0) + \exp(-n\theta)\exp(-nE_1) \geq \exp\left(-n\psi_P^*(\theta) + o(n)\right) \\
&\Rightarrow \min(E_0, E_1 + \theta) \leq \psi_P^*(\theta) \\
&\Rightarrow \text{either } E_0 \leq \psi_P^*(\theta) \text{ or } E_1 \leq \psi_P^*(\theta) - \theta,
\end{aligned}$$

proving the desired result. □

16.2 Equivalent Forms of Theorem 16.1

Alternatively, the optimal (E_0, E_1)-tradeoff can be stated in the following equivalent forms:

Theorem 16.3.

(a) *The optimal exponents are given (parametrically) in terms of* $\lambda \in [0, 1]$ *as*

$$E_0 = D(P_\lambda \| P), \quad E_1 = D(P_\lambda \| Q), \tag{16.6}$$

where the distribution P_λ[3] *is tilting of* P *along* T *given in (15.27), which moves from* $P_0 = P$ *to* $P_1 = Q$ *as* λ *ranges from* 0 *to* 1*:*

$$dP_\lambda = (dP)^{1-\lambda}(dQ)^\lambda \exp\{-\psi_P(\lambda)\}.$$

(b) *Yet another characterization of the boundary is*

$$E_1^*(E_0) = \min_{Q' : D(Q' \| P) \leq E_0} D(Q' \| Q), \quad 0 \leq E_0 \leq D(Q \| P). \tag{16.7}$$

Remark 16.3. The interesting consequence of this point of view is that it also suggests how a typical error event looks. Namely, consider an optimal hypothesis test achieving the pair of exponents (E_0, E_1). Then conditioned on the error event (under either P or Q) we have that the empirical distribution of the sample will be close to P_λ. For example, if $P = \mathrm{Bin}(m, p)$ and $Q = \mathrm{Bin}(m, q)$, then the typical error event will correspond to a sample whose empirical distribution $\hat{P}_n$ is approximately $\mathrm{Bin}(m, r)$ for some $r = r(p, q, \lambda) \in (p, q)$, and not any other distribution on $\{0, \ldots, m\}$.

Proof. The first part is verified trivially. Indeed, if we fix λ and let $\theta(\lambda) \triangleq \mathbb{E}_{P_\lambda}[T]$, then from (15.8) we have

$$D(P_\lambda \| P) = \psi_P^*(\theta),$$

whereas

$$D(P_\lambda \| Q) = \mathbb{E}_{P_\lambda}\left[\log \frac{dP_\lambda}{dQ}\right] = \mathbb{E}_{P_\lambda}\left[\log \frac{dP_\lambda}{dP}\frac{dP}{dQ}\right] = D(P_\lambda \| P) - \mathbb{E}_{P_\lambda}[T] = \psi_P^*(\theta) - \theta.$$

Also from (15.7) we know that as λ ranges in $[0, 1]$ the mean $\theta = \mathbb{E}_{P_\lambda}[T]$ ranges from $-D(P \| Q)$ to $D(Q \| P)$.

[3] This is called a geometric mixture of P and Q.

To prove the second claim (16.7), the key observation is the following: Since Q is itself a tilting of P along T (with $\lambda = 1$), the following two families of distributions,

$$dP_\lambda = \exp\{\lambda T - \psi_P(\lambda)\} \cdot dP,$$
$$dQ_{\lambda'} = \exp\{\lambda' T - \psi_Q(\lambda')\} \cdot dQ,$$

are in fact the same family with $Q_{\lambda'} = P_{\lambda'+1}$.

Now, suppose that Q^* achieves the minimum in (16.7) and that $Q^* \neq Q$, $Q^* \neq P$ (these cases should be verified separately). Note that we have not shown that this minimum is achieved, but it will be clear that our argument can be extended to the case of when Q'_n is a sequence achieving the infimum. Then, on the one hand, obviously

$$D(Q^* \| Q) = \min_{Q' :\, D(Q'\|P) \le E_0} D(Q' \| Q) \le D(P \| Q).$$

On the other hand, since $E_0 \le D(Q\|P)$ we also have

$$D(Q^* \| P) \le D(Q \| P).$$

Therefore,

$$\mathbb{E}_{Q^*}[T] = \mathbb{E}_{Q^*}\left[\log \frac{dQ^*}{dP}\frac{dQ}{dQ^*}\right] = D(Q^* \| P) - D(Q^* \| Q) \in [-D(P\|Q), D(Q\|P)]. \tag{16.8}$$

Next, we have from Corollary 15.12 that there exists a *unique* P_λ with the following three properties:[4]

$$\mathbb{E}_{P_\lambda}[T] = \mathbb{E}_{Q^*}[T],$$
$$D(P_\lambda \| P) \le D(Q^* \| P),$$
$$D(P_\lambda \| Q) \le D(Q^* \| Q).$$

Thus, we immediately conclude that the minimization in (16.7) can be restricted to Q^* belonging to the family of tilted distributions $\{P_\lambda, \lambda \in \mathbb{R}\}$. Furthermore, from (16.8) we also conclude that $\lambda \in [0, 1]$. Hence, the characterization of $E_1^*(E_0)$ given by (16.6) coincides with the one given by (16.7). □

Remark 16.4. A geometric interpretation of (16.7) is given in Figure 16.2: As λ increases from 0 to 1, or equivalently, θ increases from $-D(P\|Q)$ to $D(Q\|P)$, the optimal distribution P_λ traverses down the dotted path from P to Q. Note that there are many ways to interpolate between P and Q, for example, by taking their (arithmetic) mixture $(1-\lambda)P+\lambda Q$. In contrast, P_λ is a *geometric mixture* of P and Q, and this special path is in essence a geodesic connecting P to Q and the exponents E_0 and E_1 measure its respective distances to P and Q. Unlike Riemannian geometry,

[4] A subtlety: In Corollary 15.12 we ask that $\mathbb{E}_{Q^*}[T] \in (A, B)$. But A, B – the essential range of T – depend on the distribution under which the essential range is computed, see (15.23). Fortunately, we have $Q \ll P$ and $P \ll Q$, so the essential range is the same under both P and Q. Furthermore (16.8) implies that $\mathbb{E}_{Q^*}[T] \in (A, B)$.

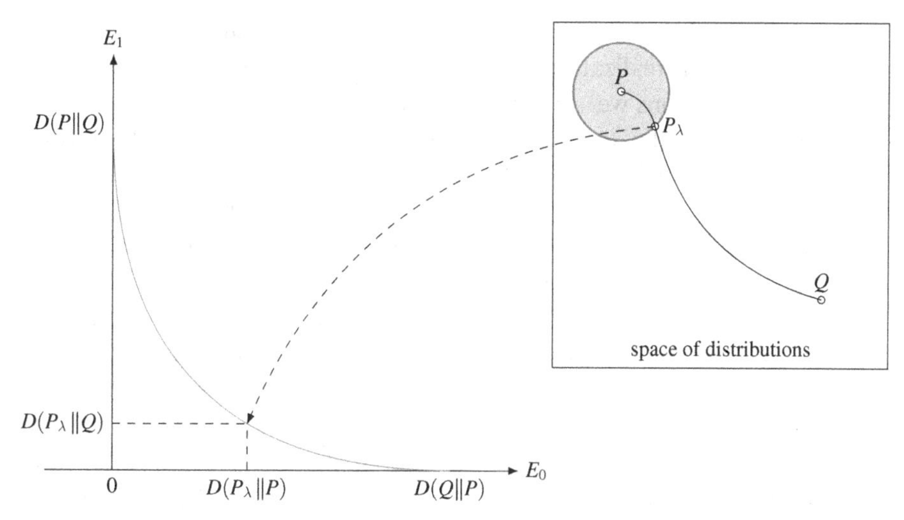

Figure 16.2 Geometric interpretation of (16.7). Here the shaded circle represents $\{Q' : D(Q'\|P) \le E_0\}$, the KL divergence "ball" of radius E_0 centered at P. The optimal $E_1^*(E_0)$ in (16.7) is given by the divergence from Q to the closest element of this ball, attained by some tilted distribution P_λ. The tilted family P_λ is the geodesic traversing from P to Q as λ increases from 0 to 1.

though, here the sum of distances to the two endpoints from an intermediate P_λ actually varies along the geodesic.

16.3* Sequential Hypothesis Testing

Review: Filtration and Stopping Time

- A sequence of nested σ-algebras $\mathcal{F}_0 \subset \mathcal{F}_1 \subset \mathcal{F}_2 \subset \cdots \subset \mathcal{F}_n \subset \cdots \subset \mathcal{F}$ is called a filtration of $\mathcal{F}$.
- A random variable τ is called a stopping time of a filtration $\mathcal{F}_n$ if (a) τ is valued in $\mathbb{Z}_+$ and (b) for every $n \ge 0$ the event $\{\tau \le n\} \in \mathcal{F}_n$.
- The σ-algebra $\mathcal{F}_\tau$ consists of all events E such that $E \cap \{\tau \le n\} \in \mathcal{F}_n$ for all $n \ge 0$.
- When $\mathcal{F}_n = \sigma\{X_1, \ldots, X_n\}$ the interpretation is that τ is a time that can be determined by causally observing the sequence X_j, and random variables measurable with respect to $\mathcal{F}_\tau$ are precisely those whose value can be determined on the basis of knowing $(X_1, \ldots, X_\tau)$.
- Let M_n be a martingale adapted to $\mathcal{F}_n$, that is, M_n is $\mathcal{F}_n$-measurable and $\mathbb{E}[M_n|\mathcal{F}_k] = M_{\min(n,k)}$. Then $\tilde{M}_n = M_{\min(n,\tau)}$ is also a martingale. If the collection $\{M_n\}$ is uniformly integrable then

$$\mathbb{E}[M_\tau] = \mathbb{E}[M_0].$$

- For more details, see [84, Chapter V].

So far we have always been working with a fixed number of observations n. However, different realizations of X^n are informative to different levels, that is, under some realizations we are very certain about declaring the true hypothesis, whereas some other realizations leave us more doubtful. In the fixed n setting, the tester is forced to take a guess in the latter case. In the sequential setting, pioneered by Wald [449], the tester is allowed to request more observations. We show in this section that the optimal test in this setting is something known as the sequential probability ratio test (SPRT) [451]. It will also be shown that the resulting tradeoff between the exponents E_0 and E_1 is much improved in the sequential setting.

We start with the concept of a *sequential test*. Informally, at each time t, upon receiving the observation X_t, a sequential test either declares H_0, declares H_1, or requests one more observation. The rigorous definition is as follows: A sequential hypothesis test consists of: (a) a stopping time τ with respect to the filtration $\{\mathcal{F}_k, k \in \mathbb{Z}_+\}$, where $\mathcal{F}_k \triangleq \sigma\{X_1, \dots, X_k\}$ is generated by the first k observations; and (b) a random variable (decision) $Z \in \{0, 1\}$ measurable with respect to $\mathcal{F}_\tau$. Each sequential test is associated with the following performance metrics:

$$\alpha = \mathbb{P}[Z = 0], \qquad \beta = \mathbb{Q}[Z = 0], \tag{16.9}$$

$$l_0 = \mathbb{E}_{\mathbb{P}}[\tau], \qquad l_1 = \mathbb{E}_{\mathbb{Q}}[\tau], \tag{16.10}$$

where under $\mathbb{P}$ (resp., $\mathbb{Q}$) the data X_i is iid from P (resp., Q).

The easiest way to see why sequential tests may be dramatically superior to fixed-sample-size tests is the following example: Consider $P = \frac{1}{2}\delta_0 + \frac{1}{2}\delta_1$ and $Q = \frac{1}{2}\delta_0 + \frac{1}{2}\delta_{-1}$. Since $P \not\perp Q$, we also have $P^n \not\perp Q^n$. Consequently, no finite-sample-size test can achieve zero error under both hypotheses. However, an obvious sequential test (waiting for the first appearance of ± 1) achieves zero error probability with a finite number of (two) observations in expectation under both hypotheses. This advantage is also clear in terms of the achievable error exponents shown in Figure 16.3.

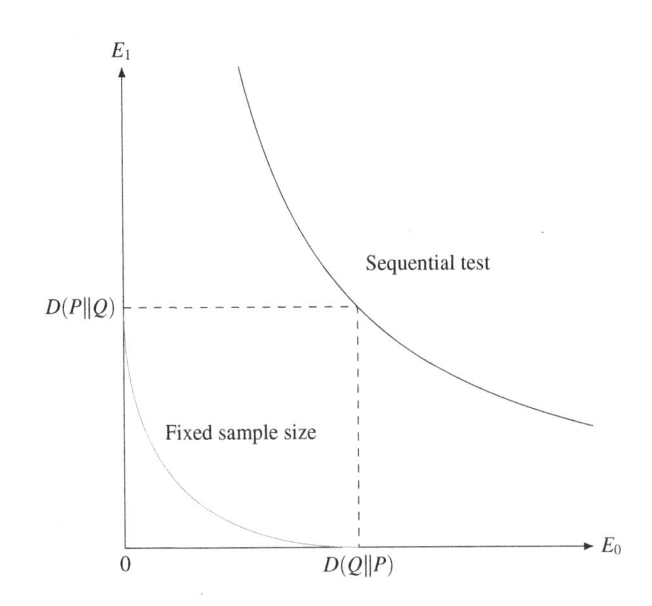

Figure 16.3 Tradeoff between type-I and type-II error exponents. The lower curve corresponds to optimal tests with fixed sample size (Theorem 16.1) and the upper curve to optimal sequential tests (Theorem 16.4).

The following result is essentially due to [451], though there it was shown only for the special case of $E_0 = D(Q\|P)$ and $E_1 = D(P\|Q)$. The version below is from [340].

Theorem 16.4. *Assume bounded LLR:*[5]

$$\left|\log \frac{P(x)}{Q(x)}\right| \le c_0, \quad \forall x,$$

where c_0 is some positive constant. Call a pair of exponents (E_0, E_1) achievable if there exist a test with $l_0, l_1 \to \infty$ and probabilities satisfying

$$\pi_{1|0} \le \exp\left(-l_0 E_0(1 + o(1))\right), \qquad \pi_{0|1} \le \exp\left(-l_1 E_1(1 + o(1))\right).$$

Then the set of achievable exponents must satisfy

$$E_0 E_1 \le D(P\|Q) D(Q\|P).$$

Furthermore, any such (E_0, E_1) is achieved by the sequential probability ratio test SPRT(A, B) *(A, B are large positive numbers) defined as follows:*

$$\tau = \inf\{n \colon S_n \ge B \text{ or } S_n \le -A\},$$

$$Z = \begin{cases} 0, & \text{if } S_\tau \ge B, \\ 1, & \text{if } S_\tau \le -A, \end{cases}$$

where

$$S_n = \sum_{k=1}^{n} \log \frac{P(X_k)}{Q(X_k)}$$

is the log–likelihood function of the first n observations.

Remark 16.5. (Interpretation of SPRT) Under the usual setup of hypothesis testing, we collect a sample of n iid observations, evaluate the LLR S_n, and compare it to the threshold to give the optimal test. Under the sequential setup, $\{S_n \colon n \ge 1\}$ is a *random walk*, which has positive (respectively negative) drift $D(P\|Q)$ (respectively $-D(Q\|P)$) under the null (respectively alternative) hypothesis! SPRT simply declares P if the random walk crosses the lower boundary B, or Q if the random walk crosses the upper boundary $-A$. See Figure 16.4 for an illustration.

Proof. As preparation we show two useful identities:

- For any stopping time with $\mathbb{E}_P[\tau] < \infty$ we have

$$\mathbb{E}_P[S_\tau] = \mathbb{E}_P[\tau] D(P\|Q), \tag{16.11}$$

  and similarly, if $\mathbb{E}_Q[\tau] < \infty$ then

$$\mathbb{E}_Q[S_\tau] = -\mathbb{E}_Q[\tau] D(Q\|P).$$

[5] This assumption is satisfied for example for a pair of fully supported discrete distributions on finite alphabets.

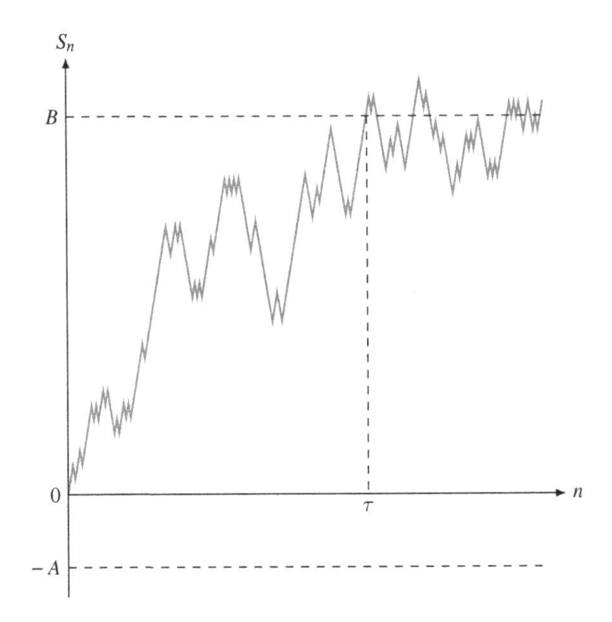

Figure 16.4 Illustration of the SPRT(A, B) test. Here, at the stopping time τ, the LLR process S_n reaches B before reaching $-A$ and the decision is $Z = 1$.

To prove these, notice that

$$M_n = S_n - nD(P\|Q)$$

is clearly a martingale w.r.t. $\mathcal{F}_n$. Consequently,

$$\tilde{M}_n \triangleq M_{\min(\tau,n)}$$

is also a martingale. Thus

$$\mathbb{E}_P[\tilde{M}_n] = \mathbb{E}_P[\tilde{M}_0] = 0,$$

or, equivalently,

$$\mathbb{E}_P[S_{\min(\tau,n)}] = \mathbb{E}_P[\min(\tau, n)]D(P\|Q). \tag{16.12}$$

This holds for every $n \geq 0$. From the boundedness assumption we have $|S_n| \leq nc_0$ and thus $|S_{\min(n,\tau)}| \leq \tau c_0$, implying that collection $\{S_{\min(n,\tau)} : n \geq 0\}$ is uniformly integrable. Thus, we can take $n \to \infty$ in (16.12) and interchange expectation and limit safely to conclude (16.11).

- Let τ be a stopping time. Recall that a random variable R is a Radon–Nikodym derivative of $\mathbb{P}$ w.r.t. $\mathbb{Q}$ on a σ-algebra $\mathcal{F}_\tau$, denoted by $\frac{d\mathbb{P}|_{\mathcal{F}_\tau}}{d\mathbb{Q}|_{\mathcal{F}_\tau}}$, if

$$\mathbb{E}_P[1_E] = \mathbb{E}_Q[R 1_E] \qquad \forall E \in \mathcal{F}_\tau. \tag{16.13}$$

We will show that it is in fact given by

$$\frac{d\mathbb{P}|_{\mathcal{F}_\tau}}{d\mathbb{Q}|_{\mathcal{F}_\tau}} = \exp\{S_\tau\}.$$

Indeed, what we need to verify is that (16.13) holds with $R = \exp\{S_\tau\}$ and an arbitrary event $E \in \mathcal{F}_\tau$, which we decompose as

$$1_E = \sum_{n\geq 0} 1_{E\cap\{\tau=n\}}.$$

By the monotone convergence theorem applied to both sides of (16.13) it is then sufficient to verify that for every n

$$\mathbb{E}_P[1_{E\cap\{\tau=n\}}] = \mathbb{E}_Q[\exp\{S_\tau\}1_{E\cap\{\tau=n\}}]. \tag{16.14}$$

This, however, follows from the fact that $E \cap \{\tau = n\} \in \mathcal{F}_n$ and $\frac{d\mathbb{P}|_{\mathcal{F}_n}}{d\mathbb{Q}|_{\mathcal{F}_n}} = \exp\{S_n\}$ by the very definition of S_n.

We now proceed to the proof. For *achievability* we apply (16.13) to infer

$$\pi_{1|0} = \mathbb{P}[S_\tau \leq -A] = \mathbb{E}_Q[\exp\{S_\tau\}1\{S_\tau \leq -A\}] \leq e^{-A}.$$

Next, we denote $\tau_0 = \inf\{n\colon S_n \geq B\}$ and observe that $\tau \leq \tau_0$, whereas the expectation of τ_0 can be bounded using (16.11) as

$$\mathbb{E}_P[\tau] \leq \mathbb{E}_P[\tau_0] = \frac{\mathbb{E}_P[S_{\tau_0}]}{D(P\|Q)} \leq \frac{B + c_0}{D(P\|Q)},$$

where in the last step we used the boundedness assumption to infer $S_{\tau_0} \leq B + c_0$. Overall,

$$l_0 = \mathbb{E}_\mathbb{P}[\tau] \leq \mathbb{E}_\mathbb{P}[\tau_0] \leq \frac{B + c_0}{D(P\|Q)}.$$

Similarly, we can show $\pi_{0|1} \leq e^{-B}$ and $l_1 \leq \frac{A+c_0}{D(Q\|P)}$. Now consider a pair of exponents E_0, E_1 at the boundary, that is, $E_0E_1 = D(P\|Q)D(Q\|P)$. Let $x = \frac{E_0}{D(P\|Q)} = \frac{D(Q\|P)}{E_1}$. Set $A = xB$ and let $B \to \infty$. From the argument above we have $\pi_{1|0} \leq e^{-xB} \leq e^{-xBl_0D(P\|Q)/(B+c_0)} = e^{-l_0E_0(1+o(1))}$ and similarly $\pi_{0|1} \leq e^{-l_1E_1(1+o(1))}$.

For the *converse* assume that (E_0, E_1) is achievable for large l_0, l_1. Recall from Section 4.6* that $D(\mathbb{P}_{\mathcal{F}_\tau}\|\mathbb{Q}_{\mathcal{F}_\tau})$ denotes the divergence between P and Q when viewed as measures on σ-algebra $\mathcal{F}_\tau$. We apply the data-processing inequality for divergence to obtain:

$$d(\mathbb{P}(Z=1)\|\mathbb{Q}(Z=1)) \leq D(\mathbb{P}_{\mathcal{F}_\tau}\|\mathbb{Q}_{\mathcal{F}_\tau}) = \mathbb{E}_P[S_\tau] \overset{(16.11)}{=} \mathbb{E}_\mathbb{P}[\tau]D(P\|Q) = l_0D(P\|Q).$$

Notice that for l_0E_0 and l_1E_1 large, we have $d(\mathbb{P}(Z = 1)\|\mathbb{Q}(Z = 1)) = l_1E_1(1 + o(1))$, therefore $l_1E_1 \leq (1 + o(1))l_0D(P\|Q)$. Similarly we can show that $l_0E_0 \leq (1 + o(1))l_1D(Q\|P)$. Thus taking $\ell_0, \ell_1 \to \infty$ we conclude

$$E_0E_1 \leq D(P\|Q)D(Q\|P).$$ □

16.4 Composite, Robust, and Goodness-of-Fit Hypothesis Testing

In this chapter we have considered the setting of distinguishing between the two alternatives, under either of which the data distribution was specified completely.

There are multiple other settings that have also been studied in the literature, which we briefly mention here for completeness.

The key departure is to replace the simple hypotheses that we started with in Chapter 14 with *composite* ones. Namely, we postulate

$$H_0\colon X_i \stackrel{\text{iid}}{\sim} P, \quad P \in \mathcal{P} \qquad \text{versus} \qquad H_1\colon X_i \stackrel{\text{iid}}{\sim} Q, \quad Q \in \mathcal{Q},$$

where $\mathcal{P}$ and $\mathcal{Q}$ are two families of distributions. In this case for a given test $Z = Z(X_1, \dots, X_n) \in \{0, 1\}$ we define the two types of error as before, but taking worst-case choices over the distribution:

$$1 - \alpha = \inf_{P \in \mathcal{P}} P^{\otimes n}[Z = 0], \qquad \beta = \sup_{Q \in \mathcal{Q}} Q^{\otimes n}[Z = 0].$$

Unlike testing simple hypotheses for which the Neyman–Pearson test is optimal (Theorem 14.11), in general there is no explicit description for the optimal test of composite hypotheses (see (32.28)). The popular choice is a generalized likelihood-ratio test (GLRT) that proposes to threshold the generalized likelihood ratio

$$T(X^n) = \frac{\sup_{P \in \mathcal{P}} P^{\otimes n}(X^n)}{\sup_{Q \in \mathcal{Q}} Q^{\otimes n}(X^n)}.$$

For examples and counterexamples of the optimality of GLRT in terms of error exponents, see, for example, [470].

Sometimes the families $\mathcal{P}$ and $\mathcal{Q}$ are small balls (in some metric) surrounding the center distributions P and Q, respectively. In this case, testing $\mathcal{P}$ against $\mathcal{Q}$ is known as *robust hypothesis testing* (since the test is robust to small deviations of the data distribution). There is a notable finite-sample optimality result in this case due to Huber [221] – see Exercise **III.31**. Asymptotically, it turns out that if $\mathcal{P}$ and $\mathcal{Q}$ are separated in the Hellinger distance, then the probability of error can be made exponentially small: see Theorem 32.8.

Sometimes in the setting of composite testing the distance between $\mathcal{P}$ and $\mathcal{Q}$ is zero. This is the case, for example, for the most famous setting of a Student t-test: $\mathcal{P} = \{\mathcal{N}(0, \sigma^2)\colon \sigma^2 > 0\}$, $\mathcal{Q} = \{\mathcal{N}(\mu, \sigma^2)\colon \mu \neq 0, \sigma^2 > 0\}$. It is clear that in this case there is no way to construct a test with $\alpha + \beta < 1$, since the data distribution under H_1 can be arbitrarily close to P_0. Here, thus, instead of minimizing the worst-case β, one tries to find a test statistic $T(X_1, \dots, X_n)$ which is (a) pivotal in the sense that its distribution under H_0 is (asymptotically) independent of the choice $P_0 \in \mathcal{P}$; and (b) consistent, in the sense that $T \to \infty$ as $n \to \infty$ under any *fixed* $Q \in \mathcal{Q}$. Optimality questions are studied by minimizing β as a function of $Q \in \mathcal{Q}$ (known as the power curve). The uniform most powerful tests are the gold standard in this area [278, Chapter 3], although besides a few classical settings (such as the one above) their existence is unknown.

In other settings, known as the *goodness-of-fit testing* [278, Chapter 14], instead of relatively low-complexity parametric families $\mathcal{P}$ and $\mathcal{Q}$ one is interested in a giant set $\mathcal{Q}$ of alternatives. For example, the simplest setting is to distinguish $H_0\colon X_i \stackrel{\text{iid}}{\sim} P_0$ versus $H_1\colon X_i \stackrel{\text{iid}}{\sim} Q, \mathrm{TV}(P_0, Q) > \delta$. If $\delta = 0$, then in this case again $\alpha + \beta = 1$ for any test and one may only ask for a statistic $T(X^n)$ with a known distribution under H_0 and $T \to \infty$ for any Q in the alternative. For $\delta > 0$ the problem is known as nonparametric detection [225, 226] and related to that of property testing [192].

16.5* Hypothesis Testing with Communication Constraints

In this section we consider a variation of the hypothesis testing problem where determination of the Stein exponent is still open, except for the special case resolved in [8]. Specifically, we still consider the case of a pair of simple iid hypotheses as in (14.14) except this time the Y sample is available to a statistician, but the X sample needs to be communicated from a remote location over a (noiseless) rate-constrained link:

$$\begin{aligned} H_0&: \ (X_1, Y_1), \ldots, (Y_n, X_n) \stackrel{\text{iid}}{\sim} P_{X,Y}, \\ H_1&: \ (X_1, Y_1), \ldots, (Y_n, X_n) \stackrel{\text{iid}}{\sim} Q_{X,Y}, \end{aligned} \tag{16.15}$$

and the tester consists of an X-compressor $W = f(X^n)$ with $W \in \{0, 1\}^{nR}$ and a decision $P_{Z|W,Y^n}$. (The illustration of the setting is given in Figure 16.5.) Our goal is to determine the dependence of the Stein exponent on the rate R, namely

$$V_\epsilon(R) \triangleq \sup\{E \colon \exists n_0, \forall n \geq n_0, \exists (f, P_{Z|X^n}) \text{ s.t. } \alpha > 1 - \epsilon, \beta < \exp(-nE)\}.$$

Exponents E satisfying constraints inside the supremum are known as ϵ-achievable exponents.

The importance of this problem is that it emphasized some new phenomenon arising in distributed statistical problems. Although, the problem is still open and the topic of characterizing Stein's exponent fell out of fashion, the tools that were developed for this problem (namely, the strong data-processing inequalities) are important and found many uses in modern distributed inference problems (see Chapter 33 and specifically Section 33.11 for more). We will discuss this after we present the main result, for which we introduce a key new concept.

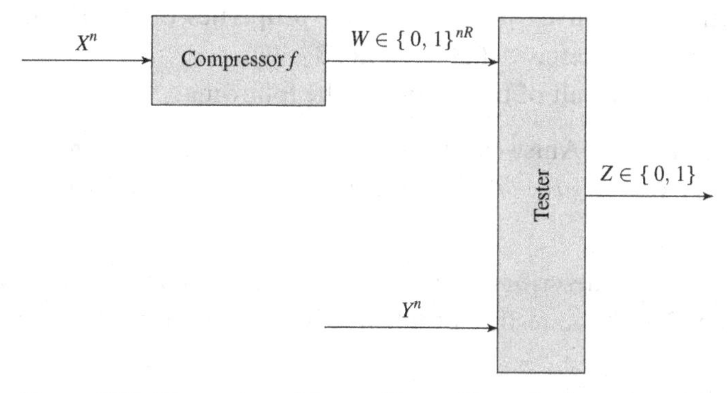

Figure 16.5 Illustration of the problem of hypothesis testing with communication constraints.

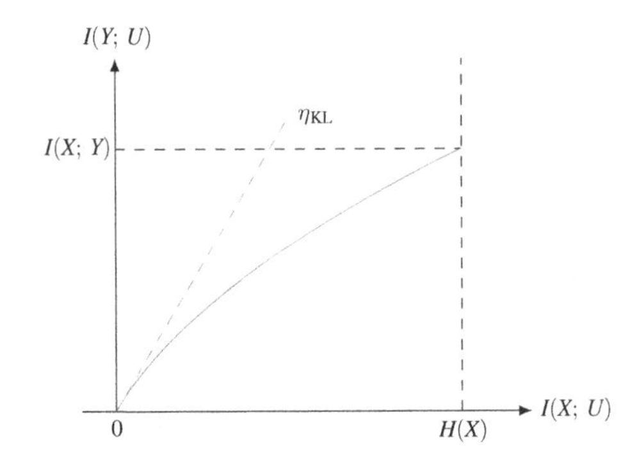

Figure 16.6 A typical F_I-curve whose slope at zero is the SDPI constant η_{KL}.

Definition 16.5. (F_I-curve[6]) Given a pair of random variables (X, Y) we define their F_I–curve as

$$F_I(t;P_{X,Y}) \triangleq \sup\{I(U;Y) \colon I(U;X) \le t, \quad U \to X \to Y\},$$

the supremum taken over all random variables U satisfying the $U \to X \to Y$ Markov relation.

Example 16.1.

A typical F_I-curve is shown in Figure 16.6. In general, computing F_I-curves is hard. An exception is the case of $X \sim \text{Ber}(1/2)$ and $Y = \mathsf{BSC}_\delta(X)$. In this case, applying MGL in Exercise **I.64** we get, in the notation of that exercise, that

$$F_I(t) = \log 2 - h(h^{-1}(\log 2 - t) * \delta),$$

achieved by taking $U \sim \text{Ber}(1/2)$ and $X = \mathsf{BSC}_p(U)$ with p chosen such that $h(p) = \log 2 - t$.

From the DPI (3.12) we know that $F_I(t) \le t$ and the F_I-curve strengthens the DPI to $I(U;X) \le F_I(I(U;Y))$ whenever $U \to X \to Y$. (Note that the roles of X and Y are not symmetric.) In general, it is not clear how to compute this function; nevertheless, in Exercise **III.32** we show that if X takes values over a finite alphabet then it is sufficient to consider $|\mathcal{U}| \le |\mathcal{X}| + 1$, and hence F_I is a value of a finite-dimensional convex program. Other properties of the F_I-curve and applications are found in Exercises **III.32** and **III.33**.

The main result of this section is the following.

Theorem 16.6. (Ahlswede–Csiszár [8]) *Suppose X, Y range over the finite alphabets and $Q_{X,Y} = P_X P_Y$ (independence testing problem). Then $V_\epsilon(R) = F_I(R)$ for all $\epsilon \in (0, 1)$.*

The setting describes the problem of detecting correlation between two sequences. When $R = 0$ the testing problem is impossible since the marginal distribution of

[6] This concept was introduced in [454], see also [136] and [346, Section 2.2] for the "P_X-independent" version.

Y is the same under both hypotheses. If only a very small communication rate is available then the sample size required will be very large (Stein exponent small).

Proof. Let us start with an upper bound. Fix a compressor W and notice that for any ϵ-achievable exponent E by Theorem 14.8 we have

$$d(1-\epsilon \| \exp\{-nV_\epsilon\}) \le D(P_{W,Y^n} \| Q_{W,Y^n}).$$

But under the conditions of the theorem we have $Q_{W,Y^n} = P_W P_{Y^n}$ and thus we obtain as in (14.18):

$$(1-\epsilon)nE \le D(P_{W,Y^n} \| P_W P_{Y^n}) + \log 2 = I(W;Y^n) + \log 2.$$

Now, from looking at Figure 16.5 we see that $W \to X^n \to Y^n$ and from Exercise **III.32** (tensorization) we know that

$$I(W;Y^n) \le nF_I\left(\frac{1}{n}I(W;X^n)\right) \le nF_I(R).$$

Thus, we have shown that for all sufficiently large n

$$E \le \frac{F_I(R)}{1-\epsilon} + \frac{\log 2}{n}.$$

This demonstrates that $\limsup_{\epsilon\to 0} V_\epsilon \le F_I(R)$. For a stronger claim of $V_\epsilon \le F_I(R)$, that is, the strong converse, see [8].

Now, for the constructive part, consider any n_1 and any compressed random variable $W_1 = f_1(X^{n_1})$ with $W_1 \in \{0,1\}^{n_1 R}$. Given blocklength n we can repeatedly send W_1 by compressing each n_1 chunk independently (for a total of n/n_1 "frames"). Then the decompressor will observe n/n_1 iid copies of W_1 and also of Y^{n_1} vector observations. Note that $D(P_{W_1,Y^{n_1}} \| Q_{W_1,Y^{n_1}}) = I(W_1;Y^{n_1})$ as above. Thus, by Theorem 14.14 we should be able to achieve $\alpha \ge 1-\epsilon$ and $\beta \le \exp\{-\frac{n}{n_1}I(W_1;Y^{n_1})\}$.

Therefore, we obtained the following lower bound (after optimizing over the choice of W_1 and blocklength n_1 that we replace by more convenient n again):

$$V_\epsilon \ge \tilde{F}_I(R) \triangleq \sup_{n,W_1}\left\{\frac{1}{n}I(W_1;Y^n) : W_1 \to X^n \to Y^n, W_1 \in \{0,1\}^{nR}\right\}. \tag{16.16}$$

This looks very similar to the definition of (tensorized) F_I-curve except that the constraint is on the cardinality of W_1 instead of the $I(W_1;X^n)$. It turns out that the two quantities coincide, that is, $\tilde{F}_I(R) = F_I(R)$. We only need to show $\tilde{F}_I(R) \ge F_I(R)$ for that.

To that end, consider any $U \to X \to Y$ and $R > I(U;X)$. We apply the covering lemma (Corollary 25.6) and Markov lemma (Proposition 25.7), where we set $A^n = X^n$, $B^n = U^n$, and $X^n = Y^n$. Overall, we get that as $n \to \infty$ there exist encoder $W_1 = f_1(X^n)$, $W_1 \in \{0,1\}^{nR}$ such that $W_1 \to X^n \to Y^n$ and

$$I(W_1;Y^n) \ge nI(U;Y) + o(n).$$

By optimizing the choice of U this proves $\tilde{F}_I(R+) \geq F_I(R)$. Since (as we showed above) $\tilde{F}_I(R+) \leq F_I(R+)$ and $F_I(R+) = F_I(R)$ (Exercise **III.32**), we conclude that $\tilde{F}_I(R) = F_I(R)$. □

The theorem shown above has interesting implications for a certain task in modern machine learning. Consider the situation where the sample size n is gigantic but the communication budget (or memory bandwidth) is constrained so that we can at most deliver k bits from terminal X to the tester. Then the rate $R = \frac{k}{n} \ll 1$ and the error probability β of an optimal test is roughly given as

$$\beta \approx 2^{-nF_I(k/n)} \approx 2^{-kF_I'(0)},$$

where we used the fact that $F_I(k/n) \approx \frac{k}{n}F_I'(0)$. We see that the error probability is decaying exponentially *with the number of communicated bits* not the sample size. In many ways, Theorem 16.6 foreshadowed various results in the last decade on distributed inference. We will get back to this topic in Chapter 33 dedicated to the strong data-processing inequality (SDPI). There is a simple relation that connects the classical results (this section) with the modern approach via SDPIs (in Chapter 33): the slope $F_I'(0)$ is precisely the SDPI constant,

$$F_I'(0) = \eta_{\mathrm{KL}}(P_X, P_{Y|X}),$$

see (33.14) for the definition of η_{KL}. In essence, SDPIs are just linearized versions of F_I-curves as illustrated in Figure 16.6.

Exercises for Part III

III.1 Let P_0 and P_1 be distributions on $\mathcal{X}$. Recall that the region of achievable pairs $(P_0[Z = 0], P_1[Z = 0])$ via randomized tests $P_{Z|X}: \mathcal{X} \to \{0, 1\}$ is denoted

$$\mathcal{R}(P_0, P_1) \triangleq \bigcup_{P_{Z|X}} (P_0[Z = 0], P_1[Z = 0]) \subseteq [0, 1]^2.$$

(a) Let $P_{Y|X}: \mathcal{X} \to \mathcal{Y}$ be a Markov kernel, which maps P_j to Q_j according to $P_j \xrightarrow{P_{Y|X}} Q_j, j = 0, 1$. Compare the regions $\mathcal{R}(P_0, P_1)$ and $\mathcal{R}(Q_0, Q_1)$. What does this say about $\beta_\alpha(P_0, P_1)$ versus $\beta_\alpha(Q_0, Q_1)$?

(b*) Prove that $\mathcal{R}(P_0, P_1) \supset \mathcal{R}(Q_0, Q_1)$ implies the existence of some $P_{Y|X}$ mapping P_0 to Q_1 and P_1 to Q_1. In other words, inclusion of $\mathcal{R}$ is equivalent to degradation or Blackwell order (see Definition 33.15).

Comment: This is the most general form of the data-processing inequality, of which all the other ones (divergence, mutual information, f-divergence, total variation, Rényi divergence, etc.) are corollaries.

III.2 Consider the following binary hypothesis testing (BHT) problem. Under both hypotheses X and Y are uniform on $\{0, 1\}$. However, under H_0, X and Y are independent, while under H_1:

$$\mathbb{P}_1[X \neq Y] = \delta < 1/2.$$

For this problem:

(a) Draw the region $\mathcal{R}(P_0, P_1)$ of achievable pairs of values $(\mathbb{P}_0[Z = 0], \mathbb{P}_1[Z = 0])$ for all randomized tests $P_{Z|XY}: \mathcal{X} \times \mathcal{Y} \to \{0, 1\}$.

(b) Find a sufficient statistic and define an equivalent BHT problem on a smaller alphabet.

(c) Let P_{F_i}, $i \in \{0, 1\}$ be the distribution of $\log \frac{P_0(X)}{P_1(X)}$ under $X \sim P_i$. What are the distributions P_{F_0}, P_{F_1}? How can you read them off of $\mathcal{R}(P_0, P_1)$?

(d) Compute the minimal probability of error in the Bayesian setup, when

$$\mathbb{P}[H_1] = 1 - \mathbb{P}[H_0] = \pi_1.$$

Identify the corresponding point on $\mathcal{R}(P_0, P_1)$.

(e) Compute the minimal probability of error in the non-Bayesian minimax setup:

$$\min \max\{\mathbb{P}_0[\text{decide } H_1], \mathbb{P}_1[\text{decide } H_0]\},$$

where the min is over the tests and the max is between the two numbers in the braces. Identify the corresponding point on $\mathcal{R}(P_0, P_1)$.

III.3 Consider distributions P and Q on $[0, 3]$ with densities of P and Q as shown below.

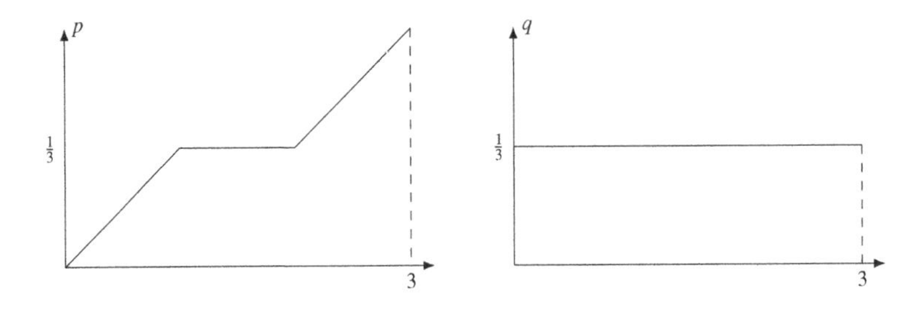

(a) Compute the expression of $\beta_\alpha(P, Q)$.
(b) Plot the region $\mathcal{R}(P, Q)$.
(c) Specify the tests achieving β_α for $\alpha = 5/6$ and $\alpha = 1/2$, respectively.

III.4 Let P be the uniform distribution on the interval $[0, 1]$. Let Q be the equal mixture of the uniform distribution on $[0, 1/2]$ and the point mass at 1.

(a) Compute the region $\mathcal{R}(P, Q)$.
(b) Explicitly describe the tests that achieve the optimal boundary $\beta_\alpha(P, Q)$.

III.5 (a) Consider the binary hypothesis test:

$$H_0 \colon X \sim \mathcal{N}(0, 1) \text{ versus } H_1 \colon X \sim \mathcal{N}(\mu, 1).$$

Compute and plot the Neyman–Pearson region $\mathcal{R}(\mathcal{N}(0, 1), \mathcal{N}(\mu, 1))$.

(b) Now suppose we have n samples and we want to test

$$H_0 \colon X_1, \ldots, X_n \overset{\text{iid}}{\sim} \mathcal{N}(0, 1) \text{ versus } H_1 \colon X_1, \ldots, X_n \overset{\text{iid}}{\sim} \mathcal{N}(\mu, 1).$$

Compute the Neyman–Pearson region $\mathcal{R}(\mathcal{N}(0, 1)^n, \mathcal{N}(\mu, 1)^n)$. As the sample size increases, describe how the region evolves and provide an interpretation. (Hint: Consider sufficient statistics.)

III.6 (a) Consider the binary hypothesis test:

$$H_0 \colon X \sim \text{Exp}(1) \text{ versus } H_1 \colon X \sim \text{Exp}(\lambda),$$

where $\text{Exp}(\lambda)$ has density $\lambda e^{-\lambda x} 1\{x \geq 0\}$. Compute the region $\mathcal{R}(\text{Exp}(1), \text{Exp}(\lambda))$. What is the optimal test for achieving β_α?

(b) Now suppose we have n samples and we want to test

$$H_0\colon X_1, \dots, X_n \stackrel{\text{iid}}{\sim} \text{Exp}(1) \text{ versus } H_1\colon X_1, \dots, X_n \stackrel{\text{iid}}{\sim} \text{Exp}(\lambda).$$

Compute the region $\mathcal{R}(\text{Exp}(1)^n, \text{Exp}(\lambda)^n)$. As the sample size increases, describe how the region evolves and provide an interpretation. What is the optimal test for achieving β_α?

III.7 (a) Prove that $\text{TV}(P, Q) = \sup_{0 \le \alpha \le 1}\{\alpha - \beta_\alpha(P, Q)\}$. Explain how to read the value $\text{TV}(P, Q)$ from the region $\mathcal{R}(P, Q)$. Does it equal half the maximal vertical segment in $\mathcal{R}(P, Q)$?

(b) (Bayesian criterion) Fix a prior $\pi = (\pi_0, \pi_1)$ such that $\pi_0 + \pi_1 = 1$ and $0 < \pi_0 < 1$. Denote the optimal average error probability by $P_e \triangleq \inf_{P_{Z|X}} \pi_0 \pi_{1|0} + \pi_1 \pi_{0|1}$. Prove that if $\pi = \left(\frac{1}{2}, \frac{1}{2}\right)$, then $P_e = \frac{1}{2}(1 - \text{TV}(P, Q))$. Find the optimal test.

(c) Find the optimal test for general prior π (not necessarily equiprobable).

(d) Show that it is sufficient to focus on a deterministic test in order to minimize the Bayesian error probability.

III.8 The function $\alpha \mapsto \beta_\alpha(P, Q)$ is monotone and thus by Lebesgue's theorem possesses a derivative

$$\beta'_\alpha \triangleq \frac{d}{d\alpha}\beta_\alpha(P, Q)$$

almost everywhere on $[0, 1]$. Prove that

$$D(P\|Q) = -\int_0^1 \log \beta'_\alpha \, d\alpha. \tag{III.1}$$

III.9 Let P, Q be distributions such that for all $\alpha \in [0, 1]$ we have

$$\beta_\alpha(P, Q) \triangleq \min_{P_{Z|X}\colon P[Z=0] \ge \alpha} Q[Z = 0] = \alpha^2.$$

Find $\text{TV}(P, Q)$, $D(P\|Q)$, and $D(Q\|P)$.

III.10 We have shown that for testing iid products and any fixed $\epsilon \in (0, 1)$:

$$\log \beta_{1-\epsilon}(P^n, Q^n) = -nD(P\|Q) + o(n), \qquad n \to \infty,$$

which is equivalent to Stein's lemma (Theorem 14.14). Show furthermore that assuming $V(P\|Q) < \infty$ we have

$$\log \beta_{1-\epsilon}(P^n, Q^n) = -nD(P\|Q) + \sqrt{nV(P\|Q)}Q^{-1}(\epsilon) + o(\sqrt{n}), \tag{III.2}$$

where $Q^{-1}(\cdot)$ is the functional inverse of $Q(x) = \int_x^\infty \frac{1}{\sqrt{2\pi}} e^{-t^2/2} dt$ and

$$V(P\|Q) \triangleq \text{Var}_P\left[\log \frac{dP}{dQ}\right].$$

III.11 (Likelihood-ratio trick) Given two distributions P and Q on $\mathcal{X}$ let us generate iid samples (X_i, Y_i) as follows: first $Y_i \sim \text{Ber}(1/2)$ and then if $Y_i = 0$ we sample $X_i \sim Q$ and otherwise $X_i \sim P$. We next train a classifier to minimize the cross-entropy loss:

$$\hat{p}^* = \underset{\hat{p}:\, \mathcal{X}\to[0,1]}{\operatorname{argmin}} \frac{1}{n}\sum_{i=1}^{n} Y_i \log\frac{1}{\hat{p}(X_i)} + (1-Y_i)\log\frac{1}{1-\hat{p}(X_i)}.$$

Show that $\frac{1-\hat{p}^*(x)}{\hat{p}^*(x)} \to \frac{dP}{dQ}(x)$ as $n \to \infty$. This trick is used in machine learning to approximate $\frac{dP}{dQ}$ for complicated high-dimensional distributions.

III.12 Prove that

$$\min_{Q_{Y^n}\in\mathcal{F}} D\left(Q_{Y^n} \middle\| \prod_{j=1}^{n} P_{Y_j}\right) = \min \sum_{j=1}^{n} D(Q_{Y_j}\|P_{Y_j})$$

whenever the constraint set $\mathcal{F}$ is marginals-based, that is,

$$Q_{Y^n} \subset \mathcal{F} \quad \iff \quad (Q_{Y_1},\ldots,Q_{Y_n}) \in \mathcal{F}'$$

for some $\mathcal{F}'$.

Conclude that in the case when $P_{Y_j} = P$ and

$$\mathcal{F} = \left\{ Q_{Y^n} \colon \mathbb{E}_Q\left[\sum_{j=1}^{n} f(Y_j)\right] \geq n\gamma \right\}$$

we have the single-letterization:

$$\min_{Q_{Y^n}} D(Q_{Y^n}\|P^n) = n \min_{Q_Y\colon\, \mathbb{E}_{Q_Y}[f(Y)]\geq\gamma} D(Q_Y\|P),$$

of which (15.15) is a special case. (Hint: Convexity of divergence.)

III.13 Fix a distribution P_X on a finite set $\mathcal{X}$ and a channel $P_{Y|X}\colon \mathcal{X} \to \mathcal{Y}$. Consider a sequence x^n with composition P_X, that is,

$$\#\{j\colon x_j = a\} = nP_X(a) \pm 1 \qquad \forall a \in \mathcal{X}.$$

Let Y^n be generated according to $P_{Y|X}^{\otimes n}(\cdot|x^n)$. Show that

$$\log \mathbb{P}\left[\sum_{j=1}^{n} f(X_j, Y_j) \geq n\gamma \middle| X^n = x^n\right] = -n \min_{\mathbb{E}_Q[f(X,Y)]\geq\gamma} D(Q_{Y|X}\|P_{Y|X}|P_X) + o(n),$$

where the minimum is over all $Q_{X,Y}$ with $Q_X = P_X$.

III.14 (Large deviations on the boundary) Recall that $A = \inf_\lambda \psi_X'(\lambda)$ and $B = \sup_\lambda \psi_X'(\lambda)$ were shown to be the boundaries of the support of P_X, for example, $B = \sup\{b\colon \mathbb{P}[X > b] > 0\}$.

(a) Show by example that $\psi_X^*(B)$ can be finite or infinite.

(b) Show by example that the asymptotic behavior of

$$\mathbb{P}\left[\frac{1}{n}\sum_{i=1}^{n} X_i \geq B\right] \tag{III.3}$$

can be quite different depending on the distribution of P_X.

(c) Compare $\Psi^*_X(B)$ and the exponent in (III.3) for your examples. Prove a general statement (you can assume that $\psi_X(\lambda) < \infty$ for all $\lambda \in \mathbb{R}$).

III.15 (Simple radar) A binary signal detector is being built. When the signal A is being sent a sequence of iid $X_j \sim \mathcal{N}(-1, 1)$ is received. When the signal B is being sent a sequence of iid $X_j \sim \mathcal{N}(+1, 1)$ is received. Given a very large number n of observations $(X_1, \dots, X_n)$ propose a detector for deciding between A and B. Consider two separate design cases:

(a) Misdetecting A for B or B for A are equally bad.
(b) Misdetecting A for B in 10^{-3} cases is acceptable, but the opposite should be avoided as much as possible.

III.16 (Small-ball probability I) Let $Z \sim \mathcal{N}(0, I_d)$. Without using the χ^2 density, show the following bounds on $\mathbb{P}[\|Z\|_2 \le \epsilon]$.

(a) Using the Chernoff bound, show that for all $\epsilon > \sqrt{d}$,

$$\mathbb{P}[\|Z\|_2 \le \epsilon] \le \left(\frac{e\epsilon^2}{d}\right)^{d/2} e^{-\epsilon^2/2}.$$

(b) Prove the lower bound

$$\mathbb{P}[\|Z\|_2 \le \epsilon] \ge \left(\frac{\epsilon^2}{2\pi d}\right)^{d/2} e^{-\epsilon^2/2}.$$

(c) Extend the results to $Z \sim \mathcal{N}(0, \Sigma)$.

See Exercise **V.30** for an example in infinite dimensions.

III.17 Consider the hypothesis testing problem:

$$H_0: X_1, \dots, X_n \overset{\text{iid}}{\sim} P = \text{Ber}(1/3),$$
$$H_1: X_1, \dots, X_n \overset{\text{iid}}{\sim} Q = \text{Ber}(2/3).$$

(a) Compute the Stein exponent.
(b) Compute the tradeoff region $\mathcal{E}$ of achievable error-exponent pairs (E_0, E_1) using the characterization $E_0(\theta) = \psi^*_P(\theta)$ and $E_1(\theta) = \psi^*_P(\theta) - \theta$. Express the optimal boundary in explicit form (eliminate the parameter).
(c) Identify the divergence-minimizing geodesic $P^{(\lambda)}$ running from P to Q, $\lambda \in [0, 1]$. Verify that $(E_0, E_1) = (D(P^{(\lambda)}\|P), D(P^{(\lambda)}\|Q)), 0 \le \lambda \le 1$ gives the same tradeoff curve.
(d) Compute the Chernoff exponent.

III.18 Let $\gamma(a, c)$ denote a Gamma distribution with shape parameter a and scale parameter c:

$$\gamma(a, c)(dx) = \frac{(cx)^{a-1}e^{-cx}}{\Gamma(a)} c\, dx.$$

Consider a hypothesis testing problem:

$$H_0 : X_1, \dots, X_n \overset{\text{iid}}{\sim} P_0 = \text{Exp}(1), \tag{III.4}$$
$$H_1 : X_1, \dots, X_n \overset{\text{iid}}{\sim} P_1 = \gamma(a, c = 1). \tag{III.5}$$

(a) Compute the Stein exponent.
(b) For $a = 3$ draw the tradeoff region $\mathcal{E}$ of achievable error-exponent pairs (E_0, E_1).
(c) Identify the divergence-minimizing geodesic P_λ running from P_0 to P_1, $\lambda \in [0, 1]$.

Hint: To simplify calculations try differentiating in u the following identity

$$\int_0^\infty x^u e^{-x} dx = \Gamma(u + 1).$$

III.19 Consider the hypothesis testing problem:

$$H_0 \colon X_1, \dots, X_n \overset{\text{iid}}{\sim} P = \mathcal{N}(0, 1),$$
$$H_1 \colon X_1, \dots, X_n \overset{\text{iid}}{\sim} Q = \mathcal{N}(\mu, 1).$$

(a) Show that the Stein exponent is $V = \frac{\log e}{2}\mu^2$.
(b) Show that the optimal tradeoff between achievable error-exponent pairs (E_0, E_1) is given by

$$E_1 = \frac{\log e}{2}(\mu - \sqrt{2E_0})^2, \quad 0 \le E_0 \le \frac{\log e}{2}\mu^2.$$

(c) Show that the Chernoff exponent is $C(P, Q) = \frac{\log e}{8}\mu^2$.

III.20 Let U_j be iid uniform on $[0, 1]$. Prove/disprove that

$$\mathbb{P}\left[\sum_{j=1}^n U_j \ge n\gamma\right], \qquad \gamma = \frac{1}{e - 1} \approx 0.582$$

converges to zero exponentially fast as $n \to \infty$. If it does then find the exponent. Repeat with $\gamma = 0.5$.

III.21 Let X_j be iid exponential with unit mean. Since the log MGF $\psi_X(\lambda) \triangleq \log \mathbb{E}[\exp\{\lambda X\}]$ does not exist for $\lambda > 1$, the large-deviations result in Theorem 15.1, namely

$$\mathbb{P}\left[\sum_{j=1}^n X_j \ge n\gamma\right] = \exp\{-n\psi_X^*(\gamma) + o(n)\}, \tag{III.6}$$

does not apply. Show (III.6) directly via the following steps:

(a) Apply the Chernoff argument directly to prove an upper bound.
(b) Fix an arbitrary $c > 0$ and prove

$$\mathbb{P}\left[\sum_{j=1}^n X_j \ge n\gamma\right] \ge \mathbb{P}\left[\sum_{j=1}^n (X_j \wedge c) \ge n\gamma\right]. \tag{III.7}$$

(c) Apply the results shown in Chapter 15 to investigate the asymptotics of the right-hand side of (III.7).

(d) Conclude the proof of (III.6) by taking $c \to \infty$.

III.22 (Hoeffding's lemma) In this exercise we prove Hoeffding's lemma (stated after Definition 4.15) and derive Hoeffding's concentration inequality. Let $X \in [-1, 1]$ with $\mathbb{E}[X] = 0$.

(a) Show that the log MGF $\psi_X(\lambda)$ satisfies $\psi_X(0) = \psi_X'(0) = 0$ and $0 \le \psi_X''(\lambda) \le 1$. (Hint: Apply Theorem 15.8(c) and the fact that the variance of any distribution supported on $[-1, 1]$ is at most 1.)

(b) By Taylor expansion, show that $\psi_X(\lambda) \le \lambda^2/2$.

(c) Applying Theorem 15.1, prove Hoeffding's inequality: Let X_i's be iid copies of X. For any $\gamma > 0$, $\mathbb{P}\left[\sum_{i=1}^n X_i \ge n\gamma\right] \le \exp(-n\gamma^2/2)$.

III.23 (Sanov's theorem for discrete $\mathcal{X}$) Let $\mathcal{X}$ be a finite set. Let $\mathcal{E}$ be a set of probability distributions on $\mathcal{X}$ with non-empty interior. Let $X^n = (X_1, \ldots, X_n)$ be iid drawn from some distribution P fully supported on $\mathcal{X}$ and let π_n denote the empirical distribution, that is, $\pi_n = \frac{1}{n}\sum_{i=1}^n \delta_{X_i}$. Our goal is to show that

$$E \triangleq \lim_{n\to\infty} \frac{1}{n} \log \frac{1}{P[\pi_n \in \mathcal{E}]} = \inf_{Q \in \mathcal{E}} D(Q\|P). \tag{III.8}$$

(a) We first assume that $\mathcal{E}$ is convex. Define the following set of joint distributions $\mathcal{E}_n \triangleq \{Q_{X^n} : Q_{X_i} \in \mathcal{E}, i = 1, \ldots, n\}$. Show that

$$\inf_{Q_{X^n} \in \mathcal{E}_n} D(Q_{X^n}\|P_{X^n}) = n \inf_{Q \in \mathcal{E}} D(Q\|P),$$

where $P_{X^n} = P^n$.

(b) Consider the conditional distribution $\tilde{P}_{X^n} = P_{X^n|\pi_n \in \mathcal{E}}$. Show that $\tilde{P}_{X^n} \in \mathcal{E}_n$.

(c) Prove the following non-asymptotic upper bound: for any convex $\mathcal{E}$,

$$P[\pi_n \in \mathcal{E}] \le \exp\Big(- n \inf_{Q \in \mathcal{E}} D(Q\|P)\Big), \quad \forall n.$$

(d) Show that for any $\mathcal{E}$:

$$P[\pi_n \in \mathcal{E}] \le \exp\Big(- n \inf_{Q \in \mathcal{E}} D(Q\|P) + o(n)\Big), \quad n \to \infty.$$

(Hint: For each $\epsilon > 0$, cover $\mathcal{E}$ by N TV balls of radius ϵ where $N = N(\epsilon)$ is finite; cf. Theorem 27.3. Apply the previous part and the union bound.)

(e) For any Q in the interior of $\mathcal{E}$, show that

$$P[\pi_n \in \mathcal{E}] \ge \exp(-nD(Q\|P) + o(n)), \quad n \to \infty.$$

(Hint: Use data processing as in the proof of the large-deviations theorem, Theorem 15.9.)

(f) Conclude (III.8) by applying the continuity of divergence on a finite alphabet (Proposition 4.8).

III.24 (Error exponents of data compression) Let X^n be iid according to P on a finite alphabet $\mathcal{X}$. Let $\epsilon^*(X^n, nR)$ denote the minimal probability of error achieved by fixed-length compressors and decompressors for X^n of compression rate R (see Definition 11.1). We know that if $R < H(P)$ then $\epsilon^*(X^n, nR) \to 0$. Here we show it converges to zero exponentially fast and find the exponent.

(a) For any sequence x^n, denote by $\hat{P}_{x^n}$ its empirical distribution and by $H(\hat{P}_{x^n})$ its empirical entropy, that is, the entropy of the empirical distribution. For example, for the binary sequence $x^n = (010110)$, the empirical distribution is Ber(1/2) and the empirical entropy is 1 bit. For each $R > 0$, define the set $T = \{x^n : H(\hat{P}_{x^n}) < R\}$. Show that $|T| \le \exp(nR)(n+1)^{|\mathcal{X}|}$.

(b) Show that for any $R > H(P)$, $\epsilon^*(X^n, nR) \le \exp(-n \inf_{Q:\, H(Q)>R} D(Q\|P))$. Specify the achievability scheme. (Hint: Use Sanov's theorem in Exercise **III.23**.)

(c) Prove that the above exponent is asymptotically optimal:

$$\limsup_{n\to\infty} \frac{1}{n} \log \frac{1}{\epsilon^*(X^n, nR)} \le \inf_{Q:\, H(Q)>R} D(Q\|P).$$

(Hint: Recall that any compression scheme for a memoryless source with rate below the entropy fails with probability tending to one. Use data-processing inequality.)

III.25 (Local KL geodesics) Recall from Section 2.6.1* the local expansion $D(\lambda Q + (1-\lambda)P\|P) = \frac{\lambda^2}{2}\chi^2(Q\|P) + o(\lambda^2)$, provided that $\chi^2(Q\|P) < \infty$. Instead of the linear mixture, consider the geometric mixture $P_\lambda \propto Q^\lambda P^{1-\lambda}$, which we argued should be called a "KL geodesic" in (15.30).

(a) Show that $D(P_\lambda\|P) = -\psi_P(\lambda) + \lambda\psi_P'(\lambda)$, where $\psi_P(\lambda) = \log \mathbb{E}_P [\exp(\lambda \log \frac{dQ}{dP})]$.

(b) State the appropriate conditions to conclude $D(P_\lambda\|P) = \frac{1}{2}\lambda^2\psi_P''(0) + o(\lambda^2)$, where $\psi_P''(0) = \mathrm{Var}_P[\log \frac{dQ}{dP}]$, which is clearly different from $\chi^2(Q\|P) = \mathrm{Var}_P[\frac{dQ}{dP}]$.

III.26 Denote by $\mathcal{N}(\mu, \sigma^2)$ the one-dimensional Gaussian distribution with mean μ and variance σ^2. Let $a > 0$. All logarithms below are natural.

(a) Show that

$$\min_{Q:\, \mathbb{E}_Q[X]\ge a} D(Q\|\mathcal{N}(0,1)) = \frac{a^2}{2}.$$

(b) Let $X_1, \ldots, X_n$ be drawn iid from $\mathcal{N}(0,1)$. Using part (a) show that

$$\lim_{n\to\infty} \frac{1}{n} \log \frac{1}{\mathbb{P}[X_1 + \cdots + X_n \ge na]} = \frac{a^2}{2}. \tag{III.9}$$

(c) Let $Q(x) = \int_x^\infty \frac{1}{\sqrt{2\pi}} e^{-t^2/2} dt$ denote the complementary CDF of the standard Gaussian distribution. Express $\mathbb{P}[X_1 + \cdots + X_n \ge na]$ in terms of the Q function. Using the fact that $Q(x) = e^{-x^2/2 + o(x^2)}$ as $x \to \infty$ (see Exercise **V.25**), re-prove (III.9).

(d) (Reverse I-projection) Let Y be a continuous random variable with zero mean and unit variance. Show that

$$\min_{\mu,\sigma} D(P_Y \| \mathcal{N}(\mu, \sigma^2)) = D(P_Y \| \mathcal{N}(0, 1)).$$

III.27 (Why temperatures equalize) Let $\mathcal{X}$ be a finite alphabet and $f\colon \mathcal{X} \to \mathbb{R}$ an arbitrary function. Let $E_{\min} = \min f(x)$.

(a) Using I-projection show that for any $E \geq E_{\min}$ the solution of

$$H^*(E) = \max\{H(X)\colon \mathbb{E}[f(X)] \leq E\}$$

is given by a Gibbs distribution (see (5.21)) $P_X(x) = \frac{1}{Z(\beta)} e^{-\beta f(x)}$ for some $\beta = \beta(E)$.

Comment: In statistical physics x is a state of the system (e.g. locations and velocities of all molecules), $f(x)$ is the energy of the system in state x, P_X is the Gibbs distribution, and $\beta = \frac{1}{T}$ is the inverse temperature of the system. In thermodynamic equilibrium, $P_X(x)$ gives the fraction of time the system spends in state x.

(b) Show that $\frac{dH^*(E)}{dE} = \beta(E)$.

(c) Next consider two functions f_0, f_1 (e.g., two types of molecules with different state–energy relations). Show that for $E \geq \min_{x_0} f_0(x_0) + \min_{x_1} f_1(x_1)$ we have

$$\max_{\mathbb{E}[f_0(X_0)+f_1(X_1)]\leq E} H(X_0, X_1) = \max_{E_0+E_1\leq E} H_0^*(E_0) + H_1^*(E_1), \qquad \text{(III.10)}$$

where $H_j^*(E) = \max_{\mathbb{E}[f_j(X)]\leq E} H(X)$.

(d) Further, show that for the optimal choice of E_0 and E_1 in (III.10) we have

$$\beta_0(E_0) = \beta_1(E_1) \qquad \text{(III.11)}$$

or equivalently that the optimal distribution P_{X_0,X_1} is given by

$$P_{X_0,X_1}(a, b) = \frac{1}{Z_0(\beta)Z_1(\beta)} e^{-\beta(f_0(a)+f_1(b))}. \qquad \text{(III.12)}$$

Remark: Equation (III.12) also just follows from part (a) by taking $f(x_0, x_1) = f_0(x_0) + f_1(x_1)$. The point here is relation (III.11): when two thermodynamical systems are brought in contact with each other, the energy distributes among them in such a way that β parameters (temperatures) equalize.

III.28 (Importance sampling [90]) Let μ and ν be two probability measures on set $\mathcal{X}$. Assume that $\nu \ll \mu$. Let $L = D(\nu \| \mu)$ and $\rho = \frac{d\nu}{d\mu}$ be the Radon–Nikodym derivative. Let $f\colon \mathcal{X} \to \mathbb{R}$ be a measurable function. We would like to estimate $\mathbb{E}_\nu f$ using data from μ.

Let $X_1, \ldots, X_n \overset{\text{iid}}{\sim} \mu$ and $I_n(f) = \frac{1}{n} \sum_{1\leq i\leq n} f(X_i)\rho(X_i)$. Prove the following.

(a) For $n \geq \exp(L+t)$ with $t \geq 0$, we have

$$\mathbb{E}\,|I_n(f) - \mathbb{E}_\nu f| \leq \|f\|_{L^2(\nu)} \left(\exp(-t/4) + 2\sqrt{\mathbb{P}_\mu(\log \rho > L + t/2)} \right).$$

(Hint: Let $h = f1\{\rho \leq \exp(L + t/2)\}$. Use the triangle inequality and bound $\mathbb{E}\,|I_n(h) - \mathbb{E}_\nu h|$, $\mathbb{E}\,|I_n(h) - I_n(f)|$, and $|\mathbb{E}_\nu f - \mathbb{E}_\nu h|$ separately.)

(b) On the other hand, for $n \leq \exp(L - t)$ with $t \geq 0$, we have

$$\mathbb{P}[(I_n(1) \geq 1 - \delta)] \leq \exp(-t/2) + \frac{\mathbb{P}_\mu(\log \rho \leq L - t/2)}{1 - \delta},$$

for all $\delta \in (0, 1)$, where 1 is the constant-1 function.
(Hint: Divide into two cases depending on whether $\max_{1 \leq i \leq n} \rho(X_i) \leq \exp(L - t/2)$.)

This shows that a sample of size $\exp(D(\nu\|\mu) + \Theta(1))$ is both necessary and sufficient for accurate estimation by importance sampling.

III.29 (M-ary hypothesis testing)[7] The following result [275] generalizes Corollary 16.2 on the best average probability of error for testing two hypotheses to multiple hypotheses.

Fix a collection of distributions $\{P_1, \ldots, P_M\}$. Conditioned on θ, which takes value i with probability $\pi_i > 0$ for $i = 1, \ldots, M$, let $X_1, \ldots, X_n \overset{\text{iid}}{\sim} P_\theta$. Denote the optimal average probability of error by $p_n^* = \inf \mathbb{P}[\hat{\theta} \neq \theta]$, where the infimum is taken over all decision rules $\hat{\theta} = \hat{\theta}(X_1, \ldots, X_n)$.

(a) Show that

$$\lim_{n\to\infty} \frac{1}{n} \log \frac{1}{p_n^*} = \min_{1 \leq i < j \leq M} C(P_i, P_j), \qquad \text{(III.13)}$$

where C is the Chernoff information defined in (16.2).

(b) It is clear that the optimal decision rule is the maximum a posteriori (MAP) rule. Does the maximum likelihood rule also achieve the optimal exponent (III.13)? Prove or disprove.

III.30 Given n observations $(X_1, Y_1), \ldots, (X_n, Y_n)$, where each observation consists of a pair of random variables, we want to test the following hypothesis:

$$H_0\colon\ (X_i, Y_i) \overset{\text{iid}}{\sim} P \otimes Q,$$
$$H_1\colon\ (X_i, Y_i) \overset{\text{iid}}{\sim} Q \otimes P,$$

where $\otimes$ denotes product distribution as usual.

(a) Show that the Stein exponent $D(P \otimes Q \| Q \otimes P)$ is equal to $D(P\|Q) + D(Q\|P)$.

(b) Show that the Chernoff exponent (Chernoff information) $C(P \otimes Q, Q \otimes P)$ is equal to $-2\log(1 - \frac{H^2(P,Q)}{2}) = -2\log \int \sqrt{dPdQ}$, where $H(P, Q)$ is the Hellinger distance – see (7.5).

[7] Not to be confused with multiple testing in the statistics literature, which refers to testing multiple pairs of binary hypotheses simultaneously.

Comment: This type of hypothesis testing problem arises in the context of community detection, where n nodes indexed by $i \in [n]$ are partitioned into two communities (labeled by $\sigma_i = +$ and $\sigma_i = -$ uniformly and independently) and the task is to classify the nodes based on the pairwise observations $W = (W_{ij}: 1 \le i < j \le n)$ which are independent conditioned on σ_i's and $W_{ij} \sim P$ if $\sigma_i = \sigma_j$ and Q otherwise. (The stochastic block model previously introduced in Exercise **I.49** corresponds to P and Q being Bernoulli.) As a means to prove the impossibility result [456], consider the setting where an oracle reveals all labels except for σ_1. Define $S_+ = \{j = 2, \ldots, n: \sigma_j = +\}$ and similarly S_-. If $\sigma_1 = +$, $\{W_{1,j}: j \in S_+\} \overset{\text{iid}}{\sim} P$ and $\{W_{1,j}: j \in S_-\} \overset{\text{iid}}{\sim} Q$ and vice versa if $\sigma_1 = -$.

III.31 (Stochastic dominance and robust LRT) Let $\mathcal{P}_0, \mathcal{P}_1$ be two families of probability distributions on $\mathcal{X}$. Suppose that there is a *least favorable pair* (LFP) $(Q_0, Q_1) \in \mathcal{P}_0 \times \mathcal{P}_1$ such that

$$Q_0[\pi > t] \ge Q_0'[\pi > t],$$
$$Q_1[\pi > t] \le Q_1'[\pi > t],$$

for all $t \ge 0$ and $Q_i' \in \mathcal{P}_i$, where $\pi = dQ_1/dQ_0$. Prove that (Q_0, Q_1) simultaneously minimizes all f-divergences between $\mathcal{P}_0$ and $\mathcal{P}_1$, that is,

$$D_f(Q_1 \| Q_0) \le D_f(Q_1' \| Q_0') \qquad \forall Q_0' \in \mathcal{P}_0, Q_1' \in \mathcal{P}_1. \tag{III.14}$$

(Hint: Interpolate between (Q_0, Q_1) and (Q_0', Q_1') and differentiate.)
Remark: For the case of two TV-balls, that is, $\mathcal{P}_i = \{Q: \mathrm{TV}(Q, P_i) \le \epsilon\}$, the existence of LFP is shown in [221], in which case $\pi = \min(c', \max(c'', \frac{dP_0}{dP_1}))$ for some $0 \le c' < c'' \le \infty$ giving the *robust likelihood-ratio test.*

III.32 Recall the F_I-curve from Definition 16.5. Suppose $\mathcal{X}$ and $\mathcal{Y}$ are finite and prove the following:

(a) (Tensorization) $F_I(nt; P_{X,Y}^{\otimes n}) = nF_I(t; P_{X,Y})$. (Hint: Theorem 11.17.)
(b) (Concavity) $t \mapsto F_I(t)$ is concave and continuous on its domain $t \in [0, H(X)]$.
(c) (Cardinality bound) It is sufficient to take $|\mathcal{U}| \le |\mathcal{X}| + 1$ in the definition of F_I. (Hint: Inspect Theorem 11.17.)
(d) Show that sup in the definition is a max and that for every t there exists a random variable U s.t. $I(U;X) = t$ and $I(U;Y) = F_I(t)$.
(e) (Strict DPI) If $P_{X,Y}(x, y) > 0$ then $F_I(t) < t$ for all $t > 0$.

III.33 Gács–Körner (GK) common information [174] between a pair of random variables (X, Y) is defined as the supremum of rates R such that for all large enough n there exist functions $f(X^n)$ and $g(Y^n)$ with $H(f(X^n)) \ge nR$ and $\mathbb{P}[f(X^n) \ne g(Y^n)] \to 0$ where $(X_i, Y_i) \overset{\text{iid}}{\sim} P_{X,Y}$ (i.e. GK common information is the maximal rate at which randomness can be extracted from two correlated sequences). Show that if $P_{X,Y}(x, y) > 0$ for all x, y then GK common information is zero. (Hint: Show that $I(Y^n; g(Y^n)) = I(Y^n; f(X^n)) + o(n)$ and apply tensorization and strict DPI from Exercise **III.32**.)

Part IV

Channel Coding

In this part we study a new type of problem. The goal of channel coding is to communicate digital information across a noisy channel. Historically, this was the first area of information theory that led to immediately and widely deployed applications. Shannon's discovery [379] of the possibility of transmitting information with vanishing error and positive (i.e. bigger than zero) rate of bits per second was quite surprising and unexpected both practically and theoretically. Our goal in this part is to understand this monumental discovery.

To explain the relation of this part to others, let us revisit what problems we have studied so far. In Part I we introduced various information measures and studied their properties irrespective of engineering applications. Then, in Part II our objective was data compression. The main object there was a single distribution P_X and the fundamental limit $\mathbb{E}[\ell(f^*(X))]$ was the minimal compression length. The main result (the "coding theorem") established the connection between the fundamental limit and an information quantity, which we can summarize as

$$\mathbb{E}[\ell(f^*(X))] \approx H(X).$$

Next, in Part III we studied binary hypothesis testing. There the main object was a pair of distributions (P, Q), the fundamental limit was the Neyman–Pearson curve $\beta_{1-\epsilon}(P^n, Q^n)$, and the main result was

$$\beta_{1-\epsilon}(P^n, Q^n) \approx \exp\{-nD(P\|Q)\},$$

again connecting an operational quantity to an information measure.

In channel coding – the topic of this part – the main object is going to be a channel $P_{Y|X}$. The fundamental limit is $M^*(\epsilon)$, the maximum number of messages that can be transmitted with probability of error at most ϵ, which we rigorously define in Chapter 17. Our main result in this part is to show the celebrated Shannon's noisy channel coding theorem:

$$\log M^*(\epsilon) \approx \max_{P_X} I(X;Y)\,.$$

We will demonstrate the possibility of sending information with high reliability and also will rigorously derive the asymptotically (and non-asymptotically!) highest achievable rates. However, we entirely omit a giant and beautiful field of *coding theory* that deals with the question of how to construct transmitters and receivers with low computational complexity. This area of science, though deeply related to the content of our book, deserves a separate dedicated treatment. We recommend reading [361] for the *sparse-graph*-based codes and [373] for an introduction to more modern *polar codes*.

The practical implications of this part are profound even without giving explicit constructions of codes. First, in the process of finding channel capacity, one needs to maximize mutual information, and the maximizing distributions reveal properties of optimal codes. In this way, the abstract mutual information results that we obtained in Part I acquire operational interpretations. For example, the water-filling solution (Exercise **I.9**) dictates how to optimally allocate power between frequency bands; the computation in Figure 3.2 suggests when to use binary modulation, Exercise **I.10** explains how much gain (in bits per second) one can expect from adding receive or transmit antennas, etc. Second, the non-asymptotic (finite-blocklength) bounds that we develop in this part are routinely used for benchmarking the performance of all newly developed codes (e.g. in the context of 4G, 5G, and 6G standardization). In all, the contents of this part have had by far the most real-world impact of all (at least at the time of writing of this book).

17 Error-Correcting Codes

In this chapter we introduce the concept of an error-correcting code (ECC). We will spend time discussing what it means for a code to have low probability of error, and what is the optimum (ML or MAP) decoder. On the special case of coding for the binary symmetric channel (BSC) we showcase the evolution of our understanding of fundamental limits from pre-Shannon to modern finite blocklength. We also briefly review the history of ECCs. We conclude with a conceptually important proof of a weak converse (impossibility) bound for the performance of ECCs.

17.1 Codes and Probability of Error

We start with a simple definition of a code.

Definition 17.1. An M-code for $P_{Y|X}$ is an encoder–decoder pair (f, g) of (possibly, randomized) functions[1]

- encoder $f\colon [M] \to \mathcal{X}$,
- decoder $g\colon \mathcal{Y} \to [M] \cup \{\mathsf{e}\}$.

In most cases f and g are deterministic functions, in which case we think of them, equivalently, in terms of codewords, codebooks, and decoding regions (see Figure 17.1 for an illustration):

- $\forall i \in [M]$: $c_i \triangleq f(i)$ are *codewords*, the collection $\mathcal{C} = \{c_1, \dots, c_M\}$ is called a *codebook*.
- $\forall i \in [M], D_i \triangleq g^{-1}(\{i\})$ is the *decoding region* for i.

Given an M-code we can define a probability space, underlying all the subsequent developments in this part. For that we chain the three objects – message W, the encoder, and the decoder – together into the following Markov chain:

$$W \xrightarrow{f} X \xrightarrow{P_{Y|X}} Y \xrightarrow{g} \hat{W}, \tag{17.1}$$

[1] For randomized encoders/decoders, we identify f and g as probability transition kernels $P_{X|W}$ and $P_{\hat{W}|Y}$.

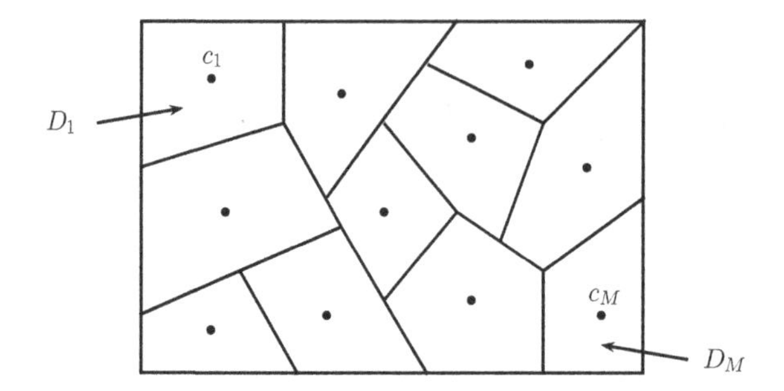

Figure 17.1 When $\mathcal{X} = \mathcal{Y}$, the decoding regions can be pictured as a partition of the space, each containing one codeword.

where we set $W \sim \text{Unif}([M])$. In the case of discrete spaces, we can explicitly write out the joint distribution of these variables as follows:

$$\text{(general)} \qquad P_{W,X,Y,\hat{W}}(m,a,b,\hat{m}) = \frac{1}{M} P_{X|W}(a|m) P_{Y|X}(b|a) P_{\hat{W}|Y}(\hat{m}|b),$$

$$\text{(deterministic } f,g) \qquad P_{W,X,Y,\hat{W}}(m,c_m,b,\hat{m}) = \frac{1}{M} P_{Y|X}(b|c_m) 1\{b \in D_{\hat{m}}\}.$$

Throughout these sections, these random variables will be referred to by their traditional names: W = original (true) message, X = (induced) channel input, Y = channel output, and $\hat{W}$ = decoded message.

Although any pair (f,g) is called an M-code, in reality we are only interested in those that satisfy certain "error-correcting" properties. To assess their quality we define the following *performance metrics*:

1 *Maximum error probability*: $P_{e,\max}(f,g) \triangleq \max_{m\in[M]} \mathbb{P}[\hat{W} \neq m | W = m]$.
2 *Average error probability*: $P_e(f,g) \triangleq \mathbb{P}[W \neq \hat{W}]$.

Note that, clearly, $P_e \le P_{e,\max}$. Therefore, the requirement of a small maximum error probability is a more stringent criterion, and offers uniform protection for all codewords. Some codes (such as linear codes, see Section 18.6) have the property of $P_e = P_{e,\max}$ by construction, but generally these two metrics could be very different.

Having defined the concept of an M-code and the performance metrics, we can finally define the *fundamental limits* for a given channel $P_{Y|X}$.

Definition 17.2. A code (f,g) is an (M,ϵ)-code for $P_{Y|X}$ if $P_e(f,g) \le \epsilon$. Similarly, an $(M,\epsilon)_{\max}$-code must satisfy $P_{e,\max} \le \epsilon$. The fundamental limits of channel coding are defined as:

$$\begin{aligned} M^*(\epsilon;P_{Y|X}) &= \max\{M\colon \exists (M,\epsilon)\text{-code}\}, \\ M^*_{\max}(\epsilon;P_{Y|X}) &= \max\{M\colon \exists (M,\epsilon)_{\max}\text{-code}\}, \\ \epsilon^*(M;P_{Y|X}) &= \inf\{\epsilon\colon \exists (M,\epsilon)\text{-code}\}, \\ \epsilon^*_{\max}(M;P_{Y|X}) &= \inf\{\epsilon\colon \exists (M,\epsilon)_{\max}\text{-code}\}. \end{aligned}$$

The argument $P_{Y|X}$ will be omitted when $P_{Y|X}$ is clear from the context.

In other words, the quantity $\log_2 M^*(\epsilon)$ gives the maximum number of bits that we can push through a noisy transformation $P_{Y|X}$, while still guaranteeing the error probability in the appropriate sense to be at most ϵ.

Example 17.1.

The channel $\mathsf{BSC}_\delta^{\otimes n}$ (recall from Example 3.6 that BSC stands for binary symmetric channel) acts between $\mathcal{X} = \{0,1\}^n$ and $\mathcal{Y} = \{0,1\}^n$, where the input X^n is contaminated by additive noise $Z^n \overset{\text{iid}}{\sim} \text{Ber}(\delta)$ independent of X^n, resulting in the channel output

$$Y^n = X^n \oplus Z^n.$$

In other words, the $\mathsf{BSC}_\delta^{\otimes n}$ channel takes a binary sequence length n and flips each bit independently with probability δ; pictorially, this is as shown below:

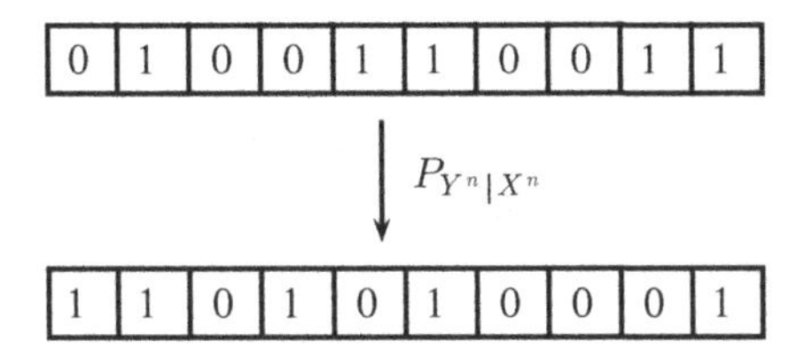

In the next section we discuss coding for the BSC channel in more detail.

17.2 Coding for Binary Symmetric Channels

To understand the problem of designing the encoders and decoders, let us consider the BSC transformation with $\delta = 0.11$ and $n = 1000$. The problem of studying $\log_2 M^*(\epsilon)$ attempts to answer what is the maximum number k of bits you can send with $P_e \leq \epsilon$? For concreteness, let us fix $\epsilon = 10^{-3}$ and discuss some of the possible ideas.

Perhaps our first attempt would be to try sending $k = 1000$ bits with one data bit mapped to one channel input position. However, a simple calculation shows that in this case we get $P_e = 1 - (1-\delta)^n \approx 1$. In other words, *uncoded transmission* does not meet our objective of small P_e and some form of coding is necessary. This incurs a fundamental tradeoff: reduce the number of bits to send (and use the freed channel inputs for sending redundant copies) in order to increase the probability of success.

So let us consider the next natural idea: *repetition coding*. We take each of the input data bits and repeat it ℓ times:

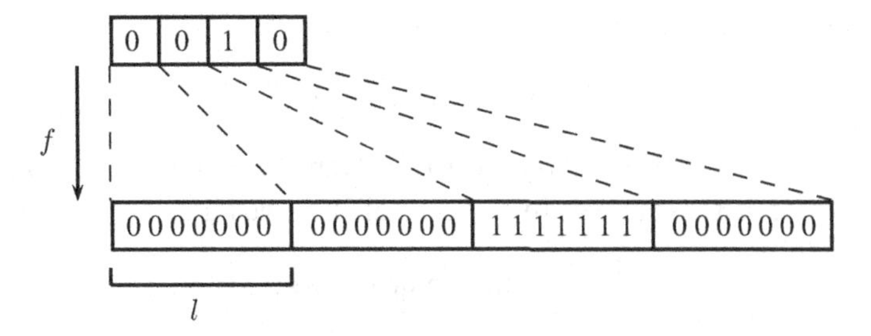

Decoding can be done by taking a majority vote inside each ℓ-block. Thus, each data bit is decoded with probability of bit error $P_b = \mathbb{P}[\text{Bin}(l, \delta) > l/2]$. However, the probability of block error of this scheme is $P_e \leq k\mathbb{P}[\text{Bin}(l, \delta) > l/2]$. (This bound is essentially tight in the current regime.) Consequently, to satisfy $P_e \leq 10^{-3}$ we must solve for k and ℓ satisfying $kl \leq n = 1000$ and also

$$k\mathbb{P}[\text{Bin}(l, \delta) > l/2] \leq 10^{-3}.$$

This gives $l = 21$, $k = 47$ bits. So we can see that using repetition coding we can send 47 data bits by using 1000 channel uses.

Repetition coding is a natural idea. It also has a very natural tradeoff: if you want better reliability, then the number ℓ needs to increase and hence the ratio $\frac{k}{n} = \frac{1}{\ell}$ should drop. Before Shannon's ground-breaking work, it was almost universally accepted that *this is fundamentally unavoidable: vanishing error probability should imply vanishing communication rate* $\frac{k}{n}$.

Before delving into optimal codes let us offer a glimpse of more sophisticated ways of injecting redundancy into the channel input n-sequence than simple repetition. For that, consider the so-called first-order Reed–Muller codes $(1, r)$. We interpret a sequence of r data bits $a_0, \ldots, a_{r-1} \in \mathbb{F}_2^r$ as a degree-one polynomial in $(r-1)$ variables:

$$a = (a_0, \ldots, a_{r-1}) \mapsto f_a(x) \triangleq \sum_{i=1}^{r-1} a_i x_i + a_0.$$

In order to transmit these r bits of data we simply evaluate $f_a(\cdot)$ at all possible values of the variables $x^{r-1} \in \mathbb{F}_2^{r-1}$. This code, which maps r bits to 2^{r-1} bits, has minimum distance $d_{\min} = 2^{r-2}$. That is, for two distinct $a \neq a'$ the number of positions in which f_a and $f_{a'}$ disagree is at least 2^{r-2}. In coding theory notation $[n, k, d_{\min}]$ we say that the first-order Reed–Muller code $(1, 7)$ is a [65, 7, 32] code. It can be shown that the optimal decoder for this code achieves over the $\text{BSC}_{0.11}^{\otimes 64}$ channel a probability of error at most 6×10^{-6}. Thus, we can use 16 such blocks (each carrying seven data bits and occupying 64 bits on the channel) over the $\text{BSC}_\delta^{\otimes 1024}$, and still have (by the union bound) overall probability of block error $P_e \lesssim 10^{-4} < 10^{-3}$. Thus, with the help of Reed–Muller codes we can send $7 \times 16 = 112$ bits in 1024 channel uses, more than doubling that of the repetition code.

Shannon's noisy channel coding theorem (Theorem 19.9) – a crown jewel of information theory – tells us that over a memoryless channel $P_{Y^n|X^n} = (P_{Y|X})^n$ of blocklength n the fundamental limit satisfies

$$\log M^*(\epsilon; P_{Y^n|X^n}) = nC + o(n) \tag{17.2}$$

as $n \to \infty$ and for arbitrary $\epsilon \in (0, 1)$. Here $C = \max_{P_{X_1}} I(X_1; Y_1)$ is the capacity of the single-letter channel. In our case of BSC we have

$$C = \log 2 - h(\delta) \approx \frac{1}{2} \text{ bit},$$

since the optimal input distribution is uniform (from symmetry) – see Section 19.3. Shannon's expansion (17.2) can be used to predict (not completely rigorously, of course, because of the $o(n)$ residual) that it should be possible to send around 500 bits reliably. As it turns out, for the blocklength $n = 1000$ this is not quite possible.

Note that computing M^* exactly requires iterating over all possible encoders and decoders – an impossible task even for small values of n. However, there exist rigorous and computationally tractable finite–blocklength bounds [335] that demonstrate for our choice of $n = 1000, \delta = 0.11$, and $\epsilon = 10^{-3}$:

$$414 \leq \log_2 M^*(\epsilon = 10^{-3}) \leq 416 \quad \text{bits.} \tag{17.3}$$

Thus we can see that Shannon's prediction is about 20% too optimistic. We will see below some such finite-length bounds. Notice, however, that while the bounds guarantee existence of an encoder–decoder pair achieving a prescribed performance, building an actual f and g implementable with modern software/hardware is a different story.

It took about 60 years after Shannon's discovery of (17.2) to construct practically implementable codes achieving that performance. The first codes that approach the bounds on $\log M^*$ are called *turbo codes* [47] (after the turbocharger engine, where the exhaust is fed back in to power the engine). This class of codes is known as *sparse-graph codes*, of which the low-density parity-check (LDPC) codes invented by Gallager are particularly well studied [361]. As a rule of thumb, these codes typically approach 80–90% of $\log M^*$ when $n \approx 10^3 - 10^4$. For shorter blocklengths in the range of $n = 100 - 1000$ there is an exciting alternative to LDPC codes: the polar codes of Arıkan [23], which are most typically used together with the list-decoding idea of Tal and Vardy [410]. And of course, the story is still evolving today as new channel models become relevant and new hardware possibilities open up.

We want to point out a subtle but very important conceptual paradigm shift introduced by Shannon's insistence on coding over many (information) bits together. Indeed, consider the situation discussed above, where we construct a powerful code with $M \approx 2^{400}$ codewords and $n = 1000$. Now, one might imagine this code as a constellation of 2^{400} points carefully arranged inside a hypercube $\{0, 1\}^{1000}$ to guarantee some degree of separation between them, see (17.6). Next, suppose one was using this code every second for the lifetime of the universe ($\approx 10^{18}$ s). Even after this laborious process one will have explored at most 2^{60} different codewords from among an overwhelmingly large codebook 2^{400}. So a natural question arises: Why did we need to carefully place all these many codewords if the majority of them will never be used by anyone? The answer is at the heart of the concept of information: to transmit information is to convey a selection of one element (W) from a collection of possibilities ($[M]$). The fact that we do not know which W will be selected forces us to a priori prepare for every one of the possibilities. This simple idea, proposed in the first paragraph of [379], is now tacitly assumed by everyone, but was one of the subtle ways in which Shannon revolutionized the scientific approach to the study of information exchange.

17.3 Optimal Decoder

Given any encoder $f\colon [M] \to \mathcal{X}$, the decoder that minimizes P_e is the *maximum a posteriori* (MAP) decoder, or equivalently, the *maximum likelihood* (ML) decoder, since the codewords are equiprobable (W is uniform):

$$g^*(y) = \operatorname*{argmax}_{m\in[M]} \mathbb{P}[W = m|Y = y]$$
$$= \operatorname*{argmax}_{m\in[M]} \mathbb{P}[Y = y|W = m]. \tag{17.4}$$

Notice that the optimal decoder is deterministic. For the special case of a deterministic encoder, where we can identify the encoder with its image $\mathcal{C}$, the minimal (MAP) probability of error for the codebook $\mathcal{C}$ can be written as

$$P_{\mathrm{e,MAP}}(\mathcal{C}) = 1 - \frac{1}{M}\sum_{y\in\mathcal{Y}} \max_{x\in\mathcal{C}} P_{Y|X}(y|x), \tag{17.5}$$

with a similar extension to non-discrete $\mathcal{Y}$.

Remark 17.1. For the case of $\mathsf{BSC}_\delta^{\otimes n}$ the MAP decoder has a nice geometric interpretation. Indeed, if $d_{\mathrm{H}}(x^n, y^n) = |\{i\colon x_i \neq y_i\}|$ denotes the Hamming distance and if f (the encoder) is deterministic with codewords $\mathcal{C} = \{c_i, i \in [M]\}$ then

$$g^*(y^n) = \operatorname*{argmin}_{m\in[M]} d_{\mathrm{H}}(c_m, y^n). \tag{17.6}$$

Consequently, the optimal decoding regions – see Figure 17.1 – become the *Voronoi cells* tessellating the Hamming space $\{0, 1\}^n$. Similarly, the MAP decoder for the AWGN channel induces a Voronoi tessellation of $\mathbb{R}^n$ – see Section 20.3.

So we have seen that the optimal decoder without loss of generality can be assumed to be deterministic. Similarly, we can represent any randomized encoder f as a function of two arguments: the true message W and an external randomness $U \perp\!\!\!\perp W$, so that $X = f(W, U)$ where this time f is a deterministic function. Then we have

$$\mathbb{P}[W \neq \hat{W}] = \mathbb{E}[\mathbb{P}[W \neq \hat{W}|U]],$$

which implies that if $P[W \neq \hat{W}] \leq \epsilon$ then there must exist some choice u_0 such that $\mathbb{P}[W \neq \hat{W}|U = u_0] \leq \epsilon$. In other words, the fundamental limit $M^*(\epsilon)$ is unchanged if we restrict our attention to deterministic encoders and decoders only.

Note, however, that neither of the above considerations apply to the maximal probability of error $P_{e,\max}$. Indeed, the fundamental limit $M^*_{\max}(\epsilon)$ does indeed require considering randomized encoders and decoders. For example, when $M = 2$ from the decoding point of view we are back to the setting of binary hypothesis testing in Part III. The optimal decoder (test) that minimizes the maximal type-I and type-II error probability, that is, $\max\{1 - \alpha, \beta\}$, will not be deterministic if $\max\{1 - \alpha, \beta\}$ is not achieved at a vertex of the Neyman–Pearson region $\mathcal{R}(P_{Y|W=1}, P_{Y|W=2})$.

17.4 Weak Converse Bound

The main focus of both the theory and the practice of channel coding lies in showing the existence of (or constructing explicit) (M, ϵ)-codes with large M and small ϵ. To understand how close the constructed code is to the fundamental limit, one needs to prove an "impossibility result" bounding M from the above or ϵ from below. Such negative results are known as "converse bounds," with the name coming from the fact that classically these bounds followed right after the positive (existential) results and were preceded with the words "Conversely, ..." The next result shows that M can never (multiplicatively) exceed capacity $\sup_{P_X} I(X;Y)$ by much.

Theorem 17.3. (Weak converse) *Any (M, ϵ)-code for $P_{Y|X}$ satisfies*

$$\log M \le \frac{\sup_{P_X} I(X;Y) + h(\epsilon)}{1 - \epsilon},$$

where $h(x) = H(\mathrm{Ber}(x))$ is the binary entropy function.

Proof. This can be derived as a one-line application of Fano's inequality (Theorem 3.12), but we proceed slightly differently with an eye toward future extensions in *meta-converse* (Section 22.3).

Consider an M-code with probability of error P_e and its corresponding probability space: $W \to X \to Y \to \hat{W}$. We want to show that this code can be used as a hypothesis test between distributions $P_{X,Y}$ and $P_X P_Y$. Indeed, given a pair (X, Y) we can sample $(W, \hat{W})$ from $P_{W,\hat{W}|X,Y} = P_{W|X} P_{\hat{W}|Y}$ and compute the binary value $Z = 1\{W \neq \hat{W}\}$. (Note that in the most interesting cases when encoder and decoder are deterministic and the encoder is injective, the value Z is a deterministic function of (X, Y).) Let us compute the performance of this binary hypothesis test under two hypotheses. First, when $(X, Y) \sim P_X P_Y$ we have that $\hat{W} \perp\!\!\!\perp W \sim \mathrm{Unif}([M])$ and therefore

$$P_X P_Y[Z = 1] = \frac{1}{M}.$$

Second, when $(X, Y) \sim P_{X,Y}$ then by definition we have

$$P_{X,Y}[Z = 1] = 1 - P_e.$$

Thus, we can now apply the data-processing inequality for divergence to conclude: Since $W \to X \to Y \to \hat{W}$, we have the following chain of inequalities (see Fano's inequality, Theorem 3.12):

$$\begin{aligned} D(P_{X,Y} \| P_X P_Y) &\overset{\text{DPI}}{\ge} d\left(1 - P_e \middle\| \frac{1}{M}\right) \\ &\ge -h(\mathbb{P}[W \neq \hat{W}]) + (1 - P_e) \log M. \end{aligned}$$

By noticing that the left-hand side is $I(X;Y) \le \sup_{P_X} I(X;Y)$ we obtain

$$\log M \le \frac{\sup_{P_X} I(X;Y) + h(P_e)}{1 - P_e},$$

and the proof is completed by checking that $p \mapsto \frac{h(p)}{1-p}$ is monotonically increasing. □

Remark 17.2. The bound can be significantly improved by considering other divergence measures in the data-processing step. In particular, we will see below how one can get a "strong" converse (explaining the term "weak" converse here as well) in Section 22.1. The proof technique is known as meta-converse; see Section 22.3.

18 Random and Maximal Coding

So far our discussion of channel coding has mostly been following the same lines as the M-ary hypothesis testing (HT) in statistics. In this chapter we introduce the key departure: the principal and most interesting goal in information theory is the design of the encoder $f\colon [M] \to \mathcal{X}$ or the codebook $\{c_i \triangleq f(i), i \in [M]\}$. Once the codebook is chosen, the problem indeed becomes that of M-ary HT and can be tackled by standard statistical tools. However, the task of choosing the encoder f has no exact analogs in statistical theory (the closest being the design of experiments). Each f gives rise to a different HT problem and the goal is to choose these M hypotheses $P_{Y|X=c_1}, \ldots, P_{Y|X=c_M}$ to ensure maximal testability. It turns out that the problem of choosing a good f will be much simplified if we adopt a suboptimal way of testing M-ary HT. Roughly speaking we will run M binary HTs testing $P_{Y|X=c_m}$ against P_Y, which tries to distinguish the channel output induced by the message m from an "average background noise" P_Y. An optimal such test, as we know from Neyman–Pearson (Theorem 14.11), thresholds the following quantity:

$$\log \frac{P_{Y|X=x}}{P_Y}.$$

This explains the central role played by information density (defined next) in this chapter. After introducing the latter we will present several results demonstrating the existence of good codes. We start with the original bound of Shannon (expressed in modern language), followed by its tightenings (the dependence texting, random coding union, and Gallager's bounds). These belong to the class of *random coding* bounds. An entirely different approach was developed by Feinstein and is called *maximal coding*. We will see that the two result in eerily similar results. Why do these two rather different methods yield similar results, which are also quite close to the best possible (i.e. "achieve capacity and dispersion")? It turns out that the answer lies in a certain submodularity property of the channel coding task. Finally, we will also discuss a more structured class of codes based on linear algebraic constructions. Similar to the case of compression it will be shown that linear codes are no worse than general codes, explaining why virtually all practical codes are linear.

While reading this chapter, we recommend also consulting Figure 22.2, in which various achievability bounds are compared for the BSC.

In this chapter it will be convenient to introduce the following *independent pairs* $(X,Y) \perp\!\!\!\perp (\overline{X},\overline{Y})$ with their joint distribution given by:

$$P_{X,Y,\overline{X},\overline{Y}}(a,b,\overline{a},\overline{b}) = P_X(a)P_{Y|X}(b|a)P_X(\overline{a})P_{Y|X}(\overline{b}|\overline{a}). \tag{18.1}$$

We will often call X the sent codeword and $\bar{X}$ the unsent codeword.

18.1 Information Density

A crucial object for the subsequent development is information density. Historically, the concept seems to originate in early works of Soviet information theorists. In a nutshell, we want to set $i(x;y) = \log \frac{P_{X,Y}(x,y)}{P_X(x)P_Y(y)}$. However, we want to make this definition sufficiently general so as to take into account continuous distributions, the possibility of $P_{X,Y} \not\ll P_X P_Y$ (in which case the value under the log can equal $+\infty$), and the possibility of the argument of the log being equal to 0. The definition below is similar to what we did in Definition 14.4 and (2.10) using the Log function, but we repeat it below for convenience.

Definition 18.1. (Information density[1]) Let $P_{X,Y} \ll \mu$ and $P_X P_Y \ll \mu$ for some dominating measure μ, and denote by $f(x,y) = \frac{dP_{X,Y}}{d\mu}$ and $\bar{f}(x,y) = \frac{dP_X P_Y}{d\mu}$ the Radon–Nikodym derivatives of $P_{X,Y}$ and $P_X P_Y$ with respect to μ, respectively. Then recalling the Log definition (2.10) we set

$$i_{P_{X,Y}}(x;y) \triangleq \mathrm{Log}\,\frac{f(x,y)}{\bar{f}(x,y)} = \begin{cases} \log \frac{f(x,y)}{\bar{f}(x,y)}, & f(x,y) > 0, \bar{f}(x,y) > 0, \\ +\infty, & f(x,y) > 0, \bar{f}(x,y) = 0, \\ -\infty, & f(x,y) = 0, \bar{f}(x,y) > 0, \\ 0, & f(x,y) = \bar{f}(x,y) = 0. \end{cases} \tag{18.2}$$

In the most common special case of $P_{X,Y} \ll P_X P_Y$ we have simply

$$i_{P_{X,Y}}(x;y) = \log \frac{dP_{X,Y}}{dP_X P_Y}(x,y),$$

with $\log 0 = -\infty$.

Notice that information density as a function depends on the underlying $P_{X,Y}$. Throughout this part, however, $P_{Y|X}$ is going to be a fixed channel (fixed by the problem at hand), and thus information density only depends on the choice of P_X. Most of the time P_X (and, correspondingly, $P_{X,Y}$) used to define information density will be apparent from the context. Thus for the benefit of the reader as well as our own, we will write $i(x;y)$ dropping the subscript $P_{X,Y}$.

We proceed to show some elementary properties of information density. The next result explains the name "information density."

[1] We remark that in machine learning (especially natural language processing) information density is also called *pointwise mutual information* (PMI) [293].

Proposition 18.2. *The expectation* $\mathbb{E}[i(X;Y)]$ *is well defined and non-negative (but possibly infinite). In any case, we have* $I(X;Y) = \mathbb{E}[i(X;Y)]$.

Proof. This follows from (2.12) and the definition of $i(x;y)$ as log-ratio. □

Being defined as log-likelihood, information density possesses the standard properties of the latter, see Theorem 14.6. However, because it is defined in terms of two variables (X, Y), there are also very useful conditional expectation versions. To illustrate the meaning of the next proposition, let us consider the case of discrete X, Y, and $P_{X,Y} \ll P_X P_Y$. Then we have *for every* x:

$$\sum_y f(x,y) P_X(x) P_Y(y) = \sum_y f(x,y) \exp\{-i(x;y)\} P_{X,Y}(x,y).$$

The general case requires a little more finesse.

Proposition 18.3. (Conditioning–unconditioning trick) *Let* $\bar{X} \perp\!\!\!\perp (X, Y)$ *be a copy of* X. *We have the following:*

1 *For any function* $f\colon \mathcal{X} \times \mathcal{Y} \to \mathbb{R}$,

$$\mathbb{E}[f(\bar{X}, Y) 1\{i(\bar{X};Y) > -\infty\}] = \mathbb{E}[f(X, Y) \exp\{-i(X;Y)\}]. \tag{18.3}$$

2 *Let* f_+ *be a non-negative function. Then for* P_X*-almost every* x *we have*

$$\mathbb{E}[f_+(\bar{X}, Y) 1\{i(\bar{X};Y) > -\infty\} | \bar{X} = x] = \mathbb{E}[f_+(X, Y) \exp\{-i(X;Y)\} | X = x]. \tag{18.4}$$

Proof. The first part (18.3) is simply a restatement of (14.5). For the second part, let us define

$$a(x) \triangleq \mathbb{E}[f_+(\bar{X}, Y) 1\{i(\bar{X};Y) > -\infty\} | \bar{X} = x],$$
$$b(x) \triangleq \mathbb{E}[f_+(X, Y) \exp\{-i(X;Y)\} | X = x].$$

We first additionally assume that f is bounded. Fix $\epsilon > 0$ and denote $S_\epsilon = \{x\colon a(x) \geq b(x) + \epsilon\}$. As $\epsilon \to 0$ we have $S_\epsilon \nearrow \{x\colon a(x) > b(x)\}$ and thus if we show $P_X[S_\epsilon] = 0$ this will imply that $a(x) \leq b(x)$ for P_X-a.e. x. The symmetric argument shows $b(x) \leq a(x)$ and completes the proof of the equality.

To show $P_X[S_\epsilon] = 0$ let us apply (18.3) to the function $f(x, y) = f_+(x, y) 1\{x \in S_\epsilon\}$. Then we get

$$\mathbb{E}[f_+(X, Y) 1\{X \in S_\epsilon\} \exp\{-i(X;Y)\}] = \mathbb{E}[f_+(\bar{X}, Y) 1\{i(\bar{X};Y) > -\infty\} 1\{X \in S_\epsilon\}].$$

Let us re-express both sides of this equality by taking the conditional expectations over Y to get:

$$\mathbb{E}[b(X) 1\{X \in S_\epsilon\}] = \mathbb{E}[a(\bar{X}) 1\{\bar{X} \in S_\epsilon\}].$$

But from the definition of S_ϵ we have

$$\mathbb{E}[b(X) 1\{X \in S_\epsilon\}] \geq \mathbb{E}[(b(\bar{X}) + \epsilon) 1\{\bar{X} \in S_\epsilon\}].$$

Recall that $X \stackrel{d}{=} \bar{X}$ and hence

$$\mathbb{E}[b(X)1\{X \in S_\epsilon\}] \geq \mathbb{E}[b(X)1\{X \in S_\epsilon\}] + \epsilon P_X[S_\epsilon].$$

Since f_+ (and therefore b) was assumed to be bounded we can cancel the common term from both sides and conclude $P_X[S_\epsilon] = 0$ as required.

Finally, to show (18.4) in full generality, given an unbounded f_+ we define $f_n(x, y) = \min(f_+(x, y), n)$. Since (18.4) holds for f_n we can take the limit as $n \to \infty$ on both sides of it:

$$\lim_{n\to\infty} \mathbb{E}[f_n(\bar{X}, Y)1\{i(\bar{X};Y) > -\infty\}|\bar{X} = x] = \lim_{n\to\infty} \mathbb{E}[f_n(X, Y)\exp\{-i(X;Y)\}|X = x].$$

By the monotone convergence theorem (for conditional expectations!) we can take the limits inside the expectations to conclude the proof. □

Corollary 18.4. (Information tails) *For P_X-almost every x we have*

$$\mathbb{P}[i(x;Y) > t] \leq \exp(-t), \tag{18.5}$$

$$\mathbb{P}[i(\overline{X};Y) > t] \leq \exp(-t). \tag{18.6}$$

Proof. Pick $f_+(x, y) = 1\{i(x;y) > t\}$ in (18.4). □

Remark 18.1. This estimate has been used by us several times before. In the hypothesis testing part we used (Corollary 14.12):

$$Q\left[\log \frac{dP}{dQ} \geq t\right] \leq \exp(-t). \tag{18.7}$$

In data compression, we used the fact that $|\{x\colon \log P_X(x) \geq t\}| \leq \exp(-t)$, which is also of the form (18.7) with Q being the counting measure.

18.2 Shannon's Random Coding Bound

In this section we present perhaps the most far-reaching technical result of Shannon. As we discussed before, a good error-correcting code is supposed to be a geometrically elegant constellation in a high-dimensional space. Its chief goal is to push different codewords as far apart as possible, so as to reduce the deleterious effects of channel noise. However, in the early 1940s there were no codes and no tools for constructing them available to Shannon. So facing the problem of understanding if error correction is even possible, Shannon decided to check if placing codewords randomly in space will somehow result in favorable geometric arrangement. To everyone's astonishment, which is still producing aftershocks today, this method not only produced reasonable codes, but in fact turned out to be optimal asymptotically (and almost optimal non-asymptotically [335]). As we mentioned earlier, the method of proving the existence of certain combinatorial objects by random selection is known as Erdös's *probabilistic method* [15], which Shannon apparently discovered independently and, perhaps, earlier.

Before going to the proof, we need to explain why we use a particular suboptimal decoder and also how information density arises in this context. First, consider, for simplicity, the case of discrete alphabets and $P_{X,Y} \ll P_X P_Y$. Then we have an equivalent expression

$$i(x;y) = \log \frac{P_{Y|X}(y|x)}{P_Y(y)}.$$

Therefore, the optimal (maximum likelihood) decoder can be written in terms of the information density:

$$g^*(y) = \operatorname*{argmax}_{m \in [M]} P_{X|Y}(c_m|y) = \operatorname*{argmax}_{m \in [M]} P_{Y|X}(y|c_m) = \operatorname*{argmax}_{m \in [M]} i(c_m;y). \tag{18.8}$$

Note that (18.8) holds regardless of the input distribution P_X used for the definition of $i(x;y)$; in particular we do not have to use the code-induced distribution $P_X = \frac{1}{M}\sum_{i=1}^{M} \delta_{c_i}$. However, if we are to threshold information density, different choices of P_X will result in different decoders, so we need to justify the choice of P_X.

To that end, recall that to distinguish between two codewords c_i and c_j, one can apply (as we learned in Part III for binary HT) the likelihood-ratio test, namely thresholding the LLR $\log \frac{P_{Y|X=c_i}}{P_{Y|X=c_j}}$. As we explained at the beginning of this chapter, a (possibly suboptimal) approach in M-ary HT is to run binary tests by thresholding each information density $i(c_i;y)$. This, loosely speaking, evaluates the likelihood of c_i against the average distribution of the other $M-1$ codewords, which we approximate by P_Y (as opposed to the more precise form $\frac{1}{M-1}\sum_{j \neq i} P_{Y|X=c_j}$). Putting these ideas together we can propose the decoder as

$$g(y) = \text{any } m \text{ s.t. } i(c_m;y) > \gamma,$$

where γ is a threshold and P_X is judiciously chosen (to maximize $I(X;Y)$ as we will see soon).

Yet another way to see why a thresholding decoder (as opposed to an ML one) is a natural idea is to simply believe the fact that for good error-correcting codes the most likely (ML) codeword has likelihood (information density) so much higher than the rest of the candidates that instead of looking for the maximum we simply can select the one (and only) codeword that exceeds a pre-specified threshold.

With these initial justifications we proceed to the main result of this section.

Theorem 18.5. (Shannon's achievability bound) *Fix a channel $P_{Y|X}$ and an arbitrary input distribution P_X. Then for every $\tau > 0$ there exists an (M,ϵ)-code with*

$$\epsilon \leq \mathbb{P}[i(X;Y) \leq \log M + \tau] + \exp(-\tau). \tag{18.9}$$

Proof. Recall that for a given codebook $\{c_1, \ldots, c_M\}$, the optimal decoder is MAP and is equivalent to maximizing information density, see (18.8). The step of maximizing $i(c_m;Y)$ makes analyzing the error probability difficult. Similar to what we did in almost lossless compression, see Theorem 11.5, the first important step for showing the achievability bound is to consider a suboptimal decoder. In Shannon's bound, we consider a threshold-based suboptimal decoder $g(y)$ as follows:

$$g(y) = \begin{cases} m, & \exists!\, c_m \text{ s.t. } i(c_m;y) \geq \log M + \tau, \\ \mathsf{e}, & \text{otherwise.} \end{cases} \tag{18.10}$$

In words, the decoder g reports m as the decoded message if and only if codeword c_m is a unique one with information density exceeding the threshold $\log M + \tau$. If there are multiple or no such codewords, then the decoder outputs a special value of e, which always results in error since $W \neq \mathsf{e}$. (We could have decreased the probability of error slightly by allowing the decoder to instead output a random message, or to choose any one of the messages exceeding the threshold, or any other clever ideas. The point, however, is that even the simplistic resolution of outputting e already achieves all qualitative goals, while simplifying the analysis considerably.)

For a given codebook $(c_1, \ldots, c_M)$, the error probability is

$$P_e(c_1, \ldots, c_M) = \mathbb{P}[\{i(c_W;Y) \leq \log M + \tau\} \cup \{\exists \bar{m} \neq W, i(c_{\bar{m}};Y) > \log M + \tau\}],$$

where W is uniform on $[M]$ and the probability space is as in (17.1).

The second (and most ingenious) step proposed by Shannon was to forgo the complicated discrete optimization of the codebook. His proposal is to generate the codebook $(c_1, \ldots, c_M)$ randomly with $c_m \sim P_X$ iid for $m \in [M]$ and then try to reason about the average $\mathbb{E}[P_e(c_1, \ldots, c_M)]$. By symmetry, this averaged error probability over all possible codebooks is unchanged if we condition on $W = 1$. Considering also the random variables $(X, Y, \bar{X})$ as in (18.1), we get the following chain:

$$\begin{aligned}
&\mathbb{E}[P_e(c_1, \ldots, c_M)] \\
&= \mathbb{E}[P_e(c_1, \ldots, c_M) | W = 1] \\
&= \mathbb{P}[\{i(c_1;Y) \leq \log M + \tau\} \cup \{\exists \bar{m} \neq 1, i(c_{\bar{m}}, Y) > \log M + \tau\} | W = 1] \\
&\leq \mathbb{P}[i(c_1;Y) \leq \log M + \tau | W = 1] + \sum_{\bar{m}=2}^{M} \mathbb{P}[i(c_{\bar{m}};Y) > \log M + \tau | W = 1] \\
&\qquad\qquad\qquad\qquad\qquad\qquad \text{(union bound)} \\
&\stackrel{(a)}{=} \mathbb{P}\left[i(X;Y) \leq \log M + \tau\right] + (M - 1)\mathbb{P}\left[i(\bar{X};Y) > \log M + \tau\right] \\
&\leq \mathbb{P}\left[i(X;Y) \leq \log M + \tau\right] + (M - 1)\exp(-(\log M + \tau)) \quad \text{(by Corollary 18.4)} \\
&\leq \mathbb{P}\left[i(X;Y) \leq \log M + \tau\right] + \exp(-\tau),
\end{aligned}$$

where the crucial step (a) follows from the fact that given $W = 1$ and $\bar{m} \neq 1$ we have

$$(c_1, c_{\bar{m}}, Y) \stackrel{d}{=} (X, \bar{X}, Y),$$

with the latter triple defined in (18.1).

The last expression does indeed conclude the proof of the existence of the (M, ϵ)-code: it shows that the average of $P_e(c_1, \ldots, c_M)$ satisfies the required bound on probability of error, and thus there must exist at least *one* choice of $c_1, \ldots, c_M$ satisfying the same bound. □

Remark 18.2. (Joint typicality) Shortly in Chapter 19, we will apply this theorem for the case of $P_X = P_{X_1}^{\otimes n}$ (the iid input) and $P_{Y|X} = P_{Y_1|X_1}^{\otimes n}$ (the memoryless channel). Traditionally, see [111], decoders in such settings were defined with the help of so-called "joint typicality." Those decoders given $y = y^n$ search for the codeword x^n (both of which are n-letter vectors) such that the empirical joint distribution is close to the true joint distribution, that is, $\hat{P}_{x^n,y^n} \approx P_{X_1,Y_1}$, where

$$\hat{P}_{x^n,y^n}(a,b) = \frac{1}{n} \cdot |\{j \in [n] \colon x_j = a, y_j = b\}|$$

is the joint empirical distribution of (x^n, y^n). This definition is used for the case when random coding is done with $c_j \sim$ uniform on the type class $\{x^n \colon \hat{P}_{x^n} \approx P_X\}$. Another alternative, "entropic typicality," see [106], is to search for a codeword with $\sum_{j=1}^n \log \frac{1}{P_{X_1,Y_1}(x_j,y_j)} \approx H(X,Y)$. We think of requirement $\{i(x^n;y^n) \geq \log M + \tau\}$ as a version of "joint typicality" that is applicable to a broader range of channels, including those that are not over product alphabets or memoryless.

18.3 Dependence Testing (DT) Bound

The following result is a slight refinement of Theorem 18.5, which results in a bound that is free from the auxiliary parameters and is provably stronger.

Theorem 18.6. (DT bound) *Fix a channel $P_{Y|X}$ and an arbitrary input distribution P_X. Then for every $M > 1$ there exists an (M, ϵ)-code with*

$$\epsilon \leq \mathbb{E}\left[\exp\left\{-\left(i(X;Y) - \log\frac{M-1}{2}\right)^+\right\}\right], \tag{18.11}$$

where $x^+ \triangleq \max(x, 0)$.

Proof. For a fixed γ, consider the following suboptimal decoder:

$$g(y) = \begin{cases} m & \text{for the smallest } m \text{ s.t. } i(c_m;y) \geq \gamma, \\ \mathsf{e} & \text{otherwise.} \end{cases}$$

Setting $\hat{W} = g(Y)$ we note that given a codebook $\{c_1, \ldots, c_M\}$, we have by union bound

$$\begin{aligned} \mathbb{P}[\hat{W} \neq j | W = j] &= \mathbb{P}[i(c_j;Y) \leq \gamma | W = j] + \mathbb{P}[i(c_j;Y) > \gamma, \exists k \in [j-1], \text{ s.t. } i(c_k;Y) > \gamma] \\ &\leq \mathbb{P}[i(c_j;Y) \leq \gamma | W = j] + \sum_{k=1}^{j-1} \mathbb{P}[i(c_k;Y) > \gamma | W = j]. \end{aligned}$$

Averaging over the randomly generated codebook, the expected error probability is upper-bounded by

$$\begin{aligned}
\mathbb{E}[P_e(c_1,\ldots,c_M)] &= \frac{1}{M}\sum_{j=1}^{M}\mathbb{P}[\hat{W}\neq j|W=j] \\
&\leq \frac{1}{M}\sum_{j=1}^{M}\left(\mathbb{P}[i(X;Y)\leq\gamma]+\sum_{k=1}^{j-1}\mathbb{P}[i(\overline{X};Y)>\gamma]\right) \\
&= \mathbb{P}[i(X;Y)\leq\gamma]+\frac{M-1}{2}\mathbb{P}[i(\overline{X};Y)>\gamma] \\
&= \mathbb{P}[i(X;Y)\leq\gamma]+\frac{M-1}{2}\mathbb{E}[\exp(-i(X;Y))1\{i(X;Y)>\gamma\}] \\
&\qquad\qquad\qquad\qquad\qquad\qquad\qquad\qquad\qquad\qquad\text{(by (18.3))} \\
&= \mathbb{E}\left[1\{i(X;Y)\leq\gamma\}+\frac{M-1}{2}\exp(-i(X;Y))1\{i(X;Y)>\gamma\}\right].
\end{aligned}$$

To optimize over γ, note the simple observation that $U1_E + V1_{E^c} \geq \min\{U,V\}$. Therefore for any x,y, $1\{i(x;y) \leq \gamma\} + \frac{M-1}{2}\exp\{-i(x;y)\}1\{i(x;y) > \gamma\} > \min(1, \frac{M-1}{2}\exp\{-i(x;y)\})$ achieved by $\gamma = \log\frac{M-1}{2}$ regardless of x,y. Thus, we continue the bounding as follows:

$$\begin{aligned}
\inf_{\gamma}\mathbb{E}[P_e(c_1,\ldots,c_M)] &\leq \inf_{\gamma}\mathbb{E}\left[1\{i(X;Y)\leq\gamma\}+\frac{M-1}{2}\exp(-i(X;Y))1\{i(X;Y)>\gamma\}\right] \\
&= \mathbb{E}\left[\min\left(1,\frac{M-1}{2}\exp(-i(X;Y))\right)\right] \\
&= \mathbb{E}\left[\exp\left\{-\left(i(X;Y)-\log\frac{M-1}{2}\right)^{+}\right\}\right]. \qquad\square
\end{aligned}$$

Remark 18.3. (Dependence testing interpretation) The RHS of (18.11) equals to $\frac{M+1}{2}$ times the minimum error probability of the following Bayesian hypothesis testing problem:

$$\begin{aligned}
&H_0\colon X,Y\sim P_{X,Y} \text{ versus } \quad H_1\colon X,Y\sim P_XP_Y \\
&\text{prior prob.: } \pi_0=\frac{2}{M+1}, \pi_1=\frac{M-1}{M+1}.
\end{aligned}$$

Note that $X,Y\sim P_{X,Y}$ and $\overline{X},Y\sim P_XP_Y$, where X is the sent codeword and $\overline{X}$ is the unsent codeword. As we know from binary hypothesis testing, the best threshold for the likelihood–ratio test (minimizing the weighted probability of error) is $\log\frac{\pi_1}{\pi_0}$, as we indeed found out.

One of the immediate benefits of Theorem 18.6 compared to Theorem 18.5 is precisely the fact that we do not need to perform a cumbersome minimization over τ in (18.9) to get the minimum upper bound in Theorem 18.5. Nevertheless, it can be shown that the DT bound is stronger than Shannon's bound with optimized τ. See also Exercise **IV.5**.

Finally, we remark (and will develop this below in our treatment of linear codes) that the DT bound and Shannon's bound both hold without change if we generate $\{c_i\}$ by any other (non-iid) procedure with a prescribed marginal *and pairwise–independent codewords* – see Theorem 18.13 below.

18.4 Feinstein's Maximal Coding Bound

The previous achievability results are obtained using *probabilistic* methods (random coding). In contrast, the following achievability bound due to Feinstein uses a *greedy* construction. One immediate advantage of Feinstein's method is that it shows the existence of codes satisfying a *maximal* probability of error criterion.[2]

Theorem 18.7. (Feinstein's lemma) *Fix a channel $P_{Y|X}$ and an arbitrary input distribution P_X. Then for every $\gamma > 0$ and for every $\epsilon \in (0,1)$ there exists an $(M,\epsilon)_{\max}$-code with*

$$M \geq \gamma(\epsilon - \mathbb{P}[i(X;Y) < \log \gamma]). \tag{18.12}$$

Remark 18.4. (Comparison with Shannon's bound) We can also interpret (18.12) differently: for any fixed M, there exists an $(M,\epsilon)_{\max}$-code that achieves the maximal error probability bounded as follows:

$$\epsilon \leq \mathbb{P}[i(X;Y) < \log \gamma] + \frac{M}{\gamma}.$$

If we take $\log \gamma = \log M + \tau$, this gives a bound of exactly the same form as Shannon's (18.9). It is rather surprising that two such different methods of proof produced essentially the same bound (modulo the difference between maximal and average probability of error). We will discuss the reason for this phenomenon in Section 18.7.

Proof. From the definition of $(M,\epsilon)_{\max}$-code, we recall that our goal is to find codewords $c_1, \ldots, c_M \in \mathcal{X}$ and disjoint subsets (decoding regions) $D_1, \ldots, D_M \subset \mathcal{Y}$, s.t.

$$P_{Y|X}(D_i|c_i) \geq 1 - \epsilon, \quad \forall i \in [M].$$

Feinstein's idea is to construct a codebook of size M in a sequential greedy manner.

For every $x \in \mathcal{X}$, associate it with a preliminary decoding region E_x defined as follows:

$$E_x \triangleq \{y \in \mathcal{Y} \colon i(x;y) \geq \log \gamma\}.$$

Notice that the preliminary decoding regions $\{E_x\}$ may be overlapping, and we will trim them into final decoding regions $\{D_x\}$, which will be disjoint. Next, we apply Corollary 18.4 and find out that there is a set $F \subset \mathcal{X}$ with two properties: (a) $P_X[F] = 1$ and (b) for every $x \in F$ we have

[2] Nevertheless, we should point out that this is not a serious advantage: from any (M,ϵ)-code we can extract an $(M',\epsilon')_{\max}$-subcode with a slightly smaller M' and larger ϵ' – see Theorem 19.4.

$$P_Y(E_x) \le \frac{1}{\gamma}. \tag{18.13}$$

We can assume that $\mathbb{P}[i(X;Y) < \log \gamma] \le \epsilon$, for otherwise the RHS of (18.12) is negative and there is nothing to prove. We first claim that there exists some $c \in F$ such that $\mathbb{P}[Y \in E_c | X = c] = P_{Y|X}(E_c|c) \ge 1 - \epsilon$. Indeed, assume (for the sake of contradiction) that $\forall c \in F$, $\mathbb{P}[i(c;Y) \ge \log \gamma | X = c] < 1 - \epsilon$. Note that since $P_X(F) = 1$ we can average this inequality over $c \sim P_X$. Then we arrive at $\mathbb{P}[i(X;Y) \ge \log \gamma] < 1 - \epsilon$, which is a contradiction.

With these preparations we construct the codebook in the following way:

1. Pick c_1 to be any codeword in F such that $P_{Y|X}(E_{c_1}|c_1) \ge 1 - \epsilon$, and set $D_1 = E_{c_1}$.
2. Pick c_2 to be any codeword in F such that $P_{Y|X}(E_{c_2} \backslash D_1 | c_2) \ge 1 - \epsilon$, and set $D_2 = E_{c_2} \backslash D_1$.

   ...
3. Pick c_M to be any codeword in F such that $P_{Y|X}(E_{c_M} \backslash \bigcup_{j=1}^{M-1} D_j | c_M) \ge 1 - \epsilon$, and set $D_M = E_{c_M} \backslash \bigcup_{j=1}^{M-1} D_j$.

We stop if a c_{M+1} codeword satisfying the requirement cannot be found. Thus, M is determined by the stopping condition:

$$\forall c \in F, \quad P_{Y|X}\left(E_c \backslash \bigcup_{j=1}^{M} D_j \middle| c \right) < 1 - \epsilon.$$

Averaging the stopping condition over $c \sim P_X$ (which is permissible due to $P_X(F) = 1$), we obtain

$$\mathbb{P}\left[i(X;Y) \ge \log \gamma \text{ and } Y \notin \bigcup_{j=1}^{M} D_j \right] < 1 - \epsilon,$$

or, equivalently,

$$\epsilon < \mathbb{P}\left[i(X;Y) < \log \gamma \text{ or } Y \in \bigcup_{j=1}^{M} D_j \right].$$

Applying the union bound to the right–hand side yields

$$\begin{aligned}
\epsilon &< \mathbb{P}[i(X;Y) < \log \gamma] + \sum_{j=1}^{M} P_Y(D_j) \\
&\le \mathbb{P}[i(X;Y) < \log \gamma] + \sum_{j=1}^{M} P_Y(E_{c_j}) \\
&\le \mathbb{P}[i(X;Y) < \log \gamma] + \frac{M}{\gamma},
\end{aligned}$$

where the last step makes use of (18.13). Evidently, this completes the proof. □

18.5 RCU and Gallager's Bound

Although the bounds we demonstrated so far will be sufficient for recovering the noisy channel coding theorem later, they are not the best possible. Namely, for a given M one can show much smaller upper bounds on the probability of error. Two such bounds are the so-called random coding union (RCU) and the Gallager's bound, which we prove here. The main new ingredient is that instead of using suboptimal (threshold) decoders as before, we will analyze the optimal maximum likelihood decoder.

Theorem 18.8. (RCU bound) *Fix a channel $P_{Y|X}$ and an arbitrary input distribution P_X. Then for every integer $M \geq 1$ there exists an (M, ϵ)-code with*

$$\epsilon \leq \mathbb{E}\left[\min\left\{1,\ (M-1)\mathbb{P}\left[i(\bar{X};Y) \geq i(X;Y) \,\middle|\, X, Y\right]\right\}\right], \tag{18.14}$$

where the joint distribution of $(X, \bar{X}, Y)$ is as in (18.1).

Proof. For a given codebook $(c_1, \ldots, c_M)$ the average probability of error for the maximum likelihood decoder, see (18.8), is upper–bounded by

$$\epsilon \leq \frac{1}{M}\sum_{m=1}^{M}\mathbb{P}\left[\bigcup_{j=1;j\neq m}^{M}\{i(c_j;Y) \geq i(c_m;Y)\} \,\middle|\, X = c_m\right].$$

Note that we do not necessarily have equality here, since the maximum likelihood decoder will resolve ties (i.e. the cases when multiple codewords maximize information density) in favor of the correct codeword, whereas in the expression above we pessimistically assume that all ties are resolved incorrectly. Now, similar to Shannon's bound in Theorem 18.5 we prove the existence of a good code by averaging the last expression over $c_j \overset{\text{iid}}{\sim} P_X$.

To that end, notice that expectations of each term in the sum coincide (by symmetry). To evaluate this expectation, let us take the $m = 1$ condition on $W = 1$ and observe that under this conditioning we have

$$(c_1, Y, c_2, \ldots, c_M) \sim P_{X,Y}\prod_{j=2}^{M}P_X.$$

With this observation in mind we have the following chain:

$$\begin{aligned}
&\mathbb{P}\left[\bigcup_{j=2}^{M}\{i(c_j;Y) \geq i(c_1;Y)\} \,\middle|\, W = 1\right] \\
&\overset{(a)}{=} \mathbb{E}_{(x,y)\sim P_{X,Y}}\left[\mathbb{P}\left[\bigcup_{j=2}^{M}\{i(c_j;Y) \geq i(c_1;Y)\} \,\middle|\, c_1 = x, Y = y, W = 1\right]\right] \\
&\overset{(b)}{\leq} \mathbb{E}\left[\min\{1, (M-1)\mathbb{P}\left[i(\bar{X};Y) \geq i(X;Y) \,\middle|\, X, Y\right]\}\right],
\end{aligned}$$

where (a) is just expressing the probability by first conditioning on the values of (c_1, Y); and (b) corresponds to applying the union bound but capping the result by

1. This completes the proof of the bound. We note that the step (b) is the essence of the RCU bound and corresponds to the self-evident fact that for any collection of events E_j we have

$$\mathbb{P}\left[\bigcup E_j\right] \le \min\left\{1, \sum_j \mathbb{P}[E_j]\right\}.$$

What makes its application clever is that we first conditioned on (c_1, Y). If we applied the union bound right from the start without conditioning, the resulting estimate on ϵ would have been much weaker (in particular, would not have led to achieving capacity). □

It turns out that Shannon's bound in Theorem 18.5 is just a weakening of (18.14) obtained by splitting the expectation according to whether or not $i(X;Y) \le \log M + \tau$ and upper–bounding $\min\{x, 1\}$ by 1 when $i(X;Y) \le \log M + \tau$ and by x otherwise. Another such weakening is the famous Gallager's bound [176], which in fact gives a tight estimate of the exponent in the decay of error probability over memoryless channels (Section 22.4*).

Theorem 18.9. (Gallager's bound) *Fix a channel $P_{Y|X}$, an arbitrary input distribution P_X, and $\rho \in [0, 1]$. Then there exists an (M, ϵ)–code such that*

$$\epsilon \le M^\rho \ \mathbb{E}\left[\left(\mathbb{E}\left[\exp \frac{i(\bar{X};Y)}{1+\rho} \,\middle|\, Y\right]\right)^{1+\rho}\right], \tag{18.15}$$

where again $(\bar{X}, Y) \sim P_X P_Y$ as in (18.1).

For a classical way of writing this bound see (22.13).

Proof. We first notice that by Proposition 18.3 applied with $f_+(x, y) = \exp\{\frac{i(x;y)}{1+\rho}\}$ and interchanged X and Y we have for P_Y-almost every y

$$\begin{aligned} \mathbb{E}\left[\exp\left\{-i(X;Y)\frac{\rho}{1+\rho}\right\} \,\middle|\, Y = y\right] &= \mathbb{E}\left[\exp\left\{i(X;\bar{Y})\frac{1}{1+\rho}\right\} \,\middle|\, \bar{Y} = y\right] \\ &= \mathbb{E}\left[\exp\left\{i(\bar{X};Y)\frac{1}{1+\rho}\right\} \,\middle|\, Y = y\right], \end{aligned} \tag{18.16}$$

where we also used the fact that $(X, \bar{Y}) \stackrel{d}{=} (\bar{X}, Y)$ under (18.1).

Now, consider the bound (18.14) and replace the min via the bound

$$\min\{t, 1\} \le t^\rho \qquad \forall t \ge 0. \tag{18.17}$$

This results in

$$\epsilon \le M^\rho \, \mathbb{E}\left[\mathbb{P}[i(\bar{X};Y) > i(X;Y)|X, Y]^\rho\right]. \tag{18.18}$$

We apply the Chernoff bound

$$\mathbb{P}\left[i(\bar{X};Y) > i(X;Y)|X, Y\right] \le \exp\left\{-\frac{1}{1+\rho} i(X;Y)\right\} \mathbb{E}\left[\exp\left\{\frac{1}{1+\rho} i(\bar{X};Y)\right\} \,\middle|\, Y\right].$$

Raising this inequality to power ρ and taking expectation $\mathbb{E}[\cdot|Y]$ we obtain

$$\mathbb{E}\left[\mathbb{P}[i(\bar{X};Y) > i(X;Y)|X,Y]^{\rho}|\, Y\right]$$
$$\leq \mathbb{E}^{\rho}\left[\exp\left\{\frac{1}{1+\rho}i(\bar{X};Y)\right\}\Big|\, Y\right]\mathbb{E}\left[\exp\left\{-\frac{\rho}{1+\rho}i(X;Y)\right\}\Big|\, Y\right].$$

The last term can be now re-expressed via (18.16) to obtain

$$\mathbb{E}\left[\mathbb{P}[i(\bar{X};Y) > i(X;Y)|X,Y]^{\rho}|Y\right] \leq \mathbb{E}^{1+\rho}\left[\exp\left\{\frac{1}{1+\rho}i(\bar{X};Y)\right\}\Big|\, Y\right].$$

Applying this estimate to (18.18) completes the proof. □

The key innovation of Gallager, namely the step (18.17), which became known as the ρ-trick, corresponds to the following version of the union bound: For any events E_j and $0 \leq \rho \leq 1$ we have

$$\mathbb{P}\left[\bigcup E_j\right] \leq \min\left\{1, \sum_j \mathbb{P}[E_j]\right\} \leq \left(\sum_j \mathbb{P}[E_j]\right)^{\rho}.$$

Now to understand properly the significance of Gallager's bound we need to first define the concept of memoryless channels (see (19.1) below). For such channels and using the iid inputs, the expression (18.15) turns, after optimization over ρ, into

$$\epsilon \leq \exp\{-nE_r(R)\},$$

where $R = \frac{\log M}{n}$ is the rate and $E_r(R)$ is the Gallager's random coding exponent. This shows that not only the error probability at a fixed rate can be made to vanish, but in fact it can be made to vanish exponentially fast in the blocklength. We will discuss such exponential estimates in more detail in Section 22.4*.

18.6 Linear Codes

So far in this chapter we have shown the existence of good error-correcting codes by doing either the random or maximal coding. The constructed codes have little structure. At the same time, most codes used in practice are so-called linear codes and a natural question arises whether restricting to linear codes leads to loss in performance. In this section we show that there exist good linear codes as well. A pleasant property of linear codes is that $P_e = P_{e,\max}$ and, therefore, bounding average probability of error (as in Shannon's bound) automatically yields control of the maximal probability of error as well.

Definition 18.10. (Linear codes) Let $\mathbb{F}_q$ denote the finite field of cardinality q (see Definition 11.7). Let the input and output spaces of the channel be $\mathcal{X} = \mathcal{Y} = \mathbb{F}_q^n$. We say a codebook $\mathcal{C} = \{c_u\colon u \in \mathbb{F}_q^k\}$ of size $M = q^k$ is a *linear code* if $\mathcal{C}$ is a k-dimensional *linear subspace* of $\mathbb{F}_q^n$.

A linear code can be equivalently described by the following:

- *Generator matrix* $G \in \mathbb{F}_q^{k \times n}$, so that the codeword for each $u \in \mathbb{F}_q^k$ is given by $c_u = uG$ (row-vector convention) and the codebook $\mathcal{C}$ is the row-span of G, denoted by $\mathrm{Im}(G)$.
- *Parity-check matrix* $H \in \mathbb{F}_q^{(n-k) \times n}$, so that each codeword $c \in \mathcal{C}$ satisfies $Hc^\top = 0$. Thus $\mathcal{C}$ is the nullspace of H, denoted by $\mathrm{Ker}(H)$. We have $HG^\top = 0$.

Example 18.1. Hamming code

The $[7,4,3]_2$ Hamming code over $\mathbb{F}_2$ is a linear code with the following generator and parity–check matrices:

$$G = \begin{bmatrix} 1 & 0 & 0 & 0 & 1 & 1 & 0 \\ 0 & 1 & 0 & 0 & 1 & 0 & 1 \\ 0 & 0 & 1 & 0 & 0 & 1 & 1 \\ 0 & 0 & 0 & 1 & 1 & 1 & 1 \end{bmatrix}, \quad H = \begin{bmatrix} 1 & 1 & 0 & 1 & 1 & 0 & 0 \\ 1 & 0 & 1 & 1 & 0 & 1 & 0 \\ 0 & 1 & 1 & 1 & 0 & 0 & 1 \end{bmatrix}.$$

In particular, G and H are of the form $G = [I;P]$ and $H = [-P^\top;I]$ (systematic codes) so that $HG^\top = 0$. The following picture helps to visualize the parity–check operation:

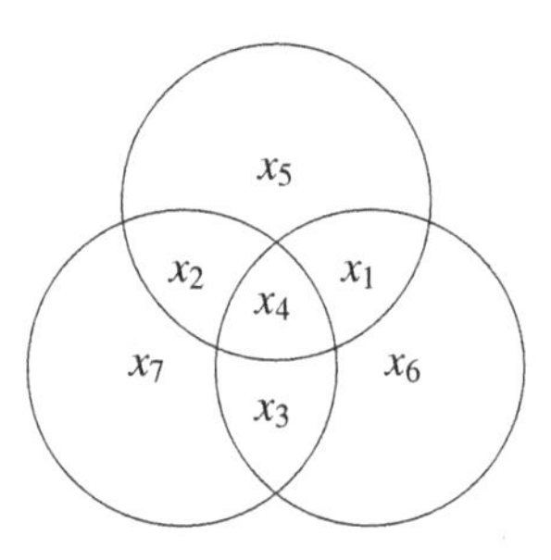

Note that all four bits in each circle (corresponding to a row of H) sum up to zero. One can verify that the minimum distance of this code is 3 bits. As such, it can correct 1 bit of error and detect 2 bits of error.

Linear codes are almost always examined with channels of additive noise, a precise definition of which is given below:

Definition 18.11. (Additive noise) A channel $P_{Y|X}$ with input and output space $\mathbb{F}_q^n$ is called additive noise if

$$P_{Y|X}(y|x) = P_Z(y - x)$$

for some random vector Z taking values in $\mathbb{F}_q^n$. In other words, $Y = X + Z$, where $Z \perp\!\!\!\perp X$.

Given a linear code and an additive noise channel $P_{Y|X}$, it turns out that there is a special "syndrome decoder" that is optimal.

Theorem 18.12. *Any $[k,n]_{\mathbb{F}_q}$ linear code over an additive noise $P_{Y|X}$ has a maximum likelihood (ML) decoder $g\colon \mathbb{F}_q^n \to \mathbb{F}_q^k$ such that:*

1. *$g(y) = y - g_{\text{synd}}(Hy^\top)$, that is, the decoder is a function of the "syndrome" $Hy^\top$ only. Here $g_{\text{synd}}: \mathbb{F}_q^{n-k} \to \mathbb{F}_q^n$, defined by $g_{\text{synd}}(s) \triangleq \operatorname{argmax}_{z:\, Hz^\top = s} P_Z(z)$, is called the "syndrome decoder," which decodes the most likely realization of the noise.*
2. *(Geometric uniformity) Decoding regions are translates of $D_0 = \operatorname{Im}(g_{\text{synd}})$: $D_u = c_u + D_0$ for any $u \in \mathbb{F}_q^k$.*
3. *$P_{e,\max} = P_e$.*

In other words, a syndrome is a sufficient statistic (Definition 3.8) for decoding a linear code.

Proof.

1. The maximum likelihood decoder for a linear code is

$$\begin{aligned} g(y) &= \operatorname*{argmax}_{c \in \mathcal{C}} P_{Y|X}(y|c) = \operatorname*{argmax}_{c:\, Hc^\top = 0} P_Z(y - c) \\ &= y - \operatorname*{argmax}_{z:\, Hz^\top = Hy^\top} P_Z(z) = y - g_{\text{synd}}(Hy^\top). \end{aligned}$$

2. For any u, the decoding region

$$\begin{aligned} D_u &= \{y\colon g(y) = c_u\} = \{y\colon y - g_{\text{synd}}(Hy^\top) = c_u\} \\ &= \{y\colon y - c_u = g_{\text{synd}}(H(y - c_u)^\top)\} = c_u + D_0, \end{aligned}$$

where we used $Hc_u^\top = 0$ and $c_0 = 0$.

3. For any u,

$$\begin{aligned} \mathbb{P}[\hat{W} \neq u | W = u] &= \mathbb{P}[g(c_u + Z) \neq c_u] = \mathbb{P}[c_u + Z - g_{\text{synd}}(Hc_u^\top + HZ^\top) \neq c_u] \\ &= \mathbb{P}[g_{\text{synd}}(HZ^\top) \neq Z]. \end{aligned}$$ □

Remark 18.5. (BSC example) As a concrete example, consider the binary symmetric channel $\mathsf{BSC}_\delta^{\otimes n}$ previously considered in Example 17.1 and Section 17.2. This is an additive-noise channel over $\mathbb{F}_2^n$, where $Y = X + Z$ and $Z = (Z_1, \ldots, Z_n) \overset{\text{iid}}{\sim} \text{Ber}(\delta)$. Assuming $\delta < 1/2$, the syndrome decoder aims to find the noise realization with the fewest number of flips that is compatible with the received codeword, namely $g_{\text{synd}}(s) = \operatorname{argmin}_{z:\, Hz^\top = s} w_H(z)$, where w_H denotes the Hamming weight. In this case elements of the image of g_{synd}, which we denoted by D_0, are known as "minimal weight coset leaders." Counting how many of them occur at each Hamming weight is a difficult open problem even for the most well-studied codes such as Reed–Muller ones. In Hamming space D_0 looks like a Voronoi region of a lattice and D_u's constitute a Voronoi tesselation of $\mathbb{F}_q^n$.

The overwhelming majority of practically used codes are in fact linear codes. Early in the history of coding, linearity was viewed as a way toward fast and *low-complexity encoding* (just binary matrix multiplication) and slightly lower complexity of the maximum likelihood decoding (via the syndrome decoder). As codes became longer and longer, though, syndrome decoding became impractical and today only those codes are used in practice for which there are fast and low-complexity (suboptimal) decoders.

Theorem 18.13. (DT bound for linear codes) *Let $P_{Y|X}$ be an additive-noise channel over $\mathbb{F}_q^n$. For all integers $k \geq 1$ there exists a linear code $f\colon \mathbb{F}_q^k \to \mathbb{F}_q^n$ with error probability:*

$$P_{e,\max} = P_e \leq \mathbb{E}\left[q^{-(n-k-\log_q(1/P_Z(Z)))^+}\right]. \tag{18.19}$$

Remark 18.6. The bound above is the same as Theorem 18.6 evaluated with $P_X = \mathrm{Unif}(\mathbb{F}_q^n)$. The analogy between Theorems 18.6 and 18.13 is the same as that between Theorems 11.5 and 11.8 (full random coding versus random linear codes).

Proof. Recall that in proving the DT bound (Theorem 18.6), we selected the codewords $c_1, \ldots, c_M \overset{\text{iid}}{\sim} P_X$ and showed that

$$\mathbb{E}[P_e(c_1, \ldots, c_M)] \leq \mathbb{P}[i(X;Y) \leq \gamma] + \frac{M-1}{2}\mathbb{P}[i(\overline{X};Y) \geq \gamma].$$

Here we will adopt the same approach and take $P_X = \mathrm{Unif}(\mathbb{F}_q^n)$ and $M = q^k$.

By Theorem 18.12 the optimal decoding regions are translational invariant, that is, $D_u = c_u + D_0, \forall u$, and therefore

$$P_{e,\max} = P_e = \mathbb{P}[\hat{W} \neq u | W = u], \quad \forall u.$$

Step 1: Random linear coding with dithering. Let codewords be chosen as

$$c_u = uG + h, \quad \forall u \in \mathbb{F}_q^k,$$

where random G and h are drawn as follows: the $k \times n$ entries of G and the $1 \times n$ entries of h are iid uniform over $\mathbb{F}_q$. We add the dithering to eliminate the special role that the all-zero codeword plays (since it is contained in any linear codebook).

Step 2: We claim that the codewords are pairwise independent and uniform, that is, $\forall u \neq u'$, $(c_u, c_{u'}) \sim (X, \overline{X})$, where $P_{X,\overline{X}}(x, \overline{x}) = 1/q^{2n}$. To see this note that

$$c_u \sim \text{uniform on } \mathbb{F}_q^n,$$
$$c_{u'} = u'G + h = uG + h + (u' - u)G = c_u + (u' - u)G.$$

We claim that $c_u \perp\!\!\!\perp G$ because, conditioned on the generator matrix $G = G_0$, $c_u \sim$ uniform on $\mathbb{F}_q^n$ due to the dithering h.

We also claim that $c_u \perp\!\!\!\perp c_{u'}$ because, conditioned on c_u, $(u' - u)G \sim$ uniform on $\mathbb{F}_q^n$.

Thus random linear coding with dithering indeed gives codewords $c_u, c_{u'}$ pairwise independent and uniformly distributed.

Step 3: Repeat the same argument in proving the DT bound for the symmetric and pairwise-independent codewords. Then we have

$$\begin{aligned}\mathbb{E}[P_e(c_1, \ldots, c_M)] &\leq \mathbb{E}\left[\exp\left\{-\left(i(X;Y) - \log\frac{M-1}{2}\right)^+\right\}\right] \\ &= \mathbb{E}\left[q^{-(i(X;Y) - \log_q((q^k-1)/2))^+}\right] \leq \mathbb{E}\left[q^{-(i(X;Y)-k)^+}\right]\end{aligned}$$

where we used $M = q^k$ and picked the base of log to be q.

Step 4: Compute $i(X;Y)$:

$$i(a;b) = \log_q \frac{P_Z(b-a)}{q^{-n}} = n - \log_q \frac{1}{P_Z(b-a)},$$

therefore

$$P_e \leq \mathbb{E}\left[q^{-(n-k-\log_q(1/P_Z(Z)))^+}\right]. \tag{18.20}$$

Step 5: Remove dithering h. We claim that there exists a linear code without dithering such that (18.20) is satisfied. The intuition is that shifting a codebook has no effect on its performance. Indeed, we can say the following:

- Before, with dithering, the encoder maps u to $uG + h$, the channel adds noise to produce $Y = uG + h + Z$, and the decoder g outputs $g(Y)$.
- Now, without dithering, we encode u to uG, the channel adds noise to produce $Y = uG+Z$, and we apply decode g' defined by $g'(Y) = g(Y+h)$.

By doing so, we "simulate" dithering at the decoder end and the probability of error remains the same as before. Note that this is possible thanks to the additivity of the noisy channel. □

We see that random coding can be done with different ensembles of codebooks. For example, we have the following.

- Shannon ensemble: $\mathcal{C} = \{c_1, \ldots, c_M\} \overset{\text{iid}}{\sim} P_X$ – fully random ensemble.
- Elias ensemble [150]: $\mathcal{C} = \{uG\colon u \in \mathbb{F}_q^k\}$, with the $k \times n$ generator matrix G drawn uniformly at random from the set of all matrices. (This ensemble is used in the proof of Theorem 18.13.)
- Gallager ensemble: $\mathcal{C} = \{c\colon Hc^\top = 0\}$, with the $(n-k) \times n$ parity-check matrix H drawn uniformly at random. Note this is not the same as the Elias ensemble.
- One issue with the Elias ensemble is that with some non-zero probability G may fail to be full rank. (It is a good exercise to find $\mathbb{P}[\text{rank}(G) < k]$ as a function of n, k, q.) If G is not full rank, then there are two identical codewords and hence $P_{e,\max} \geq 1/2$. To fix this issue, one may let the generator matrix G be uniform on the set of all $k \times n$ matrices of full (row) rank.
- Similarly, we may modify Gallager's ensemble by taking the parity-check matrix H to be uniform on all $n \times (n-k)$ full–rank matrices.

For the modified Elias and Gallager's ensembles, we could still do the analysis of random coding. A small modification would be to note that this time $(X, \bar{X})$ would have distribution

$$P_{X,\bar{X}} = \frac{1}{q^{2n} - q^n} 1\left\{X \neq X'\right\}$$

uniform on all pairs of *distinct* codewords and *not* pairwise independent.

Finally, we note that the Elias ensemble with dithering, $c_u = uG+h$, has pairwise independence property and its joint entropy $H(c_1, \ldots, c_M) = H(G)+H(h) = (nk+n) \log q$. This is significantly smaller than for Shannon's fully random ensemble that we used in Theorem 18.5. Indeed, when $c_j \stackrel{\text{iid}}{\sim} \text{Unif}(\mathbb{F}_q^n)$ we have $H(c_1, \ldots, c_M) = q^k n \log q$. An interesting question, thus, is to find

$$\min H(c_1, \ldots, c_M),$$

where the minimum is over all distributions with $P[c_i = a, c_j = b] = q^{-2n}$ when $i \neq j$ (pairwise–independent, uniform codewords). Note that $H(c_1, \ldots, c_M) \geq H(c_1, c_2) = 2n \log q$. Similarly, we may ask for (c_i, c_j) to be uniform over all pairs of *distinct* elements. In this case, the Wozencraft ensemble (see Exercise **IV.7**) for the case of $n = 2k$ achieves $H(c_1, \ldots, c_{q^k}) \approx 2n \log q$, which is essentially our lower bound.

18.7 Why Do Random and Maximal Coding Work Well?

As we will see later the bounds developed in this chapter are very tight both asymptotically and non-asymptotically. That is, the codes constructed by the apparently rather naive processes of randomly selecting codewords or greedily growing the codebook turn out to be essentially optimal in many ways. An additional mystery is that the bounds we obtained via these two rather different processes are virtually the same. These questions have puzzled researchers since the early days of information theory.

A rather satisfying reason was finally given in an elegant work of Barman and Fawzi [36]. Before going into the details, we want to vocalize explicitly the two questions we want to address:

1 Why is the greedy procedure close to optimal?
2 Why is the random coding procedure (with a simple P_X) close to optimal?

In short, we will see that the answer is that both of these methods are well known to be (almost) optimal for submodular function maximization, and this is exactly what channel coding is about.

Before proceeding, we notice that in the second question it is important to qualify that P_X is simple, since taking P_X to be supported on the optimal $M^*(\epsilon)$-achieving codebook would of course result in very good performance. However, instead we will see that choosing rather simple P_X already achieves a rather good lower bound on $M^*(\epsilon)$. More explicitly, by *simple* we mean a product distribution for the memoryless channel. Or, as an even better example to have in mind, consider an additive-noise vector channel:

$$Y^n = X^n + Z^n,$$

with addition over a product abelian group and arbitrary (even non-memoryless) noise Z^n. In this case the choice of uniform P_X in the random coding bound works, and is definitely "simple."

The key observation of [36] is *submodularity* of the function mapping a codebook $\mathcal{C} \subset \mathcal{X}$ to the $|\mathcal{C}|(1 - P_{e,\text{MAP}}(\mathcal{C}))$, where $P_{e,\text{MAP}}(\mathcal{C})$ is the probability of error under the MAP decoder (17.5). (Recall (1.8) for the definition of submodularity.) More explicitly, consider a discrete $\mathcal{Y}$ and define

$$S(\mathcal{C}) \triangleq \sum_{y \in \mathcal{Y}} \max_{x \in \mathcal{C}} P_{Y|X}(y|x), \qquad S(\emptyset) = 0.$$

It is clear that $S(\mathcal{C})$ is submodular non-decreasing as a sum of submodular non-decreasing functions (indeed, $T \mapsto \max_{x \in T} \phi(x)$ is submodular for any ϕ). On the other hand, $P_{e,\mathrm{MAP}}(\mathcal{C}) = 1 - \frac{1}{|\mathcal{C}|} S(\mathcal{C})$, and thus the search for the minimal error codebook is equivalent to maximizing the set-function S.

The question of finding

$$S^*(M) \triangleq \max_{|\mathcal{C}| \le M} S(\mathcal{C})$$

was algorithmically resolved in a ground-breaking work of [314] showing (approximate) optimality of the greedy process. Consider the following natural greedy process of constructing a sequence of good sets $\mathcal{C}_t$. Start with $\mathcal{C}_0 = \emptyset$. At each step find any

$$x_{t+1} \in \operatorname*{argmax}_{x \notin \mathcal{C}_t} S(\mathcal{C}_t \cup \{x\})$$

and set

$$\mathcal{C}_{t+1} = \mathcal{C}_t \cup \{x_{t+1}\}.$$

They showed that

$$S(\mathcal{C}_t) \ge (1 - 1/e) \max_{|\mathcal{C}|=t} S(\mathcal{C}).$$

In other words, the probability of successful (MAP) decoding for the greedily constructed codebook is at most a factor $(1 - 1/e)$ away from the largest possible probability of success among all codebooks of the same cardinality. Since we are mostly interested in success probabilities very close to 1, this result may not appear very exciting. However, a small modification of the argument yields the following (see [258, Theorem 1.5] for the proof):

Theorem 18.14. ([314]) *For any non-negative submodular set-function f and a greedy sequence $\mathcal{C}_t$ we have for all ℓ, t:*

$$f(\mathcal{C}_\ell) \ge (1 - e^{-\ell/t}) \max_{|\mathcal{C}|=t} f(\mathcal{C}).$$

Applying this to the special case of $f(\cdot) = S(\cdot)$ we obtain the result of [36]: The greedily constructed codebook $\mathcal{C}'$ with M' codewords satisfies

$$1 - P_{e,\mathrm{MAP}}(\mathcal{C}') \ge \frac{M}{M'}(1 - e^{-M'/M})(1 - \epsilon^*(M)).$$

In particular, the greedily constructed code with $M' = M2^{-10}$ achieves probability of success that is $\ge 0.9995(1 - \epsilon^*(M))$. In other words, compared to the best possible code a greedy code carrying 10 bits fewer of data suffers at most 5×10^{-4} worse probability of error. *This is a very compelling evidence for why greedy construction* *is so good.* We do note, however, that Feinstein's bound does greedy construction not with the MAP decoder, but with a suboptimal one.

Next we address the question of *random coding*. Recall that our goal is to explain how can selecting codewords uniformly at random from a "simple" distribution P_X be any good. The key idea is again contained in [314]. The set-function $S(\mathcal{C})$ can also be understood as a function with domain $\{0,1\}^{|\mathcal{X}|}$. Here is a natural extension of this function to the entire solid hypercube $[0,1]^{|\mathcal{X}|}$:

$$S_{LP}(\pi) = \sup\left\{\sum_{x,y} P_{Y|X}(y|x) r_{x,y} \colon 0 \le r_{x,y} \le \pi_x, \sum_x r_{x,y} \le 1\right\}. \tag{18.21}$$

Indeed, it is easy to see that $S_{LP}(1_{\mathcal{C}}) = S(\mathcal{C})$ and that S_{LP} is a concave function.[3]

Since S_{LP} is an extension of S it is clear that

$$S^*(M) \le S^*_{LP}(M) \triangleq \max\left\{S_{LP}(\pi) \colon 0 \le \pi_x \le 1, \sum_x \pi_x \le M\right\}. \tag{18.22}$$

In fact, we will see later in Section 22.3 that this bound coincides with the bound known as meta-converse. Surprisingly, [314] showed that the greedy construction achieves a large multiple not only of $S^*(M)$ but also of $S^*_{LP}(M)$:

$$S(\mathcal{C}_M) \ge (1 - e^{-1}) S^*_{LP}(M). \tag{18.23}$$

The importance of this result (which is specific to submodular functions $\mathcal{C} \mapsto \sum_y \max_{x \in \mathcal{C}} g(x,y)$) is that it gave one of the first integrality gap results relating the value of combinatorial optimization $S^*(M)$ and a linear program relaxation $S^*_{LP}(M)$: $(1-e^{-1}) S^*_{LP}(M) \le S^*(M) \le S^*_{LP}(M)$.

An extension of (18.23) similar to the preceding theorem can also be shown: for all M', M we have

$$S(\mathcal{C}_{M'}) \ge (1 - e^{-M'/M}) S^*_{LP}(M).$$

To connect to the concept of random coding, though, we need the following result of [36]:[4]

Theorem 18.15. *Fix $\pi \in [0,1]^{|\mathcal{X}|}$ and let $M = \sum_{x \in \mathcal{X}} \pi_x$. Let $\mathcal{C} = \{c_1, \ldots, c_{M'}\}$ with $c_j \overset{iid}{\sim} P_X(x) = \frac{\pi_x}{M}$. Then we have*

$$\mathbb{E}[S(\mathcal{C})] \ge (1 - e^{-M'/M}) S_{LP}(\pi).$$

The proof of this result trivially follows from applying the following lemma with $g(x) = P_{Y|X}(y|x)$, summing over y, and recalling the definition of S_{LP} in (18.21).

[3] There are a number of standard extensions of a submodular function f to a hypercube. The largest convex interpolant f_+, also known as Lovász extension, the least concave interpolant f_-, and multi-linear extension [80]. However, S_{LP} does not coincide with any of these and in particular is strictly larger (in general) than f_-.

[4] There are other ways of doing "random coding" to produce an integer solution from a fractional one. For example, see the multi-linear extension based one in [80].

Lemma 18.16. *Let π and C be as in Theorem 18.15. Let $g\colon \mathcal{X} \to \mathbb{R}$ be any function and denote $T(\pi, g) = \max\{\sum_x r_x g(x) : 0 \le r_x \le \pi_x, \sum_x r_x = 1\}$. Then*

$$\mathbb{E}\left[\max_{x\in C} g(x)\right] \ge (1 - e^{-M'/M})T(\pi, g).$$

Proof. Without loss of generality we take $\mathcal{X} = [m]$ and $g(1) \ge g(2) \ge \cdots \ge g(m) \ge g(m+1) \triangleq 0$. Denote for convenience $a = 1 - (1 - \frac{1}{M})^{M'} \ge 1 - e^{-M'/M}$ and $b(j) \triangleq \mathbb{P}[\{1, \dots, j\} \cap C \ne \emptyset]$. Then

$$\mathbb{P}\left[\max_{x\in C} g(x) = g(j)\right] = b(j) - b(j-1),$$

and from the summation by parts we get

$$\mathbb{E}\left[\max_{x\in C} g(x)\right] = \sum_{j=1}^{m} (g(j) - g(j+1))b(j). \tag{18.24}$$

On the other hand, denoting $c(j) = \min(\sum_{i\le j} \pi_i, 1)$, from the definition of $b(j)$ we have

$$b(j) = 1 - \left(1 - \frac{\pi_1 + \cdots + \pi_j}{M}\right)^{\ell} \ge 1 - \left(1 - \frac{c(j)}{M}\right)^{M'}.$$

From the simple inequality $(1 - \frac{x}{M})^{M'} \le 1 - ax$ (valid for any $x \in [0, 1]$) we get

$$b(j) \ge ac(j).$$

Plugging this into (18.24) we conclude the proof by noticing that $r_j = c(j) - c(j-1)$ attains the maximum in the definition of $T(\pi, g)$. □

Theorem 18.15 completes this section's goal and shows that random coding (as well as greedy/maximal coding) attains an almost optimal value of $S^*(M)$. Notice also that the random coding distribution that we should be using is the one that attains the definition of $S^*_{LP}(M)$. For input symmetric channels (such as additive-noise ones) it is easy to show that the optimal $\pi \in [0, 1]^{\mathcal{X}}$ is a constant vector, and hence the codewords are to be generated iid uniformly on $\mathcal{X}$.

19 Channel Capacity

In this chapter we apply the methods developed in the previous chapters (namely the weak converse and random/maximal coding achievability) to compute the channel capacity. This latter notion quantifies the maximal amount of (data) bits that can be reliably communicated *per single channel use* in the limit of using the channel many times. Formalizing the latter statement will require introducing the concept of a communication channel. Then for special kinds of channels (memoryless and information stable ones) we will show that computing the channel capacity reduces to maximizing the (sequence of the) mutual informations. This result, known as Shannon's noisy channel coding theorem, is the third example of a coding theorem in this book. It connects the value of an operationally defined (discrete, combinatorial) optimization problem over codebooks to that of a (convex) optimization problem over information measures. It builds a bridge between the abstraction of information measures (Part I) and a practical engineering problem of channel coding.

Information theory as a subject is sometimes accused of "asymptopia," or the obsession with asymptotic results and computing various limits. Although in this book we attempt to refrain from asymptopia, the topic of this chapter requires committing this sin *ipso facto*. After proving capacity theorems in various settings, we conclude the chapter with Shannon's separation theorem, which shows that any (stationary ergodic) source can be communicated over an (information stable) channel if and only if its entropy rate is smaller than the channel capacity. Furthermore, doing so can be done by first compressing a source to pure bits and then using channel code to match those bits to channel inputs. The fact that no performance is lost in the process of this conversion to bits had an important historical consequence in cementing *bits* as the universal currency of the digital age.

19.1 Channels and Channel Capacity

As we discussed in Chapter 17 the main information-theoretic question of data transmission is the following: How many bits can one transmit reliably if one is allowed to use a given noisy channel n times? The normalized quantity equal to the number of message bits per channel use is known as *rate*, and *capacity* refers to the highest achievable rate under a small probability of decoding error. However, what does it mean to "use a channel several times"? How do we formalize the concept of a channel use? To that end, we need to change the meaning of the term "channel." So far in this book we have used the term *channel* as a synonym of the Markov kernel (Definition 2.10). More correctly, however, this term should be used to refer to the following notion.

Definition 19.1. Fix an input alphabet $\mathcal{A}$ and an output alphabet $\mathcal{B}$. A sequence of Markov kernels $P_{Y^n|X^n}\colon \mathcal{A}^n \to \mathcal{B}^n$ indexed by the integer $n = 1, 2, \ldots$ is called a *channel*. The length of the input n is known as *blocklength*.

To give this abstract notion more concrete form one should recall Section 17.2, in which we described the BSC channel. Note, however, that despite this definition, it is customary to use the term *channel* to refer to a single Markov kernel (as we did before in this book). An even worse, yet popular, abuse of terminology is to refer to the nth element of the sequence, the kernel $P_{Y^n|X^n}$, as the n-letter channel.

Although we have not imposed any requirements on the sequence of kernels $P_{Y^n|X^n}$, one is never interested in channels at this level of generality. Most of the time the elements of the channel input $X^n = (X_1, \ldots, X_n)$ are thought of as indexed by time. That is, the X_t corresponds to the letter that is transmitted at time t inside an overall block of length n, while Y_t is the letter received at time t. The channel's action is that of "adding noise" to X_t and outputting Y_t. However, the generality of the previous definition allows one to model situations where the channel has internal state, so that the amount and type of noise added to X_t depend on the previous inputs and in principle even on the future inputs. The interpretation of t as time, however, is not exclusive. In storage (magnetic, non-volatile or flash) t indexes space. In those applications, the noise may have a rather complicated structure with transformation $X_t \to Y_t$ depending on both the "past" $X_{<t}$ and the "future" $X_{>t}$.

Almost all channels of interest satisfy one or more of the restrictions that we list next:

- A channel is called *non-anticipatory* if it has the following extension property. Under the n-letter kernel $P_{Y^n|X^n}$, the conditional distribution of the first k output symbols Y^k only depends on X^k (and not on X_{k+1}^n) and coincides with the kernel $P_{Y^k|X^k}$ (the kth element of the channel sequence). This requirement models the scenario wherein channel outputs depend causally on the inputs.
- A channel is *discrete* if $\mathcal{A}$ and $\mathcal{B}$ are finite.
- A channel is *additive-noise* if $\mathcal{A} = \mathcal{B}$ are abelian groups and $Y^n = X^n + Z^n$ for some Z^n independent of X^n (see Definition 18.11). Thus

$$P_{Y^n|X^n}(y^n|x^n) = P_{Z^n}(y^n - x^n).$$

- A channel is *memoryless* if $P_{Y^n|X^n}$ factorizes into a product distribution. Namely,

$$P_{Y^n|X^n} = \prod_{k=1}^{n} P_{Y_k|X_k}, \tag{19.1}$$

Table 19.1 Examples of DMCs.

Channel	Bipartite graph	Channel matrix
BSC_δ	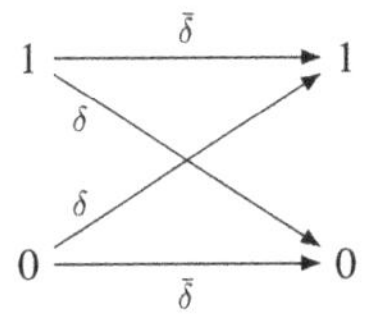 $1 \xrightarrow{\bar{\delta}} 1$, $1 \xrightarrow{\delta} 0$, $0 \xrightarrow{\delta} 1$, $0 \xrightarrow{\bar{\delta}} 0$	$\begin{bmatrix} \bar{\delta} & \delta \\ \delta & \bar{\delta} \end{bmatrix}$
BEC_δ	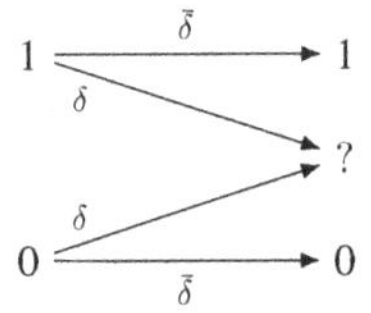 $1 \xrightarrow{\bar{\delta}} 1$, $1 \xrightarrow{\delta} ?$, $0 \xrightarrow{\delta} ?$, $0 \xrightarrow{\bar{\delta}} 0$	$\begin{bmatrix} \bar{\delta} & \delta & 0 \\ 0 & \delta & \bar{\delta} \end{bmatrix}$
Z-channel	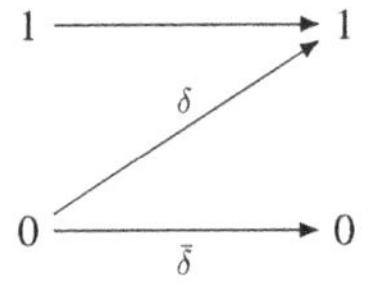 $1 \longrightarrow 1$, $0 \xrightarrow{\delta} 1$, $0 \xrightarrow{\bar{\delta}} 0$	$\begin{bmatrix} 1 & 0 \\ \delta & \bar{\delta} \end{bmatrix}$

where each $P_{Y_k|X_k}\colon \mathcal{A} \to \mathcal{B}$; in particular, $P_{Y^n|X^n}$ are compatible at different blocklengths n.

- A channel is *stationary memoryless* if (19.1) is satisfied with $P_{Y_k|X_k}$ not depending on k, denoted commonly by $P_{Y|X}$. In other words,

$$P_{Y^n|X^n} = (P_{Y|X})^{\otimes n}. \tag{19.2}$$

Thus, in discrete cases, we have

$$P_{Y^n|X^n}(y^n|x^n) = \prod_{k=1}^{n} P_{Y|X}(y_i|x_i). \tag{19.3}$$

The interpretation is that each coordinate of the transmitted codeword X^n is corrupted by noise independently with the same noise statistic.

- Stationary discrete memoryless channel (DMC): A DMC is a channel that is both discrete and stationary memoryless. It can be specified in two ways:
  - an $|\mathcal{A}| \times |\mathcal{B}|$-dimensional (row-stochastic) matrix $P_{Y|X}$ where elements specify the transition probabilities;
  - a bipartite graph with edge weight specifying the non-zero transition probabilities.

Table 19.1 lists some common binary-input binary-output DMCs: the binary symmetric channel (BSC), the binary erasure channel (BEC), and the Z-channel.

As another example, let us recall the AWGN channel in Example 3.3: the alphabets $\mathcal{A} = \mathcal{B} = \mathbb{R}$ and $Y^n = X^n + Z^n$, with $X^n \perp\!\!\!\perp Z^n \sim \mathcal{N}(0, \sigma^2 I_n)$. This channel is a non-discrete, stationary memoryless, additive-noise channel.

Having defined the notion of the channel, we can define next the operational problem that the communication engineer faces when tasked with establishing a data link across the channel. Since the channel is noisy, the data is not going to pass unperturbed and the error-correcting codes are naturally to be employed. To send one of $M = 2^k$ messages (or k data bits) with low probability of error, it is often desirable to use the shortest possible length of the input sequence. This desire explains the following definitions, which extend the fundamental limits in Definition 17.2 to involve the blocklength n.

Definition 19.2. (Fundamental limits of channel coding)

- An (n, M, ϵ)-code is an (M, ϵ)-code for $P_{Y^n|X^n}$, consisting of an encoder $f\colon [M] \to \mathcal{A}^n$ and a decoder $g\colon \mathcal{B}^n \to [M] \cup \{\mathsf{e}\}$.
- An $(n, M, \epsilon)_{\max}$-code is analogously defined for the maximum probability of error.

The (non-asymptotic) fundamental limits are

$$
\begin{aligned}
M^*(n, \epsilon) &= \max\{M\colon \exists\ (n, M, \epsilon)\text{-code}\}, && (19.4)\\
M^*_{\max}(n, \epsilon) &= \max\{M\colon \exists\ (n, M, \epsilon)_{\max}\text{-code}\},\\
\epsilon^*(n, M) &= \inf\{\epsilon\colon \exists (n, M, \epsilon)\text{-code}\}, && (19.5)\\
\epsilon^*_{\max}(n, M) &= \inf\{\epsilon\colon \exists (n, M, \epsilon)_{\max}\text{-code}\}.
\end{aligned}
$$

We will mostly focus on understanding $M^*(n, \epsilon)$ and a related quantity known as rate. Recall that blocklength n measures the amount of time or space resource used by the code. Thus, it is natural to maximize the ratio of the data transmitted to the resource used, and that leads us to the notion of the *transmission rate* defined as $R = \frac{\log_2 M}{n}$ and equal to the number of bits transmitted per channel use. Consequently, instead of studying $M^*(n, \epsilon)$ one is led to the study of $\frac{1}{n} \log M^*(n, \epsilon)$. A natural first question is to determine the first-order asymptotics of this quantity and this motivates the final definition of the section.

Definition 19.3. (Channel capacity) The *ϵ-capacity* C_ϵ and the *Shannon capacity* C are defined as follows:

$$
\begin{aligned}
C_\epsilon &\triangleq \liminf_{n\to\infty} \frac{1}{n} \log M^*(n, \epsilon);\\
C &= \lim_{\epsilon\to 0+} C_\epsilon.
\end{aligned}
$$

Channel capacity is measured in information units per channel use, for example, "bit/channel use".

The operational meaning of C_ϵ (respectively C) is the maximum achievable rate at which one can communicate through a noisy channel with probability of error at most ϵ (respectively $o(1)$). In other words, for any $R < C$, there exists an

$(n, \exp(nR), \epsilon_n)$-code, such that $\epsilon_n \to 0$. In this vein, C_ϵ and C can be equivalently defined as follows:

$$C_\epsilon = \sup\{R \colon \forall \delta > 0, \exists n_0(\delta), \forall n \geq n_0(\delta), \exists (n, \exp(n(R - \delta)), \epsilon)\text{-code}\},$$
$$C = \sup\{R \colon \forall \epsilon > 0, \forall \delta > 0, \exists n_0(\delta, \epsilon), \forall n \geq n_0(\delta, \epsilon), \exists (n, \exp(n(R - \delta)), \epsilon)\text{-code}\}.$$

The reason that capacity is defined as a large-n limit (as opposed to a supremum over n) is because we are concerned with the rate limit of transmitting large amounts of data without errors (such as in communication and storage).

The case of zero error ($\epsilon = 0$) is so different from $\epsilon > 0$ that the topic of $\epsilon = 0$ constitutes a separate subfield of its own (see the survey [253]). Introduced by Shannon in 1956 [380], the value

$$C_0 \triangleq \liminf_{n \to \infty} \frac{1}{n} \log M^*(n, 0) \tag{19.6}$$

is known as the *zero-error capacity* and represents the maximal achievable rate with no error whatsoever. Characterizing the value of C_0 is often a hard combinatorial problem. However, for many practically relevant channels it is quite trivial to show $C_0 = 0$. This is the case, for example, for the DMCs we considered before: the BSC or BEC. Indeed, for them we have $\log M^*(n, 0) = 0$ for all n, meaning transmitting any amount of information across these channels requires accepting some (perhaps vanishingly small) probability of error. Nevertheless, there are certain interesting and important channels for which C_0 is positive, see Section 23.3.1 for more.

As a function of ϵ, C_ϵ could (most generally) behave like the plot on the left-hand side below. It may have a discontinuity at $\epsilon = 0$ and may be monotonically increasing (possibly even with jump discontinuities) in ϵ. Typically, however, C_ϵ is zero at $\epsilon = 0$ and stays constant for all $0 < \epsilon < 1$ and, hence, coincides with C (see the plot on the right-hand side). In such cases we say that the *strong converse* holds (more on this later in Section 22.1).

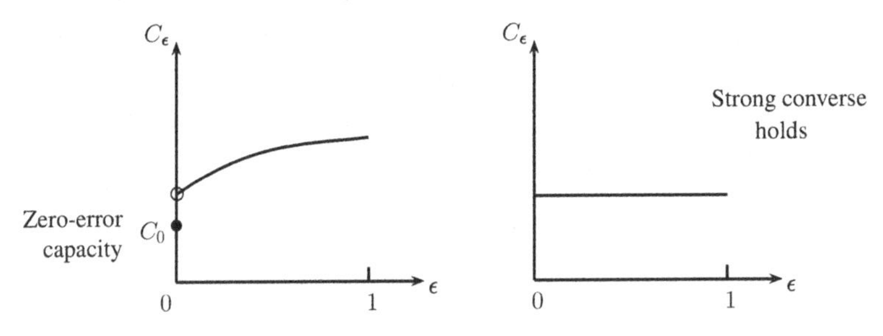

In Definition 19.3, the capacities C_ϵ and C are defined with respect to the average probability of error. By replacing M^* with $M^*_{\max}$, we can define, analogously, the capacities $C_\epsilon^{(\max)}$ and $C^{(\max)}$ with respect to the maximal probability of error. It turns out that these two definitions are equivalent, as the next theorem shows.

Theorem 19.4. *For all* $\tau \in (0, 1)$,

$$\tau M^*(n, \epsilon(1 - \tau)) \leq M^*_{\max}(n, \epsilon) \leq M^*(n, \epsilon).$$

Proof. The second inequality is obvious, since any code that achieves a maximum error probability ϵ also achieves an average error probability of ϵ.

For the first inequality, take an $(n, M, \epsilon(1-\tau))$-code, and define the error probability for the jth codeword as

$$\lambda_j \triangleq \mathbb{P}[\hat{W} \neq j | W = j].$$

Then

$$M(1-\tau)\epsilon \geq \sum \lambda_j = \sum \lambda_j 1\{\lambda_j \leq \epsilon\} + \sum \lambda_j 1\{\lambda_j > \epsilon\} \geq \epsilon |\{j: \lambda_j > \epsilon\}|.$$

Hence $|\{j: \lambda_j > \epsilon\}| \leq (1-\tau)M$. (Note that this is exactly the Markov inequality.) Now by removing those codewords[1] whose λ_j exceeds ϵ, we can extract an $(n, \tau M, \epsilon)_{\max}$-code. Finally, take $M = M^*(n, \epsilon(1-\tau))$ to finish the proof. □

Corollary 19.5. (Capacity under maximal probability of error) *One has* $C_\epsilon^{(\max)} = C_\epsilon$ *for all* $\epsilon > 0$ *such that* $C_\epsilon = C_{\epsilon-}$. *In particular,* $C^{(\max)} = C$.

Proof. Using the definition of M^* and the previous theorem, for any fixed $\tau > 0$,

$$C_\epsilon \geq C_\epsilon^{(\max)} \geq \liminf_{n\to\infty} \frac{1}{n} \log \tau M^*(n, \epsilon(1-\tau)) \geq C_{\epsilon(1-\tau)}.$$

Sending $\tau \to 0$ yields $C_\epsilon \geq C_\epsilon^{(\max)} \geq C_{\epsilon-}$. □

19.2 Shannon's Noisy Channel Coding Theorem

Now that we have the basic definitions for Shannon capacity, we define another type of capacity, and show that for a *stationary memoryless* channel, these two notions ("operational" and "information") of capacity coincide.

Definition 19.6. The *information capacity* of a channel is

$$C^{(\mathrm{I})} = \liminf_{n\to\infty} \frac{1}{n} \sup_{P_{X^n}} I(X^n; Y^n),$$

where for each n the supremum is taken over all joint distributions P_{X^n} on $\mathcal{A}^n$.

Note that information capacity $C^{(\mathrm{I})}$ so defined is not the same as the Shannon capacity C in Definition 19.3; as such, from first principles it has no direct interpretation as an operational quantity related to coding. Nevertheless, they are related by the following *coding theorems*. We start with a converse result:

Theorem 19.7. (Upper bound for C_ϵ) *For any channel,* $\forall \epsilon \in [0,1)$, $C_\epsilon \leq \frac{C^{(\mathrm{I})}}{1-\epsilon}$ *and* $C \leq C^{(\mathrm{I})}$.

[1] This operation is usually referred to as *expurgation*, which yields a smaller code by killing part of the codebook to reach a desired property.

Proof. Applying the general weak converse bound in Theorem 17.3 to $P_{Y^n|X^n}$ yields

$$\log M^*(n,\epsilon) \le \frac{\sup_{P_{X^n}} I(X^n;Y^n) + h(\epsilon)}{1-\epsilon}.$$

Normalizing this by n and taking the lim inf as $n \to \infty$, we have

$$C_\epsilon = \liminf_{n\to\infty} \frac{1}{n} \log M^*(n,\epsilon) \le \liminf_{n\to\infty} \frac{1}{n} \frac{\sup_{P_{X^n}} I(X^n;Y^n) + h(\epsilon)}{1-\epsilon} = \frac{C^{(\mathrm{I})}}{1-\epsilon}. \quad \square$$

Next we give an achievability bound:

Theorem 19.8. (Lower bound for C_ϵ) *For a stationary memoryless channel,* $C_\epsilon \ge \sup_{P_X} I(X;Y)$, *for any* $\epsilon \in (0,1]$.

Proof. Fix an arbitrary P_X on $\mathcal{A}$ and let $P_{X^n} = P_X^{\otimes n}$ be an iid product of a single-letter distribution P_X. Recall Shannon's achievability bound in Theorem 18.5 (or any other one from Chapter 18 would work just as well). From that result we know that for any n, M and any $\tau > 0$, there exists an (n, M, ϵ_n)-code with

$$\epsilon_n \le \mathbb{P}[i(X^n;Y^n) \le \log M + \tau] + \exp(-\tau).$$

Here the information density is defined with respect to the distribution $P_{X^n,Y^n} = P_{X,Y}^{\otimes n}$ and, therefore,

$$i(X^n;Y^n) = \sum_{k=1}^{n} \log \frac{dP_{X,Y}}{dP_X P_Y}(X_k, Y_k) = \sum_{k=1}^{n} i(X_k;Y_k),$$

where $i(x;y) = i_{P_{X,Y}}(x;y)$ and $i(x^n;y^n) = i_{P_{X^n,Y^n}}(x^n;y^n)$. What is important is that under P_{X^n,Y^n} the random variable $i(X^n;Y^n)$ is a sum of iid random variables with mean $I(X;Y)$. Thus, by the weak law of large numbers we have

$$\mathbb{P}[i(X^n;Y^n) < n(I(X;Y) - \delta)] \to 0$$

for any $\delta > 0$.

With this in mind, let us set $\log M = n(I(X;Y) - 2\delta)$ for some $\delta > 0$, and take $\tau = \delta n$ in Shannon's bound. Then for the error bound we have

$$\epsilon_n \le \mathbb{P}\left[\sum_{k=1}^{n} i(X_k;Y_k) \le nI(X;Y) - \delta n\right] + \exp(-\delta n) \xrightarrow{n\to\infty} 0. \tag{19.7}$$

Since the bound converges to 0, we have shown that there exists a sequence of (n, M_n, ϵ_n)-codes with $\epsilon_n \to 0$ and $\log M_n = n(I(X;Y) - 2\delta)$. Hence, for all n such that $\epsilon_n \le \epsilon$

$$\log M^*(n,\epsilon) \ge n(I(X;Y) - 2\delta).$$

And so

$$C_\epsilon = \liminf_{n\to\infty} \frac{1}{n} \log M^*(n,\epsilon) \ge I(X;Y) - 2\delta.$$

Since this holds for all P_X and all $\delta > 0$, we conclude $C_\epsilon \ge \sup_{P_X} I(X;Y)$. $\square$

The following result follows from pairing the upper and lower bounds on C_ϵ.

Theorem 19.9. (Shannon's channel coding theorem [379]) *For a stationary memoryless channel,*

$$C = C^{(\mathrm{I})} = \sup_{P_X} I(X;Y). \tag{19.8}$$

As we have mentioned several times already this result is among the most significant results in information theory. From the engineering point of view, the major surprise was that $C > 0$, that is, communication over a channel is possible with strictly positive rate for any arbitrarily small probability of error. The way to achieve this is to encode the input data jointly (i.e. over many input bits together). This is drastically different from the pre-1948 methods, which operated on a letter-by-letter basis (such as Morse code). This theoretical result gave impetus (and still gives guidance) to the evolution of practical communication systems – quite a rare achievement for an asymptotic mathematical fact.

Proof. Statement (19.8) contains two equalities. The first one follows automatically from the second and Theorems 19.7 and 19.8. To show the second equality $C^{(\mathrm{I})} = \sup_{P_X} I(X;Y)$, we note that for stationary memoryless channels $C^{(\mathrm{I})}$ is in fact easy to compute. Indeed, rather than solving a sequence of optimization problems (one for each n) and taking the limit as $n \to \infty$, the memorylessness of the channel implies that only the $n = 1$ problem needs to be solved. This type of result is known as *single-letterization* (or tensorization) in information theory and we show it formally in the following proposition, which concludes the proof. □

Proposition 19.10. (Tensorization of capacity)

- *For memoryless channels,*

$$\sup_{P_{X^n}} I(X^n;Y^n) = \sum_{i=1}^{n} \sup_{P_{X_i}} I(X_i;Y_i).$$

- *For stationary memoryless channels,*

$$C^{(\mathrm{I})} = \sup_{P_X} I(X;Y).$$

Proof. Recall that from Theorem 6.1 we know that for product kernels $P_{Y^n|X^n} = \prod P_{Y_i|X_i}$, mutual information satisfies $I(X^n;Y^n) \le \sum_{k=1}^{n} I(X_k;Y_k)$ with equality whenever X_i's are independent. Then

$$C^{(\mathrm{I})} = \liminf_{n\to\infty} \frac{1}{n} \sup_{P_{X^n}} I(X^n;Y^n) = \liminf_{n\to\infty} \sup_{P_X} I(X;Y) = \sup_{P_X} I(X;Y). \qquad \square$$

Shannon's noisy channel theorem shows that by employing codes of large blocklength, we can approach the channel capacity arbitrarily close. Given the asymptotic nature of this result (or any other asymptotic result), a natural question is understanding the price to pay for reaching capacity. This can be understood in two ways:

1. The *complexity* of achieving capacity. Is it possible to find low-complexity encoders and decoders with polynomial number of operations in the blocklength n which achieve the capacity? This question was resolved by Forney [172] who showed that this is possible in *linear* time with exponentially small error probability.

   Note that if we are content with polynomially small probability of error, for example, $P_e = O(n^{-100})$, then we can construct polynomial-time decodable codes as follows. First, it can be shown that with rate strictly below capacity, the error probability of optimal codes decays exponentially w.r.t. the blocklength. Now divide the block of length n into a shorter block of length $c \log n$ and apply the optimal code for blocklength $c \log n$ with error probability n^{-101}. Then by the union bound, the whole block has error with probability at most n^{-100}. The encoding and exhaustive-search decoding are obviously polynomial time.
2. The *speed* of achieving capacity. Suppose we are content with achieving 90% of the capacity. Then the question is how large a blocklength do we need to take in order for that to be possible? (Blocklength is directly related to latency, and, thus, equivalently we may ask how much of a delay is incurred by the desire to achieve 90% of capacity.) In other words, we want to know how fast the gap to capacity vanishes as blocklength grows. Shannon's theorem shows that the gap $C - \frac{1}{n}\log M^*(n,\epsilon) = o(1)$. The next theorem shows that under proper conditions, the $o(1)$ term is in fact $O\left(\frac{1}{\sqrt{n}}\right)$.

The main tool in the proof of Theorem 19.8 was the law of large numbers. The lower bound $C_\epsilon \geq C^{(\mathrm{I})}$ in Theorem 19.8 shows that $\log M^*(n,\epsilon) \geq nC + o(n)$ (this just restates the fact that normalizing by n and taking the lim inf must result in something $\geq C$). If instead we apply a more careful analysis using the central limit theorem (CLT), we obtain the following refined achievability result.

Theorem 19.11. *Consider a stationary memoryless channel with a capacity-achieving input distribution. Namely, $C = \max_{P_X} I(X;Y) = I(X_*;Y_*)$ is attained at P^*_X, which induces $P_{X^*Y^*} = P_{X^*}P_{Y|X}$. Assume that $V = \mathrm{Var}[i(X^*;Y^*)] < \infty$. Then*

$$\log M^*(n,\epsilon) \geq nC - \sqrt{nV}Q^{-1}(\epsilon) + o(\sqrt{n}),$$

where $Q(\cdot)$ is the complementary Gaussian CDF and $Q^{-1}(\cdot)$ is its functional inverse.

Proof. Writing the little-o notation in terms of lim inf, our goal is to show

$$\liminf_{n\to\infty} \frac{\log M^*(n,\epsilon) - nC}{\sqrt{nV}} \geq -Q^{-1}(\epsilon) = \Phi^{-1}(\epsilon),$$

where $\Phi(t)$ is the standard normal CDF.

Recall Feinstein's bound

$$\exists (n, M, \epsilon)_{\max}: \quad M \geq \beta\left(\epsilon - \mathbb{P}[i(X^n;Y^n) \leq \log\beta]\right).$$

Take $\log \beta = nC + \sqrt{nV}t$, then applying the CLT gives

$$\log M \geq nC + \sqrt{nV}t + \log\left(\epsilon - \mathbb{P}\left[\sum i(X_k;Y_k) \leq nC + \sqrt{nV}t\right]\right)$$

$$\implies \log M \geq nC + \sqrt{nV}t + \log(\epsilon - \Phi(t)) \quad \forall \Phi(t) < \epsilon$$

$$\implies \frac{\log M - nC}{\sqrt{nV}} \geq t + \frac{\log(\epsilon - \Phi(t))}{\sqrt{nV}},$$

where $\Phi(t)$ is the standard normal CDF. Taking the lim inf of both sides gives

$$\liminf_{n\to\infty} \frac{\log M^*(n,\epsilon) - nC}{\sqrt{nV}} \geq t,$$

for all t such that $\Phi(t) < \epsilon$. Finally, taking $t \nearrow \Phi^{-1}(\epsilon)$, and writing the lim inf in little-o notation completes the proof:

$$\log M^*(n,\epsilon) \geq nC - \sqrt{nV}Q^{-1}(\epsilon) + o(\sqrt{n}).$$ □

Remark 19.1. Theorem 19.9 implies that for any $R < C$, there exists a sequence of $(n, \exp(nR), \epsilon_n)$-codes such that the probability of error ϵ_n vanishes as $n \to \infty$. Examining the upper bound (19.7), we see that the probability of error actually vanishes exponentially fast, since the event in the first term is of large-deviations type (recall Chapter 15) so that both terms are exponentially small. Finding the value of the optimal exponent (or even the existence thereof) has a long history (but remains generally open) in information theory, see Section 22.4*. Recently, however, it was understood that practically more relevant, and also much easier to analyze, is the regime of fixed (non-vanishing) error ϵ, in which case the main question is to bound the speed of convergence of $R \to C_\epsilon = C$. The previous theorem shows one bound on this speed of convergence. The optimal $\frac{1}{\sqrt{n}}$ coefficient is known as channel dispersion, see Sections 22.5 and 22.6 for more. In particular, we will show that the bound on the $\sqrt{n}$ term in Theorem 19.11 is often tight.

19.3 Examples of Capacity Computation

We compute the capacities of the simple DMCs from Table 19.1 and plot them in Figure 19.1.

First, for the BSC_δ we have the following description of the input–output law:

$$Y = X + Z \mod 2, \quad Z \sim \text{Ber}(\delta) \perp\!\!\!\perp X.$$

To compute the capacity, let us notice

$$I(X;X+Z) = H(X+Z) - H(X+Z|X) = H(X+Z) - H(Z) \leq \log 2 - h(\delta)$$

with equality iff $X \sim \text{Ber}(1/2)$. Hence we have shown

$$C = \sup_{P_X} I(X;Y) = \log 2 - h(\delta).$$

More generally, recalling Example 3.7, for any additive-noise channel over a finite abelian group G, we have $C = \sup_{P_X} I(X;X+Z) = \log|G| - H(Z)$, achieved

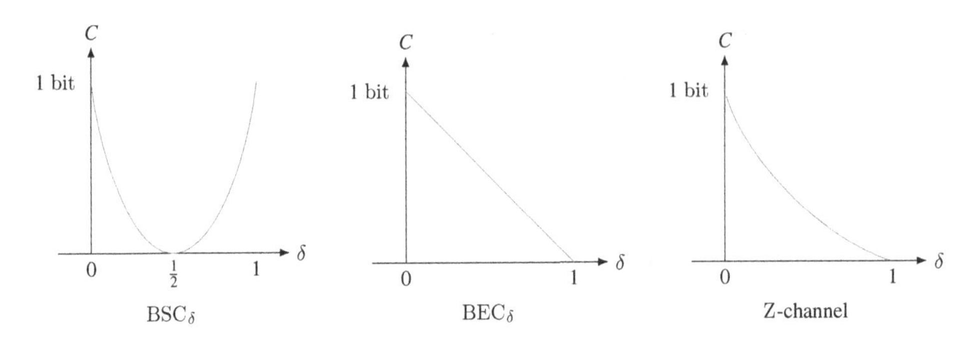

Figure 19.1 Capacities of three simple DMCs.

by $X \sim \text{Unif}(G)$. Similarly, for a group-noise channel acting over a non-abelian group G by $x \mapsto x \circ Z$, $Z \sim P_Z$ we also have capacity equal to $\log |G| - H(Z)$ and achieved by $X \sim \text{Unif}(G)$.

Next we consider the BEC_δ. This is a *multiplicative* channel. Indeed, if we equivalently redefine the input $X \in \{\pm 1\}$ and output $Y \in \{\pm 1, 0\}$, then the BEC relation can be written as

$$Y = XZ, \qquad Z \sim \text{Ber}(\delta) \perp\!\!\!\perp X.$$

To compute the capacity, we first notice that even without evaluating Shannon's formula, it is clear that $C \le 1 - \delta$ (bit), because for a large blocklength n about a fraction δ of the message is completely lost (even if the encoder knows a priori where the erasures are going to occur, the rate still cannot exceed $1 - \delta$). More formally, we notice that $\mathbb{P}[X = 1 | Y = 0] = \frac{\mathbb{P}[X=1]\delta}{\delta} = \mathbb{P}[X = 1]$ and therefore

$$I(X;Y) = H(X) - H(X|Y) = H(X) - H(X|Y = \mathsf{e}) \le (1 - \delta)H(X) \le (1 - \delta)\log 2,$$

with equality iff $X \sim \text{Ber}(1/2)$. Thus we have shown

$$C = \sup_{P_X} I(X;Y) = 1 - \delta \quad \text{bits}.$$

Finally, the Z-channel can also be thought of as a multiplicative channel with transition law

$$Y = XZ, \qquad X \in \{0, 1\} \perp\!\!\!\perp Z \sim \text{Ber}(1 - \delta),$$

so that $\mathbb{P}[Z = 0] = \delta$. For this channel if $X \sim \text{Ber}(p)$ we have

$$I(X;Y) = H(Y) - H(Y|X) = h(p(1 - \delta)) - ph(\delta).$$

Optimizing over p we get that the optimal input is given by

$$p^*(\delta) = \frac{1}{1 - \delta} \frac{1}{1 + \exp\left\{\frac{h(\delta)}{1-\delta}\right\}}.$$

The capacity-achieving input distribution $p^*(\delta)$ monotonically decreases from $\frac{1}{2}$ when $\delta = 0$ to $\frac{1}{e}$ when $\delta \to 1$. (Infamously, there is no "explanation" for this latter limiting value.) For the capacity, thus, we get

$$C = h(p^*(\delta)(1-\delta)) - p^*(\delta)h(\delta).$$

19.4* Symmetric Channels

Definition 19.12. A pair of measurable maps $f = (f_i, f_o)$ is a symmetry of $P_{Y|X}$ if

$$P_{Y|X}(f_o^{-1}(E)|f_i(x)) = P_{Y|X}(E|x),$$

for all measurable $E \subset \mathcal{Y}$ and $x \in \mathcal{X}$. Two symmetries f and g can be composed to produce another symmetry as

$$(g_i, g_o) \circ (f_i, f_o) \triangleq (g_i \circ f_i, f_o \circ g_o). \tag{19.9}$$

A symmetry group G of $P_{Y|X}$ is any collection of invertible symmetries (automorphisms) closed under the group operation (19.9).

Note that both components of an automorphism $f = (f_i, f_o)$ are bimeasurable bijections, that is f_i, f_i^{-1} and f_o, f_o^{-1} are both measurable and well-defined functions.

Naturally, every symmetry group G possesses a canonical left action on $\mathcal{X} \times \mathcal{Y}$ defined as

$$g \cdot (x, y) \triangleq (g_i(x), g_o^{-1}(y)). \tag{19.10}$$

Since the action on $\mathcal{X} \times \mathcal{Y}$ splits into actions on $\mathcal{X}$ and $\mathcal{Y}$, we will abuse notation slightly and write

$$g \cdot (x, y) \triangleq (gx, gy).$$

Let us assume in addition that our group G can be equipped with a σ-algebra $\sigma(G)$ such that the maps $h \mapsto hg$ and $h \mapsto gh$ are measurable for each $g \in G$. We say that a probability measure μ on $(G, \sigma(G))$ is a *left-invariant Haar measure* if when $H \sim \mu$ we also have $gH \sim \mu$ for any $g \in G$. (See also Exercise **V.23**.) The existence of Haar measure is trivial for finite (and compact) groups, but in general is a difficult subject. To proceed we need to make an assumption about the symmetry group G that we call regularity. (This condition is trivially satisfied whenever $\mathcal{X}$ and $\mathcal{Y}$ are finite, thus all the sophistication in these few paragraphs is only relevant to non-discrete channels.)

Definition 19.13. A symmetry group G is called *regular* if it possesses a left-invariant Haar probability measure ν such that the group action (19.10)

$$G \times \mathcal{X} \times \mathcal{Y} \to \mathcal{X} \times \mathcal{Y}$$

is measurable.

Note that under the regularity assumption the action (19.10) also defines a left action of G on $\mathcal{P}(\mathcal{X})$ and $\mathcal{P}(\mathcal{Y})$ according to

$$(gP_X)[E] \triangleq P_X[g^{-1}E], \tag{19.11}$$
$$(gQ_Y)[E] \triangleq Q_Y[g^{-1}E], \tag{19.12}$$

or, in words, if $X \sim P_X$ then $gX \sim gP_X$, and similarly for Y and gY. For every distribution P_X we define an averaged distribution $\bar{P}_X$ as

$$\bar{P}_X[E] \triangleq \int_G P_X[g^{-1}E]\nu(dg), \tag{19.13}$$

which is the distribution of random variable gX when $g \sim \nu$ and $X \sim P_X$. The measure $\bar{P}_X$ is G-invariant, in the sense that $g\bar{P}_X = \bar{P}_X$. Indeed, by the left-invariance of ν we have for every bounded function f

$$\int_G f(g)\nu(dg) = \int_G f(hg)\nu(dg) \quad \forall h \in G,$$

and therefore

$$\bar{P}_X[h^{-1}E] = \int_G P_X[(hg)^{-1}E]\nu(dg) = \bar{P}_X[E].$$

Similarly one defines $\bar{Q}_Y$:

$$\bar{Q}_Y[E] \triangleq \int_G Q_Y[g^{-1}E]\nu(dg), \tag{19.14}$$

which is also G-invariant: $g\bar{Q}_Y = \bar{Q}_Y$.

The main property of the action of G may be rephrased as follows: For arbitrary $\phi: \mathcal{X} \times \mathcal{Y} \to \mathbb{R}$ we have

$$\begin{aligned}&\int_\mathcal{X}\int_\mathcal{Y} \phi(x,y)P_{Y|X}(dy|x)(gP_X)(dx)\\ &\quad = \int_\mathcal{X}\int_\mathcal{Y} \phi(gx,gy)P_{Y|X}(dy|x)P_X(dx).\end{aligned} \tag{19.15}$$

In other words, if the pair (X,Y) is generated by taking $X \sim P_X$ and applying $P_{Y|X}$, then the pair (gX, gY) has marginal distribution gP_X but the conditional kernel is still $P_{Y|X}$. For finite $\mathcal{X}, \mathcal{Y}$ this is equivalent to

$$P_{Y|X}(gy|gx) = P_{Y|X}(y|x), \tag{19.16}$$

which may also be taken as the definition of the automorphism. In terms of the G-action on $\mathcal{P}(\mathcal{Y})$ we may also say:

$$gP_{Y|X=x} = P_{Y|X=gx} \qquad \forall g \in G, x \in \mathcal{X}. \tag{19.17}$$

It is not hard to show that for any channel and a regular group of symmetries G the capacity-achieving output distribution must be G-invariant, and the capacity-achieving input distribution can be chosen to be G-invariant. That is, the saddle-point equation

$$\inf_{P_X}\sup_{Q_Y} D(P_{Y|X}\|Q_Y|P_X) = \sup_{Q_Y}\inf_{P_X} D(P_{Y|X}\|Q_Y|P_X)$$

can be solved in the class of G-invariant distributions. Often, the action of G is transitive on $\mathcal{X}$ ($\mathcal{Y}$), in which case the capacity-achieving input (output) distribution can be taken to be uniform.

Below we systematize many popular notions of channel symmetry and explain the relationship between them.

- $P_{Y|X}$ is called input-symmetric (output-symmetric) if there exists a regular group of symmetries G acting transitively on $\mathcal{X}$ ($\mathcal{Y}$).
- An input-symmetric channel with a binary $\mathcal{X}$ is known as BMS (for binary memoryless symmetric). These channels possess a rich theory; see [361, Section 4.1] and Exercise **VI.21**.
- $P_{Y|X}$ is called weakly input-symmetric if there exists an $x_0 \in \mathcal{X}$ and a channel $T_x \colon \mathcal{B} \to \mathcal{B}$ for each $x \in \mathcal{X}$ such that $T_x \circ P_{Y|X=x_0} = P_{Y|X=x}$ and $T_x \circ P_Y^* = P_Y^*$, where P_Y^* is the capacity-achieving output distribution. In [334, Section 3.4.5] it is shown that allowing for a randomized map T_x is essential and that not all $P_{Y|X}$ are weakly input-symmetric.
- DMC $P_{Y|X}$ is a *group-noise channel* if $\mathcal{X} = \mathcal{Y}$ is a group and $P_{Y|X}$ acts by composing X with a noise variable Z:

$$Y = X \circ Z,$$

where $\circ$ is a group operation and Z is independent of X.
- DMC $P_{Y|X}$ is called *Dobrushin-symmetric* if every row of $P_{Y|X}$ is a permutation of the first one and every column of $P_{Y|X}$ is a permutation of the first one; see [131].
- DMC $P_{Y|X}$ is called *Gallager-symmetric* if the output alphabet $\mathcal{Y}$ can be split into a disjoint union of sub-alphabets such that restricted to each sub-alphabet $P_{Y|X}$ has the Dobrushin property: every row (every column) is a permutation of the first row (column); see [177, Section 4.5].
- For convenience, say that the channel is *square* if $|\mathcal{X}| = |\mathcal{Y}|$.

We demonstrate some of the relationships between these various notions of symmetry:

- Note that it is an easy consequence of the definitions that, for any input-symmetric (respectively output-symmetric) channel, all rows (respectively columns) of the channel matrix $P_{Y|X}$ are permutations of the first row (respectively column). Hence,

$$\text{input-symmetric, output-symmetric} \implies \text{Dobrushin}. \tag{19.18}$$

- Group-noise channels satisfy all other definitions of symmetry:

$$\begin{aligned} \text{group-noise} &\implies \text{square, input/output-symmetric} && (19.19) \\ &\implies \text{Dobrushin, Gallager}. && (19.20) \end{aligned}$$

- Since Gallager symmetry implies all rows are permutations of the first one, while output symmetry implies the same statement for columns, we have

$$\text{Gallager, output-symmetric} \implies \text{Dobrushin}.$$

- Clearly, not every Dobrushin-symmetric channel is square. One may wonder, however, whether every square Dobrushin channel is a group-noise channel. This is not so. Indeed, according to [391] the Latin squares that are Cayley tables are precisely the ones in which composition of two rows (as permutations) gives another row. An example of a Latin square which is not a Cayley table is the following:

$$\begin{pmatrix} 1 & 2 & 3 & 4 & 5 \\ 2 & 5 & 4 & 1 & 3 \\ 3 & 1 & 2 & 5 & 4 \\ 4 & 3 & 5 & 2 & 1 \\ 5 & 4 & 1 & 3 & 2 \end{pmatrix}. \tag{19.21}$$

Thus, by multiplying this matrix by $\frac{1}{15}$ we obtain a counterexample:

$$\text{Dobrushin, square} \not\Longrightarrow \text{group-noise.}$$

In fact, this channel is not even input-symmetric. Indeed, suppose there is $g \in G$ such that $g4 = 1$ (on $\mathcal{X}$). Then, applying (19.16) with $x = 4$ we figure out that on $\mathcal{Y}$ the action of g must be:

$$1 \mapsto 4, \quad 2 \mapsto 3, \quad 3 \mapsto 5, \quad 4 \mapsto 2, \quad 5 \mapsto 1.$$

But then we have

$$gP_{Y|X=1} = \begin{pmatrix} 5 & 4 & 2 & 1 & 3 \end{pmatrix} \cdot \frac{1}{15},$$

which by a simple inspection does not match any of the rows in (19.21). Thus, (19.17) cannot hold for $x = 1$. We conclude:

$$\text{Dobrushin, square} \not\Longrightarrow \text{input-symmetric.}$$

Similarly, if there were $g \in G$ such that $g2 = 1$ (on $\mathcal{Y}$), then on $\mathcal{X}$ it would act as

$$1 \mapsto 2, \quad 2 \mapsto 5, \quad 3 \mapsto 1, \quad 4 \mapsto 3, \quad 5 \mapsto 4,$$

which implies via (19.16) that $P_{Y|X}(g1|x)$ is not a column of (19.21). Thus:

$$\text{Dobrushin, square} \not\Longrightarrow \text{output-symmetric.}$$

- Clearly, not every input-symmetric channel is Dobrushin (e.g. BEC). One may even find a counterexample in the class of square channels:

$$\begin{pmatrix} 1 & 2 & 3 & 4 \\ 1 & 3 & 2 & 4 \\ 4 & 2 & 3 & 1 \\ 4 & 3 & 2 & 1 \end{pmatrix} \cdot \frac{1}{10}. \tag{19.22}$$

  This shows:

$$\text{input-symmetric, square} \not\Longrightarrow \text{Dobrushin.}$$

- Channel (19.22) also demonstrates:

$$\text{Gallager-symmetric, square} \not\Longrightarrow \text{Dobrushin.}$$

- Example (19.22) naturally raises the question of whether every input-symmetric channel is Gallager-symmetric. The answer is positive: by splitting $\mathcal{Y}$ into the orbits of G we see that a subchannel $\mathcal{X} \to \{\text{orbit}\}$ is input- and output-symmetric. Thus by (19.18) we have:

$$\text{input-symmetric} \Longrightarrow \text{Gallager-symmetric} \Longrightarrow \text{weakly input-symmetric.} \tag{19.23}$$

  (The second implication is evident).
- However, not all weakly input-symmetric channels are Gallager-symmetric. Indeed, consider the following channel:

$$W = \begin{pmatrix} 1/7 & 4/7 & 1/7 & 1/7 \\ 4/7 & 1/7 & 0 & 4/7 \\ 0 & 0 & 4/7 & 2/7 \\ 2/7 & 2/7 & 2/7 & 0 \end{pmatrix}. \tag{19.24}$$

  Since $\det W \neq 0$, the capacity-achieving input distribution is unique. Since $H(Y|X = x)$ is independent of x and $P_X = [1/4, 1/4, 3/8, 1/8]$ achieves uniform P_Y^* it must be the unique optimum. Clearly any permutation T_x fixes a uniform P_Y^* and thus the channel is weakly input-symmetric. At the same time it is not Gallager-symmetric since no row is a permutation of another.
- For more on the properties of weakly input-symmetric channels see [334, Section 3.4.5].

A pictorial representation of these relationships between the notions of symmetry is given schematically on Figure 19.2.

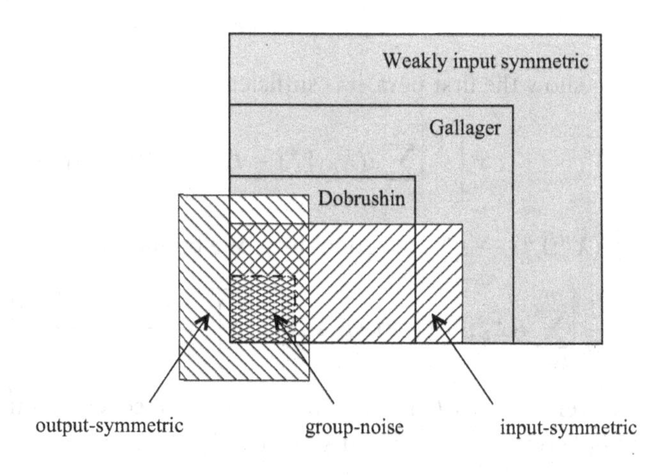

Figure 19.2 Schematic representation of inclusions of various classes of channels.

19.5* Information Stability

We saw that $C = C^{(\mathrm{I})}$ for stationary memoryless channels, but what other channels does this hold for? And what about non-stationary channels? To answer this question, we introduce the notion of *information stability*.

Definition 19.14. A channel is called *information stable* if there exists a sequence of input distributions $\{P_{X^n}, n = 1, 2, \ldots\}$ such that

$$\frac{1}{n} i(X^n;Y^n) \xrightarrow{\mathbb{P}} C^{(\mathrm{I})}.$$

For example, we can pick $P_{X^n} = (P_X^*)^n$ for stationary memoryless channels. Therefore stationary memoryless channels are information stable.

The purpose for defining information stability is the following theorem.

Theorem 19.15. *For an information stable channel, $C = C^{(\mathrm{I})}$.*

Proof. Like the stationary memoryless case, the upper bound comes from the general converse theorem, Theorem 17.3, and the lower bound uses a similar strategy as Theorem 19.8, except utilizing the definition of information stability in place of the WLLN. □

The next theorem gives conditions to check for information stability in memoryless channels which are *not* necessarily stationary.

Theorem 19.16. *A memoryless channel is information stable if there exists $\{X_k^* : k \geq 1\}$ such that both of the following hold:*

$$\frac{1}{n} \sum_{k=1}^{n} I(X_k^*;Y_k^*) \to C^{(\mathrm{I})}, \tag{19.25}$$

$$\sum_{n=1}^{\infty} \frac{1}{n^2} \mathrm{Var}[i(X_n^*;Y_n^*)] < \infty. \tag{19.26}$$

In particular, this is satisfied if

$$|\mathcal{A}| < \infty \ \ or \ \ |\mathcal{B}| < \infty. \tag{19.27}$$

Proof. To show the first part, it is sufficient to prove

$$\mathbb{P}\left[\frac{1}{n}\left|\sum_{k=1}^{n} i(X_k^*;Y_k^*) - I(X_k^*, Y_k^*)\right| > \delta\right] \to 0,$$

so that $\frac{1}{n} i(X^n;Y^n) \to C^{(\mathrm{I})}$ in probability. We bound this by Chebyshev's inequality,

$$\mathbb{P}\left[\frac{1}{n}\left|\sum_{k=1}^{n} i(X_k^*;Y_k^*) - I(X_k^*, Y_k^*)\right| > \delta\right] \leq \frac{\frac{1}{n^2}\sum_{k=1}^{n} \mathrm{Var}[i(X_k^*;Y_k^*)]}{\delta^2} \to 0,$$

where convergence to 0 follows from the Kronecker lemma (Lemma 19.17 to follow) applied with $b_n = n^2$ and $x_n = \mathrm{Var}[i(X_n^*;Y_n^*)]/n^2$.

The second part follows from the first. Indeed, notice that

$$C^{(\mathrm{I})} = \liminf_{n\to\infty} \frac{1}{n}\sum_{k=1}^{n} \sup_{P_{X_k}} I(X_k;Y_k).$$

Now select $P_{X_k^*}$ such that

$$I(X_k^*;Y_k^*) \geq \sup_{P_{X_k}} I(X_k;Y_k) - 2^{-k}.$$

(Note that each $\sup_{P_{X_k}} I(X_k;Y_k) \leq \log\min\{|\mathcal{A}|, |\mathcal{B}|\} < \infty$.) Then, we have

$$\sum_{k=1}^{n} I(X_k^*;Y_k^*) \geq \sum_{k=1}^{n} \sup_{P_{X_k}} I(X_k;Y_k) - 1,$$

and hence normalizing by n we get (19.25). We next show that for any joint distribution $P_{X,Y}$ we have

$$\mathrm{Var}[i(X;Y)] \leq 2\log^2(\min(|\mathcal{A}|, |\mathcal{B}|)). \tag{19.28}$$

The argument is symmetric in X and Y, so assume for concreteness that $|\mathcal{B}| < \infty$. Then

$$\begin{aligned}
&\mathbb{E}[i^2(X;Y)] \\
&\triangleq \int_{\mathcal{A}} dP_X(x) \sum_{y\in\mathcal{B}} P_{Y|X}(y|x)[\log^2 P_{Y|X}(y|x) + \log^2 P_Y(y) - 2\log P_{Y|X}(y|x)\cdot \log P_Y(y)] \\
&\leq \int_{\mathcal{A}} dP_X(x) \sum_{y\in\mathcal{B}} P_{Y|X}(y|x)[\log^2 P_{Y|X}(y|x) + \log^2 P_Y(y)] \qquad (19.29) \\
&= \int_{\mathcal{A}} dP_X(x) \left[\sum_{y\in\mathcal{B}} P_{Y|X}(y|x)\log^2 P_{Y|X}(y|x)\right] + \left[\sum_{y\in\mathcal{B}} P_Y(y)\log^2 P_Y(y)\right] \\
&\leq \int_{\mathcal{A}} dP_X(x) g(|\mathcal{B}|) + g(|\mathcal{B}|) \qquad (19.30) \\
&= 2g(|\mathcal{B}|),
\end{aligned}$$

where (19.29) is because $2\log P_{Y|X}(y|x) \cdot \log P_Y(y)$ is always non-negative, and (19.30) follows because each term in square brackets can be upper-bounded using the following optimization problem:

$$g(n) \triangleq \sup_{a_j\geq 0:\ \sum_{j=1}^{n} a_j = 1} \sum_{j=1}^{n} a_j \log^2 a_j. \tag{19.31}$$

Since the $x\log^2 x$ has unbounded derivative at the origin, the solution of (19.31) is always in the interior of $[0,1]^n$. Then it is straightforward to show that for $n > e$ the solution is actually $a_j = \frac{1}{n}$. For $n = 2$ it can be found directly that $g(2) = 0.5629\log^2 2 < \log^2 2$. In any case,

$$2g(|\mathcal{B}|) \leq 2\log^2 |\mathcal{B}|.$$

Finally, because of the symmetry, a similar argument can be made with $|\mathcal{B}|$ replaced by $|\mathcal{A}|$. □

Lemma 19.17. (Kronecker lemma) *If there is a sequence $0 < b_n \nearrow \infty$ and a nonnegative sequence $\{x_n\}$ such that $\sum_{n=1}^{\infty} x_n < \infty$, then*

$$\frac{1}{b_n}\sum_{j=1}^{n} b_j x_j \to 0.$$

Proof. Since b_n's are strictly increasing, we can split up the summation and bound them from above:

$$\sum_{k=1}^{n} b_k x_k \le b_m \sum_{k=1}^{m} x_k + \sum_{k=m+1}^{n} b_k x_k.$$

Now throw in the rest of the x_k's in the summation

$$\Longrightarrow \frac{1}{b_n}\sum_{k=1}^{n} b_k x_k \le \frac{b_m}{b_n}\sum_{k=1}^{\infty} x_k + \sum_{k=m+1}^{n} \frac{b_k}{b_n} x_k \le \frac{b_m}{b_n}\sum_{k=1}^{\infty} x_k + \sum_{k=m+1}^{\infty} x_k$$

$$\Longrightarrow \lim_{n\to\infty} \frac{1}{b_n}\sum_{k=1}^{n} b_k x_k \le \sum_{k=m+1}^{\infty} x_k \to 0.$$

Since this holds for any m, we can make the last term arbitrarily small. □

How to show information stability? One important class of channels with memory for which information stability can be shown easily are Gaussian channels. The complete details will be given below (see Sections 20.5* and 20.6*), but here we demonstrate a crucial fact.

For jointly Gaussian (X, Y) we always have bounded variance:

$$\mathrm{Var}[i(X;Y)] = \rho^2(X,Y)\log^2 e \le \log^2 e, \qquad \rho(X,Y) = \frac{\mathbb{E}[XY]] - \mathbb{E}[X]\mathbb{E}[Y]]}{\sqrt{\mathrm{Var}[X]\,\mathrm{Var}[Y]}}. \tag{19.32}$$

Indeed, first notice that we can always represent $Y = \tilde{X} + Z$ with $\tilde{X} = aX \perp\!\!\!\perp Z$. On the other hand, we have

$$i(\tilde{x};y) = \frac{\log e}{2}\left[\frac{\tilde{x}^2 + 2\tilde{x}z}{\sigma_Y^2} - \frac{\sigma^2}{\sigma_Y^2\sigma_Z^2}z^2\right], \qquad z \triangleq y - \tilde{x}.$$

From here by using $\mathrm{Var}[\cdot] = \mathrm{Var}[\mathbb{E}[\cdot|\tilde{X}]] + \mathrm{Var}[\cdot|\tilde{X}]$ we need to compute two terms separately:

$$\mathbb{E}[i(\tilde{X};Y)|\tilde{X}] = \frac{\log e}{2}\left[\frac{\tilde{X}^2 - \frac{\sigma_{\tilde{X}}^2}{\sigma_Z^2}}{\sigma_Y^2}\right],$$

and hence

$$\mathrm{Var}[\mathbb{E}[i(\tilde{X};Y)|\tilde{X}]] = \frac{2\log^2 e}{4\sigma_Y^4}\sigma_{\tilde{X}}^4.$$

On the other hand,

$$\mathrm{Var}[i(\tilde{X};Y)|\tilde{X}] = \frac{2\log^2 e}{4\sigma_Y^4}[4\sigma_{\tilde{X}}^2\sigma_Z^2 + 2\sigma_{\tilde{X}}^4].$$

Putting it all together we get (19.32). Inequality (19.32) justifies information stability of all sorts of Gaussian channels (memoryless and with memory), as we will see shortly.

19.6 Capacity under Bit Error Rate

In most cases of interest the space $[M]$ of messages can be given additional structure by identifying $[M] = \{0,1\}^k$, which is, of course, only possible for $M = 2^k$. In these cases, in addition to P_e and $P_{e,\max}$ every code (f, g) has another important figure of merit – the so-called *bit error rate* (BER), denoted as P_b and defined in Section 6.4:

$$P_b \triangleq \frac{1}{k}\sum_{j=1}^{k}\mathbb{P}[S_j \neq \hat{S}_j] = \frac{1}{k}\mathbb{E}[d_{\mathrm{H}}(S^k, \hat{S}^k)], \tag{19.33}$$

where we represented W and $\hat{W}$ as k-tuples: $W = (S_1, \ldots, S_k)$, $\hat{W} = (\hat{S}_1, \ldots, \hat{S}_k)$, and d_{H} denotes the Hamming distance (6.6). In words, P_b is the average fraction of errors in a decoded k-bit block.

In addition to constructing codes minimizing block error probability P_e or $P_{e,\max}$ as we studied above, the problem of minimizing the BER P_b is also practically relevant. Here, we discuss some simple facts about this setting. In particular, we will see that the capacity value for memoryless channels does not increase even if one insists only on a vanishing P_b – a much weaker criterion compared to vanishing P_e.

First, we give a bound on the average probability of error (block error rate) in terms of the bit error rate.

Theorem 19.18. *For all* (f, g), $M = 2^k \implies P_b \le P_e \le kP_b$.

Proof. Recall that $M = 2^k$ gives us the interpretation of $W = S^k$ sequence of bits,

$$\frac{1}{k}\sum_{i=1}^{k} 1\{S_i \neq \hat{S}_i\} \le 1\{S^k \neq \hat{S}^k\} \le \sum_{i=1}^{k} 1\{S_i \neq \hat{S}_i\},$$

where the first inequality is obvious and the second follows from the union bound. Taking expectation of the above expression gives the theorem. □

Next, the following pair of results is often useful for lower-bounding P_b for some specific codes.

Theorem 19.19. (Assouad's lemma) *If* $M = 2^k$ *then*

$$P_b \geq \min\{\mathbb{P}[\hat{W} = c'|W = c]\colon c, c' \in \{0,1\}^k, d_H(c,c') = 1\}.$$

Proof. Let e_i be a length-k vector that is 1 in the ith position, and 0 everywhere else. Then

$$\sum_{i=1}^{k} 1\{S_i \neq \hat{S}_i\} \geq \sum_{i=1}^{k} 1\{S^k = \hat{S}^k + e_i\}.$$

Dividing by k and taking expectation gives

$$\begin{aligned} P_b &\geq \frac{1}{k}\sum_{i=1}^{k} \mathbb{P}[S^k = \hat{S}^k + e_i] \\ &\geq \min\{\mathbb{P}[\hat{W} = c'|W = c]\colon c, c' \in \{0,1\}^k, d_H(c,c') = 1\}. \end{aligned}$$

□

Similarly, we can prove the following generalization:

Theorem 19.20. *If* $A, B \in \{0,1\}^k$ *(with arbitrary marginals!) then for every* $r \geq 1$ *we have*

$$P_b = \frac{1}{k}\mathbb{E}[d_H(A,B)] \geq \binom{k-1}{r-1} P_{r,\min}, \tag{19.34}$$

$$P_{r,\min} \triangleq \min\{\mathbb{P}[B = c'|A = c]\colon c, c' \in \{0,1\}^k, d_H(c,c') = r\}. \tag{19.35}$$

Proof. First, observe that

$$\mathbb{P}[d_H(A,B) = r|A = a] = \sum_{b\colon d_H(a,b)=r} P_{B|A}(b|a) \geq \binom{k}{r} P_{r,\min}.$$

Next, notice

$$d_H(x,y) \geq r1\{d_H(x,y) = r\}$$

and take the expectation with $x \sim A$, $y \sim B$. □

In statistics, Assouad's lemma is a useful tool for obtaining lower bounds on the minimax risk of an estimator (Section 31.2).

The following is a converse bound for channel coding under BER constraint.

Theorem 19.21. (Converse under BER) *Any* M*-code with* $M = 2^k$ *and bit error rate* P_b *satisfies*

$$\log M \leq \frac{\sup_{P_X} I(X;Y)}{\log 2 - h(P_b)}.$$

Proof. Note that $S^k \to X \to Y \to \hat{S}^k$, where $S^k \overset{\text{iid}}{\sim} \text{Ber}(1/2)$. Recall from Theorem 6.1 that for iid S^n, $\sum I(S_i;\hat{S}_i) \le I(S^k;\hat{S}^k)$. This gives us

$$\begin{aligned}
\sup_{P_X} I(X;Y) \ge I(X;Y) &\ge \sum_{i=1}^{k} I(S_i;\hat{S}_i) \\
&\ge k\frac{1}{k}\sum d\left(\mathbb{P}[S_i = \hat{S}_i]\Big\|\frac{1}{2}\right) \\
&\ge kd\left(\frac{1}{k}\sum_{i=1}^{k}\mathbb{P}[S_i = \hat{S}_i]\Big\|\frac{1}{2}\right) \\
&= kd\left(1 - P_b\Big\|\frac{1}{2}\right) = k(\log 2 - h(P_b)),
\end{aligned}$$

where the second line used Fano's inequality (Theorem 3.12) for binary random variables (or data-processing inequality for divergence), and the third line used the convexity of divergence. One should note that this last chain of inequalities is similar to the proof of Proposition 6.6. □

Pairing this bound with Proposition 19.10 shows that any sequence of codes with $P_b \to 0$ (for a memoryless channel) must have rate $R < C$. In other words, relaxing the constraint from P_e to P_b does not yield any higher rates.

Later in Section 26.3 we will see that channel coding under BER constraint is a special case of a more general paradigm known as lossy joint source channel coding so that Theorem 19.21 follows from Theorem 26.5. Furthermore, this converse bound is in fact achievable asymptotically for stationary memoryless channels.

19.7 Joint Source–Channel Coding

Now we will examine a slightly different data transmission scenario called *joint source–channel coding* (JSCC):

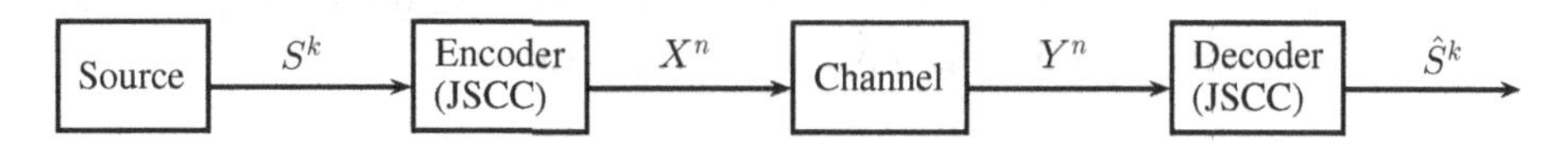

Formally, a JSCC code consists of an encoder $f\colon \mathcal{A}^k \to \mathcal{X}^n$ and a decoder $g\colon \mathcal{Y}^n \to \mathcal{A}^k$. The goal is to maximize the transmission rate $R = \frac{k}{n}$ (symbol per channel use) while ensuring the probability of error $\mathbb{P}[S^k \ne \hat{S}^k]$ is small. The fundamental limit (optimal probability of error) is defined as

$$\epsilon^*_{\text{JSCC}}(k,n) = \inf_{f,g} \mathbb{P}[S^k \ne \hat{S}^k].$$

In channel coding we are interested in transmitting M messages and all messages are born equal. Here we want to convey the source realizations which might not be equiprobable (have redundancy). Indeed, if S^k is uniformly distributed on, say,

$\{0,1\}^k$, then we are back to the channel coding setup with $M = 2^k$ under average probability of error, and $\epsilon^*_{\mathrm{JSCC}}(k,n)$ coincides with $\epsilon^*(n, 2^k)$ defined in Section 22.1.

Here, we look for a clever scheme to directly encode k symbols from $\mathcal{A}$ into a length-n channel input such that we achieve a small probability of error over the channel. This feels like a mix of two problems we have seen: compressing a source and coding over a channel. The following theorem shows that compressing and channel coding separately is optimal. This is a relief, since it implies that we do not need to develop any new theory or architectures to solve the joint source–channel coding problem. As far as the leading term in the asymptotics is concerned, the following two-stage scheme is optimal: First use the optimal compressor to eliminate all the redundancy in the source, then use the optimal channel code to add redundancy to combat the noise in the data transmission.

The result is known as a *separation theorem* since it separates the jobs of compressor and channel coder, with the two blocks interfacing in terms of bits. Note that the source can generate symbols over a very different alphabet than the channel's input alphabet. Nevertheless, the bit stream produced by the source code (compressor) is matched to the channel by the channel code. There is an even more general version of this result (Section 26.3).

Theorem 19.22. (Shannon's separation theorem) *Let the source $\{S_k\}$ be stationary memoryless on a finite alphabet with entropy H. Let the channel be stationary memoryless with finite capacity C. Then*

$$\epsilon^*_{\mathrm{JSCC}}(nR, n) \begin{cases} \to 0, & R < C/H, \\ \not\to 0, & R > C/H, \end{cases} \qquad n \to \infty.$$

The interpretation of this result is as follows: Each source symbol has information content (entropy) H bits. Each channel use can convey C bits. Therefore to reliably transmit k symbols over n channel uses, we need $kH \le nC$.

Proof. (Achievability) The idea is to separately compress our source and code it for transmission. Since this is a feasible way to solve the JSCC problem, it gives an achievability bound. This separated architecture is

$$S^k \xrightarrow{f_1} W \xrightarrow{f_2} X^n \xrightarrow{P_{Y^n|X^n}} Y^n \xrightarrow{g_2} \hat{W} \xrightarrow{g_1} \hat{S}^k,$$

where we use the optimal compressor (f_1, g_1) and optimal channel code (*maximum* probability of error) (f_2, g_2). Let W denote the output of the compressor which takes at most M_k values. Then from Corollary 11.3 and Theorem 19.9 we get:

$$\text{(from optimal compressor)} \quad \frac{1}{k}\log M_k > H + \delta \implies \mathbb{P}[\hat{S}^k \ne S^k(W)] \le \epsilon \quad \forall k \ge k_0,$$

$$\text{(from optimal channel code)} \quad \frac{1}{n}\log M_k < C - \delta \implies \mathbb{P}[\hat{W} \ne m | W = m] \le \epsilon$$

$$\forall m, \forall k \ge k_0.$$

Using both of these,

$$\mathbb{P}[S^k \neq \hat{S}^k(\hat{W})] \leq \mathbb{P}[S^k \neq \hat{S}^k, W = \hat{W}] + \mathbb{P}[W \neq \hat{W}]$$
$$\leq \mathbb{P}[S^k \neq \hat{S}^k(W)] + \mathbb{P}[W \neq \hat{W}] \leq \epsilon + \epsilon.$$

And therefore if $R(H+\delta) < C - \delta$, then $\epsilon^* \to 0$. By the arbitrariness of $\delta > 0$, we conclude the weak converse for any $R > C/H$.

(Converse) To prove the converse notice that any JSCC encoder/decoder induces a Markov chain

$$S^k \to X^n \to Y^n \to \hat{S}^k.$$

Applying data processing for mutual information:

$$I(S^k;\hat{S}^k) \leq I(X^n;Y^n) \leq \sup_{P_{X^n}} I(X^n;Y^n) = nC.$$

On the other hand, since $\mathbb{P}[S^k \neq \hat{S}^k] \leq \epsilon_n$, Fano's inequality (Theorem 3.12) yields

$$I(S^k;\hat{S}^k) = H(S^k) - H(S^k|\hat{S}^k) \geq kH - \epsilon_n \log|\mathcal{A}|^k - \log 2.$$

Combining the two gives

$$nC \geq kH - \epsilon_n \log|\mathcal{A}|^k - \log 2. \tag{19.36}$$

Since $R = \frac{k}{n}$, dividing both sides by n and sending $n \to \infty$ yields

$$\liminf_{n\to\infty} \epsilon_n \geq \frac{RH - C}{R \log|\mathcal{A}|}.$$

Therefore ϵ_n does not vanish if $R > C/H$. □

We remark that instead of using Fano's inequality we could have lower-bounded $I(S^k;\hat{S}^k)$ as in the proof of Theorem 17.3 by defining $Q_{S^k\hat{S}^k} = U_{S^k}P_{\hat{S}^k}$ (with $U_{S^k} = \mathrm{Unif}(\{0,1\}^k)$ and applying the data-processing inequality to the map $(S^k, \hat{S}^k) \mapsto 1\{S^k = \hat{S}^k\}$:

$$D(P_{S^k\hat{S}^k} \| Q_{S^k\hat{S}^k}) = D(P_{S^k} \| U_{S^k}) + D(P_{\hat{S}|S^k} \| P_{\hat{S}}|P_{S^k}) \geq d(1-\epsilon_n \| |\mathcal{A}|^{-k}).$$

Rearranging terms yields (19.36). As we discussed in Remark 17.2, replacing D with other f-divergences can be very fruitful.

In a very similar manner, by invoking Corollary 12.6 and Theorem 19.15 we obtain:

Theorem 19.23. *Let source $\{S_k\}$ be ergodic on a finite alphabet, and have entropy rate H. Let the channel have capacity C and be information stable. Then*

$$\lim_{n\to\infty} \epsilon^*_{\mathrm{JSCC}}(nR, n) \begin{cases} = 0, & R > H/C, \\ > 0, & R < H/C. \end{cases}$$

We leave the proof as an exercise for the reader.

20 Channels with Input Constraints and Gaussian Channels

In this chapter we study data transmission with constraints on the channel input. Namely, in the previous chapter the encoder for a blocklength-n code was permitted to produce arbitrary sequences of channel inputs (i.e. the range of the encoder could be all of $\mathcal{A}^n$). However, in many practical problems only a subset of $\mathcal{A}^n$ is allowed to be used. The main such example is the AWGN channel in Example 3.3. If the encoder is allowed to produce arbitrary elements of $\mathbb{R}^n$ as input, the channel capacity is infinite: $\sup_{P_X} I(X;X+Z) = \infty$ (for example, take $X \sim \mathcal{N}(0,P)$ and $P \to \infty$). That is, one can transmit arbitrarily many messages with arbitrarily small error probability by choosing elements of $\mathbb{R}^n$ with giant pairwise distance. In reality, however, allowed channel inputs are limited by the available[1] power and the encoder is only capable of using $x^n \in \mathbb{R}^n$ satisfying

$$\frac{1}{n}\sum_{t=1}^{n} x_t^2 \le P,$$

where $P > 0$ is known as the power constraint. How many bits per channel use can we transmit under this constraint on the codewords? To answer this question in general, we need to extend the setup and coding theorems to *channels with input constraints.* After doing that we will apply these results to compute capacities of various Gaussian channels (memoryless, with intersymbol interference and subject to fading).

20.1 Channel Coding with Input Constraints

We say that an (n, M, ϵ)-code satisfies the input constraint $F_n \subset \mathcal{A}^n$ if the encoder maps $[M]$ into F_n, that is, $f\colon [M] \to F_n$, as illustrated by the following diagram.

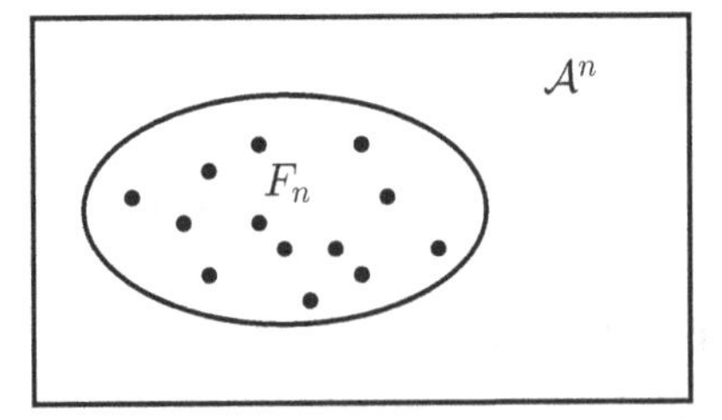

Codewords all land in a subset of $\mathcal{A}^n$

[1] Or allowed by regulatory bodies, such as the FCC in the US.

What type of constraint sets F_n are of practical interest? In the context of Gaussian channels, we have $\mathcal{A} = \mathbb{R}$. Then one often talks about the following constraints:

- Average power constraint:

$$\frac{1}{n}\sum_{i=1}^{n} |x_i|^2 \le P \quad \Leftrightarrow \quad \|x^n\|_2 \le \sqrt{nP}.$$

  In other words, codewords must lie in a ball of radius $\sqrt{nP}$.
- Peak power constraint:

$$\max_{1\le i\le n} |x_i| \le A \quad \Leftrightarrow \quad \|x^n\|_\infty \le A.$$

Notice that the second type of constraint does not introduce any new problems: we can simply restrict the input space from $\mathcal{A} = \mathbb{R}$ to $\mathcal{A} = [-A, A]$ and be back into the setting of input-unconstrained coding. The first type of constraint is known as a *separable cost constraint*. We will restrict our attention from now on to it exclusively.

Definition 20.1. A channel with a *separable* cost constraint is specified by

- input space $\mathcal{A}$ and output space $\mathcal{B}$;
- a sequence of Markov kernels $P_{Y^n|X^n}\colon \mathcal{A}^n \to \mathcal{B}^n$, indexed by the blocklength $n = 1, 2, \ldots$;
- (per-letter) cost function $\mathsf{c}\colon \mathcal{A} \to \mathbb{R} \cup \{\pm\infty\}$.

We extend the per-letter cost to n-sequences as follows:

$$\mathsf{c}(x^n) \triangleq \frac{1}{n}\sum_{k=1}^{n} \mathsf{c}(x_k).$$

We next extend the channel coding notions to such channels.

Definition 20.2.

- A code is an (n, M, ϵ, P)-code if it is an (n, M, ϵ)-code satisfying input constraint $F_n \triangleq \{x^n\colon \frac{1}{n}\sum_{k=1}^{n} \mathsf{c}(x_k) \le P\}$.
- Finite-n fundamental limits:

$$M^*(n, \epsilon, P) = \max\{M\colon \exists (n, M, \epsilon, P)\text{-code}\},$$
$$M^*_{\max}(n, \epsilon, P) = \max\{M\colon \exists (n, M, \epsilon, P)_{\max}\text{-code}\}.$$

- ϵ-capacity and Shannon capacity:

$$C_\epsilon(P) = \liminf_{n\to\infty} \frac{1}{n} \log M^*(n, \epsilon, P),$$
$$C(P) = \lim_{\epsilon\downarrow 0} C_\epsilon(P).$$

- Information capacity:

$$C^{(I)}(P) = \liminf_{n\to\infty} \frac{1}{n} \sup_{P_{X^n}:\, \mathbb{E}[\sum_{k=1}^n \mathsf{c}(X_k)]\le nP} I(X^n;Y^n).$$

- Information stability: A channel is information stable if for all (admissible) P, there exists a sequence of channel input distributions P_{X^n} such that the following two properties hold:

$$\frac{1}{n} i_{P_{X^n,Y^n}}(X^n;Y^n) \xrightarrow{\mathbb{P}} C^{(I)}(P), \tag{20.1}$$

$$\mathbb{P}[\mathsf{c}(X^n) > P + \delta] \to 0 \quad \forall \delta > 0. \tag{20.2}$$

These definitions clearly parallel those of Definitions 19.3 and 19.6 for channels without input constraints. A notable and crucial exception is the definition of the information capacity $C^{(I)}(P)$. Indeed, under input constraints instead of maximizing $I(X^n;Y^n)$ over distributions supported on F_n we extend maximization to a richer set of distributions, namely, those satisfying

$$\mathbb{E}\left[\sum_{k=1}^n \mathsf{c}(X_k)\right] \le nP.$$

This will soon be crucial for the single-letterization of $C^{(I)}(P)$.

Definition 20.3. (Admissible constraint) We say P is an admissible constraint if $\exists x_0 \in \mathcal{A}$ s.t. $\mathsf{c}(x_0) \le P$, or equivalently, $\exists P_X\colon \mathbb{E}[\mathsf{c}(X)] \le P$. The set of admissible P's is denoted by $\mathcal{D}_\mathsf{c}$, and can be in the form of either (P_0, ∞) or $[P_0, \infty)$, where $P_0 \triangleq \inf_{x\in\mathcal{A}} \mathsf{c}(x)$.

Clearly, if $P \notin \mathcal{D}_\mathsf{c}$, then there is no code (even a useless one, with one codeword) satisfying the input constraint. So from here onward we always assume $P \in \mathcal{D}_\mathsf{c}$.

Proposition 20.4. (Capacity-cost function) *Define the capacity-cost function* $\phi(P) \triangleq \sup_{P_X\colon \mathbb{E}[\mathsf{c}(X)]\le P} I(X;Y)$. *Then the following hold.*

1 *ϕ is concave and non-decreasing. The domain of ϕ is* $\operatorname{dom}\phi \triangleq \{x\colon f(x) > -\infty\} = \mathcal{D}_\mathsf{c}$.
2 *One of the following is true: $\phi(P)$ is continuous and finite on (P_0, ∞), or $\phi = \infty$ on (P_0, ∞).*

Both of these properties extend to the function $P \mapsto C^{(I)}(P)$.

Proof. In the first part all statements are obvious, except for concavity, which follows from the concavity of $P_X \mapsto I(X;Y)$. For any P_{X_i} such that $\mathbb{E}[\mathsf{c}(X_i)] \le P_i, i = 0, 1$, let $X \sim \bar\lambda P_{X_0} + \lambda P_{X_1}$. Then $\mathbb{E}[\mathsf{c}(X)] \le \bar\lambda P_0 + \lambda P_1$ and $I(X;Y) \ge \bar\lambda I(X_0;Y_0) + \lambda I(X_1;Y_1)$. Hence $\phi(\bar\lambda P_0 + \lambda P_1) \ge \bar\lambda\phi(P_0) + \lambda\phi(P_1)$. The second claim follows from the concavity of $\phi(\cdot)$.

To extend these results to $C^{(\mathrm{I})}(P)$ observe that for every n

$$P \mapsto \frac{1}{n} \sup_{P_{X^n}:\ \mathbb{E}[\mathsf{c}(X^n)] \le P} I(X^n;Y^n)$$

is concave. Then taking $\liminf_{n\to\infty}$ the same holds for $C^{(\mathrm{I})}(P)$. □

An immediate consequence is that a memoryless input is optimal for a memoryless channel with separable cost, which gives us the single-letter formula of the information capacity:

Corollary 20.5. (Single-letterization) *The information capacity of a stationary memoryless channel with separable cost is given by*

$$C^{(\mathrm{I})}(P) = \phi(P) = \sup_{\mathbb{E}[\mathsf{c}(X)] \le P} I(X;Y).$$

Proof. $C^{(\mathrm{I})}(P) \ge \phi(P)$ is obvious by using $P_{X^n} = (P_X)^{\otimes n}$. For "$\le$," fix any P_{X^n} satisfying the cost constraint. Consider the chain

$$I(X^n;Y^n) \overset{\text{(a)}}{\le} \sum_{j=1}^{n} I(X_j;Y_j) \overset{\text{(b)}}{\le} \sum_{j=1}^{n} \phi(\mathbb{E}[\mathsf{c}(X_j)]) \overset{\text{(c)}}{\le} n\phi\left(\frac{1}{n}\sum_{j=1}^{n} \mathbb{E}[\mathsf{c}(X_j)]\right) \le n\phi(P),$$

where (a) follows from Theorem 6.1; (b) from the definition of ϕ; and (c) from Jensen's inequality and the concavity of ϕ. □

20.2 Channel Capacity under Separable Cost Constraints

We start with a straightforward extension of the weak converse to the case of input constraints.

Theorem 20.6. (Weak converse)

$$C_\epsilon(P) \le \frac{C^{(\mathrm{I})}(P)}{1-\epsilon}.$$

Proof. The argument is the same as we used in Theorem 17.3. Take any (n, M, ϵ, P)-code, $W \to X^n \to Y^n \to \hat{W}$. Applying Fano's inequality and the data-processing inequality, we get

$$-h(\epsilon) + (1-\epsilon)\log M \le I(W;\hat{W}) \le I(X^n;Y^n) \le \sup_{P_{X^n}:\ \mathbb{E}[\mathsf{c}(X^n)] \le P} I(X^n;Y^n).$$

Normalizing both sides by n and taking $\liminf_{n\to\infty}$ we obtain the result. □

Next we need to extend one of the coding theorems to the case of input constraints. We do so for Feinstein's lemma (Theorem 18.7). Note that when $F = \mathcal{X}$, it reduces to the original version.

Theorem 20.7. (Extended Feinstein's lemma) *Fix a Markov kernel $P_{Y|X}$ and an arbitrary P_X. Then for any measurable subset $F \subset \mathcal{X}$, every $\gamma > 0$ and any integer $M \ge 1$, there exists an $(M, \epsilon)_{\max}$-code such that*

- *the encoder satisfies the input constraint* $f\colon [M] \to F \subset \mathcal{X}$;
- *the probability of error bound is*

$$\epsilon P_X(F) \le \mathbb{P}[i(X;Y) < \log \gamma] + \frac{M}{\gamma}.$$

Proof. Similar to the proof of the original Feinstein's lemma, define the preliminary decoding regions $E_c = \{y\colon i(c;y) \ge \log \gamma\}$ for all $c \in \mathcal{X}$. Next, we apply Corollary 18.4 and find out that there is a set $F_0 \subset \mathcal{X}$ with two properties: (a) $P_X[F_0] = 1$ and (b) for every $x \in F_0$ we have $P_Y(E_x) \le \frac{1}{\gamma}$. We now let $F' = F \cap F_0$ and notice that $P_X[F'] = P_X[F]$.

We sequentially pick codewords $\{c_1, \ldots, c_M\}$ from the set F' (!) and define the decoding regions $\{D_1, \ldots, D_M\}$ as $D_j \triangleq E_{c_j} \backslash \bigcup_{k=1}^{j-1} D_k$. The stopping criterion is that M is maximal, that is,

$$\forall x_0 \in F', \quad P_Y\left[E_{x_0} \backslash \bigcup_{j=1}^{M} D_j \middle| X = x_0\right] < 1 - \epsilon,$$

$$\Leftrightarrow \forall x_0 \in \mathcal{X}, \quad P_Y\left[E_{x_0} \backslash \bigcup_{j=1}^{M} D_j \middle| X = x_0\right] < (1-\epsilon)1[x_0 \in F'] + 1[x_0 \in F'^c].$$

Now average the last inequality over $x_0 \sim P_X$ to obtain

$$\mathbb{P}\left[\{i(X;Y) \ge \log \gamma\} \backslash \bigcup_{j=1}^{M} D_j\right] \le (1-\epsilon)P_X[F'] + P_X[F'^c] = 1 - \epsilon P_X[F].$$

From here, we complete the proof by following the same steps as in the proof of the original Feinstein's lemma (Theorem 18.7). □

Given the coding theorem we can establish a lower bound on capacity.

Theorem 20.8. (Achievability bound) *For any information stable channel with input constraints and* $P > P_0$ *we have*

$$C(P) \ge C^{(\mathrm{I})}(P). \tag{20.3}$$

Proof. Let us consider a special case of the stationary memoryless channel (the proof for a general information stable channel follows similarly). Thus, we assume $P_{Y^n|X^n} = (P_{Y|X})^{\otimes n}$.

Fix $n \ge 1$. Choose a P_X such that $\mathbb{E}[\mathsf{c}(X)] < P$. Pick $\log M = n(I(X;Y) - 2\delta)$ and $\log \gamma = n(I(X;Y) - \delta)$.

With the input constraint set $F_n = \{x^n\colon \frac{1}{n}\sum \mathsf{c}(x_k) \le P\}$, and iid input distribution $P_{X^n} = P_X^{\otimes n}$, we apply the extended Feinstein's lemma. This shows the existence of an $(n, M, \epsilon_n, P)_{\max}$-code with the encoder satisfying input constraint F_n and vanishing (maximal) error probability:

$$\epsilon_n \underbrace{P_{X^n}[F_n]}_{\to 1} \le \underbrace{\mathbb{P}[i(X^n;Y^n) \le n(I(X;Y) - \delta)]}_{\substack{\to 0 \text{ as } n\to\infty \text{ by WLLN and} \\ \text{stationary memoryless assumption}}} + \underbrace{\exp(-n\delta)}_{\to 0}.$$

Indeed, the first term is vanishing by the weak law of large numbers: since $\mathbb{E}[\mathsf{c}(X)] < P$, we have $P_{X^n}(F_n) = \mathbb{P}\left[\frac{1}{n}\sum \mathsf{c}(x_k) \leq P\right] \to 1$. Since $\epsilon_n \to 0$ this implies that for every $\epsilon > 0$ we have

$$C_\epsilon(P) \geq \frac{1}{n}\log M = I(X;Y) - 2\delta, \quad \forall \delta > 0, \forall P_X \text{ s.t. } \mathbb{E}[\mathsf{c}(X)] < P$$
$$\Rightarrow C_\epsilon(P) \geq \sup_{P_X:\, \mathbb{E}[\mathsf{c}(X)]<P} \lim_{\delta\to 0}(I(X;Y) - 2\delta)$$
$$\Rightarrow C_\epsilon(P) \geq \sup_{P_X:\, \mathbb{E}[\mathsf{c}(X)]<P} I(X;Y) = C^{(\mathrm{I})}(P-) = C^{(\mathrm{I})}(P),$$

where the last equality is from the continuity of $C^{(\mathrm{I})}$ on (P_0, ∞) by Proposition 20.4.

For a general information stable channel, we just need to use the definition to show that $P[i(X^n;Y^n) \leq n(C^{(\mathrm{I})} - \delta)] \to 0$, and the rest of the proof follows similarly. □

Theorem 20.9. (Channel capacity under cost constraint) *For an information stable channel with cost constraint and for any admissible constraint P we have*

$$C(P) = C^{(\mathrm{I})}(P).$$

Proof. The boundary case of $P = P_0$ is treated in Exercise **IV.23**, which shows that $C(P_0) = C^{(\mathrm{I})}(P_0)$ even though $C^{(\mathrm{I})}(P)$ may be discontinuous at P_0. So assume $P > P_0$ next. Theorem 20.6 shows $C_\epsilon(P) \leq \frac{C^{(\mathrm{I})}(P)}{1-\epsilon}$, thus $C(P) \leq C^{(\mathrm{I})}(P)$. On the other hand, from Theorem 20.8 we have $C(P) \geq C^{(\mathrm{I})}(P)$. □

20.3 Stationary AWGN Channel

We start our applications with perhaps the most important channel (from the point of view of communication engineering).

$Z \sim \mathcal{N}(0, \sigma^2)$

$X \longrightarrow \oplus \longrightarrow Y$

Definition 20.10. (The stationary AWGN channel) The additive white Gaussian noise (AWGN) channel is a stationary memoryless additive-noise channel with separable cost constraint: $\mathcal{A} = \mathcal{B} = \mathbb{R}$, $\mathsf{c}(x) = x^2$, and a single-letter kernel $P_{Y|X}$ given by $Y = X + Z$, as illustrated above, where $Z \sim \mathcal{N}(0, \sigma^2) \perp\!\!\!\perp X$. The n-letter kernel is given by a product extension, that is, $Y^n = X^n + Z^n$ with $Z^n \sim \mathcal{N}(0, I_n)$. When the power constraint is $\mathbb{E}[\mathsf{c}(X)] \leq P$ we say that the signal-to-noise ratio (SNR) equals $\frac{P}{\sigma^2}$. Note that our informal definition early on (Example 3.3) lacked specification of the cost constraint function, without which it was not complete.

The terminology white noise refers to the fact that the noise variables are uncorrelated across time. This makes the power spectral density of the process $\{Z_j\}$ constant in frequency (or "white"). We often drop the word "stationary" when referring to this channel. The definition we gave above more correctly should be called the *real* AWGN or $\mathbb{R}$-AWGN channel. The complex AWGN or $\mathbb{C}$-AWGN channel is defined similarly: $\mathcal{A} = \mathcal{B} = \mathbb{C}$, $\mathsf{c}(x) = |x|^2$, and $Y^n = X^n + Z^n$, with $Z^n \sim \mathcal{N}_c(0, I_n)$ being the circularly symmetric complex Gaussian.

Theorem 20.11. *For the stationary AWGN channel, the channel capacity is equal to the information capacity, and is given by:*

$$C(P) = C^{(\mathrm{I})}(P) = \frac{1}{2}\log\left(1 + \frac{P}{\sigma^2}\right) \quad \text{for } \mathbb{R}\text{-AWGN}, \tag{20.4}$$

$$C(P) = C^{(\mathrm{I})}(P) = \log\left(1 + \frac{P}{\sigma^2}\right) \quad \text{for } \mathbb{C}\text{-AWGN}.$$

Proof. By Corollary 20.5,

$$C^{(\mathrm{I})} = \sup_{P_X : \mathbb{E}X^2 \le P} I(X; X + Z).$$

Then use Theorem 5.11 (the Gaussian saddle point) to conclude $X \sim \mathcal{N}(0, P)$ (or $\mathcal{N}_c(0, P)$) is the unique capacity-achieving input distribution. □

At this point it is also instructive to revisit Section 6.2* which shows that the Gaussian capacity can in fact be derived essentially without solving the maximization of mutual information: the Euclidean rotational symmetry implies the optimal input should be Gaussian.

There is a great deal of deep knowledge embedded in the simple-looking formula of Shannon (20.4). First, from the engineering point of view we immediately see that to transmit information faster (per unit time) one needs to pay with radiating at higher power, but the payoff in transmission speed is only logarithmic. Furthermore, we can infer that The waveforms of good error-correcting codes should resemble samples of the white Gaussian process.

Second, the amount of energy spent per transmitted information bit is minimized by solving

$$\inf_{P>0} \frac{P \log 2}{C(P)} = 2\sigma^2 \log_e 2 \tag{20.5}$$

and is achieved by taking $P \to 0$. (We will discuss the notion of energy-per-bit more in Section 21.1.) Thus, we see that in order to maximize communication *rate* we need to send loud high-power waveforms. But in order to minimize energy-per-bit we need to send a very quiet "whisper" and at very low communication rate.[2] In addition, when signaling with very low power (and hence low rate), by inspecting Figure 3.2 we can see that one can restrict to just binary $\pm\sqrt{P}$ symbols (so-called binary phase shift keying (BPSK) modulation). This results in virtually no loss of capacity.

[2] This explains why, for example, deep space probes communicate with Earth via very low-rate codes and very long blocklengths.

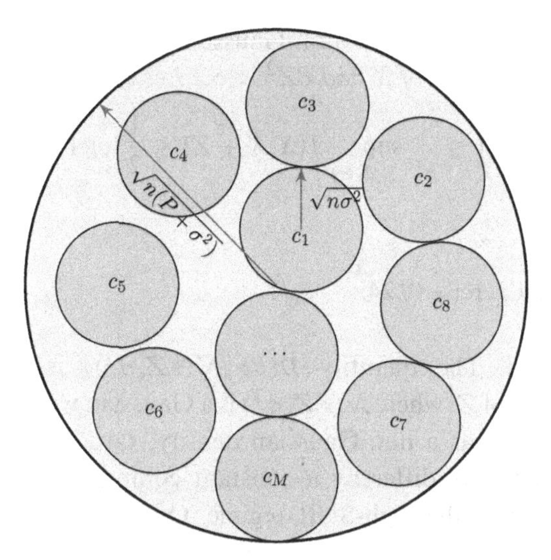

Figure 20.1 Interpretation of the AWGN capacity formula as "soft" packing.

Third, from the mathematical point of view, formula (20.4) reveals certain properties of high-dimensional Euclidean geometry as follows. Since $Z^n \sim \mathcal{N}(0, \sigma^2)$, then with high probability, $\|Z^n\|_2$ concentrates around $\sqrt{n\sigma^2}$. Similarly, due to the power constraint and the fact that $Z^n \perp\!\!\!\perp X^n$, we have $\mathbb{E}\left[\|Y^n\|^2\right] = \mathbb{E}\left[\|Y^n\|^2\right] + \mathbb{E}\left[\|Z^n\|^2\right] \le n(P + \sigma^2)$ and the received vector Y^n lies in an ℓ_2-ball of radius approximately $\sqrt{n(P+\sigma^2)}$. Since the noise can at most perturb the codeword by $\sqrt{n\sigma^2}$ in Euclidean distance, if we can pack M balls of radius $\sqrt{n\sigma^2}$ into the ℓ_2-ball of radius $\sqrt{n(P+\sigma^2)}$ centered at the origin, this yields a good codebook and decoding regions – see Figure 20.1 for an illustration. So how large can M be? Note that the volume of an ℓ_2-ball of radius r in $\mathbb{R}^n$ is given by $c_n r^n$ for some constant c_n. Then $\frac{c_n(n(P+\sigma^2))^{n/2}}{c_n(n\sigma^2)^{n/2}} = \left(1 + \frac{P}{\sigma^2}\right)^{n/2}$. Taking the log and dividing by n, we get $\frac{1}{n}\log M^* \approx \frac{1}{2}\log\left(1 + \frac{P}{\sigma^2}\right)$. This tantalizingly convincing reasoning, however, is flawed in at least two ways. (a) Computing the volume ratio only gives an upper bound on the maximal number of disjoint balls. (See Section 27.2 for an extensive discussion on this topic.) (b) Codewords need not correspond to centers of *disjoint* ℓ_2-balls. Indeed, the fact that we allow some vanishing (but non-zero) probability of error means that the $\sqrt{n\sigma^2}$ balls are slightly overlapping and Shannon's formula establishes the maximal number of such partially overlapping balls that we can pack so that they are (mostly) inside a larger ball.

Since Theorem 20.11 applies to Gaussian noise, it is natural to ask: What if the noise is non-Gaussian and how sensitive is the capacity formula $\frac{1}{2}\log(1 + \text{SNR})$ to the Gaussian assumption? Recall the Gaussian saddle-point result in Theorem 5.11 which shows that for the same variance, Gaussian noise is the worst, which shows that the capacity of any non-Gaussian noise is at least $\frac{1}{2}\log(1 + \text{SNR})$. Conversely, it turns out the increase of the capacity can be controlled by how non-Gaussian the noise is (in terms of KL divergence). The following result is due to Ihara [223].

Theorem 20.12. (Additive non-Gaussian noise) *Let Z be a real-valued random variable independent of X and $\mathbb{E}Z^2 < \infty$. Let $\sigma^2 = \operatorname{Var} Z$. Then*

$$\frac{1}{2}\log\left(1+\frac{P}{\sigma^2}\right) \le \sup_{P_X:\, \mathbb{E}X^2 \le P} I(X;X+Z) \le \frac{1}{2}\log\left(1+\frac{P}{\sigma^2}\right) + D(P_Z \| \mathcal{N}(\mathbb{E}Z,\sigma^2)). \tag{20.6}$$

Proof. See Exercise **IV.24**. □

Remark 20.1. The quantity $D(P_Z\|\mathcal{N}(\mathbb{E}Z,\sigma^2))$ is sometimes called the *non-Gaussianness* of Z, where $\mathcal{N}(\mathbb{E}Z,\sigma^2)$ is a Gaussian with the same mean and variance as Z. So if Z has a non-Gaussian density, say, Z is uniform on $[0,1]$, then the capacity can only differ by a constant compared to AWGN, which still scales as $\frac{1}{2}\log \mathrm{SNR}$ in the high-SNR regime. On the other hand, if Z is discrete, then $D(P_Z\|\mathcal{N}(\mathbb{E}Z,\sigma^2)) = \infty$ and indeed in this case one can show that the capacity is infinite because the noise is "too weak."

20.4 Parallel AWGN Channel

Definition 20.13. (Parallel AWGN) A parallel AWGN channel with L branches is a stationary memoryless channel whose single-letter kernel is defined as follows: alphabets $\mathcal{A} = \mathcal{B} = \mathbb{R}^L$; the cost $\mathsf{c}(x) = \sum_{k=1}^L |x_k|^2$; and the kernel $P_{Y^L|X^L}\colon Y_k = X_k + Z_k$, for $k = 1,\ldots,L$, and $Z_k \sim \mathcal{N}(0,\sigma_k^2)$ are independent for each branch.

Theorem 20.14. (Water-filling) *The capacity of an L-parallel AWGN channel is given by*

$$C = \frac{1}{2}\sum_{j=1}^{L} \log^+ \frac{T}{\sigma_j^2},$$

where $\log^+(x) \triangleq \max(\log x, 0)$*, and $T \ge 0$ is determined by*

$$P = \sum_{j=1}^{L} |T - \sigma_j^2|^+.$$

Proof. One can show

$$\begin{aligned}
C^{(\mathrm{I})}(P) &= \sup_{P_{X^L}:\, \sum \mathbb{E}[X_i^2] \le P} I(X^L;Y^L) \\
&\le \sup_{\sum P_k \le P, P_k \ge 0} \sum_{k=1}^{L} \sup_{\mathbb{E}[X_k^2] \le P_k} I(X_k;Y_k) \\
&= \sup_{\sum P_k \le P, P_k \ge 0} \sum_{k=1}^{L} \frac{1}{2}\log\left(1+\frac{P_k}{\sigma_k^2}\right)
\end{aligned}$$

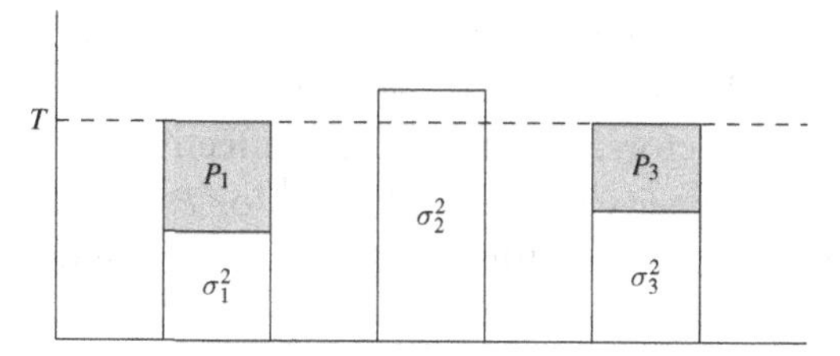

Figure 20.2 Power allocation via water-filling across three parallel channels. Here, the second branch is too noisy (σ_2 too big) for the amount of available power P and the optimal coding should discard (input zeros to) this branch altogether.

with equality if $X_k \sim \mathcal{N}(0, P_k)$ are independent. So the question boils down to the last maximization problem, known as the problem of optimal *power allocation*. Denote the Lagrangian multipliers for the constraint $\sum P_k \le P$ by λ and for the constraint $P_k \ge 0$ by μ_k. We want to solve $\max \sum \frac{1}{2}\log(1 + \frac{P_k}{\sigma_k^2}) - \mu_k P_k + \lambda(P - \sum P_k)$. The first-order condition on P_k gives that

$$\frac{1}{2}\frac{1}{\sigma_k^2 + P_k} = \lambda - \mu_k, \quad \mu_k P_k = 0,$$

and therefore the optimal solution is

$$P_k = |T - \sigma_k^2|^+, \quad T \text{ is chosen such that } P = \sum_{k=1}^{L} |T - \sigma_k^2|^+. \qquad \square$$

Figure 20.2 illustrates the water-filling solution. It has a number of practically important conclusions. First, it gives a precise recipe for how much power to allocate to different frequency bands. This solution, simple and elegant, was actually pivotal for bringing high-speed Internet to many homes (via cable modems, or ADSL): initially, before information theorists had a say, power allocations were chosen on the basis of costly and imprecise simulations of real codes. The simplicity of the water-filling scheme makes power allocation dynamic and enables instantaneous reaction to changing noise environments.

Second, there is a very important consequence for multiple-antenna (MIMO) communication. Given n_r receive antennas and n_t transmit antennas, very often one gets as a result a parallel AWGN with $L = \min(n_r, n_t)$ branches (see Exercises **I.9** and **I.10**). For a single-antenna system the capacity then scales as $\frac{1}{2}\log P$ with increasing power (Theorem 20.11), while the capacity for a MIMO AWGN channel is approximately $\frac{L}{2}\log\left(\frac{P}{L}\right) \approx \frac{L}{2}\log P$ for large P. This results in an L-fold increase in capacity at high SNR. This is the basis of a powerful technique of spatial multiplexing in MIMO, largely behind much of the advance in 4G, 5G cellular (3GPP) and post-802.11n WiFi systems.

Notice that spatial diversity (requiring both receive and transmit antennas) is different from a so-called multipath diversity (which works even if antennas are added on just one side). Indeed, if a single stream of data is sent through every parallel channel simultaneously, then the sufficient statistic would be the average of all received vectors, resulting in the effective noise level reduced by $\frac{1}{L}$ factor. This results in capacity increasing from $\frac{1}{2}\log P$ to $\frac{1}{2}\log(LP)$ – a far cry from the L-fold increase of spatial multiplexing. These exciting topics are explored in excellent textbooks [424, 269].

20.5* Non-Stationary AWGN

Definition 20.15. (Non-stationary AWGN) A non-stationary AWGN channel is a memoryless channel with single-letter alphabets $\mathcal{A} = \mathcal{B} = \mathbb{R}$ and separable cost $\mathsf{c}(x) = x^2$. The channel acts on the input vector X^n by addition $Y^n = X^n + Z^n$, where $Z_j \sim \mathcal{N}(0, \sigma_j^2)$ are independent.

Theorem 20.16. *Assume that for every $T > 0$ the following limits exist:*

$$\tilde{C}^{(\mathrm{I})}(T) = \lim_{n\to\infty} \frac{1}{n}\sum_{j=1}^{n} \frac{1}{2}\log^+ \frac{T}{\sigma_j^2},$$

$$\tilde{P}(T) = \lim_{n\to\infty} \frac{1}{n}\sum_{j=1}^{n} |T - \sigma_j^2|^+.$$

Then the capacity of the non-stationary AWGN channel is given by the parameterized form: $C(T) = \tilde{C}^{(\mathrm{I})}(T)$ with input power constraint $\tilde{P}(T)$.

Proof. Fix $T > 0$. Then it is clear from the water-filling solution in Theorem 20.14 that

$$\sup I(X^n;Y^n) = \sum_{j=1}^{n} \frac{1}{2}\log^+ \frac{T}{\sigma_j^2}, \tag{20.7}$$

where the supremum is over all P_{X^n} such that

$$\mathbb{E}[\mathsf{c}(X^n)] \le \frac{1}{n}\sum_{j=1}^{n} |T - \sigma_j^2|^+. \tag{20.8}$$

Now, by assumption, the LHS of (20.8) converges to $\tilde{P}(T)$. Thus, we have that for every $\delta > 0$

$$C^{(\mathrm{I})}(\tilde{P}(T) - \delta) \le \tilde{C}^{(\mathrm{I})}(T),$$
$$C^{(\mathrm{I})}(\tilde{P}(T) + \delta) \ge \tilde{C}^{(\mathrm{I})}(T).$$

Taking $\delta \to 0$ and invoking the continuity of $P \mapsto C^{(\mathrm{I})}(P)$, we get from Theorem 19.15 that the information capacity satisfies

$$C^{(\mathrm{I})}(\tilde{P}(T)) = \tilde{C}^{(\mathrm{I})}(T)$$

provided that the channel is information stable. Indeed, from (19.32)

$$\mathrm{Var}(i(X_j;Y_j)) = \frac{\log^2 e}{2} \frac{P_j}{P_j + \sigma_j^2} \le \frac{\log^2 e}{2}$$

and thus

$$\sum_{j=1}^{n} \frac{1}{n^2} \mathrm{Var}(i(X_j;Y_j)) < \infty.$$

From here information stability follows via Theorem 19.16. □

Non-stationary AWGN is primarily of interest due to its relationship to the additive colored Gaussian noise channel in the following section.

20.6* Additive Colored Gaussian Noise Channel

Definition 20.17. The additive colored Gaussian noise (ACGN) channel is a channel with memory defined as follows. The single-letter alphabets are $\mathcal{A} = \mathcal{B} = \mathbb{R}$ and the separable cost is $\mathsf{c}(x) = x^2$. The channel acts on the input vector X^n by addition $Y^n = X^n + Z^n$, where $\{Z_j : j \ge 1\}$ is a stationary Gaussian process with power spectral density $f_Z(\omega) \ge 0, \omega \in [-\pi, \pi]$ (recall Example 6.4).

Theorem 20.18. *The capacity of the ACGN channel with $f_Z(\omega) > 0$ for almost every $\omega \in [-\pi, \pi]$ is given by the following parametric form:*

$$C(T) = \frac{1}{2\pi} \int_{-\pi}^{\pi} \frac{1}{2} \log^+ \frac{T}{f_Z(\omega)} d\omega,$$
$$P(T) = \frac{1}{2\pi} \int_{-\pi}^{\pi} |T - f_Z(\omega)|^+ d\omega.$$

Proof. For $n \ge 1$, consider the diagonalization of the covariance matrix of Z^n:

$$\mathrm{Cov}(Z^n) = \Sigma = U^\top \widetilde{\Sigma} U, \text{ such that } \widetilde{\Sigma} = \mathrm{diag}(\sigma_1, \ldots, \sigma_n),$$

where U is an orthogonal matrix. (Since $\mathrm{Cov}(Z^n)$ is positive semidefinite this diagonalization is always possible.) Define $\widetilde{X}^n = UX^n$ and $\widetilde{Y}^n = UY^n$; the channel between $\widetilde{X}^n$ and $\widetilde{Y}^n$ is thus

$$\widetilde{Y}^n = \widetilde{X}^n + UZ^n,$$
$$\mathrm{Cov}(UZ^n) = U \cdot \mathrm{Cov}(Z^n) \cdot U^\top = \widetilde{\Sigma}.$$

Therefore we have the equivalent channel as follows:

$$\widetilde{Y}^n = \widetilde{X}^n + \widetilde{Z}^n, \quad \widetilde{Z}_j \sim \mathcal{N}(0, \sigma_j^2) \text{ independent across } j.$$

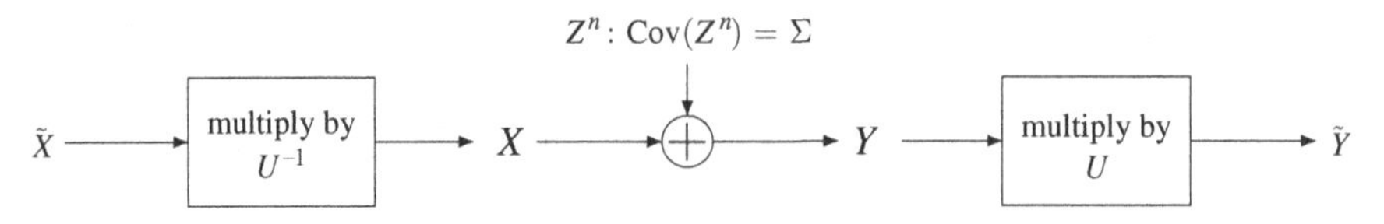

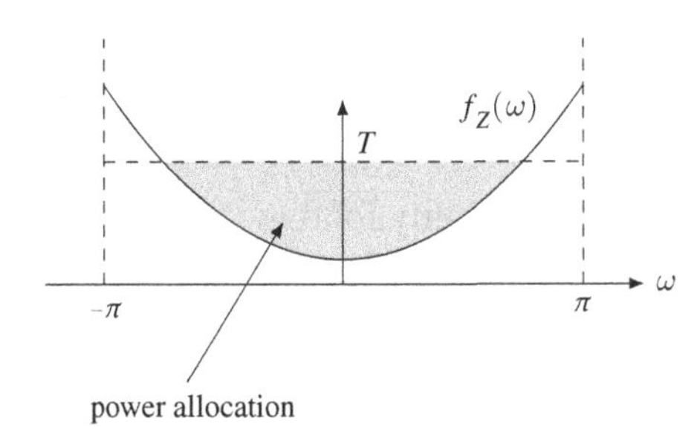

Figure 20.3 The ACGN channel: the "whitening" process used in the capacity proof and the water-filling power allocation across the spectrum.

Note that since U and $U^\top$ are orthogonal the maps $\tilde{X}^n = UX^n$ and $X^n = U^\top \tilde{X}^n$ preserve the norm $\|\tilde{X}^n\| = \|X^n\|$. Therefore, the capacities of both channels are equal: $C = \tilde{C}$. But the latter follows from Theorem 20.16. Indeed, we have that

$$\tilde{C} = \lim_{n\to\infty} \frac{1}{n} \sum_{j=1}^{n} \log^+ \frac{T}{\sigma_j^2} = \frac{1}{2\pi} \int_{-\pi}^{\pi} \frac{1}{2} \log^+ \frac{T}{f_Z(\omega)} d\omega \quad \text{(Szegö's theorem, see (6.12))}$$

$$\lim_{n\to\infty} \frac{1}{n} \sum_{j=1}^{n} |T - \sigma_j^2|^+ = P(T). \qquad \square$$

The idea used in the proof as well as the water-filling power allocation are illustrated on Figure 20.3. Note that most of the time the noise that impacts real-world systems is actually "born" white (because it is a thermal noise). However, between the place of its injection and the processing there are usually multiple circuit elements. If we model them linearly then their action can equivalently be described as the ACGN channel, since the effective noise added becomes colored. In fact, this filtering can be inserted deliberately in order to convert the actual channel into an additive-noise one. This is the content of the next section.

20.7* AWGN Channel with Intersymbol Interference

Oftentimes in wireless communication systems a signal is propagating through a rich scattering environment. Thus, reaching the receiver are multiple delayed and attenuated copies of the initial signal. Such a situation is formally called *intersymbol interference* (ISI). A similar effect also occurs when a cable modem attempts to send signals across telephone or TV wires due to the presence of various linear filters, transformers, and relays. The mathematical model for such channels is the following.

Definition 20.19. (AWGN with ISI) An AWGN channel with ISI is a channel with memory that is defined as follows: the alphabets are $\mathcal{A} = \mathcal{B} = \mathbb{R}$, and the separable cost is $\mathsf{c}(x) = x^2$. The channel law $P_{Y^n|X^n}$ is given by

$$Y_k = \sum_{j=1}^{n} h_{k-j} X_j + Z_k, \quad k = 1, \dots, n,$$

where $Z_k \overset{\text{iid}}{\sim} \mathcal{N}(0, \sigma^2)$ is white Gaussian noise, and $\{h_k, k \in \mathbb{Z}\}$ are coefficients of a discrete-time channel filter.

The coefficients $\{h_k\}$ describe the action of the environment. They are often learned by the receiver during the "handshake" process of establishing a communication link.

Theorem 20.20. *Suppose that the sequence $\{h_k\}$ is the inverse Fourier transform of a frequency response $H(\omega)$:*

$$h_k = \frac{1}{2\pi} \int_{-\pi}^{\pi} e^{i\omega k} H(\omega) d\omega.$$

Assume also that $H(\omega)$ is a continuous function on $[0, 2\pi]$. Then the capacity of the AWGN channel with ISI is given by

$$C(T) = \frac{1}{2\pi} \int_{-\pi}^{\pi} \frac{1}{2} \log^+(T|H(\omega)|^2) d\omega,$$

$$P(T) = \frac{1}{2\pi} \int_{-\pi}^{\pi} \left| T - \frac{1}{|H(\omega)|^2} \right|^+ d\omega.$$

Proof sketch. At the decoder apply the inverse filter with frequency response $\omega \mapsto \frac{1}{H(\omega)}$. The equivalent channel then becomes a stationary colored-noise Gaussian channel:

$$\tilde{Y}_j = X_j + \tilde{Z}_j,$$

where $\tilde{Z}_j$ is a stationary Gaussian process with spectral density

$$f_{\tilde{Z}}(\omega) = \frac{1}{|H(\omega)|^2}.$$

Then apply Theorem 20.18 to the resulting channel.

To make the above argument rigorous one must carefully analyze the non-zero error introduced by truncating the deconvolution filter to finite n. This would take us too much outside of the scope of this book. □

20.8* Gaussian Channels with Amplitude Constraints

We have examined some classical results of additive Gaussian noise channels. In the following, we will list some more recent results without proof.

Theorem 20.21. (Amplitude-constrained AWGN channel capacity) *For an AWGN channel $Y_i = X_i + Z_i$ with amplitude constraint $|X_i| \le A$, we denote the capacity by:*

$$C(A) = \max_{P_X:\,|X|\le A} I(X;X+Z).$$

The capacity-achieving input distribution P_X^* is discrete, with finitely many atoms on $[-A, A]$. The number of atoms is $\Omega(A)$ and $O(A^2)$ as $A \to \infty$. Moreover,

$$\frac{1}{2}\log\left(1+\frac{2A^2}{e\pi}\right) \le C(A) \le \frac{1}{2}\log\left(1+A^2\right).$$

Theorem 20.22. (Amplitude- and power-constrained AWGN channel capacity) *For an AWGN channel $Y_i = X_i + Z_i$ with amplitude constraint $|X_i| \le A$ and power constraint $\sum_{i=1}^n X_i^2 \le nP$, we denote the capacity by:*

$$C(A,P) = \max_{P_X:\,|X|\le A,\mathbb{E}|X|^2\le P} I(X;X+Z).$$

The capacity-achieving input distribution P_X^* is discrete, with finitely many atoms on $[-A, A]$. Moreover, the convergence speed of $\lim_{A\to\infty} C(A,P) = \frac{1}{2}\log(1+P)$ *is of the order $e^{-O(A^2)}$.*

For details, see [397], [344, Section III], and [144, 349] for the $O(A^2)$ bound on the number of atoms.

20.9* Gaussian Channels with Fading

So far we have assumed that the channel is either additive (as in AWGN or ACGN) or has known multiplicative gains (as in ISI). However, in practice the channel gain is a random variable. This situation is called "fading" and is often used to model urban signal propagation with multiple paths or shadowing. The received signal at time i, Y_i, is affected by multiplicative fading coefficient H_i and additive noise Z_i as follows:

$$Y_i = H_i X_i + Z_i, \quad Z_i \sim \mathcal{N}(0, \sigma^2).$$

This is illustrated in Figure 20.4.

There are two drastically different cases of fading channels, depending on the presence or absence of the dashed link in Figure 20.4. In the first case, known as the *coherent case* or the *CSIR case* (for channel state information at the receiver), the receiver is assumed to have a perfect estimate of the channel state information H_i at every time i. In other words, the channel output is effectively (Y_i, H_i). This situation occurs, for example, when there are pilot signals sent periodically and are

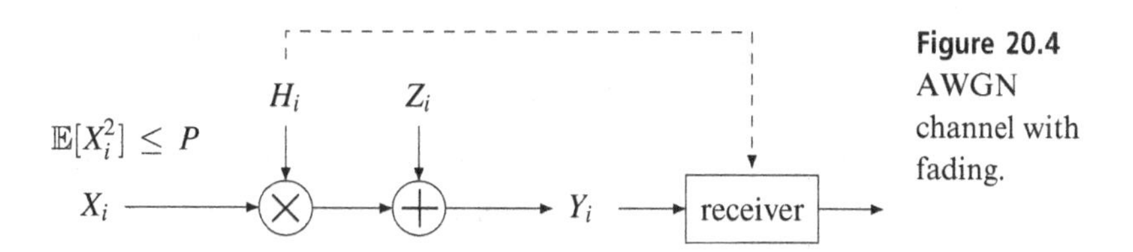

Figure 20.4 AWGN channel with fading.

used at the receiver to estimate the channel. In some cases, the index i refers to different frequencies or subchannels of an orthogonal frequency-division multiplexing (OFDM) frame.

Whenever H_j is a stationary ergodic process, we have the channel capacity given by:

$$C(P) = \mathbb{E}\left[\frac{1}{2}\log\left(1 + \frac{P|H|^2}{\sigma^2}\right)\right]$$

and the capacity-achieving input distribution is the usual $P_X = \mathcal{N}(0, P)$. Note that the capacity $C(P)$ is in the order of $\log P$ and we call the channel "energy efficient."

In the second case, known as *non-coherent* or *no-CSIR*, the receiver does not have any knowledge of the coefficients H_i's. In this case, there is no simple expression for the channel capacity. Most of the known results were shown for the case of iid H_i according to a Rayleigh distribution. In this case, the capacity-achieving input distribution is discrete [3], and the capacity scales as [416, 270]

$$C(P) = O(\log\log P), \quad P \to \infty. \tag{20.9}$$

This channel is said to be "energy inefficient" since increasing the communication rate requires dramatic expenditures in power.

Further generalization of the Gaussian channel models requires introducing multiple input and output antennas (known as MIMO channel). In this case, the single-letter input $X_i \in \mathbb{C}^{n_t}$ and the output $Y_i \in \mathbb{C}^{n_r}$ are related by

$$Y_i = H_i X_i + Z_i, \tag{20.10}$$

where $Z_i \overset{\text{iid}}{\sim} \mathcal{CN}(0, \sigma^2 I_{n_r})$, n_t and n_r are the number of transmit and receive antennas, and $H_i \in \mathbb{C}^{n_t \times n_r}$ is a matrix-valued channel gain process. For the capacity of this channel under CSIR, see Exercise **I.10**. An incredible effort in the 1990s and 2000s was invested by information-theoretic and communication-theoretic researchers to understand this channel model. Some of the highlights include:

- a beautiful transmit-diversity 2×2 code of Alamouti [10];
- generalization of Alamouti's code led to the discovery of space-time coding [418, 417];
- the result of Telatar [419] showing that under coherent fading the capacity scales as $\min(n_t, n_r)\log P$ (the coefficient appearing in front of $\log P$ is known as *pre-log* or *degrees-of-freedom* of the channel);

- the result of Zheng and Tse [476] showing a different pre-log in the scaling for the non-coherent (block-fading) case.

It is not possible to do any justice to these and many other fundamental results in MIMO communication here, unfortunately. We suggest textbook [424] as an introduction to this deep and exciting field.

21 Capacity per Unit Cost

In this chapter we will consider an interesting variation of the channel coding problem. Instead of constraining the blocklength (i.e. the number of channel uses), we will constrain the total cost incurred by the codewords. The most important special case of this problem is that of the AWGN channel and quadratic (energy) cost constraint. The standard motivation in this setting is the following. Consider a deep space probe which has a k-bit message that needs to be delivered to Earth (or to a satellite orbiting it). The duration of transmission is of little worry for the probe, but the amount of energy it has stored in its battery is the main limitation in this case. In this chapter we will learn how to study this question abstractly, how coding over a large number of bits $k \to \infty$ reduces the energy spent (per bit), and how this fundamental limit is related to communication over continuous-time channels.

21.1 Energy-per-Bit

In this chapter we will consider Markov kernels $P_{Y^\infty|X^\infty}$ acting between two spaces of infinite sequences. The prototypical example is again the AWGN channel:

$$Y_i = X_i + Z_i, \quad Z_i \sim \mathcal{N}(0, N_0/2). \tag{21.1}$$

Note that in this chapter we have denoted the noise level for Z_i to be $\frac{N_0}{2}$. There is a long tradition for such notation. Indeed, most of the noise in communication systems is white thermal noise at the receiver. The power spectral density of that noise is flat and denoted by N_0 (in joules per second per hertz). However, recall that the received signal is complex-valued and, thus, each real component has power $\frac{N_0}{2}$. Note also that thermodynamics suggests that $N_0 = kT$, where $k = 1.38 \times 10^{-23}$ is the Boltzmann constant, and T is the absolute temperature in kelvins.

In the previous chapter, we analyzed the maximum number of information messages ($M^*(n, \epsilon, P)$) that can be sent through this channel for a given n number of channel uses and under the power constraint P. We have also hinted that in (20.5) there is a fundamental minimal required cost to send each (data) bit. Here we develop this question more rigorously. Everywhere in this chapter for $v \in \mathbb{R}^\infty$

$$\|v\|_2^2 \triangleq \sum_{j=1}^{\infty} v_j^2.$$

Definition 21.1. ($(E, 2^k, \epsilon)$-code) Given a Markov kernel with input space $\mathbb{R}^\infty$ we define an $(E, 2^k, \epsilon)$-code to be an encoder–decoder pair, $f\colon [2^k] \to \mathbb{R}^\infty$ and $g\colon \mathbb{R}^\infty \to [2^k]$ (or similar randomized versions), such that for all messages $m \in [2^k]$ we have $\|f(m)\|_2^2 \le E$ and

$$\mathbb{P}[g(Y^\infty) \neq W] \le \epsilon,$$

where as usual the probability space is $W \to X^\infty \to Y^\infty \to \hat{W}$ with $W \sim \mathrm{Unif}([2^k])$, $X^\infty = f(W)$, and $\hat{W} = g(Y^\infty)$. The fundamental limit is defined to be

$$E^*(k, \epsilon) = \min\{E\colon \exists (E, 2^k, \epsilon) \text{ code}\}.$$

The operational meaning of $E^*(k,\epsilon)$ should be apparent: it is the minimal amount of energy the space probe needs to draw from the battery in order to send k bits of data.

Theorem 21.2. (Minimal "ebno"$(E_b/N_0)_{\min} = -1.6$ dB) *For the AWGN channel we have*

$$\lim_{\epsilon\to 0} \limsup_{k\to\infty} \frac{E^*(k,\epsilon)}{k} = \frac{N_0}{\log_2 e}. \tag{21.2}$$

Remark 21.1. This result, first obtained by Shannon [379], and colloquially referred to as minimal E_b/N_0 (pronounced "eebee over enzero" or "ebno"), is -1.6 dB. The latter value is simply $10\log_{10}(\frac{1}{\log_2 e}) \approx -1.592$. Achieving this value of the ebno was an ultimate quest for coding theory, first resolved by the turbo codes [47]. See [101] for a review of this long conquest. We also remark that the fundamental limit is unchanged if instead of a real-valued AWGN channel we used a $\mathbb{C}$-AWGN channel

$$Y_i = X_i + Z_i, \quad Z_i \sim \mathcal{CN}(0, N_0),$$

and energy constraint $\sum_{i=1}^{\infty} |X_i|^2 \le E$. Indeed, this channel's single input can be simply converted into a pair of inputs for the $\mathbb{R}$-AWGN channel. This doubles the blocklength, but it is anyway considered to be infinite.

Proof. We start with a lower bound (or the "converse" part). As usual, we have the working probability space

$$W \to X^\infty \to Y^\infty \to \hat{W}.$$

Then consider the following chain:

$$
\begin{aligned}
-h(\epsilon) + \bar{\epsilon}k &\le d\left((1-\epsilon)\Big\|\frac{1}{M}\right) && \text{Fano's inequality} \\
&\le I(W;\hat{W}) && \text{data processing for divergence} \\
&\le I(X^\infty;Y^\infty) && \text{data processing for mutual information} \\
&\le \sum_{i=1}^{\infty} I(X_i;Y_i) && \lim_{n\to\infty} I(X^n;U) = I(X^\infty;U) \text{ by (4.30)} \\
&\le \sum_{i=1}^{\infty} \frac{1}{2}\log\left(1+\frac{\mathbb{E}X_i^2}{N_0/2}\right) && \text{Gaussian capacity, Theorem 5.11} \\
&\le \frac{\log e}{2}\sum_{i=1}^{\infty}\frac{\mathbb{E}X_i^2}{N_0/2} && \text{linearization of log} \\
&\le \frac{E}{N_0}\log e.
\end{aligned}
$$

Thus, we have shown

$$
\frac{E^*(k,\epsilon)}{k} \ge \frac{N_0}{\log e}\left(\bar{\epsilon} - \frac{h(\epsilon)}{k}\right)
$$

and taking the double limit in $n \to \infty$ then in $\epsilon \to 0$ completes the proof.

Next, we consider the upper bound (the "achievability" part). We first give a traditional existential proof. Notice that an $(n, 2^k, \epsilon, P)$-code for the AWGN channel is also an $(nP, 2^k, \epsilon)$-code for the energy problem without time constraint. Therefore,

$$
\log_2 M^*(n,\epsilon,P) \ge k \Rightarrow E^*(k,\epsilon) \le nP.
$$

Take $k_n = \lfloor \log_2 M^*(n,\epsilon,P) \rfloor$, so that we have $\frac{E^*(k_n,\epsilon)}{k_n} \le \frac{nP}{k_n}$ for all $n \ge 1$. Taking the limit then we get

$$
\begin{aligned}
\limsup_{n\to\infty} \frac{E^*(k_n,\epsilon)}{k_n} &\le \limsup_{n\to\infty} \frac{nP}{\log M^*(n,\epsilon,P)} \\
&= \frac{P}{\liminf_{n\to\infty} \frac{1}{n}\log M^*_{\max}(n,\epsilon,P)} \\
&= \frac{P}{\frac{1}{2}\log\left(1+\frac{P}{N_0/2}\right)},
\end{aligned}
$$

where in the last step we applied Theorem 20.11. Now the above statement holds for every $P > 0$, so let us optimize it to get the best bound:

$$
\begin{aligned}
\limsup_{n\to\infty} \frac{E^*(k_n,\epsilon)}{k_n} &\le \inf_{P\ge 0} \frac{P}{\frac{1}{2}\log\left(1+\frac{P}{N_0/2}\right)} \\
&= \lim_{P\to 0} \frac{P}{\frac{1}{2}\log\left(1+\frac{P}{N_0/2}\right)} \\
&= \frac{N_0}{\log_2 e}.
\end{aligned}
\tag{21.3}
$$

Note that the fact that minimal energy-per-bit is attained at $P \to 0$ implies that in order to send information reliably at the Shannon limit of -1.6 dB, infinitely many time slots are needed. In other words, the information rate (also known as spectral efficiency) should be vanishingly small. Conversely, in order to have non-zero spectral efficiency, one necessarily has to step away from the -1.6 dB. This tradeoff is known as spectral efficiency versus energy-per-bit.

We next can give a simpler and more explicit construction of the code, not relying on the random coding implicit in Theorem 20.11. Let $M = 2^k$ and consider the following code, known as pulse-position modulation (PPM):

$$\text{PPM encoder: } \forall m, f(m) = \mathbf{c}_m \triangleq (0, 0, \ldots, \underbrace{\sqrt{E}}_{m\text{th location}}, \ldots). \tag{21.4}$$

It is not hard to derive an upper bound on the probability of error that this code achieves [338, Theorem 2]:

$$\epsilon \le \mathbb{E}\left[\min\left\{MQ\left(\sqrt{\frac{2E}{N_0}} + Z\right), 1\right\}\right], \quad Z \sim \mathcal{N}(0,1). \tag{21.5}$$

Indeed, our orthogonal codebook under a maximum likelihood decoder has probability of error equal to

$$P_e = 1 - \frac{1}{\sqrt{\pi N_0}} \int_{-\infty}^{\infty} \left[1 - Q\left(\sqrt{\frac{2}{N_0}} z\right)\right]^{M-1} e^{-(z-\sqrt{E})^2/N_0} dz, \tag{21.6}$$

which is obtained by observing that conditioned on $(W = j, Z_j)$ the events $\{\|\mathbf{c}_j + \mathbf{z}\|^2 \le \|\mathbf{c}_j + \mathbf{z} - \mathbf{c}_i\|^2\}$, $i \ne j$ are independent. A change of variables $x = \sqrt{\frac{2}{N_0}} z$ and application of the bound $1 - (1-y)^{M-1} \le \min\{My, 1\}$ weakens (21.6) to (21.5).

To see that (21.5) implies (21.3), fix $c > 0$ and condition on $|Z| \le c$ in (21.5) to relax it to

$$\epsilon \le MQ\left(\sqrt{\frac{2E}{N_0}} - c\right) + 2Q(c).$$

Recall the expansion for the Q-function [436, (3.53)]:

$$\log Q(x) = -\frac{x^2 \log e}{2} - \log x - \frac{1}{2}\log 2\pi + o(1), \quad x \to \infty. \tag{21.7}$$

Thus, choosing $\tau > 0$ and setting $E = (1+\tau)k\frac{N_0}{\log_2 e}$ we obtain

$$2^k Q\left(\sqrt{\frac{2E}{N_0}} - c\right) \to 0$$

as $k \to \infty$. Thus choosing $c > 0$ sufficiently large we obtain that $\limsup_{k\to\infty} E^*(k,\epsilon) \le (1+\tau)\frac{N_0}{\log_2 e}$ for every $\tau > 0$. Taking $\tau \to 0$ implies (21.3). □

Remark 21.2. (Simplex conjecture) The code (21.4) in fact achieves the first three terms in the large-k expansion of $E^*(k,\epsilon)$, see [338, Theorem 3]. In fact, the code can be further slightly optimized by subtracting the common center of gravity

$(2^{-k}\sqrt{E}, \ldots, 2^{-k}\sqrt{E}, \ldots)$ and rescaling each codeword to satisfy the power constraint. The resulting constellation is known as the *simplex code*. It is conjectured to be the actual optimal code achieving $E^*(k, \epsilon)$ for a fixed k and ϵ; see [105, Section 3.16] and [402] for more.

21.2 Capacity per Unit Cost

Generalizing the energy-per-bit setting of the previous section, we get the problem of *capacity per unit cost*.

Definition 21.3. Given a channel $P_{Y^\infty|X^\infty}\colon \mathcal{X}^\infty \to \mathcal{Y}^\infty$ and a cost function $\mathsf{c}\colon \mathcal{X} \to \mathbb{R}_+$, we define the (E, M, ϵ)-code to be an encoder $f\colon [M] \to \mathcal{X}^\infty$ and a decoder $g\colon \mathcal{Y}^\infty \to [M]$ s.t. (a) for every m the output of the encoder $x^\infty \triangleq f(m)$ satisfies

$$\sum_{t=1}^{\infty} \mathsf{c}(x_t) \le E, \tag{21.8}$$

and (b) the probability of error is small

$$\mathbb{P}[g(Y^\infty) \neq W] \le \epsilon,$$

where as usual we operate on the space $W \to X^\infty \to Y^\infty \to \hat{W}$ with $W \sim \mathrm{Unif}([M])$. We let

$$M^*(E, \epsilon) = \max\{M\colon (E, M, \epsilon)\text{-code}\},$$

and define capacity per unit cost as

$$C_{puc} \triangleq \lim_{\epsilon \to 0} \liminf_{E \to \infty} \frac{1}{E} \log M^*(E, \epsilon). \tag{21.9}$$

Let $C(P)$ be the capacity-cost function of the channel (in the usual sense of capacity, as defined in (20.1)). Assuming $P_0 = 0$ and $C(0) = 0$ it is not hard to show (basically following the steps of Theorem 21.2) that:

$$C_{puc} = \sup_P \frac{C(P)}{P} = \lim_{P \to 0} \frac{C(P)}{P} = \left.\frac{d}{dP}\right|_{P=0} C(P).$$

The surprising discovery of Verdú [435] is that one can avoid computing $C(P)$ and derive C_{puc} directly. This is a significant help, as for many practical channels $C(P)$ is unknown. Additionally, this gives yet another fundamental meaning to the KL divergence.

Theorem 21.4. *For a stationary memoryless channel $P_{Y^\infty|X^\infty} = \prod P_{Y|X}$ with $P_0 = \mathsf{c}(x_0) = 0$ (i.e. there is a symbol of zero cost), we have*

$$C_{puc} = \sup_{x \neq x_0} \frac{D(P_{Y|X=x} \| P_{Y|X=x_0})}{\mathsf{c}(x)}.$$

In particular, $C_{puc} = \infty$ if there exists $x_1 \neq x_0$ with $\mathsf{c}(x_1) = 0$.

Proof. Let

$$C_V = \sup_{x \neq x_0} \frac{D(P_{Y|X=x} \| P_{Y|X=x_0})}{\mathsf{c}(x)}.$$

Converse: Consider an (E, M, ϵ)-code $W \to X^\infty \to Y^\infty \to \hat{W}$. Introduce an auxiliary distribution $Q_{W,X^\infty,Y^\infty,\hat{W}}$, where a channel is a useless one:

$$Q_{Y^\infty|X^\infty} = Q_{Y^\infty} \triangleq P^{\infty}_{Y|X=x_0}.$$

That is, the overall factorization is

$$Q_{W,X^\infty,Y^\infty,\hat{W}} = P_W P_{X^\infty|W} Q_{Y^\infty} P_{\hat{W}|Y^\infty}.$$

Then, as usual we have from the data processing for divergence:

$$\begin{aligned}(1-\epsilon)\log M + h(\epsilon) &\leq d\left(1-\epsilon \middle\| \frac{1}{M}\right) \\ &\leq D\left(P_{W,X^\infty,Y^\infty,\hat{W}} \middle\| Q_{W,X^\infty,Y^\infty,\hat{W}}\right) \\ &= D(P_{Y^\infty|X^\infty} \| Q_{Y^\infty} | P_{X^\infty}) \\ &= \mathbb{E}\left[\sum_{t=1}^{\infty} d(X_t)\right], \end{aligned} \tag{21.10}$$

where we denoted for convenience $d(x) \triangleq D(P_{Y|X=x} \| P_{Y|X=x_0})$. By the definition of C_V we have

$$d(x) \leq \mathsf{c}(x) C_V.$$

Thus, continuing (21.10) we obtain

$$(1-\epsilon)\log M + h(\epsilon) \leq C_V\, \mathbb{E}\left[\sum_{t=1}^{\infty} \mathsf{c}(X_t)\right] \leq C_V \cdot E,$$

where the last step is by the cost constraint (21.8). Thus, dividing by E and taking limits we get

$$C_{puc} \leq C_V.$$

Achievability: We generalize the PPM code (21.4). For each $x_1 \in \mathcal{X}$ and $n \in \mathbb{Z}_+$ we define the encoder f as follows:

$$f(1) = (\underbrace{x_1, x_1, \ldots, x_1}_{n \text{ times}}, \underbrace{x_0, \ldots, x_0}_{n(M-1) \text{ times}}),$$

$$f(2) = (\underbrace{x_0, x_0, \ldots, x_0}_{n \text{ times}}, \underbrace{x_1, \ldots, x_1}_{n \text{ times}}, \underbrace{x_0, \ldots, x_0}_{n(M-2) \text{ times}}),$$

$$\cdots$$

$$f(M) = (\underbrace{x_0, \ldots, x_0}_{n(M-1) \text{ times}}, \underbrace{x_1, x_1, \ldots, x_1}_{n \text{ times}})$$

Now, by Stein's lemma (Theorem 14.14) there exists a subset $S \subset \mathcal{Y}^n$ with the property that

$$\mathbb{P}[Y^n \in S | X^n = (x_1, \ldots, x_1)] \geq 1 - \epsilon_1,$$
$$\mathbb{P}[Y^n \in S | X^n = (x_0, \ldots, x_0)] \leq \exp\{-nD(P_{Y|X=x_1} \| P_{Y|X=x_0}) + o(n)\}.$$

Therefore, we propose the following (suboptimal!) decoder:

$$\begin{aligned} Y^n \in S &\implies \hat{W} = 1, \\ Y_{n+1}^{2n} \in S &\implies \hat{W} = 2, \\ &\cdots. \end{aligned}$$

From the union bound we find that the overall probability of error is bounded by

$$\epsilon \leq \epsilon_1 + M \exp\{-nD(P_{Y|X=x_1} \| P_{Y|X=x_0}) + o(n)\}.$$

At the same time the total cost of each codeword is given by $n\mathsf{c}(x_1)$. Thus, taking $n \to \infty$ and after straightforward manipulations, we conclude that

$$C_{puc} \geq \frac{D(P_{Y|X=x_1} \| P_{Y|X=x_0})}{\mathsf{c}(x_1)}.$$

This holds for any symbol $x_1 \in \mathcal{X}$, and so we are free to take the supremum over x_1 to obtain $C_{puc} \geq C_V$, as required. □

21.3 Energy-per-Bit for the Fading Channel

Note that Theorem 21.4 when applied to the AWGN channel seems to be pointless: since the capacity-cost function $C(P)$ is known, it is not hard to compute $\lim_{P \to 0} \frac{C(P)}{P}$ directly. Theorem 21.4's true strength is revealed when applied to channels for which the capacity-cost function is unknown.

Specifically, we consider a stationary memoryless Gaussian channel with fading H_j unknown at the receiver (i.e. non-coherent fading channel, see Section 20.9*):

$$Y_j = H_j X_j + Z_j, \quad H_j \sim \mathcal{N}_c(0, 1) \perp\!\!\!\perp Z_j \sim \mathcal{N}_c(0, N_0).$$

(We use here a more convenient $\mathbb{C}$-valued fading channel with $H_j \sim \mathcal{N}_c$, known as Rayleigh fading.) The cost function is the usual quadratic one $\mathsf{c}(x) = |x|^2$. As we discussed previously, see (20.9), the capacity-cost function $C(P)$ is unknown in closed form, but is known to behave drastically different from the case of non-fading AWGN (i.e. when $H_j = 1$). So here Theorem 21.4 comes in handy. Let us perform a simple computation required, see (2.9):

$$\begin{aligned} C_{puc} &= \sup_{x \neq 0} \frac{D(\mathcal{N}_c(0, |x|^2 + N_0) \| \mathcal{N}_c(0, N_0))}{|x|^2} \\ &= \frac{1}{N_0} \sup_{x \neq 0} \left(\log e - \frac{\log(1 + \frac{|x|^2}{N_0})}{\frac{|x|^2}{N_0}} \right) \qquad (21.11) \\ &= \frac{\log e}{N_0}. \end{aligned}$$

Comparing with Theorem 21.2 we discover that, surprisingly, the capacity per unit cost is unaffected by the presence of fading. In other words, the random multiplicative noise which is so detrimental at high SNR, appears to be much more benign at low SNR (recall that $C_{puc} = C'(0)$ and thus computing C_{puc} corresponds to computing $C(P)$ at $P \to 0$). There is one important difference: the supremization over x in (21.11) is solved at $x = \infty$. Following the proof of the converse bound, we conclude that any code hoping to achieve optimal C_{puc} must satisfy a strange constraint:

$$\sum_t |x_t|^2 1\{|x_t| \geq A\} \approx \sum_t |x_t|^2 \quad \forall A > 0,$$

that is, the total energy expended by each codeword must be almost entirely concentrated in very large spikes. Such a coding method is called "flash signaling." Thus, we can see that unlike the non-fading AWGN (for which due to rotational symmetry all codewords can be made relatively non-spiky), the only hope of achieving full C_{puc} in the presence of fading is by signaling in short bursts of energy. Thus, while the ultimate minimal energy-per-bit is the same for the AWGN or the fading channel, the nature of optimal coding schemes is rather different.

Another fundamental difference between the two channels is revealed in the finite-blocklength behavior of $E^*(k, \epsilon)$. Specifically, we have the following different asymptotic expansions for the energy-per-bit $\frac{E^*(k,\epsilon)}{k}$:

$$\frac{E^*(k,\epsilon)}{k} = (-1.59 \text{ dB}) + \sqrt{\frac{\text{constant}}{k}} Q^{-1}(\epsilon) \quad \text{(AWGN)},$$
$$\frac{E^*(k,\epsilon)}{k} = (-1.59 \text{ dB}) + \sqrt[3]{\frac{\log k}{k}} \left(Q^{-1}(\epsilon)\right)^2 \quad \text{(non-coherent fading)}.$$

That is, we see that the speed of convergence to the Shannon limit is much slower under fading. Figure 21.1 shows this effect numerically by plotting the evaluation of (the upper and lower bounds for) $E^*(k, \epsilon)$ for the fading and non-fading channels. We see that the number of data bits k that need to be coded over for the fading channel is about a factor of 10^3 larger than for the AWGN channel. See [464] for further details.

21.4 Capacity of the Continuous-Time AWGN Channel

We now briefly discuss the topic of continuous-time channels. We would like to define the channel as acting on waveforms $x(t), t \geq 0$ by adding white Gaussian noise as follows:

$$Y(t) = X(t) + N(t),$$

where $N(t)$ is a (generalized) Gaussian process with covariance function

$$\mathbb{E}[N(t)N(s)] = \frac{N_0}{2}\delta(t - s),$$

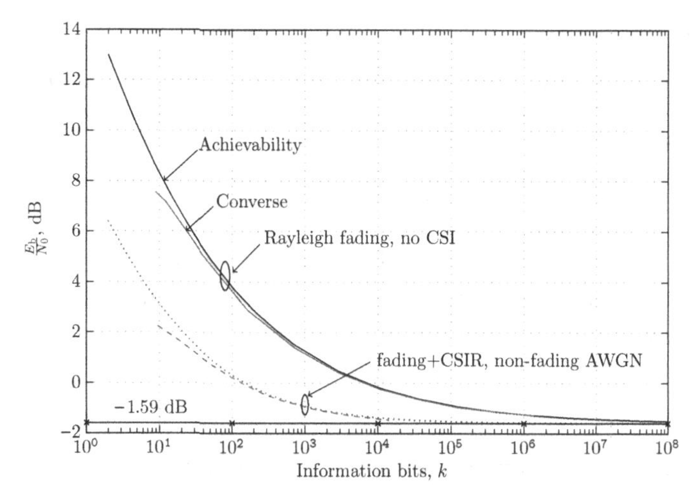

Figure 21.1 Comparing the energy-per-bit required to send a packet of k bits for different channel models (each pair of curves represent upper and lower bounds on the unknown optimal value $\frac{E^*(k,\epsilon)}{k}$). As a comparison: to get to -1.5 dB one has to code over 6×10^4 data bits when the channel is a non-fading AWGN or fading AWGN with H_j known perfectly at the receiver. For a fading AWGN without knowledge of H_j (no CSI), one has to code over at least 7×10^7 data bits to get to the same -1.5 dB. Plot generated using [398].

and δ is the Dirac δ-function. Defining the channel in this way requires careful understanding of the nature of $N(t)$ (in particular, it is not a usual stochastic process, since the random variable representing the value at a point $N(t)$ is undefined), but is preferred by engineers. Mathematicians prefer to define the continuous-time channel by introducing the standard Wiener process (Brownian motion) W_t and setting

$$Y_{int}(t) = \int_0^t X(\tau)d\tau + \sqrt{\frac{N_0}{2}} W_t,$$

where W_t is the zero-mean Gaussian process with covariance function

$$\mathbb{E}[W_s W_t] = \min(s, t).$$

Denote by $L_2([0, T])$ the space of all square-integrable functions on $[0, T]$. Let $M^*(T, \epsilon, P)$ be the maximum number of messages that can be sent through this channel such that given an encoder $f : [M] \to L_2[0, T]$ for each $m \in [M]$:

1 the wavefrom $x(t) = f(m)$ is supported on $[0, T]$ (i.e. $x(t) = 0$ for $t \notin [0, T]$)
2 the input energy is constrained to $\int_{t=0}^T x^2(t) \le TP$; and
3 the decoding error probability $P[\hat{W} \ne W] \le \epsilon$.

This is a natural extension of the previously defined $\log M^*$ functions to the continuous-time setting.

We prove the capacity result for this channel next.

Theorem 21.5. *The maximal reliable rate of communication across the continuous-time AWGN channel is* $\frac{P}{N_0}\log e$ *(per unit of time). More formally, we have*

$$\lim_{\epsilon\to 0}\liminf_{T\to\infty}\frac{1}{T}\log M^*(T,\epsilon,P) = \frac{P}{N_0}\log e. \tag{21.12}$$

Proof. Note that the space $L_2[0,T]$ has a countable basis (e.g. sinusoids). Thus, by expanding our input and output waveforms in that basis we obtain an equivalent channel model:

$$\tilde{Y}_j = \tilde{X}_j + \tilde{Z}_j,\quad \tilde{Z}_j \sim \mathcal{N}\left(0, \frac{N_0}{2}\right),$$

and energy constraint (dependent upon duration T):

$$\sum_{j=1}^{\infty} \tilde{X}_j^2 \le PT.$$

But then the problem is equivalent to the energy-per-bit for the (discrete-time) AWGN channel (see Theorem 21.2) and hence

$$\log_2 M^*(T,\epsilon,P) = k \iff E^*(k,\epsilon) = PT.$$

Thus,

$$\lim_{\epsilon\to 0}\liminf_{n\to\infty}\frac{1}{T}\log_2 M^*(T,\epsilon,P) = \frac{P}{\lim_{\epsilon\to 0}\limsup_{k\to\infty}\frac{E^*(k,\epsilon)}{k}} = \frac{P}{N_0}\log_2 e,$$

where the last step is by Theorem 21.2. □

21.5* Capacity of the Continuous-Time Band-Limited AWGN Channel

An engineer looking at the previous theorem will immediately point out an issue with the definition of an error correcting code. Namely, we allowed the waveform $x(t)$ to have bounded duration and bounded power, but did not constrain its frequency content. In practice, waveforms are also required to occupy a certain limited band of B Hz. What is the capacity of the AWGN channel subject to both the power P and the bandwidth B constraints?

Unfortunately, answering this question rigorously requires a long and delicate digression into functional analysis and prolate spheroidal functions. We thus only sketch the main result, without stating it as a rigorous theorem. For a full treatment, consider the monograph of Ihara [224].

Let us again define $M^*_{CT}(T,\epsilon,P,B)$ to be the maximum number of waveforms that can be sent with probability of error ϵ in time T, power P, and so that each waveform in addition to those two constraints also has Fourier spectrum entirely contained in $[f_c - B/2, f_c + B/2]$, where f_c is a certain "carrier" frequency.[1]

We claim that

$$C_B(P) \triangleq \lim_{\epsilon\to 0} \liminf_{n\to\infty} \frac{1}{T} \log M^*_{CT}(T, \epsilon, P, B) = B \log\left(1 + \frac{P}{N_0 B}\right). \qquad (21.13)$$

In other words, the capacity of this channel is $B \log\left(1 + \frac{P}{N_0 B}\right)$. To understand the idea of the proof, we need to recall the concept of modulation first. Every signal $X(t)$ that is required to live in the $[f_c - B/2, f_c + B/2]$ frequency band can be obtained by starting with a *complex*-valued signal $X_B(t)$ with frequency content in $[-B/2, B/2]$ and mapping it to $X(t)$ via the modulation:

$$X(t) = \mathrm{Re}(X_B(t)\sqrt{2}e^{j\omega_c t}),$$

where $\omega_c = 2\pi f_c$. Upon receiving the sum $Y(t) = X(t) + N(t)$ of the signal and the white noise $N(t)$ we may demodulate Y by computing

$$Y_B(t) = \sqrt{2}\mathrm{LPF}(Y(t)e^{j\omega_c t}),$$

where the LPF is a low-pass filter removing all frequencies beyond $[-B/2, B/2]$. The important fact is that converting from $Y(t)$ to $Y_B(t)$ does not lose information.

Overall we have the following input–output relation:

$$Y_B(t) = X_B(t) + \widetilde{N}(t),$$

where all processes are $\mathbb{C}$-valued, $\widetilde{N}(t)$ is a complex Gaussian white noise, and

$$\mathbb{E}[\widetilde{N}(t)\widetilde{N}(s)^*] = N_0\delta(t - s).$$

(Notice that after demodulation, the power spectral density of the noise is $N_0/2$ with $N_0/4$ in the real part and $N_0/4$ in the imaginary part, and after the $\sqrt{2}$ amplifier the spectral density of the noise is restored to $N_0/2$ in both real and imaginary parts.)

Next, we do Nyquist sampling to convert from continuous time to discrete time. Namely, the input waveform $X_B(t)$ is going to be represented by its values at an equispaced grid of time instants, separated by $\frac{1}{B}$. A similar representation is done to $Y_B(t)$. It is again known that these two operations do not lead to either restriction of the space of input waveforms (since every band-limited signal can be uniquely represented by its samples at Nyquist rate) or loss of the information content in $Y_B(t)$ (again, Nyquist samples represent the signal Y_B completely). Mathematically, what we have done is

[1] Here we already encounter a major issue: the waveform $x(t)$ supported on a finite interval $(0, T]$ cannot have a compactly supported spectrum. The requirements of finite duration and finite spectrum are only satisfied by the zero waveform. Rigorously, one should relax the bandwidth constraint to requiring that the signal have a vanishing out-of-band energy as $T \to \infty$. As we said, rigorous treatment of this issue led to the theory of prolate spheroidal functions [392].

$$X_B(t) = \sum_{i=-\infty}^{\infty} X_i \operatorname{sinc}_B\left(t - \frac{i}{B}\right),$$
$$Y_i = \int_{t=-\infty}^{\infty} Y_B(t) \operatorname{sinc}_B\left(t - \frac{i}{B}\right) dt,$$

where $\operatorname{sinc}_B(x) = \frac{\sin(Bx)}{x}$ and $X_i = X_B(i/B)$. After the Nyquist sampling on X_B and Y_B we get the following equivalent input–output relation:

$$Y_i = X_i + Z_i, \quad Z_i \sim \mathcal{N}_c(0, N_0), \tag{21.14}$$

where the noise $Z_i = \int_{t=-\infty}^{\infty} \widetilde{N}(t) \operatorname{sinc}_B(t - \frac{i}{B})dt$. Finally, given that $X_B(t)$ is only non-zero for $t \in (0, T]$ we see that the $\mathbb{C}$-AWGN channel (21.14) is only allowed to be used for $i = 1, \ldots, TB$. This fact is known in communication theory as "bandwidth B and duration T signal has BT complex degrees of freedom."

Let us summarize what we have obtained so far:

- After sampling the equivalent channel model is that of discrete-time $\mathbb{C}$-AWGN.
- Given time T and bandwidth B the discrete-time equivalent channel has blocklength $n = BT$.
- The power constraint in the discrete-time model corresponds to:

$$\sum_{i=1}^{BT} |X_i|^2 = \|X(t)\|_2^2 \le PT.$$

Thus the effective discrete-time power constraint becomes $P_d = \frac{P}{B}$.

Hence, we have established the following fact:

$$\frac{1}{T} \log M^*_{CT}(T, \epsilon, P, B) = \frac{1}{T} \log M^*_{\mathbb{C}\_\mathrm{AWGN}}(BT, \epsilon, P_d),$$

where $M^*_{\mathbb{C}\_\mathrm{AWGN}}$ denotes the fundamental limit of the $\mathbb{C}$-AWGN channel from Theorem 20.11. Thus, taking $T \to \infty$ we get (21.13).

Note also that in the limit of large bandwidth B the capacity formula (21.13) yields

$$C_{B=\infty}(P) = \lim_{B\to\infty} B \log\left(1 + \frac{P}{N_0 B}\right) = \frac{P}{N_0} \log e,$$

agreeing with (21.12).

22 Strong Converse, Channel Dispersion, Error Exponents, and Finite Blocklength

In previous chapters our main object of study was the fundamental limit of blocklength-n coding:

$$M^*(n,\epsilon) = \max\{M : \exists (n,M,\epsilon)\text{-code}\}.$$

Equivalently, we can define it in terms of the smallest probability of error at a given M:

$$\epsilon^*(n,M) = \inf\{\epsilon : \exists (n,M,\epsilon)\text{-code}\}.$$

What we have learned so far is that for stationary memoryless channels we have

$$\lim_{\epsilon\to 0}\liminf_{n\to\infty}\frac{1}{n}\log M^*(n,\epsilon) = C,$$

or, equivalently,

$$\limsup_{n\to\infty}\epsilon^*(n,\exp\{nR\})\begin{cases} = 0, & R < C, \\ > 0, & R > C. \end{cases}$$

These results were proved 75 years ago by Shannon. What happened in the ensuing 75 years is that we obtained much more detailed information about M^* and ϵ^*. For example, the strong converse says that in the previous limit the "> 0" can be replaced with "$= 1$." The error exponents show that the convergence of $\epsilon^*(n,\exp\{nR\})$ to zero or one happens exponentially fast (with exponents known for certain ranges of rates R). The channel dispersion refines the asymptotic description to

$$\log M^*(n,\epsilon) = nC - \sqrt{nV}Q^{-1}(\epsilon) + O(\log n).$$

Finally, the finite-blocklength information theory strives to prove the sharpest possible *computational* bounds on $\log M^*(n,\epsilon)$ at finite n, which allows evaluating real-world codes' performance taking their latency n into account. These results are surveyed in this chapter.

22.1 Strong Converse

We begin by stating the main theorem.

Theorem 22.1. *For any stationary memoryless channel with either $|\mathcal{A}| < \infty$ or $|\mathcal{B}| < \infty$ we have $C_\epsilon = C$ for $0 < \epsilon < 1$. Equivalently, for every $0 < \epsilon < 1$ we have*

$$\log M^*(n, \epsilon) = nC + o(n), \quad n \to \infty.$$

Previously in Theorem 19.7, we showed that $C \le C_\epsilon \le \frac{C}{1-\epsilon}$. Now we are asserting that equality holds for every ϵ. Our previous converse arguments (Theorem 17.3 based on Fano's inequality) showed that communication with an arbitrarily small error probability is possible only when using rate $R < C$. The strong converse shows that when communicating at any rate above capacity, $R > C$, the probability of error in fact goes to 1. (An even more detailed result of Arimoto characterizes the speed of convergence to 1 as exponential in n and gives an exact expression for the exponent.) In other words,

$$\epsilon^*(n, \exp(nR)) \to \begin{cases} 0, & R < C, \\ 1, & R > C, \end{cases} \tag{22.1}$$

where $\epsilon^*(n, M)$ is the inverse of $M^*(n, \epsilon)$ defined in (19.5).

In practice, engineers observe this effect differently. Instead of changing the coding rate, they fix a code and then allow the channel parameter (SNR for the AWGN channel, or δ for BSC_δ) to vary. This typically results in a *waterfall plot* for the probability of error:

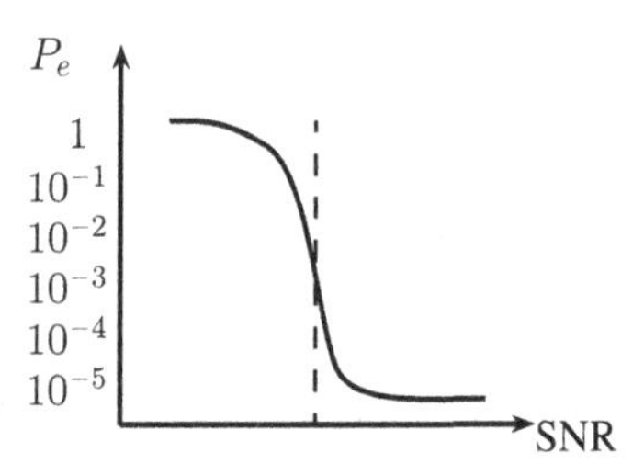

In other words, below a certain critical SNR, the probability of error quickly approaches 1, so that the receiver cannot decode anything meaningful. Above the critical SNR the probability of error quickly approaches 0 (unless there is an effect known as the *error floor*, in which case the probability of error decreases, reaches that floor value, and stays there regardless of the further SNR increase). Thus, long-blocklength codes have a threshold-like behavior of probability of error similar to (22.1). Besides changing SNR instead of rate, there is another important difference between a waterfall plot and (22.1). The former applies to only a single (perhaps rather suboptimal) code, while the latter is a statement about the best possible code for each (n, R) pair.

Proof. We will improve the method used in the proof of Theorem 17.3. Take an (n, M, ϵ)-code and consider the usual probability space

$$W \to X^n \to Y^n \to \hat{W},$$

where $W \sim \text{Unif}([M])$. Note that P_{X^n} is the empirical distribution induced by the encoder at the channel input. We denote the joint measure on $(W, X^n, Y^n, \hat{W})$ induced in this way by $\mathbb{P}$. Our goal is to replace this probability space with a different one where the true channel $P_{Y^n|X^n} = P_{Y|X}^{\otimes n}$ is replaced with an *auxiliary* channel (which is a "dummy" one in this case):

$$Q_{Y^n|X^n} = (Q_Y)^{\otimes n}.$$

We will denote the measure on $(W, X^n, Y^n, \hat{W})$ induced by this new channel by $\mathbb{Q}$. Note that for communication purposes, $Q_{Y^n|X^n}$ is a useless channel since it ignores the input and randomly picks a member of the output space according to $Y_i \overset{\text{iid}}{\sim} Q_Y$, so that X^n and Y^n are independent (under $\mathbb{Q}$). Therefore, for the probability of success under each channel we have

$$\mathbb{Q}[\hat{W} = W] = \frac{1}{M},$$
$$\mathbb{P}[\hat{W} = W] \geq 1 - \epsilon.$$

Therefore, the random variable $1\{\hat{W} = W\}$ is likely to be 1 under $\mathbb{P}$ and likely to be 0 under $\mathbb{Q}$. It thus looks like a rather good choice for a binary hypothesis test statistic distinguishing the two distributions, $P_{W,X^n,Y^n,\hat{W}}$ and $Q_{W,X^n,Y^n,\hat{W}}$. Since no hypothesis test can beat the optimal (Neyman–Pearson) test, we get the upper bound

$$\beta_{1-\epsilon}(P_{W,X^n,Y^n,\hat{W}}, Q_{W,X^n,Y^n,\hat{W}}) \leq \frac{1}{M}. \tag{22.2}$$

(Recall the definition of β from (14.3).) The likelihood ratio is a sufficient statistic for this hypothesis test, so let us compute it:

$$\frac{P_{W,X^n,Y^n,\hat{W}}}{Q_{W,X^n,Y^n,\hat{W}}} = \frac{P_W P_{X^n|W} P_{Y^n|X^n} P_{\hat{W}|Y^n}}{P_W P_{X^n|W} (Q_Y)^{\otimes n} P_{\hat{W}|Y^n}} = \frac{P_{W|X^n} P_{X^n,Y^n} P_{\hat{W}|Y^n}}{P_{W|X^n} P_{X^n} (Q_Y)^{\otimes n} P_{\hat{W}|Y^n}} = \frac{P_{X^n,Y^n}}{P_{X^n}(Q_Y)^{\otimes n}}.$$

Therefore, the inequality above becomes

$$\beta_{1-\epsilon}(P_{X^n,Y^n}, P_{X^n}(Q_Y)^{\otimes n}) \leq \frac{1}{M}, \tag{22.3}$$

where $\beta_{1-\epsilon}$ was defined in (14.3). Computing the LHS of this bound may appear to be impossible because the distribution P_{X^n} depends on the unknown code. However, it will turn out that a judicious choice of Q_Y will make knowledge of P_{X^n} unnecessary. Before presenting a formal argument, let us consider a special case of the BSC_δ channel. It will show that for symmetric channels we can select Q_Y to be the capacity-achieving output distribution (recall that it is unique by Corollary 5.5). To treat the general case later we will (essentially) decompose the channel into symmetric subchannels (corresponding to "composition" of the input).

Special case BSC_δ. So let us take $P_{Y^n|X^n} = \text{BSC}_\delta^{\otimes n}$ and for Q_Y we will take the capacity-achieving output distribution, which is simply $Q_Y = \text{Ber}(1/2)$, obtaining

$$P_{Y^n|X^n}(y^n|x^n) = P_Z^n(y^n - x^n), \quad Z^n \sim \text{Ber}(\delta)^n,$$
$$(Q_Y)^{\otimes n}(y^n) = 2^{-n}.$$

From the Neyman–Pearson lemma, the optimal hypothesis test takes the form

$$\beta_\alpha(\underbrace{P_{X^nY^n}}_{\mathbb{P}}, \underbrace{P_{X^n}(Q_Y)^{\otimes n}}_{\mathbb{Q}}) = \mathbb{Q}\left[\log \frac{P_{X^nY^n}}{P_{X^n}(Q_Y)^{\otimes n}} \geq \gamma\right]$$

$$\text{where } \alpha = \mathbb{P}\left[\log \frac{P_{X^nY^n}}{P_{X^n}(Q_Y)^{\otimes n}} \geq \gamma\right].$$

For the BSC, this becomes

$$\log \frac{P_{X^nY^n}}{P_{X^n}(P_Y^*)^n} = \log \frac{P_{Z^n}(y^n - x^n)}{2^{-n}}.$$

Notice that the effect of unknown P_{X^n} has completely disappeared, and so we can compute β_α:

$$\begin{aligned}
&\beta_\alpha(P_{X^nY^n}, P_{X^n}(Q_Y)^{\otimes n}) \\
&\quad = \beta_\alpha(\mathrm{Ber}(\delta)^{\otimes n}, \mathrm{Ber}(1/2)^{\otimes n}) \\
&\quad = \exp\{-nD(\mathrm{Ber}(\delta)\|\mathrm{Ber}(1/2)) + o(n)\} \quad \text{(by Stein's lemma: Theorem 14.14).}
\end{aligned} \tag{22.4}$$

Putting this together with our main bound (22.3), we see that any $(n, M, \epsilon)-$ code for the BSC satisfies

$$\log M \leq nD(\mathrm{Ber}(\delta)\|\mathrm{Ber}(1/2)) + o(n) = nC + o(n).$$

Clearly, this implies the strong converse for the BSC. (For a slightly different, but equivalent, proof, see Exercise **IV.32** and for the AWGN channel, see Exercise **IV.33**).

For the *general channel*, let us denote by P_Y^* the capacity-achieving output distribution. Recall that by Corollary 5.5 it is unique and by (5.1) we have for every $x \in \mathcal{A}$:

$$D(P_{Y|X=x}\|P_Y^*) \leq C. \tag{22.5}$$

This property will be very useful. We next consider two cases separately:

1 If $|\mathcal{B}| < \infty$ we take $Q_Y = P_Y^*$ and note that from (19.31) we have

$$\sum_y P_{Y|X}(y|x_0)\log^2 P_{Y|X}(y|x_0) \leq \log^2|\mathcal{B}| \quad \forall x_0 \in \mathcal{A},$$

and since $\min_y P_Y^*(y) > 0$ (without loss of generality), we conclude that for some constant $K > 0$ and for all $x_0 \in \mathcal{A}$ we have

$$\mathrm{Var}\left[\log \frac{P_{Y|X}(Y|X = x_0)}{Q_Y(Y)} \,\middle|\, X = x_0\right] \leq K < \infty.$$

Thus, if we let

$$S_n = \sum_{i=1}^n \log \frac{P_{Y|X}(Y_i|X_i)}{P_Y^*(Y_i)},$$

then we have

$$\mathbb{E}[S_n|X^n] \leq nC, \quad \mathrm{Var}[S_n|X^n] \leq Kn. \tag{22.6}$$

Hence from the Chebyshev inequality (applied conditional on X^n), we have

$$\mathbb{P}[S_n > nC + \lambda\sqrt{Kn}] \le \mathbb{P}[S_n > \mathbb{E}[S_n|X^n] + \lambda\sqrt{Kn}] \le \frac{1}{\lambda^2}. \tag{22.7}$$

2 If $|\mathcal{A}| < \infty$, then first we recall that without loss of generality the encoder can be taken to be deterministic. Then for each codeword $c \in \mathcal{A}^n$ we define its *composition* (also known as *type*) to simply be its empirical distribution:

$$\hat{P}_c(x) \triangleq \frac{1}{n}\sum_{j=1}^{n} 1\{c_j = x\}.$$

By simple counting[1] it is clear that from any (n, M, ϵ)-code, it is possible to select an (n, M', ϵ)-subcode, such that (a) all codewords have the same composition P_0; and (b) $M' > \frac{M}{(n+1)^{|\mathcal{A}|-1}}$. Note that $\log M = \log M' + O(\log n)$ and thus we may replace M with M' and focus on the analysis of the chosen subcode. Then we set $Q_Y = P_{Y|X} \circ P_0$. From now on we also assume that $P_0(x) > 0$ for all $x \in \mathcal{A}$ (otherwise just reduce $\mathcal{A}$). Let $i(x;y)$ denote the information density with respect to $P_0 P_{Y|X}$. If $X \sim P_0$ then $I(X;Y) = D(P_{Y|X}\|Q_Y|P_0) \le \log|\mathcal{A}| < \infty$ and we conclude that $P_{Y|X=x} \ll Q_Y$ for each x and thus

$$i(x;y) = \log\frac{dP_{Y|X=x}}{dQ_Y}(y).$$

From (19.28) we have

$$\mathrm{Var}\,[i(X;Y)|X] \le \mathrm{Var}[i(X;Y)] \le K < \infty.$$

Furthermore, we also have

$$\mathbb{E}\,[i(X;Y)|X] = D(P_{Y|X}\|Q_Y|P_0) = I(X;Y) \le C, \quad X \sim P_0.$$

So if we define

$$S_n = \sum_{i=1}^{n} \log\frac{dP_{Y|X=X_i}(Y_i|X_i)}{dQ_Y}(Y_i) = \sum_{i=1}^{n} i(X_i;Y_i),$$

we again first get the estimates (22.6) and then (22.7).

To proceed with (22.3) we apply the lower bound on β from (14.9):

$$\gamma\beta_{1-\epsilon}(P_{X^n,Y^n}, P_{X^n}(Q_Y)^{\otimes n}) \ge 1 - \epsilon - \mathbb{P}[S_n > \log\gamma],$$

where γ is arbitrary. We set $\log\gamma = nC + \lambda\sqrt{Kn}$ and $\lambda^2 = \frac{2}{1-\epsilon}$ to obtain (via (22.7)):

$$\gamma\beta_{1-\epsilon}(P_{X^n,Y^n}, P_{X^n}(Q_Y)^{\otimes n}) \ge \frac{1-\epsilon}{2},$$

which then implies

$$\log\beta_{1-\epsilon}(P_{X^nY^n}, P_{X^n}(Q_Y)^n) \ge -nC + O(\sqrt{n}).$$

[1] This kind of reduction from a general code to a constant-composition subcode is the essence of the method of types [115].

Consequently, from (22.3) we conclude that

$$\log M \leq nC + O(\sqrt{n}),$$

implying the strong converse. □

We note several lessons from this proof. First, we basically followed the same method as in the proof of the weak converse, except instead of invoking the data-processing inequality for divergence, we analyzed the hypothesis testing problem explicitly. Second, the bound on the variance of the information density is important. Thus, while the AWGN channel is excluded by the assumptions of the theorem, the strong converse for it does hold as well (see Exercise **IV.33**). Third, this method of proof is also known as "sphere-packing bound" for the reason that becomes clear if we do the example of the BSC slightly differently (see Exercise **IV.32**).

22.2 Stationary Memoryless Channel without Strong Converse

In the proof above we basically only used the fact that the sum of independent random variables concentrates around its expectation (we used the second moment to show that, but it could have been done more generally, when the second moment does not exist). Thus one may wonder whether the strong converse should hold for *all* stationary memoryless channels (it was shown in Theorem 22.1 only for those with finite input or output spaces). In this section we construct a counterexample.

Let the output alphabet be $\mathcal{B} = [0,1]$. The input $\mathcal{A}$ is going to be countably infinite. It will be convenient to define it as

$$\mathcal{A} = \{(j,m) : j,m \in \mathbb{Z}_+, 0 \leq j < m\}.$$

The single-letter channel $P_{Y|X}$ is defined in terms of the probability density function as

$$p_{Y|X}(y|(j,m)) = \begin{cases} a_m, & \frac{j}{m} \leq y \leq \frac{j+1}{m}, \\ b_m, & \text{otherwise}, \end{cases}$$

where a_m, b_m are chosen to satisfy

$$\frac{1}{m}a_m + \left(1 - \frac{1}{m}\right) b_m = 1, \tag{22.8}$$

$$\frac{1}{m}a_m \log a_m + \left(1 - \frac{1}{m}\right) b_m \log b_m = C, \tag{22.9}$$

where $C > 0$ is an arbitrary fixed constant. Note that for large m we have

$$a_m = \frac{mC}{\log m}\left(1 + O\left(\frac{1}{\log m}\right)\right), \tag{22.10}$$

$$b_m = 1 - \frac{C}{\log m} + O\left(\frac{1}{\log^2 m}\right). \tag{22.11}$$

It is easy to see that $P^*_Y = \mathrm{Unif}(0,1)$ is the capacity-achieving output distribution and

$$\sup_{P_X} I(X;Y) = C.$$

Thus by Theorem 19.9 the capacity of the corresponding stationary memoryless channel is C. We next show that nevertheless the ϵ-capacity can be strictly greater than C.

Indeed, fix blocklength n and consider a *single-letter* distribution P_X assigning equal weights to all atoms (j,m) with $m = \exp\{2nC\}$. It can be shown that in this case, the distribution of a single-letter information density is given by

$$i(X;Y) = \begin{cases} \log a_m, & \text{w.p. } \frac{a_m}{m} \\ \log b_m, & \text{w.p. } 1 - \frac{a_m}{m}, \end{cases} = \begin{cases} 2nC + O(\log n), & \text{w.p. } \frac{a_m}{m}, \\ O(\frac{1}{n}), & \text{w.p. } 1 - \frac{a_m}{m}. \end{cases}$$

Thus, for blocklength-n density we have

$$\frac{1}{n} i(X^n;Y^n) = \frac{1}{n}\sum_{i=1}^n i(X_i;Y_i) \stackrel{d}{=} O\left(\frac{1}{n}\right) + \left(2C + O\left(\frac{1}{n}\log n\right)\right) \cdot \mathrm{Bin}\left(n, \frac{a_m}{m}\right) \stackrel{d}{\to} 2C \cdot \mathrm{Poisson}(1/2),$$

where we used the fact that $\frac{a_m}{m} = (1 + o(1))\frac{1}{2n}$ and invoked the Poisson limit theorem for the binomial. Therefore, from Theorem 18.5 we get that for $\epsilon > e^{-1/2}$ there exist (n, M, ϵ)-codes with

$$\log M \geq 2nC(1 + o(1)).$$

In particular,

$$C_\epsilon \geq 2C \quad \forall \epsilon > e^{-1/2}.$$

22.3 Meta-converse

We have seen various ways in which one can derive upper (impossibility or converse) bounds on the fundamental limits such as $\log M^*(n,\epsilon)$. In Theorem 17.3 we used data-processing and Fano's inequalities. In the proof of Theorem 22.1 we reduced the problem to that of hypothesis testing. There are many other converse bounds that have been developed over the years. It turns out that there is a very general approach that encompasses all of them. For its versatility it is sometimes referred to as the "meta-converse."

To describe it, let us fix a Markov kernel $P_{Y|X}$ (usually, it will be the n-letter channel $P_{Y^n|X^n}$, but in the spirit of the "one-shot" approach, we avoid introducing blocklength). We are also given a certain (M,ϵ)-code and the goal is to show that there is an upper bound on M in terms of $P_{Y|X}$ and ϵ. The essence of the meta-converse is described by the following diagram:

$$W \longrightarrow X^n \xrightarrow[Q_{Y|X}]{P_{Y|X}} Y^n \longrightarrow \hat{W}$$

Here $W \to X$ and $Y \to \hat{W}$ represent the encoder and decoder of our fixed (M, ϵ)-code. The upper arrow $X \to Y$ corresponds to the actual channel, whose fundamental limits we are analyzing. The lower arrow is an *auxiliary channel* that we are free to select.

The $P_{Y|X}$ or $Q_{Y|X}$ together with P_X (distribution induced by the code) define two distributions: $P_{X,Y}$ and $Q_{X,Y}$. Consider a map $(X, Y) \mapsto Z \triangleq 1\{W \neq \hat{W}\}$ defined by the encoder and decoder pair (if decoders are randomized or $W \to X$ is not injective, we consider a Markov kernel $P_{Z|X,Y}(1|x, y) = \mathbb{P}[Z = 1|X = x, Y = y]$ instead). We have

$$P_{X,Y}[Z = 0] = 1 - \epsilon, \quad Q_{X,Y}[Z = 0] = 1 - \epsilon',$$

where ϵ and ϵ' are the average probabilities of error of the given code under $P_{Y|X}$ and $Q_{Y|X}$ respectively. This implies the following relation for the binary hypothesis testing problem of testing $P_{X,Y}$ versus $Q_{X,Y}$:

$$\beta_{1-\epsilon}(P_{X,Y}, Q_{X,Y}) \leq 1 - \epsilon'.$$

The high-level idea of the meta-converse is to select a convenient $Q_{Y|X}$, bound $1-\epsilon'$ from above (i.e. prove a converse result for the $Q_{Y|X}$), and then use the Neyman–Pearson β-function to *lift the Q-channel converse to the P-channel.*

How one chooses $Q_{Y|X}$ is a matter of art. For example, in the proof of case 2 for the general channel in Theorem 22.1 we used the trick of reducing to the constant-composition subcode. This can instead be done by taking $Q_{Y^n|X^n=c} = (P_{Y|X} \circ \hat{P}_c)^{\otimes n}$. Since there are at most $(n+1)^{|\mathcal{A}|-1}$ different output distributions, we can see that

$$1 - \epsilon' \leq \frac{(n+1)^{|\mathcal{A}|-1}}{M},$$

and bounding of β can be done similar to the case 2 proof of Theorem 22.1. For channels with $|\mathcal{A}| = \infty$ the technique of reducing to constant-composition codes is not available, but the meta-converse can still be applied. Examples include the proof of the parallel AWGN channel's dispersion [334, Theorem 78] and the study of the properties of good codes [341, Theorem 21].

However, the most common way of using the meta-converse is to apply it with the trivial channel $Q_{Y|X} = Q_Y$. We have already seen this idea in Section 22.1. Indeed, with this choice the proof of the converse for the Q-channel is trivial, because we always have: $1 - \epsilon' = \frac{1}{M}$. Therefore, we conclude that any (M, ϵ)-code must satisfy

$$\frac{1}{M} \geq \beta_{1-\epsilon}(P_{X,Y}, P_X Q_Y). \tag{22.12}$$

Or, after optimization, we obtain

$$\frac{1}{M^*(\epsilon)} \geq \inf_{P_X} \sup_{Q_Y} \beta_{1-\epsilon}(P_{X,Y}, P_X Q_Y).$$

This is a special case of the meta-converse known as the *minimax meta-converse*. It has a number of interesting properties. First, the minimax problem in question possesses a saddle point and is of convex–concave type [342]. It, thus, can be seen as a stronger version of the capacity saddle-point result for divergence in Theorem 5.4.

Second, the bound given by the minimax meta-converse coincides with the bound we obtained before via linear programming relaxation (18.22), as discovered by [296]. To see this connection, instead of writing the meta-converse as an upper bound M (for a given ϵ) let us think of it as an upper bound on $1-\epsilon$ (for a given M).

We have seen that the existence of an (M, ϵ)-code for $P_{Y|X}$ implies the existence of the (stochastic) map $(X, Y) \mapsto Z \in \{0, 1\}$, denoted by $P_{Z|X,Y}$, with the following property:

$$P_{X,Y}[Z=0] \geq 1-\epsilon \quad \text{and} \quad P_X Q_Y[Z=0] \leq \frac{1}{M} \quad \forall Q_Y.$$

That is, $P_{Z|X,Y}$ is a test of a simple null hypothesis $(X, Y) \sim P_{X,Y}$ against a composite alternative $(X, Y) \sim P_X Q_Y$ for an arbitrary Q_Y. In other words every (M, ϵ)-code must satisfy

$$1-\epsilon \leq \tilde{\alpha}(M; P_X),$$

where (we are assuming finite $\mathcal{X}, \mathcal{Y}$ for simplicity)

$$\tilde{\alpha}(M; P_X) \triangleq \sup_{P_{Z|X,Y}} \left\{ \sum_{x,y} P_{X,Y}(x, y) P_{Z|X,Y}(0|x, y) : \right.$$
$$\left. \sum_{x,y} P_X(x) Q_Y(y) P_{Z|X,Y}(0|x, y) \leq \frac{1}{M} \quad \forall Q_Y \right\}.$$

We can simplify the constraint by rewriting it as

$$\sup_{Q_Y} \sum_{x,y} P_X(x) Q_Y(y) P_{Z|X,Y}(0|x, y) \leq \frac{1}{M},$$

and further simplifying it to

$$\sum_{x} P_X(x) P_{Z|X,Y}(0|x, y) \leq \frac{1}{M}, \quad \forall y \in \mathcal{Y}.$$

Let us now replace P_X with a $\pi_x \triangleq M P_X(x), x \in \mathcal{X}$. It is clear that $\pi \in [0, 1]^{\mathcal{X}}$. Let us also replace the optimization variable with $r_{x,y} \triangleq M P_{Z|X,Y}(0|x, y) P_X(x)$. With these notational changes we obtain

$$\tilde{\alpha}(M;P_X) = \frac{1}{M}\sup\left\{\sum_{x,y} P_{Y|X}(y|x)r_{x,y}\colon 0 \le r_{x,y} \le \pi_x, \sum_x r_{x,y} \le 1\right\}.$$

It is now obvious that $\tilde{\alpha}(M;P_X) = S_{LP}(\pi)$ defined in (18.21). Optimizing over the choice of P_X (or equivalently π with $\sum_x \pi_x \le M$) we obtain

$$1-\epsilon \le \frac{1}{M}S_{LP}(\pi) \le \frac{1}{M}\sup\left\{S_{LP}(\pi)\colon \sum_x \pi_x \le M\right\} = \frac{S^*_{LP}(M)}{M}.$$

Now recall that in (18.23) we showed that a greedy procedure (essentially, the same as the one we used in the Feinstein's bound, Theorem 18.7) produces a code with probability of success

$$1-\epsilon \ge \left(1-\frac{1}{e}\right)\frac{S^*_{LP}(M)}{M}.$$

This indicates that in the regime of a fixed ϵ the bound based on the minimax meta-converse should be very sharp. This is, of course, provided we can select the best Q_Y in applying it. Fortunately, for symmetric channels, optimal Q_Y can be guessed fairly easily, see [342] for more.

22.4* Error Exponents

We have studied the question of optimal error exponents in hypothesis testing before (Chapter 16). The corresponding topic for channel coding is much harder and full of open problems.

We motivate the question by trying to understand the speed of convergence in the strong converse (22.1). If we return to the proof of Theorem 19.9, namely the step (19.7), we see that by applying the large-deviations (Theorem 15.9), we can prove that for some $\tilde{E}(R)$ and any $R < C$ we have

$$\epsilon^*(n, \exp\{nR\}) \le \exp\{-n\tilde{E}(R)\}.$$

What is the best value of $\tilde{E}(R)$ for each R? This is perhaps the most famous open question in all of channel coding. Let us elaborate.

We will treat both regimes $R < C$ and $R > C$. The *reliability function* $E(R)$ of a channel is defined as follows:

$$E(R) = \begin{cases} \lim_{n\to\infty} -\frac{1}{n}\log\epsilon^*(n,\exp\{nR\}), & R < C, \\ \lim_{n\to\infty} -\frac{1}{n}\log(1-\epsilon^*(n,\exp\{nR\})), & R > C. \end{cases}$$

We leave $E(R)$ as undefined if the limit does not exist. Unfortunately, there is no general argument showing that this limit exists. The only way to show its existence is to prove an achievability bound,

$$\liminf_{n\to\infty} -\frac{1}{n}\log\epsilon^*(n,\exp\{nR\}) \ge E_{\text{lower}}(R),$$

a converse bound,

$$\limsup_{n\to\infty} -\frac{1}{n}\log \epsilon^*(n, \exp\{nR\}) \le E_{\text{upper}}(R),$$

and conclude that the limit exists whenever $E_{\text{lower}} = E_{\text{upper}}$. It is common to abuse notation and write such pair of bounds as

$$E_{\text{lower}}(R) \le E(R) \le E_{\text{upper}}(R),$$

even though, as we said, $E(R)$ is not known to exist unless the two bounds match.

From now on we restrict our discussion to the case of a DMC. An important object to define is Gallager's E_0 function, which is nothing else than the right-hand side of Gallager's bound (18.15). For the DMC it has the following expression:

$$\begin{aligned} E_0(\rho, P_X) &= -\log \sum_{y\in\mathcal{B}}\left(\sum_{x\in\mathcal{A}} P_X(x) P_{Y|X}^{1/(1+\rho)}(y|x)\right)^{1+\rho}, \\ E_0(\rho) &= \max_{P_X} E_0(\rho, P_X), \qquad \rho \ge 0, \\ E_0(\rho) &= \min_{P_X} E_0(\rho, P_X), \qquad \rho \le 0. \end{aligned}$$

This expression is defined in terms of the single-letter channel $P_{Y|X}$. It is not hard to see that the E_0 function for the n-letter extension evaluated with $P_X^{\otimes n}$ just equals $nE_0(\rho, P_X)$, that is, it tensorizes similar to mutual information.[2] From this observation we can apply Gallager's random coding bound (Theorem 18.9) with $P_X^{\otimes n}$ to obtain

$$\epsilon^*(n, \exp\{nR\}) \le \exp\{n(\rho R - E_0(\rho, P_X))\} \quad \forall P_X, \rho \in [0,1]. \tag{22.13}$$

Optimizing the choice of P_X we obtain our first estimate on the reliability function:

$$E(R) \le E_r(R) \triangleq \sup_{\rho\in[0,1]} E_0(\rho) - \rho R.$$

An analysis, for example, [177, Section 5.6], shows that the function $E_r(R)$ is convex, decreasing, and strictly positive on $0 \le R < C$. Therefore, Gallager's bound provides a non-trivial estimate of the reliability function for the entire range of rates below capacity. At rates $R \to C$ the optimal choice of $\rho \to 0$. As R departs further away from the capacity the optimal ρ reaches 1 at a certain rate $R = R_{cr}$ known as the critical rate, so that for $R < R_{cr}$ we have $E_r(R) = E_0(1) - R$ behaving linearly. The $E_r(R)$ bound is shown on Figure 22.1 by the curve labeled "Random code ensemble."

Going to the upper bounds, taking Q_Y to be the iid product distribution in (22.12) and optimizing yields the bound [382] known as the sphere-packing bound:

[2] There is one more very pleasant analogy with mutual information: the optimization problems in the definition of $E_0(\rho)$ also tensorize. That is, the optimal distribution for the n-letter channel is just $P_X^{\otimes n}$, where P_X is optimal for a single-letter one.

$$E(R) \leq E_{sp}(R) \triangleq \sup_{\rho \geq 0} E_0(\rho) - \rho R. \tag{22.14}$$

Comparing the definitions of E_{sp} and E_r we can see that for $R_{cr} < R < C$ we have

$$E_{sp}(R) = E(R) = E_r(R),$$

thus establishing the reliability function value for high rates. However, for $R < R_{cr}$ we have $E_{sp}(R) > E_r(R)$, so that $E(R)$ remains unknown. The $E_{sp}(R)$ bound is shown on Figure 22.1 by the curve labeled "Sphere-packing (volume)."

Both upper and lower bounds have classical improvements. The random coding bound can be improved via a technique known as *expurgation* showing

$$E(R) \geq E_{ex}(R),$$

and $E_{ex}(R) > E_r(R)$ for rates $R < R_x$ where $R_x \leq R_{cr}$ is the second critical rate; see Exercise **IV.31**. The $E_{ex}(R)$ bound is shown on Figure 22.1 by the curve labeled "Typical random linear code (aka expurgation)." (See below for the explanation of the naming.)

The sphere-packing bound can also be improved at low rates by analyzing a combinatorial packing problem and showing that any code must have a pair of codewords which are close (in terms of Hellinger distance between the induced output distributions) and concluding that confusing these two leads to a lower bound on probability of error (via (16.3)). This class of bounds is known as "minimum-distance"-based bounds and several of them are shown on Figure 22.1 with the strongest labeled "MRRW + mindist," corresponding to the currently best known minimum-distance upper bound due to [300]. (This bound, also known as a linear programming or JPL bound, has not seen improvements in 60 years and it is a long-standing open problem in combinatorics to do so.)

The straight-line bound [177, Theorem 5.8.2] allows one to interpolate between any minimum-distance bound and $E_{sp}(R)$. Unfortunately, these (classical) improvements tightly bound $E(R)$ at only one additional rate point $R = 0$:

$$E(0+) = E_{ex}(0).$$

This state of affairs has remained unchanged (for a general DMC) since the foundational work of Shannon, Gallager, and Berlekamp in 1967. As far as we know, the common belief is that $E_{ex}(R)$ is in fact the true value of $E(R)$ for all rates. As we mentioned above this is perhaps one of the most famous open problems in classical information theory.

We demonstrate these bounds (with the exception of the straight-line bound) on the reliability function on Figure 22.1 for the case of the BSC_δ. For this channel, there is an interesting interpretation of the expurgated bound. To explain it, let us recall the different ensembles of random codes that we discussed in Section 18.6. In particular, we had the Shannon ensemble (as used in Theorem 18.5) and the random linear code (either Elias or Gallager ensembles, we do not need to make a distinction here).

For either ensemble, it is known [178] that $E_r(R)$ is not just an estimate, but in fact the exact value of the exponent of the average probability of error (averaged

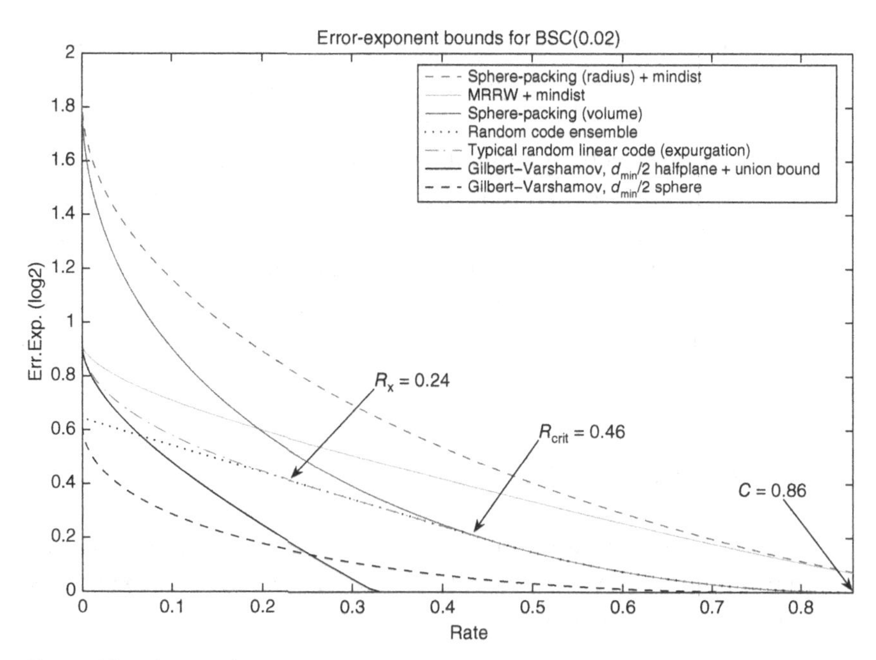

Figure 22.1 Comparison of bounds on the error exponent of the BSC. The MRRW stands for the upper bound on the minimum distance of a code [300] and Gilbert–Varshamov is a lower bound (see Theorem 27.5).

over a code in the ensemble). For either ensemble, however, for low rates the average is dominated by a few bad codes, whereas a typical (high-probability) realization of the code has a much better error exponent. For the Shannon ensemble this happens at $R < \frac{1}{2}R_x$ and for the linear ensemble it happens at $R < R_x$. Furthermore, the typical linear code in fact has error exponent exactly equal to the expurgated exponent $E_{ex}(R)$, see [34].

There is a famous conjecture in combinatorics stating that the best possible minimum pairwise Hamming distance of a code with rate R is given by the Gilbert–Varshamov bound (Theorem 27.5). If true, this would imply that $E(R) = E_{ex}(R)$ for $R < R_x$, see, for example, [284].

The most outstanding development in error exponents since 1967 was a sequence of papers starting from [284], which proposed a new technique for bounding $E(R)$ from above. Litsyn's idea was to first prove a geometric result (that any code of a given rate has a large number of pairs of codewords at a given distance) and then use de Caen's inequality to convert it into a lower bound on the probability of error. The resulting bound was very cumbersome. Thus, it was rather surprising when Barg and McGregor [35] were able to show that the new upper bound on $E(R)$ matched $E_r(R)$ for $R_{cr} - \epsilon < R < R_{cr}$ for some small $\epsilon > 0$. This, for the first time since [382] extended the range of knowledge of the reliability function. Their amazing result (together with the Gilbert–Varshamov conjecture) reinforced the belief that the typical linear codes achieve optimal error exponent in the whole range $0 \leq R \leq C$.

Regarding $E(R)$ for $R > C$ the situation is much simpler. We have

$$E(R) = \sup_{\rho \in (-1,0)} E_0(\rho) - \rho R.$$

The lower (achievability) bound here is due to Dueck [141] (see also [320]), while the harder (converse) part is by Arimoto [25]. It was later discovered that Arimoto's converse bound can be derived by a simple modification of the weak converse (Theorem 17.3): instead of applying the data-processing inequality to the KL divergence, one uses Rényi divergence of order $\alpha = \frac{1}{1+\rho}$; see [339] for details. This suggests a general conjecture that replacing Shannon information measures with Rényi ones upgrades the (weak) converse proofs to a strong converse.

22.5 Channel Dispersion

Historically, the first error-correcting codes had rather meager rates R very far from channel capacity. As we have seen in Section 22.4* the best codes at any rate $R < C$ have probability of error that behaves as

$$P_e = \exp\{-nE(R) + o(n)\}.$$

Therefore, for a while the question of non-asymptotic characterization of $\log M^*(n, \epsilon)$ and $\epsilon^*(n, M)$ was equated with establishing the sharp value of the error exponent $E(R)$. However, as codes have become better and started to have rates approaching the channel capacity, the question has changed to that of understanding the behavior of $\log M^*(n, \epsilon)$ in the regime of fixed ϵ and large n (and, thus, rates $R \to C$). It was soon discovered by [335] that the next-order terms in the asymptotic expansion of $\log M^*$ give surprisingly sharp estimates on the true value of $\log M^*$. Since then, work on channel coding has focused on establishing sharp upper and lower bounds on $\log M^*(n, \epsilon)$ for finite n (the topic of Section 22.6) and refining the classical results on asymptotic expansions, which we discuss here.

We have already seen that the strong converse (Theorem 22.1) can be stated in the asymptotic expansion form as: for every fixed $\epsilon \in (0, 1)$,

$$\log M^*(n, \epsilon) = nC + o(n), \quad n \to \infty.$$

Intuitively, though, smaller values of ϵ should make convergence to capacity slower. This suggests that the term $o(n)$ hides some interesting dependence on ϵ. What is it?

This topic has been investigated since the 1960s, see [130, 403, 335, 334], and resulted in the concept of *channel dispersion*. We first present a rigorous statement of the result and then explain its practical uses.

Theorem 22.2. *Consider one of the following channels:*

1 DMC
2 DMC with cost constraint
3 AWGN
4 Parallel AWGN

Let (X^*, Y^*) be the input–output of the channel under the capacity-achieving input distribution, and $i(x;y)$ be the corresponding (single-letter) information density. The following expansion holds for a fixed $0 < \epsilon < 1/2$ and $n \to \infty$:

$$\log M^*(n, \epsilon) = nC - \sqrt{nV} Q^{-1}(\epsilon) + O(\log n), \tag{22.15}$$

where Q^{-1} is the inverse of the complementary standard normal CDF, the channel capacity is $C = I(X^*;Y^*) = \mathbb{E}[i(X^*;Y^*)]$, and the channel dispersion[3] is $V = \mathrm{Var}[i(X^*;Y^*)|X^*]$.

Proof. The full proofs of these results are somewhat technical, even for the DMC.[4] Here we only sketch the details.

First, in the absence of cost constraints the achievability (lower bound on $\log M^*$) part has already been done by us in Theorem 19.11, where we have shown that $\log M^*(n, \epsilon) \geq nC - \sqrt{nV} Q^{-1}(\epsilon) + o(\sqrt{n})$ by refining the proof of the noisy channel coding theorem and using the CLT. Replacing the CLT with its non-asymptotic version (Berry–Esseen inequality [165, Theorem 2, Chapter XVI.5]) improves $o(\sqrt{n})$ to $O(\log n)$. In the presence of cost constraints, one is inclined to attempt to use an appropriate version of the achievability bound such as Theorem 20.7. However, for the AWGN this would require using an input distribution that is uniform on the sphere. Since this distribution is non-product, the information density ceases to be an iid sum, and CLT is harder to justify. Instead, there is a different achievability bound known as the κ–β bound [335, Theorem 25] that has become the workhorse of achievability proofs for cost-constrained channels with continuous input spaces.

The upper (converse) bound requires various special methods depending on the channel. However, the high-level idea is to always apply the meta-converse bound from (22.12) with an appropriate choice of Q_Y. Most often, Q_Y is taken as the nth power of the capacity-achieving output distribution for the channel. We illustrate the details for the special case of the BSC. In (22.4) we have shown that

$$\log M^*(n, \epsilon) \leq -\log \beta_\alpha(\mathrm{Ber}(\delta)^{\otimes n}, \mathrm{Ber}(1/2)^{\otimes n}). \tag{22.16}$$

On the other hand, Exercise **III.10** shows that

$$-\log \beta_{1-\epsilon}(\mathrm{Ber}(\delta)^{\otimes n}, \mathrm{Ber}(1/2)^{\otimes n}) = nd(\delta \| 1/2) + \sqrt{nv} Q^{-1}(\epsilon) + o(\sqrt{n}),$$

where v is just the variance of the (single-letter) log-likelihood ratio:

$$\begin{aligned} v &= \mathrm{Var}_{Z \sim \mathrm{Ber}(\delta)} \left[Z \log \frac{\delta}{\frac{1}{2}} + (1 - Z) \log \frac{1-\delta}{\frac{1}{2}} \right] = \mathrm{Var} \left[Z \log \frac{\delta}{1-\delta} \right] \\ &= \delta(1-\delta) \log^2 \frac{\delta}{1-\delta}. \end{aligned}$$

[3] There could be multiple capacity-achieving input distributions, in which case P_{X^*} should be chosen as the one that minimizes $\mathrm{Var}[i(X^*;Y^*)|X^*]$. See [335] for more details.

[4] Recently, subtle gaps in [403] and [335] in the treatment of DMCs with non-unique capacity-achieving input distributions were found and corrected in [81].

Upon inspection we notice that $v = V$, the channel dispersion of the BSC, which completes the proof of the upper bound:

$$\log M^*(n,\epsilon) \le nC - \sqrt{nV}Q^{-1}(\epsilon) + o(\sqrt{n}).$$

Improving the $o(\sqrt{n})$ to $O(\log n)$ is done by applying the Berry–Esseen inequality in place of the CLT, similar to the upper bound. Many more details on these proofs are contained in [334]. □

Remark 22.1. (Zero dispersion) We notice that $V = 0$ is entirely possible. For example, consider an additive-noise channel $Y = X + Z$ over some abelian group G with Z being uniform on some subset of G, for example, the channel in Exercise **IV.7**. Among the zero-dispersion channels there is a class of *exotic* channels [335], which for $\epsilon > 1/2$ have asymptotic expansions of the form [334, Theorem 51]:

$$\log M^*(n,\epsilon) = nC + \Theta_\epsilon(n^{1/3}).$$

The existence of this special case is why we restricted the theorem above to $\epsilon < \frac{1}{2}$.

Remark 22.2. The expansion (22.15) only applies to certain channels (as described in the theorem). If, for example, $\mathrm{Var}[i(X^*;Y^*)] = \infty$, then the theorem need not hold and there might be other stable (non-Gaussian) distributions to which the n-letter information density will converge. Also notice that in the absence of cost constraints we have

$$\mathrm{Var}[i(X^*;Y^*)|X^*] = \mathrm{Var}[i(X^*;Y^*)]$$

since, by capacity saddle point (Corollary 5.7), $\mathbb{E}[i(X^*;Y^*)|X^* = x] = C$ for P_{X^*}-almost all x.

As an example, we have the following dispersion formulas for the common channels that we have discussed so far:

$$\begin{aligned}
\mathsf{BEC}_\delta:&\quad V(\delta) = \delta\bar{\delta}\log^2 2,\\
\mathsf{BSC}_\delta:&\quad V(\delta) = \delta\bar{\delta}\log^2\frac{\bar{\delta}}{\delta},\\
\text{AWGN (real)}:&\quad V_{\text{AWGN}}(P) = \frac{P(P+2)}{2(P+1)^2}\log^2 e,\\
\text{AWGN (complex)}:&\quad V(P) = \frac{P(P+2)}{(P+1)^2}\log^2 e,\\
\text{BI-AWGN}:&\quad V(P) = \mathrm{Var}[\log(1+e^{-2P+2\sqrt{P}Z})],\quad Z \sim \mathcal{N}(0,1),
\end{aligned}$$

where for the AWGN and BI-AWGN letter P denotes the SNR. We also remind the reader that, see Example 3.4, for the BI-AWGN we have $C(P) = \log 2 - \mathbb{E}[\log(1 + e^{-2P+2\sqrt{P}Z})]$. For the parallel AWGN, see Section 20.4, we have

$$\text{parallel AWGN:}\qquad V(P, \{\sigma_j^2, j \in [L]\}) = \frac{\log^2 e}{2}\sum_{j=1}^{L}\left|1 - \left(\frac{\sigma_j^2}{T}\right)^2\right|^+,$$

where T is the optimal threshold in the *water-filling* solution, that is, $\sum_{j=1}^{L} |T - \sigma_j^2|^+ = P$. We remark that the expression for the parallel AWGN channel can be guessed by noticing that it equals $\sum_{j=1}^{L} V_{\text{AWGN}}(\frac{P_j}{\sigma_j^2})$ with $P_j = |T - \sigma_j^2|^+$ being the optimal power allocation.

What about channel dispersion for other channels? Discrete channels with memory have seen some limited success in [336], which expresses dispersion in terms of the Fourier spectrum of the information density process. The compound DMC (Exercise **IV.19**) has a much more delicate dispersion formula (and the remainder term is not $O(\log n)$, but $O(n^{1/4})$), see [343]. For non-discrete channels (other than the AWGN and Poisson) new difficulties appear in the proof of the converse part. For example, the dispersion of a (coherent) fading channel is known only if one additionally restricts the input codewords to have limited peak values, see [98, Remark 1]. In particular, the dispersion of the following *Gaussian erasure channel* is unknown:

$$Y_i = H_i(X_i + Z_i),$$

where we have $\mathcal{N}(0,1) \sim Z_i \perp\!\!\!\perp H_i \sim \text{Ber}(1/2)$ and the usual quadratic cost constraint $\sum_{i=1}^{n} x_i^2 \le nP$.

Multi-antenna (MIMO) channels (20.10) present interesting new challenges as well. For example, for coherent channels the capacity – achieving input distribution is non-unique [98]. The quasi-static channels are similar to fading channels but the $H_1 = H_2 = \cdots$, that is, the channel gain matrix in (20.10), is not changing with time. This channel model is often used to model cellular networks. By leveraging an unexpected amount of differential geometry, it was shown in [463] that they have *zero dispersion*, or more specifically:

$$\log M^*(n,\epsilon) = nC_\epsilon + O(\log n),$$

where the ϵ-capacity C_ϵ is known as outage capacity in this case (and depends on ϵ). The main implication is that C_ϵ is a good predictor of the ultimate performance limits for these practically relevant channels (better than C is for the AWGN channel, for example). But some caution must be taken in approximating $\log M^*(n,\epsilon) \approx nC_\epsilon$, nevertheless. For example, in the case where the H matrix is known at the transmitter, the same paper demonstrated that the standard water-filling power allocation (Theorem 20.14) that maximizes C_ϵ is rather suboptimal at finite n.

22.6 Finite-Blocklength Bounds and Normal Approximation

As stated earlier, direct computation of $M^*(n,\epsilon)$ by exhaustive search is doubly exponential in complexity, and thus is infeasible in most cases. However, the bounds we have developed so far can often help sandwich the unknown value pretty tightly. Less rigorously, we may also evaluate the *normal approximation* which simply suggests dropping unknown terms in the expansion (22.15):

$$\log M^*(n,\epsilon) \approx nC - \sqrt{nV}Q^{-1}(\epsilon).$$

(The $\log n$ term in (22.15) is known to be equal to $O(1)$ for the BEC, and $\frac{1}{2}\log n$ for the BSC, AWGN, and binary-input AWGN. For these latter channels, the normal approximation is typically defined with $+\frac{1}{2}\log n$ added to the previous display.)

For example, considering the $\text{BEC}_{1/2}$ channel we compute the capacity and dispersion to be $C = (1-\delta)$ and $V = \delta(1-\delta)$ (in bits and bits^2, respectively). The detailed calculation in Exercise **IV.4** leads to the following rigorous estimates:

$$213 \le \log_2 M^*(500, 10^{-3}) \le 217.$$

At the same time the normal approximation yields

$$\log M^*(500, 10^{-3}) \approx n\bar{\delta} - \sqrt{n\delta\bar{\delta}}Q^{-1}(10^{-3}) \approx 215.5 \text{ bits}.$$

We observe remarkable precision of the normal approximation, which is preserved across wide range of n, ϵ, and δ.

As another example, we consider the BSC_δ channel. We have already presented numerical results for this channel in (17.3). Here, we evaluate all the lower bounds that were discussed in Chapter 18. We show the results in Figure 22.2 together with the upper bound (22.16). We conclude that (unsurprisingly) the RCU bound is the tightest and is impressively close to the converse bound, as we have already seen in (17.3). The normal approximation (with and without the $\frac{1}{2}\log n$ term) is compared against the rigorous bounds on Figure 22.3. The excellent precision of the approximation should be contrasted with a fairly loose estimate arising from the error-exponent approximation (which coincides with the "Gallager" curve on Figure 22.2).

We can see that for the simple cases of the BEC and the BSC, the solution to the incredibly complex combinatorial optimization problem $\log M^*(n,\epsilon)$ can be

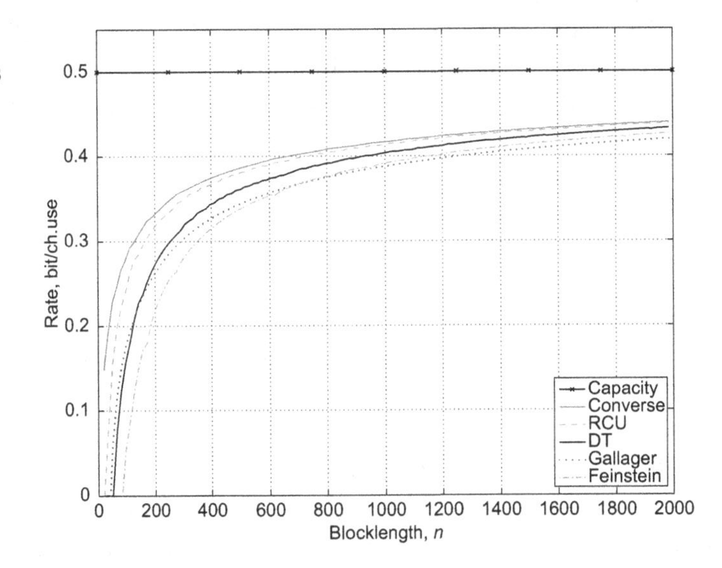

Figure 22.2 Comparing various lower (achievability) bounds on $\frac{1}{n}\log M^*(n,\epsilon)$ for the BSC_δ channel ($\delta = 0.11$, $\epsilon = 10^{-3}$).

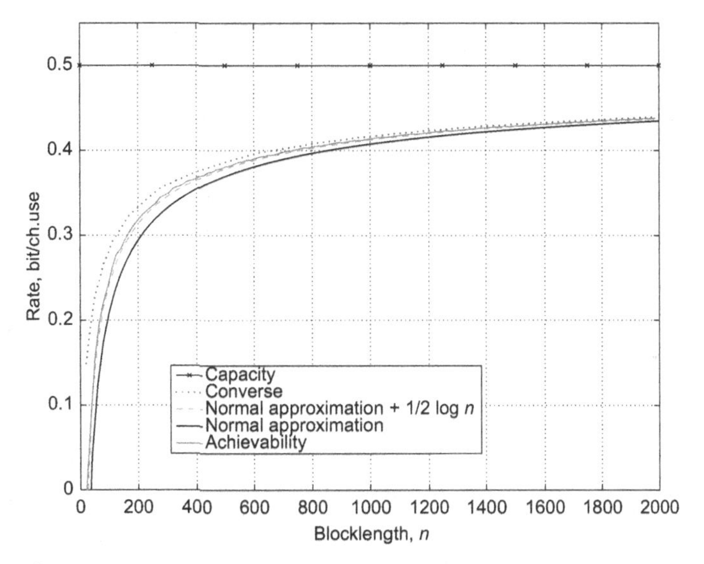

Figure 22.3 Comparing the normal approximation against the best upper and lower bounds on $\frac{1}{n}\log M^*(n,\epsilon)$ for the BSC_δ channel ($\delta = 0.11, \epsilon = 10^{-3}$).

rather well approximated by considering the first few terms in the expansion (22.15). This justifies further interest in computing channel dispersion and establishing such expansions for other channels.

22.7 Normalized Rate

Suppose we are considering two different codes. One has $M = 2^{k_1}$ and blocklength n_1 (and so, in engineering language is a k_1-to-n_1 code) and another is a k_2-to-n_2 code. How can we compare the two of them fairly? A traditional way of presenting the code performance is in terms of the "waterfall plots" showing the dependence of the probability of error on the SNR (or crossover probability) of the channel. These two codes could have waterfall plots of the following kind:

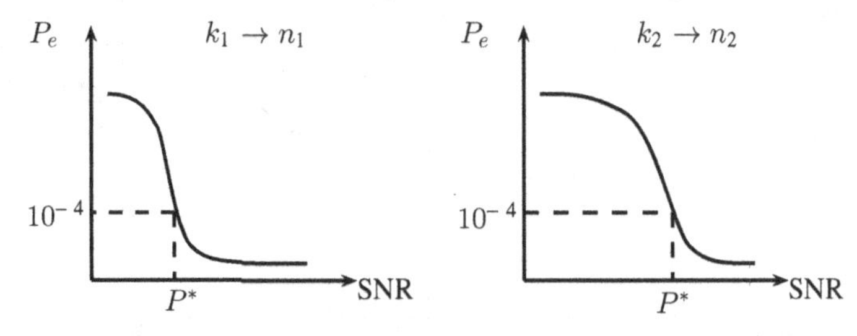

After inspecting these plots, one may believe that the $k_1 \to n_1$ code is better, since it requires a smaller SNR to achieve the same error probability. However, this ignores the fact that the rate of this code $\frac{k_1}{n_1}$ might be much smaller as well. The concept of *normalized rate* allows us to compare the codes of different blocklengths and coding rates.

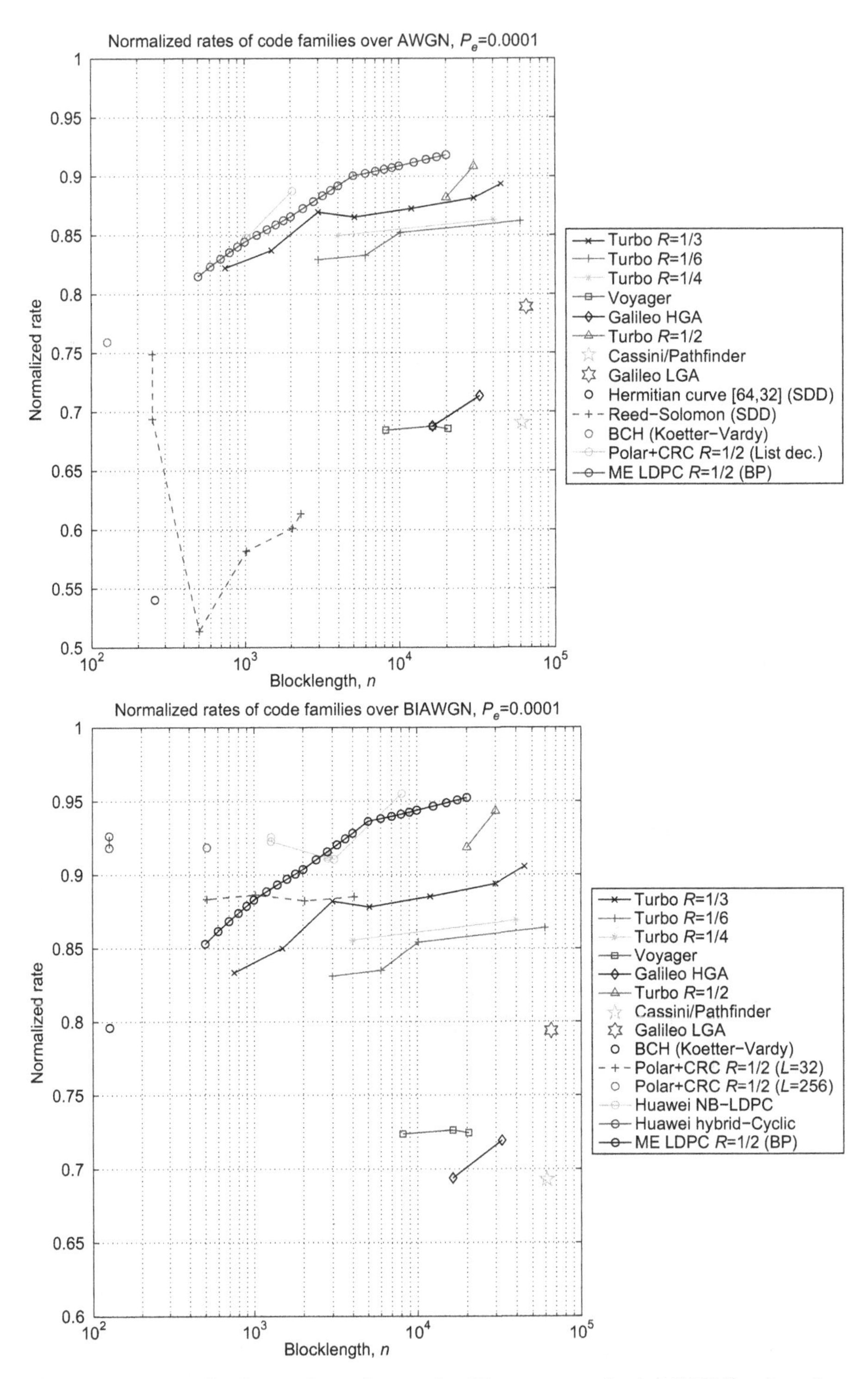

Figure 22.4 Normalized rates for various codes. Plots generated using [398] (for the color version see electronic copy).

Specifically, suppose that a $k \to n$ code is given. Fix $\epsilon > 0$ and find the value of the SNR P for which this code attains probability of error ϵ (for example, by taking a horizontal intercept at level ϵ on the waterfall plot). The normalized rate is defined as

$$R_{\text{norm}}(\epsilon) = \frac{k}{\log_2 M^*(n,\epsilon,P)} \approx \frac{k}{nC(P) - \sqrt{nV(P)}Q^{-1}(\epsilon)},$$

where $\log M^*$, capacity, and dispersion correspond to the channel over which evaluation is being made (most often the AWGN, BI-AWGN, or the fading channel). We also notice that, of course, the value of $\log M^*$ is not possible to compute exactly and thus, in practice, we use the normal approximation to evaluate it.

This idea allows us to clearly see how much different ideas in coding theory over the decades were driving the value of normalized rate upward to 1. This comparison is shown on Figure 22.4. A short summary is that at blocklengths corresponding to "data stream" channels in cellular networks ($n \sim 10^4$) the LDPC codes and non-binary LDPC codes are already achieving 95% of the information-theoretic limit. At blocklengths corresponding to "control plane" ($n \lesssim 10^3$) the polar codes and LDPC codes are at similar performance and at 90% of the fundamental limits.

23 Channel Coding with Feedback

So far we have been focusing on the paradigm for one-way communication: data are mapped to codewords and transmitted, and later decoded based on the received noisy observations. In most practical settings (except for storage), frequently the communication goes both ways so that the receiver can provide certain *feedback* to the transmitter. As a motivating example, consider the communication channel of the downlink transmission from a satellite to Earth. Downlink transmission is very expensive (power constraint at the satellite), but the uplink from Earth to the satellite is cheap which makes virtually noiseless feedback readily available at the transmitter (satellite). In general, a channel with noiseless feedback is interesting when such asymmetry exists between uplink and downlink. Even in less ideal settings, noisy or partial feedbacks are commonly available that can potentially improve the reliability or complexity of communication.

In the first half of our discussion, we shall follow Shannon to show that even with noiseless feedback nothing (in terms of capacity) can be gained in the conventional setup. In the process, we will also introduce the concept of Massey's directed information. In the second half of the chapter we examine situations where feedback is extremely helpful: low probability of error, variable transmission length, and variable transmission power.

23.1 Feedback Does Not Increase Capacity for Stationary Memoryless Channels

Definition 23.1. (Code with feedback) An (n, M, ϵ)-code with feedback is specified by the encoder–decoder pair (f, g) as follows:

- Encoder (time-varying)

$$\begin{aligned}
f_1 &: [M] \to \mathcal{A} \\
f_2 &: [M] \times \mathcal{B} \to \mathcal{A} \\
&\vdots \\
f_n &: [M] \times \mathcal{B}^{n-1} \to \mathcal{A}
\end{aligned}$$

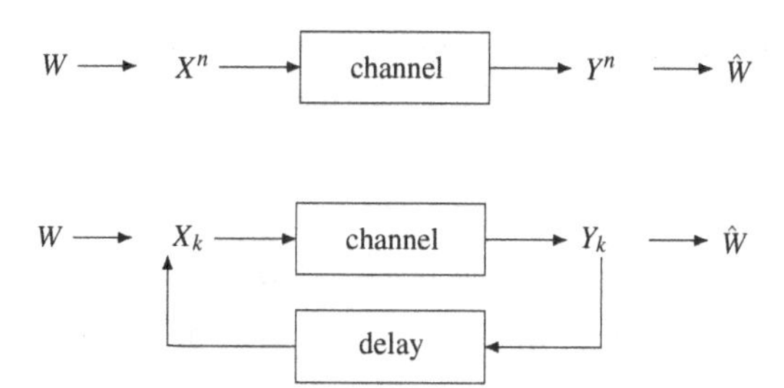

Figure 23.1 Schematic representation of coding without feedback (top) and with full noiseless feedback (bottom).

- Decoder

$$g\colon \mathcal{B}^n \to [M]$$

such that $\mathbb{P}[W \neq \hat{W}] \leq \epsilon$.

Here the symbol transmitted at time t depends on both the message and the history of received symbols (causality constraint):

$$X_t = f_t\big(W, Y_1^{t-1}\big).$$

Hence the probability space is as follows:

$$\left.\begin{aligned} &W \sim \text{ uniform on } [M], \\ &X_1 = f_1(W) \overset{P_{Y|X}}{\longrightarrow} Y_1 \\ &\vdots \\ &X_n = f_n(W, Y_1^{n-1}) \overset{P_{Y|X}}{\longrightarrow} Y_n \end{aligned}\right\} \longrightarrow \hat{W} = g(Y^n).$$

Figure 23.1 compares the settings of channel coding without feedback and with ideal full feedback:

Definition 23.2. (Fundamental limits)

$$\begin{aligned} M^*_{fb}(n, \epsilon) &= \max\{M\colon \exists (n, M, \epsilon) - \text{ code with feedback}\}, \\ C_{fb,\epsilon} &= \liminf_{n\to\infty} \frac{1}{n} \log M^*_{fb}(n, \epsilon), \\ C_{fb} &= \lim_{\epsilon\to 0} C_{fb,\epsilon} \quad \text{(Shannon capacity with feedback)}. \end{aligned}$$

Theorem 23.3. (Shannon 1956) *For a stationary memoryless channel,*

$$C_{fb} = C = C^{(\mathrm{I})} = \sup_{P_X} I(X; Y).$$

Proof. Achievability: Although it is obvious that $C_{fb} \geq C$, we wanted to demonstrate that in fact constructing codes achieving capacity with *full feedback* can be done directly, without appealing to the (much harder) problem of non-feedback codes. Let $\pi_t(\cdot) \triangleq P_{W|Y^t}(\cdot|Y^t)$ with the (random) posterior distribution after t steps.

It is clear that due to the knowledge of Y^t on both ends, the transmitter and receiver have perfectly synchronized knowledge of π_t. Now consider how the transmission progresses:

1 Initialize $\pi_0(\cdot) = \frac{1}{M}$.
2 At the $(t+1)$-th step, the encoder sends $X_{t+1} = f_{t+1}(W, Y^t)$. Note that selection of f_{t+1} is equivalent to the task of partitioning message space $[M]$ into classes $\mathcal{P}_a$, that is,

$$\mathcal{P}_a \triangleq \{j \in [M] \colon f_{t+1}(j, Y^t) = a\} \qquad a \in \mathcal{A}.$$

How to do this partitioning optimally is what we will discuss soon.
3 The channel perturbs X_{t+1} into Y_{t+1} and both parties compute the updated posterior:

$$\pi_{t+1}(j) \triangleq \pi_t(j) B_{t+1}(j), \qquad B_{t+1}(j) \triangleq \frac{P_{Y|X}(Y_{t+1}|f_{t+1}(j, Y^t))}{\sum_{a \in \mathcal{A}} \pi_t(\mathcal{P}_a)}.$$

Notice that (this is the crucial part!) the random multiplier satisfies:

$$\mathbb{E}[\log B_{t+1}(W)|Y^t] = \sum_{a \in \mathcal{A}} \sum_{y \in \mathcal{B}} \pi_t(\mathcal{P}_a) \log \frac{P_{Y|X}(y|a)}{\sum_{a \in \mathcal{A}} \pi_t(\mathcal{P}_a)} = I(\tilde{\pi}_{t+1}, P_{Y|X}), \quad (23.1)$$

where $\tilde{\pi}_{t+1}(a) \triangleq \pi_t(\mathcal{P}_a)$ is a (random) distribution on $\mathcal{A}$, induced by the encoder at the channel input in round $(t+1)$. Note that while π_t is decided before the $(t+1)$-th step, the design of partition $\mathcal{P}_a$ (and hence $\tilde{\pi}_{t+1}$) is in the hands of the encoder.

The goal of the code designer is to come up with a partitioning $\{\mathcal{P}_a \colon a \in \mathcal{A}\}$ in such a way that the speed of growth of $\pi_t(W)$ is maximal. Now, analyzing the speed of growth of a random-multiplicative process is best done by taking logs:

$$\log \pi_t(j) = \sum_{s=1}^{t} \log B_s + \log \pi_0(j).$$

Intuitively, we expect that the process $\log \pi_t(W)$ resembles a random walk starting from $-\log M$ and having a positive drift. Thus to estimate the time it takes for this process to reach value 0 we need to estimate the upward drift. Appealing to intuition and the law of large numbers (more exactly to the theory of martingales) we approximate

$$\log \pi_t(W) - \log \pi_0(W) \approx \sum_{s=1}^{t} \mathbb{E}[\log B_s].$$

Finally, from (23.1) we conclude that the best idea is to select partitioning at each step in such a way that $\tilde{\pi}_{t+1} \approx P_X^*$ (capacity-achieving input distribution) and this gives

$$\log \pi_t(W) \approx tC - \log M,$$

implying that the transmission terminates in time $\approx \frac{\log M}{C}$. The important lesson here is the following: *The optimal transmission scheme should map messages to channel inputs in such a way that the induced input distribution $P_{X_{t+1}|Y^t}$ is approximately equal to the one maximizing $I(X;Y)$.* This idea is called *posterior matching* and is explored in detail in [385].[1]

Although our argument above is not rigorous, it is not hard to make it so by an appeal to martingale convergence theory, very similar to the way we used it in Section 16.3* to analyze SPRT. We omit those details (see [385]), since the result is in principle not needed for the proof of the theorem.

Converse: We are left to show that $C_{fb} \le C^{(\mathrm{I})}$. Recall the key in proving the weak converse for channel coding without feedback: Fano's inequality plus the graphical model

$$W \to X^n \to Y^n \to \hat{W}. \tag{23.2}$$

Then

$$-h(\epsilon) + \bar{\epsilon} \log M \le I(W;\hat{W}) \le I(X^n;Y^n) \le nC^{(\mathrm{I})}.$$

With feedback the probabilistic picture becomes more complicated as Figure 23.2 demonstrates for $n = 3$ (dependence introduced by the extra squiggly arrows):

Notice that the d-separation criterion shows we no longer have a Markov chain (23.2), that is, given X^n, the W and Y^n are not independent.[2] Furthermore, the input–output relation is also no longer memoryless:

$$P_{Y^n|X^n}(y^n|x^n) \ne \prod_{j=1}^{n} P_{Y|X}(y_j|x_j)$$

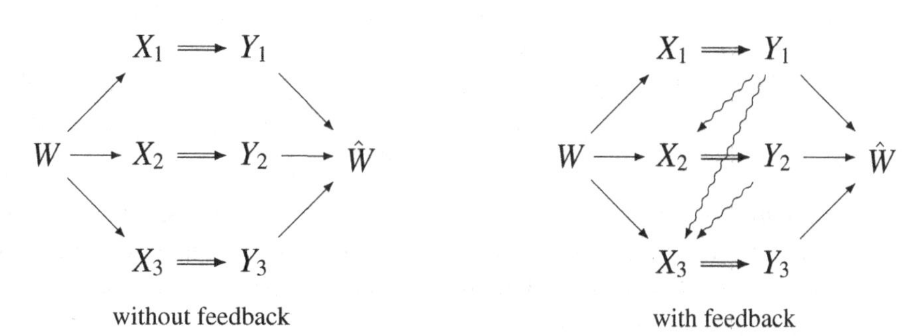

Figure 23.2 Graphical model for channel coding and $n = 3$ with and without feedback. Double arrows $\Rightarrow$ correspond to the channel links.

[1] This simple (but capacity-achieving) feedback coding scheme also helps us appreciate more fully the magic of Shannon's (non-feedback) coding theorem, which demonstrated that the (almost) optimal partitioning can be done in a way that is totally blind to actual channel outputs. That is, we can preselect partitions $\mathcal{P}_a$ that are independent of π_t (but dependent on t) and so that $\pi_t(\mathcal{P}_a) \approx P_X^*(a)$ with overwhelming probability and for almost all $t \in [n]$.

[2] For example, suppose we are transmitting $W \sim \mathrm{Ber}(1/2)$ over the BSC and set $X_1 = 0$, $X_2 = W \oplus Y_1$. Then given X_1, X_2 we see that Y_2 and W can be exactly determined from one another.

(as an example, let $X_2 = Y_1$ and thus $P_{Y_1|X_1 X_2} = \delta_{X_1}$ is a point mass). Nevertheless, there is still a large degree of independence in the channel. Namely, we have

$$(Y^{t-1}, W) \to X_t \to Y_t, \quad t = 1, \dots, n, \tag{23.3}$$

$$W \to Y^n \to \hat{W}. \tag{23.4}$$

Then

$$\begin{aligned}
-h(\epsilon) + \bar{\epsilon} \log M &\le I(W;\hat{W}) && \text{(Fano)} \\
&\le I(W;Y^n) && \text{(data processing applied to (23.4))} \\
&= \sum_{t=1}^n I(W;Y_t|Y^{t-1}) && \text{(chain rule)} \\
&\le \sum_{t=1}^n I(W, Y^{t-1};Y_t) && (I(W;Y_t|Y^{t-1}) = I(W, Y^{t-1};Y_t) \\
& && \quad -I(Y^{t-1};Y_t)) \\
&\le \sum_{t=1}^n I(X_t;Y_t) && \text{(data processing applied to (23.3))} \\
&\le nC.
\end{aligned}$$

□

In comparison with Theorem 22.2, the following result shows that, with fixed-length block coding, feedback does not even improve the speed of approaching capacity and can at most improve the third-order $\log n$ terms.

Theorem 23.4. (Dispersion with feedback [131, 337]) *For weakly input-symmetric DMC (e.g. additive noise, BSC, BEC) we have:*

$$\log M^*_{fb}(n, \epsilon) = nC - \sqrt{nV} Q^{-1}(\epsilon) + O(\log n).$$

23.2* Massey's Directed Information

In this section we show an alternative proof of the converse part of Theorem 23.3, which is more in the spirit of "channel substitution" ideas (meta-converse) that we emphasize in this book, see Sections 3.6, 17.4, and 22.3. In addition, it will also lead us to defining the concept of directed information $\vec{I}(X^n;Y^n)$ due to Massey [295]. Directed information is an important tool in the field of causal inference, though we will not go into those connections [295].

Alternative proof of Theorem 23.3 (converse). Let us revisit the steps of showing the weak converse $C \le C^{(\mathrm{I})}$, when phrased in the style of meta-converse. We take an arbitrary (n, M, ϵ)-code and define two distributions with corresponding graphical models:

$$\mathbb{P}: W \to X^n \to Y^n \to \hat{W}, \tag{23.5}$$

$$\mathbb{Q}: W \to X^n \quad Y^n \to \hat{W}. \tag{23.6}$$

We then make two key observations:

1 Under $\mathbb{Q}$, $W \perp\!\!\!\perp W$, so that $\mathbb{Q}[W = \hat{W}] = \frac{1}{M}$ while $\mathbb{P}[W = \hat{W}] \ge 1 - \epsilon$.
2 The two graphical models give the factorization:

$$P_{W,X^n,Y^n,\hat{W}} = P_{W,X^n} P_{Y^n|X^n} P_{\hat{W}|Y^n}, \qquad Q_{W,X^n,Y^n,\hat{W}} = Q_{W,X^n} Q_{Y^n} Q_{\hat{W}|Y^n}.$$

We are free to choose factors Q_{W,X^n}, Q_{Y^n}, and $Q_{\hat{W}|Y^n}$. However, as we will see soon, it is best to choose them to minimize $D(\mathbb{P}\|\mathbb{Q})$ which gives us (see the discussion of *information flow* after (4.6))

$$\min_{Q_{W,X^n},Q_{Y^n},Q_{\hat{W}|Y^n}} D(\mathbb{P}\|\mathbb{Q}) = I(X^n;Y^n) \tag{23.7}$$

and achieved by taking the factors equal to their values under $\mathbb{P}$, namely $Q_{W,X^n} = P_{W,X^n}$, $Q_{Y^n} = P_{Y^n}$, and $Q_{\hat{W}|Y^n} = P_{\hat{W}|Y^n}$. (It is a good exercise to show this by writing out the chain rule for divergence (2.26).) As we know this minimal value of $D(\mathbb{P}\|\mathbb{Q})$ measures the information flow through the links excluded in the graphical model, that is, through $X^n \to Y^n$.

From here we proceed via the data-processing inequality and tensorization of capacity for memoryless channels as follows:

$$-h(\epsilon) + \bar{\epsilon}\log M = d\left(1-\epsilon \middle\| \frac{1}{M}\right) \overset{\text{DPI}}{\leq} D(\mathbb{P}\|\mathbb{Q}) = I(X^n;Y^n) \overset{(*)}{\leq} \sum_{i=1}^{n} I(X_i;Y_i) \leq nC^{(\mathrm{I})}, \tag{23.8}$$

where the $(*)$ followed from the fact that $X^n \to Y^n$ is a memoryless channel see ((6.1)).

Let us now go back to the case of channels with feedback. There are several problems with adapting the previous argument. First, when feedback is present, $X^n \to Y^n$ is not memoryless due to the influence of the transmission protocol (for example, knowing both X_1 and X_2 affects the law of Y_1, that is, $P_{Y_1|X_1} \neq P_{Y_1|X_1,X_2}$ and also $P_{Y^n|X^n} \neq \prod_{j=1}^{n} P_{Y_j|X_j}$ even for the DMC). However, an even bigger problem is revealed by attempting to replicate the previous proof.

Suppose we again try to induce an auxiliary probability space $\mathbb{Q}$ as in (23.6). Then due to lack of a Markov chain under $\mathbb{P}$ (i.e. (23.5)), the solution of the problem (23.7) can be shown to equal this time

$$\min D(\mathbb{P}\|\mathbb{Q}) = I(W, X^n;Y^n).$$

This value can be quite a bit higher than capacity. For example, consider an extremely noisy (in fact, useless) channel $\mathsf{BSC}_{1/2}$ and a feedback transmission scheme $X_{t+1} = Y_t$. We see that $I(W, X^n;Y^n) \geq H(Y^{n-1}) = (n-1)\log 2$, whereas capacity $C = 0$. What went wrong in this case?

For the explanation, we should revisit the graphical model under $\mathbb{P}$ as shown on Figure 23.2 (right graph). When $\mathbb{Q}$ is defined as in (23.6) the value $\min D(\mathbb{P}\|\mathbb{Q}) = I(W, X^n;Y^n)$ measures the information flow through both the $\Rightarrow$ and $\rightsquigarrow$ links.

This motivates us to find a graphical model for $\mathbb{Q}$ such that $\min D(\mathbb{P}\|\mathbb{Q})$ only captures the information flow through the $\Rightarrow$ links $\{X_i \to Y_i \colon i = 1, \dots, n\}$ (and so that $\min D(\mathbb{P}\|\mathbb{Q}) \leq nC^{(\mathrm{I})}$), while still guaranteeing that $W \perp\!\!\!\perp \hat{W}$, so that $\mathbb{Q}[W = \hat{W}] = \frac{1}{M}$.

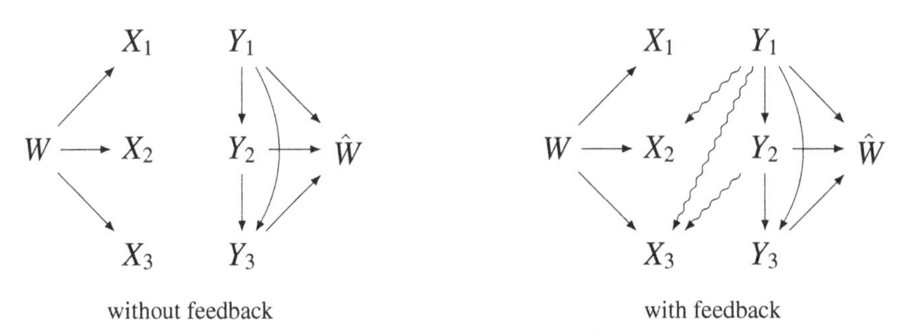

Figure 23.3 Graphical model for $n = 3$ under the auxiliary distribution Q. Compare with Figure 23.2 under the actual distribution P.

Such a graphical model is depicted on Figure 23.3 (right graph).[3] Formally, we shall restrict $Q_{W,X^n,Y^n,\hat{W}} \in \mathcal{Q}$, where $\mathcal{Q}$ is the set of distributions that can be factorized as follows:

$$Q_{W,X^n,Y^n,\hat{W}} = Q_W Q_{X_1|W} Q_{Y_1} Q_{X_2|W,Y_1} Q_{Y_2|Y_1} \cdots Q_{X_k|W,Y^{k-1}} Q_{Y_k|Y^{k-1}} \cdots Q_{X_n|W,Y^{n-1}} Q_{Y_n|Y^{n-1}} Q_{\hat{W}|Y^n}.$$

Using the d-separation criterion (see (3.11)) we can verify that for any $Q \in \mathcal{Q}$ we have $W \perp\!\!\!\perp W$: W and $\hat{W}$ are d-separated by X^n. (More directly, one can clearly see that conditioning on any fixed value of $W = w$ does affect the distributions of $X_1, \ldots, X_n$ but leaves Y^n and $\hat{W}$ unaffected.)

Notice that in the graphical model for $\mathbb{Q}$, when removing $\Rightarrow$ we also added the directional links between the Y_i's. These links serve to maximally preserve the dependence relationships between variables when $\Rightarrow$ are removed, so that $\mathbb{Q}$ could be made closer to $\mathbb{P}$, while still maintaining $W \perp\!\!\!\perp W$. We note that these links were also implicitly present in the non-feedback case (see model for $\mathbb{Q}$ in that case on the left graph in Figure 23.3).

Now since as we agreed under $\mathbb{Q}$ we still have $\mathbb{Q}[W = \hat{W}] = \frac{1}{M}$ we can use our usual data processing for divergence to conclude $d(1 - \epsilon \| \frac{1}{M}) \leq D(\mathbb{P} \| \mathbb{Q})$.

Assuming the crucial fact about this Q-graphical model that will be shown in Lemma 23.6 (to follow), we then have the following chain:

$$\begin{aligned} d\left(1 - \epsilon \middle\| \frac{1}{M}\right) &\leq \inf_{Q \in \mathcal{Q}} D\left(P_{W,X^n,Y^n,\hat{W}} \| Q_{W,X^n,Y^n,\hat{W}}\right) \\ &= \sum_{t=1}^{n} I(X_t; Y_t | Y^{t-1}) \quad \text{(Lemma 23.6)} \\ &= \sum_{t=1}^{n} \mathbb{E}_{Y^{t-1}}[I(P_{X_t|Y^{t-1}}, P_{Y|X})] \end{aligned}$$

[3] This kind of removal of one-directional links is known as *causal conditioning*.

$$
\begin{aligned}
&\leq \sum_{t=1}^{n} I(\mathbb{E}_{Y^{t-1}}[P_{X_t|Y^{t-1}}], P_{Y|X}) \quad (\text{concavity of } I(\cdot, P_{Y|X})) \\
&= \sum_{t=1}^{n} I(P_{X_t}, P_{Y|X}) \\
&\leq nC^{(\mathrm{I})}.
\end{aligned}
$$

Thus, we complete our proof:

$$
-h(\epsilon) + \bar{\epsilon} \log M \leq nC^{(\mathrm{I})} \Rightarrow \log M \leq \frac{nC + h(\epsilon)}{1-\epsilon} \Rightarrow C_{fb,\epsilon} \leq \frac{C}{1-\epsilon} \Rightarrow C_{fb} \leq C.
$$

We notice that the proof can also be adapted essentially without change to channels with cost constraints and for capacity per unit cost setting, see [338].

We now proceed to show the crucial omitted step in the above proof. Before that let us define an interesting new kind of information.

Definition 23.5. (Massey's directed information [295]) For a pair of blocklength-n random variables X^n and Y^n we define

$$
\vec{I}(X^n;Y^n) \triangleq \sum_{t=1}^{n} I(X^t;Y_t|Y^{t-1}).
$$

Note that directed information is not symmetric. As [295] (and subsequent work, e.g. [413]) shows $\vec{I}(X^n;Y^n)$ quantifies the amount of *causal* information transfer from X-process to Y-process. In the context of feedback communication a formal justification for the introduction of this concept is the following result.

Lemma 23.6. *Consider communication with feedback over a non-anticipatory channel given by a sequence of Markov kernels $P_{Y_t|X^t,Y^{t-1}}, t \in [n]$, that is, we have a probability distribution $\mathbb{P}$ on $(W, X^n, Y^n, \hat{W})$ described by factorization*

$$
P_{W,X^n,Y^n,\hat{W}} = P_W \prod_{t=1}^{n} P_{X_t|W,X^{t-1},Y^{t-1}} P_{Y_t|X^t,Y^{t-1}}. \tag{23.9}
$$

Denote by $\mathcal{Q}$ all distributions factorizing with respect to the graphical models on Figure 23.3 *(right graph), that is, those satisfying*

$$
Q_{W,X^n,Y^n,\hat{W}} = Q_W \prod_{t=1}^{n} Q_{X_k|W,Y^{k-1}} Q_{Y_k|Y^{k-1}}. \tag{23.10}
$$

Then we have

$$
\inf_{Q \in \mathcal{Q}} D(P_{W,X^n,Y^n,\hat{W}} \| Q_{W,X^n,Y^n,\hat{W}}) = \vec{I}(X^n;Y^n). \tag{23.11}
$$

In addition, if the channel is memoryless, that is, $P_{Y_t|X^t,Y^{t-1}} = P_{Y_t|X_t}$ for all $t \in [n]$, then we have

$$
\vec{I}(X^n;Y^n) = \sum_{t=1}^{n} I(X_t;Y_t|Y^{t-1}).
$$

Proof. By comparing factorizations (23.9) and (23.10) and applying the chain rule (2.26) we can immediately optimize several terms (we leave this justification as an exercise for the reader):

$$Q_{X,W} = P_{X,W},$$
$$Q_{X_t|W,Y^{t-1}} = P_{X_t|W,Y^{t-1}},$$
$$Q_{\hat{W}|Y^n} = P_{W|Y^n}.$$

From here we conclude that

$$\begin{aligned}
&\inf_{Q\in\mathcal{Q}} D(P_{W,X^n,Y^n,\hat{W}} \| Q_{W,X^n,Y^n,\hat{W}}) \\
&= \inf_{Q\in\mathcal{Q}} D(P_{Y_1|X_1} \| Q_{Y_1}|X_1) + D(P_{Y_2|X^2,Y_1} \| Q_{Y_2|Y_1}|X^2, Y_1) + \cdots \\
&\quad + D(P_{Y_n|X^n,Y^{n-1}} \| Q_{Y_n|Y^{n-1}}|X^n, Y^{n-1}) \\
&= I(X_1;Y_1) + I(X^2;Y_2|Y_1) + \cdots + I(X^n;Y_n|Y^{n-1}),
\end{aligned}$$

where in the last step we simply applied (conditional) versions of Corollary 4.2.

To prove the claim for the memoryless channels, we only need to notice that

$$I(X^t;Y_t|Y^{t-1}) = I(X_t;Y_t|Y^{t-1}) + I(X^{t-1};Y_t|Y^{t-1}, X_t),$$

and that the last term is zero. The latter can be justified via the d-separation criterion. Indeed, in the absence of channel memory every undirected path from X^{t-1} to Y_t must pass through X_t, which is a non-collider and is conditioned on. □

To summarize, we can see that Shannon's result for feedback communication can be best understood as a simple modification of the standard weak converse in channel coding: whereas in Theorem 17.3 we changed the true distribution P_{X^n,Y^n} with $P_{X^n}P_{Y^n}$ (as shown on the left of Fiure 23.3), in the case of feedback the replacement distribution retains only the forward dependence between X^n and Y^n (as shown on the right of Figure 23.3).

23.3 When Is Feedback Really Useful?

Theorems 23.3 and 23.4 state that feedback does not improve communication rate either asymptotically or for moderate blocklengths. In this section, we shall examine three cases where feedback turns out to be very useful.

23.3.1 Code with Very Small (e.g. Zero) Error Probability

Theorem 23.7. (Shannon [380]) *For any DMC $P_{Y|X}$,*

$$C_{fb,0} = \max_{P_X} \min_{y\in\mathcal{B}} \log \frac{1}{P_X(S_y)}, \tag{23.12}$$

where

$$S_y = \{x \in \mathcal{A}\colon P_{Y|X}(y|x) > 0\}$$

denotes the set of input symbols that can lead to the output symbol y.

Remark 23.1. For a stationary memoryless channel, we have

$$C_0 \overset{\text{(a)}}{\leq} C_{fb,0} \overset{\text{(b)}}{\leq} C_{fb} = \lim_{\epsilon \to 0} C_{fb,\epsilon} \overset{\text{(c)}}{=} C = \lim_{\epsilon \to 0} C_\epsilon \overset{\text{(d)}}{=} C^{(\mathrm{I})} = \sup_{P_X} I(X;Y),$$

where (a) and (b) are by definitions, (c) follows from Theorem 23.3, and (d) is due to Theorem 19.9. All capacity quantities above are defined with (fixed-length) block codes.

Remark 23.2.

1 In DMC for both zero-error capacities (C_0 and $C_{fb,0}$) only the support of the transition matrix $P_{Y|X}$, that is, whether $P_{Y|X}(b|a) > 0$ or not, matters; the values of those non-zero $P_{Y|X}(b|a)$ are irrelevant. That is, C_0 and $C_{fb,0}$ are determined by the bipartite graph representation between the input alphabet $\mathcal{A}$ and the output alphabet $\mathcal{B}$. Furthermore, C_0 (but not $C_{fb,0}$!) is a function of the *confusability graph* – a simple undirected graph on $\mathcal{A}$ with $a \neq a'$ connected by an edge iff $\exists b \in \mathcal{B}$ s.t. $P_{Y|X}(b|a)P_{Y|X}(b|a') > 0$.
2 That $C_{fb,0}$ is not a function of the confusability graph alone is easily seen from comparing the polygon channel (next example) with $L = 3$ (for which $C_{fb,0} = \log \frac{3}{2}$) and the useless channel with $\mathcal{A} = \{1, 2, 3\}$ and $\mathcal{B} = \{1\}$ (for which $C_{fb,0} = 0$). Clearly in both cases the confusability graph is the same – a triangle.
3 Oftentimes C_0 is very hard to compute, but $C_{fb,0}$ can be obtained in closed form as in (23.12). As an example, consider the following *polygon channel* (named after its confusability graph):

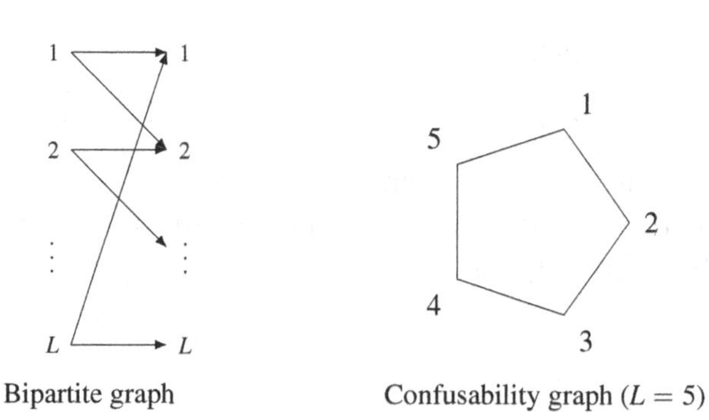

Bipartite graph Confusability graph ($L = 5$)

The following are known about the zero-error capacity C_0 of the polygon channel:

- $L = 3$: $C_0 = 0$.
- $L = 5$: $C_0 = \frac{1}{2} \log 5$. This is the famous "capacity of a pentagon" problem. For achievability, with blocklength one, one can use $\{1, 3\}$ to achieve rate 1 bit; with blocklength two, the codebook $\{(1, 1), (2, 3), (3, 5), (4, 2), (5, 4)\}$ achieves rate $\frac{1}{2} \log 5$ bits, as pointed out by Shannon in his original 1956 paper [380].

More than two decades later this was shown to be optimal by Lovász using a technique now known as semidefinite programming relaxation [287].

- Even L: $C_0 = \log \frac{L}{2}$ (Exercise **IV.36**).
- $L = 7$: $\frac{3}{5} \log 7 \leq C_0 \leq \log 3.32$. Finding the exact value for any odd $L \geq 7$ is a famous open problem in combinatorics.
- Asymptotics of odd L: Despite being unknown in general C_0 has a known asymptotic expansion: for odd L, $C_0 = \log \frac{L}{2} + o(1)$ [66].

In comparison, the zero-error capacity with feedback (Exercise **IV.36**) equals $C_{fb,0} = \log \frac{L}{2}$ for any L, which, thus, can strictly exceed C_0.

4 Notice that $C_{fb,0}$ is not necessarily equal to $C_{fb} = \lim_{\epsilon \to 0} C_{fb,\epsilon} = C$. Here is an example with

$$C_0 < C_{fb,0} < C_{fb} = C.$$

Consider a channel with the following bipartite graph representation:

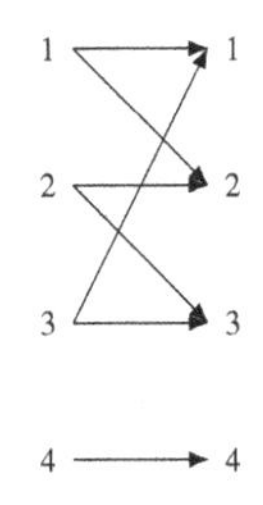

Then one can verify that

$$\begin{aligned} C_0 &= \log 2, \\ C_{fb,0} &= \max_{\delta} - \log \max\left(\tfrac{2}{3}\delta, 1-\delta\right) && (P_X^* = (\delta/3, \delta/3, \delta/3, \bar{\delta})), \\ &= \log \tfrac{5}{2} > C_0 && (\delta^* = \tfrac{3}{5}). \end{aligned}$$

On the other hand, the Shannon capacity $C = C_{fb}$ can be made arbitrarily close to $\log 4$ by picking the cross over probabilities arbitrarily close to zero, while the confusability graph stays the same.

Proof of Theorem 23.7.

1 Fix any $(n, M, 0)$-code. For each $t = 0, 1, \ldots, n$, denote the confusability set of all possible messages that could have produced the received signal $y^t = (y_1, \ldots, y_t)$ by:

$$E_t(y^t) \triangleq \{m \in [M] \colon f_1(m) \in S_{y_1}, f_2(m, y_1) \in S_{y_2}, \ldots, f_n(m, y^{t-1}) \in S_{y_t}\}.$$

Notice that zero error means no ambiguity:

$$\epsilon = 0 \Leftrightarrow \forall y^n \in \mathcal{B}^n, |E_n(y^n)| = 1 \text{ or } 0. \tag{23.13}$$

2 The key quantities in the proof are defined as follows:

$$\theta_{fb} \triangleq \min_{P_X} \max_{y \in \mathcal{B}} P_X(S_y), \quad P_X^* \triangleq \operatorname*{argmin}_{P_X} \max_{y \in \mathcal{B}} P_X(S_y).$$

The goal is to show

$$C_{fb,0} = \log \frac{1}{\theta_{fb}}.$$

By definition, we have

$$\forall P_X, \exists y \in \mathcal{B}, \text{ such that } P_X(S_y) \geq \theta_{fb}. \tag{23.14}$$

Notice that in general the minimizer P_X^* is not the capacity-achieving input distribution in the usual sense (recall Theorem 5.4). This definition also sheds light on how the encoding and decoding should proceed and serves to lower-bound the uncertainty reduction at each stage of the decoding scheme.

3 "$\leq$" (converse): Let P_{X^n} be the joint distribution of the codewords. Denote by $E_0 = [M]$ the original message set.
$t = 1$: For P_{X_1}, by (23.14), $\exists y_1^*$ such that:

$$P_{X_1}(S_{y_1^*}) = \frac{|\{m\colon f_1(m) \in S_{y_1^*}\}|}{|\{m \in [M]\}|} = \frac{|E_1(y_1^*)|}{|E_0|} \geq \theta_{fb}.$$

$t = 2$: For $P_{X_2|X_1 \in S_{y_1^*}}$, by (23.14), $\exists y_2^*$ such that:

$$P_{X_2}(S_{y_2^*}|X_1 \in S_{y_1^*}) = \frac{|\{m\colon f_1(m) \in S_{y_1^*}, f_2(m, y_1^*) \in S_{y_2^*}\}|}{|\{m\colon f_1(m) \in S_{y_1^*}\}|} = \frac{|E_2(y_1^*, y_2^*)|}{|E_1(y_1^*)|} \geq \theta_{fb}.$$

$t = n$: Continue the selection process up to y_n^* which satisfies that:

$$P_{X_n}(S_{y_n^*}|X_t \in S_{y_t^*} \text{ for } t = 1, \ldots, n-1) = \frac{|E_n(y_1^*, \ldots, y_n^*)|}{|E_{n-1}(y_1^*, \ldots, y_{n-1}^*)|} \geq \theta_{fb}.$$

Finally, by (23.13) and the above selection procedure, we have

$$\frac{1}{M} \geq \frac{|E_n(y_1^*, \ldots, y_n^*)|}{|E_0|} \geq \theta_{fb}^n.$$

Thus $\log M \leq -n \log \theta_{fb}$ and we have shown $C_{fb,0} \leq -\log \theta_{fb}$.

4 "$\geq$" (achievability): Let us construct a code that achieves $(M, n, 0)$.
As an example, consider the specific channel in Figure 23.4 with $|\mathcal{A}| = |\mathcal{B}| = 3$. As the first stage, the encoder f_1 partitions the space of all M messages into three groups of size proportional to the weight $P_X^*(a_i)$, then maps messages in each group to the corresponding a_i for $i = 1, 2, 3$. Suppose the channel outputs, say, y_1. Since in this example $S_{y_1} = \{a_1, a_2\}$, the decoder can eliminate a total number of $MP_X^*(a_3)$ candidate messages in this round. Afterwards, the "confusability set" only contains the remaining $MP_X^*(S_{y_1})$ messages. By definition of P_X^* we know that $MP_X^*(S_{y_1}) \leq M\theta_{fb}$. In the second round, the encoder f_2 partitions the remaining messages into three groups, sends the group index, and repeats.

By similar arguments, each interaction reduces the uncertainty by a factor of *at least* θ_{fb}. After n iterations, the size of the "confusability set" is upper-bounded

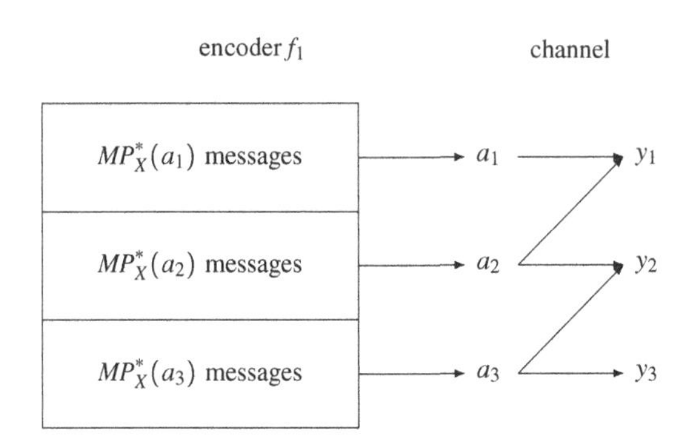

Figure 23.4 Achievability scheme for zero-error capacity with feedback.

by $M\theta^n_{fb}$, if $M\theta^n_{fb} \le 1$,[4] then zero error probability is achieved. This is guaranteed by choosing $\log M = -n\log\theta_{fb}$. Therefore we have shown that $-n\log\theta_{fb}$ bits can be reliably delivered with $n + O(1)$ channel uses with feedback, thus $C_{fb,0} \ge -\log\theta_{fb}$. □

Theorem 23.7 above shows possible advantages of feedback for zero-error communication. However, the zero-error capacity for a generic DMC (e.g. BSC_δ with $\delta \in (0, 1)$) is given by $C_0 = C_{fb,0} = 0$. Is there any advantage of feedback for such channels? Clearly for that we need to understand the behavior of $\log M^*_{fb}(n, \epsilon)$ for $\epsilon > 0$. It turns out that for weakly input-symmetric channels (Section 19.4*) we have, see Theorem 23.4,

$$\log M^*_{fb}(n,\epsilon) = nC - \sqrt{nV}Q^{-1}(\epsilon) + O(\log n),$$

and thus at least up to second order the behavior of fundamental limits is unchanged in the presence of feedback. Let us next discuss the error-exponent asymptotics (Section 22.4*) by defining

$$E_{fb}(R) \triangleq \lim_{n\to\infty} -\frac{1}{n}\log\epsilon^*_{fb}(n, \exp\{nR\}),$$

provided the limit exists and having denoted by $\epsilon^*_{fb}(n, M)$ the smallest possible error of a feedback code of blocklength n.

First, it is known that the sphere-packing bound (22.14) continues to hold in the presence of feedback [313], that is

$$E_{fb}(R) \le E_{sp}(R),$$

and thus for rates $R \in (R_{cr}, C)$ the error exponent $E_{fb}(R) = E(R)$ showing no change due to the availability of feedback. So what is the advantage of feedback then? It turns out that the error exponents do improve at rates below critical. For example, for the BEC_δ a simple transmission scheme where each bit is retransmitted until it is successfully received achieves error exponent $E_{sp}(R)$ at all rates (since the

[4] Some rounding-off errors need to be corrected in a few final steps (because P^*_X may not be closely approximable when very few messages are remaining). This does not change the asymptotics though.

probability of error here is given by $\mathbb{P}[\text{Bin}(n,\delta) > n(1 - R/\log 2)]$):

$$\text{BEC:} \qquad E_{fb}(R) = E_{sp}(R), \qquad 0 < R < C,$$

which is strictly higher than $E(R)$ for $R < R_{cr}$.

For the BSC_δ a beautiful result of Berlekamp shows that

$$\begin{aligned} E_{fb}(0+) &= -\log((1-\delta)^{1/3}\delta^{2/3} + (1-\delta)^{2/3}\delta^{1/3}) > E(0+) = E_{ex}(0) \\ &= -\frac{1}{4}\log(4\delta(1-\delta)). \end{aligned}$$

In other words, the error probability of optimal codes of size $M = \exp\{o(n)\}$ does significantly improve in the presence of feedback (at least over the BSC and BEC).

23.3.2 Code with Variable Length

Consider the example of BEC_δ channel with feedback. We can define a simple communication strategy that sends k data bits in the following way. Send each data bit repeatedly until it gets through the channel unerased. Using feedback the transmitter can know exactly when this occurs and at this point can switch to transmitting the next data bit. The expected number of channel uses for sending k bits is given by

$$\ell = \mathbb{E}[n] = \frac{k}{1-\delta}.$$

So, remarkably, we can see that by allowing variable-length transmission one can achieve capacity even by coding each data bit independently, thus completely avoiding any finite-blocklength penalty. While this cute coding scheme is special to BEC_δ, it turns out that the general fact of *zero dispersion* is universal. For that we define formally the following concept.

Definition 23.8. An (ℓ, M, ϵ) variable-length feedback (VLF) code, where ℓ is a positive real, M is a positive integer, and $0 \le \epsilon \le 1$, is defined as follows:

1. A space[5] $\mathcal{U}$ and a probability distribution P_U on it, defining a random variable U which is revealed to both transmitter and receiver before the start of transmission; that is, U acts as common randomness used to initialize the encoder and the decoder before the start of transmission.
2. A sequence of encoders $f_n \colon \mathcal{U} \times \{1, \ldots, M\} \times \mathcal{B}^{n-1} \to \mathcal{A}$, $n \ge 1$, defining channel inputs

$$X_n = f_n(U, W, Y^{n-1}), \tag{23.15}$$

where $W \in \{1, \ldots, M\}$ is the equiprobable message.
3. A sequence of decoders $g_n \colon \mathcal{U} \times \mathcal{B}^n \to \{1, \ldots, M\}$ providing the best estimate of W at time n.

[5] It can be shown that without loss of generality we can assume $|\mathcal{U}| \le 3$, see [337, Appendix].

4 A non-negative integer-valued random variable τ, a stopping time of the filtration $\mathcal{G}_n = \sigma\{U, Y_1, \ldots, Y_n\}$, which satisfies

$$\mathbb{E}[\tau] \le \ell. \tag{23.16}$$

The final decision $\hat{W}$ is computed at the time instant τ:

$$\hat{W} = g_\tau(U, Y^\tau), \tag{23.17}$$

and must satisfy

$$\mathbb{P}[\hat{W} \ne W] \le \epsilon. \tag{23.18}$$

The fundamental limit of channel coding with feedback is given by the following quantity:

$$M^*_{\text{VLF}}(\ell, \epsilon) = \max\{M \colon \exists (\ell, M, \epsilon)\text{-VLF code}\}. \tag{23.19}$$

In this language, our example above showed that for the BEC_δ we have

$$\log M^*_{\text{VLF}}(\ell, 0) \ge (1-\delta)\ell \log 2 + O(1), \qquad \ell \to \infty.$$

Notice that compared to the scheme without feedback, there is a significant improvement since the term $\sqrt{nV}Q^{-1}(\epsilon)$ in the expansion for $\log M^*(n, \epsilon)$ is now dropped. For this reason, results like this are known as *zero-dispersion* results.

It turns out that this effect is general for all DMC as long as we allow some $\epsilon > 0$ error.

Theorem 23.9. (VLF zero dispersion [337]) *For any DMC with capacity C we have*

$$\log M^*_{\text{VLF}}(l, \epsilon) = \frac{lC}{1-\epsilon} + O(\log l). \tag{23.20}$$

We omit the proof of this result, only mentioning that the achievability part relies on ideas similar to SPRT from Section 16.3*: the message keeps being transmitted until the information density $i(c_j;Y^n)$ of one of the codewords exceeds $\log M$. See [337] for details. We also mention that there is another variant of the VLF coding known as VLFT coding in which the stopping time τ, instead of being determined by the receiver, is allowed to be determined by the transmitter (see Exercise **IV.35**(d)). The expansion (23.20) continues to hold for VLFT codes as well [337].

Example 23.1.

For the channel $\text{BSC}_{0.11}$ without feedback the minimum n needed to achieve 90% of the capacity C is $n = 3000$, while there exists a VLF code with $\ell = \mathbb{E}[n] = 200$ achieving that [337]. This showcases how much feedback can improve the latency and decoding complexity.

VLF codes not only kill the dispersion term, but also dramatically improve error exponents. We have already discussed them in the context of fixed-length codes in Section 22.4* (without feedback) and at the end of the last section (with feedback). Here we mention a deep result of Burnashev [79], who showed that the optimal probability of error for VLF codes of rate R (i.e. with $\log M = \ell R$) satisfies *for every DMC*

$$\epsilon^*_{\mathrm{VLF}}(\ell, \exp\{\ell R\}) = \exp\{-\ell E_{\mathrm{VLF}}(R) + o(\ell)\},$$

where Burnashev's error exponent has a particularly simple expression:

$$E_{\mathrm{VLF}}(R) = \frac{C_1}{C}(C - R)_+ , \qquad C_1 = \max_{x_1, x_2} D(P_{Y|X=x_1} \| P_{Y|X=x_2}).$$

The simplicity of this expression when compared to the complicated (and still largely open!) situation with respect to non-feedback or fixed-length feedback error exponents is striking.

23.3.3 Codes with Variable Power

In the previous sections we discussed the advantages of feedback for the DMC. For the AWGN channel, it turns out that unlocking the potential of feedback requires relaxing the cost constraint. Recall that in Section 20.1 we postulated that every codeword $x^n \in \mathbb{R}^n$ should satisfy a fixed power constraint P, namely $\sum_{j=1}^n x_j^2 \leq nP$. It turns out that under such a power constraint one can show that feedback does not help much in any of the dispersion, finite-block length, or error-exponent senses. However, the true potential of feedback is unlocked if the power constraint is relaxed to

$$\mathbb{E}\left[\sum_{j=1}^n X_j^2\right] \leq nP,$$

where expectation here is both over the channel noise and the potential randomness employed by the transmitter in determination of X_j on the basis of the message W and $Y_1, \ldots, Y_{j-1}$. In the following, we demonstrate how to leverage this new freedom effectively.

Elias' Scheme Consider sending a standard Gaussian random variable A over the following set of AWGN channels:

$$Y_k = X_k + Z_k, \quad Z_k \sim \mathcal{N}(0, \sigma^2) \text{ iid},$$
$$\mathbb{E}[X_k^2] \leq P.$$

We assume that full noiseless feedback is available as in Figure 23.1. Note that, crucially, the power constraint is imposed *in expectation*, which does not increase the channel capacity (recall the converse in Theorem 20.6) but enables simple algorithms such as Elias' scheme below. In contrast, if we insist as in Section 20.1 that each codeword satisfies the power constraint almost surely instead of in expectation, that is, $\sum_{k=1}^n X_k^2 \leq nP$ a.s., then Elias' scheme does not work.

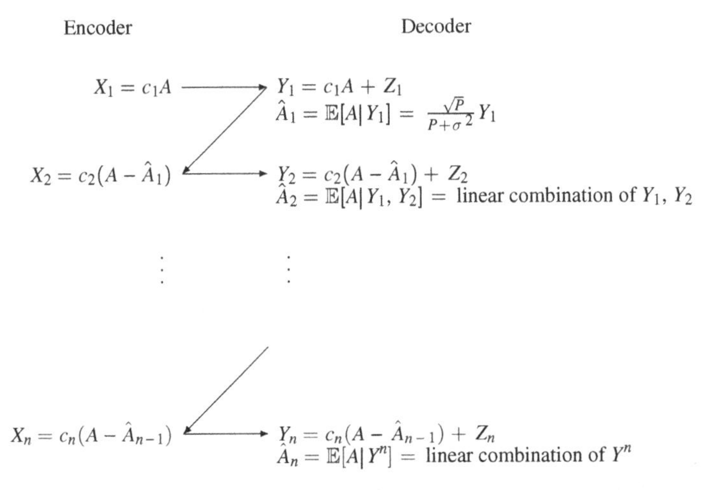

Figure 23.5 Elias' scheme for the AWGN channel with variable power. Here, each coefficient c_t is chosen such that $\mathbb{E}[X_t^2] = P$.

Using only *linear processing*, Elias' scheme proceeds according to the illustration in Figure 23.5.

According to the *orthogonality principle*, at the receiver side we have for all $t = 1, \ldots, n$,

$$A = \hat{A}_t + N_t, \quad N_t \perp\!\!\!\perp Y^t.$$

Moreover, since all operations are linear, all random variables are jointly Gaussian and hence the residual error satisfies $N_t \perp\!\!\!\perp Y^t$. Since $X_t \propto N_{t-1} \perp\!\!\!\perp Y^{t-1}$, the codeword we are sending at each time slot is independent of the history of the channel output ("innovation"), in order to maximize the information transfer.

Note that $Y^n \to \hat{A}_n \to A$, and the optimal estimator $\hat{A}_n$ (a linear combination of Y^n) is a sufficient statistic of Y^n for A thanks to Gaussianity. Then

$$\begin{aligned}
I(A;Y^n) &= I(A;\hat{A}_n, Y^n) \\
&= I(A;\hat{A}_n) + I(A;Y^n|\hat{A}_n) \\
&= I(A;\hat{A}_n) \\
&= \frac{1}{2}\log\frac{\operatorname{Var}(A)}{\operatorname{Var}(N_n)},
\end{aligned}$$

where the last equality applies $\hat{A}_n \perp\!\!\!\perp N_n$ and the Gaussian mutual information formula in Example 3.3. While $\operatorname{Var}(N_n)$ can be readily computed using standard linear MMSE results, next we determine it information-theoretically: Notice that we also have

$$\begin{aligned} I(A;Y^n) &= I(A;Y_1) + I(A;Y_2|Y_1) + \cdots + I(A;Y_n|Y^{n-1}) \\ &= I(X_1;Y_1) + I(X_2;Y_2|Y_1) + \cdots + I(X_n;Y_n|Y^{n-1}) \\ &\overset{\text{key}}{=} I(X_1;Y_1) + I(X_2;Y_2) + \cdots + I(X_n;Y_n) \\ &= \frac{n}{2}\log(1+P) = nC, \end{aligned}$$

where the key step applies $X_t \perp\!\!\!\perp Y^{t-1}$ for all t. Therefore, with Elias' scheme of sending $A \sim \mathcal{N}(0,1)$, after the nth use of the AWGN channel with feedback and expected power P, we have

$$\operatorname{Var} N_n = \operatorname{Var}(\hat{A}_n - A) = 2^{-2nC}\operatorname{Var} A = \left(\frac{P}{P+\sigma^2}\right)^n,$$

which says that the reduction of uncertainty in the estimation is exponentially fast in n.

Schalkwijk–Kailath scheme Elias' scheme can also be used to send digital data. Let $W \sim$ be uniform on the M-PAM (pulse-amplitude modulation) constellation in $[-1,1]$, that is, $\{-1, -1+\frac{2}{M}, \ldots, -1+\frac{2k}{M}, \ldots, 1\}$. In the very first step, W is sent (after scaling to satisfy the power constraint):

$$X_0 = \sqrt{P}W, \quad Y_0 = X_0 + Z_0.$$

Since Y_0 and X_0 are both known at the encoder, it can compute Z_0. Hence, to describe W it is sufficient for the encoder to describe the noise realization Z_0. This is done by employing the Elias' scheme ($n-1$ times). After $n-1$ channel uses, and the MSE estimation, the equivalent channel output is:

$$\widetilde{Y}_0 = X_0 + \widetilde{Z}_0, \quad \operatorname{Var}(\widetilde{Z}_0) = 2^{-2(n-1)C}.$$

Finally, the decoder quantizes $\widetilde{Y}_0$ to the nearest PAM point. Notice that

$$\epsilon \le \mathbb{P}\left[|\widetilde{Z}_0| > \frac{1}{2M}\right] = \mathbb{P}\left[2^{-(n-1)C}|Z| > \frac{\sqrt{P}}{2M}\right] = 2Q\left(\frac{2^{(n-1)C}\sqrt{P}}{2M}\right).$$

so that

$$\log M \ge (n-1)C + \log\frac{\sqrt{P}}{2} - \log Q^{-1}\left(\frac{\epsilon}{2}\right) = nC + O(1).$$

Hence if the rate is strictly less than capacity, the error probability decays *doubly exponentially* as n increases. More importantly, we gained an $\sqrt{n}$ term in terms of $\log M$, since for the case without feedback we have (by Theorem 22.2)

$$\log M^*(n,\epsilon) = nC - \sqrt{nV}Q^{-1}(\epsilon) + O(\log n).$$

As an example, consider $P = 1$ and then the channel capacity is $C = 0.5$ bit per channel use. To achieve error probability 10^{-3}, $2Q\left(\frac{2^{(n-1)C}}{2M}\right) \approx 10^{-3}$, so $\frac{2^{(n-1)C}}{2M} \approx 3$, and $\frac{\log M}{n} \approx \frac{n-1}{n}C - \frac{\log 8}{n}$. Notice that the capacity is achieved to within 99% in as few as $n = 50$ channel uses, whereas the best possible block codes without feedback require $n \approx 2800$ to achieve 90% of capacity.

The *take-away message* of this chapter is as follows: Feedback is best harnessed with *adaptive* strategies. Although it does not increase capacity under block coding, feedback can greatly boost reliability as well as reduce coding complexity.

Exercises for Part IV

IV.1 Consider the AWGN channel $Y^n = X^n + Z^n$, where Z_i is iid $\mathcal{N}(0,1)$ and $X_i \in [-1,1]$ (amplitude constraint). Recall that $\epsilon^*(n,2)$ denotes the optimal *average* probability of error of transmitting 1 bit of information over this channel.

(a) Express the value of $\epsilon^*(n,2)$ in terms of $Q(x) = \int_x^\infty \frac{1}{\sqrt{2\pi}} e^{-t^2/2} dt$.

(b) Compute the exponent $r = \lim_{n\to\infty} \frac{1}{n} \log \frac{1}{\epsilon^*(n,2)}$. (Hint: Exercise **V.25**.)

(c) Use asymptotics of hypothesis testing to compute r differently and check that the two values agree. (Hint: MGF of standard Gaussian $Z \sim \mathcal{N}(0,1)$ is given by $\mathbb{E}[e^{tZ}] = e^{t^2/2}$.)

IV.2 Randomized encoders and decoders may help maximal probability of error:

(a) Consider a binary asymmetric channel $P_{Y|X} \colon \{0,1\} \to \{0,1\}$ specified by $P_{Y|X=0} = \text{Ber}(1/2)$ and $P_{Y|X=1} = \text{Ber}(1/3)$. The encoder $f \colon [M] \to \{0,1\}$ tries to transmit 1 bit of information, that is, $M = 2$, with $f(1) = 0,\ f(2) = 1$. Show that the optimal decoder which minimizes the maximal probability of error is necessarily randomized. Find the optimal decoder and the optimal $P_{e,\max}$. (Hint: Recall binary hypothesis testing.)

(b) Give an example of $P_{Y|X} \colon \mathcal{X} \to \mathcal{Y}$, $M > 1$, and $\epsilon > 0$ such that there is an $(M,\epsilon)_{\max}$-code with a randomized encoder/decoder, but no such code with a deterministic encoder/decoder.

IV.3 (Lousy typist) Let $\mathcal{X} = \mathcal{Y} = \{A, S, D, F, G, H, J, K, L\}$. Let $P_{Y|X}(\alpha|\beta) = 0.1$ if α and β are neighboring letters in the keyboard, and $P_{Y|X}(\alpha|\beta) = 0$ if $\alpha \neq \beta$ and they are not neighbors. Find the smallest ϵ for which you can guarantee that a $(4,\epsilon)_{avg}$-code exists.

IV.4 (Finite-blocklength bounds for BEC). Consider a code with $M = 2^k$ operating over the blocklength-n binary erasure channel (BEC) with erasure probability $\delta \in [0,1)$.

(a) Show that regardless of the encoder–decoder pair:

$$\mathbb{P}[\text{error}|\#\text{erasures} = z] \geq \left|1 - 2^{n-z-k}\right|^+ .$$

(b) Conclude by averaging over the distribution of z that the probability of error ϵ must satisfy

$$\epsilon \geq \sum_{\ell=n-k+1}^{n} \binom{n}{\ell} \delta^\ell (1-\delta)^{n-\ell} \left(1 - 2^{n-\ell-k}\right). \tag{IV.1}$$

(c) By applying the DT bound with uniform P_X show that there exist codes with

$$\epsilon \leq \sum_{t=0}^{n} \binom{n}{t} \delta^t (1-\delta)^{n-t} 2^{-(n-t-k+1)^+}. \tag{IV.2}$$

(d) Fix $n = 500$, $\delta = 1/2$. Compute the smallest k for which the right-hand side of (IV.1) is greater than 10^{-3}.

(e) Fix $n = 500$, $\delta = 1/2$. Find the largest k for which the right-hand side of (IV.2) is smaller than 10^{-3}.

(f) Express your results in terms of lower and upper bounds on $\log M^*(500, 10^{-3})$.

IV.5 Recall that in the proof of the DT bound (Theorem 18.6) we used the decoder that outputs (for a given channel output y) the first c_m that satisfies

$$\{i(c_m; y) > \log \beta\}. \tag{IV.3}$$

One may consider the following generalization. Fix $E \subset \mathcal{X} \times \mathcal{Y}$ and let the decoder output the first c_m which satisfies

$$(c_m, y) \in E.$$

By repeating the random coding proof steps (as in the DT bound) show that the average probability of error satisfies

$$\mathbb{E}[P_e] \leq \mathbb{P}[(X, Y) \notin E] + \frac{M-1}{2} \mathbb{P}[(\bar{X}, Y) \in E],$$

where

$$P_{XY\bar{X}}(a, b, \bar{a}) = P_X(a) P_{Y|X}(b|a) P_X(\bar{a}).$$

Conclude that the optimal E is given by (IV.3) with $\beta = \frac{M-1}{2}$.

IV.6 In Section 18.6 we showed that for additive noise, random linear codes achieve the same performance as Shannon's ensemble (fully random coding). The total number of possible generator matrices is q^{nk}, which is significantly smaller than double exponential, but still quite large. Now we show that without degrading the performance, we can reduce this number to q^n by restricting to *Toeplitz* generator matrix G, that is, $G_{ij} = G_{i-1,j-1}$ for all $i, j > 1$.

Prove the following strengthening of Theorem 18.13: Let $P_{Y|X}$ be additive noise over $\mathbb{F}_q^n$. For any $1 \le k \le n$, there exists a linear code $f : \mathbb{F}_q^k \to \mathbb{F}_q^n$ with *Toeplitz* generator matrix, such that

$$P_{e,\max} = P_e \le \mathbb{E}\left[q^{-(n-k+\log_q P_{Z^n}(Z^n))^+}\right].$$

How many Toeplitz generator matrices are there?

(Hint: Analogous to the proof of Theorem 15.2, first consider random linear codewords plus random dithering, then argue that dithering can be removed without changing the performance of the codes. Show that codewords are pairwise independent and uniform.)

IV.7 (Wozencraft ensemble) Let $\mathcal{X} = \mathcal{Y} = \mathbb{F}_q^2$, a vector space of dimension two over a Galois field with q elements. A Wozencraft code of rate 1/2 is a map parameterized by $0 \ne u \in \mathbb{F}_q$ given as $a \mapsto (a, a\cdot u)$, where $a \in \mathbb{F}_q$ corresponds to the original message, multiplication is over $\mathbb{F}_q$, and $(\cdot,\cdot)$ denotes a two-dimensional vector in $\mathbb{F}_q^2$. We will show there exists u yielding a $(q,\epsilon)_{avg}$-code with

$$\epsilon \le \mathbb{E}\left[\exp\left\{-\left((X;Y) - \log\frac{q^2}{2(q-1)}\right)^+\right\}\right] \qquad \text{(IV.4)}$$

for the channel $Y = X + Z$ where X is uniform on $\mathbb{F}_q^2$, noise $Z \in \mathbb{F}_q^2$ has distribution P_Z, and

$$i(a;b) \triangleq \log\frac{P_Z(b-a)}{q^{-2}}.$$

(a) Show that the probability of error of the code $a \mapsto (av, au) + h$ is the same as that of $a \mapsto (a, auv^{-1})$.

(b) Let $\{X_a : a \in \mathbb{F}_q\}$ be a random codebook defined as

$$X_a = (aV, aU) + H,$$

with V, U uniform over non-zero elements of $\mathbb{F}_q$ and H uniform over $\mathbb{F}_q^2$, the three being jointly independent. Show that for $a \ne a'$ we have

$$P_{X_a, X_{a'}}(x_1^2, \tilde{x}_1^2) = \frac{1}{q^2(q-1)^2} 1\{x_1 \ne \tilde{x}_1, x_2 \ne \tilde{x}_2\}.$$

(c) Show that for $a \ne a'$

$$\begin{aligned}\mathbb{P}[i(X_{a'}; X_a + Z) > \log\beta] &\le \frac{q^2}{(q-1)^2}\mathbb{P}[i(\bar{X};Y) > \log\beta] \\ &\quad - \frac{1}{(q-1)^2}\mathbb{P}[i(X;Y) > \log\beta] \\ &\le \frac{q^2}{(q-1)^2}\mathbb{P}[i(\bar{X};Y) > \log\beta],\end{aligned}$$

where $P_{\bar{X}XY}(\bar{a}, a, b) = \frac{1}{q^4}P_Z(b-a)$.

(d) Conclude by following the proof of the DT bound with $M = q$ that the probability of error averaged over the random codebook $\{X_a\}$ satisfies (IV.4).

IV.8 (Universal codes) Fix finite alphabets $\mathcal{X}$ and $\mathcal{Y}$.

(a) Let $\mathcal{C}$ be a finite collection of channels $P_{Y|X}\colon \mathcal{X} \to \mathcal{Y}$. Show that for any P_X and any $R > 0$ there exists a sequence of codes (f_n, g_n) such that regardless of what DMC $P_{Y|X} \in \mathcal{C}$ is selected we have $\mathbb{P}[f_n(W) \neq g_n(Y^n)] \to 0$ as long as $I(P_X, P_{Y|X}) > R$. (Hint: Union bound over $\mathcal{C}$.)

(b) Extend the idea to show that there exists (f_n, g_n) such that for *any* DMC with $I(P_X, P_{Y|X}) > R$ we have $\mathbb{P}[f_n(W) \neq g_n(Y^n)] \to 0$. (Hint: Discretize the set of $\mathcal{X} \times \mathcal{Y}$ stochastic matrices.)

IV.9 (Information density and types) Let $P_{Y|X}\colon \mathcal{A} \to \mathcal{B}$ be a DMC and let P_X be some input distribution. Take $P_{X^nY^n} = P_{XY}^n$ and define $i(a^n;b^n)$ with respect to this $P_{X^nY^n}$.

(a) Show that $i(x^n;y^n)$ is a function of only the "joint type" $\hat{P}_{XY}$ of (x^n, y^n), which is a distribution on $\mathcal{A} \times \mathcal{B}$ defined as

$$\hat{P}_{XY}(a,b) = \frac{1}{n}\#\{i\colon x_i = a, y_i = b\},$$

where $a \in \mathcal{A}$ and $b \in \mathcal{B}$. Therefore the condition of the form $\{\frac{1}{n}i(x^n;y^n) \geq \gamma\}$ in the decoder (18.10) used in Shannon's random coding bound can be interpreted as a constraint on the joint type of (x^n, y^n).

(b) Assume also that the input x^n is such that $\hat{P}_X = P_X$. Show that

$$\frac{1}{n}i(x^n;y^n) \leq I(\hat{P}_X, \hat{P}_{Y|X}).$$

The quantity $I(\hat{P}_X, \hat{P}_{Y|X})$, sometimes written as $I(x^n \wedge y^n)$, is an *empirical mutual information*.[6] (Hint:

$$\mathbb{E}_{Q_{XY}}\left[\log \frac{P_{Y|X}(Y|X)}{P_Y(Y)}\right] = D(Q_{Y|X}\|Q_Y|Q_X) + D(Q_Y\|P_Y) - D(Q_{Y|X}\|P_{Y|X}|Q_X).)$$

IV.10 (Fitingof–Goppa universal codes) Consider a finite abelian group $\mathcal{X}$. Define the *Fitingof norm* as

$$\|x^n\|_\Phi \triangleq nH(\hat{P}_{x^n}) = nH(x_T), \quad T \sim \text{Unif}([n]),$$

where $\hat{P}_{x^n}$ is the empirical distribution of x^n.

(a) Show that $\|x^n\|_\Phi = \|-x^n\|_\Phi$ and the triangle inequality

$$\|x^n - y^n\|_\Phi \leq \|x^n\|_\Phi + \|y^n\|_\Phi.$$

[6] Invented by V. Goppa for his maximal mutual information (MMI) decoder [195]: $\hat{W} = \text{argmax}_i\, I(c_i \wedge y^n)$.

Conclude that $d_\Phi(x^n, y^n) \triangleq \|x^n - y^n\|_\Phi$ is a translation-invariant (Fitingof) metric on the set of equivalence classes in $\mathcal{X}^n$, with equivalence $x^n \sim y^n \iff \|x^n - y^n\|_\Phi = 0$.

(b) Define the Fitingof ball $B_r(x^n) \triangleq \{y^n : d_\Phi(x^n, y^n) \leq r\}$. Show that

$$\log |B_{\lambda n}(x^n)| = \lambda n + O(\log n)$$

for all $0 \leq \lambda \leq \log |\mathcal{X}|$.

(c) Show that for any product measure $P_{Z^n} = P_Z^n$ on $\mathcal{X}^n$ we have

$$\lim_{n\to\infty} P_{Z^n}[B_{\lambda n}(0^n)] = \begin{cases} 1, & H(Z) < \lambda, \\ 0, & H(Z) > \lambda. \end{cases}$$

(d) Conclude that a code $\mathcal{C} \subset \mathcal{X}^n$ with Fitingof minimal distance $d_{\min,\Phi}(\mathcal{C}) \triangleq \min_{c \neq c' \in \mathcal{C}} d_\Phi(c, c') \geq 2\lambda n$ is decodable with vanishing probability of error *on any additive-noise channel* $Y = X + Z$, as long as $H(Z) < \lambda$.

Comment: By a Feinstein-lemma-like argument it can be shown that there exist codes of size $\mathcal{X}^{n(1-\lambda)}$, such that balls of radius λn centered at codewords are almost disjoint. Such codes are universally capacity-achieving for all memoryless additive-noise channels on $\mathcal{X}$. Extension to general (non-additive) channels is done via introducing $d_\Phi(x^n, y^n) = nH(x_T|y_T)$, while extension to channels with Markov memory is done by introducing the Markov-type norm $\|x^n\|_{\Phi_1} = nH(x_T|x_{T-1})$. See [196, Chapter 3].

IV.11 A magician is performing card tricks on stage. In each round she takes a shuffled deck of 52 cards and asks someone to pick a random card N from the deck, which is then revealed to the audience. Assume the magician can prepare an arbitrary ordering of cards in the deck (before each round) and that N is distributed binomially on $\{0, \ldots, 51\}$ with mean $\frac{51}{2}$.

(a) What is the maximal number of *bits per round* that she can send over to her companion in the room (in the limit of infinitely many rounds)?

(b) Is communication possible if N were uniform on $\{0, \ldots, 51\}$? (In practice, however, nobody ever picks the top or the bottom ones.)

IV.12 (Channel with memory) Consider the additive-noise channel with $\mathcal{A} = \mathcal{B} = \mathbb{F}_2$ (Galois field of order two) and $P_{Y^n|X^n} : \mathbb{F}_2^n \to \mathbb{F}_2^n$ specified by

$$Y^n = X^n + Z^n,$$

where $Z^n = (Z_1, \ldots, Z_n)$ is a stationary Markov chain with $P_{Z_2|Z_1}(0|1) = P_{Z_2|Z_1}(1|0) = \tau$. Show information stability and find the capacity. (Hint: Your proof should work for an arbitrary stationary ergodic noise process $Z^\infty = (Z_1, \ldots)$.) Can the capacity be achieved by linear codes?

IV.13 Consider a DMC $P_{Y^n|X^n} = P_{Y|X}^{\otimes n}$, where a single-letter $P_{Y|X} : \mathcal{A} \to \mathcal{B}$ is given by $\mathcal{A} = \mathcal{B} = \{0, 1\}^7$, and

$$P_{Y|X}(\mathbf{y}|\mathbf{x}) = \begin{cases} 1 - p, & \mathbf{y} = \mathbf{x}, \\ p/7, & d_H(\mathbf{y}, \mathbf{x}) = 1, \end{cases}$$

where d_{H} stands for Hamming distance. In other words, for each 7-bit string, the channel either leaves it intact, or randomly flips exactly one bit.

(a) Compute the Shannon capacity C as a function of p and plot it.
(b) Consider the special case of $p = \frac{7}{8}$. Show that the zero-error capacity C_0 coincides with C. Moreover, show that C_0 can be achieved with blocklength $n = 1$ and give a capacity-achieving code.

IV.14 Find the capacity of the binary erasure-error channel shown below with channel matrix

where $0 \le \delta \le 1/2$.

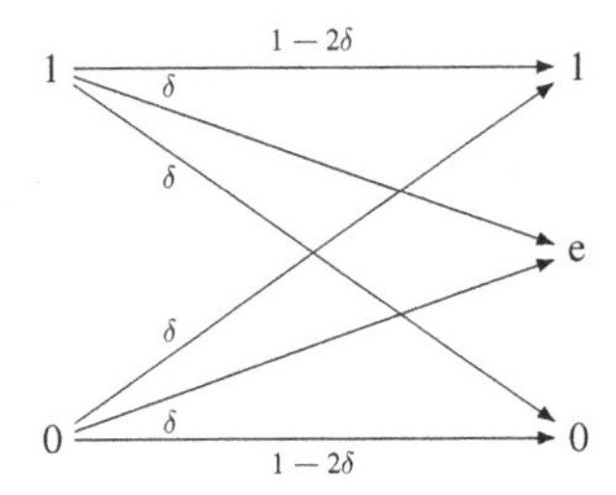

$$W = \begin{bmatrix} 1-2\delta & \delta & \delta \\ \delta & \delta & 1-2\delta \end{bmatrix},$$

IV.15 (Capacity of reordering) Routers A and B are setting up a covert communication channel in which the data is encoded in the ordering of packets. Formally, router A receives n packets, each having one of two types, Ack or Data, with probabilities p and $1-p$, respectively (and iid). It encodes k bits of secret data by reordering these packets. The network between A and B delivers packets in order with loss rate δ. *Note:* Packets have sequence numbers, so each loss is detectable by B.

What is the maximum rate of asymptotically reliable communication achievable?

IV.16 (Sum of channels) Let W_1 and W_2 denote the channel matrices of discrete memoryless channels (DMCs) $P_{Y_1|X_1}$ and $P_{Y_2|X_2}$ with capacity C_1 and C_2, respectively. The sum of the two channels is another DMC with channel matrix $\begin{bmatrix} W_1 & 0 \\ 0 & W_2 \end{bmatrix}$. Show that the capacity of the sum channel is given by

$$C = \log(\exp(C_1) + \exp(C_2)).$$

IV.17 (Product of channels) For $i = 1, 2$, let $P_{Y_i|X_i}$ be a (stationary memoryless) channel with input space $\mathcal{A}_i$, output space $\mathcal{B}_i$, and capacity C_i. Their product channel is a channel with input space $\mathcal{A}_1 \times \mathcal{A}_2$, output space $\mathcal{B}_1 \times \mathcal{B}_2$, and transition kernel $P_{Y_1Y_2|X_1X_2} = P_{Y_1|X_1}P_{Y_2|X_2}$. Show that the capacity of the product channel is given by $C = C_1 + C_2$.

IV.18 (Mixtures of DMCs) Consider two DMCs $U_{Y|X}$ and $V_{Y|X}$ with a common capacity-achieving input distribution and capacities $C_U < C_V$. Let $T = \{0, 1\}$ be uniform and consider a channel $P_{Y^n|X^n}$ that uses U if $T = 0$ and V if $T = 1$, or more formally:

$$P_{Y^n|X^n}(y^n|x^n) = \frac{1}{2}U^n_{Y|X}(y^n|x^n) + \frac{1}{2}V^n_{Y|X}(y^n|x^n). \tag{IV.5}$$

(a) Is this channel $\{P_{Y^n|X^n}\}_{n\geq 1}$ stationary? Memoryless?
(b) Show that the Shannon capacity C of this channel is not greater than C_U.
(c) Show that the maximal mutual information rate is

$$C^{(I)} = \lim_{n\to\infty} \frac{1}{n} \sup_{X^n} I(X^n;Y^n) = \frac{C_U + C_V}{2}.$$

(d) Conclude that $C < C^{(I)}$ and the strong converse does not hold.

IV.19 (Compound DMC [59]) A compound DMC is a family of DMCs with common input and output alphabets $P_{Y_s|X}: \mathcal{A} \to \mathcal{B}, s \in \mathcal{S}$. An (n, M, ϵ)-code is an encoder–decoder pair whose probability of error $\leq \epsilon$ over any channel $P_{Y_s|X}$ in the family (note that the same encoder and the same decoder are used for each $s \in \mathcal{S}$). Show that capacity is given by

$$C = \sup_{P_X} \inf_s I(X; Y_s).$$

The dispersion of the compound DMC is, however, more delicate [343].

IV.20 Consider the following (memoryless) channel. It has a side switch U that can be in positions ON and OFF. If U is on then the channel from X to Y is BSC_δ and if U is off then Y is Ber $(1/2)$ regardless of X. The receiving party sees Y but not U. A design constraint is that U should be in the ON position no more than the fraction s of all channel uses, $0 \leq s \leq 1$.

(a) One strategy is to put U into ON over the first sn time units and ignore the rest of the $(1 - s)n$ readings of Y. What is the maximal rate in bits per channel use achievable with this strategy?
(b) Can we increase the communication rate if the encoder is allowed to modulate the U switch together with the input X (while still satisfying the s-constraint on U)?
(c) Now assume nobody has access to U, which is iid Ber(s) independent of X. Find the capacity.

IV.21 Alice has n oranges and a great (essentially, infinite) number of empty trays. She wants to communicate a message to Bob by placing the oranges in (sequentially numbered) trays with at most one orange per tray. Unfortunately, before Bob gets to see the trays Eve inspects them and eats each orange independently with probability $0 < \delta < 1$. In the limit of $n \to \infty$ show that an arbitrary high rate (in bits per orange) is achievable.

Show that capacity changes to $\log_2 \frac{1}{\delta}$ bits per orange if Eve never eats any oranges but places an orange into each empty tray with probability δ (iid).

IV.22 (Non-stationary channel [106, Problem 9.12]) A train pulls out of the station at constant velocity. The received signal energy thus falls off with time as $1/i^2$. The total received signal at time i is

$$Y_i = \left(\frac{1}{i}\right) X_i + Z_i,$$

where $Z_1, Z_2, \ldots \stackrel{\text{iid}}{\sim} \mathcal{N}(0, \sigma^2)$. The transmitter cost constraint for blocklength n is $\sum_i^n |x_i^2| \le nP$. Show that the capacity C is equal to zero for this channel.

IV.23 (Capacity-cost function at the boundary) Recall from Corollary 20.5 that we have shown that for stationary memoryless channels and $P > P_0$ capacity equals $f(P)$:

$$C(P) = f(P), \tag{IV.6}$$

where

$$P_0 \triangleq \inf_{x \in \mathcal{A}} \mathsf{c}(x), \tag{IV.7}$$

$$f(P) \triangleq \sup_{X:\ \mathbb{E}[c(X)] \le P} I(X;Y). \tag{IV.8}$$

(a) If P_0 is not admissible, that is, $c(x) > P_0$ for all $x \in \mathcal{A}$, then show that $C(P_0)$ is undefined (even $M = 1$ is not possible).

(b) If there exists a unique x_0 such that $\mathsf{c}(x_0) = P_0$ show that

$$C(P_0) = f(P_0) = 0.$$

(c) If there is more than one x with $\mathsf{c}(x) = P_0$ then show that we still have

$$C(P_0) = f(P_0).$$

(d) Give an example of a channel with discontinuity of $C(P)$ at $P = P_0$. (Hint: Select a suitable cost function for the channel $Y = (-1)^Z \cdot \mathrm{sign}(X)$, where Z is Bernoulli and $\mathrm{sign}\colon \mathbb{R} \to \{-1, 0, 1\}$.)

IV.24 Consider a stationary memoryless additive *non-Gaussian* noise channel:

$$Y_i = X_i + Z_i, \qquad \mathbb{E}[Z_i] = 0, \quad \mathrm{Var}[Z_i] = 1,$$

with the standard input constraint $\sum_{i=1}^n x_i^2 \le nP$.

(a) Prove that the capacity $C(P)$ of this channel satisfies (20.6). (Hint: Use Gaussian saddle point, Theorem 5.11, and the golden formula $I(X;Y) \le D(P_{Y|X} \| Q_Y | P_X)$.)

(b) If $D(P_Z \| \mathcal{N}(0,1)) = \infty$ (Z is very non-Gaussian), then it is possible that the capacity is infinite. Let Z be ± 1 with equal probability. Show that the capacity is infinite by (a) proving the maximal mutual information is infinite; (b) giving an explicit scheme to achieve infinite capacity.

IV.25 (Input–output cost) Let $P_{Y|X}\colon \mathcal{X} \to \mathcal{Y}$ be a DMC and consider a cost function $\mathsf{c}\colon \mathcal{X} \times \mathcal{Y} \to \mathbb{R}$ (note that $\mathsf{c}(x,y) \le L < \infty$ for some L). Consider a problem of channel coding, where the error event is defined as

$$\{\text{error}\} \triangleq \{\hat{W} \neq W\} \cup \left\{ \sum_{k=1}^n \mathsf{c}(X_k, Y_k) > nP \right\},$$

where P is a fixed parameter. Define the operational capacity $C(P)$ and show it is given by

$$C^{(\mathrm{I})}(P) = \max_{P_X:\ \mathbb{E}[\mathsf{c}(X,Y)] \le P} I(X;Y)$$

for all $P > P_0 \triangleq \min_{x_0} \mathbb{E}[\mathsf{c}(X,Y)|X = x_0]$. Give a counterexample for $P = P_0$. (Hint: Do a converse directly, and for achievability reduce to an appropriately chosen cost function $\mathsf{c}'(x)$.)

IV.26 (Gauss–Markov noise) Let $\{Z_j : j = 1, 2, \ldots\}$ be a stationary ergodic Gaussian process with variance 1 such that Z_j form a Markov chain $Z_1 \to \cdots \to Z_n \to \cdots$. Consider an additive channel

$$Y^n = X^n + Z^n$$

with power constraint $\sum_{j=1}^n |x_j|^2 \le nP$. Suppose that $I(Z_1;Z_2) = \epsilon \ll 1$, then the capacity-cost function is

$$C(P) = \frac{1}{2}\log(1+P) + B\epsilon + o(\epsilon)$$

as $\epsilon \to 0$. Compute B and interpret your answer.

How does the frequency spectrum of the optimal signal change with increasing ϵ?

IV.27 A semiconductor company offers a random number generator that outputs a block of n random bits $Y_1, \ldots, Y_n$. The company wants to secretly embed a signature in every chip. To that end, it decides to encode a k-bit signature in n real numbers $X_j \in [0,1]$. Given an individual signature a chip is manufactured such that it produces the outputs $Y_j \sim \mathrm{Ber}(X_j)$. In order for the embedding to be inconspicuous the average bias p should be small:

$$\frac{1}{n}\sum_{j=1}^n \left|X_j - \frac{1}{2}\right| \le p.$$

As a function of p how many signature bits per output (k/n) can be reliably embedded in this fashion? Is there a simple coding scheme achieving this performance?

IV.28 (Capacity of sneezing) A sick student once every minute with probability p (iid) wants to sneeze. He decides to send k bits to a friend by modulating the sneezes. For that, every time he realizes he is about to sneeze he chooses to suppress the sneeze or not. A friend listens for n minutes and then tries to decode k bits.

(a) Find the capacity in bits per minute. (Hint: Think how to define the channel so that channel input at time t is not dependent on the arrival of the sneeze at time t. To rule out strategies that depend on arrivals of past sneezes, you may invoke Exercise **IV.34**.)

(b) Suppose the sender can suppress at most E sneezes and the listener can wait indefinitely ($n = \infty$). Show that the sender can transmit $C_{puc}E + o(E)$ bits reliably as $E \to \infty$ and find the capacity per unit cost C_{puc}. Curiously, $C_{puc} \ge 1.44$ bits/sneeze regardless of p. (Hint: This is similar to Exercise **IV.25**.)

(c*) Redo (a) and (b) for the case of a clairvoyant student who knows exactly when sneezes will happen in the future. (This is a simple example of a so-called Gelfand–Pinsker problem [183].)

IV.29 A data storage company is considering two options for sending its 100 Tb archive from Boston to NYC: via (physical) mail or via wireless transmission. Let us analyze these options:

(a) Given the radiated power P_t, the received power P_r at distance r for communicating at frequency f is given by $P_r = G(\frac{c}{4\pi r f})^2 P_t$, where G is antenna-to-antenna coupling gain and c is the speed of light.[7] Assuming they are transmitting between Boston and NYC compute the energy transfer coefficient η (take $G = 15$ dB and $f = 4$ GHz).

(b) The receiver's amplifier adds white Gaussian noise of power spectral density N_0 (W/Hz or J). On the basis of required energy-per-bit, compute the minimal N_0 which still makes transmission over the radio economically justified assuming optimal channel coding is done (assume 0.07$ per kWh and $20 per shipment).

(c) Compare this N_0 with the thermal noise power $N_{0,\text{thermal}} = kT$, where k is the Boltzmann constant and T is temperature in kelvins. Conclude that $T \le 10^3$ K should work.

(d) Codes that achieve Shannon's minimal E_b/N_0 in principle do not put restrictions on the receiver SNR (signal-to-noise ratio in one channel sample). However synchronization and other issues constrain this SNR to be ≥ -10 dB. Assuming communication bandwidth $B = 20$ MHz compute the minimal power (in W) required for the transmitter radio station. (Hint: Received SNR $= \frac{P_r}{BN_0}$, the answer should be a few watts.)

(e) How long will it take to send the archive at this bandwidth and SNR?

(Hint: The answer is between a few days and a few years.)

IV.30 (Optimal ϵ under ARQ) A packet of k bits is to be delivered over an AWGN channel with a given SNR. To that end, a k-to-n error-correcting code is used, whose probability of error is ϵ. The system employs automatic repeat request (ARQ) to resend the packet whenever an error occurred.[8] Suppose that the optimal k-to-n codes achieving

$$k \approx nC - \sqrt{nV}Q^{-1}(\epsilon) + \frac{1}{2}\log n$$

are available. The goal is to optimize ϵ to get the highest average throughput: ϵ too small requires excessive redundancy, ϵ too large leads to lots of retransmissions. Compute the optimal ϵ and optimal blocklength n for the following four cases: SNR = 0 dB or 20 dB; $k = 10^3$ or 10^4 bits. (This gives an idea of what ϵ you should aim for in practice.)

[7] This formula is known as the Friis transmission equation and it simply reflects the fact that the receiving antenna captures a plane wave at the effective area of $\frac{\lambda^2}{4\pi}$.

[8] Assuming there is a way for the receiver to verify whether the decoder produced the correct packet contents or not (e.g. by finding HTML tags).

IV.31 (Expurgated random coding bound)

(a) For any code $\mathcal{C}$ show the following bound on probability of error:

$$P_e(\mathcal{C}) \le \frac{1}{M} \sum_{c \ne c'} 2^{-d_B(c,c')},$$

where summation is over ordered pairs (c, c') and we recall from (16.3) the Bhattacharya distance $d_B(x^n, \tilde{x}^n) = \sum_{j=1}^n d_B(x_j, \tilde{x}_j)$ and

$$d_B(x, \tilde{x}) = -\log_2 \sum_{y \in \mathcal{Y}} \sqrt{W(y|x)W(y|\tilde{x})}.$$

(b) Fix P_X and let $E_{0,x}(\rho, P_X) \triangleq -\rho \log_2 \mathbb{E}[2^{-(1/\rho)d_B(X,X')}]$, where $X \perp\!\!\!\perp X' \sim P_X$. Show by random coding that there always exists a code $\mathcal{C}$ of rate R with

$$P_e(\mathcal{C}) \le 2^{n(E_{0,x}(1,P_X)-R)}.$$

(c) We improve the previous bound as follows. We still generate $\mathcal{C}$ by random coding. But this time we expurgate all codewords with $f(c, \mathcal{C}) > \mathrm{med}(f(c, \mathcal{C}))$, where $\mathrm{med}(\cdot)$ denotes the median and $f(c) = \sum_{c' : c' \ne c} 2^{-d_B(c,c')}$. Using the bound

$$\mathrm{med}(V) \le 2^\rho \, \mathbb{E}[V^{1/\rho}]^\rho \qquad \forall \rho \ge 1$$

show that

$$\mathrm{med}(f(c, \mathcal{C})) \le 2^{n(\rho R - E_{0,x}(\rho,P_X))}.$$

(d) Conclude that there must exist a code with rate $R - O(1/n)$ and $P_e(\mathcal{C}) \le 2^{-nE_{ex}(R)}$, where

$$E_{ex}(R) \triangleq \max_{\rho \ge 1} -\rho R + \max_{P_X} E_{0,x}(\rho, P_X).$$

IV.32 (Strong converse for BSC) In this exercise we give a combinatorial proof of the strong converse for the binary symmetric channel BSC_δ with $0 < \delta < \frac{1}{2}$.

(a) Given any $(n, M, \epsilon)_{\max}$-code with deterministic encoder f and decoder g, recall that the decoding regions $\{D_i = g^{-1}(i)\}_{i=1}^M$ form a partition of the output space. Prove that for all $i \in [M]$,

$$|D_i| \ge \sum_{j=0}^{L} \binom{n}{j},$$

where L is the largest integer such that $\mathbb{P}[\mathrm{Bin}(n, \delta) \le L] \le 1 - \epsilon$.

(b) Conclude that

$$M \le 2^{n(1-h(\delta))+o(n)}. \qquad \text{(IV.9)}$$

(c) Show that (IV.9) holds for average probability of error. (Hint: How do you go from maximal to average probability of error?)

(d) Conclude that the strong converse holds for BSC. (Hint: Argue that requiring deterministic encoder/decoder does not change the asymptotics.)

IV.33 (Strong converse for AWGN) Recall that the AWGN channel is specified by

$$Y^n = X^n + Z^n, \qquad Z^n \sim \mathcal{N}(0, I_n), \quad \mathsf{c}(x^n) = \frac{1}{n}\|x^n\|^2.$$

Prove the strong converse for the AWGN via the following steps:

(a) Let $c_i = f(i)$ and $D_i = g^{-1}(i), i = 1, \ldots, M$ be the codewords and the decoding regions of an $(n, M, P, \epsilon)_{\max}$-code. Let

$$Q_{Y^n} = \mathcal{N}(0, (1+P)I_n).$$

Show that there must exist a codeword c and a decoding region D such that

$$P_{Y^n|X^n=c}[D] \geq 1 - \epsilon, \tag{IV.10}$$

$$Q_{Y^n}[D] \leq \frac{1}{M}. \tag{IV.11}$$

(b) Show that then

$$\beta_{1-\epsilon}(P_{Y^n|X^n=c}, Q_{Y^n}) \leq \frac{1}{M}. \tag{IV.12}$$

(c) Show that the hypothesis testing problem

$$P_{Y^n|X^n=c} \quad \text{versus} \quad Q_{Y^n}$$

is equivalent to

$$P_{Y^n|X^n=Uc} \quad \text{versus} \quad Q_{Y^n},$$

where $U \in \mathbb{R}^{n\times n}$ is an orthogonal matrix. (Hint: Use the spherical symmetry of white Gaussian distributions.)

(d) Show that for any codeword c one can choose a deterministic matrix U such that

$$P_{Y^n|X^n=Uc} = \mathcal{N}(\mu, 1)^{\otimes n},$$

where mean μ will depend on $\|c\|^2$ only.

(e) Apply Stein's lemma (Theorem 14.14) to show that for a certain value $E = E(P) > 0$:

$$\beta_{1-\epsilon}(P_{Y^n|X^n=c}, Q_{Y^n}) = \exp\{-nE + o(n)\}.$$

(f) Conclude via (IV.12) that

$$\log M \leq nE + o(n) \quad \text{and, therefore,} \quad C_\epsilon \leq \frac{1}{2}\log(1+P).$$

IV.34 Consider a DMC with two outputs Y, U given by $P_{Y,U|X}$. Suppose that the receiver observes only Y, while U is (causally) fed back to the transmitter. We know that when $Y = U$ the capacity is not increased.

(a) Show that the capacity is not increased in general (even when $Y \neq U$).

(b) Suppose now that there is a cost function c and $\mathsf{c}(x_0) = 0$. Show that capacity per unit cost (with U being fed back) is still given by

$$C_V = \max_{x \neq x_0} \frac{D(P_{Y|X=x} \| P_{Y|X=x_0})}{\mathsf{c}(x)}.$$

IV.35 Consider a binary symmetric channel with crossover probability $\delta \in (0, 1)$:

$$Y = X + Z \mod 2, Z \sim \mathrm{Ber}(\delta).$$

Suppose that in addition to Y the receiver also gets to observe noise Z through a binary erasure channel with erasure probability $\delta_e \in (0, 1)$. Compute:

(a) Capacity C of the channel.
(b) Zero-error capacity C_0 of the channel.
(c) Zero-error capacity in the presence of feedback $C_{fb,0}$.
(d*) Now consider the setup when in addition to feedback also variable-length communication with feedback and termination (VLFT) is allowed. What is the zero-error capacity (in bits per average number of channel uses) in this case? (In the VLFT model, the transmitter can send a special symbol T that is received without error, but the channel dies after T has been sent; see Section 23.3.2.)

IV.36 Consider the *polygon channel* discussed in Remark 23.2, where the input and output alphabets are both $\{1, \ldots, L\}$, and $P_{Y|X}(b|a) > 0$ if and only if $b = a$ or $b = (a \bmod L) + 1$. The confusability graph is a *cycle* of L vertices. *Rigorously* prove the following:

(a) For all L, the zero-error capacity with feedback is $C_{fb,0} = \log \frac{L}{2}$.
(b) For even L, the zero-error capacity without feedback is $C_0 = \log \frac{L}{2}$.
(c) Now consider the following channel, where the input and output alphabets are both $\{1, \ldots, L\}$, and $P_{Y|X}(b|a) > 0$ if and only if $b = a$ or $b = a + 1$. In this case the confusability graph is a *path* of L vertices. Show that the zero-error capacity is given by

$$C_0 = \log \left\lceil \frac{L}{2} \right\rceil.$$

What is $C_{fb,0}$?

IV.37 (BEC with feedback) Consider the stationary memoryless binary erasure channel with erasure probability δ and noiseless feedback. Design a *fixed-blocklength* coding scheme achieving the capacity, that is, find a scheme that sends k bits over n channel uses with noiseless feedback, such that the rate $\frac{k}{n}$ approaches the capacity $1 - \delta$ when $n \to \infty$ and the maximal probability of error vanishes. Show also that for any rate $R < (1 - \delta)$ bit the error exponent matches the sphere-packing bound.

Part V

Rate-Distortion Theory and Metric Entropy

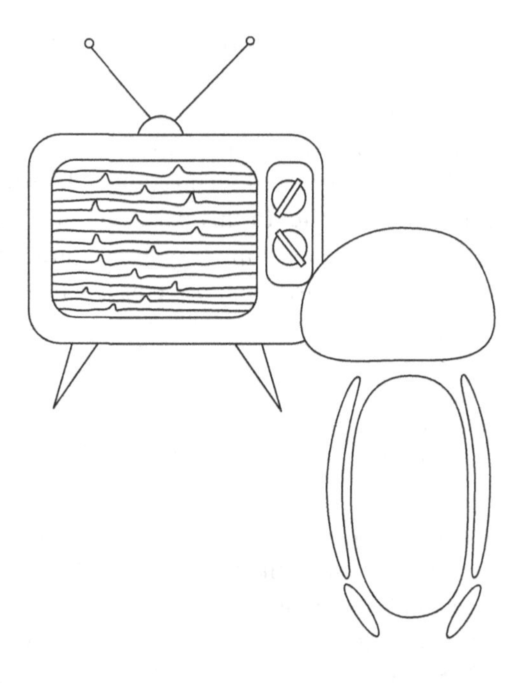

In Part II we studied lossless data compression (source coding), where the goal is to compress a random variable (source) X into a minimal number of bits so that X can be reconstructed from its compressed representation exactly (with probability either exactly 1 or close to 1) the fundamental limit was found to be given by the entropy of the source X. Clearly, this paradigm is confined to discrete random variables.

In this part we will tackle the problem of compressing continuous random variables, known as *lossy data compression*. This problem is formulated as follows. Given a random variable X, which is most likely continuous, the goal is to compress it into finitely many bits. To make such compression feasible, the decompressed version $\hat{X}$ is no longer demanded to be exactly equal to X, but only to give a reasonably faithful reconstruction of it. Mathematically, one requires that a chosen distortion metric between X and $\hat{X}$ be small on average or with high probability.

The motivations for studying lossy compression are at least two-fold.

Many natural signals (e.g. audio, images, or video) are continuously valued. As such, there is a need to represent these real-valued random variables or processes using finitely many bits, fed to downstream digital processing. We note that signals can be classified by whether their domain (time index) and whether their range (values) are continuous or discrete. In this Part, we only consider discrete time signals and thus completely ignore an important procedure, known as *sampling*, of converting from continuous to discrete time. This is schematically demonstrated in the following diagram.

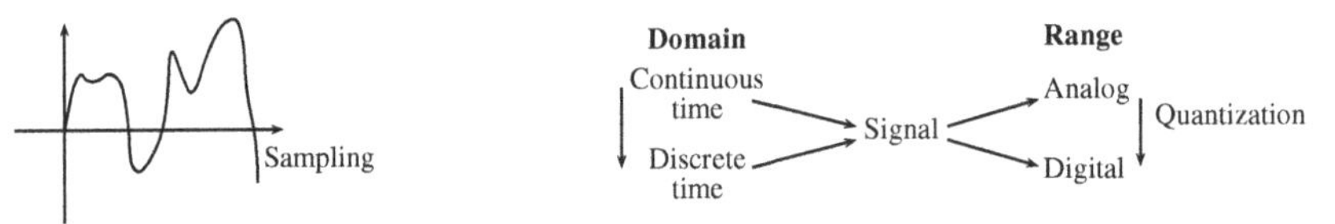

There is a lot to be gained in terms of compression rate if we allow some reconstruction errors. This is especially important in applications where certain errors (such as high-frequency components in natural audio and visual signals) are imperceptible to humans. This observation is the basis of many important compression algorithms and standards that are widely deployed in practice, including JPEG for images, MPEG for videos, and MP3 for audios.

The operation of mapping (naturally occurring) continuous/analog signals into (electronics-friendly) discrete/digital signals is known as *quantization*, which is an important subject in signal processing in its own right (see the encyclopedic survey [197]). In information theory, the study of optimal quantization is called *rate-distortion theory*, introduced by Shannon in 1959 [381]. To start, we will take a closer look at quantization next in Section 24.1, followed by the information-theoretic formulation in Section 24.2. A simple (and tight) converse bound is given in Section 24.3, with the matching achievability bound deferred to the next chapter.

In Chapter 25 we present the hard direction of the rate-distortion theorem: the random coding construction of a quantizer. This method is extended to the development of a covering lemma and soft-covering lemma, which lead to the sharp result of Cuff showing that the fundamental limit of channel simulation is given by Wyner's common information. We also derive (a strengthened form of) Han and

Verdú's results on approximating output distributions in Kullback–Leibler divergence.

Chapter 26 evaluates the rate-distortion function for Gaussian and Hamming sources. We also discuss the important foundational implication that an optimal (lossy) compressor paired with an optimal error-correcting code together form an optimal end-to-end communication scheme (known as joint source–channel coding separation principle). This principle explains why "bits" are the natural currency of the digital age.

Finally, in Chapter 27 we study Kolmogorov's *metric entropy*, which is a non-probabilistic theory of quantization for sets in metric spaces. While traditional rate distortion tries to compress samples from a fixed distribution, metric entropy tries to compress any element of the metric space. What links the two topics is that metric entropy can be viewed as a rate-distortion theory applied to the "worst-case" distribution on the space (an idea further expanded in Section 27.6). In addition to connections to the probabilistic theory of quantization in the preceding chapters, this concept has far-reaching consequences in both probability (e.g. empirical processes, small-ball probability) and statistical learning (e.g. entropic upper and lower bounds for estimation) that will be explored further in Part VI. Exercises explore applications to Brownian motion (Exercise **V.30**), random matrices (Exercise **V.29**), and more.

24 Rate-Distortion Theory

In this chapter we introduce the theory of optimal quantization. In Section 24.1 we examine the classical theory for quantization for *fixed dimension and high rate*, discussing various aspects such as uniform versus non-uniform quantization, fixed versus variable rate, quantization algorithm (of Lloyd) versus clustering, and the asymptotics of optimal quantization error. In Section 24.2 we turn to the information-theoretic formulation of quantization, known as the rate-distortion theory, that is targeted at *high dimension and fixed rate*, and the regime where the number of reconstruction points grows exponentially with dimension. Section 24.3 introduces the rate-distortion function and the main converse bound. Finally, in Section 24.4* we discuss how to relate the average distortion (upon which we focus) to excess distortion that targets a reconstruction error in high probability as opposed to in expectation.

24.1 Scalar and Vector Quantization

Before going to the information-theoretic setting, it is important to set the stage by introducing some of the classical pre-Shannon point of view on quantization. In this section and various subsections we focus on the setting where the continuous signal lives in a relatively low-dimensional space (for most of the section we only discuss scalar signals). We start with the very basic but overwhelmingly the most often used case of a scalar uniform quantization.

24.1.1 Scalar Uniform Quantization

The idea of quantizing an inherently continuous-valued signal was most explicitly expounded in the patenting of pulse-coded modulation (PCM) by A. Reeves; see [356] for some interesting historical notes. His argument was that unlike AM and FM modulation, quantized (digital) signals could be sent over long routes without the detrimental accumulation of noise. Some initial theoretical analysis of the PCM was undertaken in 1948 by Oliver, Pierce, and Shannon [319].

For a random variable $X \in [-A/2, A/2] \subset \mathbb{R}$, the scalar uniform quantizer $q_U(X)$ with N quantization points partitions the interval $[-A/2, A/2]$ uniformly:

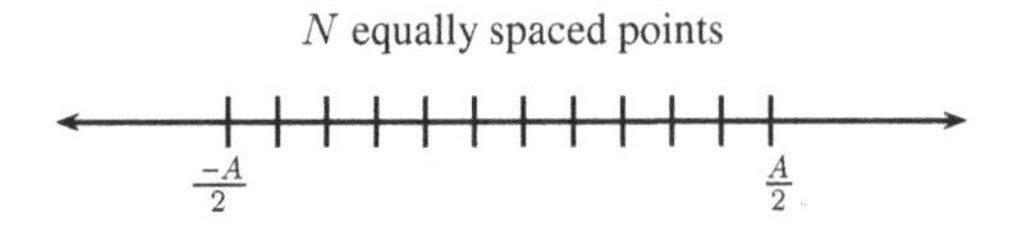

where the points are in $\{\frac{-A}{2} + \frac{kA}{N}, k = 0, \ldots, N-1\}$.

What is the *quality* (or fidelity) of this quantization? Most of the time, mean-squared error is used as the quality criterion:

$$D(N) = \mathbb{E}|X - q_U(X)|^2,$$

where D denotes the average *distortion*. Often $R = \log_2 N$ is used instead of N, so that we think about the number of bits we use for quantization instead of the number of points. To analyze this scalar uniform quantizer, we'll look at the high-rate regime ($R \gg 1$). The key idea in the high-rate regime is that (assuming a smooth density P_X), each quantization interval Δ_j looks nearly flat, so conditioned on Δ_j, the distribution is accurately approximated by a uniform distribution.

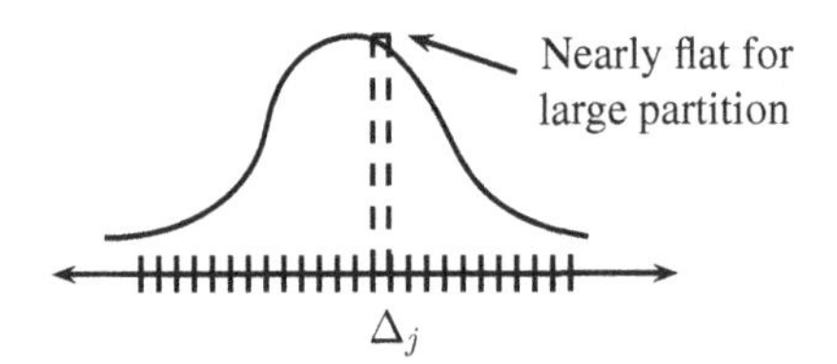

Let c_j be the jth quantization point, and Δ_j be the jth quantization interval. Here we have

$$D_U(R) = \mathbb{E}|X - q_U(X)|^2 = \sum_{j=1}^{N} \mathbb{E}[|X - c_j|^2 | X \in \Delta_j]\mathbb{P}[X \in \Delta_j] \tag{24.1}$$

$$\text{(high-rate approximation)} \quad \approx \sum_{j=1}^{N} \frac{|\Delta_j|^2}{12}\mathbb{P}[X \in \Delta_j] \tag{24.2}$$

$$= \frac{(\frac{A}{N})^2}{12} = \frac{A^2}{12}2^{-2R}, \tag{24.3}$$

where we used the fact that the variance of Unif$(-a, a)$ is $a^2/3$. How much do we gain per bit? We gain

$$\begin{aligned} 10\log_{10}\text{SNR} &= 10\log_{10}\frac{\text{var}(X)}{\mathbb{E}|X - q_U(X)|^2} \\ &= 10\log_{10}\frac{12\,\text{Var}(X)}{A^2} + (20\log_{10}2)R \\ &= \text{constant} + (6.02\text{ dB})R. \end{aligned}$$

For example, when X is uniform on $[-\frac{A}{2}, \frac{A}{2}]$, the constant is 0. Every engineer knows the rule of thumb "6 dB per bit"; adding one more quantization bit gets you

6 dB improvement in SNR. However, here we can see that this rule of thumb is valid only in the high-rate regime. (Consequently, widely articulated claims such as "16-bit PCM (CD-quality) provides 96 dB of SNR" should be taken with a grain of salt.)

The above discussion deals with X with a bounded support. When X is unbounded, it is wise to allocate the quantization points to those values that are more likely and saturate the large values at the dynamic range of the quantizer, resulting in two types of contributions to the quantization error, known as the granular distortion and overload distortion. This leads us to the question: Perhaps uniform quantization is not optimal?

24.1.2 Scalar Non-Uniform Quantization

Since our source has density p_X, a good idea may be to assign more quantization points where p_X is larger, and less where p_X is smaller, as the following picture illustrates:

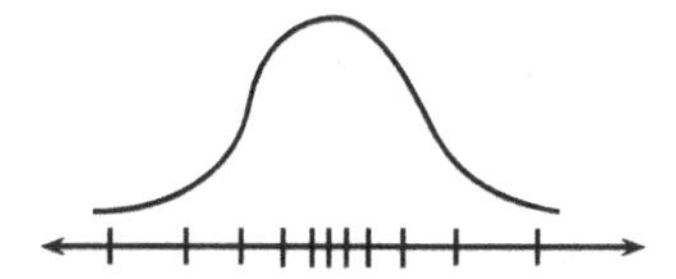

Often the way such quantizers are implemented is to take a monotone transformation of the source $f(X)$, perform uniform quantization, then take the inverse function:

$$\begin{array}{ccc} X & \xrightarrow{\;f\;} & U \\ \Big\downarrow{\scriptstyle q} & & \Big\downarrow{\scriptstyle q_U} \\ \hat{X} & \xleftarrow[f^{-1}]{} & q_U(U) \end{array} \tag{24.4}$$

that is, $q(X) = f^{-1}(q_U(f(X)))$. The function f is usually called the *compander* (compressor + expander). One of the choices of f is the CDF of X, which maps X to uniform on $[0, 1]$. In fact, this compander architecture is optimal in the high-rate regime (fine quantization) but the optimal f is not the CDF (!). We defer this discussion till Section 24.1.4.

In terms of practical considerations, for example, the human ear can detect sounds with volume as small as 0 dB, and a painful, ear-damaging sound occurs around 140 dB. Achieving this is possible because the human ear inherently uses a logarithmic companding function. Furthermore, many natural signals (such as *differences* of consecutive samples in speech or music (but not samples themselves!)) have an approximately Laplace distribution. Due to these two factors, a very popular and sensible choice for f is the μ-companding function:

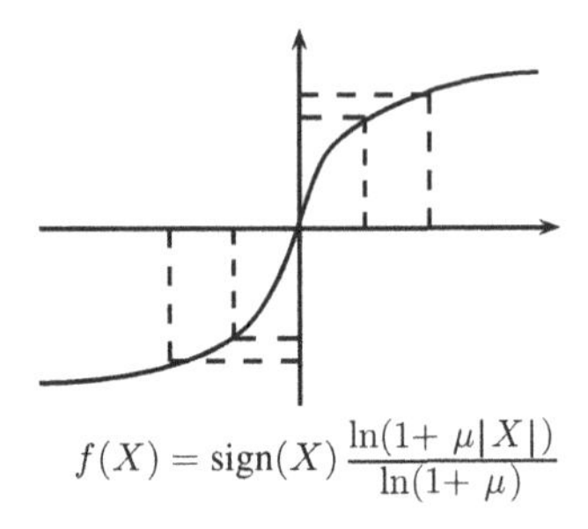

which compresses the dynamic range, uses more bits for smaller $|X|$'s, for example, $|X|$'s in the range of human hearing, and fewer quantization bits outside this region. This results in the so-called μ-law which is used in digital telecommunication systems in the US, while in Europe a slightly different compander called the A-law is used.

24.1.3 Optimal Quantizers

Now we look for the optimal scalar quantizer given R bits for reconstruction. Formally, this is

$$D_{\text{scalar}}(R) = \min_{q:\,|\text{Im}\,q|\leq 2^R} \mathbb{E}|X - q(X)|^2. \tag{24.5}$$

Intuitively, we would think that the optimal quantization regions should be contiguous; otherwise, given a point c_j, our reconstruction error will be larger. Therefore in one dimension quantizers are piecewise constant:

$$q(x) = c_j 1\left\{T_j \leq x \leq T_{j+1}\right\}$$

for some $c_j \in [T_j, T_{j+1}]$.

Example 24.1.

As a simple example, consider the one-bit quantization of $X \sim \mathcal{N}(0, \sigma^2)$. Then optimal quantization points are $c_1 = \mathbb{E}[X|X \geq 0] = \mathbb{E}[|X|] = \sqrt{\frac{2}{\pi}}\sigma$, $c_2 = \mathbb{E}[X|X \leq 0] = -\sqrt{\frac{2}{\pi}}\sigma$, with quantization error equal to $\text{Var}(|X|) = (1 - \frac{2}{\pi})\sigma^2$.

With ideas like this, in 1957 S. Lloyd developed an algorithm (now called *Lloyd's algorithm* or *Lloyd's method I*) for iteratively finding optimal quantization regions and points.[1] Suitable for both the scalar and vector cases, this method proceeds as follows: Initialized with some choice of $N = 2^k$ quantization points, the algorithm iterates between the following two steps:

[1] This work at Bell Labs remained unpublished until 1982 [285].

1 Draw the Voronoi regions around the chosen quantization points (aka minimum-distance tessellation, or set of points closest to c_j), which forms a partition of the space.
2 Update the quantization points by the centroids $\mathbb{E}[X|X \in D]$ of each Voronoi region D.

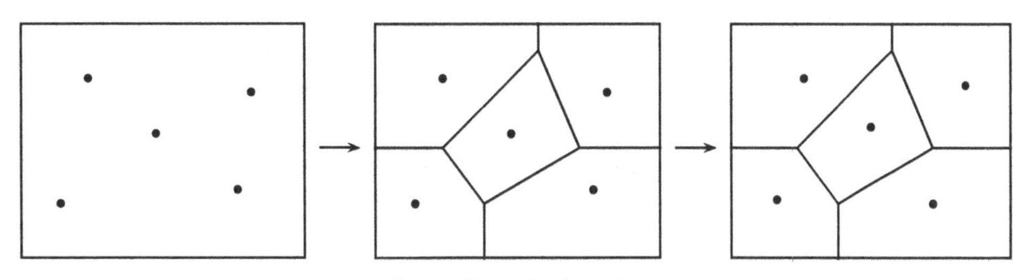

Steps of Lloyd's algorithm

Lloyd's clever observation is that the centroid of each Voronoi region is (in general) different than the original quantization points. Therefore, iterating through this procedure gives the *centroidal Voronoi tessellation* (CVT – which are very beautiful objects in their own right), which can be viewed as the fixed points of this iterative mapping. The following theorem gives the results on Lloyd's algorithm.

Theorem 24.1. (Lloyd)

1 Lloyd's algorithm always converges to a centroidal Voronoi tessellation.
2 The optimal quantization strategy is always a CVT.
3 CVT need not be unique, and the algorithm may converge to non-global optima.

Remark 24.1. The third point tells us that Lloyd's algorithm is not always guaranteed to give the optimal quantization strategy.[2] One sufficient condition for uniqueness of a CVT is the log-concavity of the density of X [171], for example Gaussians. On the other hand, even for the Gaussian P_X and $N > 3$, the optimal quantization points are not known in closed form. So it may seem to be very hard to have any meaningful theory of optimal quantizers. However, as we shall see next, when N becomes very large, the locations of optimal quantization points can be characterized. In this section we will do so in the case of fixed dimension, while for the rest of this part we will consider the regime of taking N to grow exponentially with dimension.

Remark 24.2. (k-means) A popular clustering method called *k-means* is the following: Given n data points $x_1, \ldots, x_n \in \mathbb{R}^d$, the goal is to find k centers $\mu_1, \ldots, \mu_k \in \mathbb{R}^d$ to minimize the objective function

[2] As a simple example one may consider $P_X = \frac{1}{3}\phi(x-1) + \frac{1}{3}f(x) + \frac{1}{3}f(x+1)$ where $f(\cdot)$ is a very narrow PDF, symmetric around 0. Here the CVT with centers $\pm\frac{2}{3}$ is not optimal among binary quantizers (just compare to any quantizer that quantizes two adjacent spikes to the same value).

$$\sum_{i=1}^{n} \min_{j\in[k]} \|x_i - \mu_j\|^2.$$

This is equivalent to solving the optimal vector quantization problem analogous to (24.5):

$$\min_{q:\,|\mathrm{Im}(q)|\le k} \mathbb{E}\|X - q(X)\|^2.$$

where X is distributed according to the *empirical distribution* over the dataset, namely, $\frac{1}{n}\sum_{i=1}^{n}\delta_{x_i}$. Solving the k-means problem is NP-hard in the worst case, and Lloyd's algorithm is a commonly used heuristic.

24.1.4 Fine Quantization

Following Panter and Dite [325], we now study the asymptotics of small quantization error. For this, introduce a probability density function $\lambda(x)$, which represents the density of quantization points in a given interval and allows us to approximate summations by integrals.[3] Then the number of quantization points in any interval $[a, b]$ is $\approx N\int_a^b \lambda(x)dx$. For any point x, denote the size of the quantization interval that contains x by $\Delta(x)$. Then

$$N\int_x^{x+\Delta(x)} \lambda(t)dt \approx N\lambda(x)\Delta(x) \approx 1 \implies \Delta(x) \approx \frac{1}{N\lambda(x)}.$$

With this approximation, the quality of reconstruction is

$$\begin{aligned}\mathbb{E}|X - q(X)|^2 &= \sum_{j=1}^{N} \mathbb{E}[|X - c_j|^2 | X \in \Delta_j]\mathbb{P}[X \in \Delta_j] \\ &\approx \sum_{j=1}^{N} \mathbb{P}[X \in \Delta_j]\frac{|\Delta_j|^2}{12} \approx \int p(x)\frac{\Delta^2(x)}{12}dx \\ &= \frac{1}{12N^2}\int p(x)\lambda^{-2}(x)dx.\end{aligned}$$

To find the optimal density λ that gives the best reconstruction (minimum MSE) when X has density p, we use Hölder's inequality: $\int p^{1/3} \le (\int p\lambda^{-2})^{1/3}(\int \lambda)^{2/3}$. Therefore $\int p\lambda^{-2} \ge (\int p^{1/3})^3$, with equality iff $p\lambda^{-2} \propto \lambda$. Hence the optimizer is $\lambda^*(x) = \frac{p^{1/3}(x)}{\int p^{1/3}dx}$. Therefore when[4] $N = 2^R$,

$$D_{\text{scalar}}(R) \approx \frac{1}{12}2^{-2R}\left(\int p^{1/3}(x)dx\right)^3.$$

[3] This argument can be easily made rigorous. We only need to define reconstruction points c_j as the solution to $\int_{-\infty}^{c_j} \lambda(x)\,dx = \frac{j}{N}$ (quantile).

[4] In fact when $R \to \infty$, "$\approx$" can be replaced by "$= 1 + o(1)$" as shown by Zador [468, 469].

So our optimal quantizer density in the high-rate regime is proportional to the cube root of the density of our source. This approximation is called the *Panter–Dite approximation.* For example:

- When $X \in [-\frac{A}{2}, \frac{A}{2}]$, using Hölder's inequality again $\langle 1, p^{1/3}\rangle \le \|1\|_{3/2}\|p^{1/3}\|_3 = A^{2/3}$, we have
$$D_{\text{scalar}}(R) \le \frac{1}{12}2^{-2R}A^2 = D_U(R),$$
where the RHS is the uniform quantization error given in (24.1). Therefore as long as the source distribution is not uniform, there is strict improvement. For a uniform distribution, uniform quantization is, unsurprisingly, optimal.
- When $X \sim \mathcal{N}(0, \sigma^2)$, this gives
$$D_{\text{scalar}}(R) \approx \sigma^2 2^{-2R}\frac{\pi\sqrt{3}}{2}. \tag{24.6}$$

Remark 24.3. In fact, in the *scalar* case the optimal non-uniform quantizer can be realized using the compander architecture (24.4) that we discussed in Section 24.1.2: As an exercise, use Taylor expansion to analyze the quantization error of (24.4) when $N \to \infty$. The optimal compander $f \colon \mathbb{R} \to [0,1]$ turns out to be $f(x) = \frac{\int_{-\infty}^{t} p^{1/3}(t)dt}{\int_{-\infty}^{\infty} p^{1/3}(t)dt}$ [44, 396].

24.1.5 Fine Quantization and Variable Rate

So far we have been focusing on quantization with restriction on the cardinality of the image of $q(\cdot)$. If one, however, intends to further compress the values $q(X)$ losslessly, a more natural constraint is to bound $H(q(X))$.

Koshelev [254] discovered in 1963 that in the high-rate regime uniform quantization is asymptotically optimal under the entropy constraint. Indeed, if q_Δ is a uniform quantizer with cell size Δ, then under appropriate assumptions we have (recall (2.21))

$$H(q_\Delta(X)) = h(X) - \log \Delta + o(1), \tag{24.7}$$

where $h(X) = -\int p_X(x) \log p_X(x)\, dx$ is the differential entropy of X. So a uniform quantizer with $H(q(X)) = R$ achieves

$$D = \frac{\Delta^2}{12} \approx 2^{-2R}\frac{2^{2h(X)}}{12}.$$

On the other hand, any quantizer with unnormalized point density function $\Lambda(x)$ (i.e. a smooth function such that $\int_{-\infty}^{c_j} \Lambda(x)dx = j$) can be shown to achieve (assuming $\Lambda \to \infty$ pointwise)

$$D \approx \frac{1}{12}\int p_X(x)\frac{1}{\Lambda^2(x)}dx,$$
$$H(q(X)) \approx \int p_X(x) \log \frac{\Lambda(x)}{p_X(x)}\, dx.$$

Now, from Jensen's inequality we have

$$\frac{1}{12}\int p_X(x)\frac{1}{\Lambda^2(x)}dx \geq \frac{1}{12}\exp\left\{-2\int p_X(x)\log\Lambda(x)\,dx\right\} \approx 2^{-2H(q(X))}\frac{2^{2h(X)}}{12},$$

concluding that the uniform quantizer is asymptotically optimal.

Furthermore, it turns out that for any source, even the optimal vector quantizers (to be considered next) cannot achieve distortion better than $2^{-2R}\frac{2^{2h(X)}}{2\pi e}$. That is, the maximal improvement they can gain for any iid source is 1.53 dB (or 0.255 bit/sample). This is one reason why scalar uniform quantizers followed by lossless compression is an overwhelmingly popular solution in practice.

24.2 Information-Theoretic Formulation

Before describing the mathematical formulation of optimal quantization, let us begin with two concrete examples.

Hamming Game Given 100 unbiased bits, we are asked to inspect them and scribble something down on a piece of paper that can store 50 bits at most. Later we will be asked to guess the original 100 bits, with the goal of maximizing the number of correctly guessed bits. What is the best strategy? Intuitively, it seems the optimal strategy would be to store half of the bits then randomly guess the rest, which gives 25% bit error rate (BER). However, as we will show in this chapter (Theorem 26.1), the optimal strategy amazingly achieves a BER of 11%. How is this possible? After all we are guessing independent bits and the loss function (BER) treats all bits equally.

Gaussian Example Given $(X_1, \ldots, X_n)$ drawn independently from $\mathcal{N}(0, \sigma^2)$, we are given a budget of one bit per symbol to compress, so that the decoded version $(\hat{X}_1, \ldots, \hat{X}_n)$ has a small mean-squared error $\frac{1}{n}\sum_{i=1}^n \mathbb{E}[(X_i - \hat{X}_i)^2]$.

To this end, a simple strategy is to quantize each coordinate into 1 bit. As worked out in Example 24.1, the optimal one-bit quantization error is $(1 - \frac{2}{\pi})\sigma^2 \approx 0.36\sigma^2$. In comparison, we will show later (Theorem 26.2) that there is a scheme that achieves an MSE of $\frac{\sigma^2}{4}$ per coordinate for large n; furthermore, this is optimal. More generally, given R bits per symbol, by doing optimal vector quantization in high dimensions (namely, compressing $(X_1, \ldots, X_n)$ jointly to nR bits), rate-distortion theory will tell us that when n is large, we can achieve the per-coordinate MSE:

$$D_{vec}(R) = \sigma^2 2^{-2R},$$

which, compared to the scalar quantization error (24.6), gains 4.35 dB (or 0.72 bits/sample).

The conclusions from both the Bernoulli and the Gaussian examples are rather surprising: Even when $X_1, \ldots, X_n$ are iid, there is something to be gained by quantizing these coordinates jointly. Some intuitive explanations for this high-dimensional phenomenon are as follows:

1 Applying scalar quantization componentwise results in quantization regions that are hypercubes, which may not be suboptimal for covering in high dimensions.
2 Concentration of measures effectively removes many atypical source realizations. For example, when quantizing a single Gaussian X, we need to cover a large portion of $\mathbb{R}$ in order to deal with those significant deviations of X from 0. However, when we are quantizing many $(X_1, \dots, X_n)$ together, the law of large numbers makes sure that many X_j's cannot conspire together and all produce large values. Indeed, $(X_1, \dots, X_n)$ concentrates near a sphere. As such, we may exclude large portions of the space $\mathbb{R}^n$ from consideration.

Mathematical Formulation A lossy compressor is an encoder–decoder pair (f, g) that induces the following Markov chain:

$$X \xrightarrow{f} W \xrightarrow{g} \hat{X},$$

where $X \in \mathcal{X}$ is referred to as the source, $W = f(X)$ is the compressed discrete data, and $\hat{X} = g(W)$ is the reconstruction which takes values in some alphabet $\hat{\mathcal{X}}$ that need not be the same as $\mathcal{X}$.

A *distortion metric* (or loss function) is a measurable function $d{:}\mathcal{X} \times \hat{\mathcal{X}} \to \mathbb{R} \cup \{+\infty\}$. There are various formulations of the lossy compression problem:

1 Fixed length (fixed rate), *average distortion*: $W \in [M]$, minimize $\mathbb{E}[d(X, \hat{X})]$.
2 Fixed length, *excess distortion*: $W \in [M]$, minimize $\mathbb{P}[d(X, \hat{X}) > D]$.
3 Variable length, *maximum distortion*: $W \in \{0, 1\}^*$, $d(X, \hat{X}) \leq D$ a.s., minimize the average length $\mathbb{E}[l(W)]$ or entropy $H(W)$.

In this book we focus on lossy compression with fixed length and are chiefly concerned with average distortion (with the exception of joint source–channel coding in Section 26.3 where excess distortion will be needed). The difference between average and excess distortion is analogous to that between average and high-probability risk bound in statistics and machine learning. It turns out that under mild assumptions these two formulations lead to the same asymptotic fundamental limit (see Remark 25.2). However, the speeds of convergence to that limit are very different. For excess distortion the convergence rate is $O(\frac{1}{\sqrt{n}})$, which has a rich dispersion theory [256]. The convergence under average distortion is much faster at $O(\frac{\log n}{n})$ [473]; see Exercises **V.3** and **V.4**.

As usual, of particular interest is when the source takes the form of a random vector $S^n = (S_1, \dots, S_n) \in \mathcal{S}^n$ and the reconstruction is $\hat{S}^n = (S_1, \dots, S_n) \in \hat{\mathcal{S}}^n$. We will be focusing on the so-called *separable* distortion metric defined for n-letter vectors by averaging the single-letter distortions:

$$d(s^n, \hat{s}^n) \triangleq \frac{1}{n} \sum_{i=1}^{n} d(s_i, \hat{s}_i). \tag{24.8}$$

Definition 24.2. An (n, M, D)-code consists of an encoder $f\colon \mathcal{A}^n \to [M]$ and a decoder $g\colon [M] \to \hat{\mathcal{A}}^n$ such that the average distortion satisfies $\mathbb{E}[d(S^n, g(f(S^n)))] \leq D$.

The non-asymptotic and asymptotic fundamental limits are defined as follows:

$$M^*(n, D) = \min\{M\colon \exists (n, M, D)\text{-code}\}, \tag{24.9}$$

$$R(D) = \limsup_{n\to\infty} \frac{1}{n} \log M^*(n, D). \tag{24.10}$$

Note that, for a stationary memoryless (iid) source, the large-blocklength limit in (24.10) in fact exists and coincides with the infimum over all blocklengths. This is a consequence of the average distortion criterion and the separability of the distortion metric – see Exercise **V.2**.

24.3 Converse Bounds

Now that we have the definitions, we give a (surprisingly simple) general converse.

Theorem 24.3. (General converse) *Suppose $X \to W \to \hat{X}$, where $W \in [M]$ and $\mathbb{E}[d(X, \hat{X})] \le D$. Then*

$$\log M \ge \phi_X(D) \triangleq \inf_{P_{Y|X}\colon \mathbb{E}[d(X,Y)] \le D} I(X;Y).$$

Proof.

$$\log M \ge H(W) \ge I(X;W) \ge I(X;\hat{X}) \ge \phi_X(D),$$

where the last inequality follows from the fact that $P_{\hat{X}|X}$ is a feasible solution (by assumption). □

Theorem 24.4. (Properties of ϕ_X)

(a) *ϕ_X is convex, non-increasing.*
(b) *ϕ_X continuous on (D_0, ∞), where $D_0 = \inf\{D\colon \phi_X(D) < \infty\}$.*
(c) *Suppose $\mathcal{X} = \hat{\mathcal{X}}$ and the distortion metric satisfies $d(x, x) = D_0$ for all x and $d(x, y) > D_0$ for all $x \ne y$. Then $\phi_X(D_0) = I(X;X)$.*
(d) *If d is a proper metric and $\mathcal{X}$ is a complete metric space, we have $D_0 = 0$ and $\phi_X(D_0+) = \phi_X(D_0) = I(X;X)$, where we have used the notation $f(x+)$ for the right limit of a function f at x.*
(e) *Let*

$$D_{\max} = \inf_{\hat{x}\in\hat{\mathcal{X}}} \mathbb{E}[d(X, \hat{x})].$$

Then $\phi_X(D) = 0$ for all $D > D_{\max}$. If $D_0 > D_{\max}$ then also $\phi_X(D_{\max}) = 0$.

Remark 24.4. (The role of D_0 and $D_{\max}$) By definition, $D_{\max}$ is the distortion attainable without any information. Indeed, if $D_{\max} = \mathbb{E}[d(X, \hat{x})]$ for some fixed $\hat{x}$, then this $\hat{x}$ is the "default" reconstruction of X, that is, the best estimate when we have no information about X. Therefore $D \ge D_{\max}$ can be achieved for free. This is the reason for the notation $D_{\max}$ despite that it is defined as an infimum. On the other hand, D_0 should be understood as the minimum distortion one can hope to attain. Indeed, suppose that $\hat{\mathcal{X}} = \mathcal{X}$ and d is a metric on $\mathcal{X}$. In this case, we have $D_0 = 0$, since we can choose Y to be a finitely valued approximation of X.

As an example, consider the Gaussian source with MSE distortion, namely, $X \sim \mathcal{N}(0, \sigma^2)$ and $d(x, \hat{x}) = (x - \hat{x})^2$. We will show later that $\phi_X(D) = \frac{1}{2} \log^+ \frac{\sigma^2}{D}$. In this case $D_0 = 0$ and the infimum defining it is not attained; $D_{\max} = \sigma^2$ and if $D \geq \sigma^2$, we can simply output 0 as the reconstruction which requires zero bits.

Proof.

(a) Convexity follows from the convexity of $P_{Y|X} \mapsto I(P_X, P_{Y|X})$ (Theorem 5.3).
(b) Continuity in the interior of the domain follows from convexity, since $D_0 = \inf_{P_{\hat{X}|X}} \mathbb{E}[d(X, \hat{X})] = \inf\{D \colon \phi_S(D) < \infty\}$.
(c) The only way to satisfy the constraint is to take $X = Y$.
(d) Clearly, $D_0 = d(x, x) = 0$. We also clearly have $\phi_X(D_0) \geq \phi_X(D_0+)$. Consider a sequence of Y_n such that $\mathbb{E}[d(X, Y_n)] \leq 2^{-n}$ and $I(X; Y_n) \to \phi_X(D_0+)$. By Borel–Cantelli we have with probability 1 that $d(X, Y_n) \to 0$ and hence $(X, Y_n) \to (X, X)$. Then from lower-semicontinuity of mutual information (4.28) we get $I(X; X) \leq \lim I(X; Y_n) = \phi_X(D_0+)$.
(e) For any $D > D_{\max}$ we can set $\hat{X} = \hat{x}$ deterministically. Thus $I(X; \hat{x}) = 0$. The second claim follows from continuity. □

In channel coding, the main result relates the Shannon capacity, an operational quantity, to the information capacity. Here we introduce the *information rate-distortion function* in an analogous way, which by itself is *not* an operational quantity.

Definition 24.5. The *information rate-distortion function* for a source $\{S_i\}$ is

$$R^{(\mathrm{I})}(D) = \limsup_{n \to \infty} \frac{1}{n} \phi_{S^n}(D), \quad \text{where } \phi_{S^n}(D) = \inf_{P_{\hat{S}^n|S^n} \colon \mathbb{E}[d(S^n, \hat{S}^n)] \leq D} I(S^n; \hat{S}^n).$$

The reason for defining $R^{(\mathrm{I})}(D)$ is because from Theorem 24.3 we immediately get the following:

Corollary 24.6. *For all D, $R(D) \geq R^{(\mathrm{I})}(D)$.*

Naturally, the information rate-distortion function inherits the properties of ϕ from Theorem 24.4:

Theorem 24.7. (Properties of $R^{(\mathrm{I})}$)

(a) $R^{(\mathrm{I})}(D)$ is convex, non-increasing.
(b) $R^{(\mathrm{I})}(D)$ is continuous on (D_0, ∞), where $D_0 \triangleq \inf\{D \colon R^{(\mathrm{I})}(D) < \infty\}$.
(c) Assume the same assumption on the distortion function as in Theorem 24.4(c). For stationary ergodic $\{S^n\}$, $R^{(\mathrm{I})}(D) = \mathcal{H}$ (entropy rate) or $+\infty$ if S_k is not discrete.

(d) $R^{(\mathrm{I})}(D) = 0$ *for all* $D > D_{\max}$, *where*

$$D_{\max} \triangleq \limsup_{n\to\infty} \inf_{\hat{x}^n \in \hat{\mathcal{X}}} \mathbb{E}d(X^n, \hat{x}^n).$$

If $D_0 > D_{\max}$, *then* $R^{(\mathrm{I})}(D_{\max}) = 0$ *too.*

Proof. All properties follow directly from the corresponding properties in Theorem 24.4 applied to ϕ_{S^n}. □

Next we show that $R^{(\mathrm{I})}(D)$ can be easily calculated for a stationary memoryless (iid) source without going through the multi-letter optimization problem. This parallels Corollary 20.5 for channel capacity (with separable cost function).

Theorem 24.8. (Single-letterization) *For stationary memoryless source* $S_i \overset{\text{iid}}{\sim} P_S$ *and separable distortion* d *in the sense of (24.8), we have for every* n,

$$\phi_{S^n}(D) = n\phi_S(D).$$

Thus

$$R^{(\mathrm{I})}(D) = \phi_S(D) = \inf_{P_{\hat{S}|S}:\, \mathbb{E}[d(S,\hat{S})] \le D} I(S;\hat{S}).$$

Proof. By definition we have that $\phi_{S^n}(D) \le n\phi_S(D)$ by choosing a product channel: $P_{\hat{S}^n|S^n} = P_{\hat{S}|S}^{\otimes n}$. Thus $R^{(\mathrm{I})}(D) \le \phi_S(D)$.

For the converse, for any $P_{\hat{S}^n|S^n}$ satisfying the constraint $\mathbb{E}[d(S^n, \hat{S}^n)] \le D$, we have

$$\begin{aligned}
I(S^n;\hat{S}^n) &\ge \sum_{j=1}^n I(S_j;\hat{S}_j) && (S^n \text{ independent})\\
&\ge \sum_{j=1}^n \phi_S(\mathbb{E}[d(S_j, \hat{S}_j)])\\
&\ge n\phi_S\left(\frac{1}{n}\sum_{j=1}^n \mathbb{E}[d(S_j, \hat{S}_j)]\right) && (\text{convexity of } \phi_S)\\
&\ge n\phi_S(D) && (\phi_S \text{ non-increasing}).
\end{aligned}$$

In the first step we used the crucial superadditivity property of mutual information (6.2). □

For generalization to a memoryless but non-stationary source see Exercise **V.21**.

24.4* Converting Excess Distortion to Average

Finally, we discuss how to build a compressor for average distortion if we have one for excess distortion, the former of which is our focus.

Theorem 24.9. (Excess-to-average) *Suppose that there exists* (f, g) *such that* $W = f(X) \in [M]$ *and* $\mathbb{P}[d(X, g(W)) > D] \leq \epsilon$. *Assume for some* $p \geq 1$ *and* $\hat{x}_0 \in \hat{\mathcal{X}}$ *that* $(\mathbb{E}[d(X, \hat{x}_0)^p])^{1/p} = D_p < \infty$. *Then there exists* (f', g') *such that* $W' = f'(X) \in [M+1]$ *and*

$$\mathbb{E}[d(X, g(W'))] \leq D(1-\epsilon) + D_p \epsilon^{1-1/p}. \tag{24.11}$$

Remark 24.5. This result is only useful for $p > 1$, since for $p = 1$ the right-hand side of (24.11) does not converge to D as $\epsilon \to 0$. However, a different method (as we will see in the proof of Theorem 25.1) implies that under just $D_{\max} = D_1 < \infty$ the analog of the second term in (24.11) is vanishing as $\epsilon \to 0$, albeit at an unspecified rate.

Proof. We transform the first code into the second by adding one codeword:

$$f'(x) = \begin{cases} f(x), & d(x, g(f(x))) \leq D, \\ M+1, & \text{otherwise}, \end{cases}$$

$$g'(j) = \begin{cases} g(j), & j \leq M, \\ \hat{x}_0, & j = M+1. \end{cases}$$

Then by Hölder's inequality,

$$\begin{aligned} \mathbb{E}[d(X, g'(W'))] &\leq \mathbb{E}[d(X, g(W)) | \hat{W} \neq M+1](1-\epsilon) \\ &\quad + \mathbb{E}[d(X, \hat{x}_0) \mathbf{1}\{\hat{W} = M+1\}] \\ &\leq D(1-\epsilon) + D_p \epsilon^{1-1/p}. \end{aligned}$$

□

25 Rate Distortion: Achievability Bounds

In this chapter, we prove an achievability bound and (together with the converse from the previous chapter) establish the identity $R(D) = \inf_{\hat{S}:\ \mathbb{E}[d(S,\hat{S})]\le D} I(S;\hat{S})$ for stationary memoryless sources. The key idea is again random coding, which is a probabilistic construction of quantizers by generating the reconstruction points independently from a carefully chosen distribution. Before proving this result rigorously, we first convey the main intuition in the case of Bernoulli sources by making connections to large-deviations theory (Chapter 15) and explaining how the constrained minimization of mutual information is related to optimization of the random coding ensemble. Later in Sections 25.2*–25.4*, we extend this random coding construction to establish the covering lemma and soft-covering lemma, which are at the heart of the problem of channel simulation.

We start by recalling the key concepts introduced in the last chapter:

$$R(D) = \limsup_{n\to\infty} \frac{1}{n} \log M^*(n, D) \quad \text{(rate-distortion function)},$$

$$R^{(\mathrm{I})}(D) = \limsup_{n\to\infty} \frac{1}{n} \phi_{S^n}(D) \quad \text{(information rate-distortion function)},$$

where

$$\phi_S(D) \triangleq \inf_{P_{\hat{S}|S}:\ \mathbb{E}[d(S,\hat{S})]\le D} I(S;\hat{S}), \tag{25.1}$$

$$\phi_{S^n}(D) = \inf_{P_{\hat{S}^n|S^n}:\ \mathbb{E}[d(S^n,\hat{S}^n)]\le D} I(S^n;\hat{S}^n), \tag{25.2}$$

and $d(S^n, \hat{S}^n) = \frac{1}{n}\sum_{i=1}^n d(S_i, \hat{S}_i)$ takes a separable form.

We have shown the following general converse in Theorem 24.3. For any compression scheme: $[M] \ni W \to X \to \hat{X}$ such that $\mathbb{E}[d(X, \hat{X})] \le D$, we must have $\log M \ge \phi_X(D)$, which implies in the special case of $X = S^n$, $\log M^*(n, D) \ge \phi_{S^n}(D)$ and hence, in the large-n limit, $R(D) \ge R^{(\mathrm{I})}(D)$. For a stationary memoryless source $S_i \overset{\text{iid}}{\sim} P_S$, Theorem 24.8 shows that ϕ_{S^n} single-letterizes as $\phi_{S^n}(D) = n\phi_S(D)$. As a result, we obtain the converse

$$R(D) \ge R^{(\mathrm{I})}(D) = \phi_S(D).$$

As mentioned earlier, the goal of this chapter is to show $R(D) = R^{(\mathrm{I})}(D)$.

25.1 Shannon's Rate-Distortion Theorem

The following result is (essentially) proved by Shannon in his 1959 paper [381].

Theorem 25.1. *Consider a stationary memoryless source* $S^n \overset{\text{iid}}{\sim} P_S$. *Suppose that the distortion metric d and the target distortion D satisfy:*

1. $d(s^n, \hat{s}^n)$ *is non-negative and separable;*
2. $D > D_0$, *where* $D_0 = \inf\{D\colon \phi_S(D) < \infty\}$;
3. $D_{\max}$ *is finite, that is,*

$$D_{\max} \triangleq \inf_{\hat{s}} \mathbb{E}[d(S, \hat{s})] < \infty. \tag{25.3}$$

Then

$$R(D) = R^{(\mathrm{I})}(D) = \phi_S(D) = \inf_{P_{\hat{S}|S}\colon \mathbb{E}[d(S,\hat{S})] \le D} I(S;\hat{S}). \tag{25.4}$$

Remark 25.1.

- Note that $D_{\max} < \infty$ does not require that $d(\cdot,\cdot)$ only takes values in $\mathbb{R}$. That is, Theorem 25.1 permits $d(s,\hat{s}) = \infty$.
- When $D_{\max} = \infty$, typically we have $R(D) = \infty$ for all D. Indeed, suppose that $d(\cdot,\cdot)$ is a metric (i.e. real-valued and satisfies the triangle inequality). Then, for any $x_0 \in \mathcal{A}^n$ we have

  $$d(X, \hat{X}) \ge d(X, x_0) - d(x_0, \hat{X}).$$

  Thus, for any finite codebook $\{c_1, \ldots, c_M\}$ we have $\max_j d(x_0, c_j) < \infty$ and therefore

  $$\mathbb{E}[d(X, \hat{X})] \ge \mathbb{E}[d(X, x_0)] - \max_j d(x_0, c_j) = \infty.$$

  So $R(D) = \infty$ for any finite D. This observation, however, should not be interpreted as the absolute impossibility of compressing such sources; it is just not possible with fixed-length codes. As an example, for quadratic distortion and Cauchy-distributed S, $D_{\max} = \infty$ since S has infinite second moment. But it is easy to see that[1] the *information* rate-distortion function $R^{(\mathrm{I})}(D) < \infty$ for any $D \in (0, \infty)$. In fact, in this case $R^{(\mathrm{I})}(D)$ is a hyperbola-like curve that never touches either axis. Using variable-length codes, S^n can be compressed non-trivially into W with bounded entropy (but unbounded cardinality) $H(W)$. An open question: Is $H(W) = nR^{(\mathrm{I})}(D) + o(n)$ attainable?
- We restricted the theorem to $D > D_0$ because it is possible that $R(D_0) \neq R^{(\mathrm{I})}(D_0)$. For example, consider an iid non-uniform source $\{S_j\}$ with $\mathcal{A} = \hat{\mathcal{A}}$ being a finite metric space with metric $d(\cdot,\cdot)$. Then $D_0 = 0$ and from Exercise **V.5** we have $R(D_0+) < R(D_0)$. At the same time, from Theorem 24.4(d) we know that $R^{(\mathrm{I})}$ is continuous at D_0: $R^{(\mathrm{I})}(D_0+) = \phi_S(D_0+) = \phi_S(D_0) = R^{(\mathrm{I})}(D_0)$.

[1] Indeed, if we take W to be a quantized version of S with small quantization error D and notice that the differential entropy of the Cauchy S is finite, we get from (24.7) that $R^{(\mathrm{I})}(D) \le H(W) < \infty$.

- Techniques for proving (25.4) for memoryless sources can be extended to stationary ergodic sources by making changes to the proof similar to those we have discussed in lossless compression (Chapter 12).

Before giving a formal proof, we give a heuristic derivation emphasizing the connection to large-deviations estimates from Chapter 15.

25.1.1 Intuition

Let us throw M random points $\mathcal{C} = \{c_1, \ldots, c_M\}$ into the space $\hat{\mathcal{A}}^n$ by generating them independently according to a product distribution $Q^n_{\hat{S}}$, where $Q_{\hat{S}}$ is some distribution on $\hat{\mathcal{A}}$ to be optimized. Consider the following simple coding strategy:

$$\text{encoder: } f(s^n) = \operatorname*{argmin}_{j \in [M]} d(s^n, c_j), \tag{25.5}$$

$$\text{decoder: } g(j) = c_j. \tag{25.6}$$

The basic idea is the following: Since the codewords are generated independently of the source, the probability that a given codeword is close to the source realization is (exponentially) small, say, ϵ. However, since we have many codewords, the chance that there exists a good one can be of high probability. More precisely, the probability that no good codewords exist is approximately $(1-\epsilon)^M$, which can be made close to zero provided $M \gg \frac{1}{\epsilon}$.

To explain this intuition further, consider a discrete memoryless source $S^n \overset{\text{iid}}{\sim} P_S$ and let us evaluate the excess distortion of this random code: $\mathbb{P}[d(S^n, f(S^n)) > D]$, where the probability is over all random codewords $c_1, \ldots, c_M$ and the source S^n. Define

$$P_{\text{failure}} \triangleq \mathbb{P}[\forall c \in \mathcal{C}, d(S^n, c) > D] = \mathbb{E}_{S^n}[\mathbb{P}[d(S^n, c_1) > D|S^n]^M],$$

where the last equality follows from the assumption that $c_1, \ldots, c_M$ are iid and independent of S^n. To simplify notation, let $\hat{S}^n \overset{\text{iid}}{\sim} Q^n_{\hat{S}}$ independently of S^n, so that $P_{S^n, \hat{S}^n} = P^n_S Q^n_{\hat{S}}$. Then

$$\mathbb{P}[d(S^n, c_1) > D|S^n] = \mathbb{P}[d(S^n, \hat{S}^n) > D|S^n]. \tag{25.7}$$

To evaluate the failure probability, let us consider the special case of $P_S = \text{Ber}(1/2)$ and also choose $Q_{\hat{S}} = \text{Ber}(1/2)$ to generate the random codewords, aiming to achieve a normalized Hamming distortion at most $D < \frac{1}{2}$. Since $nd(S^n, \hat{S}^n) = \sum_{i: s_i = 1}(1 - \hat{S}_i) + \sum_{i: s_i = 0} \hat{S}_i \sim \text{Bin}(n, \frac{1}{2})$ for any s^n, the conditional probability (25.7) does not depend on S^n and is given by

$$\mathbb{P}[d(S^n, \hat{S}^n) > D|S^n] = \mathbb{P}\left[\text{Bin}\left(n, \tfrac{1}{2}\right) \geq nD\right] \approx 1 - 2^{-n(1-h(D))+o(n)}, \tag{25.8}$$

where in the last step we applied large-deviations estimates from Theorem 15.9 and Example 15.1. (Note that here we actually need lower estimates on these exponentially small probabilities.) Thus, $P_{\text{failure}} = (1 - 2^{-n(1-h(D))+o(n)})^M$, which vanishes if

$M = 2^{n(1-h(D)+\delta)}$ for any $\delta > 0$.[2] As we will compute in Theorem 26.1, the rate-distortion function for $P_S = \text{Ber}(1/2)$ is precisely $\phi_S(D) = 1 - h(D)$, so we have a rigorous proof of the optimal achievability in this special case.

For a general distribution P_S (or even for $P_S = \text{Ber}(p)$ for which it is suboptimal to choose $Q_{\hat{S}}$ as Ber(1/2)), the situation is more complicated as the conditional probability (25.7) depends on the source realization S^n through its empirical distribution (type). Let T_n be the set of typical realizations whose empirical distribution is close to P_S. We have

$$\begin{aligned} P_{\text{failure}} &\approx \mathbb{P}[d(S^n, \hat{S}^n) > D | S^n \in T_n]^M \\ &= (1 - \underbrace{\mathbb{P}[d(S^n, \hat{S}^n) \le D | S^n \in T_n]}_{\approx 0, \text{ since } S^n \perp\!\!\!\perp \hat{S}^n})^M \\ &\approx (1 - 2^{-nE(Q_{\hat{S}})})^M, \end{aligned} \tag{25.9}$$

where it can be shown (using large-deviations analysis similar to information projection in Chapter 15) that

$$E(Q_{\hat{S}}) = \min_{P_{\hat{S}|S}:\, \mathbb{E}[d(S,\hat{S})] \le D} D(P_{\hat{S}|S} \| Q_{\hat{S}} | P_S). \tag{25.10}$$

Thus we conclude that for any choice of $Q_{\hat{S}}$ (from which the random codewords were drawn) and any $\delta > 0$, the above code with $M = 2^{n(E(Q_{\hat{S}})+\delta)}$ achieves vanishing excess distortion

$$P_{\text{failure}} = \mathbb{P}[\forall c \in \mathcal{C}, d(S^n, c) > D] \to 0 \text{ as } n \to \infty.$$

Finally, we optimize $Q_{\hat{S}}$ to get the smallest possible M:

$$\begin{aligned} \min_{Q_{\hat{S}}} E(Q_{\hat{S}}) &= \min_{Q_{\hat{S}}} \min_{P_{\hat{S}|S}:\, \mathbb{E}[d(S,\hat{S})] \le D} D(P_{\hat{S}|S} \| Q_{\hat{S}} | P_S) \\ &= \min_{P_{\hat{S}|S}:\, \mathbb{E}[d(S,\hat{S})] \le D} \min_{Q_{\hat{S}}} D(P_{\hat{S}|S} \| Q_{\hat{S}} | P_S) \\ &= \min_{P_{\hat{S}|S}:\, \mathbb{E}[d(S,\hat{S})] \le D} I(S; \hat{S}) \\ &= \phi_S(D), \end{aligned}$$

where the third equality follows from the variational representation of mutual information (Corollary 4.2). This heuristic derivation explains how constrained mutual information minimization arises. Below we make it rigorous using a different approach, again via random coding.

25.1.2 Proof of Theorem 25.1

Theorem 25.2. (Random coding bound of average distortion) *Fix P_X and suppose $d(x, \hat{x}) \ge 0$ for all $x, \hat{x}$. For any $P_{Y|X}$ and any $y_0 \in \hat{\mathcal{A}}$, there exists a code $X \to W \to \hat{X}$ with $W \in [M+1]$, such that $d(X, \hat{X}) \le d(X, y_0)$ almost surely, and for any $\gamma > 0$,*

[2] In fact, this argument shows that $M = 2^{n(1-h(D))+o(n)}$ codewords suffice to cover the *entire* Hamming space within distance Dn. See (27.9) and Exercise **V.26**.

$$\mathbb{E}[d(X,\hat{X})] \le \mathbb{E}[d(X,Y)] + \mathbb{E}[d(X,y_0)]e^{-M/\gamma} + \mathbb{E}[d(X,y_0)1\{i(X;Y) > \log\gamma\}].$$

Here the first and the third expectations are over $(X,Y) \sim P_{X,Y} = P_X P_{Y|X}$ and the information density $i(\cdot;\cdot)$ is defined with respect to this joint distribution (see Definition 18.1).

Some remarks are in order:

- Theorem 25.2 says that from an arbitrary $P_{Y|X}$ such that $\mathbb{E}[d(X,Y)] \le D$, we can extract a good code with average distortion D plus some extra terms which will vanish in the asymptotic regime for memoryless sources.
- The proof uses the random coding argument with codewords drawn independently from P_Y, the marginal distribution induced by the source distribution P_X, and the auxiliary channel $P_{Y|X}$. As such, $P_{Y|X}$ plays no role in the code construction and is used only in analysis (by defining a coupling between P_X and P_Y).
- The role of the deterministic y_0 is a "fail-safe" codeword (think of y_0 as the default reconstruction with $D_{\max} = \mathbb{E}[d(X,y_0)]$). We add y_0 to the random codebook for "damage control," to hedge against the (highly unlikely) event that we end up with a terrible codebook.

Proof. Similar to the intuitive argument sketched in Section 25.1.1, we apply random coding and generate the codewords randomly and independently of the source:

$$\mathcal{C} = \{c_1,\ldots,c_M\} \overset{\text{iid}}{\sim} P_Y \perp\!\!\!\perp X,$$

and add the "fail-safe" codeword $c_{M+1} = y_0$. We adopt the same encoder–decoder pair (25.5) and (25.6) and let $\hat{X} = g(f(X))$. Then by definition,

$$d(X,\hat{X}) = \min_{j\in[M+1]} d(X,c_j) \le d(X,y_0).$$

To simplify notation, let $\overline{Y}$ be an independent copy of Y (similar to the idea of introducing an unsent codeword $\overline{X}$ in channel coding – see Chapter 18):

$$P_{X,Y,\overline{Y}} = P_{X,Y} P_{\overline{Y}},$$

where $P_{\overline{Y}} = P_Y$. Recall the formula for computing the expectation of a random variable $U \in [0,a]$: $\mathbb{E}[U] = \int_0^a \mathbb{P}[U \ge u]du$. Then the average distortion is

$$\begin{aligned}
\mathbb{E}[d(X,\hat{X})] &= \mathbb{E} \min_{j\in[M+1]} d(X,c_j) \\
&= \mathbb{E}_X \mathbb{E}\Big[\min_{j\in[M+1]} d(X,c_j)\Big| X\Big] \\
&= \mathbb{E}_X \int_0^{d(X,y_0)} \mathbb{P}\Big[\min_{j\in[M+1]} d(X,c_j) > u\Big| X\Big] du \\
&\le \mathbb{E}_X \int_0^{d(X,y_0)} \mathbb{P}\Big[\cdot \min_{j\in[M]} d(X,c_j) > u\Big| X\Big] du
\end{aligned}$$

$$
\begin{aligned}
&= \mathbb{E}_X \int_0^{d(X,y_0)} \mathbb{P}[d(X,\overline{Y}) > u|X]^M du \\
&= \mathbb{E}_X \int_0^{d(X,y_0)} (1 - \mathbb{P}[d(X,\overline{Y}) \le u|X])^M du \\
&\le \mathbb{E}_X \int_0^{d(X,y_0)} (1 - \underbrace{\mathbb{P}[d(X,\overline{Y}) \le u, i(X;\overline{Y}) > -\infty|X]}_{\triangleq \delta(X,u)})^M du. \qquad (25.11)
\end{aligned}
$$

Next we upper-bound $(1 - \delta(X,u))^M$ as follows:

$$
\begin{aligned}
(1 - \delta(X,u))^M &\le e^{-M/\gamma} + |1 - \gamma\delta(X,u)|^+ &(25.12)\\
&= e^{-M/\gamma} + |1 - \gamma\mathbb{E}[\exp\{-i(X;Y)\}1\{d(X,Y) \le u\}|X]|^+ &(25.13)\\
&\le e^{-M/\gamma} + \mathbb{P}[i(X;Y) > \log\gamma|X] + \mathbb{P}[d(X,Y) > u|X], &(25.14)
\end{aligned}
$$

where

- (25.12) uses the following trick in dealing with $(1-\delta)^M$ for $\delta \ll 1$ and $M \gg 1$. First, recall the standard rule of thumb:

$$
(1-\delta)^M \approx \begin{cases} 0, & \delta M \gg 1, \\ 1, & \delta M \ll 1. \end{cases}
$$

  In order to obtain firm bounds of a similar flavor, we apply, for any $\gamma > 0$,

$$
(1-\delta)^M \le e^{-\delta M} \le e^{-M/\gamma} + (1 - \gamma\delta)_+.
$$

- (25.13) is simply a change of measure argument of Proposition 18.3. Namely we apply (18.4) with $f(x,y) = 1\{d(x,y) \le u\}$.
- For (25.14) consider the chain:

$$
\begin{aligned}
&1 - \gamma\,\mathbb{E}[\exp\{-i(X;Y)\}1\{d(X,Y) \le u\}|X] \\
&\le 1 - \gamma\,\mathbb{E}[\exp\{-i(X;Y)\}1\{d(X,Y) \le u, i(X;Y) \le \log\gamma\}|X] \\
&\le 1 - \mathbb{E}[1\{d(X,Y) \le u, i(X;Y) \le \log\gamma\}|X] \\
&= \mathbb{P}[d(X,Y) > u \text{ or } i(X;Y) > \log\gamma|X] \\
&\le \mathbb{P}[d(X,Y) > u|X] + \mathbb{P}[i(X;Y) > \log\gamma|X].
\end{aligned}
$$

Plugging (25.14) into (25.11), we have

$$
\begin{aligned}
\mathbb{E}[d(X,\hat{X})] &\le \mathbb{E}_X\left[\int_0^{d(X,y_0)} (e^{-M/\gamma} + \mathbb{P}[i(X;Y) > \log\gamma|X] + \mathbb{P}[d(X,Y) > u|X])du\right] \\
&\le \mathbb{E}[d(X,y_0)]e^{-M/\gamma} + \mathbb{E}[d(X,y_0)\mathbb{P}[i(X;Y) > \log\gamma|X]] \\
&\quad + \mathbb{E}_X \int_0^\infty \mathbb{P}[d(X,Y) > u|X]du \\
&= \mathbb{E}[d(X,y_0)]e^{-M/\gamma} + \mathbb{E}[d(X,y_0)1\{i(X;Y) > \log\gamma\}] + \mathbb{E}[d(X,Y)]. \quad \square
\end{aligned}
$$

As a side product, we have the following achievability result for excess distortion.

Theorem 25.3. (Random coding bound of excess distortion) *For any $P_{Y|X}$, there exists a code $X \to W \to \hat{X}$ with $W \in [M]$, such that for any $\gamma > 0$,*

$$\mathbb{P}[d(X,\hat{X}) > D] \le e^{-M/\gamma} + \mathbb{P}[\{d(X,Y) > D\} \cup \{i(X;Y) > \log \gamma\}].$$

Proof. Proceed exactly as in the proof of Theorem 25.2 (without using the extra codeword y_0), replace (25.11) by $\mathbb{P}[d(X,\hat{X}) > D] = \mathbb{P}[\forall j \in [M], d(X,c_j) > D] = \mathbb{E}_X[(1 - \mathbb{P}[d(X,\overline{Y}) \le D|X])^M]$, and continue similarly. □

Finally, we give a rigorous proof of Theorem 25.1 by applying Theorem 25.2 to the iid source $X = S^n \overset{\text{iid}}{\sim} P_S$ and $n \to \infty$:

Proof of Theorem 25.1. Our goal is the achievability: $R(D) \le R^{(\mathrm{I})}(D) = \phi_S(D)$.

Without loss of generality, we can assume that $D_{\max} = \mathbb{E}[d(S,\hat{s}_0)]$ is achieved at some fixed $\hat{s}_0$ – this is our default reconstruction; otherwise just take any other fixed symbol so that the expectation is finite. The default reconstruction for S^n is $\hat{s}_0^n = (\hat{s}_0, \dots, \hat{s}_0)$ and $\mathbb{E}[d(S^n, \hat{s}_0^n)] = D_{\max} < \infty$ since the distortion is separable.

Fix some small $\delta > 0$. Take any $P_{\hat{S}|S}$ such that $\mathbb{E}[d(S,\hat{S})] \le D - \delta$; such $P_{\hat{S}|S}$ exists since $D > D_0$ by assumption. Applying Theorem 25.2 to $(X,Y) = (S^n, \hat{S}^n)$ with

$$\begin{aligned}
P_X &= P_{S^n}, \\
P_{Y|X} &= P_{\hat{S}^n|S^n} = (P_{\hat{S}|S})^n, \\
\log M &= n(I(S;\hat{S}) + 2\delta), \\
\log \gamma &= n(I(S;\hat{S}) + \delta), \\
d(X,Y) &= \frac{1}{n}\sum_{j=1}^n d(S_j,\hat{S}_j), \\
y_0 &= \hat{s}_0^n,
\end{aligned}$$

we conclude that there exists a compressor $f\colon \mathcal{A}^n \to [M+1]$ and a decompressor $g\colon [M+1] \to \hat{\mathcal{A}}^n$, such that

$$\begin{aligned}
\mathbb{E}[d(S^n, g(f(S^n)))] &\le \mathbb{E}[d(S^n,\hat{S}^n)] + \mathbb{E}[d(S^n,\hat{s}_0^n)]e^{-M/\gamma} \\
&\quad + \mathbb{E}[d(S^n,\hat{s}_0^n)1\{i(S^n;\hat{S}^n) > \log\gamma\}] \\
&\le D - \delta + \underbrace{D_{\max}\, e^{-\exp(n\delta)}}_{\to 0} + \underbrace{\mathbb{E}[d(S^n,\hat{s}_0^n)1_{E_n}]}_{\to 0 \text{ (later)}},
\end{aligned} \tag{25.15}$$

where

$$E_n = \{i(S^n;\hat{S}^n) > \log\gamma\} = \left\{\frac{1}{n}\sum_{j=1}^n i(S_j;\hat{S}_j) > I(S;\hat{S}) + \delta\right\} \overset{\text{WLLN}}{\Longrightarrow} \mathbb{P}[E_n] \to 0.$$

If we can show the expectation in (25.15) vanishes, then there exists an $(n, M, \overline{D})$-code with

$$M = 2^{n(I(S;\hat{S})+2\delta)}, \quad \overline{D} = D - \delta + o(1) \le D.$$

To summarize, for any $P_{\hat{S}|S}$ such that $\mathbb{E}[d(S,\hat{S})] \le D - \delta$ we have shown that $R(D) \le I(S;\hat{S})$. Sending $\delta \downarrow 0$, we have, by continuity of $\phi_S(D)$ in (D_0, ∞) (recall Theorem 24.4), $R(D) \le \phi_S(D-) = \phi_S(D)$.

It remains to show the expectation in (25.15) vanishes. This is a simple consequence of the uniform integrability of the sequence $\{d(S^n, \hat{s}_0^n)\}$. We need the following lemma.

Lemma 25.4. *For any positive random variable U, define $g(\delta) = \sup_{H:\, \mathbb{P}[H]\le\delta} \mathbb{E}[U 1_H]$, where the supremum is over all events measurable with respect to U. Then*[3] $\mathbb{E}[U] < \infty \Rightarrow g(\delta) \xrightarrow{\delta\to 0} 0.$

Proof. For any $b > 0$, $\mathbb{E}[U1_H] \le \mathbb{E}[U1\{U > b\}] + b\delta$, where $\mathbb{E}[U1\{U > b\}] \xrightarrow{b\to\infty} 0$ by the dominated convergence theorem. Then the proof is completed by setting $b = 1/\sqrt{\delta}$. □

Now $d(S^n, \hat{s}_0^n) = \frac{1}{n}\sum_{j=1}^n U_j$, where U_j are iid copies of $U \triangleq d(S, \hat{s}_0)$. Since $\mathbb{E}[U] = D_{\max} < \infty$ by assumption, applying Lemma 25.4 yields $\mathbb{E}[d(S^n, \hat{s}_0^n) 1_{E_n}] = \frac{1}{n}\sum \mathbb{E}[U_j 1_{E_n}] \le g(\mathbb{P}[E_n]) \to 0$, since $\mathbb{P}[E_n] \to 0$. This proves the theorem. □

Remark 25.2. (Fundamental limit for excess distortion) Although Theorem 25.1 is stated for average distortion, under certain mild extra conditions, it also holds for excess distortion where the goal is to achieve $d(S^n, \hat{S}^n) \le D$ with probability arbitrarily close to one as opposed to in expectation. Indeed, the achievability proof of Theorem 25.1 is already stated in high probability. For the converse, assume in addition to (25.3) that $D_p \triangleq \mathbb{E}[d(S,\hat{s})^p]^{1/p} < \infty$ for some $\hat{s} \in \hat{\mathcal{S}}$ and $p > 1$. Applying Rosenthal's inequality [369, 235], we have $\mathbb{E}[d(S, \hat{s}^n)^p] = \mathbb{E}[(\sum_{i=1}^n d(S_i, \hat{s}))^p] \le CD_p^p$ for some constant $C = C(p)$. Then we can apply Theorem 24.9 to convert a code for excess distortion to one for average distortion and invoke the converse for the latter.

To end this section, we note that in Section 25.1.1 and in Theorem 25.1 it seems we applied different proof techniques. How come they both turn out to yield the same *tight* asymptotic result? This is because the key to both proofs is to estimate the exponent (large deviations) of the underbraced probabilities in (25.9) and (25.11), respectively. To obtain the right exponent, as we know, the key is to apply tilting (change of measure) to the distribution solving the information projection problem (25.10). When $P_{\overline{Y}} = (Q_{\hat{S}})^n = (P_{\hat{S}})^n$ with $P_{\hat{S}}$ chosen as the output distribution in the solution to rate-distortion optimization (25.1), the resulting exponent is precisely given by $2^{-I(X;Y)}$.

25.2* Covering Lemma and Joint Typicality

In this section we consider the following curious problem, a version of *channel simulation/synthesis*. We want to simulate a sequence of iid correlated strings

[3] In fact, $\Rightarrow$ is $\Leftrightarrow$.

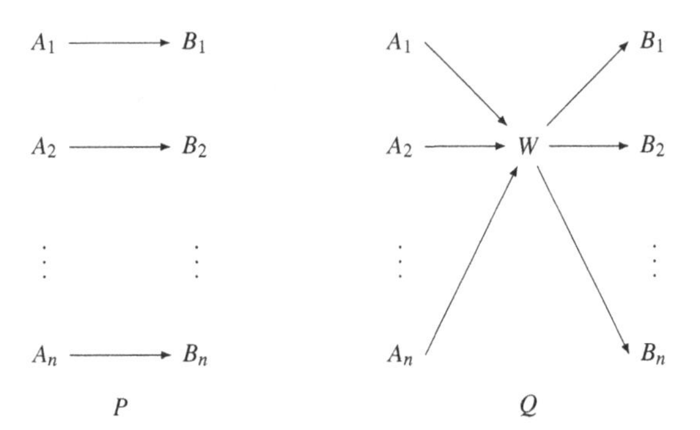

Figure 25.1 Description of the channel simulation game. The distribution P (left) is to be simulated via the distribution Q (right) at minimal rate R. Depending on the exact formulation we either require $R = I(A;B)$ (covering lemma) or $R = C(A;B)$ (soft-covering lemma).

$(A_i, B_i) \stackrel{\text{iid}}{\sim} P_{A,B}$ via a protocol we describe next. First, a sequence $A^n \stackrel{\text{iid}}{\sim} P_A$ is generated at one terminal. Then we can look at it, and produce a rate-constrained message $W \in [2^{nR}]$ which gets communicated to a remote destination (noiselessly). Upon receipt of the message, the remote decoder produces a string B^n out of it. The goal is to be able to fool the *tester* who inspects (A^n, B^n) and tries to check that it was indeed generated as $(A_i, B_i) \stackrel{\text{iid}}{\sim} P_{A,B}$. See Figure 25.1 for an illustration.

How large a rate R is required depends on how we exactly understand the requirement to "fool the tester." If the tester is fixed ahead of time (this just means that we know the set F such that $(A_i, B_i) \stackrel{\text{iid}}{\sim} P_{A,B}$ is declared whenever $(A^n, B^n) \in F$) then this is precisely the setting in which the *covering lemma* operates. In the next section we show that a higher rate $R = C(A;B)$ is required if F is not known ahead of time. We leave out the celebrated theorem of Bennett and Shor [43] which shows that rate $R = I(A;B)$ is also attainable even if F is not known, but if the encoder and decoder are given access to a source of common random bits (independent of A^n, of course).

Before proceeding, we note some simple corner cases:

1 If $R = H(A)$, we can compress A^n and send it to the "B side," who can reconstruct A^n perfectly and use that information to produce B^n through $P_{B^n|A^n}$.
2 If $R = H(B)$, the "A side" can generate B^n according to $P^n_{A,B}$ and send that B^n sequence to the "B side."
3 If $A \perp\!\!\!\perp B$, we know that $R = 0$, as the "B side" can generate B^n independently.

Our previous argument for achieving the rate distortion turns out to give a sharp answer (that $R = I(A;B)$ is sufficient) for the F-known case as follows.

Theorem 25.5. (Covering lemma) *Fix $P_{A,B}$ and let $(A_j, B_j) \stackrel{\text{iid}}{\sim} P_{A,B}$, $R > I(A;B)$. We generate a random codebook $C = \{c_1, \ldots, c_M\}$, $\log M = nR$, with each codeword c_j drawn iid from distribution P^n_B. Then as $n \to \infty$, we have for all sets F*

$$\mathbb{P}[\exists c \in \mathcal{C}\colon (A^n, c) \in F] \geq \mathbb{P}[(A^n, B^n) \in F] + \underbrace{o_R(1)}_{\text{uniform in } F}. \tag{25.16}$$

Remark 25.3. The origin of the name "covering" is from the application to a proof of Theorem 25.1. In that context we set $A = S$ and $B = \hat{S}$ to be the source and optimal reconstruction (in the sense of minimizing $R^{(I)}(D)$). Then taking $F = \{(a^n, b^n)\colon d(a^n, b^n) \leq D + \delta\}$ we see that both terms on the right-hand side of the inequality are $o(1)$. Thus, sampling 2^{nR} reconstruction points, we have *covered* the space of source realizations in such a way that with high probability we can always find a reconstruction with low distortion.

Proof. Set $\gamma > M$ and following similar arguments as in the proof of Theorem 25.2, we have

$$\begin{aligned}\mathbb{P}[\forall c \in \mathcal{C}\colon (A^n, c) \notin F] &\leq e^{-M/\gamma} + \mathbb{P}[\{(A^n, B^n) \notin F\} \cup \{i(A^n;B^n) > \log \gamma\}] \\ &= \mathbb{P}[(A^n, B^n) \notin F] + o(1) \\ \Rightarrow \mathbb{P}[\exists c \in \mathcal{C}\colon (A^n, c) \in F] &\geq \mathbb{P}[(A^n, B^n) \in F] + o(1).\end{aligned}$$

□

As we explained, the version of the covering lemma that we stated shows how to "fool the tester" by applying only one fixed test set F. However, if both A and B take values on finite alphabets then something stronger can be stated. This original version of the covering lemma [111] is what is used in sophisticated distributed compression arguments, for example, Theorem 11.17. Before stating the result we remind the reader that for two sequences a^n, b^n we denote their joint empirical distribution by

$$\hat{P}_{a^n,b^n}(\alpha, \beta) \triangleq \frac{1}{n} \sum_{i=1}^{n} 1\{a_i = \alpha, b_i = \beta\}, \qquad \alpha \in \mathcal{A}, \beta \in \mathcal{B}.$$

It is also useful to review the joint typicality discussion in Remark 18.2. In this section we say that a sequence of pairs of vectors (a^n, b^n) is *jointly typical with respect to* $P_{A,B}$ if

$$\mathrm{TV}(\hat{P}_{a^n,b^n}, P_{A,B}) = o(1).$$

Fix a distribution $P_{A,B}$ and any codebook $\mathcal{C} = \{c_1, \ldots, c_M\}$ consisting of elements $c_j \in \mathcal{B}^n$. For any fixed input string a^n we define

$$W = \operatorname*{argmin}_{j \in [M]} \mathrm{TV}(\hat{P}_{a^n,c_j}, P_{A,B}), \qquad \hat{B}^n = c_W. \tag{25.17}$$

The next corollary says that in order for us to produce a jointly typical pair $(A^n, \hat{B}^n)$ a codebook must have the rate $R > I(A;B)$ and this is optimal.

Corollary 25.6. *Fix $P_{A,B}$ on a pair of finite alphabets $\mathcal{A}, \mathcal{B}$. For any $R > I(A;B)$ we generate a random codebook $\mathcal{C} = \{c_1, \ldots, c_M\}$, $\log M = nR$, where each codeword c_j is drawn iid from distribution P_B^n. With $\hat{B}^n$ defined as in (25.17) we have that pair $(A^n, \hat{B}^n)$ is jointly typical with high probability*

$$\mathbb{E}[\mathrm{TV}(\hat{P}_{A^n,\hat{B}^n}, P_{A,B})] = o_R(1). \tag{25.18}$$

Furthermore, no codebook with rate $R < I(A;B)$ can achieve (25.18).

Proof. First, in this case $i(A^n;B^n)$ is a sum of bounded iid terms and thus the $o_R(1)$ term in (25.16) is in fact $e^{-\Omega(n)}$. Fix arbitrary $\epsilon > 0$ and apply Theorem 25.5 to

$$F = \{(a^n, b^n) \colon |\hat{P}_{a^n,b^n}(\alpha, \beta) - P_{A,B}(\alpha, \beta)| \le \epsilon\}$$

with all possible $\alpha \in \mathcal{A}$ and $\beta \in \mathcal{B}$. Then we conclude that

$$\mathbb{P}[\mathrm{TV}(P_{A^n,\hat{B}^n}, P_{A,B}) \le \epsilon] \ge 1 - |\mathcal{A}||\mathcal{B}|e^{-\Omega(n)} = 1 + o(1).$$

Due to the arbitrariness of ϵ, (25.18) follows.

This proof can also be understood combinatorially (as is done classically [111]). Indeed, the rate $R \approx I(A;B)$ is sufficient since an iid B^n ranges over about $2^{nH(B)}$ high-probability sequences (see Proposition 1.5). Applying the same proposition conditionally on a typical A^n sequence, there are around $2^{nH(B|A)}$ of B^n sequences that have the same joint distribution. Therefore, while we need $nH(B)$ bits to describe all of B^n it is intuitively clear that we only need to describe a class of B^n for each A^n sequence, and there are around $\frac{2^{nH(B)}}{2^{nH(B|A)}} = 2^{nI(A;B)}$ classes. We can represent this situation by a bipartite graph with (typical) A^n sequences on the left and (typical) B^n sequences on the right and edges corresponding to pairs having joint typicality; this graph is regular with right degrees $2^{nH(B|A)}$ and $2^{nH(A|B)}$, respectively. Thus, to convert our intuition to a rigorous proof as above we need to show that a random subset of $2^{nI(A;B)}$ right vertices covers all left vertices. (This alternative proof has the advantage of showing (25.18) conditional on the any typical A^n.)

We next proceed to proving that $R \ge I(A;B)$ is in fact necessary for simulating a jointly typical B^n. Consider an arbitrary codebook $\mathcal{C}$ such that (25.18) holds. On one hand we have

$$nR = \log M \ge I(A^n;\hat{B}^n) = H(A^n) - H(A^n|\hat{B}^n) = nH(A) - H(A^n|\hat{B}^n).$$

Thus, the proof will be complete if we show

$$H(A^n|\hat{B}^n) \le nH(A|B) + o(n).$$

To that end, let us define a random conditional type of $A^n|\hat{B}^n$ as

$$\hat{Q}(\alpha|\beta) \triangleq \frac{\#\{i \colon \hat{A}_i = \alpha, \hat{B}_i = \beta\}}{\#\{i \colon \hat{B}_i = \beta\}}.$$

For an arbitrarily small $\epsilon > 0$ we define a binary $T = 1$ if and only if for some α or β either of the following inequalities is violated:

$$\left|\frac{1}{n}\#\{i \colon \hat{B}_i = \beta\} - P_B(\beta)\right| > \epsilon,$$

$$\left|\hat{Q}(\alpha|\beta) - P_{A|B}(\alpha|\beta)\right| > \epsilon.$$

By the Markov inequality (and assuming without loss of generality that $P_B(\beta) > 0$ for all β) we get that

$$\mathbb{P}[T = 1] = o(1).$$

Thus, we have

$$H(A^n|\hat{B}^n) \le H(A^n, T|\hat{B}^n) \le \log 2 + n\log|\mathcal{A}|\mathbb{P}[T=1] + H(A^n|\hat{B}^n, T=0).$$

The first two terms are $o(n)$ so we focus on the last term. Since $\hat{Q}$ is a random variable with polynomially many possible values, see Exercise **I.2**, we have

$$H(A^n|\hat{B}^n, T=0) \le H(A^n, \hat{Q}|\hat{B}^n, T=0) \le H(A^n|\hat{Q}, \hat{B}^n, T=0) + O(\log n).$$

Let there be n_β positions with $\hat{B}_i = \beta$. Conditioned on $\hat{Q}$, the random variable A^n takes $\binom{n_\beta}{n_\beta Q(\alpha_1|\beta)\cdots n_\beta Q(\alpha_{|\mathcal{A}|}|\beta)}$ values. Since under $T=0$ we have $\hat{Q} \to P_{A|B}$ and $\frac{n_\beta}{n} \to P_B(\beta)$ as $n \to \infty$ we conclude from Proposition 1.5 and the continuity of entropy in Proposition 4.8 that

$$H(A^n|\hat{Q}, \hat{B}^n, T=0) \le n(H(A|B) + \delta)$$

for some $\delta = \delta(\epsilon) > 0$ that vanishes as $\epsilon \to 0$. □

Applications of Corollary 25.6 include distributed compression (Theorem 11.17) and hypothesis testing (Theorem 16.6). Now, those applications use the rate-constrained $\hat{B}^n$ to create a required correlation (joint typicality) not only with A^n but also with other random variables. Those applications will require the following simple observation.

Proposition 25.7. *Fix some $P_{X,A,B}$ on finite alphabets and consider a pair of random variables $(A^n, \hat{B}^n)$ which are jointly typical on average (specifically, (25.18) holds as $n \to \infty$). Given $A^n, \hat{B}^n$ suppose that X^n is generated $\sim P_{X|A,B}^{\otimes n}$. Then we have*

$$\mathbb{E}[\mathrm{TV}(\hat{P}_{X^n,A^n,\hat{B}^n}, P_{X,A,B})] = o(1).$$

And, furthermore we have

$$I(X^n;\hat{B}^n) \ge nI(X;B) + o(n). \tag{25.19}$$

Remark 25.4. (Markov lemma) This result is known as the *Markov lemma*, for example, [106, Lemma 15.8.1], because in the standard application setting one considers a joint distribution $P_{X,A,B} = P_{X,A}P_{B|A}$, that is, $X \to A \to B$. In this application, one further has $(X^n, A^n) \overset{\text{iid}}{\sim} P_{X,A}$ generated by Nature with only A^n being observed. Given A^n one constructs a jointly typical vector $\hat{B}^n$ (e.g. via the covering lemma, Corollary 25.6). Now, since with high probability (X^n, A^n) is jointly typical, it is tempting to automatically conclude that $(X^n, \hat{B}^n)$ would also be jointly typical. Unfortunately, the joint typicality relation is generally not transitive.[4] In the above result, however, what resolves this issue is the fact that X^n can be viewed as generated *after* $(A^n, \hat{B}^n)$ were already selected. Thus, viewing the process in this order we can even allow X^n to depend on $\hat{B}^n$, which is what we did. For stronger results under the classical setting of $P_{X|A,B} = P_{X|A}$ see [147, Lemma 12.1].

[4] Let $P_{X,A,B} = P_X P_A P_B$ with $P_X = P_A = P_B = \mathrm{Ber}(1/2)$. Take a^n to be any binary string in $\{0,1\}^n$ with $n/2$ ones. Set $x_j = b_j = a_j$ for $j \le n/2$ and $x_j = b_j = 1 - a_j$, otherwise. Then (x^n, a^n) and (a^n, b^n) are both jointly typical, but (x^n, b^n) is not.

Proof. Note that from condition (25.18) and the Markov inequality we get that $\mathrm{TV}(\hat{P}_{A^n,\hat{B}^n}, P_{A,B}) = o(1)$ with probability $1 - o(1)$. Fix any $a, b, x \in \mathcal{A} \times \mathcal{B} \times \mathcal{X}$ and consider $m = n\hat{P}_{A^n,\hat{B}^n}(a,b)$ coordinates $i \in [n]$ with $A_i = a, \hat{B}_i = b$. Among these there are $m' \leq m$ coordinates i that also satisfy $X_i = x$. The standard concentration estimate shows that $|m' - mP_{X|A,B}(x|a,b)| = o(m)$ with probability $1 - o(1)$. Hence, normalizing by m we obtain (from the union bound) that with probability $1 - o(1)$ we have

$$|\hat{P}_{X^n,A^n,\hat{B}^n}(x,a,b) - P_{X,A,B}(x,a,b)| = o(1).$$

This implies the first statement. Note that by summing out $a \in \mathcal{A}$ we obtain that

$$\mathbb{E}[\mathrm{TV}(\hat{P}_{X^n,\hat{B}^n}, P_{X,B})] = o(1).$$

But then repeating the steps of the second part of Corollary 25.6 we obtain $I(X^n;\hat{B}^n) \geq nI(X;B) + o(n)$, as required. □

Remark 25.5. Although in (25.19) we only proved a lower bound (which is sufficient for applications in this book), it is known that under the Markov assumption $X \to A \to B$ the inequality (25.19) holds with equality [111, Chapter 15]. This follows as a by-product of a deep entropy characterization problem for which we recommend the mentioned reference.

Let us go back to the discussion in the beginning of this section. We have learned how to "fool" the tester that uses a fixed test set F (Theorem 25.5). Then for finite alphabets we have shown that we can also "fool" the tester that computes empirical averages since

$$\frac{1}{n}\sum_{j=1}^{n} f(A_j, \hat{B}_j) \approx \mathbb{E}_{A,B\sim P_{A,B}}[f(A,B)],$$

for any bounded function f. A stronger requirement would be to demand that the joint distribution $P_{A^n,\hat{B}^n}$ fools any permutation-invariant tester, that is,

$$\sup |P_{A^n,\hat{B}^n}(F) - P^n_{A,B}(F)| \to 0,$$

where the supremum is taken over all permutation-invariant subsets $F \subset \mathcal{A}^n \times \mathcal{B}^n$. This is not guaranteed by Corollary 25.6. Indeed, note that a sufficient statistic for a permutation-invariant tester is a joint type $\hat{P}_{A^n,\hat{B}^n}$, and Corollary 25.6 does show that $\hat{P}_{A^n,\hat{B}^n} \approx P_{A,B}$ (in the sense of L_1-distance of vectors). However, it still might happen that $\hat{P}_{A^n,\hat{B}^n}$ although close to $P_{A,B}$ takes highly different values compared to those of $\hat{P}_{A^n,B^n}$. For example, if we restrict all $c \in \mathcal{C}$ to have a fixed composition P_0, the tester can easily detect the problem since the P^n_B-measure of all strings of composition P_0 cannot exceed $O(1/\sqrt{n})$. Formally, to fool the permutation-invariant tester we need to have small total variation between *the distribution* of $\hat{P}_{A^n,\hat{B}^n}$ and $\hat{P}_{A^n,B^n}$.

We conjecture, however, that nevertheless the rate $R = I(A;B)$ should be sufficient to achieve also this stronger requirement. In the next section we show that if one removes the permutation-invariance constraint, then a larger rate $R = C(A;B)$ is needed.

25.3* Wyner's Common Information

We continue discussing the channel simulation setting as in the previous section. We now want to determine the minimal possible communication rate (i.e. cardinality of $W \in [2^{nR}]$) required to have small total variation

$$\mathrm{TV}(P_{A^n,\hat{B}^n}, P^n_{A,B}) \leq \epsilon \tag{25.20}$$

between the simulated and the true output (see Figure 25.1).

Theorem 25.8. (Cuff [116]) *Let $P_{A,B}$ be an arbitrary distribution on the finite space $\mathcal{A}\times\mathcal{B}$. Consider a coding scheme where Alice observes $A^n \overset{\text{iid}}{\sim} P^n_A$, sends a message $W \in [2^{nR}]$ to Bob, who given W generates a (possibly random) sequence $\hat{B}^n$. If (25.20) is satisfied for all $\epsilon > 0$ and sufficiently large n, then we must have*

$$R \geq C(A;B) \triangleq \min_{A\to U\to B} I(A,B;U), \tag{25.21}$$

where $C(A;B)$ is known as Wyner's common information [459]. Furthermore, for any $R > C(A;B)$ and $\epsilon > 0$ there exists $n_0(\epsilon)$ such that for all $n \geq n_0(\epsilon)$ there exists a scheme satisfying (25.20).

Note that condition (25.20) guarantees that any tester (permutation-invariant or not) is fooled to believe they see the truly iid (A^n, B^n) with probability $\geq 1 - \epsilon$. However, compared to Theorem 25.5, this requires a higher communication rate since $C(A;B) \geq I(A;B)$, clearly.

Proof. Showing that Wyner's common information is a lower bound is not hard. First, since $P_{A^n,\hat{B}^n} \approx P^n_{A,B}$ (in TV) we have

$$I(A_t, \hat{B}_t; A^{t-1}, \hat{B}^{t-1}) \approx I(A_t, B_t; A^{t-1}, B^{t-1}) = 0.$$

(Here one needs to use the finiteness of the alphabet of A and B and the bounds relating $H(P) - H(Q)$ with $\mathrm{TV}(P, Q)$; see (7.20) and Corollary 6.7.) Next, for some δ depending only on ϵ and vanishing as $\epsilon \to 0$, we have

$$nR = H(W) \geq I(A^n, \hat{B}^n; W) \tag{25.22}$$
$$\geq \sum_{t=1}^{n} I(A_t, \hat{B}_t; W) - I(A_t, \hat{B}_t; A^{t-1}, \hat{B}^{t-1}) \tag{25.23}$$
$$\geq \sum_{t=1}^{n} I(A_t, \hat{B}_t; W) - \delta n \tag{25.24}$$
$$\geq nC(A;B) - 2\delta n, \tag{25.25}$$

where in the last step we used the crucial observation that

$$A_t \to W \to \hat{B}_t$$

and that, for finite alphabet, Wyner's common information $P_{A,B} \mapsto C(A;B)$ is continuous in the total variation distance on $P_{A,B}$.

To show achievability, let us notice that the problem is equivalent to constructing three random variables $(\hat{A}^n, W, \hat{B}^n)$ such that (a) $W \in [2^{nR}]$, (b) the Markov relation

$$\hat{A}^n \leftarrow W \to \hat{B}^n \tag{25.26}$$

holds, and (c) $\mathrm{TV}(P_{\hat{A}^n,\hat{B}^n}, P^n_{A,B}) \leq \epsilon/2$. Indeed, given such a triple we can use the coupling characterization of TV (7.20) and the fact that $\mathrm{TV}(P_{\hat{A}^n}, P^n_A) \leq \epsilon/2$ to extend the probability space to

$$A^n \to \hat{A}^n \to W \to \hat{B}^n$$

and $\mathbb{P}[A^n = \hat{A}^n] \geq 1 - \epsilon/2$. Again by (7.20) we conclude that $\mathrm{TV}(P_{A^n,\hat{B}^n}, P_{\hat{A}^n,\hat{B}^n}) \leq \epsilon/2$ and by the triangle inequality we conclude that (25.20) holds.

Finally, construction of the triple satisfying (a)–(c) follows from the soft-covering lemma (Corollary 25.10 next) applied with $V = (A, B)$ and W being uniform on the set of x_i's there. □

25.4* Approximation of Output Statistics and the Soft-Covering Lemma

In this section we aim to prove the remaining ingredient (the soft-covering lemma) required for the proof of Theorem 25.8. To that end, recall that in Section 7.9 we have shown that generating $X_i \overset{\text{iid}}{\sim} P_X$ and passing their empirical distribution $\hat{P}_n$ across the channel $P_{Y|X}$ results in a good approximation of $P_Y = P_{Y|X} \circ P_X$, that is,

$$D(P_{Y|X} \circ \hat{P}_n \| P_Y) \to 0.$$

A natural question is how large n should be in order for the approximation $P_{Y|X} \circ \hat{P}_n \approx P_Y$ to hold. A remarkable fact that we establish in this section is that the answer is $n \approx 2^{I(X;Y)}$, assuming $I(X;Y) \gg 1$ and certain concentration properties of $i(X;Y)$ around $I(X;Y)$. This result originated from Wyner [459] and was significantly strengthened in [212].

Here, we show a new variation of such results by strengthening our simple χ^2-information bound of Proposition 7.17 (corresponding to $\lambda = 2$).

Theorem 25.9. *Fix $P_{X,Y}$ and for any $\lambda \in \mathbb{R}$ define the Rényi mutual information of order λ,*

$$I_\lambda(X;Y) \triangleq D_\lambda(P_{X,Y} \| P_X P_Y),$$

where D_λ is the Rényi divergence, see Definition 7.24. We have for every $1 < \lambda \leq 2$

$$\mathbb{E}[D(P_{Y|X} \circ \hat{P}_n \| P_Y)] \leq \frac{1}{\lambda - 1} \log(1 + \exp\{(\lambda - 1)(I_\lambda(X;Y) - \log n)\}). \tag{25.27}$$

Proof. Since $\lambda \mapsto D_\lambda$ is non-decreasing, it is sufficient to prove an equivalent upper bound on $\mathbb{E}[D_\lambda(P_{Y|X} \circ \hat{P}_n \| P_Y)]$. From Jensen's inequality we see that

$$\mathbb{E}[D_\lambda(P_{Y|X} \circ \hat{P}_n \| P_Y)] \triangleq \frac{1}{\lambda - 1} \mathbb{E}_{X^n} \log \mathbb{E}_{Y \sim P_Y} \left[\left\{ \frac{P_{Y|X} \circ \hat{P}_n}{P_Y}(Y) \right\}^\lambda \right] \tag{25.28}$$

$$\leq \frac{1}{\lambda - 1} \log \mathbb{E}_{X^n} \mathbb{E}_{Y \sim P_Y} \left[\left\{ \frac{P_{Y|X} \circ \hat{P}_n}{P_Y}(Y) \right\}^\lambda \right] = I_\lambda(X^n;\bar{Y}),$$

where similarly to (7.56) we introduced the channel $P_{\bar{Y}|X^n} = \frac{1}{n}\sum_{i=1}^n P_{Y|X=X_i}$. To analyze $I_\lambda(X^n;\bar{Y})$ we need to bound

$$\mathbb{E}_{(X^n,\bar{Y}) \sim P_X^n \times P_Y} \left[\left\{ \frac{1}{n} \sum_i \frac{P_{Y|X}(\bar{Y}|X_i)}{P_Y(\bar{Y})} \right\}^\lambda \right]. \tag{25.29}$$

Note that conditioned on $\bar{Y}$ we get to analyze the λ-th moment of a sum of iid random variables. This puts us into a well-known setting of Rosenthal-type inequalities. In particular, we have that[5] for any iid non-negative B_j, provided $1 \leq \lambda \leq 2$,

$$\mathbb{E}\left[\left(\sum_{i=1}^n B_i \right)^\lambda \right] \leq n\,\mathbb{E}[B^\lambda] + (n\,\mathbb{E}[B])^\lambda. \tag{25.30}$$

Now using (25.30) we can overbound (25.29) by

$$\leq 1 + n^{1-\lambda}\, \mathbb{E}_{(X,\bar{Y}) \sim P_X \times P_Y} \left[\left\{ \frac{P_{Y|X}(\bar{Y}|X)}{P_Y(\bar{Y})} \right\}^\lambda \right],$$

which implies

$$I_\lambda(X^n;\bar{Y}) \leq \frac{1}{\lambda - 1} \log\left(1 + n^{1-\lambda} \exp\{(\lambda - 1) I_\lambda(X;Y)\}\right),$$

which together with (25.28) recovers the main result (25.27). □

[5] The inequality (25.30), which is known to be essentially tight [375], can be shown by applying $(a+b)^{\lambda-1} \leq a^{\lambda-1} + b^{\lambda-1}$ and Jensen's inequality to get $\mathbb{E}[B_i(B_i + \sum_{j \neq i} B_j)^{\lambda-1}] \leq \mathbb{E}[B^\lambda] + \mathbb{E}[B]$ $((n-1)\,\mathbb{E}[B])^{\lambda-1}$. Summing the left hand side over i and bounding $(n-1) \leq n$ we get (25.30).

Remark 25.6. Hayashi [217] upper–bounds the LHS of (25.27) with

$$\frac{\lambda}{\lambda-1}\log\left(1+\exp\left\{\frac{\lambda-1}{\lambda}(K_\lambda(X;Y)-\log n)\right\}\right),$$

where $K_\lambda(X;Y) = \inf_{Q_Y} D_\lambda(P_{X,Y}\|P_X Q_Y)$ is the so-called Sibson–Csiszár information, see [339]. This bound, however, does not have the right rate of convergence as $n \to \infty$, at least for $\lambda = 1$ as comparison with Proposition 7.17 reveals.

We note that [217, 212] also contain direct bounds on

$$\mathbb{E}[\mathrm{TV}(P_{Y|X} \circ \hat{P}_n, P_Y)]$$

which do not assume the existence of the λ-th moment of $\frac{P_{Y|X}}{P_Y}$ for $\lambda > 1$ and instead rely on the distribution of $i(X;Y)$. We do not discuss these bounds here, however, since for finite alphabets the next corollary is sufficient.

Corollary 25.10. (Soft-covering lemma) *Suppose* $X = (U_1, \ldots, U_d)$ *and* $Y = (V_1, \ldots, V_d)$ *are vectors with* $(U_i, V_i) \overset{\text{iid}}{\sim} P_{U,V}$ *and* $I_{\lambda_0}(U;V) < \infty$ *for some* $\lambda_0 > 1$ *(e.g. if one of U or V is over a finite alphabet). Then for any* $R > I(U;V)$ *there exists* $\epsilon > 0$, *so that for all* $d \geq 1$ *there exists* $x_1, \ldots, x_n$, $n = \lceil \exp\{dR\} \rceil$, *such that*

$$D\left(\frac{1}{n}\sum_{i=1}^{n} P_{Y|X=x_i} \middle\| P_Y\right) \leq \frac{1}{\epsilon}\exp\{-d\epsilon\}.$$

Remark 25.7. The origin of the name "soft-covering" is due to the fact that unlike the covering lemma (Theorem 25.5) which selects a single x_i (trying to make $P_{Y|X=x_i}$ as close to P_Y as possible) here we mix over n choices uniformly.

Proof. By tensorization of Rényi divergence, see Section 7.12, we have

$$I_\lambda(X;Y) = dI_\lambda(U;V).$$

For every $1 < \lambda < \lambda_0$ we have that $\lambda \mapsto I_\lambda(U;V)$ is continuous and converging to $I(U;V)$ as $\lambda \to 1$. Thus, we can find λ sufficiently close to one so that $R > I_\lambda(U;V)$. Applying Theorem 25.9 with this λ completes the proof. □

26 Evaluating the Rate-Distortion Function, and Lossy Source–Channel Separation

In previous chapters we established the main coding theorem for lossy data compression: For stationary memoryless (iid) sources and separable distortion, under the assumption that $D_{\max} < \infty$, the operational and information rate-distortion functions coincide, namely,

$$R(D) = R^{(\mathrm{I})}(D) = \inf_{P_{\hat{S}|S}:\, \mathbb{E}[d(S,\hat{S})] \le D} I(S; \hat{S}).$$

In addition, we have shown various properties of the rate-distortion function (see Theorem 24.4). In this chapter we compute the rate-distortion function for several important source distributions by evaluating this constrained minimization of mutual information. The common technique we apply to evaluate these special cases in Section 26.1 is then formalized in Section 26.2* as a saddle-point property akin to those in Sections 5.2 and 5.4* for mutual information maximization (capacity). Next we extend the paradigm of joint source–channel coding in Section 19.7 to the lossy setting; this reasoning will later be found useful in statistical applications in Part VI (see Chapter 30). Finally, in Section 26.4 we discuss several limitations, both theoretical and practical, of the classical theory for lossy compression and joint source–channel coding.

26.1 Evaluation of $R(D)$

26.1.1 Bernoulli Source

Let $S \sim \mathrm{Ber}(p)$ with Hamming distortion $d(S,\hat{S}) = 1\{S \neq \hat{S}\}$ and alphabets $\mathcal{S} = \hat{\mathcal{S}} = \{0,1\}$. Then $d(s^n,\hat{s}^n) = \frac{1}{n} d_{\mathrm{H}}(s^n,\hat{s}^n)$ is the bit error rate (fraction of erroneously decoded bits). By symmetry, we may assume that $p \le 1/2$.

Theorem 26.1.

$$R(D) = (h(p) - h(D))_+. \tag{26.1}$$

For example, when $p = 1/2$, $D = 0.11$, we have $R(D) \approx 1/2$ bits. In the Hamming game described in Section 24.2 where we aim to compress 100 bits down to 50, we indeed can do this while achieving 11% average distortion, compared to the naive scheme of storing half the string and guessing on the other half, which achieves 25% average distortion. Note that we can also get very tight non-asymptotic bounds, see Exercise **V.3**.

Proof. Since $D_{\max} = p$, in the sequel we can assume $D < p$ for otherwise there is nothing to show.

For the converse, consider any $P_{\hat{S}|S}$ such that $P[S \neq \hat{S}] \leq D \leq p \leq \frac{1}{2}$. Then

$$\begin{aligned} I(S; \hat{S}) &= H(S) - H(S|\hat{S}) \\ &= H(S) - H(S + \hat{S}|\hat{S}) \\ &\geq H(S) - H(S + \hat{S}) \\ &= h(p) - h(P[S \neq \hat{S}]) \\ &\geq h(p) - h(D). \end{aligned}$$

In order to achieve this bound, we need to saturate the above chain of inequalities, in particular, choose $P_{\hat{S}|S}$ so that the difference $S + \hat{S}$ is independent of $\hat{S}$. Let $S = \hat{S} + Z$, where $\hat{S} \sim \text{Ber}(p') \perp\!\!\!\perp Z \sim \text{Ber}(D)$, and p' is such that the convolution gives exactly $\text{Ber}(p)$, namely,

$$p' * D = p'(1 - D) + (1 - p')D = p,$$

that is, $p' = \frac{p-D}{1-2D}$. In other words, the backward channel $P_{S|\hat{S}}$ is exactly $\text{BSC}(D)$ and the resulting $P_{\hat{S}|S}$ is our choice of the forward channel $P_{\hat{S}|S}$. Then, $I(S; \hat{S}) = H(S) - H(S|\hat{S}) = H(S) - H(Z) = h(p) - h(D)$, yielding the upper bound $R(D) \leq h(p) - h(D)$. □

Remark 26.1. Here is a more general strategy (which we will later implement in the Gaussian case). Denote the optimal forward channel from the achievability proof by $P^*_{\hat{S}|S}$ and $P^*_{S|\hat{S}}$ the associated backward channel (which is $\text{BSC}(D)$). We need to show that there is no other $P_{\hat{S}|S}$ with $P[S \neq \hat{S}] \leq D$ and a smaller mutual information. Then

$$\begin{aligned} I(P_S, P_{\hat{S}|S}) &= D(P_{S|\hat{S}} \| P_S | P_{\hat{S}}) \\ &= D(P_{S|\hat{S}} \| P^*_{S|\hat{S}} | P_{\hat{S}}) + \mathbb{E}_P\left[\log \frac{P^*_{S|\hat{S}}}{P_S}\right] \\ &\geq H(S) + \mathbb{E}_P[\log D 1\{S \neq \hat{S}\} + \log \bar{D} 1\{S = \hat{S}\}] \\ &\geq h(p) - h(D), \end{aligned}$$

where the last inequality uses $P[S \neq \hat{S}] \leq D \leq \frac{1}{2}$.

Remark 26.2. By the WLLN, the distribution $P^n_S = \text{Ber}(p)^n$ concentrates near the Hamming sphere of radius np as n grows large. Recall that in proving Shannon's rate-distortion theorem, the optimal codebooks are drawn independently from $P^n_{\hat{S}} = \text{Ber}(p')^n$ with $p' = \frac{p-D}{1-2D}$. Note that $p' = 1/2$ if $p = 1/2$ but $p' < p$ if $p < 1/2$. In the latter case, the reconstruction points concentrate on a smaller sphere of radius np' and *none* of them are typical source realizations, as illustrated in Figure 26.1.

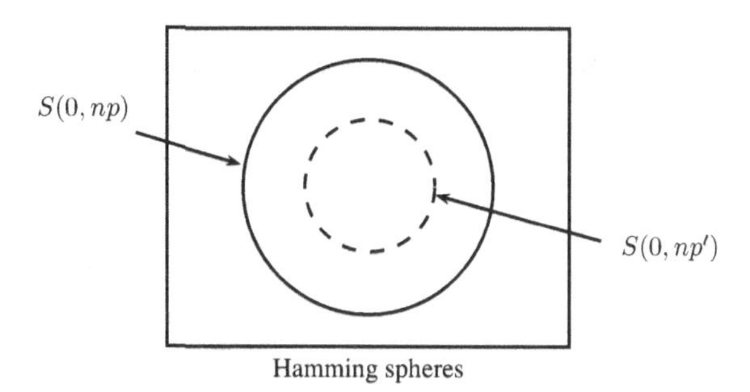

Figure 26.1 Source realizations (solid sphere) versus codewords (dashed sphere) in compressing Hamming sources.

For a generalization of Theorem 26.1 to an m-ary uniform source, see Exercise **V.6**.

26.1.2 Gaussian Source

The following results compute the Gaussian rate-distortion function for quadratic distortion in both the scalar and vector cases. (For general covariance, see Exercise **V.21**.)

Theorem 26.2. *Let $S \sim \mathcal{N}(0, \sigma^2)$ and $d(s, \hat{s}) = (s - \hat{s})^2$ for $s, \hat{s} \in \mathbb{R}$. Then*

$$R(D) = \frac{1}{2} \log^+ \frac{\sigma^2}{D}. \tag{26.2}$$

In the vector case of $S \sim \mathcal{N}(0, \sigma^2 I_d)$ and $d(s, \hat{s}) = \|s - \hat{s}\|_2^2$,

$$R(D) = \frac{d}{2} \log^+ \frac{d\sigma^2}{D}. \tag{26.3}$$

Proof. Since $D_{\max} = \sigma^2$, in the sequel we can assume $D < \sigma^2$ for otherwise there is nothing to show.

(Achievability) Choose $S = \hat{S} + Z$, where $\hat{S} \sim \mathcal{N}(0, \sigma^2 - D) \perp\!\!\!\perp Z \sim \mathcal{N}(0, D)$. In other words, the backward channel $P_{S|\hat{S}}$ is AWGN with noise power D, and the forward channel can be easily found to be $P_{\hat{S}|S} = \mathcal{N}(\frac{\sigma^2 - D}{\sigma^2} S, \frac{\sigma^2 - D}{\sigma^2} D)$. Then

$$I(S; \hat{S}) = \frac{1}{2} \log \frac{\sigma^2}{D} \implies R(D) \le \frac{1}{2} \log \frac{\sigma^2}{D}.$$

(Converse) Formally, we can mimic the proof of Theorem 26.1 replacing Shannon entropy by the differential entropy and applying the maximal entropy result from Theorem 2.8; the caveat is that for $\hat{S}$ (which may be discrete) the differential entropy may not be well defined. As such, we follow the alternative proof given in Remark 26.1. Let $P_{\hat{S}|S}$ be any conditional distribution such that $\mathbb{E}_P[(S - \hat{S})^2] \le D$. Denote the forward channel in the above achievability by $P^*_{\hat{S}|S}$. Then

$$\begin{aligned} I(P_S, P_{\hat{S}|S}) &= D(P_{S|\hat{S}} \| P^*_{S|\hat{S}} | P_{\hat{S}}) + \mathbb{E}_P\left[\log \frac{P^*_{S|\hat{S}}}{P_S}\right] \\ &\geq \mathbb{E}_P\left[\log \frac{P^*_{S|\hat{S}}}{P_S}\right] \\ &= \mathbb{E}_P\left[\log \frac{\frac{1}{\sqrt{2\pi D}} e^{-(S-\hat{S})^2/2D}}{\frac{1}{\sqrt{2\pi\sigma^2}} e^{-S^2/2\sigma^2}}\right] \\ &= \frac{1}{2}\log\frac{\sigma^2}{D} + \frac{\log e}{2}\mathbb{E}_P\left[\frac{S^2}{\sigma^2} - \frac{(S-\hat{S})^2}{D}\right] \\ &\geq \frac{1}{2}\log\frac{\sigma^2}{D}. \end{aligned}$$

Finally, for the vector case, (26.3) follows from (26.2) and the same single-letterization argument in Theorem 24.8 using the convexity of the rate-distortion function in Theorem 24.4(a). □

The interpretation of the optimal reconstruction points in the Gaussian case is analogous to that of the Hamming source previously discussed in Remark 26.2: As n grows, the Gaussian random vector concentrates on $S(0, \sqrt{n\sigma^2})$ (n-sphere in Euclidean space rather than Hamming), but each reconstruction point drawn from $\left(P^*_{\hat{S}}\right)^n$ is close to $S(0, \sqrt{n(\sigma^2 - D)})$. So again the picture is similar to Figure 26.1 of two nested spheres.

We can also understand the geometry of errors of optimal compressors. Indeed, suppose we have a sequence of quantizers $X^n \to W \to \hat{X}^n$ with $\frac{1}{n}\log M \to R(D)$. As we know, without loss of generality we may assume $\hat{X}^n = \mathbb{E}[X^n|W]$. Let us denote by $\Sigma = \mathrm{Cov}[X^n|W]$ the covariance matrix of the reconstruction errors. We know that $\frac{1}{n}\operatorname{tr}\Sigma \leq D$ by the distortion constraint. Now let us express mutual information in terms of differential entropy to obtain

$$\log M = I(X^n; W) = h(X^n) - h(X^n|W).$$

Applying the maximum entropy principle (2.19) to the second term (and taking expectation over W inside $\log\det$ via Jensen's inequality and Corollary 2.9) we obtain

$$\log M \geq \frac{n}{2}\log\sigma^2 - \frac{1}{2}\log\det\Sigma.$$

Let $\{\lambda_j : j \in [n]\}$ denote the spectrum of Σ. Dividing by n and recalling that the quantizer is optimal we get

$$\frac{1}{2}\log\frac{\sigma^2}{D} + o(1) \geq \frac{1}{n}\sum_{j=1}^{n}\frac{1}{2}\log\frac{\sigma^2}{\lambda_j}.$$

From the strict convexity of $\lambda \mapsto \frac{1}{2}\log\frac{\sigma^2}{\lambda}$ we conclude that empirical distributions of eigenvalues, that is, $\frac{1}{n}\sum_j \delta_{\lambda_j}$, must converge to a point mass, that is, to δ_D. In this sense $\Sigma \approx D \cdot I_n$ and the uncertainty regions (given the message) should be approximately spherical.

Note that the exact expression in Theorem 26.2 relies on the Gaussianity assumption of the source. How sensitive is the rate-distortion formula to this assumption? The following comparison result is a counterpart of Theorem 20.12 for channel capacity:

Theorem 26.3. *Assume that* $\mathbb{E}S = 0$ *and* $\mathrm{Var}\, S = \sigma^2$. *Consider the MSE distortion. Then*

$$\frac{1}{2}\log^+ \frac{\sigma^2}{D} - D(P_S \| \mathcal{N}(0,\sigma^2)) \le R(D) = \inf_{P_{\hat{S}|S}:\, \mathbb{E}(\hat{S}-S)^2 \le D} I(S;\hat{S}) \le \frac{1}{2}\log^+ \frac{\sigma^2}{D}.$$

Remark 26.3. A simple consequence of Theorem 26.3 is that for source distributions with a density, the rate-distortion function grows according to $\frac{1}{2}\log\frac{1}{D}$ in the low-distortion regime as long as $D(P_S \| \mathcal{N}(0,\sigma^2))$ is finite. In fact, the first inequality, known as the *Shannon lower bound* (SLB), is asymptotically tight, in the sense that

$$R(D) = \frac{1}{2}\log\frac{\sigma^2}{D} - D(P_S \| \mathcal{N}(0,\sigma^2)) + o(1), \quad D \to 0, \tag{26.4}$$

under appropriate conditions on P_S [282, 248]. Therefore, by comparing (2.21) and (26.4), we see that, for small distortion, uniform scalar quantization (Section 24.1) is in fact asymptotically optimal within $\frac{1}{2}\log(2\pi e) \approx 2.05$ bits.

Later in Section 30.1 we will apply the SLB to derive lower bounds for statistical estimation. For this we need the following general version of the SLB (see Exercise **V.22** for a proof): Let $\|\cdot\|$ be an *arbitrary* norm on $\mathbb{R}^d$ and $r > 0$. Let X be a d-dimensional continuous random vector with finite differential entropy $h(X)$. Then

$$\inf_{P_{\hat{X}|X}:\, \mathbb{E}[\|\hat{X}-X\|^r] \le D} I(X;\hat{X}) \ge h(X) + \frac{d}{r}\log\frac{d}{Dre} - \log\left(\Gamma\left(\frac{d}{r}+1\right)V\right), \tag{26.5}$$

where $V = \mathrm{vol}(\{x \in \mathbb{R}^d : \|x\| \le 1\})$ is the volume of the unit $\|\cdot\|$-ball.

Proof. Again, assume $D < D_{\max} = \sigma^2$. Let $S_G \sim \mathcal{N}(0,\sigma^2)$.

"$\le$": Use the same $P^*_{\hat{S}|S} = \mathcal{N}(\frac{\sigma^2-D}{\sigma^2}S, \frac{\sigma^2-D}{\sigma^2}D)$ as in the achievability proof of the Gaussian rate-distortion function:

$$\begin{aligned}
R(D) &\le I(P_S, P^*_{\hat{S}|S}) \\
&= I\left(S; \frac{\sigma^2-D}{\sigma^2}S + W\right) \qquad W \sim \mathcal{N}\left(0, \frac{\sigma^2-D}{\sigma^2}D\right) \\
&\le I\left(S_G; \frac{\sigma^2-D}{\sigma^2}S_G + W\right) \qquad \text{by Gaussian saddle point (Theorem 5.11)} \\
&= \frac{1}{2}\log\frac{\sigma^2}{D}.
\end{aligned}$$

"$\geq$": For any $P_{\hat{S}|S}$ such that $\mathbb{E}(\hat{S} - S)^2 \leq D$, let $P^*_{S|\hat{S}} = \mathcal{N}(\hat{S}, D)$ denote the AWGN channel with noise power D. Then

$$
\begin{aligned}
I(S;\hat{S}) &= D(P_{S|\hat{S}} \| P_S | P_{\hat{S}}) \\
&= D(P_{S|\hat{S}} \| P^*_{S|\hat{S}} | P_{\hat{S}}) + \mathbb{E}_P\left[\log \frac{P^*_{S|\hat{S}}}{P_{S_G}}\right] - D(P_S \| P_{S_G}) \\
&\geq \mathbb{E}_P\left[\log \frac{\frac{1}{\sqrt{2\pi D}} e^{-(S-\hat{S})^2/2D}}{\frac{1}{\sqrt{2\pi\sigma^2}} e^{-S^2/2\sigma^2}}\right] - D(P_S \| P_{S_G}) \\
&\geq \frac{1}{2}\log\frac{\sigma^2}{D} - D(P_S \| P_{S_G}).
\end{aligned}
$$

□

26.2* Analog of Saddle-Point Property in Rate Distortion

In the computation of $R(D)$ for the Hamming and Gaussian sources, we guessed the correct form of the rate-distortion function. In both of their converse arguments, we used the same trick to establish that any other feasible $P_{\hat{S}|S}$ gave a larger mutual information. In this section, we formalize this trick, in an analogous manner to the saddle-point property of the channel capacity (recall Theorem 5.4 and Section 5.4*). Note that typically we do not need any tricks to compute $R(D)$, since we can obtain a solution in a parametric form to the unconstrained convex optimization

$$\min_{P_{\hat{S}|S}} I(S;\hat{S}) + \lambda\, \mathbb{E}[d(S,\hat{S})].$$

In fact we have discussed in Section 5.6 iterative algorithms (Blahut–Arimoto) that compute $R(D)$. However, for peace of mind it is good to know there are some general reasons why tricks like those used in the Hamming or Gaussian cases actually are guaranteed to work.

Theorem 26.4.

1 *Suppose P_{Y^*} and $P_{X|Y^*} \ll P_X$ are such that $\mathbb{E}[d(X,Y^*)] \leq D$ and for any $P_{X,Y}$ with $\mathbb{E}[d(X,Y)] \leq D$ we have*

$$\mathbb{E}\left[\log \frac{dP_{X|Y^*}}{dP_X}(X|Y)\right] \geq I(X;Y^*). \tag{26.6}$$

Then $R(D) = I(X;Y^*)$.

2 *Suppose that $I(X;Y^*) = R(D)$. Then for any regular branch of conditional probability $P_{X|Y^*}$ and for any $P_{X,Y}$ satisfying*
- *$\mathbb{E}[d(X,Y)] \leq D$,*
- *$P_Y \ll P_{Y^*}$, and*
- *$I(X;Y) < \infty$,*

the inequality (26.6) holds.

Some remarks on the preceding theorem are as follows:

1 The first part is a sufficient condition for optimality of a given P_{X,Y^*}. The second part gives a necessary condition that is convenient to narrow down the search. Indeed, typically the set of $P_{X,Y}$ satisfying those conditions is rich enough to infer from (26.6) that

$$\log \frac{dP_{X|Y^*}}{dP_X}(x|y) = R(D) - \theta[d(x,y) - D],$$

for a positive $\theta > 0$.

2 Note that the second part is not valid without assuming $P_Y \ll P_{Y^*}$. A counterexample to this and various other erroneous (but frequently encountered) generalizations is the following: $\mathcal{A} = \{0,1\}$, $P_X = \text{Ber}(1/2)$, $\hat{\mathcal{A}} = \{0,1,0',1'\}$, and

$$d(0,0) = d(0,0') = 1 - d(0,1) = 1 - d(0,1') = 0.$$

Then $R(D) = |1-h(D)|^+$, but there exist multiple non-equivalent optimal choices of $P_{Y|X}$, $P_{X|Y}$, and P_Y.

Proof. The first part is just a repetition of the proofs in Section 26.1 for the Hamming and Gaussian cases, so we focus on the second part. Suppose there exists a counterexample $P_{X,Y}$ achieving

$$I_1 = \mathbb{E}\left[\log \frac{dP_{X|Y^*}}{dP_X}(X|Y)\right] < I^* = R(D).$$

Notice that whenever $I(X;Y) < \infty$ we have

$$I_1 = I(X;Y) - D(P_{X|Y} \| P_{X|Y^*} | P_Y),$$

and thus

$$D(P_{X|Y} \| P_{X|Y^*} | P_Y) < \infty. \tag{26.7}$$

Before going to the actual proof, we describe the principal idea. For every λ we can define a joint distribution

$$P_{X,Y_\lambda} = \lambda P_{X,Y} + (1-\lambda) P_{X,Y^*}.$$

Then, we can compute

$$\begin{aligned} I(X;Y_\lambda) &= \mathbb{E}\left[\log \frac{P_{X|Y_\lambda}}{P_X}(X|Y_\lambda)\right] = \mathbb{E}\left[\log \frac{P_{X|Y_\lambda}}{P_{X|Y^*}} \frac{P_{X|Y^*}}{P_X}(X|Y_\lambda)\right] \\ &= D(P_{X|Y_\lambda} \| P_{X|Y^*} | P_{Y_\lambda}) + \mathbb{E}\left[\frac{P_{X|Y^*}(X|Y_\lambda)}{P_X}\right] \\ &= D(P_{X|Y_\lambda} \| P_{X|Y^*} | P_{Y_\lambda}) + \lambda I_1 + (1-\lambda) I_*. \end{aligned}$$

From here we will conclude, similar to Proposition 2.20, that the first term is $o(\lambda)$ and thus for sufficiently small λ we should have $I(X;Y_\lambda) < R(D)$, contradicting the optimality of the coupling P_{X,Y^*}.

We proceed to details. For every $\lambda \in [0,1]$ define

$$\begin{aligned}
\rho_1(y) &\triangleq \frac{dP_Y}{dP_{Y^*}}(y), \\
\lambda(y) &\triangleq \frac{\lambda\rho_1(y)}{\lambda\rho_1(y)+\bar{\lambda}}, \\
P^{(\lambda)}_{X|Y=y} &= \lambda(y)P_{X|Y=y} + \bar{\lambda}(y)P_{X|Y^*=y}, \\
dP_{Y_\lambda} &= \lambda dP_Y + \bar{\lambda} dP_{Y^*} = (\lambda\rho_1(y)+\bar{\lambda})dP_{Y^*}, \\
D(y) &= D(P_{X|Y=y}\|P_{X|Y^*=y}), \\
D_\lambda(y) &= D(P^{(\lambda)}_{X|Y=y}\|P_{X|Y^*=y}).
\end{aligned}$$

Notice that

$$\text{on } \{\rho_1 = 0\}: \quad \lambda(y) = D(y) = D_\lambda(y) = 0,$$

and otherwise $\lambda(y) > 0$. By convexity of divergence

$$D_\lambda(y) \le \lambda(y)D(y)$$

and therefore

$$\frac{1}{\lambda(y)}D_\lambda(y)1\{\rho_1(y)>0\} \le D(y)1\{\rho_1(y)>0\}.$$

Notice that by (26.7) the function $\rho_1(y)D(y)$ is non-negative and P_{Y^*}-integrable. Then, applying dominated the convergence theorem we get

$$\lim_{\lambda\to 0}\int_{\{\rho_1>0\}} dP_{Y^*}\frac{1}{\lambda(y)}D_\lambda(y)\rho_1(y) = \int_{\{\rho_1>0\}} dP_{Y^*}\rho_1(y)\lim_{\lambda\to 0}\frac{1}{\lambda(y)}D_\lambda(y) = 0, \quad (26.8)$$

where in the last step we applied the result from Proposition 2.20 that

$$D(P\|Q) < \infty \qquad \Longrightarrow \qquad D(\lambda P + \bar{\lambda}Q\|Q) = o(\lambda)$$

since for each y in the set $\{\rho_1 > 0\}$ we have $\lambda(y) \to 0$ as $\lambda \to 0$.

On the other hand, notice that

$$\begin{aligned}
\int_{\{\rho_1>0\}} dP_{Y^*}\frac{1}{\lambda(y)}D_\lambda(y)\rho_1(y)1\{\rho_1(y)>0\} &= \frac{1}{\lambda}\int_{\{\rho_1>0\}} dP_{Y^*}(\lambda\rho_1(y)+\bar{\lambda})D_\lambda(y) \\
&= \frac{1}{\lambda}\int_{\{\rho_1>0\}} dP_{Y_\lambda}D_\lambda(y) \\
&= \frac{1}{\lambda}\int_{\mathcal{Y}} dP_{Y_\lambda}D_\lambda(y) = \frac{1}{\lambda}D(P^{(\lambda)}_{X|Y}\|P_{X|Y^*}|P_{Y_\lambda}),
\end{aligned}$$

where in the penultimate step we used $D_\lambda(y) = 0$ on $\{\rho_1 = 0\}$. Hence, (26.8) shows that

$$D(P^{(\lambda)}_{X|Y}\|P_{X|Y^*}|P_{Y_\lambda}) = o(\lambda)\,, \qquad \lambda \to 0.$$

Finally, since

$$P^{(\lambda)}_{X|Y} \circ P_{Y_\lambda} = P_X,$$

we have

$$I(X;Y_\lambda) = D(P^{(\lambda)}_{X|Y}\|P_{X|Y^*}|P_{Y_\lambda})+\lambda\,\mathbb{E}\left[\log\frac{dP_{X|Y^*}}{dP_X}(X|Y)\right]+\bar{\lambda}\,\mathbb{E}\left[\log\frac{dP_{X|Y^*}}{dP_X}(X|Y^*)\right]$$
$$= I^* + \lambda(I_1 - I^*) + o(\lambda),$$

contradicting the assumption

$$I(X;Y_\lambda) \geq I^* = R(D). \qquad \square$$

26.3 Lossy Joint Source–Channel Coding

Extending the lossless joint source–channel coding problem studied in Section 19.7, in this section we study its lossy version: How to transmit a source over a noisy channel such that the receiver can reconstruct the original source within a prescribed distortion?

The setup of the lossy joint source–channel coding problem is as follows. For each k and n, we are given a source $S^k = (S_1, \ldots, S_k)$ taking values on $\mathcal{S}$, a distortion metric $d \colon \mathcal{S}^k \times \hat{\mathcal{S}}^k \to \mathbb{R}$, and a channel $P_{Y^n|X^n}$ acting from $\mathcal{A}^n$ to $\mathcal{B}^n$. A lossy joint source–channel code (JSCC) consists of an encoder $f \colon \mathcal{S}^k \to \mathcal{A}^n$ and decoder $g \colon \mathcal{B}^n \to \hat{\mathcal{S}}^k$, such that the channel input is $X^n = f(S^k)$ and the reconstruction $\hat{S}^k = g(Y^n)$ satisfies $\mathbb{E}[d(S^k, \hat{S}^k)] \leq D$. By definition, we have the Markov chain

$$S^k \xrightarrow{f} X^n \xrightarrow{P_{Y^n|X^n}} Y^n \xrightarrow{g} \hat{S}^k.$$

Such a pair (f, g) is called a (k, n, D)-JSCC, which transmits k symbols over n channel uses such that the end-to-end distortion is at most D in expectation. Our goal is to optimize the encoder–decoder pair so as to maximize the transmission rate[1] (number of symbols per channel use) $R = \frac{k}{n}$. As such, we define the asymptotic fundamental limit as

$$R_{\text{JSCC}}(D) \triangleq \liminf_{n\to\infty} \frac{1}{n} \max\{k \colon \exists (k, n, D)\text{-JSCC}\}.$$

To simplify the exposition, we will focus on the JSCC for a stationary memoryless source $S^k \sim P_S^{\otimes k}$ transmitted over a stationary memoryless channel $P_{Y^n|X^n} = P_{Y|X}^{\otimes n}$ subject to a separable distortion function $d(s^k, \hat{s}^k) = \frac{1}{k}\sum_{i=1}^k d(s_i, \hat{s}_i)$.

26.3.1 Converse

The converse for the JSCC is quite simple, based on the data-processing inequality and following the weak converse of the lossless JSCC using Fano's inequality.

Theorem 26.5. (Converse)

$$R_{\text{JSCC}}(D) \leq \frac{C}{R(D)},$$

[1] Or equivalently, minimize the *bandwidth expansion factor* $\rho = \frac{n}{k}$.

where $C = \sup_{P_X} I(X;Y)$ *is the capacity of the channel and* $R(D) = \inf_{P_{\hat{S}|S}:\, \mathbb{E}[d(S,\hat{S})]\le D} I(S;\hat{S})$ *is the rate-distortion function of the source.*

The interpretation of this result is clear: Since we need at least $R(D)$ bits per symbol to reconstruct the source up to a distortion D and we can transmit at most C bits per channel use, the overall transmission rate cannot exceed $C/R(D)$. Note that the above theorem clearly holds for channels with cost constraint (Chapter 20).

Proof. Consider a (k,n,D)-code which induces the Markov chain $S^k \to X^n \to Y^n \to \hat{S}^k$ such that $\mathbb{E}[d(S^k,\hat{S}^k)] = \frac{1}{k}\sum_{i=1}^k \mathbb{E}[d(S_i,\hat{S}_i)] \le D$. Then

$$kR(D) \overset{(a)}{=} \inf_{P_{\hat{S}^k|S^k}:\, \mathbb{E}[d(S^k,\hat{S}^k)]\le D} I(S^k;\hat{S}^k) \overset{(b)}{\le} I(S^k;\hat{S}^k) \le I(X^n;Y^n) \le \sup_{P_{X^n}} I(X^n;Y^n) \overset{(c)}{=} nC,$$

where (b) applies the data-processing inequality for mutual information, and (a) and (c) follow from the respective single-letterization results for lossy compression and channel coding (Theorem 24.8 and Proposition 19.10). □

Remark 26.4. Consider the case where the source is Ber(1/2) with Hamming distortion. Then Theorem 26.5 coincides with the converse for channel coding under bit error rate P_b in (19.33):

$$R = \frac{k}{n} \le \frac{C}{1-h(P_b)},$$

which was previously given in Theorem 19.21 and proved using ad hoc techniques. In the case of a channel with cost constraints, for example, the AWGN channel with capacity $C(\mathrm{SNR}) = \frac{1}{2}\log(1+\mathrm{SNR})$, we have

$$P_b \ge h^{-1}\left(1 - \frac{C(\mathrm{SNR})}{R}\right).$$

This is often referred to as the Shannon limit in plots comparing the bit error rate of practical codes. (See, e.g. Figure. 2 from [360] for BIAWGN (binary-input) channel.) *This interpretation is erroneous*, since the P_b above refers to the bit error rate of data bits (or systematic bits), not all of the codeword bits. The latter quantity is what is typically called BER (see (19.33)) in the coding-theoretic literature.

26.3.2 Achievability via Separation

The proof strategy is similar to that of the lossless JSCC in Section 19.7 by *separately* constructing a channel coding scheme and a lossy compression scheme, as opposed to jointly optimizing the JSCC encoder–decoder pair. Specifically, first compress the data into bits then encode with a channel code; to decode, apply the channel decoder followed by the source decompressor. Under appropriate assumptions, this separately designed scheme achieves the optimal rate in Theorem 26.5.

Theorem 26.6. *For any stationary memoryless source P_S with rate-distortion function $R(D)$ satisfying Assumption 26.1 (below), and for any stationary memoryless channel $P_{Y|X}$ with capacity C,*

$$R_{\text{JSCC}}(D) = \frac{C}{R(D)}.$$

Assumption 26.1 on the source (which is rather technical and can be skipped in the first reading) is to control the distortion incurred by the channel decoder making an error. Despite this being a low-probability event, without any assumption on the distortion metric, we cannot say much about its contribution to the end-to-end average distortion. (Note that this issue does not arise in lossless JSCC.) Assumption 26.1 is trivially satisfied by bounded distortion (e.g. Hamming), and can be shown to hold more generally such as for Gaussian sources and MSE distortion.

Proof. In view of Theorem 26.5, we only prove achievability. We construct a separated compression/channel coding scheme as follows:

- Let (f_s, g_s) be a $(k, 2^{kR(D)+o(k)}, D)$-code for compressing S^k such that $\mathbb{E}[d(S^k, g_s(f_s(S^k)))] \leq D$. By Lemma 26.8 (below), we may assume that all reconstruction points are not too far from some fixed string, namely,

$$d(s_0^k, g_s(i)) \leq L \tag{26.9}$$

for all i and some constant L, where $s_0^k = (s_0, \ldots, s_0)$ is from Assumption 26.1 below.
- Let (f_c, g_c) be an $(n, 2^{nC+o(n)}, \epsilon_n)_{\max}$-code for channel $P_{Y^n|X^n}$ such that $kR(D) + o(k) \leq nC + o(n)$ and the maximal probability of error $\epsilon_n \to 0$ as $n \to \infty$. Such a code exists thanks to Theorem 19.9 and Corollary 19.5.

Let the JSCC encoder and decoder be $f = f_c \circ f_s$ and $g = g_s \circ g_c$. So the overall system is

$$S^k \xrightarrow{f_s} W \xrightarrow{f_c} X^n \longrightarrow Y^n \xrightarrow{g_c} \hat{W} \xrightarrow{g_s} \hat{S}^k.$$

Note that here we need to control the *maximal* probability of error of the channel code since, when we concatenate these two schemes, W at the input of the channel is the output of the source compressor, which need not be uniformly distributed.

To analyze the average distortion, we consider two cases depending on whether the channel decoding is successful or not:

$$\mathbb{E}[d(S^k, \hat{S}^k)] = \mathbb{E}[d(S^k, g_s(W))1\{W = \hat{W}\}] + \mathbb{E}[d(S^k, g_s(\hat{W}))1\{W \neq \hat{W}\}].$$

By assumption on our lossy code, the first term is at most D. For the second term, we have $\mathbb{P}[W \neq \hat{W}] \leq \epsilon_n = o(1)$ by assumption on our channel code. Then

$$\begin{aligned}\mathbb{E}[d(S^k, g_s(\hat{W}))1\{W \neq \hat{W}\}] &\overset{\text{(a)}}{\leq} \mathbb{E}[1\{W \neq \hat{W}\}\lambda(d(S^k, s_0^k) + d(s_0^k, g_s(\hat{W})))] \\ &\overset{\text{(b)}}{\leq} \lambda \cdot \mathbb{E}[1\{W \neq \hat{W}\}d(S^k, s_0^k)] + \lambda L \cdot \mathbb{P}[W \neq \hat{W}] \\ &\overset{\text{(c)}}{=} o(1),\end{aligned}$$

where (a) follows from the generalized triangle inequality from Assumption 26.1(a) below; (b) follows from (26.9); and in (c) we apply Lemma 25.4 that was used before to show the vanishing of the expectation in (25.15).

In all, our scheme meets the average distortion constraint. Hence we conclude that for all $R > C/R(D)$, there exists a sequence of $(k, n, D+o(1))$-JSCC codes. □

The following assumption is needed by the previous theorem:

Assumption 26.1. Fix D. There exist $\lambda \geq 0, s_0 \in \mathcal{S}$ and $\hat{s}_0 \in \hat{\mathcal{S}}$ such that:

(a) Generalized triangle inequality: $d(s, \hat{s}) \leq \lambda(d(s, \hat{s}_0) + d(s_0, \hat{s}))\ \ \forall s, \hat{s}$.
(b) $\mathbb{E}[d(S, \hat{s}_0)] < \infty$ (so that $D_{\max} < \infty$ too).
(c) $\mathbb{E}[d(s_0, \hat{S})] < \infty$ for any output distribution $P_{\hat{S}}$ achieving the rate-distortion function $R(D)$.
(d) $d(s_0, \hat{s}_0) < \infty$.

The interpretation of this assumption is that the spaces $\mathcal{S}$ and $\hat{\mathcal{S}}$ have "nice centers" s_0 and $\hat{s}_0$, in the sense that the distance between any two points is upper–bounded by a constant times the distance from the centers to each point (see figure below).

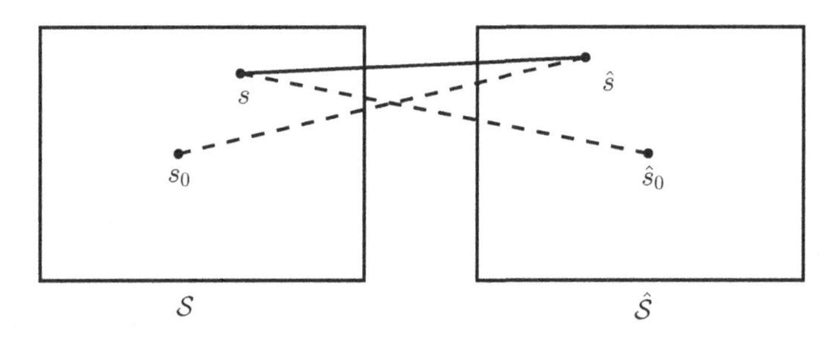

Note that Assumption 26.1 is not straightforward to verify. Next we give some more convenient sufficient conditions. First of all, Assumption 26.1 holds automatically for a bounded distortion function. In other words, for a discrete source on a finite alphabet $\mathcal{S}$, a finite reconstruction alphabet $\hat{\mathcal{S}}$, and a finite distortion function $d(s, \hat{s}) < \infty$, Assumption 26.1 is fulfilled. More generally, we have the following criterion.

Theorem 26.7. *If $\mathcal{S} = \hat{\mathcal{S}}$ and $d(s, \hat{s}) = \rho(s, \hat{s})^q$ for some metric ρ and $q \geq 1$, and $D_{\max} \triangleq \inf_{\hat{s}_0} \mathbb{E}[d(S, \hat{s}_0)] < \infty$, then Assumption 26.1 holds.*

Proof. Take $s_0 = \hat{s}_0$ that achieves a finite $D_{\max} = \mathbb{E}[d(S, \hat{s}_0)]$. (In fact, any points can serve as centers in a metric space.) Applying the triangle inequality and Jensen's inequality, we have

$$\left(\frac{1}{2}\rho(s, \hat{s})\right)^q \leq \left(\frac{1}{2}\rho(s, s_0) + \frac{1}{2}\rho(s_0, \hat{s})\right)^q \leq \frac{1}{2}\rho^q(s, s_0) + \frac{1}{2}\rho^q(s_0, \hat{s}).$$

Thus $d(s, \hat{s}) \leq 2^{q-1}(d(s, s_0) + d(s_0, \hat{s}))$. Taking $\lambda = 2^{q-1}$ verifies (a) and (b) in Assumption 26.1. To verify (c), we can apply this generalized triangle inequality to get $d(s_0, \hat{S}) \le 2^{q-1}(d(s_0, S) + d(S, \hat{S}))$. Then taking the expectation of both sides gives

$$\begin{aligned}\mathbb{E}[d(s_0, \hat{S})] &\le 2^{q-1}(\mathbb{E}[d(s_0, S)] + \mathbb{E}[d(S, \hat{S})]) \\ &\le 2^{q-1}(D_{\max} + D) < \infty. \qquad \square\end{aligned}$$

So we see that powers of distances (e.g. squared norms) satisfy Assumption 26.1. Finally, we give the lemma used in the proof of Theorem 26.6.

Lemma 26.8. *Fix a source satisfying Assumption 26.1 and an arbitrary* $P_{\hat{S}|S}$. *Let* $R > I(S; \hat{S})$, $L > \max\{\mathbb{E}[d(s_0, \hat{S})], d(s_0, \hat{s}_0)\}$, *and* $D > \mathbb{E}[d(S, \hat{S})]$. *Then, there exists a* $(k, 2^{kR}, D)$*-code such that* $d(s_0^k, \hat{s}^k) \le L$ *for every reconstruction point* $\hat{s}^k$, *where* $s_0^k = (s_0, \dots, s_0)$.

Proof. Let $\mathcal{X} = \mathcal{S}^k$, $\hat{\mathcal{X}} = \hat{\mathcal{S}}^k$, and $P_X = P_S^k$, $P_{Y|X} = P_{\hat{S}|S}^k$. We apply the achievability bound for excess distortion from Theorem 25.3 with $\gamma = 2^{k(R+I(S;\hat{S}))/2}$ to the following *non-separable* distortion function:

$$d_1(x, \hat{x}) = \begin{cases} d(x, \hat{x}), & d(s_0^k, \hat{x}) \le L, \\ +\infty, & \text{otherwise.} \end{cases}$$

For any $D' \in (\mathbb{E}[d(S, \hat{S})], D)$, there exist $M = 2^{kR}$ reconstruction points $(c_1, \dots, c_M)$ such that

$$\mathbb{P}\left[\min_{j \in [M]} d(S^k, c_j) > D'\right] \le \mathbb{P}[d_1(S^k, \hat{S}^k) > D'] + o(1),$$

where on the right-hand side $(S^k, \hat{S}^k) \sim P_{S,\hat{S}}^k$. Note that without any change in d_1-distortion we can remove all (if any) reconstruction points c_j with $d(s_0^k, c_j) > L$. Furthermore, from the WLLN we have

$$\mathbb{P}[d_1(S, \hat{S}) > D'] \le \mathbb{P}[d(S^k, \hat{S}^k) > D'] + \mathbb{P}[d(s_0^k, \hat{S}^k) > L] \to 0$$

as $k \to \infty$ (since $\mathbb{E}[d(S, \hat{S})] < D'$ and $\mathbb{E}[d(s_0, \hat{S})] < L$). Thus we have

$$\mathbb{P}\left[\min_{j \in [M]} d(S^k, c_j) > D'\right] \to 0$$

and $d(s_0^k, c_j) \le L$. Finally, by adding another reconstruction point $c_{M+1} = \hat{s}_0^k = (\hat{s}_0, \dots, \hat{s}_0)$ we get

$$\mathbb{E}\left[\min_{j \in [M+1]} d(S^k, c_j)\right] \le D' + \mathbb{E}\left[d(S^k, \hat{s}_0^k) 1\left\{\min_{j \in [M]} d(S^k, c_j) > D'\right\}\right] = D' + o(1),$$

where the last estimate follows from the same argument that shows the vanishing of the expectation in (25.15). Thus, for sufficiently large n the expected distortion is at most D, as required. $\square$

26.4 What Is Lacking in Classical Lossy Compression?

Let us discuss some issues and open problems in the classical theory of data compression and joint source-channel coding. First, for *compression* the standard results in lossless compression apply well for text files. Lossy compression theory, however, relies on the independence assumption and on separable distortion metrics. Because of that, while scalar quantization theory has been widely used in practice (in the form of analog-to-digital converters, ADCs), vector quantization (rate-distortion) theory so far has not been employed. The assumptions of rate-distortion theory can be seen to be especially problematic in the case of compressing digital images, which evidently have very strong spatial correlation compared to one-dimensional signals. (For example, the first sentence and the last in a Tolstoy novel are pretty uncorrelated. But the regions in the upper-left and bottom-right corners of one image can be strongly correlated. At the same time, the uncompressed size of the novel and the image could be easily equal.) Thus, for practicing the lossy compression of videos and images the key problem is that of coming up with a good "whitening" basis, which is an art still being refined.

For the joint source–channel coding, the separation principle has definitely been a guiding light for the entire development of digital information technology. But this now ubiquitous solution that Shannon's separation has professed led to a rather undesirable feature of dropped cellular calls (as opposed to slowly degraded quality of the old analog telephones) or "snow screen" on TV whenever the SNR falls below a certain threshold. That is, the separated systems can be very unstable, or lack *graceful degradation*. To sketch this effect consider an example of a JSCC, where the source distribution is Ber(1/2), with rate-distortion function $R(D) = 1 - h(D)$, and the channel is BSC_δ with capacity $C(\delta) = 1 - h(\delta)$. Consider two solutions:

1 A separated scheme: targeting a certain acceptable distortion level D^* we compress the source at rate $R(D^*)$. Then we can use a channel code of rate $R(D^*)$ which would achieve vanishing error as long as $R(D^*) < C(\delta)$, that is, $\delta < D^*$. Overall, this scheme has a bandwidth expansion factor $\rho = \frac{n}{k} = 1$. Note that there exist channel codes (Exercises **IV.8** and **IV.10**) that work *simultaneously* for all $\delta < \delta^* = D^*$.
2 A simple JSCC with $\rho = 1$ which transmits "uncoded" data, that is, sets $X_i = S_i$.

For large blocklengths, the achieved distortion is shown in Figure 26.2 as a function of δ. We can now see why a separated solution, though in some sense optimal, is not ideal. First, if $\delta < \delta^*$ the separated solution does achieve acceptable distortion D^*, but it does not improve if the channel improves, that is, the distortion stays constant at D^*, unlike the uncoded system. Second, and much more importantly, is a problem with $\delta > \delta^*$. In this regime, the separated scheme undergoes a catastrophic failure and distortion becomes 1/2 (that is, we observe pure noise, or "snow" in TV-speak). At the same time, the distortion of the simple "uncoded" JSCC is

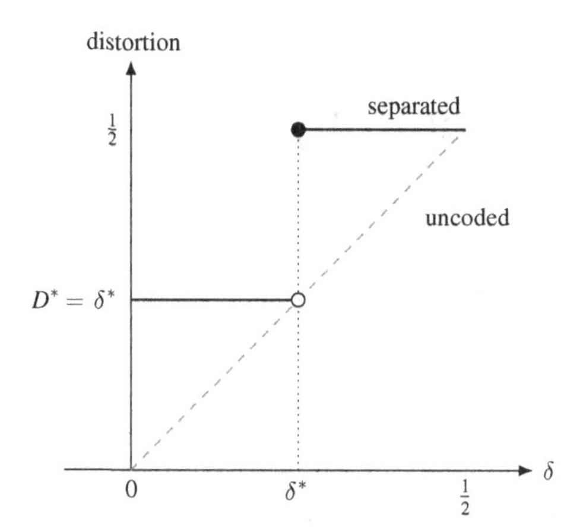

Figure 26.2 No graceful degradation of separately designed source–channel code (solid), as compared with uncoded transmission (dashed).

also deteriorating but *gracefully* so. Unfortunately, such graceful schemes are only known for very few cases, requiring $\rho = 1$ and certain "perfect match" conditions between channel noise and source (distortion metric).[2] It is a long-standing (practical and theoretical) open problem in information theory to find schemes that exhibit non-catastrophic degradation for general source–channel pairs and general ρ.

Even purely theoretically the problem of JSCC still contains many mysteries. For example, in Section 22.5 we described refined expansion of the channel coding rate as a function of block length. In particular, we have seen that convergence to channel capacity happens at the rate $\frac{1}{\sqrt{n}}$, which is rather slow. At the same time, convergence to the rate-distortion function is almost at the rate of $\frac{1}{n}$ (see Exercises **V.3** and **V.4**). Thus, it is not clear what the convergence rate of the JSCC may be. Unfortunately, sharp results here are still at a nascent stage. In fact, even for the most canonical setting of a binary source and BSC_δ channel it was only very recently shown [249] that the optimal rate $\frac{k}{n}$ converges to the ultimate limit of $\frac{C}{R(D)}$ at the speed of $\Theta(1/\sqrt{n})$ unless the Gastpar condition $R(D) = C(\delta)$ is met. Analyzing other source–channel pairs or any general results of this kind is another important *open problem*.

[2] Often informally called "Gastpar conditions" after [181].

27 Metric Entropy

In the previous chapters of this part we have discussed optimal quantization of random vectors in both fixed and high dimensions. Complementing this average-case perspective, the topic of this chapter is the deterministic (worst-case) theory of quantization. The main object of interest is the *metric entropy* of a set, which allows us to answer two key questions about: (a) covering number: the minimum number of points to cover a set up to a given accuracy; (b) packing number: the maximal number of elements of a given set with a prescribed minimum pairwise distance.

The foundational theory of metric entropy was put forth by Kolmogorov, who, together with his students, also determined the behavior of metric entropy in a variety of problems for both finite and infinite dimensions. Kolmogorov's original interest in this subject stems from Hilbert's thirteenth problem, which concerns the possibility or impossibility of representing multivariable functions as compositions of functions of fewer variables. It turns out that the theory of metric entropy can provide a surprisingly simple and powerful resolution to such problems. Over the years, metric entropy has found numerous connections to and applications in other fields such as approximation theory, empirical processes, small-ball probability, mathematical statistics, and machine learning. In particular, metric entropy will be featured prominently in Part VI of this book, wherein we discuss its applications to proving both lower and upper bounds for statistical estimation.

This chapter is organized as follows. Section 27.1 provides basic definitions and explains the fundamental connections between covering and packing numbers. These foundations are laid out by Kolmogorov and Tikhomirov in [251], which remains the definitive reference on this subject. In Section 27.2 we study metric entropy in finite-dimensional spaces and a popular approach for bounding the metric entropy known as the volume bound. To demonstrate the limitations of the volume method and the associated high-dimensional phenomenon, in Section 27.3 we discuss a few other approaches through concrete examples. Infinite-dimensional spaces are treated next for smooth functions in Section 27.4 (wherein we also discuss the application to Hilbert's thirteenth problem) and Hilbert spaces in Section 27.3.2 (wherein we also discuss the application to empirical processes). Section 27.5 gives an exposition of the connections between metric entropy and the small-ball problem in probability theory. Finally, in Section 27.6 we circle back to rate-distortion theory and discuss how it is related to metric entropy and how information-theoretic methods can be useful for the latter.

27.1 Covering and Packing

Definition 27.1. Let (V, d) be a metric space and $\Theta \subset V$.

- We say $\{v_1, \ldots, v_N\} \subset V$ is an *ϵ-covering* (or *ϵ-net*) of Θ if $\Theta \subset \bigcup_{i=1}^{N} B(v_i, \epsilon)$, where $B(v, \epsilon) \triangleq \{u \in V : d(u, v) \leq \epsilon\}$ is the (closed) ball of radius ϵ centered at v; or equivalently, $\forall \theta \in \Theta$, $\exists i \in [N]$ such that $d(\theta, v_i) \leq \epsilon$.
- We say $\{\theta_1, \ldots, \theta_M\} \subset \Theta$ is an *ϵ-packing* of Θ if $\min_{i \neq j} \|\theta_i - \theta_j\| > \epsilon$;[1] or equivalently the balls $\{B(\theta_i, \epsilon/2) : i \in [M]\}$ are disjoint.

The covering and packing of a two-dimensional set are shown in Figure 27.1.

Upon defining ϵ-covering and ϵ-packing, a natural question concerns the size of the optimal covering and packing, leading to the definition of *covering* and *packing numbers*:

$$N(\Theta, d, \epsilon) \triangleq \min\{n : \exists\, \epsilon\text{-covering of } \Theta \text{ of size } n\}, \tag{27.1}$$

$$M(\Theta, d, \epsilon) \triangleq \max\{m : \exists\, \epsilon\text{-packing of } \Theta \text{ of size } m\}, \tag{27.2}$$

with $\min \emptyset$ understood as ∞; we will sometimes abbreviate these as $N(\epsilon)$ and $M(\epsilon)$ for brevity. Similar to volume and width, covering and packing numbers provide a meaningful measure for the "massiveness" of a set. The major focus of this chapter is to understand their behavior in both finite- and infinite-dimensional spaces as well as their statistical applications.

Some remarks are in order.

- Monotonicity: $N(\Theta, d, \epsilon)$ and $M(\Theta, d, \epsilon)$ are non-decreasing and right-continuous functions of ϵ. Furthermore, both are non-decreasing in Θ with respect to set inclusion.
- Finiteness: Θ is totally bounded (e.g. compact) if $N(\Theta, d, \epsilon) < \infty$ for all $\epsilon > 0$. For Euclidean spaces, this is equivalent to Θ being bounded, namely, $\mathrm{diam}(\Theta) < \infty$ (see (5.4)).
- The logarithms of the covering and packing numbers are commonly referred to as *metric entropy*. In particular, $\log M(\epsilon)$ and $\log N(\epsilon)$ are called *ϵ-entropy* and

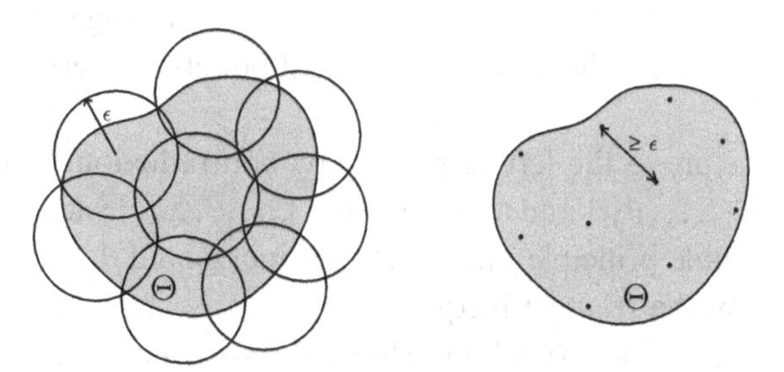

Figure 27.1 Illustration of ϵ-covering (left) and ϵ-packing (right).

[1] Notice we imposed strict inequality for convenience.

ϵ-*capacity* in [251]. Quantitative connections between metric entropy and other information measures are explored in Section 27.6.

- Widely used in the literature of functional analysis [330, 286], the notion of *entropy numbers* essentially refers to the inverse of the metric entropy: The kth entropy number of Θ is $e_k(\Theta) \triangleq \inf\{\epsilon: N(\Theta, d, \epsilon) \leq 2^{k-1}\}$. In particular, $e_1(\Theta) = \text{rad}(\Theta)$, the radius of Θ defined in (5.3).

Remark 27.1. Unlike the packing number $M(\Theta, d, \epsilon)$, the covering number $N(\Theta, d, \epsilon)$ defined in (27.1) depends implicitly on the ambient space $V \supset \Theta$, since, per Definition 27.1, an ϵ-covering is required to be a subset of V rather than Θ. Nevertheless, as Theorem 27.2 shows, this dependence on V has almost no effect on the behavior of the covering number.

As an alternative to (27.1), we can define $N'(\Theta, d, \epsilon)$ as the size of the minimal ϵ-covering of Θ that is also a subset of Θ, which is closely related to the original definition as

$$N(\Theta, d, \epsilon) \leq N'(\Theta, d, \epsilon) \leq N(\Theta, d, \epsilon/2). \tag{27.3}$$

Here, the left inequality is obvious. To see the right inequality,[2] let $\{\theta_1, \ldots, \theta_N\}$ be an $\frac{\epsilon}{2}$-covering of Θ. We can project each θ_i to Θ by defining $\theta_i' = \text{argmin}_{u \in \Theta} d(\theta_i, u)$. Then $\{\theta_1', \ldots, \theta_N'\} \subset \Theta$ constitutes an ϵ-covering. Indeed, for any $\theta \in \Theta$, we have $d(\theta, \theta_i) \leq \epsilon/2$ for some θ_i. Then $d(\theta, \theta_i') \leq d(\theta, \theta_i) + d(\theta_i, \theta_i') \leq 2d(\theta, \theta_i) \leq \epsilon$. On the other hand, the N' covering numbers need not be monotone with respect to set inclusion.

The relation between the covering and packing numbers is described by the following fundamental result.

Theorem 27.2. (Kolmogorov–Tikhomirov [251])

$$M(\Theta, d, 2\epsilon) \leq N(\Theta, d, \epsilon) \leq M(\Theta, d, \epsilon). \tag{27.4}$$

Proof. To prove the right inequality, fix a maximal packing $E = \{\theta_1, \ldots, \theta_M\}$. Then $\forall \theta \in \Theta \backslash E$, $\exists i \in [M]$, such that $d(\theta, \theta_i) \leq \epsilon$ (for otherwise we can obtain a bigger packing by adding θ). Hence E must be an ϵ-covering (which is also a subset of Θ). Since $N(\Theta, d, \epsilon)$ is the minimal size of all possible coverings, we have $M(\Theta, d, \epsilon) \geq N(\Theta, d, \epsilon)$.

We next prove the left inequality by contradiction. Suppose there exists a 2ϵ-packing $\{\theta_1, \ldots, \theta_M\}$ and an ϵ-covering $\{x_1, \ldots, x_N\}$ such that $M \geq N + 1$. Then by the pigeonhole principle, there exist distinct θ_i and θ_j belonging to the same ϵ-ball $B(x_k, \epsilon)$. By the triangle inequality, $d(\theta_i, \theta_j) \leq 2\epsilon$, which is a contradiction since $d(\theta_i, \theta_j) > 2\epsilon$ for a 2ϵ-packing. Hence the size of any 2ϵ-packing is at most that of any ϵ-covering, that is, $M(\Theta, d, 2\epsilon) \leq N(\Theta, d, \epsilon)$. □

[2] Another way to see this is from Theorem 27.2: Note that the right inequality in (27.4) yields an ϵ-covering that is included in Θ. Together with the left inequality, we get $N'(\epsilon) \leq M(\epsilon) \leq N(\epsilon/2)$.

The significance of (27.4) is that it shows that the small-ϵ behaviors of the covering and packing numbers are essentially the same. In addition, the right inequality therein, namely, $N(\epsilon) \leq M(\epsilon)$, deserves some special mention. As we will see next, it is oftentimes easier to prove negative results (lower bounds on the minimal covering or upper bounds on the maximal packing) than positive results which require explicit construction. When used in conjunction with the inequality $N(\epsilon) \leq M(\epsilon)$, these converses turn into achievability statements,[3] leading to many useful bounds on metric entropy (e.g. the volume bound in Theorem 27.3 and the Gilbert–Varshamov bound in Theorem 27.5 in the next section). Revisiting the proof of Theorem 27.2, we see that this logic actually corresponds to a *greedy construction* (greedily increase the packing until no points can be added).

27.2 Finite-Dimensional Space and Volume Bound

A commonly used method to bound metric entropy in finite dimensions is in terms of the *volume ratio*. Consider the d-dimensional Euclidean space $V = \mathbb{R}^d$ with metric given by an arbitrary norm $d(x, y) = \|x - y\|$. We have the following result.

Theorem 27.3. *Let $\|\cdot\|$ be an arbitrary norm on $\mathbb{R}^d$ and $B = \{x \in \mathbb{R}^d : \|x\| \leq 1\}$ the corresponding unit-norm ball. Then for any $\Theta \subset \mathbb{R}^d$,*

$$\left(\frac{1}{\epsilon}\right)^d \frac{\mathrm{vol}(\Theta)}{\mathrm{vol}(B)} \overset{\text{(a)}}{\leq} N(\Theta, \|\cdot\|, \epsilon) \leq M(\Theta, \|\cdot\|, \epsilon) \overset{\text{(b)}}{\leq} \frac{\mathrm{vol}(\Theta + \frac{\epsilon}{2}B)}{\mathrm{vol}(\frac{\epsilon}{2}B)} \overset{\text{(c)}}{\leq} \left(\frac{3}{\epsilon}\right)^d \frac{\mathrm{vol}(\Theta)}{\mathrm{vol}(B)},$$

where (c) holds under the extra condition that Θ is convex and contains ϵB.

Proof. To prove (a), consider an ϵ-covering $\Theta \subset \bigcup_{i=1}^{N} B(\theta_i, \epsilon)$. Applying the union bound yields

$$\mathrm{vol}(\Theta) \leq \mathrm{vol}\left(\bigcup_{i=1}^{N} B(\theta_i, \epsilon)\right) \leq \sum_{i=1}^{N} \mathrm{vol}(B(\theta_i, \epsilon)) = N\epsilon^d \mathrm{vol}(B),$$

where the last step follows from the translation invariance and scaling property of volume.

To prove (b), consider an ϵ-packing $\{\theta_1, \ldots, \theta_M\} \subset \Theta$ such that the balls $B(\theta_i, \epsilon/2)$ are disjoint. Since $\bigcup_{i=1}^{M(\epsilon)} B(\theta_i, \epsilon/2) \subset \Theta + \frac{\epsilon}{2}B$, taking the volume on both sides yields

$$\mathrm{vol}\left(\Theta + \frac{\epsilon}{2}B\right) \geq \mathrm{vol}\left(\bigcup_{i=1}^{M} B(\theta_i, \epsilon/2)\right) = M\,\mathrm{vol}\left(\frac{\epsilon}{2}B\right).$$

This proves (b).

Finally, (c) follows from the following two statements: (1) if $\epsilon B \subset \Theta$, then $\Theta + \frac{\epsilon}{2}B \subset \Theta + \frac{1}{2}\Theta$; and (2) if Θ is convex, then $\Theta + \frac{1}{2}\Theta = \frac{3}{2}\Theta$. We only prove (2). First,

[3] This is reminiscent of duality-based argument in optimization: To bound a minimization problem from above, instead of constructing an explicit feasible solution, a fruitful approach is to equate it with the dual problem (maximization) and bound this maximum from above.

$\forall \theta \in \frac{3}{2}\Theta$, we have $\theta = \frac{1}{3}\theta + \frac{2}{3}\theta$, where $\frac{1}{3}\theta \in \frac{1}{2}\Theta$ and $\frac{2}{3}\theta \in \Theta$. Thus $\frac{3}{2}\Theta \subset \Theta + \frac{1}{2}\Theta$. On the other hand, for any $x \in \Theta + \frac{1}{2}\Theta$, we have $x = y + \frac{1}{2}z$ with $y, z \in \Theta$. By the convexity of Θ, $\frac{2}{3}x = \frac{2}{3}y + \frac{1}{3}z \in \Theta$. Hence $x \in \frac{3}{2}\Theta$, implying $\Theta + \frac{1}{2}\Theta \subset \frac{3}{2}\Theta$. □

Remark 27.2. Similar to the proof of (a) in Theorem 27.3, we can start from $\Theta + \frac{\epsilon}{2}B \subset \bigcup_{i=1}^{N} B(\theta_i, \frac{3\epsilon}{2})$ to conclude that

$$(2/3)^d \le \frac{N(\Theta, \|\cdot\|, \epsilon)}{\mathrm{vol}(\Theta + \frac{\epsilon}{2}B)/\mathrm{vol}(\epsilon B)} \le 2^d.$$

In other words, the volume of the fattened set $\Theta + \frac{\epsilon}{2}$ determines the metric entropy up to constants that depend only on the dimension. We will revisit this reasoning in Section 27.5 to adapt the volumetric estimates to infinite dimensions where this fattening step becomes necessary.

Next we discuss several applications of Theorem 27.3.

Corollary 27.4. (Metric entropy of balls and spheres) *Let $\|\cdot\|$ be an arbitrary norm on $\mathbb{R}^d$. Let $B \equiv B_{\|\cdot\|} = \{x \in \mathbb{R}^d : \|x\| \le 1\}$ and $S \equiv S_{\|\cdot\|} = \{x \in \mathbb{R}^d : \|x\| = 1\}$ be the corresponding unit ball and unit sphere. Then for $\epsilon < 1$,*

$$\left(\frac{1}{\epsilon}\right)^d \le N(B, \|\cdot\|, \epsilon) \le \left(1 + \frac{2}{\epsilon}\right)^d, \tag{27.5}$$

$$\left(\frac{1}{2\epsilon}\right)^{d-1} \le N(S, \|\cdot\|, \epsilon) \le 2d\left(1 + \frac{1}{\epsilon}\right)^{d-1}, \tag{27.6}$$

where the left inequality in (27.6) holds under the extra assumption that $\|\cdot\|$ is an absolute norm (invariant to sign changes of coordinates).

Proof. For balls, the estimate (27.5) directly follows from Theorem 27.3 since $B + \frac{\epsilon}{2}B = (1 + \frac{\epsilon}{2})B$. Next we consider the spheres. Applying (b) in Theorem 27.3 yields

$$\begin{aligned} N(S, \|\cdot\|, \epsilon) \le M(S, \|\cdot\|, \epsilon) &\le \frac{\mathrm{vol}(S + \epsilon B)}{\mathrm{vol}(\epsilon B)} \le \frac{\mathrm{vol}((1+\epsilon)B) - \mathrm{vol}((1-\epsilon)B)}{\mathrm{vol}(\epsilon B)} \\ &= \frac{(1+\epsilon)^d - (1-\epsilon)^d}{\epsilon^d} \\ &= \frac{d}{\epsilon^d}\int_{-\epsilon}^{\epsilon}(1+x)^{d-1}dx \le 2d\left(1 + \frac{1}{\epsilon}\right)^{d-1}, \end{aligned}$$

where the third inequality applies $S + \epsilon B \subset ((1+\epsilon)B)\backslash((1-\epsilon)B)$ by the triangle inequality.

Finally, we prove the lower bound in (27.6) for an absolute norm $\|\cdot\|$. To this end one cannot directly invoke the lower bound in Theorem 27.3 as the sphere has zero volume. Note that $\|\cdot\|' \triangleq \|(\cdot, 0)\|$ defines a norm on $\mathbb{R}^{d-1}$. We claim that every ϵ-packing in $\|\cdot\|'$ for the unit $\|\cdot\|'$-ball induces an ϵ-packing in $\|\cdot\|$ for the unit $\|\cdot\|$-sphere. Fix $x \in \mathbb{R}^{d-1}$ such that $\|(x, 0)\| \le 1$ and define $f : \mathbb{R}_+ \to \mathbb{R}_+$ by $f(y) = \|(x, y)\|$. Using the fact that $\|\cdot\|$ is an absolute norm, it is easy to verify that f is a continuous increasing function with $f(0) \le 1$ and $f(\infty) = \infty$. By the mean value theorem, there exists y_x, such that $\|(x, y_x)\| = 1$. Finally, for any ϵ-packing

$\{x'_1, \ldots, x'_M\}$ of the unit ball $B_{\|\cdot\|'}$ with respect to $\|\cdot\|'$, setting $x'_i = (x_i, y_{x_i})$ we have $\|x'_i - x'_j\| \geq \|(x_i - x_j, 0)\| = \|x_i - x_j\|' \geq \epsilon$. This proves

$$M(S_{\|\cdot\|}, \|\cdot\|, \epsilon) \geq M(B_{\|\cdot\|'}, \|\cdot\|', \epsilon).$$

Then the left inequality of (27.6) follows from those of (27.4) and (27.5). □

Remark 27.3. Several remarks on Corollary 27.4 are in order:

(a) Using (27.5), we see that for any compact Θ with non-empty interior, we have

$$N(\Theta, \|\cdot\|, \epsilon) \asymp M(\Theta, \|\cdot\|, \epsilon) \asymp \frac{1}{\epsilon^d} \tag{27.7}$$

for small ϵ, with proportionality constants depending on both Θ and the norm. In fact, the sharp constant is also known to exist. It is shown in [251, Theorem IX] that there exists a constant τ depending only on $\|\cdot\|$ and the dimension, such that

$$M(\Theta, \|\cdot\|, 2\epsilon) = (\tau + o(1))\frac{\text{vol}(\Theta)}{\text{vol}(B)}\frac{1}{\epsilon^d}$$

holds for any Θ with positive volume. This constant τ is the *maximal sphere-packing density* in $\mathbb{R}^d$ (the proportion of the whole space covered by the balls in the packing – see [366, Chapter 1] for a formal definition); a similar result and interpretation hold for the covering number as well. Computing or bounding the value of τ is extremely difficult and remains open except for some special cases.[4] For more on this subject see the monographs [366, 99].

(b) The result (27.6) for spheres suggests that one may expect the metric entropy for a smooth manifold Θ to behave as $\left(\frac{1}{\epsilon}\right)^{\dim}$, where dim stands for the dimension of Θ as opposed to the ambient dimension. This is indeed true in many situations, for example, in the context of matrices, for the orthogonal group $O(d)$, unitary group $U(d)$, and Grassmanian manifolds [407, 408], in which case dim corresponds to the "degrees of freedom" (for example, $\dim = d(d-1)/2$ for $O(d)$). More generally, for an arbitrary set Θ, one may define the limit $\lim_{\epsilon \to 0} \frac{\log N(\Theta, \|\cdot\|, \epsilon)}{\log(1/\epsilon)}$ as its dimension (known as the *Minkowski dimension* or *box-counting dimension*). For sets of a fractal nature, this dimension can be a non-integer (e.g. $\log_2 3$ for the Cantor set).

(c) Since all norms on Euclidean space are equivalent (within multiplicative constant factors depending on dimension), the small-ϵ behavior in (27.7) holds for any norm as long as the dimension d is fixed. However, this result does not capture the full picture in *high dimensions* when ϵ is allowed to depend on d. Understanding these high-dimensional phenomena requires us to go beyond volume methods. See Section 27.3 for details.

[4] For example, it is easy to show that $\tau = 1$ for both ℓ_∞- and ℓ_1-balls in any dimension since cubes can be subdivided into smaller cubes; for the ℓ_2-ball in $d = 2$, $\tau = \frac{\pi}{\sqrt{12}}$ is the famous result of L. Fejes Tóth on the optimality of the hexagonal arrangement for circle packing [366].

Next we switch our attention to the discrete case of Hamming space. The following theorem bounds its packing number $M(\mathbb{F}_2^d, d_H, r) \equiv M(\mathbb{F}_2^d, r)$, namely, the maximal number of binary codewords of length d with a prescribed minimum distance $r+1$.[5] This is a central question in coding theory, wherein the lower and upper bounds below are known as the *Gilbert–Varshamov bound* and the *Hamming bound*, respectively.

Theorem 27.5. *For any integer* $1 \le r \le d-1$,

$$\frac{2^d}{\sum_{i=0}^{r}\binom{d}{i}} \le M(\mathbb{F}_2^d, r) \le \frac{2^d}{\sum_{i=0}^{\lfloor r/2 \rfloor}\binom{d}{i}}. \tag{27.8}$$

Proof. Both inequalities in (27.8) follow from the same argument as that in Theorem 27.3, with $\mathbb{R}^d$ replaced by $\mathbb{F}_2^d$ and volume by the counting measure (which is translation-invariant). □

Of particular interest to coding theory is the asymptotic regime of $d \to \infty$ and $r = \rho d$ for some constant $\rho \in (0,1)$. Using the asymptotics of the binomial coefficients (see Proposition 1.5), the Hamming and Gilbert–Varshamov bounds translate to

$$2^{d(1-h(\rho))+o(d)} \le M(\mathbb{F}_2^d, \rho d) \le 2^{d(1-h(\rho/2))+o(d)}.$$

Finding the exact exponent is one of the most significant open questions in coding theory. The best upper bound to date is due to McEliece, Rodemich, Rumsey and Welch [300] using the technique of linear programming relaxation.

In contrast, the corresponding covering problem in Hamming space is much simpler, as we have the following tight result:

$$N(\mathbb{F}_2^d, \rho d) = 2^{dR(\rho)+o(d)}, \tag{27.9}$$

where $R(\rho) = (1-h(\rho))_+$ is the rate-distortion function of Ber(1/2) from Theorem 26.1. Although this does not automatically follow from the rate-distortion theory, it can be shown using a similar argument – see Exercise **V.26**.

Finally, we state a lower bound on the packing number of Hamming spheres, which is needed for subsequent application in sparse estimation (Exercise **VI.12**) and useful as a basic building block for computing metric entropy in more complicated settings (Theorem 27.7).

Theorem 27.6. (Gilbert–Varshamov bound for Hamming spheres) *Denote by*

$$S_k^d = \{x \in \mathbb{F}_2^d : w_H(x) = k\} \tag{27.10}$$

the Hamming sphere of radius $0 \le k \le d$. *Then*

$$M(S_k^d, r) \ge \frac{\binom{d}{k}}{\sum_{i=0}^{r}\binom{d}{i}}. \tag{27.11}$$

[5] Recall that the packing number in Definition 27.1 is defined with a strict inequality.

In particular,

$$\log M(S_k^d, k/2) \geq \frac{k}{2} \log \frac{d}{2ek}. \tag{27.12}$$

Proof. Again (27.11) follows from the volume argument. To verify (27.12), note that for $r \leq d/2$, we have $\sum_{i=0}^{r} \binom{d}{i} \leq \exp(dh(\frac{r}{d}))$ (see Theorem 8.2 or (15.19) with $p = 1/2$). Using $h(x) \leq x \log \frac{e}{x}$ and $\binom{d}{k} \geq (\frac{d}{k})^k$, we conclude (27.12) from (27.11). □

27.3 Beyond the Volume Bound

The volume bound in Theorem 27.3 provides a useful tool for studying metric entropy in Euclidean spaces. As a result, as $\epsilon \to 0$, the covering number of any set with non-empty interior always grows exponentially in d as $\left(\frac{1}{\epsilon}\right)^d$ – see (27.7). This asymptotic result, however, has its limitations and does not apply if the dimension d is large and ϵ scales with d. In fact, one expects that there is some critical threshold of ϵ depending on the dimension d, below which the exponential asymptotics is tight, and above which the covering number can grow polynomially in d. This high-dimensional phenomenon is not fully captured by the volume method.

As a case in point, consider the maximum number of ℓ_2-balls of radius ϵ packed into the unit ℓ_1-ball, namely, $M(B_1, \|\cdot\|_2, \epsilon)$. (Recall that B_p denotes the unit ℓ_p-ball in $\mathbb{R}^d$ with $1 \leq p \leq \infty$.) We have studied the metric entropy of arbitrary norm balls under the same norm in Corollary 27.4, where the specific value of the volume was canceled from the volume ratio. Here, although ℓ_1- and ℓ_2-norms are equivalent in the sense that $\|x\|_2 \leq \|x\|_1 \leq \sqrt{d}\|x\|_2$, this relationship is too loose when d is large.

Let us start by applying the volume method in Theorem 27.3:

$$\frac{\mathrm{vol}(B_1)}{\mathrm{vol}(\epsilon B_2)} \leq N(B_1, \|\cdot\|_2, \epsilon) \leq M(B_1, \|\cdot\|_2, \epsilon) \leq \frac{\mathrm{vol}(B_1 + \frac{\epsilon}{2} B_2)}{\mathrm{vol}(\frac{\epsilon}{2} B_2)}.$$

Applying the formula for the volume of a unit ℓ_q-ball in $\mathbb{R}^d$:

$$\mathrm{vol}(B_q) = \frac{\left[2\Gamma\left(1 + \frac{1}{q}\right)\right]^d}{\Gamma\left(1 + \frac{d}{q}\right)}, \tag{27.13}$$

we get[6] $\mathrm{vol}(B_1) = 2^d/d!$ and $\mathrm{vol}(B_2) = \frac{\pi^d}{\Gamma(1+d/2)}$, which yield, by Stirling's approximation,

$$\mathrm{vol}(B_1)^{1/d} \asymp \frac{1}{d}, \quad \mathrm{vol}(B_2)^{1/d} \asymp \frac{1}{\sqrt{d}}. \tag{27.14}$$

[6] For B_1 this can be proved directly by noting that B_1 consists of 2^d disjoint "copies" of the simplex whose volume is $1/d!$ by induction on d.

Then for some absolute constant C,

$$M(B_1, \|\cdot\|_2, \epsilon) \le \frac{\mathrm{vol}(B_1 + \frac{\epsilon}{2}B_2)}{\mathrm{vol}(\frac{\epsilon}{2}B_2)} \le \frac{\mathrm{vol}((1+\frac{\epsilon\sqrt{d}}{2})B_1)}{\mathrm{vol}(\frac{\epsilon}{2}B_2)} \le \left(C\left(1+\frac{1}{\epsilon\sqrt{d}}\right)\right)^d, \tag{27.15}$$

where the second inequality follows from $B_2 \subset \sqrt{d}B_1$ by the Cauchy–Schwarz inequality. (This step is tight in the sense that $\mathrm{vol}(B_1 + \frac{\epsilon}{2}B_2)^{1/d} \gtrsim \max\{\mathrm{vol}(B_1)^{1/d}, \frac{\epsilon}{2}\mathrm{vol}(B_2)^{1/d}\} \asymp \max\{\frac{1}{d}, \frac{\epsilon}{\sqrt{d}}\}$.) On the other hand, for some absolute constant c,

$$M(B_1, \|\cdot\|_2, \epsilon) \ge \frac{\mathrm{vol}(B_1)}{\mathrm{vol}(\epsilon B_2)} = \left(\frac{1}{\epsilon}\right)^d \frac{\mathrm{vol}(B_1)}{\mathrm{vol}(B_2)} = \left(\frac{c}{\epsilon\sqrt{d}}\right)^d. \tag{27.16}$$

Overall, for $\epsilon \le \frac{1}{\sqrt{d}}$, we have $M(B_1, \|\cdot\|_2, \epsilon)^{1/d} \asymp \frac{1}{\epsilon\sqrt{d}}$; however, the lower bound trivializes and the upper bound (which is exponential in d) is loose in the regime of $\epsilon \gg \frac{1}{\sqrt{d}}$, which requires different methods than volume calculation. The following result describes the complete behavior of this metric entropy. In view of Theorem 27.2, in the proof we will go back and forth between the covering and packing numbers in the argument.

Theorem 27.7. *For $0 < \epsilon < 1$ and $d \in \mathbb{N}$,*

$$\log M(B_1, \|\cdot\|_2, \epsilon) \asymp \begin{cases} d \log \frac{e}{\epsilon^2 d}, & \epsilon \le \frac{1}{\sqrt{d}}, \\ \frac{1}{\epsilon^2}\log(e\epsilon^2 d), & \epsilon \ge \frac{1}{\sqrt{d}}. \end{cases}$$

Proof. The case of $\epsilon \le \frac{c}{\sqrt{d}}$ follows from earlier volume calculation (27.15) and (27.16). Next we focus on $\frac{1}{\sqrt{d}} \le \epsilon < 1$.

For the upper bound, we construct an ϵ-covering in ℓ_2 by quantizing each coordinate. Without loss of generality, assume that $\epsilon < 1/4$. Fix some $\delta < 1$. For each $\theta \in B_1$, there exists $x \in (\delta\mathbb{Z}^d) \cap B_1$ such that $\|x-\theta\|_\infty \le \delta$. Then $\|x-\theta\|_2^2 \le \|x-\theta\|_1\|x-\theta\|_\infty \le 2\delta$. Furthermore, x/δ belongs to the set

$$\mathcal{Z} = \left\{z \in \mathbb{Z}^d : \sum_{i=1}^d |z_i| \le k\right\} \tag{27.17}$$

with $k = \lfloor 1/\delta \rfloor$. Note that each $z \in \mathcal{Z}$ has at most k non-zeros. By enumerating the number of non-negative solutions (stars and bars calculation) and the sign pattern, we have[7] $|\mathcal{Z}| \le 2^{k\wedge d}\binom{d-1+k}{k}$. Finally, picking $\delta = \epsilon^2/2$, we conclude that $N(B_1, \|\cdot\|_2, \epsilon) \le |\mathcal{Z}| \le (2e(d+k)/k)^k$ as desired. (Note that this method also recovers the volume bound for $\epsilon \lesssim \frac{1}{\sqrt{d}}$, in which case $k \gtrsim d$.)

For the lower bound, note that $M(B_1, \|\cdot\|_2, \sqrt{2}) \ge 2d$ by considering $\pm e_1, \dots, \pm e_d$. So it suffices to consider $d \ge 8$. We construct a packing of B_1 based on a packing of the Hamming sphere. Without loss of generality, assume that

[7] By enumerating the support and counting positive solutions, it is easy to show that $|\mathcal{Z}| = \sum_{i=0}^d 2^{d-i}\binom{d}{i}\binom{k}{d-i}$.

$\epsilon > \frac{1}{4\sqrt{d}}$. Fix some $1 \le k \le d$. Applying the Gilbert–Varshamov bound in Theorem 27.6, in particular (27.12), there exists a $k/2$-packing $\{x_1, \dots, x_M\} \subset S_k^d = \{x \in \{0,1\}^d : \sum_{i=1}^d x_i = k\}$ and $\log M \ge \frac{k}{2} \log \frac{d}{2ek}$. Scale the Hamming sphere to fit the ℓ_1-ball by setting $\theta_i = x_i/k$. Then $\theta_i \in B_1$ and $\|\theta_i - \theta_j\|_2^2 = \frac{1}{k^2} d_H(x_i, x_j) \ge \frac{1}{2k}$ for all $i \ne j$. Choosing $k = \left\lfloor \frac{1}{\epsilon^2} \right\rfloor$ which satisfies $k \le d/8$, we conclude that $\{\theta_1, \dots, \theta_M\}$ is an $\frac{\epsilon}{2}$-packing of B_1 in $\|\cdot\|_2$ as desired. □

The above elementary proof can be adapted to give the following more general result (see Exercise **V.27**): Let $1 \le p < q \le \infty$. For all $0 < \epsilon < 1$ and $d \in \mathbb{N}$,

$$\log M(B_p, \|\cdot\|_q, \epsilon) \asymp_{p,q} \begin{cases} d \log \frac{e}{\epsilon^s d}, & \epsilon \le d^{-1/s}, \\ \frac{1}{\epsilon^s} \log(e\epsilon^s d), & \epsilon \ge d^{-1/s}, \end{cases} \quad \frac{1}{s} \triangleq \frac{1}{p} - \frac{1}{q}. \tag{27.18}$$

In the remainder of this section, we discuss a few generic results in connection to Theorem 27.7, in particular, metric entropy upper bounds via the *Sudakov minorization* and *Maurey's empirical method,* as well as the duality of metric entropy in Euclidean spaces.

27.3.1 Sudakov's Minoration

Theorem 27.8. (Sudakov's minoration) *Define the* Gaussian width *of* $\Theta \subset \mathbb{R}^d$ *as*[8]

$$w(\Theta) \triangleq \mathbb{E} \sup_{\theta \in \Theta} \langle \theta, Z \rangle, \quad Z \sim \mathcal{N}(0, I_d). \tag{27.19}$$

Then, for $\Theta \subset \mathbb{R}^d$,

$$w(\Theta) \gtrsim \sup_{\epsilon > 0} \epsilon \sqrt{\log M(\Theta, \|\cdot\|_2, \epsilon)}. \tag{27.20}$$

As a quick corollary, applying the volume lower bound on the packing number in Theorem 27.3 to (27.20) and optimizing over ϵ, we obtain *Urysohn's inequality:*[9]

$$w(\Theta) \gtrsim \sqrt{d} \left(\frac{\mathrm{vol}(\Theta)}{\mathrm{vol}(B_2)} \right)^{1/d} \overset{(27.14)}{\asymp} d\, \mathrm{vol}(\Theta)^{1/d}. \tag{27.21}$$

Sudakov's theorem relates the Gaussian width to the metric entropy, both of which are meaningful measures of the massiveness of a set. The important point is that the proportionality constant in (27.20) is independent of the dimension. It turns out that Sudakov's lower bound is tight up to a $\log(d)$ factor [439, Theorem 8.1.13]. The following complementary result is known as Dudley's chaining inequality (see Exercise **V.28** for a proof)

$$w(\Theta) \lesssim \int_0^\infty \sqrt{\log M(\Theta, \|\cdot\|_2, \epsilon)} d\epsilon. \tag{27.22}$$

[8] To avoid measurability difficulty, $w(\Theta)$ should be understood as $\sup_{T \subset \Theta, |T| < \infty} \mathbb{E} \max_{\theta \in T} \langle \theta, Z \rangle$.

[9] For a sharp form, see [330, Corollary 1.4], which states that for all symmetric convex Θ, $w(\Theta) \ge \mathbb{E}[\|Z\|_2] (\frac{\mathrm{vol}(\Theta)}{\mathrm{vol}(B_2)})^{1/d}$; in other words, balls minimize the Gaussian width among all symmetric convex bodies of the same volume.

Understanding the maximum of Gaussian processes is a field on its own; see the monograph [412]. In this section we focus on the lower bound (27.20) in order to develop an upper bound for metric entropy using the Gaussian width.

The proof of Theorem 27.8 relies on the following Gaussian comparison lemma of Slepian (whom we have encountered earlier in Theorem 11.13). For a self-contained proof see [89]. See also [330, Lemma 5.7, p. 70] for a simpler proof of a weaker version $\mathbb{E}\max X_i \le 2\mathbb{E}\max Y_i$, which suffices for our purposes.

Lemma 27.9. (Slepian's lemma) *Let $X = (X_1, \dots, X_n)$ and $Y = (Y_1, \dots, Y_n)$ be Gaussian random vectors. If $\mathbb{E}(Y_i - Y_j)^2 \le \mathbb{E}(X_i - X_j)^2$ for all i,j, then $\mathbb{E}\max Y_i \le \mathbb{E}\max X_i$.*

We also need the result bounding the expectation of the maximum of n Gaussian random variables.

Lemma 27.10. *Let $Z_1, \dots, Z_n$ be distributed as $\mathcal{N}(0,1)$. Then*

$$\mathbb{E}\Big[\max_{i\in[n]} Z_i\Big] \le \sqrt{2\log n}. \tag{27.23}$$

In addition, if $Z_1, \dots, Z_n \overset{\text{iid}}{\sim} \mathcal{N}(0,1)$, then

$$\mathbb{E}\Big[\max_{i\in[n]} Z_i\Big] = \sqrt{2\log n}(1 + o(1)). \tag{27.24}$$

Proof. First, let $T = \operatorname{argmax}_j Z_j$. Since Z_j are 1-sub-Gaussian (recall Definition 4.15), from Exercise **I.56** we have

$$\left|\mathbb{E}\left[\max_i Z_i\right]\right| = |\mathbb{E}[Z_T]| \le \sqrt{2I(Z^n;T)} = \sqrt{2H(T)} \le \sqrt{2\log n}.$$

Next, assume that Z_i's are iid. For any $t > 0$,

$$\begin{aligned}\mathbb{E}\left[\max_i Z_i\right] &\ge t\,\mathbb{P}\left[\max_i Z_i \ge t\right] + \mathbb{E}\left[\max_i Z_i 1\{Z_1 < 0\}1\{Z_2 < 0\}\cdots 1\{Z_n < 0\}\right]\\ &\ge t(1 - (1 - \Phi^c(t))^n) + \mathbb{E}[Z_1 1\{Z_1 < 0\}1\{Z_2 < 0\}\cdots 1\{Z_n < 0\}],\end{aligned}$$

where $\Phi^c(t) = \mathbb{P}[Z_1 \ge t]$ is the normal tail probability. The second term equals $2^{-(n-1)}\mathbb{E}[Z_1 1\{Z_1 < 0\}] = o(1)$. For the first term, recall that $\Phi^c(t) \ge \frac{t}{1+t^2}\varphi(t)$ (Exercise **V.25**). Choose $t = \sqrt{(2-\epsilon)\log n}$ for small $\epsilon > 0$ so that $\Phi^c(t) = \omega(\frac{1}{n})$ and hence $\mathbb{E}[\max_i Z_i] \ge \sqrt{(2-\epsilon)\log n}(1 + o(1))$. By the arbitrariness of $\epsilon > 0$, the lower bound part of (27.24) follows. □

Proof of Theorem 27.8. Let $\{\theta_1, \dots, \theta_M\}$ be an optimal ϵ-packing of Θ. Let $X_i = \langle \theta_i, Z\rangle$ for $i \in [M]$, where $Z \sim \mathcal{N}(0, I_d)$. Let $Y_i \overset{\text{iid}}{\sim} \mathcal{N}(0, \epsilon^2/2)$. Then

$$\mathbb{E}(X_i - X_j)^2 = (\theta_i - \theta_j)^\top \mathbb{E}[ZZ^\top](\theta_i - \theta_j) = \|\theta_i - \theta_j\|_2^2 \ge \epsilon^2 = \mathbb{E}(Y_i - Y_j)^2.$$

Then

$$\mathbb{E}\sup_{\theta\in\Theta}\langle\theta, Z\rangle \ge \mathbb{E}\max_{1\le i\le M} X_i \ge \mathbb{E}\max_{1\le i\le M} Y_i \asymp \epsilon\sqrt{\log M},$$

where the second and third steps follow from Lemma 27.9 and Lemma 27.10 respectively. □

Revisiting the packing number of the ℓ_1-ball, we apply Sudakov minorization to $\Theta = B_1$. By duality and applying Lemma 27.10,

$$w(B_1) = \mathbb{E} \sup_{x:\ \|x\|_1 \le 1} \langle x, Z \rangle = \mathbb{E}\|Z\|_\infty \le \sqrt{2 \log d}.$$

Then Theorem 27.8 gives

$$\log M(B_1, \|\cdot\|_2, \epsilon) \lesssim \frac{\log d}{\epsilon^2}. \tag{27.25}$$

When $\epsilon \gtrsim 1/\sqrt{d}$, this is much tighter than the volume bound (27.15) and almost optimal (compared to $\frac{\log(d\epsilon^2)}{\epsilon^2}$); however, when $\epsilon \asymp 1/\sqrt{d}$, (27.25) yields $d \log d$ but we know (even from the volume bound) that the correct behavior is d. In Section 27.3.3 we discuss another general approach that gives the optimal bound in this case.

27.3.2 Hilbert Ball Has Metric Entropy $\frac{1}{\epsilon^2}$

We consider a Hilbert ball $B_2 = \{x \in \mathbb{R}^\infty : \sum_i x_i^2 \le 1\}$. Under the usual metric $\ell_2(\mathbb{R}^\infty)$ this set is not compact and cannot have finite ϵ-nets for all ϵ. However, the metric of interest in many applications is often different. Specifically, let us fix some probability distribution P on B_2 s.t. $\mathbb{E}_{X\sim P}[\|X\|_2^2] \le 1$ and define

$$d_P(\theta, \theta') \triangleq \sqrt{\mathbb{E}_{X\sim P}[|\langle \theta - \theta', X \rangle|^2]}$$

for $\theta, \theta' \in B_2$. The importance of this metric is that it allows one to analyze the complexity of a class of linear functions $\theta \mapsto \langle \theta, X \rangle$ for any random variable X of unit norm and has applications in learning theory [303, 472].

Theorem 27.11. *For some universal constant c we have for all $\epsilon > 0$*

$$\log N(B_2, d_P, \epsilon) \le \frac{c}{\epsilon^2}.$$

Proof. First, we show that without loss of generality we may assume that X has all coordinates, other than the first n, zero. Indeed, take n so large that $\mathbb{E}[\sum_{j>n} X_j^2] \le \frac{\epsilon^2}{4}$. Let us denote by $\tilde{\theta}$ the vector obtained from θ by zeroing out all coordinates $j > n$. Then from Cauchy–Schwarz we see that $d_P(\theta, \tilde{\theta}) \le \frac{\epsilon}{2}$ and therefore any $\frac{\epsilon}{2}$-covering of $\tilde{B}_2 = \{\tilde{\theta} : \theta \in B_2\}$ will be an ϵ-covering of B_2. Hence, from now on we assume that the ball B_2 is in fact finite-dimensional.

We can redefine distance d_P in a more explicit way as follows:

$$d_P(\theta, \theta')^2 = (\theta - \theta')^\top \Sigma (\theta - \theta'),$$

where $\Sigma = \mathbb{E}[XX^\top]$ is a positive semidefinite matrix of second moments of $X \sim P$. Let us set $D = \sqrt{\Sigma}$ to be the symmetric square-root of Σ. To each θ let us associate $v(\theta) = D\theta$ and let $V = D(B_2)$ to be the image of B_2 under D. Note that

$d_P(\theta, \theta') = \|v(\theta) - v(\theta')\|_2$. Therefore, from the Sudakov minoration theorem (Theorem 27.8) we obtain

$$\log M(V, \|\cdot\|_2, \epsilon) \le \frac{c}{\epsilon^2} \mathbb{E}\left[\sup_{v\in V} \langle v, Z\rangle\right], \qquad Z \sim \mathcal{N}(0, I_d).$$

Since V is an ellipsoid, we can compute the supremum explicitly, indeed

$$\mathbb{E}\left[\sup_{v\in V} \langle v, Z\rangle\right] = \mathbb{E}\left[\sup_{\theta\in B_2} \langle D\theta, Z\rangle\right] = \mathbb{E}[\|DZ\|_2] \le \sqrt{\mathbb{E}[\|DZ\|_2^2]} = \sqrt{\operatorname{tr}\Sigma} \le 1. \quad \square$$

To see one simple implication of the result, recall the standard bound on empirical processes (see e.g. [430, Lemma 5.1]): By endowing any collection of functions $\{f_\theta : \theta \in \Theta\}$ with a metric $d_{\hat{P}_n}(\theta, \theta')^2 = \mathbb{E}_{\hat{P}_n}[(f_\theta(X) - f_{\theta'}(X))^2]$ we have

$$\mathbb{E}\left[\sup_\theta \mathbb{E}[f_\theta(X)] - \hat{\mathbb{E}}_n[f_\theta(X)]\right] \lesssim \mathbb{E}\left[\inf_{\delta>0} \delta + \int_\delta^\infty \sqrt{\frac{\log N(\Theta, d_{\hat{P}_n}, \epsilon)}{n}} d\epsilon\right].$$

It can be seen that when entropy behaves as ϵ^{-p} we get rate $n^{-\min(1/p, 1/2)}$ except for $p = 2$ for which the upper bound yields $n^{-1/2}\log n$. The significance of the previous theorem is that the Hilbert ball is precisely "at the phase transition" from parametric to nonparametric rate.

As a sanity check, let us take any P_X over the unit (possibly infinite-dimensional) ball B with $\mathbb{E}[X] = 0$ and let $\Theta = B$. We have

$$\mathbb{E}[\|\bar{X}_n\|] = \mathbb{E}\left[\sup_\theta \frac{1}{n}\sum_{i=1}^n \langle \theta, X_i\rangle\right] \lesssim \sqrt{\frac{\log n}{n}},$$

where $\bar{X}_n = \frac{1}{n}\sum_{i=1}^n X_i$ is the empirical mean. In this special case it is easy to bound $\mathbb{E}[\|\bar{X}_n\|] \le \sqrt{\mathbb{E}[\|\bar{X}_n\|^2]} \le \frac{1}{\sqrt{n}}$ by an explicit calculation.

27.3.3 Maurey's Empirical Method

In this sub section we discuss a powerful probabilistic method due to B. Maurey for constructing a good covering. It has found applications in approximation theory and especially that for neural nets [237, 37]. The following result gives a dimension-free bound on the cover number of convex hulls in Hilbert spaces:

Theorem 27.12. *Let H be an inner product space with the norm $\|x\| \triangleq \sqrt{\langle x, x\rangle}$. Let $T \subset H$ be a finite set, with radius $r = \operatorname{rad}(T) = \inf_{y\in H}\sup_{x\in T}\|x - y\|$ (recall (5.3)). Denote the convex hull of T by $\operatorname{co}(T)$. Then for any $0 < \epsilon \le r$,*

$$N(\operatorname{co}(T), \|\cdot\|, \epsilon) \le \binom{|T| + \lceil r^2/\epsilon^2\rceil - 2}{\lceil r^2/\epsilon^2\rceil - 1}. \tag{27.26}$$

Proof. Let $T = \{t_1, t_2, \ldots, t_m\}$ and denote the Chebyshev center of T by $c \in H$, such that $r = \max_{i\in[m]}\|c - t_i\|$. For $n \in \mathbb{Z}_+$, let

$$\mathcal{Z} = \left\{\frac{1}{n+1}\left(c + \sum_{i=1}^m n_i t_i\right) : n_i \in \mathbb{Z}_+, \sum_{i=1}^m n_i = n\right\}.$$

For any $x = \sum_{i=1}^m x_i t_i \in \text{co}(T)$ where $x_i \geq 0$ and $\sum x_i = 1$, let Z be a discrete random variable such that $Z = t_i$ with probability x_i. Then $\mathbb{E}[Z] = x$. Let $Z_0 = c$ and $Z_1, \ldots, Z_n$ be iid copies of Z. Let $\bar{Z} = \frac{1}{n+1}\sum_{i=0}^n Z_i$, which takes values in the set $\mathcal{Z}$. Since

$$
\begin{aligned}
\mathbb{E}\|\bar{Z} - x\|_2^2 &= \frac{1}{(n+1)^2}\mathbb{E}\left\|\sum_{i=0}^n (Z_i - x)\right\|^2 \\
&= \frac{1}{(n+1)^2}\left(\sum_{i=0}^n \mathbb{E}\,\|Z_i - x\|^2 + \sum_{i\neq j}\mathbb{E}\langle Z_i - x, Z_j - x\rangle\right) \\
&= \frac{1}{(n+1)^2}\sum_{i=0}^n \mathbb{E}\,\|Z_i - x\|^2 = \frac{1}{(n+1)^2}\left(\|c - x\|^2 + n\mathbb{E}[\|Z - x\|^2]\right) \\
&\leq \frac{r^2}{n+1},
\end{aligned}
$$

where the last inequality follows from that $\|c - x\| \leq \sum_{i=1}^m x_i\|c - t_i\| \leq r$ (in other words, $\text{rad}(T) = \text{rad}(\text{co}(T))$ and $\mathbb{E}[\|Z - x\|^2] \leq \mathbb{E}[\|Z - c\|^2] \leq r^2$. Set $n = \lceil r^2/\epsilon^2\rceil - 1$ so that $r^2/(n+1) \leq \epsilon^2$. There exists some $z \in N$ such that $\|z - x\| \leq \epsilon$. Therefore $\mathcal{Z}$ is an ϵ-covering of $\text{co}(T)$. Similar to (27.17), we have

$$
|\mathcal{Z}| \leq \binom{n + m - 1}{n} = \binom{m + \lceil r^2/\epsilon^2\rceil - 2}{\lceil r^2/\epsilon^2\rceil - 1}. \qquad \square
$$

We now apply Theorem 27.12 to recover the result for the unit ℓ_1-ball B_1 in $\mathbb{R}^d$ in Theorem 27.7: Note that $B_1 = \text{co}(T)$, where $T = \{\pm e_1, \ldots, \pm e_d, 0\}$ satisfies $\text{rad}(T) = 1$. Then

$$
N(B_1, \|\cdot\|_2, \epsilon) \leq \binom{2d + \lceil 1/\epsilon^2\rceil - 1}{\lceil 1/\epsilon^2\rceil - 1}, \tag{27.27}
$$

which recovers the optimal upper bound in Theorem 27.7 at both small and big scales.

27.3.4 Duality of Metric Entropy

First we define a more general notion of covering number. For $K, T \subset \mathbb{R}^d$, define the covering number of K using translates of T as

$$
N(K, T) = \min\left\{N \colon \exists x_1, \ldots, x_N \in \mathbb{R}^d \text{ such that } K \subset \bigcup_{i=1}^N T + x_i\right\}.
$$

Then the usual covering number in Definition 27.1 satisfies $N(K, \|\cdot\|, \epsilon) = N(K, \epsilon B)$, where B is the corresponding unit-norm ball.

A deep result of Artstein, Milman, and Szarek [28] establishes the following duality for metric entropy: There exist absolute constants α and β such that for any symmetric convex body K,[10]

$$\frac{1}{\beta}\log N\left(B_2, \frac{\epsilon}{\alpha}K^\circ\right) \le \log N(K, \epsilon B_2) \le \log N(B_2, \alpha\epsilon K^\circ), \tag{27.28}$$

where B_2 is the usual unit ℓ_2-ball, and $K^\circ = \{y\colon \sup_{x\in K}\langle x, y\rangle \le 1\}$ is the polar body of K.

As an example, consider $p < 2 < q$ and $\frac{1}{p} + \frac{1}{q} = 1$. By duality, $B_p^\circ = B_q$. Then (27.28) shows that $N(B_p, \|\cdot\|_2, \epsilon)$ and $N(B_2, \|\cdot\|_q, \epsilon)$ have essentially the same behavior, as verified by (27.18).

27.4 Infinite-Dimensional Space: Smooth Functions

Unlike Euclidean spaces, in infinite-dimensional spaces, the metric entropy can grow arbitrarily fast [251, Theorem XI]. Studying metric entropy in function space (for example, under shape or smoothness constraints) is an area of interest in functional analysis (see [442]), and has important applications in nonparametric statistics, empirical processes, and learning theory [139]. To gain some insight into the fundamental distinction between finite- and infinite-dimensional spaces, let us work out a concrete example, which will later be used in the application of density estimation in Section 32.4. For more general and more precise results (including some cases of equality), see [251, Sections 4 and 7]. Consider the class $\mathcal{F}(A, L)$ of all L-Lipschitz probability densities on the compact interval $[0, A]$.

Theorem 27.13. *Assume that $L, A > 0$ and $p \in [1, \infty]$ are constants. Then*

$$\log N(\mathcal{F}(A, L), \|\cdot\|_p, \epsilon) = \Theta\left(\frac{1}{\epsilon}\right). \tag{27.29}$$

Furthermore, for the sup-norm we have the sharp asymptotics:

$$\log_2 N(\mathcal{F}(A, L), \|\cdot\|_\infty, \epsilon) = \frac{LA}{\epsilon}(1 + o(1)), \quad \epsilon \to 0. \tag{27.30}$$

Proof. By replacing $f(x)$ by $\frac{1}{\sqrt{L}} f(\frac{x}{\sqrt{L}})$, we have

$$N(\mathcal{F}(A, L), \|\cdot\|_p, \epsilon) = N(\mathcal{F}(\sqrt{L}A, 1), \|\cdot\|_p, L^{\frac{1-p}{2p}}\epsilon). \tag{27.31}$$

Thus, it is sufficient to consider $\mathcal{F}(A, 1) \triangleq \mathcal{F}(A)$, the collection of 1-Lipschitz densities on $[0, A]$. Next, observe that any such density function f is bounded from above. Indeed, since $f(x) \ge (f(0) - x)_+$ and $\int_0^A f = 1$, we conclude that $f(0) \le \max\{A, \frac{A}{2} + \frac{1}{A}\} \triangleq m$.

[10] A convex body K is a compact convex set with non-empty interior. We say K is symmetric if $K = -K$.

To show (27.29), it suffices to prove the upper bound for $p = \infty$ and the lower bound for $p = 1$. Specifically, we aim to show, by explicit construction,

$$N(\mathcal{F}(A), \|\cdot\|_\infty, \epsilon) \le \frac{C}{\epsilon} 2^{\frac{A}{\epsilon}}, \tag{27.32}$$

$$M(\mathcal{F}(A), \|\cdot\|_1, \epsilon) \ge 2^{\frac{c}{\epsilon}}, \tag{27.33}$$

which imply the desired (27.29) in view of Theorem 27.2. Here and below, c, C are constants depending on A. We start with the easier (27.33). We construct a packing by perturbing the uniform density. Define a function T by $T(x) = x1\{x \le \epsilon\} + (2\epsilon - x)1\{x \ge \epsilon\} + \frac{1}{A}$ on $[0, 2\epsilon]$ and zero elsewhere. Let $n = \left\lceil \frac{A}{4\epsilon} \right\rceil$ and $a = 2n\epsilon$. For each $y \in \{0,1\}^n$, define a density f_y on $[0, A]$ such that

$$f_y(x) = \sum_{i=1}^{n} y_i T(x - 2(i-1)\epsilon), \quad x \in [0, a],$$

and linearly extend f_y to $[a, A]$ so that $\int_0^A f_y = 1$; see Figure 27.2. For sufficiently small ϵ, the resulting f_y is 1-Lipschitz since $\int_0^a f_y = \frac{1}{2} + O(\epsilon)$ so that the slope of the linear extension is $O(\epsilon)$.

Thus we conclude that each f_y is a valid member of $\mathcal{F}(A)$. Furthermore, for $y, z \in \{0,1\}^n$, we have $\|f_y - f_z\|_1 = d_H(y,z)\|T\|_1 = \epsilon^2 d_H(y,z)$. Invoking the Gilbert–Varshamov bound in Theorem 27.5, we obtain an $\frac{n}{2}$-packing $\mathcal{Y}$ of the Hamming space $\{0,1\}^n$ with $|\mathcal{Y}| \ge 2^{cn}$ for some absolute constant c. Thus $\{f_y : y \in \mathcal{Y}\}$ constitutes an $\frac{n\epsilon^2}{2}$-packing of $\mathcal{F}(A)$ with respect to the L_1-norm. This is the desired (27.33) since $\frac{n\epsilon^2}{2} = \Theta(\epsilon)$.

To construct a covering, set $J = \left\lceil \frac{m}{\epsilon} \right\rceil$, $n = \left\lceil \frac{A}{\epsilon} \right\rceil$, and $x_k = k\epsilon$ for $k = 0, \ldots, n$. Let $\mathcal{G}$ be the collection of all lattice paths (with grid size ϵ) of n steps starting from the coordinate $(0, j\epsilon)$ for some $j \in \{0, \ldots, J\}$. In other words, each element g of $\mathcal{G}$ is a continuous piecewise linear function on each subinterval $I_k = [x_k, x_{k+1})$ with slope being either $+1$ or -1. Evidently, the number of such paths is at most $(J+1)2^n = O(\frac{1}{\epsilon}2^{A/\epsilon})$. To show that $\mathcal{G}$ is an ϵ-covering, for each $f \in \mathcal{F}(A)$, we show that there exists $g \in \mathcal{G}$ such that $|f(x) - g(x)| \le \epsilon$ for all $x \in [0, A]$. This can be shown by a simple induction. Suppose that there exists g such that $|f(x) - g(x)| \le \epsilon$ for all $x \in [0, x_k]$, which clearly holds for the base case of $k = 0$. We show that g can be extended to I_k so that this holds for $k+1$. Since $|f(x_k) - g(x_k)| \le \epsilon$ and f is

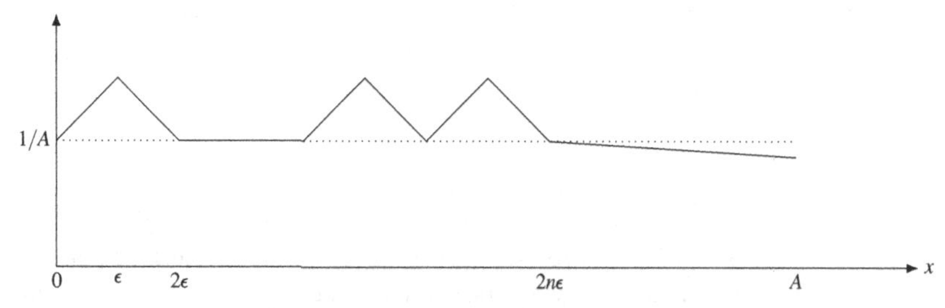

Figure 27.2 Packing that achieves (27.33). The solid line represents one such density $f_y(x)$ with $y = (1, 0, 1, 1)$. The dotted line is the density of Unif$(0, A)$.

1-Lipschitz, either $f(x_{k+1}) \in [g(x_k), g(x_k)+2\epsilon]$ or $f(x_{k+1}) \in [g(x_k)-2\epsilon, g(x_k)]$, in which case we extend g upward or downward, respectively. The resulting g satisfies $|f(x)-g(x)| \le \epsilon$ on I_k, completing the induction.

Finally, we prove the sharp bound (27.30) for $p=\infty$. The upper bound readily follows from (27.32) plus the scaling relation (27.31). For the lower bound, we apply Theorem 27.2 converting the problem to the construction of a 2ϵ-packing. Following the same idea of lattice paths, next we give an improved packing construction such that

$$M(\mathcal{F}(A), \|\cdot\|_\infty, 2\epsilon) = \Omega(\epsilon^{3/2} 2^{\frac{a}{\epsilon}}) \tag{27.34}$$

for any $a < A$. Choose any b such that $\frac{1}{A} < b < \frac{1}{A} + \frac{(A-a)^2}{2A}$. Let $a' = \epsilon \lfloor \frac{a}{\epsilon} \rfloor$ and $b' = \epsilon \left\lfloor \frac{b}{\epsilon} \right\rfloor$. Consider a density f on $[0,A]$ of the following form (see Figure 27.3): on $[0,a']$, f is a lattice path from $(0,b')$ to (a',b') that stays in the vertical range of $[b', b'+\epsilon^{1/3}]$; on $[a',A]$, f is a linear extension chosen so that $\int_0^A f = 1$. This is possible because by the 1-Lipschitz constraint we can linearly extend f so that $\int_{a'}^A f$ takes any value in the interval $[b'(A-a') - \frac{(A-a')^2}{2}, b'(A-a') + \frac{(A-a')^2}{2}]$. Since $\int_0^{a'} f = ab + o(1)$, we need $\int_{a'}^A f = 1 - \int_0^{a'} f = 1 - ab + o(1)$, which is feasible due to the choice of b. The collection $\mathcal{G}$ of all such functions constitutes a 2ϵ-packing in the sup-norm (for two distinct paths consider the first subinterval where they differ). Finally, we bound the cardinality of this packing by counting the number of such paths. This can be accomplished by standard estimates on random walks (see e.g. [164, Chapter III]). For any constant $c > 0$, the probability that a symmetric random walk on $\mathbb{Z}$ returns to zero in n (even) steps and stays in the range of $[0, n^{1+c}]$ is $\Theta(n^{-3/2})$; this implies the desired (27.34). Finally, since $a < A$ is arbitrary, the lower bound part of (27.30) follows in view of Theorem 27.2. □

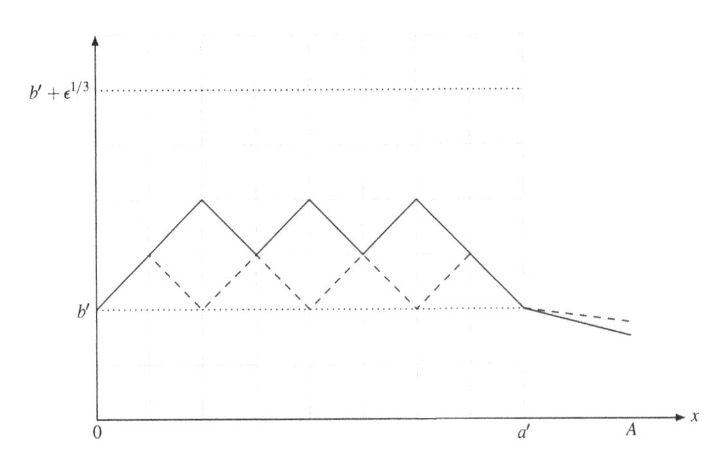

Figure 27.3 Improved packing for (27.34). Here the solid and dashed lines are two lattice paths on a grid of size ϵ starting from $(0,b')$ and staying in the range of $[b', b'+\epsilon^{1/3}]$, followed by their respective linear extensions.

The following result, due to Birman and Solomjak [57] (see [286, Section 15.6] for an exposition), is an extension of Theorem 27.13 to the more general Hölder class.

Theorem 27.14. *Fix positive constants A, L, and $d \in \mathbb{N}$. Let $\beta > 0$ and write $\beta = \ell + \alpha$, where $\ell \in \mathbb{Z}_+$ and $\alpha \in (0, 1]$. Let $\mathcal{F}_\beta(A, L)$ denote the collection of ℓ-times continuously differentiable densities f on $[0, A]^d$ whose ℓth derivative is (L, α)-Hölder continuous, namely, $\|D^{(\ell)} f(x) - D^{(\ell)} f(y)\|_\infty \le L\|x - y\|_\infty^\alpha$ for all $x, y \in [0, A]^d$. Then for any $1 \le p \le \infty$,*

$$\log N(\mathcal{F}_\beta(A, L), \|\cdot\|_p, \epsilon) = \Theta(\epsilon^{-\frac{d}{\beta}}). \tag{27.35}$$

The main message of the preceding theorem is that the entropy of the function class grows more slowly if the dimension decreases or the smoothness increases. As such, the metric entropy for very smooth functions can grow sub-polynomially in $\frac{1}{\epsilon}$. For example, Vitushkin (see [251, Eq. (129)]) showed that for the class of analytic functions on the unit complex disk D having analytic extension to a bigger disk rD for $r > 1$, the metric entropy (with respect to the sup-norm on D) is $\Theta((\log \frac{1}{\epsilon})^2)$; see [251, Sections 7 and 8] for more such results.

As mentioned at the beginning of this chapter, the conception and development of the subject of metric entropy, in particular, Theorem 27.14, are motivated by and play an important role in the study of Hilbert's thirteenth problem. In 1900, Hilbert conjectured that there exist functions of several variables which cannot be represented as a superposition (composition) of finitely many functions of fewer variables. This was disproved by Kolmogorov and Arnold in the 1950s, who showed that every continuous function of d variables can be represented by sums and superpositions of single-variable functions; however, their construction does not work if one requires the constituent functions to have specific smoothness. Subsequently, Hilbert's conjecture for smooth functions was positively resolved by Vitushkin [440], who showed that there exist functions of d variables in the β-Hölder class (in the sense of Theorem 27.14) that cannot be expressed as finitely many superpositions of functions of d' variables in the β'-Hölder class, provided $d/\beta > d'/\beta'$. The original proof of Vitushkin is highly involved. Later, Kolmogorov gave a much simplified proof by proving and applying the $\|\cdot\|_\infty$-version of Theorem 27.14. As evident in (27.35), the index d/β provides a complexity measure for the function class; this allows a proof of impossibility of superposition by an entropy comparison argument. For concreteness, let us prove the following simpler version: There exists a 1-Lipschitz function $f(x, y, z)$ of three variables on $[0, 1]^3$ that cannot be written as $g(h_1(x, y), h_2(y, z))$ where g, h_1, h_2 are 1-Lipschitz functions of two variables on $[0, 1]^2$. Suppose, for the sake of contradiction, that this is possible. Fixing an ϵ-covering of cardinality $\exp(O(\frac{1}{\epsilon^2}))$ for 1-Lipschitz functions on $[0, 1]^2$ and using it to approximate the functions g, h_1, h_2, we obtain by taking their superposition $g(h_1, h_2)$ an $O(\epsilon)$-covering of cardinality $\exp(O(\frac{1}{\epsilon^2}))$ of 1-Lipschitz functions on $[0, 1]^3$; however, this is a contradiction as any such covering must be of size $\exp(\Omega(\frac{1}{\epsilon^3}))$. For stronger and more general results along this line, see [251, Appendix I].

27.5 Metric Entropy and Small-Ball Probability

The small-ball problem in probability theory concerns the behavior of the function

$$\phi(\epsilon) \triangleq \log \frac{1}{\mathbb{P}[\|X\| \le \epsilon]}$$

as $\epsilon \to 0$, where X is a random variable taking values on some real separable Banach space $(V, \|\cdot\|)$. For example, for standard normal $X \sim \mathcal{N}(0, I_d)$ and the ℓ_2-ball, a simple large-deviations calculation (Exercise **III.16**) shows that

$$\phi(\epsilon) \asymp d \log \frac{1}{\epsilon}.$$

Of more interest is the infinite-dimensional case of Gaussian processes. For example, for the standard Brownian motion on the unit interval and the sup-norm, it is elementary to show (Exercise **V.30**) that

$$\phi(\epsilon) \asymp \frac{1}{\epsilon^2}. \tag{27.36}$$

We refer the reader to the excellent survey [280] for this field.

There is a deep connection between the small-ball probability and metric entropy, which allows one to translate results from one area to the other in fruitful ways. To identify this link, the starting point is the volume argument in Theorem 27.3. On the one hand, it is well known that there exists no analog of Lebesgue measure (translation-invariant) in infinite-dimensional spaces. As such, for functional spaces, one frequently uses a Gaussian measure. On the other hand, the "volume" argument in Theorem 27.3 and Remark 27.2 can be adapted to a measure γ that need not be translation-invariant, leading to

$$\frac{\gamma(\Theta + B(0,\epsilon))}{\max_{\theta \in V} \gamma(B(\theta, 2\epsilon))} \le N(\Theta, \|\cdot\|, \epsilon) \le M(\Theta, \|\cdot\|, \epsilon) \le \frac{\gamma(\Theta + B(0,\epsilon/2))}{\min_{\theta \in \Theta} \gamma(B(\theta, \epsilon/2))}, \tag{27.37}$$

where we recall that $B(\theta, \epsilon)$ denotes the norm ball centered at θ of radius ϵ. From here we have already seen the natural appearance of small-ball probabilities. Using properties native to the Gaussian measure, this can be further analyzed and reduced to balls centered at zero.

To be precise, let γ be a zero-mean Gaussian measure on V such that $\mathbb{E}_{X\sim\gamma}[\|X\|^2] < \infty$. Let $H \subset V$ be the reproducing kernel Hilbert space (RKHS) generated by γ and K the unit ball in H. We refer the reader to, for example, [263, Section 2] and [315, III.3.2], for the precise definition of this object.[11] For the purpose of this section, it is enough to consider the following examples (for more, see [263]).

[11] In particular, if γ is the law of a Gaussian process X on $C([0,1])$ with $\mathbb{E}[\|X\|_2^2] < \infty$, the kernel $K(s,t) = \mathbb{E}[X(s)X(t)]$ admits the eigendecomposition $K(s,t) = \sum \lambda_k \psi_k(s)\psi_k(t)$ (Mercer's theorem), where $\{\psi_k\}$ is an orthonormal basis for $L_2([0,1])$ and $\lambda_k > 0$. Then H is the closure of the span of $\{\psi_k\}$ with the inner product $\langle x, y\rangle_H = \sum_k \langle x, \psi_k\rangle\langle y, \psi_k\rangle/\lambda_k$.

- Finite dimensions. Let $\gamma = \mathcal{N}(0, \Sigma)$. Then

$$K = \{\Sigma^{1/2}x \colon \|x\|_2 \leq 1\} \tag{27.38}$$

is a rescaled Euclidean ball, with inner product $\langle x, y\rangle_H = x^\top \Sigma^{-1} y$.
- Brownian motion: Let γ be the law of the standard Brownian motion on the unit interval $[0, 1]$. Then

$$K = \left\{ f(t) = \int_0^t f'(s)ds \colon \|f'\|_2 \leq 1 \right\} \tag{27.39}$$

with inner product $\langle f, g\rangle_H = \langle f', g'\rangle \equiv \int_0^1 f'(t)g'(t)dt$.

The following fundamental result due to Kuelbs and Li [264] (see also the earlier work of Goodman [194]) describes a precise connection between the small-ball probability function $\phi(\epsilon)$ and the metric entropy of the unit Hilbert ball $N(K, \|\cdot\|, \epsilon) \equiv N(\epsilon)$.

Theorem 27.15. *For all* $\epsilon > 0$,

$$\phi(2\epsilon) - \log 2 \leq \log N\left(\frac{\epsilon}{\sqrt{2\phi(\epsilon/2)}}\right) \leq 2\phi(\epsilon/2). \tag{27.40}$$

Proof. We show that for any $\lambda > 0$,

$$\phi(2\epsilon) + \log \Phi(\lambda + \Phi^{-1}(e^{-\phi(\epsilon)})) \leq \log N(\lambda K, \epsilon) \leq \log M(\lambda K, \epsilon) \leq \frac{\lambda^2}{2} + \phi(\epsilon/2). \tag{27.41}$$

To deduce (27.40), choose $\lambda = \sqrt{2\phi(\epsilon/2)}$ and note that by scaling $N(\lambda K, \epsilon) = N(K, \epsilon/\lambda)$. Applying the normal tail bound $\Phi(-t) = \Phi^c(t) \leq e^{-t^2/2}$ (Exercise **V.25**) yields $\Phi^{-1}(e^{-\phi(\epsilon)}) \geq -\sqrt{2\phi(\epsilon)} \geq -\lambda$ so that $\Phi(\Phi^{-1}(e^{-\phi(\epsilon)}) + \lambda) \geq \Phi(0) = 1/2$.

We only give the proof in finite dimensions as the results are dimension-free and extend naturally to infinite-dimensional spaces. Let $Z \sim \gamma = \mathcal{N}(0, \Sigma)$ on $\mathbb{R}^d$ so that $K = \Sigma^{1/2} B_2$ is given in (27.38). Applying (27.37) to λK and noting that γ is a probability measure, we have

$$\frac{\gamma(\lambda K + B(0, \epsilon))}{\max_{\theta \in \mathbb{R}^d} \gamma(B(\theta, 2\epsilon))} \leq N(\lambda K, \epsilon) \leq M(\lambda K, \epsilon) \leq \frac{1}{\min_{\theta \in \lambda K} \gamma(B(\theta, \epsilon/2))}. \tag{27.42}$$

Next we further bound (27.42) using properties native to the Gaussian measure.

- For the upper bound, for any *symmetric* set $A = -A$ and any $\theta \in \lambda K$, by a change of measure

$$\begin{aligned}
\gamma(\theta + A) &= \mathbb{P}[Z - \theta \in A] \\
&= e^{-\frac{1}{2}\theta^\top \Sigma^{-1}\theta}\, \mathbb{E}[e^{\langle \Sigma^{-1}\theta, Z\rangle} 1\{Z \in A\}] \\
&\geq e^{-\lambda^2/2}\mathbb{P}[Z \in A],
\end{aligned}$$

where the last step follows from $\theta^\top \Sigma^{-1}\theta \leq \lambda^2$ and by Jensen's inequality $\mathbb{E}[e^{\langle \Sigma^{-1}\theta, Z\rangle} | Z \in A] \geq e^{\langle \Sigma^{-1}\theta, \mathbb{E}[Z|Z\in A]\rangle} = 1$, using crucially that $\mathbb{E}[Z|Z \in A] = 0$ by symmetry. Applying the above to $A = B(0, \epsilon/2)$ yields the right-hand inequality in (27.41).

- For the lower bound, recall Anderson's lemma (Lemma 28.10) stating that the Gaussian measure of a ball is maximized when centered at zero, so $\gamma(B(\theta, 2\epsilon)) \le \gamma(B(0, 2\epsilon))$ for all θ. To bound the numerator, recall the Gaussian isoperimetric inequality (see e.g. [69, Theorem 10.15]):[12]

$$\gamma(A + \lambda K) \ge \Phi(\Phi^{-1}(\gamma(A)) + \lambda). \tag{27.43}$$

Applying this with $A = B(0, \epsilon)$ proves the left-hand inequality in (27.41) and the theorem. □

The implication of Theorem 27.15 is the following. Provided that $\phi(\epsilon) \asymp \phi(\epsilon/2)$, then we should expect that approximately

$$\log N\left(\frac{\epsilon}{\sqrt{\phi(\epsilon)}}\right) \asymp \phi(\epsilon).$$

With more effort this can be made precise unconditionally (see e.g. [280, Theorem 3.3], incorporating the later improvement by [279]), leading to *very precise* connections between metric entropy and small-ball probability, for example: for fixed $\alpha > 0, \beta \in \mathbb{R}$,

$$\phi(\epsilon) \asymp \epsilon^{-\alpha}\left(\log\frac{1}{\epsilon}\right)^{\beta} \iff \log N(\epsilon) \asymp \epsilon^{-\frac{2\alpha}{2+\alpha}}\left(\log\frac{1}{\epsilon}\right)^{\frac{2\beta}{2+\alpha}}. \tag{27.44}$$

As a concrete example, consider the unit ball (27.39) in the RKHS generated by the standard Brownian motion, which is similar to a Sobolev ball.[13] Using (27.36) and (27.44), we conclude that $\log N(\epsilon) \asymp \frac{1}{\epsilon}$, recovering the metric entropy of a Sobolev ball determined in [421]. This result also coincides with the metric entropy of a Lipschitz ball in Theorem 27.14 which requires the derivative to be bounded everywhere as opposed to on average in L_2. For more applications of small-ball probability on metric entropy (and vice versa), see [264, 279].

27.6 Metric Entropy and Rate-distortion Theory

In this section we discuss a connection between metric entropy and the rate-distortion function. Note that the former is a non-probabilistic quantity whereas the latter is an information measure depending on the source distribution; nevertheless, if we consider the rate-distortion function induced by the "least favorable" source distribution, it turns out to behave similarly to the metric entropy.

[12] The connection between (27.43) and isoperimetry is that if we interpret $\lim_{\lambda\to 0}(\gamma(A + \lambda K) - \gamma(A))/\lambda$ as the surface measure of A, then among all sets with the same Gaussian measure, the half-space has the minimal surface measure.

[13] The Sobolev norm is $\|f\|_{W^{1,2}} \triangleq \|f\|_2 + \|f'\|_2$. Nevertheless, it is simple to verify a priori that the metric entropy of (27.39) and that of the Sobolev ball share the same behavior (see [264, p. 152]).

To make this precise, consider a metric space $(\mathcal{X}, d)$. For an $\mathcal{X}$-valued random variable X, denote by

$$\phi_X(\epsilon) = \inf_{P_{\hat{X}|X}:\, \mathbb{E}[d(X,\hat{X})]\leq \epsilon} I(X;\hat{X}) \tag{27.45}$$

its rate-distortion function (recall Section 24.3). Denote the *worst-case* rate-distortion function on $\mathcal{X}$ by

$$\phi_{\mathcal{X}}(\epsilon) = \sup_{P_X \in \mathcal{P}(\mathcal{X})} \phi_X(\epsilon). \tag{27.46}$$

The next theorem relates $\phi_{\mathcal{X}}$ to the covering and packing numbers of $\mathcal{X}$. The lower bound simply follows from a "Bayesian" argument, which bounds the worst case from below by the average case, akin to the relationship between minimax and Bayes risk (see Section 28.3). The upper bound was shown in [242] using the dual representation of rate-distortion functions; here we give a simpler proof via Fano's inequality.

Theorem 27.16. *For any* $0 < c < 1/2$,

$$\phi_{\mathcal{X}}(\epsilon) \leq \log N(\mathcal{X}, d, \epsilon) \leq \log M(\mathcal{X}, d, \epsilon) \leq \frac{\phi_{\mathcal{X}}(c\epsilon) + \log 2}{1-2c}. \tag{27.47}$$

Proof. Fix an ϵ-covering of $\mathcal{X}$ in d of size N. Let $\hat{X}$ denote the closest element in the covering to X. Then $d(X,\hat{X}) \leq \epsilon$ almost surely. Thus $\phi_X(\epsilon) \leq I(X;\hat{X}) \leq \log N$. Optimizing over P_X proves the left-hand inequality.

For the right-hand inequality, let X be uniformly distributed over a maximal ϵ-packing of $\mathcal{X}$. For any $P_{\hat{X}|X}$ such that $\mathbb{E}[d(X,\hat{X})] \leq c\epsilon$, let $\tilde{X}$ denote the closest point in the packing to $\hat{X}$. Then we have the Markov chain $X \to \hat{X} \to \tilde{X}$. By definition, $d(X,\tilde{X}) \leq d(\hat{X},\tilde{X}) + d(\hat{X},X) \leq 2d(\hat{X},X)$ so $\mathbb{E}[d(X,\tilde{X})] \leq 2c\epsilon$. Since either $X = \tilde{X}$ or $d(X,\tilde{X}) > \epsilon$, we have $\mathbb{P}[X \neq \tilde{X}] \leq 2c$. On the other hand, Fano's inequality (Corollary 3.13) yields $\mathbb{P}[X \neq \tilde{X}] \geq 1 - \frac{I(X;\hat{X})+\log 2}{\log M}$. In all, $I(X;\hat{X}) \geq (1-2c)\log M - \log 2$, proving the upper bound. □

Remark 27.4.

(a) Clearly, Theorem 27.16 can be extended to the case where the distortion function equals a power of the metric, namely, replacing (27.45) with

$$\phi_{X,r}(\epsilon) \triangleq \inf_{P_{\hat{X}|X}:\, \mathbb{E}[d(X,\hat{X})^r]\leq \epsilon^r} I(X;\hat{X}).$$

Then (27.47) continues to hold with $1-2c$ replaced by $1-(2c)^r$. This will be useful, for example, in the forthcoming applications where the second moment constraint is easier to work with.

(b) In the earlier literature a variant of the rate-distortion function is also considered, known as the *ϵ-entropy of* X, where the constraint is $d(X,\hat{X}) \leq \epsilon$ with probability one as opposed to in expectation (see e.g. [251, Appendix II] and [350]). With this definition, it is natural to conjecture that the maximal ϵ-entropy over all distributions on $\mathcal{X}$ coincides with the metric entropy

$\log N(\mathcal{X}, \epsilon)$; nevertheless, this need not be true (see [301, Remark, p. 1708] for a counterexample).

Theorem 27.16 points out an information-theoretic route to bound the metric entropy by the worst-case rate-distortion function (27.46).[14] Solving this maximization, however, is not easy as $P_X \mapsto \phi_X(D)$ is in general neither convex nor concave [6].[15] Fortunately, for certain spaces, one can show via a symmetry argument that the "uniform" distribution maximizes the rate-distortion function at every distortion level; see Exercise **V.24** for a formal statement. As a consequence, we have:

- For Hamming space $\mathcal{X} = \{0,1\}^d$ and Hamming distortion, $\phi_{\mathcal{X}}(D)$ is attained by $\text{Ber}(1/2)^d$. (We already knew this from Theorem 26.1 and Theorem 24.8.)
- For the unit sphere $\mathcal{X} = S^{d-1}$ and distortion function defined by the Euclidean distance, $\phi_{\mathcal{X}}(D)$ is attained by $\text{Unif}(S^{d-1})$.
- For the orthogonal group $\mathcal{X} = O(d)$ or unitary group $U(d)$ and distortion function defined by the Frobenius norm, $\phi_{\mathcal{X}}(D)$ is attained by the Haar measure. Similar statements also hold for the Grassmann manifold (collection of linear subspaces).

Next we give a concrete example by computing the rate-distortion function of $\theta \sim \text{Unif}(S^{d-1})$:

Theorem 27.17. *Let θ be uniformly distributed over the unit sphere S^{d-1}. Then for all $0 < \epsilon < 1$,*

$$(d-1)\log\frac{1}{\epsilon} - C \le \inf_{P_{\hat{\theta}|\theta}:\, \mathbb{E}[\|\hat{\theta}-\theta\|_2^2] \le \epsilon^2} I(\theta;\hat{\theta}) \le (d-1)\log\left(1+\frac{1}{\epsilon}\right) + \log(2d)$$

for some universal constant C.

Note that the random vector θ has dependent entries so we cannot invoke the single-letterization technique in Theorem 24.8. Nevertheless, we have the representation $\theta \stackrel{d}{=} Z/\|Z\|_2$ for $Z \sim \mathcal{N}(0, I_d)$, which allows us to relate the rate-distortion function of θ to that of the Gaussian found in Theorem 26.2. The resulting lower bound agrees with the metric entropy for spheres in Corollary 27.4, which scales as $(d-1)\log\frac{1}{\epsilon}$. Using similar reduction arguments (see [276, Theorem VIII.18]), one can obtain a tight lower bound for the metric entropy of the orthogonal group $O(d)$ and the unitary group $U(d)$, which scales as $\frac{d(d-1)}{2}\log\frac{1}{\epsilon}$ and $d^2\log\frac{1}{\epsilon}$, with pre-log factors commensurate with their respective degrees of freedom. As mentioned in Remark 27.3(b), these results were obtained by Szarek in [407] using a volume argument with Haar measures; in comparison, the information-theoretic approach is more elementary as we can again reduce to Gaussian rate-distortion computation.

[14] A striking parallelism between the metric entropy of Sobolev balls and the rate-distortion function of smooth Gaussian processes has been observed by Donoho in [133]. However, we cannot apply Theorem 27.16 to formally relate one to the other since it is unclear whether the Gaussian rate-distortion function is maximal.

[15] As a counterexample, consider Theorem 26.1 for the binary source.

Proof. The upper bound follows from Theorem 27.16 and Remark 27.4(a), applying the metric entropy bound for spheres in Corollary 27.4.

To prove the lower bound, let $Z \sim \mathcal{N}(0, I_d)$. Define $\theta = \frac{Z}{\|Z\|}$ and $A = \|Z\|$, where $\|\cdot\| \equiv \|\cdot\|_2$ henceforth. Then $\theta \sim \text{Unif}(S^{d-1})$ and $A \sim \chi_d$ are independent. Fix $P_{\hat{\theta}|\theta}$ such that $\mathbb{E}[\|\hat{\theta} - \theta\|^2] \leq \epsilon^2$. Since $\text{Var}(A) \leq 1$, the Shannon lower bound (Theorem 26.3) shows that the rate-distortion function of A is majorized by that of the standard Gaussian. So for each $\delta \in (0,1)$, there exists $P_{\hat{A}|A}$ such that $\mathbb{E}[(\hat{A} - A)^2] \leq \delta^2$, $I(A;\hat{A}) \leq \log \frac{1}{\delta}$, and $\mathbb{E}[A] = \mathbb{E}[\hat{A}]$. Set $\hat{Z} = \hat{A}\hat{\theta}$. Then

$$I(Z;\hat{Z}) = I(\theta, A;\hat{Z}) \leq I(\theta, A;\hat{\theta}, \hat{A}) = I(\theta;\hat{\theta}) + I(A;\hat{A}).$$

Furthermore, $\mathbb{E}[\hat{A}^2] = \mathbb{E}[(\hat{A}-A)^2] + \mathbb{E}[A^2] + 2\mathbb{E}[(\hat{A}-A)(A-\mathbb{E}[A])] \leq d + \delta^2 + 2\delta \leq d + 3\delta$. Similarly, $|\mathbb{E}[\hat{A}(\hat{A} - A)]| \leq 2\delta$ and $\mathbb{E}[\|Z - \hat{Z}\|^2] \leq d\epsilon^2 + 7\delta\epsilon + \delta$. Choosing $\delta = \epsilon$, we have $\mathbb{E}[\|Z - \hat{Z}\|^2] \leq (d+8)\epsilon^2$. Combining Theorem 24.8 with the Gaussian rate-distortion function in Theorem 26.2, we have $I(Z;\hat{Z}) \geq \frac{d}{2} \log \frac{d}{(d+8)\epsilon^2}$, so applying $\log(1+x) \leq x$ yields

$$I(\theta;\hat{\theta}) \geq (d-1)\log\frac{1}{\epsilon^2} - 4\log e.$$

□

Exercises for Part V

V.1 Let $\mathcal{S} = \hat{\mathcal{S}} = \{0, 1\}$. Consider the source X^{10} consisting of fair coin flips. Construct a simple (suboptimal) compressor achieving average Hamming distortion $\frac{1}{20}$ with 512 codewords.

V.2 Assume a separable distortion loss. Show that the minimal number of codewords $M^*(n, D)$ required to represent memoryless source X^n with average distortion D (recall (24.9)) satisfies

$$\log M^*(n_1 + n_2, D) \leq \log M^*(n_1, D) + \log M^*(n_2, D).$$

Conclude that

$$\lim_{n\to\infty} \frac{1}{n} \log M^*(n, D) = \inf_n \frac{1}{n} \log M^*(n, D). \tag{V.1}$$

That is, one can always achieve a better compression rate by using a longer blocklength. Neither claim holds for $\log M^*(n, \epsilon)$ in channel coding as defined in (19.4). Explain why this different behavior arises.

V.3 (Non-asymptotic rate distortion) Our goal is to show that the convergence to $R(D)$ happens much faster than that to capacity in channel coding. Consider binary uniform $X \sim \text{Ber}(1/2)$ with Hamming distortion.

(a) Show that there exists a lossy code $X^n \to W \to \hat{X}^n$ with M codewords and

$$\mathbb{P}[d(X^n, \hat{X}^n) > D] \leq (1 - p(nD))^M,$$

where

$$p(s) = 2^{-n} \sum_{j=0}^{s} \binom{n}{j}.$$

(b) Show that there exists a lossy code with M codewords and

$$\mathbb{E}[d(X^n, \hat{X}^n)] \leq \frac{1}{n} \sum_{s=0}^{n-1} (1 - p(s))^M . \tag{V.2}$$

(c) Show that there exists a lossy code with M codewords and

$$\mathbb{E}[d(X^n, \hat{X}^n)] \leq \frac{1}{n} \sum_{s=0}^{n-1} e^{-Mp(s)} . \tag{V.3}$$

Note: For $M \approx 2^{nR}$, numerical evaluation of (V.2) for large n is challenging. At the same time (V.3) is only slightly looser.

(d) For $n = 10, 50, 100$ and 200 compute the upper bound on $\log M^*(n, 0.11)$ via (V.3). Compare with the lower bound

$$\log M^*(n, D) \geq nR(D). \tag{V.4}$$

V.4 Continuing Exercise **V.3** use Stirling's formula and (V.3) and (V.4) to show that

$$\log M^*(n, D) = nR(D) + O(\log n).$$

Note: Thus, the optimal compression rate converges to its asymptotic fundamental limit $R(D)$ at a fast rate of $O(\log n/n)$ as opposed to $O(1/\sqrt{n})$ for channel coding (see Theorem 22.2). This result holds for most memoryless sources [473].

V.5 Let $S_j \overset{\text{iid}}{\sim} P_S$ be an iid source on a finite alphabet $\mathcal{A}$ and $P_S(a) > 0$ for all $a \in \mathcal{A}$. Suppose the distortion metric satisfies $d(x, y) = D_0 \implies x = y$. Show that $R(D_0) = \log |\mathcal{A}|$, while $R(D_0+) = H(X)$.

V.6 Consider a memoryless source X uniform on $\mathcal{X} = \hat{\mathcal{X}} = [m]$ with Hamming distortion: $d(x, \hat{x}) = 1\{x \neq \hat{x}\}$. Show that

$$R(D) = \begin{cases} \log m - h(D) - D\log(m-1), & D \leq \frac{m-1}{m}, \\ 0, & \text{otherwise.} \end{cases}$$

(Hint: Apply Fano's inequality Theorem 3.12.)

V.7 Let X and Y be random variables taking values in $\{1, 2, 3\}$ and such that $\mathbb{P}[X = Y] = \frac{1}{2}$. If Y is uniform what are the best upper and lower bounds on $I(X;Y)$ you can find?

V.8 (Erasure distortion metric) Consider Bernoulli(1/2) source $S \in \{0, 1\}$, reproduction alphabet $\hat{\mathcal{A}} = \{0, ?, 1\}$, and the distortion metric

$$d(a, \hat{a}) = \begin{cases} 0, & a = \hat{a}, \\ 1, & \hat{a} = ?, \\ \infty, & a \neq \hat{a}, \hat{a} \neq ?. \end{cases}$$

Is $D_{\max}$ finite? Find the rate-distortion function $R(D)$.

V.9 Let $X \sim \text{Unif}[\pm 1]$ and reconstruction $\hat{X} \in \mathbb{R}$ with $d(x, \hat{x}) = (x - \hat{x})^2$. Show that the rate-distortion function is given in parametric form by

$$R = \log 2 - h(p), \qquad D = 4p(1-p), \qquad p \in [0, 1/2],$$

and that for any distortion level the optimal vector quantizer only takes values $\pm(1-2p)$. (Hint: You may find Exercise **I.64**(b) useful.) Compare with the case of $\hat{X} \in \{\pm 1\}$, for which we have shown $R(D) = \log 2 - h(D/4)$, $D \in [0, 2]$.

V.10 (Product source) Consider two independent stationary memoryless sources $X \in \mathcal{X}$ and $Y \in \mathcal{Y}$ with reproduction alphabets $\hat{\mathcal{X}}$ and $\hat{\mathcal{Y}}$, distortion measures $d_1 : \mathcal{X} \times \hat{\mathcal{X}} \to \mathbb{R}_+$ and $d_2 : \mathcal{Y} \times \hat{\mathcal{Y}} \to \mathbb{R}_+$, and rate-distortion functions R_X and R_Y, respectively. Now consider the stationary memoryless product source $Z = (X, Y)$ with reproduction alphabet $\hat{\mathcal{X}} \times \hat{\mathcal{Y}}$ and distortion measure $d(z, \hat{z}) = d_1(x, \hat{x}) + d_2(y, \hat{y})$.

(a) Show that

$$I(X, Y; \hat{X}, \hat{Y}) \geq I(X; \hat{X}) + I(Y; \hat{Y}).$$

(b) Show that the rate-distortion function of Z is related to that of X and Y via the *inf-convolution*, that is,

$$R(D) = \inf_{0 \leq D_1 \leq D} R_X(D_1) + R_Y(D - D_1).$$

(c) How do you build an optimal lossy compressor for Z using optimal lossy compressors for X and Y?

V.11 (Compression with output constraints) Compute the rate-distortion function $R(D; a, p)$ of a $\text{Ber}(p)$ source, with respect to the Hamming distortion under the extra constraint that reconstruction points $\hat{X}^n$ should have average Hamming weight $\mathbb{E}[w_H(\hat{X}^n)] \leq an$, where $0 < a, p \leq \frac{1}{2}$. (Hint: Show a more general result that given two distortion metrics d_1, d_2 we have $R(D_1, D_2) = \min\{I(S; \hat{S}) : \mathbb{E}[d_i(S, \hat{S})] \leq D_i, i \in \{1, 2\}\}$.)

V.12 Commercial (mono) FM radio modulates a band-limited (15 kHz) audio signal into a radio-frequency signal of bandwidth 200 kHz. This system roughly achieves

$$\mathsf{SNR}_{\text{audio}} \approx 40 \text{ dB} + \mathsf{SNR}_{\text{channel}}$$

over the AWGN channel whenever $\mathsf{SNR}_{\text{channel}} \gtrsim 12$ dB. Thus for the 12 dB channel, we get that FM radio has distortion of 52 dB. Show that the information-theoretic limit is about 160 dB.

Hint: Assume that the input signal is low-pass filtered to a 15 kHz white, zero-mean Gaussian and use the optimal joint source–channel code (JSSC) for the given bandwidth expansion ratio and fixed $\mathsf{SNR}_{\text{channel}}$. Also recall that the SNR of the reconstruction $\hat{S}^n$ expressed in dB is defined as

$$\mathsf{SNR}_{\text{audio}} \triangleq 10\log_{10}\frac{\sum_{j=1}^{k}\mathbb{E}[S_j^2]}{\sum_{j=1}^{k}\mathbb{E}[(S_j-\hat{S}_j)^2]}.$$

V.13 Consider a memoryless Gaussian source $X\sim\mathcal{N}(0,1)$, reconstruction alphabet $\hat{\mathcal{A}}=\{\pm 1\}$, and quadratic distortion $d(a,\hat{a})=(a-\hat{a})^2$. Compute D_0, $R(D_0+)$, and $D_{\max}$. Then obtain a parametric formula for $R(D)$.

V.14 (Erokhin's rate distortion [155]) Let $d(a^k,b^k)=1\{a^k\neq b^k\}$ be a (non-separable) distortion metric for k-strings on a finite alphabet $\mathcal{S}=\hat{\mathcal{S}}$. Prove that for any source S^k we have

$$\varphi_{S^k}(\epsilon)\triangleq\min_{\mathbb{P}[S^k\neq\hat{S}^k]\le\epsilon} I(S^k;\hat{S}^k)\ge H(S^k)-\epsilon k\log|\mathcal{S}|-h(\epsilon), \tag{V.5}$$

and that the bound is tight only for S^k uniform on $\mathcal{S}^k$. Next, suppose that S^k is an iid source. Prove that

$$\phi_{S^k}(\epsilon)=(1-\epsilon)kH(S)-\sqrt{\frac{kV(S)}{2\pi}}e^{-(Q^{-1}(\epsilon))^2/2}+O(\log k),$$

where $V(S)$ is the *varentropy* (10.4). (Hint: Let $T=\hat{P}_{S^k}$ be the empirical distribution (type) of the realization S^k. Then $I(S^k;\hat{S}^k)=I(S^k,T;\hat{S}^k)=I(S^k;\hat{S}^k|T)+O(\log k)$. Denote $\epsilon_T\triangleq\mathbb{P}[S^k\neq\hat{S}^k|T]$ and given ϵ_T optimize the first term via (V.5). Then optimize the assignment ϵ_T over all $\mathbb{E}[\epsilon_T]\le\epsilon$. Also use $\mathbb{E}[Z1\{Z>c\}]=\frac{1}{\sqrt{2\pi}}e^{-c^2/2}$ for $Z\sim\mathcal{N}(0,1)$. See [255, Lemma 1] for full details.)

V.15 Consider a source $S^n\overset{\text{iid}}{\sim}\text{Ber}(1/2)$. Answer the following questions *when n is large.*

(a) Suppose the goal is to compress S^n into k bits so that one can reconstruct S^n with at most one bit of error. That is, the decoded version $\hat{S}^n$ satisfies $\mathbb{E}[d_{\text{H}}(\hat{S}^n,S^n)]\le 1$. Show that this can be done (if possible, with an explicit algorithm) with $k=n-C\log n$ bits for some constant C. Is it optimal?

(b) Suppose we are required to compress S^n into only 1 bit. Show that one can achieve (if possible, with an explicit algorithm) a reconstruction error $\mathbb{E}[d_{\text{H}}(\hat{S}^n,S^n)]\le\frac{n}{2}-C\sqrt{n}$ for some constant C. Is it optimal?

Warning: We cannot blindly apply the asymptotic rate-distortion theory to show achievability since here the distortion level changes with n. The converse, however, directly applies.

V.16 Consider a standard Gaussian vector S^n and quadratic distortion metric. We discuss *zero-rate* quantization.

(a) Let $S_{\max}=\max_{1\le i\le n}S_i$. Show that $\mathbb{E}[(S_{\max}-\sqrt{2\ln n})^2]\to 0$ when $n\to\infty$.

(b) Suppose you are given a budget of $\log_2 n$ bits. Consider the following scheme: Let i^* denote the index of the largest coordinate. The compressor stores the index i^* which costs $\log_2 n$ bits and the decompressor outputs $\hat{S}^n$ where $\hat{S}_i = \sqrt{2\ln n}$ for $i = i^*$ and $S_i = 0$ otherwise. Show that distortion in terms of mean-squared error satisfies $\mathbb{E}[\|\hat{S}^n - S^n\|_2^2] = n - 2\ln n + o(1)$ when $n \to \infty$.

(c) Show that for any compressor (using $\log_2 n$ bits) we must have $\mathbb{E}[\|\hat{S}^n - S^n\|_2^2] \geq n - 2\ln n + o(1)$.

V.17 (Noisy source coding; also remote source coding [126]) Consider the problem of compressing iid sequence X^n under separable distortion metric d. Now, however, the compressor does not have direct access to X^n but only to its noisy version Y^n obtained over a stationary memoryless channel $P_{Y|X}$ (i.e. $(X_i, Y_i) \overset{\text{iid}}{\sim} P_{X,Y}$ for a fixed $P_{X,Y}$ and the encoder is a map $f: \mathcal{Y}^n \to [M]$). Show that the rate-distortion function is

$$R(D) = \min\{I(Y;\hat{X}) \colon \mathbb{E}[d(X,\hat{X})] \leq D, X \to Y \to \hat{X}\},$$

where minimization is over all $P_{\hat{X}|Y}$. (Hint: Define $\tilde{d}(y,\hat{x}) \triangleq \mathbb{E}[d(X,\hat{x}) \mid Y = y]$.)

V.18 (Noisy/remote source coding; special case) Let $Z^n \overset{\text{iid}}{\sim} \text{Ber}(1/2)$ and $X^n = \text{BEC}_\delta(Z^n)$. The compressor is to encode X^n at rate R so that we can reconstruct Z^n with bit error rate D. Let $R(D)$ denote the optimal rate.

(a) Suppose that the locations of erasures in X^n are provided as side information to the decompressor. Show that $R(\delta/2) = \frac{\bar{\delta}}{2}$ (Hint: The compressor is very simple.)

(b) Surprisingly, the same rate is achievable without knowledge of erasures. Use Exercise **V.17** to prove $R(D) = H(\bar{\delta}/2, \bar{\delta}/2, \delta) - H(1-D-\frac{\delta}{2}, D-\frac{\delta}{2}, \delta)$ for $D \in [\frac{\delta}{2}, \frac{1}{2}]$.

V.19 (Log-loss) Consider the rate-distortion problem where the reconstruction alphabet $\hat{\mathcal{X}}_n = \mathcal{P}(\mathcal{X}^n)$ is the space of all probability mass functions on $\mathcal{X}^n$. We define two loss functions. The first one is non-separable (!),

$$d_{\text{nonsep}}(x^n, \hat{P}) = \frac{1}{n}\log\frac{1}{\hat{P}(x^n)}, \tag{V.6}$$

and the second one is separable,

$$d_{\text{sep}}(x^n, \hat{P}_{X^n}) = \frac{1}{n}\sum_{j=1}^{n}\log\frac{1}{\hat{P}_{X_j}(x_j)}.$$

Let $\{X_i \colon i \geq 1\}$ be a process with entropy rate $\mathcal{H}$.

(a) Show that any (n, M, D)-code for d_{sep} can be converted to an (n, M, D)-code for d_{nonsep}. Hence

$$R_{\text{sep}}(D) \geq R_{\text{nonsep}}(D).$$

(b) Prove that the information rate-distortion function under d_{nonsep} is given by

$$R^{(\text{I})}_{\text{nonsep}}(D) = |\mathcal{H} - D|^+ .$$

(c) Prove that if the quantizer $U = f(X^n)$ is such that the distribution of U is uniform then it is necessarily optimal for non-separable log-loss, that is, the log-cardinality of $U = R^{(\mathrm{I})}_{\mathrm{nonsep}}(D)$.

(d) Assume $\{X_i\}$ is stationary and ergodic. Show that rate $R_{i,\mathrm{nonsep}(D)}$ is achievable, that is,

$$R_{\mathrm{nonsep}}(D) = R^{(\mathrm{I})}_{\mathrm{nonsep}}(D).$$

(Hint: Use AEP (11.2).)

(e) Assume $\{X_i\}$ are iid. Show that

$$R^{(\mathrm{I})}_{\mathrm{sep}}(D) = R_{\mathrm{sep}}(D) = R_{\mathrm{nonsep}}(D) = |H(X) - D|^+.$$

(The second equality follows from the general rate-distortion theorem.)

V.20 (Information bottleneck and log-loss) Given $P_{X,Y}$ the information bottleneck (IB) [422] proposes that W is an optimal *approximate* sufficient statistic of Y for X if it solves

$$F_I(t) \triangleq \max\{I(X;W) : I(Y;W) \le t, X \to Y \to W\},$$

where the F_I-curve has been defined previously in Definition 16.5. Here we explain that IB is equivalent to noisy source coding under log-loss distortion.

(a) (Log-loss, information) Consider the rate-distortion problem (Exercise **V.19**) with $\hat{\mathcal{X}}_n = \mathcal{P}(\mathcal{X}^n)$ and the (non-separable) loss function d_{nonsep} in (V.6). Show that the solution to the noisy source coding (see Exercise **V.17**) satisfies for all n:

$$\inf\{\mathbb{E}[d_{\mathrm{nonsep}}(X^n, \hat{P}_{X^n})] : I(Y^n;\hat{P}_{X^n}) \le nR\} = n\,(H(X) - F_I(R)).$$

(b) (Log-loss, operational) Define the operational distortion-rate function as

$$D(R) \triangleq \limsup_{n\to\infty} \inf\{D : \exists (n, \exp(nR), nD)\text{-code}\}.$$

Show that

$$D(R) = H(X) - F_I(R).$$

(Hint: For achievability, restrict reconstruction to $\hat{P}_{X^n} = \prod \hat{P}_{X_i}$, which makes distortion additive, and then apply Exercise **V.17**; for the converse, use the tensorization property of the F_I-curve from Exercise **III.32**.)

V.21 (a) Let $0 \prec \Delta \preceq \Sigma$ be positive definite matrices. For $S \sim \mathcal{N}(0, \Sigma)$, show that

$$\inf_{P_{\hat{S}|S} : \mathbb{E}[(S-\hat{S})(S-\hat{S})^\top] \preceq \Delta} I(S;\hat{S}) = \frac{1}{2} \log \frac{\det \Sigma}{\det \Delta}.$$

(Hint: For achievability, consider $S = \hat{S} + Z$ with $\hat{S} \sim \mathcal{N}(0, \Sigma - \Delta) \perp\!\!\!\perp Z \sim \mathcal{N}(0, \Delta)$ and apply Example 3.5; for the converse, follow the proof of Theorem 26.2.)

(b) Prove the following extension of (26.3): Let $\sigma_1^2, \ldots, \sigma_d^2$ be the eigenvalues of Σ. Then

$$\inf_{P_{\hat{S}|S}:\, \mathbb{E}[\|S-\hat{S}\|_2^2]\le D} I(S;\hat{S}) = \frac{1}{2}\sum_{i=1}^{d} \log^{+} \frac{\sigma_i^2}{\lambda},$$

where $\lambda > 0$ is such that $\sum_{i=1}^{d} \min\{\sigma_i^2, \lambda\} = D$. This is the counterpart of the water-filling solution in Theorem 20.14.

(Hint: First, using the orthogonal invariance of distortion metric we can assume that Σ is diagonal. Next, apply the same single-letterization argument for (26.3) and solve $\min_{\sum D_i = D} \frac{1}{2}\sum_{i=1}^{d} \log^{+} \frac{\sigma_i^2}{D_i}$.)

V.22 (Shannon lower bound) Let $\|\cdot\|$ be an *arbitrary* norm on $\mathbb{R}^d$ and $r > 0$. Let X be an $\mathbb{R}^d$-valued random vector with a probability density function p_X. Denote the rate-distortion function

$$\phi_X(D) \triangleq \inf_{P_{\hat{X}|X}:\, \mathbb{E}[\|\hat{X}-X\|^r]\le D} I(X;\hat{X}).$$

Prove the Shannon lower bound (26.5), namely

$$\phi_X(D) \ge h(X) + \frac{d}{r}\log\frac{d}{Dre} - \log\left(\Gamma\left(\frac{d}{r}+1\right)V\right), \tag{V.7}$$

where the differential entropy $h(X) = \int_{\mathbb{R}^d} p_X(x)\log\frac{1}{p_X(x)}dx$ is assumed to be finite and $V = \mathrm{vol}(\{x \in \mathbb{R}^d : \|x\| \le 1\})$.

(a) Show that $0 < V < \infty$.

(b) Show that for any $s > 0$,

$$Z(s) \triangleq \int_{\mathbb{R}^d} \exp(-s\|w\|^r)dw = \Gamma\left(\frac{d}{r}+1\right)Vs^{-\frac{d}{r}}.$$

(Hint: Apply Fubini's theorem to $\int_{\mathbb{R}^d}\exp(-s\|w\|^r)dw = \int_{\mathbb{R}^d}\int_{\|w\|^r}^{\infty} s\exp(-sx)dxdw$ and use $\Gamma(x) = \int_0^{\infty} t^{x-1}e^{-t}dt$.)

(c) Show that for any feasible $P_{X|\hat{X}}$ such that $\mathbb{E}[\|X-\hat{X}\|^r] \le D$,

$$I(X;\hat{X}) \ge h(X) - \log Z(s) - sD.$$

(Hint: Define an auxiliary backward channel $Q_{X|\hat{X}}(dx|\hat{x}) = q_s(x-\hat{x})dx$, where $q_s(w) = \frac{1}{Z(s)}\exp(-s\|w\|^r)$. Then $I(X;\hat{X}) = \mathbb{E}_P[\log\frac{Q_{X|\hat{X}}}{P_X}] + D(P_{X|\hat{X}}\|Q_{X|\hat{X}}\|P_{\hat{X}})$.)

(d) Optimize over $s > 0$ to conclude (V.7).

(e) Verify that the lower bound of Theorem 26.3 is a special case of (V.7).

Note: Alternatively, the SLB can be written in the following form:

$$\phi_X(D) \ge h(X) - \sup_{P_W:\, \mathbb{E}[\|W\|^r]\le D} h(W),$$

and this entropy maximization can be solved following the argument in Example 5.2.

V.23 (Uniform distribution minimizes convex symmetric functional) Let G be a group acting on a set $\mathcal{X}$ such that each $g \in G$ sends $x \in \mathcal{X}$ to $gx \in \mathcal{X}$. Suppose G acts *transitively*, that is, for each $x, x' \in \mathcal{X}$ there exists $g \in G$ such that $gx = x'$. Let g be a random element of G with an invariant distribution, namely $h\mathsf{g} \stackrel{d}{=} \mathsf{g}$ for any $h \in G$. (Such a distribution, known as the Haar measure, exists for compact topological groups.)

(a) Show that for any $x \in \mathcal{X}$, $\mathsf{g}x$ has the same law, denoted by $\text{Unif}(\mathcal{X})$, the uniform distribution on $\mathcal{X}$.

(b) Let $f: \mathcal{P}(\mathcal{X}) \to \mathbb{R}$ be convex and G-invariant, that is, $f(P_{gX}) = f(P_X)$ for any $\mathcal{X}$-valued random variable X and any $g \in G$. Show that $\min_{P_X \in \mathcal{P}(\mathcal{X})} f(P_X) = f(\text{Unif}(\mathcal{X}))$.

V.24 (Uniform distribution maximizes rate-distortion function) Continuing the setup of Exercise **V.23**, let $d: \mathcal{X} \times \mathcal{X} \to \mathbb{R}$ be a G-invariant distortion function, that is, $d(gx, gx') = d(x, x')$ for any $g \in G$. Denote the rate-distortion function of an $\mathcal{X}$-valued X by $\phi_X(D) = \inf_{P_{\hat{X}|X}:\, \mathbb{E}[d(X,\hat{X})] \le D} I(X;\hat{X})$. Suppose that $\phi_X(D) < \infty$ for all X and all $D > 0$.

(a) Let $\phi_X^*(\lambda) = \sup_D\{\lambda D - \phi_X(D)\}$ denote the conjugate of ϕ_X. Apply Theorem 24.4 and the Fenchel–Moreau biconjugation theorem to conclude that $\phi_X(D) = \sup_\lambda\{\lambda D - \phi_X^*(\lambda)\}$.

(b) Show that

$$\phi_X^*(\lambda) = \sup_{P_{\hat{X}|X}} \{\lambda \mathbb{E}[d(X,\hat{X})] - I(X;\hat{X})\}.$$

As such, for each λ, $P_X \mapsto \phi_X^*(\lambda)$ is convex and G-invariant. (Hint: Theorem 5.3.)

(c) Apply Exercise **V.23** to conclude that $\phi_U^*(\lambda) \le \phi_X^*(\lambda)$ for $U \sim \text{Unif}(\mathcal{X})$ and that

$$\phi_X(D) \le \phi_U(D), \quad \forall D > 0.$$

V.25 (Normal tail bound.) Denote the standard normal density and tail probability by $\varphi(x) = \frac{1}{\sqrt{2\pi}} e^{-x^2/2}$ and $\Phi^c(t) = Q(t) = \int_t^\infty \varphi(x)dx$. Show that for all $t > 0$,

$$\frac{t}{1+t^2}\varphi(t) \le \Phi^c(t) \le \min\left\{\frac{\varphi(t)}{t}, e^{-t^2/2}\right\}. \tag{V.8}$$

(Hint: For $\Phi^c(t) \le e^{-t^2/2}$ apply the Chernoff bound (15.2); for the rest, note that by integration by parts $\Phi^c(t) = \frac{\varphi(t)}{t} - \int_t^\infty \frac{\varphi(x)}{x^2}dx$.)

V.26 (Covering radius in Hamming space) In this exercise we prove (27.9), namely, for any fixed $0 \le D \le 1$, as $n \to \infty$,

$$N(\mathbb{F}_2^n, d_{\text{H}}, Dn) = 2^{n(1-h(D))_+ + o(n)},$$

where $h(\cdot)$ is the binary entropy function.

(a) Prove the lower bound by invoking the volume bound in Theorem 27.3 and the large-deviations estimate in Example 15.1

(b) Prove the upper bound using probabilistic construction and a similar argument to (25.8).

(c) Show that for $D \geq \frac{1}{2}$, $N(\mathbb{F}_2^n, d_{\mathrm{H}}, Dn) \leq 2$; see Exercise **V.15a**.

V.27 (Covering ℓ_p-ball with ℓ_q-balls)

(a) For $1 \leq p < q \leq \infty$, prove the bound (27.18) on the metric entropy of the unit ℓ_p-ball with respect to the ℓ_q-norm. (Hint: For small ϵ, apply the volume calculation in (27.15) and (27.16) and the formula in (27.13); for large ϵ, proceed as in the proof of Theorem 27.7 by applying the quantization argument and the Gilbert–Varshamov bound of Hamming spheres.)

(b) What happens when $p > q$?

V.28 In this exercise we prove Dudley's chaining inequality (27.22). In view of Theorem 27.2, it is equivalent to show the following version with covering numbers:

$$w(\Theta) \lesssim \int_0^\infty \sqrt{\log N(\epsilon)} d\epsilon, \tag{V.9}$$

where $N(\epsilon) \equiv N(\Theta, \|\cdot\|_2, \epsilon)$ is the ℓ_2-covering number of Θ and $w(\Theta) = \mathbb{E}_{Z \sim \mathcal{N}(0, I_d)} \sup_{\theta \in \Theta} \langle \theta, Z \rangle$ is its Gaussian width.

(a) Show that if Θ is not totally bounded, then both sides are infinite.

(b) Next, assuming $\mathrm{rad}(\Theta) < \infty$, let T_i be the optimal ϵ_i-covering of Θ, where $\epsilon_i = \mathrm{rad}(\Theta) 2^{-i}, i \geq 0$, so $T_0 = \{\theta_0\}$, the Chebyshev center of Θ. Show that

$$\sup_{\theta \in \Theta} \langle Z, \theta \rangle \leq \langle Z, \theta_0 \rangle + \sum_{i \geq 1} M_i,$$

where $M_i \triangleq \max\{\langle Z, s_i - s_{i-1} \rangle : s_i \in T_i, s_{i-1} \in T_{i-1}\}$. (Hint: For any $\theta \in \Theta$, let θ_i denote its nearest neighbor in T_i. Then $\langle Z, \theta \rangle = \langle Z, \theta_0 \rangle + \sum_{i \geq 1} \langle Z, \theta_i - \theta_{i-1} \rangle$.)

(c) Show that $\mathbb{E}[M_i] \lesssim \epsilon_i \sqrt{\log N(\epsilon_i)}$ (Hint: $\|\theta_i - \theta_{i-1}\| \leq \|\theta_i - \theta\| + \|\theta_{i-1} - \theta\| \leq 3\epsilon_i$. Then apply Lemma 27.10 and the bounded convergence theorem.)

(d) Conclude that

$$w(\Theta) \lesssim \sum_{i \geq 0} \epsilon_i \sqrt{\log N(\epsilon_i)}$$

and compare with the integral version (V.9).

V.29 (Random matrix) Let A be an $m \times n$ matrix of iid $\mathcal{N}(0,1)$ entries. Denote its operator norm by $\|A\|_{\mathrm{op}} = \max_{v \in S^{n-1}} \|Av\|$, which is also the largest singular value of A.

(a) Show that

$$\|A\|_{\mathrm{op}} = \max_{u \in S^{m-1}, v \in S^{n-1}} \langle A, uv' \rangle. \tag{V.10}$$

(b) Let $\mathcal{U} = \{u_1, \ldots, u_M\}$ and $\mathcal{V} = \{v_1, \ldots, v_M\}$ be an ϵ-*net* for the spheres S^{m-1} and S^{n-1} respectively. Show that

$$\|A\|_{\text{op}} \le \frac{1}{(1-\epsilon)^2} \max_{u \in \mathcal{U}, v \in \mathcal{V}} \langle A, uv' \rangle.$$

(c) Apply Corollary 27.4 and Lemma 27.10 to conclude that

$$\mathbb{E}[\|A\|] \lesssim \sqrt{n} + \sqrt{m}. \tag{V.11}$$

(d) By choosing u and v in (V.10) smartly, show a matching lower bound and conclude that

$$\mathbb{E}[\|A\|] \asymp \sqrt{n} + \sqrt{m}. \tag{V.12}$$

(e) Use Sudakov minorization (Theorem 27.8) to prove a matching lower bound. (Hint: Use (27.6).)

V.30 (Small-ball probability II) In this exercise we prove (27.36). Let $\{W_t : t \ge 0\}$ be a standard Brownian motion. Show that for small ϵ,[16]

$$\phi(\epsilon) = -\log \mathbb{P}\left[\sup_{t \in [0,1]} |W_t| \le \epsilon\right] \asymp \frac{1}{\epsilon^2}.$$

(a) By rescaling space and time, show that $\mathbb{P}\left[\sup_{t\in[0,1]} |W_t| \le \epsilon\right] = \mathbb{P}\left[\sup_{t\in[0,T]} |W_t| \le 1\right] \triangleq p_T$, where $T = 1/\epsilon^2$. To show $p_T = e^{-\Theta(T)}$, there is no loss of generality to assume that T is an integer.

(b) (Upper bound) Using the independent increment property, show that $p_{T+1} \le a p_T$, where $a = \mathbb{P}[|Z| \le 1]$ with $Z \sim \mathcal{N}(0,1)$. (Hint: $g(z) \triangleq \mathbb{P}[|Z - z| \le 1]$ for $z \in [-1,1]$ is maximized at $z = 0$ and minimized at $z = \pm 1$.)

(c) (Lower bound) Again by scaling, it is equivalent to show $\mathbb{P}[\sup_{t\in[0,T]} |W_t| \le C] \ge C^{-T}$ for some constant C. Let $q_T \triangleq \mathbb{P}[\sup_{t\in[0,T]} |W_t| \le 2, \max_{t=1,\ldots,T} |W_t| \le 1]$. Show that $q_{T+1} \ge b q_T$, where $b = \mathbb{P}[|Z-1| \le 1]\mathbb{P}[\sup_{t\in[0,1]} |B_t| \le 1]$, and $B_t = B_t - tB_1$ is a Brownian bridge. (Hint: $\{W_t : t \in [0,T]\}$, $W_{T+1} - W_T$, and $\{W_{T+t} - (1-t)W_T - tW_{T+1} : t \in [0,1]\}$ are mutually independent, with the latter distributed as a Brownian bridge.)

[16] Using the large-deviations theory developed by Donsker and Varadhan, the sharp constant can be found to be $\lim_{\epsilon\to 0} \epsilon^2 \phi(\epsilon) = \frac{\pi^2}{8}$; see for example [280, Section 6.2].

Part VI

Statistical Applications

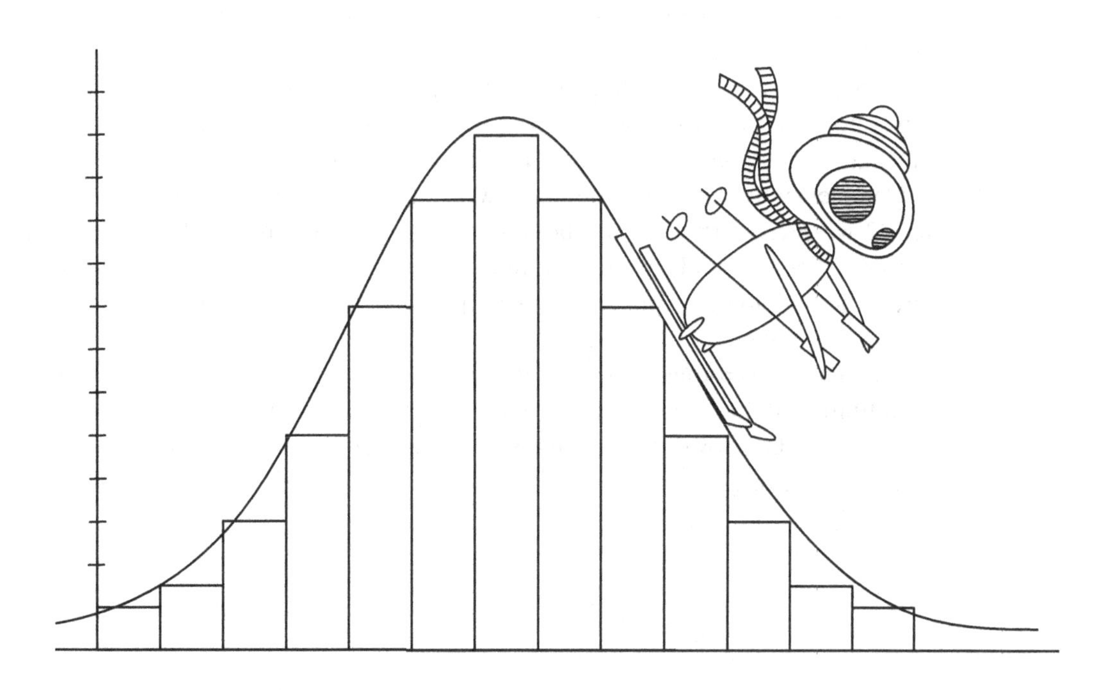

This part gives an exposition on the application of information-theoretic principles and methods in mathematical statistics; we do so by discussing a selection of topics. To start, Chapter 28 introduces the basic decision-theoretic framework of statistical estimation and the *Bayes risk* and the *minimax risk* as the fundamental limits. Chapter 29 gives an exposition of the classical large-sample asymptotics for smooth parametric models in fixed dimensions, highlighting the role of Fisher information introduced in Chapter 2. Notably, we discuss how to deduce classical lower bounds (Hammersley–Chapman–Robbins, Cramér–Rao, and van Trees) from the variational characterization and the data-processing inequality (DPI) of χ^2-divergence in Chapter 7.

Moving into high dimensions, Chapter 30 introduces the *mutual information method* for the statistical lower bound, based on the DPI for mutual information as well as the theory of capacity and rate-distortion function from Parts IV and V. This principled approach includes three popular methods for proving minimax lower bounds (Le Cam, Assouad, and Fano) as special cases, which are discussed at length in Chapter 31 drawing on results on metric entropy from Chapter 27.

Complementing the exposition on lower bounds in Chapters 30 and 31, in Chapter 32 we present three upper bounds on statistical estimation based on metric entropy. These bounds appear strikingly similar but follow from completely different methodologies. Application to nonparametric density estimation is used as a primary example.

Chapter 33 introduces *strong data processing inequalities* (SDPIs), which are quantitative strengthening of DPIs in Part I. As applications we show how to apply SDPIs to deduce lower bounds for various estimation problems on graphs or in distributed settings.

28 Basics of Statistical Decision Theory

In this chapter, we discuss the decision-theoretic framework of statistical estimation and introduce several important examples. Section 28.1 presents the basic elements of statistical experiment and statistical estimation. Section 28.3 introduces the Bayes risk (average-case) and the minimax risk (worst-case) as the respective fundamental limits of statistical estimation in the Bayesian and frequentist settings, with the latter being our primary focus in this part. We discuss several versions of the minimax theorem (and prove a simple one) that equates the minimax risk with the worst-case Bayes risk. Two variants are introduced next that extend a basic statistical experiment to either large sample size or large dimension: Section 28.4 on independent observations and Section 28.5 on tensorization of experiments. Throughout this part the Gaussian location model (GLM), introduced in Section 28.2, serves as a running example, with different focus at different places (such as the role of loss functions, parameter spaces, low versus high dimensions, etc.). In Section 28.6, we discuss a key result known as Anderson's lemma for determining the exact minimax risk of (unconstrained) GLM in any dimension for a broad class of loss functions, which provides a benchmark for various more general techniques introduced in later chapters.

28.1 Basic Setting

We start by presenting the basic elements of statistical decision theory. We refer to the classics [166, 274, 405] for a systematic treatment.

A *statistical experiment* or *statistical model* refers to a collection $\mathcal{P}$ of probability distributions (over a common measurable space $(\mathcal{X}, \mathcal{F})$). Specifically, let us consider

$$\mathcal{P} = \{P_\theta : \theta \in \Theta\}, \tag{28.1}$$

where each distribution is indexed by a *parameter* θ taking values in the *parameter space* Θ.

In the decision-theoretic framework, we play the following game: Nature picks some parameter $\theta \in \Theta$ and generates a random variable $X \sim P_\theta$. A statistician observes the data X and wants to infer the parameter θ or its certain attributes. Specifically, consider some functional $T : \Theta \to \mathcal{Y}$ and the goal is to estimate $T(\theta)$

on the basis of the observation X. Here the *estimand* $T(\theta)$ may be the parameter θ itself, or some function thereof (e.g. $T(\theta) = 1\{\theta > 0\}$ or $\|\theta\|$).

An *estimator* (decision rule) is a function $\hat{T}\colon \mathcal{X} \to \hat{\mathcal{Y}}$. Note that, similar to the rate-distortion theory in Part V, the action space $\hat{\mathcal{Y}}$ need not be the same as $\mathcal{Y}$ (e.g. $\hat{T}$ may be a confidence interval that aims to contain the scalar T). Here $\hat{T}$ can be either *deterministic*, that is, $\hat{T} = \hat{T}(X)$, or *randomized*, that is, $\hat{T}$ obtained by passing X through a conditional probability distribution (Markov transition kernel) $P_{\hat{T}|X}$, or a channel in the language of Part I. For all practical purposes, we can write $\hat{T} = \hat{T}(X, U)$, where U denotes external randomness uniform on $[0, 1]$ and independent of X.

To measure the quality of an estimator $\hat{T}$, we introduce a *loss function* $\ell\colon \mathcal{Y}\times\hat{\mathcal{Y}} \to \mathbb{R}$ such that $\ell(T, \hat{T})$ is the risk of $\hat{T}$ for estimating T. Since we are dealing with loss (as opposed to reward), all the negative (converse) results are lower bounds and all the positive (achievable) results are upper bounds. Note that X is a random variable, and so are $\hat{T}$ and $\ell(T, \hat{T})$. Therefore, to make sense of "minimizing the loss," we consider the expected risk:

$$R_\theta(\hat{T}) = \mathbb{E}_\theta[\ell(T, \hat{T})] = \int P_\theta(dx) P_{\hat{T}|X}(d\hat{t}|x)\ell(T(\theta), \hat{t}), \tag{28.2}$$

which we refer to as the *risk* of $\hat{T}$ at θ. The subscript in $\mathbb{E}_\theta$ indicates the distribution with respect to which the expectation is taken. Note that the expected risk depends on the estimator as well as the ground truth.

Remark 28.1. We note that the problem of hypothesis testing and inference can be encompassed as special cases of the estimation paradigm. As previously discussed in Section 16.4, there are three formulations for testing:

- Simple versus simple hypotheses

$$H_0\colon \theta = \theta_0 \quad \text{versus} \quad H_1: \theta = \theta_1, \qquad \theta_0 \neq \theta_1.$$

- Simple versus composite hypotheses

$$H_0\colon \theta = \theta_0 \quad \text{versus} \quad H_1: \theta \in \Theta_1, \qquad \theta_0 \notin \Theta_1.$$

- Composite versus composite hypotheses

$$H_0\colon \theta \in \Theta_0 \quad \text{versus} \quad H_1: \theta \in \Theta_1, \qquad \Theta_0 \cap \Theta_1 = \emptyset.$$

For each case one can introduce the appropriate parameter space and loss function. For example, in the last (most general) case, we may take

$$\Theta = \Theta_0 \cup \Theta_1, \quad T(\theta) = \begin{cases} 0, & \theta \in \Theta_0, \\ 1, & \theta \in \Theta_1, \end{cases} \quad \hat{T} \in \{0, 1\},$$

and use the zero–one loss $\ell(T, \hat{T}) = 1\{T \neq \hat{T}\}$ so that the expected risk $R_\theta(\hat{T}) = P_\theta\{\theta \notin \Theta_{\hat{T}}\}$ is the probability of error.

For the problem of inference, the goal is to output a confidence interval (or region) which covers the true parameter with high probability. In this case $\hat{T}$ is a subset of Θ and we may choose the loss function $\ell(\theta, \hat{T}) = 1\{\theta \notin \hat{T}\} + \lambda \cdot \text{length}(\hat{T})$ for some $\lambda > 0$, in order to balance the coverage and the size of the confidence interval.

Remark 28.2. (Randomized versus deterministic estimators) Although most of the estimators used in practice are deterministic, there are a number of reasons to consider randomized estimators:

- For certain formulations, such as the minimizing worst-case risk (minimax approach), deterministic estimators are suboptimal and it is necessary to randomize. On the other hand, if the objective is to minimize the average risk (Bayes approach), then it does not lose generality to restrict to deterministic estimators.
- The space of randomized estimators (viewed as Markov kernels) is convex which is the convex hull of deterministic estimators. This convexification is needed, for example, for the treatment of minimax theorems.

See Section 28.3 for a detailed discussion and examples.

A well-known fact is that for a convex loss function (i.e. $\hat{T} \mapsto \ell(T, \hat{T})$ is convex), randomization does not help. Indeed, for any randomized estimator $\hat{T}$, we can derandomize it by considering its conditional expectation $\mathbb{E}[\hat{T}|X]$, which is a deterministic estimator and whose risk dominates that of the original $\hat{T}$ at every θ, namely, $R_\theta(\hat{T}) = \mathbb{E}_\theta \ell(T, \hat{T}) \geq \mathbb{E}_\theta \ell(T, \mathbb{E}[\hat{T}|X])$, by Jensen's inequality.

28.2 Gaussian Location Model (GLM)

Note that, without loss of generality, all statistical models can be expressed in the parametric form of (28.1) (since we can take θ to be the distribution itself). In the statistics literature, it is customary to refer to a model as *parametric* if θ takes values in a finite-dimensional Euclidean space (so that each distribution is specified by finitely many parameters), and *nonparametric* if θ takes values in some infinite-dimensional space (e.g. density estimation or sequence model).

Perhaps the most basic parametric model is the Gaussian location model (GLM), also known as the normal mean model, which corresponds to our familiar Gaussian channel in Example 3.3. This will be our running example in this part of the book. In this model, we have

$$\mathcal{P} = \{\mathcal{N}(\theta, \sigma^2 I_d) : \theta \in \Theta\},$$

where I_d is the d-dimensional identity matrix and the parameter space $\Theta \subset \mathbb{R}^d$. Equivalently, we can express the data as a noisy observation of the unknown vector θ as

$$X = \theta + Z, \quad Z \sim \mathcal{N}(0, \sigma^2 I_d).$$

The cases of $d = 1$ and $d > 1$ refer to the univariate (scalar) and multivariate (vector) cases, respectively. (Also of interest is the case where θ is a $d_1 \times d_2$ matrix, which can be vectorized into a $d = d_1 d_2$-dimensional vector.)

The choice of the parameter space Θ represents our prior knowledges of the unknown parameter θ, for example:

- $\Theta = \mathbb{R}^d$, in which case there is no assumption on θ.
- $\Theta = \ell_p$-norm balls.
- $\Theta = \{\text{all } k\text{-sparse vectors}\} = \{\theta \in \mathbb{R}^d \colon \|\theta\|_0 \le k\}$, where $\|\theta\|_0 \triangleq |\{i \colon \theta_i \neq 0\}|$ denotes the size of the support, informally referred to as the ℓ_0-"norm."
- $\Theta = \{\theta \in \mathbb{R}^{d_1 \times d_2} \colon \operatorname{rank}(\theta) \le r\}$, the set of low-rank matrices.

By definition, more structure (smaller parameter space) always makes the estimation task easier (smaller worst-case risk), but not necessarily so in terms of computation.

For estimating θ itself (denoising), it is customary to use a loss function defined by certain norms, for example, $\ell(\theta, \hat{\theta}) = \|\theta - \hat{\theta}\|_p^\alpha$ for some $1 \le p \le \infty$ and $\alpha > 0$, where $\|\theta\|_p \triangleq (\sum |\theta_i|^p)^{\frac{1}{p}}$, with $p = \alpha = 2$ corresponding to the commonly used *quadratic loss* (squared error). Some well-known estimators include the maximum likelihood estimator (MLE)

$$\hat{\theta}_{\mathrm{ML}} = X \tag{28.3}$$

and the James–Stein estimator based on shrinkage

$$\hat{\theta}_{\mathrm{JS}} = \left(1 - \frac{(d-2)\sigma^2}{\|X\|_2^2}\right) X. \tag{28.4}$$

The choice of the estimator depends on both the objective and the parameter space. For instance, if θ is known to be sparse, it makes sense to set the smaller entries in the observed X to zero (thresholding) in order to better denoise θ (see Section 30.2).

In addition to estimating the vector θ itself, it is also of interest to estimate certain functionals $T(\theta)$ thereof, for example, $T(\theta) = \|\theta\|_p, \max\{\theta_1, \ldots, \theta_d\}$, or eigenvalues in the matrix case. In addition, the hypothesis testing problem in the GLM has been well studied. For example, one can consider detecting the presence of a signal by testing $H_0 \colon \theta = 0$ against $H_1 \colon \|\theta\| \ge \epsilon$, or testing weak signal $H_0 \colon \|\theta\| \le \epsilon_0$ versus strong signal $H_1 \colon \|\theta\| \ge \epsilon_1$, with or without further structural assumptions on θ. We refer the reader to the monograph [225] devoted to these problems.

28.3 Bayes Risk, Minimax Risk, and the Minimax Theorem

One of our main objectives in this part of the book is to understand the fundamental limit of statistical estimation, that is, to determine the performance of the best estimator. As in (28.2), the risk $R_\theta(\hat{T})$ of an estimator $\hat{T}$ for $T(\theta)$ depends on the ground truth θ. To compare the risk profiles of different estimators meaningfully requires some thought. As a toy example, Figure 28.1 depicts the risk functions of three estimators. It is clear that $\hat{\theta}_1$ is superior to $\hat{\theta}_2$ in the sense that the risk of the former is pointwise lower than that of the latter. (In statistical literature we say $\hat{\theta}_2$ is inadmissible.) However, the comparison of $\hat{\theta}_1$ and $\hat{\theta}_3$ is less clear. Although the

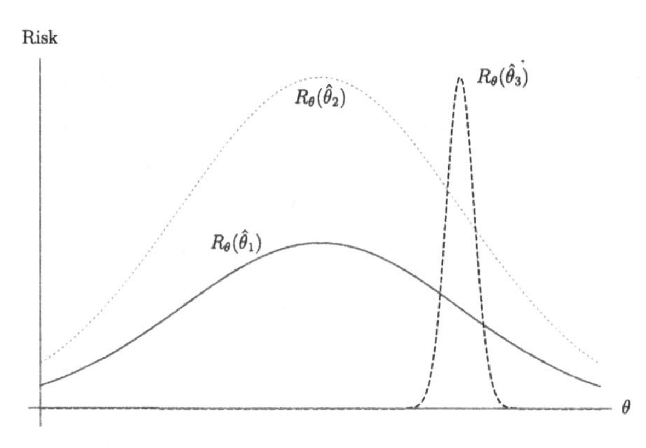

Figure 28.1 Risk profiles of three estimators.

peak risk value of $\hat{\theta}_3$ is bigger than that of $\hat{\theta}_1$, on average its risk (area under the curve) is smaller. In fact, both views are valid and meaningful, and they correspond to the worst-case (minimax) and average-case (Bayesian) approach, respectively. In the minimax formulation, we summarize the risk function into a scalar quantity, namely, the worst-case risk, and seek the estimator that minimizes this objective. In the Bayesian formulation, the objective is the average risk. Below we discuss these two approaches and their connections. For notational simplicity, we consider the task of estimating $T(\theta) = \theta$.

28.3.1 Bayes Risk

The Bayesian approach is an average-case formulation in which the statistician acts as if the parameter θ is random with a known distribution. Concretely, let π be a probability distribution (prior) on Θ. Then the *average risk* (w.r.t. π) of an estimator $\hat{\theta}$ is defined as

$$R_\pi(\hat{\theta}) = \mathbb{E}_{\theta\sim\pi}[R_\theta(\hat{\theta})] = \mathbb{E}_{\theta,X}[\ell(\theta,\hat{\theta})]. \tag{28.5}$$

Given a prior π, its *Bayes risk* is the minimal average risk, namely

$$R_\pi^* = \inf_{\hat{\theta}} R_\pi(\hat{\theta}).$$

An estimator $\hat{\theta}^*$ is called a *Bayes estimator* if it attains the Bayes risk, namely, $R_\pi^* = \mathbb{E}_{\theta\sim\pi}[R_\theta(\hat{\theta}^*)]$.

Remark 28.3. The Bayes estimator is always deterministic, a fact that holds for any loss function. To see this, note that for any randomized estimator, say $\hat{\theta} = \hat{\theta}(X, U)$, where U is some external randomness independent of X and θ, its risk is lower-bounded by

$$R_\pi(\hat{\theta}) = \mathbb{E}_{\theta,X,U}[\ell(\theta,\hat{\theta}(X,U))] = \mathbb{E}_U[R_\pi(\hat{\theta}(\cdot,U))] \geq \inf_u R_\pi(\hat{\theta}(\cdot,u)).$$

Note that for any u, $\hat{\theta}(\cdot,u)$ is a deterministic estimator. This shows that we can find a deterministic estimator whose average risk is no worse than that of the randomized estimator.

An alternative explanation of this fact is the following: Note that the average risk $R_\pi(\hat{\theta})$ defined in (28.5) is an affine function of the randomized estimator (understood as a Markov kernel $P_{\hat{\theta}|X}$), whose minimum is achieved at the extremal points. In this case the extremal points of Markov kernels are simply delta measures, which correspond to deterministic estimators.

In certain settings the Bayes estimator can be found explicitly. Consider the problem of estimating $\theta \in \mathbb{R}^d$ drawn from a prior π. Under the quadratic loss $\ell(\theta, \hat{\theta}) = \|\hat{\theta} - \theta\|_2^2$, the Bayes estimator is the conditional mean $\hat{\theta}(X) = \mathbb{E}[\theta|X]$ and the Bayes risk is the *minimum mean-squared error* (MMSE), which we previously encountered in Section 3.7* in the context of the I-MMSE relationship:

$$R_\pi^* = \mathbb{E}[\|\theta - \mathbb{E}[\theta|X]\|_2^2] = \mathbb{E}[\mathrm{Tr}(\mathrm{Cov}(\theta|X))],$$

where $\mathrm{Cov}(\theta|X = x)$ is the conditional covariance matrix of θ given $X = x$.

As a concrete example, let us consider the Gaussian location model in Section 28.2 with a Gaussian prior.

Example 28.1. Bayes risk in GLM

Consider the scalar case, where $X = \theta + Z$ and $Z \sim \mathcal{N}(0, \sigma^2)$ is independent of θ. Consider a Gaussian prior $\theta \sim \pi = \mathcal{N}(0, s)$. One can verify that the posterior distribution $P_{\theta|X=x}$ is $\mathcal{N}(\frac{s}{s+\sigma^2}x, \frac{s\sigma^2}{s+\sigma^2})$. As such, the Bayes estimator is $\mathbb{E}[\theta|X] = \frac{s}{s+\sigma^2}X$ and the Bayes risk is

$$R_\pi^* = \frac{s\sigma^2}{s+\sigma^2}. \tag{28.6}$$

Similarly, for multivariate GLM: $X = \theta + Z, Z \sim \mathcal{N}(0, I_d)$, if $\theta \sim \pi = \mathcal{N}(0, sI_d)$, then we have

$$R_\pi^* = \frac{s\sigma^2}{s+\sigma^2}d. \tag{28.7}$$

28.3.2 Minimax Risk

A common criticism of the Bayesian approach is the arbitrariness of the selected prior. A framework related to this but not discussed in this case is the empirical Bayes approach [364, 471], where one "estimates" the prior from the data instead of choosing a prior a priori. Instead, we take a frequentist viewpoint by considering the worst-case situation. The *minimax risk* is defined as

$$R^* = \inf_{\hat{\theta}} \sup_{\theta \in \Theta} R_\theta(\hat{\theta}). \tag{28.8}$$

If there exists $\hat{\theta}$ s.t. $\sup_{\theta \in \Theta} R_\theta(\hat{\theta}) = R^*$, then the estimator $\hat{\theta}$ is minimax (minimax optimal).

Finding the value of the minimax risk R^* entails proving two things, namely,

- a minimax upper bound, by exhibiting an estimator $\hat{\theta}^*$ such that $R_\theta(\hat{\theta}^*) \leq R^* + \epsilon$ for all $\theta \in \Theta$,
- a minimax lower bound, by proving that for any estimator $\hat{\theta}$, there exists some $\theta \in \Theta$, such that $R_\theta(\hat{\theta}) \geq R^* - \epsilon$,

where $\epsilon > 0$ is arbitrary. This task is frequently difficult especially in high dimensions. Instead of the exact minimax risk, it is often useful to find a constant-factor approximation Ψ, which we call *minimax rate*, such that

$$R^* \asymp \Psi, \tag{28.9}$$

that is, $c\Psi \leq R^* \leq C\Psi$ for some universal constants $c, C \geq 0$. Establishing that Ψ is the minimax rate still entails proving the minimax upper and lower bounds, albeit within multiplicative constant factors.

In practice, minimax lower bounds are rarely established according to the original definition. The next result shows that the Bayes risk is always lower than the minimax risk. Throughout this book, all lower-bound techniques essentially boil down to evaluating the Bayes risk with a sagaciously chosen prior.

Theorem 28.1. *Let $\mathcal{P}(\Theta)$ denote the collection of probability distributions on Θ. Then*

$$R^* \geq R^*_{\text{Bayes}} \triangleq \sup_{\pi \in \mathcal{P}(\Theta)} R^*_\pi. \tag{28.10}$$

(If the supremum is attained for some prior, we say it is least favorable.)

Proof. There are two (equivalent) ways to prove this fact:

1 "max $\geq$ mean": For any $\hat{\theta}$, $R_\pi(\hat{\theta}) = \mathbb{E}_{\theta\sim\pi} R_\theta(\hat{\theta}) \leq \sup_{\theta\in\Theta} R_\theta(\hat{\theta})$. Taking the infimum over $\hat{\theta}$ completes the proof.

2 "min max $\geq$ max min":

$$R^* = \inf_{\hat{\theta}} \sup_{\theta\in\Theta} R_\theta(\hat{\theta}) = \inf_{\hat{\theta}} \sup_{\pi\in\mathcal{P}(\Theta)} R_\pi(\hat{\theta}) \geq \sup_{\pi\in\mathcal{P}(\Theta)} \inf_{\hat{\theta}} R_\pi(\hat{\theta}) = \sup_{\pi} R^*_\pi,$$

where the inequality follows from the generic fact that $\min_x \max_y f(x,y) \geq \max_y \min_x f(x,y)$. □

Remark 28.4. Unlike Bayes estimators which, as shown in Remark 28.3, are always deterministic, to minimize the worst-case risk it is sometimes necessary to randomize for example in the context of hypothesis testing (Chapter 14). Specifically, consider a trivial experiment where $\theta \in \{0, 1\}$ and X is absent, so that we are forced to guess the value of θ under the zero–one loss $\ell(\theta, \hat{\theta}) = 1\{\theta \neq \hat{\theta}\}$. It is clear that in this case the minimax risk is $\frac{1}{2}$, achieved by random guessing $\hat{\theta} \sim \text{Ber}(1/2)$ but not by any deterministic $\hat{\theta}$.

As an application of Theorem 28.1, let us determine the minimax risk of the Gaussian location model under the quadratic loss function.

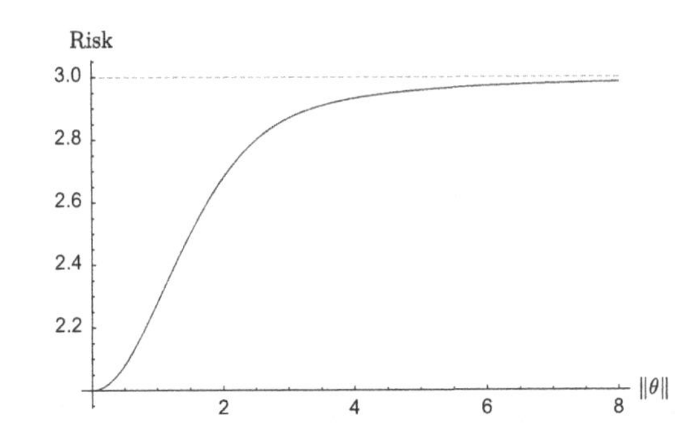

Figure 28.2 Risk of the James–Stein estimator (28.4) in dimension $d = 3$ and $\sigma = 1$ as a function of $\|\theta\|$.

Example 28.2. Minimax quadratic risk of GLM

Consider the Gaussian location model without structural assumptions, where $X \sim \mathcal{N}(\theta, \sigma^2 I_d)$ with $\theta \in \mathbb{R}^d$. We show that

$$R^* \equiv \inf_{\hat{\theta} \in \mathbb{R}^d} \sup_{\theta \in \mathbb{R}^d} \mathbb{E}_\theta[\|\hat{\theta}(X) - \theta\|_2^2] = d\sigma^2. \tag{28.11}$$

By scaling, it suffices to consider $\sigma = 1$. For the upper bound, we consider $\hat{\theta}_{\text{ML}} = X$ which achieves $R_\theta(\hat{\theta}_{\text{ML}}) = d$ for all θ. To get a matching minimax lower bound, we consider the prior $\theta \sim \mathcal{N}(0, s)$. Using the Bayes risk previously computed in (28.6), we have $R^* \geq R_\pi^* = \frac{sd}{s+1}$. Sending $s \to \infty$ yields $R^* \geq d$.

Remark 28.5. (Non-uniqueness of minimax estimators) In general, estimators that achieve the minimax risk need not be unique. For instance, as shown in Example 28.2, the MLE $\hat{\theta}_{\text{ML}} = X$ is minimax for the unconstrained GLM in any dimension. On the other hand, it is known that whenever $d \geq 3$, the risk of the James–Stein estimator (28.4) is smaller than that of the MLE everywhere (see Figure 28.2) and thus is also minimax. In fact, there exist a continuum of estimators that are minimax for (28.11) [277, Theorem 5.5].

For most of the statistical models, Theorem 28.1 in fact holds with equality; such a result is known as a *minimax theorem*. Before discussing this important topic, here is an example where minimax risk is strictly bigger than the worst-case Bayes risk.

Example 28.3.

Let $\theta, \hat{\theta} \in \mathbb{N} \triangleq \{1, 2, \ldots\}$ and $\ell(\theta, \hat{\theta}) = 1\{\hat{\theta} < \theta\}$, that is, the statistician loses one dollar if Nature's choice exceeds the statistician's guess and loses nothing otherwise. Consider the extreme case of blind guessing (i.e. no data is available, say, $X = 0$). Then for any $\hat{\theta}$ possibly randomized, we have $R_\theta(\hat{\theta}) = \mathbb{P}[\hat{\theta} < \theta]$. Thus $R^* \geq \lim_{\theta \to \infty} \mathbb{P}[\hat{\theta} < \theta] = 1$, which is clearly achievable. On the other hand, for any prior π on $\mathbb{N}$, $R_\pi(\hat{\theta}) = \mathbb{P}[\hat{\theta} < \theta]$, which vanishes as $\hat{\theta} \to \infty$. Therefore, we have $R_\pi^* = 0$. Therefore in this case $R^* = 1 > R_{\text{Bayes}}^* = 0$.

As an exercise, one can show that the minimax quadratic risk of the GLM $X \sim \mathcal{N}(\theta, 1)$ with parameter space $\theta \geq 0$ is the same as in the unconstrained case.

(This might be a bit surprising because the thresholded estimator $X_+ = \max(X, 0)$ achieves a better risk pointwise at every $\theta \geq 0$; nevertheless, just like the James–Stein estimator (see Figure 28.2), in the worst case the gain is asymptotically diminishing.)

28.3.3 Minimax and Bayes Risk: a Duality Perspective

Recall from Theorem 28.1 the inequality

$$R^* \geq R^*_{\text{Bayes}}.$$

This result can be interpreted from an optimization perspective. More precisely, R^* is the value of a convex optimization problem (primal) and R^*_{Bayes} is precisely the value of its dual program. Thus the inequality (28.10) is simply *weak duality*. If *strong duality* holds, then (28.10) is in fact an equality, in which case the minimax theorem holds.

For simplicity, we consider the case where Θ is a finite set. Then

$$R^* = \min_{P_{\hat{\theta}|X}} \max_{\theta \in \Theta} \mathbb{E}_\theta[\ell(\theta, \hat{\theta})]. \tag{28.12}$$

This is a convex optimization problem. Indeed, $P_{\hat{\theta}|X} \mapsto \mathbb{E}_\theta[\ell(\theta, \hat{\theta})]$ is affine and the pointwise supremum of affine functions is convex. To write down its dual problem, first let us rewrite (28.12) in an augmented form

$$\begin{aligned} R^* = \min_{P_{\hat{\theta}|X}, t} \quad & t \\ \text{s.t.} \quad & \mathbb{E}_\theta[\ell(\theta, \hat{\theta})] \leq t, \quad \forall \theta \in \Theta. \end{aligned} \tag{28.13}$$

Let $\pi_\theta \geq 0$ denote the Lagrange multiplier (dual variable) for each inequality constraint. The Lagrangian of (28.13) is

$$L(P_{\hat{\theta}|X}, t, \pi) = t + \sum_{\theta \in \Theta} \pi_\theta (\mathbb{E}_\theta[\ell(\theta, \hat{\theta})] - t) = \left(1 - \sum_{\theta \in \Theta} \pi_\theta\right) t + \sum_{\theta \in \Theta} \pi_\theta \mathbb{E}_\theta[\ell(\theta, \hat{\theta})].$$

By definition, we have $R^* \geq \min_{t, P_{\hat{\theta}|X}} L(\hat{\theta}, t, \pi)$. Note that unless $\sum_{\theta \in \Theta} \pi_\theta = 1$, $\min_{t \in \mathbb{R}} L(\hat{\theta}, t, \pi)$ is $-\infty$. Thus $\pi = (\pi_\theta : \theta \in \Theta)$ must be a probability measure and the dual problem is

$$\max_{\pi} \min_{P_{\hat{\theta}|X}, t} L(P_{\hat{\theta}|X}, t, \pi) = \max_{\pi \in \mathcal{P}(\Theta)} \min_{P_{\hat{\theta}|X}} R_\pi(\hat{\theta}) = \max_{\pi \in \mathcal{P}(\Theta)} R^*_\pi.$$

Hence, $R^* \geq R^*_{\text{Bayes}}$.

In summary, the minimax risk and the worst-case Bayes risk are related by convex duality, where the primal variables are (randomized) estimators and the dual variables are priors. This view can in fact be operationalized. For example, [238, 347] showed that for certain problems dualizing Le Cam's two-point lower bound (Theorem 31.1) leads to optimal minimax upper bound; see Exercise **VI.17**.

28.3.4 Minimax Theorem

Much earlier in Chapter 5 we have already seen an example of the strong minimax duality. That is, we found that capacity satisfies $C = \min_{P_X} \max_{Q_Y} D(P_{Y|X} \| Q_Y | P_X) = \max_{Q_Y} \min_{P_X} D(P_{Y|X} \| Q_Y | P_X)$, and the optimal pair (P_X^*, Q_Y^*) forms a saddle point. Here we show an example of a similar minimax theorem but for the statistical risk. Namely, we want to specify conditions that ensure (28.10) holds with equality. For simplicity, let us consider the case of estimating θ itself where the estimator $\hat{\theta}$ takes values in the action space $\hat{\Theta}$ with a loss function $\ell : \Theta \times \hat{\Theta} \to \mathbb{R}$. A very general result (see [405, Theorem 46.6]) asserts that $R^* = R^*_{\text{Bayes}}$, provided that the following conditions hold:

- The experiment is dominated, that is, $P_\theta \ll \nu$ holds for all $\theta \in \Theta$ for some ν on $\mathcal{X}$.
- The action space $\hat{\Theta}$ is a locally compact topological space with a countable base (e.g. the Euclidean space).
- The loss function is level-compact (i.e. for each $\theta \in \Theta$, $\ell(\theta, \cdot)$ is bounded from below and the sublevel set $\{\hat{\theta} : \ell(\theta, \hat{\theta}) \le a\}$ is compact for each a).

This result shows that for virtually all problems encountered in practice, the minimax risk coincides with the least favorable Bayes risk. At the heart of any minimax theorem, there is an application of the separating hyperplane theorem. Below we give a proof of a special case illustrating this type of argument.

Theorem 28.2. (Minimax theorem)

$$R^* = R^*_{Bayes}$$

in either of the following cases:

- *Θ is a finite set and the data X takes values in a finite set $\mathcal{X}$.*
- *Θ is a finite set and the loss function ℓ is bounded from below, that is,* $\inf_{\theta, \hat{\theta}} \ell(\theta, \hat{\theta}) > -\infty$.

Proof. The first case directly follows from the duality interpretation in Section 28.3.3 and the fact that strong duality holds for finite-dimensional linear programming (see for example [377, Section 7.4]).

For the second case, we start by showing that if $R^* = \infty$, then $R^*_{\text{Bayes}} = \infty$. To see this, consider the uniform prior π on Θ. Then for any estimator $\hat{\theta}$, there exists $\theta \in \Theta$ such that $R(\theta, \hat{\theta}) = \infty$. Then $R_\pi(\hat{\theta}) \ge \frac{1}{|\Theta|} R(\theta, \hat{\theta}) = \infty$.

Next we assume that $R^* < \infty$. Then $R^* \in \mathbb{R}$ since ℓ is bounded from below (say, by a) by assumption. Given an estimator $\hat{\theta}$, denote its risk vector $R(\hat{\theta}) = (R_\theta(\hat{\theta}))_{\theta \in \Theta}$. Then its average risk with respect to a prior π is given by the inner product $\langle R(\hat{\theta}), \pi \rangle = \sum_{\theta \in \Theta} \pi_\theta R_\theta(\hat{\theta})$. Define

$$S = \{R(\hat{\theta}) \in \mathbb{R}^\Theta : \hat{\theta} \text{ is a randomized estimator}\} = \text{set of all possible risk vectors},$$
$$T = \{t \in \mathbb{R}^\Theta : t_\theta < R^*, \theta \in \Theta\}.$$

Note that both S and T are convex (why?) subsets of Euclidean space $\mathbb{R}^\Theta$ and $S \cap T = \emptyset$ by definition of R^*. By the separation hyperplane theorem, there exist a non-zero $\pi \in \mathbb{R}^\Theta$ and $c \in \mathbb{R}$, such that $\inf_{s\in S} \langle \pi, s\rangle \geq c \geq \sup_{t\in T} \langle \pi, t\rangle$. Obviously, π must be componentwise positive, for otherwise $\sup_{t\in T} \langle \pi, t\rangle = \infty$. Therefore by normalization we may assume that π is a probability vector, that is, a prior on Θ. Then $R^*_{\text{Bayes}} \geq R^*_\pi = \inf_{s\in S} \langle \pi, s\rangle \geq \sup_{t\in T} \langle \pi, t\rangle \geq R^*$, completing the proof. □

28.4 Multiple Observations and Sample Complexity

Given an experiment $\{P_\theta : \theta \in \Theta\}$, consider the experiment

$$\mathcal{P}_n = \{P_\theta^{\otimes n} : \theta \in \Theta\}, \quad n \geq 1. \tag{28.14}$$

We refer to this as the *independent sampling model*, in which we observe a sample $X = (X_1, \ldots, X_n)$ consisting of independent observations drawn from P_θ for some $\theta \in \Theta \subset \mathbb{R}^d$. Given a loss function $\ell \colon \mathbb{R}^d \times \mathbb{R}^d \to \mathbb{R}^+$, the minimax risk is denoted by

$$R_n^*(\Theta) = \inf_{\hat\theta} \sup_{\theta\in\Theta} \mathbb{E}_\theta[\ell(\theta, \hat\theta)]. \tag{28.15}$$

Clearly, $n \mapsto R_n^*(\Theta)$ is non-increasing since we can always discard the extra observations. Typically, when Θ is a fixed subset of $\mathbb{R}^d$, $R_n^*(\Theta)$ vanishes as $n \to \infty$. Thus a natural question is at what rate R_n^* converges to zero. Equivalently, one can consider the *sample complexity*, namely, the minimum sample size to attain a prescribed error ϵ even in the worst case:

$$n^*(\epsilon) \triangleq \min\left\{n \in \mathbb{N} \colon R_n^*(\Theta) \leq \epsilon\right\}. \tag{28.16}$$

In the classical large-sample asymptotics (Chapter 29), the rate of convergence for the quadratic risk is usually $\Theta(\frac{1}{n})$, which is commonly referred to as the "parametric rate." In comparison, in this book we focus on understanding the dependence on the dimension and other structural parameters non-asymptotically.

As a concrete example, let us revisit the GLM in Section 28.2 with sample size n, in which case we observe $X = (X_1, \ldots, X_n) \overset{\text{iid}}{\sim} \mathcal{N}(0, \sigma^2 I_d), \theta \in \mathbb{R}^d$. In this case, the minimax quadratic risk is[1]

$$R_n^* = \frac{d\sigma^2}{n}. \tag{28.17}$$

To see this, note that in this case $\bar X = \frac{1}{n}(X_1 + \cdots + X_n)$ is a sufficient statistic (see Section 3.5) of X for θ. Therefore the model reduces to $\bar X \sim \mathcal{N}(\theta, \frac{\sigma^2}{n} I_d)$ and (28.17) follows from the minimax risk (28.11) for a single observation.

From (28.17), we conclude that the sample complexity is $n^*(\epsilon) = \left\lceil \frac{d\sigma^2}{\epsilon} \right\rceil$, which grows linearly with the dimension d. This is the common wisdom that "sample complexity scales proportionally to the number of parameters," also known as

[1] See Exercise **VI.11** for an extension of this result to nonparametric location models.

"counting the degrees of freedom." Indeed in high dimensions we typically expect the sample complexity to grow with the ambient dimension; however, the exact dependence need not be linear as it depends on the loss function and the objective of estimation. For example, consider the matrix case $\theta \in \mathbb{R}^{d\times d}$ with n independent observations in Gaussian noise. Let ϵ be a small constant. Then we have the following:

- For quadratic loss, namely, $\|\theta - \hat{\theta}\|_F^2$, we have $R_n^* = \frac{d^2}{n}$ and hence $n^*(\epsilon) = \Theta(d^2)$.
- If the loss function is $\|\theta - \hat{\theta}\|_{\text{op}}^2$, then $R_n^* \asymp \frac{d}{n}$ and hence $n^*(\epsilon) = \Theta(d)$ (Example 28.4).
- As opposed to θ itself, suppose we are content with estimating only the scalar functional $\theta_{\max} = \max\{\theta_1, \ldots, \theta_d\}$ up to accuracy ϵ, then $n^*(\epsilon) = \Theta(\log d)$ (Exercise **VI.14**).

In the last two examples, the sample complexity scales *sublinearly* with the dimension.

28.5 Tensor Product of Experiments

Tensor product is a way to define a high-dimensional model from low-dimensional models. Given statistical experiments $\mathcal{P}_i = \{P_{\theta_i} : \theta_i \in \Theta_i\}$ and the corresponding loss function ℓ_i, for $i \in [d]$, their tensor product refers to the following statistical experiment:

$$\mathcal{P} = \left\{ P_\theta = \prod_{i=1}^d P_{\theta_i} : \theta = (\theta_1, \ldots, \theta_d) \in \Theta \triangleq \prod_{i=1}^d \Theta_i \right\},$$

$$\ell(\theta, \hat{\theta}) \triangleq \sum_{i=1}^d \ell_i(\theta_i, \hat{\theta}_i), \quad \forall \theta, \hat{\theta} \in \Theta.$$

In this model, the observation $X = (X_1, \ldots, X_d)$ consists of independent (not identically distributed) $X_i \overset{\text{ind}}{\sim} P_{\theta_i}$ and the loss function takes a *separable* form, which is reminiscent of separable distortion function in (24.8). This should be contrasted with the multiple-observation model in (28.14), in which n iid observations drawn from the same distribution are given.

The minimax risk of the tensorized experiment is related to the minimax risk $R^*(\mathcal{P}_i)$ and worst-case Bayes risks $R^*_{\text{Bayes}}(\mathcal{P}_i) \triangleq \sup_{\pi_i \in \mathcal{P}(\Theta_i)} R_{\pi_i}(\mathcal{P}_i)$ of each individual experiment as follows:

Theorem 28.3. (Minimax risk of tensor product)

$$\sum_{i=1}^d R^*_{\text{Bayes}}(\mathcal{P}_i) \le R^*(\mathcal{P}) \le \sum_{i=1}^d R^*(\mathcal{P}_i). \tag{28.18}$$

Consequently, if the minimax theorem holds for each experiment, that is, $R^*(\mathcal{P}_i) = R^*_{\text{Bayes}}(\mathcal{P}_i)$, *then it also holds for the product experiment and, in particular,*

$$R^*(\mathcal{P}) = \sum_{i=1}^{d} R^*(\mathcal{P}_i). \tag{28.19}$$

Proof. The right-hand inequality of (28.18) simply follows by separately estimating θ_i on the basis of X_i, namely, $\hat{\theta} = (\hat{\theta}_1, \ldots, \hat{\theta}_d)$, where $\hat{\theta}_i$ depends only on X_i. For the left-hand inequality, consider a product prior $\pi = \prod_{i=1}^{d} \pi_i$, under which θ_i's are independent and so are X_i's. Consider any randomized estimator $\hat{\theta}_i = \hat{\theta}_i(X, U_i)$ of θ_i based on X, where U_i is some auxiliary randomness independent of X. We can rewrite it as $\hat{\theta}_i = \hat{\theta}_i(X_i, \tilde{U}_i)$, where $\tilde{U}_i = (X_{\backslash i}, U_i) \perp\!\!\!\perp X_i$. Thus $\hat{\theta}_i$ can be viewed as a randomized estimator based on X_i alone and its average risk must satisfy $R_{\pi_i}(\hat{\theta}_i) = \mathbb{E}[\ell(\theta_i, \hat{\theta}_i)] \geq R^*_{\pi_i}$. Summing over i and taking the suprema over priors π_i's yields the left-hand inequality of (28.18). □

As an example, we note that the unstructured d-dimensional GLM $\{\mathcal{N}(\theta, \sigma^2 I_d) \colon \theta \in \mathbb{R}^d\}$ with quadratic loss is simply the d-fold tensor product of the one-dimensional GLM. Since the minimax theorem holds for the GLM (see Section 28.3.4), Theorem 28.3 shows the minimax risks sum up to $\sigma^2 d$, which agrees with Example 28.2. In general, however, it is possible that the minimax risk of the tensorized experiment is less than the sum of individual minimax risks and the right-hand inequality of (28.19) can be strict. This might appear surprising since X_i only carries information about θ_i and it makes sense intuitively to estimate θ_i based solely on X_i. Nevertheless, the following is a counterexample:

Remark 28.6. Consider $X = \theta Z$, where $\theta \in \mathbb{N}$, $Z \sim \text{Ber}(1/2)$. The estimator $\hat{\theta}$ takes values in $\mathbb{N}$ as well and the loss function is $\ell(\theta, \hat{\theta}) = 1\{\hat{\theta} < \theta\}$, that is, whoever guesses the greater number wins. The minimax risk for this experiment is equal to $\mathbb{P}[Z = 0] = \frac{1}{2}$. To see this, note that if $Z = 0$, then all information about θ is erased. Therefore for any (randomized) estimator $P_{\hat{\theta}|X}$, the risk is lower-bounded by $R_\theta(\hat{\theta}) = \mathbb{P}[\hat{\theta} < \theta] \geq \mathbb{P}[\hat{\theta} < \theta, Z = 0] = \frac{1}{2}\mathbb{P}[\hat{\theta} < \theta | X = 0]$. Therefore sending $\theta \to \infty$ yields $\sup_\theta R_\theta(\hat{\theta}) \geq \frac{1}{2}$. This is achievable by $\hat{\theta} = X$. Clearly, this is a case where the minimax theorem does not hold, which is very similar to the previous Example 28.3.

Next consider the tensor product of two copies of this experiment with loss function $\ell(\theta, \hat{\theta}) = 1\{\hat{\theta}_1 < \theta_1\} + 1\{\hat{\theta}_2 < \theta_2\}$. We show that the minimax risk is strictly less than one. For $i = 1, 2$, let $X_i = \theta_i Z_i$, where $Z_1, Z_2 \overset{\text{iid}}{\sim} \text{Ber}(1/2)$. Consider the following estimator:

$$\hat{\theta}_1 = \hat{\theta}_2 = \begin{cases} X_1 \vee X_2, & X_1 > 0 \text{ or } X_2 > 0, \\ 1, & \text{otherwise}. \end{cases}$$

Then for any $\theta_1, \theta_2 \in \mathbb{N}$, averaging over Z_1, Z_2, we get

$$\mathbb{E}[\ell(\theta, \hat{\theta})] \leq \frac{1}{4}\left(1\{\theta_1 < \theta_2\} + 1\{\theta_2 < \theta_1\} + 1\right) \leq \frac{3}{4}.$$

We end this section by considering the minimax risk of GLM with non-quadratic loss. The following result extends Example 28.2:

Theorem 28.4. *Consider the Gaussian location model* $X_1, \ldots, X_n \overset{\text{iid}}{\sim} \mathcal{N}(\theta, I_d)$. *Then for* $1 \le q < \infty$,

$$\inf_{\hat\theta} \sup_{\theta \in \mathbb{R}^d} \mathbb{E}_\theta[\|\theta - \hat\theta\|_q^q] = \frac{\mathbb{E}[\|Z\|_q^q]}{n^{q/2}}, \qquad Z \sim \mathcal{N}(0, I_d).$$

Proof. Note that $\mathcal{N}(\theta, I_d)$ is a product distribution and the loss function is separable: $\|\theta - \hat\theta\|_q^q = \sum_{i=1}^d |\theta_i - \hat\theta_i|^q$. Thus the experiment is a d-fold tensor product of the one-dimensional version. By Theorem 28.3, it suffices to consider $d = 1$. The upper bound is achieved by the sample mean $\bar X = \frac{1}{n}\sum_{i=1}^n X_i \sim \mathcal{N}(\theta, \frac{1}{n})$, which is a sufficient statistic.

For the lower bound, following Example 28.2, consider a Gaussian prior $\theta \sim \pi = \mathcal{N}(0, s)$. Then the posterior distribution is also Gaussian: $P_{\theta|X} = \mathcal{N}(\mathbb{E}[\theta|X], \frac{s}{1+sn})$. The following lemma shows that the Bayes estimator is simply the conditional mean:

Lemma 28.5. *Let* $Z \sim \mathcal{N}(0,1)$. *Then* $\min_{y \in \mathbb{R}} \mathbb{E}[|y + Z|^q] = \mathbb{E}[|Z|^q]$.

Thus the Bayes risk is

$$R_\pi^* = \mathbb{E}[|\theta - \mathbb{E}[\theta|X]|^q] = \left(\frac{s}{1+sn}\right)^{q/2} \mathbb{E}|Z|^q.$$

Sending $s \to \infty$ proves the matching lower bound and completes the proof of the theorem. □

Proof of Lemma 28.5. Write

$$\mathbb{E}|y+Z|^q = \int_0^\infty \mathbb{P}\left[|y+Z|^q > c\right] dc \ge \int_0^\infty \mathbb{P}\left[|Z|^q > c\right] dc = \mathbb{E}|Z|^q,$$

where the inequality follows from the simple observation that for any $a > 0$, $\mathbb{P}[|y + Z| \le a] \le \mathbb{P}[|Z| \le a]$, due to the symmetry and unimodality of the normal density. □

28.6 Log-Concavity, Anderson's Lemma, and Exact Minimax Risk in GLM

As mentioned in Section 28.3.2, computing the exact minimax risk is frequently difficult especially in high dimensions. Nevertheless, for the special case of (unconstrained) GLM, the minimax risk is known exactly in arbitrary dimensions for a large collection of loss functions.[2] We have previously seen in Theorem 28.4 that this is possible for loss functions of the form $\ell(\theta, \hat\theta) = \|\theta - \hat\theta\|_q^q$. Examining the proof of this result, we note that the major limitation is that it only applies to separable loss functions, so that tensorization allows us to reduce the problem to one dimension. This does not apply to (and actually fails for) non-separable loss, since

[2] Another example is the multinomial model with squared error; see Exercises **VI.7** and **VI.9**.

Theorem 28.3, if applicable, dictates the risk to grow linearly with the dimension, which is not always the case. We next discuss a more general result that goes beyond separable losses.

Definition 28.6. A function $\rho\colon \mathbb{R}^d \to \mathbb{R}_+$ is called *bowl-shaped* if its sublevel set $K_c \triangleq \{x\colon \rho(x) \le c\}$ is convex and symmetric (i.e. $K_c = -K_c$) for all $c \in \mathbb{R}$.

Theorem 28.7. *Consider the d-dimensional GLM where $X_1, \dots, X_n \sim \mathcal{N}(0, I_d)$ are observed. Let the loss function be $\ell(\theta, \hat{\theta}) = \rho(\theta - \hat{\theta})$, where $\rho\colon \mathbb{R}^d \to \mathbb{R}_+$ is bowl-shaped and lower semicontinuous. Then the minimax risk is given by*

$$R^* \triangleq \inf_{\hat{\theta}} \sup_{\theta \in \mathbb{R}^d} \mathbb{E}_\theta[\rho(\theta - \hat{\theta})] = \mathbb{E}\rho\left(\frac{Z}{\sqrt{n}}\right), \qquad Z \sim \mathcal{N}(0, I_d).$$

Furthermore, the upper bound is attained by $\bar{X} = \frac{1}{n}\sum_{i=1}^n X_i$.

The following corollary extends Theorem 28.4 to arbitrary norms.

Corollary 28.8. *Let $\rho(\cdot) = \|\cdot\|^q$ for some $q > 0$, where $\|\cdot\|$ is an arbitrary norm on $\mathbb{R}^d$. Then*

$$R^* = \frac{\mathbb{E}\|Z\|^q}{n^{q/2}}. \tag{28.20}$$

Example 28.4.

Some applications of Corollary 28.8:

- For $\rho = \|\cdot\|_2^2$, $R^* = \frac{1}{n}\mathbb{E}\|Z\|^2 = \frac{d}{n}$, which has been shown in (28.17).
- For $\rho = \|\cdot\|_\infty$, $\mathbb{E}\|Z\|_\infty \asymp \sqrt{\log d}$ (Lemma 27.10) and $R^* \asymp \sqrt{\frac{\log d}{n}}$.
- For a matrix $\theta \in \mathbb{R}^{d\times d}$, let $\rho(\theta) = \|\theta\|_{\mathrm{op}}$ denote the operator norm (maximum singular value). It has been shown in Exercise **V.29** that $\mathbb{E}\,\|Z\|_{\mathrm{op}} \asymp \sqrt{d}$ and so $R^* \asymp \sqrt{\frac{d}{n}}$; for $\rho(\cdot) = \|\cdot\|_{\mathrm{F}}$, $R^* \asymp \frac{d}{\sqrt{n}}$.

We can also phrase the result of Corollary 28.8 in terms of the sample complexity $n^*(\epsilon)$ as defined in (28.16). For example, for $q = 2$ we have $n^*(\epsilon) = \lceil \mathbb{E}[\|Z\|^2]/\epsilon \rceil$. The above examples show that the scaling of $n^*(\epsilon)$ with dimension depends on the loss function, and the "rule of thumb" that the sampling complexity is proportional to the number of parameters need not always hold.

Finally, for the high-probability (as opposed to average) risk bound, consider $\rho(\theta - \hat{\theta}) = 1\{\|\theta - \hat{\theta}\| > \epsilon\}$, which is lower semicontinuous and bowl-shaped. Then the exact expression $R^* = \mathbb{P}\left[\|Z\| \ge \epsilon\sqrt{n}\right]$. This result is stronger since the sample mean is optimal simultaneously for all ϵ, so that integrating over ϵ recovers (28.20).

Proof of Theorem 28.7. We only prove the lower bound. We bound the minimax risk R^* from below by the Bayes risk R^*_π with the prior $\pi = \mathcal{N}(0, sI_d)$:

$$
\begin{aligned}
R^* \geq R^*_\pi &= \inf_{\hat\theta} \mathbb{E}_\pi[\rho(\theta - \hat\theta)] \\
&= \mathbb{E}\left[\inf_{\hat\theta} \mathbb{E}[\rho(\theta - \hat\theta)|X]\right] \\
&\overset{\text{(a)}}{=} \mathbb{E}[\mathbb{E}[\rho(\theta - \mathbb{E}[\theta|X])|X]] \\
&\overset{\text{(b)}}{=} \mathbb{E}\left[\rho\left(\sqrt{\frac{s}{1+sn}}Z\right)\right],
\end{aligned}
$$

where (a) follows from the crucial Lemma 28.9 below; and (b) uses the fact that $\theta - \mathbb{E}[\theta|X] \sim \mathcal{N}(0, \frac{s}{1+sn} I_d)$ under the Gaussian prior. Since $\rho(\cdot)$ is lower semicontinuous, sending $s \to \infty$ and applying Fatou's lemma, we obtain the matching lower bound:

$$
R^* \geq \lim_{s\to\infty} \mathbb{E}\left[\rho\left(\sqrt{\frac{s}{1+sn}}Z\right)\right] \geq \mathbb{E}\left[\rho\left(\frac{Z}{\sqrt{n}}\right)\right]. \qquad \square
$$

The following lemma establishes the conditional mean as the Bayes estimator under the Gaussian prior for all bowl-shaped losses, extending the previous Lemma 28.5 in one dimension:

Lemma 28.9. (Anderson [21]) *Let $X \sim \mathcal{N}(0, \Sigma)$ for some $\Sigma \succ 0$ and $\rho\colon \mathbb{R}^d \to \mathbb{R}_+$ be a bowl-shaped loss function. Then*

$$
\min_{y\in\mathbb{R}^d} \mathbb{E}[\rho(y + X)] = \mathbb{E}[\rho(X)].
$$

In order to prove Lemma 28.9, it suffices to consider ρ being indicator functions. This is done in the next lemma, which we prove later.

Lemma 28.10. *Let $K \in \mathbb{R}^d$ be a symmetric convex set and $X \sim \mathcal{N}(0, \Sigma)$. Then* $\max_{y\in\mathbb{R}^d} \mathbb{P}(X + y \in K) = \mathbb{P}(X \in K)$.

Proof of Lemma 28.9. Denote the sublevel set set $K_c = \{x \in \mathbb{R}^d\colon \rho(x) \leq c\}$. Since ρ is bowl-shaped, K_c is convex and symmetric, which satisfies the conditions of Lemma 28.10. So,

$$
\begin{aligned}
\mathbb{E}[\rho(y + x)] &= \int_0^\infty \mathbb{P}(\rho(y + x) > c)dc \\
&= \int_0^\infty (1 - \mathbb{P}(y + x \in K_c))dc \\
&\geq \int_0^\infty (1 - \mathbb{P}(x \in K_c))dc \\
&= \int_0^\infty \mathbb{P}(\rho(x) \geq c)dc \\
&= \mathbb{E}[\rho(x)].
\end{aligned}
$$

Hence, $\min_{y\in\mathbb{R}^d} \mathbb{E}[\rho(y + x)] = \mathbb{E}[\rho(x)]$. $\square$

Before going into the proof of Lemma 28.10, we need the following definition.

Definition 28.11. A measure μ on $\mathbb{R}^d$ is said to be *log-concave* if

$$\mu(\lambda A + (1-\lambda)B) \geq \mu(A)^\lambda \mu(B)^{1-\lambda}$$

for all measurable $A, B \subset \mathbb{R}^d$ and any $\lambda \in [0,1]$.

The following result, due to Prékopa [351], characterizes the log-concavity of measures in terms of that of its density function; see also [362] (or [179, Theorem 4.2]) for a proof.

Theorem 28.12. *Suppose that μ has a density f with respect to the Lebesgue measure on $\mathbb{R}^d$. Then μ is log-concave if and only if f is log-concave.*

Example 28.5.

The following are examples of log-concave measures.

- Lebesgue measure: Let $\mu = \mathrm{vol}$ be the Lebesgue measure on $\mathbb{R}^d$, which satisfies Theorem 28.12 ($f \equiv 1$). Then

$$\mathrm{vol}(\lambda A + (1-\lambda)B) \geq \mathrm{vol}(A)^\lambda \mathrm{vol}(B)^{1-\lambda}, \tag{28.21}$$

which implies[3] the *Brunn–Minkowski inequality*:

$$\mathrm{vol}(A+B)^{1/d} \geq \mathrm{vol}(A)^{1/d} + \mathrm{vol}(B)^{1/d}. \tag{28.22}$$

- Gaussian distribution: Let $\mu = \mathcal{N}(0, \Sigma)$, with a log-concave density f since $\log f(x) = -\frac{p}{2}\log(2\pi) - \frac{1}{2}\log\det(\Sigma) - \frac{1}{2}x^\top \Sigma^{-1} x$ is concave in x.

Proof of Lemma 28.10. By Theorem 28.12, the distribution of X is log-concave. Then

$$\begin{aligned}
\mathbb{P}[X \in K] &\overset{\text{(a)}}{=} \mathbb{P}\Big[X \in \frac{1}{2}(K+y) + \frac{1}{2}(K-y)\Big] \\
&\overset{\text{(b)}}{\geq} \sqrt{\mathbb{P}[X \in K-y]\mathbb{P}[X \in K+y]} \\
&\overset{\text{(c)}}{=} \mathbb{P}[X + y \in K],
\end{aligned}$$

where (a) follows from $\frac{1}{2}(K+y) + \frac{1}{2}(K-y) = \frac{1}{2}K + \frac{1}{2}K = K$ since K is convex; (b) follows from the definition of log-concavity in Definition 28.11 with $\lambda = \frac{1}{2}$, $A = K - y = \{x - y : x \in K\}$ and $B = K + y$; and (c) follows from $\mathbb{P}[X \in K+y] = \mathbb{P}[X \in -K - y] = \mathbb{P}[X + y \in K]$ since X has a symmetric distribution and K is symmetric ($K = -K$). □

[3] Applying (28.21) to $A' = \mathrm{vol}(A)^{-1/d}A$, $B' = \mathrm{vol}(B)^{-1/d}B$ (both of which have unit volume), and $\lambda = \mathrm{vol}(A)^{1/d}/(\mathrm{vol}(A)^{1/d} + \mathrm{vol}(B)^{1/d})$ yields (28.22).

29 Classical Large-Sample Asymptotics

In this chapter we give an overview of the classical large-sample theory in the setting of iid observations in Section 28.4 focusing again on the minimax risk (28.15). These results pertain to smooth parametric models in fixed dimensions, with the sole asymptotics being the sample size going to infinity. The main result is that, under suitable conditions, the minimax squared error of estimating θ based on $X_1, \ldots, X_n \stackrel{\text{iid}}{\sim} P_\theta$ satisfies

$$\inf_{\hat{\theta}} \sup_{\theta \in \Theta} \mathbb{E}_\theta \left[\|\hat{\theta} - \theta\|_2^2 \right] = \frac{1 + o(1)}{n} \sup_{\theta \in \Theta} \operatorname{Tr} J_F^{-1}(\theta), \tag{29.1}$$

where $J_F(\theta)$ is the Fisher information matrix introduced in (2.32) in Chapter 2. This is an asymptotic characterization of the minimax risk with a sharp constant. In later chapters, we will proceed to high dimensions where such precise results are difficult and rare.

Throughout this chapter, we focus on the quadratic risk and assume that Θ is an open set of the Euclidean space $\mathbb{R}^d$. While reading this chapter, the reader is advised to consult Exercise **VI.7** to understand the minimax risk in a simple setting of estimating the parameter of a Bernoulli model.

29.1 Statistical Lower Bound from Data Processing

In this section we derive several statistical lower bounds from a data-processing argument. Specifically, we will take a comparison-of-experiment approach by comparing the actual model with a perturbed model. The performance of a given estimator can then be related to the f-divergence via the data-processing inequality and the variational representation (Chapter 7).

We start by discussing the Hammersley–Chapman–Robbins lower bound which implies the well-known Cramér–Rao lower bound. Because these results are restricted to unbiased estimators, we will also discuss their Bayesian versions; in particular, the Bayesian Cramér–Rao lower bound is responsible for proving the lower bound in (29.1). We focus on explaining how these results can be anticipated from information-theoretic reasoning and postpone the exact statement and assumption of the Bayesian Cramér–Rao bound to Section 29.2.

29.1.1 Hammersley–Chapman–Robbins (HCR) Lower Bound

The following result due to [210, 87] is a direct consequence of the variational representation of χ^2-divergence in Section 7.13, which relates it to the mean and variance of test functions.

Theorem 29.1. (HCR lower bound) *The quadratic loss of any estimator* $\hat{\theta}$ *at* $\theta \in \Theta \subset \mathbb{R}^d$ *satisfies*

$$R_\theta(\hat{\theta}) = \mathbb{E}_\theta[(\hat{\theta} - \theta)^2] \geq \mathrm{Var}_\theta(\hat{\theta}) \geq \sup_{\theta' \neq \theta} \frac{(\mathbb{E}_\theta[\hat{\theta}] - \mathbb{E}_{\theta'}[\hat{\theta}])^2}{\chi^2(P_{\theta'} \| P_\theta)}. \tag{29.2}$$

Proof. Let $\hat{\theta}$ be a (possibly randomized) estimator based on X. Fix $\theta' \neq \theta \in \Theta$. Denote by P and Q the probability distribution when the true parameter is θ or θ', respectively. That is, $P_X = P_\theta$ and $Q_X = P_{\theta'}$. Then

$$\chi^2(P_X \| Q_X) \geq \chi^2(P_{\hat{\theta}} \| Q_{\hat{\theta}}) \geq \frac{(\mathbb{E}_\theta[\hat{\theta}] - \mathbb{E}_{\theta'}[\hat{\theta}])^2}{\mathrm{Var}_\theta(\hat{\theta})}, \tag{29.3}$$

where the first inequality applies the data-processing inequality (Theorem 7.4) and the second inequality applies the variational representation (7.91) of the χ^2-divergence. □

Next we apply Theorem 29.1 to an unbiased estimator $\hat{\theta}$ that satisfies $\mathbb{E}_\theta[\hat{\theta}] = \theta$ for all $\theta \in \Theta$. Then

$$\mathrm{Var}_\theta(\hat{\theta}) \geq \sup_{\theta' \neq \theta} \frac{(\theta - \theta')^2}{\chi^2(P_{\theta'} \| P_\theta)}.$$

Lower-bounding the supremum by the limit of $\theta' \to \theta$ and recalling the asymptotic expansion of χ^2-divergence from Theorem 7.22 in terms of the Fisher information, we get, under the regularity conditions in Theorem 7.22, the celebrated Cramér–Rao (CR) lower bound [108, 355]:

$$\mathrm{Var}_\theta(\hat{\theta}) \geq \frac{1}{J_F(\theta)}. \tag{29.4}$$

A few more remarks are as follows:

- Note that the HCR lower bound in Theorem 29.1 is based on the χ^2-divergence. For a version based on Hellinger distance which also implies the CR lower bound, see Exercise **VI.5**.
- Both the HCR and the CR lower bounds extend to the multivariate case as follows. Let $\hat{\theta}$ be an unbiased estimator of $\theta \in \Theta \subset \mathbb{R}^d$. Assume that its covariance matrix $\mathrm{Cov}_\theta(\hat{\theta}) = \mathbb{E}_\theta[(\hat{\theta} - \theta)(\hat{\theta} - \theta)^\top]$ is positive definite. Fix $a \in \mathbb{R}^d$. Applying Theorem 29.1 to $\langle a, \hat{\theta} \rangle$, we get

$$\chi^2(P_\theta \| P_{\theta'}) \geq \frac{\langle a, \theta - \theta' \rangle^2}{a^\top \mathrm{Cov}_\theta(\hat{\theta}) a}.$$

Optimizing over a yields[1]

$$\chi^2(P_\theta \| P_{\theta'}) \geq (\theta - \theta')^\top \operatorname{Cov}_\theta(\hat{\theta})^{-1}(\theta - \theta').$$

Sending $\theta' \to \theta$ and applying the asymptotic expansion $\chi^2(P_\theta \| P_{\theta'}) = (\theta - \theta')^\top J_F(\theta)(\theta - \theta')(1 + o(1))$ (see Remark 7.13), we get the multivariate version of the CR lower bound:

$$\operatorname{Cov}_\theta(\hat{\theta}) \succeq J_F^{-1}(\theta). \tag{29.5}$$

- For a sample of n iid observations, by the additivity property (2.36), the Fisher information matrix is equal to $nJ_F(\theta)$. Taking the trace on both sides, we conclude that the squared error of any unbiased estimators satisfies

$$\mathbb{E}_\theta[\|\hat{\theta} - \theta\|_2^2] \geq \frac{1}{n}\operatorname{Tr}(J_F^{-1}(\theta)).$$

This is already very close to (29.1), except for the fundamental restriction of unbiased estimators.

29.1.2 Bayesian CR and HCR

The drawback of the HCR and CR lower bounds is that they are confined to unbiased estimators. For the minimax settings in (29.1), there is no sound reason to restrict to unbiased estimators; in fact, it is often wise to trade bias with variance in order to achieve a smaller overall risk.

Next we discuss a lower bound, known as the Bayesian Cramér–Rao (BCR) lower bound [188] or the van Trees inequality [434], for a Bayesian setting that applies to *all* estimators; to apply to the minimax setting, in view of Theorem 28.1, one just needs to choose an appropriate prior. The exact statement and the application to minimax risk are postponed till the next section. Here we continue the previous line of thinking and derive it from the data-processing argument.

Fix a prior π on Θ and a (possibly randomized) estimator $\hat{\theta}$. Then we have the Markov chain $\theta \to X \to \hat{\theta}$. Consider two joint distributions for (θ, X).

- Under Q, θ is drawn from π and $X \sim P_\theta$ conditioned on θ;
- Under P, θ is drawn from $T_\delta \pi$, where T_δ denotes the pushforward of shifting by δ, that is, $T_\delta \pi(A) = \pi(A - \delta)$, and $X \sim P_{\theta - \delta}$ conditioned on θ.

Similar to (29.3), applying data processing and variational representation of χ^2-divergence yields

$$\chi^2(P_{\theta X} \| Q_{\theta X}) \geq \chi^2(P_{\theta\hat{\theta}} \| Q_{\theta\hat{\theta}}) \geq \chi^2(P_{\theta-\hat{\theta}} \| Q_{\theta-\hat{\theta}}) \geq \frac{(\mathbb{E}_P[\theta - \hat{\theta}] - \mathbb{E}_Q[\theta - \hat{\theta}])^2}{\operatorname{Var}_Q(\hat{\theta} - \theta)}.$$

Note that by design, $P_X = Q_X$ and thus $\mathbb{E}_P[\hat{\theta}] = \mathbb{E}_Q[\hat{\theta}]$; on the other hand, $\mathbb{E}_P[\theta] = \mathbb{E}_Q[\theta] + \delta$. Furthermore, $\mathbb{E}_\pi[(\hat{\theta} - \theta)^2] \geq \operatorname{Var}_Q(\hat{\theta} - \theta)$. Since this

[1] For $\Sigma \succ 0$, $\sup_{x \neq 0} \frac{\langle x, y \rangle^2}{x^\top \Sigma x} = y^\top \Sigma^{-1} y$, attained at $x = \Sigma^{-1} y$.

applies to any estimators, we conclude that the Bayes risk R_π^* (and hence the minimax risk) satisfies

$$R_\pi^* \triangleq \inf_{\hat{\theta}} \mathbb{E}_\pi[(\hat{\theta} - \theta)^2] \geq \sup_{\delta \neq 0} \frac{\delta^2}{\chi^2(P_{X\theta} \| Q_{X\theta})}, \tag{29.6}$$

which is referred to as the *Bayesian HCR* lower bound in comparison with (29.2).

Similar to the deduction of the CR lower bound from the HCR, we can further lower-bound this supremum by evaluating the small-δ limit. First note the following chain rule for the χ^2-divergence:

$$\chi^2(P_{X\theta} \| Q_{X\theta}) = \chi^2(P_\theta \| Q_\theta) + \mathbb{E}_Q\left[\chi^2(P_{X|\theta} \| Q_{X|\theta}) \cdot \left(\frac{dP_\theta}{dQ_\theta}\right)^2\right].$$

Under suitable regularity conditions in Theorem 7.22, again applying the local expansion of χ^2-divergence yields

- $\chi^2(P_\theta \| Q_\theta) = \chi^2(T_\delta \pi \| \pi) = (J(\pi) + o(1))\delta^2$, where $J(\pi) \triangleq \int \frac{\pi'^2}{\pi}$ is the Fisher information of the prior (see Example 2.7);
- $\chi^2(P_{X|\theta} \| Q_{X|\theta}) = [J_F(\theta) + o(1)]\delta^2$.

Thus from (29.6) we get

$$R_\pi^* \geq \frac{1}{J(\pi) + \mathbb{E}_{\theta \sim \pi}[J_F(\theta)]}. \tag{29.7}$$

We conclude this section by revisiting the Gaussian location model (GLM) in Example 28.1.

Example 29.1.

Let $X^n = (X_1, \ldots, X_n) \overset{\text{iid}}{\sim} \mathcal{N}(\theta, 1)$ and consider the prior $\theta \sim \pi = \mathcal{N}(0, s)$. To apply the Bayesian HCR bound (29.6), note that

$$\begin{aligned}
\chi^2(P_{\theta X^n} \| Q_{\theta X^n}) &\overset{(a)}{=} \chi^2(P_{\theta \bar{X}} \| Q_{\theta \bar{X}}) \\
&= \chi^2(P_\theta \| Q_\theta) + \mathbb{E}_Q\left[\left(\frac{dP_\theta}{dQ_\theta}\right)^2 \chi^2(P_{\bar{X}|\theta} \| Q_{\bar{X}|\theta})\right] \\
&\overset{(b)}{=} e^{\delta^2/s} - 1 + e^{\delta^2/s}(e^{n\delta^2} - 1) \\
&= e^{\delta^2(n + \frac{1}{s})} - 1,
\end{aligned}$$

where (a) follows from the sufficiency of $\bar{X} = \frac{1}{n}\sum_{i=1}^n X_i$; and (b) is by $Q_\theta = \mathcal{N}(0, s)$, $Q_{\bar{X}|\theta} = \mathcal{N}(\theta, \frac{1}{n})$, $P_\theta = \mathcal{N}(\delta, s)$, $P_{\bar{X}|\theta} = \mathcal{N}(\theta - \delta, \frac{1}{n})$, and the fact (7.43) for Gaussians. Therefore,

$$R_\pi^* \geq \sup_{\delta \neq 0} \frac{\delta^2}{e^{\delta^2(n + \frac{1}{s})} - 1} = \lim_{\delta \to 0} \frac{\delta^2}{e^{\delta^2(n + \frac{1}{s})} - 1} = \frac{s}{sn + 1}.$$

In view of the Bayes risk found in Example 28.1, we see that in this case the Bayesian HCR and Bayesian Cramér–Rao lower bounds are *exact*.

29.2 Bayesian CR Lower Bounds and Extensions

In this section we give the rigorous statement of the Bayesian Cramér–Rao lower bound and discuss its extensions and consequences. For the proof, we take a more direct approach as opposed to the data-processing argument in Section 29.1 based on asymptotic expansion of the χ^2-divergence.

Theorem 29.2. (BCR lower bound) *Let π be a differentiable prior density on the interval $[\theta_0, \theta_1]$ such that $\pi(\theta_0) = \pi(\theta_1) = 0$ and*

$$J(\pi) \triangleq \int_{\theta_0}^{\theta_1} \frac{\pi'(\theta)^2}{\pi(\theta)} d\theta < \infty. \tag{29.8}$$

Let $P_\theta(dx) = p_\theta(x)\mu(dx)$, where the density $p_\theta(x)$ is differentiable in θ for μ-almost every x. Assume that for π-almost every θ,

$$\int \mu(dx)\partial_\theta p_\theta(x) = 0. \tag{29.9}$$

Then the Bayes quadratic risk $R_\pi^* \triangleq \inf_{\hat{\theta}} \mathbb{E}[(\theta - \hat{\theta})^2]$ satisfies

$$R_\pi^* \geq \frac{1}{\mathbb{E}_{\theta\sim\pi}[J_F(\theta)] + J(\pi)}. \tag{29.10}$$

Proof. In view of Remark 28.3, it loses no generality to assume that the estimator $\hat{\theta} = \hat{\theta}(X)$ is deterministic. For each x, integration by parts yields

$$\int_{\theta_0}^{\theta_1} d\theta(\hat{\theta}(x) - \theta)\partial_\theta(p_\theta(x)\pi(\theta)) = \int_{\theta_0}^{\theta_1} p_\theta(x)\pi(\theta)d\theta.$$

Integrating both sides over $\mu(dx)$ yields

$$\mathbb{E}[(\hat{\theta} - \theta)V(\theta, X)] = 1,$$

where $V(\theta, x) \triangleq \partial_\theta(\log(p_\theta(x)\pi(\theta))) = \partial_\theta \log p_\theta(x) + \partial_\theta \log \pi(\theta)$ and the expectation is over the joint distribution of (θ, X). Applying Cauchy–Schwarz, we have $\mathbb{E}[(\hat{\theta} - \theta)^2]\mathbb{E}[V(\theta, X)^2] \geq 1$. The proof is completed by noting that $\mathbb{E}[V(\theta, X)^2] = \mathbb{E}[(\partial_\theta \log p_\theta(X))^2] + \mathbb{E}[(\partial_\theta \log \pi(\theta))^2] = \mathbb{E}[J_F(\theta)] + J(\pi)$, thanks to the assumption (29.9). □

The multivariate version of Theorem 29.2 is the following.

Theorem 29.3. (Multivariate BCR) *Consider a product prior density $\pi(\theta) = \prod_{i=1}^d \pi_i(\theta_i)$ over the box $\prod_{i=1}^d [\theta_{0,i}, \theta_{1,i}]$, where each π_i is differentiable on $[\theta_{0,i}, \theta_{1,i}]$ and vanishes on the boundary. Assume that for π-almost every θ,*

$$\int \mu(dx)\nabla_\theta p_\theta(x) = 0. \tag{29.11}$$

Then

$$R_\pi^* \triangleq \inf_{\hat{\theta}} \mathbb{E}_\pi[\|\hat{\theta} - \theta\|_2^2] \geq \mathrm{Tr}((\mathbb{E}_{\theta\sim\pi}[J_F(\theta)] + J(\pi))^{-1}), \tag{29.12}$$

where the Fisher information matrices are given by $J_F(\theta) = \mathbb{E}_\theta[\nabla_\theta \log p_\theta(X) \nabla_\theta \log p_\theta(X)^\top]$ *and* $J(\pi) = \text{diag}(J(\pi_1), \ldots, J(\pi_d))$.

Proof. Fix an estimator $\hat{\theta} = (\hat{\theta}_1, \ldots, \hat{\theta}_d)$ and a non-zero $u \in \mathbb{R}^d$. For each $i, k = 1, \ldots, d$, integration by parts yields

$$\int_{\theta_{0,i}}^{\theta_{1,i}} (\hat{\theta}_k(x) - \theta_k) \partial_{\theta_i}(p_\theta(x)\pi(\theta)) d\theta_i = 1\{k = i\} \int_{\theta_{0,i}}^{\theta_{1,i}} p_\theta(x)\pi(\theta) d\theta_i.$$

Integrating both sides over $\prod_{j \neq i} d\theta_j$ and $\mu(dx)$, multiplying by u_i, and summing over i, we obtain

$$\mathbb{E}[(\hat{\theta}_k(X) - \theta_k)\langle u, \nabla \log(p_\theta(X)\pi(\theta))\rangle] = \langle u, e_k \rangle,$$

where e_k denotes the kth standard basis. Applying Cauchy–Schwarz and optimizing over u yields

$$\mathbb{E}[(\hat{\theta}_k(X) - \theta_k)^2] \geq \sup_{u \neq 0} \frac{\langle u, e_k \rangle^2}{u^\top \Sigma u} = \Sigma_{kk}^{-1},$$

where $\Sigma \equiv \mathbb{E}[\nabla \log(p_\theta(X)\pi(\theta)) \nabla \log(p_\theta(X)\pi(\theta))^\top] = \mathbb{E}_{\theta \sim \pi}[J_F(\theta)] + J(\pi)$, thanks to (29.11). Summing over k completes the proof of (29.12). □

Several remarks are in order:

- The above versions of the BCR bound assume a prior density that vanishes at the boundary. If we choose a uniform prior, the same derivation leads to a similar lower bound known as the Chernoff–Rubin–Stein inequality (see Exercise **VI.4**), which also suffices for proving the optimal minimax lower bound in (29.1).
- For the purpose of the lower bound, it is advantageous to choose a prior density with the minimum Fisher information. The optimal density with a compact support is known to be a squared cosine density [219, 427]:

$$\min_{g \text{ on } [-1,1]} J(g) = \pi^2,$$

attained by

$$g(u) = \cos^2 \frac{\pi u}{2}. \tag{29.13}$$

- Suppose the goal is to estimate a smooth *functional* $T(\theta)$ of the unknown parameter θ, where $T\colon \mathbb{R}^d \to \mathbb{R}^s$ is differentiable with $\nabla T(\theta) = (\frac{\partial T_i(\theta)}{\partial \theta_j})$ its $s \times d$ Jacobian matrix. Then under the same conditions as in Theorem 29.3, we have the following Bayesian Cramér–Rao lower bound for functional estimation:

$$\inf_{\hat{T}} \mathbb{E}_\pi[\|\hat{T}(X) - T(\theta)\|_2^2] \geq \text{Tr}(\mathbb{E}[\nabla T(\theta)](\mathbb{E}[J_F(\theta)] + J(\pi))^{-1}\mathbb{E}[\nabla T(\theta)]^\top), \tag{29.14}$$

where the expectation on the right-hand side is over $\theta \sim \pi$.

As a consequence of the BCR bound, we prove the lower-bound part for the asymptotic minimax risk in (29.1).

Theorem 29.4. *Assume that $\theta \mapsto J_F(\theta)$ is continuous. Denote the minimax squared error $R_n^* \triangleq \inf_{\hat{\theta}} \sup_{\theta \in \Theta} \mathbb{E}_\theta[\|\hat{\theta} - \theta\|_2^2]$, where $\mathbb{E}_\theta$ is taken over $X_1, \ldots, X_n \overset{\text{iid}}{\sim} P_\theta$. Then as $n \to \infty$,*

$$R_n^* \geq \frac{1 + o(1)}{n} \sup_{\theta \in \Theta} \operatorname{Tr} J_F^{-1}(\theta). \tag{29.15}$$

Proof. Fix $\theta \in \Theta$. Then for all sufficiently small δ, $B_\infty(\theta, \delta) = \theta + [-\delta, \delta]^d \subset \Theta$. Let $\pi_i(\theta_i) = \frac{1}{\delta} g(\frac{\theta - \theta_i}{\delta})$, where g is the prior density in (29.13). Then the product distribution $\pi = \prod_{i=1}^d \pi_i$ satisfies the assumptions of Theorem 29.3. By the scaling rule of Fisher information (see (2.35)), $J(\pi_i) = \frac{1}{\delta^2} J(g) = \frac{\pi^2}{\delta^2}$. Thus $J(\pi) = \frac{\pi^2}{\delta^2} I_d$.

It is known that (see [68, Theorem 2, Appendix V]) the continuity of $\theta \mapsto J_F(\theta)$ implies (29.11). So we are ready to apply the BCR bound in Theorem 29.3. Lower-bounding the minimax by the Bayes risk and also applying the additivity property (2.36) of Fisher information, we obtain

$$R_n^* \geq \frac{1}{n} \cdot \operatorname{Tr} \left(\left(\mathbb{E}_{\theta \sim \pi}[J_F(\theta)] + \frac{\pi^2}{n\delta^2} I_d \right)^{-1} \right).$$

Finally, choosing $\delta = n^{-1/4}$ and applying the continuity of $J_F(\theta)$ in θ, the desired (29.15) follows. $\square$

Similarly, for estimating a smooth functional $T(\theta)$, applying (29.14) with the same argument yields

$$\inf_{\hat{T}} \sup_{\theta \in \Theta} \mathbb{E}_\theta[\|\hat{T} - T(\theta)\|_2^2] \geq \frac{1 + o(1)}{n} \sup_{\theta \in \Theta} \operatorname{Tr}(\nabla T(\theta) J_F^{-1}(\theta) \nabla T(\theta)^\top). \tag{29.16}$$

29.3 Maximum Likelihood Estimator and Asymptotic Efficiency

Theorem 29.4 shows that in a small neighborhood of each parameter θ, the best estimation error is at best $\frac{1}{n}(\operatorname{Tr} J_F^{-1}(\theta) + o(1))$ when the sample size n grows; this is known as the *information bound* as determined by the Fisher information matrix. Estimators achieving this bound are called *asymptotic efficient*. A cornerstone of the classical large-sample theory is the asymptotic efficiency of the *maximum likelihood estimator* (MLE). Rigorously stating this result requires a lengthy list of technical conditions, and an even lengthier one is needed to make the error uniform so as to attain the minimax lower bound in Theorem 29.4. In this section we give a sketch of the asymptotic analysis of MLE, focusing on the main ideas and how Fisher information emerges from the likelihood optimization.

Suppose we observe a sample $X^n = (X_1, X_2, \ldots, X_n) \overset{\text{iid}}{\sim} P_{\theta_0}$, where θ_0 stands for the true parameter. The MLE is defined as

$$\hat{\theta}_{\text{MLE}} \in \arg\max_{\theta \in \Theta} L_\theta(X^n), \tag{29.17}$$

where

$$L_\theta(X^n) = \sum_{i=1}^n \log p_\theta(X_i)$$

is the total log-likelihood and $p_\theta(x) = \frac{dP_\theta}{d\mu}(x)$ is the density of P_θ with respect to some common dominating measure μ. For a discrete distribution P_θ, the MLE can also be written as the KL projection[2] of the empirical distribution $\hat{P}_n$ to the model class: $\hat{\theta}_{\text{MLE}} \in \arg\min_{\theta\in\Theta} D(\hat{P}_n \| P_\theta)$.

The main intuition why MLE works is as follows. Assume that the model is identifiable, namely, $\theta \mapsto P_\theta$ is injective. Then for any $\theta \neq \theta_0$, we have by positivity of the KL divergence (Theorem 2.3)

$$\mathbb{E}_{\theta_0}\left[L_\theta - L_{\theta_0}\right] = \mathbb{E}_{\theta_0}\left[\sum_{i=1}^n \log \frac{p_\theta(X_i)}{p_{\theta_0}(X_i)}\right] = -nD(P_{\theta_0}\|P_\theta) < 0.$$

In other words, $L_\theta - L_{\theta_0}$ is an iid sum with a negative mean and thus negative with high probability for large n. From here the consistency of MLE follows upon assuming appropriate regularity conditions, among which is Wald's integrability condition $\mathbb{E}_{\theta_0}[\sup_{\|\theta-\theta_0\|\leq\epsilon} \log \frac{p_\theta}{p_{\theta_0}}(X)] < \infty$ [450, 455].

Assuming more conditions one can obtain the asymptotic normality and efficiency of the MLE. This follows from the local quadratic approximation of the log-likelihood function. Define $V(\theta, x) \triangleq \nabla_\theta \log p_\theta(x)$ (score) and $H(\theta, x) \triangleq \nabla_\theta^2 \log p_\theta(x)$. By Taylor expansion,

$$L_\theta = L_{\theta_0} + (\theta-\theta_0)^\top \left(\sum_{i=1}^n V(\theta_0, X_i)\right) + \frac{1}{2}(\theta-\theta_0)^\top \left(\sum_{i=1}^n H(\theta_0, X_i)\right)(\theta-\theta_0) + o(n(\theta-\theta_0)^2). \tag{29.18}$$

Recall from Section 2.6.2* that, under suitable regularity conditions, we have

$$\mathbb{E}_{\theta_0}[V(\theta_0, X)] = 0, \quad \mathbb{E}_{\theta_0}[V(\theta_0, X)V(\theta_0, X)^\top] = -\mathbb{E}_{\theta_0}[H(\theta_0, X)] = J_F(\theta_0).$$

Thus, by the central limit theorem and the weak law of large numbers, we have

$$\frac{1}{\sqrt{n}}\sum_{i=1}^n V(\theta_0, X_i) \xrightarrow{\text{d}} \mathcal{N}(0, J_F(\theta_0)), \quad \frac{1}{n}\sum_{i=1}^n H(\theta_0, X_i) \xrightarrow{\mathbb{P}} -J_F(\theta_0).$$

Substituting these quantities into (29.18), we obtain the following stochastic approximation of the log-likelihood:

$$L_\theta \approx L_{\theta_0} + \left\langle \sqrt{nJ_F(\theta_0)}Z, \theta-\theta_0 \right\rangle - \frac{n}{2}(\theta-\theta_0)^\top J_F(\theta_0)(\theta-\theta_0),$$

where $Z \sim \mathcal{N}(0, I_d)$. Maximizing the right-hand side yields:

$$\hat{\theta}_{\text{MLE}} \approx \theta_0 + \frac{1}{\sqrt{n}} J_F(\theta_0)^{-1/2} Z.$$

[2] Note that this is the reverse of the information projection studied in Section 15.3.

From this asymptotic normality, we can obtain $\mathbb{E}_{\theta_0}[\|\hat{\theta}_{\text{MLE}} - \theta_0\|_2^2] \leq \frac{1}{n}(\operatorname{Tr} J_F(\theta_0)^{-1} + o(1))$, and for smooth functionals by Taylor-expanding T at θ_0 (delta method), $\mathbb{E}_{\theta_0}[\|T(\hat{\theta}_{\text{MLE}}) - T(\theta_0)\|_2^2] \leq \frac{1}{n}(\operatorname{Tr}(\nabla T(\theta_0) J_F(\theta_0)^{-1} \nabla T(\theta_0)^\top) + o(1))$, matching the information bounds (29.15) and (29.16).

Of course, the above heuristic derivation requires additional assumptions to justify (for example, Cramér's condition, see [168, Theorem 18] and [376, Theorem 7.63]). Even stronger assumptions are needed to ensure the error is uniform in θ in order to achieve the minimax lower bound in Theorem 29.4; see, for example, Theorem 34.4 (and also Chapters 36 and 37) of [68] for the exact conditions and statements. A more general and abstract theory of MLE and the attainment of information bound were developed by Hájek and Le Cam; see [209, 274].

Despite its wide applicability and strong optimality properties, the methodology of MLE is not without limitations. We conclude this section with some remarks along this line.

- MLE may not exist even for simple parametric models. For example, consider $X_1, \ldots, X_n$ drawn iid from the location-scale mixture of two Gaussians $\frac{1}{2}\mathcal{N}(\mu_1, \sigma_1^2) + \frac{1}{2}\mathcal{N}(\mu_2, \sigma_2^2)$, where $(\mu_1, \mu_2, \sigma_1, \sigma_2)$ are unknown parameters. Then the likelihood can be made arbitrarily large by setting for example $\mu_1 = X_1$ and $\sigma_1 \to 0$.
- MLE may be inconsistent; see [376, Example 7.61] and [167] for examples, both in a one-dimensional parametric family.
- In high dimensions, it is possible that MLE fails to achieve the minimax rate (Exercise **VI.15**).

29.4 Application: Estimating Discrete Distributions and Entropy

As an application in this section we consider the concrete problems of estimating a discrete distribution or its property (such as Shannon entropy) based on iid observations. Of course, the asymptotic theory developed in this chapter applies only to the classical setting of fixed alphabet and large sample size. Along the way, we will also discuss extensions a to large alphabet and what may go wrong with the classical theory.

Throughout this section, let $X_1, \ldots, X_n \overset{\text{iid}}{\sim} P \in \mathcal{P}_k$, where $\mathcal{P}_k \equiv \mathcal{P}([k])$ denotes the collection of probability distributions over $[k] = \{1, \ldots, k\}$. We first consider the estimation of P under the squared loss.

Theorem 29.5. *For fixed k, the minimax squared error of estimating P satisfies*

$$R_{\text{sq}}^*(k, n) \triangleq \inf_{\widehat{P}} \sup_{P \in \mathcal{P}_k} \mathbb{E}[\|\widehat{P} - P\|_2^2] = \frac{1}{n}\left(\frac{k-1}{k} + o(1)\right), \quad n \to \infty. \tag{29.19}$$

Proof. Let $P = (P_1, \ldots, P_k)$ be parameterized, as in Example 2.6, by $\theta = (P_1, P_2, \ldots, P_{k-1})$ and $P_k = 1 - P_1 - \cdots - P_{k-1}$. Then $P = T(\theta)$, where

$T\colon \mathbb{R}^{k-1} \to \mathbb{R}^k$ is an affine functional so that $\nabla T(\theta) = \left[\begin{smallmatrix} I_{k-1} \\ -\mathbf{1}^\top \end{smallmatrix}\right]$, with $\mathbf{1}$ being the all-ones (column) vector.

The Fisher information matrix and its inverse have been calculated in (2.37) and (2.38): We have $J_F^{-1}(\theta) = \mathrm{diag}(\theta) - \theta\theta^\top$ and

$$\nabla T(\theta) J_F^{-1}(\theta) \nabla T(\theta)^\top = \begin{bmatrix} \mathrm{diag}(\theta) - \theta\theta^\top & -P_k\theta \\ -P_k\theta^\top & P_k(1-P_k). \end{bmatrix}$$

Thus $\mathrm{Tr}(\nabla T(\theta) J_F^{-1}(\theta) \nabla T(\theta)^\top) = \sum_{i=1}^k P_i(1-P_i) = 1 - \sum_{i=1}^k P_i^2$, which achieves its maximum $1 - \frac{1}{k}$ at the uniform distribution. Applying the functional form of the BCR bound in (29.16), we conclude $R^*_{\mathrm{sq}}(k,n) \geq \frac{1}{n}(1 - \frac{1}{k} + o(1))$.

For the upper bound, consider the MLE, which in this case coincides with the empirical distribution $\hat{P} = (\hat{P}_i)$ (Exercise **VI.8**). Note that $n\hat{P}_i = \sum_{j=1}^n 1\{X_j = i\} \sim \mathrm{Bin}(n, P_i)$. Then for any P, $\mathbb{E}[\|\hat{P} - P\|_2^2] = \frac{1}{n}\sum_{i=1}^k P_i(1-P_i) \leq \frac{1}{n}(1 - \frac{1}{k})$. □

Some remarks on Theorem 29.5 are in order:

- In fact, for any k, n, we have the precise result: $R^*_{\mathrm{sq}}(k,n) = \frac{1-1/k}{(1+\sqrt{n})^2}$ – see Exercise **VI.7h**. This can be shown by considering a Dirichlet prior (13.16) and applying the corresponding Bayes estimator, which is an additively smoothed empirical distribution (Section 13.5).
- Note that $R^*_{\mathrm{sq}}(k,n)$ does not grow with the alphabet size k; this is because squared loss is too weak for estimating probability vectors. More meaningful loss functions include the f-divergences in Chapter 7, such as the total variation, KL divergence, and χ^2-divergence. These minimax rates are worked out in Exercises **VI.8** and **VI.10**, for both small and large alphabets, and they indeed depend on the alphabet size k. For example, the minimax KL risk satisfies $\Theta(\frac{k}{n})$ for $k \leq n$ and grows as $\Theta(\log \frac{k}{n})$ for $k \gg n$. This agrees with the rule of thumb that consistent estimation requires the sample size to scale faster than the dimension.

As a final application, let us consider the classical problem of *entropy estimation* in information theory and statistics [305, 128, 215], where the goal is to estimate the Shannon entropy, a non-linear functional of P. The following result follows from the functional BCR lower bound (29.16) and analyzing the MLE (in this case the empirical entropy) [39].

Theorem 29.6. *For fixed k, the minimax quadratic risk of entropy estimation satisfies*

$$R^*_{\mathrm{ent}}(k,n) \triangleq \inf_{\widehat{H}} \sup_{P \in \mathcal{P}_k} \mathbb{E}[(\widehat{H}(X_1,\ldots,X_n) - H(P))^2] = \frac{1}{n}\left(\max_{P \in \mathcal{P}_k} V(P) + o(1)\right),\quad n \to \infty,$$

where $H(P) = \sum_{i=1}^k P_i \log \frac{1}{P_i} = \mathbb{E}[\log \frac{1}{P(X)}]$ *and* $V(P) = \mathrm{Var}[\log \frac{1}{P(X)}]$ *are the Shannon entropy and varentropy (see (10.4)) of P.*

Let us analyze the result of Theorem 29.6 and see how it extends to large alphabets. It can be shown that[3] $\max_{P \in \mathcal{P}_k} V(P) \asymp \log^2 k$, which suggests that

[3] Indeed, $\max_{P \in \mathcal{P}_k} V(P) \leq \log^2 k$ for all $k \geq 3$ [335, Eq. (464)]. For the lower bound, consider $P = (\frac{1}{2}, \frac{1}{2(k-1)}, \ldots \frac{1}{2(k-1)})$.

$R^*_{\text{ent}} \equiv R^*_{\text{ent}}(k,n)$ may satisfy $R^*_{\text{ent}} \asymp \frac{\log^2 k}{n}$ even when the alphabet size k grows with n; however, this result only holds for sufficiently small alphabet. In fact, back in Lemma 13.2 we have shown that for the empirical entropy which achieves the bound in Theorem 29.6, its bias is on the order of $\frac{k}{n}$, which is no longer negligible on large alphabets. Using techniques of polynomial approximation [457, 233], one can reduce this bias to $\frac{k}{n \log k}$ and further show that consistent entropy estimation is only possible if and only if $n \gg \frac{k}{\log k}$ [429], in which case the minimax rate satisfies

$$R^*_{\text{ent}} \asymp \left(\frac{k}{n \log k}\right)^2 + \frac{\log^2 k}{n}.$$

In summary, one needs to exercise caution extending classical large-sample results to high dimensions, especially when bias becomes the dominating factor.

30 Mutual Information Method

In this chapter we describe a strategy for proving statistical lower bound we call the *mutual information method* (MIM), which entails comparing the amount of information data provides with the minimum amount of information needed to achieve a certain estimation accuracy. Similar to Section 29.2, the main information-theoretical ingredient is the data-processing inequality, this time for mutual information as opposed to f-divergences.

Here is the main idea of the MIM: We fix some prior π on Θ and we aim to lower-bound the Bayes risk R_π^* of estimating $\theta \sim \pi$ on the basis of X with respect to some loss function ℓ. Let $\hat{\theta}$ be an estimator such that $\mathbb{E}[\ell(\theta,\hat{\theta})] \leq D$. Then we have the Markov chain $\theta \to X \to \hat{\theta}$. Applying the data-processing inequality (Theorem 3.7), we have

$$\inf_{P_{\tilde{\theta}|\theta}:\, \mathbb{E}\ell(\theta,\tilde{\theta})\leq D} I(\theta;\tilde{\theta}) \leq I(\theta;\hat{\theta}) \leq I(\theta;X). \tag{30.1}$$

Note that:

- The leftmost quantity can be interpreted as the minimum amount of information required to achieve a given estimation accuracy. This is precisely the rate-distortion function $\phi(D) \equiv \phi_\theta(D)$ (recall Section 24.3).
- The rightmost quantity can be interpreted as the amount of information provided by the data about the latent parameter. Sometimes it suffices to further upper-bound it by the capacity of the channel $P_{X|\theta}$ by maximizing over all priors (Chapter 5):

$$I(\theta;X) \leq \sup_{\pi\in\mathcal{P}(\Theta)} I(\theta;X) \triangleq C. \tag{30.2}$$

Therefore, we arrive at the following lower bound on the Bayes and hence the minimax risks:

$$R_\pi^* \geq \phi^{-1}(I(\theta;X)) \geq \phi^{-1}(C). \tag{30.3}$$

The reasoning behind the mutual information method is reminiscent of the converse proof for joint source–channel coding in Section 26.3. As such, the argument here retains the flavor of "source–channel separation," in that the lower bound in (30.1) depends only on the prior (source) and the loss function, while the capacity upper bound (30.2) depends only on the statistical model (channel).

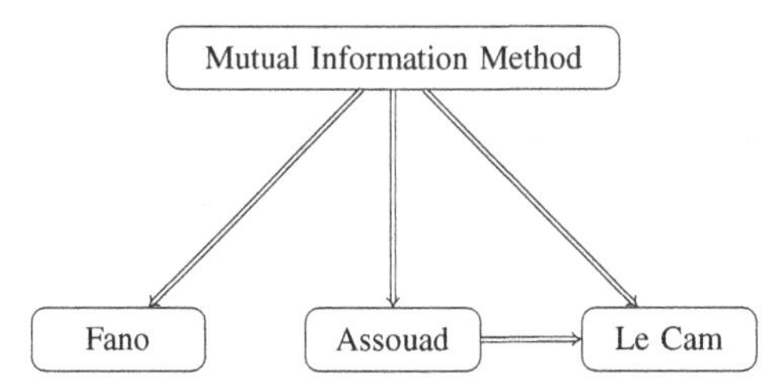

Figure 30.1 The three lower-bound techniques as consequences of the mutual information method.

In the next few sections, we discuss a sequence of examples to illustrate the MIM and its execution:

- Denoising a vector in Gaussian noise, where we will compute the exact minimax risk.
- Denoising a sparse vector, where we determine the sharp minimax rate.
- Community detection, where the goal is to recover a dense subgraph planted in a bigger Erdös–Rényi graph.

In the next chapter we will discuss three popular approaches for proving statistical lower bounds, namely, *Le Cam's method*, *Assouad's lemma*, and *Fano's method*. As illustrated in Figure 30.1, all three follow from the mutual information method, corresponding to different choice of prior π for θ, namely, the uniform distribution over a two-point set $\{\theta_0, \theta_1\}$, the hypercube $\{0,1\}^d$, and a packing (recall Section 27.1). While these methods are highly useful in determining the minimax rate for many problems, they are often loose with constant factors compared to the MIM. In the last section of this chapter, we discuss the problem of how and when non-trivial estimation is achievable by applying the MIM; for this purpose, none of the three methods in the next chapter works.

30.1 GLM Revisited and the Shannon Lower Bound

Consider the d-dimensional GLM, where we observe $X = (X_1, \dots, X_n) \overset{\text{iid}}{\sim} \mathcal{N}(\theta, I_d)$ and $\theta \in \Theta$ is the parameter. Denote by $R^*(\Theta)$ the minimax risk with respect to the quadratic loss $\ell(\theta, \hat{\theta}) = \|\hat{\theta} - \theta\|_2^2$.

First, let us consider the unconstrained model where $\Theta = \mathbb{R}^d$. Estimating using the sample mean $\bar{X} = \frac{1}{n}\sum_{i=1}^n X_i \sim \mathcal{N}(\theta, \frac{1}{n} I_d)$, we achieve the upper bound $R^*(\mathbb{R}^d) \leq \frac{d}{n}$. This turns out to be the exact minimax risk, as shown in Example 28.2 by computing the Bayes risk for Gaussian priors. Next we apply the mutual information method to obtain the same matching lower bound without evaluating the Bayes risk. Again, let us consider $\theta \sim \mathcal{N}(0, sI_d)$ for some $s > 0$. We know from the Gaussian rate-distortion function (Theorem 26.2) that

$$\phi(D) = \inf_{P_{\hat{\theta}|\theta}\colon \mathbb{E}[\|\hat{\theta}-\theta\|_2^2]\le D} I(\theta;\hat{\theta}) = \begin{cases} \frac{d}{2}\log\frac{sd}{D}, & D < sd, \\ 0, & \text{otherwise.} \end{cases}$$

Using the sufficiency of $\bar{X}$ and the formula for the Gaussian channel capacity (see Theorem 5.11 or Theorem 20.11), the mutual information between the parameter and the data can be computed as

$$I(\theta;X) = I(\theta;\bar{X}) = \frac{d}{2}\log(1+sn).$$

It then follows from (30.3) that $R^*_\pi \ge \frac{sd}{1+sn}$, which in fact matches the exact Bayes risk in (28.7). Sending $s \to \infty$ we recover the result in (28.17), namely

$$R^*\left(\mathbb{R}^d\right) = \frac{d}{n}. \tag{30.4}$$

In the above unconstrained GLM, we are able to compute everything in closed form when applying the mutual information method. Such exact expressions are rarely available in more complicated models in which case various bounds on the mutual information will prove useful. Next, let us consider the GLM with bounded means, where the parameter space $\Theta = B(\rho) = \{\theta\colon \|\theta\|_2 \le \rho\}$ is the ℓ_2-ball of radius ρ centered at zero. In this case there is no known closed-form formula for the minimax quadratic risk even in one dimension.[1] Nevertheless, the next result determines the sharp minimax rate, which characterizes the minimax risk up to universal constant factors.

Theorem 30.1. (Bounded GLM)

$$R^*(B(\rho)) \asymp \frac{d}{n} \wedge \rho^2. \tag{30.5}$$

Remark 30.1. Comparing (30.5) with (30.4), we see that if $\rho \gtrsim \sqrt{d/n}$, it is rate-optimal to ignore the bounded-norm constraint; if $\rho \lesssim \sqrt{d/n}$, we can discard all observations and estimate by zero, because data do not provide a better resolution than the prior information.

Proof. The upper bound $R^*(B(\rho)) \le \frac{d}{n} \wedge \rho^2$ follows from considering the estimator $\hat{\theta} = \bar{X}$ and $\hat{\theta} = 0$. To prove the lower bound, we apply the mutual information method with a uniform prior $\theta \sim \mathrm{Unif}(B(r))$, where $r \in [0, \rho]$ is to be optimized. The mutual information can be upper-bounded using the AWGN capacity as follows:

$$I(\theta;X) = I(\theta;\bar{X}) \le \sup_{P_\theta\colon \mathbb{E}[\|\theta\|_2^2]\le r} I\left(\theta;\theta + \frac{1}{\sqrt{n}}Z\right) = \frac{d}{2}\log\left(1+\frac{nr^2}{d}\right) \le \frac{nr^2}{2}, \tag{30.6}$$

where $Z \sim \mathcal{N}(0, I_d)$. Alternatively, we can use Corollary 5.8 to bound the capacity (as information radius) by the KL diameter, which yields the same bound within constant factors:

[1] It is known that there exists some ρ_0 depending on d/n such that, for all $\rho \le \rho_0$, the uniform prior over the sphere of radius ρ is *exactly* least favorable (see [82] for $d = 1$ and [48] for $d > 1$).

$$I(\theta;X) \le \sup_{P_\theta:\,\|\theta\|\le r} I\left(\theta;\theta+\frac{1}{\sqrt{n}}Z\right) \le \max_{\theta,\theta'\in B(r)} D(\mathcal{N}(\theta,I_d/n)\|\mathcal{N}(\theta,I_d/n)) = 2nr^2. \tag{30.7}$$

For the lower bound, due to the lack of a closed-form formula for the rate-distortion function for a uniform distribution over Euclidean balls, we apply the *Shannon lower bound* (SLB) from Section 26.1. Since θ has an isotropic distribution, applying Theorem 26.3 yields

$$\inf_{P_{\hat\theta|\theta}:\,\mathbb{E}\|\theta-\hat\theta\|^2\le D} I(\theta;\hat\theta) \ge h(\theta) + \frac{d}{2}\log\frac{2\pi ed}{D} \ge \frac{d}{2}\log\frac{cr^2}{D},$$

for some universal constant c, where the last inequality is because for $\theta \sim \text{Unif}(B(r))$, $h(\theta) = \log\text{vol}(B(r)) = d\log r + \log\text{vol}(B(1))$ and the volume of a unit Euclidean ball in d dimensions satisfies (recall (27.14)) $\text{vol}(B(1))^{1/d} \asymp \frac{1}{\sqrt{d}}$.

Finally, applying (30.3) yields $\frac{1}{2}\log\frac{cr^2}{R^*} \le \frac{nr^2}{2}$, that is, $R^* \ge cr^2e^{-nr^2/d}$. Optimizing over r and using the fact that $\sup_{0<x<1} xe^{-ax} = \frac{1}{ea}$ if $a\ge 1$ and e^{-a} if $a<1$, we have

$$R^* \ge \sup_{r\in[0,\rho]} cr^2e^{-nr^2/d} \asymp \frac{d}{n}\wedge\rho^2. \qquad \square$$

As a final example, let us consider a non-quadratic loss $\ell(\theta,\hat\theta) = \|\theta-\hat\theta\|^r$, the rth power of an arbitrary norm on $\mathbb{R}^d$. Recall that we have determined in Corollary 28.8 the exact minimax risk using Anderson's lemma, namely,

$$\inf_{\hat\theta}\sup_{\theta\in\mathbb{R}^d} \mathbb{E}_\theta[\|\hat\theta-\theta\|^r] = n^{-r/2}\mathbb{E}[\|Z\|^r], \quad Z\sim\mathcal{N}(0,I_d). \tag{30.8}$$

In order to apply the mutual information method, consider again a Gaussian prior $\theta\sim\mathcal{N}(0,sI_d)$. Suppose that $\mathbb{E}[\|\hat\theta-\theta\|^r]\le D$. By the data-processing inequality,

$$\frac{d}{2}\log(1+ns) \ge I(\theta;X) \ge I(\theta;\hat\theta) \ge \frac{d}{2}\log(2\pi es) - \log\left\{V_{\|\cdot\|}\left(\frac{Dre}{d}\right)^{d/r}\Gamma\left(1+\frac{d}{r}\right)\right\},$$

where the last inequality follows from the general SLB (26.5). Rearranging terms and sending $s\to\infty$ yields

$$\begin{aligned}\inf_{\hat\theta}\sup_{\theta\in\mathbb{R}^d} \mathbb{E}_\theta[\|\hat\theta-\theta\|^r] &\ge \frac{d}{re}\left(\frac{2\pi e}{n}\right)^{r/2}\left[V_{\|\cdot\|}\Gamma\left(1+\frac{d}{r}\right)\right]^{-r/d} \asymp n^{-r/2}V_{\|\cdot\|}^{-r/d} \\ &\gtrsim \left(\frac{d}{\sqrt{n}\mathbb{E}[\|Z\|_*]}\right)^r,\end{aligned} \tag{30.9}$$

where the middle inequality applies Stirling's approximation $\Gamma(x)^{1/x}\asymp x$ for $x\to\infty$, and the right applies Urysohn's volume inequality (27.21), with $\|x\|_* = \sup\{\langle x,y\rangle:\,\|y\|\le 1\}$ denoting the dual norm of $\|\cdot\|$.

To evaluate the tightness of the lower bound from the SLB in comparison with the exact result (30.8), consider the example of $r=2$ and the ℓ_q-norm $\|x\|_q = \left(\sum_{i=1}^d |x_i|^q\right)^{1/q}$ with $1\le q\le\infty$. Recall the volume of a unit ℓ_q-ball given in (27.13). In the special case of $q=2$, the (first) lower bound in (30.9) is in fact exact and coincides with (30.4). For general $q \in [1, \infty)$, (30.9) gives the tight minimax rate $\frac{d^{2/q}}{n}$; however, for $q = \infty$, the minimax lower bound we get is $1/n$, independent of the dimension d. In comparison, from (30.8) we get the sharp rate $\frac{\log d}{n}$, since $\mathbb{E}\|Z\|_\infty \asymp \sqrt{\log d}$ (see Lemma 27.10). We will revisit this example in Section 31.4 and show how to obtain the optimal dependence on the dimension.

Remark 30.2. (SLB versus the volume method) Recall the connection between the rate-distortion function and the metric entropy in Section 27.6. As we have seen in Section 27.2, a common lower bound for metric entropy is via the volume bound. In fact, the SLB can be interpreted as a volume-based lower bound to the rate-distortion function. To see this, consider $r = 1$ and let θ be uniformly distributed over some compact set Θ, so that $h(\theta) = \log \mathrm{vol}(\Theta)$ (Theorem 2.7(a)). Applying Stirling's approximation, the lower bound in (26.5) becomes $\log \frac{\mathrm{vol}(\Theta)}{\mathrm{vol}(B_{\|\cdot\|}(c\epsilon))}$ for some constant c, which has the same form as the volume ratio in Theorem 27.3 for metric entropy. We will see later in Section 31.4 that in statistical applications, applying the SLB yields basically the same lower bound as applying Fano's method to a packing obtained from the volume bound, although the SLB does not rely explicitly on a packing.

30.2 GLM with Sparse Means

In this section we consider the problem of denoising a sparse vector. Specifically, consider again the Gaussian location model $\mathcal{N}(\theta, I_d)$ where the mean vector θ is known to be k-sparse, taking values in the "ℓ_0-ball"

$$B_0(k) = \{\theta \in \mathbb{R}^d \colon \|\theta\|_0 \le k\}, \quad k \in [p],$$

where $\|\theta\|_0 = |\{i \in [d] \colon \theta_i \neq 0\}|$ is the number of non-zero entries of θ, indicating the sparsity of θ. Our goal is to characterize the minimax quadratic risk

$$R_n^*(B_0(k)) = \inf_{\hat{\theta}} \sup_{\theta \in B_0(k)} \mathbb{E}_\theta \|\hat{\theta} - \theta\|_2^2.$$

Next we prove an optimal lower bound applying MIM. (For a different proof using Fano's method in Section 31.4, see Exercise **VI.12**.)

Theorem 30.2.

$$R_n^*(B_0(k)) \gtrsim \frac{k}{n} \log \frac{ed}{k}. \tag{30.10}$$

A few remarks are in order:

Remark 30.3.

(a) The lower bound (30.10) turns out to be tight, achieved by the maximum likelihood estimator

$$\hat{\theta}_{\mathrm{MLE}} = \arg \min_{\|\theta\|_0 \le k} \|\bar{X} - \theta\|_2, \tag{30.11}$$

which is equivalent to keeping the k entries from $\bar{X}$ with the largest magnitude and setting the rest to zero, or the following hard-thresholding estimator $\hat{\theta}^\tau$ with an appropriately chosen τ (see Exercise **VI.13**):

$$\hat{\theta}_i^\tau = X_i 1\{|X_i| \geq \tau\}. \tag{30.12}$$

(b) Sharp asymptotics: For sublinear sparsity $k = o(d)$, we have $R_n^*(B_0(k)) = (2 + o(1))\frac{k}{n} \log \frac{d}{k}$ (Exercise **VI.13**); for linear sparsity $k = (\eta + o(1))d$ with $\eta \in (0, 1)$, $R_n^*(B_0(k)) = (\beta(\eta) + o(1))d$ for some constant $\beta(\eta)$. For the latter and more refined results, we refer the reader to the monograph [236, Chapter 8].

Proof. First, note that $B_0(k)$ is a union of linear subspaces of $\mathbb{R}^d$ and thus homogeneous. Therefore by scaling, we have

$$R_n^*(B_0(k)) = \frac{1}{n} R_1^*(B_0(k)) \triangleq \frac{1}{n} R^*(k, d). \tag{30.13}$$

Thus it suffices to consider $n = 1$. Denote the observation by $X = \theta + Z$.

Next, note the following oracle lower bound:

$$R^*(k, d) \geq k,$$

which is the optimal risk given the extra information of the support of θ, in view of (30.4). Thus to show (30.10), below it suffices to consider $k \leq d/4$.

We now apply the mutual information method. Recall from (27.10) that S_k^d denotes the Hamming sphere, namely,

$$S_k^d = \{b \in \{0, 1\}^d : w_H(b) = k\},$$

where $w_H(b)$ denotes the Hamming weights of b. Let b be uniformly distributed over S_k^d and let $\theta = \tau b$, where $\tau = \sqrt{\log \frac{d}{k}}$. Given any estimator $\hat{\theta} = \hat{\theta}(X)$, define an estimator $\hat{b} \in \{0, 1\}^d$ for b by

$$\hat{b}_i = \begin{cases} 0, & \hat{\theta}_i \leq \tau/2, \\ 1, & \hat{\theta}_i > \tau/2, \end{cases} \quad i \in [d].$$

Thus the Hamming loss of $\hat{b}$ can be related to the squared loss of $\hat{\theta}$ as

$$\|\theta - \hat{\theta}\|_2^2 \geq \frac{\tau^2}{4} d_H(b, \hat{b}). \tag{30.14}$$

Let $\mathbb{E} d_H(b, \hat{b}) = \delta k$. Assume that $\delta \leq \frac{1}{4}$, for otherwise, we are done.

Note the following Markov chain $b \to \theta \to X \to \hat{\theta} \to \hat{b}$ and thus, by the data-processing inequality of mutual information,

$$I(b;\hat{b}) \leq I(\theta;X) \leq \frac{d}{2} \log\left(1 + \frac{k\tau^2}{d}\right) \leq \frac{k\tau^2}{2} = \frac{k}{2} \log \frac{d}{k},$$

where the second inequality follows from the fact that $\|\theta\|_2^2 = k\tau^2$ and the Gaussian channel capacity.

Conversely,

$$\begin{aligned}
I(\hat{b};b) &\geq \min_{\mathbb{E}d_{\mathrm{H}}(b,\hat{b})\leq\delta d} I(\hat{b};b) \\
&= H(b) - \max_{\mathbb{E}d_{\mathrm{H}}(b,\hat{b})\leq\delta k} H(b|\hat{b}) \\
&\geq \log\binom{d}{k} - \max_{\mathbb{E}w_{\mathrm{H}}(b\oplus\hat{b})\leq\delta k} H(b\oplus\hat{b}) \\
&= \log\binom{d}{k} - dh\left(\frac{\delta k}{d}\right),
\end{aligned} \tag{30.15}$$

where the last step follows from Exercise **I.7**.

Combining the lower and upper bounds on the mutual information and using $\binom{d}{k} \geq (\frac{d}{k})^k$, we get $dh(\frac{\delta k}{d}) \geq \frac{k}{2}k\log\frac{d}{k}$. Since $h(p) \leq -p\log p + p$ for $p \in [0,1]$ and $k/d \leq \frac{1}{4}$ by assumption, we conclude that $\delta \geq ck/d$ for some absolute constant c, completing the proof of (30.10) in view of (30.14). □

30.3 Community Detection

As another application of the mutual information method, let us consider the following statistical problem of detecting a single hidden community in random graphs, also known as the *planted dense subgraph model*. (The reader should compare this with the stochastic block model with two communities introduced in Exercise **I.49**.) Let C^* be drawn uniformly at random from all subsets of $[n]$ of cardinality $k \geq 2$. Let G denote a random graph on the vertex set $[n]$, such that for each $i \neq j$, they are connected independently with probability p if both i and j belong to C^*, and with probability q otherwise. Assuming that $p > q$, the set C^* represents a densely connected community, which forms an Erdös–Rényi graph $\mathrm{ER}(k,p)$ planted in the bigger $\mathrm{ER}(n,q)$ graph. Upon observing G, the goal is to reconstruct C^* as accurately as possible. In particular, given an estimator $\hat{C} = \hat{C}(G)$, we say it achieves *almost exact recovery* if $\mathbb{E}|C\Delta\hat{C}| = o(k)$. The following result gives a necessary condition in terms of the parameters (p,q,n,k):

Theorem 30.3. *Assume that k/n is bounded away from one. If almost exact recovery is possible, then*

$$d(p\|q) \geq \frac{2+o(1)}{k-1}\log\frac{n}{k}. \tag{30.16}$$

Remark 30.4. In addition to Theorem 30.3, another necessary condition is that

$$d(p\|q) = \omega\left(\frac{1}{k}\right), \tag{30.17}$$

which can be shown by a reduction to testing the membership of two nodes given the rest. It turns out that conditions (30.16) and (30.17) are optimal, in the sense that almost exact recovery can be achieved (via maximum likelihood) provided that

(30.17) holds and $d(p\|q) \geq \frac{2+\epsilon}{k-1} \log \frac{n}{k}$ for any constant $\epsilon > 0$. For details, we refer the reader to [208].

Proof. Suppose $\hat{C}$ achieves almost exact recovery of C^*. Let $\xi^*, \hat{\xi} \in \{0,1\}^n$ denote their indicator vectors, respectively, for example, $\xi_i^* = 1\{i \in C^*\}$ for each $i \in [n]$. Then $\mathbb{E}[d_H(\xi, \hat{\xi})] = \epsilon_n k$ for some $\epsilon_n \to 0$. Applying the mutual information method as before, we have

$$I(G;\xi^*) \geq I(\hat{\xi};\xi^*) \overset{(a)}{\geq} \log \binom{n}{k} - nh\left(\frac{\epsilon_n k}{n}\right) \overset{(b)}{\geq} k \log \frac{n}{k}(1+o(1)),$$

where (a) follows in the same manner as (30.15) did from Exercise **I.7**; and (b) is due to the assumption that $k/n \leq 1 - c$ for some constant c.

On the other hand, the mutual information between the hidden community and the graph can be upper-bounded as:

$$I(G;\xi^*) \overset{(a)}{=} \min_Q D(P_{G|\xi^*}\|Q|P_{\xi^*}) \overset{(b)}{\leq} D(P_{G|\xi^*}\|\mathrm{Ber}(q)^{\otimes\binom{n}{2}}|P_{\xi^*}) \overset{(c)}{=} \binom{k}{2} d(p\|q),$$

where (a) is by the variational representation of mutual information in Corollary 4.2; (b) follows from choosing Q to be the distribution of the Erdös–Rényi graph $\mathrm{ER}(n,q)$; and (c) is by the tensorization property of KL divergence for product distributions (Theorem 2.16(d)). Combining the last two displays completes the proof. □

30.4 Estimation Better than Chance

Instead of characterizing the rate of convergence of the minimax risk to zero as the amount of data grows, suppose we are in a regime where this is impossible due to limited sample size, poor signal-to-noise ratio, or high dimensionality; instead, we are concerned with the modest goal of achieving an estimation error strictly better than the trivial error (without data). In the context of clustering, this is known as weak recovery or correlated recovery, where the goal is to achieve not a vanishing misclassification rate but one strictly better than randomly guessing the labels. It turns out that MIM is particularly suited for this regime. (In fact, we will see in the next chapter that all three popular further relaxations of MIM fall short due to the loss of constant factors.)

As an example, let us continue the setting of Theorem 30.1, where the goal is to estimate a vector in a high-dimensional unit-ball based on noisy observations. Since the radius of the parameter space is one, the trivial squared error equals one. The following theorem shows that in high dimensions, non-trivial estimation is achievable if and only if the sample size n grows proportionally with the dimension d; otherwise, when $d \gg n \gg 1$, the optimal estimation error is $1 - \frac{n}{d}(1+o(1))$.

Theorem 30.4. (Bounded GLM continued) *Suppose $X_1, \dots, X_n \overset{\text{iid}}{\sim} \mathcal{N}(\theta, I_d)$, where θ belongs to B, the unit ℓ_2-ball in $\mathbb{R}^d$. Then for some universal constant C_0,*

$$e^{-\frac{n+C_0}{d-1}} \le \inf_{\hat{\theta}} \sup_{\theta \in B} \mathbb{E}_\theta[\|\hat{\theta} - \theta\|^2] \le \frac{d}{d+n}.$$

Proof. Without loss of generality, assume that the observation is $X = \theta + \frac{Z}{\sqrt{n}}$, where $Z \sim \mathcal{N}(0, I_d)$. For the upper bound, applying the shrinkage estimator[2] $\hat{\theta} = \frac{1}{1+d/n} X$ yields $\mathbb{E}[\|\hat{\theta} - \theta\|^2] \le \frac{d}{n+d}$.

For the lower bound, we apply MIM as in the proof of Theorem 30.1 with the prior $\theta \sim \text{Unif}(S^{d-1})$. We still apply the AWGN capacity in (30.6) to get $I(\theta;X) \le n/2$. (Here the constant $1/2$ is important and so the diameter-based bound (30.7) is too loose.) For the rate-distortion function of a spherical uniform distribution, applying Theorem 27.17 yields $I(\theta;\hat{\theta}) \ge \frac{d-1}{2} \log \frac{1}{\mathbb{E}[\|\hat{\theta}-\theta\|^2]} - C$. Thus the lower bound on $\mathbb{E}[\|\hat{\theta} - \theta\|^2]$ follows from the data-processing inequality. □

A similar phenomenon also occurs in the problem of estimating a discrete distribution P on k elements based on n iid observations, which has been studied in Section 29.4 for a small alphabet in the large-sample asymptotics and extended in Exercises **VI.7–VI.10** to large alphabets. In particular, consider the total variation loss, which is at most one. Exercise **VI.10f** shows that the TV error of any estimator is $1 - o(1)$ if $n \ll k$; conversely, Exercise **VI.10b** demonstrates an estimator $\hat{P}$ such that $\mathbb{E}[\chi^2(P\|\hat{P})] \le \frac{k-1}{n+1}$. Applying the joint range (7.32) between TV and χ^2 and Jensen's inequality, we have

$$\mathbb{E}[\text{TV}(P, \hat{P})] \le \begin{cases} \frac{1}{2}\sqrt{\frac{k-1}{n+1}}, & n \ge k-2, \\ \frac{k-1}{k+n}, & n \le k-2, \end{cases}$$

which is bounded away from one whenever $n = \Omega(k)$. In summary, non-trivial estimation in total variation is possible if and only if n scales at least proportionally with k.

Finally, let us mention the problem of *correlated recovery* in the stochastic block model (see Exercise **I.49**), which refers to estimating the community labels better than chance. The optimal information-theoretic threshold of this problem can be established by bounding the appropriate mutual information; see Section 33.9 for the Gaussian version (spiked Wigner model).

[2] This corresponds to the Bayes estimator (Example 28.1) when we choose $\theta \sim \mathcal{N}(0, \frac{1}{d} I_d)$, which is approximately concentrated on the unit sphere for large d.

31 Lower Bounds via Reduction to Hypothesis Testing

In this chapter we study three commonly used techniques for proving minimax lower bounds, namely, *Le Cam's method*, *Assouad's lemma*, and *Fano's method*. Compared to the results in Chapter 29 geared toward large-sample asymptotics in smooth parametric models, the approach here is more generic, less tied to mean-squared error, and applicable in non-asymptotic settings such as nonparametric or high-dimensional problems.

The common rationale of all three methods is reducing statistical estimation to hypothesis testing. Specifically, to lower-bound the minimax risk $R^*(\Theta)$ for the parameter space Θ, the first step is to notice that $R^*(\Theta) \geq R^*(\Theta')$ for any subcollection $\Theta' \subset \Theta$, and Le Cam, Assouad, and Fano's methods amount to choosing Θ' to be a two-point set, a hypercube, or a packing, respectively. In particular, Le Cam's method reduces the estimation problem to binary hypothesis testing. This method is perhaps the easiest to evaluate; however, the disadvantage is that it is frequently loose in estimating high-dimensional parameters. To capture the correct dependence on the dimension, both Assouad's and Fano's method rely on reduction to testing multiple hypotheses.

As illustrated in Figure 30.1, all three methods in fact follow from the common principle of the mutual information method (MIM) in Chapter 30, corresponding to different choice of priors. The limitation of these methods, compared to the MIM, is that, due to the looseness in constant factors, they are ineffective for certain problems such as estimation better than chance discussed in Section 30.4.

31.1 Le Cam's Two-Point Method

Theorem 31.1. *Suppose the loss function $\ell\colon \Theta \times \Theta \to \mathbb{R}_+$ satisfies the following α-triangle inequality for some $\alpha > 0$: For all $\theta_0, \theta_1, \theta \in \Theta$,*

$$\ell(\theta_0, \theta_1) \leq \alpha(\ell(\theta_0, \theta) + \ell(\theta_1, \theta)). \tag{31.1}$$

Then

$$\inf_{\hat\theta} \sup_{\theta \in \Theta} \mathbb{E}_\theta \ell(\theta, \hat\theta) \geq \sup_{\theta_0, \theta_1 \in \Theta} \frac{\ell(\theta_0, \theta_1)\ell(\theta_1, \theta_0)}{\alpha(\ell(\theta_0, \theta_1) + \ell(\theta_1, \theta_0))}(1 - \mathrm{TV}(P_{\theta_0}, P_{\theta_1})). \tag{31.2}$$

Proof. Fix $\theta_0, \theta_1 \in \Theta$. Given any estimator $\hat{\theta}$, let us convert it into the following (randomized) test:

$$\tilde{\theta} = \begin{cases} \theta_0 & \text{with probability } \frac{\ell(\theta_1,\hat{\theta})}{\ell(\theta_0,\hat{\theta})+\ell(\theta_1,\hat{\theta})}, \\ \theta_1 & \text{with probability } \frac{\ell(\theta_0,\hat{\theta})}{\ell(\theta_0,\hat{\theta})+\ell(\theta_1,\hat{\theta})}. \end{cases}$$

By the α-triangle inequality, we have

$$\ell(\theta_0,\theta_1)\mathbb{P}_{\theta_0}[\tilde{\theta} \neq \theta_0)] = \ell(\theta_0,\theta_1)\mathbb{E}_{\theta_0}\left[\frac{\ell(\theta_0,\hat{\theta})}{\ell(\theta_0,\hat{\theta})+\ell(\theta_1,\hat{\theta})}\right] \leq \alpha\mathbb{E}_{\theta_0}[\ell(\hat{\theta},\theta_0)],$$

and similarly for θ_1. Consider the prior $\pi = \frac{\ell(\theta_1,\theta_0)}{\ell(\theta_0,\theta_1)+\ell(\theta_1,\theta_0)}\delta_{\theta_0} + \frac{\ell(\theta_0,\theta_1)}{\ell(\theta_0,\theta_1)+\ell(\theta_1,\theta_0)}\delta_{\theta_1}$ and let $\theta \sim \pi$. Taking expectation on both sides yields the following lower bound on the Bayes risk:

$$\mathbb{E}_\pi[\ell(\hat{\theta},\theta)] \geq \frac{\ell(\theta_0,\theta_1)\ell(\theta_1,\theta_0)}{\alpha(\ell(\theta_0,\theta_1)+\ell(\theta_1,\theta_0))}(\mathbb{P}_{\theta_0}[\tilde{\theta} \neq \theta_0] + \mathbb{P}_{\theta_1}[\tilde{\theta} \neq \theta_1]).$$

Applying the lower bound on the minimum total probability of error in binary hypothesis testing (Theorem 7.7) concludes the proof. □

Remark 31.1. As an example where the bound (31.2) is tight (up to constants), consider a binary hypothesis testing problem with $\Theta = \{\theta_0, \theta_1\}$ and the Hamming loss $\ell(\theta,\hat{\theta}) = 1\{\theta \neq \hat{\theta}\}$, where $\theta, \hat{\theta} \in \{\theta_0,\theta_1\}$ and $\alpha = 1$. Then the left-hand side is the minimax probability of error, and the right-hand side is the optimal average probability of error (see (7.19)). These two quantities can coincide (for example, for the Gaussian location model).

Another special case of interest is the quadratic loss $\ell(\theta,\hat{\theta}) = \|\theta - \hat{\theta}\|_2^2$, where $\theta, \hat{\theta} \in \mathbb{R}^d$, which satisfies the α-triangle inequality with $\alpha = 2$. In this case, the leading constant $\frac{1}{4}$ in (31.2) makes sense, because in the extreme case of TV $= 0$ where P_{θ_0} and P_{θ_1} cannot be distinguished, the best estimate is simply $\frac{\theta_0+\theta_1}{2}$. In addition, the inequality (31.2) can be deduced based on the properties of f-divergences and their joint range (Chapter 7). To this end, abbreviate P_{θ_i} as P_i for $i = 0, 1$ and consider the prior $\pi = \frac{1}{2}(\delta_{\theta_0} + \delta_{\theta_1})$. Then the Bayes estimator (posterior mean) is $\frac{\theta_0 dP_0 + \theta_1 dP_1}{dP_0 + dP_1}$ and the Bayes risk is given by

$$\begin{aligned} R_\pi^* &= \frac{\|\theta_0 - \theta_1\|^2}{2}\int \frac{dP_0 dP_1}{dP_0 + dP_1} \\ &= \frac{\|\theta_0 - \theta_1\|^2}{4}(1 - \mathrm{LC}(P_0,P_1)) \geq \frac{\|\theta_0 - \theta_1\|^2}{4}(1 - \mathrm{TV}(P_0,P_1)), \end{aligned}$$

where $\mathrm{LC}(P_0,P_1) = \int \frac{(dP_0 - dP_1)^2}{dP_0 + dP_1}$ is the Le Cam divergence defined in (7.7) and satisfies LC $\leq$ TV.

Example 31.1.

As a concrete example, consider the one-dimensional GLM with sample size n. By considering the sufficient statistic $\bar{X} = \frac{1}{n}\sum_{i=1}^n X_i$, the model is simply $\{\mathcal{N}(\theta, \frac{1}{n}) : \theta \in \mathbb{R}\}$. Applying Theorem 31.1 yields

$$R^* \geq \sup_{\theta_0,\theta_1 \in \mathbb{R}} \frac{1}{4}|\theta_0 - \theta_1|^2 \left\{1 - \mathrm{TV}\left(N\left(\theta_0, \frac{1}{n}\right), N\left(\theta_1, \frac{1}{n}\right)\right)\right\}$$

$$\stackrel{(a)}{=} \frac{1}{4n} \sup_{s>0} s^2(1 - \mathrm{TV}(\mathcal{N}(0,1), \mathcal{N}(s,1))) \stackrel{(b)}{=} \frac{c}{n}, \tag{31.3}$$

where (a) follows from the shift and scale invariance of the total variation; in (b) $c \approx 0.083$ is some absolute constant, obtained by applying the formula $\mathrm{TV}(\mathcal{N}(0,1), \mathcal{N}(s,1)) = 2\Phi(\frac{s}{2}) - 1$ from (7.40). On the other hand, we know from Example 28.2 that the minimax risk equals $\frac{1}{n}$, so the two-point method is rate-optimal in this case.

In the above example, for two points separated by $\Theta(\frac{1}{\sqrt{n}})$, the corresponding hypothesis cannot be tested with vanishing probability of error so that the resulting estimation risk in squared error cannot be smaller than $\frac{1}{n}$. This convergence rate is commonly known as the "parametric rate," which we have studied in Chapter 29 for smooth parametric families focusing on the Fisher information as the sharp constant. More generally, the $\frac{1}{n}$ rate is not improvable for models with locally quadratic behavior:

$$H^2(P_{\theta_0}, P_{\theta_0+t}) \asymp t^2, \quad t \to 0. \tag{31.4}$$

(Recall that Theorem 7.23 gives a sufficient condition for this behavior.) Indeed, pick θ_0 in the interior of the parameter space and set $\theta_1 = \theta_0 + \frac{1}{\sqrt{n}}$, so that $H^2(P_{\theta_0}, P_{\theta_1}) = \Theta(\frac{1}{n})$ thanks to (31.4). By Theorem 7.8, we have $\mathrm{TV}(P_{\theta_0}^{\otimes n}, P_{\theta_1}^{\otimes n}) \leq 1 - c$ for some constant c and hence Theorem 31.1 yields the lower bound $\Omega(1/n)$ for the squared error. Furthermore, later we will show that the same locally quadratic behavior in fact guarantees the achievability of the $1/n$ rate; see Corollary 32.12.

Example 31.2.

As a different example, consider the family $\mathrm{Unif}(0,\theta)$. Note that as opposed to the quadratic behavior (31.4), we have

$$H^2(\mathrm{Unif}(0,1), \mathrm{Unif}(0,1+t)) = 2(1 - 1/\sqrt{1+t}) \asymp t.$$

Thus an application of Theorem 31.1 yields an $\Omega(1/n^2)$ lower bound. This rate is achieved not by the empirical mean estimator (which only achieves $1/n$ rate), but by the maximum likelihood estimator $\hat{\theta} = \max\{X_1, \ldots, X_n\}$. Other types of behavior in t, and hence the rates of convergence, can occur even in compactly supported location families – see Example 7.1.

The limitation of Le Cam's two-point method is that it does not capture the correct dependence on the dimensionality. To see this, let us revisit Example 31.1 for d dimensions.

Example 31.3.

Consider the d-dimensional GLM in Corollary 28.8. Again, it is equivalent to consider the reduced model $\{\mathcal{N}(\theta, \frac{1}{n}) : \theta \in \mathbb{R}^d\}$. We know from Example 28.2 (see also Theorem 28.4) that for quadratic risk $\ell(\theta, \hat{\theta}) = \|\theta - \hat{\theta}\|_2^2$, the exact minimax risk is $R^* = \frac{d}{n}$ for any d and n. Let us compare this with the best two-point lower bound. Applying Theorem 31.1 with $\alpha = 2$,

$$
\begin{aligned}
R^* &\geq \sup_{\theta_0, \theta_1 \in \mathbb{R}^d} \frac{1}{4}\|\theta_0 - \theta_1\|_2^2 \left\{1 - \mathrm{TV}\left(N\left(\theta_0, \frac{1}{n}I_d\right), N\left(\theta_1, \frac{1}{n}I_d\right)\right)\right\} \\
&= \sup_{\theta \in \mathbb{R}^d} \frac{1}{4n}\|\theta\|_2^2 \left\{1 - \mathrm{TV}\left(N(0, I_d), N(\theta, I_d)\right)\right\} \\
&= \frac{1}{4n} \sup_{s > 0} s^2 (1 - \mathrm{TV}(\mathcal{N}(0,1), \mathcal{N}(s,1))),
\end{aligned}
$$

where the second step applies the shift and scale invariance of the total variation; in the last step, by the rotational invariance of isotropic Gaussians, we can rotate the vector θ to align with a coordinate vector (say, $e_1 = (1, 0, \ldots, 0)$) which reduces the problem to one dimension, namely,

$$
\begin{aligned}
\mathrm{TV}(\mathcal{N}(0, I_d), \mathcal{N}(\theta, I_d)) &= \mathrm{TV}(\mathcal{N}(0, I_d), \mathcal{N}(\|\theta\| e_1, I_d)) \\
&= \mathrm{TV}(\mathcal{N}(0, 1), \mathcal{N}(\|\theta\|, 1)).
\end{aligned}
$$

Comparing the above display with (31.3), we see that the best Le Cam two-point lower bound in d dimensions coincides with that in one dimension.

Let us mention in passing that although Le Cam's two-point method is typically suboptimal for estimating a high-dimensional parameter θ, for functional estimation in high dimensions (e.g. estimating a scalar functional $T(\theta)$), Le Cam's method is much more effective and sometimes even optimal. The subtlety is that as opposed to testing a pair of simple hypotheses $H_0 : \theta = \theta_0$ versus $H_1 : \theta = \theta_1$, we need to test $H_0 : T(\theta) = t_0$ versus $H_1 : T(\theta) = t_1$, both of which are composite hypotheses and require a sagacious choice of priors. See Exercise **VI.14** for an example.

31.2 Assouad's Lemma

From Example 31.3 we see that Le Cam's two-point method effectively only perturbs one out of d coordinates, leaving the remaining $d-1$ coordinates unexplored; this is the source of its suboptimality. In order to obtain a lower bound that scales with the dimension, it is necessary to randomize all d coordinates. Our next topic, Assouad's lemma, is an extension in this direction.

Theorem 31.2. (Assouad's lemma) *Assume that the loss function ℓ satisfies the α-triangle inequality (31.1). Suppose Θ contains a subset $\Theta' = \{\theta_b : b \in \{0,1\}^d\}$ indexed by the hypercube, such that $\ell(\theta_b, \theta_{b'}) \geq \beta \cdot d_{\mathrm{H}}(b, b')$ for all b, b' and some $\beta > 0$. Then*

$$\inf_{\hat{\theta}} \sup_{\theta \in \Theta} \mathbb{E}_\theta \ell(\theta, \hat{\theta}) \geq \frac{\beta d}{4\alpha} \left(1 - \max_{d_{\mathrm{H}}(b,b')=1} \mathrm{TV}(P_{\theta_b}, P_{\theta_{b'}}) \right). \tag{31.5}$$

Proof. We lower-bound the Bayes risk with respect to the uniform prior over Θ'. Given any estimator $\hat{\theta} = \hat{\theta}(X)$, define $\hat{b} \in \operatorname{argmin} \ell(\hat{\theta}, \theta_b)$. Then for any $b \in \{0,1\}^d$,

$$\beta d_{\mathrm{H}}(\hat{b}, b) \leq \ell(\theta_{\hat{b}}, \theta_b) \leq \alpha(\ell(\theta_{\hat{b}}, \hat{\theta}_b) + \ell(\hat{\theta}, \theta_b)) \leq 2\alpha \ell(\hat{\theta}, \theta_b).$$

Let $b \sim \mathrm{Unif}(\{0,1\}^d)$ and we have $b \to \theta_b \to X$. Then

$$\begin{aligned}
\mathbb{E}[\ell(\hat{\theta}, \theta_b)] &\geq \frac{\beta}{2\alpha} \mathbb{E}[d_{\mathrm{H}}(\hat{b}, b)] \\
&= \frac{\beta}{2\alpha} \sum_{i=1}^{d} \mathbb{P}\left[\hat{b}_i \neq b_i\right] \\
&\geq \frac{\beta}{4\alpha} \sum_{i=1}^{d} (1 - \mathrm{TV}(P_{X|b_i=0}, P_{X|b_i=1})),
\end{aligned}$$

where the last step is again by Theorem 7.7, just like in the proof of Theorem 31.1. Each total variation can be upper-bounded as follows:

$$\begin{aligned}
\mathrm{TV}(P_{X|b_i=0}, P_{X|b_i=1}) &\overset{(a)}{=} \mathrm{TV}\left(\frac{1}{2^{d-1}} \sum_{b:\, b_i=1} P_{\theta_b}, \frac{1}{2^{d-1}} \sum_{b:\, b_i=0} P_{\theta_b} \right) \\
&\overset{(b)}{\leq} \max_{d_{\mathrm{H}}(b,b')=1} \mathrm{TV}(P_{\theta_b}, P_{\theta_{b'}}),
\end{aligned}$$

where (a) follows from the Bayes rule, and (b) follows from the convexity of total variation (Theorem 7.5). This completes the proof. □

Example 31.4.

Let us continue the discussion of the d-dimensional GLM in Example 31.3. Consider the quadratic loss first. To apply Theorem 31.2, consider the hypercube $\theta_b = \epsilon b$, where $b \in \{0,1\}^d$. Then $\|\theta_b - \theta'_b\|_2^2 = \epsilon^2 d_{\mathrm{H}}(b, b')$. Applying Theorem 31.2 yields

$$\begin{aligned}
R^* &\geq \frac{\epsilon^2 d}{4} \left\{ 1 - \max_{b,b' \in \{0,1\}^d, d_{\mathrm{H}}(b,b')=1} \mathrm{TV}\left(N\left(\epsilon b, \frac{1}{n} I_d\right), N\left(\epsilon b', \frac{1}{n} I_d\right) \right) \right\} \\
&= \frac{\epsilon^2 d}{4} \left\{ 1 - \mathrm{TV}\left(N\left(0, \frac{1}{n}\right), N\left(\epsilon, \frac{1}{n}\right) \right) \right\},
\end{aligned}$$

where the last step applies (7.11) for f-divergence between product distributions that only differ in one coordinate. Setting $\epsilon = \frac{1}{\sqrt{n}}$ and by the scale invariance of TV, we get the desired $R^* \gtrsim \frac{d}{n}$.

Next, let's consider the loss function $\|\theta_b - \theta'_b\|_\infty$. In the same setup, we only have $\|\theta_b - \theta'_b\|_\infty \geq \frac{\epsilon}{d} d_{\mathrm{H}}(b, b')$. Then Assouad's lemma yields $R^* \gtrsim \frac{1}{\sqrt{n}}$, which does not depend on d. In fact, $R^* \asymp \sqrt{\frac{\log d}{n}}$ as shown in Corollary 28.8. In a subsequent section, we will discuss Fano's method which can resolve this deficiency.

31.3 Assouad's Lemma from the Mutual Information Method

One can integrate Assouad's idea into the mutual information method. Consider the Bayesian setting of Theorem 31.2, where $b^d = (b_1, \ldots, b_d) \overset{\text{iid}}{\sim} \text{Ber}(1/2)$. From the rate-distortion function of the Bernoulli source (Section 26.1.1), we know that for any $\hat{b}^d$ and $\tau > 0$ there is some $\tau' > 0$ such that

$$I(b^d;X) \leq d(1-\tau)\log 2 \quad \Longrightarrow \quad \mathbb{E}[d_{\text{H}}(\hat{b}^d, b^d)] \geq d\tau'. \tag{31.6}$$

Here τ' is related to τ by $\tau \log 2 = h(\tau')$. Thus, using the same "hypercube embedding $b \to \theta_b$," the bound similar to (31.5) will follow once we can bound $I(b^d;X)$ away from $d \log 2$.

Can we use the pairwise total variation bound in (31.5) to do that? Yes! Notice that thanks to the independence of b_i's we have[1]

$$I(b_i;X|b^{i-1}) = I(b_i;X, b^{i-1}) \leq I(b_i;X, b_{\setminus i}) = I(b_i;X|b_{\setminus i}).$$

Applying the chain rule leads to the upper bound

$$\begin{aligned} I(b^d;X) &= \sum_{i=1}^d I(b_i;X|b^{i-1}) \leq \sum_{i=1}^d I(b_i;X|b_{\setminus i}) \\ &\leq d \log 2 \max_{d_{\text{H}}(b,b')=1} \text{TV}(P_{X|b^d=b}, P_{X|b^d=b'}), \end{aligned} \tag{31.7}$$

where in the last step we used the fact that whenever $B \sim \text{Ber}(1/2)$,

$$I(B;X) \leq \text{TV}(P_{X|B=0}, P_{X|B=1}) \log 2, \tag{31.8}$$

which follows from (7.39) by noting that the mutual information is expressed as the Jensen–Shannon divergence as $2I(B;X) = \text{JS}(P_{X|B=0}, P_{X|B=1})$. Combining (31.6) and (31.7), the mutual information method implies the following version of Assouad's lemma: Under the assumption of Theorem 31.2,

$$\inf_{\hat{\theta}} \sup_{\theta \in \Theta} \mathbb{E}_\theta \ell(\theta, \hat{\theta}) \geq \frac{\beta d}{4\alpha} \cdot f\left(\max_{d_{\text{H}}(\theta,\theta')=1} \text{TV}(P_\theta, P_{\theta'}) \right), \quad f(t) \triangleq h^{-1}\left(\frac{(1-t)\log 2}{2} \right), \tag{31.9}$$

where $h^{-1}\colon [0, \log 2] \to [0, 1/2]$ is the inverse of the binary entropy function. Note that (31.9) is slightly weaker than (31.5). Nevertheless, as seen in Example 31.4, Assouad's lemma is typically applied when the pairwise total variation is bounded away from one by a constant, in which case (31.9) and (31.5) differ by only a constant factor.

[1] Equivalently, this also follows from the convexity of the mutual information in the channel (see Theorem 5.3).

In all, we may summarize Assouad's lemma as a convenient method for bounding $I(b^d;X)$ away from the full entropy (d bits) on the basis of distances between $P_{X|b^d}$ and adjacent b^d's.

31.4 Fano's Method

In this section we discuss another method for proving the minimax lower bound by reduction to multiple hypothesis testing. To this end, assume that the loss function is a metric. The idea is to consider an ϵ-packing (Chapter 27) of the parameter space, namely, a finite collection of parameters whose minimum separation is ϵ. Suppose we can show that given data one cannot reliably distinguish these hypotheses. Then the best estimation error is at least proportional to ϵ. The impossibility of testing is often shown by applying Fano's inequality (Corollary 3.13), which bounds the probability of error of testing in terms of the mutual information in Section 3.6. As such, we refer to this program as *Fano's method*. The following is a precise statement.

Theorem 31.3. *Let d be a metric on Θ. Fix an estimator $\hat{\theta}$. For any $T \subset \Theta$ and $\epsilon > 0$,*

$$\mathbb{P}\left[d(\theta,\hat{\theta}) \geq \frac{\epsilon}{2}\right] \geq 1 - \frac{C(T) + \log 2}{\log M(T,d,\epsilon)}, \tag{31.10}$$

where $C(T) \triangleq \sup I(\theta;X)$ is the capacity of the channel from θ to X with input space T, with the supremum taken over all distributions (priors) on T. Thus for any $r > 0$,

$$\inf_{\hat{\theta}} \sup_{\theta \in \Theta} \mathbb{E}_\theta[d(\theta,\hat{\theta})^r] \geq \sup_{T \subset \Theta, \epsilon > 0} \left(\frac{\epsilon}{2}\right)^r \left(1 - \frac{C(T) + \log 2}{\log M(T,d,\epsilon)}\right). \tag{31.11}$$

Proof. It suffices to show (31.10). Fix $T \subset \Theta$. Consider an ϵ-packing $T' = \{\theta_1, \ldots, \theta_M\} \subset T$ such that $\min_{i \neq j} d(\theta_i, \theta_j) \geq \epsilon$. Let θ be uniformly distributed on this packing and $X \sim P_\theta$ conditioned on θ. Given any estimator $\hat{\theta}$, construct a test by rounding $\hat{\theta}$ to $\tilde{\theta} = \operatorname{argmin}_{\theta \in T'} d(\hat{\theta}, \theta)$. By the triangle inequality, $d(\theta,\tilde{\theta}) \leq 2d(\theta,\hat{\theta})$. Thus $\mathbb{P}[\theta \neq \tilde{\theta}] \leq \mathbb{P}[d(\theta,\hat{\theta}) \geq \epsilon/2]$. On the other hand, applying Fano's inequality (Corollary 3.13) yields

$$\mathbb{P}[\theta \neq \tilde{\theta}] \geq 1 - \frac{I(\theta;X) + \log 2}{\log M}.$$

The proof of (31.10) is completed by noting that $I(\theta;X) \leq C(T)$. □

In applying Fano's method, since it is often difficult to evaluate the capacity $C(T)$, it is useful to recall from Theorem 5.9 that $C(T)$ coincides with the KL radius of the set of distributions $\{P_\theta : \theta \in T\}$, namely, $C(T) = \inf_Q \sup_{\theta \in T} D(P_\theta \| Q)$. As such, choosing any Q leads to an upper bound on the capacity. As an application, we revisit the d-dimensional GLM in Corollary 28.8 under the ℓ_q-loss ($1 \leq q \leq \infty$), with the particular focus on the dependence on the dimension. (For a different application in the sparse setting see Exercise **VI.12**.)

Example 31.5.

Consider the GLM with sample size n, where $P_\theta = \mathcal{N}(\theta, I_d)^{\otimes n}$. Taking natural logs here and below, we have

$$D(P_\theta \| P_{\theta'}) = \frac{n}{2}\|\theta - \theta'\|_2^2;$$

in other words, KL neighborhoods are ℓ_2-balls. As such, let us apply Theorem 31.3 to $T = B_2(\rho)$ for some $\rho > 0$ to be specified. Then $C(T) \le \sup_{\theta \in T} D(P_\theta \| P_0) = \frac{n\rho^2}{2}$. To bound the packing number from below, we apply the volume bound in Theorem 27.3,

$$M(B_2(\rho), \|\cdot\|_q, \epsilon) \ge \frac{\rho^d \mathrm{vol}(B_2)}{\epsilon^d \mathrm{vol}(B_q)} \ge \left(\frac{c_q \rho d^{1/q}}{\epsilon\sqrt{d}}\right)^d$$

for some constant c_q, where the last step follows from the volume formula (27.13) for ℓ_q-balls. Choosing $\rho = \sqrt{d/n}$ and $\epsilon = \frac{c_q}{e^2}\rho d^{1/q-1/2}$, an application of Theorem 31.3 yields the minimax lower bound

$$R_q \equiv \inf_{\hat\theta} \sup_{\theta\in\mathbb{R}^d} \mathbb{E}_\theta[\|\hat\theta - \theta\|_q] \ge C_q \frac{d^{1/q}}{\sqrt{n}} \tag{31.12}$$

for some constant C_q depending on q. This is the same lower bound as that in (30.9) obtained via the mutual information method plus the Shannon lower bound (which is also volume-based).

For any $q \ge 1$, (31.12) is rate-optimal since we can apply the MLE $\hat\theta = \bar{X}$. (Note that at $q = \infty$, the constant C_q is still finite since $\mathrm{vol}(B_\infty) = 2^d$.) However, for the special case of $q = \infty$, (31.12) does not depend on the dimension at all, as opposed to the correct dependence $\sqrt{\log d}$ shown in Corollary 28.8. In fact, previously in Example 31.4 the application of Assouad's lemma yields the same suboptimal result. So is it possible to fix this looseness with Fano's method? It turns out that the answer is "yes" and the suboptimality is due to the volume bound on the metric entropy, which, as we have seen in Section 27.3, can be ineffective if ϵ scales with dimension. Indeed, if we apply the tight bound of $M(B_2, \|\cdot\|_\infty, \epsilon)$ in (27.18),[2] with $\epsilon = \sqrt{\frac{c\log d}{n}}$ and $\rho = \sqrt{c'\frac{\log d}{n}}$ for some absolute constants c, c', we do get $R_\infty \gtrsim \sqrt{\frac{\log d}{n}}$ as desired.

We end[2] this section with some comments regarding the application of Theorem 31.3:

- It is sometimes convenient to further bound the KL radius by the KL diameter, since $C(T) \le \mathrm{diam}_{\mathrm{KL}}(T) \triangleq \sup_{\theta,\theta'\in T} D(P_{\theta'}\|P_\theta)$ (see Corollary 5.8). This suffices for Example 31.5.
- In Theorem 31.3 we actually lower-bound the global minimax risk by that restricted on a parameter subspace $T \subset \Theta$ for the purpose of controlling the mutual information, which is often difficult to compute. For the GLM considered in Example 31.5, the KL divergence is proportional to the squared ℓ_2-distance and T

[2] In fact, in this case we can also choose the explicit packing $\{\epsilon e_1, \ldots, \epsilon e_d\}$.

is naturally chosen to be a Euclidean ball. For other models such as the covariance model (Exercise **VI.16**) wherein the KL divergence is more complicated, the KL neighborhood T needs to be chosen carefully. Later in Section 32.4 we will apply the same Fano's method to the infinite-dimensional problem of estimating smooth density.

32 Entropic Bounds for Statistical Estimation

So far our discussion on information-theoretic methods have been mostly focused on statistical lower bounds (impossibility results), with matching upper bounds obtained on a case-by-case basis. In this chapter, we will discuss three information-theoretic upper bounds for statistical estimation. These three results apply to different loss functions and are obtained using completely different means. However, they take on exactly the same form involving the appropriate metric entropy of the model.

Specifically, suppose that we observe $X_1, \ldots, X_n$ drawn independently from a distribution P_θ for some unknown parameter $\theta \in \Theta$, and the goal is to produce an estimate $\hat{P}$ for the true distribution P_θ. We have the following entropic minimax upper bounds:

- KL loss (Yang–Barron [465]):

$$\inf_{\hat{P}} \sup_{\theta \in \Theta} \mathbb{E}_\theta[D(P_\theta \| \hat{P})] \lesssim \inf_{\epsilon > 0} \left\{ \epsilon^2 + \frac{1}{n} \log N_{\mathrm{KL}}(\mathcal{P}, \epsilon) \right\}. \tag{32.1}$$

- Hellinger loss (Le Cam–Birgé [274, 53]):

$$\inf_{\hat{P}} \sup_{\theta \in \Theta} \mathbb{E}_\theta[H^2(P_\theta, \hat{P})] \lesssim \inf_{\epsilon > 0} \left\{ \epsilon^2 + \frac{1}{n} \log N_H(\mathcal{P}, \epsilon) \right\}. \tag{32.2}$$

- Total variation loss (Yatracos [466]):

$$\inf_{\hat{P}} \sup_{\theta \in \Theta} \mathbb{E}_\theta[\mathrm{TV}^2(P_\theta \| \hat{P})] \lesssim \inf_{\epsilon > 0} \left\{ \epsilon^2 + \frac{1}{n} \log N_{\mathrm{TV}}(\mathcal{P}, \epsilon) \right\}. \tag{32.3}$$

Here $N(\mathcal{P}, \epsilon)$ refers to the metric entropy (see Chapter 27) of the model class $\mathcal{P} = \{P_\theta : \theta \in \Theta\}$ under various distances, which we will formalize along the way.

In particular, we will see that these methods achieve minimax optimal rates for the classical problem of *density estimation* under smoothness constraints. To place these results in the bigger context, we remind the reader that we have already discussed modern methods of density estimation based on machine learning ideas (Examples 4.2 and 7.5). However, those methods, beautiful and empirically successful, are not known to achieve optimality over any reasonable classes. The metric entropy methods as above, though, could and should be used to derive fundamental limits for the classes which are targeted by machine learning methods. Thus, there is a rich field of modern applications, which this chapter will hopefully encourage the reader to explore.

We note that there are other entropic upper bounds for statistical estimation, notably, MLE and other M-estimators. These require a different type of metric entropy (bracketing entropy, which is akin to metric entropy under the sup-norm) and the style of analysis is more related in spirit to the theory of empirical processes (e.g. Dudley's entropy integral (27.22)). We refer the reader to the monographs [333, 430, 432] for details. In this chapter we focus on more information-theoretic style results.

32.1 Yang–Barron's Construction

Let $\mathcal{P} = \{P_\theta : \theta \in \Theta\}$ be a parametric family of distributions on the space $\mathcal{X}$. Given $X^n = (X_1, \ldots, X_n) \overset{\text{iid}}{\sim} P_\theta$ for some $\theta \in \Theta$, we obtain an estimate $\hat{P} = \hat{P}(\cdot|X^n)$, which is a distribution depending on X^n. The loss function is the KL divergence $D(P_\theta \| \hat{P})$.[1] The average risk is thus

$$\mathbb{E}_\theta D(P_\theta \| \hat{P}) = \int D(P_\theta \| \hat{P}(\cdot|X^n)) P^{\otimes n}(dx^n).$$

If the family has a common dominating measure μ, the problem is equivalent to estimating the density $p_\theta = \frac{dP_\theta}{d\mu}$, commonly referred to as the problem of *density estimation* in the statistics literature.

Our objective is to prove the upper bound (32.1) for the minimax KL risk,

$$R^*_{\mathrm{KL}}(n) \triangleq \inf_{\hat{P}} \sup_{\theta \in \Theta} \mathbb{E}_\theta D(P_\theta \| \hat{P}), \tag{32.4}$$

where the infimum is taken over all estimators $\hat{P} = \hat{P}(\cdot|X^n)$ which is a distribution on $\mathcal{X}$; in other words, we allow *improper* estimates in the sense that $\hat{P}$ can step outside the model class $\mathcal{P}$. Indeed, the construction we will use in this section (such as predictive density estimators (Bayes) or their mixtures) need not be a member of $\mathcal{P}$. Later we will see in Sections 32.2 and 32.3 that for total variation and Hellinger loss we can always restrict to *proper* estimators;[2] however, these loss functions are weaker than the KL divergence.

The main result of this section is the following.

Theorem 32.1. *Let C_n denote the capacity of the channel $\theta \mapsto X^n \sim P_\theta^{\otimes n}$, namely*

$$C_n = \sup I(\theta; X^n), \tag{32.5}$$

[1] Note the asymmetry in this loss function. Alternatively the loss $D(\hat{P} \| P)$ is typically infinite in nonparametric settings, because it is impossible to estimate the support of the true density exactly.

[2] This is in fact a generic observation: Whenever the loss function satisfies an approximate triangle inequality, any improper estimate can be converted to a proper one by its projection on the model class whose risk is inflated by no more than a constant factor.

where the supremum is over all distributions (priors) of θ taking values in Θ. Denote by

$$N_{\text{KL}}(\mathcal{P},\epsilon) \triangleq \min\left\{N\colon \exists Q_1,\ldots,Q_N \text{ s.t. } \forall\theta\in\Theta, \exists i\in[N], D(P_\theta\|Q_i)\le\epsilon^2\right\} \quad (32.6)$$

the KL covering number for the class $\mathcal{P}$. Then

$$R^*_{\text{KL}}(n) \le \frac{C_{n+1}}{n+1} \quad (32.7)$$

$$\le \inf_{\epsilon>0}\left\{\epsilon^2 + \frac{1}{n+1}\log N_{\text{KL}}(\mathcal{P},\epsilon)\right\}. \quad (32.8)$$

Conversely,

$$\sum_{t=0}^{n} R^*_{\text{KL}}(t) \ge C_{n+1}. \quad (32.9)$$

Note that the capacity C_n is precisely the redundancy (13.10) which governs the minimax regret in universal compression; the fact that it bounds the KL risk can be attributed to a generic relation between individual and cumulative risks which we explain later in Section 32.1.4. As explained in Chapter 13, it is in general difficult to compute the exact value of C_n even for models as simple as Bernoulli ($P_\theta = \text{Ber}(\theta)$). This is where (32.8) comes in: one can use metric entropy and tools from Chapter 27 to bound this capacity, leading to useful (and even optimal) risk bounds. We discuss two types of applications of this result.

Finite-Dimensional Models Consider a family $\mathcal{P} = \{P_\theta\colon \theta\in\Theta\}$ of smooth parameterized densities, where $\Theta\subset\mathbb{R}^d$ is some compact set. Suppose that the KL divergence behaves like a squared norm, namely, $D(P_\theta\|P_{\theta'}) \asymp \|\theta-\theta'\|^2$ for any $\theta,\theta'\in\Theta$ and some norm $\|\cdot\|$ on $\mathbb{R}^d$. (For example, for the GLM with $P_\theta = \mathcal{N}(\theta, I_d)$, we have $D(P_\theta\|P_{\theta'}) = \frac{1}{2}\|\theta-\theta'\|_2^2$.) In this case, the KL covering number inherits the usual behavior of metric entropy in finite-dimensional space (see Theorem 27.3 and Corollary 27.4) and we have

$$N_{\text{KL}}(\mathcal{P},\epsilon) \lesssim \left(\frac{1}{\epsilon}\right)^d.$$

Then (32.8) yields

$$C_n \lesssim \inf_{\epsilon>0}\left\{n\epsilon^2 + d\log\frac{1}{\epsilon}\right\} \asymp d\log n, \quad (32.10)$$

which is consistent with the typical asymptotics of redundancy $C_n = \frac{d}{2}\log n + o(\log n)$ (recall (13.24) and (13.25)).

Applying the upper bound (32.7) or (32.8) yields

$$R^*_{\text{KL}}(n) \lesssim \frac{d\log n}{n}.$$

As compared to the usual parametric rate of $\frac{d}{n}$ in d dimensions (e.g. GLM), this upper bound is suboptimal only by a logarithmic factor.

Infinite-Dimensional Models Similar to the results in Section 27.4, for nonparametric models $N_{\mathrm{KL}}(\epsilon)$ typically grows super-polynomially in $\frac{1}{\epsilon}$ and, in turn, the capacity C_n grows super-logarithmically. In fact, whenever we have $C_n = n^\alpha \mathrm{polylog}(n)$ for some $\alpha > 0$ where $(\log n)^{c_0} \le \mathrm{polylog}(n) \le (\log n)^{c_1}$ for some absolute c_0, c_1, Theorem 32.1 shows the minimax KL rate satisfies

$$R^*_{\mathrm{KL}}(n) = n^{\alpha-1}\mathrm{polylog}(n), \tag{32.11}$$

which easily follows from combining (32.7) and (32.8) – see (32.27) for details. For concrete examples, see Section 32.4 for the application of estimating smooth densities.

Next, we explain the intuition behind and the proof of Theorem 32.1.

32.1.1 Bayes Risk as Conditional Mutual Information and Capacity Bound

To gain some insight, let us start by considering the Bayesian setting with a prior π on Θ. Conditioned on $\theta \sim \pi$, the data $X^n = (X_1, \ldots, X_n) \overset{\text{iid}}{\sim} P_\theta$.[3] Any estimator, $\hat{P} = \hat{P}(\cdot|X^n)$, is a distribution on $\mathcal{X}$ depending on X^n. As such, $\hat{P}$ can be identified with a *conditional distribution*, say, $Q_{X_{n+1}|X^n}$, and we shall do so henceforth. For convenience, let us introduce an (unseen) observation X_{n+1} that is drawn from the same P_θ and independent of X^n conditioned on θ. In this light, the role of the estimator is to *predict* the distribution of the unseen X_{n+1}.

The following lemma shows that the Bayes KL risk equals the conditional mutual information and the Bayes estimator is precisely $P_{X_{n+1}|X^n}$ (with respect to the joint distribution induced by the prior), known as the *predictive density estimator* in the statistics literature.

Lemma 32.2. *The Bayes risk for prior π is given by*

$$R^*_{\mathrm{KL,Bayes}}(\pi) \triangleq \inf_{\hat{P}} \int \pi(d\theta) P_\theta^{\otimes n}(dx^n) D(P_\theta \| \hat{P}(\cdot|x^n)) = I(\theta; X_{n+1}|X^n),$$

where $\theta \sim \pi$ and $(X_1, \ldots, X_{n+1}) \overset{\text{iid}}{\sim} P_\theta$ conditioned on θ. The Bayes estimator achieving this infimum is given by $\hat{P}_{Bayes}(\cdot|x^n) = P_{X_{n+1}|X^n = x^n}$. If each P_θ has a density p_θ with respect to some common dominating measure μ, the Bayes estimator has density

$$\hat{p}_{\mathrm{Bayes}}(x_{n+1}|x^n) = \frac{\int \pi(d\theta) \prod_{i=1}^{n+1} p_\theta(x_i)}{\int \pi(d\theta) \prod_{i=1}^{n} p_\theta(x_i)}. \tag{32.12}$$

[3] Throughout this chapter, we continue to use the conventional notation P_θ for a parametric family of distributions and use π to stand for the distribution of θ.

Proof. The Bayes risk can be computed as follows:

$$\begin{aligned}
\inf_{Q_{X_{n+1}|X^n}} \mathbb{E}_{\theta,X^n}\left[D(P_\theta \| Q_{X_{n+1}|X^n})\right] &= \inf_{Q_{X_{n+1}|X^n}} D(P_{X_{n+1}|\theta} \| \hat{P}_{X_{n+1}|X^n} | P_{\theta,X^n}) \\
&= \mathbb{E}_{X^n}\left[\inf_{Q_{X_{n+1}|X^n}} D(P_{X_{n+1}|\theta} \| \hat{P}_{X_{n+1}|X^n} | P_{\theta|X^n})\right] \\
&\overset{(a)}{=} \mathbb{E}_{X^n}\left[D(P_{X_{n+1}|\theta} \| P_{X_{n+1}|X^n} | P_{\theta|X^n})\right] \\
&= D(P_{X_{n+1}|\theta} \| P_{X_{n+1}|X^n} | P_{\theta,X^n}) \\
&\overset{(b)}{=} I(\theta;X_{n+1}|X^n).
\end{aligned}$$

Here (a) follows from the variational representation of mutual information (Theorem 4.1 and Corollary 4.2); and (b) invokes the definition of the conditional mutual information (Section 3.4) and the fact that $X^n \to \theta \to X_{n+1}$ form a Markov chain, so that $P_{X_{n+1}|\theta,X^n} = P_{X_{n+1}|\theta}$. In addition, the Bayes optimal estimator is given by $P_{X_{n+1}|X^n}$. □

Note that the operational meaning of $I(\theta;X_{n+1}|X^n)$ is the information provided by one extra observation about θ having already obtained n observations. In most situations, since X^n will have already allowed θ to be consistently estimated as $n \to \infty$, the additional usefulness of X_{n+1} is vanishing. This is made precise by the following result.

Lemma 32.3. (Diminishing marginal utility in information) $n \mapsto I(\theta;X_{n+1}|X^n)$ *is a decreasing sequence. Furthermore,*

$$I(\theta;X_{n+1}|X^n) \le \frac{1}{n+1} I(\theta;X^{n+1}). \tag{32.13}$$

Proof. In view of the chain rule for mutual information (Theorem 3.7), $I(\theta;X^{n+1}) = \sum_{i=1}^{n+1} I(\theta;X_i|X^{i-1})$, (32.13) follows from the monotonicity. To show the latter, let us consider a "sampling channel" where the input is θ and the output is X sampled from P_θ. Let $I(\pi)$ denote the mutual information when the input distribution is π, which is a concave function in π (Theorem 5.3). Then

$$I(\theta;X_{n+1}|X^n) = \mathbb{E}_{X^n}[I(P_{\theta|X^n})] \le \mathbb{E}_{X^{n-1}}[I(P_{\theta|X^{n-1}})] = I(\theta;X_n|X^{n-1}),$$

where the inequality follows from Jensen's inequality, since $P_{\theta|X^{n-1}}$ is a mixture of $P_{\theta|X^n}$. □

Lemma 32.3 allows us to prove the converse bound (32.9): Fix any prior π. Since the minimax risk dominates any Bayes risk (Theorem 28.1), in view of Lemma 32.2, we have

$$\sum_{t=0}^{n} R^*_{\mathrm{KL}}(t) \ge \sum_{t=0}^{n} I(\theta;X_{t+1}|X^t) = I(\theta;X^{n+1}).$$

Recall from (32.5) that $C_{n+1} = \sup_{\pi \in \mathcal{P}(\Theta)} I(\theta;X^{n+1})$. Optimizing over the prior π yields (32.9).

Now suppose that the minimax theorem holds for (32.4), so that $R^*_{\mathrm{KL}} = \sup_{\pi \in \mathcal{P}(\Theta)} R^*_{\mathrm{KL,Bayes}}(\pi)$. Lemma 32.2 then allows us to express the minimax risk as the conditional mutual information maximized over the prior π:

$$R^*_{\mathrm{KL}}(n) = \sup_{\pi \in \mathcal{P}(\Theta)} I(\theta; X_{n+1}|X^n).$$

Thus Lemma 32.3 implies the desired

$$R^*_{\mathrm{KL}}(n) \leq \frac{1}{n+1} C_{n+1}.$$

Next, we prove this directly without going through the Bayesian route or assuming the minimax theorem. The main idea, due to Yang and Barron [465], is to consider a Bayes estimator (of the form (32.12)) but analyze it in the *worst case*. Fix an arbitrary joint distribution $Q_{X^{n+1}}$ on $\mathcal{X}^{n+1}$, which factorizes as $Q_{X^{n+1}} = \prod_{i=1}^{n+1} Q_{X_i|X^{i-1}}$. (This joint distribution is an auxiliary object used only for constructing an estimator.) For each i, the conditional distribution $Q_{X_i|X^{i-1}}$ defines an estimator taking the sample X^i of size i as the input. Taking their Cesàro mean results in the following estimator operating on the full sample X^n:

$$\hat{P}(\cdot|X^n) \triangleq \frac{1}{n+1} \sum_{i=1}^{n+1} Q_{X_i|X^{i-1}}. \tag{32.14}$$

Let us bound the worst-case KL risk of this estimator. *Fix* $\theta \in \Theta$ and let X^{n+1} be drawn independently from P_θ so that $P_{X^{n+1}} = P_\theta^{\otimes(n+1)}$. Taking expectations with this law, we have

$$\begin{aligned}
\mathbb{E}_\theta[D(P_\theta \| \hat{P}(\cdot|X^n))] &= \mathbb{E}\left[D\left(P_\theta \middle\| \frac{1}{n+1} \sum_{i=1}^{n+1} Q_{X_i|X^{i-1}} \right) \right] \\
&\overset{(a)}{\leq} \frac{1}{n+1} \sum_{i=1}^{n+1} D(P_\theta \| Q_{X_i|X^{i-1}} | P_{X^{i-1}}) \\
&\overset{(b)}{=} \frac{1}{n+1} D(P_\theta^{\otimes(n+1)} \| Q_{X^{n+1}}),
\end{aligned}$$

where (a) and (b) follow from the convexity (Theorem 5.1) and the chain rule for KL divergence (Theorem 2.16(c)). Taking the supremum over $\theta \in \Theta$ bounds the worst-case risk as

$$R^*_{\mathrm{KL}}(n) \leq \frac{1}{n+1} \sup_{\theta \in \Theta} D(P_\theta^{\otimes(n+1)} \| Q_{X^{n+1}}).$$

Optimizing over the choice of $Q_{X^{n+1}}$, we obtain

$$R^*_{\mathrm{KL}}(n) \leq \frac{1}{n+1} \inf_{Q_{X^{n+1}}} \sup_{\theta \in \Theta} D(P_\theta^{\otimes(n+1)} \| Q_{X^{n+1}}) = \frac{C_{n+1}}{n+1},$$

where the last identity applies Theorem 5.9 of Kemperman, completing the proof of (32.7). Furthermore, Theorem 5.9 asserts that the optimal $Q_{X^{n+1}}$ exists and is given uniquely by the capacity-achieving output distribution $P^*_{X^{n+1}}$. Thus the above minimax upper bound can be attained by taking the Cesàro average of $P^*_{X_1}, P^*_{X_2|X_1}, \ldots, P^*_{X_{n+1}|X^n}$, namely,

$$\hat{P}^*(\cdot|X^n) = \frac{1}{n+1}\sum_{i=1}^{n+1} P^*_{X_i|X^{i-1}}. \tag{32.15}$$

Note that in general this is an *improper* estimate as it steps outside the class $\mathcal{P}$.

In the special case where the capacity-achieving input distribution π^* exists, the capacity-achieving output distribution can be expressed as a mixture over product distributions as $P^*_{X^{n+1}} = \int \pi^*(d\theta)P_\theta^{\otimes(n+1)}$. Thus the estimator $\hat{P}^*(\cdot|X^n)$ is in fact the *average of Bayes estimators* (32.12) under prior π^* for sample sizes ranging from 0 to n.

Finally, as will be made clear in the next section, in order to achieve the further upper bound (32.8) in terms of the KL covering numbers, namely $R^*_{\mathrm{KL}}(n) \le \epsilon^2 + \frac{1}{n+1}\log N_{\mathrm{KL}}(\mathcal{P}, \epsilon)$, it suffices to choose the following $Q_{X^{n+1}}$ as opposed to the exact capacity-achieving output distribution: Pick an ϵ-KL cover $Q_1, \ldots, Q_N$ for $\mathcal{P}$ of size $N = N_{\mathrm{KL}}(\mathcal{P}, \epsilon)$, choose π to be the uniform distribution, and define $Q_{X^{n+1}} = \frac{1}{N}\sum_{j=1}^N Q_j^{\otimes(n+1)}$ – this was the original construction in [465]. In this case, applying the Bayes rule (32.12), we see that the estimator is in fact a convex combination $\hat{P}(\cdot|X^n) = \sum_{j=1}^N w_j Q_j$ of the centers $Q_1, \ldots, Q_N$, with data-driven weights given by

$$w_j = \frac{1}{n+1}\sum_{i=1}^{n+1} \frac{\prod_{t=1}^{i-1} Q_j(X_t)}{\sum_{\ell=1}^N \prod_{t=1}^{i-1} Q_\ell(X_t)}.$$

Again, except for the extraordinary case where $\mathcal{P}$ is convex and the centers Q_j belong to $\mathcal{P}$, the estimate $\hat{P}(\cdot|X^n)$ is improper.

32.1.2 Capacity Upper Bound via KL Covering Numbers

As explained earlier, finding the capacity C_n requires solving the difficult optimization problem in (32.5). In this subsection we prove (32.8) which bounds this capacity by metric entropy. Conceptually speaking, both metric entropy and capacity measure the complexity of a model class. The following result, which applies to a more general setting than (32.5), makes precise their relations.

Theorem 32.4. *Let $\mathcal{Q} = \{P_{B|A=a} : a \in \mathcal{A}\}$ be a collection of distributions on some space $\mathcal{B}$ and denote the capacity $C = \sup_{P_A \in \mathcal{P}(\mathcal{A})} I(A;B)$. Then*

$$C = \inf_{\epsilon>0}\{\epsilon^2 + \log N_{\mathrm{KL}}(\mathcal{Q}, \epsilon)\}, \tag{32.16}$$

where N_{KL} is the KL covering number defined in (32.6).

Proof. Fix ϵ and let $N = N_{\mathrm{KL}}(\mathcal{Q}, \epsilon)$. Then there exist $Q_1, \ldots, Q_N$ that form an ϵ-KL cover, such that for any $a \in \mathcal{A}$ there exists $i(a) \in [N]$ such that $D(P_{B|A=a}\|Q_{i(a)}) \le \epsilon^2$. Fix any P_A. Then

$$
\begin{aligned}
I(A;B) = I(A,i(A);B) &= I(i(A);B) + I(A;B|i(A)) \\
&\leq H(i(A)) + I(A;B|i(A)) \leq \log N + \epsilon^2,
\end{aligned}
$$

where the last inequality follows from the fact that $i(A)$ takes at most N values and, by applying Theorem 4.1,

$$
I(A;B|i(A)) \leq D\left(P_{B|A} \| Q_{i(A)} | P_{i(A)}\right) \leq \epsilon^2.
$$

For the lower bound, note that if $C = \infty$, then in view of the upper bound above, $N_{\mathrm{KL}}(\mathcal{Q}, \epsilon) = \infty$ for any ϵ and (32.16) holds with equality. If $C < \infty$, Theorem 5.9 shows that C is the KL radius of $\mathcal{Q}$, namely, there exists P_B^* such that $C = \sup_{P_A \in \mathcal{P}(\mathcal{A})} D(P_{B|A} \| P_B^* | P_A) = \sup_{a \in \mathcal{A}} D(P_{B|A=a} \| P_B^* | P_A)$. In other words, $N_{\mathrm{KL}}(\mathcal{Q}, \sqrt{C+\delta}) = 1$ for any $\delta > 0$. Sending $\delta \to 0$ proves the equality of (32.16). □

Next we specialize Theorem 32.4 to our statistical setting (32.5) where the input A is θ and the output B is $X^n \overset{\text{iid}}{\sim} P_\theta$. Recall that $\mathcal{P} = \{P_\theta : \theta \in \Theta\}$. Let $\mathcal{P}_n \triangleq \{P_\theta^{\otimes n} : \theta \in \Theta\}$. By tensorization of KL divergence (Theorem 2.16(d)), $D(P_\theta^{\otimes n} \| P_{\theta'}^{\otimes n}) = nD(P_\theta \| P_{\theta'})$. Thus

$$
N_{\mathrm{KL}}(\mathcal{P}_n, \epsilon) \leq N_{\mathrm{KL}}\left(\mathcal{P}, \frac{\epsilon}{\sqrt{n}}\right).
$$

Combining this with Theorem 32.4, we obtain the following upper bound on the capacity C_n in terms of the KL metric entropy of the (single-letter) family $\mathcal{P}$:

$$
C_n \leq \inf_{\epsilon > 0} \left\{ n\epsilon^2 + \log N_{\mathrm{KL}}(\mathcal{P}, \epsilon) \right\}. \tag{32.17}
$$

This proves (32.8), completing the proof of Theorem 32.1.

32.1.3 Bounding Capacity and KL Covering Number Using Hellinger Entropy

Recall that in order to deduce from (32.9) concrete bounds on the minimax KL risk, such as (32.11), one needs to have matching upper and lower bounds on the capacity C_n. Although Theorem 32.4 characterizes capacity in terms of KL covering number, working with the latter is not convenient as it is not a metric so that results developed in Chapter 27 such as Theorem 27.2 do not apply. Next, we give bounds on the KL covering number and the capacity C_n using metric entropy with respect to the Hellinger distance, which are tight up to logarithmic factors under mild conditions.

Theorem 32.5. *Let $\mathcal{P} = \{P_\theta : \theta \in \Theta\}$ and $M_{\mathrm{H}}(\epsilon) \equiv M(\mathcal{P}, H, \epsilon)$ the Hellinger packing number of the set $\mathcal{P}$, see (27.2). Then C_n defined in (32.5) satisfies*

$$
C_n \geq \min\left(\frac{\log e}{2} n\epsilon^2, \log M_{\mathrm{H}}(\epsilon)\right) - \log 2. \tag{32.18}
$$

Proof. The idea of the proof is simple. Given a packing $\theta_1, \ldots, \theta_M \in \Theta$ with pairwise distances $H^2(Q_i, Q_j) \geq \epsilon^2$ for $i \neq j$, where $Q_i \equiv P_{\theta_i}$, we know that one can test $Q_i^{\otimes n}$ versus $Q_j^{\otimes n}$ with error $e^{-n\epsilon^2/2}$; see Theorems 7.8 and 32.8. Then by the union bound, if $Me^{-n\epsilon^2/2} < \frac{1}{2}$, we can distinguish these M hypotheses with error $< \frac{1}{2}$. Let $\theta \sim \mathrm{Unif}(\theta_1, \dots, \theta_M)$. Then from Fano's inequality we get $I(\theta;X^n) \gtrsim \log M$.

To get sharper constants, though, we will proceed via the inequality shown in Exercise **I.58**. In the notation of that exercise we take $\lambda = 1/2$ and from Definition 7.24 we get that

$$D_{1/2}(Q_i, Q_j) = -2\log(1 - \tfrac{1}{2}H^2(Q_i, Q_j)) \geq H^2(Q_i, Q_j)\log e \geq \epsilon^2 \log e, \qquad i \neq j.$$

By the tensorization property (7.79) for Rényi divergence, $D_{1/2}(Q_i^{\otimes n}, Q_j^{\otimes n}) = nD_{1/2}(Q_i, Q_j)$ and we get by Exercise **I.58**

$$\begin{aligned} I(\theta;X^n) &\geq -\sum_{i=1}^{M} \frac{1}{M} \log\left(\sum_{j=1}^{M} \frac{1}{M} \exp\left\{-\frac{n}{2} D_{1/2}(Q_i, Q_j)\right\}\right) \\ &\overset{(a)}{\geq} -\sum_{i=1}^{M} \frac{1}{M} \log\left(\frac{M-1}{M} e^{-n\epsilon^2/2} + \frac{1}{M}\right) \\ &\geq -\sum_{i=1}^{M} \frac{1}{M} \log\left(e^{-n\epsilon^2/2} + \frac{1}{M}\right) = -\log\left(e^{-n\epsilon^2/2} + \frac{1}{M}\right), \end{aligned}$$

where in (a) we used the fact that pairwise distances are all $\geq n\epsilon^2$ except when $i = j$. Finally, since $\frac{1}{A} + \frac{1}{B} \leq \frac{2}{\min(A,B)}$ we conclude the result. □

Note that, from the joint range (7.33) that $D(P\|Q) \geq H^2(P, Q)$, a different (weaker) lower bound on the KL risk also follows from Section 32.2.4 below.

Next we proceed to the converse of Theorem 32.5. The KL and Hellinger covering numbers always satisfy

$$N_{\mathrm{KL}}(\mathcal{P}, \epsilon) \geq N_{\mathrm{H}}(\epsilon) \equiv N(\mathcal{P}, H, \epsilon). \tag{32.19}$$

We next show that, assuming that the class $\mathcal{P}$ has a finite radius in Rényi divergence, (32.19) and hence the capacity bound in Theorem 32.5 are tight up to logarithmic factors. Later in Section 32.4 we will apply these results to the class of smooth densities, which has a finite χ^2-radius (by choosing the uniform distribution as the center).

Theorem 32.6. *Suppose that the family $\mathcal{P}$ has a finite D_λ radius for some $\lambda > 1$, that is,*

$$R_\lambda(\mathcal{P}) \triangleq \inf_U \sup_{P\in\mathcal{P}} D_\lambda(P\|U) < \infty, \tag{32.20}$$

where D_λ is the Rényi divergence of order λ (see Definition 7.24). Then there exist ϵ_0 and c depending only on λ and R_λ, such that for all $\epsilon \leq \epsilon_0$,

$$N_{\mathrm{KL}}\left(\mathcal{P}, c\epsilon\sqrt{\log\frac{1}{\epsilon}}\right) \leq N_{\mathrm{H}}(\epsilon) \tag{32.21}$$

and, consequently,

$$C_n \le \inf_{\epsilon \le \epsilon_0} \left\{ cn\epsilon^2 \log \frac{1}{\epsilon} + \log N_{\mathrm{H}}(\epsilon) \right\}. \tag{32.22}$$

Proof. Let $Q_1, \dots, Q_M$ be an ϵ-covering of $\mathcal{P}$ such that for any $P \in \mathcal{P}$, there exists $i \in [M]$ such that $H^2(P, Q_i) \le \epsilon^2$. Fix an arbitrary U and let $P_i = \epsilon^2 U + (1-\epsilon^2) Q_i$. Applying Exercise **I.59** yields

$$D(P \| P_i) \le 24\epsilon^2 \left(\frac{2\lambda}{\lambda - 1} \log \frac{1}{\epsilon} + D_\lambda(P \| U) \right).$$

Optimizing over U to approach (32.20) proves (32.21). Finally, (32.22) follows from applying (32.21) to (32.17). □

32.1.4 General Bounds between Cumulative and Individual (One-Step) Risks

In summary, we can see that the beauty of the Yang–Barron method lies in two ideas:

- Instead of directly studying the risk $R^*_{\mathrm{KL}}(n)$, (32.7) relates it to a cumulative risk C_n.
- The cumulative risk turns out to be equal to a capacity, which can be conveniently bounded in terms of covering numbers.

In this subsection we want to point out that while the second step is very special to KL (log-loss), the first idea is generic. Namely, we have the following relationship between individual risk (also known as batch loss) and cumulative risk (also known as online loss), which were previously introduced in Section 13.6 in the context of universal compression.

Proposition 32.7. (Online-to-batch conversion) *Fix a loss function $\ell\colon \mathcal{P}(\mathcal{X}) \times \mathcal{P}(\mathcal{X}) \to \bar{\mathbb{R}}$ and a class Π of distributions on $\mathcal{X}$. Define the cumulative and one-step minimax risks as follows:*[4]

$$C_n = \inf_{\{\hat{P}_t(\cdot)\}} \sup_{P \in \Pi} \mathbb{E}\left[\sum_{t=1}^n \ell(P, \hat{P}_t(X^{t-1})) \right], \tag{32.23}$$

$$R^*_n = \inf_{\hat{P}(\cdot)} \sup_{P \in \Pi} \mathbb{E}[\ell(P, \hat{P}(X^{n-1}))], \tag{32.24}$$

where both infima are over measurable (possibly randomized) estimators $\hat{P}_t\colon \mathcal{X}^{t-1} \to \mathcal{P}(\mathcal{X})$, and the expectations are over $X_i \overset{\text{iid}}{\sim} P$ and the randomness of the estimators. Then we have

$$nR^*_n \le C_n \le C_{n-1} + R^*_n \le \sum_{t=1}^n R^*_t. \tag{32.25}$$

[4] Note that for KL loss, C_n and R^*_n coincide with AvgReg^*_n and $\mathrm{BatchReg}^*_n$ defined in (*13.34*) and (*13.35*), respectively.

Thus, if the sequence $\{R_n^*\}$ *satisfies* $R_n^* \asymp \frac{1}{n}\sum_{t=1}^{n-1} R_t^*$ *then* $C_n \asymp nR_n^*$. *Conversely, if* $n^{\alpha_-} \lesssim C_n \lesssim n^{\alpha_+}$ *for all n and some* $\alpha_+ \geq \alpha_- > 0$, *then*

$$n^{(\alpha_- - 1)\alpha_+/\alpha_-} \lesssim R_n^* \lesssim n^{\alpha_+ - 1}. \tag{32.26}$$

Remark 32.1. The meaning of the above is that $R_n^* \approx \frac{1}{n}C_n$ within either constant or polylogarithmic factors, for most cases of interest.

Proof. To show the first inequality in (32.25), given predictors $\{\hat{P}_t(X^{t-1}) : t \in [n]\}$ for C_n, consider a randomized predictor $\hat{P}(X^{n-1})$ for R_n^* that equals each of the $\hat{P}_t(X^{t-1})$ with equal probability. The second inequality follows from interchanging $\sup_P$ and $\sum_t$ via:

$$\sup_{P\in\Pi} \mathbb{E}\left[\sum_{t=1}^{n} \ell(P, \hat{P}_t(X^{t-1}))\right] \leq \sup_{P\in\Pi} \mathbb{E}\left[\sum_{t=1}^{n-1} \ell(P, \hat{P}_t(X^{t-1}))\right] + \sup_{P\in\Pi} \mathbb{E}[\ell(P, \hat{P}_n(X^{n-1}))].$$

(In other words, C_n is bounded by using the C_{n-1}-optimal online learner for the first $n-1$ rounds and the R_n^*-optimal batch learner for the last round.) The third inequality in (32.25) follows from the second by induction and $C_1 = R_1^*$.

To derive (32.26) notice that the upper bound on R_n^* follows from (32.25). For the lower bound, notice that the sequence R_n^* is non-increasing and hence we have for any $n < m$

$$C_m \leq \sum_{t=1}^{m-1} R_t^* \leq \sum_{t=1}^{n-1} \frac{C_t}{t} + (m-n)R_n^*. \tag{32.27}$$

Setting $m = an^{\alpha_+/\alpha_-}$ with some appropriate constant a yields the lower bound. □

32.2 Pairwise Comparison à la Le Cam–Birgé

When we proved the lower bound in Theorem 31.3, we applied the reasoning that if an ϵ-packing of the parameter space Θ cannot be tested, then $\theta \in \Theta$ cannot be estimated with more than precision ϵ, thereby establishing a minimax *lower bound* in terms of the KL metric entropy. Conversely, we can ask the following question:

Is it possible to construct an estimator based on tests, and produce a minimax *upper bound* in terms of the metric entropy?

For Hellinger loss, the answer is "yes," although the metric entropy involved is with respect to the Hellinger distance not KL divergence. The basic construction is due to Le Cam and further developed by Birgé. The main idea is as follows: Fix an ϵ-covering $\{P_1, \ldots, P_N\}$ of the set of distributions $\mathcal{P}$. Given n observations drawn from $P \in \mathcal{P}$, let us test which ball P belongs to; this allows us to estimate P up to Hellinger loss ϵ. This can be realized by a pairwise comparison argument of testing the (composite) hypothesis $P \in B(P_i, \epsilon)$ versus $P \in B(P_j, \epsilon)$. This program can be further refined to involve the local entropy of the model.

32.2.1 Composite Hypothesis Testing and Hellinger Distance

Recall the problem of composite hypothesis testing introduced in Section 16.4. Let $\mathcal{P}$ and $\mathcal{Q}$ be two (*not* necessarily convex) classes of distributions. Given iid observations $X_1, \ldots, X_n$ drawn from some distribution P, we want to test, according to some decision rule $\phi = \phi(X_1, \ldots, X_n) \in \{0, 1\}$, whether $P \in \mathcal{P}$ (indicated by $\phi = 0$) or $P \in \mathcal{Q}$ (indicated by $\phi = 1$). By the minimax theorem, the optimal error is given by the total variation between the worst-case mixtures:

$$\min_{\phi}\left\{\sup_{P\in\mathcal{P}} P(\phi = 1) + \sup_{Q\in\mathcal{Q}} Q(\phi = 0)\right\} = 1 - \mathrm{TV}(\mathrm{co}(\mathcal{P}^{\otimes n}), \mathrm{co}(\mathcal{Q}^{\otimes n})), \qquad (32.28)$$

wherein the notations are explained as follows:

- $\mathcal{P}^{\otimes n} \triangleq \{P^{\otimes n} : P \in \mathcal{P}\}$ consists of all n-fold products of distributions in $\mathcal{P}$.
- $\mathrm{co}(\cdot)$ denotes the convex hull, that is, the set of all mixtures. For example, for a parametric family, $\mathrm{co}(\{P_\theta : \theta \in \Theta\}) = \{P_\pi : \pi \in \mathcal{P}(\Theta)\}$, where $P_\pi = \int P_\theta \pi(d\theta)$ is the mixture under the mixing distribution π, and $\mathcal{P}(\Theta)$ denotes the collection of all distributions (priors) on Θ.

The optimal test that achieves (32.28) is the likelihood ratio given by the worst-case mixtures, that is, the closest[5] pair of mixtures (P_n^*, Q_n^*) such that $\mathrm{TV}(P_n^*, Q_n^*) = \mathrm{TV}(\mathrm{co}(\mathcal{P}^{\otimes n}), \mathrm{co}(\mathcal{Q}^{\otimes n}))$.

The exact result (32.28) is unwieldy as the RHS involves finding the least favorable priors over the n-fold product space. However, there are several known examples where much simpler and explicit results are available. In the case when $\mathcal{P}$ and $\mathcal{Q}$ are TV-balls around P_0 and Q_0, Huber [221] showed that the minimax optimal test has the form

$$\phi(x^n) = 1\left\{\sum_{i=1}^{n} \min\left(c', \max\left(c'', \log\frac{dP_0}{dQ_0}(X_i)\right)\right) > t\right\}.$$

(See also Exercise **III.31**.) However, there are few other examples where minimax optimal tests are known explicitly. Fortunately, as was shown by Le Cam, there is a general "single-letter" upper bound in terms of the Hellinger separation between $\mathcal{P}$ and $\mathcal{Q}$. It is the consequence of the more general tensorization property of Rényi divergence in Proposition 7.25 (of which Hellinger is a special case).

Theorem 32.8.

$$\min_{\phi}\left\{\sup_{P\in\mathcal{P}} P(\phi = 1) + \sup_{Q\in\mathcal{Q}} Q(\phi = 0)\right\} \le e^{-\frac{n}{2}H^2(\mathrm{co}(\mathcal{P}),\mathrm{co}(\mathcal{Q}))}. \qquad (32.29)$$

Remark 32.2. For the case when $\mathcal{P}$ and $\mathcal{Q}$ are Hellinger balls of radius r around P_0 and Q_0 (which arises in the proof of Theorem 32.9 below), respectively, Birgé [56] constructed an explicit test. Namely, under the assumption $H(P_0, Q_0) > 2.01r$,

[5] In case the closest pair does not exist, we can replace it by an infimizing sequence.

there is a test of the form $\phi(x^n) = 1\left\{\sum_{i=1}^n \log \frac{\alpha+\beta\psi(X_i)}{\beta+\alpha\psi(X_i)} > t\right\}$ attaining error $e^{-\Omega(nr^2)}$, where $\psi(x) = \sqrt{\frac{dP_0}{dQ_0}(x)}$ and $\alpha, \beta > 0$ depend only on $H(P_0, Q_0)$.

Proof. We start by restating the special case of Proposition 7.25:

$$1 - \frac{1}{2}H^2\left(\text{co}\left(\bigotimes_{i=1}^n \mathcal{P}_i\right), \text{co}\left(\bigotimes_{i=1}^n \mathcal{Q}_i\right)\right) \le \prod_{i=1}^n \left(1 - \frac{1}{2}H^2(\text{co}(\mathcal{P}_i), \text{co}(\mathcal{Q}_i))\right). \tag{32.30}$$

Then from (32.28) we get

$$\begin{aligned} 1 - \text{TV}(\text{co}(\mathcal{P}^{\otimes n}), \text{co}(\mathcal{Q}^{\otimes n})) &\overset{\text{(a)}}{\le} 1 - \frac{1}{2}H^2(\text{co}(\mathcal{P}^{\otimes n}), \text{co}(\mathcal{Q}^{\otimes n})) \\ &\overset{\text{(b)}}{\le} \left(1 - \frac{1}{2}H^2(\text{co}(\mathcal{P}), \text{co}(\mathcal{Q}))\right)^n \\ &\le \exp\left(-\frac{n}{2}H^2(\text{co}(\mathcal{P}), \text{co}(\mathcal{Q}))\right), \end{aligned}$$

where (a) follows from (7.22); and (b) follows from (32.30). □

In the sequel we will apply Theorem 32.8 to two disjoint Hellinger balls (both are convex).

32.2.2 Hellinger Guarantee on Le Cam–Birgé Pairwise Comparison Estimator

The idea of constructing an estimator based on pairwise tests is due to Le Cam ([274], see also [431, Section 10]) and Birgé [53]. We are given n iid observations $X_1, \dots, X_n$ generated from P, where $P \in \mathcal{P}$ is the distribution to be estimated. Here let us emphasize that $\mathcal{P}$ need *not* be a convex set. Let the loss function between the true distribution P and the estimated distribution $\hat{P}$ be their squared Hellinger distance, that is,

$$\ell(P, \hat{P}) = H^2(P, \hat{P}).$$

Then, we have the following result:

Theorem 32.9. (Le Cam–Birgé) *Denote by $N_{\text{H}}(\mathcal{P}, \epsilon)$ the ϵ-covering number of the set $\mathcal{P}$ under the Hellinger distance (see (27.1)). Let ϵ_n be such that*

$$n\epsilon_n^2 \ge \log N_{\text{H}}(\mathcal{P}, \epsilon_n) \vee 1.$$

Then there exists an estimator $\hat{P} = \hat{P}(X_1, \dots, X_n)$ taking values in $\mathcal{P}$ such that for any $t \ge 1$,

$$\sup_{P \in \mathcal{P}} P[H(P, \hat{P}) > 4t\epsilon_n] \lesssim e^{-t^2} \tag{32.31}$$

and, consequently,

$$\sup_{P \in \mathcal{P}} \mathbb{E}_P[H^2(P, \hat{P})] \lesssim \epsilon_n^2 \tag{32.32}$$

Proof of Theorem 32.9. It suffices to prove the high-probability bound (32.31). Abbreviate $\epsilon = \epsilon_n$ and $N = N_{\mathrm{H}}(\mathcal{P}, \epsilon_n)$. Let $P_1, \dots, P_N$ be a maximal ϵ-packing of $\mathcal{P}$ under the Hellinger distance, which also serves as an ϵ-covering (see Theorem 27.2). Thus, $\forall i \neq j$,

$$H(P_i, P_j) \geq \epsilon,$$

and for $\forall P \in \mathcal{P}, \exists i \in [N]$, s.t.

$$H(P, P_i) \leq \epsilon.$$

Let $B(P, \epsilon) = \{Q \colon H(P, Q) \leq \epsilon\}$ denote the ϵ-Hellinger ball centered at P. Crucially, the *Hellinger ball is convex* thanks to the convexity of *squared* Hellinger distance as an f-divergence; see Theorem 7.5. (In contrast, recall from (7.6) that the Hellinger distance itself is not convex.) Indeed, for any $P', P'' \in B(P, \epsilon)$ and $\alpha \in [0, 1]$,

$$H^2(\bar{\alpha} P' + \alpha P'', P) \leq \bar{\alpha} H^2(P', P) + \alpha H^2(P'', P) \leq \epsilon^2.$$

Next, consider the following *pairwise comparison problem*, where we test two Hellinger balls (composite hypothesis) against each other:

$$H_i \colon P \in B(P_i, \epsilon) \qquad \text{versus} \qquad H_j \colon P \in B(P_j, \epsilon)$$

for all $i \neq j$, s.t. $H(P_i, P_j) \geq \delta = 4\epsilon$.

Since both $B(P_i, \epsilon)$ and $B(P_j, \epsilon)$ are convex, applying Theorem 32.8 yields a test $\psi_{ij} = \psi_{ij}(X_1, \dots, X_n)$, with $\psi_{ij} = 0$ corresponding to declaring $P \in B(P_i, \epsilon)$, and $\psi_{ij} = 1$ corresponding to declaring $P \in B(P_j, \epsilon)$, such that $\psi_{ij} = 1 - \psi_{ji}$ and the following large-deviations bound holds: for all i, j, s.t. $H(P_i, P_j) \geq \delta$,

$$\sup_{P \in B(P_i, \epsilon)} P(\psi_{ij} = 1) \leq e^{-(n/8) H(P_i, P_j)^2}, \tag{32.33}$$

where we used the triangle inequality of Hellinger distance: for any $P \in B(P_i, \epsilon)$ and any $Q \in B(P_j, \epsilon)$,

$$H(P, Q) \geq H(P_i, P_j) - 2\epsilon \geq H(P_i, P_j)/2 \geq 2\epsilon.$$

For $i \in [N]$, define the random variable

$$T_i \triangleq \begin{cases} \max_{j \in [N]} H^2(P_i, P_j) & \text{s.t. } \psi_{ij} = 1, \quad H(P_i, P_j) > \delta; \\ 0, & \text{no such } j \text{ exists.} \end{cases}$$

Basically, T_i records the maximum distance from P_i to those P_j outside the δ-neighborhood of P_i that is confusable with P_i given the present sample. Our density estimator is defined as

$$\hat{P} = P_{i^*}, \quad \text{where} \quad i^* \in \operatorname*{argmin}_{i \in [N]} T_i. \tag{32.34}$$

Now for the proof of correctness, assume that $P \in B(P_1, \epsilon)$. The intuition is that we should expect, typically, that $T_1 = 0$, and furthermore, $T_j \geq \delta^2$ for all j such that

$H(P_1, P_j) \geq \delta$. Note that by the definition of T_i and the symmetry of the Hellinger distance, for any pair i, j such that $H(P_i, P_j) \geq \delta$, we have

$$\max\{T_i, T_j\} \geq H(P_i, P_j).$$

Consequently,

$$\begin{aligned} H(\hat{P}, P_1)1\{H(\hat{P}, P_1) \geq \delta\} &= H(P_{i_*}, P_1)1\{H(P_{i_*}, P_1) \geq \delta\} \\ &\leq \max\{T_{i_*}, T_1\}1\{\max\{T_{i_*}, T_1\} \geq \delta\} = T_1 1\{T_1 \geq \delta\}, \end{aligned}$$

where the last equality follows from the definition of i_* as a global minimizer in (32.34). Thus, for any $t \geq 1$,

$$\begin{aligned} P[H(\hat{P}, P_1) \geq t\delta] &\leq P[T_1 \geq t\delta] \\ &\leq N(\epsilon)e^{-2n\epsilon^2 t^2} \qquad (32.35) \\ &\lesssim e^{-t^2}, \qquad (32.36) \end{aligned}$$

where (32.35) follows from (32.33) and (32.36) uses the assumption that $n\epsilon^2 \geq 1$ and $N \leq e^{n\epsilon^2}$. □

32.2.3 Refinement Using Local Entropy

Just like Theorem 32.1, while they are often tight for nonparametric problems with super-logarithmically metric entropy, for finite-dimensional models a direct application of Theorem 32.9 results in a slack by a log factor. For example, for a d-dimensional parametric family, for example, the Gaussian location model or its finite mixtures, the metric entropy usually behaves as $\log N_{\mathrm{H}}(\epsilon) \asymp d \log \frac{1}{\epsilon}$. Thus when $n \gtrsim d$, Theorem 32.9 entails choosing $\epsilon_n^2 \asymp \frac{d}{n} \log \frac{n}{d}$, which falls short of the parametric rate $\mathbb{E}[H^2(\hat{P}, P)] \lesssim \frac{d}{n}$ which is typically achievable.

As usual, such a log factor can be removed using the local entropy argument. To this end, define the local Hellinger entropy:

$$N_{\mathrm{loc}}(\mathcal{P}, \epsilon) \triangleq \sup_{P \in \mathcal{P}} \sup_{\eta \geq \epsilon} N_{\mathrm{H}}(B(P, \eta) \cap \mathcal{P}, \eta/2). \qquad (32.37)$$

Theorem 32.10. (Le Cam–Birgé: local entropy version) *Let ϵ_n be such that*

$$n\epsilon_n^2 \geq \log N_{\mathrm{loc}}(\mathcal{P}, \epsilon_n) \vee 1. \qquad (32.38)$$

Then there exists an estimator $\hat{P} = \hat{P}(X_1, \ldots, X_n)$ taking values in $\mathcal{P}$ such that for any $t \geq 2$,

$$\sup_{P \in \mathcal{P}} P[H(P, \hat{P}) > 4t\epsilon_n] \leq e^{-t^2} \qquad (32.39)$$

and hence

$$\sup_{P \in \mathcal{P}} \mathbb{E}_P[H^2(P, \hat{P})] \lesssim \epsilon_n^2. \qquad (32.40)$$

Remark 32.3. (Doubling dimension) Suppose that for some $d > 0$, $\log N_{\text{loc}}(\mathcal{P}, \epsilon) \leq d \log \frac{1}{\epsilon}$ holds for all sufficiently small ϵ; this is the case for finite-dimensional models where the Hellinger distance is comparable with the vector norm by the usual volume argument (Theorem 27.3). Then we say the *doubling dimension* (also known as the *Le Cam dimension* [431]) of $\mathcal{P}$ is at most d; this terminology comes from the fact that the local entropy concerns covering Hellinger balls using balls of half the radius. Then Theorem 32.10 shows that it is possible to achieve the "parametric rate" $O(\frac{d}{n})$. In this sense, the doubling dimension serves as the effective dimension of the model $\mathcal{P}$.

Lemma 32.11. *For any $P \in \mathcal{P}$, $\eta \geq \epsilon$, and $k \geq \mathbb{Z}_+$,*

$$N_{\text{H}}(B(P, 2^k\eta) \cap \mathcal{P}, \eta/2) \leq N_{\text{loc}}(\mathcal{P}, \epsilon)^k. \tag{32.41}$$

Proof. We proceed by induction on k. The base case of $k = 0$ follows from the definition (32.37). For $k \geq 1$, assume that (32.41) holds for $k - 1$ for all $P \in \mathcal{P}$. To prove it for k, we construct a cover of $B(P, 2^k\eta) \cap \mathcal{P}$ as follows: first cover it with $2^{k-1}\eta$-balls, then cover each ball with $\eta/2$-balls. By the induction hypothesis, the total number of balls is at most

$$N_{\text{H}}(B(P, 2^k\eta) \cap \mathcal{P}, 2^{k-1}\eta) \cdot \sup_{P' \in \mathcal{P}} N_{\text{H}}(B(P', 2^{k-1}\eta) \cap \mathcal{P}, \eta/2) \leq N_{\text{loc}}(\epsilon) \cdot N_{\text{loc}}(\epsilon)^{k-1},$$

completing the proof. □

We now prove Theorem 32.10:

Proof. We analyze the same estimator (32.34) following the proof of Theorem 32.9, except that the estimate (32.35) is improved as follows: Define the Hellinger shell $A_k \triangleq \{P \in \mathcal{P} \colon 2^k\delta \leq H(P_1, P) < 2^{k+1}\delta\}$ and $G_k \triangleq \{P_2, \ldots, P_N\} \cap A_k$. Recall that $\delta = 4\epsilon$. Given $t \geq 2$, let $\ell = \lfloor \log_2 t \rfloor$ so that $2^\ell \leq t < 2^{\ell+1}$. Then

$$\begin{aligned} P[T_1 \geq t\delta] &\leq \sum_{k \geq \ell} P[2^k\delta \leq T_1 < 2^{k+1}\delta] \\ &\overset{\text{(a)}}{\leq} \sum_{k \geq \ell} |G_k| e^{-(n/8)(2^k\delta)^2} \\ &\overset{\text{(b)}}{\leq} \sum_{k \geq \ell} N_{\text{loc}}(\epsilon)^{k+3} e^{-2n\epsilon^2 4^k} \\ &\overset{\text{(c)}}{\lesssim} e^{-4^\ell} \leq e^{-t^2}, \end{aligned}$$

where (a) follows from (32.33); (c) follows from the assumption that $\log N_{\text{loc}}(\epsilon) \leq n\epsilon^2$ and $k \geq \ell \geq \log_2 t \geq 1$; and (b) follows from the following reasoning: since $\{P_1, \ldots, P_N\}$ is an ϵ-packing, we have

$$|G_k| \leq M(A_k, \epsilon) \leq N(A_k, \epsilon/2) \leq N(B(P_1, 2^{k+1}\delta) \cap \mathcal{P}, \epsilon/2) \leq N_{\text{loc}}(\epsilon)^{k+3},$$

where the first and the last inequalities follow from Theorem 27.2 and Lemma 32.11 respectively. □

As an application of Theorem 32.10, we show that parametric rate (namely, dimension divided by the sample size) is achievable for models with locally quadratic behavior, such as those smooth parametric models (see Section 7.11 and in particular Theorem 7.23).

Corollary 32.12. *Consider a parametric family $\mathcal{P} = \{P_\theta : \theta \in \Theta\}$, where $\Theta \subset \mathbb{R}^d$ and $\mathcal{P}$ is totally bounded in Hellinger distance. Suppose that there exists a norm $\|\cdot\|$ and constants t_0, c, C such that for all $\theta_0, \theta_1 \in \Theta$ with $\|\theta_0 - \theta_1\| \le t_0$,*

$$c\|\theta_0 - \theta_1\| \le H(P_{\theta_0}, P_{\theta_1}) \le C\|\theta_0 - \theta_1\|. \tag{32.42}$$

Then there exists an estimator $\hat{\theta}$ based on $X_1, \dots, X_n \overset{\text{iid}}{\sim} P_\theta$, such that

$$\sup_{\theta \in \Theta} \mathbb{E}_\theta[H^2(P_\theta, P_{\hat{\theta}})] \lesssim \frac{d}{n}.$$

Proof. It suffices to bound the local entropy $\mathcal{N}_{\text{loc}}(\mathcal{P}, \epsilon)$ in (32.37). Fix $\theta_0 \in \Theta$. Indeed, for any $\eta > t_0$, we have $\mathcal{N}_{\text{H}}(B(P_{\theta_0}, \eta) \cap \mathcal{P}, \eta/2) \le \mathcal{N}_{\text{H}}(\mathcal{P}, t_0) \lesssim 1$. For $\epsilon \le \eta \le t_0$,

$$\begin{aligned} \mathcal{N}_{\text{H}}(B(P_{\theta_0}, \eta) \cap \mathcal{P}, \eta/2) &\overset{(a)}{\le} \mathcal{N}_{\|\cdot\|}(B_{\|\cdot\|}(\theta_0, \eta/c), \eta/(2C)) \\ &\overset{(b)}{\le} \frac{\text{vol}(B_{\|\cdot\|}(\theta_0, \eta/c + \eta/(2C)))}{\text{vol}(B_{\|\cdot\|}(\theta_0, \eta/(2C)))} = \left(1 + \frac{2C}{c}\right)^d, \end{aligned}$$

where (a) and (b) follow from (32.42) and Theorem 27.3 respectively. This shows that $\log \mathcal{N}_{\text{loc}}(\mathcal{P}, \epsilon) \lesssim d$, completing the proof by applying Theorem 32.10. □

32.2.4 Lower Bound Using Local Hellinger Packing

It turns out that under certain regularity assumptions we can prove an almost matching lower bound (typically within a logarithmic term) on the Hellinger risk. First we define the local packing number as follows:

$$\begin{aligned} M_{\text{loc}}(\epsilon) &\equiv M_{\text{loc}}(\mathcal{P}, H, \epsilon) \\ &= \max\left\{M : \exists R, P_1, \dots, P_M \in \mathcal{P} : H(P_i, R) \le \epsilon, H(P_i, P_j) \ge \frac{\epsilon}{2} \quad \forall i \ne j\right\}. \end{aligned}$$

The local packing number is related to the global one $M(\epsilon) \equiv M(\mathcal{P}, H, \epsilon)$ by the following general lemma that holds for any metric. This result shows that the local and global packing numbers behave similarly as long as the growth is super-polynomial in $1/\epsilon$ (e.g. for those nonparametric classes considered in Section 27.4).

Lemma 32.13.

$$\frac{M(\epsilon/2)}{M(\epsilon)} \le M_{\text{loc}}(\epsilon) \le M(\epsilon).$$

Proof. The upper bound is obvious. For the lower bound, let $P_1, \dots, P_M$ be a maximal ϵ-packing for $\mathcal{P}$ with $M = M(\epsilon)$. Let $Q_1, \dots, Q_{M'}$ be a maximal $\epsilon/2$-packing for $\mathcal{P}$ with $M' = M(\epsilon/2)$. Partition $E = \{P_1, \dots, P_M\}$ into Voronoi cells centered at each Q_i, namely, $E_i \triangleq \{P_j : H(P_j, Q_i) = \min_{k \in [M]} H(P_k, Q_i)\}$ (with ties broken arbitrarily), so that $E_1, \dots, E_{M'}$ are disjoint and $E = \bigcup_{i\in[M']} E_i$. Thus $\max|E_i| \geq M/M'$. Finally, note that each $E_i \subset B(Q_i, \epsilon)$ because E is also an ϵ-covering. □

Note that unlike the definition of N_{loc} in (32.37) we are not taking the supremum over the scale $\eta \geq \epsilon$. For this reason, we cannot generally apply Theorem 27.2 to conclude that $N_{\text{loc}}(\epsilon) \geq M_{\text{loc}}(\epsilon)$. In all instances known to us we have $\log N_{\text{loc}} \asymp \log M_{\text{loc}}$, in which case the following general result provides a minimax lower bound that matches the upper bound in Theorem 32.10 up to logarithmic factors.

Theorem 32.14. *Suppose that the D_λ radius $R_\lambda(\mathcal{P})$ of the family $\mathcal{P}$ is finite for some $\lambda > 1$; see (32.20). There exist constants $c = c(\lambda)$ and $\epsilon < \epsilon_0(\lambda)$ such that whenever n and $\epsilon < \epsilon_0$ are such that*

$$c(\lambda)n\epsilon^2\left(\log\frac{1}{\epsilon^2} + R_\lambda(\mathcal{P})\right) + 2\log 2 < \log M_{\text{loc}}(\epsilon), \tag{32.43}$$

any estimator $\hat{P} = \hat{P}(\cdot; X^n)$ must satisfy

$$\sup_{P\in\mathcal{P}} \mathbb{E}_P[H^2(P, \hat{P})] \geq \frac{\epsilon^2}{32},$$

where $\mathbb{E}_P$ is taken with respect to $X^n \overset{\text{iid}}{\sim} P$.

Proof. Let $M = M_{\text{loc}}(\mathcal{P}, \epsilon)$. From the definition there exists an $\epsilon/2$-packing $P_1, \dots, P_M$ in some Hellinger ball $B(R, \epsilon)$.

Let $\theta \sim \text{Unif}([M])$ and $X^n \overset{\text{iid}}{\sim} P_\theta$ conditioned on θ. Then from Fano's inequality in the form of Theorem 31.3 we get

$$\sup_{P\in\mathcal{P}} \mathbb{E}[H^2(P, \hat{P})] \geq \left(\frac{\epsilon}{4}\right)^2\left(1 - \frac{I(\theta; X^n) + \log 2}{\log M}\right).$$

It remains to show that

$$\frac{I(\theta; X^n) + \log 2}{\log M} \leq \frac{1}{2}. \tag{32.44}$$

To that end for an arbitrary distribution U define

$$Q = \epsilon^2 U + (1 - \epsilon^2)R.$$

We first notice that from Exercise **I.59** we have that for all $i \in [M]$

$$D(P_i\|Q) \leq 8(H^2(P_i, R) + 2\epsilon^2)\left(\frac{\lambda}{\lambda - 1}\log\frac{1}{\epsilon^2} + D_\lambda(P_i\|U)\right)$$

provided that $\epsilon < 2^{-5\lambda/(2(\lambda-1))} \triangleq \epsilon_0$. Since $H^2(P_i, R) \leq \epsilon^2$, by optimizing U (as the D_λ-center of $\mathcal{P}$) we obtain

$$\inf_U \max_{i\in[M]} D(P_i\|Q) \leq 24\epsilon^2\left(\frac{\lambda}{\lambda - 1}\log\frac{1}{\epsilon^2} + R_\lambda\right) \leq \frac{c(\lambda)}{2}\epsilon^2\left(\log\frac{1}{\epsilon^2} + R_\lambda\right).$$

By Theorem 4.1 we have

$$I(\theta;X^n) \le \max_{i\in[M]} D(P_i^{\otimes n}\|Q^{\otimes n}) \le \frac{nc(\lambda)}{2}\epsilon^2\left(\log\frac{1}{\epsilon^2}+R_\lambda\right).$$

This final bound and condition (32.43) then imply (32.44) and the statement of the theorem. □

Finally, we mention that for *sufficiently regular* models wherein the KL divergence and the squared Hellinger distances are comparable, the upper bound in Theorem 32.10 based on local entropy gives the *exact* minimax rate. Models of this type include the GLM and more generally Gaussian mixture models with bounded centers in arbitrary dimensions [232].

Corollary 32.15. *Assume that*

$$H^2(P,P') \asymp D(P\|P'), \quad \forall P,P' \in \mathcal{P}.$$

Then

$$\inf_{\hat{P}}\sup_{P\in\mathcal{P}} \mathbb{E}_P[H^2(P,\hat{P})] \asymp \epsilon_n^2,$$

where ϵ_n was defined in (32.38).

Proof. By assumption, KL neighborhoods coincide with Hellinger balls up to constant factors. Thus the lower bound follows from applying Fano's method in Theorem 31.3 to a Hellinger ball of radius $O(\epsilon_n)$. □

32.3 Yatracos' Class and Minimum-Distance Estimator

In this section we prove (32.3), the third entropy upper bound on statistical risk. Paralleling the result (32.1) of Yang–Barron (for KL divergence) and (32.2) of Le Cam–Birgé (for Hellinger distance), the following result bounds the minimax total variation risk using the metric entropy of the parameter space in total variation:

Theorem 32.16. (Yatracos [466]) *There exists a universal constant C such that the following holds. Let $X_1,\dots,X_n \overset{\text{iid}}{\sim} P \in \mathcal{P}$, where $\mathcal{P}$ is a collection of distributions on a common measurable space $(\mathcal{X},\mathcal{E})$. For any $\epsilon > 0$, there exists a proper estimator $\hat{P} = \hat{P}(X_1,\dots,X_n) \in \mathcal{P}$, such that*

$$\sup_{P\in\mathcal{P}} \mathbb{E}_P[\mathrm{TV}(\hat{P},P)^2] \le C\left(\epsilon^2 + \frac{1}{n}\log N(\mathcal{P},\mathrm{TV},\epsilon)\right). \tag{32.45}$$

For a loss function that is a distance, a natural idea for obtaining a proper estimator is the *minimum-distance estimator*. In the current context, we compute the minimum-distance projection of the empirical distribution on the model class $\mathcal{P}$:[6]

$$P_{\text{min-dist}} = \operatorname*{argmin}_{P\in\mathcal{P}} \mathrm{TV}(\hat{P}_n, P),$$

[6] Here and below, if the minimizer does not exist, we can replace it by an infimizing sequence.

where $\hat{P}_n = \frac{1}{n}\sum_{i=1}^{n}\delta_{X_i}$ is the empirical distribution. However, since the empirical distribution is discrete, this strategy does not make sense if the elements of $\mathcal{P}$ have densities. The reason for this degeneracy is because the total variation distance is too strong. The key idea is to replace TV, which compares two distributions over all measurable sets, by a proxy, which only inspects a "low-complexity" family of sets.

To this end, let $\mathcal{A} \subset \mathcal{E}$ be a finite collection of measurable sets to be specified later. Define a pseudo-distance

$$\mathsf{dist}(P,Q) \triangleq \sup_{A\in\mathcal{A}} |P(A) - Q(A)|. \tag{32.46}$$

(Note that if $\mathcal{A} = \mathcal{E}$, then this is just TV.) One can verify that dist satisfies the triangle inequality. As a result, the estimator

$$\tilde{P} \triangleq \operatorname*{argmin}_{P\in\mathcal{P}} \mathsf{dist}(P, \hat{P}_n), \tag{32.47}$$

as a minimizer, satisfies

$$\mathsf{dist}(\tilde{P}, P) \le \mathsf{dist}(\tilde{P}, \hat{P}_n) + \mathsf{dist}(P, \hat{P}_n) \le 2\mathsf{dist}(P, \hat{P}_n). \tag{32.48}$$

The min-max form of the estimator (32.47) is remiscient of GAN (Example 7.5), with the indicators in the collection $\mathcal{A}$ being the discriminators.

In addition, applying the binomial tail bound (Example 15.1) and the union bound, we have

$$\mathbb{E}[\mathsf{dist}(P, \hat{P}_n)^2] \le \frac{C_0 \log|\mathcal{A}|}{n} \tag{32.49}$$

for some absolute constant C_0.

The main idea of Yatracos [466] boils down to the following choice of $\mathcal{A}$: Consider an ϵ-covering $\{Q_1, \ldots, Q_N\}$ of $\mathcal{P}$ in TV. Define the set

$$A_{ij} \triangleq \left\{x\colon \frac{dQ_i}{d(Q_i+Q_j)}(x) \ge \frac{dQ_j}{d(Q_i+Q_j)}(x)\right\}$$

and the collection (known as the *Yatracos class*)

$$\mathcal{A} \triangleq \{A_{ij}\colon i \ne j \in [N]\}. \tag{32.50}$$

Then the corresponding dist approximates the TV on $\mathcal{P}$, in the sense that

$$\mathsf{dist}(P,Q) \le \mathrm{TV}(P,Q) \le \mathsf{dist}(P,Q) + 4\epsilon, \qquad \forall P,Q \in \mathcal{P}. \tag{32.51}$$

To see this, we only need to justify the upper bound. For any $P, Q \in \mathcal{P}$, there exist $i,j \in [N]$, such that $\mathrm{TV}(P,Q_i) \le \epsilon$ and $\mathrm{TV}(Q,Q_j) \le \epsilon$. By the key observation that $\mathsf{dist}(Q_i,Q_j) = \mathrm{TV}(Q_i,Q_j)$, we have

$$\begin{aligned}
\mathrm{TV}(P,Q) &\le \mathrm{TV}(P,Q_i) + \mathrm{TV}(Q_i,Q_j) + \mathrm{TV}(Q_j,Q)\\
&\le 2\epsilon + \mathsf{dist}(Q_i,Q_j)\\
&\le 2\epsilon + \mathsf{dist}(Q_i,P) + \mathsf{dist}(P,Q) + \mathsf{dist}(Q,Q_j)\\
&\le 4\epsilon + \mathsf{dist}(P,Q).
\end{aligned}$$

Finally, we analyze the estimator (32.47) with $\mathcal{A}$ given in (32.50). Applying (32.51) and (32.48) yields

$$\begin{aligned}\mathrm{TV}(\tilde{P}, P) &\le \mathsf{dist}(P, \tilde{P}) + 4\epsilon \\ &\le 2\mathsf{dist}(P, \hat{P}_n) + 4\epsilon.\end{aligned}$$

Squaring both sizes, taking expectation, and applying (32.49), we have

$$\mathbb{E}[\mathrm{TV}(\tilde{P}, P)^2] \le 32\epsilon^2 + 8\mathbb{E}[\mathsf{dist}(P, \hat{P}_n)^2] \le 32\epsilon^2 + \frac{8C_0 \log|N|}{n}.$$

Choosing the optimal TV covering completes the proof of (32.45).

Remark 32.4. (Robust version) Note that Yatracos' scheme idea works even if the model is mis-specified, that is, when the data generating distribution P is outside (but close to) $\mathcal{P}$. Indeed, denote $Q^* = \mathrm{argmin}_{Q\in\{Q_i\}} \mathrm{TV}(P, Q)$ and notice that

$$\mathsf{dist}(Q^*, \hat{P}_n) \le \mathsf{dist}(Q^*, P) + \mathsf{dist}(P, \hat{P}_n) \le \mathrm{TV}(P, Q^*) + \mathsf{dist}(P, \hat{P}_n),$$

since $\mathsf{dist}(Q, Q') \le \mathrm{TV}(Q, Q')$ for any pair of distributions. Then we have

$$\begin{aligned}\mathrm{TV}(\tilde{P}, P) &\le \mathrm{TV}(\tilde{P}, Q^*) + \mathrm{TV}(Q^*, P) = \mathsf{dist}(\tilde{P}, Q^*) + \mathrm{TV}(Q^*, P) \\ &\le \mathsf{dist}(\tilde{P}, \hat{P}_n) + \mathsf{dist}(\hat{P}_n, P) + \mathsf{dist}(P, Q^*) + \mathrm{TV}(Q^*, P) \\ &\le \mathsf{dist}(Q^*, \hat{P}_n) + \mathsf{dist}(\hat{P}_n, P) + 2\mathrm{TV}(P, Q^*) \\ &\le 2\mathsf{dist}(P, \hat{P}_n) + 3\mathrm{TV}(P, Q^*).\end{aligned}$$

Since $3\mathrm{TV}(P, Q^*) \le 3\epsilon + 3\min_{P'\in\mathcal{P}} \mathrm{TV}(P, P')$ we can see that the estimator also works for the "mis-specified case." Surprisingly, the multiplier 3 is not improvable if the estimator is required to be proper (inside $\mathcal{P}$), see [70].

32.4 Density Estimation over Hölder Classes

In this section we discuss the classical problem of nonparametric density estimation under a smoothness constraint. Following Theorem 27.14, for brevity denote by $\mathcal{F} \equiv \mathcal{F}_\beta(1, 1)$ the collection of β-Hölder densities f on the unit cube $[0, 1]^d$ for some constant d. (In this case the parameter is simply the density f, so we shall refrain from writing a parameterized form.) Given $X_1, \ldots, X_n \overset{\text{iid}}{\sim} f \in \mathcal{F}$, an estimator of the unknown density f is a function $\hat{f}(\cdot) = \hat{f}(\cdot;X_1, \ldots, X_n)$. Let us first consider the conventional quadratic risk $\|f - \hat{f}\|_2^2 = \int_0^1 (f(x) - \hat{f}(x))^2$ $\|f - \hat{f}\|_2^2 = \int_0^1 (f(x) - \hat{f}(x))^2 dx$. Then we will state the results for Hellinger, KL, and total variation risks.

Theorem 32.17. *Given $X_1, \ldots, X_n \overset{\text{iid}}{\sim} f \in \mathcal{F}$, the minimax quadratic risk over $\mathcal{F}$ satisfies*

$$R^*_{L_2}(n;\mathcal{F}) \triangleq \inf_{\hat{f}} \sup_{f\in\mathcal{F}} \mathbb{E}\,\|f - \hat{f}\|_2^2 \asymp n^{-2\beta/(2\beta+d)}. \tag{32.52}$$

Capitalizing on the metric entropy of smooth densities studied in Section 27.4, we will prove this result by applying the entropic upper bound in Theorem 32.1 and the minimax lower bound based on Fano's inequality in Theorem 31.3. However, Theorem 32.17 pertains to the L_2 rather than KL risk. This can be fixed by a simple reduction.

Lemma 32.18. *Let $\mathcal{F}'$ denote the collection of $f \in \mathcal{F}$ which is bounded from below by 1/2. Then*

$$R^*_{L_2}(n;\mathcal{F}') \le R^*_{L_2}(n;\mathcal{F}) \le 4R^*_{L_2}(n;\mathcal{F}').$$

Proof. The left-hand inequality follows because $\mathcal{F}' \subset \mathcal{F}$. For the right-hand inequality, we apply a simulation argument. Fix some $f \in \mathcal{F}$ and we observe $X_1, \dots, X_n \overset{\text{iid}}{\sim} f$. Let us sample $U_1, \dots, U_n$ independently and uniformly from $[0,1]^d$. Define

$$Z_i = \begin{cases} U_i & \text{w.p. } \frac{1}{2}, \\ X_i & \text{w.p. } \frac{1}{2}. \end{cases}$$

Then $Z_1, \dots, Z_n \overset{\text{iid}}{\sim} g \triangleq \frac{1}{2}(1+f) \in \mathcal{F}'$. Let $\hat{g}$ be an estimator that achieves the minimax risk $R^*_{L_2}(n;\mathcal{F}')$ on $\mathcal{F}'$. Consider the estimator $\hat{f} = 2\hat{g} - 1$. Then $\|f - \hat{f}\|_2^2 = 4\|g - \hat{g}\|_2^2$. Taking the supremum over $f \in \mathcal{F}$ proves $R^*_{L_2}(n;\mathcal{F}) \le 4R^*_{L_2}(n;\mathcal{F}')$. □

Lemma 32.18 allows us to focus on the subcollection $\mathcal{F}'$, where each density is lower-bounded by 1/2. In addition, each β-Hölder density in $\mathcal{F}$ is also upper-bounded by an absolute constant. Therefore, the KL divergence and squared L_2-distance are in fact equivalent on $\mathcal{F}'$, that is,

$$D(f\|g) \asymp \|f - g\|_2^2, \quad f, g \in \mathcal{F}', \tag{32.53}$$

as shown by the following lemma:

Lemma 32.19. *Suppose both $f = \frac{dP}{d\mu}$ and $g = \frac{dQ}{d\mu}$ are upper- and lower-bounded by absolute constants c and C respectively. Then*

$$\frac{1}{4C}\int d\mu (f-g)^2 \le H^2(P,Q) \le D(P\|Q) \le \chi^2(P\|Q) \le \frac{1}{c}\int d\mu (f-g)^2.$$

Proof. For the upper bound, applying (7.34), $D(P\|Q) \le \chi^2(P\|Q) = \int d\mu \frac{(f-g)^2}{g} \le \frac{1}{c}\int d\mu (f-g)^2$. For the lower bound, applying (7.33), $D(P\|Q) \ge H^2(P,Q) = \int d\mu \frac{(f-g)^2}{(\sqrt{f}+\sqrt{g})^2} \ge \frac{1}{4C}\int d\mu (f-g)^2$. □

We now prove Theorem 32.17:

Proof. In view of Lemma 32.18, it suffices to consider $R^*_{L_2}(n;\mathcal{F}')$. For the upper bound, we have

$$\begin{aligned} R^*_{L_2}(n;\mathcal{F}') &\overset{(a)}{\asymp} R^*_{\mathrm{KL}}(n;\mathcal{F}') \\ &\overset{(b)}{\lesssim} \inf_{\epsilon>0}\left\{\epsilon^2 + \frac{1}{n}\log N_{\mathrm{KL}}(\mathcal{F}',\epsilon)\right\} \\ &\overset{(c)}{\asymp} \inf_{\epsilon>0}\left\{\epsilon^2 + \frac{1}{n}\log N(\mathcal{F}',\|\cdot\|_2,\epsilon)\right\} \\ &\overset{(d)}{\asymp} \inf_{\epsilon>0}\left\{\epsilon^2 + \frac{1}{n\epsilon^{d/\beta}}\right\} \asymp n^{-2\beta/(2\beta+d)}, \end{aligned}$$

where both (a) and (c) apply (32.53), so that both the risk and the metric entropy are equivalent for KL and L_2-distance; (b) follows from Theorem 32.1; and (d) applies the metric entropy (under L_2) of the Lipschitz class from Theorem 27.14 and the fact that the metric entropy of the subclass $\mathcal{F}'$ is at most that of the full class $\mathcal{F}$.

For the lower bound, we apply Fano's inequality. Applying Theorem 27.14 and the relation between covering and packing numbers in Theorem 27.2, we have $\log N(\mathcal{F},\|\cdot\|_2,\epsilon) \asymp \log M(\mathcal{F},\|\cdot\|_2,\epsilon) \asymp \epsilon^{-d/\beta}$. Fix ϵ to be specified and let $f_1,\ldots,f_M$ be an ϵ-packing in $\mathcal{F}$, where $M \geq \exp(C\epsilon^{-d/\beta})$. Then $g_1,\ldots,g_M$ is an $\frac{\epsilon}{2}$-packing in $\mathcal{F}'$, with $g_i = (f_i+1)/2$. Applying Fano's inequality in Theorem 31.3, we have

$$R^*_{L_2}(n;\mathcal{F}') \gtrsim \epsilon^2\left(1 - \frac{C'_n}{\log M}\right), \tag{32.54}$$

where C'_n is the capacity (or KL radius) from $f \in \mathcal{F}'$ to $X_1,\ldots,X_n \overset{\text{iid}}{\sim} f$. Using (32.17) and Lemma 32.19, we have

$$\begin{aligned} C'_n &\leq \inf_{\epsilon>0}(n\epsilon^2 + \log N_{\mathrm{KL}}(\mathcal{F}',\epsilon)) \asymp \inf_{\epsilon>0}(n\epsilon^2 + \log N(\mathcal{F}',\|\cdot\|_2,\epsilon)) \\ &\lesssim \inf_{\epsilon>0}(n\epsilon^2 + \epsilon^{-d/\beta}) \asymp n^{d/(2\beta+d)}. \end{aligned}$$

Thus choosing $\epsilon = cn^{-\beta/(2\beta+d)}$ for sufficiently small c ensures $C'_n \leq \frac{1}{2}\log M$ and hence $R^*_{L_2}(n;\mathcal{F}') \gtrsim \epsilon^2 \asymp n^{-2\beta/(2\beta+d)}$. □

Remark 32.5. The above lower-bound proof, based on Fano's inequality and the intuition that small mutual information implies large estimation error, requires us to upper-bound the capacity C'_n of the subcollection $\mathcal{F}'$. On the other hand, as hinted earlier in (32.11) (and shown next), the risk is expected to be proportional to $\frac{C'_n}{n}$, which suggests one should lower-bound the capacity using metric entropy. Indeed, this is possible: Applying Theorem 32.5,

$$\begin{aligned} C'_n &\gtrsim \min\{n\epsilon^2, \log M(\mathcal{F}',H,\epsilon)\} - 2 \\ &\asymp \min\{n\epsilon^2, \log M(\mathcal{F}',\|\cdot\|_2,\epsilon)\} - 2 \\ &\asymp \min\{n\epsilon^2, \epsilon^{-d/\beta}\} - 2 \asymp n^{d/(2\beta+d)}, \end{aligned}$$

where we picked the same ϵ as in the previous proof. So $C_n' \asymp n^{d/(2\beta+d)}$. Finally, applying the online-to-batch conversion (32.26) in Proposition 32.7 (or equivalently, combining (32.7) and (32.9)) yields $R^*_{\mathrm{KL}}(n;\mathcal{F}') \asymp \frac{C_n'}{n} \asymp n^{-2\beta/(2\beta+d)}$.

Remark 32.6. Note that the above proof of Theorem 32.17 relies on the entropic risk bound (32.1), which, though rate-optimal, is not attained by a computationally efficient estimator. (The same criticism also applies to (32.2) and (32.3) for Hellinger and total variation.) To remedy this, for the squared loss, a classical idea is to apply the kernel density estimator (KDE) – see Section 7.9. Specifically, one computes the convolution of the empirical distribution $\hat{P}_n = \frac{1}{n}\sum_{i=1}^n \delta_{X_i}$ with a kernel function $K(\cdot)$ whose shape and bandwidth are chosen according to the smooth constraint. For example, for Lipschitz densities, the optimal rate in Theorem 32.17 can be attained by a box kernel $K(\cdot) = \frac{1}{2h}1\{|\cdot| \le h\}$ with bandwidth $h = n^{-1/3}$ (see e.g. [425, Section 1.2]).

The classical literature of density estimation is predominantly concerned with the L_2 loss, mainly due to the convenient quadratic nature of the loss function that allows bias-variance decomposition and facilitates the analysis of KDE. However, the L_2-distance between densities has no clear operational meaning. Next we consider the three f-divergence losses introduced at the beginning of this chapter. Paralleling Theorem 32.17, we have

$$R^*_{\mathrm{TV}}(n;\mathcal{F}) \triangleq \inf_{\hat{f}} \sup_{f\in\mathcal{F}} \mathbb{E}\,\mathrm{TV}(f,\hat{f}) \asymp n^{-\beta/(2\beta+d)}, \tag{32.55}$$

$$R^*_{H_2}(n;\mathcal{F}) \triangleq \inf_{\hat{f}} \sup_{f\in\mathcal{F}} \mathbb{E}\,H^2(f,\hat{f}) \asymp n^{-\beta/(\beta+d)}, \tag{32.56}$$

$$n^{-\beta/(\beta+d)} \lesssim R^*_{\mathrm{KL}}(n;\mathcal{F}) \triangleq \inf_{\hat{f}} \sup_{f\in\mathcal{F}} \mathbb{E}\,D(f\|\hat{f}) \asymp n^{-\beta/(\beta+d)}(\log n)^{d/(\beta+d)}. \tag{32.57}$$

For TV loss, the upper bound follows from the L_2 rates in Theorem 32.17 and $\|f-\hat{f}\|_1 \le \|f-\hat{f}\|_2$ by Cauchy–Schwarz; alternatively, we can also apply Yatracos' estimator from Theorem 32.16. The matching lower bound can be shown using the same argument based on Fano's method as the metric entropy under L_1-distance behaves the same (Theorem 27.14).

Recall that for L_2/L_1 the rate is derived by considering a subclass $\mathcal{F}'$, which has the same estimation rate, but on which $L_p \asymp H \asymp \mathrm{KL}$, see Lemma 32.18. It was thus surprising when Birgé [54] found the Hellinger rate on the full family $\mathcal{F}$ to be different.

To derive the H^2 result (32.56), first note that neither upper nor lower bound follow from the generic comparison inequality $\frac{H^2}{2} \le \mathrm{TV} \le H$ in (7.22). Instead, what works is comparing entropy numbers via the first of these inequalities. Specifically, we have

$$\log N(\mathcal{F},H,\epsilon) \le \log N(\mathcal{F},\mathrm{TV},\epsilon^2/2) \lesssim \epsilon^{-2d/\beta}, \tag{32.58}$$

where in the last step we invoked Theorem 27.14. Combining this with the Le Cam–Birgé method (Theorem 32.9) proves the upper-bound part of (32.56).[7]

The lower bound follows from a similar argument as in the proof of Theorem 32.17, although the construction is more involved. Below $c_0, c_1, \ldots$ are absolute constants. Fix a small ϵ and consider the subcollection $\mathcal{F}' = \{f \in \mathcal{F} : f \geq \epsilon\}$ of densities lower-bounded by ϵ. We first construct a Hellinger packing of $\mathcal{F}'$. Applying the same argument as in Lemma 32.13 yields an L_2-packing in an L_∞-local ball: there exist $f_0, f_1, \ldots, f_M \in \mathcal{F}$, such that $\|f_i - f_j\|_2 \geq c_0\epsilon$ for all $i \neq j$, $\|f_i - f_0\|_\infty \leq \epsilon$ for all i, and $M \geq M(\mathcal{F}, \|\cdot\|_2, c_0\epsilon)/M(\mathcal{F}, \|\cdot\|_\infty, \epsilon) \geq \exp(c_1\epsilon^{-d/\beta})$, the last step applying Theorem 32.17 for sufficiently small c_0. Let $h_i = f_i - f_0$ and define f_i by $f_i(x) = \epsilon + h_i(2x)$ for $x \in [0, \frac{1}{2}]^d$ and extend f_i smoothly elsewhere so that it is a valid density in $\mathcal{F}'$. Then $f_1, \ldots, f_M$ form a Hellinger $\Omega(\sqrt{\epsilon})$-packing of $\mathcal{F}'$, since $H^2(f_i, f_j) \geq \int_{[0,\frac{1}{2}]^d} \frac{(f_i - f_j)^2}{(\sqrt{f_i} + \sqrt{f_j})^2} \geq c_2\epsilon$. (This construction also shows that the metric entropy bound (32.58) is tight.) It remains to bound the capacity C_n' of $\mathcal{F}'$ as a function of n and ϵ. Note that for any $f, g \in \mathcal{F}'$, we have as in Lemma 32.19 $D(f\|g) \leq \chi^2(f\|g) \leq \|f - g\|_2^2/\epsilon$. Thus $N_{\mathrm{KL}}(\mathcal{F}', \delta^2/\epsilon) \leq N(\mathcal{F}', \|\cdot\|_2, \delta)$. Applying (32.17) and Lemma 32.19, $C_n' \lesssim \inf_{\delta>0}(n\delta^2/\epsilon + \delta^{-d/\beta}) \asymp (n/\epsilon)^{d/(2\beta+d)}$. Applying Fano's inequality as in (32.54) yields an $\Omega(\epsilon)$ lower bound in squared Hellinger, provided $\log M \geq 2C_n'$. This is achieved by choosing $\epsilon = c_3 n^{-\beta/(\beta+d)}$, completing the proof of (32.56).

For KL loss, the lower bound of (32.57) follows from (32.56) because $D \geq H^2$. For the upper bound, applying (32.7) in Theorem 32.1, we have $R^*_{\mathrm{KL}}(n; \mathcal{F}) \leq \frac{C_{n+1}}{n+1}$, where C_n is the capacity (32.5) of the channel between f and $X^n \overset{\text{iid}}{\sim} f \in \mathcal{F}$. This capacity can be bounded, in turn, using Theorem 32.6 via the Hellinger entropy. Applying (32.58) in conjunction with (32.22), we obtain $C_n \leq \inf_\epsilon(n\epsilon^2 \log\frac{1}{\epsilon} + \epsilon^{-2d/\beta}) \asymp (n\log n)^{d/(d+\beta)}$, proving the upper bound (32.57).[8] To the best of our knowledge, resolving the logarithmic gap in (32.57) remains open.

[7] Comparing (32.56) with (32.52), we see that the Hellinger rate coincides with the L_2 rate upon replacing the smoothness parameter β by $\beta/2$. Note that Hellinger distance is the L_2 distance between root densities. For $\beta \leq 1$, one can indeed show that $\sqrt{f}$ is $\beta/2$-Hölder continuous, which explains the result in (32.56). However, this interpretation fails for general β. For example, Glaeser [191] constructs an infinitely differentiable f such that $\sqrt{f}$ has points with arbitrarily large second derivative.

[8] This capacity bound is tight up to logarithmic factors. Note that the construction in the proof of the lower bound in (32.56) shows that $\log M(\mathcal{F}, H, \epsilon) \gtrsim \epsilon^{-2d/\beta}$, which, via Theorem 32.5, implies that $C_n \geq n^{d/(d+\beta)}$.

33 Strong Data-Processing Inequality

In this chapter we explore statistical implications of the following effect. For any Markov chain

$$U \to X \to Y \to V \tag{33.1}$$

we know from the data-processing inequality (DPI, Theorem 3.7) that

$$I(U;Y) \le I(U;X), \qquad I(X;V) \le I(Y;V).$$

However, something stronger can often be said. Namely, if the Markov chain (33.1) factors through a *known* noisy channel $P_{Y|X}:\mathcal{X} \to \mathcal{Y}$, then oftentimes we can prove strong data-processing inequalities (SDPI):

$$I(U;Y) \le \eta I(U;X), \qquad I(X;V) \le \eta^{(p)} I(Y;V),$$

where coefficients $\eta = \eta(P_{Y|X}), \eta^{(p)}(P_{Y|X}) < 1$ only depend on the channel and not the (generally unknown or very complex) $P_{U,X}$ or $P_{Y,V}$. The coefficients η and $\eta^{(p)}$ approach 0 for channels that are very noisy (for example, η is always up to a constant factor equal to the Hellinger-squared diameter of the channel).

The purpose of this chapter is two-fold. First, we want to introduce the general properties of the SDPI coefficients. Second, we want to show how SDPIs help prove sharp lower (impossibility) bounds on statistical estimation questions. The flavor of the statistical problems in this chapter is different from the rest of the book in that here the information about the unknown parameter θ is either more "thinly spread" across a high-dimensional vector of observations than in classical $X = \theta + Z$ type of models (see, spiked Wigner and tree-coloring examples), or distributed across different terminals (as in correlation and mean estimation examples).

We point out that SDPIs are an area of current research and multiple topics are not covered by our brief exposition here. For more, we recommend surveys [346] and [353], of which the latter explores the functional-theoretic side of SDPIs and their close relation to logarithmic Sobolev inequalities – a topic we entirely omit in our book.

33.1 Computing a Boolean Function with Noisy Gates

A boolean function with n inputs is defined as $f:\{0,1\}^n \to \{0,1\}$. Note that a boolean function can be described as a network of primitive logic gates of the three kinds illustrated in Figure 33.1.

In 1938 Shannon showed any boolean function f can be represented with primitive logic gates [378] from the top row of Figure 33.1. In the 1950s John von Neumann was laying the groundwork for digital computers, and he was bothered by the following question. Since real physical (and biological) networks are composed of imperfect elements, can we compute any boolean function f if the constituent basis gates are in fact noisy? His model of the δ-noisy gate (bottom row of Figure 33.1) was to take a primitive noiseless gate and apply (mod 2) additive noise to the output.

In this case, we have a network of noisy gates, and such a network necessarily has noisy (non-deterministic) output. Therefore, when we say that a noisy gate circuit C computes f we require the existence of some $\epsilon_0 = \epsilon_0(\delta)$ (that cannot depend on f) such that

$$\mathbb{P}[C(x_1,\ldots,x_n) \neq f(x_1,\ldots,x_n)\big) \leq \frac{1}{2} - \epsilon_0, \tag{33.2}$$

where $C(x_1,\ldots,x_n)$ is the output of the noisy circuit with inputs $x_1,\ldots,x_n$. If we build the circuit according to the classical (Shannon) methods, we would obviously have catastrophic error accumulation so deep circuits necessarily have $\epsilon_0 \to 0$. At the same time, von Neumann was bothered by the fact that evidently our brains operate with very noisy gates and yet are able to carry out very long computations without mistakes. His thoughts culminated in the following ground-breaking result.

Theorem 33.1. (von Neumann [444]) *There exists $\delta^* > 0$ such that for all $\delta < \delta^*$ it is possible to compute every boolean function f via δ-noisy 3-majority gates.*

Von Neumann's original estimate $\delta^* \approx 0.087$ was subsequently improved by Pippenger. The main (still open) question of this area is to find the largest δ^* for which the above theorem holds.

The condition in (33.2) implies the output should be correlated with the inputs. This requires the mutual information between the inputs (if they are random) and

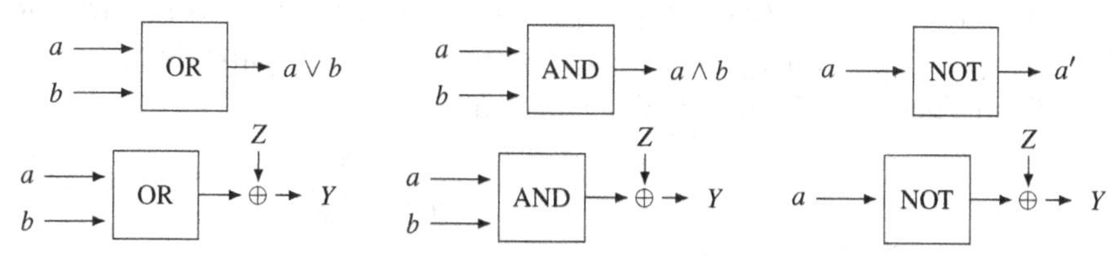

Figure 33.1 Basic building blocks of any boolean circuit. Top row shows the classical (Shannon) noiseless gates. Bottom row shows noisy (von Neumann) gates. Here $Z \sim \text{Ber}(\delta)$ is assumed to be independent of the inputs.

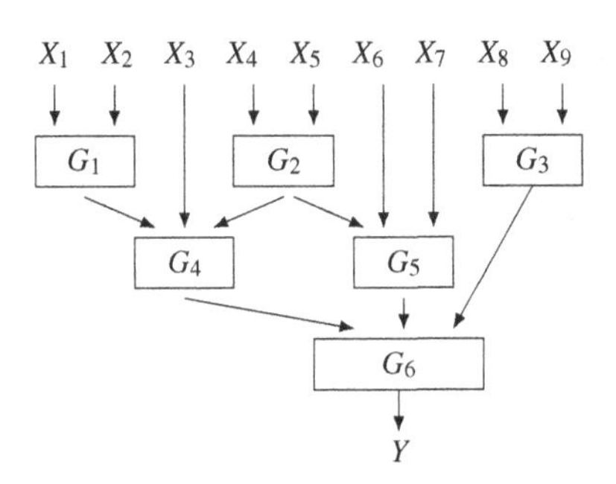

Figure 33.2 An example of a nine-input boolean circuit.

the output to be greater than zero. We now give a theorem of Evans and Schulman that gives an upper bound to the mutual information between any of the inputs and the output. We will prove the theorem in Section 33.3 as a consequence of the more general *directed information percolation* theory.

Theorem 33.2. ([158]) *Suppose one has an n-input noisy boolean circuit composed of gates with at most K inputs and with noise components having at most δ probability of error. Then, the mutual information between any input X_i and output Y is upper-bounded as*

$$I(X_i;Y) \leq \left(K(1-2\delta)^2\right)^{d_i} \log 2,$$

where d_i is the minimum length between X_i and Y (i.e. the minimum number of gates required to be passed through until reaching Y).

Theorem 33.2 implies that noisy computation is only possible for $\delta < \frac{1}{2} - \frac{1}{2\sqrt{K}}$. This is the best known threshold. To illustrate this result consider the circuit given in Figure 33.2. That circuit has nine inputs and is composed of gates with at most three inputs. The three-input gates are G_4, G_5, and G_6. The minimum distance between X_3 and Y is $d_3 = 2$, and the minimum distance between X_5 and Y is $d_5 = 3$. If G_i's are δ-noisy gates, we can invoke Theorem 33.2 between any input and the output.

Now, the main conceptual implication of Theorem 33.2 is in demonstrating that some circuits are not computable with δ-noisy gates unless δ is sufficiently small. For instance, take $f(X_1,\dots,X_n) = \mathrm{XOR}(X_1,\dots,X_n)$. Note that function f depends essentially on every input X_i, since $\mathrm{XOR}(X_1,\dots,X_n) = \mathrm{XOR}\left(\mathrm{XOR}(X_1,\dots,X_{i-1},X_{i+1},\dots,X_n),X_i\right)$. Thus, any circuit that ignores any one of the inputs X_i will not be able to satisfy requirement (33.2). Since we are composing our circuit via K-input gates, this implies that there must exist at least one input X_i with $d_i \geq \frac{\log n}{\log K}$ (indeed, going from Y up we are to make K-ary choice at each gate and thus at height d we can at most reach K^d inputs). Now as $n \to \infty$ we see from Theorem 33.2 that $I(X_i;Y) \to 0$ unless

$$\delta \leq \delta^*_{ES} = \frac{1}{2} - \frac{1}{2\sqrt{K}}.$$

As we argued $I(X_i;Y) \to 0$ is incompatible with satisfying (33.2). Hence the value of δ^*_{ES} (the Evans–Schulman estimate) gives an (at present, the tightest) upper bound for the noise limit under which reliable computation with K-input gates is possible.

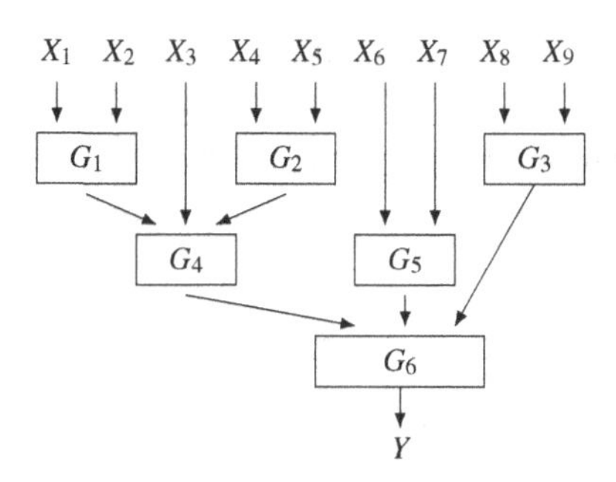

Figure 33.3 An example of a nine-input boolean formula.

Computation with formulas Note that the graph structure given in Figure 33.2 contains some undirected loops. A *formula* is a type of boolean circuit that does not contain any undirected loops unlike the case in Figure 33.2. In other words, for a formula the underlying graph structure forms a tree. For example, removing one of the outputs of G_2 in Figure 33.2 we obtain a formula as given in Figure 33.3.

For computation with formulas much stronger results are available. For example, for any odd K, the threshold is exactly known from [157, Theorem 1]. Specifically, it is shown there that we can compute reliably any boolean function f that is represented with a formula composed of K-input δ-noisy gates (with K odd) if $\delta < \delta_f^*$, and no such computation is possible for $\delta > \delta_f^*$, where

$$\delta_f^* = \frac{1}{2} - \frac{2^{K-1}}{K\binom{K-1}{(K-1)/2}}.$$

Since every formula is also a circuit, we of course have $\delta_f^* < \delta_{ES}^*$, so that there is no contradiction with Theorem 33.2. However, comparing the thresholds gives us the ability to appreciate the tightness of Theorem 33.2 for general boolean circuits. Indeed, for large K we have the approximation

$$\delta_f^* \approx \frac{1}{2} - \frac{\sqrt{\pi/2}}{2\sqrt{K}}, \quad K \gg 1,$$

whereas the Evans–Schulman estimate is $\delta_{ES}^* \approx \frac{1}{2} - \frac{1}{2\sqrt{K}}$. We can thus see that it has at least the right rate of convergence to $1/2$ for large K.

33.2 Strong Data-Processing Inequality

Definition 33.3. (Contraction coefficient for $P_{Y|X}$) For a fixed conditional distribution (or kernel) $P_{Y|X}$, define

$$\eta_f(P_{Y|X}) = \sup \frac{D_f(P_Y \| Q_Y)}{D_f(P_X \| Q_X)}, \tag{33.3}$$

where $P_Y = P_{Y|X} \circ P_X$, $Q_Y = P_{Y|X} \circ Q_X$, and the supremum is over all pairs (P_X, Q_X) satisfying $0 < D_f(P_X \| Q_X) < \infty$.

Recall that the data-processing inequality (DPI) in Theorem 7.4 states that $D_f(P_X\|Q_X) \geq D_f(P_Y\|Q_Y)$. The concept of the strong DPI introduced above quantifies the multiplicative decrease between the input and output f-divergences.

We note that in general $\eta_f(P_{Y|X})$ is hard to compute. However, total variation is an exception.

Theorem 33.4. ([127]) $\eta_{\mathrm{TV}} = \sup_{x\neq x'} \mathrm{TV}(P_{Y|X=x}, P_{Y|X=x'})$.

Proof. We consider two directions separately.

- $\eta_{\mathrm{TV}} \geq \sup_{x_0\neq x_0'} \mathrm{TV}(P_{Y|X=x_0}, P_{Y|X=x_0'})$:
  This case is obvious. Take $P_X = \delta_{x_0}$ and $Q_X = \delta_{x_0'}$.[1] Then from the definition of η_{TV}, we have $\eta_{\mathrm{TV}} \geq \mathrm{TV}(P_{Y|X=x_0}, P_{Y|X=x_0'})$ for any x_0 and x_0', $x_0 \neq x_0'$.

- $\eta_{\mathrm{TV}} \leq \sup_{x_0\neq x_0'} \mathrm{TV}(P_{Y|X=x_0}, P_{Y|X=x_0'})$:
  Define $\tilde{\eta} \triangleq \sup_{x_0\neq x_0'} \mathrm{TV}(P_{Y|X=x_0}, P_{Y|X=x_0'})$. We consider the discrete alphabet case for simplicity. Fix any P_X, Q_X and $P_Y = P_{Y|X} \circ P_X$, $Q_Y = P_{Y|X} \circ Q_X$. Observe that for any $E \subseteq \mathcal{Y}$

  $$P_{Y|X=x_0}(E) - P_{Y|X=x_0'}(E) \leq \tilde{\eta}\,1\{x_0 \neq x_0'\}. \tag{33.4}$$

  Now suppose there are random variables X_0 and X_0' having some marginals P_X and Q_X respectively. Consider any coupling $\pi_{X_0,X_0'}$ with marginals P_X and Q_X respectively. Then averaging (33.4) and taking the supremum over E, we obtain

  $$\sup_{E\subseteq\mathcal{Y}} P_Y(E) - Q_Y(E) \leq \tilde{\eta}\pi[X_0 \neq X_0'].$$

  Now the left-hand side equals $\mathrm{TV}(P_Y, Q_Y)$ by Theorem 7.7(a). Taking the infimum over couplings π the right-hand side evaluates to $\mathrm{TV}(P_X, Q_X)$ by Theorem 7.7(b). □

Example 33.1. η_{TV} of a binary symmetric channel

The η_{TV} of the BSC_δ is given by

$$\eta_{\mathrm{TV}}(\mathsf{BSC}_\delta) = \mathrm{TV}(\mathrm{Ber}(\delta), \mathrm{Ber}(1-\delta)) = |1-2\delta|.$$

We sometimes want to relate η_f to the f-information (Section 7.8) instead of f-divergence. This relation is given in the following theorem.

Theorem 33.5.

$$\eta_f(P_{Y|X}) = \sup_{P_{UX}:\, U\to X\to Y} \frac{I_f(U;Y)}{I_f(U;X)}.$$

[1] Recall that δ_{x_0} is the probability distribution with $\mathbb{P}(X = x_0) = 1$.

Recall that for any Markov chain $U \to X \to Y$, DPI states that $I_f(U;Y) \le I_f(U;X)$ and Theorem 33.5 gives the stronger bound

$$I_f(U;Y) \le \eta_f(P_{Y|X}) I_f(U;X). \tag{33.5}$$

Proof. First, notice that for any u_0, we have $D_f(P_{Y|U=u_0} \| P_Y) \le \eta_f D_f(P_{X|U=u_0} \| P_X)$. Averaging the above expression over any P_U, we obtain

$$I_f(U;Y) \le \eta_f I_f(U;X).$$

Second, fix $\tilde{P}_X$, $\tilde{Q}_X$ and let $U \sim \mathrm{Ber}(\lambda)$ for some $\lambda \in [0,1]$. Define the conditional distribution $P_{X|U}$ as $P_{X|U=1} = \tilde{P}_X$, $P_{X|U=0} = \tilde{Q}_X$. Take $\lambda \to 0$, then (see [346] for technical subtleties)

$$I_f(U;X) = \lambda D_f(\tilde{P}_X \| \tilde{Q}_X) + o(\lambda),$$
$$I_f(U;Y) = \lambda D_f(\tilde{P}_Y \| \tilde{Q}_Y) + o(\lambda).$$

The ratio $\frac{I_f(U;Y)}{I_f(U;X)}$ will then converge to $\frac{D_f(\tilde{P}_Y \| \tilde{Q}_Y)}{D_f(\tilde{P}_X \| \tilde{Q}_X)}$. Thus, optimizing over $\tilde{P}_X$ and $\tilde{Q}_X$ we can get the ratio of I_f's arbitrarily close to η_f. □

We next state some of the fundamental properties of contraction coefficients.

Theorem 33.6. *In the statements below η_f (and others) correspond to $\eta_f(P_{Y|X})$ for some fixed $P_{Y|X}$ from $\mathcal{X}$ to $\mathcal{Y}$.*

(a) For any f, $\eta_f \le \eta_{TV}$.

(b) $\eta_{\mathrm{KL}} = \eta_{H^2} = \eta_{\chi^2}$. More generally, for any operator-convex and twice continuously differentiable f we have $\eta_f = \eta_{\chi^2}$.

(c) η_{χ^2} equals the squared maximal correlation: Denote by $\rho(A,B) \triangleq \frac{\mathrm{Cov}(A,B)}{\sqrt{\mathrm{Var}[A]\,\mathrm{Var}[B]}}$ the correlation coefficient between scalar random variables A and B. Then $\eta_{\chi^2} = \sup_{P_X,f,g} \rho^2(f(X),g(Y))$, where the supremum is over all distributions P_X on $\mathcal{X}$, and all functions $f\colon \mathcal{X} \to \mathbb{R}$ and $g\colon \mathcal{Y} \to \mathbb{R}$.

(d) For binary-input channels, denote $P_0 = P_{Y|X=0}$ and $P_1 = P_{Y|X=1}$. Then

$$\eta_{\mathrm{KL}} = \mathrm{LC}_{\max}(P_0,P_1) \triangleq \sup_{0<\beta<1} \mathrm{LC}_\beta(P_0 \| P_1),$$

where (recall $\bar{\beta} \triangleq 1-\beta$)

$$\mathrm{LC}_\beta(P\|Q) = D_f(P\|Q), \quad f(x) = \bar{\beta}\beta\frac{(1-x)^2}{\bar{\beta}x+\beta}$$

is the Le Cam divergence of order β (recall (7.7) for $\beta = 1/2$).

(e) Consequently,

$$\frac{1}{2}H^2(P_0,P_1) \le \eta_{\mathrm{KL}} \le H^2(P_0,P_1) - \frac{H^4(P_0,P_1)}{4}. \tag{33.6}$$

(f) If a binary-input channel $P_{Y|X}$ is also input-symmetric (or binary memoryless symmetric, see Section 19.4), then $\eta_{\mathrm{KL}}(P_{Y|X}) = I_{\chi^2}(X;Y)$ for $X \sim \mathrm{Ber}(1/2)$.*

(g) *For any channel $P_{Y|X}$, the supremum in (33.3) can be restricted to P_X, Q_X with a common binary support. In other words, $\eta_f(P_{Y|X})$ coincides with that of the least contractive binary subchannel. Consequently, from (e) we conclude that*

$$\frac{1}{2}\mathrm{diam}_{H^2} \le \eta_{\mathrm{KL}}(P_{Y|X}) = \mathrm{diam}_{\mathrm{LC}_{\max}} \le \mathrm{diam}_{H^2} - \frac{\mathrm{diam}_{H^2}}{4}$$

(in particular $\eta_{\mathrm{KL}} \asymp \mathrm{diam}_{H^2}$) where $\mathrm{diam}_{H^2}(P_{Y|X}) = \sup_{x,x' \in \mathcal{X}} H^2(P_{Y|X=x}, P_{Y|X=x'})$ and $\mathrm{diam}_{\mathrm{LC}_{\max}} = \sup_{x,x'} \mathrm{LC}_{\max}(P_{Y|X=x}, P_{Y|x=x'})$ are the squared Hellinger and Le Cam diameters of the channel.

Proof. Most proofs in full generality can be found in [346]. For (a) one first shows that $\eta_f \le \eta_{\mathrm{TV}}$ for the so-called $\mathcal{E}_\gamma$ divergence, an f-divergence corresponding to $f(x) = |x - \gamma|_+ - |1 - \gamma|_+$, which is not hard to believe since $\mathcal{E}_\gamma$ is piecewise linear and reduces to TV when $\gamma = 1$. Then the general result follows from the fact that any convex function f can be approximated (as $N \to \infty$) in the form

$$\sum_{j=1}^{N} a_j|x - c_j|_+ + a_0 x + c_0.$$

For (b) see [93, Theorem 1] and [97, Proposition II.6.13 and Corollary II.6.16]. The idea of this proof is as follows:

- $\eta_{\mathrm{KL}} \ge \eta_{\chi^2}$ by restricting to local perturbations. Recall that KL divergence behaves locally as χ^2 – Proposition 2.21.
- Using the identity $D(P\|Q) = \int_0^\infty \chi^2(P\|Q_t)dt$ where $Q_t = \frac{tP+Q}{1+t}$, we have

$$D(P_Y\|Q_Y) = \int_0^\infty \chi^2(P_Y\|Q_{Y_t})dt \le \eta_{\chi^2}\int_0^\infty \chi^2(P_X\|Q_{X_t})dt = \eta_{\chi^2} D(P_X\|Q_X).$$

For (c), we fix Q_X (and thus $Q_{X,Y} = Q_X P_{Y|X}$). If $g = \frac{dP_X}{dQ_X}$ then $Tg(y) = \frac{dP_Y}{dQ_Y} = \mathbb{E}_{Q_{X|Y}}[g(X)|Y = y]$ is a linear operator. Then $\eta_{\chi^2}(P_{Y|X})$ is nothing but the maximal singular value (spectral norm squared) of $T\colon L_2(Q_X) \to L_2(Q_Y)$ when restricted to a linear subspace $\{h\colon \mathbb{E}_{Q_X}[h] = 0\}$. The adjoint of T is $T^*h(x) = \mathbb{E}_{P_{Y|X}}[h(Y)|X = x]$. The spectral norms of an operator and its adjoint coincide and the spectral norm of T^* is precisely the squared maximal correlation. These two facts together yield the result. (See Theorem 33.12(c) which strengthens this result for a fixed P_X.)

Part (d) follows from the definition of $\eta_{\chi^2} = \sup_{\alpha,\beta} \frac{\chi^2(\alpha P_1 + \bar{\alpha} P_0 \| \beta P_1 + \bar{\beta} P_0)}{\chi^2(\mathrm{Ber}(\alpha)\|\mathrm{Ber}(\beta))}$ and some algebra.

Next, (e) follows from bounding (via Cauchy–Schwarz, etc.) $\mathrm{LC}_{\max}$ in terms of H^2; see [346, Appendix B].

Part (f) follows from the fact that every BMS channel can be represented as $X \mapsto Y = (Y_\Delta, \Delta)$ where $\Delta \in [0, 1/2]$ is independent of X and $Y_\delta = \mathsf{BSC}_\delta(X)$. In other words, every BMS channel is a mixture of BSCs; see [361, Section 4.1]. Thus, we have for any $U \to X \to Y = (Y_\Delta, \Delta)$ and $\Delta \perp\!\!\!\perp (U, X)$ the following chain:

$$I(U;Y) = I(U;Y|\Delta) \le \mathbb{E}_{\delta\sim P_\Delta}[(1 - 2\delta)^2 I(U;X|\Delta = \delta)] = \mathbb{E}[(1 - 2\Delta)^2]I(U;X),$$

where we used the fact that $I(U;X|\Delta = \delta) = I(U;X)$ and Example 33.2 below.
For (g) see Exercise **VI.20**. □

Example 33.2. Computing $\eta_{\mathrm{KL}}(\mathsf{BSC}_\delta)$

Consider the BSC_δ channel. In Example 33.1 we computed η_{TV}. Here we have $\mathrm{diam}_{H^2} = 2-4\sqrt{\delta(1-\delta)}$ and thus from the bound (33.6) we get $\eta_{\mathrm{KL}} \le (1-2\delta)^2$. On the other hand taking $U = \mathrm{Ber}(1/2)$ and $P_{X|U} = \mathrm{Ber}(\alpha)$ we get

$$\eta_{\mathrm{KL}} \ge \frac{I(U;Y)}{I(U;X)} = \frac{\log 2 - h(\alpha + (1-2\alpha)\delta)}{\log 2 - h(\alpha)} \to (1-2\delta)^2, \qquad \alpha \to \frac{1}{2}.$$

Thus we have shown for BSC_δ that

$$\eta_{\mathrm{KL}} = \eta_{H^2} = \eta_{\chi^2} = (1-2\delta)^2.$$

This example has the following consequence for the KL divergence geometry.

Proposition 33.7. *Consider any distributions P_0 and P_1 on $\mathcal{X}$ and let us consider the interval in $\mathcal{P}(\mathcal{X})$: $P_\lambda = \lambda P_1 + (1-\lambda)P_0$ for $\lambda \in [0,1]$. Then divergence (with respect to the midpoint) behaves subquadratically:*

$$D(P_\lambda \| P_{1/2}) + D(P_{1-\lambda} \| P_{1/2}) \le (1-2\lambda)^2 (D(P_0 \| P_{1/2}) + D(P_1 \| P_{1/2})).$$

The same statement holds with D replaced by χ^2 (and any other D_f satisfying Theorem 33.6(b)).

Proof. Let $X \sim \mathrm{Ber}(1/2)$ and $Y = \mathsf{BSC}_\lambda(X)$. Let $U \leftarrow X \rightarrow Y$ be defined with $U \sim P_0$ if $X = 0$ and $U \sim P_1$ if $X = 1$. Then

$$I_f(U;Y) = \frac{1}{2} D_f(P_\lambda \| P_{1/2}) + \frac{1}{2} D_f(P_\lambda \| P_{1/2}).$$

Thus, applying SDPI (33.5) completes the proof. □

Remark 33.1. Let us introduce $d_{\mathrm{JS}}(P,Q) = \sqrt{\mathrm{JS}(P,Q)}$ and $d_{\mathrm{LC}} = \sqrt{\mathrm{LC}(P,Q)}$ as the Jensen–Shannon (7.8) and Le Cam (7.7) metrics. Then the proposition can be rewritten as

$$d_{\mathrm{JS}}(P_\lambda, P_{1-\lambda}) \le |1-2\lambda| d_{\mathrm{JS}}(P_0, P_1),$$
$$d_{\mathrm{LC}}(P_\lambda, P_{1-\lambda}) \le |1-2\lambda| d_{\mathrm{LC}}(P_0, P_1).$$

Notice that for any metric $d(P,Q)$ on $\mathcal{P}(\mathcal{X})$ that is induced from the norm on the vector space $\mathcal{M}(\mathcal{X})$ of all signed measures (such as TV), we must necessarily have $d(P_\lambda, P_{1-\lambda}) = |1-2\lambda| d(P_0, P_1)$. Thus, the $\eta_{\mathrm{KL}}(\mathsf{BSC}_\lambda) = (1-2\lambda)^2$ which yields the inequality is rather natural.

33.3 Directed Information Percolation

In this section, we are concerned with the amount of information decay experienced in a directed acyclic graph (DAG) $G = (V, E)$. In the following context the vertex set V refers to a set of vertices v, each associated with a random variable X_v and the edge set E refers to a set of directed edges whose configuration allows us to factorize the joint distribution of X_V (Section 3.4). Throughout the section, we consider Shannon mutual information, that is, $f = x \log x$. Let us give a detailed example below.

Example 33.3.

Suppose we have a graph $G = (V, E)$ as follows:

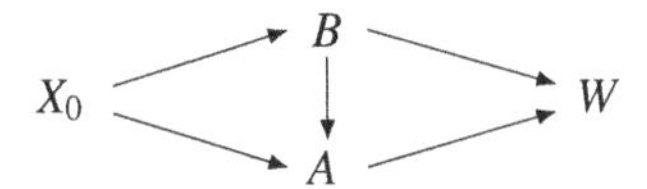

This means that we have a joint distribution factorizing as

$$P_{X_0,A,B,W} = P_{X_0} P_{B|X_0} P_{A|B,X_0} P_{W|A,B}.$$

Then every node has a channel from its parents to itself, for example W corresponds to a noisy channel $P_{W|A,B}$, and we can define $\eta \triangleq \eta_{\mathrm{KL}}(P_{W|A,B})$. Now, prepend another random variable $U \sim \mathrm{Ber}(\lambda)$ at the beginning; the new graph $G' = (V', E')$ is shown below:

B

$U \longrightarrow X_0 \qquad\qquad W$

A

We want to verify the relation

$$I(U;B, W) \le \bar{\eta} I(U;B) + \eta I(U;A, B). \tag{33.7}$$

Recall that from the chain rule we have $I(U;B, W) = I(U;B) + I(U;W|B) \ge I(U;B)$. Hence, if (33.7) is correct, then $\eta \to 0$ implies $I(U;B, W) \approx I(U;B)$ and symmetrically $I(U;A, W) \approx I(U;A)$. Therefore for small δ, observing W, A or W, B does not give any advantage over observing solely A or B, respectively.

Observe that G' forms a Markov chain $U \to X_0 \to (A, B) \to W$, which allows us to factorize the joint distribution over E' as

$$P_{U,X_0,A,B,W} = P_U P_{X_0|U} P_{A,B|X_0} P_{W|A,B}.$$

Now consider the joint distribution conditioned on $B = b$, that is, $P_{U,X_0,A,W|B=b}$. We claim that the conditional Markov chain $U \to X_0 \to A \to W|B = b$ holds. Indeed, given B and A, X_0 is independent of W, that is, $P_{X_0|A,B} P_{W|A,B} = P_{X_0,W|AB}$, from which follows the mentioned conditional Markov chain. Using the conditional Markov chain, SDPI gives us for any b,

$$I(U;W|B=b) \le \eta I(U;A|B=b).$$

Averaging over b and adding $I(U;B)$ to both sides we obtain

$$I(U;W,B) \le \eta I(U;A|B) + I(U;B)$$
$$= \eta I(U;A,B) + \bar{\eta} I(U;B).$$

From the characterization of η_f in Theorem 33.5 we conclude

$$\eta_{\mathrm{KL}}(P_{W,B|X_0}) \le \eta \cdot \eta_{\mathrm{KL}}(P_{A,B|X_0}) + (1-\eta) \cdot \eta_{\mathrm{KL}}(P_{B|X_0}). \tag{33.8}$$

Now, we provide another example which has in some sense an analogous setup to Example 33.3.

Example 33.4. Percolation

Take the graph $G = (V, E)$ in Example 33.3 with a small modification. See Figure 33.4. Now, suppose X, A, B, W are some cities and the edge set E represents the roads between these cities. Let R be a random variable denoting the state of the road connecting to W with $\mathbb{P}(R \text{ is open}) = \eta$ and $\mathbb{P}[R \text{ is closed}] = \bar{\eta}$. For any $Y \in V$, let the event $\{X \to Y\}$ indicate that one can drive from X to Y. Then

$$\mathbb{P}[X \to B \text{ or } W] = \eta \mathbb{P}[X \to A \text{ or } B] + \bar{\eta} \mathbb{P}[X \to B]. \tag{33.9}$$

Observe the resemblance between (33.8) and (33.9).

We will now give a theorem that relates η_{KL} to the percolation probability on a DAG under the following setting: Consider a DAG $G = (V, E)$.

- All edges are open.
- Every vertex is open with probability $p(v) = \eta_{\mathrm{KL}}(P_{X_v|X_{\mathrm{Pa}(v)}})$ where $\mathrm{Pa}(v)$ denotes the set of parents of v.

Under this model, for two subsets $T, S \subset V$ we define $\mathrm{perc}[T \to S] = \mathbb{P}[\exists \text{ open path } T \to S]$.

Note that $P_{X_v|X_{\mathrm{Pa}(v)}}$ describe the stochastic recipe for producing X_v based on its parent variables. We assume that in addition to a DAG we also have been given all these constituent channels (or at least bounds on their η_{KL} coefficients).

Theorem 33.8. ([346]) *Let $G = (V, E)$ be a DAG and let 0 be a node with in-degree equal to zero (i.e. a source node). Note that for any $0 \not\ni S \subset V$ we can inductively stitch together constituent channels $P_{X_v|X_{Pa(v)}}$ and obtain $P_{X_S|X_0}$. Then we have*

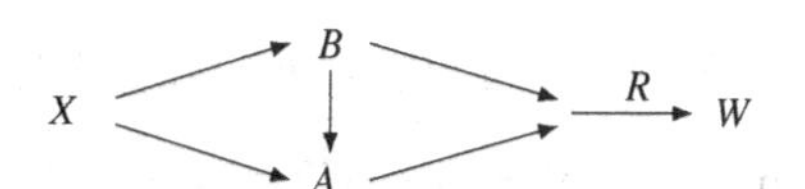

Figure 33.4 Illustration for Example 33.4.

$$\eta_{\mathrm{KL}}(P_{X_S|X_0}) \leq \mathrm{perc}(0 \to \mathrm{S}). \tag{33.10}$$

Proof. For convenience let us denote $\eta(T) = \eta_{\mathrm{KL}}(P_{X_T|X_0})$ and $\eta_v = \eta_{\mathrm{KL}}(P_{X_v|X_{\mathrm{Pa}(v)}})$. The proof follows from an induction on the size of G. The statement is clear for $|V(G)| = 1$ since $S = \emptyset$ or $S = \{X_0\}$. Now suppose the statement is already shown for all graphs smaller than G. Let v be the node with out-degree 0 in G. If $v \notin S$ then we can exclude it from G and the statement follows from the induction hypothesis. Otherwise, define $S_A = \mathrm{Pa}(v) \setminus S$ and $S_B = S \setminus \{v\}$, $A = X_{S_A}, B = X_{S_B}, W = X_v$. (If $0 \in A$ then we can create a fake $0'$ with $X_{0'} = X_0$ and retain $0' \in A$ while moving 0 out of A. So without loss of generality, $0 \notin A$.) Prepending arbitrary U to the graph as $U \to X_0$, the joint DAG of random variables (X_0, A, B, W) is then given by precisely the graph in (33.7). Thus, we obtain from (33.8) the estimate

$$\eta(S) \leq \eta_v \eta(S_A \cup S_B) + (1 - \eta_v)\eta_{\mathrm{KL}}(S_B). \tag{33.11}$$

From the induction hypothesis $\eta(S_A \cup S_B) \leq \mathrm{perc}(0 \to S_A)$ and $\eta(S_B) \leq \mathrm{perc}(0 \to S_B)$ (they live on a graph $G \setminus \{v\}$). Thus, from the computation in (33.9) we see that the right-hand side of (33.11) is precisely $\mathrm{perc}(0 \to S)$ and thus $\eta(S) \leq \mathrm{perc}(S)$ as claimed. □

We are now in position to complete the postponed proof.

Proof of Theorem 33.2. First observe the noisy boolean circuit is a form of DAG. Since the gates are δ-noisy, the contraction coefficients of constituent channels η_v in the DAG can be bounded by $(1-2\delta)^2$. Thus, in the percolation question all vertices are open with probability $(1 - 2\delta)^2$.

From SDPI, for each i, we have $I(X_i;Y) \leq \eta_{\mathrm{KL}}(P_{Y|X_i})H(X_i)$. From Theorem 33.8, we know $\eta_{\mathrm{KL}}(P_{Y|X_i}) \leq \mathrm{perc}(X_i \to Y)$. We now want to upper-bound $\mathrm{perc}(X_i \to Y)$. Recall that the minimum distance between X_i and Y is d_i. For any path π of length $\ell(\pi)$ from X_i to Y, therefore, the probability that it will be open is $\leq (1 - 2\delta)^{2\ell(\pi)}$. We can thus bound

$$\mathrm{perc}(X_i \to Y) \leq \sum_{\pi:\, X_i \to Y} (1 - 2\delta)^{2\ell(\pi)}. \tag{33.12}$$

Let us now build paths backward starting from Y, which allows us to represent paths $X \to Y_i$ as vertices of a K-ary tree with root Y_i. By labeling all vertices on a K-ary tree corresponding to paths $X \to Y_i$ we observe two facts: the labeled set V is prefix-free (two labeled vertices are never in ancestral relation) and the depth of each labeled set is at least d_i. It is easy to see that $\sum_{u \in V} c^{\mathrm{depth}(u)} \leq (Kc)^{d_i}$ provided $Kc \leq 1$ and attained by taking V to be the set of all vertices in the tree at depth d_i. We conclude that whenever $K(1-2\delta)^2 \leq 1$ the right-hand side of (33.12) is bounded by $(K(1 - 2\delta)^2)^{d_i}$, which concludes the proof by upper-bounding $H(X_i) \leq \log 2$ as

$$I(X_i;Y) \leq \eta_{\mathrm{KL}}(P_{Y|X_i})H(X_i) \leq K^{d_i}(1 - 2\delta)^{2d_i} \log 2.$$ □

As another (simple) application of Theorem 33.8 we show the following.

Corollary 33.9. *Consider a channel $P_{Y|X}$ and its n-letter memoryless extension $P_{Y|X}^{\otimes n}$. Then we have*

$$\eta_{\mathrm{KL}}(P_{Y|X}^{\otimes n}) \leq 1 - (1 - \eta_{\mathrm{KL}}(P_{Y|X}))^n \leq n\eta_{\mathrm{KL}}(P_{Y|X}).$$

The first inequality can be sharp for some channels. For example, it is sharp when $P_{Y|X}$ is a binary or q-ary erasure channel (defined below in Example 33.6). This fact is proven in [346, Theorem 17].

Proof. The graph here consists of n parallel edges $X_i \to Y_i$. Theorem 33.8 shows that $\eta_{\mathrm{KL}}(P_{Y|X}^{\otimes n}) \leq \mathrm{perc}(\{X_1, \ldots, X_n\} \to \{Y_1, \ldots, Y_n\})$. The latter simply equals $1 - (1 - \eta)^n$ where $\eta = \eta(P_{Y|X})$ is the probability of an edge being open. □

We conclude the section with a more sophisticated application of Theorem 33.8, emphasizing how it can yield stronger bounds when compared to Theorem 33.2.

Example 33.5.

Suppose we have the topological restriction on the placement of gates (namely that the inputs to each gate should be from nearest neighbors to the right), resulting in the following circuit of two-input δ-noisy gates.

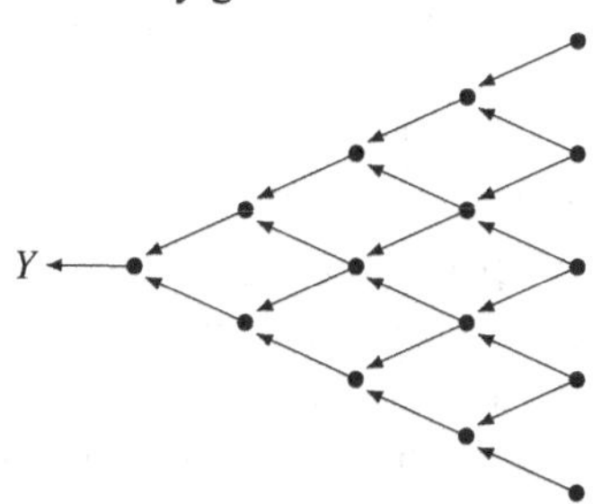

Note that each gate may be a simple passthrough (i.e. serve as router) or a constant output. Theorem 33.2 states that if $(1 - 2\delta)^2 < \frac{1}{2}$, then noisy computation within arbitrary topology is not possible. Theorem 33.8 improves this to $(1 - 2\delta)^2 < p_c$, where p_c is the oriented site percolation threshold for the particular graph we have. Namely, if each vertex is open with probability $p < p_c$ then with probability 1 the connected component emanating from any given node (and extending to the right) is finite. For the example above the site percolation threshold is estimated as $p_c \approx 0.705$ (so-called Stavskaya automata).

33.4 Input-Dependent SDPI; Mixing of Markov Chains

Previously we have defined the contraction coefficient $\eta_f(P_{Y|X})$ as the maximum contraction of an f-divergence over all input distributions. We now define an analogous concept for a fixed input distribution P_X.

Definition 33.10. (Input-dependent contraction coefficient) For any input distribution P_X, Markov kernel $P_{Y|X}$, and convex function f, we define

$$\eta_f(P_X, P_{Y|X}) \triangleq \sup \frac{D_f(Q_Y \| P_Y)}{D_f(Q_X \| P_X)},$$

where $P_Y = P_{Y|X} \circ P_X$, $Q_Y = P_{Y|X} \circ Q_X$, and the supremum is over Q_X satisfying $0 < D_f(Q_X \| P_X) < \infty$.

We refer to $\eta_f(P_X, P_{Y|X})$ as the input-dependent contraction coefficient, in contrast with the input-independent contraction coefficient $\eta_f(P_{Y|X})$. It is obvious that

$$\eta_f(P_X, P_{Y|X}) \leq \eta_f(P_{Y|X}),$$

but as we will see below the inequality is often strict and the difference can lead to significant improvements in applications (Example 33.10). In Theorem 33.6 we have seen that for most interesting f's we have $\eta_f(P_{Y|X}) = \eta_{\chi^2}(P_{Y|X})$. Unfortunately, for the input-dependent version this is not true: we only have a one-sided comparison, namely for any twice continuously differentiable f with $f''(1) > 0$ (in particular for KL divergence) it holds that [346, Theorem 2]

$$\eta_{\chi^2}(P_X, P_{Y|X}) \leq \eta_f(P_X, P_{Y|X}). \tag{33.13}$$

For example, for jointly Gaussian X, Y, we in fact have $\eta_{\chi^2} = \eta_{\mathrm{KL}}$ (see Example 33.7 next); however, in general we only have $\eta_{\chi^2} < \eta_{\mathrm{KL}}$ (see [19] for an example). Thus, unlike the input-independent case, here the choice of f is very important. A general rule is that $\eta_{\chi^2}(P_X, P_{Y|X})$ is the easiest to bound and by (33.13) it contracts the fastest. However, for various reasons other f are more useful in applications. Consequently, the theory of input-dependent contraction coefficients is much more intricate (see [201] for many recent results and references). In this section we try to summarize some important similarities and distinctions between $\eta_f(P_X, P_{Y|X})$ and $\eta_f(P_{Y|X})$.

First, just as in Theorem 33.5 we can similarly prove a mutual information characterization of $\eta_{\mathrm{KL}}(P_X, P_{Y|X})$ as follows [353, Theorem V.2]:

$$\eta_{\mathrm{KL}}(P_X, P_{Y|X}) = \sup_{P_{U|X}:\, U \to X \to Y} \frac{I(U;Y)}{I(U;X)}.$$

In particular, we see that $\eta_{\mathrm{KL}}(P_X, P_{Y|X})$ is also a slope of the F_I-curve (see Definition 16.5):

$$\eta_{\mathrm{KL}}(P_X, P_{Y|X}) = \left.\frac{d}{dt}\right|_{t=0+} F_I(t; P_{X,Y}). \tag{33.14}$$

(Indeed, from Exercise **III.32** we know $F_I(t)$ is concave and thus $\sup_{t \geq 0} \frac{F_I(t)}{t} = F_I'(0)$.)

The next property of an input-dependent SDPI emphasizes the key difference compared to its input-independent counterpart. Recall that Corollary 33.9 (and

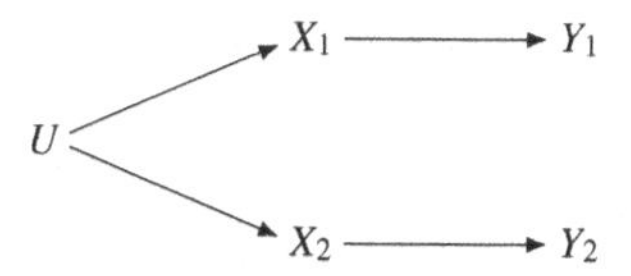

Figure 33.5 Illustration for the probability space in the proof of Proposition 33.11.

the discussion thereafter) shows that generally $\eta_{\mathrm{KL}}(P_{Y|X}^{\otimes n}) \to 1$ exponentially fast. At the same time, $\eta_{\mathrm{KL}}(P_X^{\otimes n}, P_{Y|X}^{\otimes n})$ stays constant.

Proposition 33.11. (Tensorization of η_{KL})

$$\eta_{\mathrm{KL}}(P_X^{\otimes n}, P_{Y|X}^{\otimes n}) = \eta_{\mathrm{KL}}(P_X, P_{Y|X}).$$

In particular, if $(X_i, Y_i) \overset{\mathrm{iid}}{\sim} P_{X,Y}$, *then* $\forall P_{U|X^n}$

$$I(U;Y^n) \leq \eta_{\mathrm{KL}}(P_X, P_{Y|X}) I(U;X^n).$$

Note that not all η_f satisfy tensorization. We will show below (Theorem 33.12) that η_{χ^2} does satisfy it. On the other hand, $\eta_{\mathrm{TV}}(P_X^{\otimes n}, P_{Y|X}^{\otimes n}) \to 1$ exponentially fast (which follows from (7.21)).

Proof. This result is implied by (33.14) and Exercise **III.32**, but a simple direct proof is useful. Without loss of generality (by induction) it is sufficient to prove the proposition for $n = 2$. It is always useful to keep in mind the diagram in Figure 33.5. Let $\eta = \eta_{\mathrm{KL}}(P_X, P_{Y|X})$, then

$$\begin{aligned} I(U;Y_1, Y_2) &= I(U;Y_1) + I(U;Y_2|Y_1) \\ &\leq \eta\,[I(U;X_1) + I(U;X_2|Y_1)] && (33.15)\\ &= \eta\,[I(U;X_1) + I(U;X_2|X_1) + I(U;X_1|Y_1) - I(U;X_1|Y_1, X_2)] && (33.16)\\ &\leq \eta\,[I(U;X_1) + I(U;X_2|Y_1)] && (33.17)\\ &= \eta I(U;X_1, X_2), \end{aligned}$$

where (33.15) is due to the fact that, conditioned on Y_1, $U \to X_2 \to Y_2$ is still a Markov chain, (33.16) is because $U \to X_1 \to Y_1$ is a Markov chain, and (33.17) follows from the fact that $X_2 \to U \to X_1$ is a Markov chain even when conditioned on Y_1. □

As an example, let us analyze the erasure channel.

Example 33.6.

($\eta_{\mathrm{KL}}(P_X, P_{Y|X})$ for erasure channel) We define EC_τ as the following channel

$$Y = \begin{cases} X & \text{w.p. } 1-\tau, \\ ? & \text{w.p. } \tau. \end{cases}$$

Consider an arbitrary $U \to X \to Y$ and define an auxiliary random variable $B = 1\{Y = ?\}$. We have

$$I(U;Y) = I(U;Y,B) = \underbrace{I(U;B)}_{=0,\text{ since } B \perp\!\!\!\perp U} + I(U;Y|B) = (1-\tau)I(U;X),$$

where the last equality is due to the fact that $I(U;Y|B=1) = 0$ and $I(U;Y|B=0) = I(U;X)$. By the mutual information characterization of $\eta_{\mathrm{KL}}(P_X, P_{Y|X})$, we have $\eta_{\mathrm{KL}}(P_X, \mathsf{EC}_\tau) = 1-\tau$. Note that by tensorization we also have $\eta_{\mathrm{KL}}(P_X^{\otimes n}, \mathsf{EC}_\tau^{\otimes n}) = 1-\tau$. However, for non-product input measure μ the study of $\eta_{\mathrm{KL}}(\mu, \mathsf{EC}_\tau^{\otimes n})$ is essentially as hard as the study of mixing of Glauber dynamics for μ – see Exercise **VI.26**.

Another interesting observation about the erasure channel is that even for a uniform input distribution we may have

$$\eta_{\mathrm{KL}}(P_Y, P_{X|Y}) > \eta_{\mathrm{KL}}(P_X, P_{Y|X}) = 1-\tau$$

(at least for $\tau > 0.74$, see Example 33.14). Thus, η_{KL} is not symmetric even for such a simple channel. This is in contrast with η_{χ^2}, as we show next.

Among the input-dependent η_f the most elegant is the theory of η_{χ^2}. The properties hold for general $P_{X,Y}$ but we only state it for the finite case for simplicity.

Theorem 33.12. (Properties of $\eta_{\chi^2}(P_X, P_{Y|X})$) *Consider finite $\mathcal{X}$ and $\mathcal{Y}$. Then, we have:*

(a) (Spectral characterization) Let $M_{x,y} = \frac{P_{X,Y}(x,y)}{\sqrt{P_X(x)P_Y(y)}}$ be an $|\mathcal{X}| \times |\mathcal{Y}|$ matrix. Let $1 = \sigma_1(M) \geq \sigma_2(M) \geq \cdots \geq 0$ be the singular values of M, that is, $\sigma_j(M) = \sqrt{\lambda_j(M^\top M)}$. Then $\eta_{\chi^2}(P_X, P_{Y|X}) = \sigma_2^2(M)$.

(b) (Symmetry) $\eta_{\chi^2}(P_X, P_{Y|X}) = \eta_{\chi^2}(P_Y, P_{X|Y})$.

(c) (Maximal correlation) $\eta_{\chi^2}(P_X, P_{Y|X}) = \sup_{g_1,g_2} \rho^2(g_1(X), g_2(Y))$, where the supremum is over all functions $g_1 : \mathcal{X} \to \mathbb{R}$ and $g_2 : \mathcal{Y} \to \mathbb{R}$.

(d) (Tensorization) $\eta_{\chi^2}(P_X^{\otimes n}, P_{Y|X}^{\otimes n}) = \eta_{\chi^2}(P_X, P_{Y|X})$.

Proof. We focus on the spectral characterization which implies the rest. Denote by $\mathbb{E}_{X|Y}$ a linear operator that acts on function g as $\mathbb{E}_{X|Y}\, g(y) = \sum_x P_{X|Y}(x|y) g(x)$. For any Q_X let $g(x) = \frac{Q_X(x)}{P_X(x)}$ then we have $\frac{Q_Y(y)}{P_Y(y)} = \mathbb{E}_{X|Y}\, g$. Therefore, we have

$$\eta_{\chi^2}(P_X, P_{Y|X}) = \sup_g \frac{\mathrm{Var}_{P_Y}[\mathbb{E}_{X|Y}\, g]}{\mathrm{Var}_{P_X}[g]}$$

with supremum over all $g \geq 0$ and $\mathbb{E}_{P_X}[g] = 1$. We claim that this supremum is also equal to

$$\eta_{\chi^2}(P_X, P_{Y|X}) = \sup_h \frac{\mathbb{E}_{P_Y}[\mathbb{E}_{X|Y}^2\, h]}{\mathbb{E}_{P_X}[h^2]},$$

taken over all h with $\mathbb{E}_{P_X} h = 0$. Indeed, for any such h we can take $g = 1 + \epsilon h$ for some sufficiently small ϵ (to satisfy $g \geq 0$) and, conversely, for any g we can set $h = g - 1$. Finally, let us reparameterize $\phi_x \triangleq \sqrt{P_X(x)} h(x)$ in which case we get

$$\eta_{\chi^2}(P_X, P_{Y|X}) = \sup_{\phi} \frac{\phi^\top M^\top M \phi}{\phi^\top \phi},$$

where ϕ ranges over all vectors in $\mathbb{R}^{\mathcal{X}}$ that are orthogonal to the vector ψ with $\psi_x = \sqrt{P_X(x)}$. Finally, we notice that the largest singular value of M corresponds to the singular vector ψ and thus restricting $\phi \perp \psi$ results in recovering the second-largest singular vector.

Symmetry follows from noticing that matrix M is replaced by $M^\top$ when we interchange X and Y. The maximal correlation characterization follows from the fact that $\sup_{g_2} \frac{\mathbb{E}[g_1(X)g_2(Y)]}{\sqrt{\mathrm{Var}[g_2(Y)]}}$ is attained at $g_2 = \mathbb{E}_{X|Y} g_1$. Tensorization follows from the fact that the singular values of the Kronecker product $M^{\otimes n}$ are just products of (all possible) n-tuples of singular values of M. □

Example 33.7. (SDPI constants of joint Gaussian)

Let X, Y be jointly Gaussian with correlation coefficient ρ. Then

$$\eta_{\chi^2}(P_X, P_{Y|X}) = \eta_{\mathrm{KL}}(P_X, P_{Y|X}) = \rho^2.$$

Indeed, it is well known that the maximal correlation of X and Y is simply $|\rho|$. (This can be shown by finding the eigenvalues of the (Mehler) kernel defined in Theorem 33.12(a); see e.g. [267].) Applying Theorem 33.12(c) yields $\eta_{\chi^2}(P_X, P_{Y|X}) = \rho^2$.

Next, in view of (33.13), it suffices to show $\eta_{\mathrm{KL}} \le \rho^2$, which is a simple consequence of the entropy-power inequality (EPI). Without loss of generality, let us consider $Y = X + Z$, where $X \sim P_X = \mathcal{N}(0,1)$ and $Z \sim \mathcal{N}(0, \sigma^2)$. Then $P_Y = \mathcal{N}(0, 1+\sigma^2)$ and $\rho^2 = \frac{1}{1+\sigma^2}$. Let $\tilde{X}$ have finite second moment and finite differential entropy and set $\tilde{Y} = \tilde{X} + Z$. Applying Lieb's EPI (3.36) with $U_1 = \tilde{X}$, $U_2 = Z/\sigma$, and $\cos^2\alpha = \frac{1}{1+\sigma^2}$, we obtain

$$h(\tilde{Y}) \ge \frac{1}{1+\sigma^2} h(\tilde{X}) + \frac{\sigma^2}{2(1+\sigma^2)} \log(2\pi e) + \frac{1}{2}\log(1+\sigma^2),$$

which implies $D(P_{\tilde{Y}} \| P_Y) \le \frac{1}{1+\sigma^2} D(P_{\tilde{X}} \| P_X)$ as desired.

Before proceeding to statistical applications we mention a very important probabilistic application.

Example 33.8. Mixing of Markov chains

One area in which input-dependent contraction coefficients have found a lot of use is in estimating the mixing time (the time to convergence to equilibrium) of Markov chains. Indeed, suppose $K = P_{Y|X}$ is a kernel for a time-homogeneous Markov chain $X_0 \to X_1 \to \cdots$ with stationary distribution π (i.e. $K = P_{X_{t+1}|X_t}$). Then for any initial distribution q, SDPI gives the following bound:

$$D_f(qP^n \| \pi) \le \eta_f(\pi, K)^n D_f(q\|\pi),$$

showing exponential decrease of D_f provided that $\eta_f(\pi, K) < 1$. For most chains of interest the TV version is useless, but χ^2 and KL are rather effective (the two known as the *spectral gap* and *modified log-Sobolev inequality* methods). For example, for reversible Markov chains, we have [124, Proposition 3]

$$\chi^2(P_{X_n}\|\pi) \le \gamma_*^{2n}\chi^2(P_{X_0}\|\pi), \tag{33.18}$$

where γ_* is the absolute spectral gap of P. See Exercise **VI.19**. The most efficient modern method for bounding η_{KL} is known as *spectral independence*, see Exercise **VI.26**.

33.5 Application: Broadcasting and Coloring on Trees

Consider an infinite b-ary tree $G = (\mathcal{V}, \mathcal{E})$. We assign a random variable X_v for each $v \in \mathcal{V}$. These random variables X_v's are defined on the same alphabet $\mathcal{X}$. In this model, the joint distribution is induced by the distribution on the root vertex π, that is, $X_\rho \sim \pi$, and the edge kernel $P_{X'|X}$, that is, $\forall(p, c) \in \mathcal{E}, P_{X_c|X_p} = P_{X'|X}$, as shown:

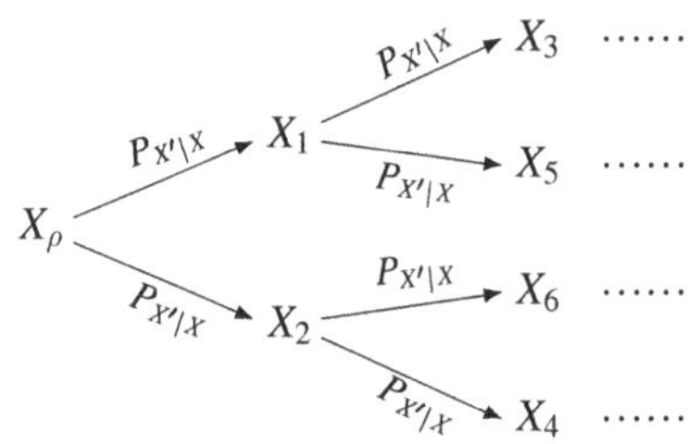

To simplify our discussion, we will assume that π is a reversible measure on kernel $P_{X'|X}$, that is,

$$P_{X'|X}(a|b)\pi(b) = P_{X'|X}(b|a)\pi(a). \tag{33.19}$$

By a standard result on Markov chains, this also implies that π is a stationary distribution of the reversed Markov kernel $P_{X|X'}$.

This model, known as *broadcasting on trees*, turns out to be rather deep. It first arose in statistical physics as a simplification of the Ising model on lattices (trees are called Bethe lattices in physics) [63]. Then, it was found to be closely related to a problem of phylogenetic reconstruction in computational biology [307] and almost simultaneously appeared in random constraint satisfaction problems [262] and sparse-graph coding theory. Our own interest was triggered by a discovery of a certain equivalence between reconstruction on trees and *community detection* in the stochastic block model [308, 119].

We make the following observations:

- We can think of this model as a broadcasting scenario, where the root broadcasts its message X_ρ to the leaves through noisy channels $P_{X'|X}$. The condition (33.19) here is only made to avoid defining the reverse channel. In general, one only requires that π is a stationary distribution of $P_{X'|X}$, in which case (33.21) should be replaced with $\eta_{\mathrm{KL}}(\pi, P_{X|X'})b < 1$.

- Under the assumption (33.19), the joint distribution of this tree can also be written as a Gibbs distribution

$$P_{X_{\mathrm{all}}} = \frac{1}{Z}\exp\left(\sum_{(p,c)\in\mathcal{E}} f(X_p, X_c) + \sum_{v\in\mathcal{V}} g(X_v)\right), \tag{33.20}$$

where Z is the normalization constant, and $f(x_p, x_c) = f(x_c, x_p)$ is symmetric. When $\mathcal{X} = \{0, 1\}$, this model is known as the Ising model (on a tree). Note, however, that not every measure factorizing as (33.20) (with symmetric f) can be written as a broadcasting process for some P and π.

The *broadcasting on trees* is an inference problem in which we want to reconstruct the root variable X_ρ given the observations $X_{L_d} = \{X_v : v \in L_d\}$, with $L_d = \{v : v \in \mathcal{V}, \mathrm{depth}(v) = d\}$. A natural question is when it is possible for any algorithm to reconstruct the root variable better than random guessing. The following theorem provides a necessary condition in terms on the branching factor b and the contraction coefficient of the kernel $P_{X'|X}$.

Theorem 33.13. *Consider the broadcasting problem on an infinite b-ary tree* ($b > 1$), *with root distribution π and edge kernel $P_{X'|X}$. If π is a reversible measure of $P_{X'|X}$ such that*

$$\eta_{\mathrm{KL}}(\pi, P_{X'|X})b < 1, \tag{33.21}$$

then $I(X_\rho; X_{L_d}) \to 0$ as $d \to 0$.

Proof. For every $v \in L_1$, we define the set $L_{d,v} = \{u : u \in L_d, v \in \mathrm{ancestor}(u)\}$. We can upper-bound the mutual information between the root vertex and leaves at depth d:

$$I(X_\rho; X_{L_d}) \le \sum_{v\in L_1} I(X_\rho; X_{L_{d,v}}).$$

For each term in the summation, we consider the Markov chain

$$X_{L_{d,v}} \to X_v \to X_\rho.$$

Due to our assumption on π and $P_{X'|X}$, we have $P_{X_\rho|X_v} = P_{X'|X}$ and $P_{X_v} = \pi$. By the definition of the contraction coefficient, we have

$$I(X_{L_{d,v}}; X_\rho) \le \eta_{\mathrm{KL}}(\pi, P_{X'|X}) I(X_{L_{d,v}}; X_v).$$

Observe that because $P_{X_v} = \pi$ and all edges have the same kernel, then $I(X_{L_{d,v}}; X_v) = I(X_{L_{d-1}}; X_\rho)$. This gives us the inequality

$$I(X_\rho; X_{L_d}) \le \eta_{\mathrm{KL}}(\pi, P_{X'|X}) b I(X_\rho; X_{L_{d-1}}),$$

which implies

$$I(X_\rho; X_{L_d}) \le (\eta_{\mathrm{KL}}(\pi, P_{X'|X})b)^d H(X_\rho).$$

Therefore if $\eta_{\mathrm{KL}}(\pi, P_{X'|X})b < 1$ then $I(X_\rho; X_{L_d}) \to 0$ exponentially fast as $d \to \infty$. □

Note that a weaker version of this theorem (non-reconstruction when $\eta_{\mathrm{KL}}(P_{X'|X})b \le 1$) is implied by the directed information percolation theorem. The k-coloring example (see below) demonstrates that this strengthening is essential; see [203] for details.

Example 33.9. Broadcasting on BSC tree

Consider a broadcasting problem on a b-ary tree with vertex alphabet $\mathcal{X} = \{0, 1\}$, edge kernel $P_{X'|X} = \mathsf{BSC}_\delta$, and $\pi = \mathrm{Ber}(1/2)$. Note that the uniform distribution is a reversible measure for BSC_δ. In Example 33.2, we calculated $\eta_{\mathrm{KL}}(\mathsf{BSC}_\delta) = (1 - 2\delta)^2$. Therefore, using Theorem 33.13, we can deduce that if

$$b(1 - 2\delta)^2 < 1,$$

then no inference algorithm can recover the root nodes better than random guessing as the depth of the tree goes to infinity. When $b(1 - 2\delta)^2 > 1$, non-trivial reconstruction is achieved by majority vote of the leaf values. (See also Exercise **VI.22**). This result is originally proved in [63].

Example 33.10. *k*-coloring on a tree

Given a b-ary tree, we assign a k-coloring $X_{v_{\mathrm{all}}}$ by sampling uniformly from the ensemble of all valid k-colorings. For this model, we can define a corresponding inference problem, namely given all the colors of the leaves at a certain depth, that is, X_{L_d}, determine the color of the root node, that is, X_ρ.

This problem can be modeled as a broadcasting problem on a tree where the root distribution π is given by the uniform distribution on k colors, and the edge kernel $P_{X'|X}$ is defined as

$$P_{X'|X}(a|b) = \begin{cases} \frac{1}{k-1}, & a \neq b, \\ 0, & a = b. \end{cases}$$

It can be shown (see Exercise **VI.24**) that $\eta_{\mathrm{KL}}(\mathrm{Unif}, P_{X'|X}) = \frac{1}{k \log k(1+o(1))}$ for large k. By Theorem 33.13, this implies that if $b < k \log k(1 + o(1))$ then reliable reconstruction of the root node is not possible. This result is originally proved in [394] and [50].

The other direction $b > k \log k(1 + o(1))$ can be shown by observing that if $b > k \log k(1 + o(1))$ then the probability of the children of a node taking all available colors (except its own) is close to 1. Thus, an inference algorithm can always determine the color of a node by finding a color that is not assigned to any of its children. Similarly, when $b > (1+\epsilon)k \log k$ even observing $(1-\epsilon)$ fraction of the node's children is sufficient to reconstruct its color exactly. Proceeding recursively from the bottom up, such a reconstruction algorithm will succeed with high probability. In this regime with positive probability (over the leaf variables) the posterior distribution of the root color is a point mass (deterministic). This effect is known as "freezing" of the root given the boundary.

We may also consider another reconstruction method which simply computes the majority of the leaves, that is, $\hat{X}_\rho = j$ for the color j that appears the most among the leaves. This method gives success probability strictly above $\frac{1}{k}$ if and only if $d > (k-1)^2$, by a famous result of Kesten and Stigum [245]. While the threshold is suboptimal, the method is quite *robust* in the sense that it also works if we only have access to a small fraction ϵ of the leaves (and the rest are replaced by erasures).

Let us now consider $\eta_{\chi^2}(\mathrm{Unif}, P_{X'|X})$. The transition matrix is symmetric with eigenvalues $\{1, -\frac{1}{k-1}\}$ and thus from Theorem 33.12 we have that

$$\eta_{\chi^2}(\mathrm{Unif}, P_{X'|X}) = \frac{1}{(k-1)^2} \ll \eta_{\mathrm{KL}}(\mathrm{Unif}, P_{X'|X}) = \frac{1}{k \log k(1+o(1))}.$$

Thus if Theorem 33.13 could be shown with I_{χ^2} instead of I_{KL} we would be able to show non-reconstruction for $d < (k-1)^2$, contradicting the result of the previous paragraph. What goes wrong is that I_{χ^2} fails to be subadditive, see (7.47). However, it is locally subadditive (when e.g. $I_{\chi^2}(X;A) \ll 1$) by [202, Lemma 26]. Thus, the argument in Theorem 33.13 can be repeated for the case where the leaves are observed through a very noisy channel (for example, an erasure channel leaving only ϵ fraction of the leaves). Consequently, the robust reconstruction threshold for coloring exactly equals $d = (k-1)^2$. See [228] for more on robust reconstruction thresholds.

33.6 Application: Distributed Correlation Estimation

As an application of the SDPI and its tensorization property, consider the problem of correlation estimation under communication constraints. Suppose Alice observes $\{X_i\}_{i\geq 1} \overset{\text{iid}}{\sim} \mathrm{Ber}(1/2)$ and Bob observes $\{Y_i\}_{i\geq 1} \overset{\text{iid}}{\sim} \mathrm{Ber}(1/2)$ such that the (X_i, Y_i) are iid with $\mathbb{E}[X_i Y_i] = \rho \in [-1,1]$. The goal is for Bob to send W to Alice with $H(W) = B$ bits and for Alice to estimate $\hat{\rho} = \hat{\rho}(X^\infty, W)$ with minimax squared error:

$$R^*(B) = \inf_{W,\hat{\rho}} \sup_{\rho} \mathbb{E}[(\rho - \hat{\rho})^2].$$

Notice that in this problem we are not sample-limited (each party has infinitely many observations), but communication-limited (only B bits can be exchanged).

Here is a trivial attempt to solve it. Notice that if Bob sends $W = (Y_1, \ldots, Y_B)$ then the optimal estimator is $\hat{\rho}(X^\infty, W) = \frac{1}{B}\sum_{i=1}^B X_i Y_i$ which has minimax error $\frac{1}{B}$, hence $R^*(B) \leq \frac{1}{B}$. Surprisingly, this can be improved.

Theorem 33.14. ([207]) *The optimal rate when $B \to \infty$ is given by*

$$R^*(X^\infty, W) = \frac{1+o(1)}{2\ln 2} \cdot \frac{1}{B}.$$

Proof. Fixing $P_{W|Y^\infty}$, we get the following decomposition:

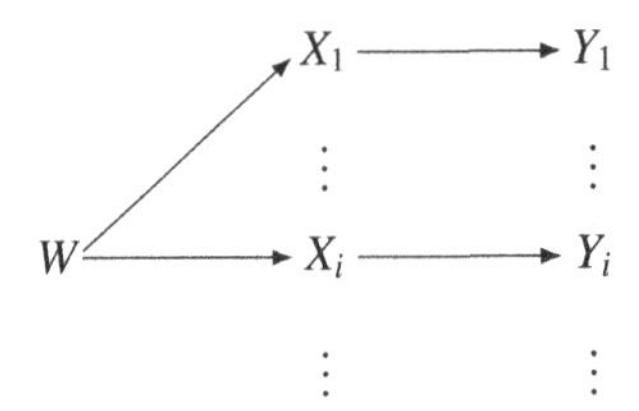

Note that once the messages W are fixed we have a parameter estimation problem $\{Q_\rho : \rho \in [-1,1]\}$ where Q_ρ is a distribution of (X^∞, W) when A^∞, B^∞ are ρ-correlated. Since we minimize mean-squared error, we know from the van Trees inequality (Theorem 29.2)[2] that $R^*(B) \geq \frac{1+o(1)}{\min_\rho J_F(\rho)} \geq \frac{1+o(1)}{J_F(0)}$ where $J_F(\rho)$ is the Fisher information of the family $\{Q_\rho\}$.

Recall that we also know from the local approximation (Section 2.6.2*) that

$$D(Q_\rho \| Q_0) = \frac{\rho^2 \log e}{2} J_F(0) + o(\rho^2).$$

Furthermore, notice that under $\rho = 0$ we have X^∞ and W independent and thus

$$\begin{aligned} D(Q_\rho \| Q_0) &= D(P^\rho_{X^\infty,W} \| P^0_{X^\infty,W}) \\ &= D(P^\rho_{X^\infty,W} \| P^\rho_{X^\infty} \times P^\rho_W) \\ &= I(W;X^\infty) \\ &\leq \rho^2 I(W;Y^\infty) \\ &\leq \rho^2 B \log 2, \end{aligned}$$

where the penultimate inequality follows from Example 33.7 and Proposition 33.11. Hence $J_F(0) \leq (2\ln 2)B + o(1)$ which in turns implies the theorem. For full details and the extension to interactive communication between Alice and Bob, see [207].

We turn to the upper bound next. First, notice that by taking blocks of $m \to \infty$ consecutive bits and setting $\tilde{X}_i = \frac{1}{\sqrt{m}} \sum_{j=(i-1)m}^{im-1} X_j$ and similarly for $\tilde{Y}_i$, Alice and Bob can replace ρ-correlated bits with ρ-correlated standard Gaussians $(\tilde{X}_i, \tilde{Y}_i) \overset{\text{iid}}{\sim} \mathcal{N}(0, \begin{pmatrix} 1 & \rho \\ \rho & 1 \end{pmatrix})$. Next, fix some very large N and let

$$W = \operatorname*{argmax}_{1 \leq j \leq N} Y_j.$$

(See Exercise **V.16** for a motivation behind this idea.) From standard concentration results we know that $\mathbb{E}[Y_W] = \sqrt{2 \ln N}(1 + o(1))$ (Lemma 27.10) and $\mathrm{Var}[Y_W] = O(\frac{1}{\ln N})$. Therefore, knowing W Alice can estimate

$$\hat{\rho} = \frac{X_W}{\mathbb{E}[Y_W]}.$$

This is an unbiased estimator and $\mathrm{Var}_\rho[\hat{\rho}] = \frac{1-\rho^2+o(1)}{2\ln N}$. Finally, setting $N = 2^B$ completes the argument. □

[2] This requires some technical justification about smoothness of the Fisher information $J_F(\rho)$.

33.7 Channel Comparison: Degradation, Less Noisy, More Capable

It turns out that the η_{KL} coefficient is intimately related to the concept of less noisy partial order on channels. We define several such partial orders together.

Definition 33.15. (Partial orders on channels) Let $P_{Y|X}$ and $P_{Z|X}$ be two channels.

- We say that $P_{Y|X}$ is a *degradation* of $P_{Z|X}$, denoted by $P_{Y|X} \leq_{\mathrm{deg}} P_{Z|X}$, if there exists $P_{Y|Z}$ such that $P_{Y|X} = P_{Y|Z} \circ P_{Z|X}$.
- We say that $P_{Z|X}$ is *less noisy* than $P_{Y|X}$, denoted by $P_{Y|X} \leq_{\mathrm{ln}} P_{Z|X}$, if for every $P_{U,X}$ on the following Markov chain

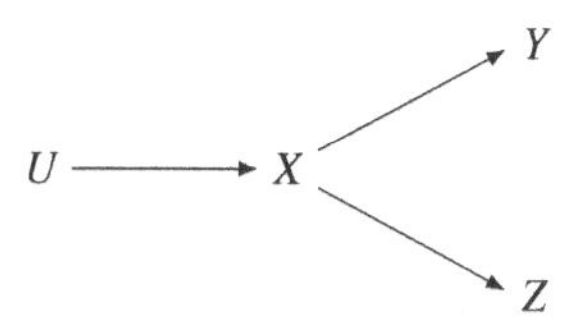

  we have $I(U;Y) \leq I(U;Z)$.
- We say that $P_{Z|X}$ is *more capable* than $P_{Y|X}$, denoted $P_{Y|X} \leq_{\mathrm{mc}} P_{Z|X}$, if for any P_X we have $I(X;Y) \leq I(X;Z)$.

We make some remarks on these definitions and refer to [346] for proofs:

- $P_{Y|X} \leq_{\mathrm{deg}} P_{Z|X} \implies P_{Y|X} \leq_{\mathrm{ln}} P_{Z|X} \implies P_{Y|X} \leq_{\mathrm{mc}} P_{Z|X}$. Counterexamples for reverse implications can be found in [111, Problem 15.11].
- The less noisy relation can be defined equivalently in terms of the divergence, namely $P_{Y|X} \leq_{\mathrm{ln}} P_{Z|X}$ if and only if for all P_X, Q_X we have $D(Q_Y\|P_Y) \leq D(Q_Z\|P_Z)$. We refer to [291, Sections I.B, II.A] and [346, Section 6] for alternative useful characterizations of the less noisy order.
- For BMS channels (see Section 19.4*) it turns out that among all channels with a given $I_{\chi^2}(X;Y) = \eta$ (with $X \sim \mathrm{Ber}(1/2)$) the BSC and BEC are the minimal and maximal elements in the poset of $\leq_{\mathrm{ln}}$; see Exercise **VI.21** for details.

Proposition 33.16. *$\eta_{\mathrm{KL}}(P_{Y|X}) \leq 1 - \tau$ if and only if $P_{Y|X} \leq_{\mathrm{ln}} \mathsf{EC}_\tau$, where EC_τ was defined in Example 33.6.*

Proof. For EC_τ we always have

$$I(U;Z) = (1-\tau)I(U;X).$$

By the mutual information characterization of η_{KL} we have

$$I(U;Y) \leq (1-\tau)I(U;X).$$

Combining these two inequalities gives us

$$I(U;Y) \leq I(U;Z).$$

□

This proposition gives us an intuitive interpretation of contraction coefficient as the worst erasure channel that still dominates the channel.

Proposition 33.17. (Tensorization of less noisy and more capable) *If for all* $i \in [n], P_{Y_i|X_i} \leq_{\text{ln}} P_{Z_i|X_i}$, *then* $\bigotimes_{i\in[n]} P_{Y_i|X_i} \leq_{\text{ln}} \bigotimes_{i\in[n]} P_{Z_i|X_i}$.[3] *If for all* $i \in [n], P_{Y_i|X_i} \leq_{\text{mc}} P_{Z_i|X_i}$, *then* $\bigotimes_{i\in[n]} P_{Y_i|X_i} \leq_{\text{mc}} \bigotimes_{i\in[n]} P_{Z_i|X_i}$.

Proof. By induction it is sufficient to consider $n = 2$ only. Consider the following directed graphical model (cf. Section 3.4):

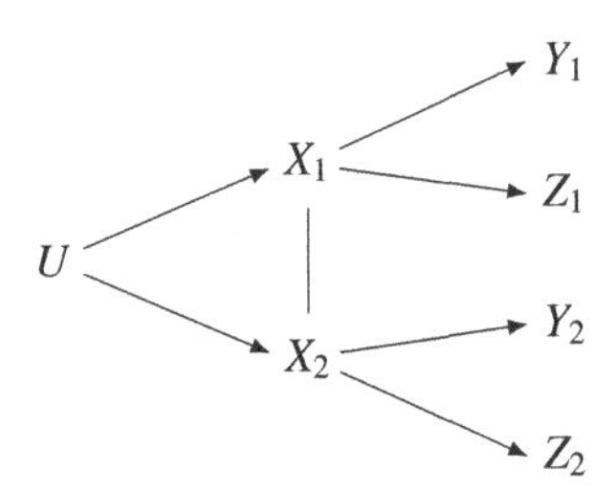

We have the following inequalities:

$$\begin{aligned} I(U;Y_1,Y_2) &= I(U;Y_1) + I(U;Y_2|Y_1) \\ &\leq I(U;Y_1) + I(U;Z_2|Y_1) \\ &= I(U;Y_1,Z_2). \end{aligned}$$

Hence $I(U;Y_1,Y_2) \leq I(U;Y_1,Z_2)$ for any $P_{X_1,X_2,U}$. Applying the same argument replacing Y_1 with Z_1 we get $I(U;Y_1,Z_2) \leq I(U;Z_1,Z_2)$, completing the proof.

For the second claim, notice that

$$\begin{aligned} I(X^2;Y^2) &= I(X_2;Y^2) + I(X_1;Y^2|X_2) \\ &\overset{(a)}{=} I(X_2;Y_1) + I(X_2;Y_2|Y_1) + I(X_1;Y_1|X_2) \\ &\leq I(X_2;Y_1) + I(X_2;Z_2|Y_1) + I(X_1;Z_1|X_2) \\ &\overset{(b)}{=} I(X_2;Y_1,Z_2) + I(X_1;Z^2|X_2) \\ &= I(X_2;Z_2) + I(X_2;Y_1|Z_2) + I(X_1;Z^2|X_2) \\ &\leq I(X_2;Z_2) + I(X_2;Z_1|Z_2) + I(X_1;Z^2|X_2) = I(X^2;Z^2), \end{aligned}$$

where the equalities are just applications of the chain rule (and in (a) and (b) we also notice that conditioned on X_2 the Y_2 or Z_2 are non-informative) and both inequalities are applications of the most capable relation to the conditional distributions. For example, for every y we have $I(X_2;Y_2|Y_1 = y) \leq I(X_2;Z_2|Y_1 = y)$ and hence we can average over $y \sim P_{Y_1}$. □

[3] We remind the reader that $\bigotimes P_{Y_i|X_i}$ refers to the product (memoryless) channel with $x^n \mapsto Y^n \sim \prod_i P_{Y_i|X_i=x_i}$.

33.8 Undirected Information Percolation

In this section we study the problem of inference on an undirected graph. Consider an undirected graph $G = (\mathcal{V}, \mathcal{E})$. We assign a random variable X_v on the alphabet $\mathcal{X}$ to each vertex v. For each $e = (u, v) \in \mathcal{E}$, we assign Y_e sampled according to the kernel $P_{Y_e|X_e}$ with $X_e = (X_u, X_v)$. The goal is to estimate the value of X_v's given the value of Y_e's. As a visual illustration consider the following graph:

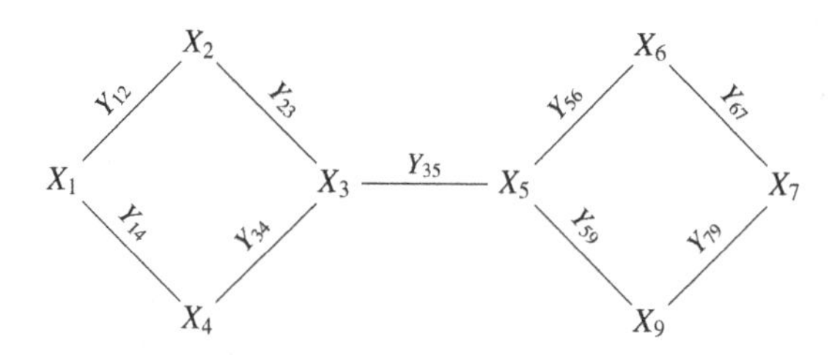

Example 33.11. Community detection

Consider a complete graph with n vertices, that is, K_n, and the random variables X_v representing the membership of each vertex to one of the m communities. We assume that X_v is sampled uniformly from $[m]$ and independent of the other vertices. The observation $Y_{u,v}$ is defined as

$$Y_{uv} \sim \begin{cases} \text{Ber}(a/n), & X_u = X_v, \\ \text{Ber}(b/n), & X_u \neq X_v. \end{cases}$$

This is the the stochastic block model with m communities (see Exercise **I.49** for $m = 2$.)

Example 33.12. $\mathbb{Z}_2$ synchronization

For any graph G, we sample X_v uniformly from $\{-1, +1\}$ and $Y = \text{BSC}_\delta(X_u X_v)$ if u and v are connected in G.

Example 33.13. Spiked Wigner model

We consider the problem of estimating the value of vector $(X_i)_{i\in[n]}$ given the observation $(Y_{ij})_{i,j\in[n],i\leq j}$. The X_i's and Y_{ij}'s are related by

$$Y_{ij} = \sqrt{\frac{\lambda}{n}} X_i X_j + W_{ij},$$

where $\mathbf{X} = (X_1, \ldots, X_n)^\top$ is sampled uniformly from $\{\pm 1\}^n$ and $W_{i,j} = W_{j,i} \overset{\text{iid}}{\sim} \mathcal{N}(0, 1)$, so that $\mathbf{W}$ forms a Wigner matrix (symmetric Gaussian matrix). This model can also be written in matrix form as

$$\mathbf{Y} = \sqrt{\frac{\lambda}{n}} \mathbf{X}\mathbf{X}^\top + \mathbf{W}$$

as a rank-one perturbation of a Wigner matrix $\mathbf{W}$, hence the name of the model. It is used as a probabilistic model for principal component analysis.

This problem can also be treated as a problem of inference on an undirected graph. In this case, the underlying graph is a complete graph, and we assign X_i to the ith vertex. Under this model, the edge observation is given by $Y_{ij} = \mathsf{BIAWGN}_{\lambda/n}(X_i X_j)$, see Example 3.4.

Although seemingly different, these problems share the following common characteristics, namely:

Assumption 33.1.

- Each X_v is uniformly distributed.
- Defining an auxiliary random variable $B = 1\{X_u \neq X_v\}$ for any edge $e = (u, v)$, the following Markov chain holds:

$$(X_u, X_v) \to B \to Y_e.$$

In other words, the observation on each edge only depends on whether the random variables on its endpoints are equal.

Due to Assumption 33.1, the reconstructed X_v's are symmetric up to any permutation on $\mathcal{X}$. In the case of alphabet $\mathcal{X} = \{-1, +1\}$, this implies that for any realization σ then $P_{X_{\text{all}}|Y_{\text{all}}}(\sigma|b) = P_{X_{\text{all}}|Y_{\text{all}}}(-\sigma|b)$. Consequently, our reconstruction metric also needs to accommodate this symmetry. For $\mathcal{X} = \{-1, +1\}$, this leads to the use of $\frac{1}{n}|\sum_{i=1}^{n} X_i \hat{X}_i|$ as our reconstruction metric (overlap).

Our main theorem for the undirected inference problem can be seen as the analog of the information percolation theorem for DAG (Theorem 33.8). However, instead of controlling the contraction coefficient, the percolation probability is used to directly control the conditional mutual information between any subsets of vertices in the graph.

Before stating our main theorem, we will need to define the corresponding percolation model on an undirected graph. For any undirected graph $G = (\mathcal{V}, \mathcal{E})$ we define a percolation model on this graph as follows:

- Every edge $e \in \mathcal{E}$ is open with probability $\eta_{\mathrm{KL}}(P_{Y_e|X_e})$, independent of the other edges.
- For any $v \in \mathcal{V}$ and $S \subset \mathcal{V}$, we define $v \leftrightarrow S$ as the event that there exists an open path from v to any vertex in S.
- For any $S_1, S_2 \subset \mathcal{V}$, we define the function $\mathrm{perc}_u(S_1, S_2)$ as

$$\mathrm{perc}_u(S_1, S_2) \triangleq \sum_{v \in S_1} P(v \leftrightarrow S_2).$$

Notice that this function is different from the percolation function for information percolation in DAG (Theorem 33.8). Most importantly, this function is not equivalent to the exact percolation probability. Instead, it is an upper bound on the percolation probability by union bounding with respect to S_1. Hence, it is natural that this function is not symmetric, that is, $\mathrm{perc}_u(S_1, S_2) \neq \mathrm{perc}_u(S_2, S_1)$.

Theorem 33.18. (Undirected information percolation [348]) *Consider an inference problem on undirected graph $G = (\mathcal{V}, \mathcal{E})$. For any $S_1, S_2 \subset \mathcal{V}$, the following holds:*

$$I(X_{S_1};X_{S_2}|Y) \le \mathrm{perc}_u(S_1, S_2) \log |\mathcal{X}|.$$

Instead of proving Theorem 33.18 in its full generality, we will prove the theorem under Assumption 33.1. The main step of the proof utilizes the fact that we can upper-bound the mutual information of any channel by its less noisy counterpart.

Theorem 33.19. *Consider the problem of inference on undirected graph $G = (\mathcal{V}, \mathcal{E})$ with $X_1, \dots, X_n$ not necessarily independent. If $P_{Y_e|X_e} \le_{\mathrm{ln}} P_{Z_e|X_e}$ for every $e \in \mathcal{E}$, then for any $S_1, S_2 \subset \mathcal{V}$ and $E \subset \mathcal{E}$*

$$I(X_{S_1};Y_E|X_{S_2}) \le I(X_{S_1};Z_E|X_{S_2}).$$

Proof. From our assumption and the tensorization property of less noisy ordering (Proposition 33.17), we have $P_{Y_E|X_{S_1},X_{S_2}} \le_{\mathrm{ln}} P_{Z_E|X_{S_1},X_{S_2}}$. This implies that for σ as a valid realization of X_{S_2} we will have

$$\begin{aligned} I(X_{S_1};Y_E|X_{S_2} = \sigma) &= I(X_{S_1}, X_{S_2};Y_E|X_{S_2} = \sigma) \le I(X_{S_1}, X_{S_2};Z_E|X_{S_2} = \sigma) \\ &= I(X_{S_1};Z_E|X_{S_2} = \sigma). \end{aligned}$$

As this inequality holds for all realization of X_{S_2}, then the following inequality also holds

$$I(X_{S_1};Y_E|X_{S_2}) \le I(X_{S_1};Z_E|X_{S_2}).$$

□

Proof of Theorem 33.18. We only give a proof under Assumption 33.1 and only for the case $S_1 = \{i\}$. For the full proof (that proceeds by induction and does not leverage the less noisy idea), see [348]. We have the equalities

$$I(X_i;X_{S_2}|Y_E) = I(X_i;X_{S_2}, Y_E) = I(X_i;Y_E|X_{S_2}), \tag{33.22}$$

where the first inequality is due to the fact $B_E \perp\!\!\!\perp X_i$ under Assumption 33.1, and the second inequality is due to $X_i \perp\!\!\!\perp X_{S_2}$ under Assumption 33.1.

Due to our previous result, if $\eta_{\mathrm{KL}}(P_{Y_e|X_e}) = 1 - \tau$ then $P_{Y_e|X_e} \le_{\mathrm{ln}} P_{Z_e|X_e}$ where $P_{Z_e|X_e} = \mathsf{EC}_\tau$. By the tensorization property, this ordering also holds for the channel $P_{Y_E|X_E}$, thus we have

$$I(X_i;Y_E|X_{S_2}) \le I(X_j;Z_E|X_{S_2}).$$

Let us define another auxiliary random variable $D = 1\{i \leftrightarrow S_2\}$, namely the indicator that there is an open path from i to S_2. Notice that D is fully determined by Z_E. By the same argument as in (33.22), we have

$$\begin{aligned}
I(X_i;Z_E|X_{S_2}) &= I(X_i;X_{S_2}|Z_E) \\
&= I(X_i;X_{S_2}|Z_E, D) \\
&= (1 - \mathbb{P}[i \leftrightarrow S_2])\, \underbrace{I(X_i;X_{S_2}|Z_E, D=0)}_{0} \\
&\quad + \mathbb{P}[i \leftrightarrow S_2]\, \underbrace{I(X_i;X_{S_2}|Z_E, D=1)}_{\leq \log|\mathcal{X}|} \\
&\leq \mathbb{P}[i \leftrightarrow S_2] \log|\mathcal{X}| \\
&= \mathrm{perc}_u(i, S_2) \log|\mathcal{X}|. \qquad \square
\end{aligned}$$

33.9 Application: Spiked Wigner Model

As an application we show how the undirected information percolation concept allows us to derive a converse result for the spiked Wigner model, which we described in Example 33.13. To restate the problem, we are given an observation

$$\mathbf{Y} = \sqrt{\frac{\lambda}{n}}\mathbf{X}\mathbf{X}^\top + \mathbf{W},$$

where $\mathbf{W}$ is a Gaussian Wigner matrix and $\mathbf{X} = (X_1, \ldots, X_n)^\top$ consists of iid uniform ± 1 entries. As in many modern problems (see Section 30.4) our goal here is not to recover $\mathbf{X}$ perfectly, but only to outperform random guess ("weak recovery"). That is, we will be content with finding an estimator with non-trivial overlap:

$$\frac{1}{n}\mathbb{E}\left[\left|\sum_{i=1}^{n} X_i \hat{X}_i\right|\right] \geq \epsilon_0 > 0 \tag{33.23}$$

as $n \to \infty$. Now, it turns out that in such spiked models there is a so-called *BBP phase transition*, first discovered in [29, 327]. Specifically, the eigenvalues of $\frac{1}{\sqrt{n}}\mathbf{W}$ are well known to follow *Wigner's semicircle law* supported on the interval $(-2, 2)$. At the same time the rank-one matrix $\frac{\sqrt{\lambda}}{n}\mathbf{X}\mathbf{X}^\top$ has only one non-zero eigenvalue equal to $\sqrt{\lambda}$. It turns out that for $\lambda < 1$ the effect of this "spike" is undetectable and the spectrum of $\mathbf{Y}/\sqrt{n}$ is unaffected. For $\lambda > 1$ the top eigenvalue of $\mathbf{Y}/\sqrt{n}$ moves above the edge of the semicircle law to $\lambda + \frac{1}{\lambda} > 2$. Furthermore, computing the top eigenvector and taking the sign of its coordinates achieves a correlated recovery of the true $\mathbf{X}$ in the sense of (33.23). Note, however, that inability to change the spectrum (when $\lambda < 1$) does not imply that reconstruction of $\mathbf{X}$ is impossible by other means. Nevertheless, in this section, we will show that indeed for $\lambda \leq 1$ no method can achieve (33.23). Thus, together with the mentioned spectral algorithm for $\lambda > 1$ we may conclude that $\lambda^* = 1$ is the critical threshold separating the two phases of the problem.

Theorem 33.20. *Consider the spiked Wigner model. If* $\lambda \le 1$, *then for any sequence of estimators* $\hat{X}^n(Y)$,

$$\frac{1}{n}\mathbb{E}\left[\left|\sum_{i=1}^{n} X_i\hat{X}_i\right|\right] \to 0 \tag{33.24}$$

as $n \to \infty$.

Proof. By Cauchy–Schwarz, for (33.24) it suffices to show

$$\sum_{i\neq j}\mathbb{E}[X_iX_j\hat{X}_i\hat{X}_j] = o(n^2).$$

Next, it is clear that we can simplify the task of maximizing (over $\hat{X}^n$) by allowing separate estimation of each X_iX_j by $\hat{T}_{i,j}$, that is,

$$\max_{\hat{X}^n}\sum_{i\neq j}\mathbb{E}[X_iX_j\hat{X}_i\hat{X}_j] \le \sum_{i\neq j}\max_{\hat{T}_{i,j}}\mathbb{E}[X_iX_j\hat{T}_{i,j}].$$

The latter maximization is solved by the MAP decoder:

$$\hat{T}_{i,j}(\mathbf{Y}) = \arg\max_{\sigma\in\{\pm 1\}}\mathbb{P}[X_iX_j = \sigma\,|\mathbf{Y}].$$

Since each $X_i \sim \mathrm{Ber}(1/2)$ it is easy to see that

$$I(X_i;X_j|\mathbf{Y}) \to 0 \quad\iff\quad \max_{\hat{T}_{i,j}}\mathbb{E}[X_iX_j\hat{T}_{i,j}] \to 0.$$

(For example, we may notice $I(X_i;X_j|\mathbf{Y}) = I(X_i,X_j;\mathbf{Y}) \ge I(X_iX_j;\mathbf{Y})$ and apply Fano's inequality.) Thus, from the symmetry of the problem it is sufficient to prove $I(X_1;X_2|Y) \to 0$ as $n \to \infty$.

By using the undirected information percolation theorem, we have

$$I(X_2;X_1|\mathbf{Y}) \le \mathrm{perc}_u(\{1\},\{2\}).$$

Now, to compute perc_u we need to compute the probability of having an open edge, which in our case simply equals $\eta_{\mathrm{KL}}(\mathsf{BIAWGN}_{\lambda/n})$. From Theorem 33.6 we know the latter equals $I_{\chi^2}(X;Y)$ where $X \sim \mathrm{Ber}(1/2)$ and $Y = \mathsf{BIAWGN}_{\lambda/n}(X)$. A short computation shows thus

$$\eta_{\mathrm{KL}}(\mathsf{BIAWGN}_{\lambda/n}) = \frac{\lambda}{n}(1+o(1)).$$

Suppose that $\lambda < 1$. In this case, we can overbound $\frac{\lambda+o(1)}{n}$ by $\frac{\lambda'}{n}$ with $\lambda' < 1$. The percolation random graph then is equivalent to the Erdös–Rényi random graph with n vertices and λ'/n edge probability, that is, $\mathrm{ER}(n,\lambda'/n)$. Using this observation, the inequality can be rewritten as

$$I(X_2;X_1|\mathbf{Y}) \le P(\text{vertices 1 and 2 are connected in } \mathrm{ER}(n,\lambda'/n)).$$

A classical result in random graph theory is that the largest connected component in $\mathrm{ER}(n,\lambda'/n)$ contains $O(\log n)$ vertices if $\lambda' < 1$ [154]. This implies that the probability that two specific vertices are connected is $o(1)$, hence $I(X_2;X_1|\mathbf{Y}) \to 0$ as $n \to \infty$.

To treat the critical case of $\lambda = 1$ we need slightly more refined information about $\eta_{\mathrm{KL}}(\mathsf{BIAWGN}_{\lambda/n})$ and about the behavior of the giant component of $\mathrm{ER}(n, \frac{1+o(1)}{n})$ graph; see [348] for full details. □

Remark 33.2. (Dense–sparse correspondence) The proof above changes the underlying structure of the graph. Namely, instead of dealing with a complete graph, the information percolation method replaced it with an Erdös–Rényi random graph. Moreover, if η_{KL} is small enough, then the underlying percolation graph tends to be very sparse and has a locally tree-like structure. This demonstrates a ubiquitous and actively studied effect in modern statistics: dense inference (such as spiked Wigner model, sparse regression, sparse PCA, etc.) with very weak signals ($\eta_{\mathrm{KL}} \approx 0$) is similar to sparse inference (broadcasting on trees) with moderate signals ($\eta_{\mathrm{KL}} \in (\epsilon, 1-\epsilon)$). The information percolation method provides a certain bridge between these two worlds, perhaps partially explaining why the results in these two worlds often parallel one another. (For example, results on optimality and phase transitions for belief propagation (sparse inference) often parallel those for approximate message passing (AMP, dense inference).) We do want to caution, however, that the reduction given by the information percolation method is not generally tight (spiked Wigner being a lucky exception). For example [348], for correlated recovery in the stochastic block model with k communities and edge probability a/n and b/n it yields an impossibility result $(\sqrt{a} - \sqrt{b})^2 \le \frac{k}{2}$, weaker than the best known upper bounds of [203].

33.10 Strong Data-Post-Processing Inequality (Post-SDPI)

For applications in distributed estimation the following version of the SDPI is useful.

Definition 33.21. (Post-SDPI constant) Given a conditional distribution $P_{Y|X}$, define the input-dependent and input-free contraction coefficients respectively as

$$\eta_{\mathrm{KL}}^{(p)}(P_X, P_{Y|X}) = \sup_{P_{U|Y}} \left\{ \frac{I(U;X)}{I(U;Y)} : X \to Y \to U \right\}$$

and

$$\eta_{\mathrm{KL}}^{(p)}(P_{Y|X}) = \sup_{P_X, P_{U|Y}} \left\{ \frac{I(U;X)}{I(U;Y)} : X \to Y \to U \right\}.$$

To get characterization in terms of KL divergence we simply notice that

$$\eta_{\mathrm{KL}}^{(p)}(P_X, P_{Y|X}) = \eta_{\mathrm{KL}}(P_Y, P_{X|Y}), \tag{33.25}$$

$$\eta_{\mathrm{KL}}^{(p)}(P_{Y|X}) = \sup_{P_X} \eta_{\mathrm{KL}}(P_Y, P_{X|Y}), \tag{33.26}$$

where $P_Y = P_{Y|X} \circ P_X$ and $P_{X|Y}$ is the conditional distribution corresponding to $P_X P_{Y|X}$. From (33.25) and Proposition 33.11 we also get the tensorization property for input-dependent post-SDPI:

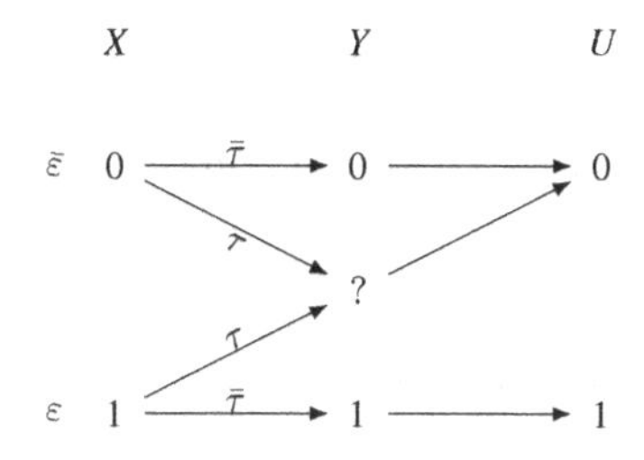

Figure 33.6 Post-SDPI coefficient of BEC equal to 1.

$$\eta_{\mathrm{KL}}^{(p)}(P_X^{\otimes n}, (P_{Y|X})^{\otimes n}) = \eta_{\mathrm{KL}}^{(p)}(P_X, P_{Y|X}). \tag{33.27}$$

It is easy to see that by the data-processing inequality, $\eta_{\mathrm{KL}}^{(p)}(P_{Y|X}) \leq 1$. Unlike the η_{KL} coefficient, $\eta_{\mathrm{KL}}^{(p)}$ can equal to 1 even for a noisy channel $P_{Y|X}$.

Example 33.14. $\eta_{\mathrm{KL}}^{(p)} = 1$ for erasure channels

Let $P_{Y|X} = \mathsf{BEC}_\tau$ and $X \to Y \to U$ be defined as on Figure 33.6. Then we can compute $I(Y;U) = H(U) = h(\varepsilon\bar{\tau})$ and $I(X;U) = H(U) - H(U|X) = h(\varepsilon\bar{\tau}) - \varepsilon h(\tau)$ hence

$$\begin{aligned}\eta_{\mathrm{KL}}^{(p)}(P_{Y|X}) &\geq \frac{I(X;U)}{I(Y;U)} \\ &= 1 - h(\tau)\frac{\varepsilon}{h(\varepsilon\bar{\tau})}.\end{aligned}$$

This last term tends to 1 when ε tends to 0 hence

$$\eta_{\mathrm{KL}}^{(p)}(\mathsf{BEC}_\tau) = 1,$$

even though Y is not a one-to-one function of X.

Note that this example also shows that even for an input-constrained version of $\eta_{\mathrm{KL}}^{(p)}$ the natural conjecture $\eta_{\mathrm{KL}}^{(p)}(\mathrm{Unif}, \mathrm{BMS}) = \eta_{\mathrm{KL}}(\mathrm{BMS})$ is *incorrect*. Indeed, by taking $\varepsilon = \frac{1}{2}$, we have that $\eta_{\mathrm{KL}}^{(p)}(\mathrm{Unif}, \mathsf{BEC}_\tau) > 1 - \tau$ for $\tau \to 1$.

Nevertheless, the post-SDPI constant is often non-trivial, most importantly for the BSC:

Theorem 33.22.

$$\eta_{\mathrm{KL}}^{(p)}(\mathsf{BSC}_\delta) = (1 - 2\delta)^2.$$

To prove this theorem, the following lemma is useful.

Lemma 33.23. *Consider a pair of binary random variable X, Y such that for all $x, y \in \{0, 1\}$ we have*

$$P_{X,Y}(x, y) = f(x)\left(\frac{\delta}{1-\delta}\right)^{1\{x \neq y\}} g(y)$$

for some functions f and g. Then $\eta_{\mathrm{KL}}(P_{Y|X}) \leq (1 - 2\delta)^2$.

Proof. From (33.6) we know that for binary-input channels

$$\eta_{\mathrm{KL}}(P_{Y|X}) \le H^2(P_{Y|X=0}, P_{Y|X=1}) - \frac{H^4(P_{Y|X=0}, P_{Y|X=1})}{4}.$$

If we let $\phi = \frac{g(0)}{g(1)}$, then we have $P_{Y|X=0} = \mathrm{Ber}\left(\frac{\lambda}{\phi+\lambda}\right)$ and $P_{Y|X=1} = \mathrm{Ber}\left(\frac{1}{1+\phi\lambda}\right)$ and a simple computation shows that

$$\max_{\phi} H^2(P_{Y|X=0}, P_{Y|X=1}) - \frac{H^4(P_{Y|X=0}, P_{Y|X=1})}{4} \overset{\phi=1}{=} (1-2\delta)^2.$$

Now observe that $P_{X,Y}$ in Theorem 33.22 satisfies the property of the lemma with X and Y exchanged, hence $\eta_{\mathrm{KL}}(P_Y, P_{X|Y}) \le (1-2\delta)^2$ which implies that $\eta_{\mathrm{KL}}^{(p)}(P_{Y|X}) = \sup_{P_X} \eta_{\mathrm{KL}}(P_Y, P_{X|Y}) \le (1-2\delta)^2$ with equality if P_X is uniform. □

Theorem 33.24. (Post-SDPI for BI-AWGN) *Let $0 \le \epsilon \le 1$ and consider the channel $P_{Y|X}$ with $X \in \{\pm 1\}$ given by*

$$Y = \epsilon X + Z, \qquad Z \sim \mathcal{N}(0,1).$$

Then for any $\pi \in (0,1)$ taking $P_X = \mathrm{Ber}(\pi)$ we have for some absolute constant K the estimate

$$\eta_{\mathrm{KL}}^{(p)}(P_X, P_{Y|X}) \le K \frac{\epsilon^2}{\pi(1-\pi)}.$$

Proof. In this proof we assume all information measures are with respect to base e. First, notice that

$$v(y) \triangleq P[X=1|Y=y] = \frac{1}{1 + \frac{1-\pi}{\pi} e^{-2y\epsilon}}.$$

Then, the optimization defining $\eta_{\mathrm{KL}}^{(p)}$ can be written as

$$\eta_{\mathrm{KL}}^{(p)}(P_X, P_{Y|X}) \le \sup_{Q_Y} \frac{d(\mathbb{E}_{Q_Y}[v(Y)] \| \pi)}{D(Q_Y \| P_Y)}. \tag{33.28}$$

From (7.34) we have

$$\eta_{\mathrm{KL}}^{(p)}(P_X, P_{Y|X}) \le \frac{1}{\pi(1-\pi)} \sup_{Q_Y} \frac{(\mathbb{E}_{Q_Y}[v(Y)] - \pi)^2}{D(Q_Y \| P_Y)}. \tag{33.29}$$

To proceed, we need to introduce a new concept. The T_1-transportation inequality, first introduced by K. Marton, for the measure P_Y states the following: For every Q_Y we have for some $c = c(P_Y)$

$$W_1(Q_Y, P_Y) \le \sqrt{2cD(Q_Y \| P_Y)}, \tag{33.30}$$

where $W_1(Q_Y, P_Y)$ is the 1-Wasserstein distance defined as

$$\begin{aligned} W_1(Q_Y, P_Y) &= \sup\{\mathbb{E}_{Q_Y}[f] - \mathbb{E}_{P_Y}[f] \colon f \text{ 1-Lipschitz}\} \\ &= \inf\{\mathbb{E}[|A-B|] \colon A \sim Q_Y, B \sim P_Y\}. \end{aligned} \tag{33.31}$$

The constant $c(P_Y)$ in (33.30) has been characterized in [64, 125] in terms of the properties of P_Y. One such estimate is the following:

$$c(P_Y) \le \frac{2}{\delta} \sup_{k\ge 1} \left(\frac{G(\delta)}{\binom{2k}{k}} \right)^{1/k},$$

where $G(\delta) = \mathbb{E}[e^{\delta(Y-Y')^2}]$ where $Y, Y' \overset{\text{iid}}{\sim} P_Y$. Using the estimate $\binom{2k}{k} \ge \frac{4^k}{\sqrt{\pi(k+1/2)}}$ and the fact that $\frac{1}{k}\ln(k+1/2) \le \frac{1}{2}$ we get a further bound

$$c(P_Y) \le \frac{2}{\delta} G(\delta) \frac{\pi\sqrt{e}}{4} \le \frac{6G(\delta)}{\delta}.$$

Next notice that $Y - Y' \overset{\text{d}}{=} B_\epsilon + \sqrt{2}Z$ where $B_\epsilon \perp\!\!\!\perp Z \sim \mathcal{N}(0,1)$, B_ϵ is symmetric, and $|B_\epsilon| \le 2\epsilon$. Thus, we conclude that for any $\delta < 1/4$ we have $\bar{c} \triangleq \frac{6}{\delta}\sup_{\epsilon\le 1} G(\delta) < \infty$. In the end, we have inequality (33.30) with constant $c = \bar{c}$ that holds uniformly for all $0 \le \epsilon \le 1$.

Now, notice that $\left|\frac{d}{dy}v(y)\right| \le \frac{\epsilon}{2}$ and therefore v is $\frac{\epsilon}{2}$-Lipschitz. From (33.30) and (33.31) we obtain then

$$\left|\mathbb{E}_{Q_Y}[v(Y)] - \mathbb{E}_{P_Y}[v(Y)]\right| \le \frac{\epsilon}{2}\sqrt{2\bar{c}D(Q_Y\|P_Y)}.$$

Squaring this inequality and plugging back into (33.29) completes the proof. □

Remark 33.3. Notice that we can also compute the exact value of $\eta_{\mathrm{KL}}^{(p)}(P_X, P_{Y|X})$ by noticing the following. From (33.28) it is evident that among all measures Q_Y with a given value of $\mathbb{E}_{Q_Y}[v(Y)]$ we are interested in the one minimizing $D(Q_Y\|P_Y)$. From Theorem 15.11 we know that such Q_Y is given by $dQ_Y = e^{bv(y)-\psi_V(b)}dP_Y$, where $\psi_V(b) \triangleq \ln \mathbb{E}_{P_Y}[e^{bv(Y)}]$. Thus, by defining the convex dual $\psi_V^*(\lambda)$ we can get the exact value in terms of the following single-variable optimization:

$$\eta_{\mathrm{KL}}^{(p)}(P_X, P_{Y|X}) = \sup_{\lambda\in[0,1]} \frac{d(\lambda\|\pi)}{\psi_V^*(\lambda)}.$$

Numerically, for $\pi = 1/2$ it turns out that the optimal value is $\lambda \to \frac{1}{2}$, justifying our overbounding of d by χ^2, and surprisingly giving

$$\eta_{\mathrm{KL}}^{(p)}(\mathrm{Ber}(1/2), P_{Y|X}) = 4\,\mathbb{E}_{P_Y}[\tanh^2(\epsilon Y)] = \eta_{\mathrm{KL}}(P_{Y|X}),$$

where in the last equality we used Theorem 33.6(f)).

33.11 Application: Distributed Mean Estimation

We want to estimate $\theta \in [-1,1]^d$ and we have m machines observing $Y_i = \theta + \sigma Z_i$ where $Z_i \sim \mathcal{N}(0, I_d)$ independently for $i = 1, \ldots, m$. They can send a total of B bits to a remote estimator which produces $\hat{\theta}$ with the goal of minimizing the worst-case risk $\sup_\theta \mathbb{E}[\|\theta - \hat{\theta}\|^2]$. If we denote the messages by $U_i \in \mathcal{U}_i$, then we have the communication constraint $\sum_i \log_2 |\mathcal{U}_i| \le B$ and the diagram is the following:

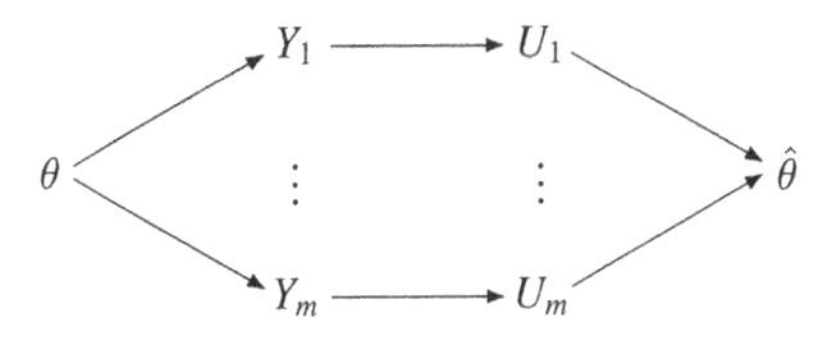

Finally, denote the minimax risk of estimation by

$$R^*(m, d, \sigma^2, B) = \inf_{U_1,\dots,U_m,\hat{\theta}} \sup_{\theta} \mathbb{E}[\|\theta - \hat{\theta}\|^2].$$

We begin with some simple observations:

- Without the constraint $\theta \in [-1, 1]^d$, we could take $\theta \sim \mathcal{N}(0, bI_d)$ and from rate distortion quickly conclude that estimating θ within risk R requires communicating at least $\frac{d}{2} \log \frac{bd}{R}$ bits, which diverges as $b \to \infty$. Thus, restricting the magnitude of θ is necessary in order for it to be estimable with finitely many bits communicated.
- Without communication constraint, it is easy to establish that $R^*(m, d, \sigma^2, \infty) = \mathbb{E}[\|\frac{\sigma}{m} \sum_i Z_i\|^2] = \frac{d\sigma^2}{m}$ by taking $U_i = Y_i$ and $\hat{\theta} = \frac{1}{m} \sum_i U_i$, which matches the minimax risk (28.17) in the non-distributed setting.
- In order to achieve a risk of order $\frac{d}{m}$ we can apply a crude quantizer as follows. Let $U_i = \mathrm{sign}(Y_i)$ (coordinatewise sign). This yields $B = md$ and it is easy to show that the achievable risk is $O_\sigma(\frac{d}{m})$. Indeed, notice that by taking $V = \frac{1}{m} \sum_{i=1}^m U_i$ we see that each coordinate V_j, $j \in [d]$, estimates (within $O_p(\frac{1}{\sqrt{m}})$) quantities $\Phi(\theta_j/\sigma)$ with Φ denoting the CDF of $\mathcal{N}(0,1)$. Since Φ has derivative bounded away from 0 on $[-1/\sigma, 1/\sigma]$, we get that the estimate $\hat{\theta}_j \triangleq \sigma\Phi^{-1}(V_j)$ will have mean-squared error of $O(1/m)$ (with a poor dependence on σ, though), which gives overall error $O(d/m)$ as claimed.
  Our main result below shows that the previous simple strategy is order-optimal in terms of communicated bits. This simplifies the proofs of [137, 73].
- We remark that these results are also implicitly contained in the long line of work in the information-theoretic literature on the so-called *Gaussian CEO problem*. We recommend consulting [156]; in particular, Theorem 3 there implies the $B \gtrsim dm$ lower bound we show below. However, the Gaussian CEO (chief executive officer) work uses a lot more sophisticated machinery (the entropy-power inequality and related results), while our SDPI proof is more elementary.

Our goal is to show that $R^* \lesssim \frac{d}{m}$ implies $B \gtrsim md$.

Notice, first of all, that this is completely obvious for $d = 1$. Indeed, if $B \le \tau m$ then less than τm machines are communicating anything at all, and hence $R^* \ge \frac{K}{\tau m}$ for some universal constant K (which is not 1 because θ is restricted to $[-1, 1]$). However, for $d \gg 1$ it is not clear whether each machine is required to communicate $\Omega(d)$ bits. Perhaps sending $\ll d$ single-bit measurements taken in different orthogonal bases could work? Hopefully, this (incorrect) guess demonstrates why the following result is interesting and non-trivial.

Theorem 33.25. *There exists a constant* $c_1 > 0$ *such that if* $R^*(m,d,\sigma^2,B) \le \frac{d\epsilon^2}{9}$ *then* $B \ge \frac{c_1 d}{\epsilon^2}$.

Proof. Let $X \sim \mathrm{Unif}(\{\pm 1\}^d)$ and set $\theta = \epsilon X$. Given an estimate $\hat{\theta}$ we can convert it into an estimator of X via $\hat{X} = \mathrm{sign}(\hat{\theta})$ (coordinatewise). Then, clearly

$$\mathbb{E}[d_{\mathrm{H}}(X,\hat{X})]\frac{\epsilon^2}{4} \le \mathbb{E}[\|\hat{\theta}-\theta\|^2] \le \frac{d\epsilon^2}{9}.$$

Thus, we have an estimator of X within Hamming distance $\frac{4}{9}d$. From rate distortion (Theorem 26.1) we conclude that $I(X;\hat{X}) \ge cd$ for some constant $c > 0$. On the other hand, from the standard DPI we have

$$cd \le I(X;\hat{X}) \le I(X;U_1,\ldots,U_m) \le \sum_{j=1}^{m} I(X;U_j), \tag{33.32}$$

where we also applied Theorem 6.1. Next we estimate $I(X;U_j)$ via $I(Y_j;U_j)$ by applying the post-SDPI. To do this we need to notice that the channel $X \to Y_j$ for each j is just a memoryless extension of the binary-input AWGN channel with SNR ϵ. Since each coordinate of X is uniform, we can apply Theorem 33.24 (with $\pi = 1/2$) together with tensorization (33.27) to conclude that

$$I(X;U_j) \le 4K\epsilon^2 I(Y_j;U_j) \le 4K\epsilon^2 \log|\mathcal{U}_j|.$$

Together with (33.32) we thus obtain

$$cd \le I(X;\hat{X}) \le 4K\epsilon^2 B\log 2. \tag{33.33}$$

□

We notice that in this short section we only considered a *non-interactive* setting in the sense that the message U_i is produced by machine i independently and without consulting anything except its private measurement Y_i. More generally, we could allow machines to communicate their bits over a public broadcast channel, so that each communicated bit is seen by all other machines. We could still restrict the total number of bits sent by all machines to be B and ask for the best possible *interactive* estimation rate. While [137, 73] claim lower bounds that apply to this setting, those bounds contain subtle errors (see [5, 4] for details). There are lower bounds applicable to non-interactive settings but they are weaker by certain logarithmic terms. For example, [5, Theorem 4] shows that to achieve risk $\lesssim d\epsilon^2$ one needs $B \gtrsim \frac{d}{\epsilon^2\log(dm)}$ in the limited interactive setting where U_i may depend on U_1^{i-1} but there are no other interactions (i.e. the ith machine sends its entire message at once instead of sending part of it and waiting for others to broadcast theirs before completing its own transmission, as permitted by the fully interactive protocol).

Exercises for Part VI

VI.1 Let $X_1, \ldots, X_n \overset{\text{iid}}{\sim} \text{Exp}(\exp(\theta))$, where θ follows the Cauchy distribution π with parameter s, whose PDF is given by $p(\theta) = \frac{1}{\pi s(1+\theta^2/s^2)}$ for $\theta \in \mathbb{R}$. Show that the Bayes risk

$$R_\pi^* \triangleq \inf_{\hat{\theta}} \mathbb{E}_{\theta\sim\pi} \mathbb{E}(\hat{\theta}(X^n) - \theta)^2$$

satisfies $R_\pi^* \geq \frac{2s^2}{2ns^2+1}$.

VI.2 (System identification) Let $\theta \in \mathbb{R}$ be an unknown parameter of a dynamical system:

$$X_t = \theta X_{t-1} + Z_t, \quad Z_t \overset{\text{iid}}{\sim} \mathcal{N}(0,1), \quad X_0 = 0.$$

Learning the parameters of dynamical systems is known as "system identification." Denote the law of $(X_1, \ldots, X_n)$ parametrized by θ by P_θ.

(a) Compute $D(P_\theta \| P_{\theta_0})$. (Hint: The chain rule saves a lot of effort.)

(b) Show that the Fisher information is given by

$$J_F(\theta) = \sum_{1\leq t\leq n-1} \theta^{2t-2}(n-t).$$

(c) Argue that the hardest regime for system identification is when $\theta \approx 0$, and that instability ($|\theta| > 1$) is in fact helpful.

VI.3 (Linear regression) Consider the model

$$Y = X\beta + Z,$$

where the design matrix $X \in \mathbb{R}^{n\times d}$ is known and $Z \sim \mathcal{N}(0, I_n)$. Define the minimax mean-squared error of estimating the regression coefficient $\beta \in \mathbb{R}^d$ based on X and Y as follows:

$$R_{\text{est}}^* = \inf_{\hat{\beta}} \sup_{\beta\in\mathbb{R}^d} \mathbb{E}\|\hat{\beta} - \beta\|_2^2. \tag{VI.1}$$

(a) Show that if $\text{rank}(X) < d$, then $R^*_{\text{est}} = \infty$.
(b) Show that if $\text{rank}(X) = d$, then

$$R^*_{\text{est}} = \text{tr}((X^\top X)^{-1})$$

and identify which estimator achieves the minimax risk.
(c) As opposed to the estimation error in (VI.1), consider the *prediction error*:

$$R^*_{\text{pred}} = \inf_{\hat{\beta}} \sup_{\beta \in \mathbb{R}^d} \mathbb{E}\|X\hat{\beta} - X\beta\|_2^2. \tag{VI.2}$$

Redo (a) and (b) by finding the value of R^*_{pred} and identify the minimax estimator. Explain intuitively why R^*_{pred} is always finite even when d exceeds n.

VI.4 (Chernoff–Rubin–Stein lower bound) Let $X_1, \ldots, X_n \overset{\text{iid}}{\sim} P_\theta$ and $\theta \in [-a, a]$.
(a) State the appropriate regularity conditions and prove the following minimax lower bound:

$$\inf_{\hat{\theta}} \sup_{\theta \in [-a,a]} \mathbb{E}_\theta[(\theta - \hat{\theta})^2] \geq \min_{0<\epsilon<1} \max\left\{\epsilon^2 a^2, \frac{(1-\epsilon)^2}{n\bar{J}_F}\right\},$$

where $\bar{J}_F = \frac{1}{2a}\int_{-a}^{a} J_F(\theta)d\theta$ is the average Fisher information. (Hint: Consider the uniform prior on $[-a, a]$ and proceed as in the proof of Theorem 29.2 by applying integration by parts.)
(b) Simplify the above bound and show that

$$\inf_{\hat{\theta}} \sup_{\theta \in [-a,a]} \mathbb{E}_\theta[(\theta - \hat{\theta})^2] \geq \frac{1}{(a^{-1} + \sqrt{n\bar{J}_F})^2}. \tag{VI.3}$$

(c) Assuming the continuity of $\theta \mapsto J_F(\theta)$, show that the above result also leads to the optimal local minimax lower bound in Theorem 29.4 obtained from Bayesian Cramér–Rao:

$$\inf_{\hat{\theta}} \sup_{\theta \in [\theta_0 \pm n^{-1/4}]} \mathbb{E}_\theta[(\theta - \hat{\theta})^2] \geq \frac{1 + o(1)}{nJ_F(\theta_0)}.$$

Note: (VI.3) is an improvement of the inequality given in [92, Lemma 1] without proof and credited to Rubin and Stein.

VI.5 In this exercise we give a Hellinger-based lower bound analogous to the χ^2-based HCR lower bound in Theorem 29.1. Let $\hat{\theta}$ be an unbiased estimator for $\theta \in \Theta \subset \mathbb{R}$.
(a) For any $\theta, \theta' \in \Theta$, show that [387]

$$\frac{1}{2}(\text{Var}_\theta(\hat{\theta}) + \text{Var}_{\theta'}(\hat{\theta})) \geq \frac{(\theta - \theta')^2}{4}\left(\frac{1}{H^2(P_\theta, P_{\theta'})} - 1\right). \tag{VI.4}$$

(Hint: For any c, $\theta - \theta' = \int(\hat{\theta} - c)(\sqrt{p_\theta} + \sqrt{p_{\theta'}})(\sqrt{p_\theta} - \sqrt{p_{\theta'}})$. Apply Cauchy–Schwarz and optimize over c.)

(b) Show that

$$H^2(P_\theta, P_{\theta'}) \le \frac{1}{4}(\theta - \theta')^2 \bar{J}_F, \tag{VI.5}$$

where $\bar{J}_F = \frac{1}{\theta'-\theta}\int_\theta^{\theta'} J_F(u)du$ is the average Fisher information.

(c) State the needed regularity conditions and deduce the Cramér–Rao lower bound from (VI.4) and (VI.5) with $\theta' \to \theta$.

(d) Extend the previous parts to the multivariate case.

VI.6 (Bayesian distribution estimation) Let $\{P_\theta : \theta \in \Theta\}$ be a family of distributions on $\mathcal{X}$ with a common dominating measure μ and density $p_\theta(x) = \frac{dP_\theta}{d\mu}(x)$. Given a sample $X^n = (X_1, \ldots, X_n) \overset{\text{iid}}{\sim} P_\theta$ for some $\theta \in \Theta$, the goal is to estimate the data-generating distribution P_θ by some estimator $\hat{P}(\cdot) = \hat{P}(\cdot;X^n)$ with respect to some loss function $\ell(P, \hat{P})$. Suppose we are in a Bayesian setting where θ is drawn from a prior π. Let's find the form of the Bayes estimator and the Bayes risk.

(a) For convenience, let X_{n+1} denote a test data point (unseen) drawn from P_θ and independent of the observed data X^n. Convince yourself that every estimator $\hat{P}$ can be formally identified as a conditional distribution $Q_{X_{n+1}|X^n}$.

(b) Consider the KL loss $\ell(P, \hat{P}) = D(P\|\hat{P})$. Using Corollary 4.2, show that the Bayes estimator minimizing the average KL risk is the posterior (conditional mean), that is, its μ-density is given by

$$q_{X_{n+1}|X^n}(x_{n+1}|x^n) = \frac{\mathbb{E}_{\theta\sim\pi}[\prod_{i=1}^{n+1} p_\theta(x_i)]}{\mathbb{E}_{\theta\sim\pi}[\prod_{i=1}^{n} p_\theta(x_i)]}. \tag{VI.6}$$

(c) Conclude that the Bayes KL risk equals $I(\theta;X_{n+1}|X^n)$. Compare with the conclusion of Exercise **II.19** and the KL risk interpretation of batch regret in (13.35).

(d) Now, consider the χ^2 loss $\ell(P, \hat{P}) = \chi^2(P\|\hat{P})$. Using (I.12) in Exercise **I.45** show that the optimal risk is given by

$$\inf_{\hat{P}} \mathbb{E}_{\theta,X^n}[\chi^2(P_\theta\|\hat{P})] = \mathbb{E}_{X^n}\left[\left(\int \mu(dx_{n+1})\sqrt{\mathbb{E}_\theta[p_\theta(x_{n+1})^2|X^n]}\right)^2\right] - 1 \tag{VI.7}$$

attained by

$$q_{X_{n+1}|X^n}(x_{n+1}|x^n) \propto \sqrt{\mathbb{E}_\theta[p_\theta(x_{n+1})^2|X^n = x^n]}. \tag{VI.8}$$

(e) Now, consider the reverse-χ^2 loss $\ell(P, \hat{P}) = \chi^2(\hat{P}\|P)$, a weighted quadratic loss. Using (I.13) in Exercise **I.45** show that the optimal risk is attained by

$$q_{X_{n+1}|X^n}(x_{n+1}|x^n) \propto \left(\mathbb{E}_\theta[p_\theta(x_{n+1})^{-1}|X^n = x^n]\right)^{-1}. \tag{VI.9}$$

(f) Consider the discrete alphabet $[k]$ and $X^n \overset{\text{iid}}{\sim} P$, where $P = (P_1, \ldots, P_k)$ is drawn from the Dirichlet$(\alpha, \ldots, \alpha)$ prior. Applying previous results (with μ the counting measure), show that the Bayes estimator for the KL loss and reverse-χ^2 loss is given by the Krichevsky–Trofimov add-β estimator (Section 13.5)

$$\widehat{P}(j) = \frac{n_j + \beta}{n + k\beta}, \qquad n_j \triangleq \sum_{i=1}^{n} 1\{X_i = j\}, \tag{VI.10}$$

where $\beta = \alpha$ for KL and $\beta = \alpha - 1$ for reverse-χ^2 (assuming $\alpha \geq 1$).

(Hint: The posterior is $(P_1, \ldots, P_k)|X^n \sim \text{Dirichlet}(n_1 + \alpha, \ldots, n_k + \alpha)$ and $P_j|X^n \sim \text{Beta}(n_j + \alpha, n - n_j + (k-1)\alpha)$.)

(g) For the χ^2 loss, show that the Bayes estimator is

$$\widehat{P}(j) = \frac{\sqrt{(n_j + \alpha)(n_j + \alpha + 1)}}{\sum_{j=1}^{k} \sqrt{(n_j + \alpha)(n_j + \alpha + 1)}}. \tag{VI.11}$$

VI.7 (Coin flips) Given $X_1, \ldots, X_n \overset{\text{iid}}{\sim} \text{Ber}(\theta)$ with $\theta \in \Theta = [0, 1]$, we aim to estimate θ with respect to the quadratic loss function $\ell(\theta, \hat{\theta}) = (\theta - \hat{\theta})^2$. Denote the minimax risk by R_n^*.

(a) Use the empirical frequency $\hat{\theta}_{\text{emp}} = \bar{X}$ to estimate θ. Compute and plot the risk $R_\theta(\hat{\theta})$ and show that

$$R_n^* \leq \frac{1}{4n}.$$

(b) Compute the Fisher information of $P_\theta = \text{Ber}(\theta)^{\otimes n}$ and $Q_\theta = \text{Bin}(n, \theta)$. Explain why they are equal.

(c) Invoke the Bayesian Cramér–Rao lower bound from Theorem 29.2 to show that

$$R_n^* = \frac{1 + o(1)}{4n}.$$

(d) Notice that the risk of $\hat{\theta}_{\text{emp}}$ is maximized at 1/2 (fair coin), which suggests that it might be possible to hedge against this situation with the following randomized estimator:

$$\hat{\theta}_{\text{rand}} = \begin{cases} \hat{\theta}_{\text{emp}}, & \text{with probability } \delta, \\ \frac{1}{2}, & \text{with probability } 1 - \delta. \end{cases} \tag{VI.12}$$

Find the worst-case risk of $\hat{\theta}_{\text{rand}}$ as a function of δ. Optimizing over δ, show the improved upper bound:

$$R_n^* \leq \frac{1}{4(n+1)}.$$

(e) As discussed in Remark 28.3, the randomized estimator can always be improved if the loss is convex; so we should average out the randomness in (VI.12) by considering the estimator

$$\hat{\theta}^* = \mathbb{E}[\hat{\theta}_{\text{rand}}|X] = \bar{X}\delta + \frac{1}{2}(1 - \delta). \tag{VI.13}$$

Optimizing over δ to minimize the worst-case risk, find the resulting estimator $\hat{\theta}^*$ and its risk, show that it is constant (independent of θ), and conclude that

$$R_n^* \leq \frac{1}{4(1+\sqrt{n})^2}.$$

(f) Next we show $\hat{\theta}^*$ found in part (e) is exactly minimax and hence

$$R_n^* = \frac{1}{4(1+\sqrt{n})^2}.$$

Consider the following prior Beta(a, b) with density:

$$\pi(\theta) = \frac{\Gamma(a+b)}{\Gamma(a)\Gamma(b)}\theta^{a-1}(1-\theta)^{b-1}, \quad \theta \in [0, 1],$$

where $\Gamma(a) \triangleq \int_0^\infty x^{a-1}e^{-x}dx$. Show that if $a = b = \frac{\sqrt{n}}{2}$, $\hat{\theta}^*$ coincides with the Bayes estimator for this prior, which is therefore least favorable.

(Hint: work with the sufficient statistic $S = X_1 + \ldots + X_n$.)

(g) Show that the least favorable prior is not unique; in fact, there is a continuum of them. (Hint: Consider the Bayes estimator $\mathbb{E}[\theta|X]$ and show that it only depends on the first $n+1$ moments of π.)

(h) (k-ary alphabet) Suppose $X_1, \ldots, X_n \overset{\text{iid}}{\sim} P$ on $[k]$. Show that for any k, n, the minimax squared risk of estimating P in Theorem 29.5 is exactly

$$R_{\text{sq}}^*(k, n) = \inf_{\widehat{P}} \sup_{P \in \mathcal{P}_k} \mathbb{E}[\|\widehat{P} - P\|_2^2] = \frac{1}{(\sqrt{n}+1)^2}\frac{k-1}{k}, \qquad \text{(VI.14)}$$

achieved by the add-$\frac{\sqrt{n}}{k}$ estimator. (Hint: For the lower bound, show that the Bayes estimators for the squared loss and the KL loss coincide, then apply (VI.10) in Exercise **VI.6**.)

VI.8 (Distribution estimation in TV) Continuing (VI.14), we show that the minimax rate for estimating P with respect to the total variation loss is

$$R_{\text{TV}}^*(k, n) \triangleq \inf_{\hat{P}} \sup_{P \in \mathcal{P}_k} \mathbb{E}_P[\text{TV}(\hat{P}, P)] \asymp \sqrt{\frac{k}{n}} \wedge 1, \qquad \forall k \geq 2, n \geq 1. \quad \text{(VI.15)}$$

(a) Show that the MLE coincides with the empirical distribution.

(b) Show that the MLE achieves the RHS of (VI.15) within constant factors. (Hint: Either apply (7.58) plus Pinsker's inequality, or directly use the variance of empirical frequencies.)

(c) Establish the minimax lower bound. (Hint: Apply Assouad's lemma, or Fano's inequality (with volume method or explicit construction of packing), or the mutual information method directly.)

VI.9 (Distribution estimation under reverse-χ^2) Consider estimating a discrete distribution P on $[k]$ in reverse-χ^2 divergence from $X^n \overset{\text{iid}}{\sim} P$, which is a weighted version of the quadratic loss in (VI.14). We show that the minimax risk is given by

$$R^*_{\mathrm{rev}\chi^2}(k,n) \triangleq \inf_{\hat{P}} \sup_{P\in\mathcal{P}([k])} \mathbb{E}_P[\chi^2(\hat{P}\|P)] = \frac{k-1}{n}.$$

(a) Show that taking $\hat{P}(j) = \frac{1}{n}\sum_{i=1}^n 1\{X_i = j\}$ to be the empirical distribution we always have $\mathbb{E}[\chi^2(\hat{P}\|P)] = \frac{k-1}{n}$.
(b) Given $P \sim \mathrm{Dirichlet}(\alpha,\dots,\alpha)$ show that when $\alpha = 1$ the Bayes optimal estimator is precisely the empirical distribution. (Hint: See (VI.9).)
(c) Conclude that the uniform prior on $\mathcal{P}([k])$ is least favorable and the empirical distribution is exactly minimax optimal.

VI.10 (Distribution estimation in KL and χ^2) Consider estimating a discrete distribution P on $[k]$ in KL and χ^2-divergence, which are examples of an unbounded loss (KL loss is also known as *cross-entropy* or *log-loss* in machine learning). Since these divergences are not symmetric, we define (recall reverse-χ^2 has been addressed in Exercise **VI.9**)

$$R^*_{\mathrm{KL}}(k,n) \triangleq \inf_{\hat{P}} \sup_{P\in\mathcal{P}_k} \mathbb{E}_P[D(P\|\hat{P})], \quad R^*_{\mathrm{revKL}}(k,n) \triangleq \inf_{\hat{P}} \sup_{P\in\mathcal{P}_k} \mathbb{E}_P[D(\hat{P}\|P)],$$

$$R^*_{\chi^2}(k,n) \triangleq \inf_{\hat{P}} \sup_{P\in\mathcal{P}_k} \mathbb{E}_P[\chi^2(P\|\hat{P})].$$

We have (up to universal constant factors) for all k,n:

$$R^*_{\mathrm{KL}}(k,n) \asymp R^*_{\mathrm{revKL}}(k,n) \asymp \log\left(1+\frac{k}{n}\right) \asymp \begin{cases} \frac{k}{n}, & k \le 1.1n, \\ \log\frac{k}{n}, & k > 1.1n, \end{cases} \tag{VI.16}$$

$$R^*_{\chi^2}(k,n) \asymp \frac{k}{n}. \tag{VI.17}$$

(a) Show that the empirical distribution, optimal for the TV loss (Exercise **VI.8**), implies the claimed upper bound for the reverse KL loss.

(Hint: See (7.58).) Show, on the other hand, that for KL and χ^2 it results in infinite loss.

(b) To show the upper bound for χ^2, consider the add-α estimator $\hat{P}$ in (VI.10) with $\alpha = 1$. Show that

$$\mathbb{E}[\chi^2(P\|\hat{P})] \le \frac{k-1}{n+1}.$$

Using (7.34) conclude the upper-bound part of (VI.16). (Hint: $\mathbb{E}_{N\sim\mathrm{Bin}(n,p)}[\frac{1}{N+1}] = \frac{1}{(n+1)p}(1-\bar{p}^{n+1})$.)
(c) Show that in the small alphabet regime of $k \lesssim n$, all lower bounds follow from (VI.15).
(d) Next assume $k \ge 4n$. Consider a Dirichlet$(\alpha,\dots,\alpha)$ prior in (13.16). Applying (VI.11) and (VI.7) for the Bayes χ^2 risk and choosing $\alpha = n/k$, show the lower bound $R^*_{\chi^2}(k,n) \gtrsim \frac{k}{n}$.
(e) Consider the prior under which P is uniform over a support set S chosen uniformly at random from all s-subsets of $[k]$, where $s < k$ is to be specified. Applying (VI.6), show that for this prior the Bayes estimator for KL loss takes a natural form:

$$\hat{P}_j = \begin{cases} \frac{1}{s}, & i \in \hat{S}, \\ \frac{1-\hat{s}/s}{k-\hat{s}}, & i \notin \hat{S}, \end{cases}$$

where $\hat{S} = \{i: n_i \geq 1\}$ is the support of the empirical distribution and $\hat{s} = |\hat{S}|$.

(f) Choosing $s = \sqrt{nk}$, conclude $\mathbb{E}[\mathrm{TV}(P, \hat{P})] \geq 1 - 2\sqrt{\frac{n}{k}}$. (Hint: Show that $\mathrm{TV}(P, \hat{P}) \geq (1 - \frac{\hat{s}}{s})(1 - \frac{s}{k})$ and $\hat{s} \leq n$.)

(g) Using (7.31), show that $\mathbb{E}[D(\hat{P}\|P)], \mathbb{E}[D(P\|\hat{P})] \geq \Omega(\log \frac{k}{n})$, concluding the lower bound in (VI.16). (Hint: (7.31) is convex in TV.)

Note: For $k \asymp 1$, [259] found that the best add-α estimator has $\alpha^* \approx 0.509$ (unlike $\alpha = 1/2$ optimal for cumulative loss in Section 13.5) and it achieves loss $\alpha^* \frac{k-1+o(1)}{n}$. In this regime, the optimal risk is only slightly smaller and equals $R^*_{\mathrm{KL}}(k,n) = \frac{k-1+o(1)}{2n}$, which was shown in [72, 71] via deep results in polynomial approximation (the optimal estimator is the add-c estimator but with c varying according to the empirical count in each bin). For $k \gg n$, Paninski [324] showed $R^*_{\mathrm{KL}}(k,n) = \log \frac{k}{n}(1 + o(1))$ by a careful analysis of the Dirichlet prior.

VI.11 (Nonparametric model) In this exercise we consider some nonparametric extensions of the Gaussian location model and the Bernoulli model. Observing $X_1, \ldots, X_n \overset{\text{iid}}{\sim} P$ for some $P \in \mathcal{P}$, where $\mathcal{P}$ is a collection of distributions on the real line, our goal is to estimate the *mean* of the distribution P: $\mu(P) \triangleq \int xP(dx)$, which is a linear functional of P. Denote the minimax quadratic risk by

$$R^*_n = \inf_{\hat{\mu}} \sup_{P \in \mathcal{P}} \mathbb{E}_P[(\hat{\mu}(X_1, \ldots, X_n) - \mu(P))^2].$$

(a) Let $\mathcal{P}$ be the class of distributions (which need not have a density) on the real line with variance at most σ^2. Show that $R^*_n = \frac{\sigma^2}{n}$.

(b) Let $\mathcal{P} = \mathcal{P}([0,1])$, the collection of all probability distributions on $[0,1]$. Show that $R^*_n = \frac{1}{4(1+\sqrt{n})^2}$. (Hint: For the upper bound, using the fact that, for any $[0,1]$-valued random variable Z, $\mathrm{Var}(Z) \leq \mathbb{E}[Z](1 - \mathbb{E}[Z])$, mimic the analysis of the estimator (VI.13) in Exercise **VI.7e**.)

VI.12 Prove Theorem 30.2 using Fano's method. (Hint: Apply Theorem 31.3 with $T = \epsilon \cdot S^d_k$, where S^d_k denotes the Hamming sphere of radius k in d dimensions. Choose ϵ appropriately and apply the Gilbert–Varshamov bound for the packing number of S^d_k in Theorem 27.6.)

VI.13 (Sharp minimax rate in sparse denoising) Continuing Theorem 30.2, in this exercise we determine the sharp minimax risk for denoising a high-dimensional sparse vector. In the notation of (30.13), we show that, for the d-dimensional GLM model $X \sim \mathcal{N}(\theta, I_d)$, the following minimax risk satisfies, as $d \to \infty$ and $k/d \to 0$,

$$R^*(k,d) \triangleq \inf_{\hat{\theta}} \sup_{\|\theta\|_0 \leq k} \mathbb{E}_\theta[\|\hat{\theta} - \theta\|_2^2] = (2 + o(1))k \log \frac{d}{k}. \tag{VI.18}$$

(a) We first consider 1-sparse vectors and prove

$$R^*(1,d) \triangleq \inf_{\hat\theta} \sup_{\|\theta\|_0\le 1} \mathbb{E}_\theta[\|\hat\theta-\theta\|_2^2] = (2+o(1))\log d, \qquad d\to\infty. \tag{VI.19}$$

For the lower bound, consider the prior π under which θ is uniformly distributed over $\{\tau e_1,\dots,\tau e_d\}$, where e_i's denote the standard basis. Let $\tau = \sqrt{(2-\epsilon)\log d}$. Show that for any $\epsilon>0$, the Bayes risk is given by

$$\inf_{\hat\theta} \mathbb{E}_{\theta\sim\pi}[\|\hat\theta-\theta\|_2^2] = \tau^2(1+o(1)), \qquad d\to\infty.$$

(Hint: Either apply the mutual information method, or directly compute the Bayes risk by evaluating the conditional mean and conditional variance.)

(b) Demonstrate an estimator $\hat\theta$ that achieves the RHS of (VI.19) asymptotically. (Hint: Consider the hard-thresholding estimator (30.13) or the MLE (30.11).)

(c) To prove the lower-bound part of (VI.18), prove the generic result

$$R^*(k,d) \ge kR^*\left(1,\frac{d}{k}\right),$$

and then apply (VI.19). (Hint: Consider a prior of d/k blocks each of which is 1-sparse.)

(d) Similar to the 1-sparse case, demonstrate an estimator $\hat\theta$ that achieves the RHS of (VI.18) asymptotically.

Note: For both the upper and lower bounds, the normal tail bound in Exercise **V.25** is helpful.

VI.14 Consider the following functional estimation problem in GLM. Observing $X\sim\mathcal{N}(\theta,I_d)$, we intend to estimate the maximal coordinate of θ: $T(\theta)=\theta_{\max}\triangleq\max\{\theta_1,\dots,\theta_d\}$. Prove the minimax rate:

$$\inf_{\hat T}\sup_{\theta\in\mathbb{R}^d} \mathbb{E}_\theta(\hat T-\theta_{\max})^2 \asymp \log d. \tag{VI.20}$$

(a) Prove the upper bound by considering $\hat T = X_{\max}$, the plug-in estimator with the MLE.

(b) For the lower bound, consider two hypotheses:

$$H_0: \theta=0, \qquad H_1:\theta\sim \mathrm{Unif}\{\tau e_1,\tau e_2,\dots,\tau e_d\},$$

where e_i's are the standard bases and $\tau>0$. Then under H_0, $X\sim P_0=\mathcal{N}(0,I_d)$; under H_1, $X\sim P_1=\frac{1}{d}\sum_{i=1}^d\mathcal{N}(\tau e_i,I_d)$. Show that $\chi^2(P_1\|P_0)=\frac{e^{\tau^2}-1}{d}$. (Hint: Exercise **I.48**.)

(c) Applying the joint range (7.32) (or (7.38)) to bound $\mathrm{TV}(P_0,P_1)$, conclude the lower-bound part of (VI.20) via Le Cam's method (Theorem 31.1).

(d) By improving both the upper and lower bounds prove the sharp version:

$$\inf_{\hat{T}} \sup_{\theta \in \mathbb{R}^d} \mathbb{E}_\theta (\hat{T} - \theta_{\max})^2 = \left(\frac{1}{2} + o(1)\right) \log d, \quad d \to \infty. \tag{VI.21}$$

VI.15 (Suboptimality of MLE in high dimensions [55]) Consider the d-dimensional GLM: $X \sim \mathcal{N}(\theta, I_d)$, where θ belongs to the parameter space

$$\Theta = \{\theta \in \mathbb{R}^d : |\theta_1| \leq d^{1/4}, \|\theta_{\setminus 1}\|_2 \leq 2(1 - d^{-1/4}|\theta_1|)\}$$

with $\theta_{\setminus 1} \equiv (\theta_2, \ldots, \theta_d)$. For the square loss, prove the following for sufficiently large d.

(a) The minimax risk is bounded:

$$\inf_{\hat{\theta}} \sup_{\theta \in \Theta} \mathbb{E}_\theta [\|\hat{\theta} - \theta\|_2^2] \lesssim 1.$$

(b) The worst-case risk of the maximum likelihood estimator

$$\theta_{\mathrm{MLE}} \triangleq \operatorname*{argmin}_{\tilde{\theta} \in \Theta} \|X - \tilde{\theta}\|_2$$

is unbounded:

$$\sup_{\theta \in \Theta} \mathbb{E}_\theta [\|\hat{\theta}_{\mathrm{MLE}} - \theta\|_2^2] \gtrsim \sqrt{d}.$$

VI.16 (Covariance model) Let $X_1, \ldots, X_n \overset{\text{i.i.d.}}{\sim} \mathcal{N}(0, \Sigma)$, where Σ is a $d \times d$ covariance matrix. Let us show that the minimax quadratic risk of estimating Σ using $X_1, \ldots, X_n$ satisfies

$$\inf_{\hat{\Sigma}} \sup_{\|\Sigma\|_F \leq r} \mathbb{E}[\|\hat{\Sigma} - \Sigma\|_F^2] \asymp \left(\frac{d}{n} \wedge 1\right) r^2, \qquad \forall r > 0, d, n \in \mathbb{N}, \tag{VI.22}$$

where $\|\hat{\Sigma} - \Sigma\|_F^2 = \sum_{ij} (\hat{\Sigma}_{ij} - \Sigma_{ij})^2$.

(a) Show that unlike the location model, without restricting to a compact parameter space for Σ, the minimax risk in (VI.22) is infinite.

(b) Consider the sample covariance matrix $\hat{\Sigma} = \frac{1}{n}\sum_{i=1}^n X_i X_i^\top$. Show that

$$\mathbb{E}[\|\hat{\Sigma} - \Sigma\|_F^2] = \frac{1}{n}\left(\|\Sigma\|_F^2 + \operatorname{Tr}(\Sigma)^2\right)$$

and use this to deduce the minimax upper bound in (VI.22).

(c) To prove the minimax lower bound, we can proceed in several steps. Show that for any positive semidefinite (PSD) $\Sigma_0, \Sigma_1 \succeq 0$, the KL divergence satisfies

$$D(\mathcal{N}(0, I_d + \Sigma_0) \| \mathcal{N}(0, I_d + \Sigma_1)) \leq \frac{1}{2}\|\Sigma_0^{1/2} - \Sigma_1^{1/2}\|_F^2, \tag{VI.23}$$

where I_d is the $d \times d$ identity matrix. (Hint: Apply (2.8).)

(d) Let $B(\delta)$ denote the Frobenius ball of radius δ centered at the zero matrix. Let $\mathsf{PSD} = \{X : X \succeq 0\}$ denote the collection of $d \times d$ PSD matrices. Show that

$$\frac{\mathrm{vol}(B(\delta) \cap \mathsf{PSD})}{\mathrm{vol}(B(\delta))} = \mathbb{P}[Z \succeq 0], \tag{VI.24}$$

where Z is a random matrix distributed according to the Gaussian orthogonal ensemble (GOE), that is, Z is symmetric with independent diagonals $Z_{ii} \overset{\text{iid}}{\sim} \mathcal{N}(0,2)$ and off-diagonals $Z_{ij} \overset{\text{iid}}{\sim} \mathcal{N}(0,1)$.

(e) Show that $\mathbb{P}[Z \succeq 0] \geq c^{d^2}$ for some absolute constant c.[4]

(f) Prove the following lower bound on the packing number on the set of PSD matrices:

$$M(B(\delta) \cap \mathsf{PSD}, \|\cdot\|_{\mathrm{F}}, \epsilon) \geq \left(\frac{c'\delta}{\epsilon}\right)^{d^2/2} \tag{VI.25}$$

for some absolute constant c'. (Hint: Use the volume bound and the result of parts (d) and (e).)

(g) Complete the proof of the lower bound of (VI.22). (Hint: Without loss of generality, we can consider $r \asymp \sqrt{d}$ and aim for the lower bound $\Omega(\frac{d^2}{n} \wedge d)$. Take the packing from (VI.25) and shift by the identity matrix I. Then apply Fano's method and use (VI.23).)

VI.17 For a family of probability distributions $\mathcal{P}$ and a functional $T : \mathcal{P} \to \mathbb{R}$ define its χ^2-modulus of continuity as

$$\delta_{\chi^2}(t) = \sup_{P_1, P_2 \in \mathcal{P}} \{T(P_1) - T(P_2) : \chi^2(P_1 \| P_2) \leq t\}.$$

When the functional T is affine, and continuous, and $\mathcal{P}$ is compact[5] it can be shown [347] that

$$\frac{1}{7}\delta_{\chi^2}(1/n)^2 \leq \inf_{\hat{T}_n} \sup_{P \in \mathcal{P}} \mathbb{E}_{X_i \overset{\text{iid}}{\sim} P}(T(P) - \hat{T}_n(X_1, \ldots, X_n))^2 \leq \delta_{\chi^2}(1/n)^2. \tag{VI.26}$$

Consider the following problem (interval censored model): A lab conducts experiments with n mice. In the ith mouse a tumour develops at time $A_i \in [0,1]$ with $A_i \overset{\text{iid}}{\sim} \pi$ where π is a PDF on $[0,1]$ bounded by $\frac{1}{2} \leq \pi \leq 2$ pointwise. For each i the existence of a tumour is checked at another random time $B_i \overset{\text{iid}}{\sim} \mathrm{Unif}(0,1)$ with $B_i \perp\!\!\!\perp A_i$. Given observations $X_i = (1\{A_i \leq B_i\}, B_i)$ one is trying to estimate $T(\pi) = \pi[A \leq 1/2]$. Show that

$$\inf_{\hat{T}_n} \sup_{\pi} \mathbb{E}[(T(\pi) - \hat{T}_n(X_1, \ldots, X_n))^2] \asymp n^{-2/3}.$$

VI.18 (Comparison between contraction coefficients)
Prove (33.13) for $\eta_f = \eta_{\mathrm{KL}}$.
(Hint: Use local behavior of f-divergences (Proposition 2.21).)

[4] Getting the exact exponent is a difficult result (see [26]). Here we only need some crude estimate.

[5] Both under the same, but otherwise arbitrary, topology on $\mathcal{P}$.

VI.19 (Spectral gap and χ^2-contraction of Markov chains)
In this exercise we prove (33.18). Let $P = (P(x,y))$ denote the transition matrix of a time-reversible Markov chain with finite state space $\mathcal{X}$ and stationary distribution π, so that $\pi(x)P(x,y) = \pi(y)P(y,x)$ for all $x, y \in \mathcal{X}$. It is known that the $k = |\mathcal{X}|$ eigenvalues of P satisfy $1 = \lambda_1 \geq \lambda_2 \geq \cdots \geq \lambda_k \geq -1$. Define by $\gamma_* \triangleq \max\{\lambda_2, |\lambda_k|\}$ the absolute spectral gap.

(a) Show that

$$\chi^2(P_{X_1}\|\pi) \leq \chi^2(P_{X_0}\|\pi)\gamma_*^2,$$

from which (33.18) follows.

(b) Conclude that for any initial state x,

$$\chi^2(P_{X_n|X_0=x}\|\pi) \leq \frac{1-\pi(x)}{\pi(x)}\gamma_*^{2n}.$$

(c) Compute γ_* for the BSC_δ channel and compare with the η_{χ^2} contraction coefficients.

For a continuous-time version, see [124].

VI.20 (Input-independent contraction coefficient is achieved by binary inputs [322])
Let $K\colon \mathcal{X} \to \mathcal{Y}$ be a Markov kernel with countable $\mathcal{X}$. Prove that for all f-divergence, we have

$$\eta_f(K) = \sup_{\substack{P,Q\colon |\mathrm{supp}(P)\cup\mathrm{supp}(Q)|\leq 2\\ 0<D_f(P\|Q)<\infty}} \frac{D_f(K\circ P\|K\circ Q)}{D_f(P\|Q)}.$$

(Hint: Define the function

$$L_\lambda(P,Q) = D_f(K\circ P\|K\circ Q) - \lambda D_f(P\|Q)$$

and prove that $\hat{Q} \mapsto L_\lambda(\frac{P}{Q}\hat{Q}, \hat{Q})$ is convex on the set

$$\{\hat{Q} \in \mathcal{P}(\mathcal{X})\colon \mathrm{supp}(\hat{Q}) \subseteq \mathrm{supp}(Q), \tfrac{P}{Q}\hat{Q} \in \mathcal{P}(\mathcal{X})\}.)$$

VI.21 (BMS channel comparison [372, 368]) Below, $X \sim \mathrm{Ber}(1/2)$ and $P_{Y|X}$ is an input-symmetric channel (BMS). It turns out that BSC and BEC are extremal for various partial orders. Prove the following statements.

(a) If $I_{\mathrm{TV}}(X;Y) = \frac{1}{2}(1-2\delta)$, then

$$\mathsf{BSC}_\delta \leq_{\mathrm{deg}} P_{Y|X} \leq_{\mathrm{deg}} \mathsf{BEC}_{2\delta}.$$

(b) If $I(X;Y) = C$, then

$$\mathsf{BSC}_{h^{-1}(\log 2 - C)} \leq_{\mathrm{mc}} P_{Y|X} \leq_{\mathrm{mc}} \mathsf{BEC}_{1-C/\log 2}.$$

(c) If $I_{\chi^2}(X;Y) = \eta$, then

$$\mathsf{BSC}_{1/2-\sqrt{\eta}/2} \leq_{\mathrm{ln}} P_{Y|X} \leq_{\mathrm{ln}} \mathsf{BEC}_{1-\eta}.$$

(Hint: Apply Exercise **I.64**.)

VI.22 (Broadcasting on trees with BSC [204]) We have seen in Example 33.9 that broadcasting on trees (BOT) with BSC_δ has non-reconstruction when $b(1-2\delta)^2 < 1$. In this exercise we prove the achievability bound (known as the Kesten–Stigum bound [245]) using channel comparison.

We work with an infinite b-ary tree with BSC_δ edge channels. Let ρ be the root and L_k be the set of nodes at distance k to ρ. Let M_k denote the channel $X_\rho \to X_{L_k}$.

In the following, assume that $b(1-2\delta)^2 > 1$.

(a) Prove that there exists $\tau < \frac{1}{2}$ such that

$$\mathsf{BSC}_\tau \le_{\ln} \mathsf{BSC}_\tau^{\otimes b} \circ M_1.$$

(Hint: Use Exercise **VI.21**.)

(b) Prove $\mathsf{BSC}_\tau \le_{\ln} M_k$ by induction on k. Conclude that reconstruction holds.

(Hint: Use tensorization of less noisy ordering.)

VI.23 (Broadcasting on a 2D grid) Consider the following broadcasting model on a 2D grid:

- nodes are labeled with (i,j) for $i,j \in \mathbb{Z}$;
- $X_{i,j} = 0$ when $i < 0$ or $j < 0$;
- $X_{0,0} \sim \mathrm{Ber}(1/2)$;
- $X_{i,j} = f_{i,j}(X_{i-1,j} \oplus Z_{i,j,1}, X_{i,j-1} \oplus Z_{i,j,2})$ for $i,j \ge 0$ and $(i,j) \neq (0,0)$, where $Z_{i,j,k} \overset{\text{iid}}{\sim} \mathrm{Ber}(\delta)$, and $f_{i,j}$ is any function $\{0,1\} \times \{0,1\} \to \{0,1\}$.

Let $L_n = \{(n-i, i) \colon 0 \le i \le n\}$ be the set of nodes at level n. Let p_c be directed *bond* percolation threshold from $(0,0)$ to L_n for $n \to \infty$. Apply Theorem 33.8 to show that for $(1-2\delta)^2 < p_c$ we have

$$\lim_{n\to\infty} I(X_{0,0}; X_{L_n}) = 0.$$

It is known that $p_c \approx 0.645$ (e.g. [231]).

Note: We could also use Theorem 33.8 differently. Above we replaced each directed edge by a BSC_δ. We could instead consider channels $(X_{i-1,j}, X_{i,j-1}) \mapsto X_{i,j}$ and relate the contraction coefficient to the *directed site percolation* threshold p_c'. This would yield a non-reconstruction whenever

$$1 - 2\delta + 4\delta^3 - 2\delta^4 - 2\delta(1-\delta)\sqrt{\delta(1+\delta)(1-\delta)(2-\delta)} < p_c'.$$

Since $p_c' \approx 0.705$ it turns out the bond percolation result is stronger.

VI.24 (Input-dependent contraction coefficient for coloring channel [203]) Fix an integer $q \ge 3$ and let $\mathcal{X} = [q]$. Consider the following coloring channel $K \colon \mathcal{X} \to \mathcal{X}$:

$$K(y|x) = \begin{cases} 0, & y = x, \\ \frac{1}{q-1}, & y \neq x. \end{cases}$$

Let π be the uniform distribution on $\mathcal{X}$.

(a) Compute $\eta_{\mathrm{KL}}(\pi, K)$.

(b) Conclude that there exists a function $f(q) = (1 - o(1))q \log q$ as $q \to \infty$ such that for all $d < f(q)$, BOT with the coloring channel on a d-ary tree has non-reconstruction.

Note: This bound is tight up to the first order: there exists a function $g(q) = (1 + o(1))q \log q$ such that for all $d > g(q)$, BOT with coloring channel on a d-ary tree has reconstruction.

VI.25 ([203]) Fix an integer $q \geq 2$ and let $\mathcal{X} = [q]$. Let $\lambda \in [-\frac{1}{q-1}, 1]$ be a real number. Let us define a special kind of q-ary symmetric channel, known as the Potts channel, by taking $P_\lambda \colon \mathcal{X} \to \mathcal{X}$ as

$$P_\lambda(y|x) = \begin{cases} \lambda + \frac{1-\lambda}{q}, & y = x, \\ \frac{1-\lambda}{q}, & y \neq x. \end{cases}$$

Prove that

$$\eta_{\mathrm{KL}}(P_\lambda) = \frac{q\lambda^2}{(q-2)\lambda + 2}.$$

VI.26 (Spectral independence [20]) We say that a probability distribution $\mu = \mu_{X^n}$ supported on $[q]^n$ is c-pairwise independent if for every $T \subset [n]$, $\sigma_T \in [q]^\top$, the conditional measure $\mu^{(\sigma_T)} \triangleq \mu_{X_{T^c}|X_T=\sigma_T}$ satisfies for every $\nu_{X_{T^c}}$,

$$\sum_{i \neq j \in T^c} D\left(\nu_{X_i,X_j} \| \mu^{(\sigma_T)}_{X_i,X_j}\right) \geq \left(2 - \frac{c}{n - |T| - 1}\right) \sum_{i \in T^c} D\left(\nu_{X_i} \| \mu^{(\sigma_T)}_{X_i}\right).$$

Prove that for such a measure μ we have

$$\eta_{\mathrm{KL}}(\mu, \mathrm{EC}_\tau^{\otimes n}) \leq 1 - \tau^{c+1},$$

where EC_τ is the erasure channel, see Example 33.6 (Hint: Define $f(\tau) = D(\mathrm{EC}_\tau^{\otimes n} \circ \nu \| \mathrm{EC}_\tau^{\otimes n} \circ \mu)$ and prove $f''(\tau) \geq \frac{c}{\tau} f'(\tau)$.)

Remark: Applying the above with $\tau = \frac{1}{n}$ shows that a Markov chain G_τ known as (small-block) Glauber dynamics for μ is mixing in $O(n^{c+1} \log n)$ time. Indeed, G_τ consists of first applying $\mathrm{EC}_\tau^{\otimes n}$ and then "imputing" erasures in the set S from the conditional distribution $\mu_{X_S|X_{S^c}}$. It is also known that c-pairwise independence is implied (under some additional conditions on μ and $q = 2$) by the uniform boundedness of the operator norms of the covariance matrices of all $\mu^{(\sigma_T)}$ (see [91] for details). Thus a hard question of bounding $\eta_{\mathrm{KL}}(\mu, G_\tau)$ is first reduced to $\eta_{\mathrm{KL}}(\mu, \mathrm{EC}_\tau^{\otimes n})$ and then to the study of the spectrum of a covariance matrix.

References

[1] E. Abbe and E. B. Adserà, "Subadditivity beyond trees and the chi-squared mutual information," in *2019 IEEE International Symposium on Information Theory (ISIT)*. IEEE, 2019, pp. 697–701. (p. 135)

[2] M. C. Abbott and B. B. Machta, "A scaling law from discrete to continuous solutions of channel capacity problems in the low-noise limit," *Journal of Statistical Physics*, vol. 176, no. 1, pp. 214–227, 2019. (p. 250)

[3] I. Abou-Faycal, M. Trott, and S. Shamai, "The capacity of discrete-time memoryless Rayleigh-fading channels," *IEEE Transactions on Information Theory*, vol. 47, no. 4, pp. 1290–1301, 2001. (p. 417)

[4] J. Acharya, C. L. Canonne, Y. Liu, Z. Sun, and H. Tyagi, "Interactive inference under information constraints," *IEEE Transactions on Information Theory*, vol. 68, no. 1, pp. 502–516, 2021. (p. 675)

[5] J. Acharya, C. L. Canonne, Z. Sun, and H. Tyagi, "Unified lower bounds for interactive high-dimensional estimation under information constraints," *Advances in Neural Information Processing Systems*, vol. 36, no. 36, pp. 51133–51165, 2023. (p. 675)

[6] R. Ahlswede, "Extremal properties of rate distortion functions," *IEEE Transactions on Information Theory*, vol. 36, no. 1, pp. 166–171, 1990. (p. 556)

[7] R. Ahlswede, B. Balkenhol, and L. Khachatrian, "Some properties of fix free codes," in *Proceedings of the First INTAS International Seminar on Coding Theory and Combinatorics*, Thahkadzor, Armenia, 1996, pp. 20–33. (pp. 211 and 352)

[8] R. Ahlswede and I. Csiszár, "Hypothesis testing with communication constraints," *IEEE Transactions on Information Theory*, vol. 32, no. 4, pp. 533–542, 1986. (pp. 329, 330, and 331)

[9] R. Ahlswede and J. Körner, "Source coding with side information and a converse for degraded broadcast channels," *IEEE Transactions on Information Theory*, vol. 21, no. 6, pp. 629–637, 1975. (p. 229)

[10] S. M. Alamouti, "A simple transmit diversity technique for wireless communications," *IEEE Journal on Selected Areas in Communications*, vol. 16, no. 8, pp. 1451–1458, 1998. (p. 417)

[11] P. H. Algoet and T. M. Cover, "A sandwich proof of the Shannon–McMillan–Breiman theorem," *The Annals of Probability*, vol. 16, no. 2, pp. 899–909, 1988. (p. 235)

[12] C. D. Aliprantis and K. C. Border, *Infinite Dimensional Analysis: a Hitchhiker's Guide*, 3rd ed. Springer, 2006. (p. 130)

[13] N. Alon, "On the number of subgraphs of prescribed type of graphs with a given number of edges," *Israel Journal of Mathematics*, vol. 38, no. 1-2, pp. 116–130, 1981. (p. 161)

[14] N. Alon and A. Orlitsky, "A lower bound on the expected length of one-to-one codes," *IEEE Transactions on Information Theory*, vol. 40, no. 5, pp. 1670–1672, 1994. (p. 201)

[15] N. Alon and J. H. Spencer, *The Probabilistic Method*, 3rd ed. Wiley, 2008. (pp. 210 and 360)

[16] P. Alquier, "User-friendly introduction to PAC-Bayes bounds," *arXiv preprint arXiv:2110.11216*, 2021. (p. 84)

[17] S.-I. Amari and H. Nagaoka, *Methods of Information Geometry*, vol. 191. American Mathematical Society, 2007. (pp. 39 and 312)

[18] G. Aminian, Y. Bu, L. Toni, M. R. Rodrigues, and G. Wornell, "Characterizing the generalization error of Gibbs algorithm with symmetrized KL information," *arXiv preprint arXiv:2107.13656*, 2021. (p. 190)

[19] V. Anantharam, A. Gohari, S. Kamath, and C. Nair, "On maximal correlation, hypercontractivity, and the data processing inequality studied by Erkip and Cover," *arXiv preprint arXiv:1304.6133*, 2013. (p. 654)

[20] N. Anari, K. Liu, and S. O. Gharan, "Spectral independence in high-dimensional expanders and applications to the hardcore model," *SIAM Journal on Computing*, Special Section, pp. FOCS20–1, 2021. (p. 689)

[21] T. W. Anderson, "The integral of a symmetric unimodal function over a symmetric convex set and some probability inequalities," *Proceedings of the American Mathematical Society*, vol. 6, no. 2, pp. 170–176, 1955. (p. 586)

[22] A. Antos and I. Kontoyiannis, "Convergence properties of functional estimates for discrete distributions," *Random Structures & Algorithms*, vol. 19, no. 3-4, pp. 163–193, 2001. (p. 138)

[23] E. Arıkan, "Channel polarization: A method for constructing capacity-achieving codes for symmetric binary-input memoryless channels," *IEEE Transactions on Information Theory*, vol. 55, no. 7, pp. 3051–3073, 2009. (p. 353)

[24] S. Arimoto, "An algorithm for computing the capacity of arbitrary discrete memoryless channels," *IEEE Transactions on Information Theory*, vol. 18, no. 1, pp. 14–20, 1972. (p. 102)

[25] S. Arimoto, "On the converse to the coding theorem for discrete memoryless channels (corresp.)," *IEEE Transactions on Information Theory*, vol. 19, no. 3, pp. 357–359, 1973. (p. 444)

[26] G. B. Arous and A. Guionnet, "Large deviations for Wigner's law and Voiculescu's non-commutative entropy," *Probability Theory and Related Fields*, vol. 108, no. 4, pp. 517–542, 1997. (p. 686)

[27] S. Artstein, K. Ball, F. Barthe, and A. Naor, "Solution of Shannon's problem on the monotonicity of entropy," *Journal of the American Mathematical Society*, vol. 17, no. 4, pp. 975–982, 2004. (p. 64)

[28] S. Artstein, V. Milman, and S. J. Szarek, "Duality of metric entropy," *Annals of Mathematics*, vol. 159, no. 3, pp. 1313–1328, 2004. (p. 548)

[29] J. Baik, G. Ben Arous, and S. Péché, "Phase transition of the largest eigenvalue for nonnull complex sample covariance matrices," *The Annals of Probability*, vol. 33, no. 5, pp. 1643–1697, 2005. (p. 668)

[30] A. V. Banerjee, "A simple model of herd behavior," *The Quarterly Journal of Economics*, vol. 107, no. 3, pp. 797–817, 1992. (pp. 135 and 183)

[31] Z. Bar-Yossef, T. S. Jayram, R. Kumar, and D. Sivakumar, "An information statistics approach to data stream and communication complexity," *Journal of Computer and System Sciences*, vol. 68, no. 4, pp. 702–732, 2004. (p. 184)

[32] B. Bárány and I. Kolossváry, "On the absolute continuity of the Blackwell measure," *Journal of Statistical Physics*, vol. 159, pp. 158–171, 2015. (pp. 111 and 352)

[33] B. Bárány, M. Pollicott, and K. Simon, "Stationary measures for projective transformations: the Blackwell and Furstenberg measures," *Journal of Statistical Physics*, vol. 148, pp. 393–421, 2012. (p. 111)

[34] A. Barg and G. D. Forney, "Random codes: Minimum distances and error exponents," *IEEE Transactions on Information Theory*, vol. 48, no. 9, pp. 2568–2573, 2002. (p. 443)

[35] A. Barg and A. McGregor, "Distance distribution of binary codes and the error probability of decoding," *IEEE Transactions on Information Theory*, vol. 51, no. 12, pp. 4237–4246, 2005. (p. 443)

[36] S. Barman and O. Fawzi, "Algorithmic aspects of optimal channel coding," *IEEE Transactions on Information Theory*, vol. 64, no. 2, pp. 1038–1045, 2017. (pp. 374, 375, and 376)
[37] A. R. Barron, "Universal approximation bounds for superpositions of a sigmoidal function," *IEEE Transactions on Information Theory*, vol. 39, no. 3, pp. 930–945, 1993. (p. 546)
[38] P. L. Bartlett and S. Mendelson, "Rademacher and Gaussian complexities: Risk bounds and structural results," *Journal of Machine Learning Research*, vol. 3, pp. 463–482, November, 2002. (p. 87)
[39] G. Basharin, "On a statistical estimate for the entropy of a sequence of independent random variables," *Theory of Probability & Its Applications*, vol. 4, no. 3, pp. 333–336, 1959. (p. 597)
[40] A. Beck, *First-Order Methods in Optimization*. SIAM, 2017. (p. 92)
[41] A. Beirami and F. Fekri, "Fundamental limits of universal lossless one-to-one compression of parametric sources," in *Information Theory Workshop (ITW)*. IEEE, 2014, pp. 212–216. (p. 251)
[42] C. H. Bennett, "Notes on Landauer's principle, reversible computation, and Maxwell's Demon," *Studies in History and Philosophy of Modern Physics*, vol. 34, no. 3, pp. 501–510, 2003. (p. xix)
[43] C. H. Bennett, P. W. Shor, J. A. Smolin, and A. V. Thapliyal, "Entanglement-assisted classical capacity of noisy quantum channels," *Physical Review Letters*, vol. 83, no. 15, p. 3081, 1999. (p. 510)
[44] W. R. Bennett, "Spectra of quantized signals," *Bell System Technical Journal*, vol. 27, no. 3, pp. 446–472, 1948. (p. 495)
[45] P. Bergmans, "A simple converse for broadcast channels with additive white Gaussian noise (corresp.)," *IEEE Transactions on Information Theory*, vol. 20, no. 2, pp. 279–280, 1974. (p. 65)
[46] J. M. Bernardo, "Reference posterior distributions for Bayesian inference," *Journal of the Royal Statistical Society: Series B (Methodological)*, vol. 41, no. 2, pp. 113–128, 1979. (p. 254)
[47] C. Berrou, A. Glavieux, and P. Thitimajshima, "Near Shannon limit error-correcting coding and decoding: Turbo-codes. 1," in *Proceedings of ICC'93-IEEE International Conference on Communications*, vol. 2. IEEE, 1993, pp. 1064–1070. (pp. 353 and 420)
[48] J. C. Berry, "Minimax estimation of a bounded normal mean vector," *Journal of Multivariate Analysis*, vol. 35, no. 1, pp. 130–139, 1990. (p. 601)
[49] D. P. Bertsekas, A. Nedić, and A. E. Ozdaglar, *Convex Analysis and Optimization*. Athena Scientific, 2003. (p. 93)
[50] N. Bhatnagar, J. Vera, E. Vigoda, and D. Weitz, "Reconstruction for colorings on trees," *SIAM Journal on Discrete Mathematics*, vol. 25, no. 2, pp. 809–826, 2011. (p. 660)
[51] A. Bhatt, B. Nazer, O. Ordentlich, and Y. Polyanskiy, "Information-distilling quantizers," *IEEE Transactions on Information Theory*, vol. 67, no. 4, pp. 2472–2487, 2021. (p. 193)
[52] A. Bhattacharyya, "On a measure of divergence between two statistical populations defined by their probability distributions," *Bulletin of the Calcutta Mathematical Society*, vol. 35, pp. 99–109, 1943. (p. 117)
[53] L. Birgé, "Approximation dans les espaces métriques et théorie de l'estimation," *Zeitschrift für Wahrscheinlichkeitstheorie und Verwandte Gebiete*, vol. 65, no. 2, pp. 181–237, 1983. (pp. xxii, 617, and 629)
[54] L. Birgé, "On estimating a density using Hellinger distance and some other strange facts," *Probability Theory and Related Fields*, vol. 71, no. 2, pp. 271–291, 1986. (p. 640)
[55] L. Birgé, "Model selection via testing : an alternative to (penalized) maximum

likelihood estimators", *Annales de l'I.H.P. Probabilités et statistiques*, vol. 42, no. 3, pp. 273–325, 2006. (p. 685)

[56] L. Birgé, "Robust tests for model selection," in *From Probability to Statistics and Back: High-Dimensional Models and Processes–A Festschrift in Honor of Jon A. Wellner, IMS Collections*, Vol. 9. Institute of Mathematical Statistics, 2013, pp. 47–64. (p. 628)

[57] M. Š. Birman and M. Solomjak, "Piecewise-polynomial approximations of functions of the classes," *Mathematics of the USSR-Sbornik*, vol. 2, no. 3, p. 295, 1967. (p. 551)

[58] N. Blachman, "The convolution inequality for entropy powers," *IEEE Transactions on Information Theory*, vol. 11, no. 2, pp. 267–271, 1965. (p. 187)

[59] D. Blackwell, L. Breiman, and A. Thomasian, "The capacity of a class of channels," *The Annals of Mathematical Statistics*, vol. 30, no. 24, pp. 1229–1241, 1959. (p. 477)

[60] D. H. Blackwell, "The entropy of functions of finite-state Markov chains," in *Transactions of the First Prague Conference on Information Theory, Statistical Decision Functions, Random Processes*, pp. 13–20, 1956. (p. 111)

[61] R. E. Blahut, "Hypothesis testing and information theory," *IEEE Transactions on Information Theory*, vol. 20, no. 4, pp. 405–417, 1974. (p. 294)

[62] R. Blahut, "Computation of channel capacity and rate-distortion functions," *IEEE Transactions on Information Theory*, vol. 18, no. 4, pp. 460–473, 1972. (p. 102)

[63] P. M. Bleher, J. Ruiz, and V. A. Zagrebnov, "On the purity of the limiting Gibbs state for the Ising model on the Bethe lattice," *Journal of Statistical Physics*, vol. 79, no. 1, pp. 473–482, 1995. (pp. 658 and 660)

[64] S. G. Bobkov and F. Götze, "Exponential integrability and transportation cost related to logarithmic Sobolev inequalities," *Journal of Functional Analysis*, vol. 163, no. 1, pp. 1–28, 1999. (p. 673)

[65] S. Bobkov and G. P. Chistyakov, "Entropy power inequality for the Rényi entropy," *IEEE Transactions on Information Theory*, vol. 61, no. 2, pp. 708–714, 2015. (pp. 26 and 352)

[66] T. Bohman, "A limit theorem for the Shannon capacities of odd cycles I," *Proceedings of the American Mathematical Society*, vol. 131, no. 11, pp. 3559–3569, 2003. (p. 462)

[67] H. F. Bohnenblust, "Convex regions and projections in Minkowski spaces," *Annals of Mathematics*, vol. 39, no. 2, pp. 301–308, 1938. (p. 96)

[68] A. Borovkov, *Mathematical Statistics*. CRC Press, 1999. (pp. xxii, 141, 594, and 596)

[69] S. Boucheron, G. Lugosi, and P. Massart, *Concentration Inequalities: A Nonasymptotic Theory of Independence*. Oxford University Press, 2013. (pp. 86, 151, 307, and 554)

[70] O. Bousquet, D. Kane, and S. Moran, "The optimal approximation factor in density estimation," in *Conference on Learning Theory*. PMLR, 2019, pp. 318–341. (p. 637)

[71] D. Braess and T. Sauer, "Bernstein polynomials and learning theory," *Journal of Approximation Theory*, vol. 128, no. 2, pp. 187–206, 2004. (p. 683)

[72] D. Braess, J. Forster, T. Sauer, and H. U. Simon, "How to achieve minimax expected Kullback–Leibler distance from an unknown finite distribution," in *Algorithmic Learning Theory*. Springer, 2002, pp. 380–394. (p. 683)

[73] M. Braverman, A. Garg, T. Ma, H. L. Nguyen, and D. P. Woodruff, "Communication lower bounds for statistical estimation problems via a distributed data processing inequality," in *Proceedings of the Forty-Eighth Annual ACM Symposium on Theory of Computing*. ACM, 2016, pp. 1011–1020. (pp. 674 and 675)

[74] L. M. Bregman, "Some properties of nonnegative matrices and their

permanents," *Soviet Mathematics Doklady*, vol. 14, no. 4, pp. 945–949, 1973. (p. 162)

[75] L. Breiman, "The individual ergodic theorem of information theory," *The Annals of Mathematical Statistics*, vol. 28, no. 3, pp. 809–811, 1957. (p. 235)

[76] L. Brillouin, *Science and Information Theory,* 2nd ed. Academic Press, 1962. (p. xvii)

[77] L. D. Brown, "Fundamentals of statistical exponential families with applications in statistical decision theory," in *Lecture Notes–Monograph Series*, S. S. Gupta, Ed. Institute of Mathematical Statistics, 1986, vol. 9. (pp. 302, 314, and 315)

[78] P. Bühlmann and S. van de Geer, *Statistics for High-Dimensional Data: Methods, Theory and Applications.* Springer Science & Business Media, 2011. (p. xxii)

[79] M. V. Burnashev, "Data transmission over a discrete channel with feedback. Random transmission time," *Problemy Peredachi Informatsii*, vol. 12, no. 4, pp. 10–30, 1976. (p. 467)

[80] G. Calinescu, C. Chekuri, M. Pal, and J. Vondrák, "Maximizing a monotone submodular function subject to a matroid constraint," *SIAM Journal on Computing*, vol. 40, no. 6, pp. 1740–1766, 2011. (p. 376)

[81] M. X. Cao and M. Tomamichel, "Comments on 'Channel coding rate in the finite blocklength regime': On the quadratic decaying property of the information rate function," *IEEE Transactions on Information Theory*, vol. 69, no. 9, 2023. (p. 445)

[82] G. Casella and W. E. Strawderman, "Estimating a bounded normal mean," *The Annals of Statistics*, vol. 9, no. 4, pp. 870–878, 1981. (p. 601)

[83] O. Catoni, "PAC-Bayesian supervised classification: the thermodynamics of statistical learning," *Lecture Notes–Monograph Series, IMS*, vol. 1277. Institute of Mathematical Statistics, 2007. (p. 84)

[84] E. Çinlar, *Probability and Stochastics.* Springer, 2011. (pp. 19, 29, 30, 31, 81, and 323)

[85] N. N. Cencov, *Statistical Decision Rules and Optimal Inference*, Translations of Mathematical Monographs, vol. 53. American Mathematical Society, 2000. (pp. 39, 311, 312, and 314)

[86] N. Cesa-Bianchi and G. Lugosi, *Prediction, Learning, and Games.* Cambridge University Press, 2006. (p. 257)

[87] D. G. Chapman and H. Robbins, "Minimum variance estimation without regularity assumptions," *The Annals of Mathematical Statistics*, vol. 22, no. 4, pp. 581–586, 1951. (p. 589)

[88] Z. Chase and S. Lovett, "Approximate union closed conjecture," *arXiv preprint arXiv:2211.11689*, 2022. (p. 192)

[89] S. Chatterjee, "An error bound in the Sudakov-Fernique inequality," *arXiv preprint arXiv:0510424*, 2005. (p. 544)

[90] S. Chatterjee and P. Diaconis, "The sample size required in importance sampling," *The Annals of Applied Probability*, vol. 28, no. 2, pp. 1099–1135, 2018. (p. 341)

[91] Z. Chen, K. Liu, and E. Vigoda, "Optimal mixing of Glauber dynamics: Entropy factorization via high-dimensional expansion," in *Proceedings of the 53rd Annual ACM SIGACT Symposium on Theory of Computing.* ACM, 2021, pp. 1537–1550. (p. 689)

[92] H. Chernoff, "Large-sample theory: Parametric case," *The Annals of Mathematical Statistics*, vol. 27, no. 1, pp. 1–22, 1956. (p. 678)

[93] M. Choi, M. Ruskai, and E. Seneta, "Equivalence of certain entropy contraction coefficients," *Linear Algebra and Its Applications*, vol. 208, pp. 29–36, 1994. (p. 648)

[94] N. Chomsky, "Three models for the description of language," *IRE Transactions on Information Theory*, vol. 2, no. 3, pp. 113–124, 1956. (p. 196)

[95] B. S. Clarke and A. R. Barron, "Information-theoretic asymptotics of Bayes methods," *IEEE Transactions on

Information Theory, vol. 36, no. 3, pp. 453–471, 1990. (p. 254)

[96] B. S. Clarke and A. R. Barron, "Jeffreys' prior is asymptotically least favorable under entropy risk," *Journal of Statistical Planning and Inference*, vol. 41, no. 1, pp. 37–60, 1994. (p. 254)

[97] J. E. Cohen, J. H. B. Kempermann, and G. Zbăganu, *Comparisons of Stochastic Matrices with Applications in Information Theory, Statistics, Economics and Population*. Springer, 1998. (p. 648)

[98] A. Collins and Y. Polyanskiy, "Coherent multiple-antenna block-fading channels at finite blocklength," *IEEE Transactions on Information Theory*, vol. 65, no. 1, pp. 380–405, 2018. (p. 447)

[99] J. H. Conway and N. J. A. Sloane, *Sphere Packings, Lattices and Groups*. Springer Science & Business Media, 1999, vol. 290. (p. 539)

[100] M. Costa, "A new entropy power inequality," *IEEE Transactions on Information Theory*, vol. 31, no. 6, pp. 751–760, 1985. (p. 64)

[101] D. J. Costello and G. D. Forney, "Channel coding: The road to channel capacity," *Proceedings of the IEEE*, vol. 95, no. 6, pp. 1150–1177, 2007. (p. 420)

[102] T. A. Courtade, "Monotonicity of entropy and Fisher information: a quick proof via maximal correlation," *Communications in Information and Systems*, vol. 16, no. 2, pp. 111–115, 2017. (p. 64)

[103] T. A. Courtade, "A strong entropy power inequality," *IEEE Transactions on Information Theory*, vol. 64, no. 4, pp. 2173–2192, 2017. (p. 64)

[104] T. M. Cover, "Universal data compression and portfolio selection," in *Proceedings of 37th Conference on Foundations of Computer Science*. IEEE, 1996, pp. 534–538. (p. xx)

[105] T. M. Cover and B. Gopinath, *Open Problems in Communication and Computation*. Springer Science & Business Media, 2012. (p. 423)

[106] T. M. Cover and J. A. Thomas, *Elements of Information Theory,* 2nd ed. Wiley-Interscience, 2006. (pp. xvii, xxi, 65, 212, 218, 363, 477, and 513)

[107] H. Cramér, "Über eine eigenschaft der normalen verteilungsfunktion," *Mathematische Zeitschrift*, vol. 41, no. 1, pp. 405–414, 1936. (p. 101)

[108] H. Cramér, *Mathematical Methods of Statistics*. Princeton University Press, 1946. (p. 589)

[109] I. Csiszár, "Information-type measures of difference of probability distributions and indirect observation," *Studia Scientiarum Mathematicarum Hungarica*, vol. 2, pp. 229–318, 1967. (p. 115)

[110] I. Csiszár and J. Körner, "Graph decomposition: a new key to coding theorems," *IEEE Transactions on Information Theory*, vol. 27, no. 1, pp. 5–12, 1981. (p. 47)

[111] I. Csiszár and J. Körner, *Information Theory: Coding Theorems for Discrete Memoryless Systems*. Academic Press, 1981. (pp. xvii, xxi, 363, 511, 512, 514, and 663)

[112] I. Csiszár and P. C. Shields, "Information theory and statistics: A tutorial," *Foundations and Trends in Communications and Information Theory*, vol. 1, no. 4, pp. 417–528, 2004. (pp. 104 and 251)

[113] I. Csiszár and G. Tusnády, "Information geometry and alternating minimization problems," *Statistics & Decision, Supplement Issue* vol. 1, 1984. (pp. 102 and 103)

[114] I. Csiszár, "I-divergence geometry of probability distributions and minimization problems," *The Annals of Probability*, pp. 146–158, 1975. (pp. 308 and 316)

[115] I. Csiszár and J. Körner, *Information Theory: Coding Theorems for Discrete Memoryless Systems*, 2nd ed. Cambridge University Press, 2011. (pp. xxi, 12, 218, and 435)

[116] P. Cuff, "Distributed channel synthesis," *IEEE Transactions on Information Theory*, vol. 59, no. 11, pp. 7071–7096, 2013. (p. 515)

[117] M. Cuturi, "Sinkhorn distances: Lightspeed computation of optimal transport," in *Advances in Neural Information Processing Systems*, 26, pp. 2292–2300, 2013. (p. 105)
[118] L. Davisson, R. McEliece, M. Pursley, and M. Wallace, "Efficient universal noiseless source codes," *IEEE Transactions on Information Theory*, vol. 27, no. 3, pp. 269–279, 1981. (p. 272)
[119] A. Decelle, F. Krzakala, C. Moore, and L. Zdeborová, "Asymptotic analysis of the stochastic block model for modular networks and its algorithmic applications," *Physical Review E*, vol. 84, no. 6, p. 066106, 2011. (p. 658)
[120] A. Dembo and O. Zeitouni, *Large Deviations Techniques and Applications*. Springer, 2009. (pp. 297 and 312)
[121] A. P. Dempster, N. M. Laird, and D. B. Rubin, "Maximum likelihood from incomplete data via the EM algorithm," *Journal of the Royal Statistical Society. Series B (Methodological)*, vol. 39, no. 1, pp. 1–22, 1977. (p. 103)
[122] P. Diaconis and L. Saloff-Coste, "Logarithmic Sobolev inequalities for finite Markov chains," *The Annals of Applied Probability*, vol. 6, no. 3, pp. 695–750, 1996. (pp. 132 and 194)
[123] P. Diaconis and D. Freedman, "Finite exchangeable sequences," *The Annals of Probability*, vol. 8, no. 4, pp. 745–764, 1980. (p. 189)
[124] P. Diaconis and D. Stroock, "Geometric bounds for eigenvalues of Markov chains," *The Annals of Applied Probability*, vol. 1, no. 1, pp. 36–61, 1991. (pp. 658 and 687)
[125] H. Djellout, A. Guillin, and L. Wu, "Transportation cost-information inequalities and applications to random dynamical systems and diffusions," *The Annals of Probability*, vol. 32, no. 3B, pp. 2702–2732, 2004. (p. 673)
[126] R. Dobrushin and B. Tsybakov, "Information transmission with additional noise," *IRE Transactions on Information Theory*, vol. 8, no. 5, pp. 293–304, 1962. (p. 563)
[127] R. L. Dobrushin, "Central limit theorem for nonstationary Markov chains, I," *Theory of Probability & Its Applications*, vol. 1, no. 1, pp. 65–80, 1956. (p. 646)
[128] R. L. Dobrushin, "A simplified method of experimentally evaluating the entropy of a stationary sequence," *Theory of Probability & Its Applications*, vol. 3, no. 4, pp. 428–430, 1958. (p. 597)
[129] R. L. Dobrushin, "A general formulation of the fundamental theorem of Shannon in the theory of information," *Uspekhi Matematicheskikh Nauk*, vol. 14, no. 6, pp. 3–104, 1959. English translation in *Eleven Papers in Analysis: Nine Papers on Differential Equations, Two on Information Theory*, American Mathematical Society Translations: Series 2, Volume 33, 1963. (p. 83)
[130] R. L. Dobrushin, "Mathematical problems in the Shannon theory of optimal coding of information," in *Proceedings of the 4th Berkeley Symposium on Mathematics, Statistics, and Probability*, vol. 1, Berkeley, CA, USA, 1961, pp. 211–252. (p. 444)
[131] R. L. Dobrushin, "Asymptotic bounds on error probability for transmission over DMC with symmetric transition probabilities," *Theory of Probability & Its Applications*, vol. 7, pp. 283–311, 1962. (pp. 391 and 456)
[132] J. Dong, A. Roth, and W. J. Su, "Gaussian differential privacy," *Journal of the Royal Statistical Society Series B: Statistical Methodology*, vol. 84, no. 1, pp. 3–37, 2022. (p. 184)
[133] D. L. Donoho, "Wald lecture I: Counting bits with Kolmogorov and Shannon," *Note for the Wald Lectures, IMS Annual Meeting*, July 1997. (p. 556)
[134] M. D. Donsker and S. S. Varadhan, "Asymptotic evaluation of certain Markov process expectations for large time. IV," *Communications on Pure and Applied Mathematics*, vol. 36, no. 2, pp. 183–212, 1983. (p. 72)
[135] J. L. Doob, *Stochastic Processes*. Wiley, 1953. (p. 234)
[136] F. du Pin Calmon, Y. Polyanskiy, and Y. Wu, "Strong data processing

inequalities for input constrained additive noise channels," *IEEE Transactions on Information Theory*, vol. 64, no. 3, pp. 1879–1892, 2017. (p. 330)

[137] J. C. Duchi, M. I. Jordan, M. J. Wainwright, and Y. Zhang, "Optimality guarantees for distributed statistical estimation," *arXiv preprint arXiv:1405.0782*, 2014. (pp. 674 and 675)

[138] J. Duda, "Asymmetric numeral systems: entropy coding combining speed of Huffman coding with compression rate of arithmetic coding," *arXiv preprint arXiv:1311.2540*, 2013. (p. 247)

[139] R. M. Dudley, *Uniform Central Limit Theorems*. Cambridge University Press, 1999, no. 63. (pp. 87 and 548)

[140] G. Dueck, "The strong converse to the coding theorem for the multiple-access channel," *Journal of Combinatorics, Information & System Sciences*, vol. 6, no. 3, pp. 187–196, 1981. (p. 190)

[141] G. Dueck and J. Körner, "Reliability function of a discrete memoryless channel at rates above capacity (corresp.)," *IEEE Transactions on Information Theory*, vol. 25, no. 1, pp. 82–85, 1979. (p. 444)

[142] N. Dunford and J. T. Schwartz, *Linear Operators, Part 1: General Theory*. Wiley, 1988. (p. 81)

[143] R. Durrett, *Probability: Theory and Examples*, 4th ed. Cambridge University Press, 2010. (p. 125)

[144] A. Dytso, S. Yagli, H. V. Poor, and S. S. Shitz, "The capacity achieving distribution for the amplitude constrained additive Gaussian channel: An upper bound on the number of mass points," *IEEE Transactions on Information Theory*, vol. 66, no. 4, pp. 2006–2022, 2019. (p. 416)

[145] H. G. Eggleston, *Convexity*, Cambridge Tracts in Mathematics and Mathematical Physics, vol. 47. Cambridge University Press, 1958. (p. 129)

[146] A. El Alaoui and A. Montanari, "An information-theoretic view of stochastic localization," *IEEE Transactions on Information Theory*, vol. 68, no. 11, pp. 7423–7426, 2022. (p. 194)

[147] A. El Gamal and Y.-H. Kim, *Network Information Theory*. Cambridge University Press, 2011. (pp. xxi and 513)

[148] R. Eldan, "Taming correlations through entropy-efficient measure decompositions with applications to mean-field approximation," *Probability Theory and Related Fields*, vol. 176, no. 3-4, pp. 737–755, 2020. (p. 194)

[149] P. Elias, "The efficient construction of an unbiased random sequence," *The Annals of Mathematical Statistics*, vol. 43, no. 3, pp. 865–870, 1972. (p. 167)

[150] P. Elias, "Coding for noisy channels," *IRE Convention Record*, vol. 3, pp. 37–46, 1955. (p. 373)

[151] E. O. Elliott, "Estimates of error rates for codes on burst-noise channels," *Bell System Technical Journal*, vol. 42, pp. 1977–1997, Sep. 1963. (p. 111)

[152] D. M. Endres and J. E. Schindelin, "A new metric for probability distributions," *IEEE Transactions on Information theory*, vol. 49, no. 7, pp. 1858–1860, 2003. (p. 117)

[153] P. Erdös, "Some remarks on the theory of graphs," *Bulletin of the American Mathematical Society*, vol. 53, no. 4, pp. 292–294, 1947. (p. 217)

[154] P. Erdös and A. Rényi, "On random graphs, I," *Publicationes Mathematicae (Debrecen)*, vol. 6, pp. 290–297, 1959. (p. 669)

[155] V. Erokhin, "ε-entropy of a discrete random variable," *Theory of Probability & Its Applications*, vol. 3, no. 1, pp. 97–100, 1958. (p. 562)

[156] K. Eswaran and M. Gastpar, "Remote source coding under Gaussian noise: Dueling roles of power and entropy power," *IEEE Transactions on Information Theory*, vol. 65, no. 7, pp. 4486–4498, 2019. (p. 674)

[157] W. Evans and L. J. Schulman "On the maximum tolerable noise of k-input gates for reliable computation by formulas," *IEEE Transactions on Information Theory*, vol. 49, no. 11, pp. 3094–3098, 2003. (p. 645)

[158] W. S. Evans and L. J. Schulman, "Signal propagation and noisy circuits," *IEEE Transactions on Information Theory*, vol. 45, no. 7, pp. 2367–2373, 1999. (p. 644)

[159] M. Falahatgar, A. Orlitsky, V. Pichapati, and A. Suresh, "Learning Markov distributions: Does estimation trump compression?" in *2016 IEEE International Symposium on Information Theory (ISIT)*. IEEE, 2016, pp. 2689–2693. (p. 259)

[160] M. Feder, "Gambling using a finite state machine," *IEEE Transactions on Information Theory*, vol. 37, no. 5, pp. 1459–1465, 1991. (p. 266)

[161] M. Feder, N. Merhav, and M. Gutman, "Universal prediction of individual sequences," *IEEE Transactions on Information Theory*, vol. 38, no. 4, pp. 1258–1270, 1992. (p. 261)

[162] M. Feder and Y. Polyanskiy, "Sequential prediction under log-loss and misspecification," in *Conference on Learning Theory*. PMLR, 2021, pp. 1937–1964. (pp. 177 and 263)

[163] A. A. Fedotov, P. Harremoës, and F. Topsøe, "Refinements of Pinsker's inequality," *IEEE Transactions on Information Theory*, vol. 49, no. 6, pp. 1491–1498, 2003. (p. 131)

[164] W. Feller, *An Introduction to Probability Theory and Its Applications*, 3rd ed. Wiley, 1970, vol. I. (p. 550)

[165] W. Feller, *An Introduction to Probability Theory and Its Applications*, 2nd ed. Wiley, 1971, vol. II. (p. 445)

[166] T. S. Ferguson, *Mathematical Statistics: A Decision Theoretic Approach*. Academic Press, 1967. (p. 571)

[167] T. S. Ferguson, "An inconsistent maximum likelihood estimate," *Journal of the American Statistical Association*, vol. 77, no. 380, pp. 831–834, 1982. (p. 596)

[168] T. S. Ferguson, *A Course in Large Sample Theory*. CRC Press, 1996. (p. 596)

[169] R. A. Fisher, "The logic of inductive inference," *Journal of the Royal Statistical Society*, vol. 98, no. 1, pp. 39–82, 1935. (p. xvii)

[170] B. M. Fitingof, "The compression of discrete information," *Problemy Peredachi Informatsii*, vol. 3, no. 3, pp. 28–36, 1967. (p. 247)

[171] P. Fleisher, "Sufficient conditions for achieving minimum distortion in a quantizer," *IEEE International Convention Record*, pp. 104–111, 1964. (p. 493)

[172] G. D. Forney, "Concatenated codes," *MIT RLE Technical Report*, 440, 1965. (p. 386)

[173] E. Friedgut and J. Kahn, "On the number of copies of one hypergraph in another," *Israel Journal of Mathematics*, vol. 105, pp. 251–256, 1998. (p. 161)

[174] P. Gács and J. Körner, "Common information is far less than mutual information," *Problems of Control and Information Theory*, vol. 2, no. 2, pp. 149–162, 1973. (p. 343)

[175] A. Galanis, D. Štefankovič, and E. Vigoda, "Inapproximability of the partition function for the antiferromagnetic Ising and hard-core models," *Combinatorics, Probability and Computing*, vol. 25, no. 4, pp. 500–559, 2016. (p. 75)

[176] R. G. Gallager, "A simple derivation of the coding theorem and some applications," *IEEE Transactions on Information Theory*, vol. 11, no. 1, pp. 3–18, 1965. (p. 368)

[177] R. G. Gallager, *Information Theory and Reliable Communication*. Wiley, 1968. (pp. xvii, xxi, 391, 441, and 442)

[178] R. G. Gallager, "The random coding bound is tight for the average code (corresp.)," *IEEE Transactions on Information Theory*, vol. 19, no. 2, pp. 244–246, 1973. (p. 442)

[179] R. Gardner, "The Brunn-Minkowski inequality," *Bulletin of the American Mathematical Society*, vol. 39, no. 3, pp. 355–405, 2002. (p. 587)

[180] A. M. Garsia, *Topics in Almost Everywhere Convergence*. Markham, 1970. (p. 239)

[181] M. Gastpar, B. Rimoldi, and M. Vetterli, "To code, or not to code: Lossy

source-channel communication revisited," *IEEE Transactions on Information Theory*, vol. 49, no. 5, pp. 1147–1158, 2003. (p. 533)

[182] I. M. Gel'fand, A. N. Kolmogorov, and A. M. Yaglom, "On the general definition of the amount of information," *Doklady Akademii Nauk SSSR*, vol. 11, pp. 745–748, 1956. (p. 71)

[183] S. I. Gelfand and M. Pinsker, "Coding for channels with random parameters," *Problems in Control and Information Theory*, vol. 9, no. 1, pp. 19–31, 1980. (p. 480)

[184] Y. Geng and C. Nair, "The capacity region of the two-receiver Gaussian vector broadcast channel with private and common messages," *IEEE Transactions on Information Theory*, vol. 60, no. 4, pp. 2087–2104, 2014. (p. 109)

[185] G. L. Gilardoni, "On a Gel'fand–Yaglom–Peres theorem for f-divergences," *arXiv preprint arXiv:0911.1934*, 2009. (p. 154)

[186] G. L. Gilardoni, "On Pinsker's and Vajda's type inequalities for Csiszár's-divergences," *IEEE Transactions on Information Theory*, vol. 56, no. 11, pp. 5377–5386, 2010. (p. 133)

[187] E. N. Gilbert, "Capacity of burst-noise channels," *Bell System Technical Journal*, vol. 39, pp. 1253–1265, Sep. 1960. (p. 111)

[188] R. D. Gill and B. Y. Levit, "Applications of the van Trees inequality: a Bayesian Cramér-Rao bound," *Bernoulli*, vol. 1, no. 1–2, pp. 59–79, 1995. (p. 590)

[189] J. Gilmer, "A constant lower bound for the union-closed sets conjecture," *arXiv preprint arXiv:2211.09055*, 2022. (p. 192)

[190] C. Giraud, *Introduction to High-Dimensional Statistics*. Chapman and Hall/CRC, 2014. (p. xxii)

[191] G. Glaeser, "Racine carrée d'une fonction différentiable," *Annales de l'Institut Fourier*, vol. 13, no. 2, pp. 203–210, 1963. (p. 641)

[192] O. Goldreich, *Introduction to Property Testing*. Cambridge University Press, 2017. (p. 329)

[193] I. Goodfellow, J. Pouget-Abadie, M. Mirza, B. Xu, D. Warde-Farley, S. Ozair, A. Courville, and Y. Bengio, "Generative adversarial nets," in *Advances in Neural Information Processing Systems*, 27, 2014. (pp. 149 and 150)

[194] V. Goodman, "Characteristics of normal samples," *The Annals of Probability*, vol. 16, no. 3, pp. 1281–1290, 1988. (p. 553)

[195] V. D. Goppa, "Nonprobabilistic mutual information with memory," *Problems in Control Information Theory*, vol. 4, pp. 97–102, 1975. (p. 474)

[196] V. D. Goppa, "Codes and information," *Russian Mathematical Surveys*, vol. 39, no. 1, p. 87, 1984. (p. 475)

[197] R. M. Gray and D. L. Neuhoff, "Quantization," *IEEE Transactions on Information Theory*, vol. 44, no. 6, pp. 2325–2383, 1998. (p. 486)

[198] R. M. Gray, *Entropy and Information Theory*. Springer, 1990. (p. xxi)

[199] U. Grenander and G. Szegö, *Toeplitz Forms and Their Applications*, 2nd ed. Chelsea, 1984. (p. 114)

[200] L. Gross, "Logarithmic Sobolev inequalities," *American Journal of Mathematics*, vol. 97, no. 4, pp. 1061–1083, 1975. (pp. 107 and 194)

[201] Y. Gu, "Channel comparison methods and statistical problems on graphs," Ph.D. dissertation, MIT, Cambridge, MA, 2023. (p. 654)

[202] Y. Gu and Y. Polyanskiy, "Uniqueness of BP fixed point for the Potts model and applications to community detection," in *Conference on Learning Theory (COLT)*, pp. 837–884. PMLR, 2023. (pp. 135 and 661)

[203] Y. Gu and Y. Polyanskiy, "Non-linear log-Sobolev inequalities for the Potts semigroup and applications to reconstruction problems," *Communications in Mathematical Physics*, vol. 404, pp. 769–831, 2023. Also *arXiv preprint*, 2022. (pp. 660, 670, 688, and 689)

[204] Y. Gu, H. Roozbehani, and Y. Polyanskiy, "Broadcasting on trees

near criticality," in *2020 IEEE International Symposium on Information Theory (ISIT)*. IEEE, 2020, pp. 1504–1509. (p. 688)

[205] D. Guo, S. Shamai (Shitz), and S. Verdú, "Mutual information and minimum mean-square error in Gaussian channels," *IEEE Transactions on Information Theory*, vol. 51, no. 4, pp. 1261–1283, Apr. 2005. (p. 59)

[206] D. Guo, Y. Wu, S. S. Shamai, and S. Verdú, "Estimation in Gaussian noise: Properties of the minimum mean-square error," *IEEE Transactions on Information Theory*, vol. 57, no. 4, pp. 2371–2385, 2011. (p. 63)

[207] U. Hadar, J. Liu, Y. Polyanskiy, and O. Shayevitz, "Communication complexity of estimating correlations," in *Proceedings of the 51st Annual ACM SIGACT Symposium on Theory of Computing*. ACM, 2019, pp. 792–803. (pp. 661 and 662)

[208] B. Hajek, Y. Wu, and J. Xu, "Information limits for recovering a hidden community," *IEEE Transactions on Information Theory*, vol. 63, no. 8, pp. 4729–4745, 2017. (p. 606)

[209] J. Hájek, "Local asymptotic minimax and admissibility in estimation," in *Proceedings of the Sixth Berkeley Symposium on Mathematical Statistics and Probability*, vol. 1, 1972, pp. 175–194. (p. 596)

[210] J. M. Hammersley, "On estimating restricted parameters," *Journal of the Royal Statistical Society. Series B (Methodological)*, vol. 12, no. 2, pp. 192–240, 1950. (p. 589)

[211] T. S. Han, *Information-Spectrum Methods in Information Theory*. Springer Science & Business Media, 2003. (pp. xix and xxi)

[212] T. S. Han and S. Verdú, "Approximation theory of output statistics," *IEEE Transactions on Information Theory*, vol. 39, no. 3, pp. 752–772, 1993. (pp. 516 and 518)

[213] Y. Han, S. Jana, and Y. Wu, "Optimal prediction of Markov chains with and without spectral gap," *IEEE Transactions on Information Theory*, vol. 69, no. 6, pp. 3920–3959, 2023. (pp. 259 and 260)

[214] P. Harremoës and I. Vajda, "On pairs of f-divergences and their joint range," *IEEE Transactions on Information Theory*, vol. 57, no. 6, pp. 3230–3235, Jun. 2011. (pp. 115 and 129)

[215] B. Harris, "The statistical estimation of entropy in the non-parametric case," in *Topics in Information Theory*, I. Csiszár and P. Elias, Eds. Springer Netherlands, 1975, vol. 16, pp. 323–355. (p. 597)

[216] D. Haussler and M. Opper, "Mutual information, metric entropy and cumulative relative entropy risk," *The Annals of Statistics*, vol. 25, no. 6, pp. 2451–2492, 1997. (pp. xxii, 190, and 191)

[217] M. Hayashi, "General nonasymptotic and asymptotic formulas in channel resolvability and identification capacity and their application to the wiretap channel," *IEEE Transactions on Information Theory*, vol. 52, no. 4, pp. 1562–1575, 2006. (p. 518)

[218] W. Hoeffding, "Asymptotically optimal tests for multinomial distributions," *The Annals of Mathematical Statistics*, vol. 36, no. 2, pp. 369–401, 1965. (p. 294)

[219] P. J. Huber, "Fisher information and spline interpolation," *The Annals of Statistics*, vol. 2, no. 5, pp. 1029–1033, 1974. (p. 593)

[220] P. J. Huber, *Robust Statistics*. Wiley-Interscience, 1981. (pp. 151 and 152)

[221] P. J. Huber, "A robust version of the probability ratio test," *The Annals of Mathematical Statistics*, pp. 1753–1758, 1965. (pp. 328, 343, and 628)

[222] I. A. Ibragimov and R. Z. Khas'minski, *Statistical Estimation: Asymptotic Theory*. Springer, 1981. (pp. xxii and 143)

[223] S. Ihara, "On the capacity of channels with additive non-Gaussian noise," *Information and Control*, vol. 37, no. 1, pp. 34–39, 1978. (p. 409)

[224] S. Ihara, *Information Theory for Continuous Systems*, vol. 2, World Scientific, 1993. (p. 428)

[225] Y. I. Ingster and I. A. Suslina, *Nonparametric Goodness-of-Fit Testing under Gaussian Models*. Springer, 2003. (pp. 134, 187, 329, and 574)

[226] Y. I. Ingster, "Minimax testing of nonparametric hypotheses on a distribution density in the L_p metrics," *Theory of Probability & Its Applications*, vol. 31, no. 2, pp. 333–337, 1987. (p. 329)

[227] S. Janson, "Random regular graphs: asymptotic distributions and contiguity," *Combinatorics, Probability and Computing*, vol. 4, no. 4, pp. 369–405, 1995. (p. 188)

[228] S. Janson and E. Mossel, "Robust reconstruction on trees is determined by the second eigenvalue," *The Annals of Probability*, vol. 32, no. 3B, pp. 2630–2649, 2004. (p. 661)

[229] E. T. Jaynes, *Probability Theory: The Logic of Science*. Cambridge University Press, 2003. (p. 254)

[230] T. S. Jayram, "Hellinger strikes back: A note on the multi-party information complexity of AND," in *International Workshop on Approximation Algorithms for Combinatorial Optimization*, 2009, pp. 562–573. (p. 185)

[231] I. Jensen and A. J. Guttmann, "Series expansions of the percolation probability for directed square and honeycomb lattices," *Journal of Physics A: Mathematical and General*, vol. 28, no. 17, p. 4813, 1995. (p. 688)

[232] Z. Jia, Y. Polyanskiy, and Y. Wu, "Entropic characterization of optimal rates for learning Gaussian mixtures," in *Conference on Learning Theory (COLT)*. PMLR, 2023. (p. 635)

[233] J. Jiao, K. Venkat, Y. Han, and T. Weissman, "Minimax estimation of functionals of discrete distributions," *IEEE Transactions on Information Theory*, vol. 61, no. 5, pp. 2835–2885, 2015. (p. 598)

[234] C. Jin, Y. Zhang, S. Balakrishnan, M. J. Wainwright, and M. I. Jordan, "Local maxima in the likelihood of Gaussian mixture models: Structural results and algorithmic consequences," in *Advances in Neural Information Processing Systems*, vol. 29, 2016, pp. 4116–4124. (p. 104)

[235] W. B. Johnson, G. Schechtman, and J. Zinn, "Best constants in moment inequalities for linear combinations of independent and exchangeable random variables," *The Annals of Probability*, vol. 13, no. 1, pp. 234–253, 1985. (p. 509)

[236] I. Johnstone, *Gaussian Estimation: Sequence and Wavelet Models*, 2011, available at www-stat.stanford.edu/~imj/. (p. 604)

[237] L. K. Jones, "A simple lemma on greedy approximation in Hilbert space and convergence rates for projection pursuit regression and neural network training," *The Annals of Statistics*, vol. 20, no. 1, pp. 608–613, 1992. (p. 546)

[238] A. B. Juditsky and A. S. Nemirovski, "Nonparametric estimation by convex programming," *The Annals of Statistics*, vol. 37, no. 5A, pp. 2278–2300, 2009. (p. 579)

[239] H. Jung, "Über den kleinsten Kreis, der eine ebene Figur einschliesst," *J. Reine Angew. Math.*, vol. 137, no. 4, pp. 310–313, 1910. (p. 96)

[240] S. M. Kakade, K. Sridharan, and A. Tewari, "On the complexity of linear prediction: Risk bounds, margin bounds, and regularization," in *Advances in Neural Information Processing Systems*, vol. 21, 2008. (p. 87)

[241] S. Kamath, A. Orlitsky, D. Pichapati, and A. Suresh, "On learning distributions from their samples," in *Conference on Learning Theory*, pp. 1066–1100, PMLR 2015. (p. 259)

[242] T. Kawabata and A. Dembo, "The rate-distortion dimension of sets and measures," *IEEE Transactions on Information Theory*, vol. 40, no. 5, pp. 1564–1572, Sep. 1994. (p. 555)

[243] M. Keane and G. O'Brien, "A Bernoulli factory," *ACM Transactions on Modeling and Computer Simulation*, vol. 4, no. 2, pp. 213–219, 1994. (p. 172)

[244] J. Kemperman, "On the Shannon capacity of an arbitrary channel," in *Indagationes Mathematicae (Proceedings)*, vol. 77, no. 2. North-Holland, 1974, pp. 101–115. (p. 97)

[245] H. Kesten and B. P. Stigum, "Additional limit theorems for indecomposable multidimensional Galton-Watson processes," *The Annals of Mathematical Statistics*, vol. 37, no. 6, pp. 1463–1481, 1966. (pp. 661 and 688)

[246] D. P. Kingma and M. Welling, "Auto-encoding variational Bayes," *arXiv preprint arXiv:1312.6114*, 2013. (pp. 77 and 78)

[247] D. P. Kingma, M. Welling, "An introduction to variational autoencoders," *Foundations and Trends in Machine Learning*, vol. 12, no. 4, pp. 307–392, 2019. (p. 78)

[248] T. Koch, "The Shannon lower bound is asymptotically tight," *IEEE Transactions on Information Theory*, vol. 62, no. 11, pp. 6155–6161, 2016. (p. 523)

[249] Y. Kochman, O. Ordentlich, and Y. Polyanskiy, "A lower bound on the expected distortion of joint source-channel coding," *IEEE Transactions on Information Theory*, vol. 66, no. 8, pp. 4722–4741, 2020. (p. 533)

[250] A. Kolchinsky and B. D. Tracey, "Estimating mixture entropy with pairwise distances," *Entropy*, vol. 19, no. 7, p. 361, 2017. (p. 190)

[251] A. N. Kolmogorov and V. M. Tikhomirov, "ε-entropy and ε-capacity of sets in function spaces," *Uspekhi Matematicheskikh Nauk*, vol. 14, no. 2, pp. 3–86, 1959. Reprinted in Shiryayev, A. N., ed. *Selected Works of AN Kolmogorov: Volume III: Information Theory and the Theory of Algorithms*, Vol. 27, Springer Netherlands, 1993, pp. 86–170. (pp. 534, 536, 539, 548, 551, and 555)

[252] I. Kontoyiannis and S. Verdú, "Optimal lossless data compression: Non-asymptotics and asymptotics," *IEEE Transactions on Information Theory*, vol. 60, no. 2, pp. 777–795, 2014. (p. 200)

[253] J. Körner and A. Orlitsky, "Zero-error information theory," *IEEE Transactions on Information Theory*, vol. 44, no. 6, pp. 2207–2229, 1998. (p. 382)

[254] V. Koshelev, "Quantization with minimal entropy," *Problemy Peredachi Informatsii*, vol. 14, pp. 151–156, 1963. (p. 495)

[255] V. Kostina, Y. Polyanskiy, and S. Verdú, "Variable-length compression allowing errors," *IEEE Transactions on Information Theory*, vol. 61, no. 8, pp. 4316–4330, 2015. (p. 562)

[256] V. Kostina and S. Verdú, "Fixed-length lossy compression in the finite blocklength regime," *IEEE Transactions on Information Theory*, vol. 58, no. 6, pp. 3309–3338, 2012. (p. 497)

[257] O. Kosut and L. Sankar, "Asymptotics and non-asymptotics for universal fixed-to-variable source coding," *IEEE Transactions on Information Theory*, vol. 63, no. 6, pp. 3757–3772, 2017. (p. 251)

[258] A. Krause and D. Golovin, "Submodular function maximization," *Tractability*, vol. 3, pp. 71–104, 2014. (p. 375)

[259] R. Krichevskiy, "Laplace's law of succession and universal encoding," *IEEE Transactions on Information Theory*, vol. 44, no. 1, pp. 296–303, 1998. (p. 683)

[260] R. Krichevsky, "A relation between the plausibility of information about a source and encoding redundancy," *Problems of Information Transmission*, vol. 4, no. 3, pp. 48–57, 1968. (p. 249)

[261] R. Krichevsky and V. Trofimov, "The performance of universal encoding," *IEEE Transactions on Information Theory*, vol. 27, no. 2, pp. 199–207, 1981. (p. 255)

[262] F. Krzakała, A. Montanari, F. Ricci-Tersenghi, G. Semerjian, and L. Zdeborová, "Gibbs states and the set of solutions of random constraint satisfaction problems," *Proceedings of the National Academy of Sciences*, vol. 104, no. 25, pp. 10318–10323, 2007. (p. 658)

[263] J. Kuelbs, "A strong convergence theorem for Banach space valued random variables," *The Annals of Probability*, vol. 4, no. 5, pp. 744–771, 1976. (p. 552)

[264] J. Kuelbs and W. V. Li, "Metric entropy and the small ball problem for Gaussian measures," *Journal of Functional Analysis*, vol. 116, no. 1, pp. 133–157, 1993. (pp. 553 and 554)

[265] S. Kullback, *Information Theory and Statistics*. Dover, 1968. (p. xxi)

[266] C. Külske and M. Formentin, "A symmetric entropy bound on the non-reconstruction regime of Markov chains on Galton-Watson trees," *Electronic Communications in Probability*, vol. 14, pp. 587–596, 2009. (p. 135)

[267] H. O. Lancaster, "Some properties of the bivariate normal distribution considered in the form of a contingency table," *Biometrika*, vol. 44, no. 1/2, pp. 289–292, 1957. (p. 657)

[268] R. Landauer, "Irreversibility and heat generation in the computing process," *IBM Journal of Research and Development*, vol. 5, no. 3, pp. 183–191, 1961. (p. xix)

[269] A. Lapidoth, *A Foundation in Digital Communication*. Cambridge University Press, 2017. (p. 412)

[270] A. Lapidoth and S. M. Moser, "Capacity bounds via duality with applications to multiple-antenna systems on flat-fading channels," *IEEE Transactions on Information Theory*, vol. 49, no. 10, pp. 2426–2467, 2003. (p. 417)

[271] B. Laurent and P. Massart, "Adaptive estimation of a quadratic functional by model selection," *The Annals of Statistics*, vol. 28, no. 5, pp. 1302–1338, 2000. (p. 86)

[272] S. L. Lauritzen, *Graphical Models*. Clarendon, 1996. (p. 51)

[273] L. Le Cam, "Convergence of estimates under dimensionality restrictions," *The Annals of Statistics*, vol. 1, no. 1, pp. 38–53, 1973. (p. xxii)

[274] L. Le Cam, *Asymptotic Methods in Statistical Decision Theory*. Springer, 1986. (pp. 117, 133, 571, 596, 617, and 629)

[275] C. C. Leang and D. H. Johnson, "On the asymptotics of m-hypothesis Bayesian detection," *IEEE Transactions on Information Theory*, vol. 43, no. 1, pp. 280–282, 1997. (p. 342)

[276] K. Lee, Y. Wu, and Y. Bresler, "Near optimal compressed sensing of sparse rank-one matrices via sparse power factorization," *IEEE Transactions on Information Theory*, vol. 64, no. 3, pp. 1666–1698, 2018. (p. 556)

[277] E. L. Lehmann and G. Casella, *Theory of Point Estimation*, 2nd ed. Springer, 1998. (pp. xxii and 578)

[278] E. Lehmann and J. Romano, *Testing Statistical Hypotheses*, 3rd ed. Springer, 2005. (pp. 278 and 328)

[279] W. V. Li and W. Linde, "Approximation, metric entropy and small ball estimates for Gaussian measures," *The Annals of Probability*, vol. 27, no. 3, pp. 1556–1578, 1999. (p. 554)

[280] W. V. Li and Q.-M. Shao, "Gaussian processes: inequalities, small ball probabilities and applications," *Handbook of Statistics*, vol. 19, pp. 533–597, 2001. (pp. 552, 554, and 568)

[281] E. H. Lieb, "Proof of an entropy conjecture of Wehrl," *Communications in Mathematical Physics*, vol. 62, no. 1, pp. 35–41, 1978. (p. 64)

[282] T. Linder and R. Zamir, "On the asymptotic tightness of the Shannon lower bound," *IEEE Transactions on Information Theory*, vol. 40, no. 6, pp. 2026–2031, 1994. (p. 523)

[283] R. S. Liptser, F. Pukel'sheim, and A. N. Shiryaev, "Necessary and sufficient conditions for contiguity and entire asymptotic separation of probability measures," *Russian Mathematical Surveys*, vol. 37, no. 6, p. 107, 1982. (p. 126)

[284] S. Litsyn, "New upper bounds on error exponents," *IEEE Transactions on Information Theory*, vol. 45, no. 2, pp. 385–398, 1999. (p. 443)

[285] S. Lloyd, "Least squares quantization in PCM," *IEEE Transactions on Information Theory*, vol. 28, no. 2, pp. 129–137, 1982. (p. 492)

[286] G. G. Lorentz, M. v. Golitschek, and Y. Makovoz, *Constructive Approximation: Advanced Problems*. Springer, 1996. (pp. 536 and 551)

[287] L. Lovász, "On the Shannon capacity of a graph," *IEEE Transactions on Information Theory*, vol. 25, no. 1, pp. 1–7, 1979. (p. 462)

[288] D. J. MacKay, *Information Theory, Inference and Learning Algorithms*. Cambridge University Press, 2003. (p. xxi)

[289] M. Madiman and P. Tetali, "Information inequalities for joint distributions, with interpretations and applications," *IEEE Transactions on Information Theory*, vol. 56, no. 6, pp. 2699–2713, 2010. (p. 18)

[290] M. Mahoney, "Large text compression benchmark," Aug. 2021, www .mattmahoney.net/dc/text.html. (p. 246)

[291] A. Makur and Y. Polyanskiy, "Comparison of channels: Criteria for domination by a symmetric channel," *IEEE Transactions on Information Theory*, vol. 64, no. 8, pp. 5704–5725, 2018. (p. 663)

[292] B. Mandelbrot, "An informational theory of the statistical structure of language," *Communication Theory*, vol. 84, pp. 486–502, 1953. (pp. 205, 207, and 208)

[293] C. Manning and H. Schutze, *Foundations of Statistical Natural Language Processing*. MIT Press, 1999. (p. 358)

[294] J. Massey, "On the fractional weight of distinct binary *n*-tuples (corresp.)," *IEEE Transactions on Information Theory*, vol. 20, no. 1, pp. 131–131, 1974. (p. 158)

[295] J. Massey, "Causality, feedback and directed information," in *Proceedings of the International Symposium on Information Theory and Its Applications(ISITA-90)*, 1990, pp. 303–305. (pp. 456 and 459)

[296] W. Matthews, "A linear program for the finite block length converse of Polyanskiy–Poor–Verdú via nonsignaling codes," *IEEE Transactions on Information Theory*, vol. 58, no. 12, pp. 7036–7044, 2012. (p. 439)

[297] H. H. Mattingly, M. K. Transtrum, M. C. Abbott, and B. B. Machta, "Maximizing the information learned from finite data selects a simple model," *Proceedings of the National Academy of Sciences*, vol. 115, no. 8, pp. 1760–1765, 2018. (p. 250)

[298] A. Maurer, "A note on the PAC Bayesian theorem," *arXiv preprint cs/0411099*, 2004. (p. 84)

[299] D. A. McAllester, "Some PAC-Bayesian theorems," in *Proceedings of the Eleventh Annual Conference on Computational Learning Theory*, 1998, pp. 230–234. (pp. 84 and 88)

[300] R. McEliece, E. Rodemich, H. Rumsey, and L. Welch, "New upper bounds on the rate of a code via the Delsarte-Macwilliams inequalities," *IEEE Transactions on Information Theory*, vol. 23, no. 2, pp. 157–166, 1977. (pp. 442, 443, and 540)

[301] R. J. McEliece and E. C. Posner, "Hide and seek, data storage, and entropy," *The Annals of Mathematical Statistics*, vol. 42, no. 5, pp. 1706–1716, 1971. (p. 556)

[302] B. McMillan, "The basic theorems of information theory," *The Annals of Mathematical Statistics*, vol. 24, no. 2, pp. 196–219, 1953. (p. 235)

[303] S. Mendelson, "Rademacher averages and phase transitions in Glivenko-Cantelli classes," *IEEE Transactions on Information Theory*, vol. 48, no. 1, pp. 251–263, 2002. (p. 545)

[304] N. Merhav and M. Feder, "Universal prediction," *IEEE Transactions on Information Theory*, vol. 44, no. 6, pp. 2124–2147, 1998. (p. 261)

[305] G. A. Miller, "Note on the bias of information estimates," in *Information Theory in Psychology: Problems and Methods*, vol. 2, H. Quastler, Ed. Free Press, 1955, pp. 95–100. (p. 597)

[306] M. Mitzenmacher, "A brief history of generative models for power law and lognormal distributions," *Internet*

Mathematics, vol. 1, no. 2, pp. 226–251, 2004. (pp. 205 and 208)

[307] E. Mossel, "Phase transitions in phylogeny," *Transactions of the American Mathematical Society*, vol. 356, no. 6, pp. 2379–2404, 2004. (p. 658)

[308] E. Mossel, J. Neeman, and A. Sly, "Reconstruction and estimation in the planted partition model," *Probability Theory and Related Fields*, vol. 162, no. 3-4, pp. 431–461, 2015. (pp. 188 and 658)

[309] E. Mossel and Y. Peres, "New coins from old: computing with unknown bias," *Combinatorica*, vol. 25, no. 6, pp. 707–724, 2005. (p. 172)

[310] J. Mourtada, "Exact minimax risk for linear least squares, and the lower tail of sample covariance matrices," *The Annals of Statistics*, vol. 50, no. 4, pp. 2157–2178, 2022. (p. 87)

[311] X. Mu, L. Pomatto, P. Strack, and O. Tamuz, "From Blackwell dominance in large samples to Rényi divergences and back again," *Econometrica*, vol. 89, no. 1, pp. 475–506, 2021. (pp. 145 and 184)

[312] N. Mukhanova, "Illustrator with a focus on children's books," https://muhanovs.com/nastya, 2023. (p. xvi)

[313] B. Nakiboğlu, "The sphere packing bound via Augustin's method," *IEEE Transactions on Information Theory*, vol. 65, no. 2, pp. 816–840, 2018. (p. 464)

[314] G. L. Nemhauser, L. A. Wolsey, and M. L. Fisher, "An analysis of approximations for maximizing submodular set functions–I," *Mathematical Programming*, vol. 14, no. 1, pp. 265–294, 1978. (pp. 375 and 376)

[315] J. Neveu, *Mathematical Foundations of the Calculus of Probability*. Holden-Day, 1965. (p. 552)

[316] M. E. Newman, "Power laws, Pareto distributions and Zipf's law," *Contemporary Physics*, vol. 46, no. 5, pp. 323–351, 2005. (p. 205)

[317] M. Okamoto, "Some inequalities relating to the partial sum of binomial probabilities," *Annals of the Institute of Statistical Mathematics*, vol. 10, no. 1, pp. 29–35, 1959. (p. 307)

[318] R. I. Oliveira, "The lower tail of random quadratic forms with applications to ordinary least squares," *Probability Theory and Related Fields*, vol. 166, pp. 1175–1194, 2016. (p. 87)

[319] B. Oliver, J. Pierce, and C. Shannon, "The philosophy of PCM," *Proceedings of the IRE*, vol. 36, no. 11, pp. 1324–1331, 1948. (p. 489)

[320] Y. Oohama, "On two strong converse theorems for discrete memoryless channels," *IEICE Transactions on Fundamentals of Electronics, Communications and Computer Sciences*, vol. 98, no. 12, pp. 2471–2475, 2015. (p. 444)

[321] OpenAI, "GPT-4 technical report," *arXiv preprint arXiv:2303.08774*, 2023. (pp. xix, 110, 196, and 259)

[322] O. Ordentlich and Y. Polyanskiy, "Strong data processing constant is achieved by binary inputs," *IEEE Transactions on Information Theory*, vol. 68, no. 3, pp. 1480–1481, 2022. (p. 687)

[323] D. Ornstein, "Bernoulli shifts with the same entropy are isomorphic," *Advances in Mathematics*, vol. 4, no. 3, pp. 337–352, 1970. (pp. xix and 231)

[324] L. Paninski, "Variational minimax estimation of discrete distributions under KL loss," in *Advances in Neural Information Processing Systems*, vol. 17, 2004. (p. 683)

[325] P. Panter and W. Dite, "Quantization distortion in pulse-count modulation with nonuniform spacing of levels," *Proceedings of the IRE*, vol. 39, no. 1, pp. 44–48, 1951. (p. 494)

[326] M. Pardo and I. Vajda, "About distances of discrete distributions satisfying the data processing theorem of information theory," *IEEE Transactions on Information Theory*, vol. 43, no. 4, pp. 1288–1293, 1997. (p. 121)

[327] S. Péché, "The largest eigenvalue of small rank perturbations of Hermitian random matrices," *Probability Theory and Related*

Fields, vol. 134, pp. 127–173, 2006. (p. 668)

[328] Y. Peres, "Iterating von Neumann's procedure for extracting random bits," *The Annals of Statistics*, vol. 20, no. 1, pp. 590–597, 1992. (p. 167)

[329] M. S. Pinsker, "Optimal filtering of square-integrable signals in Gaussian noise," *Problemy Peredachi Informatsii*, vol. 16, no. 2, pp. 52–68, 1980. (p. xxii)

[330] G. Pisier, *The Volume of Convex Bodies and Banach Space Geometry*. Cambridge University Press, 1999. (pp. 536, 543, and 544)

[331] J. Pitman, "Probabilistic bounds on the coefficients of polynomials with only real zeros," *Journal of Combinatorial Theory, Series A*, vol. 77, no. 2, pp. 279–303, 1997. (p. 306)

[332] E. Plotnik, M. J. Weinberger, and J. Ziv, "Upper bounds on the probability of sequences emitted by finite-state sources and on the redundancy of the Lempel-Ziv algorithm," *IEEE Transactions on Information Theory*, vol. 38, no. 1, pp. 66–72, 1992. (p. 266)

[333] D. Pollard, *Empirical Processes: Theory and Applications*, NSF-CBMS Regional Conference Series in Probability and Statistics, vol. 2. IMS and ASA, 1990. (p. 618)

[334] Y. Polyanskiy, "Channel coding: non-asymptotic fundamental limits," Ph.D. dissertation, Princeton University, Princeton, NJ, 2010. (pp. 109, 391, 393, 438, 444, and 446)

[335] Y. Polyanskiy, H. V. Poor, and S. Verdú, "Channel coding rate in the finite blocklength regime," *IEEE Transactions on Information Theory*, vol. 56, no. 5, pp. 2307–2359, 2010. (pp. 353, 360, 444, 445, 446, and 597)

[336] Y. Polyanskiy, H. V. Poor, and S. Verdú, "Dispersion of the Gilbert–Elliott channel," *IEEE Transactions on Information Theory*, vol. 57, no. 4, pp. 1829–1848, 2011. (p. 447)

[337] Y. Polyanskiy, H. V. Poor, and S. Verdú, "Feedback in the non-asymptotic regime," *IEEE Transactions on Information Theory*, vol. 57, no. 4, pp. 4903–4925, 2011. (pp. 456, 465, and 466)

[338] Y. Polyanskiy, H. V. Poor, and S. Verdú, "Minimum energy to send k bits with and without feedback," *IEEE Transactions on Information Theory*, vol. 57, no. 8, pp. 4880–4902, 2011. (pp. 422 and 459)

[339] Y. Polyanskiy and S. Verdú, "Arimoto channel coding converse and Rényi divergence," in *Proceedings of the Forty-eighth Annual Allerton Conference on Communication, Control, and Computing*, 2010, pp. 1327–1333. (pp. 121, 444, and 518)

[340] Y. Polyanskiy and S. Verdú, "Binary hypothesis testing with feedback," in *Information Theory and Applications Workshop (ITA)*, 2011. (p. 325)

[341] Y. Polyanskiy and S. Verdú, "Empirical distribution of good channel codes with non-vanishing error probability," *IEEE Transactions on Information Theory*, vol. 60, no. 1, pp. 5–21, 2014. (p. 438)

[342] Y. Polyanskiy, "Saddle point in the minimax converse for channel coding," *IEEE Transactions on Information Theory*, vol. 59, no. 5, pp. 2576–2595, 2012. (pp. 439 and 440)

[343] Y. Polyanskiy, "On dispersion of compound DMCs," in *2013 51st Annual Allerton Conference on Communication, Control, and Computing (Allerton)*. IEEE, 2013, pp. 26–32. (pp. 447 and 477)

[344] Y. Polyanskiy and Y. Wu, "Peak-to-average power ratio of good codes for Gaussian channel," *IEEE Transactions on Information Theory*, vol. 60, no. 12, pp. 7655–7660, 2014. (p. 416)

[345] Y. Polyanskiy and Y. Wu, "Wasserstein continuity of entropy and outer bounds for interference channels," *IEEE Transactions on Information Theory*, vol. 62, no. 7, pp. 3992–4002, 2016. (pp. 60 and 64)

[346] Y. Polyanskiy and Y. Wu, "Strong data-processing inequalities for channels and Bayesian networks," in *Convexity and Concentration*, The IMA Volumes in

Mathematics and its Applications, vol. 161, E. Carlen, M. Madiman, and E. M. Werner, Eds. Springer, 2017, pp. 211–249. (pp. 330, 642, 647, 648, 651, 653, 654, and 663)

[347] Y. Polyanskiy and Y. Wu, "Dualizing Le Cam's method for functional estimation, with applications to estimating the unseens," *arXiv preprint arXiv:1902.05616*, 2019. (pp. 579 and 686)

[348] Y. Polyanskiy and Y. Wu, "Application of the information-percolation method to reconstruction problems on graphs," *Mathematical Statistics and Learning*, vol. 2, no. 1, pp. 1–24, 2020. (pp. 667 and 670)

[349] Y. Polyanskiy and Y. Wu, "Self-regularizing property of nonparametric maximum likelihood estimator in mixture models," *arXiv preprint arXiv:2008.08244*, 2020. (p. 416)

[350] E. C. Posner and E. R. Rodemich, "Epsilon entropy and data compression," *The Annals of Mathematical Statistics*, vol. 42, no. 6, pp. 2079–2125, 1971. (p. 555)

[351] A. Prékopa, "Logarithmic concave measures with application to stochastic programming," *Acta Scientiarum Mathematicarum*, vol. 32, pp. 301–316, 1971. (p. 587)

[352] J. Radhakrishnan, "An entropy proof of Bregman's theorem," *Journal of Combinatorial Theory, Series A*, vol. 77, no. 1, pp. 161–164, 1997. (p. 162)

[353] M. Raginsky, "Strong data processing inequalities and ϕ-Sobolev inequalities for discrete channels," *IEEE Transactions on Information Theory*, vol. 62, no. 6, pp. 3355–3389, 2016. (pp. 642 and 654)

[354] M. Raginsky and I. Sason, "Concentration of measure inequalities in information theory, communications, and coding," *Foundations and Trends in Communications and Information Theory*, vol. 10, no. 1-2, pp. 1–246, 2013. (p. xxii)

[355] C. R. Rao, "Information and the accuracy attainable in the estimation of statistical parameters," *Bulletin of the Calcutta Mathematical Society*, vol. 37, pp. 81–91, 1945. (p. 589)

[356] A. H. Reeves, "The past, present and future of PCM," *IEEE Spectrum*, vol. 2, no. 5, pp. 58–62, 1965. (p. 489)

[357] A. Rényi, "On measures of entropy and information," in *Proceedings of the 4th Berkeley Symposium on Mathematics, Statistics, and Probability*, vol. 1, Berkeley, CA, 1961, pp. 547–561. (p. 12)

[358] A. Rényi, "On the dimension and entropy of probability distributions," *Acta Mathematica Hungarica*, vol. 10, no. 1–2, 1959. (p. 28)

[359] Z. Reznikova and B. Ryabko, "Analysis of the language of ants by information-theoretical methods," *Problemy Peredachi Informatsii*, vol. 22, no. 3, pp. 103–108, 1986. English translation: http://reznikova.net/R-R-entropy-09.pdf. (p. 8)

[360] T. J. Richardson, M. A. Shokrollahi, and R. L. Urbanke, "Design of capacity-approaching irregular low-density parity-check codes," *IEEE Transactions on Information Theory*, vol. 47, no. 2, pp. 619–637, 2001. (p. 528)

[361] T. Richardson and R. Urbanke, *Modern Coding Theory*. Cambridge University Press, 2008. (pp. xxii, 63, 346, 353, 391, and 648)

[362] Y. Rinott, "On convexity of measures," *The Annals of Probability*, vol. 4, no. 6, pp. 1020–1026, 1976. (p. 587)

[363] J. J. Rissanen, "Fisher information and stochastic complexity," *IEEE Transactions on Information Theory*, vol. 42, no. 1, pp. 40–47, 1996. (p. 263)

[364] H. Robbins, "An empirical Bayes approach to statistics," in *Proceedings of the Third Berkeley Symposium on Mathematical Statistics and Probability, Vol. 1: Contributions to the Theory of Statistics*. Regents of the University of California, 1956. (p. 576)

[365] R. W. Robinson and N. C. Wormald, "Almost all cubic graphs are Hamiltonian," *Random Structures & Algorithms*, vol. 3, no. 2, pp. 117–125, 1992. (p. 188)

[366] C. Rogers, *Packing and Covering*, Cambridge Tracts in Mathematics and Mathematical Physics. Cambridge University Press, 1964. (p. 539)

[367] H. Roozbehani and Y. Polyanskiy, "Algebraic methods of classifying directed graphical models," *arXiv preprint arXiv:1401.5551*, 2014. (p. 182)

[368] H. Roozbehani and Y. Polyanskiy, "Low density majority codes and the problem of graceful degradation," *arXiv preprint arXiv:1911.12263*, 2019. (pp. 194 and 687)

[369] H. P. Rosenthal, "On the subspaces of $L^p (p > 2)$ spanned by sequences of independent random variables," *Israel Journal of Mathematics*, vol. 8, no. 3, pp. 273–303, 1970. (p. 509)

[370] D. Russo and J. Zou, "Controlling bias in adaptive data analysis using information theory," in *Artificial Intelligence and Statistics*. PMLR, 2016, pp. 1232–1240. (pp. 90 and 190)

[371] I. N. Sanov, "On the probability of large deviations of random magnitudes," *Matematicheskii Sbornik*, vol. 84, no. 1, pp. 11–44, 1957. (p. 312)

[372] E. Şaşoğlu, "Polar coding theorems for discrete systems," Ph.D. Dissertation, EPFL, Lausanne, 2011. (p. 687)

[373] E. Şaşoğlu, "Polarization and polar codes," *Foundations and Trends in Communications and Information Theory*, vol. 8, no. 4, pp. 259–381, 2012. (p. 346)

[374] I. Sason and S. Verdú, "f-divergence inequalities," *IEEE Transactions on Information Theory*, vol. 62, no. 11, pp. 5973–6006, 2016. (p. 132)

[375] G. Schechtman, "Extremal configurations for moments of sums of independent positive random variables," in *Banach Spaces and Their Applications in Analysis*. de Gruyter, 2011, pp. 183–192. (p. 517)

[376] M. J. Schervish, *Theory of Statistics*. Springer, 1995. (p. 596)

[377] A. Schrijver, *Theory of Linear and Integer Programming*. Wiley, 1998. (p. 580)

[378] C. E. Shannon, "A symbolic analysis of relay and switching circuits," *Electrical Engineering*, vol. 57, no. 12, pp. 713–723, 1938. (p. 643)

[379] C. E. Shannon, "A mathematical theory of communication," *Bell System Technical Journal*, vol. 27, pp. 379–423 and 623–656, 1948. (pp. xvii, 41, 196, 217, 235, 346, 353, 385, and 420)

[380] C. E. Shannon, "The zero error capacity of a noisy channel," *IRE Transactions on Information Theory*, vol. 2, no. 3, pp. 8–19, 1956. (pp. 382, 460, and 461)

[381] C. E. Shannon, "Coding theorems for a discrete source with a fidelity criterion," *IRE National Convention Record*, vol. 4, pp. 142–163, 1959. (pp. 486 and 503)

[382] C. E. Shannon, R. G. Gallager, and E. R. Berlekamp, "Lower bounds to error probability for coding on discrete memoryless channels I," *Information and Control*, vol. 10, pp. 65–103, 1967. (pp. 441 and 443)

[383] J. Shawe-Taylor and R. C. Williamson, "A PAC analysis of a Bayesian estimator," in *Proceedings of the Tenth Annual Conference on Computational Learning Theory*, 1997, pp. 2–9. (p. 84)

[384] O. Shayevitz, "On Rényi measures and hypothesis testing," in *2011 IEEE International Symposium on Information Theory Proceedings*. IEEE, 2011, pp. 894–898. (p. 184)

[385] O. Shayevitz and M. Feder, "Optimal feedback communication via posterior matching," *IEEE Transactions on Information Theory*, vol. 57, no. 3, pp. 1186–1222, 2011. (p. 455)

[386] A. N. Shiryaev, *Probability-1*, Graduate Texts in Mathematics, no. 95. Springer, 2016. (p. 126)

[387] G. Simons and M. Woodroofe, "The Cramér–Rao inequality holds almost everywhere," in *Recent Advances in Statistics: Papers in Honor of Herman Chernoff on his Sixtieth Birthday*. Academic Press, pp. 69–93. (p. 678)

[388] Y. G. Sinai, "On the notion of entropy of a dynamical system," *Doklady of Russian Academy of Sciences*, vol. 124, no. 3, 1959, pp. 768–771. (pp. xix and 231)

[389] R. Sinkhorn, "A relationship between arbitrary positive matrices and doubly stochastic matrices," *The Annals of*

Mathematical Statistics, vol. 35, no. 2, pp. 876–879, 1964. (p. 105)

[390] M. Sion, "On general minimax theorems," *Pacific Journal of Mathematics*, vol. 8, no. 1, pp. 171–176, 1958. (p. 93)

[391] M.-K. Siu, "Which Latin squares are Cayley tables?" *American Mathematical Monthly*, vol. 98, no. 7, pp. 625–627, 1991. (p. 392)

[392] D. Slepian and H. O. Pollak, "Prolate spheroidal wave functions, Fourier analysis and uncertainty–I," *Bell System Technical Journal*, vol. 40, no. 1, pp. 43–63, 1961. (p. 429)

[393] D. Slepian and J. Wolf, "Noiseless coding of correlated information sources," *IEEE Transactions on Information Theory*, vol. 19, no. 4, pp. 471–480, 1973. (p. 225)

[394] A. Sly, "Reconstruction of random colourings," *Communications in Mathematical Physics*, vol. 288, no. 3, pp. 943–961, 2009. (p. 660)

[395] A. Sly and N. Sun, "Counting in two-spin models on d-regular graphs," *The Annals of Probability*, vol. 42, no. 6, pp. 2383–2416, 2014. (p. 75)

[396] B. Smith, "Instantaneous companding of quantized signals," *Bell System Technical Journal*, vol. 36, no. 3, pp. 653–709, 1957. (p. 495)

[397] J. G. Smith, "The information capacity of amplitude and variance-constrained scalar Gaussian channels," *Information and Control*, vol. 18, pp. 203–219, 1971. (p. 416)

[398] Spectre, "SPECTRE: Short packet communication toolbox," 2015, GitHub repository, https://github.com/yp-mit/spectre. (pp. 427 and 450)

[399] R. Speer, J. Chin, A. Lin, S. Jewett, and L. Nathan, "Luminosoinsight/wordfreq: v2.2," 2018, available at https://doi.org/10.5281/zenodo.1443582. (p. 206)

[400] A. J. Stam, "Some inequalities satisfied by the quantities of information of Fisher and Shannon," *Information and Control*, vol. 2, no. 2, pp. 101–112, 1959. (pp. 64, 187, and 194)

[401] A. J. Stam, "Distance between sampling with and without replacement," *Statistica Neerlandica*, vol. 32, no. 2, pp. 81–91, 1978. (pp. 188 and 189)

[402] M. Steiner, "The strong simplex conjecture is false," *IEEE Transactions on Information Theory*, vol. 40, no. 3, pp. 721–731, 1994. (p. 423)

[403] V. Strassen, "Asymptotische Abschätzungen in Shannon's Informationstheorie," in *Transactions of the 3rd Prague Conference on Information Theory*, Prague, 1962, pp. 689–723. (pp. 444 and 445)

[404] V. Strassen, "The existence of probability measures with given marginals," *The Annals of Mathematical Statistics*, vol. 36, no. 2, pp. 423–439, 1965. (p. 122)

[405] H. Strasser, *Mathematical Theory of Statistics: Statistical Experiments and Asymptotic Decision Theory*. de Gruyter, 1985. (pp. 571 and 580)

[406] J. Suzuki, "Some notes on universal noiseless coding," *IEICE Transactions on Fundamentals of Electronics, Communications and Computer Sciences*, vol. 78, no. 12, pp. 1840–1847, 1995. (p. 254)

[407] S. Szarek, "Nets of Grassmann manifold and orthogonal groups," in *Proceedings of Banach Space Workshop*. University of Iowa Press, 1982, pp. 169–185. (pp. 539 and 556)

[408] S. Szarek, "Metric entropy of homogeneous spaces," *Banach Center Publications*, vol. 43, no. 1, pp. 395–410, 1998. (p. 539)

[409] W. Szpankowski and S. Verdú, "Minimum expected length of fixed-to-variable lossless compression without prefix constraints," *IEEE Transactions on Information Theory*, vol. 57, no. 7, pp. 4017–4025, 2011. (p. 202)

[410] I. Tal and A. Vardy, "List decoding of polar codes," *IEEE Transactions on Information Theory*, vol. 61, no. 5, pp. 2213–2226, 2015. (p. 353)

[411] M. Talagrand, "The Parisi formula," *Annals of Mathematics*, vol. 163, no. 1, pp. 221–263, 2006. (p. 63)

[412] M. Talagrand, *Upper and Lower Bounds for Stochastic Processes*. Springer, 2014. (p. 544)

[413] T. Tanaka, P. M. Esfahani, and S. K. Mitter, "LQG control with minimum directed information: Semidefinite programming approach," *IEEE Transactions on Automatic Control*, vol. 63, no. 1, pp. 37–52, 2017. (p. 459)

[414] W. Tang and F. Tang, "The Poisson binomial distribution – old & new," *Statistical Science*, vol. 38, no. 1, pp. 108–119, 2023. (p. 306)

[415] T. Tao, "Szemerédi's regularity lemma revisited," *Contributions to Discrete Mathematics*, vol. 1, no. 1, pp. 8–28, 2006. (pp. 127 and 193)

[416] G. Taricco and M. Elia, "Capacity of fading channel with no side information," *Electronics Letters*, vol. 33, no. 16, pp. 1368–1370, 1997. (p. 417)

[417] V. Tarokh, H. Jafarkhani, and A. R. Calderbank, "Space-time block codes from orthogonal designs," *IEEE Transactions on Information Theory*, vol. 45, no. 5, pp. 1456–1467, 1999. (p. 417)

[418] V. Tarokh, N. Seshadri, and A. R. Calderbank, "Space-time codes for high data rate wireless communication: Performance criterion and code construction," *IEEE Transactions on Information Theory*, vol. 44, no. 2, pp. 744–765, 1998. (p. 417)

[419] E. Telatar, "Capacity of multi-antenna Gaussian channels," *European Transactions on Telecommunications*, vol. 10, no. 6, pp. 585–595, 1999. (pp. 178 and 417)

[420] E. Telatar, "Wringing lemmas and multiple descriptions," 2016, unpublished draft. (p. 190)

[421] V. N. Temlyakov, "On estimates of ϵ-entropy and widths of classes of functions with a bounded mixed derivative or difference," *Doklady Akademii Nauk*, vol. 301, no. 2, pp. 288–291, 1988. (p. 554)

[422] N. Tishby, F. C. Pereira, and W. Bialek, "The information bottleneck method," *arXiv preprint physics/0004057*, 2000. (p. 564)

[423] F. Topsøe, "Some inequalities for information divergence and related measures of discrimination," *IEEE Transactions on Information Theory*, vol. 46, no. 4, pp. 1602–1609, 2000. (p. 133)

[424] D. Tse and P. Viswanath, *Fundamentals of Wireless Communication*. Cambridge University Press, 2005. (pp. xxii, 412, and 418)

[425] A. B. Tsybakov, *Introduction to Nonparametric Estimation*. Springer, 2009. (pp. xxi, xxii, 132, and 640)

[426] B. P. Tunstall, "Synthesis of noiseless compression codes," Ph.D. Dissertation, Georgia Institute of Technology, 1967. (p. 197)

[427] E. Uhrmann-Klingen, "Minimal Fisher information distributions with compact-supports," *Sankhyā: The Indian Journal of Statistics, Series A*, vol. 57, no. 3, pp. 360–374, 1995. (p. 593)

[428] I. Vajda, "Note on discrimination information and variation (corresp.)," *IEEE Transactions on Information Theory*, vol. 16, no. 6, pp. 771–773, 1970. (p. 131)

[429] G. Valiant and P. Valiant, "Estimating the unseen: an $n/\log(n)$-sample estimator for entropy and support size, shown optimal via new CLTs," in *Proceedings of the 43rd Annual ACM Symposium on Theory of Computing*, 2011, pp. 685–694. (p. 598)

[430] S. van de Geer, *Empirical Processes in M-Estimation*. Cambridge University Press, 2000. (pp. 87, 546, and 618)

[431] A. van der Vaart, "The statistical work of Lucien Le Cam," *The Annals of Statistics*, vol. 30, no. 3, pp. 631–682, 2002. (pp. 629 and 632)

[432] A. W. van der Vaart and J. A. Wellner, *Weak Convergence and Empirical Processes*. Springer, 1996. (pp. 87 and 618)

[433] T. Van Erven and P. Harremoës, "Rényi divergence and Kullback-Leibler

divergence," *IEEE Transactions on Information Theory*, vol. 60, no. 7, pp. 3797–3820, 2014. (p. 145)

[434] H. L. Van Trees, *Detection, Estimation, and Modulation Theory*. Wiley, 1968. (p. 590)

[435] S. Verdú, "On channel capacity per unit cost," *IEEE Transactions on Information Theory*, vol. 36, no. 5, pp. 1019–1030, 1990. (p. 423)

[436] S. Verdú, *Multiuser Detection*. Cambridge University Press, 1998. (p. 422)

[437] S. Verdú, "Information theory, part I," draft (personal communication), 2017. (p. xv)

[438] S. Verdú and D. Guo, "A simple proof of the entropy-power inequality," *IEEE Transactions on Information Theory*, vol. 52, no. 5, pp. 2165–2166, 2006. (p. 64)

[439] R. Vershynin, *High-Dimensional Probability: An Introduction with Applications in Data Science*. Cambridge University Press, 2018. (pp. 86 and 543)

[440] A. G. Vitushkin, "On the 13th problem of Hilbert," *Doklady Akademii Nauk SSSR*, vol. 95, no. 4, pp. 701–704, 1954. (p. 551)

[441] A. G. Vitushkin, "On Hilbert's thirteenth problem and related questions," *Russian Mathematical Surveys*, vol. 59, no. 1, p. 11, 2004. (p. xix)

[442] A. G. Vitushkin, *Theory of the Transmission and Processing of Information*. Pergamon, 1961. (p. 548)

[443] J. von Neumann, "Various techniques used in connection with random digits," in *Monte Carlo Method*, National Bureau of Standards, Applied Math Series, no. 12, pp. 36–38, 1951. (p. 166)

[444] J. von Neumann, "Probabilistic logics and the synthesis of reliable organisms from unreliable components," in *Automata Studies*, Vol. 34, C. E. Shannon and J. McCarthy, Eds. Princeton University Press, 1956, pp. 43–98. (p. 643)

[445] D. Von Rosen, "Moments for the inverted Wishart distribution," *Scandinavian Journal of Statistics*, vol. 15, no. 2, pp. 97–109, 1988. (p. 275)

[446] V. G. Vovk, "Aggregating strategies," *Proceedings of the Third Annual Workshop on Computational Learning Theory*, 1990, pp. 371–383. (pp. xx and 274)

[447] M. J. Wainwright, *High-Dimensional Statistics: A Non-Asymptotic Viewpoint*. Cambridge University Press, 2019. (p. xxii)

[448] M. J. Wainwright and M. I. Jordan, "Graphical models, exponential families, and variational inference," *Foundations and Trends in Machine Learning*, vol. 1, no. 1–2, pp. 1–305, 2008. (pp. 75 and 76)

[449] A. Wald, "Sequential tests of statistical hypotheses," *The Annals of Mathematical Statistics*, vol. 16, no. 2, pp. 117–186, 1945. (p. 324)

[450] A. Wald, "Note on the consistency of the maximum likelihood estimate," *The Annals of Mathematical Statistics*, vol. 20, no. 4, pp. 595–601, 1949. (p. 595)

[451] A. Wald and J. Wolfowitz, "Optimum character of the sequential probability ratio test," *The Annals of Mathematical Statistics*, vol. 19, no. 3, pp. 326–339, 1948. (pp. 324 and 325)

[452] M. M. Wilde, *Quantum Information Theory*. Cambridge University Press, 2013. (p. xxi)

[453] R. J. Williams, "Simple statistical gradient-following algorithms for connectionist reinforcement Learning," *Machine learning*, vol. 8, pp. 229–256, 1992. (p. 78)

[454] H. Witsenhausen and A. Wyner, "A conditional entropy bound for a pair of discrete random variables," *IEEE Transactions on Information Theory*, vol. 21, no. 5, pp. 493–501, 1975. (p. 330)

[455] J. Wolfowitz, "On Wald's proof of the consistency of the maximum likelihood estimate," *The Annals of Mathematical Statistics*, vol. 20, no. 4, pp. 601–602, 1949. (p. 595)

[456] Y. Wu and J. Xu, "Statistical problems with planted structures: Information-theoretical and computational limits," in

Information-Theoretic Methods in Data Science, Y. Eldar and M. Rodrigues, Eds. Cambridge University Press, 2020. Also *arXiv preprint arXiv:1806.00118*. (p. 343)

[457] Y. Wu and P. Yang, "Minimax rates of entropy estimation on large alphabets via best polynomial approximation," *IEEE Transactions on Information Theory*, vol. 62, no. 6, pp. 3702–3720, 2016. (p. 598)

[458] A. Wyner and J. Ziv, "A theorem on the entropy of certain binary sequences and applications–I," *IEEE Transactions on Information Theory*, vol. 19, no. 6, pp. 769–772, 1973. (pp. 193 and 194)

[459] A. Wyner, "The common information of two dependent random variables," *IEEE Transactions on Information Theory*, vol. 21, no. 2, pp. 163–179, 1975. (pp. 515 and 516)

[460] A. Wyner, "On source coding with side information at the decoder," *IEEE Transactions on Information Theory*, vol. 21, no. 3, pp. 294–300, 1975. (p. 229)

[461] Q. Xie and A. R. Barron, "Minimax redundancy for the class of memoryless sources," *IEEE Transactions on Information Theory*, vol. 43, no. 2, pp. 646–657, 1997. (p. 254)

[462] A. Xu and M. Raginsky, "Information-theoretic analysis of generalization capability of learning algorithms," in *Advances in Neural Information Processing Systems*, vol. 30, 2017. (p. 90)

[463] W. Yang, G. Durisi, T. Koch, and Y. Polyanskiy, "Quasi-static multiple-antenna fading channels at finite blocklength," *IEEE Transactions on Information Theory*, vol. 60, no. 7, pp. 4232–4265, 2014. (p. 447)

[464] W. Yang, G. Durisi, and Y. Polyanskiy, "Minimum energy to send k bits over multiple-antenna fading channels," *IEEE Transactions on Information Theory*, vol. 62, no. 12, pp. 6831–6853, 2016. (p. 426)

[465] Y. Yang and A. R. Barron, "Information-theoretic determination of minimax rates of convergence," *The Annals of Statistics*, vol. 27, no. 5, pp. 1564–1599, 1999. (pp. xxii, 617, 622, and 623)

[466] Y. G. Yatracos, "Rates of convergence of minimum distance estimators and Kolmogorov's entropy," *The Annals of Statistics*, vol. 13, no. 2, pp. 768–774, 1985. (pp. 617, 635, and 636)

[467] S. Yekhanin, "Improved upper bound for the redundancy of fix-free codes," *IEEE Transactions on Information Theory*, vol. 50, no. 11, pp. 2815–2818, 2004. (p. 211)

[468] P. L. Zador, "Development and evaluation of procedures for quantizing multivariate distributions," Ph.D. Dissertation, Stanford University, 1963. (p. 494)

[469] P. L. Zador, "Asymptotic quantization error of continuous signals and the quantization dimension," *IEEE Transactions on Information Theory*, vol. 28, no. 2, pp. 139–149, 1982. (p. 494)

[470] O. Zeitouni, J. Ziv, and N. Merhav, "When is the generalized likelihood ratio test optimal?" *IEEE Transactions on Information Theory*, vol. 38, no. 5, pp. 1597–1602, 1992. (p. 328)

[471] C.-H. Zhang, "Compound decision theory and empirical Bayes methods," *The Annals of Statistics*, vol. 31, no. 2, pp. 379–390, 2003. (p. 576)

[472] T. Zhang, "Covering number bounds of certain regularized linear function classes," *Journal of Machine Learning Research*, vol. 2, pp. 527–550, 2002. (p. 545)

[473] Z. Zhang, E. Yang, and V. Wei, "The redundancy of source coding with a fidelity criterion," *IEEE Transactions on Information Theory*, vol. 43, no. 1, pp. 71–91, 1997. (pp. 497 and 560)

[474] Z. Zhang and R. W. Yeung, "A non-Shannon-type conditional inequality of information quantities," *IEEE Transactions on Information Theory*, vol. 43, no. 6, pp. 1982–1986, 1997. (p. 16)

[475] Z. Zhang and R. W. Yeung, "On characterization of entropy function via information inequalities," *IEEE Transactions on Information Theory*, vol. 44, no. 4, pp. 1440–1452, 1998. (p. 16)

[476] L. Zheng and D. N. C. Tse, "Communication on the Grassmann manifold: A geometric approach to the noncoherent multiple-antenna channel," *IEEE Transactions on Information Theory*, vol. 48, no. 2, pp. 359–383, 2002. (p. 418)

[477] N. Zhivotovskiy, "Dimension-free bounds for sums of independent matrices and simple tensors via the variational principle," *Electronic Journal of Probability*, vol. 29 pp. 1–28, 2024. (pp. 85 and 86)

[478] W. Zhou, V. Veitch, M. Austern, R. P. Adams, and P. Orbanz, "Non-vacuous generalization bounds at the ImageNet scale: a PAC-Bayesian compression approach," in *International Conference on Learning Representations (ICLR)*, 2018. (p. 90)

[479] G. Zipf, *Selective Studies and the Principle of Relative Frequency in Language*. Harvard University Press, 1932. (pp. 205 and 206)

Index

Printed in the United States
by Baker & Taylor Publisher Services